Spectral Distributions in Nuclei and Statistical Spectroscopy

Spectral Distributions in Nuclei and Statistical Spectroscopy

Edited by

V K B Kota
Physical Research Laboratory, Ahmedabad, India

Rizwan ul Haq
Laurentian University, Canada

Foreword by Daniel Koltun

World Scientific

NEW JERSEY • LONDON • SINGAPORE • BEIJING • SHANGHAI • HONG KONG • TAIPEI • CHENNAI

Published by

World Scientific Publishing Co. Pte. Ltd.
5 Toh Tuck Link, Singapore 596224
USA office: 27 Warren Street, Suite 401-402, Hackensack, NJ 07601
UK office: 57 Shelton Street, Covent Garden, London WC2H 9HE

British Library Cataloguing-in-Publication Data
A catalogue record for this book is available from the British Library.

Cover Illustration:
A schematic drawing of the spectrum of a heavy nucleus, with handwritten notes, has been taken from lectures given by J.B. French at the Summer Institute of Theoretical Physics held at Panjab University, Chandigarh, India in May 1971.

SPECTRAL DISTRIBUTIONS IN NUCLEI AND STATISTICAL SPECTROSCOPY

ISBN-13 978-981-4287-38-8
ISBN-10 981-4287-38-5

Printed in Singapore by B & Jo Enterprise Pte Ltd

This book is dedicated to the memory of J. Bruce French

"The striking features of spectral distribution methods are their wide applicability and the connections which they display, and make use of, between statistical behavior, unitary and other symmetries and their associated geometries, and information content and propagation. As for the future it might be good for us not to think of the methods we are discussing as forming a separate domain with its own special tricks, devices, methods and assumptions. It is probably wise instead to think of the whole subject as forming a sub-domain of statistical mechanics in which special attention is paid to the nature of the model space."

Foreword

This volume contains two components, as explained in the following Preface: First, Chapter 1 is a survey of the subject of Statistical Spectroscopy, particularly as the subject applies to atomic nuclei. Second is a collection of 46 papers published between 1966 and 2006 on the subject covered in the survey. Many of these papers, but not all, were written by J. Bruce French and his many collaborators, who were students and/or research colleagues of his at the University of Rochester during those 40 years or so. Both of the authors of this volume, R.U. Haq and V.K.B. Kota, began their interactions with French around 1980. They have dedicated this volume to the memory of J. Bruce French, who contributed so much to the subject matter of the book.

Bruce French pretty much invented the forms of statistical spectroscopy presented in this volume. It was exciting for me to talk with Bruce or his visitors informally as the theory was emerging and taking form. Bruce came to Rochester in 1950 and developed an interest in nuclear spectra and the shell model theory of lighter mass nuclei. He gradually developed the point of view that as shell model theory of nuclear spectra advanced in complexity, with many unknown parameters for interactions, little would be learned by comparison of models with experimental data, in the long run. His line of thought led to statistical spectroscopy, where complex nuclei at higher excitation energy could be successfully characterized by a small number of parameters. Statistical theory of fluctuations in spectra had already been developed earlier, based on random matrix theory. Bruce's program was to deal with a theory of averages. What has come out of this program is generally called the theory of Spectral Distributions.

So, how should the reader use this volume? If you are new to the subject, start with the Table of Contents, and then turn to the beginning of the introductory Chapter 1.1, where the ideas to be used are explained and the language developed. If you are interested in how the subject developed historically, turn to the earliest published papers listed in the Contents. If you are familiar with the statistical arguments in Chapter 1, go to the discussions of applications to problems like symmetry breaking in nuclei. If you know what you want - go for it!

Daniel Koltun
February 26, 2010

Preface

In the book entitled "Statistical Theories of Spectra: Fluctuations" published in 1965, Charles E. Porter wrote - "A discussion of fluctuations presented a natural vehicle for developing the general ideas and including a number of relevant papers all within a single volume. It is quite likely that a similar treatment of the average properties of spectra would require at least one volume of the size of this one for each property separately; a sequence of such volumes would be a natural extrapolation to the present one".

We have attempted here to fill this gap of spectral averages in statistical spectroscopy and used the same format as in Porter of an introductory review followed by a collection of original papers. Starting with the seminal work of J. Bruce French in the mid-sixties, we trace the development and applications of spectral distributions in nuclei over the past forty-five years or so. While the focus is on nuclei, included also is some discussion regarding applications to atoms, mesoscopic systems and interacting bosons.

The lead overview article starts with the typical spectrum of a heavy nucleus describing the various domains and the different methods traditionally used to study them. Statistical spectroscopy with its two complementary aspects, one dealing with averages and the other with fluctuations around this average is introduced as a unifying approach. The methods used, spectral distributions for the former and random matrix ensembles for the latter are discussed and the basic principles leading to a simplification of the complicated many-body problem are summarized in the form of five general laws. We then consider in some detail the theory and applications of distributions of energy levels (densities), expectation values and transition strengths as well as the role played by group symmetries. It is now recognized that the two-body random matrix ensemble and quantum chaos form the basis of spectral distributions in nuclei. Since our aim here is to deal primarily with averages, for a complete picture of statistical spectroscopy, this monograph needs to be supplemented by reviews on fluctuations and random matrix ensembles: Brody et al (1981); Zelevinsky et al (1996); Kota (2001); Weidenmüller and Mitchell (2009).

We have strived to include specifically those papers which reflect the development of new concepts and/or applications of spectral distributions in nuclei. Other important papers are referred to and included in the Bibliography. With a few notable exceptions, we have, in general, not included review articles or review talks given at conferences. Even though we have tried to be as objective as possible, the choice of included papers may have been, nonetheless, influenced by our own bias. We therefore apologize to those who might feel that their contribution has been inadvertently omitted. It should also be mentioned that other books that complement the material presented here are: Parikh (1978); Wong (1986); and Karazija (1991, in Russian) on spectral distributions; Porter (1965) and Mehta (2004) on random matrices; and Haake (2001) on chaos.

It is our belief that an appreciation of the concepts and general applicability will lead to further development of the subject and define new directions in quantum chaos, random matrix theory, and statistical mechanics of finite quantum systems such as nuclei, atoms, quantum dots, etc. The book should be of interest to the expert for a deeper understanding and to the graduate student for an introduction to the subject of statistical spectroscopy.

This book is our tribute to Bruce French for initiating the subject of spectral distributions in nuclei, for his far reaching contributions to physics in general and nuclear physics in particular, and also to express our gratitude for the several decades of scientific mentorship and research collaboration. In writing the review we have borrowed from unpublished manuscripts and research reports where we had been co-authors.

We are indebted to R. Arvieu, O. Bohigas, J.P. Draayer, J.M.G. Gomez, K. Kar, D. Majumdar, A. Pandey, J.C. Parikh, V. Potbhare, C. Quesne, R. Sahu, S. Tomsovic, S.S.M. Wong, and A. Zuker for the innumerable discussions and collaboration over an extended period of time. We are also grateful to Ashok Das for his continued interest in the book. We thank Dan Koltun for the hospitality shown to us during our numerous visits to the University of Rochester and are indebted to him for writing the Foreword. We also thank Manan Vyas for her assistance in preparing several figures of Chapter 1. Copyright permission from the American Physical Society, the American Institute of Physics, the National Research Council Research Press, Elsevier Science, Springer-Verlag, and World Scientific to reproduce the 46 original papers included in this work is acknowledged. Finally, we would like to thank Lakshmi Narayanan of World Scientific for her constant help with the entire process.

R.U. Haq and V.K.B. Kota

Contents

Foreword vii

Preface ix

1. Fluctuation-free Nuclear Spectroscopy 1

1.1 Introduction . . . 1
1.1.1 Spectral distributions and fluctuations . . . 4
1.1.2 Average-Fluctuation separation . . . 6
1.1.3 Random matrix ensembles . . . 9
1.1.4 Fluctuations and quantum chaos . . . 12
1.1.5 Laws of statistical spectroscopy . . . 15
1.2 Statistical Methods in Finite-dimensional Model Spaces . . . 16
1.2.1 Moments and distributions . . . 16
1.2.2 Central Limit Theorems for NIP and IP systems . . . 24
1.2.3 Partitioning and partial densities . . . 31
1.3 Expectation Value Distributions . . . 37
1.3.1 Five methods for moment expansions for expectation values . . . 37
1.3.2 Gram-Charlier and Edgeworth expansions . . . 41
1.3.3 Partitioning, convergence and multi-variate expectation values . . . 43
1.4 Unitary Decompositions, Geometry and Configuration Traces: State and Spin Cutoff Densities . . . 47
1.4.1 Unitary decompositions and renormalization of the NIP Hamiltonian . . . 48
1.4.2 Geometry, norms and correlation coefficients . . . 54
1.4.3 State density and variances of IP spreading Gaussians . . 57
1.4.4 Moments for spin cutoff densities . . . 59
1.5 Nuclear Level Densities . . . 65
1.5.1 Level densities using spherical configurations . . . 65
1.5.2 Level densities using convolution forms . . . 68

1.6 Group Symmetries and Trace Propagation 76
- 1.6.1 Introduction 77
- 1.6.2 Classification of group symmetries 78
- 1.6.3 Propagation of fixed symmetry trace matrix 79
- 1.6.4 Casimir propagation 81
- 1.6.5 Propagation with I.B. operators 84
- 1.6.6 Trace of the trace matrix 86
- 1.6.7 Fixed-J traces 87
- 1.6.8 Reduced traces 93
- 1.6.9 Regular structures with random interactions 96

1.7 Bivariate Distributions for Transition Strengths 97
- 1.7.1 Preliminaries 97
- 1.7.2 Bivariate distributions 100
- 1.7.3 NIP strength densities: Convolution form 104
- 1.7.4 Bivariate Gaussian form for NIP strength densities 108
- 1.7.5 Spherical configuration NIP strength densities 110
- 1.7.6 Unitary configuration NIP strength densities 113
- 1.7.7 IP strength densities: Convolution form 113

1.8 Applications to Atoms, Mesoscopic Systems and Interacting Bosons 120

Bibliography 123

Original Papers 145

2. Combinatorial Methods for Spectroscopic Averaging and Spectral Moments
J. B. French
Phys. Lett. **23**, *248–250 (1966)* 146

3. Measures for Group-Symmetry Breaking in Nuclei
J. B. French
Phys. Lett. B **26**, *75–77 (1967)* 149

4. Validity of Random Matrix Theories for Many-Particle Systems
J. B. French and S. S. M. Wong
Phys. Lett. B **33**, *449–452 (1970)* 152

5. Spectral Distributions in Nuclei
J. B. French and K. F. Ratcliff
Phys. Rev. C **3**, *94–117 (1971)* 156

6. Applications of Spectral Distributions in Nuclear Spectroscopy
K. F. Ratcliff
Phys. Rev. C **3**, *117–143 (1971)* 180

7. Distribution Methods for Nuclear Energies, Level Densities and Excitation Strengths
F. S. Chang, J. B. French, and T. H. Thio
Ann. Phys. (N. Y.) **66**, *137–188 (1971)* 207

8. Two-Body Random Hamiltonian and Level Density
O. Bohigas and J. Flores
Phys. Lett. B **34**, *261–263 (1971)* 259

9. Spin-Isospin SU(4) Symmetry in the (ds) Shell
J. B. French and J. C. Parikh
Phys. Lett. B **35**, *1–4 (1971)* 262

10. Angular-Momentum Dependence of the Density of States
J. N. Ginnochio
Phys. Rev. Lett. **31**, *1260–1263 (1973)* 266

11. Occupation Probabilities and Spin Cut-Off Factors
F. S. Chang and J. B. French
Phys. Lett. B **44**, *135–138 (1973)* 270

12. Recent Results on Energy-Level Fluctuations
O. Bohigas, J. Flores, J. B. French, M. J. Giannoni, P. A. Mello, and S. S. M. Wong
Phys. Rev. C **10**, *1551–1553 (1974)* 274

13. Statistical Properties of Many-Particle Spectra
K. K. Mon and J. B. French
Ann. Phys. (N. Y.) **95**, *90–111 (1975)* 277

14. Group Theory and Propagation of Operator Averages
C. Quesne
J. Math. Phys. **16**, *2427–2431 (1975)* 299

15. Strength Distributions and Statistical Spectroscopy. I. General Theory
J. P. Draayer, J. B. French, and S. S. M. Wong
Ann. Phys. (N. Y.) **106**, *472–502 (1977)* 304

16. Symmetries and Statistical Behavior in Fermion Systems
J. B. French and J. P. Draayer
Group Theoretical Methods in Physics, eds. W. Beiglbock, A. Bohm, and E. Takasugi (Springer, Berlin), 394–407 (1979) 335

17. Special Topics in Spectral Distributions
J. B. French
Moment Methods in Many Fermion Systems, eds. B. J. Dalton, S. M. Grimes, J. P. Vary, and S. A. Williams (Plenum, New York), 91–108 (1980) 349

18. Shape of Eigenvalue Distribution for Bosons in Scalar Space
V. K. B. Kota and V. Potbhare
Phys. Rev. C **21**, *2637–2642 (1980)* 367

19. Fluctuation Properties of Nuclear Energy Levels: Do Theory and Experiment Agree?
R. U. Haq, A. Pandey, and O. Bohigas
Phys. Rev. Lett. **48**, *1086–1089 (1982)* 373

20. Shell Model and Spectral Distribution Study of SU(4) Symmetry Mixing in the S-D Shell
R. U. Haq and C. Quesne
Phys. Lett. B **114**, *91–94 (1982)* 377

21. Electromagnetic Sum Rules by Spectral Distribution Methods
T. R. Halemane and J. B. French
Phys. Rev. C **25**, *2029–2058 (1982)* 381

22. Nuclear Level Densities and Partition Functions with Interactions
J. B. French and V. K. B. Kota
Phys. Rev. Lett. **51**, *2183–2186 (1983)* 411

23. Fluctuation Properties of Nuclear Energy Levels and Widths: Comparison of Theory with Experiment
O. Bohigas, R. U. Haq, and A. Pandey
Nuclear Data for Science and Technology, ed. K. H. Böckhoff (Reidel, Dordrecht), 809–813 (1983) 415

24. Characterization of Chaotic Quantum Spectra and Universality of Level Fluctuation Laws
O. Bohigas, M. J. Giannoni, and C. Schmit
Phys. Rev. Lett. **52**, *1–4 (1984)* 420

25. Bivariate Distributions in Statistical Spectroscopy Studies: Fixed-J Level Densities, Fixed-J Averages and Spin Cut-Off Factors
V. K. B. Kota
Z. Phys. A **315**, *91–98 (1984)* 424

26. Higher-Order Correlations in Spectra of Complex Systems
O. Bohigas, R. U. Haq, and A. Pandey
Phys. Rev. Lett. **54**, *1645–1648 (1985)* 432

27. Propagation Coefficients for Fixed-isospin (T, T_z) Average and Related Spectroscopic Sum Rules Masao Nomura
J. Math. Phys. **27**, *536–543 (1986)* 436

28. Symmetries and Quantum Chaos: Time-Reversal Invariance in the Nucleon-Nucleon Interaction
J. B. French, V. K. B. Kota, A. Pandey, and S. Tomsovic
Phys. Rev. Lett. **58**, *2400–2403 (1987)* 444

29. Compound-Nuclear Tests of Time Reversal Invariance in the Nucleon-Nucleon Interaction
J. B. French, A. Pandey, and J. Smith
Tests of Time Reversal Invariance in Neutron Physics, eds. N. R. Roberson, C. R. Gould, and J. D. Bowman (World Scientific, Singapore), 80–99 (1987) 448

30. Statistical Properties of Many-Particle Spectra VI. Fluctuation Bounds on N-N T-Noninvariance
J. B. French, V. K. B. Kota, A. Pandey, and S. Tomsovic
Ann. Phys. (N. Y.) **181**, *235–260 (1988)* 468

31. "Universality" of Gaussian Orthogonal Ensemble Fluctuations: the Two-Body Random Ensemble and Shell Model Spectra
G. J. H. Laberge and R. U. Haq
Can. J. Phys. **68** *301–312 (1990)* 494

32. Influence of Saturation Properties on Shell-Model Calculations
A. Abzouzi, E. Caurier, and A. P. Zuker
Phys. Rev. Lett. **66**, *1134–1137 (1991)* 506

33. Bivariate Distributions in Statistical Spectroscopy Studies: IV. Interacting Particle Gamow-Teller Strength Densities and β-decay Rates of fp-shell Nuclei for Presupernova Stars
V. K. B. Kota and D. Majumdar
Z. Phys. A **351**, *377–383 (1995)* 510

34. Application of Spectral Averaging Theory in Large Shell Model Spaces: Analysis of Level Density Data of fp-shell Nuclei
V. K. B. Kota and D. Majumdar
Nucl. Phys. A **604**, *129–162 (1996)* 517

35. Gamow-Teller Strengths in fp Shell
S. Sarkar and K. Kar
Phys. Lett. B **387**, *227–232 (1996)* 551

36. Strength Functions and Spreading Widths of Simple Shell Model Configurations
N. Frazier, B.A. Brown, and V. Zelevinsky
Phys. Rev. C **54**, *1665–1674 (1996)* 557

37. Shell-Model Tests of the Bimodal Partial State Densities in a 2×2 Partitioned Embedded Random Matrix Ensemble
V. K. B. Kota, D. Majumdar, R. Haq, and R. J. Leclair
Can. J. Phys. **77**, *893–901 (1999)* 567

38. Level Densities for Nuclei with $20 \leq A \leq 41$
P. Huang, S. M. Grimes, and T. N. Massey
Phys. Rev. C **62**, *024002/1-12 (2000)* 576

39. Statistical Theory of Parity Nonconservation in Compound Nuclei
S. Tomsovic, M. B. Johnson, A. C. Hayes, and J. D. Bowman
Phys. Rev. C **62**, *054607/1-17 (2000)* 588

40. Theory for Matrix Elements of One-body Transition Operators in the Quantum Chaotic Domain of Interacting Particle Systems
V. K. B. Kota and R. Sahu
Phys. Rev. E **62**, *3568–3571 (2000)* 605

41. Nonuniversal Behavior of the k-body Embedded Gaussian Unitary Ensemble of Random Matrices
L. Benet, T. Rupp, and H. A. Weidenmüller
Phys. Rev. Lett. **87**, *010601/1-4 (2001)* 609

42. Convergence of Moment Expansions for Expectation Values with Embedded Random Matrix Ensembles and Quantum Chaos
V. K. B. Kota
Ann. Phys. (N. Y.) **300**, *58–77 (2003)* 613

43. Spin- and Parity-Dependent Nuclear Level Densities and the Exponential Convergence Method
M. Horoi, J. Kaiser, and V. Zelevinsky
Phys. Rev. C **67**, *054309/1-7 (2003)* 633

44. Regularities with Random Interactions in Energy Centroids defined by Group Symmetries
V. K. B. Kota
Phys. Rev. C **71**, *041304(R)/1-5 (2005)* 640

45. SU(N) Wigner-Racah Algebra for the Matrix of Second Moments of Embedded Gaussian Unitary Ensemble of Random Matrices
V. K. B. Kota
J. Math. Phys. **46**, *033514/1-9 (2005)* 645

46. Underlying Symmetries of Realistic Interactions and the Nuclear Many Body Problem
K. D. Sviratcheva, J. P. Draayer, and J. P. Vary
Phys. Rev. C **73**, *034324/1-12 (2006)* 654

47. Nuclear Spectroscopy in the Chaotic Domain: Level Densities
J. B. French, S. Rab, J. F. Smith, R. U. Haq, and V. K. B. Kota
Can. J. Phys. **84**, *677–706 (2006)* 666

Chapter 1

Fluctuation-free Nuclear Spectroscopy

1.1 Introduction

We start with a typical spectrum of a heavy nucleus as shown schematically in Fig. 1.1. This complex spectrum may be broadly classified into four distinct spectral regions: (D1) ground-state domain beginning at the ground state energy and extending to about 2 MeV excitation energy; (D2) close-lying bound states in the excitation energy range of approximately (2-6) MeV; (D3) slow-neutron resonances in the energy window (6-6.002) MeV; (D4) overlapping levels for excitation energies greater than 6.002 MeV.

Region D1, extraordinarily rich in experimental data, has been studied in exquisite detail through the shell model and other microscopic models to determine ground state energies, low-lying spectra, transition strengths and the goodness of symmetries among various other quantities of interest. This conventional spectroscopy with a lot of structure and a broad domain has been very successful in the study of the low-lying states of light and intermediate nuclei. However, because of severe limitations on the size of the spectroscopic space that can be considered, the range of applicability gets limited and does not provide answers to a large class of problems. Further, because of the restricted model spaces defined by the single particle (sp) orbits and applicable exact symmetries, it is generally regarded that such detailed spectroscopy could only be valid at relatively low energies or close to the yrast line.

Very little is known about region D2 which may contain an extremely large number of levels. With increasing excitation energy, the level spacing decreases to the extent that it becomes more or less impossible to distinguish all the levels up to 6 MeV in a heavy nucleus. However, starting at the excitation energy at which neutron decay becomes feasible and extending over a small range (a few keV), it becomes possible to identify individual levels by bombardment with slow neutrons. The resonances, resulting from the strong reflection of the slow neutrons at the surface by the deep and narrow nuclear potential well are sharp because of the spreading of the target plus neutron giant-resonance state over a large number of eigenstates. In this region D3, neutron resonance spectroscopy has led to the

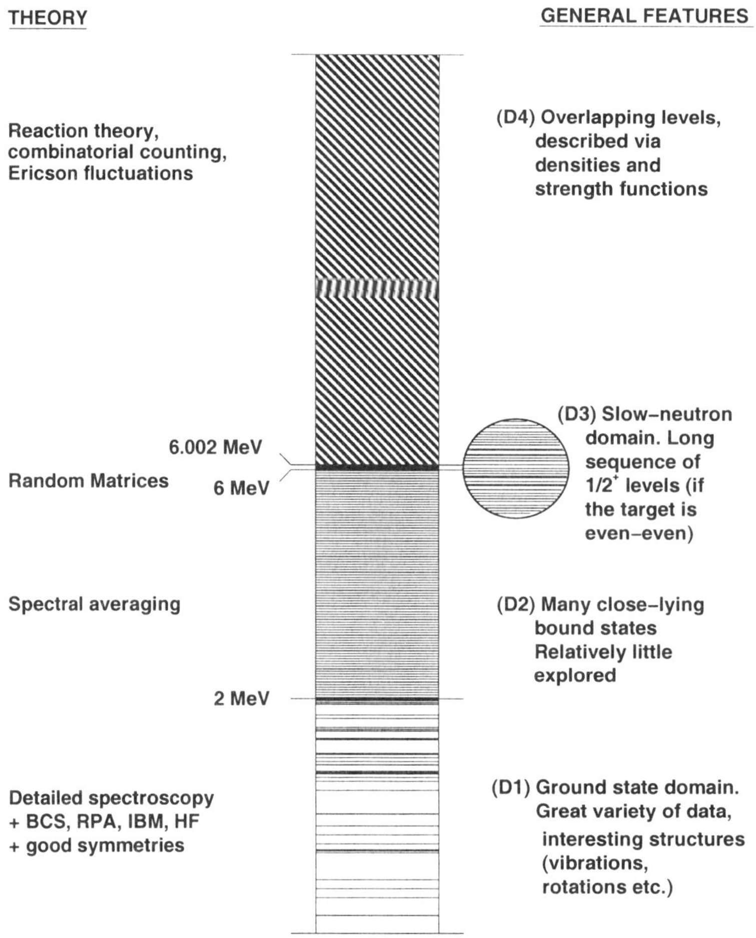

Fig. 1.1 Typical spectrum of a heavy nucleus such as ^{169}Er. Figure adapted from [French and Kota (1989b)].

measurement of complete spectra containing hundreds of individually resolved and measured neutron resonances. Further, advances in charged particle spectroscopy have made possible the measurement of similar spectra for proton resonances.

The importance of D3 has been emphasized by Bohr and Mottelson (1973): *However, the whole development of nuclear physics has been decisively influenced by the existence of a small window, in the region of the neutron binding energy, within which the slow neutron reactions provide a probe of enormously greater resolving power. Already the earliest experiments with slow neutrons revealed, very unexpectedly, a dense spectrum of resonances. This discovery led to the strong coupling between the motion of the incident neutron and many degrees of freedom of the target. The coupling gives rise to the formation of a compound system with a lifetime very long compared with the one-particle periods.*

For a very interesting account of the ideas surrounding the compound nucleus [Bohr (1936)], we refer the reader to Mottelson (1985): *It was Niels Bohr who saw first and most clearly that these experimental discoveries concerning the interactions of neutrons with nuclei demanded a radical revision in the basic picture of nuclear dynamics. He recognized that the assumed single particle motion, copied from atomic physics, was being falsified and he suggested in its stead an idealization which focused on the many body features and the strong coupling of all the different degrees of freedom of the nuclear system - the compound nucleus. ... The study of radioactivity had shown that the lowest excited states in heavy nuclei of the order of a fraction of 1 MeV and Bohr assumed that these excitations represent some sort of collective vibration of the whole nucleus. With increasing excitation energy an increasing number of different vibrational modes can be excited and the different possibilities for partitioning the total excitation energy between the different modes leads to an enormous increase in the total number of excited states. All of these quantum states can be resonantly excited by an incoming neutron thus accounting for the dense spacing and the narrowness of the levels observed in the selective absorption phenomena ... the level scheme above and below the neutron separation energy (~ 10 MeV) are not significantly different .... Only at higher energies will the neutron emission probability contribute appreciably to the width of the individual levels and lead eventually to a smearing out of the spectrum.* Therefore, at excitation energies beyond the neutron resonance energy, the strong reflection weakens and the resonances broaden resulting in overlapping levels in region D4.

The characteristic nature of the different regions has largely defined the quantities of interest studied. In the ground-state region because of the importance of the detailed nature of the Hamiltonian, one has not been interested in averaged properties such as level densities. In the neutron-threshold region and at higher excitation energies, the study of averages and fluctuations about them has been of major interest. Further, the mathematical approaches and the underlying assumptions vary greatly from region to region. For example, we go from a detailed Hamiltonian in D1, to random matrices in D3, to using just sp energies in D4. The disconcerting feature is that there has been little overlap between the assumptions made and the methods used in the various regions. For example, the parameters for high-excitation theory have not been evaluated in terms of more fundamental quantities and conversely, the great body of level density and similar data has not been used to study or test the effective interaction. In fact there has been relatively little interaction between those involved in the ground-state region D1 and those interested in the higher energy regions D2, D3 and D4.

Statistical spectroscopy, based on statistical laws operating in model spaces, unifies the very different approaches used in the ground state and higher energy regions and makes clear the connection between the domains arising from the effective nuclear interaction. It works in the model spaces of conventional spectroscopy but because it does not deal with the construction and diagonalization of Hamiltonian matrices, it is not constrained by the dimensionality of these spaces. Further, the statistical methods are applicable in a very wide range of circumstances. The aim is to deal with general features of complex nuclei keeping in mind that the statistical behavior observed at high excitations extends right down to the ground state domain. It then seems appropriate to use the same methods for all parts of the spectrum and indeed over the entire periodic table.

1.1.1 *Spectral distributions and fluctuations*

The origins of statistical nuclear theory can perhaps be traced back to Bethe's [Bethe (1936)] derivation of level density. His calculation based on statistical mechanics of essentially non-interacting particles (NIP) in an unbound sp energy spectrum is an example of a "true" spectral distribution. For a general discussion see Bloch (1969).

We refer to the discussion in [Bohr and Mottelson (1969)] for Bethe's derivation of NIP state density. An alternate derivation follows from the recognition that it is essentially the Hardy-Ramanujan partition formula [Hardy and Ramanujan (1918)]. Here we reproduce this derivation following French (1980b). Except at extremely high energies we can consider the NIP system as an almost degenerate Fermi gas and therefore the many particle level density depends only on the sp density at the Fermi surface. Now replacing the true sp spectrum by a uniform spectrum with unit spacing [later we will renormalize the energies E by the inverse of the level spacing of the sp spectrum at the Fermi surface], the calculation reduces to combinatorial counting [Van Lier and Uhlenbeck (1937)]. The sp levels are denoted by 1, 2, ... and for each distribution of the given number of particles, the occupied orbits are denoted by s. Then the m particle (identical fermions) density $g_m(E^*)$ at excitation energy E^* is [French (1980b)],

$$\begin{aligned} g_m(E^*) &= \text{no of solutions of } E = \sum_{occ} s\,; \quad m = \sum_{occ} 1 \\ &= \text{no of partitions of } E \text{ into } m \text{ unequal parts} \\ &= \text{no of partitions of } E^* \text{ into } \leq m \text{ parts} \\ &= \text{no of partitions of } E^* \\ &\xrightarrow{E^*>1} \frac{1}{\sqrt{48}\,E^*} \exp\left[2\sqrt{\frac{\pi^2 E^*}{6}}\right] . \end{aligned} \tag{1.1}$$

In the final step use is made of the asymptotic Hardy-Ramanujan partition formula. If we change the sp level spacing to $[g(\epsilon_F)]^{-1}$, we have $a = \pi^2 g(\epsilon_F)/6$ the Bethe parameter. Then we have, denoting E^* by E, the well known Bethe form $(1/\sqrt{48}\,E)\exp(2\sqrt{aE})$. For proton-neutron (p,n) systems, the density follows by convolution of the proton and neutron densities each represented by the Bethe form. By adding to this the back shifting parameter (essentially due to pairing interaction), we recover Eq. (1.117) given ahead. It is important to note that the significance of adding oscillatory corrections to the Bethe form, first worked out by Rademacher (1938), has been only recently recognized [Leboeuf and Roccia (2006)]. Note that the Bethe density is a state density and the J decomposition, giving level densities, follows by applying locally the Central Limit Theorem (CLT) for the J_z operator; see [Bethe (1937)] and Sec. 1.6.7. Although Bethe density is for NIP systems, treating a as a free parameter and taking into account, phenomenologically or otherwise, the interaction effects through various modification, it is applied to data analysis; see [Goriely, Hilaire and Koning (2008)] for most recent results.

Clearly then, the starting point in statistical spectroscopy is the density of states $\rho(E)$ arising from a Hamiltonian H acting in a spectroscopic space of m particles (nucleons) distributed in N sp states. This density may be regarded as composed of two distinct parts: an average or smooth density and fluctuations around this average. We illustrate with a shell model example of ^{24}Mg considered as 8 nucleons in the $(2s1d)$-shell with $J = 2, T = 0$. The dimensionality of the space is 1206 and the eigenvalue distribution is shown in Fig. 1.2. The exact distribution is well approximated by a Gaussian indicating that the smooth (average) density has a simple characteristic shape describable in terms of a few parameters (dimensionality, centroid, variance and perhaps one or two shape parameters for added accuracy). Further, departures from the smooth density are small indicating a rigidity extending over the entire spectrum.
The exact density may then be written as

$$\rho(E) = \rho_s(E) + \rho_f(E) \tag{1.2}$$

where $\rho_s(E)$ and $\rho_f(E)$ refer to the smooth and fluctuating parts respectively. The basic questions are:

(1) Is there a clear separation between the average behavior and fluctuations? This separation would allow treating the two phenomena by different methods, spectral distributions for the average behavior and random matrix ensembles for the fluctuations.
(2) What is the nature and magnitude of fluctuations? The magnitude provides an estimate of the irreducible errors (limitations) implicit in spectral distributions and the nature of the fluctuations is linked to the amount of information carried by them. Other important questions are: (i) how do the calculated measures used to define fluctuations agree with experiment; (ii) are the fluctuations universal and if so what is the origin of the universality; (iii) what mechanisms affect fluctuations and what limits may be imposed by agreement with data?

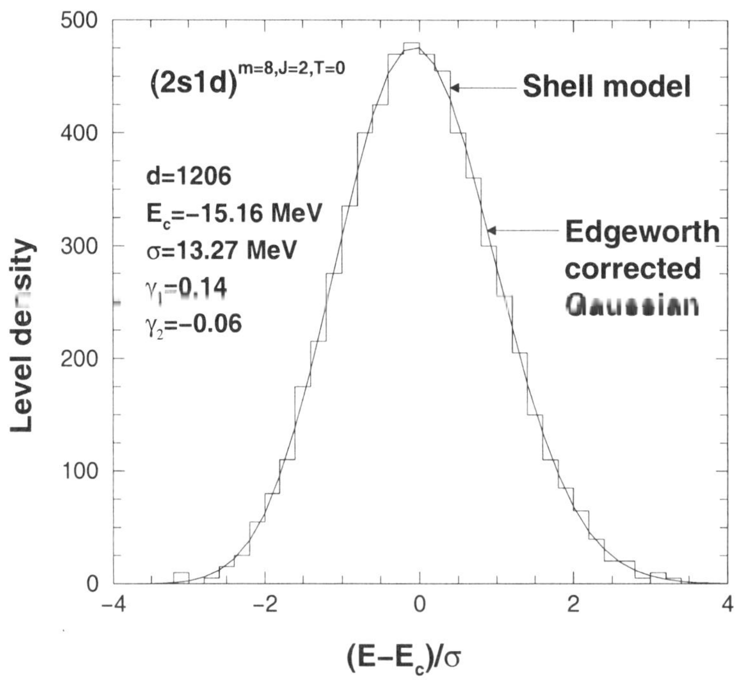

Fig. 1.2 Eigenvalue density for $(2s1d)^{m=8,J=2,T=0}$ shell model example. The exact shell model distribution (histogram) is well approximated by an Edgeworth corrected Gaussian [Eq. (1.17)] with the shape parameters skewness (γ_1) and excess (γ_2). Values of the matrix dimension (d), energy centroid (E_c), spectrum width (σ) and γ_1 and γ_2 are shown. The distributions are normalized to the matrix dimension. The eigenvalues are from [Soyeur and Zuker (1972); Leclair (1994)].

Comparing Bethe density with the shell model Gaussian densities, firstly it should be recognized that the Gaussian densities are indeed only partial densities. Therefore Bethe density follows by adding Gaussian densities appropriately. Statistical spectroscopy clarifies this as discussed in Sec. 1.5.

In spectral averaging we deal with a given Hamiltonian H and rely (derived using random matrix theory or otherwise) on the operation of CLTs to generate local energy averages. The aim is to calculate for H and other operators of the system, fluctuation-free values for quantities of physical interest such as densities or partition functions, expectation values of operators in physical states, transition strengths, etc. These emerge as explicit functions of the defining matrix elements and determine much of the statistical behavior of the system. Spectral distributions draw on the techniques of conventional spectroscopy, make connections with statistical methods and thus bring out new simplicities in even very complex systems.

1.1.2 *Average-Fluctuation separation*

The separation of information into two distinct parts, averages and fluctuations provides a physical basis for statistical spectroscopy of finite quantum systems with interactions such as nuclei, atoms and molecules. The decoupling arises due to the action of CLTs. It has been shown by Mon and French (1975*)[1] that the eigenvalue density generated by a 2-body Hamiltonian in a many-body space gets smoother as more particles are added to the system and converges to a Gaussian distribution. The density is then describable in terms of low-order moments of the Hamiltonian, defined by traces of powers of H, and all information contained in higher-order moments, being of little consequence, is washed away by the CLT. This CLT smoothing action does not affect the spectral fluctuations when renormalized according to the local density since as particle number increases, the local spacing must rapidly decrease.

If we represent the exact density in terms of excitations built upon a specific shape, a normal mode decomposition, a sharp separation implies that in the power spectrum we have: (i) a few excitations with wavelengths comparable to the spectrum span that take care of the secular variation, (ii) no excitations of intermediate wavelengths because of the CLT and (iii) short wavelength excitations of the order of the mean spacing responsible for fluctuations. This is indeed seen in shell model examples [French (1980c*); Brody *et al.* (1981); Leclair, Haq, Kota and Chavda (2008)] and in random matrix ensembles, generated by random two-body interactions, for many particle (fermions or bosons) systems [Laberge and Haq (1990*); Patel, Desai, Potbhare and Kota (2000); Leclair, Haq, Kota and Chavda (2008)]. A shell model example demonstrating average-fluctuation separation is shown in Fig. 1.3 (a first attempt in this direction is due to Chang and Zuker (1972)).

Consider a simple system of m interacting spinless fermions in N single particle states with $m << N$. An ensemble of random two-body interactions then generates the many particle ensemble averaged eigenvalue density to be a Gaussian (see Fig. 1.4 ahead) just as in the shell model example shown in Fig. 1.2. Then, with bar denoting ensemble average, $\overline{\rho(E)} = \rho_{\mathcal{G}}(E)$. It is possible to expand $\rho(E)$ in terms of $\rho_{\mathcal{G}}(E)$ and the orthonormal polynomials [Hermite polynomials He_ζ] defined by it. The expansion coefficients S_ζ are, in principle, related to higher moments of the state density $\rho(E)$ and each ζ term represents an excitation 'mode', the wavelength of the modes being proportional to ζ^{-1}. The S_ζ change from member to member of the ensemble and generate fluctuations. The distribution function $F(E)$, the integrated version of $\rho(E)$, is given by $F(E) = d\int_{-\infty}^{E} \rho(E')dE'$ where d is the dimensionality. Deviation of a given level with energy E from its smoothed (with respect to the ensemble) counterpart $\overline{E}$ gives the level motion. In terms of $F(E)$ and the local mean spacing $\overline{D(E)}$, we have $\delta E = E - \overline{E} = \left[F(E) - \overline{F(E)}\right] \overline{D(E)}$. Then the variance of the level motion is given by the ensemble average of $\frac{(\delta E)^2}{\overline{D(E)}^2}$ and the expansion for it in the dilute limit defined by $m \rightarrow \infty$, $N \rightarrow \infty$ and

[1]references with asterisk indicate that these papers have been included in the book

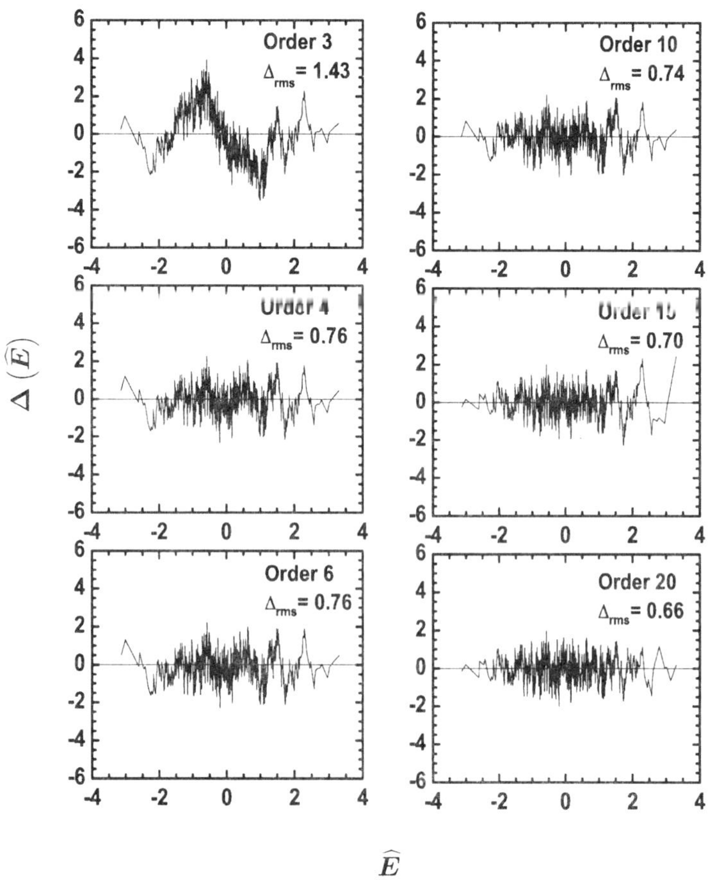

Fig. 1.3 Level motion $\Delta(\widehat{E})$ vs $\widehat{E}$ for ^{24}Mg shell model example (see Fig. 1.2) with 8 nucleons in the $(2s1d)$ shell with angular momentum $J = 2$ and isospin $T = 0$. The matrix dimension is 1206. The $\Delta(E) = F(E) - \overline{F(E)}$ for a given order ζ_0 is obtained by using the Gram-Charlier expansion $\overline{\rho(E)} = \rho_{\mathcal{G}}(E)\ \{1 + \sum_{\zeta\geq 3}^{\zeta_0} (\zeta!)^{-1}\ S_\zeta\ He_\zeta(\widehat{E})\}$ for the smooth density of states. As the spectrum is discrete, S_ζ are determined, for a given ζ_0, by minimizing $\sum_{i=1}^{d}\{\Delta(E_i)\}^2$ with respect to S_ζ. These optimized S_ζ, for a given ζ_0 order, give $\Delta(E)$ vs E, i.e. level motion. Note that $\widehat{E}$, the standardized E and $\Delta(\widehat{E})$ have no units. See [Leclair, Haq, Kota and Chavda (2008); Leclair (1994)] for details.

$m/N \to 0$ is

$$
\begin{aligned}
\frac{\overline{(\delta E)^2}}{\overline{D(E)}^2} &= \binom{N}{m}^2 \binom{m}{2}^2 [\rho_{\mathcal{G}}(E)]^2 \\
&\quad \times \left\{ \sum_{\zeta\geq 1} (\zeta!)^{-2}\, 2\zeta \binom{m}{2}^{2-\zeta} \binom{N}{2}^{-2} \left[He_{\zeta-1}(\widehat{E})\right]^2 \right\} \\
&\xrightarrow{\hat{E}=0} \frac{1}{\pi} \binom{N}{m}^2 \binom{m}{2} \binom{N}{2}^{-2} \\
&\quad \times \left\{ 1 + \frac{1}{12}\binom{m}{2}^{-2} + \frac{1}{320}\binom{m}{2}^{-4} + \ldots \right\} .
\end{aligned} \tag{1.3}
$$

Here $\widehat{E}$ is the standardized E (zero centered with unit variance). Thus, as ζ increases, deviations in $\overline{(\delta E)^2}$ from the leading term rapidly go to zero due to the $\binom{m}{2}^{-2r}$, $r = 1, 2, \ldots$ terms in Eq. (1.3). There will be no change until $\zeta \sim m/2$ thereby defining separation. Beyond this, as argued by French [Mon and French (1975*); Brody *et al.* (1981)], for $\zeta >> m/2$ for fermions the deviations grow, i.e. fluctuations set in and they will tend to that of GOE. All this is well confirmed by the shell model results in Fig. 1.3. Details of the derivation of Eq. (1.3) are given in [Mon and French (1975*); Brody *et al.* (1981)]

1.1.3 *Random matrix ensembles*

Fluctuations deal with deviations from local uniformity. As a model for describing fluctuations, the Gaussian Orthogonal Ensemble (GOE) of random matrices introduced by Wigner [Wigner (1955); Porter (1965)] has been very successful (along with the Gaussian Unitary Ensemble GUE and Gaussian Symplectic Ensemble GSE). The aim here is not to calculate individual fluctuations but to understand the general nature of the fluctuation patterns and to calculate physically relevant statistical measures such as spacing distributions, number variances, spectral correlation functions etc.

The GOE corresponds to an ensemble of asymptotically large real symmetric matrices which apart from rotational and time reversal invariance have no other specific features. The GOE matrix for a many-particle system describes simultaneous interactions between all particles because of the statistical independence of the matrix elements. Realistic Hamiltonians are, in general, two-body in nature. To keep the generality of GOE on the one hand and to conform to reasonable Hamiltonians, such as those used in shell model calculations, on the other, statistical extensions of the shell model have been proposed. A matrix ensemble of random two-body Hamiltonians, with shell model angular momentum J (and isospin T) symmetry, called the two-body random ensemble (TBRE) was first introduced by French and Wong (1970*, 1971) and Bohigas and Flores (1971a*,b). More generically, as stated by French [French (1980c*)]: *For a better a priori model we can choose an ensemble of k-body interactions ($k = 2$ is an interesting case) by generating a GOE in k-particle space and using it in the space of m-particles. For most purposes the resulting embedded GOE (or EGOE) is very difficult to deal with but by good luck, we can use it to study the questions we have posed and the answers are different from, and much more enlightening than, those which would come from GOE.* Denoting the rank of the interaction by k, in an m-particle space for spinless fermion systems, we have GOE for $m = k$ and the embedded GOE, EGOE(k) for $m >> k$ [French (1973)]. For $k = 2$, $m >> 2$ and H preserving the shell model J symmetry we have TBRE or EGOE(2)-J. (similarly with JT symmetry). The ensemble averaged level density for GOE is semi-circular while for TBRE or EGOE(2) it is close to Gaussian. The reasons which have to do with the rank of the Hamiltonian (for m particles GOE implies that the rank of H is m) have been established in [Wong and French (1972); Gervois (1972); Mon and French (1975*); Benet, Rupp and Weidenmüller (2001a*)]. A numerical example showing semi-circle to Gaussian transition is given in Fig. 1.4.

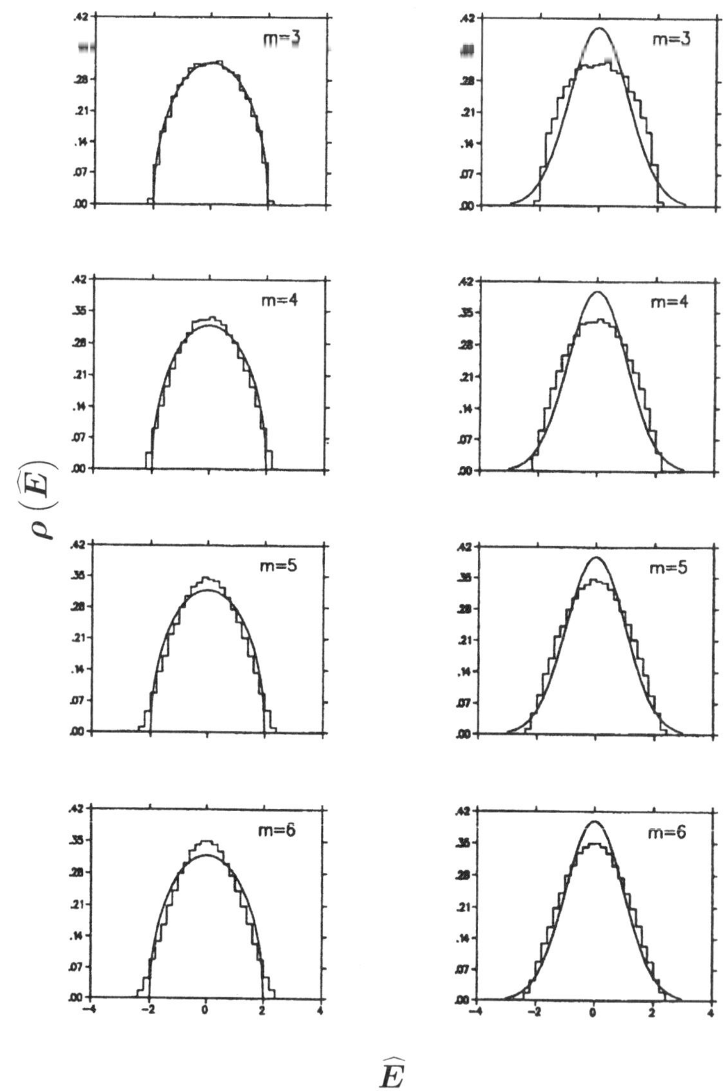

Fig. 1.4 Eigenvalue densities (histograms) for a 20 member two-body ensemble for spinless fermions with the number (m) of fermions changing from $m = 4$ to 6. Each density is given twice, compared on the left with a semi-circle, and on the right with a Gaussian. Given the eigenvalues E, the x-axis is the standardized variable ($\widehat{E}$) and the y-axis is the eigenvalue density normalized to unity. We see a clear transition from semi-circle to Gaussian as m increases. Figure taken from [Leclair (1994)]. Similar results with m fixed (say $m = 6$) and the rank k of the interaction varied (from $2 - 6$) exhibiting a Gaussian to semi-circle transition are given in [Brody *et al.* (1981); Leclair (1994)].

It should be clear that for level densities, GOE is an unphysical ensemble. The state density near the ground state as given by the semi-circle has the wrong curvature as compared to the experimentally observed density. For a historical perspective, it is worthwhile recalling the interesting discussion regarding the differences between GOE and EGOE at the Albany conference [Garg (1972)]. Commenting on the work by French and Wong and Bohigas and Flores, Wigner in his introductory talk stated: *The principal difference between the model considered by these authors, and the Wishart model considered earlier, is that a great many matrix elements of the new model's Hamiltonian are 0. These include the matrix elements which connect independent particle states which differ in more than two orbits. If I understood them right, French and Wong mention a case in which less than 10 percent of all the matrix elements of a 50 by 50 matrix are different from 0. In the usual ensemble, that is the Wishart ensemble, the probability of the vanishing, or even approximate vanishing, of so many matrix elements is practically zero. ... The first conclusion of the new theory is that the density of the levels, as function of the energy, is a Gaussian rather than the semicircle characteristic of the Wishart and related ensembles.* In the discussion following the talk French says: *Perhaps I should make something clear about the zeros in matrices. It is indeed true that if one uses a picture or a representation in space which is simple, a Hartree-Fock kind of thing, of describing it in configurations. Then when you put in two body interactions, you do get a fair number of zeros. I don't think you get as many as has been suggested. But, that we regard as not the real essence of the thing, that is done. The real essence instead, is that constraints on all the matrix elements are imposed by the condition that we have a two-body interaction. And we have, for example, verified that if you take large matrices and put them and think of them as being in block form and impose only the zeros corresponding to the fact that you cannot move two particles from here to there, that is much more likely to give you a semicircular form than to have a Gaussian form. If now, however you impose a further condition that the matrix elements have to satisfy a large number of constraints then the distribution becomes Gaussian. In other words, I don't think that one should put too much stress on inserting zeros. That doesn't really convert you from semicircular to Gaussian.*

The transition from semi-circle to Gaussian in EGOE(k) has been studied mathematically in great detail by Mon and French (1975*) and more recently by Benet, Rupp and Weidenmüller (2001a*,b). The Mon and French (1975*) method is based on evaluating the ensemble averaged moments of the Hamiltonian in the two limits $m = k$ and $m >> k$. By choosing matrix elements from a zero centered distribution, all odd-moments vanish. For even-order moments, the binary (pairwise) correlations dominate, higher order correlations being smaller by a factor $1/N$ (N is number of sp states) and go to zero in the limit as N goes to infinity. The binary correlations are of two kinds, linked and unlinked. For $m >> k$ all terms contribute while for $m = k$, the linked correlations go to zero. This then gives either the moments of a Gaussian or a semicircle. Going beyond this, Benet, Rupp and Weidenmüller (2001b) proved that the semi-circle to the Gaussian transition point is $m = 2k$.

Let us remark that EGOEs with group symmetries provide a complete statistical description, including both spectral averages and fluctuations, of interacting finite many particle systems such as nuclei. However, if the focus is on local fluctuations in a given spectrum, GOE is sufficient. The EGOEs generate forms for spectral distributions for various observables such as density of states, occupancies, transition strengths, strength sums and so on, although in practice we need to apply corrections to the EGOE forms. More significantly, EGOEs generate correlations between many particle states with different quantum numbers including particle number (spectra of different nuclei and or with different J or JT values) and these cross correlations will be zero in a GOE description [Papenbrock and Weidenmüller (2006, 2007); Kota, Chavda and Sahu (2006b); Kota (2006a); Vyas and Kota (2009)]. Experimental tests of this important feature are not yet available. For details of EGOEs with group symmetries, we refer the readers to the reviews [Brody *et al.* (1981); Kota (2001); Benet and Weidenmüller (2003); Papenbrock and Weidenmüller (2007); Vyas, Kota and Chavda (2010)] and do not discuss these any further since our focus here is on spectral averages.

1.1.4 *Fluctuations and quantum chaos*

The theoretical framework for studying fluctuations is provided by random matrix theories through appropriate ensembles introduced by Wigner and others. Fluctuation measures are derived from the set of k-order correlation functions by averaging over the Hamiltonian ensemble [Dyson (1970); Mehta (1971)]. These ensembles have been shown to have proper ergodic behavior [Pandey (1979)] so that the results of ensemble averaging apply to individual spectra. The 2-point fluctuations [French, Mello and Pandey (1978)] are dominated by short range von Neumann-Wigner level repulsion [von Neumann and Wigner (1929)] and Dyson-Mehta long-range order [Dyson and Mehta (1963)]. Similarly, the strength fluctuations follow the Porter-Thomas distribution [Porter and Thomas (1956)]. The best experimental evidence of GOE fluctuations has been provided by nuclear data based on slow neutron resonances in heavy nuclei and proton resonances in intermediate nuclei [Lynn (1968); Mughabghab (1984); Bilpuch, Lane, Mitchell and Moses (1976)]. The essential requirement of the theory is to have a complete set of levels with the same quantum numbers (spin, parity, etc.) thereby ensuring that there are no missing or spurious (those having different quantum numbers) levels. By combining all the available high quality data into a nuclear data ensemble (NDE) and introducing new spectral measures, Haq, Pandey and Bohigas (1982*); Bohigas, Haq and Pandey (1983*, 1985*); Lombardi, Bohigas and Seligman (1994) have found remarkably close agreement between the predictions of random matrix theory and experiment confirming

Wigner's suggestion [Wigner (1961)] that: *the Hamiltonian which governs the behavior of a complicated system is a random symmetric matrix with no particular properties except for its symmetric nature.*

Besides the nucleus, evidence of GOE fluctuations comes from a variety of quantum systems and beyond [Haake (2001); Mehta (2004)]. The same fluctuations are found in shell model spectra and also in many numerical EGOE spectra; see Fig. 1.5. However, there are open questions about the fluctuation properties and ergodicity of EGOEs [Flores, Horoi, Müller and Seligman (2001)] and in fact a major gap is that the two-point function is not yet available even for the EGOE(2) for spinless fermion systems [Mon and French (1975*); Benet, Rupp and Weidenmüller (2001b); Srednicki (2002); Pluhar, Weidenmüller, Papenbrock and Tithof (2009)]. It is worthwhile to recall that historically during and after the Albany conference (1971), confusion

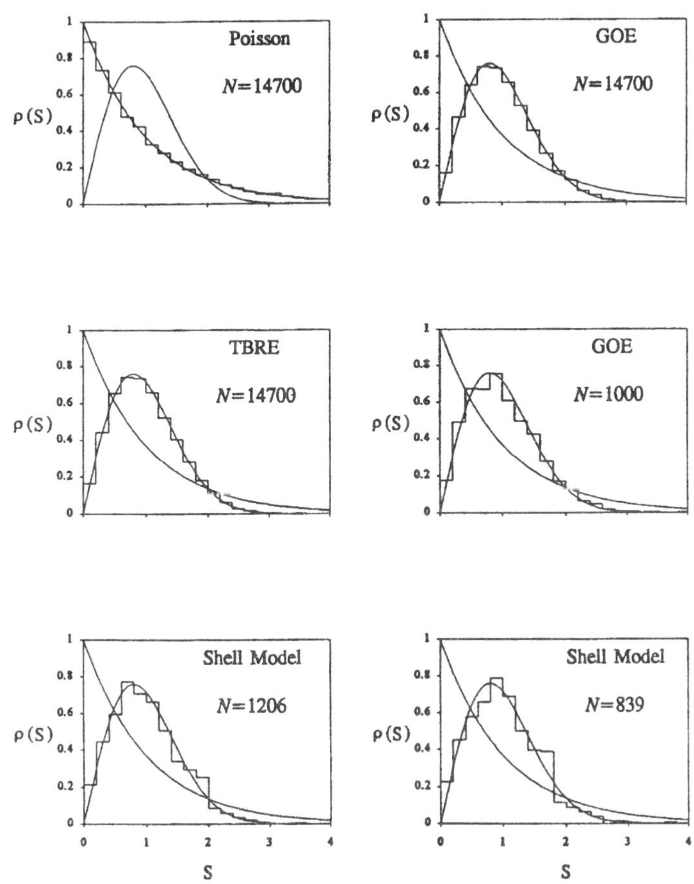

Fig. 1.5 Nearest-neighbor spacing distributions $P(S)$ (histograms) for two shell model examples with matrix dimensions 1206 and 839, one GOE matrix with dimension 1000 and three 50 member ensembles (GOE, TBRE and Poisson) each with dimension 294. The total number (N) of eigenvalues for each example is indicated in the figure. Each example is compared to Poisson ($\exp -S$) and Wigner ($[\pi S/2] \exp -\pi S^2/4$) spacing distributions (smooth curves). Figure taken from [Leclair (1994)].

arose regarding the possibility of distinguishing between GOE and TBRE through the study of spacing distributions. This was based on some preliminary work. However, it was established soon thereafter [Bohigas *et al.* (1974*); French (1973)] that both GOE and TBRE give essentially the same spacing distributions and are not dependent on the rank k of the interaction. Further, with a view to understand why fluctuations are unaffected by specific features of the Hamiltonian describing various physical systems, Pandey (1981) has shown that the perturbation of a non-random matrix corresponding to a given Hamiltonian by a random GOE matrix leads very quickly to GOE fluctuations. Also, there exist arguments that two-body ensembles in general contain local GOE structure giving rise to GOE fluctuations [French, Kota, Pandey and Tomsovic (1988a); Papenbrock and Weidenmüller (2005)].

Fluctuation patterns are stable under a wide class of changes in the system Hamiltonian. But under the gradual breaking of a good symmetry they change from one pattern to another. Of fundamental interest is the breaking of time-reversal invariance (TRI) under which the GOE pattern changes to the more rigid GUE, the model for time-reversal non-invariance (TRNI). The problem then is to determine from data a value or an upper bound to the TRNI nucleon-nucleon interaction. The first step is to calculate the GOE to GUE transition curve for the fluctuation measures and thereby determine the RMS value of the symmetry breaking matrix element in the complex system. The final step is to determine from this the magnitude of the V(TRNI) part of the nucleon-nucleon interaction. This analysis has been done in detail by French, Kota, Pandey and Tomsovic (1985, 1987*, 1988b*); French, Pandey and Smith (1987*). A similar study has been carried out for parity breaking by Tomsovic, Johnson, Hayes and Bowman (2000*).

The fact that such a wide variety of systems show the same fluctuation patters as those of a parameter-free theory suggests the existence of a universal law. The belief in the universality of GOE fluctuations has been strengthened by the connection between chaos in classical systems and fluctuation properties of their quantum analogs. This led Bohigas, Giannoni and Schmit (1984*) to conjecture that: *the fluctuation properties of generic quantum systems with (without) time-reversal symmetry, which in the classical limit are fully chaotic, coincide with those of GOE (GUE).* This link between fluctuations and chaos has provided a much deeper understanding of the fluctuations specially for the nucleus [Seligman and Nishioka (1986); Bohigas and Weidenmüller (1988); Weidenmüller and Mitchell (2009)] and has also led to a study of the relationship between chaos and statistical spectroscopy [Zelevinsky, Brown, Frazier and Horoi (1996); Flambaum, Gribakina, Gribakin and Ponomarev (1999); Kota (2001)]. As Papenbrock and Weidenmüller (2007) state: *We speak of chaos in quantum systems if the statistical properties of the eigenvalue spectrum coincide with predictions of random-matrix theory. Chaos is a typical feature of atomic nuclei and other self bound Fermi systems.* Similarly Berry and Tabor (1977) conjectured that, with certain exceptions, completely integrable systems should lead to Poisson fluctuations which are much larger than GOE fluctuations. All this is well summarized by Altshuler in the abstract of the colloquium he gave in memory of J.B. French in Rochester in 2004: *Classical dynamical systems can be separated into two classes - integrable and chaotic. For quantum systems this distinction manifests itself, e.g. in spectral statistics. Roughly speaking integrability leads to Poisson distribution for the energies while chaos implies Wigner-Dyson statistics of levels, which are characteristic for the ensemble of random matrices ... the onset of chaotic behavior for a rather broad class of systems can be understood as a delocalization in the space of quantum numbers that characterize the original integrable system.* It thus appears that chaos ensures that the description of observables in terms of only the smooth component (spectral average) is adequate. It is also clear that apart from the symmetries implicit in the definition of the relevant ensemble, fluctuations carry little or no information about the system, the exceptional case being that of weakly-broken symmetries such as time-reversal, parity and isospin.

1.1.5 *Laws of statistical spectroscopy*

Statistical spectroscopy deals with spectroscopy in terms of the complete set of correlation functions of various orders (k-point functions). The 1-point function defines the state density and by parametric differentiation or otherwise we obtain occupancies, spin cutoff factors and expectation values more generally. There is a natural extension to 2-point correlation functions which give a theory for transition strengths and symmetry breaking (time-reversal, parity, isospin, etc.). Moreover, the ($k > 2$)-point functions carry essentially no information so that we have an economical structure.

We end this Section with the five general laws [French and Kota (1982)]:

(1) There is, in the model space, *a macroscopic simplicity* deriving from the action of the CLT. The smoothed eigenvalue density is close to a characteristic form, usually Gaussian, describable therefore in terms of a small number of low-order Hamiltonian traces (moments).

(2) There is a *microscopic simplicity* corresponding to a remarkable spectral rigidity [Dyson and Mehta (1963)], which extends over the whole spectrum and ensures that the fluctuations are small and for the most part carry little information.

(3) There is indeed, as implied above, usually a *sharp separation* between the secular behavior of the spectrum and its fluctuations, so that the two can be treated separately and by different methods. This separation also arises from CLT action.

(4) There is a *propagation of information* (i.e. of traces) throughout the set of model subspaces defined by N (the number of single-particle states) and the symmetries that label the subspaces. This enables us to express either exactly or approximately, depending upon the symmetries involved, the many-particle traces as linear combinations of the few-particle input traces.

(5) The ensembles that one uses have a *strong ergodic behavior.*

Similar laws apply to expectation values of operators and transition strength distributions, i.e. for all spectroscopic observables generating spectral distribution theory.

Statistical nuclear spectroscopy was initiated by French during 1966-67 with a series of five papers on trace propagation [Banerjee and French (1966); French (1966a*, 1967b*); Chang, French and Ratcliff (1966); French and Hsu (1967)]. It should be mentioned that early papers dealing with trace propagation by Moszkowski (1962); Layzer (1963); Pasachoff (1968) were in the context of atomic physics. A first detailed account of spectral distributions in nuclei was given by French and Ratcliff (1971*); Ratcliff (1971*); Chang, French and Thio (1971*). In addition, there are articles by French (1967a, 1969, 1972, 1974, 1975); French and Draayer (1979*); French (1980a,b,c, 1983a,b, 1984) published in various conference proceedings describing different aspects of statistical spectroscopy. There are also three review articles [French and Kota (1982); Manfredi (1986); Kota and Kar (1989)] and books [Parikh (1978); Wong (1986); Karazija (1991); Koltun and Das (1995)] on the subject.

We now give a preview. In Sec. 1.2 we discuss spectral distributions and the operation of central limit theorems in many-particle spaces for both non-interacting and interacting particle systems. This leads naturally to the distributions of expectation values as given in Sec. 1.3. In Sec. 1.4 we discuss the significance of the unitary group $U(N)$ and its various subgroups in a space defined by N sp states. The unitary decomposition of the Hamiltonian clarifies the role of the various parts of H and leads to expressions for state and spin cutoff densities. Nuclear level densities are discussed in Sec. 1.5. The importance and consequences of group symmetries is introduced in Sec. 1.6 and the general principles of trace propagation discussed. Distributions for transition strengths and their applications to symmetry breaking are dealt with in Sec. 1.7. Finally, in Sec. 1.8, we discuss briefly applications of spectral distributions and EGOE for atoms, mesoscopic systems and interacting bosons.

1.2 Statistical Methods in Finite-dimensional Model Spaces

We begin with those aspects of distributions which do not depend on the nature and structure of the model space though we shall often arrange things in a way best adapted to the many particle spaces which we come to later.

1.2.1 *Moments and distributions*

Given that various operators G (one of which is the Hamiltonian H) are defined in the m particle model space; it implies that, for $\psi_\alpha \in m$, $G\psi_\alpha$ is also a vector in m though this relationship does not necessarily hold for subspaces of m. The eigenvectors of H are denoted by $\psi_{r,i}$ where $i = 1, 2, \ldots, g_r$ distinguishes between degenerate states. The eigenvalues are E_r with $E_r < E_{r+1}$ and $r = 1, 2, \ldots, l$, and the model space dimensionality is $d = \sum g_r$. The microscopic state density $I(x)$, its normalized counter part $\rho(x)$ and the distribution function $F(x)$ are:

$$\begin{aligned} I^H(x) \Leftrightarrow I(x) = d\rho(x) &= \sum_{r=1}^{l} g_r \delta(x - E_r) = \sum_{r,i} \langle r\, i \,|\delta(H-x)|\, r\, i\rangle \\ &= \langle\langle \delta(H-x)\rangle\rangle^m = d\,\langle \delta(H-x)\rangle^m \ , \\ \int_{-\infty}^{+\infty} I(x)dx = d\ ; &\quad \int_{-\infty}^{+\infty} \rho(x)dx = 1\ , \\ F(x) = \int_{-\infty}^{x} \rho(z)dz\ ; &\quad F(-\infty) = 0,\ \ F(+\infty) = 1. \end{aligned} \tag{1.4}$$

We have written things for finite d; no difficulties arise for $d \to \infty$. We often drop the label which describes the variable when it is explicitly understood as in I, ρ, and F. $F(x)$ is a staircase function with jumps $d^{-1}g_r$ at eigenvalues E_r. Thus

$$F(x) = F(E_r) = d^{-1} \sum_{k=1}^{r} g_k\ , \quad E_r \le x < E_{r+1}\ . \tag{1.5}$$

We shall usually write things as if the degeneracy $g_r = 1$; any necessary modifications will be obvious. The moments M_p of ρ and the characteristic function ϕ are,

$$\begin{aligned} M_p &= \textstyle\int \rho(x) x^p dx = d^{-1} \sum_r g_r (E_r)^p = \langle H^p\rangle^m\ , \\ \phi(t) &= \textstyle\int \rho(x)\exp(itx)dx = \sum_p \frac{(it)^p}{p!} M_p = \langle \exp(itH)\rangle^m\ , \\ M_0 &= 1,\ \ ,M_1 = \xi,\ \ M_2 - (M_1)^2 = \sigma^2,\ \ \phi(0) = 1\ . \end{aligned} \tag{1.6}$$

For a Gaussian density $\rho_{\mathcal{G}}(x)$, the characteristic function is

$$\rho_{\mathcal{G}}(x) = \frac{1}{\sqrt{2\pi}\,\sigma} \exp -\frac{1}{2}\left(\frac{x-\xi}{\sigma}\right)^2 \Rightarrow \phi_{\mathcal{G}}(t;\xi,\sigma^2) = \exp\left(it\xi - \frac{\sigma^2 t^2}{2}\right)\ . \tag{1.7}$$

The characteristic function (which always exists) uniquely determines the distribution. The moments on the other hand do not in all generality determine it and indeed they may not even exist. But they do so for most of the cases which we encounter, correspondingly, as they do, for finite-dimensional model spaces and for operators with bounded eigenvalues. The centroid ξ fixes the location of the distribution and the width σ defines the scale. The translation invariant central moments (taken about the centroid as origin) are $\mathcal{M}_p$. we have $\mathcal{M}_0 = 1$, $\mathcal{M}_1 = 0$, $\mathcal{M}_2 = \sigma^2$. The set of M_p $(p \le s)$ fixes the $\mathcal{M}_p$ $(p \le s)$, and vice versa, by the homogeneous expressions,

$$\mathcal{M}_p = \sum_r (-1)^r \binom{p}{r} M_{p-r}\xi^r; \quad M_p = \sum_r \binom{p}{r} \mathcal{M}_{p-r}\xi^r\ . \tag{1.8}$$

The reduced central moments $\mu_p = \mathcal{M}_p/\sigma^p$ are $1, 0, 1$ for $p = 0, 1, 2$. Since the energy E and centroid ξ can occur only as $(E - \xi)$ in the density and since σ fixes the scale and the total integral of $\rho(E)$ is unity, we see that $\rho(x)$ must have the form

$$\rho(x) = \sigma^{-1}\eta(\widehat{x}); \;\; \rho(x)dx = \eta(\widehat{x})d\widehat{x}; \;\; \widehat{x} = (x - \xi)/\sigma \tag{1.9}$$

where $\widehat{x}$ is the standardized form of x, i.e. x measured with respect to ξ as origin and renormalized to unit variance. It follows then that

$$\frac{\partial \rho(x)}{\partial \xi} = -\frac{\partial \rho(x)}{\partial x}; \;\; \frac{\partial \rho(x)}{\partial \sigma^2} = -\frac{\partial}{\partial x}\left[\frac{x-\xi}{2\sigma^2}\rho(x)\right] . \tag{1.10}$$

Moreover, if $P_\nu^{(\rho)}(x)$ and $P_\nu^{(\eta)}(x)$, with $\nu = 0, 1, \ldots$, are orthonormal polynomials defined respectively with $\rho(x)$ and $\eta(x)$ as weight functions we have $P_\nu^{(\rho)}(x) = P_\nu^{(\eta)}(\widehat{x})$. Several distributions are discussed later. Most of the distributions that we deal with are continuous rather than discrete; we encounter several of them in the $d \to \infty$ limit or as a result of otherwise smoothing a discrete distribution. The polynomial excitation functions (exact definition given later) oscillate about zero and are therefore not true probability densities.

Since for $p \geq 3$, μ_p are translation and scale invariant, they determine the shape of the distribution. In general terms we can describe homogeneous combinations of the $\mu_{p \geq 3}$ as shape parameters and write them as S_ν with $\nu \geq 3$; it will sometimes be convenient to write $\xi = S_1$, $\sigma^2 = S_2$. A particular set of distribution parameters, the reduced cumulants $k_p = K_p/\sigma^p$, which are non-trivial shape parameters for $p \geq 3$ will be of major interest. Just as the moments enter into the Taylor expansion of $\phi(t)$, the cumulants K_p enter into the expansion of its logarithm,

$$\log \phi(t) = \sum_{p=1}^{\infty} \frac{(it)^p}{p!} K_p; \;\; \phi(t) = \exp\left(\sum \frac{K_p (it)^p}{p!}\right) . \tag{1.11}$$

From Eq. (1.6) we see that $\log \phi(t) = i\xi t - \sigma^2 t^2/2 + \ldots$ which, combined with Eq. (1.11) gives $K_1 = \xi$, $K_2 = \sigma^2$ (as long as these moments exist). From Eq. (1.7) it follows that the Gaussian cumulants K_p vanish for $p > 2$, which is obviously a defining property of the Gaussian distribution. More generally, using the two expansion forms in Eqs. (1.6) and (1.11) one finds [Kendall and Stuart (1969); Stuart and Ord (1987)] that, when $p \geq 2$, the reduced cumulants k_p are given in terms of the reduced central moments μ_p. The shape parameter $k_3 = \gamma_1$ is called skewness and $k_4 = \gamma_2$ the excess. Broadly speaking $k_3 > 0$ defines a distribution which extends more in the $(x > \xi)$ domain than in $(x < \xi)$, and $k_4 > 0$ a distribution more sharply peaked than the Gaussian. Finally, the expression for the reduced cumulants in terms of the reduced central moments is,

$$k_P = P! \sum_{[P']} \left[\left\{ \prod_{i=1}^{\ell} \left(\frac{\mu_{p_i}}{P_i!}\right)^{S_i} \frac{1}{S_i!} \right\} (-1)^{S-1} \, (S-1)! \right] \tag{1.12}$$

and this gives,

$$\begin{aligned}
&k_1 = \xi/\sigma\,, \quad k_2 = 1\,, \\
&k_3 = \gamma_1(m) = \mu_3(m)\,, \quad k_4 = \gamma_2(m) = \mu_4(m) - 3\,, \\
&k_5 = \mu_5(m) - 10k_3(m)\,, \\
&k_6 = \mu_6(m) - 15k_4(m) - 10[k_3(m)]^2 - 15\,.
\end{aligned} \tag{1.13}$$

In Eq. (1.12) $[P']$ are all partitions of P such that $P_i \geq 2$ for all P_i.

The elementary CLT, valid with only very minor restrictions on the individual distributions, derives from the fact that the densities convolute under the addition of independent variables so that their Fourier transforms multiply, and the logarithms and hence the cumulants, add. Though the variables we encounter will be neither additive nor independent there will, nonetheless, be CLTs operating in our model spaces so that the fundamental description of a distribution according to its location, scale and shape will be of major importance to us. The CLT action will largely determine the shape, usually a Gaussian, and we will therefore be able to deal with phenomena in terms of centroids and variances (locations and scales) along with a few low-order corrections to accommodate deviations from a Gaussian. On the other hand, we shall occasionally encounter distributions for which the second central moment does not really define the physical extent in a satisfactory way; this is particularly true for a bimodal or a multimodal one, in which small "pieces" of the density occur a great distance away from the dominant piece. We shall encounter such distributions only when the "chaos" [term used by French probably for the first time in nuclear structure; see for example French (1975); Garg (1972)] which is generated by strong interactions, and which is responsible for the success of the statistical procedure, is modified by the presence of relatively weak admixings, e.g. between particles in orbits which are very far apart in energy. It will be important for us to recognize and take account of this feature when it is present; see the discussion on level densities.

Turning now to polynomial expansions for the densities, we refer to Cramer (1946); Feller (1971); Kendall and Stuart (1969); Stuart and Ord (1987). There are two problems of immediate interest. The first is to find an adequate expansion for the density in terms of the asymptotic density and polynomials defined by it. Given that density, the second problem (which we leave to the next subsection) is that of deriving the first order corrections to it when we add a small term to the Hamiltonian; with the solution to this (the problem of "linear response") we shall be able to derive expectation values and sum rules. Only in the very simplest "textbook" cases do we have any prospect of evaluating a complete set of moments; usually we must settle for a small number of the low-order moments. These of course place constraints on the distribution [Kendall and Stuart (1969), p.87] in accordance with the so called "Principle of Moments", and they determine inequalities, as for example Chebyshev inequality [Kendall and Stuart (1969); Stuart and Ord (1987)] on various quantities defined by the distribution; see also Wheeler and Gordon

(1970). These inequalities however are inadequate for our purpose. But we are saved by the CLT generation of a close-to-Gaussian smoothed density. Then it will turn out that, "to within fluctuations", calculable (low order) shape corrections will give adequate results for particle numbers and Hamiltonians of interest. As a consequence, we seek expansions of $\rho(x)$ around an asymptotic shape which we take to be Gaussian (other forms will arise in special cases; the modifications needed or the extensions will be straightforward). The general nature of this expansion will be in terms of a sequence of "polynomial excitations" of the asymptotic density (see ahead for GC, ED and CF expansions). Given a density ρ with central moments $\mathcal{M}_r$, it is possible to write orthonormal polynomials $P_\mu(x)$,

$$\begin{aligned} &P_0(x) = 1, \quad P_1(x) = \frac{(x-\xi)}{\sigma} = \widehat{x}\,, \\ &P_2(x) = \frac{\mathcal{M}_2^2\widehat{x}^2 - \mathcal{M}_3\sigma\widehat{x} - \mathcal{M}_2^2}{[\mathcal{M}_2(\mathcal{M}_2\mathcal{M}_4 - \mathcal{M}_3^2 - \mathcal{M}_2^3)]^{1/2}} = \frac{\widehat{x}^2 - \gamma_1\widehat{x} - 1}{[2+\gamma_2-\gamma_1^2]^{1/2}}\,. \end{aligned} \tag{1.14}$$

Eq. (1.45) ahead gives the general formula. Note that, in order to be useful, the first few terms of the polynomial expansion must give a satisfactory representation of the density. Compatible with this are the facts that our interest is in fluctuation free densities and (as discussed in Sec. 1.1) there is a wide gap in the spectrum between the few long wavelength excitations of present interest and the short wavelength ones which describe the fluctuations. It follows then that a finite expansion will often be useful even when the complete expansion is not formally point convergent. This is what we call "convergence to within fluctuations". In fact statistical methods are valid only for strong interactions; when they are weak perturbation theory is appropriate. The situation is somewhat more complex when the interactions are of intermediate strength.

For illustration and later use, we discuss three expansions for a univariate density function. Given the standardized variable $\widehat{x}$ and the corresponding Gaussian density it was argued by Edgeworth that $\eta(\widehat{x})$,

$$\eta(\widehat{x}) = \exp\left\{\sum_{\nu\geq 3}(-1)^\nu\frac{k_\nu}{\nu!}\frac{\partial^\nu}{\partial\widehat{x}^\nu}\right\}\eta_{\mathcal{G}}(\widehat{x})\,; \quad \eta_{\mathcal{G}}(\widehat{x}) = \frac{1}{\sqrt{2\pi}}\exp-\frac{\widehat{x}^2}{2} \tag{1.15}$$

is a "true and unique law that represents the frequency curve of a magnitude that depends on a number of independent elements" [Bowley (1972)]. If the number Υ of such events varies, then $k_\nu \propto \Upsilon^{-\nu/2}$, and that the appropriate method of approximation is by truncating the ν series. The k_ν are reduced cumulants and $k_\nu(m) = k_\nu(1)/\Upsilon^{(\nu/2)-1}$ if the independent variables are similarly distributed. Thus on the one hand Edgeworth argument is a statement of the CLT and, on the other, an argument that uniformity with regard to Υ (particle number m in our examples) is a good guide to a method of approximation. Expanding the exponential in Eq. (1.15) and collecting all terms that behave as $\Upsilon^{-P/2}$, $P = 1, 2, \ldots$, a compact form for the

Edgeworth (ED) expansion is,

$$\eta_{ED}(\widehat{x}) = \eta_{\mathcal{G}}(\widehat{x}) \left\{ 1 + \sum_{P=1}^{\infty} \sum_{[P]} \left[\prod_{i=1}^{\ell} \left(\frac{k_{P_i+2}}{(P_i+2)!} \right)^{S_i} \frac{1}{(S_i\,!)} \right] He_{P+2S}(\widehat{x}) \right\} . \quad (1.16)$$

Here $[P] = [P_1^{S_1}, P_2^{S_2} \ldots P_\ell^{S_\ell}]$ is a partition of the integer P such that $P_1 \geq P_2 \geq \ldots \geq P_\ell > 0$, $S_i > 0$, $S = \sum_{i=1}^{\ell} S_i$ and $\sum_{i=1}^{\ell} P_i S_i = P$. Feller (1971) has shown that if the moments $\mu_3, \mu_4, \ldots, \mu_r$ exist and $|\phi(t)|^s$ is integrable for some $s \geq 1$, then $\eta(\widehat{x})$ exists for $m > s$ and the ED representation given by Eq. (1.16) is asymptotically convergent, i.e. as $m \to \infty$ the series converges to the first $r = P+1$ terms. The ED expansion to order $P = 6$ is, with $k'_\nu = k_\nu/\nu!$,

$$\begin{aligned}
\eta_{ED}(\widehat{x}) = \eta_{\mathcal{G}}(\widehat{x}) \Bigg\{ & 1 + [k'_3 He_3(\widehat{x})] + \left[k'_4 He_4(\widehat{x}) + \frac{(k'_3)^2}{2!} He_6(\widehat{x}) \right] \\
& + \left[k'_5 He_5(\widehat{x}) + k'_3 k'_4 He_7(\widehat{x}) + \frac{(k'_3)^3}{3!} He_9(\widehat{x}) \right] \\
& + \left[k'_6 He_6(\widehat{x}) + \left(\frac{(k'_4)^2}{2!} + k'_3 k'_5 \right) He_8(\widehat{x}) + \frac{(k'_3)^2 k'_4}{2!} He_{10}(\widehat{x}) + \frac{(k'_3)^4}{4!} He_{12}(\widehat{x}) \right] \\
& + [k'_7 He_7(\widehat{x}) + (k'_3 k'_6 + k'_4 k'_5)\, He_9(\widehat{x}) \\
& + \left(\frac{(k'_3)^2 k'_5}{2!} + \frac{k'_3 (k'_4)^2}{2!} \right) He_{11}(\widehat{x}) \frac{(k'_3)^3 k'_4}{3!} He_{13}(\widehat{x}) + \frac{(k'_3)^5}{5!} He_{15}(\widehat{x})] \\
& + \Bigg[k'_8 He_8(\widehat{x}) + \left(\frac{(k'_5)^2}{2!} + k'_3 k'_7 + k'_4 k'_6 \right) He_{10}(\widehat{x}) \\
& + \left(\frac{(k'_4)^3}{3!} + \frac{(k'_3)^2 k'_6}{2!} + k'_3 k'_4 k'_5 \right) He_{12}(\widehat{x}) \\
& + \left(\frac{(k'_3)^2 (k'_4)^2}{2!2!} + \frac{(k'_3)^3 k'_5}{3!} \right) He_{14}(\widehat{x}) + \frac{(k'_3)^4 k'_4}{4!} He_{16}(\widehat{x}) + \frac{(k'_3)^6}{6!} He_{18}(\widehat{x}) \Bigg] \Bigg\} .
\end{aligned} \quad (1.17)$$

The Hermite polynomials $He_r(\widehat{x})$ satisfy the recursion equation $He_{r+1}(\widehat{x}) = \widehat{x}\, He_r(\widehat{x}) - r\, He_{r-1}(\widehat{x})$ and explicit expressions for the lowest 6 polynomials are,

$$\begin{aligned}
He_0(\widehat{x}) &= 1 \\
He_1(\widehat{x}) &= \widehat{x} \\
He_2(\widehat{x}) &= \widehat{x}^2 - 1 \\
He_3(\widehat{x}) &= \widehat{x}^3 - 3\widehat{x} \\
He_4(\widehat{x}) &= \widehat{x}^4 - 6\widehat{x}^2 + 3 \\
He_5(\widehat{x}) &= \widehat{x}^5 - 10\widehat{x}^3 + 15\widehat{x} \\
He_6(\widehat{x}) &= \widehat{x}^6 - 15\widehat{x}^4 + 45\widehat{x}^2 - 15 \,.
\end{aligned} \quad (1.18)$$

It should be noted that the centroid and width of $\rho(x)$ and $\rho_{\mathcal{G}}(x)$ that correspond to $\eta_{ED}(\widehat{x})$ and $\eta_{\mathcal{G}}(\widehat{x})$ respectively are identical in the above ED expansion.

The second expansion commonly used is the Gram-Charlier (GC) expansion where the density $\eta(\widehat{x})$ is expanded in terms of polynomial excitations of a given density $\eta_0(\widehat{x})$, i.e. in terms of the polynomials that are orthonormal with η_0 as weight function. For standard GC, $\eta_0 = \eta_{\mathcal{G}}$ and then [Stuart and Ord (1987); Cramer (1946)],

$$\eta_{GC}(\widehat{x}) = \eta_{\mathcal{G}}(\widehat{x}) \left\{ 1 + \sum_{\nu \geq 3} \frac{C_\nu}{\nu!} He_\nu(\widehat{x}) \right\} \; ; \quad C_\nu = \left\langle He_\nu(\widehat{K}) \right\rangle^m \; . \tag{1.19}$$

In Eq. (1.19), the density $\eta(\widehat{x})$ is for the eigenvalues of an operator $\widehat{K}$ (in most applications $\widehat{K} = H$). Thus the shape parameters here are Hermite polynomials The GC expansion truncated to include only $\gamma_1 = k_3$ and $\gamma_2 = k_4$ corrections is,

$$\eta_{GC}(\widehat{x}) = \eta_{\mathcal{G}}(\widehat{x}) \left\{ 1 + \frac{\gamma_1}{6} He_3(\widehat{x}) + \frac{\gamma_2}{24} He_4(\widehat{x}) \right\} \; . \tag{1.20}$$

Observe that, formally the GC and ED series are rearrangements of one another, but of course they do not in general coincide to finite order. There is some evidence that the ED series, which should be favored a priori for systems with an operative CLT, and which to the same order as GC requires calculation of the same parameters, gives in fact a better agreement with numerical calculations [Haq and Wong (1979); Zirnbauer and Brink (1981); French, Kota and Smith (1989)].

Instead of expanding the density in terms of an asymptotic density (often assumed to be Gaussian), it is sometimes useful to consider an expansion of the variable in terms of a Gaussian variable following the principle used in the ED expansion. This gives the Cornish-Fisher (CF) expansion [Kendall and Stuart (1969); Stuart and Ord (1987)] for the density. Including only γ_1 and γ_2 corrections, the CF expansion is [Kendall and Stuart (1969); Kota, Potbhare and Shenoy (1986)],

$$\begin{aligned} \eta_{CF}(\widehat{x}) = & \frac{1}{\sqrt{2\pi}} \left[1 - \frac{\gamma_1}{3} \widehat{x} - \frac{\gamma_2}{8} (\widehat{x}^2 - 1) + \frac{\gamma_1^2}{36} (12\widehat{x}^2 - 7) \right] \\ & \times \exp \left\{ -\frac{1}{2} \left(\widehat{x} - \frac{\gamma_1}{6} (\widehat{x}^2 - 1) - \frac{\gamma_2}{24} (\widehat{x}^3 - 3\widehat{x}) + \frac{\gamma_1^2}{36} (4\widehat{x}^3 - 7\widehat{x}) \right)^2 \right\} \; . \end{aligned} \tag{1.21}$$

The truncation of ED, GC or CF expansions to a finite number of correction terms (commonly only those that include γ_1 and γ_2) may give a negative density particularly in the distribution tail, a region which is important for example in locating the ground state. We can usually get around this with partitioning of the space which also generates new information about the partitioning symmetry; this is discussed in more detail ahead. Experience indicates that partitioning is needed if the ground state (GS) is 3.5σ or more below the centroid, this corresponds to a dimensionality of a few thousand. The same problem, but less severe may arise with partitioning. For completeness we mention that, the domain of validity (giving positive density for all values of the variable) for CF representation is very much larger than that for ED and GC expansions. For avoiding negative densities there are some non-canonical forms (not expansions) suggested by some authors and we will consider these at the end of this Section. Before going further, let us mention that in most of the discussion ahead the variable x corresponds to energies (E) and therefore from now on the centroid ξ is referred to as the energy centroid E_c and the width as the spectrum width σ.

In order to apply the moment expansions, it is essential to understand the origin of the CLT for NIP and IP systems and we consider this next. *It is difficult to exaggerate the importance of convolutions in many branches of mathematics* [Feller (1971), p.143]; including, we might add, communication and information theory, and in physics and other domains as well. Indeed, the essential content of the elementary CLT, which, with extensions, will dominate everything that we do, is that the m-fold convolution of a function $\rho_1(x)$ with itself, $\rho_m = \rho_1 \otimes \rho_1 \otimes \ldots \rho_1$, becomes Gaussian for large m; the minor restrictions on ρ for the validity of CLT [Feller (1971)] are always satisfied in the model spaces which we shall encounter. The Gaussian form of the distribution derives from the fact that, if the parameters of ρ_1 are $E_c(1)$, $\sigma^2(1)$, $k_{\nu\geq 3}(1)$, then for ρ_m they are $mE_c(1)$, $m\sigma^2(1)$, $k_\nu(1)/m^{((\nu/2)-1)}$. Thus we have not only the Gaussian asymptotic result but, in the simple case at least, full information about the approach to that form as m increases; we shall regard this as an essential part of the CLT. Described in different terms, the CLT states that the probability density for the sum of m independent random variables is the m-fold convolution of the separate densities, and therefore, asymptotically ($m \to \infty$) Gaussian. Of course a function of such a sum of variables will in general have an asymptotic distribution which is different from Gaussian (the square of the sum for example would give a χ^2 distribution but these forms derive also from the CLT). With m particles distributed over a finite set of N sp states, forming thereby a *spectroscopic space* of dimension $d = \binom{N}{m}$, we can, if the particles are non-interacting, think of each random variable as the energy of the state occupied by the corresponding particle. These variables are independent if the fermion "blocking effects" (or effects of the opposite kind in the case of bosons) can be ignored. If then we take the sp probability density ρ_1 as uniform (probability N^{-1} for each state), the m-particle density reproduces the m-particle spectrum. The scale of the distribution, as measured by $\sigma(m)$, increases as $m^{1/2}$ while the number of states increases more rapidly, as N^m (or with fermions as $\binom{N}{m} \stackrel{N\gg m}{\longrightarrow} N^m/m!$) so that the level spacing decreases rapidly with m. This of course is one aspect of the smoothing which must take place if the discrete spectrum is to be well approximated by Gaussian. Moreover, if the spectrum is bounded from above or below, or both, so will be its multiple convolution so that we may expect departures from Gaussian to show up first in the tails of the distribution. With fermions our random variables are not really independent because of the Pauli effects which, for example, limit the particle number to $m << N$. Except in the case of exotic sp spectra these effects are relatively mild and do not destroy the CLT operation. Far more important and difficult to deal with are the interactions between particles, as a result of which the random variables, even if they can be defined as above, no longer add to give the total energy. Since direct convolution methods are then not available, we shall instead proceed by considering moments and polynomial expansions.

Besides dealing with the effects of statistics and interactions we shall, by partitioning methods, be able to deal also with symmetries, both exact and broken. In fact we will find quite remarkable connections between symmetries, statistical behavior and information content. The possibilities of doing these things will derive from the fact that, with the interactions and spaces of nuclear physics (perhaps also for atoms), more general CLTs will be found to operate and to produce essentially Gaussian spectra. See Sec. 1.1 for some examples and also the papers included in this book. The basic questions then concern the class of Hamiltonians which produce an asymptotic Gaussian spectrum, and whether or not the approach to Gaussian is governed by some type of generalized convolution. There are several ways in which the first, or both, of these questions can be studied: (i) By evaluating, for a given H, the asymptotic spectral distribution and non-asymptotic corrections to it. This is not yet a general procedure, although it is used in some special cases as discussed earlier; (ii) By introducing ensembles of Hamiltonians and constructing the ensemble averaged spectrum, taking for granted that an ergodic property will give an essential equality between the ensemble spectrum and the smoothed spectrum of H [Mon and French (1975*)]; (iii) By evaluating the low-order moments, in particular up to order 4, and thereby the skewness and excess. We shall see that this is quite practical in some cases [Potbhare (1978)] but very difficult in others, depending on the symmetries specified; (iv) By shell model calculations with realistic Hamiltonians; see [Brody *et al.* (1981); Zelevinsky, Brown, Frazier and Horoi (1996); Kota (2001)]. However, we begin with the case of NIP systems.

1.2.2 *Central Limit Theorems for NIP and IP systems*

Given the NIP Hamiltonian $h = \sum_{\alpha=1}^{N} \epsilon_\alpha n_\alpha$, with α being a spherical j-orbit with energy ϵ_α and the number operator n_α, as we encounter in nuclear shell model and similarly with extensions to other type of orbits such as for example Nilsson levels [Patel, Desai and Potbhare (1999)], we consider partitioning of the sp space into ℓ spherical orbits α, β, ..., each having degeneracy N_i so that N is partitioned into $N \rightarrow (N_1, N_2, \ldots, N_\ell)$. Then distributing particles in these orbits will give m particle spherical configurations $\mathbf{m} = [m_1, m_2, \ldots, m_\ell]$ where m_i is the number of particles in the i-th spherical orbit. The exact NIP state density can then be written easily using spherical configurations $\mathbf{m}$,

$$I^{h,m}(E) = \langle\langle \delta(h - E) \rangle\rangle^m = \sum_{\mathbf{m}} d(\mathbf{m}) \delta(E - \epsilon(\mathbf{m})) \; ; \quad \epsilon(\mathbf{m}) = \sum_{\alpha} m_\alpha \epsilon_\alpha \; . \quad (1.22)$$

However, in large spaces it is more useful to consider unitary orbits, a combination of spherical orbits with the same parity. Then for L-unitary orbits it is possible to group spherical orbits into $\boldsymbol{\alpha}$, $\boldsymbol{\beta}$, ... each having degeneracy (i.e. the number of sp states) $\mathcal{N}_i$ so that $N = \sum_{i=1}^{L} \mathcal{N}_i$ and then define configurations

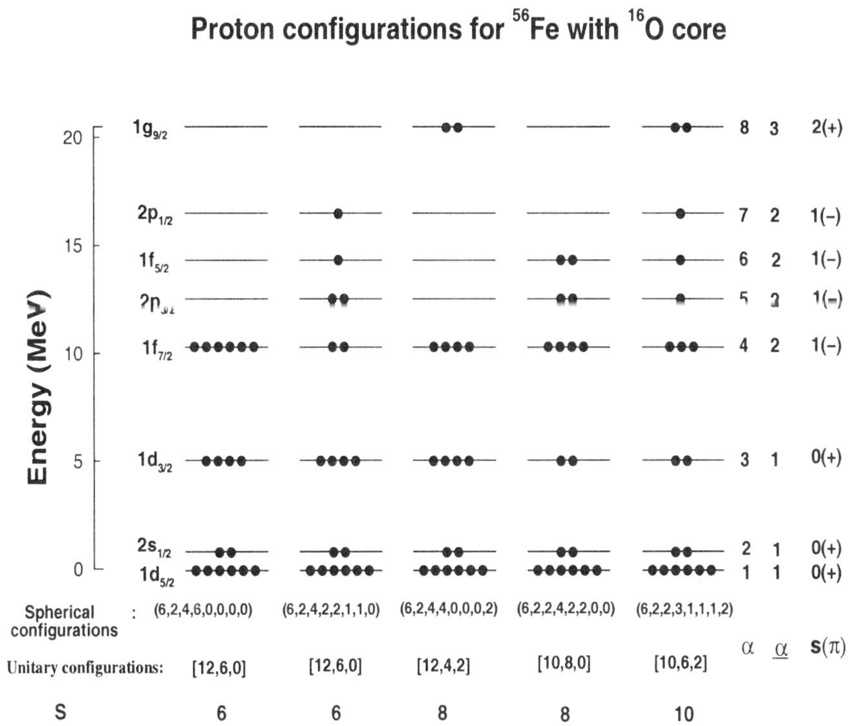

Fig. 1.6 An example for spherical (α) and unitary orbits ($\boldsymbol{\alpha}$) and the corresponding spherical and unitary configurations. The active orbits appropriate for protons in ^{56}Fe are shown. Also shown are the s label for marking distant orbits, parity (π) and the total S value.

$[\mathbf{m}] = [\mathbf{m}_1, \mathbf{m}_2, \ldots, \mathbf{m}_L]$; see Fig. 1.6. Then

$$I^{h,m}(E) = \sum_{[\mathbf{m}]} I^{h,[\mathbf{m}]}(E) \; ; \quad I^{h,[\mathbf{m}]}(E) = \langle\langle \delta(h-E) \rangle\rangle^{[\mathbf{m}]} \; . \tag{1.23}$$

For each of the unitary configuration densities $I^{h,[\mathbf{m}]}(E)$ there is moreover a convolution factoring giving [French *et al.* (2006*)],

$$\begin{aligned} I_N^{h,[\mathbf{m}]}(E) &= \langle\langle \delta(h-E) \rangle\rangle_N^{[\mathbf{m}]} = \left\langle\!\left\langle \delta(\{\textstyle\sum_{\boldsymbol{\alpha}=1}^{L} h_{\boldsymbol{\alpha}}\} - E) \right\rangle\!\right\rangle_N^{[\mathbf{m}]} \\ &= \int dx \left\langle\!\left\langle \delta(\{\textstyle\sum_{\boldsymbol{\alpha}=1}^{L-1} h_{\boldsymbol{\alpha}}\} - x) \right\rangle\!\right\rangle_{N-\mathcal{N}_L}^{[\mathbf{m}]-m_L} \langle\langle \delta(h_L + x - E) \rangle\rangle_{\mathcal{N}_L}^{\mathbf{m}_L} \\ &= I_{N-\mathcal{N}_L}^{h-h_L,[\mathbf{m}]-m_L} \otimes I_{\mathcal{N}_L}^{h,m_L}[E] = I_{\mathcal{N}_1}^{h_1,m_1} \otimes I_{\mathcal{N}_2}^{h_2,m_2} \otimes \ldots \otimes I_{\mathcal{N}_L}^{h_L,m_L}[E] \; . \end{aligned} \tag{1.24}$$

In Eq. (1.24) $N - \mathcal{N}_L$ contain all unitary unitary orbits except the L-th orbit and $[\mathbf{m}]-m_L$ is the unitary configuration obtained after dropping all the m_L particles occupying the L-th unitary orbit. Convolutions in Eq. (1.24) are carried out simply by adding low-order cumulants. An alternative procedure would convolute by $(L-1)$-fold integration involving the separate single-orbit densities. In the (p, n) formalism a further convolution of proton and neutron densities, each given by Eq. (1.24), will give the total NIP density. For applying Eq. (1.24) we need single orbit cumulants.

Without loss of generality let us consider a single unitary orbit with N sp energies ϵ_i, then $h = \sum_{i=1}^{N} \epsilon_i n_i$ and h is in single state representation (rather than using orbits α). The p-th moment of h is easy to write down as $n_i^r = n_i$ for $r \geq 1$. Then $n_i n_j n_k \ldots n_\ell$ with $i \neq j \neq k \ldots \neq \ell$ will be an $\ell-$body operator. Now applying the elementary but important trace result,

$$\langle \mathcal{O}(k) \rangle^m = \binom{m}{k} \langle \mathcal{O}(k) \rangle^k \tag{1.25}$$

for any k-body operator, we get $\langle n_i n_j n_k \ldots n_\ell \rangle^m = \binom{N}{\ell}^{-1} \binom{m}{\ell}$. This gives,

$$\begin{aligned} M_p(m) &= \sum_{[p]} d_{[p]} \, \langle \epsilon^{p_1} \wedge \epsilon^{p_2} \wedge \ldots \wedge \epsilon^{p_q} \rangle \; N^q \binom{N}{q}^{-1} \binom{m}{q} ; \\ d_{[p]} &= \frac{p!}{\prod p_i! \, \prod s_j!} \, . \end{aligned} \tag{1.26}$$

In Eq. (1.26) the summation is over unordered partitions $[p]$ of p and $d_{[p]}$, with s_k the number of k's in a given partition (for example [332111] gives $s_1 = 3$, $s_2 = 1$ and $s_3 = 2$), is a multinomial coefficient. With $\sum'$ indicating an ℓ-fold sum with $i_1 \neq i_2 \ldots \neq i_\ell$,

$$\langle \epsilon^{p_1} \wedge \epsilon^{p_2} \wedge \ldots \wedge \epsilon^{p_\ell} \rangle = N^{-\ell} \sum{}' \, (\epsilon_{i_1})^{p_1} \, (\epsilon_{i_2})^{p_2} \ldots (\epsilon_{i_\ell})^{p_\ell} \; . \tag{1.27}$$

The sums in Eq. (1.27), i.e. $\langle \epsilon^{p_1} \wedge \epsilon^{p_2} \wedge \ldots \wedge \epsilon^{p_q} \rangle$, are "augmented symmetric functions". An essential result is that they can be written as sums of products of "power sums" $(r) = \sum \epsilon_i^r = N M_r(1)$, and tables are available for this expansion [David, Kendall and Barton (1966)]. Thus, in a slightly different context, all the NIP results are well known to statisticians. Starting with zero centered sp energies, so that $\sum_i \epsilon_i = 0$, using Eqs. (1.26), (1.27) and the expansion of augmented symmetric functions in terms of power sums, it is possible to write the m-particle central moments $\mathcal{M}_p(m)$ in terms of one particle moments $\mathcal{M}_r(1)$ for any p. The structure of the NIP moments is now describable as follows. Given the sp energies ϵ_i, $M_1(1) = \epsilon = N^{-1} \sum_i \epsilon_i = N^{-1} \sum_\alpha N_\alpha \epsilon_\alpha$ and $M_1(m) = m\epsilon(1)$. Then $\mathcal{M}_p(1) = N^{-1} \sum_i (\epsilon_i - \epsilon)^p = N^{-1} \sum_\alpha N_\alpha (\epsilon_\alpha - \epsilon)^p$. With this we can write $\mathcal{M}_p(m)$ as,

$$\mathcal{M}_p(m) = \sum_{[p]} R_{[p]}(m, N) \, \mathcal{M}_{[p]}(1) \;\; ; \;\; \mathcal{M}_{[p]}(1) = \prod_i \mathcal{M}_{p_i}(1) \, . \tag{1.28}$$

In Eq. (1.28) only those partitions of p with every $p_i \neq 1$ will contribute. The final results for the propagating polynomials $R_{[p]}(m, N)$ for $p = 2, 3, 4, 5, 6$ are given in Table 1.1. Before going further, for illustration we consider an alternative derivation of the results for $p = 2, 3$ and 4 where we do not use Eq. (1.26). Firstly, from the basic trace equation (1.25) we have $\mathcal{M}_2(m) = a_0 + a_1 m + a_2 m^2$ and the parameters a_i follow from the values of $\mathcal{M}_2(m)$ for $m = 0, N$ and 1. Note that $\mathcal{M}_2(0) = 0$ and $\mathcal{M}_2(N) = 0$. Then we have $\mathcal{M}_2(m) = [m(N-m)/(N-1)] \, \mathcal{M}_2(1)$. Similarly

Table 1.1 Propagating polynomials $R_{[p]}$ for the central moments $\mathcal{M}_p(m)$, $p \geq 2$. Note that $\mathcal{M}_p(m) = \sum_{[p]=[p_1,p_2,\ldots]} R_{[p]} \prod_i \mathcal{M}_{p_i}(1)$ where $[p]$ is a partition of p and the partitions with any $p_i = 1$ will not appear in the summation. In the table $g(p) = m(N-m)[(N-1)(N-2)\ldots(N-p+1)]^{-1}$. From [French *et al.* (2006*)].

p	$[p]$	$R_{[p]}$
2	[2]	$g(2)$
3	[3]	$g(3)(N-2m)$
4	[4]	$g(4)[N(N+1)-6m(N-m)]$
	[22]	$g(4)[3(m-1)(N-m-1)N]$
5	[5]	$g(5)\{(N-2m)[N(N+5)-12m(N-m)]\}$
	[32]	$g(5)[(N-2m)10N(m-1)(N-m-1)]$
6	[6]	$g(6)[120m^2(N-m)^2-30N(N+3)m(N-m)+N(N+1)(N^2+15N-4)]$
	[42]	$g(6)\{15N(m-1)(N-m-1)[(N+4)(N-1)-6m(N-m)]\}$
	[33]	$g(6)\{10N(m-1)(N-m-1)[(N^2-N+4)-4m(N-m)]\}$
	[222]	$g(6)[15N^2(m-1)(m-2)(N-m-1)(N-m-2)]$

$\mathcal{M}_3(m)$ is a 3rd order polynomial in m. Now using the basic traces $\mathcal{M}_3(0) = 0$, $\mathcal{M}_3(N) = 0$, $\mathcal{M}_3(N-1) = -\mathcal{M}_3(1)$ we have easily $\mathcal{M}_3(m) = [m(N-m)(N-2m)/(N-1)(N-2)]\mathcal{M}_3(1)$. Finally for $p = 4$, $\mathcal{M}_4(m)$ is a 4-th order polynomial and we can use the basic traces $\mathcal{M}_4(0) = 0$, $\mathcal{M}_4(N) = 0$, $\mathcal{M}_4(N-1) = \mathcal{M}_4(1)$ and $\mathcal{M}_4(2)$. Elementary algebra, by putting 2 fermions in N sp states with energies ϵ_i, gives $\mathcal{M}_4(2) = \sum_{i \neq j}(\epsilon_i+\epsilon_j)^4 = [(2N-16)/(N-1)]\mathcal{M}_4(1)+[6N/(N-1)]\{\mathcal{M}_2(1)\}^2$. Then the formula for $\mathcal{M}_4(m)$ follows easily as given in Table 1.1.

Given the central moments listed in Table 1.1, it is possible to examine the asymptotic structure of the reduced central moments $\mu_p = \mathcal{M}_p/\sigma^p$ and then that of the cumulants. In the asymptotic limit defined by $N \to \infty$, $m \to \infty$ and $m/N \to 0$ (also called the dilute limit), it is easy to see from Table 1.1 that $\sigma^2(m) \to m\sigma^2(1)$, $\mu_3(m) \to \mu_3(1)/\sqrt{m}$, $\mu_4(m) \to 3$, $\mu_5(m) \to 10\mu_3(m)$ and $\mu_6(m) \to 15$. Using Eqs. (1.26) and (1.27), it is seen that in general, the binary correlations dominate with even moments coming from $[p] = [2^\nu]$ and odd moments from $[2^{\nu-1}, 3]$,

$$\begin{aligned} \mu_{2\nu}(m) &\to (2\nu-1)!! \,, \\ \mu_{2\nu+1}(m) &\to \tfrac{1}{3}\nu\,(2\nu+1)!!\,\mu_3(m) \,. \end{aligned} \tag{1.29}$$

Therefore asymptotically the odd moments will vanish [as $\mu_3(m) \to \frac{1}{\sqrt{m}}\mu_3(1)$] and even moments will be those of a Gaussian. Thus *NIP Hamiltonians in general will give Gaussian state densities*. The exception occurs only for singular spectra with unusual values for $\gamma_1(1)$ and $\gamma_2(1)$. An example being the pairing NIP spectrum with $\epsilon_1 = 0$ and $\epsilon_i = \Delta$ for $i = 2, 3, \ldots, N$; see ahead. Dominance of binary correlations that give Eq. (1.29) will be seen later to generate Gaussians even with interactions. Going beyond Eq. (1.29), it is possible to write the exact expressions, taking fully into account Pauli effects, for the reduced cumulants k_r, $r = 3-6$

and they give information about the departures from Gaussian form; they can be taken into account in practice for example by using Edgeworth expansion. Using Table 1.1 and relationship between k_r and μ_r we obtain,

$$
\begin{aligned}
k_3(m) = \gamma_1(m) &= \frac{N-2m}{r(3)}\, X^{1/2}\, k_3(1)\,, \\
k_4(m) = \gamma_2(m) &= \frac{X}{r(4)} \left\{ \left[N(N+1) - 6\frac{N-1}{X} \right] k_4(1) + 6N \left[1 - \frac{1}{X} \right] \right\}, \\
k_5(m) &= \frac{N-2m}{r(5)}\, X^{3/2} \left\{ \left[N(N+5) - 12\frac{N-1}{X} \right] k_5(1) + 60N \left[1 - \frac{1}{X} \right] k_3(1) \right\}, \\
k_6(m) &= \frac{60X^2Y}{(N-1)r(6)} \left[\frac{4(N-1)(2N+1)}{X} - (N-1)(N^2+9N-4) \right] \\
&+ \frac{30X^2Y}{r(6)} \left[\frac{36(N-1)}{X} - (7N^2+9N-4) \right] k_4(1) \\
&+ \frac{60X^2Y}{(N-2)r(6)} \left[\frac{2(N-1)(7N-15)}{X} - N(N-2)(3N+1) \right] k_3^2(1) \\
&+ \frac{X^2}{r(6)} \left[\frac{120(N-1)^2}{X^2} - \frac{30N(N+3)(N-1)}{X} + N(N+1)(N^2+15N-4) \right] k_6(1)\,.
\end{aligned}
\tag{1.30}
$$

In Eq. (1.30), $X = (N-1)/m(N-m)$, $Y = N(m-1)(N-m-1)/(N-1)$ and $r(p) = (N-2)(N-3)\ldots(N-p+1)$. Therefore asymptotically, $\gamma_1(m) \to \gamma_1(1)/\sqrt{m}$ and $\gamma_2(m) \to \gamma_2(1)/m$.

Whenever new coherent states occur far from equilibrium, the application of probability theory, as implied by the counting of the number of complexions, breaks down. Though we are not concerned here with non-equilibrium statistical mechanics, this remark of Prigogine (1980) is relevant when we ask about exceptions to the Gaussian law derived above. We have taken for granted there that $\gamma_2(1) \sim O(1)$. Let us consider large N, m and $\gamma_2(1) = N$. Then with fixed m and $N \to \infty$ we have the canonical result (from the exact formula for $\gamma_2(m)$), that $\gamma_2(m) = N/m$, while, with fixed $m/N = 1/2$ say, we have $\gamma_2(m) \to -2$, both non-Gaussian results. The trivial explanation is that a $\gamma_2(1) \sim O(N)$ spectrum is of such an exotic nature that it would require far more than N convolutions (i.e. particles) to produce a good Gaussian spectrum, this of course being ruled out for fermions. In general these "singular" spectra are those in which essentially all the "weight" is carried by a fraction $\sim N^{-1}$ of the levels. They are characterized by $\mu_{2\nu} \sim O(N^{\nu-1})$, whereas for non-singular spectra such as Gaussian, uniform, χ_r^2 etc, $\mu_{2\nu} \sim O(1)$ (the distinction of course is in the N dependence for large N, not in the ν dependence). The extreme example is the pairing spectrum $(O^{N-1},1)$, for which obviously $M_p = N^{-1}$ so that $\gamma_2(1) \xrightarrow{N\to\infty} N$, its largest possible value. For this case $\gamma_1(1) \to N^{1/2}$ [there are symmetrical singular spectra for which the odd cumulants vanish, an example is $(\pm 1, O^{N-2})$]. Taking the pairing spectrum as an example, and noting that only a single energy eigenvalue is non-zero, we see that binary associations play no role whatsoever in the correlated sums. The only contributing partition is $[p,0,0,\ldots]$ and then $M_p(m) = m/N \overset{m \ll N}{\longrightarrow} \mathcal{M}_p(m)$, which of course also defines a singular spectrum. It is important to note that there are analogs to singular spectra for interacting particle systems and they are connected to collective behavior. A trivial example is the pairing interaction which we consider later.

We have seen that in the NIP case a non-singular sp spectrum generates an asymptotic Gaussian density while a singular spectrum generates another singular spectrum. At first sight we might not expect a general result like this with interacting particle (IP) systems since energies are not additive so that there is no convolution. Thus for example $H = h^2$ with h non-singular and traceless is (1+2) body, the two body part describing interactions. Its asymptotic spectrum is obviously χ_1^2 $[\propto x^{-1/2}\exp(-x/2)]$ which is far from Gaussian. But χ_1^2 derives directly from the CLT action generated by h, for which there is a convolution. Thus in some circumstances there might indeed be an operative CLT in IP systems. For a first understanding of this [French and Draayer (1979*); Ginocchio (1980)] we proceed by analogy with h^2. Taking the dilute limit ($N \to \infty$, $m \to \infty$, $m/N \to 0$), to eliminate uninteresting blocking effects, consider the spectrum of

$$H = \sum_{\alpha=1}^{\ell} h_\alpha^2 \; ; \; h_\alpha = \sum_{i,j=1}^{N} \epsilon_{ij}(\alpha) a_i^\dagger a_j \; ; \; \epsilon_{ij}(\alpha) = \epsilon_{ji}^*(\alpha) \; ; \; \sum_i \epsilon_{ii}(\alpha) = \langle\langle h_\alpha \rangle\rangle^1 = 0 \, , \tag{1.31}$$

the sum of squares of traceless one body operators which we take to have non-singular spectra. The binary correlation result discussed before giving $\mu_{2\nu} = (2\nu - 1)!!$ implies that operators effectively commute in the dilute limit. It then follows that

$$M_{2\nu}(m) \overset{dilute\ limit}{\longrightarrow} \sum_{\nu_1 \ldots \nu_\ell} (2\nu : \nu_1 \ldots \nu_\ell) \left\langle h_1^{\nu_1} h_2^{\nu_2} \ldots h_l^{\nu_\ell} \right\rangle^m \tag{1.32}$$

in which the traces indicated can be written as a sum of ν-fold products of bi-linear traces $\langle h_\alpha h_\beta \rangle^1$. As $\langle h_\alpha h_\beta \rangle^m$ is a second order polynomial in m and the traces for $m = 0, 1, N$ will be zero for one-body operators uncorrelated in the one particle space (note that $\langle h_\alpha h_\beta \rangle^N = \langle h_\alpha \rangle^N \langle h_\beta \rangle^N$),

$$\begin{aligned} \langle h_\alpha h_\beta \rangle^1 &= 0 = \langle h_\alpha h_\beta \rangle^m \quad \alpha \neq \beta \\ \left\langle h_\alpha^2 \right\rangle^m &= \sigma_\alpha^2 = m \left\langle h_\alpha^2 \right\rangle^1 = m\,\sigma_\alpha^2(1) \end{aligned} \tag{1.33}$$

so that the trace in Eq. (1.32) decomposes into an ℓ-fold product of traces $\langle h_s^{\nu_s} \rangle^m$ as a result of which, $M_{2\nu}(m)$ defines a convolution of the ℓ separate h_α^2 densities, which, for large enough ℓ becomes Gaussian in general. When the h_α variances are equal, $\sigma_\alpha^2 = \sigma^2$, the asymptotic spectrum has a χ_ℓ^2 form

$$\rho(x) = 2^{\ell/2} \sigma^{\ell-2} [\Gamma(\ell/2)]^{-1} x^{\ell/2-1} \exp\left(-x/2\sigma^2\right) \, . \tag{1.34}$$

The $\ell = 1$ case is trivially derivable, as indicated above. The $\ell = 3$ distribution was derived by Bethe (1937) who, in his J-decomposition of the level density, treated J_z

as a random variable whose resultant Gaussian distribution then gives Maxwellian for J itself. The $Q \cdot Q$ interaction $\ell = 5$ was first studied by Nomura (1972c) who derived the skewness and excess but did not give the density. A numerical example for $\ell = 8$ using the quadratic Casimir invariant of $SU(3)$ was given by French and Draayer (1979*). Making a natural extension of Eq. (1.34), and relying on the forgiving nature of the CLT, we would argue that any H which is decomposable into l non-singular terms of comparable magnitude whose actions are in some appropriate sense "independent" would give a density close to χ^2_ℓ which tends to Gaussian for large ℓ. It is not yet established if this leads to normality with realistic nuclear interactions. To re-emphasize that nuclear shell model with realistic interactions generates Gaussian densities, we show, in addition to Fig. 1.2, in Fig. 1.7 a larger numerical example.

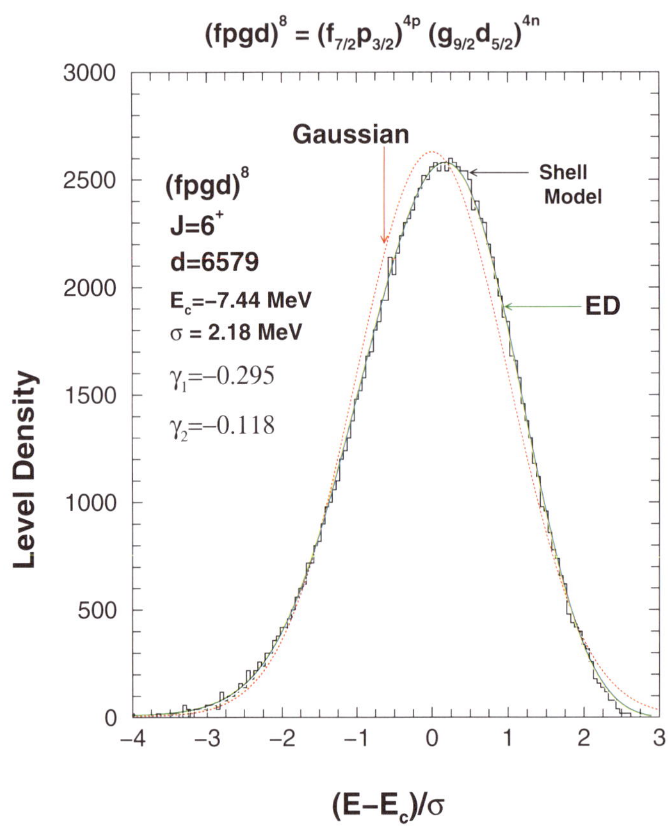

Fig. 1.7 Shell model level density for $(fpgd)^{m=8,J=6^+}$ system compared with a Gaussian and Edgeworth corrected Gaussian (ED). The ED form is in good agreement with the calculated density. Note that $(fpgd)^{m=8}$ corresponds to a proton-neutron system $\left[f_{7/2}p_{3/2}\right]^{m_p=4}\left[g_{9/2}d_{5/2}\right]^{m_n=4}$. The matrix dimension d, energy centroid E_c, spectrum width σ, γ_1 and γ_2 are shown. The calculated results are shown as a histogram constructed with bin size of 0.05. Note that the ground state is -3.98σ below the centroid. The effective interaction and other shell model details are given in [Zuker, Ndeuna, Nowacki and Caurier (2001)]. Shell model eigenvalues provided by A.P. Zuker.

For a generic two-body (or k-body with $k \geq 2$) interaction, an appropriate approach to derive the CLT result is to consider an ensemble of random interactions and examine the ensemble averaged m particle space eigenvalue density. If the ensemble is ergodic, the ensemble averaged results apply to individual members. An analytically tractable ensemble is the embedded GOE or EGOE for spinless fermion systems where the Hamiltonian in two particle (k particle for k-body H) spaces is represented by GOE (assuming we are dealing with a TRI H) and then the m fermion H matrix is generated by using the direct product structure of the m particle states. Following Mon and French, applying effective commutation of operators for $k << m$, it is seen that for EGOE, by counting the binary correlations, in the dilute limit there is CLT Gaussian form. Corrections to the Gaussian moments can be derived. For example, $\gamma_2 \sim -k^2/m$, $k_6 \sim k^3(6k-1)/m^2$ and $k_8 \sim -4k^5(23k-9)/m^3$ [Mon and French (1975*); Kota (2001)]. Since Mon and French (1975*) is included, we refer the reader to this for details of the binary correlation theory for EGOE. Also Nomura (1978) has derived results for the density for the mid shell region, i.e. $m = N/2$ and $m, N - m >> k$, using shell model methods and Gervois (1972) has derived some results for ensembles with two-body interactions using diagrammatic methods. Further, as seen in numerical calculations [Wong and French (1972)] and proved recently by Benet, Rupp and Weidenmüller (2001a*), (2001b), there is a transition from the GOE semi-circle form to a Gaussian, as m increases from k with the transition point at $m = 2k$. Finally, exact analytical results for EGOE follow from the Wigner-Racah algebra of $U(N)$, the algebra that generates the embedding and this has been recognized only recently [Kota (2005b*, 2007a)].

1.2.3 *Partitioning and partial densities*

In the discussion so far we have assumed that we are dealing with a finite set of sp states. However, for real systems (for example nuclei) the sp space has to be expanded as the excitation energy increases. Then the state density keeps growing and takes the well known Bethe form. Therefore the densities of the previous subsections are, in fact, partial densities. To be precise, let us consider the decomposition of the m particle space into subspaces defined by Γ each with $d(\Gamma)$ number of states. Then $m \to \sum \Gamma$ and $d = \sum_\Gamma d(\Gamma)$. Going back to the definition for the total density $I(E)$ we have the remarkable (exact) decomposition, not just for NIP [see Eq. (1.23)] but also for IP systems, of the density into sum of partial densities $I^\Gamma(E)$,

$$\begin{aligned} I^{m,H}(E) &= \langle\langle \delta(H-E) \rangle\rangle^m = \sum_\Gamma \langle\langle \delta(H-E) \rangle\rangle^\Gamma = I^{\Gamma,H}(E) \\ \Rightarrow \rho^{m,H}(E) &= \sum_\Gamma \frac{d(\Gamma)}{d} \rho^{\Gamma,H}(E) \,. \end{aligned} \tag{1.35}$$

It is easy to see that $I^{\Gamma}(E)$ is nothing but the distribution of the strength of each eigenstate in a Γ state averaged over all its sub-states. Thus it represents an intensity distribution. It is important to recognize that for IP systems, the moments $M_p(\Gamma) = \langle H^p \rangle^{\Gamma}$ of $I^{\Gamma,H}(E)$ involve excitations of states from Γ to other Γ subspaces. To see this explicitly consider the second moment. Then the variance involves internal $\Gamma \to \Gamma$ and external $\Gamma \to \Gamma'$, $\Gamma' \neq \Gamma$ variances. The external variances measure the amount of breaking of the symmetry that generates the Γ subspaces. To understand the relationship between the Bethe density and the Gaussian densities, consider the NIP density decomposed into partial densities defined by unitary orbit configurations as given by Eq. (1.23). For light nuclei we can group unitary orbits such that we can construct densities defined over spaces with fixed oscillator excitations $S\hbar\omega$. Similarly for heavy nuclei an appropriate choice of unitary orbits is made such that distant configurations, that admix weakly (when interactions are turned on), are easily identifiable. Now the total density is seen to be a sum of displaced (partial) Gaussian densities one for each S value. The total, as it should, fits extremely well the Bethe form; see Fig. 1.8 for a numerical example and further discussion. The first calculation of this type is due to Smith (1987). Let us point out that there are many calculations of NIP level densities, with and without the use of moment methods, in large spaces but without considering the S decomposition; see for example [Haq and Wong (1980); Jacquemin and Kataria (1986)]. It is should be clear that moment methods generate a theory for level densities for IP systems and we will return to this in Sec. 1.5. Let us add that the role of partial densities in statistical spectroscopy is well recognized; see for example [Chang, French and Thio (1971*); Dalton, Grimes, Vary and Williams (1980); Nishioka, Weidenmüller and Yoshida (1988)] and the discussion ahead.

It is assumed that the CLT Gaussian form derived before for interacting particle systems applies to partial densities as well [see for example French (1967a); Chang, French and Thio (1971*); Dalton, Grimes, Vary and Williams (1980); Parikh (1978); Wong (1986)]. This has its basis in one plus two-body random ensembles [EGOE(1+2)]. To establish this a large number of numerical calculations have been carried out using EGOE(1+2) for spinless fermion systems consisting of random two-body interactions (V) represented by GOE in two particle spaces plus a mean-field one-body part (h) just as in the shell model. In the absence of interactions, the density will be a sum of spikes (delta functions) at the h energies, one for each configuration defined by distributing the m particles in the sp states defined by h. Thus we have partial densities or strength functions (defined by h basis) that are just delta functions. As the interaction is turned on, the spikes spread due to the mixing generated by V and the strength functions are found to quickly acquire the Breit-Wigner (BW) shape. When we further increase the strength λ of V, measured in units of the average spacing (Δ) of the sp states, it is found that level fluctuations start changing from Poisson to GOE at a value $\lambda = \lambda_c$ [Aberg

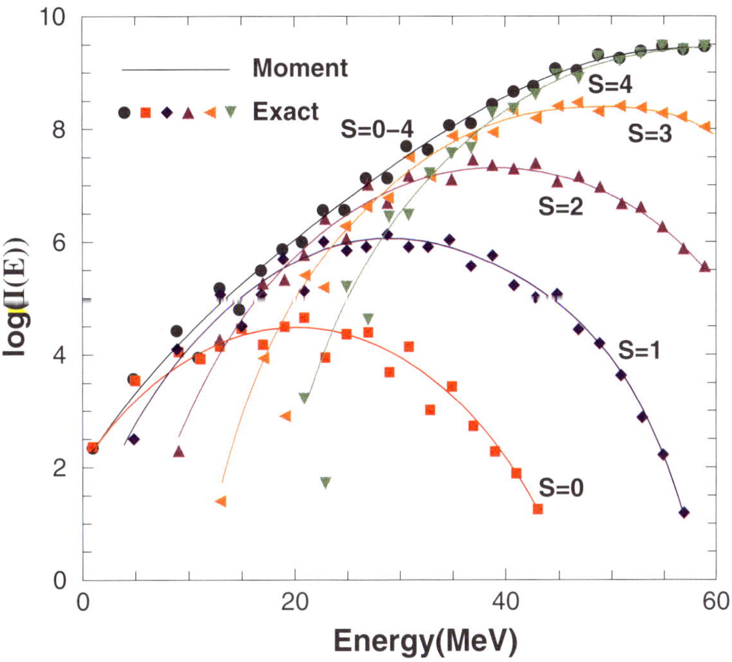

Fig. 1.8 NIP state-density for a system of twelve protons ($m_p = 12$) and twelve neutrons ($m_n = 12$) in sixteen spherical orbits with $S\hbar\omega = 0 - 4\hbar\omega$. The sp orbits are ($\alpha = 1 - 16$) $1s_{1/2}$, $1p_{3/2}$, $1p_{1/2}$, $1d_{5/2}$, $2s_{1/2}$, $1d_{3/2}$, $1f_{7/2}$, $2p_{3/2}$, $1f_{5/2}$, $2p_{1/2}$, $1g_{9/2}$, $2d_{5/2}$, $1g_{7/2}$, $3s_{1/2}$, $2d_{1/2}$, $1h_{11/2}$ and the corresponding sp energies (in MeV) are 0.0, 8.076, 12.206, 16.388, 20.596, 21.851, 24.752, 29.431, 31.208, 32.201, 33.116, 38.213, 40.383, 41.297, 42.238, 41.454 respectively. The sp energies follow from Seeger energies for neutrons [Hillman and Grover (1969)]. The spherical orbits are divided into eight unitary orbits ($\boldsymbol{\alpha}$ = 1-8) and they are [$1s_{1/2}$], [$1p_{3/2}$, $1p_{1/2}$], [$1d_{5/2}$, $2s_{1/2}$, $1d_{3/2}$], [$1f_{7/2}$], [$2p_{3/2}$, $1f_{5/2}$, $2p_{1/2}$], [$1g_{9/2}$], [$2d_{5/2}$, $1g_{7/2}$, $3s_{1/2}$, $2d_{3/2}$], and [$1h_{11/2}$]. The corresponding s values and parities ($s(\pi)$) are 0(+), 1(−), 2(+), 3(−), 3(−), 4(+), 4(+) and 5(−) respectively (in the above choice, parity is uniquely defined for a given s; in principle one can take $s = 4$ for $1h_{11/2}$ orbit but then one has to decompose the space according to s and π). The exact results are obtained by direct counting [i.e. by using spherical configurations ($\mathbf{m}_p$, $\mathbf{m}_n$)]. Exact counting results are compared with those obtained using 4-moment Edgeworth expansion for the unitary configuration densities. The agreements between exact and the moment method are excellent both for the total state density and also for partial S-densities. Also, the total density fits well upto 45 MeV with the back shifted Fermi gas form with $a = 3.75$ MeV $^{-1}$ and $\Delta = -4.68$ MeV (thus, beyond 45 MeV we need to include partial densities for $S = 5, 6$ etc.). The results in the figure are from [Kota and Majumdar (1992a); Majumdar (1994)]. Similar results are also valid [Kota and Majumdar (1992a); Majumdar (1994)] for spin cutoff and occupancy densities (see ahead for the definition of expectation value densities).

(1990); Jacquod and Shepelyansky (1997)]. As we further increase λ, the BW form for strength functions make a transition to the Gaussian form and this is the so called strong coupling limit. Thus as λ increases from zero, after a small value λ_0, the partial densities will be of BW shape [Georgeot and Shepelyansky (1997)] and after it reaches a value $\lambda = \lambda_F$, transition to Gaussian form sets in. Note that the transition to GOE fluctuations start much before, $\lambda_c << \lambda_F$. Thus there is a region, called the BW region, with GOE fluctuations and partial densities that are close to BW form ($\lambda_c < \lambda < \lambda_F$) and a region, called the Gaussian region, where fluctuations are of GOE type and strength functions close to Gaussian form. This

BW to Gaussian transition was investigated using shell model by Lewenkopf and Zelevinsky (1994); Frazier, Brown and Zelevinsky (1996*), 3 level Lipkin model by Wang, Izrailev and Casati (1998), by Flambaum and Izrailev (1997) and Kota and Sahu (2001a), (2001b) using EGOE(1+2) for fermions and finally for boson systems with random interactions by Chavda, Potbhare and Kota (2004). A EGOE(1+2) example is shown in Fig. 1.9. The BW region appears to be relevant for atoms [Flambaum, Gribakina, Gribakin and Ponomarev (1999); Angom, Ghosh and Kota (2004); Angom and Kota (2005)] and the Gaussian region for nuclei. Moreover, it is also established that there is a thermodynamic region defined by $\lambda_t \gg \lambda_F$ where different definitions of entropy, temperature etc. coincide and in this region strength functions will be Gaussian, independent of the basis [Kota and Sahu (2002); Horoi, Zelevinsky and Brown (1995); Jacquod and Varga (2002)]. This strong result is assumed to be valid in the discussion ahead, i.e. in deriving spectral distributions in nuclei. See [Kota (2003*); Angom, Ghosh and Kota (2004); Vyas, Kota and Chavda (2010)] for further discussion on the chaos markers λ_c, λ_F and λ_t. Before going further let us recall, briefly, the standard derivation of the BW form [see Bohr and Mottelson (1969)].

Consider a basis state $|k\rangle$ in the m particle space for a Hamiltonian H. With $|k\rangle = \sum_E C_k^E |E\rangle$, the strength function $F_k(E) = \sum_{E'} \left|C_k^E\right|^2 \delta(E-E')$. Note that $\int dE F_k(E) = \rho(E)$ and therefore $F_k(E)$ are nothing but partial densities and sometimes they are also called local density of states. Let us consider N basis states and denote the eigenvalues of H by E. To derive the form for $|C_k^E|^2$, we denote the first basis state as $|k\rangle$, with energy $\overline{E_k} = \int_{-\infty}^{+\infty} F_k(E) E dE$, and this state couples to the remaining $N-1$ states $|\nu\rangle$. The diagonalization of the $(N-1)\times(N-1)$ sub-matrix gives the eigenvalues ϵ_ν and eigenvectors $|B_\nu\rangle$, $\nu = 1, 2, \ldots, N-1$. Now the eigenvalue equation for the H matrix in the $|k\rangle \oplus |B_\nu\rangle$ basis can be solved giving a coupled equation for the eigenvalues E and this involves ϵ_ν and the matrix elements $V_{k\nu} = V_{\nu k}$. Similarly we can write an expression for $|C_k^E|^2$. To proceed further the following assumptions are made (this is also referred to as the standard model). (i) The background spectrum, defined by the $N-1 \times N-1$ sub-matrix, is dense and rigid, and therefore it can be replaced by a uniform spectrum with mean spacing $\overline{D}$, i.e. $\epsilon_\nu = \overline{n}\overline{D}$, $-\infty \leq \overline{n} \leq +\infty$. (ii) The coupling matrix elements $V_{k\nu}^2$ are uncorrelated with energies ϵ_ν and weakly fluctuate around the mean value. Then $(V_{k\nu})^2 = \overline{(V_{k\nu})^2} = \mathcal{V}^2$. (iii) The mixing is 'sufficiently'strong so that $V_{k\nu}^2 \gg \overline{D}^2$. These finally lead to the BW form for $F_k(E)$,

$$F_{k:BW}(E) = \frac{1}{2\pi} \frac{\Gamma}{(E-\overline{E_k})^2 + \dfrac{\Gamma^2}{4}} \;; \quad \Gamma = \frac{2\pi\mathcal{V}^2}{\overline{D}} \,. \tag{1.36}$$

Here, Γ is the spreading width. Note that the assumptions of the standard model are no longer valid in the "strong coupling" limit, i.e. when the spreading width

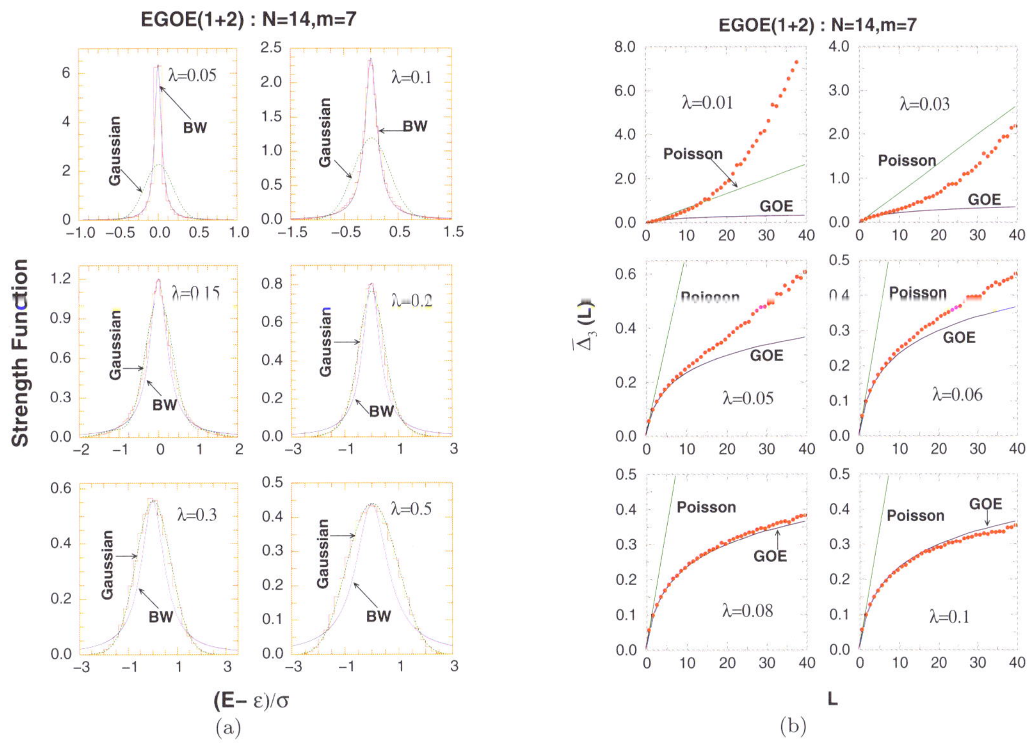

Fig. 1.9 (a) Strength function $F_k(E)$ for EGOE(1+2) for various values of the interaction strength λ in $\{H\} = h(1) + \lambda\{V(2)\}$ for a system of 7 fermions in 14 single particle states; the matrix dimension is 3432. The single particle energies are $\epsilon_i = (i + 1/i), i = 1, 2, \ldots, 14$. The histograms are EGOE(1+2) results for the strength functions, continuous curves are BW fit and the dotted curves are Gaussian for $\lambda \leq 0.1$ and Edgeworth corrected Gaussian for $\lambda > 0.1$. In constructing the strength functions, $|C_k^E|^2$ are summed over the basis states $|k\rangle$ in the energy window $\hat{E}_k \pm \Delta$ and then the ensemble averaged $F_{\hat{E}_k}(\hat{E})$ vs $\hat{E}$ is constructed as a histogram; Δ is chosen as 0.025 for $\lambda \leq 0.1$ and beyond this $\Delta = 0.1$. Here $\hat{E}_k = (E_k - \epsilon_H)/\sigma_H$ and in the figure $\hat{E}_k = 0$. Note the BW to Gaussian transition for $\lambda_{F_k} \sim 0.2$. (b) Dyson-Mehta $\overline{\Delta}_3$ statistic for level fluctuations. The $\Delta_3(L)$ statistic for overlapping intervals of length $L \leq 40$ are compared with Poisson and GOE values. For $\lambda \sim 0.06$, there is Poisson to GOE transition in the $\overline{\Delta}_3$ statistic. Results are from [Kota and Sahu (2001a)].

grows larger than the energy interval ΔE where the level density can be considered as approximately constant.

For situations intermediate to BW and Gaussian it is possible to use an extended t-distribution [Angom, Ghosh and Kota (2004)],

$$\rho_{t-ex}(E:\alpha,\beta) = \frac{(\alpha\beta)^{\alpha-\frac{1}{2}}\,\Gamma(\alpha)}{\sqrt{\pi}\,\Gamma(\alpha-\dfrac{1}{2})}\,\frac{1}{[(E-E_0)^2+\alpha\beta]^{\alpha}}\;, \quad \alpha \geq 1\,. \tag{1.37}$$

Here $\Gamma(-)$ are Γ-functions and α and β are parameters. Eq. (1.37) gives BW for $\alpha = 1$ and Gaussian for $\alpha \to \infty$ (this can be easily checked using Stirling's approximation). As required, it is normalized to unity for any positive value of the continuous parameter α. For $2\alpha - 1$ an integer, $\rho_{t-ex}(E)$ gives the so called

Student's t-distribution [Stuart and Ord (1987)] which is well known in statistics. A non-canonical correction to BW has been suggested recently following the large γ_1 and γ_2 values observed for shell model partial densities in many examples [Kota, Potbhare and Shenoy (1986); Terán and Johnson (2006a)], and this is called modified BW (MBW). The MBW form is defined by [Terán and Johnson (2006b)],

$$\rho_{MBW}(E) = \frac{\mathcal{N}}{W^3} \frac{(E - E_{min})^2 (E_{max} - E)^2}{(E - E_0)^2 + W^2} . \tag{1.38}$$

Here $\mathcal{N}$ is the normalization factor and E_{min} and E_{max} are the end points of the eigenvalues, $E_{min} \le E \le E_{max}$. The four parameters $(E_0, E_{min}, E_{max}, W)$ can be determined from the centroid, variance, γ_1 and γ_2. It should be noted that in the situation that the partial densities are close to Gaussian, it is possible to use GC, ED or CF expansions. However, they can give negative densities near the ground state region (sometimes it is possible to recognize this and correct it by an adhoc procedure). One significant aspect of MBW is that although this form is not derived from any CLT related principle, it is always positive definite. In a similar attempt, Grimes and Massey (1995) suggested the maximum entropy (maxent) method for constructing densities, with moments up to a given order being the same for the approximated and the given density, that are always positive definite. In this approach [Mead and Papanicolaou (1984)], the function

$$S(\rho) = -\int_a^b [\rho(E) \ln \rho(E) \; - \; \rho(E)] \, dE + \sum_{n=0}^{r} \lambda_n \left[\int_a^b E^n \rho(E) dE - M_n \right] \tag{1.39}$$

is maximized, where M_n are the moments of the density function $\rho(E)$ defined in the range $a - b$. Then $\delta S = -\int_a^b (\delta\rho \ln \rho + \delta\rho - \delta\rho) dE \, + \, \sum_{n=0}^{r} \int_a^b \lambda_n E^n \delta\rho dE$. Now $\delta S / \delta \rho = 0$ gives $\rho_r(E) = \exp - \sum_{n=0}^{r} \lambda_n \, E^n$. The parameters λ_n are determined by the moments μ_n. The form as advocated by Grimes and Massey (1995) for $\rho(E)$ is,

$$\rho_{maxent}(E) = \exp(a + bE + cE^2 + dE^3 + eE^4) \tag{1.40}$$

and the parameters (a, b, c, d, e) in Eq. (1.40) are fixed (with the help of an appropriate least square fit procedure) by the normalization condition and the first four moments. Finally, yet another positive definite form suggested, first by Zuker (2001) and based on Eq. (1.29) with $\mu_3(m) \neq 0$, is the continuous binomial $I_{Z-b}(E)$ (with $Z - b$ denoting the continuous binomial due to Zuker),

$$\begin{aligned} I_{Z-b}(E) = d_0 \sqrt{\frac{8}{\pi n}} \exp \Bigg\{ &-(n-1) \left[\frac{E}{E_{max}} \ln \frac{E}{E_{max}} + \frac{E_{max} - E}{E_{max}} \ln \frac{E_{max} - E}{E_{max}} \right] \\ &+ n \frac{E}{E_{max}} \ln \lambda \Bigg\} . \end{aligned} \tag{1.41}$$

In Eq. (1.41), E is the excitation energy (i.e. the energy measured with respect to the ground state). Solving $d = (1 + \lambda)^n$ with d (putting $d_0 = 1$) is the dimension and $\gamma_1 = (1 - \lambda)/\sqrt{n\lambda}$ one can obtain (n, λ) in Eq. (1.41). Alternatively solving

$\gamma_1 = (1-\lambda)/\sqrt{n\lambda}$ and $\gamma_2 = (1+\lambda^2-4\lambda)/n\lambda$ one can obtain (n,λ) and then multiplying $I(E)$ with a constant gives the correct dimension. Finally the exact centroid and width are taken into account by moving the center and scaling. In the limit $n \to \infty$ and $\lambda \to 1$, $I_{Z-b}(E)$ goes to a Gaussian.

1.3 Expectation Value Distributions

Fluctuation free expectation values $K(E) = \langle K \rangle^E$ of an operator K in the H eigen states are encountered for example in calculating occupation probabilities, electromagnetic moments, in the studies of symmetries (where K might be a function of Casimir operators), in calculating spin cutoff factors or J decomposition of state densities, in evaluating strength sums for excitations, for example Gamow Teller (GT) strength sums that are important in beta decay rates calculations and so on. We shall also encounter the expectation value of the Hamiltonian in the eigenstates of another operator, and cases where neither operator is the Hamiltonian. For generality, let us consider the expectation value $\left\langle \widehat{K}_1 \right\rangle^{K_2}$ of an operator $\widehat{K}_1$ in the eigenstate of another operator $\widehat{K}_2$ with eigenvalues K_2. Taking degeneracies in the $\widehat{K}_2$ spectrum into account, the expectation value of $\widehat{K}_1$ is defined by

$$\left\langle \widehat{K}_1 \right\rangle^{K_2} = \left[I^{\widehat{K}_2}(K_2)\right]^{-1} \sum_\alpha \left\langle K_2\alpha \mid \widehat{K}_1 \mid K_2\alpha \right\rangle . \tag{1.42}$$

Then,

$$\left\langle \widehat{K}_1 \right\rangle^{K_2} I^{\widehat{K}_2}(K_2) = \left\langle \left\langle \widehat{K}_1\, \delta(\widehat{K}_2 - K_2) \right\rangle \right\rangle^m = I^{\widehat{K}_2}_{\widehat{K}_1}(K_2) . \tag{1.43}$$

We will see ahead the significance of the second equality in Eq. (1.43). Our goal is to derive statistical laws for expectation values.

1.3.1 *Five methods for moment expansions for expectation values*

Expansion of the δ-function in Eq. (1.43) in terms of orthonormal polynomials P^R_μ defined by some given density say ρ^R immediately produces a moment theory for expectation values. Firstly,

$$\begin{aligned} &\delta(\widehat{K}_2 - K_2) = \rho^R(K_2) \sum_\mu P^R_\mu(\widehat{K}_2)\, P^R_\mu(K_2) \,; \\ &\textstyle\int P^R_\mu(x)\, P^R_\nu(x)\, \rho^R(x)\, dx = \delta_{\mu\nu} \,. \end{aligned} \tag{1.44}$$

Note that, given $\rho(x)$ one can define a set of orthonormal polynomials $P_\mu(x)$ such that $\int_{-\infty}^{+\infty} P_\mu(x) P_\nu(x) \rho(x) dx = \delta_{\mu\nu}$. Explicit form of the polynomials $P_\mu(x)$ in terms of the moments M_i of $\rho(x)$ is [Szego (1939)],

$$P_\mu(x) = [D_\mu D_{\mu-1}]^{-1/2} \begin{vmatrix} 1 & M_1 & - - - & M_\mu \\ M_1 & M_2 & - - - & M_{\mu+1} \\ - & - & - - - & - \\ - & - & - - - & - \\ M_{\mu-1} & M_\mu & - - - & M_{2\mu-1} \\ 1 & x & x^2 - - & x^\mu \end{vmatrix} ;$$

$$D_\mu = \begin{vmatrix} 1 & M_1 & & M_\mu \\ M_1 & M_2 & - - - & M_{\mu+1} \\ - & - & - - - & - \\ - & - & - - - & - \\ M_\mu & M_{\mu+1} & - - - & M_{2\mu} \end{vmatrix} . \tag{1.45}$$

For example the general form for $P_1(x)$ and $P_2(x)$ is given in Eq. (1.14) and the Hermite polynomials $He_\mu(\widehat{x}) = (-1)^\mu \eta_\mathcal{G}^{-1}(\widehat{x}) \dfrac{\partial^\mu}{\partial \widehat{x}^\mu} \eta_\mathcal{G}(\widehat{x})$ are orthogonal polynomials with respect to the standard Gaussian $\eta_\mathcal{G}(\widehat{x})$. Substituting in Eq. (1.43) the expansion for the δ-function given in Eq. (1.44), we obtain

$$\left\langle \widehat{K}_1 \right\rangle^{K_2} = \frac{\rho^R(K_2)}{\rho^{\widehat{K}_2}(K_2)} \sum_{\mu=0}^{\infty} \left\langle \widehat{K}_1 P_\mu^R(\widehat{K}_2) \right\rangle^m P_\mu^R(K_2) \,. \tag{1.46}$$

Note that the ratio of densities before the summation will be unity if we choose $\rho^R(K_2) = \rho^{\widehat{K}_2}(K_2)$ and then the polynomials are those defined by the $\widehat{K}_2$-density of states,

$$\left\langle \widehat{K}_1 \right\rangle^{K_2} = \sum_{\mu=0}^{\infty} \left\langle \widehat{K}_1 P_\mu^{\widehat{K}_2}(\widehat{K}_2) \right\rangle^m P_\mu^{\widehat{K}_2}(K_2) \,. \tag{1.47}$$

This has been used in many studies; see [Draayer, French, Potbhare and Wong (1975a,b); Draayer *et al.* (1975c); Draayer, French and Wong (1977a*,b); Halemane and French (1982a*); Kar (1981a); Sarkar and Kar (1988, 1989)]. Draayer, French and Wong (1977a*) describe the polynomial theory and Halemane and French (1982a*) discuss applications to electromagnetic strengths. However, Eq. (1.46) is valid for any ρ^R. As discussed before, for H representable by EGOE(2), the density of states $\rho(x)$ is close to Gaussian. For the wide class of other K operators which also have a Gaussian spectrum, we can expect that $H_\alpha = H + \alpha K$ is also a Gaussian so that the smoothed $\rho(x)$ and $\rho_\alpha(x)$ can differ only in centroid and variance. For much wider range of operators K (including operators whose spectrum is by no means Gaussian) the same will be true to lowest order in α, which is all we need for our present purposes. Then the change in a given E is $E \to E_\alpha = [\sigma_{H_\alpha}/\sigma_H](E + \alpha \langle K \rangle^m)$; here E is with respect to E_c, i.e. $E = \widetilde{E} = E - E_c$. Now $[\partial E_\alpha / \partial \alpha]_{\alpha=0} = \langle K \rangle^m + \left\langle \widetilde{K}\widetilde{H} \right\rangle^m \widetilde{E}/\sigma_H^2(m)$. Note that $\widetilde{H}$ is defined by subtracting from H, the m particle trace generating part of H and similarly for $\widetilde{K}$.

The first order perturbation theory gives $[\partial E_\alpha/\partial\alpha]_{\alpha=0} = \langle K\rangle^E$. Therefore, under the deformation $H \to H_\alpha = H + \alpha K$, if there is no change in the shape of the state density in the limit of small α, expectation values will be linear in energy,

$$
\begin{aligned}
\langle K\rangle^E &= \langle K\rangle^m + \left\langle \widetilde{K}\widetilde{H}\right\rangle^m \frac{[E - E_c(m)]}{\sigma_H^2(m)} \\
&= \langle K\rangle^m + \zeta_{K-H}\, \frac{\sigma_K(m)[E - E_c(m)]}{\sigma_H(m)} \,.
\end{aligned}
\tag{1.48}
$$

This CLT result, with ζ the correlation coefficient, is the same as the sum of the first two terms in the polynomial expansion given by Eq. (1.47). The procedure followed above tells us little or nothing about the convergence of linearity of the polynomial expansion. Therefore we consider the linear statistical response in greater detail.

The second procedure for expectation values, that of studying the response of the system to a deformation of $\widehat{K}_2$, is more general. It includes the first as a special case and extends also, as we shall see later, to problems involving a non-linear response. In fact this was proposed first by Chang and French (1973a), and applied [Chang and French (1973b*)] to occupation probabilities and spin cutoff factors, before the polynomial expansion given by Eq. (1.47) was introduced. It is easy to show that under the deformation $\widehat{K}_2 \to \widehat{K}_2(\alpha) = \widehat{K}_2 + \alpha\widehat{K}_1$ with $\rho^{\widehat{K}_2}(x) \to \rho^{\widehat{K}_2(\alpha)}(x)$, we have to first order in α,

$$
\begin{aligned}
\rho^{\widehat{K}_2(\alpha)}(x) &= \left\langle \delta(\widehat{K}_2 + \alpha\widehat{K}_1 - x)\right\rangle^m \longrightarrow \rho^{\widehat{K}_2}(x) - \alpha\frac{\partial}{\partial x}\left\langle \widehat{K}_1\delta(\widehat{K}_2 - x)\right\rangle^m \\
&= \rho^{\widehat{K}_2}(x) - \alpha\frac{\partial}{\partial x}\left[\rho^{\widehat{K}_2}(x)\left\langle \widehat{K}_1\right\rangle^x\right] .
\end{aligned}
\tag{1.49}
$$

Here the evaluation of $\left\langle \widehat{K}_1\delta(\widehat{K}_2 - x)\right\rangle^m$ follows using the $\widehat{K}_2$ basis. Then, differentiating with respect to the parameter and integrating with respect to the variable, we have [Chang and French (1973a)]

$$
\left\langle \widehat{K}_1\right\rangle^{K_2} \rho^{\widehat{K}_2}(K_2) = -\int_{-\infty}^{K_2} \left.\frac{\partial\rho^{\widehat{K}_2(\alpha)}(x)}{\partial\alpha}\right|_{\alpha=0} dx = -\lim_{\alpha\to 0}\frac{\partial}{\partial\alpha}F^\alpha(K_2) \,. \tag{1.50}
$$

Here $F^\alpha(K_2)$ is the distribution function in the eigenvalues of $\widehat{K}_2 + \alpha\widehat{K}_1$. For $\rho^{\widehat{K}_2}(x)$ defined by a set of parameters S_ν, $\frac{\partial\rho^{\widehat{K}_2(\alpha)}(x)}{\partial\alpha} = \sum_\nu \frac{\partial\rho^{\widehat{K}_2}(x)}{\partial S_\nu}\frac{\partial S_\nu}{\partial\alpha}$. For example, in the ED representation, we can choose S_1 to be the centroid, S_2 to be the variance and $S_\nu = k_\nu$ for $\nu \geq 3$. The S_1 plus S_2 part gives a linear form (in K_2) for expectation values of $\widehat{K}_1$, i.e. Eq. (1.48). Similarly with the GC expansion, it is possible to choose S_1 and S_2 as the centroid and variance respectively while $S_\nu = C_\nu$, $\nu \geq 3$. Depending on the representation chosen for $\rho^{\widehat{K}_2}(x)$, we get different expansions for expectation values; see ahead for examples. If we take all terms into account, all expansions will give the same result. However, if we truncate them, the results will be different. Now applying the earlier results, i.e. Eq. (1.10), for the change in the centroid ξ and the variance σ^2 of ρ_α, the CLT result is recovered. Thus, we see that the CLT approximation to $K(x)$ will be of general validity provided that, starting at the point H in the operator space, we are able to move a small distance in "almost" any direction without modifying the smoothed shape to first order in α. With $K = H$ this is so, but, on the other hand $K = H^2$ (whose expectation value is quadratic in x, and which by itself would generate a non-Gaussian spectrum) defines an exceptional direction in the operator space, as would other operators closely correlated with a power of H which would similarly require further terms in Eq. (1.48). Under $H \to H + \alpha K$, deformations of the shape give the corrections to the CLT result. We can relate the $\mu \geq 2$ terms in the polynomial expansion given by Eq. (1.47) to the shape deformations and can therefore relate the convergence of the expansion to that of the state density [Draayer, French and Wong (1977a*)]. From this discussion it is easily seen that for occupancies $\langle n_\alpha \rangle^E$ the polynomial expansion works very well giving an essentially linear form with respect to E and the slope of $[\langle n_\alpha \rangle^E - \langle n_\alpha \rangle^m]/\sigma_{n_\alpha}(m)$ vs $\widehat{E}$ is the correlation coefficient $\left\langle \widetilde{n_\alpha} \widetilde{H} \right\rangle^m / [\sigma_{n_\alpha}(m)\, \sigma_H(m)]$. Note that in the covariance term $\left\langle \widetilde{n_\alpha} \widetilde{H} \right\rangle^m$ only the one-body part (defining induced sp energies, see Sec. 1.4) generated by H will contribute and hence occupancy data probes this piece of H. The data, analyzed with a variety of interactions, for $(2s1d)$ shell nuclei is well described by the polynomial theory; see [Potbhare and Pandya (1976); Sarkar, Kar and Kota (1987)].

For $\widehat{K}_1$ operators that are positive definite, $\left\langle \widehat{K}_1 \right\rangle^{K_2}$ is always positive and then the $I_{\widehat{K}_1}^{\widehat{K}_2}(K_2)$ function is positive for all K_2 values. Hence it can be treated as a probability distribution. Therefore it is possible to construct both the $I^{\widehat{K}_2}(K_2)$ state density with moments $M_P^{\widehat{K}_2} = \left\langle \widehat{K}_2^P \right\rangle^m$ and the $I_{\widehat{K}_1}^{\widehat{K}_2}(K_2)$ expectation value density defined by the moments $M_P^{\widehat{K}_2}(\widehat{K}_1) = \left\langle \widehat{K}_1 \widehat{K}_2^P \right\rangle^m / \left\langle \widehat{K}_1 \right\rangle^m$, as GC, ED, CF or some other representation via the lower order moments. Note that the expectation values are now given as the ratio of two densities [French, Kota, Pandey and Tomsovic (1988b*); Kota (2001)] and this is the third procedure,

$$\left\langle \widehat{K}_1 \right\rangle^{K_2} = \frac{I_{\widehat{K}_1}^{\widehat{K}_2}(K_2)}{I^{\widehat{K}_2}(K_2)} = \left\langle \widehat{K}_1 \right\rangle^m \frac{\rho_{\widehat{K}_1}^{\widehat{K}_2}(K_2)}{\rho^{\widehat{K}_2}(K_2)} , \tag{1.51}$$

where

$$\rho_{\widehat{K}_1}^{\widehat{K}_2}(K_2) = \frac{\left\langle \widehat{K}_1\, \delta(\widehat{K}_2 - K_2) \right\rangle^m}{\left\langle \widehat{K}_1 \right\rangle^m} . \tag{1.52}$$

In general for Hamiltonians that are representable by EGOE, the state density $\rho^{\widehat{K}_2}(K_2)$ will be a Gaussian (with $\widehat{K}_2 = \widehat{H}$) and in many situations the bivariate transition strength density generated by a transition operator $\mathcal{O}$ will be a bivariate Gaussian (see Sec. 1.7 for detailed discussion). Then the expectation values of $\widehat{K}_1 = \mathcal{O}^\dagger \mathcal{O}$ will be a ratio of Gaussians, i.e. the two densities in Eq. (1.51) will be

Gaussians (in practice with Edgeworth corrections). Recognizing that $\langle \mathcal{O}^\dagger \mathcal{O} \rangle^E$ is the (non-energy weighted) strength sum, it is easily seen that Eq. (1.51) is appropriate for strength sums [French, Kota, Pandey and Tomsovic (1988b*)]. Indeed Eq. (1.51) has been applied for example, successfully for GT strength sums needed for astrophysical applications; see [Sarkar and Kar (1996*)].

In the situation that two operators $\widehat{K}_1$ and $\widehat{K}_2$ commute, it is possible to extend Eq. (1.50) to $\left\langle \widehat{K}_1^t \right\rangle^{K_1}$ for any t by replacing the first order derivative w.r.t. α to t-th order derivative. Firstly, note that with $M_p^\alpha = \left\langle (\widehat{K}_2 + \alpha \widehat{K}_1)^p \right\rangle$,

$$t! \binom{p+t}{t} \left\langle \widehat{K}_1 \right\rangle M_p^{\widehat{K}_2}(\widehat{K}_1^t) = \lim_{\alpha \to 0} \frac{\partial^t}{\partial \alpha^t} M_{p+t}^\alpha \,. \tag{1.53}$$

Now, extending the definition of the distribution function $F(x) = \int_{-\infty}^x \rho(y)dy$ to multiple-integrated function $F^{(t)}(x) = \int_{-\infty}^x F^{(t-1)}(y)dy$ with $F^{(1)} = F$, will give the fourth procedure [French, Kota and Smith (1989)],

$$\left\langle \widehat{K}_1^t \right\rangle^{K_2} = \left[\rho^{\widehat{K}_2}(K_2) \right]^{-1} (-1)^t \lim_{\alpha \to 0} \frac{\partial^t}{\partial \alpha^t} F^{\alpha,(t)}(K_2) \,. \tag{1.54}$$

Note that the function F in Eq. (1.54) is for the eigenvalues of the operator $\widehat{K}_2 + \alpha \widehat{K}_1$.

Alternatively, it is also possible to consider the bivariate distribution $\rho_{biv}(K_1, K_2)$ and the expectation value of $\widehat{K}_1^P$ or $\widehat{K}_2^P$ is then nothing but the conditional P-th moment of this bivariate density. This gives the fifth procedure [Kota (1984*)],

$$\left\langle \widehat{K}_1^P \right\rangle^{K_2} = [\rho_2(K_2)]^{-1} \int K_1^P \, \rho_{biv}(K_1, K_2) \, dK_1 \,. \tag{1.55}$$

Again if we represent the bivariate density as a bivariate GC or ED or CF etc. we get different expansions for the expectation values. A good example for Eq. (1.55) is fixed-J moments of H and it will give expansions in powers of $J(J+1)$ (see Sec. 1.6 ahead). It is useful to point out that the methods 1 and 2 apply for expectation values of any operator $\widehat{K}_1$, while the third method only to positive definite operators and the fourth and fifth methods only when $\left[\widehat{K}_1 \,,\, \widehat{K}_2 \right]_- = 0$.

1.3.2 *Gram-Charlier and Edgeworth expansions*

To illustrate the expansions for expectation values, let us consider GC and ED expansions with $\widehat{K}_1 = K$, $\widehat{K}_2 = H$ and $K_2 = E$. Using the parametric derivative method defined by Eq. (1.50), we obtain an expansion for the expectation values $\langle K \rangle^E$ by choosing for example $S_1 = E_c$, $S_2 = \sigma^2$ and $S_\nu = C_\nu$ for $\nu \geq 3$ where C_ν are the expansion coefficients in the GC expansion; see Eq. (1.19). It is easily seen that the E_c and σ^2 terms give $\langle K \rangle^m$ and $\left\langle \widetilde{K} \widehat{H} \right\rangle^m \widehat{E}$. Note that $\widetilde{K} = K - \langle K \rangle^m$, $\widetilde{H} = H - \langle H \rangle^m$, $\widehat{H} = \widetilde{H}/\sigma(m)$ and $\widehat{E} = (E - \langle H \rangle^m)/\sigma(m)$ where $\sigma(m) = \{ \left\langle \widetilde{H}^2 \right\rangle^m \}^{1/2}$.

Under CLT action as $\rho(E) \to \rho_{\mathcal{G}}(E)$, expectation values $\langle K \rangle^E \to K_{CLT}(E)$ where

$$\begin{aligned} K_{CLT} &= \langle K \rangle^m + \left\langle \widetilde{K}\widehat{H} \right\rangle^m \widehat{H} \\ \Rightarrow \langle K \rangle^E \stackrel{CLT}{\longrightarrow} & K_{CLT}(E) = \langle K \rangle^m + \left\langle \widetilde{K}\widehat{H} \right\rangle^m \widehat{E} \; . \end{aligned} \tag{1.56}$$

The corrections to CLT arise from S_ν with $\nu \geq 3$. Using the GC expansion we have $\frac{\partial \rho}{\partial S_\nu} = \frac{1}{\nu!} He_\nu(\widehat{E})\rho_{\mathcal{G}}(\widehat{E})$. Similarly

$$\begin{aligned} \left.\frac{\partial S_\nu}{\partial \alpha}\right|_{\alpha=0} &= \frac{\partial}{\partial \alpha} \left\langle He_\nu(\widehat{H}(\alpha)) \right\rangle^m \Big|_{\alpha=0} \\ &= \nu \left\langle He_{\nu-1}(\widehat{H}) \left.\frac{\partial \widehat{H}(\alpha)}{\partial \alpha}\right|_{\alpha=0} \right\rangle^m = \frac{1}{\sigma}\left[K - K_{CLT}\right] \; ; \\ \left.\frac{\partial \widehat{H}(\alpha)}{\partial \alpha}\right|_{\alpha=0} &= \frac{\partial}{\partial \alpha} \left\{ \frac{H + \alpha K - \langle H + \alpha K \rangle^m}{[\langle (H+\alpha K)^2 \rangle^m - [\langle (H+\alpha K) \rangle^m]^2]^{1/2}} \right\}\Bigg|_{\alpha=0} . \end{aligned} \tag{1.57}$$

This finally gives,

$$\begin{aligned} \langle K \rangle^E &= K_{CLT}(E) + \Delta[K(E)] \; ; \\ \Delta[K(E)] &= \frac{\eta_{\mathcal{G}}(\widehat{E})}{\eta(\widehat{E})} \sum_{\nu \geq 2} \frac{1}{\nu!} \left\langle [K - K_{CLT}] He_\nu(\widehat{H}) \right\rangle^m He_\nu(\widehat{E}) \; . \end{aligned} \tag{1.58}$$

For the ED expansion, $S_\nu = k_\nu$ for $\nu \geq 3$; see Eqs. (1.16) and (1.17). Then,

$$\begin{aligned} \Delta[K(E)] = \frac{\eta_{\mathcal{G}}(\widehat{E})}{\eta(\widehat{E})} \sum_{P=1}^{\infty} \sum_{[P]} & \left\{ \left[\prod_{i=1}^{\ell} \left(\frac{k_{P_i+2}}{(P_i+2)!} \right)^{S_i} \frac{1}{(S_i\,!)} \right] \right. \\ & \left. \times \sum_{i=1}^{\ell} S_i \left(k_{P_i+2}\right)^{-1} He_{P+2S-1}(\widehat{x}) \left[\frac{\partial k_{P_i+2}(\alpha)}{\partial \alpha} \right]_{\alpha=0} \right\} , \end{aligned} \tag{1.59}$$

where

$$\begin{aligned} \left.\frac{\partial k_P(\alpha)}{\partial \alpha}\right|_{\alpha=0} = P! \sum_{[P']} & \left[\left\{ \prod_{i=1}^{\ell} \left(\frac{\mu_{p_i}}{P_i!} \right)^{S_i} \frac{1}{S_i!} \right\} (-1)^S \, (S-1)! \right. \\ & \left. \times \left\{ \sum_{\substack{i=1 \\ P_i \neq 2}}^{\ell} \left(\mu_{p_i}\right)^{-1} S_i P_i \left\langle [K - K_{CLT}] \widehat{H}^{P_i - 1} \right\rangle^m \right\} \right] . \end{aligned} \tag{1.60}$$

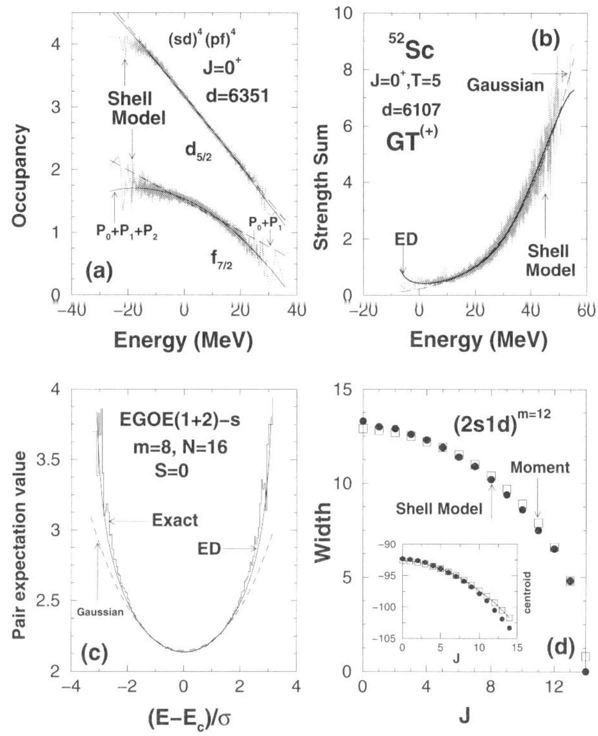

Fig. 1.10 Examples for expectation values calculated using spectral distribution methods 1, 3 and 5 as described in the text. (a) $f_{7/2}$ and $d_{5/2}$ occupancies in a $(2s1d)^{m_p=4}(2p1f)^{m_n=4}$ space with $J = 0^+$. To avoid overlap between the figures, results for $d_{5/2}$ are displaced by one unit. Exact shell model results are compared with the polynomial expansion upto 1st and second order (method #1). The second order corrections are important for $f_{7/2}$ occupancies. For details see [Gómez *et al.* (2001)]. (b) $\text{GT}^{(+)}$ strength sum originating from $J = 0^+, T = 5$ levels (energies in the figure are for these levels) in ^{52}Sc. Exact shell model results are compared with ratio of Gaussians and ratio of Edgeworth corrected Gaussians (method #3). Figure constructed from the results in [Gómez *et al.* (2001)]; see also [Kota *et al.* (1999)]. (c) Pair expectation value $\langle PP^\dagger \rangle^E$ vs E for a one plus two-body EGOE with spin degree of freedom for 8 fermions in 16 sp states (or 8 levels). The Hamiltonian $H = h(1) + \lambda\{V(2)\}$ where the sp energies defining $h(1)$ are $\epsilon_i = i + 1/i$ for the i-th level and $\{V(2)\}$ is a GOE in two particle space. Results are shown for levels with total spin $S = 0$. Energy of the level is denoted by E in the figure and E_c and σ are the energy centroid and the spectrum width respectively. Exact results are compared with ED and Gaussian curves as in (b). The ED results are in good agreement with exact results. All the results are from [Vyas, Kota and Chavda (2009)]. (d) Fixed-J centroid $E_c(J)$ and width $\sigma(J)$ vs J in $(2s1d)^{m=12}$ space calculated (open squares) using method #5. See [Haq (1975); Kota (1984*); Kota and Kar (2002)] for details.

We show in Fig. 1.10 comparison of expectation values calculated using methods 1-4 (in some cases using ED expansion) with exact results.

1.3.3 *Partitioning, convergence and multi-variate expectation values*

Just as with state densities, in large spaces (and in small spaces for better accuracy) it is essential to consider partitioning of the m particle spaces. Then the expectation value density decomposes (exactly) into the corresponding densities defined over the subspaces,

$$\begin{aligned} I_{\widehat{K}_1}^{\widehat{K}_2}(K_2) = \left\langle\left\langle \widehat{K}_1 \delta(\widehat{K}_2 - K_2) \right\rangle\right\rangle^m &= \sum_{\Gamma \in m} \left\langle\left\langle \widehat{K}_1 \delta(\widehat{K}_2 - K_2) \right\rangle\right\rangle^\Gamma \\ &= \sum_\Gamma I_{\widehat{K}_1}^{\widehat{K}_2;\Gamma}(K_2) \; . \end{aligned} \tag{1.61}$$

Now, for the partial expectation value densities $I_{\widehat{K}_1}^{\widehat{K}_2;\Gamma}(K_2)$, defined by the traces $\left\langle\left\langle \widehat{K}_1 (\widehat{K}_2)^p \right\rangle\right\rangle^\Gamma$, it is possible to apply the methods 1-4 depending on the nature of $\widehat{K}_1$ and the Γ. Sometimes this may result in a very simple expression. For example for $\widehat{K}_1 = n_\alpha$, if we choose spherical configurations, we have the simple expression,

$$\langle n_\alpha \rangle^E = \sum_{\mathbf{m}} \frac{I^{\mathbf{m}}(E)}{I(E)} \, m_\alpha(\mathbf{m}) \; . \tag{1.62}$$

where $m_\alpha(\mathbf{m})$ is number of particles in the orbit α for the configuration $\mathbf{m}$. Further discussion and applications of this result for $(2s1d)$ shell nuclei is given in [Potbhare and Pandya (1976); Potbhare and Tressler (1991)], for $(2p1f)$ shell nuclei in [Kota and Potbhare (1979); Kota (1991a)] and for $(pf_{5/2}g_{9/2})$ shell nuclei in [Kota, Pandya and Potbhare (1982); Naik, Ramkumar, Shenoy and Tressler (1985)]. Similarly, partitioning the space into Wigner's spin-isospin $SU(4)$ subspaces will give a simple expression for GT strength sums (also for $M1$ strength sums) [Kar (1981b, 1983)]. Note that the GT operator $\sigma\tau$ is a generator of the $SU(4)$ algebra.

Applying the methods 1-5 with or without partitioning, a variety of expansions for expectation values of an operator can be derived and they differ from each other when truncated to a finite number of terms. Thus the issue of convergence of these expansions is extremely important. In many situations, operation of two-body ensembles and quantum chaos in many particle spaces will determine the appropriate expansion for a given operator $\widehat{K}_1$. For detailed discussion see Kota (2003*). Let us add that in some situations as in $(2p1f)$ shell for GT strength sums, Eq. (1.61) with spherical configuration partitioning and polynomial expansion gives a simple formula as the correlation coefficient [see Eq. (1.48) for definition] here is very small [Sarkar and Kar (1996*); Kota (2003*)] and therefore results in an expression similar to Eq. (1.62). Interestingly the resulting formula for the GT strength sums is the same as that derived by Macfarlane (1986) using quite a different approach.

Two important remarks at this stage are: (i) linear response theory when extended to second order derivative of $\rho(E)$ allows one to study inverse energy weighted sums [Halemane (1981b); Kota, Potbhare, Shenoy and Tressler (1989)]; (ii) extensions to bivariate and multivariate expectation values, though complicated, are important and it is possible to derive moment expansions for them. An example for (ii) is fixed-(E, J) occupancies $\langle n_\alpha \rangle^{E,J}$ and also for fixed-(E, J, T). Here one has to deal with polynomials in two or more variables [Kota and Potbhare (1985)].

To illustrate the usefulness of multivariate distributions, let us consider a multivariate Gaussian in R number of variables $\widehat{x}_i$. We will write $\widehat{x}_i$ as a row matrix $[\widehat{x}]$ with the matrix element in i-th column being $\widehat{x}_i$. Similarly $[\rho]$ denotes the correlation matrix, $\rho_{ij} = \langle \widehat{x}_i \widehat{x}_j \rangle$. Note that $\rho_{ii} = 1$ for all i and $\rho_{ij} = \rho_{ji}$. With $[\Lambda] = [\rho]^{-1}$, the multivariate Gaussian is defined by [Cramer (1946)]

$$\eta_{\mathcal{G}}([\widehat{x}]) = \frac{|\Lambda|^{1/2}}{(2\pi)^{n/2}} \exp -\frac{1}{2} \left([\widehat{x}] \, [\Lambda] \, [\widehat{x}]^T \right) . \tag{1.63}$$

Here $[\widehat{x}]^T$ is the transpose of the matrix $[\widehat{x}]$. Now consider the conditional density $\eta_{\mathcal{G}}(\widehat{x}_1/\widehat{x}_2, \widehat{x}_3, \ldots, \widehat{x}_R)$ in x_1. Then Eq. (1.63) gives,

$$\begin{aligned} \eta_{\mathcal{G}}(\widehat{x}_1/\widehat{x}_2, \widehat{x}_3, \ldots, \widehat{x}_R) &= \frac{\eta_{\mathcal{G}}(\widehat{x}_1, \widehat{x}_2, \widehat{x}_3, \ldots, \widehat{x}_R)}{\int \eta_{\mathcal{G}}(\widehat{x}_1, \widehat{x}_2, \widehat{x}_3, \ldots, \widehat{x}_R) d\widehat{x}_1} \\ &= \frac{1}{\sqrt{2\pi}\, \Lambda_{11}^{-1/2}} \exp - \frac{\left(\widehat{x}_1 + \sum\limits_{k=2}^{R} \frac{\Lambda_{1k}}{\Lambda_{11}} \widehat{x}_k \right)^2}{2\Lambda_{11}^{-1}} . \end{aligned} \tag{1.64}$$

Here the exponential form is easy to recognize and the normalization follows from $\int \eta_{\mathcal{G}}(\widehat{x}_1/\widehat{x}_2, \widehat{x}_3, \ldots, \widehat{x}_R) d\widehat{x}_1 = 1$. Thus the conditional density is a Gaussian with the centroid and variance given by,

$$\begin{aligned} E_c^{\widehat{X}_1}(\widehat{x}_2, \widehat{x}_3, \ldots, \widehat{x}_R) &= \left\langle \widehat{X}_1 \right\rangle^{\widehat{x}_2, \widehat{x}_3, \ldots, \widehat{x}_R} = -\sum_{k=2}^{R} \frac{\Lambda_{1k}}{\Lambda_{11}} \widehat{x}_k \\ \left[\sigma^{\widehat{X}_1}(\widehat{x}_2, \widehat{x}_3, \ldots, \widehat{x}_R) \right]^2 &= \left\langle \widehat{X}_1^2 \right\rangle^{\widehat{x}_2, \widehat{x}_3, \ldots, \widehat{x}_R} - \left\{ E_c^{\widehat{X}_1}(\widehat{x}_2, \widehat{x}_3, \ldots, \widehat{x}_R) \right\}^2 = \Lambda_{11}^{-1} \end{aligned} \tag{1.65}$$

If we have a situation where the correlation matrix is of the simple form with $\rho_{1j} = \rho_{j1}$ non-zero, $\rho_{ii} = 1$ and all other ρ_{ij} are zero, the conditional centroid and variance will take a simple form. Firstly, we have the determinant $|\rho| = (1 - \sum_{k=2}^{R} \rho_{1k}^2)$, $\Lambda_{11} = 1/|\rho|$ and $\Lambda_{1k} = \Lambda_{k1} = -\rho_{1k}/|\rho|$ for $k \neq 1$. Then we have, using Eq. (1.65)

$$\begin{aligned} E_c^{\widehat{X}_1}(\widehat{x}_2, \widehat{x}_3, \ldots, \widehat{x}_R) &= \sum_{k=2}^{R} \rho_{1k} \widehat{x}_k , \\ \left[\sigma^{\widehat{X}_1}(\widehat{x}_2, \widehat{x}_3, \ldots, \widehat{x}_R) \right]^2 &= \left(1 - \sum_{k=2}^{R} \rho_{1k}^2 \right) . \end{aligned} \tag{1.66}$$

As an application of Eq. (1.66), consider the example where $\widehat{X}_1 = H$ and $\widehat{X}_i = n_{i-1}$ for $i > 1$ where n_i are number operators for the orbits i. The eigenvalues of the number operators define configurations and they label the eigenvectors of $h(1) = \sum_i \epsilon_i n_i$. Let us say that the degeneracy of each orbit is N_i so that $\sum_i N_i = N$.

Similarly $n = \sum_i n_i$ gives total number of particles. Then we have,

$$\begin{aligned}\widetilde{n}_i &= n_i - n\frac{N_i}{N} \,, \\ \langle \widetilde{n}_i \widetilde{n}_j \rangle^m &= \frac{m(N-m)}{N(N-1)} \langle \widetilde{n}_i \widetilde{n}_j \rangle^1 \\ &= \frac{m(N-m)}{N(N-1)} N_i \left(\delta_{ij} - \frac{N_j}{N} \right) .\end{aligned} \tag{1.67}$$

The correlation coefficient giving $\rho_{n_i n_j}$ is

$$\begin{aligned}\rho_{n_i n_j} &= \frac{\langle \widetilde{n}_i \widetilde{n}_j \rangle^m}{[\langle \widetilde{n}_i \widetilde{n}_i \rangle^m \langle \widetilde{n}_j \widetilde{n}_j \rangle^m]^{1/2}} \\ &= -\sqrt{\frac{N_i N_j}{(N-N_i)(N-N_j)}} \quad \text{for} \quad i \neq j \\ &\longrightarrow 0 \quad \text{for} \quad N_i << N, \;\; N_j << N, \\ \langle \widetilde{n}_i \widetilde{n}_i \rangle^m &= \sigma^2_{n_i}(m) \\ &\longrightarrow \frac{m(N-m)}{N(N-1)} N_i \left(1 - \frac{N_i}{N} \right) .\end{aligned} \tag{1.68}$$

For H, a one plus two-body operator, as discussed ahead in detail in Sec. 1.4, it decomposes into a scalar ($\nu = 0$) part, an effective one-body ($\nu = 1$) part and the remaining irreducible two-body ($\nu = 2$) part; the $\nu = 1$ part is defined by the m dependent traceless sp energies $\lambda_i(m)$; $H^{\nu=1} = \sum_i \lambda_i(m) n_i$. Then in the $\left\langle \widetilde{n}_i \widetilde{H} \right\rangle^m$ trace, only the $\nu = 1$ part of H will contribute. This gives,

$$\left\langle \widetilde{n}_i \widetilde{H} \right\rangle^m = \frac{m(N-m)}{N(N-1)} \langle \langle \widetilde{n}_i H^{\nu=1} \rangle \rangle^1 = \frac{m(N-m)}{N(N-1)} \lambda_i(m) N_i \,. \tag{1.69}$$

Clearly the structure of ρ_{ij} is the same as that of the special situation considered just above Eq. (1.66). Therefore, with m_i giving the eigenvalues of n_i,

$$\begin{aligned}\left\langle \widehat{H} \right\rangle^{m_1, m_2, \ldots m_K} &= \sum_{r=1}^{K} \frac{\left\langle \widetilde{n}_r \widehat{H} \right\rangle^m}{\sigma_{n_r}(m)} \frac{\widetilde{m}_r}{\sigma_{n_r}(m)} \\ &= \frac{1}{\sigma_H(m)} \sum_{r=1}^{K} \lambda_r(m) m_r \\ \Rightarrow \langle H \rangle^{m_1, m_2, \ldots m_K} &= \langle H \rangle^m + \sum_{r=1}^{K} m_r \lambda_r(m) \,.\end{aligned} \tag{1.70}$$

Similarly,

$$
\begin{aligned}
&\left\langle \widehat{H}^2 \right\rangle^{m_1,m_2,\ldots m_K} - \left[\left\langle \widehat{H} \right\rangle^{m_1,m_2,\ldots m_K} \right]^2 = 1 - \sum_{r=1}^{K} \rho^2_{n_r H} \\
&= 1 - \frac{1}{\sigma^2_H(m)} \sum_{r=1}^{K} \left[\frac{m(N-m)}{N(N-1)} N_r \right]^{-1} \left[\frac{m(N-m)}{N(N-1)} \lambda_r(m) N_r \right]^2 \\
&- \frac{1}{\sigma^2_H(m)} \left\{ \sigma^2_H(m) - \sum_{r=1}^{K} \frac{m(N-m)}{N(N-1)} N_r \lambda^2_r(m) \right\} \; ; \quad K - R - 1 \, .
\end{aligned}
\tag{1.71}
$$

Thus the conditional centroids depend only on the $\nu = 1$ part (i.e. the mean-field producing part) of H and the variance is m-particle H variances minus the variance of the configuration centroid distribution (this is the variance generated by the $\nu = 2$ part). Thus the variances are a constant, independent of the configurations $(m_1, m_2, \ldots, m_K)$. Therefore the assumption that $\rho(E, m_1, m_2, \ldots, m_K)$ is a multivariate Gaussian leads to (i) configuration centroids with contributions only from the $H^{\nu=1}$ part and (ii) constant configuration variances (derive only from the $H^{\nu=2}$ part). Compact formulas for the centroids and variances follow in the dilute limit defined by $N_r/N \to 0$ for all r (this also assumes that K, the number of orbits is large). All these structures are seen in nuclear shell model examples [French and Kota (1983*)]. An important outcome of Eqs. (1.70) and (1.71) is that the bivariate distribution $\rho(E, E_k)$, with H generating E and $H_k = H^{\nu=1}$ generating E_k, will be close to a Gaussian with the correlation coefficient $\zeta = \sigma_{H_k}/\sigma_H$. Recently, this result has been exploited to derive simple formulas for the chaos measures, number of principle components (NPC) and information entropy, for wavefunctions expanded in the mean-field basis [Kota and Sahu (2001a)]. Finally, the bivariate Gaussian form for $\rho(E, E_k)$ also implies strong correlation between the diagonal matrix elements of H in the mean-field basis (i.e. E_k) and the eigenvalues E. Recently, there have been detailed investigations of this result for many systems [Shen, Zhao, Arima and Yoshinaga (2008b); Yoshinaga, Arima, Shen and Zhao (2009)].

1.4 Unitary Decompositions, Geometry and Configuration Traces: State and Spin Cutoff Densities

Operator traces or defining moments propagate from few particle to many particle spaces and some examples for NIP have already been discussed. To go further, it is useful to consider the unitary tensor decomposition of the Hamiltonian and other operators [French and Kota (1989a)] and recognize that CLT defines a geometry, i.e. norms and correlation coefficients for operators. In addition, unitary decompositions simplify trace propagation equations for the moments of the distributions that are needed in applications. We now discuss these aspects of statistical spectroscopy with state and level densities as our reference. For this, trace propagation involves only particle numbers, hence it is largely possible to avoid group theory and simply use combinatorial methods.

1.4.1 *Unitary decompositions and renormalization of the NIP Hamiltonian*

We take the Hamiltonian $H = C(0) + h(1) + V(2) \equiv C + h + V$ operating in a model space generated by distributing m particles over N sp states which are labeled by i and which form a set of spherical, or often harmonic oscillator (h.o.) orbits; we take for granted that the space is large enough to accommodate the $S\hbar\omega$ excitations of interest. Then with i denoting the sp states and α the sp orbits, the NIP part $h(1) = \sum_i \epsilon_i(1)n_i = \sum_\alpha \epsilon_\alpha(1)n_\alpha$. Similarly, the two-body matrix elements defining $V(2)$ are V_{ijkl} in single state representation and $V^{JT}_{\alpha\beta\gamma\delta}$ for spherical orbits with JT coupling; with $|(\alpha\beta)JT\rangle$ denoting normalized two-particle states, we have $V^{JT}_{\alpha\beta\gamma\delta} = \langle(\alpha\beta)JT \mid V(2) \mid (\gamma\delta)JT\rangle$. The relationship between the two-body matrix elements in isospin formalism and the (p,n) formalism, for charge-independent V, is [French (1966b)],

$$
\begin{aligned}
V^{J}_{\alpha\beta\gamma\delta:pn} &= (2\zeta_{\alpha\beta}\zeta_{\gamma\delta})^{-1}\{V^{J,0}_{\alpha\beta\gamma\delta} + V^{J,1}_{\alpha\beta\gamma\delta}\};\ \zeta_{\alpha\beta} = (1+\delta_{\alpha\beta})^{-1/2} \\
V^{J}_{\alpha\beta\gamma\delta:pp} &= V^{J}_{\alpha\beta\gamma\delta:nn} = V^{J,1}_{\alpha\beta\gamma\delta}\ .
\end{aligned}
\tag{1.72}
$$

Partitioning the sp space into h.o. orbits and the model space into interacting subspaces labeled by (S,π) where π denotes parity, gives a very simple structure for the state density $I(E)$. Finer partitioning will usually be needed for adequate NIP accuracy, for example, as in Sec. 1.2.2., into spherical orbits α and spherical configurations $\mathbf{m}$ or L unitary orbits $\boldsymbol{\alpha}$ and unitary configurations $[\mathbf{m}]$.

Given V, it can be decomposed into a part $V^{[0]}$ that produces spherical configuration centroids and the remaining part $\boldsymbol{V}$ that produces the spreading of the configurations,

$$
V = V^{[0]} + \boldsymbol{V}\ . \tag{1.73}
$$

Partitioning into unitary configurations is sometimes needed also for the density (spreading function) $\rho^{\boldsymbol{V}}$ with $S > 0$. Note that the symbol N is always the dimensionality of the significant sp space. For further discussion we introduce the symbols $\mathcal{X}_{ab}$ and $\hat{\mathcal{X}}_{abc\ldots}$ where

$$
\begin{aligned}
&\mathcal{X}_{ab} = \mathcal{X}_a\left(\mathcal{X}_b - \delta_{ab}\right)/(1+\delta_{ab})\ , \quad \hat{\mathcal{X}}_{abc\ldots} = \mathcal{X}_a\left(\mathcal{X}_b - \delta_{ba}\right)\left(\mathcal{X}_c - \delta_{ca} - \delta_{cb}\right)\ldots \\
&\mathcal{X} = n, N, m, m^X\ ; \quad a, b\ldots = \alpha, \beta, \ldots \ \text{or}\ \boldsymbol{\alpha}, \boldsymbol{\beta}, \ldots\ .
\end{aligned}
\tag{1.74}
$$

The $V^{[0]}$ that yields as eigenvalues the centroids of all spherical configurations $\mathbf{m}$ is,

$$
\begin{aligned}
&V^{[0]} = \sum V_{\alpha\beta} n_{\alpha\beta}\ ; \quad V_{\alpha\beta} = (N_{\alpha\beta})^{-1} \sum_J (2J+1)\, V^J_{\alpha\beta\alpha\beta} \\
&V^{[0],J}_{\alpha\beta\gamma\delta} = V_{\alpha\beta}\delta_{\alpha\gamma}\delta_{\beta\delta}\ ; \quad \boldsymbol{V}^J_{\alpha\beta\gamma\delta} = V^J_{\alpha\beta\gamma\delta} - V_{\alpha\beta}\delta_{\alpha\gamma}\delta_{\beta\delta}\ ; \quad \boldsymbol{V}_{\alpha\beta} = 0\ .
\end{aligned}
\tag{1.75}
$$

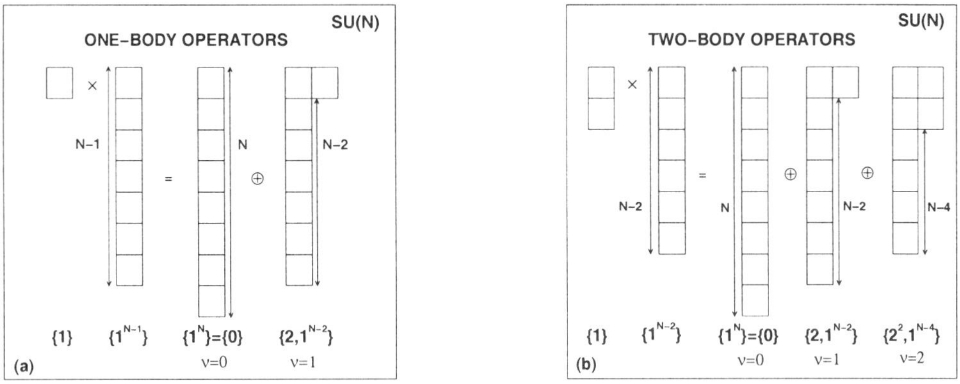

Fig. 1.11 Young tableaux for various tensor parts of (a) one-body and (b) two-body operators with respect to SU(N). See [Vincent (1967); Chang, French and Thio (1971*)] for details.

The vanishing of every $\boldsymbol{V}_{\alpha\beta}$ [as long as we ignore purely radial excitations (e.g. 1s $\rightarrow$ 2s) which is appropriate; see Chang, French and Thio (1971*) for the general case] follows directly from the $V_{\alpha\beta}$ construction. Thus the spreading interaction $\boldsymbol{V}$ is traceless in every spherical configuration. By making use of J-averaging at the outset we have lost information about angular momentum. However, we shall recover the J-dependence of the state density (and of other quantities) by introducing spin cutoff factors and taking account of $(J^2 - \boldsymbol{V})$ and $(J^2 - \boldsymbol{V}^2)$ correlations.

A number-conserving k-body operator $F(k)$ which acts in an (N, m) space can be represented as a sum of $SU(N)$ tensors $F^{\nu}(k)$ which are ν-body operators ($\nu = 0, 1, \ldots k$) promoted to particle-rank k by the unitary-scalar (ν- preserving) operator $\binom{n-\nu}{k-\nu}$ (see Fig. 1.11 for $k = 1, 2$ examples),

$$F(k) = \sum_{\nu=0}^{k} F^{\nu}(k) = \sum_{\nu=0}^{k} \binom{n-\nu}{k-\nu} F^{\nu}(\nu) \,. \tag{1.76}$$

Since only the $\nu = 0$ part contributes to m-particle traces, we have the important relation,

$$\langle F(k) \rangle^m = \langle F^0(k) \rangle^m = \binom{m}{k} \langle F(k) \rangle^k \,. \tag{1.77}$$

An essential property of a Hermitian F^k, or anti-hermitian G^k, is that, under the hole-particle transformation ($a_i \leftrightarrow a_i^{\dagger}$; $i = 1 \ldots N$)

$$F^k \rightarrow F^{k,X} = (-1)^k F^k; \quad G^k \rightarrow G^{k,X} = (-1)^{k+1} G^k \tag{1.78}$$

where X denotes the "hole-particle adjoint" under which for example $n \rightarrow (N - n)$, $n_{\alpha} \rightarrow N_{\alpha} - n_{\alpha}$, $\binom{n}{2} \rightarrow \binom{N-n}{2}$. If not specified otherwise, all number-conserving operators will be assumed Hermitian. Using Eq. (1.76) it is easy to write the trace-orthogonality relation

$$\left\langle F^{\nu} G^{\nu'} \right\rangle^m = \binom{m}{\nu} \binom{N-m}{\nu} \binom{N-\nu}{\nu}^{-1} \left\langle F^{\nu} G^{\nu'} \right\rangle^{\nu} \delta_{\nu\nu'} \,. \tag{1.79}$$

We have here also the propagation rule for the trace of the product of two tensors, and hence for their squared norm $(| F |_m)^2 = \langle F^\dagger F \rangle^m - | \langle F \rangle^m |^2$ which decomposes without cross-terms in ν and also for the correlation coefficient between them. Taking $\nu' = 0$ in Eq. (1.79) gives the result that non-scalar ($\nu \neq 0$) tensors are traceless. The ν-decomposition is trivial for a one-body (number-conserving) operator K since it involves only a shift of the origin to the spherical centroid as indicated for K^1 in Eq. (1.80) below. Moreover, an easy manipulation of Eq. (1.75) gives for K and also for a two-body operator $F = \binom{n}{2} F^0 + (n-1)F^1 + F^2$ the results,

$$\begin{aligned}
&K^0 = \langle K \rangle^1 \;;\quad K^1 = K - \langle K \rangle^1 n = N^{-1}\{n^X K - nK^X\} \\
&F^0 = \langle F \rangle^2 \;;\quad F^1 = (N-2)^{-1}\{F - F^X + \tfrac{1}{2}(N-1)(N-2n)\langle F \rangle^2\} \\
&F^2 = (N-2)^{-1}\{n^X - 1)F + (n-1)F^X - \tfrac{1}{2}N(n-1)(n^X-1)\langle F \rangle^2\} \;.
\end{aligned} \tag{1.80}$$

Recall that F^X is the hole-particle adjoint of F. Since F is two-body, F^X is a two-hole operator or, equivalently a (0+1+2)-body operator, i.e. $F^X = F^X(0) + F^X(1) + F^X(2)$. Decomposing the F^1 equation in Eq. (1.80) by particle rank k then gives $F^X(2) = F$ (which would follow also from ignoring the $a, a^\dagger$ anti-commutators which, under particle-hole transformation, generate $F^X(1)$ and $F^X(2)$ and $F^X(0) = \binom{N}{2} \langle F \rangle^2$ whose genesis should be obvious).

We now consider the unitary decomposition of $V^{[0]}$. Since $V^{[0]}$ defined in Eq. (1.75) is a function only of (commuting) spherical-orbit number operators, unitary decomposition and behavior under hole-particle transformations follow immediately from those of n, n_α and $n_{\alpha\beta}$. Its $\nu = 0$ part will shift the primary spectrum by an amount proportional to $\binom{n}{2}$ while the $\nu = 1$ part will renormalize the spectrum by adding a zero-centered spectrum with magnitude proportional to $(n-1)$; both represent n-dependent changes in the sp energies. The $\nu = 2$ part, quadratic in the n-operators, cannot be reduced to NIP form. It could be incorporated into the $\nu = 2$ spreading interaction but, as we shall see, its effects are so small that it is best to ignore it. Let the spherical orbits labeled by $\alpha, \beta, \gamma, \delta$ belong to an N-dimensional orbit (i.e. a single unitary orbit). We have, from Eq. (1.80) with $n = \sum n_\alpha$,

$$\begin{aligned}
n_\alpha^0 &= N^{-1} N_\alpha n \;;\quad n_\alpha^1 = n_\alpha - N^{-1} N_\alpha n \\
n_{\alpha\beta}^X &= n_{\alpha\beta} + N_{\alpha\beta} - (N_\alpha n_\beta + N_\beta n_\alpha) + \delta_{\alpha\beta}(N_\alpha + 1)n_\alpha \\
n_{\alpha\beta}^0 &= \langle n_{\alpha\beta} \rangle^2 \binom{n}{2} = \binom{N}{2}^{-1} N_{\alpha\beta} \binom{n}{2} \\
n_{\alpha\beta}^1 &= (n-1)(N-2)^{-1}[(N_\alpha - \delta_{\alpha\beta})n_\beta + (N_\beta - \delta_{\alpha\beta})n_\alpha] \\
&\quad \times (1 - \tfrac{1}{2}\delta_{\alpha\beta}) - 2N^{-1}(N-2)^{-1} N_{\alpha\beta} n(n-1)
\end{aligned} \tag{1.81}$$

where $n_{\alpha\beta}^X$ derives from $n_\alpha^X = (N_\alpha - n_\alpha)$ and $n_{\alpha\beta}^2$ follows by subtraction. Here $n_{\alpha\beta}^0$, being scalar, must be a multiple of $\binom{n}{2}$, the coefficient following from $\langle n_{\alpha\beta} \rangle^2 = \binom{N}{2}^{-1} N_{\alpha\beta}$. The $\nu = 1, 2$ terms can be derived by observing that $\sum n_{\alpha\beta}^\nu = n_{\alpha\beta}$ while $(n_{\alpha\beta}^\nu)^X = (-1)^\nu n_{\alpha\beta}^\nu$. Then keeping in mind that $\sum_{\alpha\beta}$ is always restricted by $\alpha \leq \beta$

(or equivalently by $\alpha \geq \beta$ if the quantity being summed is symmetric in α, β), we have

$$\begin{aligned}
V^{[0]} &= \sum V^{[0],\nu} = \sum V_{\alpha\beta} n^{\nu}_{\alpha\beta} \\
V^{[0],0} &= \sum V_{\alpha\beta} n^0_{\alpha\beta} = \binom{N}{2}^{-1} \binom{n}{2} \sum N_{\alpha\beta} V_{\alpha\beta} \\
V^{[0],1} &= (n-1) \sum \epsilon^1_\alpha(2) n_\alpha \; ; \quad \epsilon^1_\alpha(2) = (N-2)^{-1} \textstyle\sum_\beta (N_\beta - \delta_{\alpha\beta})(V_{\alpha\beta} - \langle V \rangle^2) \\
V^{[0],2} &= V^{[0]} - V^{[0],0} - V^{[0],1} \; .
\end{aligned} \tag{1.82}$$

The second line of Eq. (1.82) gives the number-dependent shift of the sp spectrum, while the third, in its first form, displays the effective sp nature of $V^{[0],1}$; the notation here is that the $\epsilon^1_\alpha(2)$, the induced sp energies, define the $\nu = 1$ sp operator deriving from the 2-body V. Note that $\sum N_\alpha \epsilon^1_\alpha(2) = 0$. Now for the decomposition of $\boldsymbol{V}$ we know that $\boldsymbol{V}^{\nu=0} = 0$. Also we find then $V^X(1) = -\sum N_{\alpha\beta} \boldsymbol{V}_{\alpha\beta} (N_\alpha^{-1} n_\alpha + N_\beta^{-1} n_\beta)$ which vanishes because $\boldsymbol{V}_{\alpha\beta} = 0$, this confirming the $\nu = 2$ irreducibility of $\boldsymbol{V}$.

For the primary NIP Hamiltonian $C(0) + h(1)$, first note that $C(0) = C^{[0],0}$. Similarly the decomposition of $h(1) = h^{[0]} = \sum \epsilon_\alpha(1) n_\alpha$ is,

$$\begin{aligned}
h(1) &= h^{[0],0} + h^{[0],1} \\
h^{[0],0} &= n\bar{\epsilon} \; ; \qquad \bar{\epsilon} = N^{-1} \sum N_\alpha \epsilon_\alpha(1) \\
h^{[0],1} &= \sum \epsilon^1_\alpha(1) n_\alpha \, , \qquad \epsilon^1_\alpha(1) = \epsilon_\alpha(1) - \bar{\epsilon} \; .
\end{aligned} \tag{1.83}$$

Then, dropping $V^{[0],2}$, we have for its renormalization and the residual interaction

$$\begin{aligned}
\boldsymbol{C} &= C(0) + h^{[0],0} + V^{[0],0} \\
\boldsymbol{h} &= h^{[0],1} + V^{[0],1} = \sum \{\epsilon^1_\alpha(1) + (n-1)\epsilon^1_\alpha(2)\} n_\alpha = \sum \zeta_\alpha(n) n_\alpha \\
\boldsymbol{V} &= V - V^{[0]} \; .
\end{aligned} \tag{1.84}$$

Next, we give the $V^{[0]}$ decomposition when (α, β) in $V_{\alpha\beta}$ belong to different unitary orbits. Let α, β belong to distinct unitary orbits labeled respectively by u, v with dimensionality N_u, N_v. Then with $\nu_1 = 0, 1$ referring to u and $\nu_2 = 0, 1$ to v, with $n_u = \sum n_\alpha, n_v = \sum n_\beta$, we find immediately

$$\begin{aligned}
n^0_\alpha &= N_u^{-1} N_\alpha n_u \; ; \quad n^1_\alpha = n_\alpha - n^0_\alpha \; ; \quad n^0_\beta = N_v^{-1} N_\beta n_v \; ; \quad n^1_\beta = n_\beta - n^0_\beta \\
V^{[0]} &= \sum V_{\alpha\beta} n_\alpha n_\beta = \sum V_{\alpha\beta} (n^{\nu_1}_\alpha)(n^{\nu_2}_\beta) \\
\left\langle V^{[0]} \right\rangle^{1,1} &= \langle V \rangle^{1,1} = (N_u N_v)^{-1} \sum N_\alpha N_\beta V_{\alpha\beta} \\
V^{[0],0,0} &= \langle V \rangle^{1,1} n_u n_v \\
V^{[0],1,0} &= n_v \textstyle\sum_\alpha \epsilon^{1,0}_\alpha(2) n_\alpha \; ; \quad \epsilon^{1,0}_\alpha(2) = N_v^{-1} \displaystyle\sum_\beta N_\beta \{V_{\alpha\beta} - \langle V \rangle^{1,1}\} \\
V^{[0],0,1} &= n_u \sum_\beta \epsilon^{0,1}_\beta(2) n_\beta \; ; \quad \epsilon^{0,1}_\beta(2) = N_u^{-1} \sum_\alpha N_\alpha \{V_{\alpha\beta} - \langle V \rangle^{1,1}\}
\end{aligned} \tag{1.85}$$

and of course $V^{[0],1,1}$ and the spreading interaction $V^{1,1}_{\alpha\beta}$ follow by subtraction. Note now that the NIP Hamiltonian and the sets of $V^{[0],\nu}$ (or $V^{[0],\nu_1,\nu_2}$) operators, all of which commute with each other, determine the $(C(0) + h(1) + V^{[0]})$ eigenvalues (the IP configuration centroids), but only if $V^{[0],2}$ (or $V^{[0],1,1}$) can be dropped (probably true in most of the interesting cases) do they constitute an effective NIP Hamiltonian.

For the single-orbit contributions to $\boldsymbol{C}$ and $\boldsymbol{h}$, from $(h^{[0],0} + V^{[0],0})$ and $V^{[0],1}$ respectively, we assign a label u to the single h.o. orbit involved in Eq. (1.82) so that $N \to N_u$ and $n \to n_u$, and then sum over u as in Eq. (1.86) below. For the two-orbit contributions, from $V^{[0],0,0}$ and $(V^{[0],1,0} + V^{[0],0,1})$ we sum over u, v with $u < v$ in Eq. (1.85). Then, with $H = \boldsymbol{C} + \boldsymbol{h} + \boldsymbol{V}$ we have explicit forms for the back-shifting parameter and the sp (or mean-field) renormalization,

$$
\begin{aligned}
\boldsymbol{C} &= C(0) + \sum_u n_u N_u^{-1} \sum_{\alpha\in u} N_\alpha \epsilon_\alpha(1) + \sum_{u\leq v} n_{uv} N_{uv}^{-1} \sum_{\alpha\in u,\,\beta\in v} N_{\alpha\beta} V_{\alpha\beta} \\
\boldsymbol{h} &= h^{[0],\mathbf{1}} + V^{[0],\mathbf{1}} = \sum_u \sum_{\alpha\in u} \left\{\epsilon^1_\alpha(1) + (n_u - 1)\epsilon^1_\alpha(2)\right\} n_\alpha \\
&\quad + \sum_{u<v} \left\{ n_v \sum_{\alpha\in u} \epsilon^{1,0}_\alpha(2) n_\alpha + n_u \sum_{\beta\in v} \epsilon^{0,1}_\beta(2) n_\beta \right\}
\end{aligned}
\tag{1.86}
$$

where, in the third term of $\boldsymbol{C}$ we have combined the one-orbit ($u = v$) and two-orbit results. Observe that, in $\boldsymbol{h}$, the $\epsilon^1_\alpha(2)$, $\epsilon^{1,0}_\alpha(2)$ are respectively the sp energies induced in spherical-orbit $\alpha \in u$ and $\beta \in v$ by the $V_{\alpha\beta}$ interactions respectively in u and v. Note that $\mathbf{1} = (1, 0, 0 \ldots) + (0, 1, 0, 0, \ldots) + \ldots$ where the position of 1 indicates the rank $\nu = 1$ with respect to the corresponding unitary orbit.

In the (p, n) formalism, the NIP Hamiltonian is $[C + h(1)] = C(0) + h_p(1) + h_n(1)$ and we have three interactions. V_{pp} renormalizes $C(0) + h_p(1)$ precisely as in Eq. (1.86), in which u, v now refer only to proton h.o. orbits and similarly V_{nn} renormalizes $C(0) + h_n(1)$. Since in this formalism proton and neutron orbits must be regarded as distinct (even if otherwise identical) the V_{pn} renormalizations show up only in the two-orbit terms of Eq. (1.86) the $(u \leq v)$ summation for $\boldsymbol{C}$ and $(u < v)$ for $\boldsymbol{h}$ both being replaced by $u \neq v$, and u ranging over the proton orbits and v over the neutron orbits. The $\epsilon^{1,0}_\alpha$ terms renormalize h_p and the $\epsilon^{0,1}_\alpha$ renormalize h_n so that

$$
\begin{aligned}
H &= \left\{C(0) + h_p^{[0],0} + h_n^{[0],0} + V_{pp}^{[0],0} + V_{nn}^{[0],0} + V_{pn}^{[0],0,0}\right\} + \left\{h_p^{[1],1} + h_n^{[1],1}\right\} \\
&\quad + \left\{V_{pp}^{[0],1} + V_{pn}^{[0],1,0} + V_{nn}^{1,0} + V_{pn}^{[0],0,1}\right\} + \left\{V_{pp}^{[0],2} + V_{nn}^{[0],2} + V_{pn}^{[0],1,1}\right\} \\
&\quad + \left\{\mathbf{V}_{pp} + \mathbf{V}_{nn} + \mathbf{V}_{pn}\right\} \\
&= \boldsymbol{C} + h^{[0],\mathbf{1}} + V^{[0],\mathbf{1}} + V^{[0],\mathbf{2}} + \boldsymbol{V} \\
&= \boldsymbol{C} + \boldsymbol{h} + V^{[0],\mathbf{2}} + \boldsymbol{V} \simeq \boldsymbol{C} + \boldsymbol{h} + \boldsymbol{V}
\end{aligned}
\tag{1.87}
$$

where the $\boldsymbol{V}$ operators are defined as before ($= V - V^{[0]}$) and the final form involves dropping the $V^{[0]}$ terms of *total* unitary rank 2 i.e. $V^{[0],\mathbf{2}} = V_{pp}^{[0],2} + V_{nn}^{[0],2} + V_{pn}^{[0],1,1}$. We remark that the NIP renormalizations defined by Eq. (1.86) and its (p,n) extension defined by Eq. (1.87) are not surprising in form. They represent a type of averaging often attempted by intuitive procedures. But the unitary group technique is simple. It displays the correction term $V^{[0],2}$ of Eq. (1.82) and its (p,n) extension $V^{[0],1,1}$, which are lost in an intuitive procedure. It also gives methods for evaluating norms in many-particle spaces which can be used to test the approximation of dropping the $V^{[0],\mathbf{2}}$ term in Eq. (1.87) so that the last form in this equation is good.

Before turning to this [French and Kota (1983*); Kota and Majumdar (1992b, 1996*)], we will consider a numerical example of interaction renormalization. In the level density calculations of ^{28}Si for example, one may consider 2 h.o. orbits $(2s1d)$ and $(2p1f)$ by eliminating the lowest two h.o. orbits $1s$ and $1p$ which are relatively inactive in the energy domain of up to 15 MeV excitation and compensate this by a Hamiltonian renormalization, thus simplifying the ^{28}Si description (from $10 \rightarrow 7$ spherical orbits). Broadly speaking, in doing this we must take account of the three parts $(C, h, \mathbf{V})$ of H insofar as these, before renormalization, involve interactions between active nucleons and those in the orbits to be discarded. See Abzouzi, Caurier and Zuker (1991*) for a related treatment for the renormalization and parametrization of the mean field, and their influence on shell-model calculations. For C, the renormalizations will follow in the (p,n) case from Eqs. (1.85) and (1.87). If $u = 1, 2$ denotes respectively the $(1s, 1p)$ h.o. orbits and $v = 3, 4$ the $(2s1d), (2p1f)$ orbits we have $\overline{V_{uv}} = N_{uv}^{-1} \sum N_{\alpha\beta} V_{\alpha\beta}$ for the average interaction energy for a particle in u and one in v; here α labels the spherical orbits of u, and β those of v. Then we find for the $(A = 17)$ $(2s1d) - (2p1f)$ separation generated by the renormalization,

$$\begin{aligned} \Delta = {} & 2(\overline{V_{14}}(pp) - \overline{V_{13}}(pp)) + 6(\overline{V_{24}}(pp) - \overline{V_{23}}(pp)) \\ & + 2(\overline{V_{14}}(np) - \overline{V_{13}}(np)) + 6(\overline{V_{24}}(np) - \overline{V_{23}}(np)) \,. \end{aligned} \tag{1.88}$$

Because the $(2s1d)$ interactions with $1s$ and $1p$ are stronger than the $(2p1f)$ interactions, the effective $(2p1f)$ sp energies will move upward by an amount Δ relative to $(2s1d)$. The h.o. (unitary orbits) $\overline{V_{uv}}$ interactions, for Kuo's 10-orbit interaction [Ratcliff (1971*)] are $\overline{V_{13}} = -0.87$, $\overline{V_{14}} = -0.59$, $\overline{V_{23}} = -0.77$, $\overline{V_{24}} = -0.45$ MeV for pp (or nn) while for pn the corresponding energies are -2.38, -1.52, -2.11, -1.42 MeV. These give rise to individual shifts -23.78 MeV for $(2s1d)$ interactions with the filled h.o. orbits #1,2 and -15.95 for the $(2p1f)$ interactions. Thus $\Delta = 7.83$ MeV and if the original $(V = 0)$ separation had been 7 MeV [from the kinetic energy only; see Ratcliff (1971*)] the final would now be 14.8 MeV which is a reasonable $\hbar\omega$ value.

1.4.2 Geometry, norms and correlation coefficients

Given a set of operators $\mathcal{O}$ preserving a symmetry defined by an algebra say G, the $\mathcal{O}$ do not connect spaces labeled by different irreducible representations (irreps) Γ of G. Then the part of $\mathcal{O}$ that produces $\langle \mathcal{O} \rangle^{\Gamma}$ is the trace equivalent operator $\mathcal{O}_{TE}$. With $\widetilde{\mathcal{O}} = \mathcal{O} - \mathcal{O}_{TE}$, the fixed-$\Gamma$ space norm of $\mathcal{O}$ is,

$$|\mathcal{O}|_{\Gamma} = \left\{ \left\langle \widetilde{\mathcal{O}}^{\dagger} \widetilde{\mathcal{O}} \right\rangle^{\Gamma} \right\}^{1/2} . \tag{1.89}$$

For example, for the effective interactions of $(2s1d)$ and $(2p1f)$ shell nuclei one can consider fixed m, or mT or (m_p, m_n) (also mJ, mJT) norms. An important property is that the tensorial decomposition of $\mathcal{O}$ with respect to G will be an orthogonal decomposition with respect to the norm defined by Eq. (1.89). Given $\mathcal{O}^{\lambda}$, the different tensorial parts of $\mathcal{O}$ with respect to G, then

$$|\mathcal{O}|^2_{\Gamma} = \sum_{\lambda} \left| \mathcal{O}^{\lambda} \right|^2_{\Gamma} . \tag{1.90}$$

With norms having the proper meaning of 'length', it is also possible to define correlation coefficients ζ,

$$\zeta_{\mathcal{O}_1, \mathcal{O}_2}(\Gamma) = \frac{\left\langle \widetilde{\mathcal{O}_1}^{\dagger} \widetilde{\mathcal{O}_2} \right\rangle^{\Gamma}}{|\mathcal{O}_1|_{\Gamma} \, |\mathcal{O}_2|_{\Gamma}} . \tag{1.91}$$

It is easy to see that ζ defined by Eq. (1.91) has all the properties of the correlation coefficient, i.e. ζ gives the cosine of the angle, in the space defined by the irrep Γ, between the operators $\mathcal{O}_1$ and $\mathcal{O}_2$. It is important to repeat that $\mathcal{O}_1$ and $\mathcal{O}_2$ should preserve the symmetry defining the irreps Γ. Norms and correlation coefficients defined here derive their significance from the action of CLTs as squares of norms of Hamiltonians are nothing but spectral variances. The unitary geometry defined by norms and correlation coefficients allows us to make a meaningful comparison and analysis of operators. With this it is possible to generate simple approximations to a given operator, know the importance of its various parts, estimate the deviations of the spectra generated by one operator from that of another and also deduce information on various sum rule quantities, which are determined, in the CLT limit, by the correlation coefficients and so on. Among various operators available, the effective interactions acquire special importance due to the fact that they are often given as a large numerical set of matrix elements. Usually they are analyzed by comparing the two-body matrix elements. The norms and correlation coefficients defined by Eq. (1.89) and (1.91) fill this gap. They are well defined and are easily calculable in many particle spaces, thereby allowing one to see how various relationships evolve with particle number and isospin (in principle also with J).

The CLT defined geometry was recognized first by French [French (1967b*, 1972); Chang, French and Thio (1971*)]. Given H, it is seen from Eq. (1.90) that the norms of various tensorial parts provide information about their relative importance as the total norm is the spectral width. For example, analysis of $(2s1d)$ shell interactions have shown that the $\nu = 1$ and $\nu = 2$ parts [with respect to $U(24)$] contribute equally to the norms of Hamiltonians and therefore both are important [Kota (1979a)]. Decomposition with respect to $SU(4)$ tensors showed that the parts that are scalar and one-body with respect to $SU(4)$ generate about 90-95% of $|H|$ [Chakraborty, Kota and Parikh (1980a)]. Also comparing $(2s1d)$ shell interactions via the norm of their $\nu = 1$ part [with respect to $U(24)$] showed that they differ significantly in this part and hence produce quite different results for occupancies [Potbhare (1977)]. Norms of irreducible tensor parts with respect to spherical configuration subgroup [Chang, French and Thio (1971*)], quasi-spin group [Bohigas, Quesne and Arvieu (1968)], isospin [Kota (1979a)] and Hartree-Fock basis [Satyan and Parikh (1979)] have also studied. It is possible to use norms to define a measure for the amount of H that goes into the generation of the mean-field when H is decomposed into tensors with respect to a sub-algebra G of $U(N)$ [Chang, French and Thio (1971*); Satyan and Parikh (1979); Kota (1979a); Chakraborty, Kota and Parikh (1980a)]. In addition, norms or spectral variances can be used to define new variational methods [Satyan and Parikh (1976)]. Comparing effective interactions via correlation coefficients in many particle spaces showed that the correlations coefficients between realistic interactions is ~ 0.9 for both $(2s1d)$ and $(2p1f)$ shell examples.

Another important application of the unitary geometry is that it is possible to expand a given operator in terms of a given set of operators $\mathcal{O}_\alpha$ with good physical meaning (pairing, quadrupole-quadrupole, group Casimir operators etc.) and so on. For example, consider $H = \sum_\alpha C_\alpha \mathcal{O}_\alpha + X = H_M + X$. Then the expansion coefficients C_α can be determined using the condition that the norm of X is minimum or the correlation between H and H_M is maximum. The goodness of the approximation $H \sim H_M$ is determined by ζ_{H-H_M}. This is used in the analysis of $(2s1d)$ and $(2p1f)$ shell effective interactions in terms of group Casimirs [Halemane, Kar and Draayer (1978); Countee, Draayer, Halemane and Kar (1981); Sviratcheva, Draayer and Vary (2006*, 2007)] and in terms of the pairing plus quadrupole-quadrupole interaction [Kota, Pandya and Potbhare (1980)]. It has also been used in establishing that the collective model [Bohr and Mottelson (1975)] Hamiltonians are dominantly two-body in character for axially symmetric nuclei whereas asymmetric shapes require three-body terms [Draayer and Rosensteel (1982)]. Let us point out that it is possible to study the energy dependence of C_α [Chang, French and Thio (1971*); Chakraborty, Kota and Parikh (1980a); Kota and Potbhare (1982)]. Finally, recent attempts by Sviratcheva, Draayer and Vary (2006*, 2007) brought new insights into $(2p1f)$ shell interactions in terms of the proton-neutron pairing $Sp(4)$ algebra. Norms and correlation coefficients have also been used in symplectic no-core shell model studies [Draayer, *et al.* (2008)].

In order to establish that the $V^{[0],\mathbf{2}}$ part is small compared to the $\{h^{[0],\mathbf{1}} + V^{[0],\mathbf{1}}\}$, the unitary norm which gives a measure of the size of an operator is employed; $h^{[0],\mathbf{0}}$ + $V^{[0],\mathbf{0}}$ behaves as a scalar (number) with respect to unitary configurations and hence this part need not be considered in the calculation of norms. Firstly, given L_p proton unitary orbits and L_n neutron unitary orbits, we can arrange them such that in the total $L = L_p + L_n$ unitary orbits, the first L_p orbits are proton orbits and the remaining are neutron orbits. Then a proton-neutron unitary configuration is $[\mathbf{m}] = [\mathbf{m}_1, \mathbf{m}_2, \ldots, \mathbf{m}_{L_p}, \ldots, \mathbf{m}_L]$. For a fixed $[\mathbf{m}]$, $h^{[0],\mathbf{1}}$ is defined by the sp energies $\epsilon_\alpha^{\mathbf{1}}$ and $V^{[0],\mathbf{1}}$ by $[\mathbf{m}]$ dependent sp energies $\zeta_\alpha^{\mathbf{1}}([\mathbf{m}])$; explicit forms for them follow easily from Eqs. (1.82)-(1.86). Then, $\boldsymbol{h}$ is defined by the sp energies $\xi_\alpha^{\mathbf{1}}([\mathbf{m}]) = \epsilon_\alpha^{\mathbf{1}} + \zeta_\alpha^{\mathbf{1}}([\mathbf{m}])$. This can be termed as 'algebraic mean-field'.

Propagation equations for the squares of fixed unitary configuration norms of $\boldsymbol{h} = h^{[0],\mathbf{1}} + V^{[0],\mathbf{1}}$ and $V^{[0],\mathbf{2}}$ are [they follow from Eqs. (1.76), (1.77), (1.79) and (1.89)],

$$
\begin{aligned}
|\boldsymbol{h}|^2_{[\mathbf{m}]} &= \sum_{\boldsymbol{\alpha}} \frac{m_{\boldsymbol{\alpha}} m_{\boldsymbol{\alpha}}^X}{\hat{N}_{\boldsymbol{\alpha}\boldsymbol{\alpha}}} \sum_{\alpha\in\boldsymbol{\alpha}} N_\alpha \left\{\xi_\alpha^{\mathbf{1}}([\mathbf{m}])\right\}^2 \;; \quad \xi_\alpha^{\mathbf{1}}([\mathbf{m}]) = \epsilon_\alpha^{\mathbf{1}} + \zeta_\alpha^{\mathbf{1}}([\mathbf{m}]) \\
\left|V^{[0],\mathbf{2}}\right|^2_{[\mathbf{m}]} &= \sum_{\boldsymbol{\alpha}\geq\boldsymbol{\beta}} \frac{\hat{m}_{\boldsymbol{\alpha}\boldsymbol{\beta}} \hat{m}_{\boldsymbol{\alpha}\boldsymbol{\beta}}^X}{\hat{N}_{\boldsymbol{\alpha}\boldsymbol{\beta}\boldsymbol{\alpha}\boldsymbol{\beta}}} \sum_{\alpha\in\boldsymbol{\alpha},\,\beta\in\boldsymbol{\beta}} \left[(V_{\alpha\beta}^{[0],\mathbf{2}})^2 N_{\alpha\beta}(1+\delta_{\alpha\beta})\right] [1+\delta_{\boldsymbol{\alpha}\boldsymbol{\beta}}]^{-1} \;.
\end{aligned}
\tag{1.92}
$$

It is essential to recognize that the different $\boldsymbol{\nu}$ tensor parts of $h + V^{[0]}$ are orthogonal to each other with respect to the fixed-$[\mathbf{m}]$ norm defined above. As a result $\left|\boldsymbol{h} + V^{[0],\mathbf{2}}\right|^2_{[\mathbf{m}]} = |\boldsymbol{h}|^2_{[\mathbf{m}]} + \left|V^{[0],\mathbf{2}}\right|^2_{[\mathbf{m}]}$. Therefore $V^{[0],\mathbf{2}}$ can be neglected if $\left|V^{[0],\mathbf{2}}\right|_{[\mathbf{m}]} << \left|\boldsymbol{h} + V^{[0],\mathbf{2}}\right|_{[\mathbf{m}]}$. Also, the first equality in Eq. (1.92) gives $\left|h^{[0],\mathbf{1}}\right|^2_{[\mathbf{m}]}$, $\left|V^{[0],\mathbf{1}}\right|^2_{[\mathbf{m}]}$ and the correlation coefficient between these two parts of $\boldsymbol{h}$. Finally, fixed-S norms in terms of fixed-$[\mathbf{m}]$ norms, are given by,

$$
|K|^2_S = \left[\sum_{[\mathbf{m}]\in S} d([\mathbf{m}])\right]^{-1} \left[\sum_{[\mathbf{m}]\in S} d([\mathbf{m}]) |K|^2_{[\mathbf{m}]}\right]
\tag{1.93}
$$

where $K = h^{[0],\mathbf{1}}$, $V^{[0],\mathbf{1}}$, $\boldsymbol{h}$ or $V^{[0],\mathbf{2}}$. Tests of smallness of $V^{[0],\mathbf{2}}$ are given in Table 1.2. The examples for the norms shown in the Table are: (i) $(2s1d)$ shell with proton $(2s1d)$ and neutron $(2s1d)$ shells as the two unitary orbits both having $s = 0$; (ii) $(2p1f)$ shell with proton $(2p1f)$ and neutron $(2p1f)$ shells as the two unitary orbits both having $s = 0$; (iii) $(2s1d), (2p1f)$ shell with $(2s1d)$ $[s = 0]$ and $(2p1f)$ $[s = 1]$ for protons and neutrons as four unitary orbits. Given in order are the norms of: primary traceless sp Hamiltonian $h^{[0],\mathbf{1}}$, the induced field-producing interaction $V^{[0],\mathbf{1}}$, total traceless effective NIP Hamiltonian $\boldsymbol{h}$, the non-NIP part from the interaction $V^{[0],\mathbf{2}}$. It is seen that the errors generated by ignoring the $V^{[0],\mathbf{2}}$ term are negligible (3% or less). This result is verified in many other examples in different shell model spaces [Kota and Majumdar (1992b); Majumdar (1994)]. It is also seen that in general the correlation coefficient between $h^{[0],\mathbf{1}}$ and $V^{[0],\mathbf{1}}$ is positive and large ($\gtrsim 0.9$).

Table 1.2 Norms in MeV for the various Hamiltonian terms, decomposed with respect to harmonic oscillator orbits, as described in the text. The last column gives the percentage change in $|h^{[0],\mathbf{1}} + V^{[0]}|$ when $V^{[0],\mathbf{2}}$ is set to zero. For $(2s1d)$ examples, PW interaction [Preedom and Wildenthal (1972)] with ^{17}O sp energies and for $(2p1f)$ examples the MWH2 interaction [Sharma and Bhatt (1973); McGrory, Wildenthal and Halbert (1970)] with ^{41}Ca sp energies have been used. For the $(2s1d) - (2p1f)$ examples, Kuo's 10-orbit interaction [Ratcliff (1971*)], omitting the $1s$ and $1p$ orbits is used and the single-particle energies corresponding to the 7-orbits $(1d_{5/2}, 1d_{3/2}, 2s_{1/2}, 1f_{7/2}, 1f_{5/2}, 2p_{3/2}$ and $2p_{1/2}$ are (−4.15, 0.93, −3.28, 6.55, 12.74, 8.57 and 10.58) MeV respectively.

sp space	(m_p, m_n, s)	$\|h^{[0],\mathbf{1}}\|$	$\|V^{[0],\mathbf{1}}\|$	$\|\boldsymbol{h}\|$	$\|V^{[0],\mathbf{2}}\|$	Error
$(2s1d)$	(2,2,0)	4.41	1.19	5.48	0.92	1.41%
	(4,6,0)	5.75	4.55	10.01	1.66	1.37%
	(8,8,0)	5.58	7.33	12.55	1.55	0.77%
$(2p1f)$	(4,4,0)	7.09	0.35	7.42	0.93	0.78%
	(6,6,0)	8.12	0.63	8.72	1.24	1.01%
	(10,10,0)	8.86	1.20	9.99	1.49	1.11%
$(2s1d) - (2p1f)$	(4,4,0)	5.58	0.56	6.04	1.66	3.8%
	(4,4,1)	5.96	0.77	6.50	1.62	3.1%
	(4,4,2)	6.27	0.85	6.84	1.57	2.6%
	(6,6,0)	5.92	0.94	6.72	1.88	3.9%
	(6,6,1)	6.43	1.28	7.39	1.95	3.5%
	(6,6,2)	6.86	1.45	7.91	2.00	3.2%

1.4.3 *State density and variances of IP spreading Gaussians*

With $V^{[0],\mathbf{2}}$ negligible so that the only significant role of $V^{[0]}$ is the NIP spectrum shift and renormalization, and without loss of generality adding $\boldsymbol{C}$ to $\boldsymbol{h}$ so that $\boldsymbol{C}+\boldsymbol{h} \rightarrow \boldsymbol{h}$, I^H takes a convolution form provided the spreading widths are constant, i.e. same for all spherical (or unitary) configurations. Then, with S decomposition and unitary configurations $[\mathbf{m}]$, we have

$$I^H(E) = \sum_S I^{\boldsymbol{h},S} \otimes \rho_{\mathcal{G}}^{\boldsymbol{V},S}[E] = \sum_S \sum_{[\mathbf{m}]\in S} I^{\boldsymbol{h},[\mathbf{m}]} \otimes \rho_{\mathcal{G}}^{\boldsymbol{V},[\mathbf{m}]}[E] \,. \tag{1.94}$$

A structure similar to Eq. (1.94) was identified by Haq and Wong (1982) by introducing an energy dependent spreading width. The convolution form given by Eq. (1.94) was first recognized by French and Kota (1983*). For establishing and applying Eq. (1.94) we need the variances of the spreading function $\rho_{\mathcal{G}}^{\boldsymbol{V},[\mathbf{m}]}$. Note that we have already discussed in Sec. 1.2 the moment method construction of $I^{\boldsymbol{h},[\mathbf{m}]}$.

Let us start with spreading variances in the situation with a single unitary orbit $\boldsymbol{\alpha}$. As $\boldsymbol{V}$ is of type $\nu = 2$, as seen from Eq. (1.79), for m particles in the orbit $\boldsymbol{\alpha}$ with dimension N, $\sigma^2(m) = \left\langle \boldsymbol{V}^2 \right\rangle^m = [\hat{m}_{\boldsymbol{\alpha}\boldsymbol{\alpha}} \hat{m}^X_{\boldsymbol{\alpha}\boldsymbol{\alpha}} / \hat{N}_{\boldsymbol{\alpha}\boldsymbol{\alpha}\boldsymbol{\alpha}\boldsymbol{\alpha}}] \left\langle\left\langle \boldsymbol{V}^2 \right\rangle\right\rangle^2$. With spherical orbits $\alpha, \beta, \ldots$ in $\boldsymbol{\alpha}$, the trace $\left\langle\left\langle \boldsymbol{V}^2 \right\rangle\right\rangle^2$ is sum of the squares of matrix elements connecting all states in two-particle $(\alpha\beta)$ configurations with states in $(\gamma\delta)$ configurations. With the notation $\boldsymbol{V}^2(\alpha\beta \rightarrow \gamma\delta) = \sum_{i\in(\alpha\beta), j\in(\gamma\delta)} |\langle(\alpha\beta)i \mid \boldsymbol{V} \mid (\gamma\delta)j\rangle|^2$, the spreading variance $\sigma^2(m)$ is,

$$\sigma^2(m) = \frac{\hat{m}_{\boldsymbol{\alpha\alpha}}\hat{m}^X_{\boldsymbol{\alpha\alpha}}}{\hat{N}_{\boldsymbol{\alpha\alpha\alpha\alpha}}} \sum_{(\alpha\beta),(\gamma\delta)} \boldsymbol{V}^2(\alpha\beta \to \gamma\delta) \; . \tag{1.95}$$

Extension of Eq. (1.95) for fixed-$\mathbf{m}$ variances $\sigma^2(\mathbf{m}) = \left\langle \boldsymbol{V}^2 \right\rangle^{\mathbf{m}}$ is immediate as $\boldsymbol{V}$ is $\boldsymbol{\nu} = 2$ with respect to the direct sum group generating $\mathbf{m}$. For $\sigma^2(\mathbf{m})$, there will be a propagator for each $\boldsymbol{V}^2(\alpha\beta \to \gamma\delta)$. This follows by taking the propagator $[m_{\boldsymbol{\alpha\alpha}} m^X_{\boldsymbol{\alpha\alpha}} / N_{\boldsymbol{\alpha\alpha\alpha\alpha}}]$ inside the summation in Eq. (1.95) and changing $\boldsymbol{\alpha}$ appropriately. The propagator, by inspection, is simply $\hat{m}_{\alpha\beta}\hat{m}^X_{\gamma\delta}/\hat{N}_{\alpha\beta\gamma\delta}$. Then,

$$\sigma^2(\mathbf{m}) = \left\langle \boldsymbol{V}^2 \right\rangle^{\mathbf{m}} = \sum_{(\alpha\beta),(\gamma\delta)} \hat{m}_{\alpha\beta}\, \hat{m}^X_{\gamma\delta} \left(\hat{N}_{\alpha\beta\gamma\delta}\right)^{-1} \boldsymbol{V}^2(\alpha\beta \to \gamma\delta) \; . \tag{1.96}$$

For a formal derivation, see [Chang, French and Thio (1971*); Hecht (1975); Wong (1986); Nomura (1975)]. It should be remarked that the delta-functions in $m_{\alpha\beta}$ etc. are due to Pauli blocking. The expression for $\sigma^2(\mathbf{m})$ goes over easily, by similar intuitive arguments, to $\sigma^2([\mathbf{m}])$. For (p,n) unitary configurations $[\mathbf{m}] = [\mathbf{m}_p, \mathbf{m}_n]$, in the notation described earlier, the spreading variance $\sigma^2([\mathbf{m}])$ of $\rho_{\mathcal{G}}^{\boldsymbol{V},[\mathbf{m}]}$, with appropriate Pauli-blocking terms, is (for J-coupled two-body matrix elements)

$$\begin{aligned} \sigma^2([\mathbf{m}]) = \left\langle \boldsymbol{V}^2 \right\rangle^{[\mathbf{m}]} &= \sum_{\boldsymbol{\alpha},\boldsymbol{\beta},\boldsymbol{\gamma},\boldsymbol{\delta}} \left[\hat{m}_{\boldsymbol{\alpha\beta}}\hat{m}^{\times}_{\boldsymbol{\gamma\delta}}\right] \left(\hat{N}_{\boldsymbol{\alpha\beta\gamma\delta}}\right)^{-1} \\ &\times \frac{1}{4} \sum_{J;\alpha\in\boldsymbol{\alpha},\, \beta\in\boldsymbol{\beta},\, \gamma\in\boldsymbol{\gamma},\, \delta\in\boldsymbol{\delta}} (2J+1) \left\{\boldsymbol{V}^J_{\alpha\beta\gamma\delta}\right\}^2 (1+\delta_{\alpha\beta})(1+\delta_{\gamma\delta}) \; . \end{aligned} \tag{1.97}$$

Now the fixed-S spreading variances are simply (with $[\mathbf{m}] = [\mathbf{m}_p, \mathbf{m}_n]$),

$$\sigma^2(m_p, m_n : S) = \left[\sum_{[\mathbf{m}]\in S} d([\mathbf{m}])\right]^{-1} \left[\sum_{[\mathbf{m}]\in S} d([\mathbf{m}])\sigma^2_{[\mathbf{m}]}\right] \; . \tag{1.98}$$

As an illustrative example of the application of Eqs. (1.97) and (1.98), consider three unitary orbits with $s = -1, 0$ and 1 with degeneracies N_1, N_2 and N_3 respectively and m active particles with $s = -1$ orbit completely filled. Numbering the orbits as 1,2,3 (instead of the canonical $-1, 0, +1$ as in [French *et al.* (2006*)]) we see that the $s = 0$ configuration is $[N_1, m, 0]$ while the $s = 1$ configurations are $C_1 = [N_1 - 1, m+1, 0]$ and $C_2 = [N_1, m-1, 1]$ of dimensionalities $d_1 = N_1\binom{N_2}{m+1}$ and $d_2 = N_3\binom{N_2}{m-1}$; $d = d_1 + d_2$. We consider here S conserving $\mathbf{V}$. The partial traces (or variances) $\boldsymbol{V}^2(\boldsymbol{\alpha\beta} \to \boldsymbol{\gamma\delta})$ now derive from transitions both within and between the two configurations, the latter two being equal to within a dimensionality ratio. The fact that $\mathbf{V}$ conserves s implies that the orbital numbers for contributing transitions satisfy $\boldsymbol{\alpha}+\boldsymbol{\beta}=\boldsymbol{\gamma}+\boldsymbol{\delta}$. Then,

$$
\begin{aligned}
\sigma_A^2(m,s=0) &= m(m-1)(N_2-m)(N_2-m-1) \\
&\quad \times \{N_2(N_2-1)(N_2-2)(N_2-3)\}^{-1}\boldsymbol{V}^2(22\to 22)\;, \\
\sigma_A^2(m,s=1) &= d_1 d^{-1}\left[\frac{(m+1)(N_2-m-1)}{N_1N_2(N_2-1)}\boldsymbol{V}^2(2:12\to 12)\right. \\
&\quad + \frac{m(m+1)(N_2-m-1)(N_2-m-2)}{N_2(N_2-1)(N_2-2)(N_2-3)}\boldsymbol{V}^2(2:22\to 22) \\
&\quad \left. + \frac{m(m+1)}{N_1N_2(N_2-1)}\boldsymbol{V}^2(2:22\to 13)\right] \\
&\quad + d_2 d^{-1}\left[\frac{(m-1)(m-2)(N_2-m+1)(N_2-m)}{N_2(N_2-1)(N_2-2)(N_2-3)}\boldsymbol{V}^2(2:22\to 22)\right. \\
&\quad + \frac{(m-1)(N_2-m+1)}{N_2(N_2-1)N_3}\boldsymbol{V}^2(2:23\to 23) \\
&\quad \left. + \frac{(N_2-m+1)(N_2-m)}{N_2(N_2-1)N_3}\boldsymbol{V}^2(2:13\to 22)\right]\;.
\end{aligned}
\tag{1.99}
$$

where the expression in the first bracket [ ] defines the C_1 spreading variance, decomposed into an internal (the first two terms) and an external variance (the third term) which connects C_1 with C_2; similarly for the C_2 variance defined by the second bracket.

We end this Section with a brief discussion regarding the constancy of variances. For the fluctuation in the spherical-configuration variance, we define $\Delta(\mathbf{m}) = \mid \sigma^2(\mathbf{m}) - \sigma_A^2 \mid /2\sigma_A^2$ and evaluate its (dimensionality-weighted) RMS average $\Delta(m)$ which is given in terms of the variance of the configuration variances by $\Delta(m) = Var^{1/2}(\sigma^2(\mathbf{m}))/2\sigma_A^2$. Then for $\Delta << 1$ we can write for the RMS average and deviation that $\sigma(\mathbf{m}) = \sigma_A\{1 \pm \Delta(m)\}$. Table 1 of [French *et al.* (2006*)] gives one good example for constancy of variances. Constancy of variances is also discussed in [Lougheed and Wong (1975b); Parikh (1978); Wong (1986); Kota and Majumdar (1992b)]. We see from Eq. (1.97) that $\sigma^2([\mathbf{m}])$ is a sum of, say K with K being usually very large, $\boldsymbol{V}^2$ matrix elements. Chaos leading to EGOE(2) representation for $\boldsymbol{V}$ implies that the variance distribution will be a χ_K^2 distribution [Flambaum, Gribakin and Izrailev (1996)]. Then $\Delta \sim 1/\sqrt{K}$ and the fluctuations will be small. Thus constancy of variances appears to be a result of quantum chaos and EGOE(2) behavior of $\boldsymbol{V}$ [Flambaum, Gribakin and Izrailev (1996); Zelevinsky, Brown, Frazier and Horoi (1996); Kota (2001)].

1.4.4 *Moments for spin cutoff densities*

Spin cutoff factors $\sigma_J^2(E) = \langle J_z^2\rangle^E$ decompose state densities $I^H(E)$ in Eq. (1.94) into level densities and they are equal to, as discussed in Sec. 1.3, the ratio of spin cutoff densities and state densities,

$$\sigma_J^2(E) = I^H_{J_z^2}(E)/I^H(E) \; . \tag{1.100}$$

Extending Eq. (1.94), with (p, n) unitary configurations $[\mathbf{m}]$, the spin cutoff densities can be written in a convolution form,

$$I^H_{J_z^2}(E) = \sum_{S} \sum_{[\mathbf{m}] \in S} I^{\boldsymbol{h},[\mathbf{m}]}_{J_z^2} \otimes \rho^{\boldsymbol{V},[\mathbf{m}]}_{J_z^2:\mathcal{G}}[E] \; . \tag{1.101}$$

In order to apply Eq. (1.101), we construct the NIP spin cutoff density $I^{\boldsymbol{h},[\mathbf{m}]}_{J_z^2}$, using its first four moments, as a Gaussian with Edgeworth corrections. We first discuss the results for these NIP moments and then consider the variances of the spreading functions $\rho^{\boldsymbol{V},[\mathbf{m}]}_{J_z^2:\mathcal{G}}$. It should be noted that in some situations it is also useful to carry out an isospin decomposition of the level density [Jensen (1977)].

As J_z is a one-body operator with $\boldsymbol{\nu} = \mathbf{1}$, the convolution form [Eq. (1.24)] for the one-body operator $\boldsymbol{h}(\lambda) = \boldsymbol{h} + \lambda J_z$ gives $K_r[\boldsymbol{h}(\lambda)] = \sum_{\boldsymbol{\alpha}} K_r[\boldsymbol{h}_{\boldsymbol{\alpha}}(\lambda)]$; with $\boldsymbol{\alpha}$ the unitary orbits, $\boldsymbol{h}(\lambda) = \sum_{\boldsymbol{\alpha}} h_{\boldsymbol{\alpha}}(\lambda)$. Now, applying Eq. (1.13) for K_2 to K_6 of $I^{\boldsymbol{h}(\lambda)}$ and carrying out double parametric differentiation, i.e. using Eq. (1.53), allows us to write the traces $\left\langle J_z^2 \boldsymbol{h}^p \right\rangle^{[\mathbf{m}]}$, $p \leq 4$ in terms of single orbit traces $\langle (\boldsymbol{h}_{\boldsymbol{\alpha}})^p \rangle^{m_{\boldsymbol{\alpha}}}$ and $\left\langle J_z^2 (\boldsymbol{h}_{\boldsymbol{\alpha}})^p \right\rangle^{m_{\boldsymbol{\alpha}}}$. The final results are (using $\langle J_z \boldsymbol{h}^p \rangle = 0$) [French, Kota and Smith (1989)],

$$\begin{aligned}
\left\langle J_z^2 \right\rangle^{[\mathbf{m}]} &= \sum_{\boldsymbol{\alpha}} \left\langle J_z^2 \right\rangle^{m_{\boldsymbol{\alpha}}} = \sum_{\boldsymbol{\alpha}} \frac{m_{\boldsymbol{\alpha}} m_{\boldsymbol{\alpha}}^X}{3\hat{N}_{\boldsymbol{\alpha}\boldsymbol{\alpha}}} \sum_{\alpha \in \boldsymbol{\alpha}} j_\alpha (j_\alpha + 1)(2j_\alpha + 1) \; , \\
\left\langle J_z^2 \boldsymbol{h} \right\rangle^{[\mathbf{m}]} &= \sum_{\boldsymbol{\alpha}} \left\langle J_z^2 \boldsymbol{h}_{\boldsymbol{\alpha}} \right\rangle^{m_{\boldsymbol{\alpha}}} \; , \\
\left\langle J_z^2 \boldsymbol{h}^2 \right\rangle^{[\mathbf{m}]} &= \left\langle J_z^2 \right\rangle^{[\mathbf{m}]} \left\langle \boldsymbol{h}^2 \right\rangle^{[\mathbf{m}]} \\
&\quad + \sum_{\boldsymbol{\alpha}} \left\{ \left\langle J_z^2 \boldsymbol{h}_{\boldsymbol{\alpha}}^2 \right\rangle^{m_{\boldsymbol{\alpha}}} - \left\langle J_z^2 \right\rangle^{m_{\boldsymbol{\alpha}}} \left\langle \boldsymbol{h}_{\boldsymbol{\alpha}}^2 \right\rangle^{m_{\boldsymbol{\alpha}}} \right\} \; , \\
\left\langle J_z^2 \boldsymbol{h}^3 \right\rangle^{[\mathbf{m}]} &= \left\langle J_z^2 \right\rangle^{[\mathbf{m}]} \left\langle \boldsymbol{h}^3 \right\rangle^{[\mathbf{m}]} + 3 \left\langle J_z^2 \boldsymbol{h} \right\rangle^{[\mathbf{m}]} \left\langle \boldsymbol{h}^2 \right\rangle^{[\mathbf{m}]} \\
&\quad + \sum_{\boldsymbol{\alpha}} \left\{ \left\langle J_z^2 \boldsymbol{h}_{\boldsymbol{\alpha}}^3 \right\rangle^{m_{\boldsymbol{\alpha}}} - 3 \left\langle J_z^2 \boldsymbol{h}_{\boldsymbol{\alpha}} \right\rangle^{m_{\boldsymbol{\alpha}}} \left\langle \boldsymbol{h}_{\boldsymbol{\alpha}}^2 \right\rangle^{m_{\boldsymbol{\alpha}}} - \left\langle J_z^2 \right\rangle^{m_{\boldsymbol{\alpha}}} \left\langle \boldsymbol{h}_{\boldsymbol{\alpha}}^3 \right\rangle^{m_{\boldsymbol{\alpha}}} \right\} \; , \\
\left\langle J_z^2 \boldsymbol{h}^4 \right\rangle^{[\mathbf{m}]} &= \left\langle J_z^2 \right\rangle^{[\mathbf{m}]} \left\langle \boldsymbol{h}^4 \right\rangle^{[\mathbf{m}]} + 6 \left\langle J_z^2 \boldsymbol{h}^2 \right\rangle^{[\mathbf{m}]} \left\langle \boldsymbol{h}^2 \right\rangle^{[\mathbf{m}]} \\
&\quad + 4 \left\langle J_z^2 \boldsymbol{h} \right\rangle^{[\mathbf{m}]} \left\langle \boldsymbol{h}^3 \right\rangle^{[\mathbf{m}]} - 6 \left\langle J_z^2 \right\rangle^{[\mathbf{m}]} \left[\left\langle \boldsymbol{h}^2 \right\rangle^{[\mathbf{m}]} \right]^2 \\
&\quad + \sum_{\boldsymbol{\alpha}} \left\{ \left\langle J_z^2 \boldsymbol{h}_{\boldsymbol{\alpha}}^4 \right\rangle^{m_{\boldsymbol{\alpha}}} - \left\langle J_z^2 \right\rangle^{m_{\boldsymbol{\alpha}}} \left\langle \boldsymbol{h}_{\boldsymbol{\alpha}}^4 \right\rangle^{m_{\boldsymbol{\alpha}}} - 6 \left\langle J_z^2 \boldsymbol{h}_{\boldsymbol{\alpha}}^2 \right\rangle^{m_{\boldsymbol{\alpha}}} \left\langle \boldsymbol{h}_{\boldsymbol{\alpha}}^2 \right\rangle^{m_{\boldsymbol{\alpha}}} \right. \\
&\quad \left. - 4 \left\langle J_z^2 \boldsymbol{h}_{\boldsymbol{\alpha}} \right\rangle^{m_{\boldsymbol{\alpha}}} \left\langle \boldsymbol{h}_{\boldsymbol{\alpha}}^3 \right\rangle^{m_{\boldsymbol{\alpha}}} + 6 \left\langle J_z^2 \right\rangle^{m_{\boldsymbol{\alpha}}} \left[\left\langle \boldsymbol{h}_{\boldsymbol{\alpha}}^2 \right\rangle^{m_{\boldsymbol{\alpha}}} \right]^2 \right\} \; .
\end{aligned} \tag{1.102}$$

With $\langle \boldsymbol{h}^p \rangle^{[\mathbf{m}]}$ and $\langle h_{\boldsymbol{\alpha}} \rangle^{m_{\boldsymbol{\alpha}}}$ already known via Eq. (1.24) and Table 1.1, all we need, for applying Eq. (1.102) are single unitary orbit traces $\left\langle J_z^2 (\boldsymbol{h}_{\boldsymbol{\alpha}})^p \right\rangle^{m_{\boldsymbol{\alpha}}}$.

For the single orbit traces, to simplify the notation, we replace $\boldsymbol{h}$ by h^1 and $(m_{\boldsymbol{\alpha}}, N_{\boldsymbol{\alpha}})$ by (m, N). For the deformed one-body Hamiltonian $h^1(\lambda) = h^1 + \lambda J_z$, the central moments are $\mathcal{M}_p^\lambda(m) = \left\langle h^1(\lambda)\right\rangle^m$. Using Table 1.1, $\mathcal{M}_p^\lambda(m)$ in terms of $\mathcal{M}_p^\lambda(1)$ are,

$$\mathcal{M}_p^\lambda(m) = \sum_{[p]} R_{[p]}(m,N) \prod_i \mathcal{M}_{p_i}^\lambda(1) \; . \tag{1.103}$$

The partitions $[p]$ and the functions $R_{[p]}(m,N)$ are listed in Table 1.1 for $p \leq 6$. By double parametric differentiation on both sides of Eq. (1.103) and taking the $\lambda \to 0$ limit, it is seen that,

$$\left\langle J_z^2 (h^1)^{p-2}\right\rangle^m = \sum_{[p]} R_{[p]}(m,N) \left\{ \sum_i \left[\frac{p_i(p_i-1)}{p(p-1)} \left\langle J_z^2 (h^1)^{p_i-2}\right\rangle^1 \prod_{j\neq i} \left\langle (h^1)^{p_j}\right\rangle^1 \right] \right\} . \tag{1.104}$$

The one particle traces in Eq. (1.104) are,

$$\begin{aligned} \left\langle j_z^2 (h^1)^p\right\rangle^1 &= \frac{1}{3N} \sum_\alpha j_\alpha(j_\alpha+1)(2j_\alpha+1)(\epsilon_\alpha^1)^p \, , \\ \left\langle (h^1)^p\right\rangle^1 &= \frac{1}{N} \sum_\alpha (2j_\alpha+1)(\epsilon_\alpha^1)^p \, . \end{aligned} \tag{1.105}$$

Thus we have complete results for one-orbit traces for any p and hence for $\left\langle J_z^2 \boldsymbol{h}^p\right\rangle^{[\mathbf{m}]}$ via Eq. (1.102). However, the moments $M_p^{[\mathbf{m}]}(J_z) = \left\langle J_z^2 \boldsymbol{h}^p\right\rangle^{[\mathbf{m}]} / \left\langle J_z^2\right\rangle^{[\mathbf{m}]}$ are non-central moments of $I_{J_z^2}^{h,[\mathbf{m}]}$. In order to convert these into cumulants we need to use Eq. (3.42) of Kendall and Stuart (1969),

$$\begin{aligned} K_1 &= M_1 \, , \\ K_2 &= M_2 - (M_1)^2 \, , \\ K_3 &= M_3 - 3M_2M_1 + 2\,(M_1)^2 \, , \\ K_4 &= M_4 - 4M_3M_1 - 3\,(M_2)^2 + 12M_2\,(M_1)^2 - 6\,(M_1)^4 \, . \end{aligned} \tag{1.106}$$

A test of the method given by Eqs. (1.102), (1.104), (1.105), (1.106) and Table 1.1 for constructing $I_{J_z^2}^h(E) = \sum_s \sum_{[\mathbf{m}]\in s} I_{J_z^2}^{h,[\mathbf{m}]}(E)$ in large spaces, is shown in Fig. 1.12. Note that the exact results follow easily by using spherical configurations $\mathbf{m}$, then $\left\langle J_z^2\right\rangle^{\mathbf{m}} = \sum_\alpha [m_\alpha(N_\alpha - m_\alpha)/3(N_\alpha - 1)]\, j_\alpha(j_\alpha+1)$. As seen from Fig. 1.12, clearly the results obtained with the Edgeworth representation for $I_{J_z^2}^{h,[\mathbf{m}]}$ are in good agreement with exact results (up to 60 MeV excitation in the present example). Further tests, including those for occupancies, are given in [Kota and Majumdar (1992a)].

For the spin cutoff spreading Gaussians $\rho_{J_z^2:\mathcal{G}}^{\boldsymbol{V},[\mathbf{m}]}$, we need the centroids $\epsilon_{J_z^2:\boldsymbol{V}}([\mathbf{m}])$ and variances $\sigma^2_{J_z^2:\boldsymbol{V}}([\mathbf{m}])$. Before giving the general expressions which are complex, we consider, for illustrating the structures, first the situation with a single unitary orbit [say $\boldsymbol{\alpha}$ with spherical orbits α, and then $N = \sum(2j_\alpha+1)$] for identical nucleons.

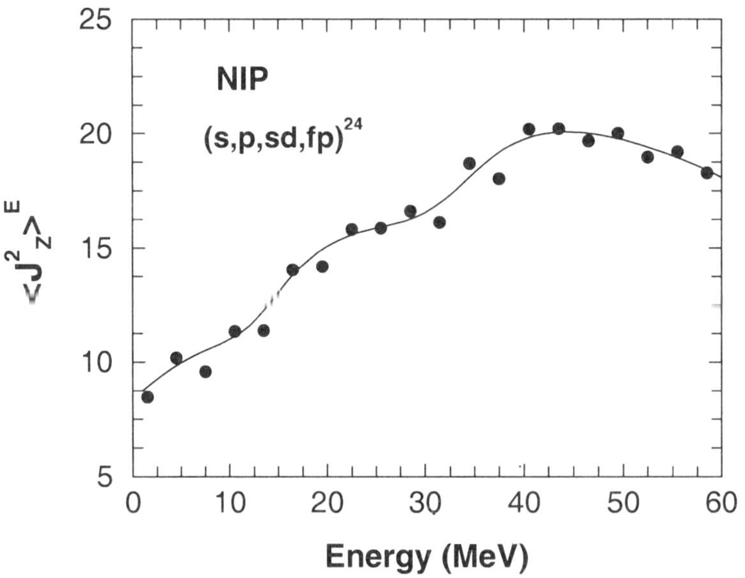

Fig. 1.12 Spin cutoff factors $\langle J_z^2 \rangle^E$ for 24 non-interacting nucleons in (s, p, sd, fp) space. All spherical configurations with positive parity up to $4\hbar\omega$ excitation are included in the exact calculation and results averaged over 3MeV bin (filled circles) are compared with those obtained using ${}^1s_{1/2}$, $({}^1p_{3/2}, {}^1p_{1/2})$, $({}^1d_{5/2}, {}^2s_{1/2}, {}^1d_{5/2})$, ${}^1f_{7/2}$ and $({}^2p_{3/2}, {}^1f_{5/2}, {}^2p_{1/2})$ as unitary orbits with $s = 0$, 1, 2, 3 and 3 respectively; the minimum s value is 28 $\hbar\omega$. The sp energies for the 10 orbits are 0.0, 8.1, 12.2, 16.4, 20.6, 21.9, 24.8, 29.4, 31.2 and 32.2 respectively and they derive from Seeger energies [Hillman and Grover (1969)]. Results are from [Kota (1991b)].

Here we need to start with the unitary decomposition of the J^2 operator. With $q_\alpha = j_\alpha(j_\alpha+1)$, $q = N^{-1}\sum(2j_\alpha+1)\, q_\alpha = \langle J^2 \rangle^1$, $\boldsymbol{J} = (J^2)^{\nu=2}$ and using Eqs. (1.79) - (1.82), we have

$$
\begin{aligned}
(J^2)^{\nu=0} &= n(N-n)(N-1)^{-1} q \,, \\
(J^2)^{\nu=1} &= (N-2n)(N-2)^{-1} \sum_\alpha \{q_\alpha - q\} n_\alpha = \sum_\alpha \epsilon_\alpha^{1:J^2}(n)\, n_\alpha \,, \\
\boldsymbol{J} &\leftrightarrow \boldsymbol{J}^J_{\alpha\beta\alpha\beta} \\
&= \left[J(J+1) - (N-4)(N-2)^{-1}\{q_\alpha + q_\beta\} - 2N\{(N-1)(N-2)\}^{-1} q \right] .
\end{aligned}
\tag{1.107}
$$

Note that the off-diagonal two-particle matrix elements of $\boldsymbol{J}$ are all zero in the $|(\alpha\beta)J\rangle$ basis. Also from Eq. (1.107) we have, $\langle J^2 \rangle^m = m(N-m)(N-1)^{-1} q$. This and Eq. (1.95) immediately yield a simple formula for the spin cutoff centroids,

$$
\begin{aligned}
\epsilon_{J_z^2:\boldsymbol{V}}(m) &= \{\langle J^2 \rangle^m\}^{-1} \langle J^2 \boldsymbol{V} \rangle^m = \{\langle J^2 \rangle^m\}^{-1} \langle \boldsymbol{J}\boldsymbol{V} \rangle^m = \Delta_1 \,, \\
\langle \boldsymbol{J}\boldsymbol{V} \rangle^m &= \frac{\hat{m}_{\boldsymbol{\alpha\alpha}} \hat{m}^X_{\boldsymbol{\alpha\alpha}}}{4\hat{N}_{\boldsymbol{\alpha\alpha\alpha\alpha}}} \sum_{\alpha,\beta,\gamma,\delta \in \boldsymbol{\alpha}} (2J+1)\, \boldsymbol{J}^J_{\alpha\beta\gamma\delta} \boldsymbol{V}^J_{\alpha\beta\gamma\delta} (1+\delta_{\alpha\beta})(1+\delta_{\gamma\delta}) \,.
\end{aligned}
\tag{1.108}
$$

As $\langle \boldsymbol{V} \rangle^m = 0$, the Δ_1 in Eq. (1.108) is in fact the centroid shift. Similarly the variance is,

$$\sigma^2_{J_z^2:\boldsymbol{V}}(m) = \left\langle \boldsymbol{V}^2 \right\rangle^m + \left\{ \left\langle J^2 \right\rangle^m \right\}^{-1} \left\langle (J^2)^{\nu=1} \boldsymbol{V}^2 \right\rangle^m + \left\{ \left\langle J^2 \right\rangle^m \right\}^{-1} \left\langle \boldsymbol{J} \boldsymbol{V}^2 \right\rangle^m - (\Delta_1)^2 \,. \tag{1.109}$$

Propagation equation for the first term depends only on $\left\langle (J^2)^{\nu=1} \boldsymbol{V}^2 \right\rangle^2$ and this follows from Eq. (1.78); note that the trace here is a polynomial of 5th degree in m with the traces for $m = 0, 1, N, N-1$ being zero and those for $m = 2$ and $m = N-2$ differing only in phase. Hence,

$$\begin{aligned} \left\langle (J^2)^{\nu=1} \boldsymbol{V}^2 \right\rangle^m = & \frac{\hat{m}_{\boldsymbol{\alpha\alpha}} \hat{m}^X_{\boldsymbol{\alpha\alpha}} (m^X - m)}{2 \hat{N}_{\boldsymbol{\alpha\alpha\alpha\alpha\alpha}}} \\ & \times \sum_{\alpha,\beta,\gamma,\delta} (2J+1) \epsilon^{1:J^2}_\alpha (m) \left[\boldsymbol{V}^J_{\alpha\beta\gamma\delta} \right]^2 (1+\delta_{\alpha\beta})(1+\delta_{\gamma\delta}) \,. \end{aligned} \tag{1.110}$$

Finally, we need the propagation equation for $\left\langle \boldsymbol{J} \boldsymbol{V}^2 \right\rangle^m$. This will be a polynomial of degree 6 in m and the operator $\boldsymbol{J}\boldsymbol{V}^2$ is particle-hole symmetric. Therefore the trace will propagate from the $m = 2$ and 3 particle traces. The 3-particle traces of $\boldsymbol{J}\boldsymbol{V}^2$ will be the sum of the traces of its 2 and 3-body parts. These two parts are easy to identify by transforming the $\boldsymbol{J}$ and $\boldsymbol{V}$ into single state representation and then writing them in normal order form. This exercise, though tedious but straightforward, will give finally,

$$\begin{aligned} \left\langle \boldsymbol{J} \boldsymbol{V}^2 \right\rangle^m = & \frac{\hat{m}_{\boldsymbol{\alpha\alpha\alpha}} \hat{m}^X_{\boldsymbol{\alpha\alpha\alpha}}}{\hat{N}_{\boldsymbol{\alpha\alpha\alpha\alpha\alpha\alpha}}} \mathbf{A} + \frac{\hat{m}_{\boldsymbol{\alpha\alpha}} \hat{m}^X_{\boldsymbol{\alpha\alpha\alpha\alpha}} + \hat{m}_{\boldsymbol{\alpha\alpha\alpha\alpha}} \hat{m}^X_{\boldsymbol{\alpha\alpha}}}{\hat{N}_{\boldsymbol{\alpha\alpha\alpha\alpha\alpha\alpha}}} \mathbf{B} \,, \\ \mathbf{A} = & \sum_{\alpha,\beta,\gamma,\delta,\tau,\theta \in \boldsymbol{\alpha}} (-1)^{j_\alpha - j_\gamma - \Delta} (2\Delta+1)^{-1/2} \mathcal{C}^\Delta_{\alpha\gamma\beta\delta}(\boldsymbol{J}) \mathcal{C}^\Delta_{\tau\theta\beta\delta}(\boldsymbol{V}) \mathcal{C}^\Delta_{\tau\theta\gamma\alpha}(\boldsymbol{V}) \,, \\ \mathbf{B} = & \frac{1}{8} \sum_{\alpha,\beta,\gamma,\delta,\tau,\theta,\in \boldsymbol{\alpha}} (2J+1) \boldsymbol{J}^J_{\alpha\beta\gamma\delta} \boldsymbol{V}^J_{\gamma\delta\tau\theta} \boldsymbol{V}^J_{\tau\theta\alpha\beta} (1+\delta_{\alpha\beta})(1+\delta_{\gamma\delta})(1+\delta_{\tau\theta}) \,, \end{aligned} \tag{1.111}$$

$$\begin{aligned} \mathcal{C}^\Delta_{\alpha\beta\gamma\delta}(\mathbf{X}) = & \sum_J (-1)^{j_\beta + j_\gamma + J + \Delta} (2\Delta+1)^{1/2} (2J+1) \begin{Bmatrix} j_\alpha & j_\beta & J \\ j_\delta & j_\gamma & \Delta \end{Bmatrix} \\ & \times \left[(1+\delta_{\alpha\beta})(1+\delta_{\gamma\delta}) \right]^{1/2} \mathbf{X}^J_{\alpha\beta\gamma\delta} \;;\; \mathbf{X} = \boldsymbol{J} \quad \text{or} \quad \boldsymbol{V} \,. \end{aligned}$$

In Eq. (1.111), $\{- - -\}$ is a $6-j$ symbol.

Extension of the propagation Eqs. (1.108), (1.110) and (1.111) to the unitary configurations $[\mathbf{m}]$ follow from the principle used in going from Eq. (1.95) to Eq. (1.97). To apply this, first the J^2 operator, using the formulation given in Sec. 1.4.1, is decomposed into unitary tensors with rank $\boldsymbol{\nu} = \mathbf{0} + \mathbf{1} + \mathbf{2}$. From now on we denote $(J^2)^{\mathbf{2}}$ by $\boldsymbol{J}$ and this should not be confused with $\boldsymbol{J}$ used in Eq. (1.107). Similarly the $[\mathbf{m}]$ dependent sp energies defining $(J^2)^{\mathbf{1}}$ are denoted by $\epsilon^{\mathbf{1},J^2}_\alpha([\mathbf{m}])$. Firstly, the extension of Eq. (1.108) from $\langle \boldsymbol{J}\boldsymbol{V} \rangle^m$ to $\langle \boldsymbol{J}\boldsymbol{V} \rangle^{[\mathbf{m}]}$ is quite simple and this follows from

Eq. (1.97) by replacing $(\boldsymbol{V}^J_{\alpha\beta\gamma\delta})^2$ by $(\boldsymbol{J}^J_{\alpha\beta\gamma\delta})(\boldsymbol{V}^J_{\alpha\beta\gamma\delta})$. For extending Eq. (1.110), we can follow a similar procedure [alternatively, a formal method given in Wong (1986) can be used]. Then we take the propagator inside the summation and replace $\hat{m}_{\boldsymbol{\alpha}\boldsymbol{\alpha}}$ by $\hat{m}_{\boldsymbol{\alpha}\boldsymbol{\beta}}$ etc. The only complication is in associating the unitary orbit indices $\boldsymbol{\alpha}$ with the spherical orbit indices α. This is best done using a computer programme and by comparing the trace result with exact shell model calculations. The Rochester-Oak Ridge shell model code [French, Halbert, McGrory and Wong (1969)] for $(2s1d)$ shell nuclei has been used for testing the final formula. Carrying out this exercise, the following formula for $\langle (J^2)^1\boldsymbol{V}^2\rangle^{[\mathrm{m}]}$ was obtained,

$$\begin{aligned}\langle (J^2)^1\boldsymbol{V}^2\rangle^{[\mathrm{m}]} = \frac{1}{2} \sum_{\boldsymbol{\alpha\beta\gamma\delta}} \left[\hat{N}^{-1}_{\boldsymbol{\alpha\beta\gamma\delta\alpha}}\hat{m}_{\boldsymbol{\alpha\beta}}\hat{m}^X_{\boldsymbol{\gamma\delta\alpha}} - \hat{N}^{-1}_{\boldsymbol{\gamma\delta\alpha\alpha\beta}}\hat{m}_{\boldsymbol{\gamma\delta\alpha}}\hat{m}^X_{\boldsymbol{\beta\alpha}}\right] \\ \times \sum_{J;(\alpha\beta\gamma\delta)'} (2J+1)\epsilon^{1,J^2}_{\alpha}([\mathrm{m}]) \left[\boldsymbol{V}^J_{\alpha\beta\gamma\delta}\right]^2 (1+\delta_{\alpha\beta})(1+\delta_{\gamma\delta}) \,.\end{aligned} \tag{1.112}$$

Here $(\alpha\beta\gamma\delta)'$ implies that the sum is restricted to $\alpha \in \boldsymbol{\alpha}$, $\beta \in \boldsymbol{\beta}$, $\gamma \in \boldsymbol{\gamma}$ and $\delta \in \boldsymbol{\delta}$. Proceeding in the same manner, we have obtained the propagation formula for $\langle \boldsymbol{J}\boldsymbol{V}^2\rangle^{[\mathrm{m}]}$. The association of the unitary orbits to spherical orbit indices is much more complex here and we are guided by the expression given in the book by

Table 1.3 Centroid shifts and spreading widths for the spin cutoff density spreading Gaussians and the spreading widths for the state density spreading Gaussians (all in MeV). Details of the interactions, the unitary orbits, S-values etc. are given in Table 1.2 and in the text following Eq. (1.93).

sp space	(m_p, m_n, S)	$\epsilon_{J_z^2:V}$	$\sigma_{J_z^2:V}$	σ_V
$(2s1d)$	(2,2,0)	−0.518	4.085	5.200
	(3,5,0)	−0.775	7.697	8.364
	(4,4,0)	−0.814	8.015	8.681
	(6,6,0)	−0.913	9.081	9.831
	(6,8,0)	−0.860	8.518	9.254
	(8,10,0)	−0.626	6.317	6.875
$(2p1f)$	(3,7,0)	−0.213	5.582	5.905
	(5,10,0)	−0.293	7.028	7.430
	(6,10,0)	−0.315	7.435	7.846
	(8,12,0)	−0.332	7.783	8.211
	(10,10,0)	−0.344	8.136	8.565
	(14,14,0)	−0.296	6.789	7.151
$(2s1d) - (2p1f)$	(3,5,0)	−1.030	6.711	7.287
	(3,5,1)	−0.972	7.954	8.365
	(3,5,2)	−0.950	8.751	9.076
	(4,4,0)	−1.077	6.993	7.567
	(4,4,1)	−1.017	8.212	8.624
	(4,4,2)	−0.995	8.993	9.318
	(6,6,0)	−1.222	7.933	8.560
	(6,6,1)	−1.219	9.583	10.017
	(6,6,2)	−1.253	10.866	11.325

Wong (1986) for $\langle \boldsymbol{VUW} \rangle^{\mathbf{m}}$. Then the final result is,

$$\begin{aligned}
\langle \boldsymbol{JV}^2 \rangle^{[\mathbf{m}]} &= \sum_{\boldsymbol{\alpha},\boldsymbol{\beta},\boldsymbol{\gamma},\boldsymbol{\delta},\boldsymbol{\tau},\boldsymbol{\theta}} \Big[\Big\{ \hat{m}_{\alpha\beta\theta} \hat{m}^X_{\gamma\delta\tau} \hat{N}^{-1}_{\alpha\beta\theta\gamma\delta\tau} \Big\} \mathbf{A}_{\boldsymbol{\alpha},\boldsymbol{\beta},\boldsymbol{\gamma},\boldsymbol{\delta},\boldsymbol{\tau},\boldsymbol{\theta}} \\
&\quad + \Big\{ \hat{m}_{\alpha\beta} \hat{m}^X_{\gamma\delta\tau\theta} \hat{N}^{-1}_{\alpha\beta\gamma\delta\tau\theta} + \hat{m}_{\tau\theta\gamma\delta} \hat{m}^X_{\alpha\beta} \hat{N}^{-1}_{\tau\theta\gamma\delta\alpha\beta} \Big\} \Big] \mathbf{B}_{\boldsymbol{\alpha},\boldsymbol{\beta},\boldsymbol{\gamma},\boldsymbol{\delta},\boldsymbol{\tau},\boldsymbol{\theta}} \; ; \\
\mathbf{A}_{\boldsymbol{\alpha},\boldsymbol{\beta},\boldsymbol{\gamma},\boldsymbol{\delta},\boldsymbol{\tau},\boldsymbol{\theta}} &= \sum_{\Delta;(\alpha\beta\gamma\delta\tau\theta)'} (-1)^{j_\alpha - j_\gamma - \Delta} (2\Delta+1)^{-1/2} \mathcal{C}^{\Delta}_{\alpha\gamma\beta\delta}(\boldsymbol{J}) \mathcal{C}^{\Delta}_{\tau\theta\beta\delta}(\boldsymbol{V}) \mathcal{C}^{\Delta}_{\tau\theta\gamma\alpha}(\boldsymbol{V}) \, , \\
\mathbf{B}_{\boldsymbol{\alpha},\boldsymbol{\beta},\boldsymbol{\gamma},\boldsymbol{\delta},\boldsymbol{\tau},\boldsymbol{\theta}} &= \frac{1}{8} \sum_{\Delta;(\alpha\beta\gamma\delta\tau\theta)'} (2J+1) \boldsymbol{J}^J_{\alpha\beta\gamma\delta} \boldsymbol{V}^J_{\gamma\delta\tau\theta} \boldsymbol{V}^J_{\tau\theta\alpha\beta} (1+\delta_{\alpha\beta})(1+\delta_{\gamma\delta})(1+\delta_{\tau\theta}) \, .
\end{aligned} \tag{1.113}$$

Here $(\alpha\beta\gamma\delta\tau\theta)'$ implies that the sums are restricted to $\alpha \in \boldsymbol{\alpha}$, $\beta \in \boldsymbol{\beta}$, $\gamma \in \boldsymbol{\gamma}$, $\delta \in \boldsymbol{\delta}$, $\tau \in \boldsymbol{\tau}$ and $\theta \in \boldsymbol{\theta}$. Similarly, the function $\mathcal{C}()$ is already defined in Eq. (1.111). Using Eqs. (1.97), (1.112) and (1.113) one can calculate the centroids and variances of $\rho^{\boldsymbol{V},[\mathbf{m}]}_{J_z^2:\mathcal{G}}$ and hence the spin cutoff factors via Eq. (1.101). Results for the spin cutoff centroid shifts $\epsilon_{J_z^2:\boldsymbol{V}}([\mathbf{m}])$ and the spin cutoff widths $\sigma_{J_z^2:\boldsymbol{V}}([\mathbf{m}])$ are given in Table 1.3 for some $(2s1d)$, $(2p1f)$ and $(2s1d)-(2p1f)$ examples. These spaces and other related details are same as those in Table 1.2. Also $\sigma_{J_z^2:\boldsymbol{V}}$ are compared with $\sigma_{\boldsymbol{V}}$ in the Table. The spin cutoff widths are in general smaller by about 10% compared to $\sigma_{\boldsymbol{V}}$ and the centroid shifts are ~ -0.5 MeV.

1.5 Nuclear Level Densities

With Eq. (1.94) and the spin decomposition using spin cutoff factors calculated via Eq. (1.101) and (1.100), we have a spectral distribution theory, for nuclear level densities where the interaction effects are introduced via convolutions (by Gaussians) of the NIP partial densities defined over unitary configurations. Before discussing its applications, we briefly describe the more conventional approach based on spherical configuration partial densities.

1.5.1 *Level densities using spherical configurations*

Following the discussion in Sec. 1.2, a convenient method for calculating level densities, from the shell model point of view, is to first construct partial state densities $I^{\mathbf{m}}(E)$ defined over spherical configuration $(\mathbf{m})$ spaces as Gaussian or Edgeworth corrected Gaussians with $\gamma_1(\mathbf{m})$ and $\gamma_2(\mathbf{m})$ corrections. Next, the angular momentum decomposition of the total state density $I(E) = \sum_{\mathbf{m}} I^{\mathbf{m}}(E)$ is carried out using spin cutoff factors [Chang, French and Thio (1971*); Grimes, Poppe, Wong and Dalton (1978); Baldridge and Dalton (1978)] so that total level densities or fixed-J densities can be calculated and compared with data. It is important to note that the decomposition of the total space into (p, n) configurations $(\mathbf{m}_p, \mathbf{m}_n)$ automatically generates the decomposition of the level densities into densities with positive and negative parity. Understanding parity ratios for level densities is an important problem; see for example [Grimes (1988); Mocelj et. al. (2007); Grimes and Voinov (2009)].

To apply the spectral distribution procedure it is necessary to locate the ground state or an appropriate excited (reference) state in the theoretical calculations. A standard approach is to employ the so called Ratcliff procedure [Ratcliff (1971*)]. For the ground state angular momentum J_{GS} [in the (p,n) formalism], the ground state energy E_{GS} is given by

$$\frac{(2J_{GS}+1)}{2} = \int_{-\infty}^{E_{GS}} I(E)dE \; . \tag{1.114}$$

Note that the integral in Eq. (1.114) gives the total number of states $(2J_{GS}+1)$ upto the ground state. However, we take only half of this due to the curvature effect near the ground state; see [Langhoff (1980)] for a rigorous theoretical explanation. To avoid the necessarily large departures in the density near the ground state from its CLT smoothed form, instead of using E_{GS}, a high enough (from complete spectroscopy) reference energy state (E_{Ref}) is used. With total number of states upto E_{Ref} being N_{Ref} and the angular momentum J_{Ref} of the reference state, it is simple to modify Eq. (1.114) to

$$N_{Ref} - \frac{(2J_{Ref}+1)}{2} = \int_{-\infty}^{E_{Ref}} I(E)dE \; . \tag{1.115}$$

For a Gaussian $I(E)$, Eq. (1.114) can be inverted [Abramowtiz and Stegun (1964)]. For 0^+ ground states this gives

$$(E_{GS} - E_c) \simeq -\sigma \sqrt{\ln(2d)} \; . \tag{1.116}$$

Some phenomenological modifications of Eq. (1.116) have been given recently by Yoshinaga, Arima and Zhao (2006); Shen, Zhao, Arima and Yoshinaga (2008a).

Starting with the analysis by Chang, French and Thio (1971*), there have been several studies of binding energies and spectra of nuclei using Eq. (1.116) and also applications to occupancies and spin cutoff factors [Dalton, Vary and Baldridge (1977); Dalton, Baldridge and Vary (1979); Wong and Chang (1982); Margetan, Klar and Vary (1983)]. In many analyses the focus has been to test the methods using exact shell model results. With large scale shell model describing lower $(2p1f)$ shell nuclei extremely well [Caurier *et al.* (2005)], more recent efforts following earlier work by Haq and Parikh (1976), have focused on $(2p1f)$ shell comparisons [Kar *et al.* (1997); Choubey, Kar, Gómez and Manfredi (1998)]; see also [Delfini and Brussaard (1985); Sviratcheva, Draayer and Vary (2006*, 2007)]. An important extension of Eq. (1.114) is in determining the spectrum end (the integration limits will be from maximum energy to $+\infty$) and this will determine the upper bound [Hsu (1967)] on transition strength sums [non-energy weighted sum rule (NEWSR)] generated by a transition operator $\mathcal{O}$. Replacing H by $\mathcal{O}\cdot\mathcal{O}$ for electromagnetic transition operators

$\mathcal{O}$, Halemane [Halemane (1982); Halemane and French (1982a*)] calculated upper bounds for NEWSR of isoscalar and isovector E2, E4, M1, M3 and M5 transitions from the ground states in $(2s1d)$ shell nuclei. In addition, the polynomial expansion theory has been applied to NEWSR and LEWSR for electromagnetic transitions in $(2s1d)$ shell nuclei in detail and compared with data by Halemane [Halemane, Abbas and Zamick (1981); Halemane and French (1982a*,b)].

The first attempt to calculate nuclear level densities using spectral distributions (i.e. partitioning with $\mathbf{m}$ and spin cutoff factors) was by Chang, French and Thio (1971*) and Ayik and Ginocchio (1974b). Grimes and collaborators [Grimes, Bloom, Vanach and Hausman Jr (1983); Strohmaier, Grimes and Satyanarayana (1987); Strohmaier and Grimes (1988)] made a detailed and systematic analysis of data for ^{20}Ne, ^{24}Mg and ^{28}Si. For a more comprehensive account of the analysis of nuclei with $20 \leq A \leq 40$ see [Huang, Grimes and Massey (2000*); Grimes (2002); Al-Quraishi, Grimes, Massey and Resler (2003)]. Instead of using a corrected Gaussian form for the partial densities, an attempt has been made [Nabi, Johnson and Ormand (2001); Johnson, Nabi and Ormand (2001); Johnson and Terán (2005)] to employ the binomial form defined by Eq. (1.41). Exact propagation methods for calculating $\gamma_1(\mathbf{m})$ and $\gamma_2(\mathbf{m})$ have been developed by several groups [Li and Koltun (1973); Ginocchio (1973a); Ayik and Ginocchio (1974a); Potbhare (1975); Chang and Wong (1978)] and computer codes for these have been published [Chang and Wong (1979, 1980); Chang, Draayer and Wong (1982); Chang, Vincent and Wong (1984)]. However, as stated recently by Terán and Johnson (2006a) following the calculations they made for some $(2p1f)$ shell nuclei: "For a realistic application, one must include thousands of configurations ... Further more, higher moments take more and more computational effort. The configuration centroids and widths (first and second moments, respectively) take CPU-minutes, while configuration asymmetries (third moments) take CPU-hours or days and configuration fourth moments even longer; in our example of ^{54}Fe it takes about 43 min on a single 2.4GHz Pentium to compute all the second moments, but 9 days to compute the third moments". Thus exact methods are not practical. This was known before and an alternative was suggested and used for calculating third and fourth configuration moments by Grimes, Bloom, Hausman Jr and Dalton (1979); Bloom, Grimes and Hausman Jr (1980); Strohmaier, Grimes and Bloom (1985) - this is called 'random representative vectors method'where a Monte-Carlo sampling is used. Another approach is to look for analytical approximations [French (1980c*); Tomsovic (1986)]. Higher order traces are needed for applications in beta decay and double beta decay matrix elements calculations [Kota and Majumdar (1995b*); Kota (2007b)].

An alternative to the decomposition using spin cutoff factors is to construct directly $\mathbf{m}JT$ [or $(\mathbf{m}_p\mathbf{m}_n)J$] Gaussian densities using the $\langle H\rangle^{\mathbf{m}JT}$ and $\langle H^2\rangle^{\mathbf{m}JT}$ traces. This has been done by Jacquemin (1973); Lougheed and Wong (1975a,b); Verbaarschot and Brussaard (1981). See Sec. 1.6.6 ahead for more details. Recently, along with a new method for determining the ground state [Horoi, Volya and Zelevinsky (1999)], Horoi *et al.* [Horoi, Kaiser and Zelevinsky (2003*); Horoi, Ghita and Zelevinsky (2004)] suggested that one can use this method for realistic nuclear level density calculations. Continuing with this they have also developed a method for removal of center-of-mass spurious states from level densities (this is needed at higher excitation energies that involve multi-shell excitations); see [Horoi and Zelevinsky (2007a); Verbaarschot, Timmer and Brussaard (1982)]. More interestingly, further investigation of the method based on exact fixed-J moments has shown that the exact moment approach is impractical for $(2p1f)$ shell and beyond. As Horoi, Ghita and Zelevinsky (2005) state. "The calculation of the fixed-J positive-parity widths in the middle of the fp-shell could take days on a single modern workstation. However, if one includes the sd-shell or/and the gsd-shell in order to calculate the fixed-J negative parity widths for some cases of interest, the calculations may become impossible even when using the most powerful modern computer. In contrast, the all-averaged configuration centroids and widths can be calculated .... It will thus be interesting to investigate if a method of calculating the fixed-J density of states using the spin-cutoff factor could be an efficient and accurate alternative".

1.5.2 *Level densities using convolution forms*

In this Section we consider applications of the level density theory given by the convolution form discussed in Sec. 1.4, i.e. Eq. (1.94) along with the corresponding convolution form for spin cutoff densities. First, we discuss some important aspects of the formalism and then consider two examples as applications. The first example illustrates the calculation of level densities and spin cutoff factors starting from a given spectroscopic space and a known effective interaction. The second example deals with the determination of the Hamiltonian parameters from observed level densities in the neutron resonance domain and information propagation from one region of the nuclear spectrum to another.

The specific problem is that of calculating the density given the Hamiltonian or, conversely, of extracting from the density the information about the Hamiltonian which it carries. For effective interactions such as the surface-delta, which are used for heavy nuclei and do not renormalize the sp spectrum, a resonance-region density accurately determines, for a given sp spectrum, a magnitude parameter of the interaction; for other interactions it gives a combination of magnitude and renormalization parameters. In general terms our results should tell us new things about the propagation of information up through the spectrum, as represented by the relationship between spectra in the ground-state domain and densities at high excitations.

The formalism is as presented in Sec. 1.4 and we show in Fig. 1.13 the structure of the theory. We start with the Hamiltonian H written as a sum of four parts. The zero-body operator $C(0)$ may be regarded as the internal energy of the inactive

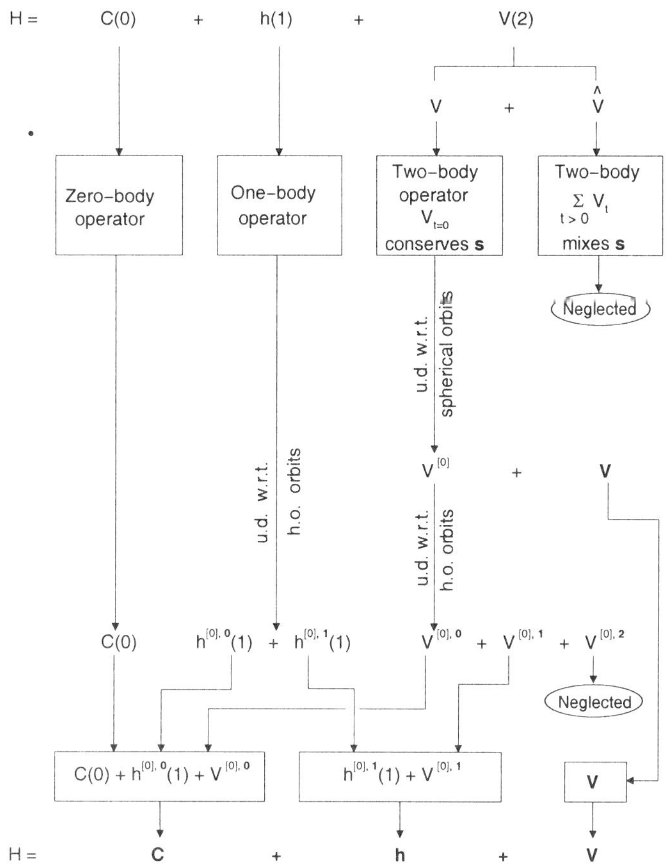

Fig. 1.13 Schematic structure of the theory showing the renormalization of the mean field and spreading interaction via unitary group decompositions of the Hamiltonian. In the figure *u.d.* stands for unitary decomposition and *h.o.* for harmonic oscillator. See text for details.

nucleons, which along with the one-body $h(1)$ defines the primary mean field in which the active nucleons find themselves. For the indicated t-decomposition $(V + \hat{V})$ of the two-body interaction, we introduce h.o. orbits which we label by Λ_r, the higher orbits being modified as usual for spin orbit effects. Thus labeling and ordering the h.o. orbits by Λ_r with $r = 0, \pm1, \pm2, \ldots$, with Λ_0 so defined that $(\Lambda_0)^m$ is the ground-state configuration with $S = 0$, and $(\Lambda_{-1})^{-1}(\Lambda_0)^{m+1}$ and $(\Lambda_0)^{m-1}(\Lambda_1)^1$ have $S = 1$, and in general $S = \sum |r| m_r$ where m_r counts particles when $r > 0$ and holes when $r < 0$; we take $m = 0$ for a ground state configuration with only filled h.o. orbits. For light nuclei and small S, the even-S configurations will then give states of natural parity (+ for $(2s1d)^m$ and $(-1)^m$ for $(2p1f)^m$), and the odd-S of opposite parity. In heavy nuclei all many-particle h.o. configurations will have essentially equal numbers of positive- and negative-parity states.

The V_t with $t > 0$ is that part of V which directly connects only configurations which differ by $t\hbar\omega$ in h.o. energy. Then $\hat{V}$ will, on the average, connect only distant states, generating weak admixing for which the variance is not a proper measure. It is clear that the strong interactions between close configurations and weak interactions between distant ones need to be treated differently as first suggested by Mon and French (1975*): "It is known by this time that the statistical methods and more conventional perturbation methods are complementary in that the simplest version of the first are valid only when the interactions are strong enough to generate a kind of *chaos*, a requirement that is quite opposite to the conditions needed for the validity of perturbation methods". A preliminary schematic treatment of this problem is given in French and Kota (1983*) and considered in some detail by Kota, Majumdar, Haq and Leclair (1999). The model analyzed is a simple partitioned EGOE. It is useful to note that while a general partitioned GOE is solvable [Pluhar and Weidenmüller (1988)], even a simple partitioned EGOE presents huge challenges; see also [Touchard, Haq and Arvieu (1997); Kota (2001)]. These suggest that it is a sensible approximation to neglect $\hat{V}$. We do that in what we call our first "structural approximation" and we have then a set of major-shell configurations which interact only if their h.o. energies are equal; this of course does not at all imply that particles in different major shells are treated as non-interacting.

The next question is whether $V^{[0]}$ simply renormalizes the mean field. In general it has a non-renormalizing ($\nu = 2$) part in the decomposition of $V^{[0]} = \sum V^{[0],\nu}$ according to a unitary group defined by a h.o. orbit or by a unitary-orbit subgroup of it. However, this term is negligible when we use compact unitary orbits (zero with spherical orbits) so that we can, though this is not necessary, usually drop it. The centroid generator $V^{[0]}$ then simply renormalizes the NIP spectrum. As shown in Fig. 1.13, the $V^{[0]}$ renormalization of the NIP Hamiltonian gives an effective NIP Hamiltonian $C + h$. The spectrum consists of a set of degenerate "spikes", one for each spherical configuration and these spikes do not move when the remaining interaction $V - V^{[0]} = \boldsymbol{V}$ is turned on. Under $\boldsymbol{V}$ the states of different spikes admix and the spikes spread with a spreading variance $\left\langle \boldsymbol{V}^2 \right\rangle^{\mathbf{m}}$. A result of major significance is that the spreading widths are essentially the same for all spherical-orbit configurations of the same h.o. excitation S. The $S = 0$ width fluctuation is about $3 - 4\%$ and as shown in Fig. 1.14 is of no consequence except possibly in the ground state domain where it could be corrected by using the true individual variances. But the sum of densities with the same variance and shape is simply the convolution of the centroid distribution with the characteristic shape function so that we have for fixed s the important result $\rho^H = \rho^C \otimes \rho^h \otimes \rho^{\boldsymbol{V}}$. From a practical point of view the use of unitary orbits instead of spherical orbits in calculating the NIP density is recommended. Though this is not essential, compact orbit grouping causes negligible error as shown in Fig. 1.14 and greatly reduces the number of configurations. Going beyond $S = 0$, for higher energies other orbits with $s > 0$ gradually enter, the appropriate group being then the direct-product group $U(N_1) \otimes U(N_2)$which defines separate averaging in the h.o. orbits. It is the gradual entry of the product groups as the excitation energy increases which generates an effective

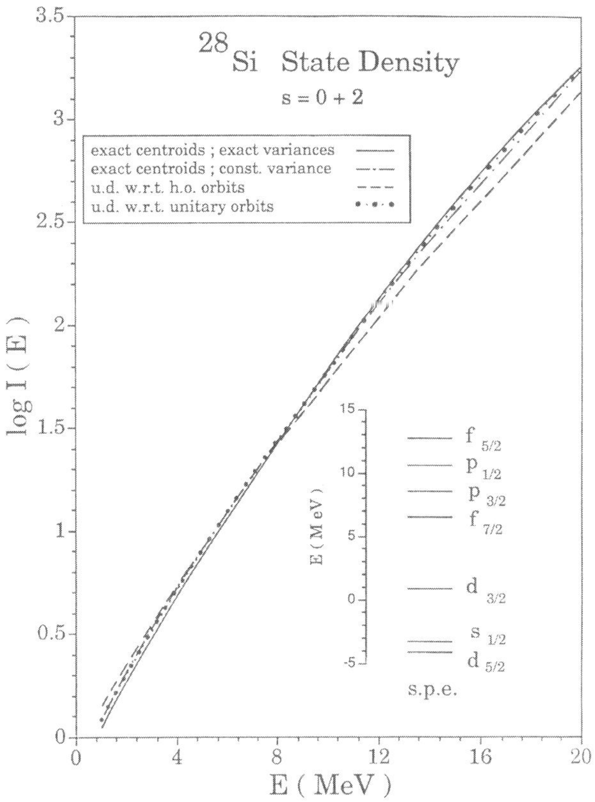

Fig. 1.14 Positive parity ($S = 0 + 2$) IP state density for 6 neutrons and 6 protons in the $(2s1d) - (2p1f)$-subspace using a truncated Kuo interaction. Results corresponding to various structural approximations are shown. In the figure, $I(E)$ is in units of MeV^{-1}, *u.d.* stands for unitary decomposition and *h.o.* for harmonic oscillator. See text for details.

$N(E)$ and a density which contains a back-shifted Bethe NIP form as a convolution factor.

Finally, the procedure for calculating state densities is as follows: (1) By considering the maximum excitation energy of interest, determine the set of active spherical orbits to be dealt with. (2) Choose the single-particle spectrum and the interaction V (perhaps leaving magnitude parameters, not specified a priori, to be determined by calculating densities in the neutron or proton resonance region). Note that enlarging the set of active orbits and re-calculating will yield information about the renormalization of the mean field and the spreading interaction. (3) Discard the $t > 0$ part of the interaction. (4) Construct the centroid shift operator, $V^{[0]}$, and make a decomposition into $\sum V^{[0],\nu}$; discard $V^{[0],2}$ and use the $\nu = 0$, 1 parts to renormalize the mean field. (5) Form unitary orbits and unitary configurations for the (S, π) values of interest. (6) Calculate the low-order configuration-density cumulants, defining thereby (via Edgeworth corrections to the Gaussian) the configuration partial densities, whose summation for fixed (S, π), yields the total NIP

(S,π) density; see Fig. 1.8. (7) Choose the reference energies. (8) Calculate the fixed-(S,π) spreading variances $\langle \boldsymbol{V}^2 \rangle^{m,S}$. This gives the corresponding spreading function, convolution with NIP yields the final (S,π) density. (9) Similarly calculate spin cutoff factors via spin cutoff densities. (10) Use the spin cutoff factors for the J-decomposition of the state density to obtain level densities.

We describe calculations which test the first three structural approximations, namely that fluctuations in spreading variances can be ignored except perhaps near the ground state, that unitary instead of spherical orbits may be used, and that $V^{[0],2}$, the non-renormalizing part of $V^{[0]}$, may also be ignored if the decomposition is based on sufficiently compact unitary orbits. We consider ^{28}Si and calculate the IP density for 6 neutrons and 6 protons in the $S = (0{+}2)$ subspace of the $(2s1d)-(2p1f)$ space using a truncated [i.e. acting only in $(2s1d) - (2p1f)$ space and dropping the $V_{t>0}$ part] Kuo interaction [Ratcliff (1971*)]. We take the experimental reference point as the $J = 3^+$ state at 8.59 MeV, at which the distribution function $F(E)$ has a value 79.5 (in the sum in Eq. (1.115) we include half the degeneracy of the 3^+ state). The four calculations (see Fig. 1.14) are characterized by different treatments of $V^{[0]} = \sum V^{[0],\nu}$ and $\boldsymbol{V}$ which are treated without approximation in the first. The second calculation ignores the fluctuations in $\langle \boldsymbol{V}^2 \rangle^{\mathbf{m}}$, replacing the traces by $\langle \boldsymbol{V}^2 \rangle^{m}$, their properly-weighted average. We conclude from Fig. 1.14 that the variance fluctuations can be ignored so that we do in fact encounter a convolution factoring. In the third calculation we use the direct-product h.o. group [factors defined by $(2s1d)$ and $(2p1f)$] and discard all terms in which $\{\nu(2s1d) + \nu(2p1f)\} = 2$; for some purposes this h.o. decomposition will be inadequate since for example at 20 MeV it gives a 30% error in the $S = (0+2)$ spectrum. The cure, of course, is to decompose $V^{[0]}$ by h.o. subgroups defined by more compact unitary groups. This is the procedure used for the final calculation in which the $(2s1d)$ decomposition is $(d_{5/2}, s_{1/2}) + (d_{3/2})$ whereas the $(2p1f)$ decomposition is still treated as a single orbit (satisfactory because the $S = 2$ density is not a major part of the total). Before turning to applications let us mention that a more detailed discussion of the structure of the theory is given in [Kota and Majumdar (1996*); French *et al.* (2006*)].

Employing the sp spectrum as shown in Fig. 1.15 and the truncated Kuo interaction with other details as described above, state density and spin cutoff factors are calculated for ^{28}Si. The average variances for $S = 0$ and $S = 2$ spaces being 73 and 128 MeV^2 respectively. These two variances define the Gaussian spreading functions $\rho^{\boldsymbol{V},S}(E)$. The NIP density is constructed using the same unitary orbits as above and the 4-moment Edgeworth expansion provides accurate representation of the partial densities. By convoluting I_{NIP} with $\rho^{\boldsymbol{V}}$, (this is achieved by adding cumulants, up to fourth order, of the two functions and again using Edgeworth representation for the partial densities) we have obtained the state density function. The results are shown in Fig. 1.15. The calculated $S = 0 + 2$ density is in good agreement with

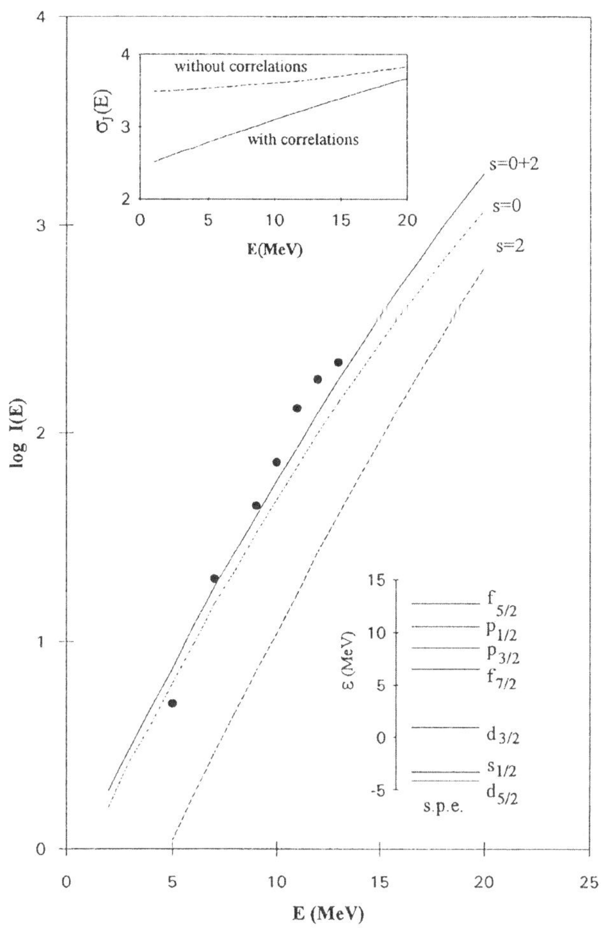

Fig. 1.15 State density and spin cutoff factor for positive parity states in ^{28}Si. Note that $I(E)$ is in units of MeV^{-1}. The filled circles represent experimental observation and continuous curves are calculated results. The calculations are from [Haq (1995)] and [Kota (1997)] and the figure is from [Haq (1995)]. See text for details.

data upto about 15 MeV excitation; The $S = 4$ contribution starts around 20 MeV. Note that $S = 0$ corresponds to configurations arising from $[(2s1d)^{6p}(2s1d)^{6n}]$ and similarly those arising from $[(2s1d)^{4p}(2s1d)^{6n}(2p1f)^{2p}]$, $[(2s1d)^{6p}(2s1d)^{4n}(2p1f)^{2n}]$ and $[(2s1d)^{5p}(2s1d)^{5n}(2p1f)^{1p}(2p1f)^{1n}]$ belong to $S = 2$. Also shown in the figure is the spin cutoff factor $\sigma_J(E)$ calculated with and without $J_z^2 - \boldsymbol{V}$ and $J_z^2 - \boldsymbol{V}^2$ correlations. It is clear that these correlations have a pronounced effect, both in terms of the magnitude and energy dependence, and should not be neglected. With the state density and spin cutoff factor calculated, the level density can be easily obtained. It is important to note that the calculated state density follows closely the standard back shifted Fermi gas (BSFG) form,

$$I_{BSFG}(E) = \frac{\sqrt{\pi}}{12} \frac{\exp\left(2\sqrt{a(E-\Delta)}\right)}{a^{1/4}\,(E-\Delta)^{5/4}} \tag{1.117}$$

where the level density parameter a and the back shifting (due to pairing) parameter Δ are in MeV^{-1} and MeV respectively. A fit of the calculated density to Eq. (1.117) gives [Kota (1997)], $a = 2.236$ MeV^{-1} and $\Delta = 1.265$ MeV. Similarly a fit of the spin cutoff parameter to the Fermi gas form $\chi_J\sqrt{a(E-\Delta)}$ gives $\chi_J = 2.927$. Alternatively a fit to the form $A(E-\Delta)+B$ due to the polynomial theory given in Sec. 1.3, gives $A = 0.426$ MeV^{-1} and B=5.84. In conclusion, the ^{28}Si example shows that the convolution result for state and spin cutoff densities along with unitary orbit decompositions can be used in the analysis of level density data starting with an appropriate set of sp orbits and effective interaction. A systematic analysis of level density data has been carried out for $(2p1f)$ shell nuclei ^{55}Mn, ^{56}Fe, ^{59,60}Co, ^{60,62}Ni and ^{63,65}Cu. These results are reported in [Kota and Majumdar (1996*)].

In our second application we consider the neutron resonance spectrum of ^{168}Er. The resonance domain starts at excitation energy 7.771 MeV and the resonance spin window is $J^\pi = (3^+, 4^+)$ as the ground state spin of ^{167}Er is $J = 7/2^+$. The density of resonance levels is $\sim 2.5 \times 10^5$ levels/MeV. Using the mean spacing (or level density) of the experimentally observed resonances, it is possible to deduce the magnitude of the effective nucleon-nucleon interaction whose form is more or less well known from the studies of the levels near the ground state. It is then possible to interpolate the level density from the low-energy region to the resonance region and perhaps beyond. More specifically, we use the surface delta interaction (SDI) [Brussaard and Glaudemans (1977)] and deduce its strength G by fitting the resonance density. The sp spectrum, with lower four levels for protons and the upper five for neutrons is shown in Fig. 1.16 and the sp energies are zero deformed Nilsson energies [Bohr and Mottelson (1975)]. With this ^{168}Er has 10 valence protons and 10 valence neutrons and the entire space is taken to be $S = 0$. With $G = 1$ the average spreading variance is 183.6 MeV^2. For the $V^{[0]}$ decomposition we choose 4 unitary orbits: $(d_{5/2}, d_{3/2}, s_{1/2})$ and $h_{11/2}$ for protons and $(h_{9/2}, p_{3/2}, f_{5/2}, p_{1/2})$ and $i_{13/2}$ for neutrons. This gives the effective one-body Hamiltonian and then I_{NIP} is calculated using 6 unitary orbits: $d_{5/2}$, $h_{11/2}$, $(d_{3/2}, s_{1/2})$, $h_{9/2}$, $i_{13/2}$ and $(p_{3/2}, f_{5/2}, p_{1/2})$. The experimental $J^\pi = 12^+$ level at 1.94 MeV excitation with $F(E) = 278.5$ is taken as the reference state. Starting with an initial value of G, the state and spin cutoff densities are calculated and from them the density of levels for the resonance spins $(3^+, 4^+)$ at the resonance energy 7.771 MeV is obtained. The value of G is varied and it is found that $G = 0.15$ MeV gives the observed resonance density; see Fig. 1.16. It is pleasing that the deduced G value is close to the often quoted value 25/A MeV, obtained from the spectroscopy of low-lying levels. The state density and spin cutoff factors for this G, interpolating from the reference energy to the resonance energy and beyond, are shown in Fig. 1.16. Once again we notice the strong effect of correlations in $\sigma_J(E)$. It should be noted that Fig. 1.16 shows information propagation downwards in energy to (1-2) MeV excitation from the resonance energy as generated by spectral distribution theory. The calculated state density fits quite well the BSFG form [Kota (1997)] giving (a, Δ) in MeV^{-1}

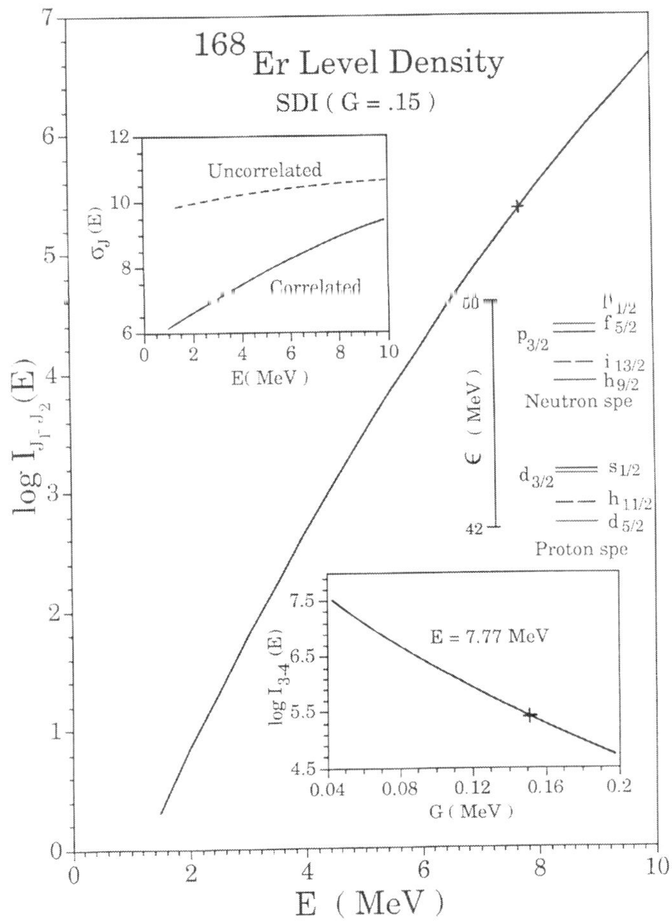

Fig. 1.16 ^{168}Er level density in the spin window $(3^+, 4^+)$ using the sp energies shown with a surface delta interaction (SDI). The strength G of the interaction is determined by matching the density at the resonance energy (point marked +). Note the sensitive dependence of the density on G (inset). Also shown is the spin cutoff factor with (continuous line) and without (dashed line) correlations. Note that the dimensionality of the complete model space is 2.5×10^{14} and the spin cutoff factor determines the density for other J values. The density $I(E)$ is in units of MeV^{-1}. Results are from [Kota (1991b); Haq (1995); French (1995); Kota (1997)] and the figure adapted from [French (1995)].

to be $(18.981, 0.511)$ and these values are compatible with the standard results $a \sim A/10$ and $|\Delta| \sim 1$. Similarly for spin cutoff factors the parameters (A, B, χ_J) with A in MeV^{-1} are $(8.11, 16.97, 6.25)$. The analysis of determining the interaction strength parameter G and interpolating the low-energy region to the far separated resonance region has been done for a large number of heavy nuclei as described in [French *et al.* (2006*); Gharghar (2008)].

The level density theory based on Eq. (1.94), the information content, renormalization and separation of the interaction it displays and finally the solution (when available) to the more general partitioned EGOE obviously have implications for effective interaction theory. There have been significant attempts to apply spectral distribution methods to effective interaction theory [Chang (1978); Chang and Draayer (1979); Chang and Vincent (1981); Darema-Rogers and Vincent (1978); Vary, Belehrad and Dalton (1979)]. It should be added that shell model (similarly in other models) truncation schemes based on configuration centroids and widths have been suggested in Lougheed and Wong (1975b); Lougheed, Ng and Wong (1976); Wong (1978, 1986); Horoi, Brown and Zelevinsky (1994). Similarly, the so-called monopole correction [Abzouzi, Caurier and Zuker (1991*); Caurier *et al.* (2005)] is based on adjusting the average two-body matrix elements defining configuration (p, n) or isospin centroids.

1.6 Group Symmetries and Trace Propagation

Results of simple propagation discussed in Sec. 1.4 can be obtained without much attention paid to $U(N)$ algebra for scalar (m particle) averages and $U(N) \supset U(N_i)\oplus$ direct sum subgroup algebra for configuration averages. The simplicity in these examples derives from the fact that the propagators for these involve just particle numbers. This approach can be extended for fixed isospin traces [French (1969)]. Let us consider fixed-T variances of a two-body H in a single j shell. Then the propagation equation is

$$\sigma^2(m,T) = a_0 + a_1 m + a_2 m^2 + a_3 m^3 + a_4 m^4 + a_5 T^2 + a_6 m T^2 + a_7 m^2 T^2 + a_8 (T^2)^2 \, .$$

We can solve for the nine a_i using the input variances for (m, T) with $m = 0 - 4$. However, an optimal net is to choose,

$$\begin{aligned}(m,T) : \,& (0,0), (1,1/2), (2,0), (2,1), (2\Omega,0), (2\Omega-1,1/2), \\ & (2\Omega-2,0), (2\Omega-2,1), (\Omega,\Omega/2) \; ; \quad \Omega = (2j+1) \, .\end{aligned}$$

Note that, as we have a single j-shell, the inputs $\sigma^2(m,T)$ are trivially zero for the points #1, #2, #5 and #6 and #9. It is then easy to solve for the parameters a_i in terms of just the 2-particle variances $\sigma^2(2,0)$ and $\sigma^2(2,1)$ and generate the propagation equation. The final result was given first by French (1969) and derived later using many other methods by several others. It is in fact instructive to plot m vs T and see the optimal net and points to where one can propagate the variances. The optimum net eliminates the need to calculate 3 and 4 particle variances for propagation. However, this simplification will not extend when we have a multi-j shell situation. It is here that one sees the usefulness of group theory and the Wigner-Racah algebra of the higher groups involved allows one to write the trace propagation equation just in terms of one and two-body matrix elements of H for $\sigma^2(m,T)$. In the next three subsections, we discuss briefly the general theory of trace propagation and for details we refer the readers to [French and Draayer (1979*); Draayer, French and Kota (1982); Draayer and Rosensteel (1985); Kota and Kar (1989)].

1.6.1 *Introduction*

This Section deals with information and its propagation in model spaces organized and partitioned by irreps of group symmetries. A close relationship between group symmetries and spectral averaging is one strong reason for partitioning. The introduction of N, the number of sp states, automatically introduces the group $U(N)$ of unitary transformations in the sp space, and, with $U(N)$, all of its subgroups and subgroup chains. When we deal with an infinite sp spectrum, at a given excitation energy E (or energy region) the value of effective N will be fixed by the number of sp states that are active (neither empty or filled). The number of sp states N is significant because distributing m particles (assumed to be fermions, for most part, in the following discussion and extensions to bosons is straight forward) over these N states generates the unique antisymmetric $U(N)$ irrep $\{1^m\}$. Here we are using Young tableaux notation. Furthermore, the k-body operators of the system decompose into $(k+1)$ $U(N)$ irreps $\{2^\nu 1^{N-2\nu}\}$, a decomposition, already discussed in Sec. 1.4, is of major physical interest. Going beyond this, each $U(N)$ irrep decomposes into a (direct) sum of G irreps, with G being a subgroup, i.e. $U(N) \supset G$, with a corresponding decomposition of the m particle eigenvalue density. This decomposition is of obvious consequence if G is a good symmetry (by which we mean that every H-eigenstate belongs to a G irrep). For a broken symmetry it leads, among other things, to measures for the goodness of G. Decomposition by shell model spherical configurations is a good example of this.

Measurable quantities, being invariant under basis transformations, are representable as traces of operators in the model space or its subspaces. Then the input information is defined by a certain set of simple traces (two-particle ones for a two-body operator and so forth). If the subspaces are defined by symmetries, the information which survives the CLT action "propagates" from the defining subspaces throughout the entire subspace lattice. In many cases, depending on the symmetries involved, the propagation is particularly simple, the many-particle subspace traces being linear combinations of the defining-subspace traces which follow directly from the elementary matrix elements; in other cases further features of the model subspaces come into play, giving rise to certain modifications, but preserving the essential simplicity of the propagation. The really important result is that trace propagation (simple or modified) occurs only when the subspace decomposition is defined by symmetries. This propagation then defines the interesting information (apart from the little if any carried by fluctuations), gives methods for calculating it in complex spaces, and establishes a close connection between phenomena in related systems (same N, different m) such as nuclei in the same region of the periodic table. Since the significant information consists of low-order traces, the higher-order ones being filtered away, we see that the cooperation between CLT and symmetries gives rise to remarkable simplicities which form the basis for our fluctuation-free treatment of complicated spectroscopic systems.

Obviously, a pertinent question is "What propagates and how?" In proceeding towards a proper mathematical answer, it is useful to first identify and label in a general but simple and useful way the types of symmetries encountered in (nuclear) spectroscopy. It is also important to identify and distinguish the different kinds of averages that are of interest in statistical spectroscopy. In this regard, what is meant by propagation must be clearly defined since from one perspective, i.e. shell model or interacting boson model (IBM), everything propagates. As already mentioned, trace evaluation and propagation is crucial to the whole subject of statistical spectroscopy.

1.6.2 *Classification of group symmetries*

For a given model space and system Hamiltonian H, the most important aspect of any group is the commutation properties of its generators C_α with H. If H commutes with all group generators, i.e. $[H, C_\alpha] = 0$ for all α, H is itself a group invariant or combination of these, and the symmetry is called **exact** symmetry. Then each eigenstate of H belongs to degenerate irreps of the group, i.e. there will be degenerate multiplets. Simplest examples are $SO_J(3)$ and $SU_T(2)$ generating angular momentum and isospin respectively in the shell model for nucleons and similarly $SO_L(3)$ generating angular momentum in the *sd* interacting boson model (*sd*IBM); see [Arima and Iachello (1987)] for details of *sd*IBM.

A more common situation is when H does not commute with the generators of the group but is expressible solely in terms of the group generators. The simplest example is the H of a standard shell model theory with N sp states. By definition it conserves particle number m and hence has no matrix elements connecting different $U(N)$ irreps; the $U(N)$ irrep for m particles is $\{1^m\}$. Another examples is the $U(6)$ of *sd*IBM. Then the group symmetry is called **conserved** symmetry; H, although not a group scalar (except in the trivial case where the conserved symmetry is also exact), has no matrix elements between different irreps of the group so that every H eigenstate belongs to a single irrep. The degeneracy found for the exact symmetry is lifted and there is internal spreading but no external (i.e. other irrep) mixing. Thus $[H, C_\alpha] \neq 0$ but $H = H(C_\alpha)$. Other examples of conserved symmetry are, in the exact dynamical symmetry limits, the $U(5)$, $SU(3)$ and $O(6)$ of *sd*IBM.

Observe that H itself defines all of the exact and conserved symmetries. When, H is not expressible solely in terms of group generators and therefore has symmetry breaking parts which connect different irreps, we call the symmetry an **imposed** (or broken) symmetry. Here $[H, C_\alpha] \neq 0$ and there will be external (one irrep with another) mixing. Many symmetries in nuclear spectroscopy are of this type. Wigner's [Wigner (1937)] spin-isospin $SU(4) \supset SU_S(2) \otimes SU_T(2)$, pairing $SU(2)$ and $SO(5)$ [Talmi (1993)], Elliott's [Elliott (1958)] rotational $SU(3)$ etc. in shell model and, for transitional nuclei, the $U(5) \supset SO(5)$, $SU(3)$ and $SO(6) \supset SO(5)$ of *sd*IBM are common examples. As stated in the beginning, an interesting question for imposed symmetries is about their goodness or usefulness, i.e. about the extent and type of the irrep admixtures in individual H eigenstates. Weak symmetry breaking can be treated by perturbation theory and strong symmetry breaking statistically. In the latter case, as long as the symmetry admixing is reasonably localized in energy, the symmetry may still be of great interest, shell model spherical configurations being a major example.

In any particular physically interesting problem all three of these symmetry types usually occur. Exact symmetries are used to partition a space into disjoint subspaces, parity and angular momentum being prime examples for nuclei. In most examples $SU_J(3)$ is realized as the final member of a group chain that starts with a largest group that is conserved. For example $U(24)$ for $2s1d$ nuclei, $U(40)$ for $2p1f$ nuclei, $U(6)$ for sdIBM etc. Then the intermediate groups, for example the supermultiplet $U(\Omega) \otimes SU(4)$ in $U(24)$ (with $\Omega = 6$) and in $U(40)$ (with $\Omega = 10$), are imposed symmetries. Thus in many situations we have conserved $\rightarrow$ imposed $\rightarrow$ exact. We will consider some examples in the discussion ahead. We now turn to three types of traces and their propagation.

1.6.3 *Propagation of fixed symmetry trace matrix*

Let us consider the group-subgroup structure $G \supset S$ and assume that G is a conserved symmetry algebra for H. Then the basis states are

$$\left| \begin{matrix} G & S & \\ F\ \beta & f & \mu \end{matrix} \right\rangle \tag{1.118}$$

where F and f are respectively, irreps of G and S, β is the multiplicity of f in the reduction $F \rightarrow f$ and μ labels all the states of the single irrep f of S. Now, the fixed symmetry, i.e. fixed S, trace of a given operator $\mathcal{O}$ is defined by

$$C_{\beta,\beta'}(F,f) = \langle\langle \mathcal{O} \rangle\rangle^{F,\beta,\beta',f} = \sum_{\mu} \langle F,\beta' f,\mu \mid \mathcal{O} \mid F,\beta f,\mu \rangle \tag{1.119}$$

a uniform sum over the basis states of two equivalent but, if $\beta' \neq \beta$, different irreps of S. As the notation suggests, for each (F,f) pair the set of these traces $C_{\beta,\beta'}(F,f)$ forms a matrix. The dimensionality of this matrix is $d(F \rightarrow f)$, the number of times f occurs in F; note that $d(F) = d(F \rightarrow f)d(f)$ where $d(f)$ is number of μ labels. We shall see that the matrix of traces for any (F,f) pair propagates. Firstly, it is important to recognize that only the scalar part, with respect to S, of $\mathcal{O}$ can contribute to the traces $C_{\beta,\beta'}(F,f)$ and this is due to the uniform sum over μ, i.e. over all states of a given f. In terms of the defining space (F_0,f_0), the trace generating part $\mathcal{O}_{TE}$ is, with $\psi(F_0\beta_0 f_0\mu_0)$ a creation operator for the state $|F_0\beta_0 f_0\mu_0\rangle$,

$$\begin{aligned} \mathcal{O}_{TE} &= \sum_{F_0,\beta_0,\beta_0',f_0} \frac{\langle\langle \mathcal{O} \rangle\rangle^{F_0,\beta_0,\beta_0',f_0}}{d(F_0 \rightarrow f_0)} \sum_{\mu_0} \psi(F_0\beta_0' f_0\mu_0)\psi^\dagger(F_0\beta_0 f_0\mu_0) \\ &= \sum_{F_0,\beta_0,\beta_0',f_0} \langle\langle \mathcal{O} \rangle\rangle^{F_0,\beta_0,\beta_0',f_0}\ P(F_0,\beta_0,\beta_0',f_0)\ . \end{aligned} \tag{1.120}$$

The projection operators Ps in Eq. (1.120) are clearly S scalars in G and then,

$$\begin{aligned}\langle\langle \mathcal{O}\rangle\rangle^{F,\beta,\beta',f} &= \langle\langle \mathcal{O}_{TE}\rangle\rangle^{F,\beta,\beta',f} \\ &= \sum_{F_0,\beta_0,\beta_0',f_0} \langle\langle \mathcal{O}\rangle\rangle^{F_0,\beta_0,\beta_0',f_0} \langle\langle P(F_0,\beta_0,\beta_0',f_0)\rangle\rangle^{F,\beta,\beta',f} .\end{aligned} \tag{1.121}$$

Thus we have the remarkable phenomenon of 'matrix propagation', the elements of the matrix of traces, defined by Eq. (1.119), are expressible in terms of the corresponding defining space traces with the traces of the projection operators P propagating the trace information [Quesne (1975b*); French and Draayer (1979*)].

In group theory, the minimal set of S scalars in G is called $G \supset S$ integrity basis (I.B.). Then all other S scalars in G will be polynomials in the I.B. operators. A good example for I.B. operators is $SU(3) \supset SO(3)$ discussed in detail by Judd, Miller Jr., Patera and Winterniz (1974); see also Gilmore and Draayer (1985). Obviously the $\mathcal{O}_{TE}$ or the Ps will be polynomials in the $G \supset S$ I.B. operators. The technical problems are: (i) identify the I.B. operators; (ii) determine their traces. However, when the integrity basis reduces to the set of Casimir operators of G and S, we shall call the propagation as 'simple'or 'Casimir'propagation. It is simple because the eigenvalues of Casimir operators are known functions of the irrep labels and the invariants of a group act as a multiple of identity in any of its irreps. The integrity basis for shell model $U(N)$ is just $\{\hat{n}\}$ and for isospin $U(N) \supset U(N/2) \otimes SU_T(2)$ is $\{\hat{n}, T^2\}$. In general this happens only when the $G \supset S$ reduction is multiplicity free, i.e. when no irrep of S occurs more than once in an irrep of G. The $G \supset S$ reduction is then called 'canonical'and all trace matrices are one dimensional.

Even when the integrity basis does not reduce to the set of Casimir invariants, there can still be simple propagation for special operators and/or special classes of G irreps. This happens for example when the particle rank of an operator is less than the rank of non-Casimir I.B. scalars. Examples are centroids generated by a $(1+2)$-body operator for $U(24) \supset [U(6) \supset SU(3)] \otimes SU(4)$ traces in $2s1d$ shell [Parikh (1972)] and $U(4\Omega) \supset U(\Omega) \otimes [SU(4) \supset SU_S(2) \otimes SU_T(2)]$ traces for any Ω [Haq and Parikh (1974)]. The sdIBM $U(6) \supset SU(3)$ is an example where the integrity basis includes non-Casimir scalars but for all G irreps of physical interest, i.e. symmetric ones, each $SU(3)$ irrep occurs here at most once and there is Casimir basis $\{\hat{n}, C_2, C_3\}$ where C_2 and C_3 are the second and third order invariants of $SU(3)$. Thus the $SU(3)$ traces here have simple propagation [Kota (1981)]. Let us point out that a simple counting of the defining space irreps of S in G and the Casimir operators will confirm if there is Casimir propagation are not; see [French and Draayer (1979*); Kota and Kar (1989)] for details. Many examples, from shell model [$SU(3)$, $U(\Omega \otimes SU(4)$, $U(\Omega \otimes \{SU(4) \supset SU(2) \otimes SU(2)\}$, $Sp(\Omega)$, configuration seniority, configuration isospin etc.], for major shell centroids in the symplectic $Sp(6,R)$ collective model, from various interacting boson models [$U(5)$, $SU(3)$ and $SO(6)$ limits of sdIBM, $SU(3)$ and $U(1) \oplus U(5) \oplus U(9)$ limits of sdgIBM, some of the symmetry limits of *sdpf*IBM, IBM-3 and IBM-4 etc.] and also for averages over many-quasi-particle subspaces, for Casimir propagation, are given in [Nissimov, Arvieu and Bohigas (1972); Nomura (1972a, 1974); Quesne and Spitz (1974, 1978); Parikh (1978); Kota (1981); Draayer and Rosensteel (1983b); Wong (1986); Nomura (1987a, 1990)] and more recent results in [Kota (2005a*); Zhao *et al.* (2005); Kota (2006a,b)]. For illustration, we consider in some detail one example for Casimir propagation and one where we need non-Casimir I.B. operators.

1.6.4 *Casimir propagation*

Consider m fermions distributed in Ω number of sp levels (denoted by $i, j, \ldots$) each with spin $s = 1/2$. The total number of sp states is $N = 2\Omega$. With $a^\dagger_{i\frac{1}{2}m_s}$ denoting one particle creation operator and $\tilde{a}_{i\frac{1}{2}m_s} = (-1)^{\frac{1}{2}-m_s} a_{i\frac{1}{2}-m_s}$, the $4\Omega^2$ number of operators, in spin coupled representation, $u^r_\mu(i,j) = (a^\dagger_i a^\dagger_j)^r_\mu = \left(a^\dagger_{i\frac{1}{2}} \tilde{a}_{j\frac{1}{2}}\right)^r_\mu$, $r = 0,\ 1$ generate the $U(2\Omega)$ algebra. Its irreps are represented by the particle number m. The $2\Omega(\Omega-1)$ number of operators $V^r_\mu(i,j)$,

$$V^r_\mu(i,j) = \sqrt{(-1)^{r+1}}\left[u^r_\mu(i,j) - (-1)^r u^r_\mu(j,i)\right] \ ; \quad i > j\ , \ \ r = 0,\ 1$$

along with $u^1_\mu(i,i)$ form the $Sp(2\Omega)$ sub-algebra. The quadratic Casimir operators of the $U(2\Omega)$ and $Sp(2\Omega)$ algebras are,

$$\begin{aligned}
C_2[U(2\Omega)] &= \sum_{i,j,r} u^r(i,j) \cdot u^r(j,i) \\
C_2[Sp(2\Omega)] &= 2\sum_i u^1(i,i) \cdot u^1(i,i) + \sum_{i>j,r} V^r(i,j) \cdot V^r(i,j) \\
&\Rightarrow C_2[U(2\Omega)] - C_2[Sp(2\Omega)] = 4\ PP^\dagger - \hat{n}
\end{aligned} \tag{1.122}$$

where $P = \frac{1}{\sqrt{2}}\sum_i (a^\dagger_i a^\dagger_j)^0$ is the pairing operator and $P^\dagger = -\frac{1}{\sqrt{2}}(\tilde{a}_j \tilde{a}_i)^0$. It is also seen that the operators P, $P^\dagger$ and $P_0 = \hat{n} - \Omega$ form $SU(2)$ quasi-spin (Q) algebra. The P_0 eigenvalue $M_Q = (m-\Omega)/2$ and therefore $Q = (\Omega - v)/2$. For $m \leq \Omega$, v take values $v = m, m-2, \ldots, 0$ or 1. The eigenvalues of the pairing Hamiltonian $H_P = P\ P^\dagger$ are given by,

$$\langle H_P \rangle^{m,v,S} = E_p(m,v,S) = \frac{1}{4}(m-v)(2\Omega + 2 - m - v)\ . \tag{1.123}$$

With respect to the $U(2\Omega)$ algebra all m nucleon states behave as the totally anti-symmetric irrep $\{1^m\}$ and $\langle C_2[U(2\Omega)]\rangle^{\{1^m\}} = m(2\Omega + 1 - m)$. Then Eq. (1.123)

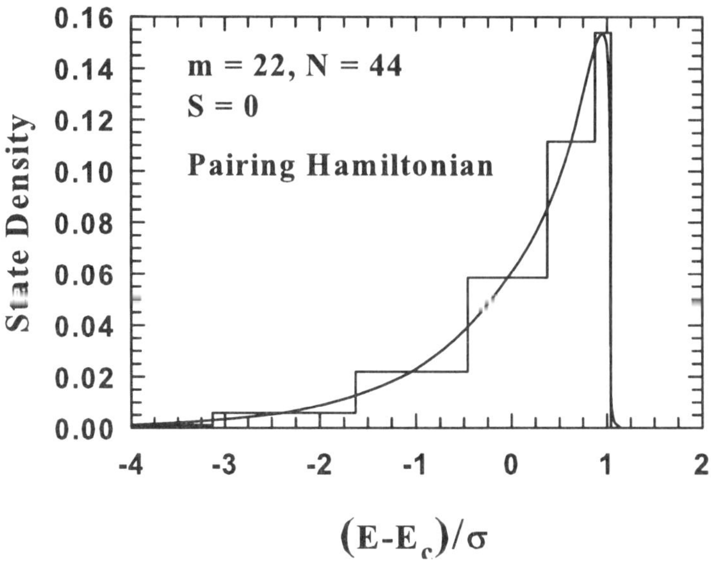

Fig. 1.17 Eigenvalue density for the pairing Hamiltonian $H = -H_p$ for a system of 22 fermions in $\Omega = 22$ orbits ($N = 44$) with total spin $S = 0$. For this system, the spectrum spread is 132 (note that $v_{max} = 22$), energy centroid $E_c \sim 5.7$ and spectrum width $\sigma \sim 6$. The normalized density $\rho_{(-H_p)}(E : v) = [d(m, S)\ \Delta E]^{-1}\ D(m, v, S)$ with $\Delta E = E(m, v+1, S) - E(m, v-1, S) = \Omega - v + 1$. In the histogram, $\rho(\widehat{E})$ for a given $\widehat{E} = (E - E_c)/\sigma$ is plotted with $\widehat{E}$ as center with width given by $\Delta\widehat{E} = \Delta E/\sigma$. The smooth curve obtained by joining the mid-points is shown as a visual guide. Figure taken with permission from an unpublished paper by Manan Vyas.

gives $\langle C_2[Sp(2\Omega)]\rangle^{m,v} = 2v(\Omega+1-\frac{v}{2})$. Comparing this with the formula [Wybourne (1974)] for $\langle C_2[Sp(2\Omega)]\rangle^{\langle g\rangle}$ with $\langle g\rangle$ being any irrep of $Sp(2\Omega)$, it is easily seen that the seniority quantum number v corresponds to the totally antisymmetric irrep $\langle 1^v\rangle$ of $Sp(2\Omega)$. Thus $Sp(2\Omega)$ corresponds to $SU(2)$ algebra generated by (P, $P^\dagger$, P_0), with the m particle spin S generated by the v free particles. Then, for a given m and S we have

$$v = m,\ m-2,\ \ldots,\ 2S\ , \quad (m \leq \Omega)\ . \tag{1.124}$$

With fixed-(m, S) dimension $d(m, S) = \frac{2S+1}{\Omega+1}\binom{\Omega+1}{\frac{m}{2}+S+1}\binom{\Omega+1}{\frac{m}{2}-S}$, the dimension of fixed-(m, v, S) space is given by $D(m, v, S) = d(v, S) - d(v-2, S)$. Using $D(m, v, S)$ and Eqs. (1.123), (1.124), it is easy to construct the eigenvalue density for the pairing operator H_P over fixed-(m, S) spaces. An example is shown in Fig. 1.17 and clearly, as stated before (see also [Ginocchio (1980)]), the pairing Hamiltonian generates a highly skewed distribution.

We now consider fixed (m, v, S) traces and show that their propagation is simple. The number of (m, v, S) irreps are 5 for m up to 2 and there are 5 Casimir operators of maximum body rank 2. These being 1, $\hat{n}$, $\binom{\hat{n}}{2}$, H_P and $\hat{S}^2$. Therefore the energy centroids over fixed (m, v, S) spaces propagate for any general $(0 + 1 + 2)$-body Hamiltonian and with $\mathcal{S}^2 = S(S+1)$, $X(m, S) = m(m+2) - 4S(S+1)$ and

$Y(m,S) = m(m-2) - 4S(S+1)$, are given by

$$\begin{aligned}
&\langle H(0+1+2)\rangle^{m,v,S} = E_c(m,v,S) = a_0 + a_1 m + a_2\tbinom{m}{2} + a_3\mathcal{S}^2 + a_4 E_p(m,v,S) \\
&\Rightarrow E_c(m,v,S) = \frac{1}{2}(m-1)(m-2)\; E_c(0,0,0) - m(m-2)\; E_c(1,1,\frac{1}{2}) \\
&+ \frac{1}{8\Omega}\left[E_p(m,v,S) - \Omega X(m,S)\right] E_c(2,2,0) + \frac{1}{\Omega}\; E_p(m,v,S)\; E_c(2,0,0) \\
&+ \frac{1}{8}\left[4m(m-2) - Y(m,S)\right]\; E_c(2,2,1)\;.
\end{aligned} \tag{1.125}$$

Even more remarkable is that, for $m \leq 4$, the number of (m,v,S) irreps is 14 and the available simple scalars $\mathcal{O}$ of maximum body rank 4 are also exactly 14. These are 1, $\hat{n}$, $\binom{\hat{n}}{2}$, $\binom{\hat{n}}{3}$, $\binom{\hat{n}}{4}$, H_P, $\hat{n}H_P$, $\binom{\hat{n}}{2}H_P$, $[H_P]^2$, $H_P\hat{S}^2$, $\hat{S}^2$, $\hat{n}\hat{S}^2$, $\binom{\hat{n}}{2}\hat{S}^2$ and $(\hat{S}^2)^2$. Therefore, the spectral variances over (m,v,S) spaces propagate simply and they will be linear combinations of the eigenvalues of the 14 operators above. The constants in the expansion will follow from the variances for $m \leq 4$. The final propagation formula is,

$$\begin{aligned}
&\left\langle [H(0+1+2)]^2\right\rangle^{m,v,S} = \tfrac{1}{24}(m-1)(m-2)(m-3)(m-4)\;\left\langle H^2\right\rangle^{0,0,0} \\
&- \tfrac{1}{6}m(m-2)(m-3)(m-4)\;\left\langle H^2\right\rangle^{1,1,\frac{1}{2}} \\
&+ \tfrac{1}{16\Omega}(m-3)(m-4)\left[\Omega X(m,S) - 8E_p(m,v,S)\right]\;\left\langle H^2\right\rangle^{2,2,0} \\
&+ \tfrac{1}{2\Omega}(m-3)(m-4)E_p(m,v,S)\;\left\langle H^2\right\rangle^{2,0,0} \\
&+ \tfrac{1}{16}(m-3)(m-4)[3m(m-2) + 4\mathcal{S}^2]\;\left\langle H^2\right\rangle^{2,2,1} \\
&- \tfrac{(m-2)(m-4)}{12(\Omega-1)}\left[(\Omega-1)X(m,S) - 12E_p(m,v,S)\right]\;\left\langle H^2\right\rangle^{3,3,\frac{1}{2}} \\
&- \tfrac{1}{\Omega-1}(m-2)(m-4)E_p(m,v,S)\;\left\langle H^2\right\rangle^{3,1,\frac{1}{2}} \\
&- \tfrac{1}{12}(m-2)(m-4)[m(m-4) + 4\mathcal{S}^2]\;\left\langle H^2\right\rangle^{3,3,\frac{3}{2}} \\
&+ \tfrac{1}{192(\Omega-2)(\Omega-1)}\Big[96\left\{E_p(m,v,S)\right\}^2 + 24\big\{-(\Omega-1)m^2 + 2(\Omega+1)m \\
&+ 4(\Omega-1)\mathcal{S}^2 - 4(\Omega+2)\big\}\; E_p(m,v,S) \\
&+ (\Omega-1)(\Omega-2)X(m,S)Y(m,S)\Big]\;\left\langle H^2\right\rangle^{4,4,0} \\
&+ \frac{1}{8\Omega(\Omega-2)}E_p(m,v,S) \\
&\times\left[\Omega\{m(m-2) - 4\mathcal{S}^2 + 8\} - 8\left\{E_p(m,v,S) + m - 2\right\}\right]\;\left\langle H^2\right\rangle^{4,2,0} \\
&+ \frac{1}{2\Omega(\Omega-1)}E_p(m,v,S)\left\{E_p(m,v,S) + m - \Omega - 2\right\}\;\left\langle H^2\right\rangle^{4,0,0} \\
&+ \frac{\left[3m(m-6) + 4\mathcal{S}^2 + 24\right]}{128(\Omega-2)} \\
&\times\;\left[(\Omega-2)X(m,S) - 16E_p(m,v,S)\right]\;\left\langle H^2\right\rangle^{4,4,1} \\
&+ \frac{1}{8(\Omega-2)}E_p(m,v,S)\left[3m(m-6) + 4\mathcal{S}^2S + 24\right]\;\left\langle H^2\right\rangle^{4,2,1} \\
&+ \frac{1}{384}\left[16(\mathcal{S}^2)^2 + 40m^2\mathcal{S}^2 - 240m\mathcal{S}^2 + 288\mathcal{S}^2\right. \\
&\left.+ 5m(m-2)(m-4)(m-6)\right]\;\left\langle H^2\right\rangle^{4,4,2}\;.
\end{aligned} \tag{1.126}$$

The propagation equations Eq. (1.125) and Eq. (1.126) have been derived and used in analyzing pairing correlations generated by EGOE(1+2) for fermions with spin degree of freedom [Vyas, Kota and Chavda (2009)].

1.6.5 *Propagation with I.B. operators*

There are several physically important situations where in order to deal with matrix propagation one needs non-Casimir I.B. scalars. For example, details of I.B. construction and determination of the matrix elements of I.B. operators are discussed in [Quesne (1976)] for spin-isospin SU(4) in shell model and in [Draayer and Rosensteel (1983a); Rosensteel and Draayer (1985)] for $Sp(6,R) \supset SU(3)$ in the $Sp(6,R)$ model. For illustration, with current interest in 3-body interactions, consider the example of energy centroid $\overline{E_{m,(\lambda\mu),T}}$ of $U_{sd}(18) \supset U(6) \otimes [SU_T(3) \supset SO_T(3)]$ in the isospin invariant sdIBM with three body interactions; see [Kota (2006a)] for details. Note first that the $SU_T(3)$ irreps are $(\lambda,\mu) = ((f_1 - f_2), (f_2 - f_3))$ where $m = f_1 + f_2 + f_3$, $f_1 \geq f_2 \geq f_3 \geq 0$ and the $(\lambda\mu) \to T$ reductions follow from Elliott's rules [Kota (2005a*)]. Counting of number of irreps $(\lambda\mu)T$ for $m \leq 3$ shows that besides the operators 1, $\hat{n}$, $\hat{C}_2$, $\hat{C}_3$ and $\hat{T}^2$, we need one extra $SO_T(3)$ scalar in $SU_T(3)$ and it should be 3-body. The $SU(3) \supset SO(3)$ I.B. indeed contains, besides the Casimir operators, a 3-body operator X_3 and a 4-body operator X_4 [Judd, Miller Jr., Patera and Winterniz (1974)]. The X_3 operator given in [Draayer and Rosensteel (1985)], called X_3^{Dr} is related to the X_3^{Judd} given in [Judd, Miller Jr., Patera and Winterniz (1974)] by the relation $X_3^{Judd} = -\frac{5}{\sqrt{10}} X_3^{Dr}$. Then the average of X_3^{Judd} is [Draayer and Rosensteel (1985)],

$$\left\langle X_3^{Judd} \right\rangle^{(\lambda\mu)T} = (-1)^{1-\delta_{\mu,0}} \; \frac{[3 - 4T(T+1)]\sqrt{T(T+1)}}{\sqrt{(2T-1)(2T+3)}} \times \sqrt{C_2(\lambda\,\mu)} \; \frac{\sum_k \langle (\lambda\mu)kT \;\; (11)2 \mid\mid (\lambda\mu)kT \rangle_{\rho=1}}{\sum_{k'} \; 1} \,. \tag{1.127}$$

In Eq. (1.127), $\langle - - - \rangle$ are $SU_T(3) \supset SO_T(3)$ reduced Wigner coefficients and programmes to calculate them are available [Akiyama and Draayer (1973)]. Eq. (1.127) gives the average of X_3^{Judd} for (00)0, (10)1, (20)0, (20)2, (01)1, (30)1, (30)3, (11)1 and (11)2 as 0, 5, 0, 21, −5, 9, 54, 0 and 0 respectively. These are verified with the results in Table 1 and Eq. (44) of [Judd, Miller Jr., Patera and Winterniz (1974)]. Using these, a propagation formula for $\overline{E_{m,(\lambda\mu),T}}$ generated by a (0 + 1 + 2 + 3)-body H is derived and the result is [with $T^2 = T(T+1)$, $C_2(\lambda\mu) = \lambda^2 + \mu^2 + \lambda\mu + 3(\lambda + \mu)$, $C_3(\lambda\mu) = (\lambda - \mu)(\lambda + 2\mu + 3)(2\lambda + \mu + 3)/9$ and

$$X_3((\lambda\mu)T) = \langle X_3^{Judd} \rangle^{(\lambda\mu)T}],$$

$$\begin{aligned}
&\overline{E_{m,(\lambda\mu),T}} = \langle H(0+1+2+3)\rangle^{m,(\lambda\mu),T} \\
&= \left[1 - \tfrac{11m}{6} + m^2 - \tfrac{m^3}{6}\right] \overline{E_{0,(00),0}} \\
&+ \left[3m - \tfrac{5m^2}{2} - \tfrac{m^3}{2}\right] \overline{E_{1,(10),1}} \\
&+ \left[-\tfrac{m}{2} + \tfrac{m^2}{3} + \tfrac{C_2(\lambda\,\mu)}{3} - \tfrac{T^2}{2} - \tfrac{m^3}{18} - \tfrac{mC_2(\lambda\,\mu)}{9} + \tfrac{mT^2}{6}\right] \overline{E_{2,(20),0}} \\
&+ \left[-\tfrac{5m}{2} + \tfrac{5m^2}{3} + \tfrac{C_2(\lambda\,\mu)}{6} + \tfrac{T^2}{2} - \tfrac{5m^3}{18} - \tfrac{mC_2(\lambda\,\mu)}{18} - \tfrac{mT^2}{6}\right] \overline{E_{2,(20),2}} \\
&+ \left[\tfrac{3m}{2} - \tfrac{C_2(\lambda\,\mu)}{2} - \tfrac{m^3}{6} + \tfrac{mC_2(\lambda\,\mu)}{6}\right] \overline{E_{2,(01),1}} \\
&\left[\tfrac{7m}{9} - \tfrac{7m^2}{18} - \tfrac{7C_2(\lambda\,\mu)}{90} - \tfrac{7T^2}{30} + \tfrac{7m^3}{162} + \tfrac{7mC_2(\lambda\,\mu)}{270} + \tfrac{7mT^2}{90}\right. \\
&\left. - \tfrac{C_3(\lambda\,\mu)}{90} + \tfrac{X_3((\lambda\mu)T)}{45}\right] \overline{E_{3,(30),3}} \\
&+ \left[\tfrac{m}{3} - \tfrac{m^2}{6} - \tfrac{C_2(\lambda\,\mu)}{5} + \tfrac{7T^2}{30} + \tfrac{m^3}{54} + \tfrac{mC_2(\lambda\,\mu)}{15} - \tfrac{7mT^2}{90} + \tfrac{C_3(\lambda\,\mu)}{15}\right. \\
&\left. - \tfrac{X_3((\lambda\mu)T)}{45}\right] \overline{E_{3,(30),1}} \\
&+ \left[-\tfrac{m}{3} + \tfrac{C_2(\lambda\,\mu)}{6} - \tfrac{T^2}{12} + \tfrac{m^3}{27} - \tfrac{mT^2}{18} - \tfrac{C_3(\lambda\,\mu)}{6} + \tfrac{X_3((\lambda\mu)T)}{18}\right] \overline{E_{3,(11),1}} \\
&+ \left[-\tfrac{5m}{9} + \tfrac{C_2(\lambda\,\mu)}{6} + \tfrac{T^2}{12} + \tfrac{5m^3}{81} - \tfrac{2mC_2(\lambda\,\mu)}{27} + \tfrac{mT^2}{18} + \tfrac{C_3(\lambda\,\mu)}{18}\right. \\
&\left. - \tfrac{X_3((\lambda\mu)T)}{18}\right] \overline{E_{3,(11),2}} \\
&+ \left[\tfrac{m}{9} + \tfrac{m^2}{18} - \tfrac{C_2(\lambda\,\mu)}{18} + \tfrac{m^3}{162} - \tfrac{mC_2(\lambda\,\mu)}{54} + \tfrac{C_3(\lambda\,\mu)}{18}\right] \overline{E_{3,(00),0}}\ .
\end{aligned} \tag{1.128}$$

This formula has been applied [Kota (2006a)] to study regularities generated by random interactions; see discussion in Sec. 1.6.9.

Calculation of centroids and variances by trace propagation allows for construction of partial densities $I^{\{f\}}(E)$ defined over group irreps $\{f\}$ as Gaussians and then the relative intensity of $\{f\}$ in the eigenstate with energy E is $I^{\{f\}}(E)/I(E)$. This enables the study of goodness of group symmetries. The best studied example is $SU(4)$ supermultiplet symmetry in $(2s1d)$ shell nuclei. Analysis of goodness of $SU(4)$ has been carried out both using the Casimir propagation for SU(4) [French and Parikh (1971*); Parikh and Wong (1972); Parikh (1973)] and also propagation using I.B. operators for $SU(4)-ST$ [Haq and Quesne (1982a*,b)]. It has been clearly established that $SU(4)$ is a broken symmetry and the Gaussian partial densities give intensities of $SU(4)$ irreps close to those of exact shell model results. The conclusion that $SU(4)$ is a broken symmetry for $(2s1d)$ shell nuclei is further supported by admixings calculated using $SU(4)$ partial variances [Hecht and Draayer (1974); Chakraborty, Kota and Parikh (1980a)]. Further, Parikh has established, following the spectral distribution results for $SU(4)$ mixing, that the Franzini-Radicati mass relationship is not a good measure for studying the goodness of supermultiplet $SU(4)$ symmetry; see [Chakraborty, Kota and Parikh (1980b)].

In addition to $SU(4)$, seniority [defined by $Sp(N)$ algebra] admixtures in the low-lying states of various isotopes of nuclei such as Ni, Sn etc. have been studied by Quesne and collaborators [Quesne (1973); Quesne, Salmon and Spitz (1977); Quesne and Spitz (1977); Spitz and Quesne (1976); Quesne and Spitz (1978)] using Gaussian partial densities defined over seniority or configuration seniority subspaces. The corresponding centroid and variance propagation equations are given in [Quesne and Spitz (1974); Quesne (1975a); Quesne and Spitz (1978)]. Moreover, as the pairing interaction [or the quadratic Casimir invariant of the $Sp(N)$ algebra] generates highly skewed eigenvalue distribution (see Sec. 1.6.4), further investigation of the Gaussian form for fixed seniority partial densities is called for. It will be useful to derive the trace propagation results for the skewness and excess parameters of these densities.

Before moving onto other traces, let us emphasize that the occurrence of a multiplicity, regardless of the simplicity or complexity of G and S algebras, introduces a matrix of traces whose propagation requires the introduction of non-invariant I.B. scalars.

1.6.6 *Trace of the trace matrix*

In cases when the multiplicity is high and the integrity basis large, the significance of the individual traces of the trace matrix is lost and the centroid, variance etc. of the matrix emerge as the quantities having statistical significance and physical importance. For the $G \supset S$ geometry, the centroid or the trace of the trace matrix, called fixed symmetry trace, is defined by

$$\langle\langle \mathcal{O} \rangle\rangle^{F,f} = \sum_{\beta,\mu} \langle F, \beta, f, \mu \mid \mathcal{O} \mid F, \beta, f, \mu \rangle = \sum_{\beta} \langle\langle \mathcal{O} \rangle\rangle^{F,\beta,\beta,f} \; . \tag{1.129}$$

The sum running over the $d(F \to f)$ multiplicity labels. The prime example here is fixed-J traces encountered for example in the spin decomposition of level densities. The multiplicity $d(F \to f)$ is then the dimensionality of the fixed-J subspaces with μ labeling the $(2J+1)$ M sub-states. Whereas the matrix of traces always propagates in terms of the set of $G \supset S$ I.B. operators, there is no such simple rule for fixed symmetry traces defined by Eq. (1.129). In general, $\langle\langle \mathcal{O} \rangle\rangle^{F,f}$ cannot be determined in terms of the corresponding traces in the defining space of $\mathcal{O}$. Only under very special circumstances or trivial situations do the fixed symmetry traces propagate. One special case is $U(N) \supset U(N/r) \otimes U(r)$ with $U(N)$ irreps being symmetric or antisymmetric. Then in the direct product subgroup, the $U(r)$ irreps uniquely define the $U(N/r)$ irreps. For shell model $r = 2$ is isospin and $r = 4$ is $SU(4)$ symmetry. Similar structures appear in IBM-3 and IBM-4 models [Kota (2006b)]. The trivial situation is when $\mathcal{O}$ itself is a polynomial in the Casimir operators of G and S.

Although fixed symmetry traces (defined w.r.t an algebra S) do not propagate, it is possible to deal with them in a variety of ways: (i) using a multivariate distribution in the eigenvalues of H and the eigenvalues of the Casimir invariants of S (provided they commute with H) and then generating an expansion; (ii) by using a recursive method; (iii) by going to a subgroup of S where simple propagation, though cumbersome, is possible; (iv) seeking asymptotic expansions using group theory; (v) EGOEs with appropriate symmetries. Below we will consider (i)-(iii) with fixed-J traces as example. We refer to [French (1980c*)] for (iv) and [Mon (1973); Tomsovic (1986)] for (v).

1.6.7 *Fixed-J traces*

With m fermions occupying sp orbits $(j_1, j_2, \ldots, j_r)$, the many particle states can be written as $|m, \alpha_J, J, M\rangle$; $M = -J, -J+1, \ldots, J-1, J$ and $\alpha_J = 1, 2, \ldots, D(m, J)$. For J scalar Hamiltonians, the $(2J+1)$ M-states will be degenerate. Then the Hamiltonian matrix dimension for a given J is $D(m, J)$. Although it is possible to calculate $D(m, J)$ exactly, for large m and several j-orbits, i.e. for $r >> 1$, the $D(m, J)$ follow statistical laws (this is a part of statistical group theory [Cleary and Wybourne (1971); Hirst and Wybourne (1986); Kota, Mayya and Castilho Alcarás (2008)]) as discussed in [Bethe (1936, 1937); Chang, French and Thio (1971*)] in the context of level densities while introducing spin cutoff factors.

We consider a simple derivation of the statistical law for $D(m, J)$. Starting with the J_z operator, which is a one-body operator, we can write down the sp spectrum for the J_z operator whose eigenvalues in the $|j_i m_i\rangle$ basis are m_i, the largest j_i value being j_{max}. With $s = 2j_{max} + 1$, the eigenvalues are $m_{z_1} = j_{max}$, $m_{z_2} = j_{max} - 1$, …, $m_{z_{s-1}} = -j_{max} + 1$, $m_{z_s} = -j_{max}$ each with a degeneracy $d(m_{z_r}) = \sum_{j_i \geq |m_{z_r}|} 1$. Distributing the m fermions in these sp orbits will give the configurations $\widetilde{m(m_z)} = (m_1(m_z), m_2(m_z), \ldots)$ where $m_i(m_z)$ is number of particles in the m_{z_i} orbit. Generating all possible $\widetilde{m(m_z)}$ configurations such that $M = \sum_{i=1}^{s} m_{z_i}\; m_i(m_z)$, the fixed-$M$ dimension $\mathcal{D}(m, M) = \sum_{\widetilde{m(m_z)} \in M} \prod_{i=1}^{s} \binom{d(m_{z_i})}{m_i(m_z)}$. However, this is nothing but the m-particle eigenvalue density for the NIP Hamiltonian J_z. As discussed before, NIP densities approach Gaussian form for large enough m and a reasonably uniform sp spectrum. Therefore,

$$
\begin{aligned}
\mathcal{D}(m, M) &= d(m)\rho(m, M) = \frac{d(m)}{\sigma_{J_z}(m)}\,\eta(\widehat{M}) \\
&\quad \rightarrow \mathcal{D}_{\mathcal{G}}(m, M) = d(m)\rho_{\mathcal{G}}(m, M) = \frac{d(m)}{\sigma_{J_z}(m)}\,\eta_{\mathcal{G}}(\widehat{M})\;, \\
\eta_{\mathcal{G}}(\widehat{M}) &= \frac{1}{\sqrt{2\pi}} \exp -\frac{\widehat{M}^2}{2}\;,\;\; \widehat{M} = M/\sigma_{J_z}(m)\,,\;\; \sigma^2_{J_z}(m) = \left\langle J_z^2 \right\rangle^m\;,\;\; d(m) = \binom{N}{m}\,.
\end{aligned}
\tag{1.130}
$$

Note that here we have introduced the densities $\rho(m, M)$, $\rho_{\mathcal{G}}(m, M)$, $\eta_{\mathcal{G}}(\widehat{M})$, the standardized variable $\widehat{M} = \frac{M}{\sigma_{J_z}(m)}$ and the variance $\sigma^2_{J_z}(m)$ of the J_z operator ($\langle J_z \rangle^m = 0$). Now the difference equation

$$D(m, J) = \mathcal{D}(m, M = J) - \mathcal{D}(m, M = J + 1) \tag{1.131}$$

will give the statistical law for fixed-J dimensions $D(m, J)$. A compact (standard) formula follows by applying the approximation,

$$D(m, J) \sim \left.\frac{\partial \mathcal{D}(m, M)}{\partial M}\right|_{M=J+1/2} , \tag{1.132}$$

Then we have easily using Eq. (1.130),

$$D(m, J) = d(m) \frac{(2J + 1)}{\sqrt{8\pi}\,\sigma^3_{J_z}(m)} \exp - \frac{(J + 1/2)^2}{2\sigma^2_{J_z}(m)} = d(m)\; C_J(m) \,. \tag{1.133}$$

We will use the $C_J(m)$ function in Eq. (1.133) later. In many examples Eq. (1.133) has been verified to be fairly accurate. As $\langle J_z^r \rangle^m = 0$ for r odd, the first two Edgeworth corrections to $\mathcal{D}_{\mathcal{G}}(m, M)$ come from the reduced cumulants $k_4(m, J_z) = \gamma_2(m, J_z)$ and $k_6(m, J_z)$. Then Eq. (1.17) gives

$$\begin{aligned} \mathcal{D}(m, M) = \mathcal{D}_{\mathcal{G}}(m, M) \Bigg\{ 1 &+ \left[\frac{k_4(m, J_z)}{24} He_4(\widehat{M}) \right] \\ &+ \left[\frac{k_6(m, J_z)}{720} He_6(\widehat{M}) + \frac{1}{2} \left(\frac{k_4(m, J_z)}{4!} \right)^2 He_8(\widehat{M}) \right] \Bigg\} \,. \end{aligned} \tag{1.134}$$

Propagation formulas for $\sigma^2(m, J_Z) = \sigma^2_{J_z}(m)$, $k_4(m, J_z)$ and $k_6(m, J_z)$ follow from Table 1.1. It is useful to remark that Eq. (1.133) also follows from the fact that $J^2 = J_x^2 + J_y^2 + J_z^2$, i.e. it is a sum of the squares of three one-body operators and hence $D(m, J)$ is χ^2_3. However, to calculate corrections to $D(m, J)$ it is much better to start with $\mathcal{D}_{\mathcal{G}}(m, M)$ as we can use the Edgeworth expansion which has known mathematical convergence properties and hence is physically more meaningful. As an aside, it is important to recognize that via Eq. (1.134) we have generated a statistical formula for the multiplicities of $SO(3)$ irreps J in the m particle anti-symmetric irrep of $U(N)$ with the basic association $m = 1 \to J = j_1 \oplus j_2 \oplus \ldots j_r$ and $N = \sum_{\nu=1}^{r} (2j_\nu + 1)$. Extension of Eq. (1.134) to bosons is also available [Kota (2006b)]. There have been attempts to extend this 'statistical group theory'result to Wigner's supermultiplet $SU(4)$ [Bloch (1954)] and Elliott's SU(3) [Kanestrom (1966)]. However, only recently a complete result for $SU(3)$ irreps has been derived [Kota, Mayya and Castilho Alcarás (2008)]. Having understood the structure of the distribution of fixed-J (also fixed-M) dimensions, we now consider fixed-J level densities.

Formal definition of normalized fixed-(E, M) density of states, which is a bivariate density, is

$$\rho^{H,m}(E, M) = \langle \delta(H - E)\delta(J_z - M) \rangle^m = \{d(m)\}^{-1} \langle\langle \delta(H - E)\delta(J_z - M) \rangle\rangle^m \,. \tag{1.135}$$

An important point here is that it is possible to define a bivariate density for the eigenvalues of any two operators that commute. The $\rho(E,M)$ density can be decomposed into marginal and conditional densities,

$$\begin{aligned}\rho^{H,m}(E,M) &= \{d(m)\}^{-1}\langle\langle\delta(H-E)\rangle\rangle^{M} = [\mathcal{D}(m,M)/d(m)]\,\langle\delta(H-E)\rangle^{M} \\ &= \rho(E \mid M)\,\rho(M)\;; \\ \rho(E \mid M) &= \langle\delta(H-E)\rangle^{M}\;,\quad \rho(M) = [\mathcal{D}(m,M)/d(m)] = \langle\delta(J_z-M)\rangle^{m}\;.\end{aligned} \tag{1.136}$$

Similarly,

$$\begin{aligned}\rho^{H,m}(E,M) &= \{d(m)\}^{-1}\langle\langle\delta(J_z-M)\rangle\rangle^{E} = [I(E)/d(m)]\,\langle\delta(J_z-M)\rangle^{E} \\ &= \rho(M \mid E)\,\rho(E)\;; \\ \rho(M \mid E) &= \langle\delta(J_z-M)\rangle^{E}\;,\quad \rho(E) = [I(E)/d(m)] = \langle\delta(H-E)\rangle^{m}\;.\end{aligned} \tag{1.137}$$

Thus we can deal directly with the bivariate distribution $\rho(E,M)$ or the marginals $\rho(E)$, $\rho(M)$ and the conditionals $\rho(E \mid M)$, $\rho(M \mid E)$ that are univariate distributions.

The operators H and J_z whose eigenvalues are E and M respectively, commute and therefore the bivariate moments of $\rho^{H,m}(E,M)$ are just $M_{rs}(m) = \langle H^r J_z^s\rangle^m$; note that nuclear effective Hamiltonians are all J invariant. For simplicity we drop m in $M_{rs}(m)$ and similarly in $k_{rs}(m)$. Some important results are: (i) $M_{rs}(m) = 0$ for s odd and therefore all $k_{rs}(m) = 0$ for s odd; (ii) the marginal densities $\rho(E)$ and $\rho(M)$ are close to Gaussian, the first one is a result of the fact that nuclear Hamiltonians can be represented by two-body random matrix ensembles giving $k_{40}(m) \sim -4/m$ and the second because J_z is a one-body operator giving $k_{04}(m) \sim -1/m$; (iii) the correlation coefficient $\zeta(m) = k_{11}(m) = 0$ and hence the bivariate Gaussian in (E,M) is just $\rho_{\mathcal{G}}(E)\rho_{\mathcal{G}}(M)$; (iv) random matrix assumptions show that $k_{22}(m) \sim -2/3m$; (v) as $k_{rs}(m) \sim 1/m$ for $r+s=4$, one can further assume that $k_{rs}(m) \sim 1/[m^{(r+s-2)/2}]$. With (i)-(v) it is possible to use bivariate ED expansion for $\rho(E,M)$. Details of the expansion are given in [Kota (1984*)]. The fixed-J densities, or level densities $I_\ell^{m,J}(E)$ where the $(2J+1)$ degeneracy is not counted, follow from fixed-M densities via the difference equation, an extension of Eq. (1.131),

$$\begin{aligned}&(2J+1)^{-1}\langle\langle\delta(H-E)\rangle\rangle^{m,J} = \langle\langle\delta(H-E)\rangle\rangle^{m,M=J} - \langle\langle\delta(H-E)\rangle\rangle^{m,M=J+1} \\ &\Longrightarrow I_\ell^{m,J}(E) = d(m)[\rho^m(E,M=J) - \rho^m(E,M=J+1)] \\ &\Longrightarrow [D(m,J)/d(m)]\rho^{m,J}(E) \simeq -\left.\frac{\partial\rho^m(E,M)}{\partial M}\right|_{M=J+1/2}\;.\end{aligned} \tag{1.138}$$

More importantly, the ED expansion gives simple expressions for $E_c(m,J) = \langle H\rangle^{m,J}$ and $\sigma^2(m,J) = \langle H^2\rangle^{m,J} - (\langle H\rangle^{m,J})^2$. Note that the fixed-$J$ averages of a J invariant operator $\mathcal{O}$ follow from fixed-M averages using,

$$\begin{aligned}
\langle\mathcal{O}\rangle^{m,J} &= \frac{\langle\langle\mathcal{O}\rangle\rangle^{m,M=J} - \langle\langle\mathcal{O}\rangle\rangle^{m,M=J+1}}{\mathcal{D}(m,M=J) - \mathcal{D}(m,M=J+1)} \\
&\simeq \left[-\left.\frac{\partial\mathcal{D}(m,M)}{\partial M}\right|_{M=J+1/2}\right]^{-1}\left[-\left.\frac{\partial\langle\langle\mathcal{O}\rangle\rangle^{M}}{\partial M}\right|_{M=J+1/2}\right] .
\end{aligned} \tag{1.139}$$

These give expansions for $d(m,J)$ and fixed-J moments of H spectra in powers of $J(J+1)$ with coefficients involving only fixed-m traces. Note that

$$\langle H^p\rangle^{m,M} = [\rho^m(M)]^{-1}\int E^p\rho^m(E,M)dE\ . \tag{1.140}$$

For example, using the bivariate Edgeworth corrected Gaussian representation for $\rho(E,M)$ and Eqs. (1.140) and (1.139), expansions in powers of $J(J+1)$, the only J scalar, for fixed-J centroids and variances are derived. Truncating these expansions to second order, we have [Kota, Vyas and Mayya (2008)]

$$\begin{aligned}
& D(m,J) = [d(m)(2J+1)/\sqrt{8\pi}\sigma^3_{J_z}(m)]\ \exp -\frac{(J+1/2)^2}{2\sigma^2_{J_z}(m)} \\
& \times\left[1+\tfrac{K_{04}(m)}{24}\left\{\left[\tfrac{J(J+1)}{\sigma^2_{J_z}(m)}\right]^2 - 10\tfrac{J(J+1)}{\sigma^2_{J_z}(m)}+15\right\}\right]\ ; \\
& \frac{E_c(m,J)-E_c(m)}{\sigma(m)} \\
& =\simeq \left[-\frac{3k_{12}(m)}{2}\right] + \frac{k_{12}(m)}{2}\frac{J(J+1)}{\sigma^2_{J_z}(m)} + \left\{\frac{k_{14}(m)}{24} - \frac{k_{04}(m)k_{12}(m)}{6}\right\}\frac{[J(J+1)]^2}{\sigma^4_{J_z}(m)}\ , \\
& \frac{\sigma^2(m,J)}{\sigma^2(m)} \simeq \left[1-\frac{3k_{22}(m)}{2}\right] \\
& + \left[\frac{k_{22}(m)}{2}\right]\frac{J(J+1)}{\sigma^2_{J_z}(m)} + \left\{\frac{k_{24}(m)}{24} - \frac{k_{22}(m)k_{04}(m)}{6}\right\}\left[\frac{J(J+1)}{\sigma^2_{J_z}(m)}\right]^2 .
\end{aligned} \tag{1.141}$$

Here $E_c(m) = \langle H\rangle^m$, $\sigma^2(m) = \sigma^2_H(m) = \langle H^2\rangle^m - [E_c(m)]^2$ and $k_{rs}(m)$ are the reduced cumulants defined by the $H - J_z$ bivariate moments $M_{rs}(m) = \langle H^r J_z^s\rangle^m$. Note that in the expansions in Eq. (1.141), fixed-J averages are written in terms of fixed-m averages, i.e. the J averages are projected from total m space averages. The price one pays here is that for a lower order J trace we need higher order m traces., For a single j-shell, it is possible to write analytical expressions for all the $k_{rs}(m)$ cumulants in Eq. (1.141) using the fixed-m trace propagation formulas given in [Nomura (1972b); Wong (1986)] and this has been reported in [Kota, Vyas and Mayya (2008)].

Instead of dealing directly with the bivariate $\rho(E, M)$ density, one may equivalently consider the decomposition $\rho(E, M) = \rho(E \mid M)\, \rho(M)$. The marginal density $\rho(M)$ is the fixed-M dimensionality and may be calculated exactly or approximately as described before. The conditional density $\rho(E|M)$ is defined by the conditional moments $\langle H^p \rangle^M$. To the extent that the conditional density is close to a Gaussian, all we need are the lower order conditional moments of H. Expansions for these traces follow from the methods described in Sec. 1.3 with $\widehat{K}_2 = J_z$. Instead of constructing $\rho(E|J)$, the quantity of interest in nuclear spectroscopy, using Eq. (1.138), an alternative is to convert fixed-M moments into fixed-J moments using Eq. (1.139) and then construct $\rho(E|J)$ either as an ED or GC expansion. This approach has been used in [Ginocchio (1973b*); Ayik and Ginocchio (1974b); Ginocchio and Yen (1975)]. The relative merits of the various expansions for $\langle H^p \rangle^{m,J}$ have been studied by many groups [Ginocchio and Yen (1975); Baldridge and Dalton (1978); Haq and Wong (1979)]. Although the J^2 density is not a Gaussian, it is possible to use the methods discussed in Sec. 1.3 with $\widehat{K}_2 = J^2$ [Halemane (1981a)]. However, convergence properties of the resulting expansions are not known. Instead of calculating the fixed-M or J moments approximately, it is also possible to calculate them exactly. There are two methods available for this.

The first method, explored by Jacquemin and collaborators [Jacquemin (1973); Jacquemin and Spitz (1979a,b); Jacquemin (1981); Jacquemin, Auger and Quesne (1982)], follows from the m_{z_i} orbits introduced in the beginning of this subsection. Given a k-body operator $\mathcal{O}(k)$, propagation equation for the trace $\langle\langle \mathcal{O}(k) \rangle\rangle^{m,M}$ follows by generating the configurations $\widetilde{m(m_z)}$ and $\widetilde{k(m_z)}$. As configuration traces propagate, we have,

$$\langle\langle \mathcal{O}(k) \rangle\rangle^{m,M} = \sum_{\widetilde{m(m_z)} \in M} \; \sum_{\widetilde{k(m_z)}} \langle\langle \mathcal{O}(k) \rangle\rangle^{\widetilde{k(m_z)}} \; P[\widetilde{k(m_z)} \, : \, \widetilde{m(m_z)}] \, . \tag{1.142}$$

Jacquemin derived recursion equations for the propagators Ps. Verbaarschot and Brussaard, using a generalized Wick's theorem and a generating function method, simplified the propagators further and also derived explicit expressions for the input traces in terms of sp energies and two-body matrix elements for H and H^2 propagation [Verbaarschot and Brussaard (1981); Verbaarschot (1982); Verbaarschot and Brussaard (1984)]. The main difficulty with Eq. (1.142), though it is simple, is that there will be too many propagators as m_{z_i} orbits are employed. The number of propagators increases further for fixed $\mathbf{m} - J$ traces that are needed in practice. Therefore the method has been used so far only for $(2s1d)$ shell.

The second method, giving directly fixed-J traces, is due to French, Halbert, McGrory and Wong (1969) and used for calculations by Mugambi (1970). However, a more systematic development and applications are given in [Lougheed and Wong (1975a,b); Wong and Lougheed (1978); Wong (1986)]. The principle of this method is to derive recursion relations for traces. For an m-particle trace of a k-body operator $\mathcal{O}(k) = A(k)B(k)$ we have,

$$\begin{aligned} \langle\langle A(k)B(k)\rangle\rangle^m &= \langle\langle B(k)A(k)\rangle\rangle^{m-k} \\ &= \langle\langle A(k)B(k)\rangle\rangle^{m-k} + \langle\langle [B(k)\,,\, A(k)]_-\rangle\rangle^{m-k} \;. \end{aligned} \tag{1.143}$$

As the commutator $[B(k)\,,\, A(k)]_-$ is a $(k-1,\, k-2,\, \ldots,\, 0)$-body operator, Eq. (1.143) is recursive both in particle number and in the body rank of the operator. Thus, if we know the traces of a general one-body operator, Eq. (1.143) will allow us to calculate the traces for a 2-body operator and so on. Though Eq. (1.143) is unnecessary for m-particle (scalar) traces, the important point is that it extends to fixed-J traces. For $A(k)$ a spherical tensor of rank σ and denoted by $A^\sigma(k)$ and similarly $B(k)$ by $B^\lambda(k)$, we obtain

$$\begin{aligned} &\langle\langle ||\, \left[A^\sigma(k)\; B^\lambda(k)\right]^\nu \,||\rangle\rangle^{m,J} \\ &= \sum_{\omega,J'} (-1)^{\nu-\omega} \left[\frac{(2J+1)}{(2J'+1)}\right]^{1/2} U(J\sigma J\lambda; J'\nu)\, U(J'\lambda J'\sigma; J\omega) \\ &\times \left\{(-1)^{\lambda+\sigma-\omega} \langle\langle ||\, \left[A^\sigma(k)\; B^\lambda(k)\right]^\omega \,||\rangle\rangle^{m-k,J'} \right. \\ &\left. + \left\langle\left\langle ||\, \left[B^\lambda(k)\;\; A^\sigma(k)\right]_-^\omega \,||\right\rangle\right\rangle^{m-k,J'} \right\} \;. \end{aligned} \tag{1.144}$$

Again there is recursion with respect to particle number and the rank of the interaction. Starting with (p,n) spherical configurations $(m_1, m_2, \ldots)$ and denoting the angular momentum of the m_i particles as τ_i, we have $(\tilde{m}, \tilde{\tau}, \tilde{J})$ configurations with $J_i = \tau_i \oplus J_{i-1}$; $J_1 = \tau_1$ and $J_n = J$. Decomposing a given operator into product of operators with each acting in a single j-orbit of proton or neutron, the fixed-(m,J) trace calculation reduces to the calculation of traces of a general one-body operator for identical particles in a single j-shell; Ginocchio and Yen (1972) tabulated these for $j = 1/2, 3/2, \ldots, 15/2$. See [Lougheed and Wong (1975a); Ng and Wong (1976); Wong (1986); Horoi, Kaiser and Zelevinsky (2003*)] for computer programmes and $(2s1d)$ shell applications of this method. Because of the induction involved, one has to calculate, store and retrieve a large number of traces. This method is computationally extensive and hence so far has not been applied beyond $(2s1d)$ shell.

Instead of using $\rho(E,M)$ directly or the decomposition $\rho(E|M)\rho(M)$, a third alternative for constructing fixed-J densities is to consider the decomposition $\rho(E,M) = \rho(M \mid E)\, \rho(E)$ and then covert it into J densities. This method is used in almost all calculations of level densities as in Secs. 1.4 and 1.5. The marginal density $\rho(E)$ is nothing but the state density and it can be constructed by using methods described in Sec. 1.5. The conditional density $\rho(M|E)$ is defined by the conditional moments $\langle J_z^p\rangle^E$ and the lowest moment is the energy dependent spin cutoff factor $\sigma_J^2(E) = \left\langle J_z^2\right\rangle^E$, expectation value of the J_z^2 operator. Extending the CLT argument used for constructing $\rho(M)$ and applying it locally, one can assume that $\rho(M|E)$ is a Gaussian, its centroid $\langle J_z\rangle^E = 0$ and variance is the spin cutoff factor. Then applying Eq. (1.132) we obtain

$$\rho(E,J) = \rho(E)\frac{(2J+1)}{\sqrt{8\pi}\ \sigma_J^3(E)}\ \exp - \frac{(J+1/2)^2}{2\sigma_J^2(E)} = \rho(E)\ C_J(E)\ . \tag{1.145}$$

The factor $C_J(E)$ gives the angular momentum decomposition of state densities into level densities. Note that the most probable spin at a given energy E is $\sim \sigma_J(E)$. To the lowest approximation, Eq. (1.48) gives,

$$\begin{aligned} \sigma_J^2(E) &= \left\langle J_z^2 \right\rangle^E = \left\langle J_z^2 \right\rangle^m \left[1 + \zeta_{H,J_z^2}(m)\frac{\sigma(m,J_z^2)}{\langle J_z^2 \rangle^m}\ \widehat{E}\right] \ ; \\ \zeta_{H,J_z^2}(m) &= \frac{\left\langle \widehat{H} J_z^2 \right\rangle^m}{\sigma(m,J_z^2)}\ , \quad \sigma^2(m,J_z^2) = \left\langle (J_z^2 - \left\langle J_z^2 \right\rangle^m)^2 \right\rangle^m \ . \end{aligned} \tag{1.146}$$

The correlation coefficient in Eq. (1.146) is found to be ~ -0.15 for $(2s1d)$ shell nuclei. Thus $\sigma_J(E)$ depends weakly on E; Bethe's theory [Bethe (1936, 1937)] assumes $\sigma_J(E)$ to be energy independent. A more accurate calculation of spin cutoff factors follows from applying Eqs. (1.100) and (1.101); see for example [French *et al.* (2006*)].

1.6.8 *Reduced traces*

Symmetry breaking of an imposed symmetry S can be measured by the partial variances $\sigma^2(f \to f')$ connecting two different irreps (f, f') of S as these traces will be zero if the symmetry S is a conserved (or exact) symmetry,

$$\sigma^2(f \to f') = [D(f)]^{-1} \sum_{\lambda,\lambda',\alpha,\alpha'} \left\langle \lambda, f, \alpha \mid \widetilde{H}^2 \mid \lambda', f', \alpha' \right\rangle \ . \tag{1.147}$$

Here we using a slightly different notation compared to Eq. (1.119). The α are labels of the subgroups of S and λ represents all other labels. For example for shell model $SU(4)$, f are $SU(4)$ irreps, λ correspond to $U(N/4)$ labels and α are $SU(4)$ subgroup labels; note that $SU(4)$ multiplicities are all one. Similarly for isospin decomposition, α are M_T labels and λ are all other labels. In identical particle seniority decomposition $U(2\Omega) \supset Sp(2\Omega)$, the labels λ are not needed. The $D(f)$ in Eq. (1.147) is the total dimension of the space with the irrep f fixed. We now show that partial variances involve traces of reduced matrix elements and the discussion below follows [Hecht and Draayer (1974); Hecht (1975)].

Given H, it is possible to carry out a unitary decomposition $H = \sum_{\lambda_0 f_0 \alpha_0} C(\lambda_0 f_0 \alpha_0)\ T^{\lambda_0 f_0 \alpha_0}$ with respect to $G \supset S$. Now consider

$$X = \sum_{\lambda,\lambda',\alpha,\alpha'} \left\langle \lambda f \alpha \mid T^{\lambda_0 f_0 \alpha_0} \mid \lambda' f' \alpha' \right\rangle \left\langle \lambda' f' \alpha' \mid \left[T^{\lambda_0' f_0' \alpha_0'}\right]^\dagger \mid \lambda f \alpha \right\rangle \ . \tag{1.148}$$

Using the following properties, with $d(f)$ denoting the dimension of the irreps f (i.e. number of states within a given f and this is number of α labels),

$$
\begin{aligned}
&\left(T^{\lambda_0 f_0 \alpha_0}\right)^\dagger = (-1)^{\eta(\alpha_0)}\, T^{\lambda_0^* f_0^* \alpha_0^*} \\
&\langle f_1\alpha_1 f_2\alpha_2 \mid f_3\alpha_3\rangle_\rho = (-1)^{\phi(f_1 f_2 f_3)}\, \langle f_2\alpha_2 f_1\alpha_1 \mid f_3\alpha_3\rangle_\rho \\
&\langle f_1\alpha_1 f_2\alpha_2 \mid f_3\alpha_3\rangle_\rho = (-1)^{\eta(\alpha_2)}(-1)^{\phi(f_1 f_2 f_3)} \sqrt{\frac{d(f_3)}{d(f_1)}}\, \langle f_3\alpha_3 f_2^*\alpha_2^* \mid f_1\alpha_1\rangle_\rho \\
&(-1)^{\eta(\alpha_0)} \left\langle \lambda f\alpha \mid T^{\lambda_0^* f_0^* \alpha_0^*} \mid \lambda' f'\alpha'\right\rangle = \left\langle \lambda' f'\alpha' \mid T^{\lambda_0 f_0 \alpha_0} \mid \lambda f\alpha\right\rangle \;,
\end{aligned}
\tag{1.149}
$$

and the orthonormal properties of S space Wigner coefficients $\langle f_1\alpha_1 f_2\alpha_2 \mid f_3\alpha_3\rangle_\rho$, where ρ is the multiplicity label for $f_1 \otimes f_2 \to f_3$,

$$
\sum_{\alpha_1,\alpha_3} \langle f_1\alpha_1 f_2\alpha_2 \mid f_3\alpha_3\rangle_\rho \left\langle f_1\alpha_1 f_2'\alpha_2' \mid f_3\alpha_3\right\rangle_{\rho'} = [d(f_3)/d(f_2)]\delta_{\rho\rho'}\delta_{f_2 f_2'}\delta_{\alpha_2\alpha_2'} \tag{1.150}
$$

along with the Wigner-Eckart theorem

$$
\left\langle \lambda f\alpha \mid T^{\lambda_0 f_0 \alpha_0} \mid \lambda' f'\alpha'\right\rangle = \sum_\rho \langle f'\alpha'\; f_0\alpha_0 \mid f\alpha\rangle_\rho \left\langle \lambda f \mid\mid T^{\lambda_0 f_0} \mid\mid \lambda' f'\right\rangle_\rho \tag{1.151}
$$

will allow us to write the partial variances in terms of reduced matrix elements. Eq. (1.150) is obtained by changing the order in the Wigner coefficients to $f_3 \times f_1 \to f_2$ by applying the 2nd and 3rd equalities in Eq. (1.149). Then, firstly X in Eq. (1.148) is

$$
X = d(f) \sum_{\lambda,\lambda',\rho} \left\langle \lambda f \mid\mid T^{\lambda_0 f_0} \mid\mid \lambda' f'\right\rangle_\rho \left\langle \lambda f \mid\mid T^{\lambda_0' f_0} \mid\mid \lambda' f'\right\rangle_\rho \delta_{\alpha_0\alpha_0'} \;. \tag{1.152}
$$

Now the result for the partial variances is,

$$
\begin{aligned}
&\sigma^2(f \to f') = \sum_{\lambda_0,\lambda_0',f_0} \left[\{d(f_0)\}^{-1} \sum_{\alpha_0} C(\lambda_0 f_0\alpha_0)C^*(\lambda_0' f_0\alpha_0)\right] \sigma^2_{\lambda_0,\lambda_0',f_0}(f \to f')\;, \\
&\sigma^2_{\lambda_0,\lambda_0',f_0}(f \to f') = [d(f)/D(f)] \sum_{\lambda,\lambda',\rho} \left\langle \lambda f \mid\mid T^{\lambda_0 f_0} \mid\mid \lambda' f'\right\rangle_\rho \left\langle \lambda f \mid\mid T^{\lambda_0' f_0} \mid\mid \lambda' f'\right\rangle_\rho \;.
\end{aligned}
\tag{1.153}
$$

Similarly using the U or Racah coefficients of S we have,

$$
\begin{aligned}
&\left\langle \lambda f \mid\mid \left[T^{\lambda_0 f_0} \times T^{(\lambda_0')^*(f_0')^*}\right]^{F_0\rho_0} \mid\mid \lambda f\right\rangle_{\rho_0'} = \sum_{\lambda',f'} (-1)^{\phi(f_0\,(f_0')^*\,F_0)+\phi(f'\,f_0'\,f)} \\
&\times U(f(f_0')^* f f_0; f'_{\rho_1\rho_2} F_{0:\rho_0\rho_0'}) \sqrt{\frac{d(f)}{d(f')}} \left\langle \lambda f \mid\mid T^{\lambda_0 f_0} \mid\mid \lambda' f'\right\rangle_{\rho_1} \left\langle \lambda f \mid\mid T^{\lambda_0' f_0'} \mid\mid \lambda' f'\right\rangle_{\rho_2}
\end{aligned}
\tag{1.154}
$$

and this equation can be inverted by using the orthonormal properties of the U-coefficients,

$$\begin{gathered}\sum_{F_0,\rho_0,\rho_0'} (-1)^{\phi(f_0\,(f_0')^*\,F_0)}\; U(f(f_0')^* f f_0; f'_{\rho_1\rho_2} F_{0:\rho_0\rho_0'}) \\ \times \left\langle \lambda f \,||\, \left[T^{\lambda_0 f_0} \times T^{(\lambda_0')^*(f_0')^*}\right]^{F_0\rho_0} \,||\, \lambda f \right\rangle_{\rho_0'} \\ = \sum_{\lambda'} (-1)^{\phi(f'\,f_0'\,f)} \sqrt{\frac{d(f)}{d(f')}} \left\langle \lambda f \,||\, T^{\lambda_0 f_0} \,||\, \lambda' f' \right\rangle_{\rho_1} \left\langle \lambda f \,||\, T^{\lambda_0' f_0'} \,||\, \lambda' f' \right\rangle_{\rho_2} .\end{gathered} \tag{1.155}$$

The phase factor on the r.h.s can be written in terms of a U-coefficient,

$$\begin{gathered} U(f(f_0')^* f f_0'; f'_{\rho\rho}[0]) = (-1)^{\phi(f' f_0' f)} \sqrt{\frac{d(f')}{d(f) d(f_0')}} \\ \Rightarrow (-1)^{\phi(f' f_0' f)} \sqrt{\frac{d(f')}{d(f)}} = \sqrt{d(f_0')}\; U(f(f_0')^* f f_0'; f'_{\rho\rho}\,[0]) .\end{gathered} \tag{1.156}$$

With this we finally obtain

$$\begin{gathered}\sigma^2_{\lambda_0,\lambda_0',f_0}(f \to f') = \sum_{F_0,\rho,\rho_0,\rho_0'} (-1)^{\phi(f_0 f_0^* F_0)} \sqrt{\frac{d(f_0) d^2(f)}{D^2(f)}}\; U(f f_0^* f f_0; f'_{\rho_a\rho_a}[0]) \\ \times U(f(f_0)^* f f_0; f'_{\rho\rho} F_{0:\rho_0\rho_0'}) \left\langle\left\langle f \,||\, \left[T^{\lambda_0 f_0} \times T^{(\lambda_0')^*(f_0)^*}\right]^{F_0\rho_0} \,||\, f \right\rangle\right\rangle_{\rho_0'} .\end{gathered} \tag{1.157}$$

Note that ρ_a in Eq. (1.157) can be any one of the ρ_a labels. The reduced matrix elements trace $\langle\langle || - - -- || \rangle\rangle$ in Eq. (1.157) involves a sum over all λ. It is easy to recognize that in some situations, it is possible to write the reduced traces in terms of Wigner coefficients for $G \supset S$ and Racah coefficients for the algebras G and S; see [Hecht (1974); Hecht and Draayer (1974); Hecht (1975); Quesne and Spitz (1978); Kota (1979b)].

We close this subsection with the following comments:

(1) For variances $\sigma^2(f) = \sum_{f'} \sigma^2(f \to f')$, it is easy to see that, via the sum rule for the U coefficients, only $F_0 = 0$ will contribute in Eq. (1.157).

(2) Reduced traces appear when we deal with transition strength sums generated by electromagnetic transitions (also for GT strength sums). The operators (for example $E2$ and $M1$) here are by definition tensors with respect to both J and T. Hence one has to deal with JT reduced traces for evaluating strength sums using spectral averaging methods [Draayer, French and Wong (1977a*,b); Verbaarschot (1982); Verbaarschot and Brussaard (1984)].

(3) For reduced traces it is possible to consider extended I.B. defined for tensor operators. This has been discussed in [Kopsky (1979)] but not much has been explored till now.

(4) It is useful to mention that only for fixed isospin (T) reduced traces, the propagation equations are available in literature and they were given in the early years of the subject of trace propagation by French (1969). However, later, Nomura developed a new formulation [Nomura (1985a,b,c)] and gave a derivation of the propagation equation [Nomura (1986*)].

(5) It is important to note from Eqs. (1.155) and (1.157) that operator traces can be written in terms of Wigner and Racah coefficients for the group structures involved. Using this result, compact formulas (without the need to construct matrices for 3 and 4 particle systems) for partial variances for fixed $SU(4)$ [Hecht and Draayer (1974)], for fixed identical particle seniority [Quesne and Spitz (1978); Quesne (1975a); Kota (1979b)] and also for the $SU(3)$ and $SO(6)$ symmetry limits of sdIBM [Kota (1981)] have been derived.

(6) Even for the well known scalar and configuration averages it can be shown that the propagators can be written in terms of Wigner and Racah coefficients [Hecht (1975)]. This fact has been exploited recently to derive some exact results for two-body random matrix ensembles [Kota (2005b*, 2007a); Vyas and Kota (2009)]. For example for scalar, i.e. $U(N)$, averages, the tensorial decomposition of H, a p-body operator, with respect to $U(N)$ is [Hecht (1975)]

$$H = \sum C_{\{1^p\}\{1^{N-p}\}\{f_0\}} \, T^{\{1^p\}\{1^{N-p}\}\{f_0\}}$$

where $\{f_0\} = \{2^x 1^{N-2x}\}$, $x = 0, 1 \ldots, p$. Then $U(N)$ trace of $HH^\dagger$ will involve reduced traces,

$$\begin{aligned}
\left\langle\left\langle \left[T^{\{1^p\}\{1^{N-p}\}\{f_0\}}\right]^2 \right\rangle\right\rangle^m &= \left[\frac{m!}{(m-p)!}\right]^2 \frac{(N-p)!p!}{N!} \frac{1}{d(f_0)} \\
&\times \left[\frac{U(\{1^m\}\{1^{N-p}\}\{1^m\}\{1^p]\}; \{1^{m-p}\}\{f_0\})}{U(\{1^m\}\{1^{N-p}\}\{1^m\}\{1^p]\}; \{1^{m-p}\}\{0\})}\right]^2 \\
&= \left[\frac{m!(m-x)!(N-m)!}{(m-p)!(m-p)!(N-m-x)!}\right] \left[\frac{p!p!(N-p-x)!}{(p-x)!N!}\right] .
\end{aligned} \tag{1.158}$$

For configuration averages we need to consider the tensor operators $T^{\{1^p\}\{1^{N-h}\}\{f_0\}}$ with $p = h$ and also $p \neq h$. Finally, Nomura (1987b) has shown that scalar or $U(N)$ averages can be written in terms of symmetric group characters.

Finally, we discuss briefly a novel application of trace propagation.

1.6.9 *Regular structures with random interactions*

We have already mentioned at various places that the random matrix ensembles generated by random interactions acting in many particle Hilbert spaces form the basis for statistical spectroscopy. Operation of these ensembles could in fact be used as a definition for many-body quantum chaos. A new paradigm, first recognized by Johnson, Bertsch and Dean (1998), is that random interactions generate a variety of regular structures with a high probability. For example random two-body interactions generate, for even-even nuclei, ground states with spin 0^+ with high probability. Similarly Bijker and Frank (2000) found that random interactions in IBM generate vibrational and rotational structures with high probability. Mulhall, Volya and Zelevinsky (2000) introduced the idea of geometric chaos, Zhao, Arima and Yoshinaga (2004) developed a phenomenological prescription and Papenbrock and Weidenmüller (2004) examined spectral widths, for understanding the origin of regularities generated by random interactions. A very large number of studies that followed between 1998-2007 are reviewed in [Zelevinsky and Volya (2004); Zhao, Arima and Yoshinaga (2004); Papenbrock and Weidenmüller (2007)]. More recently it has been pointed out that that symmetries emerge as k in EGOE(k) increases [Volya (2008)]. Spectral distributions with exact or approximate trace propagation for centroids, variances and other higher order moments (or operator traces) opened a new window to investigate regularities generated by random interactions as traces can be calculated for each member of the two-body or a k-body ensemble. For example using fixed-T centroids and variances, [Kota and Kar (2002)] explained the regularities found by Jacquod and Stone (2000, 2001) in interacting spin systems. Similarly, regularities in energy centroids and spectral variance, using propagation equations in a variety of examples, have been studied [Kota (2005a*); Zhao *et al.* (2005); Kota (2006a); Kota, Vyas and Mayya (2008)]. For example, for fermions in a single j and bosons in a single ℓ shell, propagation formulas for energy centroids generated by 3-body Hamiltonians are derived using Eq. (1.141) and applied [Zhao *et al.* (2005)] to establish 'robustness of regularities for energy centroids in the presence of random interactions'. Similarly it is shown in [Kota, Vyas and Mayya (2008)] that the approach based on spectral distributions and trace propagation is equivalent to the 'geometric chaos'approach of Mulhall, Volya and Zelevinsky (2000). Further exploration of the connections between these two should provide interesting results. It is also of interest to examine the relationship between spectral distributions and the shell model Monte Carlo methods [Koonin, Dean and Langanke (1997)] that start with the partition function.

1.7 Bivariate Distributions for Transition Strengths

1.7.1 *Preliminaries*

The one-point density calculations discussed in Sec. 1.5 in general can be extended to the ($k > 1$)-point correlation functions, but it is highly probable that, in the chaotic domain, very little information can be extracted when $k > 2$ [an exception is for spectral fluctuations from which a fairly good $k = 3$ and relatively less certain $k = 4$ information have been extracted by Bohigas, Haq and Pandey (1985*)]. Thus extending one-point density results to two-point functions should give an essentially complete statistical spectroscopy theory. In this Section we turn to the two-point density which gives rise to transition strengths and symmetry breaking [Draayer, French and Wong (1977a*); French, Kota, Pandey and Tomsovic (1988b*)]. Both of these phenomena involve the same formal structure, specifically that of bivariate distributions, in the initial and final Hamiltonian eigenenergies x and y, of strengths $|\langle y \mid \mathcal{O} \mid x\rangle|^2$ generated by the transition $|x\rangle \stackrel{\mathcal{O}}{\rightarrow} |y\rangle$. Here $\mathcal{O}$ is a transition operator which acts in the Hamiltonian space. Before proceeding further some remarks are in order. The first is that by specifying the (m, N) space (as in Secs. 1.3- 1.5) we restrict ourselves to number conserving transitions, ruling out, for example, particle transfer processes. However, it is elementary to remove this restriction and we shall do so ahead. The second is that for an NIP system (or for an effective one-body h which has no $\nu > 1$ part) and an arbitrary transition operator, the problem of the strength distribution is exactly solved as discussed in Sec. 1.7.2 ahead. Then for a heuristic derivation of the corresponding IP system (i.e. one with a $\nu = 2$ interaction), it is reasonable to assume that "the spreading of the strength follows that of the density, i.e. that when a pair (initial and final) of h-eigenstates spreads among the H-eigenstates as $\boldsymbol{V}$ is turned on, the initial strength connecting them spreads in proportion to the product of the admixing intensities". This conjecture, which has its basis in the operation of two-body ensembles [French, Kota, Pandey and Tomsovic (1988b*)], i.e. H and $\mathcal{O}$ are typical members of such ensembles, leads to a bivariate Gaussian strength distribution which can be verified or, if necessary, improved upon by taking account of low-order Edgeworth corrections to the bivariate Gaussian; it has been verified in this way for the case where $\mathcal{O}$ is represented by a "featureless ensemble" [French, Kota, Pandey and Tomsovic (1988b*); Kota and Sahu (1998)] and also for E2 and M1 operators [Gómez *et al.* (2004)]. We shall make use of that and consider also the case of a uniquely specified $H = \boldsymbol{h}+\boldsymbol{V}$, a correction however being called for when the transition operator is strongly correlated with the Hamiltonian. The third remark is that, although we shall start with an elementary NIP system with no specific symmetries, it is necessary to impose the symmetries in the final result and this is possible. Finally, just as with the state density problem, spreading due to interactions between configurations which are far apart must be ignored, or treated by perturbation theory (which may not be entirely satisfactory), or by a multimodal extension of the statistical method [this requires further studies of the partitioned EGOE, see Kota, Majumdar, Haq and Leclair (1999)].

Following closely Sec. 1.2, we write the bivariate strength density $\mathrm{I}_{\mathcal{O}}^{H,m_i,m_f}(x,y)$, its marginal densities $\mathrm{I}_{\mathcal{O}}^{H,m_i}(x)$ and $\mathrm{I}_{\widetilde{\mathcal{O}}}^{H,m_f}(y)$, its bivariate moments $\widetilde{M_{pq}}$, the normalized bivariate moments $M_{pq} = \widetilde{M_{pq}}/\widetilde{M_{00}}$ and the corresponding central moments $\mathcal{M}_{pq}$, its marginal centroids $\epsilon_x = M_{10}$, $\epsilon_y = M_{01}$, marginal variances $\sigma_x^2 = M_{20} - (\epsilon_x)^2$, $\sigma_y^2 = M_{02} - (\epsilon_y)^2$, and correlation coefficient $\zeta = \mathcal{M}_{11}/(\sigma_x\sigma_y)$. Similarly, the normalized bivariate density, dropping the superscripts and subscripts, is $\rho(x,y)$ and the density in terms of standard variables $\widehat{x} = (x - \epsilon_x)/\sigma_x$ and $\widehat{y} = (y - \epsilon_y)/\sigma_y$ is $\eta(\widehat{x},\widehat{y})$. The moments of $\eta(\widehat{x},\widehat{y})$ are the reduced moments

$\mu_{rs} = \mathcal{M}_{rs}/\sigma_x^r \sigma_y^s$. Firstly,

$$\begin{aligned}
I_{\mathcal{O}}^{H,m_i,m_f}(x,y) &= I^{H,m_f}(y) |\langle m_f, y \mid \mathcal{O} \mid m_i, x\rangle|^2 I^{H,m_i}(x) \\
&= \langle\langle \mathcal{O}^\dagger \delta(H-y)\mathcal{O}\delta(H-x)\rangle\rangle^{m_i} \\
\mathrm{I}_{\mathcal{O}}^{H,m_i}(x) &= \int I_{\mathcal{O}}^{H,m_i,m_f}(x,y)dy = \left\langle\left\langle \mathcal{O}^\dagger \mathcal{O}\delta(H-x)\right\rangle\right\rangle^{m_i} \\
\mathrm{I}_{\mathcal{O}}^{H,m_f}(y) &= \int I_{\mathcal{O}}^{H,m_i,m_f}(x,y)dx = \left\langle\left\langle \mathcal{O}^\dagger \delta(H-y)\mathcal{O}\right\rangle\right\rangle^{m_i} \\
\widetilde{M_{pq}} &= \int I_{\mathcal{O}}^{H,m_i,m_f}(x,y)x^p y^q \, dxdy = \left\langle\left\langle \mathcal{O}^\dagger H^q \mathcal{O} H^p\right\rangle\right\rangle^{m_i} \\
M_p(x) &= \sum_y |\langle m_f, y \mid \mathcal{O} \mid m_i, x\rangle|^2 \, y^p = \left\langle\left\langle \mathcal{O}^\dagger H^p \mathcal{O}\right\rangle\right\rangle^{m_i,x} / I^{H,m_i}(x) \\
M_p(y) &= \sum_x |\langle m_f, y \mid \mathcal{O} \mid m_i, x\rangle|^2 \, x^p = \left\langle\left\langle \mathcal{O} H^p \mathcal{O}^\dagger\right\rangle\right\rangle^{m_f,y} / I^{H,m_i}(y) \, .
\end{aligned} \tag{1.159}$$

The delta function formalism used here facilitates the recognition of convolutions and extensions to the IP systems (from NIP). We see that, at fixed x, $M_p(x)/M_0(x)$ is the p-th moment of a normalized mono-variate density so that the total strength originating at $x = E$ is $M_0(E)$ while its centroid and variance are $\epsilon(E) = M_1(E)/M_0(E)$ and $\sigma^2(E) = M_2(E)/M_0(E) - [\epsilon(E)]^2$; similarly for higher-order moments both for NIP and IP. For a chaotic system, it is profitable to proceed first with the NIP system and then incorporate the interactions via convolutions, a procedure well demonstrated for state densities and expectation values.

Finally, we stress that, though every strength moment may be regarded as the outcome of a sum rule, only rarely, if at all, could we expect to make effective use of high-order sum rules in IP systems; this is because, on the one hand, operators $\mathcal{O}^\dagger H^p \mathcal{O}$ of high-particle rank are called for, while on the other hand [French (1966b)] 'most model wavefunctions promise more than they can deliver', and give many-particle correlations which are inadequate for dealing with operators of high particle rank. This problem is alleviated by the fact that the CLTs operating in the spaces do specify, to within fluctuations, the smoothed form of the strength distribution and confirm that high-order moments carry little useful information. But, beyond that, even for low-order sum rules it may not be feasible to measure strengths over the whole significant energy range [for example because only an exothermic or endothermic reaction is available for a sum rule which requires both, see Draayer, French and Wong (1977a*,b)]. However, the techniques used make feasible the theoretical evaluation of such low-order partial sums, since we evaluate the strength itself. For a detailed account of sum rules see French (1966b). We mention that partitioning of the many-particle spaces will usually be called for (the interaction between subspaces however being taken into account), the fundamental angular momentum, parity and isospin symmetries must be considered (other symmetries also may be of specific interest), and that we must be able to evaluate correlation coefficients of the bivariate strength density.

1.7.2 *Bivariate distributions*

We briefly review some properties of bivariate distributions and then turn to NIP and IP strength distributions. Given the marginal centroids (ϵ_i, ϵ_f), widths (σ_i, σ_f) and the bivariate correlation coefficient ζ_{biv}, a normalized bivariate Gaussian is defined by,

$$\rho_{biv-\mathcal{G}}(E_i, E_f) = \rho_{biv-\mathcal{G}}(E_i, E_f; \epsilon_i, \epsilon_f, \sigma_i, \sigma_f, \zeta_{biv}) = \frac{1}{2\pi\sigma_i\sigma_f\sqrt{(1-\zeta_{biv}^2)}}$$
$$\times \exp -\frac{1}{2(1-\zeta_{biv}^2)}\left\{\left(\frac{E_i-\epsilon_i}{\sigma_i}\right)^2 - 2\zeta_{biv}\left(\frac{E_i-\epsilon_i}{\sigma_i}\right)\left(\frac{E_f-\epsilon_f}{\sigma_f}\right)+\left(\frac{E_f-\epsilon_f}{\sigma_f}\right)^2\right\} . \tag{1.160}$$

When the reduced bivariate cumulants k_{rs} for $r+s \geq 3$ vanish, the bivariate moments define bivariate Gaussian and, in terms of the standardized variables $\widehat{x} = (E_i - \epsilon_i)\sigma_i$ and $\widehat{y} = (E_f - \epsilon_f)/\sigma_f$, it is

$$\eta_{\mathcal{G}}(\widehat{x}, \widehat{y}) = \frac{1}{2\pi\sqrt{(1-\zeta^2)}}\, exp\left\{-\frac{\widehat{x}^2 - 2\zeta\widehat{x}\widehat{y} + \widehat{y}^2}{2(1-\zeta^2)}\right\} . \tag{1.161}$$

The bivariate Edgeworth expansion for any bivariate distribution $\eta(\widehat{x}, \widehat{y})$ follows from,

$$\eta(\widehat{x}, \widehat{y}) = \exp\left\{\sum_{r+s\geq 3}(-1)^{r+s}\frac{k_{rs}}{r!s!}\frac{\partial^r}{\partial\widehat{x}^r}\frac{\partial^s}{\partial\widehat{y}^s}\right\}\eta_{\mathcal{G}}(\widehat{x}, \widehat{y}) . \tag{1.162}$$

Assuming that the bivariate reduced cumulants k_{rs} behave as $k_{r+s} \propto \Upsilon^{-(r+s-2)/2}$ where Υ is a system parameter, and collecting all the terms that behave as $\Upsilon^{-P/2}$, $P = 1, 2, \ldots$, in the expansion of Eq. (1.162), the bivariate ED expansion is generated [Kota (1984*)] as,

$$\eta_{ED}(\widehat{x}, \widehat{y}) = \eta_{\mathcal{G}}(\widehat{x}, \widehat{y})\left\{1 + \sum_{P=1}^{\infty}\left[\sum_{[P]}\sum_{\{\prod_{i=1}^m (p_iq_i)^{\pi_i}\}}\right.\right.$$
$$\left.\left.\left[\prod_{i=1}^m\left\{\frac{k_{p_iq_i}}{p_i!q_i!}\right\}^{\pi_i}\frac{1}{\pi_i!}\right] He_{\sum p_i\pi_i, \sum q_i\pi_i}(\widehat{x}, \widehat{y})\right]\right\} . \tag{1.163}$$

In Eq. (1.163), $[P]$ are the partitions of the integer P and the corresponding bipartitions $\prod_{i=1}^m (p_iq_i)^{\pi_i}$ are defined as follows. With $[P] = [P_1, P_2, \ldots, P_\ell]$, $P_1 \geq P_2 \geq \ldots \geq P_\ell > 0$, generate all possible $[(p_1q_1)(p_2q_2)\ldots(p_\ell q_\ell)]$ such that $p_i + q_i = P_i + 2$ and $p_i, q_i \geq 0$. If (p_iq_i) is repeated π_i times, the bipartition is written as $\prod_{i=1}^{m}(p_i q_i)^{\pi_i} = [(p_1 q_1)^{\pi_1}(p_2 q_2)^{\pi_2} \ldots (p_m q_m)^{\pi_m}]$. Note that $\sum_{i=1}^{m} \pi_i = \ell$. The explicit form to order $P = 2$ is

$$\begin{aligned}
\eta_{biv-ED}(\widehat{x},\widehat{y}) = \Bigg\{ 1 &+ \left(\frac{k_{30}}{6} He_{30}(\widehat{x},\widehat{y}) + \frac{k_{21}}{2} He_{21}(\widehat{x},\widehat{y}) \right. \\
&+ \left. \frac{k_{12}}{2} He_{12}(\widehat{x},\widehat{y}) + \frac{k_{03}}{6} He_{03}(\widehat{x},\widehat{y}) \right) \\
&+ \left(\left\{ \frac{k_{40}}{24} He_{40}(\widehat{x},\widehat{y}) + \frac{k_{31}}{6} He_{31}(\widehat{x},\widehat{y}) \right. \right. \\
&+ \left. \frac{k_{22}}{4} He_{22}(\widehat{x},\widehat{y}) + \frac{k_{13}}{6} He_{13}(\widehat{x},\widehat{y}) + \frac{k_{04}}{24} He_{04}(\widehat{x},\widehat{y}) \right\} \\
&+ \left\{ \frac{k_{30}^2}{72} He_{60}(\widehat{x},\widehat{y}) + \frac{k_{30}k_{21}}{12} He_{51}(\widehat{x};\widehat{y}) \right. \\
&+ \left[\frac{k_{21}^2}{8} + \frac{k_{30}k_{12}}{12} \right] He_{42}(\widehat{x},\widehat{y}) \\
&+ \left[\frac{k_{30}k_{03}}{36} + \frac{k_{12}k_{21}}{4} \right] He_{33}(\widehat{x},\widehat{y}) \\
&+ \left[\frac{k_{12}^2}{8} + \frac{k_{21}k_{03}}{12} \right] He_{24}(\widehat{x},\widehat{y}) + \frac{k_{12}k_{03}}{12} He_{15}(\widehat{x},\widehat{y}) \\
&+ \left. \left. \frac{k_{03}^2}{72} He_{06}(\widehat{x},\widehat{y}) \right\} \right) \Bigg\} \eta_{\mathcal{G}}(\widehat{x},\widehat{y}) \,.
\end{aligned} \tag{1.164}$$

The bivariate Hermite polynomials $He_{m_1 m_2}(\widehat{x},\widehat{y})$ in Eq. (1.164) are generated by

$$He_{m_1 m_2}(\widehat{x},\widehat{y}) = [\eta_{\mathcal{G}}(\widehat{x},\widehat{y})]^{-1} \, (-1)^{m_1+m_2} \frac{\partial^{m_1}}{\partial \widehat{x}^{m_1}} \frac{\partial^{m_2}}{\partial \widehat{y}^{m_2}} \eta_{\mathcal{G}}(\widehat{x},\widehat{y}) \tag{1.165}$$

and satisfy the recursion relation,

$$\begin{aligned}
(1-\zeta^2) He_{m_1+1,m_2}(\widehat{x},\widehat{y}) = (\widehat{x} - \zeta\widehat{y}) He_{m_1,m_2}(\widehat{x},\widehat{y}) - m_1 He_{m_1-1,m_2}(\widehat{x},\widehat{y}) \\
+ m_2 \zeta He_{m_1,m_2-1}(\widehat{x},\widehat{y}) \;\; .
\end{aligned} \tag{1.166}$$

The polynomials $He_{m_1 m_2}$ with $m_1 + m_2 \leq 3$ are

$$\begin{aligned}
He_{00}(\widehat{x},\widehat{y}) &= 1 \\
He_{10}(\widehat{x},\widehat{y}) &= (\widehat{x} - \zeta\widehat{y}) \,/\, \left(1 - \zeta^2\right) \\
He_{20}(\widehat{x},\widehat{y}) &= \frac{(\widehat{x} - \zeta\widehat{y})^2}{(1-\zeta^2)^2} - \frac{1}{(1-\zeta^2)} \\
He_{11}(\widehat{x},\widehat{y}) &= \frac{(\widehat{x} - \zeta\widehat{y})(\widehat{y} - \zeta\widehat{x})}{(1-\zeta^2)^2} + \frac{\zeta}{1-\zeta^2} \\
He_{30}(\widehat{x},\widehat{y}) &= \frac{(\widehat{x} - \zeta\widehat{y})^3}{(1-\zeta^2)^3} - 3\frac{(\widehat{x} - \zeta\widehat{y})}{(1-\zeta^2)^2} \\
He_{21}(\widehat{x},\widehat{y}) &= \frac{(\widehat{x} - \zeta\widehat{y})^2(\widehat{y} - \zeta\widehat{x})}{(1-\zeta^2)^3} + \frac{2\zeta(\widehat{x} - \zeta\widehat{y})}{(1-\zeta^2)^2} - \frac{(\widehat{y} - \zeta\widehat{x})}{(1-\zeta^2)^2} \,.
\end{aligned} \tag{1.167}$$

Note that $He_{m_1m_2}(\widehat{x},\widehat{y}) = He_{m_2m_1}(\widehat{y},\widehat{x})$. For completeness, we give here the bivariate cumulants K_{rs} for $r+s \leq 6$ in terms of the central moments $\mathcal{M}_{r'+s'}$ [Kendall and Stuart (1969)],

$$
\begin{aligned}
K_{30} &= \mathcal{M}_{30} \\
K_{21} &= \mathcal{M}_{21} \\
K_{40} &= \mathcal{M}_{40} - 3\mathcal{M}_{20}^2 \\
K_{31} &= \mathcal{M}_{31} - 3\mathcal{M}_{20}\mathcal{M}_{11} \\
K_{22} &= \mathcal{M}_{22} - \mathcal{M}_{20}\mathcal{M}02 - 2\mathcal{M}_{11}^2 \\
K_{50} &= \mathcal{M}_{50} - 10\mathcal{M}_{30}\mathcal{M}_{20} \\
K_{41} &= \mathcal{M}_{41} - 4\mathcal{M}_{30}\mathcal{M}_{11} - 6\mathcal{M}_{21}\mathcal{M}_{20} \\
K_{32} &= \mathcal{M}_{32} - \mathcal{M}_{30}\mathcal{M}_{02} - 6\mathcal{M}_{21}\mathcal{M}_{11} - 3\mathcal{M}_{20}\mathcal{M}_{12} \\
K_{60} &= \mathcal{M}_{60} - 15\mathcal{M}_{40}\mathcal{M}_{20} - 10\mathcal{M}_{30}^2 + 30\mathcal{M}_{20}^3 \\
K_{51} &= \mathcal{M}_{51} - 5\mathcal{M}_{40}\mathcal{M}_{11} - 10\mathcal{M}_{31}\mathcal{M}_{20} - 10\mathcal{M}_{30}\mathcal{M}_{21} + 30\mathcal{M}_{20}^2\mathcal{M}_{11} \\
K_{42} &= \mathcal{M}_{42} - \mathcal{M}_{40}\mathcal{M}_{02} - 8\mathcal{M}_{31}\mathcal{M}_{11} - 4\mathcal{M}_{30}\mathcal{M}_{12} - 6\mathcal{M}_{22}\mathcal{M}_{20} \\
&\quad -6\mathcal{M}_{21}^2 + 6\mathcal{M}_{20}^2\mathcal{M}_{02} + 24\mathcal{M}_{20}\mathcal{M}_{11}^2 \\
K_{33} &= \mathcal{M}_{33} - 3\mathcal{M}_{31}\mathcal{M}_{02} - \mathcal{M}_{30}\mathcal{M}_{03} - 9\mathcal{M}_{22}\mathcal{M}_{11} - 9\mathcal{M}_{21}\mathcal{M}_{12} \\
&\quad -3\mathcal{M}_{20}\mathcal{M}_{13} + 18\mathcal{M}_{20}\mathcal{M}_{11}\mathcal{M}_{02} + 12\mathcal{M}_{11}^3 \;.
\end{aligned}
\tag{1.168}
$$

Note that $K_{rs} \to K_{sr}$ with $\mathcal{M}_{r's'} \to \mathcal{M}_{s'r'}$. Similarly, $K_{20} = \sigma_{20}^2 = \mathcal{M}_{20}$ and $K_{02} = \sigma_{02}^2 = \mathcal{M}_{02}$. The reduced cumulants $k_{rs} = K_{rs}/[\{K_{20}\}^{r/2}\{K_{02}\}^{s/2}]$.

The conditional densities ρ_{12} and ρ_{21} are

$$
\rho_{12}(x_1|x_2) = \rho(x_1,x_2)/\rho_2(x_2)\;, \quad \rho_{21}(x_2|x_1) = \rho(x_1,x_2)/\rho_1(x_1)\;. \tag{1.169}
$$

where $\rho_1(x_1)$ and $\rho_2(x_2)$ are the marginal densities. The conditional moments $M_p(x_2)$ of ρ_{12} (similarly $M_p(x_1)$ of ρ_{21}) are,

$$
\begin{aligned}
M_p(x_2) &= \int_{-\infty}^{\infty} x_1^p\, \rho_{12}(x_1|x_2)\, dx_1 \\
&= \int_{-\infty}^{\infty} x_1^p\, \rho(x_1,x_2)\, dx_1 \;/\; \int_{-\infty}^{\infty} \rho(x_1,x_2)\, dx_1 \;.
\end{aligned}
\tag{1.170}
$$

An important result is that for the bivariate Gaussian defined by Eq. (1.161) the corresponding conditional densities will be Gaussians,

$$
\eta_{12:\mathcal{G}}(\widehat{x}|\widehat{y}) = \frac{1}{\sqrt{2\pi(1-\zeta^2)}}\, \exp\left\{-\frac{(\widehat{x}-\zeta\widehat{y})^2}{2(1-\zeta^2)}\right\} \;. \tag{1.171}
$$

Therefore, for a bivariate Gaussian, the conditional centroids are linear in the fixed variable with a slope given by the correlation coefficient. Similarly, the conditional width is a constant defined by the correlation coefficient and independent of the fixed variable. These results have been verified in many shell model examples for transition strengths ($E2$, $M1$ and GT) and form the basis for considering the bivariate Gaussian form as a good starting point for transition strength densities. Initial shell model tests are given in [Draayer, French and Wong (1977b); French, Kota, Pandey and Tomsovic (1988b*)] and further tests in [Kota and Sahu (1998); Kota *et al.* (1999); Gómez *et al.* (2001, 2004)]. Two remarks are appropriate here. Firstly, when the correlation coefficient is small, which will however not be so for most of shell model transition strength examples, the (exact) expansion

$$\eta_{\mathcal{G}}(\widehat{x},\widehat{y}) = \eta_{\mathcal{G}}(\widehat{x})\eta_{\mathcal{G}}(\widehat{y}) \sum_{\mu=0}^{\infty} \zeta^{\mu}\, P_{\mu}(\widehat{x}) P_{\mu}(\widehat{y})\,. \tag{1.172}$$

is useful. Here the $P_\mu(\widehat{x})$ are orthonormal (Hermite) polynomials defined by the corresponding density as weight function. The double polynomial theory for strengths (good when ζ is small) has been investigated in considerable detail [Draayer, French, Potbhare and Wong (1975a,b); Draayer, French and Wong (1977a*,b); Wong and Zhao (1982); Wong (1986)]. Secondly, under linear transformation of variables the form $\eta_{\mathcal{G}}(\widehat{x},\widehat{y})$ is unchanged, but not the parameters. Just as a mono-variate density is reduced to standard form ($\epsilon = 0, \sigma^2 = 1$) by using $z = (x - \epsilon_x)/\sigma_x$ so also is the case with the bivariate form where $z_1 = (x - \epsilon_x)/\sigma_x$ and $z_2 = (y - \epsilon_y)/\sigma_y$, in which the ζ has its original value, or with $(z_1 + z_2)/\sqrt{2(1+\zeta)}$ and $(z_1 - z_2)/\sqrt{2(1-\zeta)}$ which gives the standard form with zero correlation.

In the discussion of transition strengths, just as with the univariate t-distribution interpolating BW and Gaussian forms for strength functions, it is useful to consider the bivariate t-distribution [Kota, Chavda and Sahu (2006a)],

$$\begin{array}{l}
\rho_{biv-t}(E_i, E_f; \epsilon_i, \epsilon_f, \sigma_1, \sigma_2, \zeta; \nu) = \frac{1}{2\pi\sigma_1\sigma_2\sqrt{1-\zeta^2}} \\
\times \left[1 + \frac{1}{\nu(1-\zeta^2)}\left\{\left(\frac{E_i-\epsilon_i}{\sigma_1}\right)^2 - 2\zeta\left(\frac{E_i-\epsilon_i}{\sigma_1}\right)\left(\frac{E_f-\epsilon_f}{\sigma_2}\right) + \left(\frac{E_f-\epsilon_f}{\sigma_2}\right)^2\right\}\right]^{-\frac{\nu+2}{2}}\,, \quad \nu \geq 1\,.
\end{array} \tag{1.173}$$

Properties of ρ_{biv-t} are given in [Johnson and Kotz (1972); Hutchinson and Lai (1990)]. Most important is that for $\nu = 1$, ρ_{biv-t} gives a bivariate Cauchy (i.e. bivariate BW) distribution [Hutchinson and Lai (1990); Gideon and Rothan (2004)] and as $\nu \to \infty$, ρ_{biv-t} becomes a bivariate Gaussian. Secondly, the marginal distributions of ρ_{biv-t} are easily seen to be univariate t-distributions, with ν degrees of freedom, independent of ζ with univariate Cauchy (BW) distribution for $\nu = 1$ and Gaussian as $\nu \to \infty$. Thus, as the parameter ν changes from 1 to ∞, ρ_{biv-t} changes from bivariate BW to bivariate Gaussian. Assuming that the transition strength densities follow strength functions (they change from BW to Gaussian), ρ_{biv-t} defined by Eq. (1.173) has the correct limiting form and the intermediate shapes (for fixed ζ) are determined by the ν parameter. In Eq. (1.173), ϵ_i and ϵ_f are the centroids of the two marginals of ρ_{biv-t}. However, σ_1 and σ_2 will approach the marginal widths σ_i and σ_f only in the limit $\nu \to \infty$, i.e. for the bivariate Gaussian. In fact, the second central moments $\mu_{20} = \sigma_i^2$ and $\mu_{02} = \sigma_f^2$ are related to

σ_1^2 and σ_2^2 by $\mu_{20} = \frac{\nu}{\nu-2}\,\sigma_1^2$ and $\mu_{02} = \frac{\nu}{\nu-2}\,\sigma_2^2$ for $\nu > 2$. However, ζ remains to be the bivariate correlation coefficient. Exceptions will occur for $\nu \leq 2$ and then the spreading widths of the marginal densities define σ_1, σ_2 and other parameters. Finally, the fourth order cumulants for ρ_{biv-t} are $k_{40} = 6(3\nu-4)/(\nu-2)(\nu-4)$, $k_{31} = \zeta k_{40}$ and $k_{22} = (1/3)(1+2\zeta^2)k_{40}$.

1.7.3 *NIP strength densities: Convolution form*

Let us first define the transition operator $\mathcal{O}(k', k)$ for a system of m fermions in N sp states. We denote by A_{μ_i} the operator that creates a fermion in the μ_i-th sp state and similarly $B_{\mu_i} = (A_{\mu_i})^\dagger$ is the corresponding annihilation operator. With these, a k' particle creation operator $\phi_\mu(k')$ and a k particle annihilation operator $\phi_\nu^\dagger(k)$ are,

$$\begin{aligned} &\phi_\mu(k') = A_\mu = A_{\mu_1}A_{\mu_2}\ldots A_{\mu_{k'}}\;; \quad \mu_1 < \mu_2 < \ldots \mu_{k'} \\ &\phi_\nu^\dagger(k) = (A_{\nu_1}A_{\nu_2}\ldots A_{\nu_k})^\dagger = B_\nu = B_{\nu_k}B_{\nu_{k-1}}\ldots B_{\nu_1}; \quad \nu_1 < \nu_2 < \ldots \nu_k\;. \end{aligned} \tag{1.174}$$

In Eq. (1.174), $\mu = \mu_1, \mu_2, \ldots, \mu_{k'}$ and similarly $\nu = \nu_1, \nu_2, \ldots, \nu_k$. Now any general (k', k)-body operator (for $k' = k$, it is called just k-body operator and, for $k \neq k'$ it is number non-conserving) is defined as,

$$\mathcal{O}(k', k) = \sum_{\mu,\nu} C_{\mu,\nu}^{\mathcal{O}(k',k)}\; \phi_\mu(k')\phi_\nu^\dagger(k)\;. \tag{1.175}$$

Then its hermitian adjoint is,

$$\mathcal{O}^\dagger(k', k) = \sum_{\mu,\nu} \left[C_{\mu,\nu}^{\mathcal{O}(k',k)}\right]^*\; \phi_\nu(k)\phi_\mu^\dagger(k')\;. \tag{1.176}$$

In Eqs. (1.175,1.176), the $C_{\mu,\nu}$ are the structure factors that define the transition operator $\mathcal{O}(k', k)$; see ahead for examples. For later use, let us define,

$$\begin{aligned} \Phi_{\mu\nu} &= \phi_\mu(k')\phi_\nu^\dagger(k) = A_\mu B_\nu\;, \\ \Phi_{\mu\nu}^\dagger &= \phi_\nu(k)\phi_\mu^\dagger(k') = A_\nu B_\mu\;. \end{aligned} \tag{1.177}$$

With these, the NIP transition strength density can be written as a trace,

$$\mathrm{I}^{h,m:\mathcal{O}(k',k)}(x,y) = \left\langle\left\langle [\mathcal{O}(k',k)]^\dagger\; \delta(h-y)\; \mathcal{O}(k',k)\; \delta(h-x)\right\rangle\right\rangle_N^m\;. \tag{1.178}$$

Using the configurations $(m_1, m_2, \ldots, m_N)$ where m_1 is number of fermions (0 or 1) in state #1, m_2 is number of fermions in state #2 etc., it is easily seen that, for $k \neq k'$, there can only be $|C_{\mu\nu}|^2$ terms in $\mathrm{I}^{h,m}(x,y)$. Note that the $C_{\mu\mu}$ terms will appear for transition operators with $k = k'$; see [Kota and Majumdar (1995a)] for an example. Simplifying Eq. (1.178) using Eqs. (1.175, 1.177), we get

$$\begin{aligned} \mathrm{I}^{h,m:\mathcal{O}(k',k)}(x,y) &= \sum_{\mu,\nu} \left|C_{\mu\nu}^{\mathcal{O}(k',k)}\right|^2 \mathrm{I}_{\mu,\nu}^{h,m}(x,y)\;, \\ \mathrm{I}_{\mu,\nu}^{h,m}(x,y) &= \left\langle\left\langle \Phi_{\mu\nu}^\dagger \delta(h-y)\Phi_{\mu\nu}\delta(h-x)\right\rangle\right\rangle_N^m\;. \end{aligned} \tag{1.179}$$

For further simplification of $\mathrm{I}^{h,m}(x,y)$ we need some basic results which are discussed next.

Given the number operator for the rth sp state $n_r = A_r B_r$, first we have $(n_r)^p = n_r$ for $p > 0$ as the eigenvalue of n_r is 0 or 1. Then $e^{it\epsilon_r n_r} = 1 + n_r\{e^{it\epsilon_r} - 1\}$. Similarly $A_t n_r = (n_r - \delta_{tr})A_t$, $A_r n_r = (n_r - 1)A_r$, $B_t n_s = (n_s + \delta_{st})B_t$ and $B_s n_s = (n_s + 1)B_s$. Now, given $f(n_r)$, a polynomial in n_r, we have $f(n_r) = \sum_p c_p (n_r)^p = A_0 + B_0 n_r$. This gives the important relations,

$$\begin{aligned} A_r f(n_r) &= f(n_r - 1)A_r \;, \quad B_r f(n_r) = f(n_r + 1)B_r \\ A_s f(n_r) &= f(n_r)A_s \;, \quad B_s f(n_r) = f(n_r)B_s \;, \quad \text{for } r \neq s \;. \end{aligned} \tag{1.180}$$

Eq. (1.180) gives an important identity for $A_\mu B_\nu \delta(h-y)$ where $h = \sum_i \epsilon_i n_i$. Note that $B_\nu h = [\sum_{i\in\nu} \epsilon_i(n_i+1) + \sum_{i\notin\nu} \epsilon_i n_i]B_\nu = [h+\epsilon_\nu]B_\nu$. Here $\epsilon_\nu = \sum_{i=\nu_1,\nu_2,\ldots,\nu_k} \epsilon_i$. Similarly $\epsilon_\mu = \sum_{i=\mu_1,\mu_2,\ldots,\mu_{k'}} \epsilon_i$ and $A_\mu h = [h - \epsilon_\mu]A_\mu$. Therefore,

$$A_\mu B_\nu \delta(h-y) = \delta(h - y + \epsilon_\nu - \epsilon_\mu)A_\mu B_\nu \;. \tag{1.181}$$

It is easy to see that,

$$\begin{aligned} \mathrm{I}^{h,m}_{\mu,\nu}(x,y) &= \langle\langle A_\nu B_\mu \delta(h-y) A_\mu B_\nu \delta(h-x)\rangle\rangle^m_N \\ &= \langle\langle \delta(h - y + \epsilon_\mu - \epsilon_\nu) A_\nu B_\mu A_\mu B_\nu \delta(h-x)\rangle\rangle^m_N \;. \end{aligned} \tag{1.182}$$

In the trace in Eq. (1.182) only configurations with centroid $\sum_i \epsilon_i = x$ survive because of $\delta(h-x)$. With this we pull $\delta(h - y + \epsilon_\mu - \epsilon_\nu)$ outside the trace giving,

$$\mathrm{I}^{h,m}_{\mu,\nu}(x,y) = \delta(x - y + \epsilon_\mu - \epsilon_\nu)\, \langle\langle A_\nu B_\mu A_\mu B_\nu \delta(h-x)\rangle\rangle^m_N \;. \tag{1.183}$$

For further simplification of $\mathrm{I}^{h,m}_{\mu,\nu}$, it is useful to consider the Venn diagrams involving the sp states in μ and ν. From Fig. 1.18 we have $\mu = \overline{\mu} + \overline{\zeta}$ and $\nu = \overline{\nu} + \overline{\zeta}$, i.e. the states in $\overline{\zeta}$ are common to both μ and ν. Then,

$$\begin{aligned} A_\nu B_\mu A_\mu B_\nu &= A_{\overline{\zeta}} A_{\overline{\nu}} B_\mu A_\mu B_{\overline{\zeta}} B_{\overline{\nu}} \\ &= A_{\overline{\zeta}} A_{\overline{\nu}} B_{\overline{\mu}} B_{\overline{\zeta}} A_{\overline{\mu}} A_{\overline{\zeta}} B_{\overline{\zeta}} B_{\overline{\nu}} \\ &= B_{\overline{\mu}} A_{\overline{\mu}} A_{\overline{\zeta}} B_{\overline{\zeta}} A_{\overline{\zeta}} B_{\overline{\zeta}} A_{\overline{\nu}} B_{\overline{\nu}} \\ &= B_{\overline{\mu}} A_{\overline{\mu}} A_\nu B_\nu \\ &= \prod_{\mu_i\in\mu-(\mu\cap\nu)} (1 - n_{\mu_i}) \prod_{\nu_i\in\nu} n_{\nu_i} \;. \end{aligned} \tag{1.184}$$

In the fourth equality in Eq. (1.184) we have used the results $A_{\overline{\zeta}} B_{\overline{\zeta}} = \prod_{i\in\overline{\zeta}} n_i$ and $n_i^2 = n_i$. Then the last equality is obvious. Fig. 1.18 shows $\mu - (\mu\cap\nu)$ that appears in Eq. (1.184). Applying Eq. (1.184) to Eq. (1.183) gives,

$$\mathrm{I}^{h,m}_{\mu,\nu}(x,y) = \delta(x - y + \epsilon_\mu - \epsilon_\nu) \left\langle\left\langle \prod_{\mu_i\in\mu-(\mu\cap\nu)} (1 - n_{\mu_i}) \prod_{\nu_i\in\nu} n_{\nu_i} \delta(h-x) \right\rangle\right\rangle^m_N \;. \tag{1.185}$$

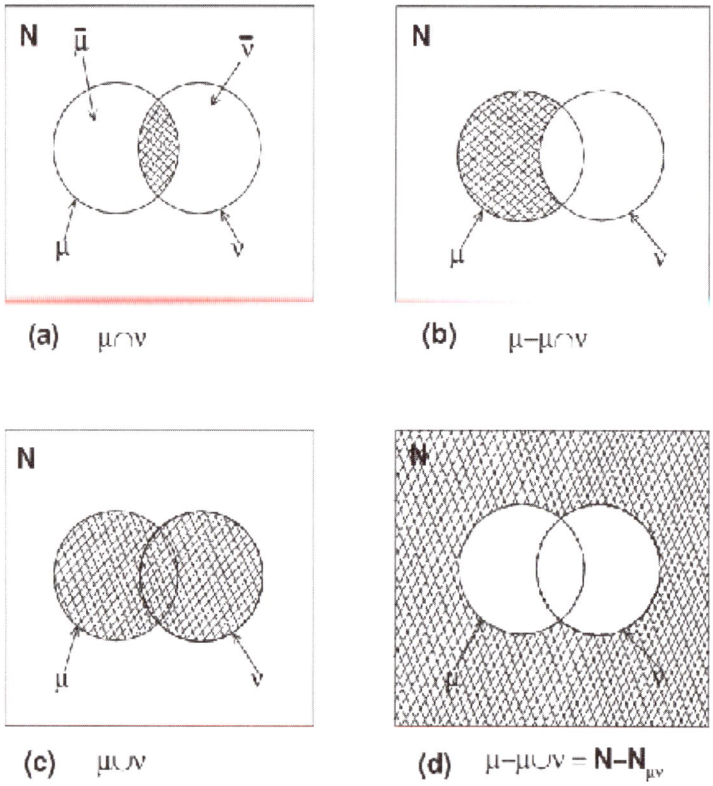

Fig. 1.18 Venn diagrams for the total number of sp states N and for the sp states in μ and ν. Shaded portions in the figure are: (a) $\overline{\zeta} = \mu \cap \nu$; (b) $\mu - \mu \cap \nu$; (c) $\mu \cup \nu$; (d) $N - \mu \cup \nu = N - N_{\mu\nu}$.

To proceed further we consider the trace $\langle\langle n_i \delta(h-x)\rangle\rangle^m_N$. The action of n_i on any configuration $\mathbf{m} = (m_1, m_2, \ldots, m_N)$ is $n_i \left|\mathbf{m}\right\rangle = 0$ if $m_i = 0$ and 1 if $m_i = 1$. Therefore in the trace we need consider only those $\mathbf{m}$ that have $m_i = 1$. This is easily achieved by distributing $m-1$ particles in $N - 1_i$ sp states. Here $N - 1_i$ denotes $(N-1)$ sp states with the ith state removed. Then,

$$\langle\langle n_i \delta(h-x)\rangle\rangle^m_N = \langle\langle \delta(h - [x - \epsilon_i])\rangle\rangle^{m-1}_{N-1_i} \tag{1.186}$$

and similarly

$$\langle\langle (1-n_i)\delta(h-x)\rangle\rangle^m_N = \langle\langle \delta(h-x)\rangle\rangle^m_{N-1_i} \quad . \tag{1.187}$$

Here we have used the results $(1-n_i)\left|\mathbf{m}\right\rangle = 0$ if $m_i = 1$ and 1 if $m_i = 0$. Hence for a non-zero result all the m particles should be in $N - 1_i$ states. Using Eqs. (1.186) (1.187) we have,

$$\begin{aligned}
&\left\langle\left\langle \prod_{\mu_i \in \mu - (\mu\cap\nu)} (1-n_{\mu_i}) \prod_{\nu_i \in \nu} n_{\nu_i} \delta(h-x) \right\rangle\right\rangle^m_N \\
&= \langle\langle \delta(h - x + \epsilon_\nu)\rangle\rangle^{m-k}_{N-\nu-(\mu-\mu\cap\nu)} \\
&= \langle\langle \delta(h - x + \epsilon_\nu)\rangle\rangle^{m-k}_{N-\mu\cup\nu} \quad .
\end{aligned} \tag{1.188}$$

Figure 1.18 shows $\mu \cup \nu$ and $N - \mu \cup \nu$. Using the simpler notation $N_{\mu\nu} = \mu \cup \nu$ and noting that the trace on the R.H.S. of Eq. (1.188) is an NIP state density, gives the compact result

$$\begin{aligned} \mathrm{I}^{h,m}_{\mu,\nu}(x,y) &= \delta(x-y+\epsilon_\mu-\epsilon_\nu)\,\langle\langle\delta(h-x+\epsilon_\nu)\rangle\rangle^{m-k}_{N-N_{\mu\nu}} \\ &= \delta(x-y+\epsilon_\mu-\epsilon_\nu)\,I^{h,m-k}_{N-N_{\mu\nu}}(x-\epsilon_\nu)\ . \end{aligned} \tag{1.189}$$

Introducing the basic bivariate density $S^{h,(k',k)}_{\mu\nu}$ in the defining (k',k) space and an auxiliary m-particle bivariate density $T^{h,m}_{N'}$ where,

$$\begin{aligned} S^{h,(k',k)}_{\mu\nu}(u,v) &= \delta(u-\epsilon_\nu)\delta(v-\epsilon_\mu) \\ T^{h,m}_{N'}(u,v) &= \delta(u-v)I^{h,m}_{N'}(u)\ , \end{aligned} \tag{1.190}$$

it is possible to write $\mathrm{I}^{h,m}_{\mu,\nu}(x,y)$ in a convolution form,

$$\begin{aligned} \mathrm{I}^{h,m:\mathcal{O}(k',k)}(x,y) &= \sum_{\mu,\nu}\left|C^{\mathcal{O}(k',k)}_{\mu\nu}\right|^2 \mathrm{I}^{h,m}_{\mu,\nu}(x,y)\ ; \\ \mathrm{I}^{h,m}_{\mu,\nu}(x,y) &= S^{h,(k',k)}_{\mu\nu}\otimes T^{h,m-k}_{N-N_{\mu\nu}}[x,y]\ . \end{aligned} \tag{1.191}$$

It is easily seen that Eq. (1.189) reduces to the convolution form for $\mathrm{I}^{h,m}_{\mu,\nu}(x,y)$ given in Eq. (1.191).

Eq. (1.191) is exact and reduces to a simpler form under the approximation $N - N_{\mu\nu} \sim N$. Then the (μ,ν) summation in Eq. (1.191) can be carried out independent of the T density. This gives a pleasing form for $\mathrm{I}^{h,m}(x,y)$,

$$\begin{aligned} \mathrm{I}^{h,m:\mathcal{O}(k',k)}(x,y) &= S^{h,\mathcal{O}(k',k)}\otimes T^{h,m-k}_{N}[x,y]\ ; \\ S^{h,\mathcal{O}(k',k)}(u,v) &= \sum_{\mu,\nu}\left|C^{\mathcal{O}(k',k)}_{\mu\nu}\right|^2 \delta(u-\epsilon_\nu)\delta(v-\epsilon_\mu)\ . \end{aligned} \tag{1.192}$$

Note that $S^{h,\mathcal{O}(k',k)}$ is the bivariate strength density generated by $\mathcal{O}$ in the defining (k',k) space. It is often easy to construct the NIP state density T either exactly or approximately. The approximation leading to Eq. (1.192) is not yet well tested; see however [French, Pandey and Smith (1987*)]. Note that Eq. (1.186) gives a method for calculating NIP occupancy and spin cutoff densities.

For $k' = k$, when $k << m$ only a small fraction of the particles ($\sim k/m$) will be transferred by $\mathcal{O}$ and therefore, in the h diagonal basis, the matrix elements of $\mathcal{O}$ usually lie in a narrow diagonal band; this is seen in NIP [Kota and Majumdar (1995a)] and also in IP examples [French, Kota, Pandey and Tomsovic (1988b*)]. For $k = 0$ or $k' = 0$, appropriate for single and two nucleon transfer, Eq. (1.191) reduces to a simple form,

$$\begin{aligned} \mathrm{I}^{h,m:\mathcal{O}(k',0)}(x,y) &= \sum_{\mu}\left|C^{\mathcal{O}(k',0)}_{\mu,0}\right|^2 \delta(x-y+\epsilon_\mu)\,T^{h,m}_{N-N_\mu}(x)\ , \\ \mathrm{I}^{h,m:\mathcal{O}(0,k)}(x,y) &= \sum_{\nu}\left|C^{\mathcal{O}(0,k)}_{0,\nu}\right|^2 \delta(x-y-\epsilon_\nu)\,T^{h,m-k}_{N-N_\nu}(x-\epsilon_\nu)\ . \end{aligned} \tag{1.193}$$

In the (p, n) formalism, the $C_{\mu\nu}$ coefficients carry the p, n indices. Similarly, they carry JM and/or TM_T indices where appropriate.

Although Eq. (1.191) is exact, it is cumbersome to apply in practice as the formulation uses single state representation. For real nuclei, in the shell model description, nucleons will be in spherical sp orbits and the transition operators will have a definite spherical tensorial rank. In this situation, if we use spherical configurations **m**, exact results follow by applying Eq. (1.185) or otherwise. We will turn to these by considering three examples: (i) $k' = 1$, $k = 0$ or $k' = 0$, $k = 1$; (ii) $k' = k = 1$; (iii) $k' = k = 2$. In all the discussion we will be using implicitly, with conventions as given before, the (p, n) formalism. Before turning to these, it is worthwhile discussing the asymptotic form for the NIP strength densities.

1.7.4 *Bivariate Gaussian form for NIP strength densities*

In order to understand the general form of the NIP strength densities, a plausible model is to assume that the matrix in the defining space for $\mathcal{O}$ is a GOE and this amounts to assuming that $|C_{\mu\nu}|^2$ are zero center Gaussian variables with unit variance. Applying Eq. (1.192) and carrying out an ensemble average (then $\overline{|C_{\mu\nu}|^2} = 1$), the S density in Eq. (1.192) reduces to product of NIP state densities in the initial and final energies. As bivariate cumulants add under convolution, we have easily,

$$K_{pq}(m) = K_{p+q}(m - k') + K_p(k')\delta_{q0} + K_q(k)\delta_{p0} \ . \tag{1.194}$$

Here K_r are NIP state density cumulants; note that the first term is from the T density and the last two from the defining space S density. Applying the dilute limit with large m, we have $K_r(m) = mK_r(1)$; here we are essentially ignoring Pauli corrections which are small in the limit $N, m >> 1$ and $m/N \to 0$. Then $\sigma_{20}^2(m) = mK_2(1)$ and $\sigma_{02}^2 = (m - k' + k)K_2(1)$. With these, the reduced cumulants k_{pq} are

$$k_{pq}(m) = \frac{\{(m - k')\,k_{p+q}(1) + k'\,k_p(1)\delta_{q0} + k\,k_q(1)\delta_{p0}\}}{\{m^{p/2}(m + k - k')^{q/2}\}} \ . \tag{1.195}$$

Important results that follow from Eq. (1.195) are that the bivariate correlation coefficient of ζ,

$$\zeta = \frac{(m - k')}{\sqrt{m(m + k - k')}} = 1 - \frac{k + k'}{2m} + O(1/m^2) \tag{1.196}$$

goes to unity and the cumulants k_{pq} for $p + q \geq 3$ go to zero in the dilute limit giving a bivariate Gaussian form. Note that [with $\gamma_1(1) = 0$ and $\gamma_2(1) \to -1$ for sp energies drawn from a GOE] the k_{pq} for $p+q=4$ are,

$$\begin{aligned}
k_{40} &= \frac{1}{m}\,\gamma_2(1) \\
k_{31} &= \frac{m-k'}{m^{3/2}(m+k-k')^{1/2}}\,\gamma_2(1) \longrightarrow \frac{\gamma_2(1)}{m}\left[1-\frac{k+k'}{m}+O(\frac{1}{m^2}\right] \\
k_{22} &= \frac{m-k'}{m(m+k-k')}\,\gamma_2(1) \longrightarrow \frac{\gamma_2(1)}{m}\left[1-\frac{k}{m}+O(\frac{1}{m^2}\right] \\
k_{13} &= \frac{m-k'}{m^{1/2}(m+k-k')^{3/2}}\,\gamma_2(1) \longrightarrow \frac{\gamma_2(1)}{m}\left[1-\frac{3k-k'}{2m}+O(\frac{1}{m^2}\right] \\
k_{04} &= \frac{1}{m+k-k'}\,\gamma_2(1) \longrightarrow \frac{\gamma_2(1)}{m}\left[1-\frac{k-k'}{m}+O(\frac{1}{m^2}\right] .
\end{aligned} \tag{1.197}$$

As the distribution will be singular for $\zeta=1$, in reality we have to consider finite m effects, i.e. the cumulants correction to the bivariate Gaussian form. To see this important feature more clearly, we consider the bivariate moments/cumulants in the variables $X=(E_i+E_f)$ and $Y=(E_i-E_f)$, i.e. in the center of mass and relative variables. Then clearly $\sigma_X^2=2(1+\zeta)$ and $\sigma_Y^2=2(1-\zeta)$. These and the k_{rs} given by Eq. (1.197) will determine the reduced central moments $\mathcal{M}_{pq}^{XY}$ and the corresponding cumulants k_{pq}^{XY} for $p+q=4$. As an example, consider $\mathcal{M}_{40}^{XY}$,

$$\begin{aligned}
\mathcal{M}_{40}^{XY} &= \frac{\langle (X+Y)^4\rangle}{4(1+\zeta)^2} \\
&= \frac{\langle X^4+Y^4+4X^3Y+4XY^3+6X^2Y^2\rangle}{4(1+\zeta)^2} \\
&= \frac{k_{40}+k_{04}+6+4(k_{31}+k_{13}+6\zeta)+6(k_{22}+1+2\zeta^2)}{4(1+\zeta)^2}\,, \\
\Longrightarrow k_{40}^{XY} &= \frac{(k_{40}+k_{04})+4(k_{31}+k_{13})+6k_{22}}{4(1+\zeta)^2} \\
&= \frac{\gamma_2(1)}{m}+O(\frac{1}{m^2})\,.
\end{aligned} \tag{1.198}$$

Similarly we have,

$$\begin{aligned}
\mathcal{M}_{31}^{XY} &= \frac{(k_{40}-k_{04})+2(k_{31}-k_{13})}{4(1+\zeta)^{3/2}(1-\zeta)^{1/2}} \rightarrow k_{31}^{XY} = \frac{\gamma_2(1)}{m^{3/2}}\left[\frac{3(k-k')}{8(k+k')^{1/2}}\right]+O(\frac{1}{m^{5/2}}) \\
\mathcal{M}_{22}^{XY} &= \frac{(k_{40}+k_{04})-2k_{22}}{4(1-\zeta^2)}+1 \rightarrow k_{22}^{XY} = \frac{\gamma_2(1)}{4m}+O(\frac{1}{m^2}) \\
\mathcal{M}_{13}^{XY} &= \frac{(k_{40}-k_{04})-2(k_{31}-k_{13})}{4(1+\zeta)^{1/2}(1-\zeta)^{3/2}} \rightarrow k_{13}^{XY} = \frac{\gamma_2(1)}{m^{1/2}}\left[\frac{3(k'-k)}{2(k+k')^{3/2}}\right]+O(\frac{1}{m^{3/2}}) \\
\mathcal{M}_{04}^{XY} &= \frac{(k_{40}+k_{04})-4(k_{31}+k_{13})+6k_{22}}{4(1-\zeta)^2}+3 \rightarrow k_{04}^{XY} = \frac{\gamma_2(1)}{k+k'}+O(\frac{1}{m})\,.
\end{aligned} \tag{1.199}$$

The cumulants k_{pq}^{XY} with $p+q=4$ show that there will be departures from a true Gaussian connected with the narrowly constrained Y variable; $k_{04}^{XY}=\gamma_2(1)/(k+k')$

will go to zero only for sufficiently large values of $(k+k')$. Thus strength densities will never be perfect bivariate Gaussians and this is unlike the situation with state densities that approach Gaussian form in the asymptotic limit. However, this minor deviation found even for IP systems may perhaps not be of much consequence.

1.7.5 *Spherical configuration NIP strength densities*

1.7.5.1 $k'=1,\ k=0$ *or* $k'=0,\ k=1$ *operators*

Consider a single particle creation operator $\mathcal{O} = a^\dagger_{j_\alpha m_\alpha}$, a $k'=1$, $k=0$ operator; note that α also includes p or n index and all configurations $\mathbf{m}$ in this Section refer to (p,n) configurations. Now the transition strength density is,

$$
\begin{aligned}
I^{h,m:a^\dagger_{j_\alpha}}(E_i,E_f) &= (2j_\alpha+1)^{-1} \sum_{m_\alpha} \left|\left\langle E_f \mid a^\dagger_{j_\alpha m_\alpha} \mid E_i \right\rangle\right|^2\ I^{m+1}(E_f) I^m(E_i) \\
&= \frac{1}{N_\alpha} \sum_{\mathbf{m}_i,\gamma_i,\mathbf{m}_f,\gamma_f,m_\alpha} \left|\left\langle \mathbf{m}_f\gamma_f \mid a^\dagger_{j_\alpha m_\alpha} \mid \mathbf{m}_i\gamma_i \right\rangle\right|^2\ \delta(E_i-\epsilon(\mathbf{m}_i))\delta(E_f-\epsilon(\mathbf{m}_f)) \\
&= \sum_{\mathbf{m}} \left(1-\frac{m_\alpha(\mathbf{m})}{N_\alpha}\right)\ d(\mathbf{m})\ \delta(E_i-\epsilon(\mathbf{m}))\ \delta(\epsilon(E_f-\epsilon[\mathbf{m}_f=\mathbf{m}\times 1^+_\alpha])\ .
\end{aligned}
\tag{1.200}
$$

In Eq. (1.200), N_α is the degeneracy of orbit α and $m_\alpha(\mathbf{m})$ gives the number of particles in orbit α for the m-particle configuration $\mathbf{m}$. Note that $m_f = m+1$ and $\mathbf{m}_f$ in the last step is uniquely defined by $\mathbf{m}$ and α; $\mathbf{m}_f = \mathbf{m}\times 1^\alpha$, i.e. m_α of $\mathbf{m}$ changes to $m_\alpha+1$. Also it is useful to note that, symbolically, $|\langle \mathbf{m}_f \mid -- \mid \mathbf{m}_i\rangle|^2 = \langle \mathbf{m} \mid aa^\dagger \mid \mathbf{m}\rangle = \langle (1-a^\dagger a)\rangle$ and this gives the final result. Extension of Eq. (1.200) for $\mathcal{O} = a_{j_\alpha m_\alpha}$ is immediate,

$$
I^{h,m:a_{j_\alpha}}(E_i,E_f) = \sum_{\mathbf{m}} \frac{m_\alpha(\mathbf{m})}{N_\alpha}\ d(\mathbf{m})\ \delta(E_i-\epsilon(\mathbf{m}))\ \delta(E_f-\epsilon[\mathbf{m}_f:\mathbf{m}=\mathbf{m}_f\times 1^+_\alpha])\ .
\tag{1.201}
$$

1.7.5.2 $(k'=1,k=1)$ *or one-body transition operators*

A general one-body transition operator $\mathcal{O}^k_\mu$ is defined by,

$$
\begin{aligned}
\mathcal{O}^k_\mu &= \sum \epsilon^k_{\alpha\beta}\ \left(a^\dagger_{\alpha_x}\tilde{a}_{\beta_y}\right)^k_\mu = \sum \epsilon^k_{\alpha_x\beta_y}\ \mathcal{O}^k_{\alpha_x\beta_y;\mu} \\
(\mathcal{O}^k_\mu)^\dagger &= \sum (-1)^{j_\alpha-j_\beta}\ \epsilon^k_{\alpha_x\beta_y}\ (-1)^\mu\ \left(a^\dagger_{\beta_y}\tilde{a}_{\alpha_y}\right)^k_{-\mu}\ .
\end{aligned}
\tag{1.202}
$$

The (x,y) in Eq. (1.202) label p and n. For identical nucleon systems, $x=y=p$ or n. For electromagnetic operators, $\mathcal{O}$ is the sum of the $\mathcal{O}$ operators for p and n, i.e. $x=y$ with $x=p$ and n in the sum in Eq. (1.202). For the GT(β^-) operator, $x=p$ and $y=n$ and similarly for GT(β^+) operator, $x=n$ and $y=p$. As an example for $\epsilon^1_{\alpha_x\beta_y}$ Eq. (1.202) we consider the GT(β^-) operator. Let us say that for the nuclei under consideration, protons are in single particle (sp) orbits j^p and similarly neutrons in j^n. Using the usual assumption that the radial part of the sp states are those of the harmonic oscillator, the proton sp states are completely specified by $(\mathbf{n}^p, \ell^p, j^p)$ with $\mathbf{n}^p$ denoting oscillator radial quantum number so that for an oscillator shell $\mathcal{N}^p$, $2\mathbf{n}^p + \ell^p = \mathcal{N}^p$. Similarly the neutron sp states are $(\mathbf{n}^n, \ell^n, j^n)$. Then $\mathrm{GT}(\beta^-) = (\sigma\tau^+)^1_\mu$ where,

$$
\begin{gathered}
(\sigma\tau^+)^1_\mu = \sum_{j_1^p, j_2^n} \epsilon^1_{j_1^p, j_2^n} \left(a^\dagger_{j_1^p} \tilde{a}_{j_2^n}\right)^1_\mu , \\
\epsilon^1_{(\mathbf{n}_1^p \ell_1^p j_1^p),(\mathbf{n}_2^n \ell_2^n j_2^n)} \stackrel{j_1^p = j_2^n}{\longrightarrow} \delta_{\ell_1^p \ell_2^n} \, \delta_{\mathbf{n}_1^p \mathbf{n}_2^n} \sqrt{\frac{2j_1^p+1}{3}} \, \frac{j_1^p(j_1^p+1) + 3/4 - \ell_1^p(\ell_1^p+1)}{\sqrt{j_1^p(j_1^p+1)}} , \\
\epsilon^1_{(\mathbf{n}_1^p \ell_1^p j_1^p),(\mathbf{n}_2^n \ell_2^n j_2^n)} \stackrel{j_1^p \neq j_2^n}{\longrightarrow} \delta_{\ell_1^p \ell_2^n} \, \delta_{\mathbf{n}_1^p \mathbf{n}_2^n} (\pm) \sqrt{\frac{8\ell_1^p(\ell_1^p+1)}{3(2\ell_1^p+1)}} .
\end{gathered}
\tag{1.203}
$$

In Eq. (1.203) we have $\pm$ for $j_1^p = \ell_1^p \mp \frac{1}{2}$ and $j_2^n = \ell_1^p \pm \frac{1}{2}$. The exact NIP strength density, using spherical configurations, is

$$
\begin{aligned}
\mathrm{I}^{h,m:\mathcal{O}^k}(E_i, E_f) = {} & (2k+1)^{-1} \sum_{\mathbf{m}_i, \mathbf{m}_f, \mu} \; \sum_{\gamma_i \in \mathbf{m}_i, \gamma_f \in \mathbf{m}_f} \\
& \left| \left\langle \mathbf{m}_f \gamma_f \mid \mathcal{O}^k_\mu \mid \mathbf{m}_i \gamma_i \right\rangle \right|^2 \delta(E_i - \epsilon(\mathbf{m}_i)) \, \delta(E_f - \epsilon(\mathbf{m}_f)) .
\end{aligned}
\tag{1.204}
$$

After some algebra, using Eq. (1.202), we obtain a simple expression,

$$
\begin{aligned}
\mathrm{I}^{h,m:\mathcal{O}^k}(E_i, E_f) = {} & \sum_{\mathbf{m}} d(\mathbf{m}) \sum_{\alpha,\beta} \left\{\epsilon^k_{\alpha\beta}\right\}^2 \frac{m_\beta(\mathbf{m})(N_\alpha - m_\alpha(\mathbf{m}))}{N_\beta(N_\alpha - \delta_{\alpha\beta})} \\
& \times \delta(E_i - \epsilon(\mathbf{m})) \, \delta\left(E_f - \epsilon\left[\mathbf{m}_f = \mathbf{m} \times (1^\dagger_\alpha 1_\beta)\right]\right) .
\end{aligned}
\tag{1.205}
$$

Note that $\mathbf{m}_f$ is obtained from $\mathbf{m}$ configuration by removing a particle from the orbit β and creating a particle in the orbit α. Eq. (1.205) extends easily to (p,n) spherical configurations; see [Kota and Majumdar (1995a)] for further discussion and examples.

1.7.5.3 *($k' = 2, k = 2$) or two-body transition operators*

Consider the two-body transition operator, the double Gamow-Teller (dGT) operator $\mathcal{O}^{dGT}(2)$,

$$
\mathcal{O}^{dGT}(2) = \sqrt{3} \left[(\sigma\tau^+) \times (\sigma\tau^+)\right]^0 . \tag{1.206}
$$

With the normalized two-particle creation operator $A^J_\mu(j_1 j_2) = (1 + \delta_{j_1 j_2})^{-1/2} (a^\dagger_{j_1} a^\dagger_{j_2})^J_\mu$, the transition operator $\mathcal{O}^{dGT}(2))$ in standard form, with V's denoting two-body matrix elements, is

$$
\mathcal{O}^{dGT}(2)) = \sum_{j_1^p \geq j_2^p \,;\, j_3^n \geq j_4^n \,;\, J_0} V^{J_0}_{j_1^p j_2^p : j_3^n j_4^n}(dGT) \sum_\mu A^{J_0}_\mu(j_1^p j_2^p) \left\{A^{J_0}_\mu(j_3^n j_4^n)\right\}^\dagger .
\tag{1.207}
$$

Applying the $9-j$ transformation to $(a^\dagger_{j_1^p}\tilde{a}_{j_2^n})^1 \cdot (a^\dagger_{j_3^p}\tilde{a}_{j_4^n})^1$, we obtain

$$\begin{aligned} V^{J_0}_{j_1^p j_2^p : j_3^n j_4^n}(dGT) = {} & v^{J_0}_{j_1^p j_2^p : j_3^n j_4^n} - (-1)^{j_1^p + j_2^p + J_0}(1-\delta_{j_1^p j_2^p})\, v^{J_0}_{j_2^p j_1^p : j_3^n j_4^n} \\ & - (-1)^{j_3^n + j_4^n + J_0}(1-\delta_{j_3^n j_4^n})\, v^{J_0}_{j_1^p j_2^p : j_4^n j_3^n} \\ & + (-1)^{j_1^p + j_2^p + j_3^n + j_4^n}(1-\delta_{j_1^p j_2^p})(1-\delta_{j_3^n j_4^n})\, v^{J_0}_{j_2^p j_1^p : j_4^n j_3^n} \end{aligned} \tag{1.208}$$

where $v^{J_0}_{j_1^p j_2^p : j_3^n j_4^n}$ are,

$$\begin{aligned} v^{J_0}_{j_1^p j_2^p : j_3^n j_4^n} = {} & -3\, \epsilon^1_{j_1^p, j_3^n}\, \epsilon^1_{j_2^p, j_4^n} \begin{Bmatrix} j_1^p & j_3^n & 1 \\ j_4^n & j_2^p & J_0 \end{Bmatrix} \\ & \times (-1)^{j_2^p + j_3^n + J_0} \sqrt{\left(1+\delta_{j_1^p j_2^p}\right)\left(1+\delta_{j_3^n j_4^n}\right)} \;. \end{aligned} \tag{1.209}$$

Note that $\mathcal{O}^{dDGT}(2)$ behaves just like $\boldsymbol{V}$ with respect to spherical configurations, i.e. the two-body transition operator does not contribute to spherical configuration centroids, $\langle \mathcal{O}^{DGT}(2)\rangle^{\mathbf{m}} = 0$ for all $\mathbf{m}$. The exact transition strength density is easy to write by using the fixed-$\mathbf{m}$ variance propagation formula Eq. (1.97). Firstly, the strength density is [Kota (2007b)]

$$\begin{aligned} & \mathrm{I}^{h,m:\mathcal{O}^{dGT}(2)}(E_i, E_f) = \sum_{(\widetilde{m_p},\widetilde{m_n})_i\,,\,(\widetilde{m_p},\widetilde{m_n})_f} \left|\left\langle\left\langle (\widetilde{m_p},\widetilde{m_n})_f \mid \mathcal{O}^{dGT}(2) \mid (\widetilde{m_p},\widetilde{m_n})_i \right\rangle\right\rangle\right|^2 \\ & \quad \times \delta(E_i - \epsilon[(\widetilde{m_p},\widetilde{m_n})_i])\; \delta(E_f - [(\widetilde{m_p},\widetilde{m_n})_f]) \;; \\ & \left|\langle\langle (\widetilde{m_p},\widetilde{m_n})_f \mid \mathcal{O}^{dGT}(2) \mid (\widetilde{m_p},\widetilde{m_n})_i\rangle\rangle\right|^2 \\ & \quad = \sum_{\alpha\in(\widetilde{m_p},\widetilde{m_n})_i,\; \beta\in(\widetilde{m_p},\widetilde{m_n})_f} \left|\left\langle (\widetilde{m_p},\widetilde{m_n})_f\; \beta \mid \mathcal{O}^{dGT}(2) \mid (\widetilde{m_p},\widetilde{m_n})_i\; \alpha \right\rangle\right|^2 \;. \end{aligned} \tag{1.210}$$

Propagation formula for the trace in Eq. (1.210) is,

$$\begin{aligned} & \left|\langle\langle (\widetilde{m_p},\widetilde{m_n})_f \mid \mathcal{O}^{dGT}(2) \mid (\widetilde{m_p},\widetilde{m_n})_i\rangle\rangle\right|^2 = \{d[(\widetilde{m_p},\widetilde{m_n})_i]\} \\ & \times \sum_{\alpha,\beta,\gamma,\delta} \frac{m_n^i(\alpha)[m_n^i(\beta)-\delta_{\alpha\beta}][N_p(\gamma)-m_p^i(\gamma)][N_p(\delta)-m_p^i(\delta)-\delta_{\gamma\delta}]}{N_n(\alpha)[N_n(\beta)-\delta_{\alpha\beta}]N_p(\gamma)[N_p(\delta)-\delta_{\gamma\delta}]} \\ & \times \sum_{J_0} \left[V^{J_0}_{\gamma^p\delta^p\alpha^n\beta^n}(dGT)\right]^2 (2J_0+1) \;, \\ & (\widetilde{m_p},\widetilde{m_n})_f = (\widetilde{m_p},\widetilde{m_n})_i \times \left(1^+_{\gamma_p} 1^+_{\delta_p} 1_{\alpha_n} 1_{\beta_n}\right) \;. \end{aligned} \tag{1.211}$$

Note that in Eq. (1.211) the final configuration is defined by removing one neutron from orbit α and another from β and then adding one proton in orbit γ and another in orbit δ. Also, $N_p(\alpha)$ is the degeneracy of the proton orbit α and similarly $N_n(\alpha)$ for the neutron orbit α. Note that the $V^{J_0}(dGT)$ are given by Eq. (1.208) and

(1.209). Extension of Eqs. (1.210, 1.211) is straightforward for any two-body transition operator that is of $\boldsymbol{V}$ type [French, Kota, Pandey and Tomsovic (1988b*); Tomsovic (1986); Kota and Kar (1989); Tomsovic, Johnson, Hayes and Bowman (2000*)].

1.7.6 *Unitary configuration NIP strength densities*

In general, though the formulation is simple, construction of strength densities in spherical configuration representation will be computationally prohibitive and the structure of strength densities will not be transparent in this formulation. Therefore, just as with level densities, it is important to consider strength densities in a unitary configuration representation. For $k'=1$, $k=0$ or $k'=0$, $k=1$ operators (i.e. for particle creation or annihilation operators) it is easy to write the expression for NIP strength densities using unitary configurations. For $k'=k=1$ or one-body transition operators, the NIP strength density generated by a one-body operator $\mathcal{O}^k_\mu$ can be decomposed into unitary configuration densities $\mathrm{I}^{[\mathbf{m}_i],[\mathbf{m}_f]}$,

$$\mathrm{I}^{h,m:\mathcal{O}^k}(E_i,E_f) = \sum_{[\mathbf{m}_i],\,[\mathbf{m}_f]} \mathrm{I}^{h,[\mathbf{m}_i],[\mathbf{m}_f]:\mathcal{O}^k}(E_i,E_f)\;; \tag{1.212}$$

and the unitary configuration densities are,

$$\begin{aligned}\mathrm{I}^{h,[\mathbf{m}_i][\mathbf{m}_f]:\mathcal{O}^k}(E_i,E_f) = (2k+1)^{-1} & \sum_{\mathbf{m}_i\in[\mathbf{m}_i]\,,\,\mathbf{m}_f\in[\mathbf{m}_f]}\;\sum_{\mu}\;\sum_{\gamma_i\in\mathbf{m}_i\,,\gamma_f\in\mathbf{m}_f} \\ & \left|\left\langle \mathbf{m}_f\gamma_f \mid \mathcal{O}^k_\mu \mid \mathbf{m}_i\gamma_i\right\rangle\right|^2\,\delta(E_i-\epsilon(\mathbf{m}_i))\,\delta(E_f-\epsilon(\mathbf{m}_f))\;.\end{aligned} \tag{1.213}$$

Then the bivariate moments of $\mathrm{I}^{[\mathbf{m}_i],[\mathbf{m}_f]}$ are,

$$\widetilde{M_{PQ}}([\mathbf{m}_i],[\mathbf{m}_f]) = (2k+1)^{-1}\sum_{\mu}\left\langle(\mathcal{O}^k_\mu)^\dagger P([\mathbf{m}_f])h^Q\mathcal{O}^k_\mu h^P\right\rangle^{[\mathbf{m}_i]} \tag{1.214}$$

and this can be verified using $\sum_\gamma |[\mathbf{m}]\gamma\rangle = \sum_{\mathbf{m}\in[\mathbf{m}];\gamma'}|\mathbf{m}\gamma'\rangle$ and the definition given by Eq. (1.213). Note that $P([\mathbf{m}_f])$ in Eq. (1.214) is a projection operator allowing only $[\mathbf{m}_f]$ in that position. The normalized moments $M_{PQ}([\mathbf{m}_i],[\mathbf{m}_f]) = \widetilde{M_{PQ}}([\mathbf{m}_i],[\mathbf{m}_f])/\widetilde{M_{00}}([\mathbf{m}_i],[\mathbf{m}_f])$ with $P+Q\leq 2$ define the bivariate Gaussian form for the $\mathrm{I}^{[\mathbf{m}_i],[\mathbf{m}_f]}$ densities. Propagation formulas for these are given in detail in [Kota and Majumdar (1995a)]. Using these it is possible to construct the unitary configuration strength densities as bivariate Gaussians and some numerical tests are given in [Kota and Majumdar (1995a)]; see Fig. 1.19. Finally, for $k'=k=2$ or two-body transition operators (for example for neutrinoless double beta decay) it is possible to construct $\mathrm{I}^{h:[\mathbf{m}_i],[\mathbf{m}_f])}$ densities.

1.7.7 *IP strength densities: Convolution form*

Returning to $H=\boldsymbol{h}+\boldsymbol{V}$, it is reasonable to postulate that interactions ($\boldsymbol{V}$), just as with state densities, generate spreading of the NIP strength densities via a bivariate

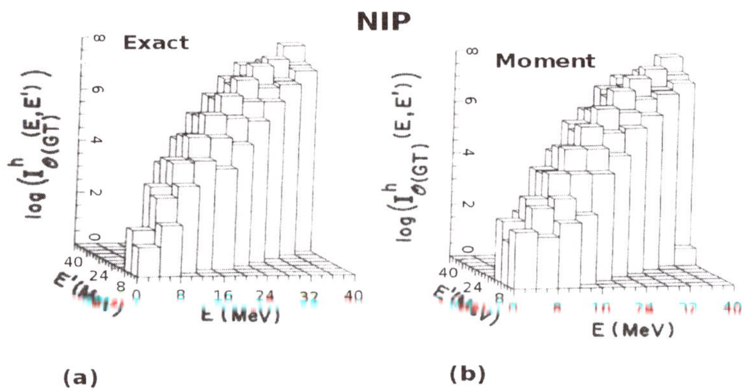

Fig. 1.19 3D histogram for NIP strength densities. (a) Exact results (i.e. NIP strength densities obtained using spherical configurations). (b) Result obtained using Eq. (1.212), i.e. unitary configurations with bivariate Edgeworth representation for the partial strength densities. Calculations are in $(2s1d)-(2p1f)$ space with $(2s1d)$ and $(2p1f)$ orbits having $s=0$ and 1 and the unitary orbits are $1d_{5/2}$, $(2s_{1/2},1d_{3/2})$, $1f_{7/2}$ and $(2p_{3/2},2p_{1/2},1f_{5/2})$. Results are from [Kota and Majumdar (1995a)].

convolution,

$$\mathrm{I}^{H,m:\mathcal{O}}(E_i,E_f)=\mathrm{I}^{\boldsymbol{h},m:\mathcal{O}}\otimes\rho_{biv}^{\boldsymbol{V},m:\mathcal{O}}[E_i,E_f] \tag{1.215}$$

where ρ_{biv} is a normalized density. In applying Eq. (1.215) a distinction needs to be made between sufficiently large and strongly interacting spaces and large spaces that necessarily (as required by shell model stability) contain weakly interacting subspaces. Firstly, in IP spaces that are strongly interacting, for example $(2s1d)$ and $(2p1f)$ shell spaces for nuclei such as ^{24}Mg and ^{56}Fe respectively, we can treat the entire $\mathrm{I}^{H,m:\mathcal{O}}$ as a single bivariate distribution with moments given by $M_{PQ}=\left\langle\mathcal{O}^\dagger H^Q\mathcal{O}H^P\right\rangle^m$. Representing a k-body H by EGOE(k) and a t-body $\mathcal{O}$ by another independent EGOE(t), it is seen that $\rho_{biv}^{H,m:\mathcal{O}}$ approaches bivariate Gaussian form. The most significant parameter, the bivariate correlation coefficient ζ_{biv} in the dilute limit is given by,

$$\zeta_{biv}=\frac{\left\langle\mathcal{O}^\dagger(t)H(k)\mathcal{O}(t)H(k)\right\rangle^m}{\left\langle\mathcal{O}^\dagger(t)\mathcal{O}(t)\right\rangle^m\left\langle H(k)H(k)\right\rangle^m}\overset{EGOE}{\longrightarrow}\binom{m-t}{k}\binom{m}{k}^{-1}-1\,. \tag{1.216}$$

Similarly the fourth order cumulants are $k_{40}=k_{04}=\binom{m-k}{k}\binom{m}{k}^{-1}-1$, $k_{31}=k_{13}=\zeta_{biv}k_{40}$ and $k_{22}=\zeta_{biv}^2\{\binom{m-k-t}{k}\binom{m-t}{k}^{-1}-1\}$. Then, for k fixed and $m\to\infty$, the bivariate cumulants of order 4 and higher (odd cumulants vanish by definition for EGOE) tend to zero giving the bivariate Gaussian form. Eq. (1.216) leads to the important result that for EGOE, $\zeta_{biv}\to 1$ and hence the strengths will be concentrated close to the diagonal, i.e. along $E_i\sim E_f$. As a consequence, the strength distribution will be narrow and the strengths close to $E_i\sim E_f$ will be enhanced. Of course the extreme limit of $\zeta_{biv}=1$ is unphysical. On the other hand for GOE ($k=m$) we have $\zeta_{biv}=0$ giving strengths to be constant (independent of E_i and E_f) as expected. Note that the strengths are locally averaged strengths with fluctuations following Porter-Thomas (P-T) form, evidence for P-T form for IP systems is demonstrated with numerical examples in [Brody *et al.* (1981); Kota and Sahu (1998); Kota (2001)]. In realistic nuclear shell model examples we find $\zeta_{biv} \sim 0.6-0.9$. Extensions of Eq. (1.216) for (m_p, m_n) systems useful in practical applications is available in [Tomsovic (1986)]. For a more detailed discussion of EGOE formulation and shell model tests see [French, Kota, Pandey and Tomsovic (1988b*); Tomsovic (1986); Kota (2001)]. It is useful to add that the shell model results for electromagnetic strengths and strength sums as reported in [Draayer, French and Wong (1977b); Verbaarschot and Brussaard (1985)] do reveal that the transition strength densities are close to a bivariate Gaussian form. Finally, in Fig. 1.20 are shown two shell model examples demonstrating the applicability of the bivariate Gaussian form for the chaos measure NPC in transition strengths. Given a transition operator $\mathcal{O}$ and the corresponding transition strengths $|\langle E_f \mid \mathcal{O} \mid E\rangle|^2$, the normalized strength $\mathcal{R}(E,E_f) = \{\langle E \mid \mathcal{O}^\dagger\mathcal{O} \mid E\rangle\}^{-1}\; |\langle E_f \mid \mathcal{O} \mid E\rangle|^2$. Then the chaos measures NPC for strength distributions is $(\text{NPC})_E = \{\sum_{E_f} \{\mathcal{R}(E,E_f)\}^2\}^{-1}$. Note that: (i) smallness of NPC signifies presence of collective states [Draayer, French and Wong (1977a*)]; (ii) NPC can be used as a measure for defining the region of onset of chaos in the spectrum [Gómez *et al.* (2001)]; (iii) it is useful in the study of rotational damping [Sheikh and Sun (2004)]. The average-fluctuation separation in strengths, strength fluctuations following P-T law and the bivariate Gaussian form for the smoothed strength densities, give a simple formula for $(\text{NPC})_E$ [Kota and Sahu (1998)],

$$
\begin{gathered}
(\text{NPC})_E \stackrel{EGOE}{\longrightarrow} \frac{d_f}{3}\left\{\hat{\sigma}\sqrt{1-\zeta^2}X\;\exp-\left(\frac{\hat{\sigma}\zeta\hat{E}+\hat{\Delta}}{X}\right)^2\right\}\;; \\
\hat{\sigma} = \sigma_2/\sigma_f,\;\; \hat{\Delta} = (\epsilon_2-\epsilon_f)/\sigma_f,\;\; \hat{E} = (E-\epsilon_1)/\sigma_1,\; X = \left[2-(\hat{\sigma})^2(1-\zeta^2)\right]^{1/2}\,.
\end{gathered}
\tag{1.217}
$$

Here ϵ_f and σ_f^2 are the centroid and variance of $I^{m_f}(E_f)$ and d_f is the dimension of the E_f space. Similarly, (ϵ_1, ϵ_2) are the centroids and (σ_1, σ_2) the variances of the marginal densities and ζ is the correlation coefficient. The crucial factor that determines the structure of $(\text{NPC})_E$ is the bivariate correlation coefficient ζ. As $\zeta \to 0$ we recover the GOE result $(\text{NPC})_E = d_f/3$ independent of E. For shell model examples, as ζ is large, NPC is very different from GOE.

In the situation where the interaction strength, measured in units of average spacing of the sp levels, is smaller than the critical strength required for generating a Gaussian form for strength functions, it is plausible that, just as the strength functions will be close BW form, the transition strength densities will be close to bivariate BW form. Thus it is important to consider a bivariate distribution that interpolates bivariate BW and Gaussian forms. The bivariate t-distribution given by Eq. (1.173) generates this interpolation and appears to describe strength densities in intermediate situations. This has been adequately demonstrated in [Kota, Chavda

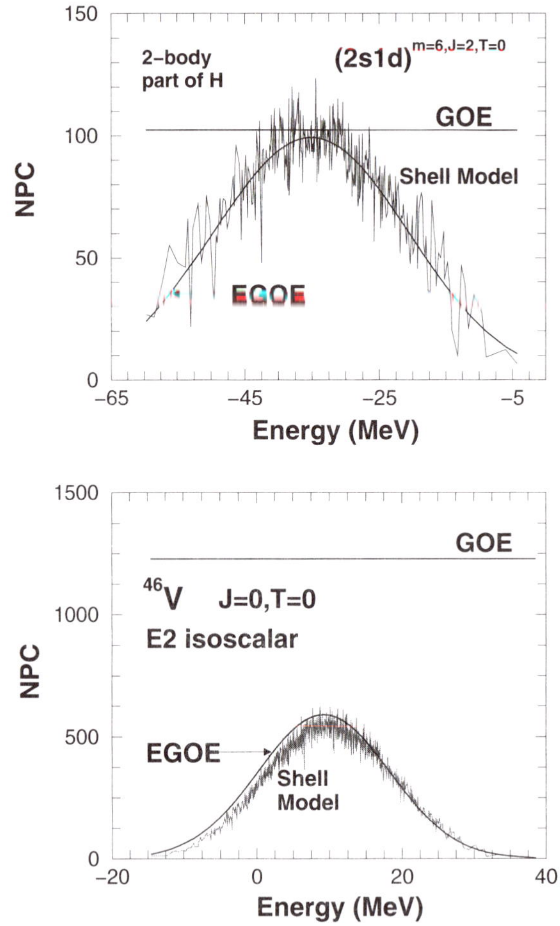

Fig. 1.20 Number of principle components (NPC) in transition strengths. In the $(2s1d)$ example, the $\nu = 2$ part (which is two-body) of H is used as the transition operator [Kota and Sahu (1998)]. For the $(2p1f)$ example, the $E2$ isoscalar operator (which is one-body) is used [Gómez *et al.* (2004)].

and Sahu (2006a)] using shell model and EGOE examples. The interpolating form applies to atoms [Angom, Ghosh and Kota (2004); Angom and Kota (2003)].

In large spaces, as discussed in Secs. 1.7.5 and 1.7.6, decomposition of $\mathrm{I}^{H,m:\mathcal{O}}$ is essential both for accuracy and for identifying weakly interacting subspaces. Then Eq. (1.215) applies locally. With unitary configurations, this gives

$$\mathrm{I}^{H,m_i;\mathcal{O}}(E_i, E_f) = \sum_{[\mathbf{m}_i],[\mathbf{m}_f]} \mathrm{I}^{\boldsymbol{h},[\mathbf{m}_i],[\mathbf{m}_f]:\mathcal{O}} \otimes \rho_{biv-\mathcal{G}}^{\boldsymbol{V},[\mathbf{m}_i],[\mathbf{m}_f]:\mathcal{O}}[E_i, E_f] \ . \tag{1.218}$$

The marginal centroids and variances of the spreading Gaussian $\rho_{biv-\mathcal{G}}$, verified using several one-body operators [French, Kota, Pandey and Tomsovic (1988b*); Kota and Majumdar (1995b*); Tomsovic (1986)], are well approximated by the corresponding state density centroids and variances. Similarly the correlation coefficient

$$\zeta_{biv} \sim \left\langle \mathcal{O}^\dagger \boldsymbol{V} \mathcal{O} \boldsymbol{V} \right\rangle / \left\langle \mathcal{O}^\dagger \mathcal{O} \right\rangle \langle \boldsymbol{V} \boldsymbol{V} \rangle$$

can be treated, with estimates given by Eq. (1.216) and its (p, n) extension, as a constant. Then the marginal centroids of ρ_{biv} will be zero and the variances will be unitary configuration variances. On the other hand, with spherical configurations, we have

$$\begin{aligned} |\langle E_f \mid \mathcal{O} \mid E_i \rangle|^2 = \sum_{\mathbf{m}_i, \mathbf{m}_f} & \frac{I_{\mathcal{G}}^{m_i}(E_i)\, I_{\mathcal{G}}^{m_f}(E_f)}{I^{m_i}(E_i)\, I^{m_f}(E_f)} |\langle \mathbf{m}_f \mid \mathcal{O} \mid \mathbf{m}_i \rangle|^2 \\ & \times \frac{\rho_{biv-\mathcal{G}}^{\mathbf{m}_i, \mathbf{m}_f : \mathcal{O}}(E_i, E_f \,;\, \epsilon_i, \epsilon_f, \sigma_i, \sigma_f, \zeta)}{\rho_{\mathcal{G}}^{\mathbf{m}_i}(E_i)\, \rho_{\mathcal{G}}^{\mathbf{m}_f}(E_f)} \;; \end{aligned} \tag{1.219}$$

$$|\langle \mathbf{m}_f \mid \mathcal{O} \mid \mathbf{m}_i \rangle|^2 = [d(\mathbf{m}_i) d(\mathbf{m}_f)]^{-1} \sum_{\alpha \in \mathbf{m}_i,\, \beta \in \mathbf{m}_f} |\langle \mathbf{m}_f \beta \mid \mathcal{O} \mid \mathbf{m}_i \alpha \rangle|^2 \;,$$

$$I^{\mathbf{m}_r}(E_r) = \langle\langle \delta(H - E_r) \rangle\rangle^{\mathbf{m}_r} = d(\mathbf{m}_r) \rho^{\mathbf{m}_r}(E_r) \;,$$

$$\epsilon_r = \langle H \rangle^{\mathbf{m}_r} \;,\; \sigma_r^2 = \left\langle (H - \epsilon_r)^2 \right\rangle^{\mathbf{m}_r} \;,\; r = i, j \;.$$

Eq. (1.219) has been used in many applications described ahead.

Given a one-body transition operator $\mathcal{O} = \sum_{\alpha,\beta} \epsilon_{\alpha\beta}\, a_\alpha^\dagger a_\beta$, it is possible to simplify Eq. (1.219), assuming the operation of chaos (or EGOE), to a form that involves occupation probabilities. The steps involved are: (i) evaluating $\mid \langle \mathbf{m}_f \mid \mathcal{O} \mid \mathbf{m}_i \rangle \mid^2$ using Eq. (1.205) and this gives the $\langle \hat{n}_\beta (1 - \hat{n}_\alpha) \rangle^{\mathbf{m}_i}$ term (after ignoring $\delta_{\alpha\beta}$ corrections) where $\hat{n}_\alpha = n_\alpha / N_\alpha$; (ii) replacing it by $\langle \hat{n}_\beta (1 - \hat{n}_\alpha) \rangle^{E_i}$ and this is valid in the chaotic domain where occupancies vary slowly; (iii) assuming constant spectral widths, i.e. σ_r^2 in Eq. (1.219) do not depend on $\mathbf{m}_r$ so that $\sigma_i^2 \to \overline{\sigma_i}^2$ and $\sigma_f^2 \to \overline{\sigma_f}^2$; (iv) converting the sum over $\mathbf{m}_i$ into an integral. Then we have the pleasing form [Kota and Sahu (2000*)],

$$\begin{aligned} & |\langle E_f \mid \mathcal{O} \mid E_i \rangle|^2 \\ & \quad = \sum_{\alpha,\beta} |\epsilon_{\alpha\beta}|^2 \; \langle \hat{n}_\beta (1 - \hat{n}_\alpha) \rangle^{E_i}\, \overline{D(E_f)}\, \mathcal{F}(\Delta = E_f - E_i + \epsilon_\beta - \epsilon_\alpha, \overline{\sigma_i}, \overline{\sigma_f}, \zeta_{biv}) \end{aligned} \tag{1.220}$$

where

$$\mathcal{F}(\Delta, \overline{\sigma_i}, \overline{\sigma_f}, \zeta) = \frac{1}{\sqrt{2\pi \left(\overline{\sigma_i}^2 + \overline{\sigma_f}^2 - 2\zeta\, \overline{\sigma_i}\, \overline{\sigma_f} \right)}} \exp - \frac{\Delta^2}{2 \left(\overline{\sigma_i}^2 + \overline{\sigma_f}^2 - 2\zeta\, \overline{\sigma_i}\, \overline{\sigma_f} \right)} \;. \tag{1.221}$$

Eq. (1.221) extends simply to the situation $\rho_{biv-\mathcal{G}} \rightarrow \rho_{biv-t}$ with $\mathcal{F}$ changing to [Kota, Chavda and Sahu (2006a)]

$$\mathcal{F} = \frac{\Gamma(\frac{\nu+1}{2})}{\sqrt{\pi}\Gamma(\frac{\nu}{2})} \frac{1}{\sqrt{\nu(\sigma_1^2+\sigma_2^2-2\zeta\sigma_1\sigma_2)}} \left[1+\frac{\Delta^2}{\nu(\sigma_1^2+\sigma_2^2-2\zeta\sigma_1\sigma_2)}\right]^{-\frac{\nu+1}{2}} . \tag{1.222}$$

Note that, for $\nu > 2$, σ_1 and σ_2 are related (as discussed in Sec. 1.7.2) to the marginal variances μ_{20} and μ_{02}. More remarkable is the fact that Eq. (1.220) reduces to the result derived by Flambaum *et al.* employing the BW form for the strength functions [Flambaum, Gribakina, Gribakin and Kozlov (1994); Flambaum, Gribakina, Gribakin and Ponomarev (1999)]. Then,

$$\mathcal{F}(\Delta, \overline{\Gamma_i}, \overline{\Gamma_f})_{BW} = \frac{1}{2\pi}\frac{\overline{\Gamma_i}+\overline{\Gamma_f}}{\Delta^2+\left(\overline{\Gamma_i}+\overline{\Gamma_f}\right)^2/4} \tag{1.223}$$

where $\overline{\Gamma_i}$ and $\overline{\Gamma_f}$ are the average BW spreading widths for the basis states over the initial and final many-particle states respectively. It is well understood that Eq. (1.223) under estimates the transition matrix elements since it ignores the effects due to the bivariate correlation coefficient ζ_{biv} [Kota, Chavda and Sahu (2006a)].

Spectral distribution theory for squares of matrix elements, as given by Eqs. (1.216)-(1.223) has been employed in a wide variety of application involving both symmetry breaking studies and transition strengths calculations. Briefly:

- With level fluctuations in the neutron.resonance region, a chaotic region of the nuclear spectra, following GOE, the analysis of experimental data determines a bound on the mean squared many particle matrix element of the time reversal non-invariant (TRNI) part of the nucleon-nucleon interaction in heavy nuclei such as ^{232}Th and ^{239}U. Reduction of this to a bound on α, the relative strength of the $\nu = 2$ parts of the TRNI and TRI effective nucleon-nucleon interaction involves calculation of the mean squared matrix element of the $\nu = 2$ part of the TRI effective interaction. This is a locally averaged transition strength generated by a two-body operator in H eigenstates with $E_i \sim E_f$. Applying the theory given by Eq. (1.219) gives a bound $\alpha \simeq 2 \times 10^{-3}$. The bounds are given in [French, Kota, Pandey and Tomsovic (1988b*); French, Pandey and Smith (1987*)].
- TRIPLE collaboration measured, by scattering polarized neutrons from compound nuclei, weak spreading widths due to parity non-conserving (PNC) interactions in nuclei in the mass A $\sim$ 230 and A $\sim$ 100 region [Mitchell, Bowman and Weidenmüller (1999); Mitchell, Bowman, Penttilä and Sharapov (2001)]. The spreading widths can be converted into PNC root mean squared matrix elements M. Using the Desplanques-Donoghue-Holstein estimates of the PNC meson coupling constants, and applying the theory given by Eq. (1.219), Tomsovic, Johnson, Hayes and Bowman (2000*) showed that the calculated M value is about one third of the experimental value for ^{238}U and about a factor 1.7

smaller for Pd isotopes. Similarly [Smith *et al.* (2001)], the observed spreading width for ^{117}Sn relative to the nearby nuclides confirmed the predictions of statistical spectroscopy.

- Single particle (proton or neutron) removal and addition centroids and widths for several $(2s1d)$ shell nuclei are calculated using realistic interactions and Eq. (1.219). Results of this first application to one particle transfer have been reported in [Potbhare and Tressler (1991)].
- Gamow-Teller (GT) strength distributions for A > 60 neutron excess nuclei are of interest in presupernova evolution calculations. The β-decay rates for these nuclei play a very important role in determining the structure of the core of massive presupernova stars and hence on their subsequent evolution towards gravitational collapse and supernova explosion phases. Fixing the bivariate correlation coefficient ζ_{biv} in Eq. (1.219) using experimental β-decay half lives and the calculated GT strength distributions, gives $\zeta_{biv} \sim 0.67$, β-decay rates for a number of A > 60 nuclei are tabulated in [Kota and Majumdar (1995b*); Majumdar and Kar (2007)] at presupernova matter densities $\rho = 10^7 - 10^9$ gm/cc, temperatures $T = (3-5) \times 10^9\ {}^oK$ and electron fractions $Y_e = 0.43 - 0.5$. These studies used the surface delta and pairing plus quadrupole-quadrupole interactions. The first study of beta decay matrix elements using spectral distribution theory is due to Kar (1981a); Kar and Wong (1981). Also Kar and his collaborators developed and applied a simpler version of the spectral distribution theory to calculate beta decay rates for $(2p1f)$-shell nuclei for presupernova evolution [Kar, Sarkar and Ray (1991, 1994); Sutaria and Ray (1995)], beta decay rates for nuclei with $115 < A < 140$ for r-process nucleosynthesis [Kar, Chakravarti and Manfredi (2006)] and also capture rates of solar neutrinos in Ga detectors [Kar, Goswami and Raychaudhuri (1995); Sarkar and Kar (1995)]. Note that spectral distributions have also been applied to study astrophysical partition functions [Wong, Zhao and Zhu (1980)].
- Giant dipole resonance (GDR) widths and GDR absorption cross sections, using the spectral distribution theory, have been studied by Majumdar, Kar and Ansari (1997, 1998). For example, using the giant dipole transition operator, data for GDR built on ground state in ^{76}Se is well reproduced. The correlation coefficient ζ_{biv} has been determined to be ~ 0.87 and the calculated GDR width ~ 9 MeV is close to the experimental value. It will be interesting to extend this analysis to the data [Kusnezov, Alhassid and Snover (1998)] on giant dipole resonance widths in hot rotating nuclei.
- In a novel application in atomic physics, the strongly enhanced low-energy electron recombination observed in Au^{25+} was studied by Flambaum *et al.* [Flambaum, Gribakina, Gribakin and Harabati (2002)] using the theory given by Eqs. (1.220) and (1.223) for transition strengths generated by one-body operators; the operator involved here is the dipole operator. Similarly recombination of low energy electrons with U^{28+} has been studied by Gribakin, Sahoo and Dzuba (2004).

1.8 Applications to Atoms, Mesoscopic Systems and Interacting Bosons

Our focus so far has been almost exclusively on nuclei. However, spectral distributions and the two-body random matrix ensembles that generate them are expected to apply more generally to many fermion and boson quantum systems. Before concluding and for completeness we discuss briefly their application to atoms, mesoscopic systems and interacting boson.

The initial papers on statistical properties of atomic spectra, with emphasis on spacing distributions and their comparison with GOE results, are due to Porter, Rosenzweig and Trees; these papers are reprinted in [Porter (1965)]. Empirical studies on the nature of level density arising from a definite configuration partial level density in several atomic configurations of small dimension, by Parikh (1978a) and Cowan (1981) showed that the level density is close to a Gaussian. A more detailed analysis later confirmed that fixed-J densities in many situations are indeed close to Gaussian. For example in [Bancewicz and Karwowski (1991)], statistical properties of the $(3d+4s)^5 4p$ spectrum of CrI have been studied. They find that fixed-J densities are close to Gaussians with third and fourth moment correction and the smooth spectra extracted from these are close to the exact results. Thus spectral distribution methods based on low order moments are important in statistical atomic spectroscopy. Low order traces (or moments) are also used in deriving envelopes of spectra in transition arrays [Bauche and Bauche-Arnoult (1992); Bauche, Bauche-Arnoult and Wyart (1994)]. With close to Gaussian forms operating in many atomic systems, propagation equations for centroids and variances generated by two-body operators over fixed-m and also (m, S) spaces have been derived by several authors; see for example [Karwowski and Bancewicz (1987)]. It is important to add that they follow, in principle, from the results in [French (1969); Hecht and Draayer (1974); Wong (1986)] derived in the nuclear context. This was explicitly shown by Nomura [Nomura (1988)]. Early reviews on statistical properties of atomic spectra are due to Bauche, Bauche-Arnoult and Klapisch (1988); Bauche and Bauche-Arnoult (1990) and a book on this subject (in Russian) is due to Karazija (1991). Following these developments in the 80's, Karwowski and collaborators [Rajadell, Planelles and Karwowski (1993); Rajadell,Planelles, Karwowski and Mas (1995); Karwowski, Rajadell, Planelles and Mas (1995); Planelles, Rajadell, Karwowski and Mas (1996)] have developed in detail a formalism for evaluating 3rd and 4th moments (even higher ones), for one plus two-body operators, defined over fixed-(Ω, m, S) spaces where S is the total (m particle) electron spin and Ω is the number of orbital degrees of freedom. They employed a diagrammatic approach based on Hugenholtz-like diagrams. The results given here can, in principle, be generated from those given in [Ayik and Ginocchio (1974a); Wong (1986)] by considering traces over spaces defined by fixed number of protons and neutrons (i.e. using 2-orbit configuration averages) and then generating fixed-T (for many electron systems T will go over to S) averages by a subtraction procedure. An overview of the formalism for higher moments and their possible applications has been given by Karwowski (1994). Identifying the diagrams that contribute in the dilute limit, it is shown that in fixed-(m, S) spaces the asymptotic behavior of the moments is same as in Eq. (1-27). Further the spin dependence on the moments is weak [Planelles, Rajadell and Karwowski (1997)]. Thus the densities approach a Gaussian form for interacting fermion systems with spin degree of freedom thereby enlarging the domain of applicability of the results of Mon and French (1975*).

A new direction in statistical theory of atomic spectra was initiated in 1994 with the detailed (shell model type) analysis of the Hamiltonian matrices of large dimension for the CeI atom by Flambaum, Gribakina, Gribakin and Kozlov (1994). They constructed H matrices for $J = 4^{\pm}$, with dimension $d = 260$ for odd parity and 276 for even parity levels in CeI and analyzed the matrix structure, strength functions, transition strengths, etc. It is observed that the strength functions (partial densities) are close to BW rather than Gaussian. Starting with this they developed a theory for the distribution of (smoothed) transition strengths and occupation probabilities [Flambaum, Gribakina, Gribakin and Kozlov (1994); Flambaum, Gribakina and Gribakin (1998); Flambaum, Gribakina, Gribakin and Ponomarev (1999)]. They used ideas based on 'chaos'[Flambaum, Gribakin and Izrailev (1996); Flambaum and Izrailev (1997); Flambaum and Gribakin (2000)] and their results also follow from appropriate two-body random matrix ensembles as shown in [Kota and Sahu (2000*); Kota (2003*)]. Later, continuing with the Lanthanide atoms, it was shown by Angom and Kota (2003); Angom, Ghosh and Kota (2004); Angom and Kota (2005) that, as we go from CeI to SmI (through PmI, PrI and NdI), there is a transition in strength functions from BW to Gaussian. Therefore, statistical theory based on BW for lighter Lanthanide and Gaussian form for heavier Lanthanide atoms will be applicable. As mentioned earlier (see also [Kota (2003*)]), random two-body interactions in the presence of a mean-field generate, in the regime where level fluctuations follow GOE, BW to Gaussian transition in strength functions. It was also proved, as discussed in Sec. 1.7.7 and in [Kota and Sahu (2000*)], that the theory for transition strengths given in [Flambaum, Gribakina, Gribakin and Ponomarev (1999)] converges to the theory given by Eq. (1.219). There are now attempts to analyze the eigenvalues and eigenstates of Actinide atoms, in particular, configuration mixing between doubly excited states and multiply excited states in Au^{24+} has been studied in [Gribakin and Sahoo (2003)] and this mixing accounts for the large electron recombination rate in Au^{25+}.

Going beyond atomic physics, spectral distribution methods have been applied to the description of vibronic spectra of molecules and solids. In [Bielinska-Waz and Karwowski (1995)], expressions for envelopes of electronic bands are derived. Also some applications in crystal field theory have been presented by Yueng and Newman (1986) and for disordered systems by Parikh (1983). In the last ten years or so it has been shown that random two-body interactions in the presence of a mean-field, i.e. EGOE(1+2) for spinless fermion systems and for fermions with spin are needed to describe some of the properties of mesoscopic systems such as quantum dots and small metallic grains. These include, for example, conductance peak spacing distributions in quantum dots and small metallic grains, suppression of ground-state magnetization in finite-size systems and odd-even staggering in ground state energies of small metallic grains [Alhassid (2000); Jacquod and Stone (2000); Jacquod and Varga (2002); Papenbrock, Kaplan and Bertsch (2002); Alhassid, Weidenmüller and Wobst (2005); Schmidt and Alhassid (2008)]. It is of interest to note that many of these results could be obtained using trace propagation and Gaussian form for state and expectation value densities [Kota and Kar (2002); Vyas, Kota and Chavda (2009)]. It thus appears that the applicability of two-body random matrix ensembles (and trace propagation) extends from femtoscience to nanoscience [Alhassid (2005)].

Although the discussion in this book has been mainly regarding fermion systems, it is important to mention that there are several studies on spectral distributions for non-interacting as well as interacting boson systems. For example, for a system of m bosons in N sp states, in the dense limit defined by $m \to \infty$, $N \to \infty$ and $m/N \to 0$, the state density approaches a Gaussian. This is shown in [Kota and Potbhare (1980*)] by deriving propagation equations, for both NIP and IP boson systems, for the first four moments. The propagation equations show interesting symmetries between dilute fermion and dense boson systems [Kota and Potbhare (1980*); Nomura (1975); Kota (1979b)]. Boson systems have also been investigated by Manfredi and his collaborators [Bortingnon, Francesca and Manfredi (1977); Bortingnon, Manfredi and Spitz (1978); Manfredi and Salmaso (1980); Manfredi (1981, 1984, 1985, 1986)] and Nomura [Nomura (1975, 1985a,b, 1990)]. More importantly, by extending the formulation in [Kota and Potbhare (1980*)], it is shown [Patel, Desai, Potbhare and Kota (2000)] that in the dense limit IP boson systems exhibit average-fluctuation separation in energy levels. Moreover, random matrix ensembles with random two-body interactions in presence of a mean-field for boson systems in the dense limit are also being explored [Asaga, Benet, Rupp and Weidenmüller (2002); Chavda, Potbhare and Kota (2003, 2004)] with the possibility that these ensembles may be useful in the study of Bose condensates. As Asaga, Benet, Rupp and Weidenmüller (2002) state: "In an atomic trap, bosonic atoms occupy partly degenerate single particle states. The interaction will lift the degeneracy. A random matrix approach should reveal the generic features of the resulting system".

Bibliography

Abramowtiz, M. and Stegun, I. A. (eds.) (1964). *Handbook of Mathematical Functions, NBS Applied Mathematics Series*, Vol. 55 (U.S. Govt. Printing Office, Washington D.C.).

Aberg, S. (1990). Onset of chaos in rapidly rotating nuclei, *Phys. Rev. Lett.* **64**, pp. 3119–3122.

Abzouzi, A., Caurier, E. and Zuker, A. P. (1991). Influence of saturation properties on shell-model calculations, *Phys. Rev. Lett.* **66**, pp. 1134–1137.

Akiyama, Y. and Draayer, J. P. (1973). A users guide to fortran programs for Wigner and Racah coefficients of SU_3, *Comp. Phys. Commu.* **5**, pp. 405–415.

Alhassid, Y. (2000). The statistical theory of quantum dots, *Rev. Mod. Phys.* **72**, pp. 895–968 .

Alhassid, Y. (2005). Mesoscopic fluctuations in quantum dots, nanoparticles and nuclei, *AIP Conf. Proc.* **777**, pp. 250–269.

Alhassid, Y., Weidenmüller, H. A. and Wobst, A. (2005). Disordered mesoscopic systems with interactions: Induced two-body ensembles and the Hartree-Fock approach, *Phys. Rev. B* **72**, pp. 045318/1-19.

Al-Quraishi, S. I., Grimes, S. M., Massey, T. N. and Resler, D. A. (2003). Level densities for $20 \leq A \leq 110$, *Phys. Rev. C* **67**, pp. 015803/1-7.

Angom, D., Ghosh, S. and Kota, V. K. B. (2004). Strength functions, entropies and duality in weakly to strongly interacting fermion systems, *Phys. Rev. E* **70**, pp. 016209/1-11.

Angom, D. and Kota, V. K. B. (2003). Signatures of Two-body Random Matrix Ensembles in Sm I, *Phys. Rev. A* **67**, pp. 052508/1-4.

Angom, D. and Kota, V. K. B. (2005). Chaos and localization in the wavefunctions of complex atoms NdI, PmI and SmI, *Phys. Rev. A* **71**, pp. 042504/1-14.

Arima, A. and Iachello, F. (1987). *The Interacting boson model* (Cambridge University Press, New York).

Asaga, T., Benet, L., Rupp, T. and Weidenmüller, H. A. (2002). Spectral properties of the k-body embedded Gaussian ensembles of random matrices for bosons, *Ann. Phys. (N.Y.)*, **298**, pp. 229–247.

Ayik, S. and Ginocchio, J. N. (1974a). Configuration averaging of operators, *Nucl. Phys. A* **221**, pp. 285–298.

Ayik, S. and Ginocchio, J. N. (1974b). Shell model level densities for light nuclei, *Nucl. Phys. A* **234**, pp. 13–32.

Baldridge, W. J. and Dalton, B. J. (1978). Shell-model studies for the ^{132}Sn region. II. Exact and statistical results for multi-proton cases, *Phys. Rev. C* **18**, pp. 539–548.

Bancewicz, M. and Karwowski, J. (1991). Statistical analysis of the $(3d + 4s)^5 4p$ spectrum of Cr I, *Phys. Rev. A* **44**, pp. 3054–3059.

Banerjee, H. and French, J. B. (1966). Spectroscopic averaging, *Phys. Lett.* **23**, pp. 245–248.

Bauche, J. and Bauche-Arnoult, C. (1990). Statistical properties of atomic spectra, *Comp. Phys. Rep.* **12**, pp. 3–28.

Bauche, J. and Bauche-Arnoult, C. (1992). Statistical approach to the spectra of plasmas, *Physica Scripta* **T40**, pp. 58–64.

Bauche, J., Bauche-Arnoult, C. and Klapisch, M. (1988). Transition arrays in the spectra of ionized atoms, *Adv. Atom. Mol. Phys.* **23**, pp. 131–193.

Bauche, J., Bauche-Arnoult, C. and Wyart, J. -F. (1994). Intensity distribution in atomic transition arrays, *J. Quant. Spec. Rad. Tran.* **51**, pp. 41–47.

Berry, M. V. and Tabor, M. (1977). Level clustering in the regular spectrum, *Proc. R. Soc. Lond. A* **356**, pp. 375–394.

Bethe, H. A. (1936). An attempt to calculate the number of energy levels of a heavy nucleus, *Phys. Rev.* **50**, pp. 332–341.

Bethe, H. A. (1937). Nuclear physics B. nuclear dynamics: Theoretical, *Rev. Mod. Phys.* **9**, pp. 69–244.

Benet, L., Rupp, T. and Weidenmüller, H. A. (2001a). Nonuniversal behavior of the k-body embedded Gaussian unitary ensemble of random matrices, *Phys. Rev. Lett.* **87**, pp. 010601/1-4.

Benet, L., Rupp, T. and Weidenmüller, H. A. (2001b). Spectral properties of the k-body embedded Gaussian ensembles of random matrices, *Ann. Phys. (N.Y.)* **292**, pp. 67–94.

Benet, L. and Weidenmüller, H. A. (2003). Review of the k-body embedded ensembles of Gaussian random matrices, *J. Phys. A* **36**, pp. 3569–3594.

Bielinska-Waz, D. and Karwowski, J. (1995). Statistical theory of vibronic spectra: Envelopes of the electronic bands, *Phys. Rev. A* **52**, pp. 1067–1071.

Bijker, R. and Frank, A. (2000). Band structure from random interactions, *Phys. Rev. Lett.* **84**, pp. 420–422.

Bilpuch, A. M., Lane, G. E., Mitchell, G. E. and Moses, J. D. (1976). Fine structure of analogue states, *Phys. Rep.* **28**, pp 145–244.

Bloch, C. (1954). Theory of Nuclear level density, *Phys. Rev.* **93**, pp. 1094–1106.

Bloch, C. (1969). Statistical Nuclear Theory, in C. DeWitt and V. Gillet (eds), *Physique Nucleaire* (Gordon and Breach, New York), pp. 303–411.

Bloom, S. D., Grimes, S. M. and Hausman Jr, R. F. (1980). A Monte Carlo method for calculating strength functions in many-fermion systems, *Phys. Lett. B* **93**, pp. 227–231.

Bohigas, O. and Flores, J. (1971a). Two-body random Hamiltonian and level density, *Phys. Lett. B* **34**, pp. 261–263.

Bohigas, O. and Flores, J. (1971b). Spacing and individual eigenvalue distributions of two-body random hamiltonians *Phys. Lett. B* **35**, pp. 383–387.

Bohigas, O. and Flores, J., French, J. B., Giannoni, M. J., Mello, P. A. and Wong, S. S. M. (1974). Recent Results on Energy-Level Fluctuations, *Phys. Rev. C* **10**, pp. 1551–1553.

Bohigas, O., Giannoni, M. J. and Schmit, C. (1984). Characterization of chaotic quantum spectra and universality of level fluctuation laws, *Phys. Rev. Lett.* **52**, pp. 1–4.

Bohigas, O., Haq, R. U. and Pandey, A. (1983). Fluctuation properties of nuclear energy levels and widths: Comparison of theory and experiment, in K.H. Böckhoff (ed.), *Nuclear Data for Science and Technology* (Reidel, Dordrecht), pp. 809–813.

Bohigas, O., Haq, R. U. and Pandey, A. (1985). Higher-Order Correlations in Spectra of Complex Systems, *Phys. Rev. Lett.* **54**, pp. 1645–1648.

Bohigas, O., Quesne, C. and Arvieu, R. (1968). Generalized seniority breaking of two body interactions, *Phys. Lett. B* **26**, pp. 562–567.

Bohigas, O. and Weidenmüller, H. A. (1988). Aspects of chaos in nuclear physics, *Ann. Rev. Nucl. Part. Sci.* **39**, pp. 421–453.

Bohr, N. (1936). Neutron Capture and Nuclear Constitution, *Nature* **137**, pp. 344–348.

Bohr, A. and Mottelson, B. R. (1969). *Nuclear Structure: Single-Particle Motion*, Vol. I (Benjamin, New York).

Bohr, A. and Mottelson, B. R. (1973). The many facets of nuclear structure, *Ann. Rev. Nucl. Sci.* **23**, pp. 363–393.

Bohr, A. and Mottelson, B. R. (1975). *Nuclear Structure: Collective Motion*, Vol. II (Benjamin, New York).

Bortingnon, P. F., Francesca, M. D. and Manfredi, V. R. (1977). The spectral-distribution method in the space of M-bosons, *Lettere Al Nuovo Cimento* **19**, pp. 15–19.

Bortingnon, P. F., Manfredi, V. R. and Spitz, S. (1978). Scalar averages in the boson space, *Lettere Al Nuovo Cimento* **23**, pp. 219–222.

Bowley, A.L. (1972). *F.Y. Edgeworth's Contributions to Mathematical Statistics* (Clifton, New Jersey).

Brody, T. A., Flores, J., French, J. B., Mello, P. A., Pandey, A. and Wong, S. S. M. (1981). Random matrix physics: spectrum and strength fluctuations, *Rev. Mod. Phys.* **53**, pp. 385–479.

Brussaard, P. J. and Glaudemans, P. W. M. (1977). *Shell Model Applications in Nuclear Spectroscopy* (North Holland,, Amsterdam, 1977).

Caurier, E., Martínez-Pinedo, Nowacki, G. F., Poves, A. and Zuker, A. P. (2005). The shell model as a unified view of nuclear structure, *Rev. Mod. Phys.* **77**, pp. 427–488.

Chakraborty, M., Kota, V. K. B. and Parikh, J. C. (1980a). Unitary decomposition of Hamiltonian operators II. SU(4) irreducible tensors, norms and their energy variation and symmetry breaking, *Ann. Phys. (N.Y.)* **127**, pp. 413–435

Chakraborty, M., Kota, V. K. B. and Parikh, J. C. (1980b). What does Franzini-Radicati mass relationship imply about the supermultiplet symmetry?, *Phys. Rev. Lett.* **45**, pp. 1073–1076.

Chang, B. D. (1978). Spectral distribution methods in effective interaction theory, *Nucl. Phys. A* **304**, pp. 127–140.

Chang, B. D. and Draayer, J. P. (1979). Orthogonal polynomial expansions for effective interactions in the modified Lipkin model, *Phys. Lett. B* **80**, pp. 332–334.

Chang, B. D., Draayer, J. P. and Wong, S. S. M. (1982). A system to generate fortran programs for calculating configuration traces of angular momentum coupled product operators, *Comp. Phys. Comm.* **28**, pp. 41–60; (1984) **35**, pp. C833–C834.

Chang, B. D. and Vincent, C. M. (1981). Rigorous bounds on the energy-averaged effective hamiltonian, *Phys. Rev. C* **24**, pp. 2267–2278.

Chang, B. D. and Wong, S. S. M. (1978). Scalar and configuration traces of operators in large spectroscopic spaces, *Nucl. Phys. A* **294**, pp. 19–40.

Chang, B. D. and Wong, S. S. M. (1979). A program to generate closed basic diagrams for product operators, *Comp. Phys. Comm.* **18**, pp. 35–61; (1984) **35**, pp. 575.

Chang, B. D. and Wong, S. S. M. (1980). A program to evaluate closed diagrams algebraically for angular momentum coupled product operators, *Comp. Phys. Comm.* **20**, pp. 191–211; (1984) **35**, pp. C629–C630.

Chang, B. D., Vincent, C. M. and Wong, S. S. M. (1984). Automated algebra for higher order perturbation diagrams with angular momentum coupling, *Phys. Rev. C* **30**, pp. 1055–1062.

Chang, F. S. and French, J. B. (1973a). Energy dependence of expectation values in many-particle spectroscopy, *Phys. Lett. B* **44**, pp. 131–134.

Chang, F. S. and French, J. B. (1973b). Occupation probabilities and spin cut-off factors, *Phys. Lett. B* **44**, pp. 135–138.

Chang, F. S., French, J. B. and Ratcliff, K. F. (1966). Centroid energies and widths of nuclear configurations, *Phys. Lett. B* **23**, pp. 251–254.

Chang, F. S., French, J. B. and Thio, T. H. (1971). Distribution methods for nuclear energies, level densities and excitation strengths, *Ann. Phys (N.Y.)* **66**, pp. 137–188.

Chang, F. S. and Zuker, A. (1972). Validity of spectral distribution methods in spectroscopy, *Nucl. Phys. A* **198**, pp. 417–429.

Chavda, N. D., Potbhare, V. and Kota, V. K. B. (2003). Statistical Properties of dense interacting boson systems with One plus two-body random matrix ensembles, *Phys. Lett. A* **311**, pp. 331–339.

Chavda, N. D., Potbhare, V. and Kota, V. K. B. (2004). Strength functions for interacting bosons in a mean-field with random two-body interactions, *Phys. Lett. A* **326**, pp. 47–54.

Choubey, S., Kar, K., Gómez, J. M. G. and Manfredi, V. R. (1998). Spectral distribution studies with a modified Kuo-Brown interaction in the upper half of the fp shell, *Phys. Rev. C* **58**, pp. 597–600.

Cleary, J. G. and Wybourne, B. G. (1971). Statistical group theory and the distribution of angular momentum states, *J. Math. Phys.* **12**, pp. 45–52.

Countee, C. R., Draayer, J. P., Halemane, T. R. and Kar, K. (1981). Shell-model studies of statistical approximations to model interactions, *Nucl. Phys. A* **356**, pp. 1–14.

Cowan, R. D. (1981). *The Theory of Atomic Structure and Spectra*, (University of California Press, Berkeley), pp. 616–631.

Cramer, H. (1946). *Mathematical Methods of Statistics*, (Princeton University Press, Princeton, New Jersey).

Dalton, B. J., Baldridge, W. J. and Vary, J. P. (1979). Realistic effective interactions for the *sd* shell: Global and discrete nuclear properties *Phys. Rev. C* **20**, pp. 1908–1916.

Dalton, B. J., Grimes, S. M., Vary, J. P. and Williams, S. A. (eds.) (1980). *Theory and Applications of Moment Methods in Many Fermion Systems* (Plenum, New York).

Dalton, B. J., Vary, J. P. and Baldridge, W. J. (1977). Binding energies of *sd*-shell nuclei with a realistic effective hamiltonian, *Phys. Rev. Lett.* **38**, pp. 1348–1351.

Darema-Rogers, F. and Vincent, C. M. (1978). Energies and bounds from perturbative approximations to the Bloch-Horowitz effective hamiltonian, *Phys. Rev. C* **17**, pp. 1461–1479.

David, F. N., Kendall, M. G. and Barton, D. E. (1966). *Symmetric function and Allied Tables* (Cambridge University Press, London).

Delfini, M. G. and Brussaard, P. G. (1985). Spectral distributions in ^{64}Cu, *J. Phys. G* **11**, pp. 113–117.

Draayer, J. P., Dytrych, T., Sviratcheva, K. D., Bahri, C. and Vary, J. P. (2008). Symplectic no-core shell model, *Int. J. Mod. Phys. E* **17** (Suppl), pp. 133–145.

Draayer, J. P., French, J. B. and Kota, V. K. B. (1982). Group symmetries and information propagation (unpublished).

Draayer, J. P., French, J. B., Potbhare, V. and Wong, S. S. M. (1975a). Polynomial expansions for excitation strengths, *Phys. Lett. B* **55**, pp. 263–266.

Draayer, J. P., French, J. B., Potbhare, V. and Wong, S. S. M. (1975b). $E2$ polynomial strengths: A comparison with shell-model results, *Phys. Lett. B* **55**, pp. 349–353.

Draayer, J. P., French, J. B., Prasad, M., Potbhare, V. and Wong, S. S. M. (1975c). Spectral averaging for single-nucleon-transfer sum rules, *Phys. Lett. B* **57**, pp. 130–134.

Draayer, J. P., French, J. B., and Wong, S. S. M. (1977a). Spectral distributions and statistical spectroscopy. I. General theory, *Ann. Phys. (N.Y.)* **106**, pp. 472–502.

Draayer, J. P., French, J. B., and Wong, S. S. M. (1977b). Spectral distributions and statistical spectroscopy. II. Shell-model comparisons, *Ann. Phys. (N.Y.)* **106**, pp. 503–524.

Draayer, J. P. and Rosensteel, G. (1982). Two-body potentials in the collective model, *Nucl. Phys. A* **386**, pp. 189–199.

Draayer, J. P. and Rosensteel, G. (1983a). Centroids and widths in the symplectic collective model, *Phys. Lett. B* **124**, pp. 281–283.

Draayer, J. P. and Rosensteel, G. (1983b). Major shell centroids in the symplectic collective model , *Phys. Lett. B* **125**, pp. 237–239.

Draayer, J. P. and Rosensteel, G. (1985). $U(3) \longrightarrow R(3)$ integrity-basis spectroscopy, *Nucl. Phys. A* **439**, pp. 61–85.

Dyson, F. J. and Mehta, M. L. (1963). Statistical theory of the energy levels of complex systems IV, *J. Math. Phys.* **4**, pp. 701–712.

Dyson, F. J. (1970). Correlations between eigenvalues of a random matrix, *Comm. Math. Phys.* **19**, pp. 235–250.

Elliott, J. P. (1958). Collective motion in the nuclear shell model. I. Classification schemes for states of mixed configurations, *Proc. Roy. Soc. London (Series A)* **245**, pp. 128–145.

Feller, W. (1971). *An Introduction to probability theory and its applications*, Vol. II (Wiley, New York).

Flambaum, V. V. and Gribakin, G. F. (2000). Statistical theory of finite Fermi systems with chaotic excited eigenstates, *Phil. Mag. B* **80**, pp. 2143–2173.

Flambaum, V. V., Gribakina, A. A., Gribakin, G. F. (1998). Statistics of electromagnetic transitions as a signature of chaos in many electron atoms, *Phys. Rev. A* **58**, pp. 230–237.

Flambaum, V. V., Gribakina, A. A., Gribakin, G. F. and Harabati, C. (2002). Electron recombination with multicharged ions via chaotic many-electron states, *Phys. Rev. A* **66**, pp. 012713/1-7.

Flambaum, V. V., Gribakina, A. A., Gribakin, G. F. and Kozlov, M. G. (1994). Structure of compound states in the chaotic spectrum of the Ce atom: Localization properties, matrix elements, and enhancement of weak perturbations, *Phys. Rev. A* **50**, pp. 267–296.

Flambaum, V. V., Gribakina, A. A., Gribakin, G. F. and Ponomarev, I. V. (1999). Quantum chaos in many-body systems: what can we learn from the Ce atom, *Physica D* **131**, pp. 205–220.

Flambaum, V. V., Gribakin, G. F., Izrailev, F. M. (1996). Correlations within eigenvectors and transition amplitudes in the two-body random interaction model, *Phys. Rev. E* **53**, pp. 5729–5741.

Flambaum, V. V. and Izrailev, F. M. (1997). Statistical theory of finite Fermi systems based on the structure of chaotic eigenstates, *Phys. Rev. E* **56**, pp. 5144–5159.

Flores, J., Horoi, M., Müller, M. and Seligman, T. H. (2001). Spectral statistics of the two-body random ensemble revisited, *Phys. Rev. E* **63**, pp. 026204/1-7.

Frazier, N., Brown, B. A. and Zelevinsky, V. (1996). Strength functions and spreading widths of simple shell model configurations, *Phys. Rev. C* **54**, pp. 1665–1674.

French, J. B. (1966a). Combinatorial methods for spectroscopic averaging and spectral moments, *Phys. Lett.* **23**, pp. 248–250.

French, J. B. (1966b). Multipole and sum rule methods in spectroscopy, in C. Bloch (ed.), *Proc. Int. School of Physics (Enrico Fermi Course 36)* (Academic Press, New York), pp. 278–393.

French, J. B. (1967a). Spectral distributions in nuclei, in A. Hossain, Harun-ar-Rashid and M.Islam (eds.), *Nuclear Structure* (North Holland, Amsterdam), pp. 85–123.

French, J. B. (1967b). Measures for group-symmetry breaking in nuclei, *Phys. Lett. B* **26**, pp. 75–77.

French, J. B. (1969). Isospin distributions in nuclei, in D. H. Wilkinson (ed.), *Isospin in Nuclear Physics* (North Holland, Amsterdam), pp. 259–295.

French, J. B. (1972). Nuclear symmetries and distributions, in D. J. Rowe, L.E.H. Trainor, S.S.M. Wong and T.W. Donnelly (eds.), *Dynamic Structure of Nuclear States* (Univ. of. Toronto Press, Toronto), pp. 154–204.

French, J. B. (1973). Analysis of distant-neighbour spacing distributions for k-body interaction ensembles, *Rev. Mex. Phys.* **22**, pp. 221–229.

French, J. B. (1974). Distributions, fluctuations and symmetries in many-particle spectroscopy, *Rev. Mex. Phys.* **23**, pp. 189–219.

French, J.B. (1975). Relevant aspects of statistical spectroscopy, in B. R. Barrett (ed.), *Effective interactions and operators in nuclei*, Lecture notes in Physics 40 (Springer, Berlin), pp. 191–206.

French, J. B. (1980a). Statistical spectroscopy, in G. F. Bertsch and D. Kurath (eds.), *Nuclear Spectroscopy* (Springer, Berlin), pp. 180–239.

French, J. B. (1980b). Elementary principles of spectral distributions, in B. J. Dalton, S. M. Grimes, J. P. Vary and S. A. Williams (ed.), *Theory and Applications of Moment Methods in Many Fermion Systems* (Plenum, New York), pp. 91–108.

French, J. B. (1980c). Special topics in spectral distributions, in B. J. Dalton, S. M. Grimes, J. P. Vary and S. A. Williams (ed.), *Theory and Applications of Moment Methods in Many Fermion Systems* (Plenum, New York), pp. 91–108.

French, J. B. (1983a). Statistical Nuclear Spectroscopy, *Nucl. Phys. A* **396**, pp. 87c–104c.

French, J. B. (1983b). General principles of statistical spectroscopy, in K. Abraham, K. Allaart and A. E. L. Dieperink (eds.), *Symmetries in Nuclear Structure* (Plenum, New York), pp. 177–202.

French, J. B. (1984). Spectral averaging and partition functions, in J. S. Dehesa, J.M.G. Gómez and A. Polls (eds.), *Mathematical and Computational Methods in Nuclear Physics* (Springer, Berlin), pp. 100–120.

French, J. B. (1995). Elementary aspects of nuclear quantum chaos, in Sudarshan, E. C. G. (ed.), *A Gift of Prophecy - Essays in Celebration of the Life of R.E. Marshak* (World Scientific, Singapore), pp. 156–167.

French, J. B. and Draayer, J. P. (1979). Symmetries and statistical behavior in Fermion systems, in W. Beiglbock, A. Bohm and E. Takasugi (eds.), *Group Theoretical Methods in Physics* (Springer, Berlin), pp. 394–407.

French, J. B., Halbert, E. C., McGrory, J. B. and Wong, S. S. M. (1969). Complex spectroscopy, in M. Baranger and E. Vogt (ed.), *Advances in Nuclear Physics*, Vol. 3 (Plenum, New York), pp. 193–257.

French, J. B. and Hsu, L. S. (1967). Strength distributions, *Phys. Lett. B* **25**, pp. 75–78.

French, J. B. and Kota, V. K. B. (1982). Statistical spectroscopy, *Ann. Rev. Nucl. Part. Sci.* **32**, pp. 35–64.

French, J. B. and Kota, V. K. B. (1983). Nuclear level densities and partition functions with interactions, *Phys. Rev. Lett.* **51**, pp. 2183–2186.

French, J. B. and Kota, V. K. B. (1989a). Nuclear state densities and smoothed partition functions: I. General Theory, Technical Report UR-1116, University of Rochester, Rochester, New York.

French, J. B. and Kota, V. K. B. (1989b). State densities and spectrum fluctuations: Information propagation in complex nuclei, *Pramana-J. Phys.* **32**, pp. 459–473.

French, J. B., Kota, V. K. B., Pandey, A. and Tomsovic, S. (1985). Bound on time-reversal non-invariance in the nuclear Hamiltonian, *Phys. Rev. Lett.* **54**, pp. 2313–2316.

French, J. B., Kota, V. K. B., Pandey, A. and Tomsovic, S. (1987). Symmetries and Quantum Chaos: time reversal invariance in the nucleon-nucleon interaction, *Phys. Rev. Lett.* **58**, pp. 2400–2403.

French, J. B., Kota, V. K. B., Pandey, A. and Tomsovic, S. (1988a). Statistical properties of many - particle spectra V: Fluctuations and symmetries, *Ann. Phys. (N.Y.)* **181**, pp. 198–234.

French, J. B., Kota, V. K. B., Pandey, A. and Tomsovic, S. (1988b). Statistical properties of many - particle spectra VI: Fluctuation bounds on N-N T-noninvariance, *Ann. Phys. (N.Y.)* **181**, pp. 235–260.

French, J. B., Kota V. K. B. and Smith, J. F. (1989) Nuclear state densities and smoothed partition functions: II. Expectation values, Technical Report UR-1122(ER40425-245), University of Rochester, Rochester, New York.

French, J. B., Mello, P. A. and Pandey, A. (1978). Statistical properties of many-particle spectra II: Two-point correlations and fluctuations, *Ann. Phys. (N.Y.)* **113**, pp. 277–293.

French, J. B., Pandey, A. and Smith, J. F. (1987). Compound-nuclear tests of time reversal invariance in the nucleon-nucleon interaction, in N. R. Roberson, C. R. Gould and J. D. Bowman (eds.), *Tests of Time Reversal Invariance in Neutron Physics* (World Scientific, Singapore), pp. 80–99.

French, J. B. and J.C.Parikh, J. C. (1971). Spin-isospin SU(4) symmetry in the (ds) shell, *Phys. Lett. B* **35**, pp. 1–4.

French, J. B., Rab, S., Smith, J. F., Haq, R. U. and Kota, V. K. B. (2006). Nuclear Spectroscopy in the Chaotic Domain: Level Densities, *Can. J. Phys.* **84**, pp. 677–706.

Fréňch, J. B. and Ratcliff, K. F. (1971). Spectral distributions in nuclei, *Phys. Rev. C* **3**, pp. 94–117.

French, J. B. and Wong, S. S. M. (1970). Validity of random matrix theories for many-particle systems, *Phys. Lett. B* **33**, pp. 449–452.

French, J. B. and Wong, S. S. M. (1971). Some random-matrix level and spacing distributions for fixed-particle-rank interactions, *Phys. Lett. B* **35**, pp. 5–7.

Garg, J. B. (ed.) (1972). *Statistical Properties of Nuclei* (Plenum, New York).

Georgeot, B. and Shepelyansky, D. L. (1997). Breit-Wigner width and inverse participation ratio in finite interacting Fermi systems, *Phys. Rev. Lett.* **79**, pp. 4365–4368.

Gervois, A. (1972). Level densities for random one- and two-body potentials, *Nucl. Phys. A* **184**, pp. 507–532.

Gharghar, F. (2008). Nuclear level density, M. Sc. Thesis, Laurentian University, Sudbury, Ontario, Canada.

Gideon, R. A. and Rothan, A. M. (2004). Cauchy regression and confidence intervals for the slope, **http://www.math.umt.edu/gideon/cauchyreg.pdf**.

Gilmore, R. and Draayer, J. P. (1985). Dynamical group chains and integrity bases, *J. Math. Phys.* **26**, pp. 3053–3067.

Ginnochio, J. N. (1973a). Operator Averages in a shell-model basis, *Phys. Rev. C* **8**, pp. 135–145.

Ginnochio, J. N. (1973b). Angular-momentum dependence of the density of states, *Phys. Rev. Lett.* **31**, pp. 1260–1263.

Ginocchio, J. N. (1980). On the averages of operators in finite fermion systems, in B. J. Dalton, S. M. Grimes, J. P. Vary and S. A. Williams (ed.) *Theory and Applications of Moment Methods in Many Fermion Systems* (Plenum, New York), pp. 109–127.

Ginocchio, J. N. and Yen, M. M. (1972). Average one-body operators with fixed angular momentum and seniority, *Nucl. Phys. A* **195**, pp. 129–145.

Ginocchio, J. N. and Yen, M. M. (1975). The dependence of shell model state densities on angular momentum, *Nucl. Phys. A* **239**, pp. 365–396.

Gómez, J. M. G., Kar, K., Kota, V. K. B., Molina, R. A. and Retamosa, J. (2004). Number of principal components and localization length in E2 and M1 transition strengths in ^{46}V, *Phys. Rev. C* **69**, pp. 057302/1-4.

Gómez, J. M. G., Kar, K., Kota, V. K. B., Retamosa, J. and Sahu, R. (2001). Transition strengths and quantum chaos in large shell-model spaces, *Phys. Rev. C* **64**, pp. 034305/1-11.

Goriely, S., Hilaire, S. and Koning, A. J. (2008). Improved microscopic nuclear level densities within the Hartree-Fock-Bogoliubov plus combinatorial method, *Phys. Rev. C* **78**, pp. 064307/1-14.

Gribakin, G. F. and Sahoo, S. (2003). Mixing of dielectronic and multiply excited states in electron-ion recombination: a study of Au24+, *J. Phys. B: At. Mol. Opt. Phys.* **36**, pp. 3349–3370.

Gribakin, G. F., Sahoo, S. and Dzuba, V (2004). Recombination of low energy electrons with U^{28+}, *arXiv:physics/0401157V1.*

Grimes, S. M. (1988). Problems in calculating parity ratios for nuclear level densities with the Fermi gas model, *Phys. Rev. C* **38**, pp. 2362–2365.

Grimes, S.M. (2002). Nuclear level densities, *J. Nucl. Sci. Tech. Suppl* **2**, pp. 709–713.

Grimes, S. M., Bloom, S. D., Hausman Jr, H. F. and Dalton, B. J. (1979). Method for calculating operator traces, *Phys. Rev. C* **19**, pp. 2378–2386.

Grimes, S. M., Bloom, S. D., Vanach, H. K. and Hausman Jr, H. F. (1983). Spectral distribution calculation of the level density and spin-cutoff parameters of ^{28}Si, *Phys. Rev. C* **27**, pp. 2893–2901.

Grimes, S. M. and Massey, T. N. (1995). New expansion technique for spectral distribution calculations, *Phys. Rev. C* **51**, pp. 606–610.

Grimes, S. M., Poppe, C. H., Wong, C. and Dalton, B. J. (1978). Spin cutoff parameters for ^{28}Si, ^{29}Si, and ^{30}P: Comparison of measurements with spectral distribution calculations, *Phys. Rev. C* **18**, pp. 1100–1109.

Grimes, S. M. and Voinov, A. (2009). Nuclear level densities, *AIP Conf. Proc.* **1090**, pp. 79–83.

Haake, F. (2001). *Quantum Signatures of Chaos*, 2nd edn. (Springer, Berlin).

Halemane, T. R. (1981a). A method for evaluation of fixed-J traces, *Z. Phys. A* **301**, pp. 177–181.

Halemane, T. R. (1981b). A statistical approach to perturbation theory and inverse-energy-weighted sum rules, *J. Math. Phys.* **22**, pp. 2961–2967.

Halemane, T. R. (1982). Upper bounds for electromagnetic excitations, *Phys. Lett. B* **108**, pp. 256–260.

Halemane, T. R., Abbas , A. and Zamick, L. (1981). Magnetic dipole energy-weighted sum rules with a delta interaction, *J. Phys. G* **7**, pp. 1639–1645.

Halemane, T. R. and French, J. B. (1982a). Electromagnetic sum rules by spectral distribution methods, *Phys. Rev. C* **25**, pp. 2029–2058.

Halemane, T. R. and French, J. B. (1982b). Isovector magnetic dipole sum rules in (ds)-shell *Z. Phys. A* **304**, pp. 363–366.

Halemane, T. R., Kar, K. and Draayer, J. P. (1978). Simple approximations and symmmetry measures for model interactions, *Nucl. Phys. A* **311**, pp. 301–323.

Haq, R. U. (1975). Applications of distribution methods to symmetries in nuclei, Ph.D. Thesis, Gujarat University, Ahmedabad, India.

Haq, R. U. (1995). From chaos to spectroscopy, in D.S. Koltun and A. Das (ed.), *From Spectroscopy to Chaos* (World Scientific, Singapore), pp. 89–104

Haq, R. U., Pandey, A. and Bohigas, O. (1982). Fluctuation properties of nuclear energy levels: Do theory and experiment agree?, *Phys. Rev. Lett.* **48**, pp. 1086–1089.

Haq, R. U. and Parikh, J. C. (1974). Space symmetry in light nuclei II. SU(4) spin-isospin averages, *Nucl. Phys. A* **220**, pp. 349–366.

Haq, R. U. and Parikh, J. C. (1976). Statistical spectroscopy for nuclei in the fp shell, *Nucl. Phys. A* **273**, pp. 410–428.

Haq, R. U. and Quesne, C. (1982a). Shell model and spectral distribution study of SU(4) symmetry mixing in the $s-d$ shell, *Phys. Lett. B* **114**, pp. 91–94.

Haq, R. U. and Quesne, C. (1982b). Spin-isospin $SU(4)$ symmetry breaking in statistical spectroscopy , *Physica A* **114**, pp. 327–330.

Haq, R. U. and Wong, S. S. M. (1979). Fixed angular momentum moments and level densities, *Nucl. Phys. A* **327**, pp. 314–334.

Haq, R. U. and Wong, S. S. M. (1980). Level density in large spectroscopic spaces, *Phys. Lett. B* **93**, pp. 357–362.

Haq, R. U. and Wong, S. S. M. (1982). Nuclear level densities, *Can. J. Phys.* **60**, pp. 1502–1509.

Hardy, C. H. and Ramanujan, S. (1918). Asymptotic formulae in combinatory analysis, *Proc. London. Math. Soc.* **17**, pp. 75–115.

Hecht, K. T. (1974). Summation relation for $U(N)$ Racah coefficients, *J. Math. Phys.* **15**, pp. 2148–2156.

Hecht, K. T. (1975). A simple class of $U(N)$ Racah coefficients and their applications, *Comm. Math. Phys.* **41**, pp. 135–156.

Hecht, K. T. and Draayer, J. P. (1974). Spectral distributions and the breaking of isospin and supermultiplet symmetries in nuclei, *Nucl. Phys. A* **223**, pp. 285–319.

Hillman, M. and Grover, J. R. (1969). Shell-Model Combinatorial Calculations of Nuclear Level Densities, *Phys. Rev.* **185**, pp. 1303–1319.

Hirst, M. G. and Wybourne, B. G. (1986). Statistical group theory and the distribution of angular momentum states: II, *J. Phys. A* **19**, pp. 1545–1549.

Horoi, M., Brown, B. A. and Zelevinsky, V. (1994). Truncation method for shell model calculations, *Phys. Rev. C* **50**, pp. R2274–R2277.

Horoi, M., Ghita, M. and Zelevinsky, V. (2004). Fixed spin and parity nuclear level density for restricted shell model configurations, *Phys. Rev. C* **69**, pp. 041307(R)/1-4.

Horoi, M., Ghita, M. and Zelevinsky, V. (2005). Comparison of approaches for spin- and parity-dependent shell model nuclear level density, *Nucl. Phys. A* **758**, pp. 142c–145c.

Horoi, M., Kaiser, J. and Zelevinsky, V. (2003). Spin- and Parity-dependent nuclear level densities and the exponential convergence method, *Phys. Rev. C* **67**, pp. 054309/1-7.

Horoi, M., Volya, A. and Zelevinsky, V. (1999). Chaotic wave functions and exponential convergence of low-lying energy eigenvalues, *Phys. Rev. Lett.* **82**, pp. 2064–2067.

Horoi, M. and Zelevinsky, V. (2007a). Exact removal of center-of-mass spurious states from level density, *Phys. Rev. Lett.* **98**, pp. 262503/1-4.

Horoi, M and Zelevinsky, V. (2007b). Pairing phase transitions in nuclear wave functions, *Phys. Rev. C* **75**, pp. 054303/1-11.

Horoi, M., Zelevinsky, V. and Brown, B. A. (1995). Chaos vs thermalization in the nuclear shell model, *Phys. Rev. Lett.* **74**, pp. 5194–5197.

Hsu, L. S. (1967). Upper bounds for electromagnetic transition rates, *Phys. Lett. B* **25**, pp. 588–592.

Huang, P., Grimes, S. M. and Massey, T.N. (2000). Level densities for nuclei with $20 \leq$ A ≤ 41, *Phys. Rev. C* **62**, pp. 024002/1-12.

Hutchinson, T. P. and Lai, C. D. (1990). *Continuous Bivariate Distributions, Emphasizing Applications* (Rumsby scientific publishing, Adelaide).

Jacquemin, C. (1973). A method of calculation of J-fixed shell model spectral distributions, *Phys. Lett. B* **43**, pp. 253–258.

Jacquemin, C. (1981). Fixed J spectral distributions in large shell model spaces II. Fixed J configuration centroids and widths, *Z. Phys. A* **303**, pp. 135–146.

Jacquemin, C., Auger, G. and Quesne, C. (1982). Fixed J spectral distributions in large shell model spaces III. Fixed J quasiparticle centroids and widths , *Z. Phys. A* **309**, pp. 55–64.

Jacquemin, C. and Kataria, S. K. (1986). Exact calculation of level densities for many non-interacting fermions in large model spaces, *Z. Phys. A* **324**, pp. 261–269.

Jacquemin, C. and Spitz, S. (1979a). A method of exact calculation of fixed-J configuration centroids, *J. Phys. G* **5**, pp. L95–L99.

Jacquemin, C. and Spitz, S. (1979b). Fixed J spectral distributions in large shell model spaces I. Scalar centroids and widths , *Z. Phys. A* **290**, pp. 251–260.

Jacquod, Ph. and Shepelyansky, D. L. (1997). Emergence of Quantum chaos in finite interacting Fermi systems, *Phys. Rev. Lett.* **79**, pp. 1837–1840.

Jacquod, Ph. and Stone, A. D. (2000). Suppression of Ground-State Magnetization in Finite-Size Systems due to Off-Diagonal Interaction Fluctuations, *Phys. Rev. Lett.* **84**, pp. 3938–3941.

Jacquod, Ph. and Stone, A. D. (2001). Ground state magnetization for interacting fermions in a disordered potential: Kinetic energy, exchange interaction, and off-diagonal fluctuations, *Phys. Rev. B* **64**, pp. 214416/1-26.

Jacquod, Ph. and Varga, I. (2002). Duality between the weak and strong interaction limits of deformed fermionic two-body random ensembles, *Phys. Rev. Lett.* **89**, pp. 134101/1-4.

Jensen, A. S. (1977). Isospin dependence of the nuclear level density, *Phys. Lett. B* **68**, pp. 105–107.

Johnson, C. W., Bertsch, G. F. and Dean, D. J. (1998). Orderly spectra from random interactions, *Phys. Rev. Lett.* **80**, pp. 2749–2753.

Johnson, N. L. and Kotz, S. (1972). *Distributions in Statistics: Continuous Multivariate Distributions* (John Wiley and Sons, New York).

Johnson, C. W., Nabi, J. U. and Ormand, W. E. (2001). Statistical models of nuclear level densities, *arXiv:nucl-th/0105041-v1.*

Johnson, C. W. and Terán, E. (2005). The nuclear density of states and the role of the residual interaction. *arXiv:nucl-th/0512008-v1.*

Judd, B. R., Miller Jr., W., Patera, J. and Winterniz, P. (1974). Complete set of commuting operators and $O(3)$ scalars in the enveloping algebra of $SU(3)$, *J. Math. Phys.* **15**, pp. 1787–1799.

Kanestrom, I. (1966). Some features of the calculated nuclear level density, *Nucl. Phys.* **83**, pp. 380–400.

Kar, K. (1981a). A statistical method for the study of beta decay giant resonances, *Nucl. Phys. A* **368**, pp. 285–318.

Kar, K. (1981b). Gamow-Teller decay strength in the Wigner supermultiplet scheme, *Phys. Rev. C* **28**, pp. 2446–2451.

Kar, K. (1983). M1 strengths and sum rules in the Wigner supermultiplet scheme, *J. Phys. G* **9**, pp. 735–744.

Kar, K., Chakravarti, S. and Manfredi, V. R. (2006). Beta decay rates for nuclei with $115 < A < 140$ for r-process nucleosynthesis, *Pramana-J. Phys.* **67**, pp. 363–368.

Kar, K., Goswami, S. and Raychaudhuri, A. (1995). Neutrino mass and mixing: The impact of excited state contributions on capture rates of Ga detectors, *Int. J. Mod. Phys. A* **10**, pp. 133–143.

Kar, K., Sarkar, S., Gómez, J.M.G., Manfredi, V. R. and Salasnich, L. (1997). Spectral distribution studies of fp shell nuclei with a modified Kuo-Brown interaction, *Phys. Rev. C* **55**, pp. 1260–1265.

Kar, K., Sarkar, S. and Ray, A. (1991). Beta decay rates of fp shell nuclei in advanced evolutionary stages of presupernova stars, *Phys. Lett. B* **261**, pp. 217–220.

Kar, K., Sarkar, S. and Ray. A. (1994). β decay rates of fp shell nuclei with $A > 60$ in massive stars at the presupernova stage, *The Astrophysical Journal* **434**, pp. 662–683.

Kar, K. and Wong, S. S. M. (1981). A shell model study of the Gamow-Teller giant resonance, *Phys. Lett. B* **98**, pp. 393–397.

Karazija, R. (1991). *Sums of atomic quantities and mean characteristics of spectra* (Mokslas publishers, Vilnius), in Russian.

Karwowski, J. (1994). Statistical theory of spectra, *Int. J. Quantum Chem.* **51**, pp. 425–437.

Karwowski, J. and Bancewicz, M. (1987). The first two moments of energy level distributions in N-electron spin-adopted model spaces, *J. Phys. A: Math. Gen.* **20**, pp. 6309–6320.

Karwowski, J., Rajadell, F., Planelles, J. and Mas. V. (1995). The first four moments of spectral density distribution of an N-electron hamiltonian matrix defined in an antisymmetric and spin-adapted model space, *Atomic Data and Nuclear Data Tables* **61**, pp. 177–232.

Kendall, M. G. and Stuart, A. (1969). *Advanced Theory of Statistics*, Vol. 1, 3rd edn. (Hafner Publishing Company, N.Y.).

Koltun, D. S. and Das, A. (eds.) (1995). *From Spectroscopy to Chaos* (World Scientific, Singapore).

Koonin, S. E., Dean, D. J. and Langanke, K. (1997). Shell model Monte Carlo methods, *Phys. Rep.*, **278**, pp. 1–77.

Kopsky, V. (1979). Extended integrity bases of finite groups, *J. Phys. A* **12**, pp. 429–443.

Kota, V. K. B. (1979a). Sizes of effective single particle fields for $s-d$ shell effective interactions, *Phys. Rev. C* **20**, pp. 347–356.

Kota, V. K. B. (1979b). A symmetry for the widths of the eigenvalue spectra of boson and fermion systems, *J. de Physique Lettres (Paris)* **40**, pp. L579–L582.

Kota, V. K. B. (1981). Studies on the goodness of IBA group symmetries: Centroids, widths and partial widths for irreducible representations of IBA group symmetries, *Ann. Phys. (N.Y.)* **134**, pp. 221–258.

Kota, V. K. B. (1984). Bivariate distributions in statistical spectroscopy studies I: Fixed-J level densities, fixed-J averages and spin cut-off factors, *Z. Phys. A* **315**, pp. 91–98.

Kota, V. K. B. (1991a). Single nucleon transfer strengths and sum rules in the interacting boson model and spectral averaging theory, in N.G. Puttaswamy (ed.), *Direct nuclear reactions - Proceedings of the international seminar* (Indian Academy of Sciences, Bangalore, India), pp. 321–339.

Kota, V. K. B. (1991b). Level densities, expectation values and strengths with interactions: Convolution forms and applications, in R.W. Hoff (ed.), *Capture Gamma Ray Spectroscopy*, Vol. 238 (AIP conference proceedings, New York), pp. 685–693.

Kota, V. K. B. (1997). Recent Developments in Statistical Nuclear Spectroscopy: Level Densities for ^{28}Si, ^{168}Er, ^{181}W, and ^{234}U nuclei, in G. Shanmugam and N. Arunachalam (ed.), *Recent Trends in Nuclear Structure Physics* (Published by M.S. University, Tirunelveli, India, 1997), pp. I2.1–I2.19.

Kota, V. K. B. (2001). Embedded random matrix ensembles for complexity and chaos in finite interacting particle systems, *Physics Reports* **347**, pp. 223–288.

Kota, V. K. B. (2003). Convergence of moment expansions for expectation values with embedded random matrix ensembles and quantum chaos, *Ann. Phys. (N.Y.)* **306**, pp. 58–77.

Kota, V. K. B. (2005a). Regularities with random interactions in energy centroids defined by group symmetries, *Phys. Rev. C* **71**, pp. 041304(R)/1-5.

Kota, V. K. B. (2005b). SU(N) Wigner-Racah algebra for the matrix of second moments of embedded Gaussian unitary ensemble of random matrices, *J. Math. Phys.* **46**, pp. 033514/1-9.

Kota, V. K. B. (2006a). Two-body ensembles with group symmetries for chaos and regular structures, *Int. J. Mod. Phys. E* **15**, pp. 1869–1883.

Kota, V. K. B. (2006b). Group Theoretical and Statistical Properties of Interacting Boson Models of Atomic Nuclei: Recent Developments, in A.V. Ling (ed.), *Focus on Boson Research* (Nova Science Publishers Inc., New York), pp. 57–105.

Kota, V. K. B. (2007a). $U(2\Omega) \supset U(\Omega) \otimes SU(2)$ Wigner-Racah algebra for embedded Gaussian unitary ensemble of random matrices with spin, *J. Math. Phys.* **48**, pp. 053304/1-9 .

Kota, V. K. B. (2007b). Nuclear Models and Statistical Spectroscopy for Double Beta Decay, in V.K.B. Kota and Utpal Sarkar (eds.), *Neutrinoless Double Beta Decay* (NAROSA, New Delhi), pp. 63–76.

Kota, V. K. B., Chavda, N. D. and Sahu, R. (2006a). Bivariate t-distribution for transition matrix elements in Breit-Wigner to Gaussian domains of interacting particle systems, *Phys. Rev. E* **73**, pp. 047203/1-4.

Kota, V. K. B., Chavda, N. D. and Sahu, R. (2006b). One plus two-body random matrix ensemble with spin: analysis using spectral variances, *Phys. Lett. A* **359**, pp. 381–389.

Kota, V. K. B. and Kar, K. (1989). Spectral distributions in nuclei: General principles and applications, *Pramana-J. Phys.* **32**, pp. 647–692.

Kota, V. K. B. and Kar, K. (2002). Group symmetries in two-body random matrix ensembles generating order out of complexity, *Phys. Rev. E* **64**, pp. 026130/1-7.

Kota, V. K. B. and Majumdar, D. (1992a). Tests of moment methods for spin cut-off and occupancy densities for non-interacting particle systems, Technical Report PRL-TH/92-12, Physical Research Laboratory, Ahmedabad, India.

Kota, V. K. B. and Majumdar, D. (1992b). Unitary decomposition of spherical configuration scalar part of nuclear hamiltonians all across the periodic table, Technical Report PRL-TH/92-21, Physical Research Laboratory, Ahmedabad, India.

Kota, V. K. B. and Majumdar, D. (1995a). Bivariate distributions in statistical spectroscopy studies: III. Non interacting particle strength densities for one-body transition operators, *Z. Phys. A* **351**, pp. 365–376.

Kota, V. K. B. and Majumdar, D. (1995b). Bivariate distributions in statistical spectroscopy studies: IV. Interacting particle Gamow-Teller strength densities and β-decay rates of fp-shell nuclei for presupernova stars, *Z. Phys. A* **351**, pp. 377–383.

Kota, V. K. B. and Majumdar, D. (1996). Application of spectral averaging theory in large shell model spaces: Analysis of level density data of fp-shell nuclei, *Nucl. Phys. A* **604**, pp. 129–162.

Kota, V. K. B., Majumdar, D., Haq, R. U. and Leclair, R. J. (1999). Shell-model tests of the bimodal partial state densities in a 2×2 partitioned embedded random matrix ensemble, *Can. J. Phys.* **77**, pp. 893–901.

Kota, V. K. B., Mayya, K. B. K. and Castilho Alcarás, J. A. (2008). Statistical law for multiplicities of $SU(3)$ irreps (λ, μ) in the plethysm $\{\eta\} \overset{3}{\otimes} \{m\} \to (\lambda, \mu)$, *J. Phys. A: Math. Theor.* **42**, pp. 145201/1-20.

Kota, V. K. B., Pandya, S. P. and Potbhare, V. (1980). A comparative study of effective interactions and correlations with pairing and quadrupole operators, *Nucl. Phys. A* **349**, pp. 397–415.

Kota, V. K. B., Pandya, S. P. and Potbhare, V. (1982). Occupancies and structural changes in upper $f-p$ shell nuclei, *Phys. Rev. C* **25**, pp. 1667–1670.

Kota, V. K. B. and Potbhare, V. (1979). Ground state proton-neutron occupancies in the $f-p$ shell, *Nucl. Phys. A* **331**, pp. 93–116.

Kota, V. K. B. and Potbhare, V. (1980). Shape of eigenvalue distribution for bosons in scalar space, *Phys. Rev. C* **21**, pp. 2637–2642.

Kota, V. K. B. and Potbhare, V. (1982). Energy dependence of the strengths of pairing and quadrupole - quadrupole components of the nuclear effective interactions, *Z. Phys. A* **309**, pp. 159–162.

Kota, V. K. B. and Potbhare, V. (1985). Bivariate distributions in statistical spectroscopy studies: II. Orthogonal polynomials in two variables and fixed-E, J expectation values, *Z. Phys. A* **322**, pp. 129–136.

Kota, V. K. B., Potbhare, V. and Shenoy, P. (1986). Representation of shell-model density of states, *Phys. Rev. C* **34**, pp. 2330–2333.

Kota, V. K. B., Potbhare, V., Shenoy, P. and Tressler, N. (1989). Inverse energy weighted sum rules, *Pramana J. Phys.* **32**, pp. 507–513.

Kota, V. K. B. and Sahu, R. (1998). Information Entropy and Number of Principal Components in Shell Model Transition Strength Distributions, *Phys. Lett. B* **429**, pp. 1–6.

Kota, V. K. B. and Sahu, R. (2000). Theory for matrix elements of one-body transition operators in the quantum chaotic domain of interacting particle systems, *Phys. Rev. E* **62**, pp. 3568–3571.

Kota, V. K. B. and Sahu, R. (2001a). Structure of wavefunctions in (1+2)-body random matrix ensembles, *Phys. Rev. E* **64**, pp. 016219/1-8.

Kota, V. K. B. and Sahu, R. (2001b). Breit-Wigner to Gaussian transition in strength functions, *arXiv:nucl-th/0006079.*

Kota, V. K. B. and Sahu, R. (2002). Single particle entropy in (1+2)-body random matrix ensembles, *Phys. Rev. E* **66**, pp. 037103/1-4.

Kota, V. K. B., Sahu, R., Kar, K., Gómez, J. M. G. and Retamosa, J. (1999). Transition Strength Sums and Quantum Chaos, *Phys. Rev. C* **60**, pp. 051306/1-4.

Kota, V. K. B., Vyas, M. and Mayya, K. B. K. (2008). Spectral distribution analysis of random interactions with J-symmetry and its extensions, *Int. J. Mod. Phys. E* **17** (Suppl), pp. 318–333.

Kusnezov, D., Alhassid, Y. and Snover, K. A. (1998). Scaling properties of the giant dipole resonance widths in hot rotating nuclei, *Phys. Rev. Lett.* **81**, pp. 542–545.

Langhoff, P. W. (1980). Stieltjes-Tchebycheff moment-theory approach to photoeffect studies in Hilbert space, in B. J. Dalton, S. M. Grimes, J. P. Vary and S. A. Williams (eds.) *Theory and Applications of Moment Methods in many Fermion Systems* (Plenum, New York), pp. 191–212.

Laberge, G. J. H. and Haq, R. U. (1990). "Universality" of Gaussian orthogonal ensemble fluctuations: the two-body random ensemble and shell model spectra, *Can. J. Phys.* **68**, pp. 301–312.

Layzer, D. (1963). Proof of Moszkowski's formula for the variance of term energies in an electronic configuration of the Form ℓ^n, *Phys. Rev.* **132**, pp. 2125–2127.

Leboeuf, P. and Roccia, J. (2006). Level density of a Fermi gas: Average growth and fluctuations, *Phys. Rev. Lett.* **97**, pp. 010401/1-4.

Leclair, R. J. (1994).Information content of quantum spectra, M.Sc. Thesis, Laurentian University, Sudbury, Ontario, Canada.

Leclair, R. J., Haq, R. U., Kota, V. K. B. and Chavda, N. D. (2008). Power spectrum analysis of the average-fluctuation density separation in interacting particle systems, *Phys. Lett. A* **372**, pp. 4373–4378.

Lewenkopf, C. H. and Zelevinsky, V. G. (1994). Single and multiple giant resonances: Counterplay of collective and chaotic dynamics, *Nucl. Phys. A* **569**, pp. 183c–194c.

Li, S. Y. and Koltun, D. S. (1973). (unpublished: private communication).

Lombardi, M., Bohigas, O. and Seligman, T. H. (1994). New evidence of GOE statistics for compound nuclear resonances, *Phys. Lett. B* **324**, pp. 263–266.

Lougheed, G. D., Ng, W. Y. and Wong, S. S. M. (1976). Truncation of the shell model vector space, *Phys. Lett. B* **64**, pp. 242–246.

Lougheed, G. D. and Wong, S. S. M. (1975a). Preliminary results on fixed-JT averaging method for many-particle spectroscopy, *Nucl. Phys. A* **243**, pp. 215–228.

Lougheed, G. D. and Wong, S. S. M. (1975b). Constant width approximations for large spectroscopic calculations, *Nucl. Phys. A* **246**, pp. 29–42.

Lynn, J. E. (1968). *The theory of neutron resonance reactions* (Oxford University Press, New York).

Macfarlane, M. H. (1986). Gamow-Teller sum rules and ground-state correlations, *Phys. Lett. B* **182**, pp. 265–268.

Majumdar, D. (1994). Nuclear Level Densities and Other Statistical Quantities with Interactions in Spectral Averaging Theory: Systematic Studies and Applications, Ph. D. Thesis, M.S. University of Baroda, Vadodara, India.

Majumdar, D., Kar, K. and Ansari, A. (1997). Giant dipole resonance in spectral averaging theory, *J. Phys. G* **23**, pp. L41–L47.

Majumdar, D., Kar, K. and Ansari, A. (1998). Giant dipole resonance in spectral averaging theory and analysis of experimental data for the ^{76}Se nucleus, *J. Phys. G* **24**, pp. 2103–2109.

Majumdar, D. and Kar, K. (2007). Beta decay rates of nuclei with $65 < A < 75$ for presupernova and supernova evolution, *Pramana-J.Phys.* **68**, pp. 423–433.

Manfredi,V. R. (1981). The spectral-distribution method in boson space, *Il Nuovo Cimento A* **64**, pp. 101–113.

Manfredi,V. R. (1984). Level density fluctuations of interacting bosons, *Lettere Al Nuovo Cimento* **40**, pp. 135–139.

Manfredi,V. R. (1985). Global and local properties of interacting bosons, *Il Nuovo Cimento A* **86**, pp. 219–230.

Manfredi,V. R. (1986). Old and new results in statistical nuclear spectroscopy, *La Rivista del Nuovo Cimento* **9**, No. 9, pp. 1–31.

Manfredi,V. R. and Salmaso, G. (1980). The level density of interacting bosons, *Lettere Al Nuovo Cimento* **28**, pp. 569–572.

Margetan, F. J., Klar, A. and Vary, J. P. (1983). Nuclear binding energies from moment methods: Harmonic oscillator Hamiltonian, *Phys. Rev. C* **27**, pp. 852–861.

McGrory, J. B., Wildenthal, B. H. and Halbert, E. C. (1970). Shell-model structure of $^{42-50}$Ca, *Phys. Rev. C* **2**, pp. 186–212.

Mead, L. R. and Papanicolaou, N. (1984). Maximum entropy in the problem of moments, *J. Math. Phys.* **25**, pp. 2404–2417.

Mehta, M. L. (1971). A note on correlations between eigenvalues of a random matrix, *Comm. Math. Phys.* **20**, pp. 245–250.

Mehta, M. L. (2004). *Random Matrices*, 3rd edition (Elsevier B.V., The Netherlands).

Mitchell, G. E., Bowman, J. D., Penttilä, S. I. and Sharapov, E. I. (2001). Parity violation in compound nuclei: experimental methods and recent results, *Phys. Rep.* **354**, pp. 157–241.

Mitchell, G. E., Bowman, J. D. and Weidenmüller, H. A. (1999). Parity violation in the compound nucleus, *Rev. Mod. Phys.* **71**, pp. 445–457.

Mocelj, D., Rauscher, T., Martnez-Pinedo, G., Langanke, K., Pacearescu,L., Faessler, A., Thielemann, F. -K. and Alhassid, Y. (2007). Large-scale prediction of the parity distribution in the nuclear level density and application to astrophysical reaction rates, *Phys. Rev. C* **75**, pp. 045805/1-15.

Mon, K. K. (1973). Ensemble-averaged eigenvalue distributions for k-body interactions, B.A. Dissertation, Princeton University, Princeton, USA.

Mon, K. K. and French, J. B. (1975). Statistical properties of many-particle spectra, *Ann. Phys(N.Y.)* **95**, pp. 90–111.

Moszkowski, S. A. (1962). On the Energy Distribution of Terms and Line Arrays in Atomic Spectra, *Prog. Theor. Phys.* **28**, pp. 1–23.

Mottelson, B. (1985). Some Periods in the History of Nuclear Physics with Particular Emphasis on the Compound Nucleus in R. Broglia, G. Hageman and B. Herskind (ed.), *Nuclear Structure 1985* (North-Holland, Amsterdam), pp. 3–24.

Mugambi, P. E. (1970). Theory of nuclear densities with residual interactions, Ph.D. thesis, University of Rochester, Rochester, New York, USA.

Mughabghab, S. F. (1984). *Neutron resonance parameters and thermal cross sections*, Vol. 1 (Academic Press, New York).

Mulhall, D., Volya, A. and Zelevinsky, V. (2000). Geometric chaoticity leads to ordered spectra for randomly interacting fermions, *Phys. Rev. Lett.* **85**, pp. 4016–4019.

Nabi, J. U., Johnson, C. W., and Ormand, W. E. (2001). Reliable calculations of level densities using statistical spectroscopy , *arXiv:nucl-th/0111067-v1.*

Naik, S., Ramkumar, S., Shenoy, P. and Tressler, N. (1985). Ground state occupancies of $f-p-g$ shell nuclei, *Pramana-J.Phys.* **25**, pp. 251–258.

Ng, W. Y. and Wong, S. S. M. (1976). A comparison of the fixed-JT averaging and projected Hartree-Fock methods for nuclear structure studies, *Can. J. Phys.* **54**, pp. 2367–2372.

Nishioka, H., Weidenmüller, H. A. and Yoshida, S. (1988). Partial level densities in precompound reactions, *Phys. Lett. B* **203**, pp. 1–3.

Nissimov, H., Arvieu, R. and Bohigas, O. (1972). Extension of the spectral distribution method to many-quasiparticle subspaces, *Nucl. Phys. A* **190**, pp. 514–532.

Nomura, M. (1972a). Sum rules for many-particle matrix elements in the seniority scheme, *Prog. Theo. Phys.* **47**, pp. 1858–1877.

Nomura, M. (1972b). Reduction relations for the third and the fourth moments of energy spectra in a many-particle configuration, *Prog. Theo. Phys.* **48**, pp. 110–132.

Nomura, M. (1972c). Distribution of energy spectra in a large j shell, *Prog. Theo. Phys.* **48**, pp. 442–458.

Nomura, M. (1974). A calculation method for a high moment of energy spectra, *Prog. Theo. Phys.* **51**, pp. 489–505.

Nomura, M. (1975). Spectral distribution method in the J-scheme, *Ann. Phys. (N.Y.)* **91**, pp. 83–105.

Nomura, M. (1978). Study of gaussian distribution in nuclear spectroscopy, *Prog. Theo. Phys.* **59**, pp. 1199–1212.

Nomura, M. (1985a). Extension of Wick's theorem by means of the state operator formalism, *J. Math. Phys.* **26**, pp. 732–737.

Nomura, M. (1985b). A reduction relation for an n-body trace of an operator product consisting of coupled C+'s and C's, *J. Math. Phys.* **26**, pp. 738–742.

Nomura, M. (1985c). On the propagation of the operator average in truncated space, *J. Math. Phys.* **26**, pp. 965–969.

Nomura, M. (1986). Propagation coefficients for fixed-isospin (T, T_z) average and related spectroscopic sum rules, *J. Math. Phys.* **27**, pp. 536–543.

Nomura, M. (1987a). The sum rule for spectroscopic factors in the seniority scheme of identical particles, *J. Math. Phys.* **28**, pp. 128–133.

Nomura, M. (1987b). A reduction relation for the trace over a finite set of many-fermion states with a definite symmetry $U(N)$, *Phys. Lett. A* **126**, pp. 1–4.

Nomura, M. (1988). Spin-adapted reduced hamiltonians in view of the spectral-distribution method, *Phys. Rev. A* **37**, pp. 2709–2711.

Nomura, M. (1990). Novel correspondence of the fixed-seniority and the fixed-isospin averages in their reduction formulas, *J. Math. Phys.* **31**, pp. 351–353.

Pandey, A. (1979). Statistical properties of many-particle spectra III: Ergodic behavior in random-matrix ensembles, *Ann. Phys. (N.Y.)* **119**, pp. 170–191.

Pandey, A. (1981). Statistical properties of many-particle spectra IV: New ensembles by Stieltjes transform methods ensembles, *Ann. Phys. (N.Y.)* **119**, pp. 170–191.

Papenbrock, P., Kaplan, L. and Bertsch, G. F. (2002). Odd-even binding effect from random two-body interactions, *Phys. Rev. B* **65**, pp. 235120/1-8.

Papenbrock, P. and Weidenmüller, H. A. (2004). Distribution of spectral widths and preponderance of spin-0 ground states in nuclei, *Phys. Rev. Lett.* **93**, pp. 132503/1-4.

Papenbrock, P. and Weidenmüller, H. A. (2005).Origin of chaos in the spherical nuclear shell model: Role of symmetries, *Nucl. Phys. A* **757**, pp. 422–438.

Papenbrock, P. and Weidenmüller, H. A. (2006). Two-body random ensemble in nuclei, *Phys. Rev. C* **73**, pp. 014311/1-15.

Papenbrock, P. and Weidenmüller, H. A. (2007). Random matrices and chaos in nuclear spectra, *Rev. Mod. Phys.* **79**, pp. 997–1013.

Parikh, J. C. (1972). Mixing of $SU(3)$ symmetry for nuclei in the ds-shell, *Phys. Lett. B* **41**, pp. 468–472.

Parikh, J. C. (1973). Space symmetry in light nuclei I. Application to the $2s-1d$ shell, *Ann. Phys. (N.Y.)* **76**, pp. 202–235.

Parikh, J. C. (1978). *Group Symmetries in Nuclear Structure* (Plenum, New York).

Parikh, J. C. (1978a). Statistical properties of atomic energy levels I: Partial level density, *J. Phys. B* **11**, pp. 1881–1887.

Parikh, J. C. (1983). Random matrix theory and the statistical mechanics of disordered systems, *Pramana-J. Phys.* **20**, pp. 467–476.

Parikh, J. C. and Wong, S. S. M. (1972). A study of some measures for group symmetry breaking in nuclei, *Nucl. Phys. A*, **182**, pp. 593–605.

Pasachoff, J. M. (1966). Proof of a conjecture by Moszkowski for the variances of spectral-line distributions, *Phys. Rev.* **150**, pp. 47–50.

Patel, K., Desai, M. S. and Potbhare, V. (1999). Spectral averaging methods and Nilsson model, *Phys. Rev. C* **59**, pp. 1199–1202.

Patel, K., Desai, M. S., Potbhare, V. and Kota, V. K. B. (2000). Average-fluctuations separation in energy levels in dense interacting boson systems, *Phys. Lett. A* **275**, pp. 330–337.

Planelles, J., Rajadell, F., Karwowski, J., and Mas. V. (1996). A diagrammatic approach to statistical spectroscopy of many-fermion hamiltonians, *Phys. Rep.* **267**, pp. 161–194.

Planelles, J., Rajadell, F. and Karwowski, J. (1997). Spectral density distribution moments of N-electron hamiltonians in the low-density limit, *J. Phys. A* **30**, pp. 2181–2196.

Pluhar, Z. and Weidenmüller, H. A. (1988). Approximation for shell-model level densities, *Phys. Rev. C* **38**, pp. 1046–1057.

Pluhar, Z., Weidenmüller, H. A., Papenbrock, T. and Tithof, J. (2009). Chaos in fermionic many-body systems and metal-insulator transition, *arXiv:0911.0316v1 [nlin.CD]*.

Porter, C. E. (1965). *Statistical Theories of Spectra: Fluctuations* (Academic Press, New York).

Porter, C. E. and Thomas, R. G. (1956). Fluctuations of nuclear reaction widths, *Phys. Rev.* **104**, pp. 483–491.

Potbhare, V. (1975). Spectral averaging methods in nuclear physics, Ph. D. Thesis, University of Rochester, Rochester, New York, USA.

Potbhare, V. (1977). Correlation coefficients of effective interactions in the $s-d$ shell, *Nucl. Phys. A* **289**, pp. 373–380.

Potbhare, V. (1978). Shape of spectrum with ensemble of 2-body random Hamiltonians, *Pramana-J. Phys.* **11**, pp. 205–208.

Potbhare, V. and Pandya, S. P. (1976). Ground-state occupancies and effective interactions in $s-d$ shell nuclei, *Nucl. Phys. A* **256**, pp. 253–270.

Potbhare, V. and Tressler, N. (1991). Single-nucleon transfer sum-rules in the $2s1d$ shell, *Nucl. Phys. A* **530**, pp. 171–186.

Prigogine, I. (1980). *From Being to Becoming: Time and Complexity in the Physical Sciences* (W. H. Freeman and Company, New York), pp. 89.

Preedom, B. M. and Wildenthal, B. H. (1972). Shell-model calculations for ^{22}Na and ^{22}Ne, *Phys. Rev. C* **6**, pp. 1633–1644.

Quesne,C. (1973). Study of the validity of the Gaussian approximation for the nuclear spectral distributions in a solvable model, *Phys. Lett. B* **43**, pp. 463–466.

Quesne,C. (1975a). Partial widths of fixed generalized seniority spectral distributions, *Phys. Lett. B* **57**, pp. 422–424.

Quesne,C. (1975b). Group theory and propagation of operator averages, *J. Math. Phys.* **16**, pp. 2427–2431.

Quesne,C. (1976). $SU(2) \times SU(2)$ scalars in the enveloping algebra of $SU(4)$, *J. Math. Phys.* **17**, pp. 1452–1467.

Quesne,C., Salmon, Y. and Spitz, S. (1977). New statistical measures of symmetry breaking in the ground state region of nuclei , *J. de Physique Lettres (Paris)* **38**, pp. 237–240.

Quesne,C. and Spitz, S. (1974). Spectral distributions of mixed configurations of identical nucleons in the seniority scheme I. Generalized seniority scheme, *Ann. Phys. (N.Y.)* **85**, pp. 115–151.

Quesne,C. and Spitz, S. (1977). Seniority four admixtures in the low-lying 0^+ states of even-mass tin and lead nuclei , *J. de Physique Lettres (Paris)* **38**, pp. 337–342.

Quesne,C. and Spitz, S. (1978). Spectral distributions of mixed configurations of identical nucleons in the seniority scheme II. Configuration-seniority scheme, *Ann. Phys. (N.Y.)* **112**, pp. 304–327.

Rademacher, H. (1938). On the Partition Function $p(n)$, *Proc. London Math. Soc.* **S2-43**, pp. 241–254.

Ratcliff, K. F. (1971). Application of spectral distributions in nuclear Spectroscopy, *Phys. Rev. C* **3**, pp. 117–143.

Rajadell, F., Planelles, J. and Karwowski, J. (1993), Traces of powers of the hamiltonian operator in finite-dimensional antisymmetric model spaces, *J. Phys. A* **26**, pp. 3883–3901.

Rajadell, F., Planelles, J., Karwowski, J. and Mas, V. (1995). Spectral density distribution of an N-electron hamiltonian in a finite-dimensional and spin-adapted model space, *Int. J. Quant. Chem.* **56**, pp. 71–79.

Rosensteel, G. and Draayer, J. P. (1985). Analytic expressions for energy centroids and widths in the microscopic collective model, *Nucl. Phys. A* **436**, pp. 445–457.

Sarkar, S. and Kar, K. (1988). Gamow-Teller strength sums for nuclei in the fp shell, *J. Phys. G* **14**, pp. L123–L129.

Sarkar, S. and Kar, K. (1989). Gamow-Teller and M1 strength sums for *sd* shell nuclei by spectral distribution methods, *Phys. Rev. C* **40**, pp. 1826–1829.

Sarkar, S. and Kar, K. (1995). Effect of inclusion of $1g$ orbit on the Gamow-Teller strengths for the capture rates of solar neutrinos in Ga detectors, *J. Phys. G* **21**, pp. 411–418.

Sarkar, S. and Kar, K. (1996). Gamow-Teller strengths in fp shell, *Phys. Lett. B* **387**, pp. 227–232.

Sarkar, S., Kar, K. and Kota, V. K. B. (1987) Spectral distribution calculations using Wildenthal's universal *sd* interaction, *Phys. Rev. C* **36**, pp. 2700–2703.

Satyan, V. and Parikh, J. C. (1976). Minimization of energy and of energy variance for Slater determinants in nuclei, *Phys. Rev. C* **14**, pp. 1198–1206.

Satyan, V. and Parikh, J. C. (1979). Unitary decomposition of Hamiltonian operators I: Structure and sizes in Hartree-Fock basis, *Ann. Phys.(N.Y.)* **119**, pp. 285–304.

Schmidt, S. and Alhassid, Y. (2008). Mesoscopic competition of superconductivity and ferromagnetism: Conductance peak statistics for metallic grains, *Phys. Rev. Lett.* **101**, pp. 207003/1-4.

Seligman, T. H. and Nishioka, H. (eds.) (1986). *Quantum chaos and statistical nuclear physics*, Lecture notes in physics 263 (Springer, Berlin).

Sharma, S. and Bhatt, K. H. (1973). E2 Transition systematics in the 1f-2p shell: A constraint on the effective interaction, *Phys. Rev. Lett.* **30**, pp. 620–623.

Sheikh, J. A. and Sun, Y. (2004). Chaos and rotational damping in particle-rotor model, *Nucl. Phys. A* **733**, pp. 67–84 (2004).

Shen, J. J., Zhao, Y. M., Arima, A. and Yoshinaga, N. (2008a). Lowest eigenvalue of random hamiltonians, *Phys. Rev. C* **77**, pp. 054312/1-5.

Shen, J. J., Zhao, Y. M., Arima, A. and Yoshinaga, N. (2008b). Strong correlation between eigenvalues and diagonal matrix elements, *Phys. Rev. C* **78**, pp. 044305/1-9.

Smith, D. A. et al., (2001). Parity violation in neutron resonances of ^{117}Sn, *Phys. Rev. C* **64**, pp. 015502/1-9.

Smith III, J. F. (1987). Statistical nuclear spectroscopy: New methods and applications; densities and related functions for interacting-particle nuclei, Ph.D. Thesis, University of Rochester, Rochester, New York, USA.

Soyeur, M. and Zuker, A. P. (1972). Structure of ^{28}Si, some answers and their problems, *Phys. Lett. B* **41**, pp. 135–142.

Spitz, S. and Quesne, C. (1976). Configuration-seniority spectral distributions in ^{62}Ni, *J. de Physique Lettres (Paris)* **37**, pp. 3–6.

Srednicki, M. (2002). Spectral statistics of the k-body random interaction model, *Phys. Rev. E* **66**, pp. 046138/1-8.

Strohmaier, B. and Grimes, S. M. (1988). Spectral distribution calculation of the level density of ^{20}Ne, *Z. Phys. A* **329**, pp. 431–440.

Strohmaier, B., Grimes, S. M. and Bloom, S. D. (1985). Comparison of Lanczos methods and moment methods in calculating nuclear level densities, *Phys. Rev. C* **32**, pp. 1397–1405.

Strohmaier, B., Grimes, S.M. and Satyanarayana, H. (1987). Spectral distribution calculation of the level density of ^{24}Mg, *Phys. Rev. C* **36**, pp. 1604–1610.

Stuart, A. and Ord, J. K. (1987). *Kendall's Advanced Theory of Statistics: Distribution Theory, 5th edition of Vol. 1* (Oxford University Press, New York).

Sutaria, F. K. and Ray, A. (1995). Gamow-Teller strength distributions for nuclei in presupernova stellar cores, *Phys. Rev. C* **52**, pp. 3460–3469.

Sviratcheva, K. D., Draayer, J. P. and Vary, J. P. (2006). Underlying symmetries of realistic interactions and the nuclear many-body problem, *Phys. Rev. C* **73**, pp. 034324/1-12.

Sviratcheva, K. D., Draayer, J. P. and Vary, J. P. (2007). Global properties of fp-shell interactions in many-nucleon systems, *Nucl. Phys. A* **786**, pp. 31–46.

Szego, G. (1939). *Orthogonal Polynomials* (American Mathematical Society, Providence, Rhode Island).

Talmi, I. (1993). *Simple models of complex nuclei: the shell model and interacting boson model* (Harwood Academic Publishers, Chur, Switzerland).

Terán, E. and Johnson, C. W. (2006a). Behavior of shell model configuration moments, *Phys. Rev. C* **73**, pp. 024303/1-8.

Terán, E. and Johnson, C. W. (2006b). Simple models for shell-model configuration densities, *Phys. Rev. C* **74**, pp. 067302/1-4.

Tomsovic, S. (1986). Bounds on the time-reversal non-invariant nucleon-nucleon interaction derived from transition-strength fluctuations, Ph.D. Thesis, University of Rochester, Rochester, New York, USA.

Tomsovic, S., Johnson, M. B., Hayes, A. C. and Bowman, J. D. (2000). Statistical theory of parity nonconservation in compound nuclei, *Phys. Rev. C* **62**, pp. 054607/1-17.

Touchard, J., Haq, R. U. and Arvieu, R. (1997). An ensemble of random particle-hole matrices with collective eigenstates, *Z. Phys. A*, **282**, pp. 191–201.

Van Lier, C. and Uhlenbeck,G. E. (1937). On the statistical calculation of the density of the energy levels of the nuclei, *Physica* **4**, pp. 531–542.

Vary, J. P., Belehrad, R. and Dalton, B. J. (1979). Nuclear properties with realistic Hamiltonians through spectral distribution theory, *Nucl. Phys. A* **328**, pp. 526–536.

Verbaarschot, J. J. M. (1982). On spectral averages in nuclear spectroscopy, Ph. D. Thesis, Fysisch Laboratorium, Rijksuniversiteit Utrecht, Utrecht, The Netherlands.

Verbaarschot, J. J. M. and Brussaard, P. J. (1981). A new method to evaluate operator moments for fixed angular momentum, *Phys. Lett. B* **102**, pp. 201–208.

Verbaarschot, J. J. M. and Brussaard, P. J. (1984). Fixed-JT averages of electromagnetic operators, *Nucl. Phys. A* **423**, pp. 77–92.

Verbaarschot, J. J. M. and Brussaard, P. J. (1985). Distribution of electromagnetic transition amplitudes, *Z. Phys. A* **321**, pp. 125–135.

Verbaarschot, J. J. M., Timmer, G. A. and Brussaard, P. J. (1982). On the systematic energy difference between states of positive and negative parity, *Nucl. Phys. A* **378**, pp. 280–292.

Vincent, C. M. (1967). Group classification of many-body interactions, *Phys. Rev.* **163**, pp. 1044–1050.

Volya, A. (2008). Emergence of symmetry from random n-body interactions, *Phys. Rev. Lett.* **100**, pp. 162501/1-4.

von Neumann, J. and Wigner, E. (1929). Behaviour of eigenvalues in adiabatic processes, *Z. Physik* **30**, pp. 467–470.

Vyas, M., Kota, V. K. B. and Chavda, N. D. (2009). One plus two-body random matrix ensembles with spin: Results for pairing correlations, *Phys. Lett. A* **373**, pp. 1434–1443.

Vyas, M. and Kota, V. K. B. (2009). Random matrix ensembles with random interactions: Results for EGUE(2)-$SU(4)$, *Pramana-J.Phys.* **73**, pp. 521–531.

Vyas, M., Kota, V. K. B. and Chavda, N. D. (2010). Transitions in eigenvalue and wavefunction structure in (1+2)-body random matrix ensembles with spin, Phys. Rev. E, in press.

Wang, W., Izrailev, F. M. and Casati, G. (1998). Structure of eigenstates and local spectral density of states: A three orbital schematic shell model, *Phys. Rev. E* **57**, pp. 323–339.

Weidenmüller, H. A. and Mitchell, G. E. (2009). Random matrices and chaos in nuclear physics: Nuclear structure, *Rev. Mod. Phys.* **81**, pp. 539–589.

Wheeler, J. C. and Gordon, R. G. (1970). Bounds for averages using moment constraints, in Baker, G. (ed.), *The Pade Approximation in Theoretical Physics* (Academic Press, New York), pp. 99–128.

Wigner, E. P. (1937). On the consequences of the symmetry of the nuclear Hamiltonian on the spectroscopy of nuclei, *Phys. Rev.* **51**, pp. 106–119.

Wigner, E. P. (1955). Characteristic vectors of bordered matrices with infinite dimensions, *Ann. Math.* **62**, pp. 548–564.

Wigner, E. P. (1961). The Probability of the Existence of a Self Reproducing Unit. Reproduced from *The Logic of Personal Knowledge: Essays in Honor of Michael Polanyi, Chapter 19 (London: Routledge and Kegan Paul Lts., 1961)*, in Moore, W.J. and Scriven, M. (ed.), *Symmetries and Reflections: Scientific Essays of E.P. Wigner* (Indiana University Press, Bloomington, 1967, reprinted by Ox Bow Press, Connecticut, 1979), pp. 200–208.

Wong, S. S. M. (1978). Truncation of the nuclear shell-model space using spectral strength distribution, *Nucl. Phys. A* **295**, pp. 275–288.

Wong, S. S. M. (1986). *Nuclear Statistical Spectroscopy* (Oxford University Press, New York)

Wong, S. S. M. and Chang, B. D. (1982). Binding energy, occupancy and spin-cutoff factor calculated by statistical spectroscopy with configuration-isospin skewness and excess terms, *Nucl. Phys.* **A389**, pp. 324–340.

Wong, S. S. M. and French, J. B. (1972). Level-density fluctuations and two-body versus multi-body interactions, *Nucl. Phys. A* **198**, pp. 188–208.

Wong, S. S. M. and Lougheed, G. D. (1978). A study of ^{28}Si using shell-model and statistical spectroscopic methods, *Nucl. Phys. A* **295**, pp. 289–300.

Wong, S. S. M. and Zhao, E. (1982). A study of excitation strength distribution with shell-model examples, *Nucl. Phys. A* **385**, pp. 1–13.

Wong, S. S. M., Zhao, E. and Zhu, X. (1980). Astrophysical partition function, *The Astrophysical Journal* **242**, pp. L173–L177.

Wybourne, B. G. (1974). *Classical groups for physicists* (Wiley Interscience, New York).

Yoshinaga, Y., Arima, A. and Zhao, Y. M. (2006). Lowest bound of energies for random interactions and the origin of spin-zero ground state dominance in even-even nuclei, *Phys. Rev C* **73**, pp. 017303/1-4.

Yoshinaga, Y., Arima, A., Shen, J. J. and Zhao, Y. M. (2009). Correlation between eigenvalues and sorted diagonal elements of a large dimensional matrix, *Phys. Rev. C* **79**, pp. 017301/1-4.

Yueng, Y. Y. and Newman, D. J. (1986). High order crystal field invariants and the determination of lanthanide crystal field parameters, *J. Chem. Phys.* **84**, pp. 4470–4473.

Zelevinsky, V., Brown, B. A., Frazier, N. and Horoi, M. (1996). The nuclear shell model as a testing ground for many-body quantum chaos, *Phys. Rep.* **276**, pp. 85–176.

Zelevinsky, V. and Volya, A. (2004). Nuclear structure, random interactions and mesoscopic physics, *Phys. Rep.* **391**, pp. 311–352.

Zhao, Y. M., Arima, A. and Yoshinaga, N. (2004). Regularities of many-body systems interacting by a two-body random ensemble, *Phys. Rep.* **400**, pp. 1–66.

Zhao, Y. M., Arima, A., Yoshida, N., Ogawa, K., Yoshinaga, N. and Kota, V. K. B. (2005). Robustness of regularities for energy centroids in the presence of random interactions, *Phys. Rev. C* **72**, pp. 064314/1-8.

Zirnbauer, M. R. and Brink, B. M. (1981). Distribution of angular momenta in shell-model configurations, *Z. Phys. A* **301**, pp. 237–242.

Zuker, A. P. (2001). Binomial level densities, *Phys. Rev. C* **64**, pp. 021303(R)/1-4.

Zuker, A. P., Ndeuna, L. W., Nowacki, F. and Caurier, E. (2001). Canonical form of Hamiltonian matrices, *Phys. Rev. C* **64**, pp. 021304(R)/1-4.

Original Papers

Volume 23, number 4 PHYSICS LETTERS 24 October 1966

COMBINATORIAL METHODS FOR SPECTROSCOPIC AVERAGING AND SPECTRAL MOMENTS *

J. B. FRENCH
Department of Physics and Astronomy, University of Rochester, Rochester, New York

Received 21 September 1966

The scalar and fixed-partition scalar averagings of an operator are treated by combinatorial methods which exploit the fact that the averages are polynomials in occupation numbers. The results are of particular value for studying the low-order moments of spectral distributions.

This paper will give combinatorial methods for the scalar and fixed-partition scalar averagings [1] of an operator. The combinatorial method is to some extent an alternative to the density-matrix method of the preceding paper. In some cases one method is preferrable, in other cases the other. The particular advantages of the combinatorial method are its transparent economy with respect to the input information and the fact that it may be used for operators which are not readily available in normal or antinormal form (indeed they may not even be completely specified). Though we shall write things in general form, our particular interest will be in the fixed-partition averages of low powers of the system Hamiltonian.

Consider to begin with, the (scalar) averaging of an operator T over all the m-particle states of a system of degeneracy N. Let T be of mixed particle rank, but let us agree that it has no rank larger than v. Then from the fact that, for an operator of definite particle rank k, $\langle T(k)\rangle^m$ is a polynomial of order k in the variable m, we see

* Supported in part by the U.S. Atomic Energy Commission.

that $\langle T\rangle^m$ is also a polynomial of order v. Let us emphasize this by writing $\langle T\rangle^m = P^{(v)}(m)$. But then it is clear that $\langle T\rangle^m$ can be expressed, for $m \geqslant v$, in terms of the $(v+1)$ quantities $\langle T\rangle^t$ where $t = 0 \ldots v$. In terms of the polynomials we have

$$P^{(v)}(m) = \sum_{t=0}^{v} \Lambda_t^v(m)\, P^{(v)}(t)\,. \qquad (1)$$

To evaluate the "propagation" coefficients Λ is an easy combinatorial problem: we find *

$$\Lambda_t^v(m) = (-1)^{t-v}\frac{(m-v)}{(m-t)}\binom{m}{v}\binom{v}{t} = \qquad (2)$$

$$= (-1)^{t-v}\binom{m}{t}\binom{m-t-1}{m-v-1}$$

and thus

$$\langle T\rangle^m = \sum_{t=0}^{v} (-1)^{t-v}\binom{m}{t}\binom{m-t-1}{m-v-1}\langle T\rangle^t\,. \qquad (3)$$

We are free here to take v as large as we wish; the form with the smallest v (= maximum particle rank of T) is of maximum economy in the sense that the number of pieces of input data $\langle T\rangle^t$ has then its smallest value, namely $(v+1)$. Observe too that, since the product of polynomials is itself a polynomial (of higher order), the same propagation equation holds for the product of averages, provided only that we take v not less than the *combined* maximum particle ranks. Then

$$\langle T\rangle^m\langle U\rangle^m\langle V\rangle^m \ldots = \sum_{t=0}^{v} \Lambda_t^v(m)\langle T\rangle^t\langle U\rangle^t\langle V\rangle^t \ldots$$

$$\langle TUV\ldots\rangle^m = \sum_{t=0}^{v} \Lambda_t^v(m)\langle TUV\ldots\rangle^t\,. \qquad (4)$$

It is clear then that the maximum-economy (smallest v) equation for the moments of an operator is valid also for the cumulants. This important result will be seen to apply also to the extended propagation equations which we write below.

As it stands, eq. (3) applied to the p'th moment of the Hamiltonian (assumed of maximum rank 2) requires as input the averages over states with up to $2p$ particles. A very great simplication results if we "symmetrize" the input data with respect to particles and holes. There are two points here.

* The resultant eq. (3) is intimately connected with the well known Sylvester Sieve equation [2]. Note also that for $m \leqslant v$ eq. (2) gives $\Lambda_t^v(m) = \delta_{mt}$ as is obviously necessary.

The first is that, if we are considering the *calculation* of spectral moments, then from the t-particle calculations we can quite easily by a hole $\rightleftharpoons$ particle transformation produce the t-hole results, while if we are taking the averages from data the few-hole results are more likely to the available than the many-particle results. The second point is that the averages for different t's are "linearly independent" and we can evaluate $\langle T\rangle^m$ in terms of the averages for any $(v+1)$ t-values whatever.

By writing the polynomials as the sum of two parts one of which vanishes for $m = 0, 1 \ldots v_1$ the other for $m = N-v_2, \ldots N$ (where as always N is the system degeneracy) we find the appropriate generalization of eq. (3) to be (with $v = v_1 + v_2 + 1$).

$$\begin{aligned}\langle T\rangle^m = &\binom{N-m}{v_2+1}\sum_{t=0}^{v_1}(-1)^{t-v_1}\times\\ &\times\binom{m}{t}\binom{m-t-1}{m-v_1-1}\binom{N-t}{v_2+1}^{-1}\langle T\rangle^t +\\ &+\binom{m}{v_1+1}\sum_{t=N-v_2}^{N}(-1)^{t-v_2+N}\times\\ &\times\binom{N-m}{N-t}\binom{t-1-m}{N-m-v_2-1}\binom{t}{v_1+1}^{-1}\langle T\rangle^t\,.\end{aligned} \qquad (5)$$

Now for the p'th moment of a (0 + 1 + 2)-body Hamiltonian we could take‡ $v_1 = v_2 + 1 = p$. Since the complexity of t-hole spectroscopy is exactly that of t-particle spectroscopy we see then that, as far as the moments are concerned, the hole $\rightleftharpoons$ particle symmetrization has effectively reduced the complexity of H to that of a (0 + 1)-body operator, a great simplification indeed. Observe now the general result that the traces of $(H)^p$ for t-particle and t-hole systems with $t = 0, \ldots p$ determine (in fact slightly overdetermine) the p'th spectral moments for arbitrary numbers of particles. Generalizations of this result, involving partial traces, are valid for finer averagings.

As minor applications of these equations we

‡ If instead we take the less economical set $v_1 = v_2 = p$ the two terms of eq. (5) are related by a hole $\rightleftharpoons$ particle transformation and may be combined into one by measuring occupancies of states with respect to the half-filled system. We can maintain simultaneously the maximum economy and the formal advantages of this symmetry by dealing separately with the "h $\rightleftharpoons$ p symmetric" and "h $\rightleftharpoons$ p antisymmetric" averages $\langle (H)^p\rangle^m \pm \langle (H)^p\rangle^{N-m}$, this coming about because the antisymmetric average is represented by a polynomial of order $(2p-1)$.

take $v_1 = 1$, $v_2 = 0$ and $v_1 = 2$, $v_2 = 1$, in eq. (5) to produce respectively convenient forms for the spectrum centroid energy $E(m) = \langle H\rangle^m$ and the square of the dispersion $\sigma^2(m) = \langle H^2\rangle^m - \{\langle H\rangle^m\}^2$. We have immediately

$$E(m) = \frac{m(m-1)}{N(N-1)}E(N) - \frac{(N-m)(m-1)}{N}E(0) + \frac{(N-m)m}{(N-1)}E(1)\,,$$

$$\sigma^2(m) = \frac{m(m-1)(m-2)(N-m)(N-m-1)}{(N-1)(N-2)(N-3)} \times \left\{\frac{\sigma^2(N-1)}{(N-m-1)} - \frac{(N-3)}{(m-1)}\sigma^2(1) + \frac{(N-1)}{2(m-2)}\sigma^2(2)\right\}. \tag{6}$$

The energy result is trivial. A special case of the dispersion result has been given previously by Moszkowski [3] and Layzer [4].

The evaluation of the low-order scalar moments does tell us something about the spectra but not very much. The fixed-partition moments however would tell us a great deal, and it is therefore important to extend the above results to the fixed-partition case. Let us divide the N single-particle states into l sets, which we may think of as orbits. Then a simple extension of the arguments used above informs us that the corresponding fixed-partition average of an operator of maximum total rank v is a polynomial of order v in the variables (occupation numbers) $m_1 \dots m_l$, being then defined by any $\binom{l+v}{v}$ of its linearly independent values. On solving the combinatorial problem which extends eq. (3) to this case we find the result

$$\langle T\rangle^{m_1 \dots m_l} = \sum_{\substack{t_1=0 \\ (t \leq v)}}^{v} \cdots \sum_{t_l=0}^{v} (-1)^{t-v} \times \binom{m-t-1}{m-v-1}\binom{m_1}{t_1}\cdots\binom{m_l}{t_l}\langle T\rangle^{t_1 \dots t_l}\,, \tag{7}$$

where $m = \Sigma m_i$, $t = \Sigma t_i$. It is easy to see that this is a solution of maximum economy, and is therefore a valid solution to the fixed-partition problem *. The symmetrization with respect to holes and particles is not however a trivial operation in the fixed-partition case. An inelegant way to carry it out is to use eq. (7) to eliminate unwanted sets $[t_1 \dots t_l]$ in favour of sets involving small $(N_i - m_i)$.

We shall later in a more detailed paper discuss various features of combinatorial methods, give a "canonical" procedure for the symmetrization, and make some applications for which the combinatorial method is particularly suitable. It should be observed that neither the method of the present paper nor that of the preceding one is suitable for high-order spectral moments. We hope also to discuss that problem.

The author acknowledges many helpful conversations with D. S. Koltun and K. F. Ratcliff.

References

1. H. Banerjee and J. B. French, Physics Letters 23 (1966) 245.
2. H. J. Ryser, Combinatorial mathematics (John Wiley, New York, 1963).
3. S. A. Moszkowski, Progr. Theoret. Phys. (Kyoto) 28 (1962) 1.
4. D. Layzer, Phys. Rev. 132 (1963) 2125.

* Note the rather remarkable form. One of the factors in the kernel of eq. (3) is extended multiplicatively while, in the other factor the extension is additive with respect to the (m_i, t_i) variables. The form in which *both* factors are extended multiplicatively, the $(t \leq v)$ restriction being then removed, is also a solution (one which ignores the correlations between the maximum ranks in different orbits) but it is not of maximum economy and is in fact of no real interest. We remark too that the structure of eq. (7) can be made clear in a "non-combinatorial" way by using the same "net Green's function" method as used for isospin averaging in the previous paper and indeed the equation can be easily derived that way.

* * * * *

Volume 26B, number 2 PHYSICS LETTERS 25 December 1967

MEASURES FOR GROUP-SYMMETRY BREAKING IN NUCLEI *

J. B. FRENCH **

Department of Physics and Astronomy, University of Rochester, Rochester, New York, USA

and Instituto de Física, Universidad de México, México, D.F.

Received 6 November 1967

An orthonal separation of the Hamiltonian operator into parts which preserve and break a given symmetry is introduced, along with norms for measuring the separate components. The procedure takes full account of the restrictions imposed by the exact symmetries.

We discuss the question of the goodness of group symmetries in nuclei. The aim is to produce measures for the symmetry breaking which will, among other things, enable us to decide upon the conditions under which a given group symmetry is good, and whether, when these are not met, the symmetry is still good enough for the group to be of interest. We proceed by defining a measure for an operator, this in a standard way implying a definition of operator orthonormality. There are two important requirements on this measure. It must, in the usual mathematical sense, be a norm, and moreover one which is compatible with the unitary norm used for the state vectors of the system. We can then be certain that the norm of the sum of two operators is not greater than the sum of their norms and that the norm of the product of two operators or of an operator and a state vector will not be greater than the product of *their* norms; without this requirement, conclusions drawn from the values of the norms could be very misleading. The second requirement is that it must be possible to evaluate the norm even in very complicated spaces. Both of these requirements are met by the *Euclidean norm* which, for an operator O given as a transformation on a vector space S is defined by $(\| O \|_S)^2 = \mathrm{Tr}(O^+O) = d(s)\langle O^+O\rangle^S$ where $\langle\ \rangle^S$ is an average over the states of S and d(s) is the dimensionality of S. Then, dropping the dimensionality factor, we may say that O_α, O_β are orthonormal in S if $\langle O_\alpha O_\beta\rangle^S = \delta_{\alpha\beta}$.

The vector spaces in which we shall be interested involve states of m particles distributed over N single-particle states ($N = 24$ for the nuclear (ds)-shell). It follows then that the group U(N) of transformations among these states really dominates the whole thing, and our concern is with the goodness of the symmetries for a group $\mathcal{G}$ which is a subgroup of U(N). Besides these symmetries there are also the "exact" symmetries whose existence has a profound effect on the question which we are considering. For the present, the only exact symmetries which we shall tolerate are particle number, isospin (in cases where it is relevant), and angular momentum, the parity being irrelevant in the cases which we now consider. Let it be understood then that $\mathrm{U}(N) \supset \mathrm{U}(\frac{1}{2}N) \times \mathrm{U}(2)$ (isospin) $\supset \mathrm{R}_3$ (angular momentum) and that we have a Hamiltonian operator H. We are interested in the extent to which the eigenstates of the system are describable in terms of irreducible representations of the group $\mathcal{G}$. We shall agree that H is Hermitian, scalar with respect to isospin and angular momentum, and of maximum particle rank 2. Our aim is to make an orthogonal decomposition of H into parts which preserve and violate the symmetry of $\mathcal{G}$ and then to measure the separate parts. It is immediately clear that we should measure things separately in each of the subspaces S in which the exact symmetries are specified.

We introduce now the following vector spaces of operators; (i) $\boldsymbol{H}$, the space of all H operators compatible with the states conditions (the dimensionality of $\boldsymbol{H}$ equals the number of linearly independent defining matrix elements, e.g. 67 for the ds-shell); and $\boldsymbol{H}$(s) its representation in the "exact-symmetry subspace" S; (ii) $\boldsymbol{G}$, a basis for which is the set of Casimir operators of $\mathcal{G}$,

* Supported in part by the U.S. Atomic Energy Commission.

** N.S.F. Senior Postdoctoral Fellow.

Volume 26B, number 2 PHYSICS LETTERS 25 December 1967

and G(s), its representations; (iii) X, the space of H operators which have no symmetry-breaking matrix elements in * $\{N\}$, and X(s). A basis for X is supplied by bilinear forms (scalar products) of the infinitesimal operators of $\mathcal{G}$ together with those further operators of H which behave like multiples of unity in every irreducible representation of $\mathcal{G}$; in the present case a basis for the latter subspace of X is given by the operators 1, n, $\binom{n}{1}$ and T^2. Alternatively X is defined by $X \subset H$ and $[X,G] = 0$; (iv) $\tilde{X}$, defined by the fact ** that its representation $\tilde{X}$ (s) in any S is the set of all H_α (s) for which $[H_\alpha(\mathrm{s}), G(\mathrm{s})] = 0$. It is easy to see that $\tilde{X} \supset X$. It follows now that any element H_α (s) of H(s) which is not an element of $\tilde{X}$(s) necessarily has symmetry-breaking matrix elements in S, i.e. matrix elements in S connecting two inequivalent irreducible representations of $\mathcal{G}$. A necessary and sufficient condition then that there be no symmetry breaking in S is that $H(\mathrm{s}) \in \tilde{X}(\mathrm{s})$, while if $H(\mathrm{s}) = \tilde{X}(\mathrm{s})$ no Hamiltonian can induce a symmetry breaking. These essential results follow from the fact that the space S decomposes into subspaces giving irreducible representation of $\mathcal{G}$, and, in a basis adapted to this decomposition (in which we should, in the case of multiplicities, combine together the spaces giving equivalent representations of $\mathcal{G}$) every G_β in G is represented by a diagonal matrix which decomposes into a set of multiples of unity. A matrix then, which fails to commute with a matrix of this sort, must have at least one non-diagonal element connecting vectors which give different diagonal elements for G_β and therefore having different symmetries. Of course the converse fact that if an operator commutes with G in S it has no symmetry breaking matrix elements in S follows directly from the fact that the operators of G uniquely label the irreducible representations.

We have now an appropriate formal orthogonal decomposition of H(s) into H_p(s), a symmetry-preserving part which lies in $\tilde{X}$(s) (and the collection of which for all S defines the abstract operator H_p), and H_b(s), symmetry-breaking ‡, which is orthogonal to H_p(s). If $\tilde{X}_\alpha$(s) be an orthonormal basis for $\tilde{X}$(s) then

$$H(\mathrm{s}) = \tilde{X}(\mathrm{s}) + H_\perp(\mathrm{s})$$

$$\langle H_\perp(\mathrm{s})\tilde{X}(\mathrm{s})\rangle^\mathrm{S} = \langle H_\perp \tilde{X}\rangle^\mathrm{S} = 0$$

$$H(\mathrm{s}) = H_\mathrm{p}(\mathrm{s}) + H_\mathrm{b}(\mathrm{s})$$

$$H_\mathrm{p}(\mathrm{s}) = \sum_\alpha \langle \tilde{X}_\alpha H\rangle^\mathrm{S}\, \tilde{X}_\alpha(\mathrm{s}) \tag{1}$$

$$\langle H_\mathrm{p}(\mathrm{s})H_\mathrm{b}(\mathrm{s})\rangle^\mathrm{S} = \langle H_\mathrm{p}H_\mathrm{b}\rangle^\mathrm{S} = 0$$

$$(\|H\|_\mathrm{s})^2 = (\|H_\mathrm{p}\|_\mathrm{s})^2 + (\|H_\mathrm{b}\|_\mathrm{s})^2$$

$$(\|H_\mathrm{p}\|_\mathrm{s})^2 = \langle H_\mathrm{p}^2\rangle^\mathrm{S} = \sum_\alpha \langle H\tilde{X}_\alpha\rangle^\mathrm{S}\langle \tilde{X}_\alpha H\rangle^\mathrm{S}.$$

Before proceeding we must remember that $\tilde{X}$(s) contains, as a one-dimensional subspace, multiples of the unit operator in S, deriving in part from the trivial 4-dimensional subspace of X mentioned above. Since the unit operator does not affect the states of S, the component of H(s) in this subspace should be removed by a further orthogonal splitting (or alternatively by using the dispersion instead of the square of H_p in evaluating the norms), and discarded. Let us assume

* The N-space $\{N\}$ is the 2^N-dimensional space of all m-fermion states ($m = 0, 1, \ldots, N$). $\{N;m\}$, $\{N;m,T\}$, $\{N;m,T,J\}$ etc. are subspaces with specified symmetries as indicated while, in the present case, $\mathrm{S} = \{N;m,T,J\}$. Defining an operator by its second-quantized form, which we may loosely label as "abstract", is fully equivalent to specifying it as a transformation on the N-space.

** The point here is that $\tilde{X}$ (s) defined by $[\tilde{X}(\mathrm{s}), G(\mathrm{s})] = 0$ may not coincide with X(s), the representation in S of X defined by $[X,G] = 0$. An example of this arises from the fact that, in the one-shell case, the two-body part of $[H,G]$ vanishes and thus $[H,G]$ is a space of three-body operators; thus all symmetries are good for the two-particle system, an obvious result which would however not be produced if we were to measure the components of H in the spaces produced by the "abstract" separation of H, thereby ignoring the distinction between X and $\tilde{X}$. This behaviour corresponds to the fact that even a representation which is separately faithful for H and G need not be faithful for $[H,G]$.

‡ It must however be understood that H_b has both symmetry-breaking and non-breaking matrix elements, this coming about because of the restriction on the decomposition of H into subspaces. One effect of this restriction is that a symmetry-breaking matrix element "generates" an accompanying operator in $\tilde{X}$. Things would be different if we had allowed a *general* decomposition of H (into parts, in particular, with unrestricted particle ranks) for then, as is easily seen, the resultant H_b would have only symmetry-breaking matrix elements. Its norm, which would take account of all such elements of H, and only of these, would, at least in principle, give a better account of the symmetry breaking. Its evaluation, which would be much more difficult, might best be considered via a group averaging method applied to calculate the partial widths of representations (partial in the sense that one considers the width of one representation due to interactions with another). On the one hand however, the formal theory of partial widths is, except for some special groups, in a rudimentary state, while, on the other hand, an evaluation of the partial widths in a space S would tell us much more than we need for the present purpose.

that this has been done, the resultant H then becoming H'. Our criterion now for the goodness of the symmetry in the subspace S will then be that

$$\zeta(s) = \frac{\|H_b\|_s}{\|H'\|_s} \ll 1 \qquad (2)$$

and, supplemented by some experience in cases which are soluble in detail, the set of numerical values of $\zeta(s)$ should then give quite detailed information about the symmetry. The condition under which a symmetry will be exactly good in a subspace S is simply that $\zeta(s) = 0$, while the condition that it be good for all states is that $\zeta(s) = 0$ for all S, or equivalently that $H \in \tilde{X}$. It should be clear also that of the symmetry is good in any representation which is faithful for H and for $[H,G]$ then it is necessarily good for all representations; an example of this, which has been known for some time, is that the single-shell symplectic symmetry is necessarily good if it is good for the three-particle system. These exact-symmetry results are of course independent of which norm we use.

For the construction of the $\tilde{X}$ space we need the explicit abstract form of G, which may be regarded as available, and the abstract commutator space $[H,G]$ whose construction is then straightforward. But we must also be able to identify the subspace of $[H,G]$ which is zero in S; we already know that this subspace contains the space X whose explicit structure is known, so that the question here is that of finding the "extra parts of H" which, in part because of the restrictions imposed on the system by the exact symmetries, are pushed into the symmetry-preserving operator space. In some cases, as in the trivial ($m = 2$) one-shell case mentioned above, the identification will be immediate. If, in others, we can convince ourselves that the dimensionalities of X(s) and $\tilde{X}$(s) are equal, then we know that $X(s) = \tilde{X}(s)$. A quite general procedure uses the fact that an operator vanishes in S if and only if its S-norm vanishes. The identification problem then becomes one of evaluating the $[H,G]$ norms and examining the metric matrix defined by them.

The whole problem of measuring the symmetry breaking thus reduces to one of evaluating subspace norms of H and of $[H,G]$. Available for this is the method of spectral averaging, recently discussed in some detail [1], which treats the propagation throughout a lattice of subspaces of $\{N\}$ of traces of operators in these subspaces, exactly what we need for evaluating norms. As things stand at present the propagation is well understood for the $\{N; m, T\}$ spaces, so that if we lower our standards a little and use norms in these larger-than-minimal subspaces * we may achieve an explicit numerical solution to the problem for a given group and even draw some *general* conclusions about the way in which the symmetry breaking varies throughout the subspaces.

We shall shortly report a detailed application of the present method and, in the future, hope to apply it, and others related to it, in a more or less systematic study of group symmetries in nuclei. To a considerable extent the present work grows from earlier work done by Ginocchio and the author [3], which however did not make use of an adequate norm and which paid little attention to the exact symmetries. The author acknowledges the cooperation of Dr. H. Banerjee in the early stages of this work, and some valuable comments from Dr. D. S. Koltun.

References

1. H. Banerjee and J. B. French, Phys. Letters 23 (1966) 245;
J. B. French, in: Nuclear Structure, eds. A. Hossain et al. (North-Holland Publishing Co., 1967) p. 85, 124.
2. e.g. C. A. Levinson, Phys. Rev. 132 (1963) 2184.
3. J. N. Ginocchio, University of Rochester Thesis (1964). Nucl. Phys. 63 (1965) 449.

* The $\{N; m, T, J\}$ norms are much more difficult to calculate though it appears now that a solution is at hand. Without this norm it is not feasible to perform the elimination of the "J^2 part" of H (in the $H \rightarrow H'$ transformation). One can however do this approximately by using $(H - \alpha J^2)$ as the Hamiltonian and then carrying out a variation with respect to α, a procedure which has been often used in studies of rotational structures [2].

* * * * *

Volume 33B, number 7 PHYSICS LETTERS 7 December 1970

VALIDITY OF RANDOM MATRIX THEORIES FOR MANY-PARTICLE SYSTEMS*

J. B. FRENCH
Department of Physics and Astronomy, University of Rochester, Rochester, New York, USA

and

S. S. M. WONG
Department of Physics, University of Toronto, Toronto, Canada
and Department of Physics and Astronomy, University of Rochester, Rochester, New York, USA

Received 19 October 1970

The assumption of statistical independence of matrix elements leads to ensembles of Hamiltonians which involve simultaneous interactions between many particles. Then, for example, the resultant (semicircular) spectra are quite different from the Gaussian spectra found for ensembles properly restricted to involve only two-body interactions.

In this paper we consider certain aspects of the random-matrix theory of spectra [1-3], asking in particular whether the ensemble of Hamiltonians which results from the assumption of statistical independence of the matrix elements is a proper one. Our conclusion will be that the ensemble is by no means appropriate for many-particle system in which the predominant interactions involve only two particles. As a result the "semi-circular" (or "semi-elliptical") eigenvalue distribution, which results when the independence condition is supplemented by certain conditions on the matrix-element distributions [4], is not the characteristic distribution to be encountered in nuclear or atomic spectroscopy. The characteristic distribution is in fact normal (Gaussian), as has become clear recently [5-6] from examining the results of spectroscopic calculations, and will be demonstrated in this paper for an ensemble of H's which *is* appropriate. Of course random matrix calculations have been concerned less with simple eigenvalue distributions** than with more subtle quantities (spacing distributions and so on), but considerable uncertainty would seem to attach to any results which rely on the assumption of statistical independence of matrix elements.

The independence assumption greatly extends the space of the Hamiltonians which one would normally admit. To take an example, the most generaly (d,s)-shell Hamiltonian satisying the usual requirements of symmetry and particle rank has 67 matrix elements, so that these Hamiltonians form a 67-dimensional vector space; this is in strong contrast to the 2×10^7-dimensional space encountered in the random matrix treatment of the $J = 3$, $T = 1$ states of $(d, s)^{12}$. The difference between the two cases is moreover not just one of dimensionalities, for the particle vector spaces have an (anti-symmetrized) direct-product structure, the 10-particle space for example being simply the 10-fold product of the 1-particle space with itself. The conventional H is then defined in a (0 + 1 + 2)-particle factor space of this multi-particle space; in this circumstance we may expect that, by the operation of a Central Limit Theorem, the many-particle system will tend to have a characteristic spectrum even if no ensemble average is taken. For one-body Hamiltonians in spaces with a large number of fermions (particles and holes both) and a large number of single-particle states, this spectrum is close to normal [8,7] and it is not then surprising (especially if we think of the Hamiltonian defined in a factor space) that normality is found also with two-body interactions#. The direct-

* Supported in part by the U.S. Atomic Energy Commission and the National Research Council of Canada.

** This seems in part on the ground that the random matrix densities are not in agreement with experiment; this is true but one must still distinguish between the unbounded level densities of experiment and the bounded densities (the eigenfunctions lying in a finite vector space) with which we deal here and which are in fact partial level densities.

See footnote on next page.

Volume 33B, number 7 PHYSICS LETTERS 7 December 1970

product structure of the space being ignored in the random matrix calculations, it is likewise not surprising that the spectrum which emerges is altogether different.

Fig. 1a gives the random-matrix spectrum for a set of 100 50-dimensional matrices, the ensemble being one in which the individual matrix elements H_{ij} ($i \leqslant j$) are, independently, uniformly distributed over the interval (-1,1). This ensemble satisfies the Wigner conditions [4] for a semi-circular spectrum and, as found in similar cases by other authors [2], this is found here also. We now "realize" the 50-dimensional space in terms of the states of several particles constrained to a set of single-particle states, thereby introducing a direct-product structure. At the same time we restrict H to be a (0 + 1 + 2)-body operator and to have the standard invariances. The realization which we choose is $(f_{7/2}, f_{5/2})$ for identical particles, which, for 7 particles ($\equiv$ atomic f^7) with $J = \frac{7}{2}$, has dimensionality 50. There are 21 two-particle matrix elements, as well as a zero-body energy and two one-body energies (one for each j-orbit), so that the properly restricted H resides in a space of 24 dimensions instead of the 1275 of the conventional theory. Fig. 1b shows the spectrum for an ensemble of 100 Hamiltonians with independent *two-body* matrix elements chosen uniformly in (-1,1) interval, the zero- and one-body parameters being chosen zero (a choice which should have no significant effect on the results); the Hamiltonian space is now 21-dimensional. We see an excellent normal spectrum**.

We should consider whether the spectrum shapes are characteristic for typical members

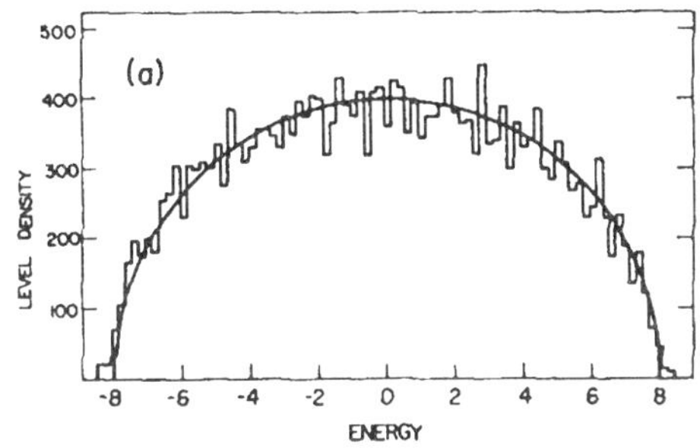

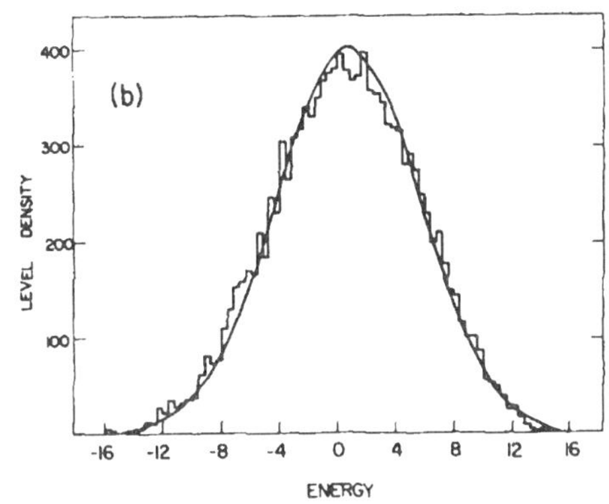

Fig. 1. The cumulative spectra for two 100-member sets of 50-dimensional matrices. (a) is for a conventional ensemble while (b) is for an ensemble of two-body operators acting in the 7-particle space f^7 with $T = J = \frac{7}{2}$.

of the Hamiltonian ensembles or whether they emerge only after ensemble averaging. It is convenient for this purpose to define finite-dimensional semi-circular and normal spectra, and a correlation function which measures how close in shape (ignoring energy translations and scale expansions) two spectra are to each other. For the spectra we use a procedure of Ratcliff [6]; if $f(E)$ is the frequency function, then the (cumulative) distribution function

$$F(E) = \int_{-\infty}^{E} f(E')\,\mathrm{d}E' \;,$$

where

$$F(\infty) = d \;,$$

the matrix dimensionality, is a multi-step function with unit jumps at the eigenvalues. A natural correspondence then with any continuous distribution function is to define the eigenvalue as those points where the function takes on half-integral values,

\# For a one-body H we may choose a single-particle basis which diagonalizes H, and then the total energy is a sum of single-particle energies to which, if Pauli principle effects are ignored, we may apply the usual limit theorem. For two-body interactions, things are by no means well understood. It can be shown that, in spaces with a large number N of single-particle states, the dominant part of a given Hamiltonian will, with high probability, tend, as the fermion number increases, to a number-dependent one-body H (an operator of lower than maximum symmetry with respect to the group $U(N)$ of transformations in the single-particle space), so that the same result should emerge in this case; but the results are clearly more general than that.

** A similar analysis, with similar results, has been made for sets of 500 50-dimensional traceless fixed-strength matrices, in which extra conditions have been imposed that the centroid of the eigenvalues be zero and the trace of H^2 have a fixed value. The resultant spectra are reproduced in ref. [7].

Volume 33B, number 7 PHYSICS LETTERS 7 December 1970

$\frac{1}{2}, \frac{3}{2}, \frac{5}{2}, \ldots$, the standard spectra being then defined by $F(E_j) = j - \frac{1}{2}$, with $j = 1, 2, \ldots$.

For the correlation function we introduce, for a given spectrum labelled by α, the d-dimensional "vector" $E^\alpha = (E_1^\alpha \ldots E_d^\alpha)$ where the E_i^α are ordered $(E_i^\alpha \leqslant E_{i+1}^\alpha)$; then, choosing the centroid

$$\xi^\alpha = d^{-1} \sum_i E_i^\alpha$$

as a new energy zero, and renormalizing the new vector by a positive scale change, so that its length, defined for a vector F as $|F| = \{\sum_i E^2\}^{1/2}$ is unity, we end up with the unit vector

$$\omega^\alpha = \frac{E^\alpha - \xi^\alpha \mathbf{1}}{|E^\alpha - \xi^\alpha \mathbf{1}|}$$

which describes the spectrum "shape". The correlation between two spectra, α, β, may now be considered in terms of the scalar product of the two unit vectors, $\rho(\alpha, \beta) = (\omega^\alpha \cdot \omega^\beta)$, a unit value for which implies that the unit spectra are identical (the original spectra being then identical to within translation and expansion).

Since the operation of replacing a spectrum by its traceless fixed-strength equivalent has no effect on its shape, we use, instead of the 100-member sets of fig. 1, the 500-member sets of such spectra referred to in the footnote above. For the general random matrices we find that in every one of the 500 cases the shapes are closer to semi-circular than to normal; for the ensemble of two-body Hamiltonians 435 cases are closer to normal than to semi-circular while 65 are closer to semi-circular. The distribution of the correlations for each ensemble with each of the characteristic spectra is shown in fig. 2. The distributions of the correlations are very narrow for the general ensemble, indicating that the individual spectra are essentially identical in shape; considerably larger fluctuations in the spectrum shapes are found for the restricted ensemble.

We conclude with a few general remarks. It should be clear that any spectrum shape whatever can be realized within the general random ensemble (not true at all for the restricted ensemble), and, in particular, so can the closer-to-normal spectra produced by the restricted H's. The fact that we see no such case tells us, as we would expect from the dimensionality arguments given at the beginning, that our chances of finding a permissible Hamiltonian in the random ensembles are very small indeed. The conclusion then is that the conventional ensemble consists largely of H's involving multi-particle interactions, and one might in fact be inclined to guess that the maximum weight attaches to H's in which all the

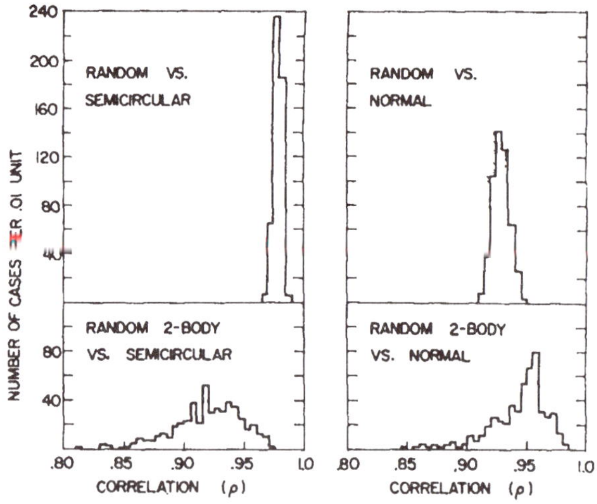

Fig. 2. The distribution of the correlation coefficients, with respect to the semi-circular and normal spectra, for a conventional fixed-strength ensemble and a fixed-strength two-body-interaction ensemble. The ensembles contain 500 50-dimensional matrices. The correlation between the two characteristic spectra is 0.93.

particles interact simultaneously, a situation far removed from that usually considered. We see here a failure of the common belief that it may be safe to use an ensemble of Hamiltonians as long as the great majority of H's in the ensemble produce the same results for the things of interest; the great majority of matrices in the conventional ensemble do produce the same spectra but the very small minority of *permissible* Hamiltonians in the set produce something quite different. On the other hand the evidence is that, if we could deal analytically with the restricted ensemble, we might do well*.

It is proposed to give shortly, for the same realization of the space, the results for ensembles of k-body operators ($k = 3, 4, \ldots 7$) this as a way of studying the transition from one characteristic spectrum to the other. Besides that it is hoped to present results of an analytical study, done via unitary-group decompositions of the Hamiltonian, of the kinds of interactions implied by the usual matrix ensemble.

We mention finally that O. Bohigas and J. G. Flores [9] have been independently carrying out

* We almost certainly would for spectra, for which however quite different methods not requiring ensemble averaging are available anyway.

calculations of a very similar variety and from the same point of view. The first-named author acknowledges recent very interesting discussions with them.

References

[1] E.P.Wigner, SIAM Review 9 (1967) 1.
[2] Statistical theories of spectra: Fluctuations, ed. C.E.Porter, (Academic Press, New York, 1965).
[3] M.L.Mehta, Random matrices and the statistical theory of energy levels (Academic Press, New York, 1967).
[4] E.P.Wigner, Ann.of Math.67 (1958) 325. This article is reprinted in ref.[2].
[5] F.S.Chang, J.B.French and K.F.Ratcliff, Phys. Letters 23 (1966) 251.
[6] K.F.Ratcliff, Phys.Rev., to be published.
[7] F.S.Chang, J.B.French and T.H.Thio, submitted to Annals of Physics.
[8] F.C.Williams, private communication.
[9] O.Bohigas and J.G.Flores, private communication.

* * * * *

PHYSICAL REVIEW C VOLUME 3, NUMBER 1 JANUARY 1971

Spectral Distributions in Nuclei*

J. B. French

Department of Physics and Astronomy, University of Rochester, Rochester, New York 14627
and Instituto de Física, Universidad de México, México, D. F.

and

K. F. Ratcliff†

Department of Physics and Astronomy, University of Rochester, Rochester, New York 14627
and Department of Physics, Massachusetts Institute of Technology, Cambridge, Massachusetts 02139

(Received 8 May 1970)

Given a finite many-particle vector space, which may be of large dimensionality, and a Hamiltonian which operates inside it, we may investigate the general properties of the system by a method of expanding preselected basis states in terms of the Hamiltonian eigenfunctions. The corresponding intensity, considered as a function of the Hamiltonian eigenvalues, is a spectral distribution which may be studied via its energy moments; we consider such distributions averaged over subsets of basis states. The properties of these averaged distributions are discussed, as well as methods for applying them in general studies of complicated systems; methods are given for evaluating the moments in some significant cases. Detailed formal results are given for configuration distributions in the spherical shell model.

1. INTRODUCTION

There has been steady progress in the attempt to give, in the domain of the conventional shell model, a microscopic description of nuclear structure. But there remain great difficulties. For one thing there are severe limitations on the dimensionalities of the vector spaces in which one can work, and it is clear that, in many or all nuclear problems, much of the interesting physics will "lie outside" the spaces which can be handled in the most conventional way. One attempts to get around this by introducing effective Hamiltonians (and effective transition operators for various processes), determining such operators either directly, from experimental data[1] or by theoretical considerations, usually grounded in perturbation theory, which begin with the free nucleon-nucleon interaction.[2] Besides this problem, and of course related to it, one often does not know how to choose the subspace, compatible with a given dimensionality limit, which is optimal for considering a problem at hand, and an *a posteriori* recourse to comparison with experiment is often not satisfactory. There has been much progress here too, especially via Hartree-Fock theory and its various extensions[3] which achieve a degree of self-consistency in the choice of single-particle states which underly the subspace in which one works.

The aim of the theoretical considerations mentioned above has been almost exclusively that of simplifying a nuclear many-body problem so that a detailed ("microscopic") solution becomes feasible in the domain of conventional spectroscopy. In this pair of papers, and in others to follow, we suggest methods in which conventional spectroscopy plays a much less prominent role, and in which some of the limitations and ambiguities of the standard methods are missing. These methods do not, for the most part, consider the detailed spectrum of the Hamiltonian operator or the detailed strengths for various processes. They deal instead with certain distributions associated with these quantities, examining these in terms of their low-order moments ("spectral" or "strength" moments) with respect to energy, isospin, angular momentum, and so on.[4] They are not limited by the dimensionalities of vector spaces, since they involve neither complete matrix-element calculations nor matrix diagonalization. They do not solve the problems associated with detailed spectroscopy, but deal instead with more general questions. An essential feature of these spectral-moment methods is that, at the beginning when we deal with very low-order moments, we throw away almost all the "information" implicit in a given problem; then as we proceed (taking higher and higher moments, or more or less equivalently, as we shall see, dealing with finer and finer group structures) we gradually regain the lost information. While it is *in principle* possible by proceeding in this way to reproduce exactly the results of a conventional microscopic calculation, that would in fact be feasible only in quite trivial cases and it does not at all represent our view as to a reasonable way of proceeding.

It is in fact our belief that, in many cases, low-order moments of various distributions are more significant quantities than the individual energies and transition strengths which come from detailed calculations. The low-order moments depend essentially on interactions between small numbers of particles, interactions which we can indeed expect to be reasonably treated by our nuclear models; the moments, being defined as averages over sets of states, are properly insensitive to minor variations in the model or interaction parameters. On the other hand, the experimentalist usually deals with and measures detailed quantities, so that the use of spectral-moment methods would then, in many cases, inhibit a *direct* comparison of experiment and theory. We will see in fact that a low-moment spectral distribution can sometimes give a surprisingly good account of the detailed spectra; and when we deal later with excitation strengths rather than Hamiltonian spectra it will turn out that the moments of the appropriate distributions are much more closely connected to experimental measurement. Nonetheless, there will be a burden upon us to demonstrate how, direct comparisons with experiment being often ruled out, we can still make progress via spectral-moment methods.

From the mathematical standpoint, the calculation of spectra moments can make good use of a number of concepts and methods of linear algebra together with elementary notions borrowed from a less familiar domain, in particular the theory of partially-ordered finite sets. These are combined with the usual second-quantization procedures used in many-fermion problems. It will become clear that representations of groups or chains of groups are essential ingredients, and indeed the entire method has everywhere a group-theoretical foundation. Parenthetically then it will turn out, because of this, that spectral-moment methods are ideally suited to a study of the goodness of group symmetries in many-particle systems, an important subject in which progress has been very slow. Although group theory underlies the whole thing, we can consider several important distributions without explicit reference to groups, and in order to make things as clear and simple as possible, we shall, in the first two papers do exactly that. We shall, moreover, discuss the essential ideas in considerable detail even though they may already be familiar to many people.

All the present work concerns itself with the behavior of an m-particle system, described in terms of a finite vector space, a basis for which is given by the states of m fermions distributed over a finite number of single-particle states. We shall describe such a space as a *spectroscopic space*, a term which will remind us that the things we do are still related to the standard operations of spectroscopy. However, we shall ask questions of such a sort and answer them in such a way that the dimensionality of these spaces will not be limiting. And eventually we would in fact hope to deal with spaces which have a more complicated basis.

In Sec. 2 of this paper we introduce the idea of spectral distributions and the moments which characterize them, and give also a brief account of the way in which they may be used. Section 3 will deal, as an illustrative example, with the "scalar" distribution in which one averages over all the states for a fixed number of particles. The theory of spectral moments leads to the discussion of averaging over sets of states as given in Sec. 4. Section 5 contains derivations of certain trace operators and the law of propagation of averages of the configuration distribution. In Sec. 6 the centroid energies and dispersions of the complete scalar and configuration distributions are derived in terms of the one- and two-body matrix elements of the nuclear shell-model Hamiltonian. Various properties of the resulting moment formulas are then investigated in Sec. 7.

2. SPECTRAL DISTRIBUTIONS

Suppose that the problems in which we are interested are describable in terms of a finite set of N single-particle states: For the nuclear (ds) shell in an isospin formalism we have, for example, $N=24$. These N states form a basis for a vector space which we shall call the *orbital space* and denote by $\{N\}$. The states for m fermions distributed over these states[5] form an $\binom{N}{m}$-dimensional vector space $[\binom{N}{m}$ being a binomial coefficient$]$ which we shall label as $\vec{\mathrm{m}}$ or $(\vec{\mathrm{m}})$ or $(N; \vec{\mathrm{m}})$, the last of these when we need to indicate N explicitly. The vector space for any number of fermions $(m=0, 1 \ldots N)$ distributed over $\{N\}$ we shall indicate as $\vec{\mathfrak{N}}$. We have then $\vec{\mathfrak{N}} = \sum \vec{\mathrm{m}}$. The dimensionality of $\vec{\mathfrak{N}}$ is $\sum_{m=0}^{N} \binom{N}{m} = 2^N$. More generally, we may decompose $\vec{\mathfrak{N}}$ in other ways, introducing subspaces $\vec{\alpha}$ or $(\vec{\mathrm{m}}, \vec{\alpha})$, the latter symbol indicating that $\vec{\alpha}$ is also a subspace of $\vec{\mathrm{m}}$. We assume that the Hamiltonian H is a linear operator in $\vec{\mathfrak{N}}$, this implying that H operating on any vector in $\vec{\mathfrak{N}}$ yields another vector in $\vec{\mathfrak{N}}$. Then of course H^p is the pth power[6] of a transformation of $\vec{\mathfrak{N}}$ and is itself a transformation of $\vec{\mathfrak{N}}$.

Let $\{\phi_\alpha(m)\}$ denote an orthonormal basis set in $\vec{\mathrm{m}}$. The standard spectroscopic problem is to compute and diagonalize the ϕ-matrix representation of H, thus producing the eigenvalues E_i and eigenvectors $\psi_i(m)$ as linear combinations of the basis functions $\phi_\alpha(m)$.

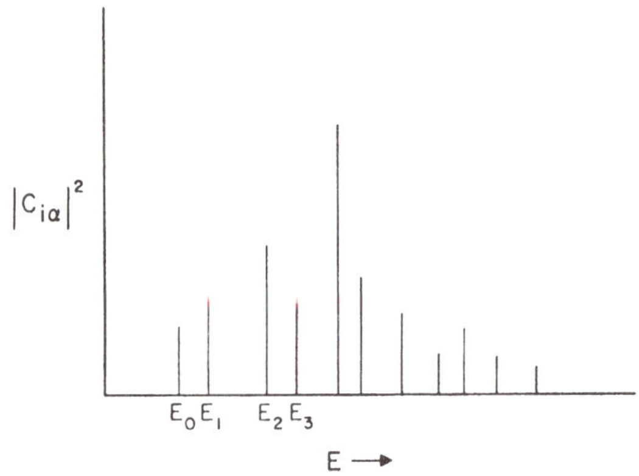

FIG. 1. Schematic representation of the spectral distribution of a vector ϕ_α over the eigenvectors of H.

$$H\psi_i(m) = E_i\psi_i(m)\,, \tag{1}$$

$$\psi_i(m) = \sum_\alpha C_{i\alpha}\phi_\alpha(m)\,. \tag{2}$$

Equally well, one may consider the basis vectors to be expanded in terms of the eigenvectors of H

$$\phi_\alpha(m) = \sum_i C_{i\alpha}{}^*\psi_i(m)\,. \tag{3}$$

Figure 1 gives a schematic picture of the distribution of the intensity of the vector $\phi_\alpha(m)$ over the eigenvectors of H. We refer to this as the "spectral distribution" of $\phi_\alpha(m)$. Such a distribution is completely characterized by its energy moments where the pth moment $M_p(m, \alpha)$ is defined by

$$M_p(m, \alpha) = \sum_i (E_i)^p\,|C_{i\alpha}|^2\,. \tag{4}$$

Using the definition of the expansion coefficients it is easily seen that $M_p(m, \alpha)$ may be rewritten as

$$M_p(m, \alpha) = \langle\phi_\alpha(m) H^p\,\phi_\alpha(m)\rangle\;, \tag{5}$$

where H^p denotes the pth power of the Hamiltonian. Equation (5) tells us that the distribution of ϕ_α over the eigenvectors of H is completely determined by expectation values of powers of H in the state ϕ_α, and that knowledge of the detailed eigenvalues and eigenvectors of H is unnecessary for this purpose.

The zeroth moment $M_0(m, \alpha)$ is unity, which merely expresses the normalization of the ϕ_α. The first moment $M_1(m, \alpha)$, which we shall also call $\mathcal{E}(m, \alpha)$ is the *centroid energy* of the distribution. Going beyond this it is more convenient to introduce either the pth *central moment* [defined for $p \geq 2$ as the pth moment of $H - \mathcal{E}(m, \alpha)$], or the pth cumulant, which we shall turn to later. The non-negative-definite variance of the distribution is denoted by $\sigma^2(m, \alpha)$ and is defined as the second central moment, i.e.,

$$\sigma^2(m, \alpha) = \langle\phi_\alpha(m)[H - \mathcal{E}(m, \alpha)]^2\phi_\alpha(m)\rangle$$
$$= M_2(m, \alpha) - [M_1(m, \alpha)]^2\,. \tag{6}$$

The positive square root of the variance is the "width" $\sigma(m, \alpha)$ of the distribution. The skewness or asymmetry of the distribution about its centroid is measured by the third central moment $\langle\phi_\alpha(m) \times [H - \mathcal{E}(m, \alpha)]^3\phi_\alpha(m)\rangle$, and similarly for higher moments.

In the event that $\phi_\alpha(m)$ is an eigenvector of H then $\mathcal{E}(m, \alpha)$ is the corresponding eigenvalue, and all central moments of order two or more will vanish. A nonvanishing value for $\sigma(m, \alpha)$ may be called a pure "mixing" width, since it implies that $\phi_\alpha(m)$ is actually distributed over a number of eigenvectors of H. In general, the smaller the value of $\sigma(m, \alpha)$ the better is the approximation that $\phi_\alpha(m)$ is an eigenvector of H with $\mathcal{E}(m, \alpha)$ the approximate eigenvalue.

There are as many single-state distributions of the kind discussed above as there are vectors in the vector space, and thus, of course, no individual one is, on an *a priori* basis, of any particular interest. So we next proceed to the notion of distributions averaged over a set $(\vec{m}, \vec{\alpha}) \equiv \vec{\alpha}$ of the m-particle states. The moments of this distribution are then defined as

$$M_p(\vec{m}, \vec{\alpha}) = d^{-1}(\vec{m}, \vec{\alpha}) \sum_{\alpha\in\vec{\alpha}} M_p(m, \alpha)$$
$$= d^{-1}(\vec{m}, \vec{\alpha}) \sum_{\alpha\in\vec{\alpha}} \langle\phi_\alpha(m) H^p\,\phi_\alpha(m)\rangle\;, \tag{7}$$

where $d(\vec{m}, \vec{\alpha})$ is the dimensionality of the $\vec{\alpha}$ subspace of $(N; \vec{m})$.

From the purely formal standpoint now, the set $\vec{\alpha}$ must be chosen in such a way that its moments (at least as far as $p = 2$) can be evaluated. We rule out evaluation by means of an explicit construction of the individual matrix elements, since our major interest is in large spaces where such construction is not feasible; hence we require new techniques which are not to be found among those of conventional spectroscopy. Not surprisingly, it will turn out that to produce such techniques it will be advantageous for the subsets to define representations of groups or chains of groups.

From the standpoint of physics it should be clear that if the resultant averaged distributions are narrow then they are interesting, and, besides that, the groups which define $\vec{\alpha}$ then correspond to approximately good symmetries for the problem. But, as we shall see, there is much more to it than that, and, in particular, wide distributions are not necessarily uninteresting. Among the groups which we shall study in detail in early papers are the $U(N)$ group (which dominates almost everything), the direct-sum unitary group which defines configurations, and the isospin, angular momentum, symplectic, and space-symmetry groups.

It follows immediately from Eq. (7) that the $\vec{\alpha}$ centroid energy is simply the average of the centroid energies for the individual states,

$$\mathcal{E}(\vec{m}, \vec{\alpha}) = d^{-1}(\vec{m}, \vec{\alpha}) \sum_{\alpha \in \vec{\alpha}} \mathcal{E}(m, \alpha) . \tag{8}$$

Things are different for higher-order quantities; in particular we have for the variance

$$\begin{aligned}\sigma^2(\vec{m}, \vec{\alpha}) &\equiv M_2(\vec{m}, \vec{\alpha}) - [M_1(\vec{m}, \vec{\alpha})]^2 \\ &= d^{-1}(\vec{m}, \vec{\alpha}) \sum_{\alpha \in \vec{\alpha}} \sigma^2(m, \alpha) \\ &\quad + \tfrac{1}{2} d^{-2}(\vec{m}, \vec{\alpha}) \sum_{\alpha, \alpha' \in \vec{\alpha}} [\mathcal{E}(m, \alpha) - \mathcal{E}(m, \alpha')]^2 \\ &= d^{-1}(\vec{m}, \vec{\alpha}) \sum_{\alpha \in \vec{\alpha}} \{\sigma^2(m, \alpha) \\ &\quad + [\mathcal{E}(m, \alpha) - \mathcal{E}(\vec{m}, \vec{\alpha})]^2\} . \end{aligned} \tag{9}$$

The energy term in Eq. (9) expresses a new contribution to the width which enters when averaged distributions are considered. Thus, even if every state of $\vec{\alpha}$ were an eigenstate of H, in which case there would be zero mixing widths for all the states, the width of the averaged distribution would still not vanish, since, in general, the eigenvectors would not be degenerate. We remark also that Eq. (9), applied to the Hamiltonian eigenstate representation [in which $\sigma^2(m, \alpha) = 0$], informs us that $\sigma^2(\vec{m}, \vec{\alpha})$ is a measure connected with the spacing distribution of the eigenvalues of H. Thus, though we make no statistical assumptions, we do have here a first contact with the statistical theory of spectra.[7]

The distribution moments $M_p(\vec{m}, \vec{\alpha})$, for a basis subset $(\vec{m}, \vec{\alpha})$, of the space $\vec{m}$ which we have introduced above are not really adequate for our purposes. In dealing with second moments, for example, it will be important to know what part of the width comes from intermediate states in $(\vec{m}, \vec{\alpha})$ and what part from states outside $(\vec{m}, \vec{\alpha})$. Therefore we should divide the width into an *internal*, or self-width, and an *external* width, according to

$$\begin{aligned} M_2(\vec{\alpha}) &= M_2^{\mathrm{i}}(\vec{\alpha}) + M_2^{\mathrm{e}}(\vec{\alpha}) , \\ M_2^{\mathrm{i}}(\vec{\alpha}) &= d^{-1}(\vec{\alpha}) \sum_{\alpha, \alpha' \in \alpha} \langle \alpha | H | \alpha' \rangle \langle \alpha' | H | \alpha \rangle \\ &= d^{-1}(\vec{\alpha}) \sum_{\alpha \in \vec{\alpha}} \langle \alpha | H P(\vec{\alpha}) H | \alpha \rangle , \\ M_2^{\mathrm{e}}(\vec{\alpha}) &= d^{-1}(\vec{\alpha}) \sum_{\substack{\alpha \in \vec{\alpha} \\ \beta \notin \vec{\alpha}}} \langle \alpha | H | \beta \rangle \langle \beta | H | \alpha \rangle \\ &= d^{-1}(\vec{\alpha}) \sum_{\alpha \in \vec{\alpha}} \langle \alpha | H [1 - P(\vec{\alpha})] H | \alpha \rangle , \\ \sigma^2(\vec{\alpha}) &= \sigma^2{}_{\mathrm{i}}(\vec{\alpha}) + \sigma^2{}_{\mathrm{e}}(\vec{\alpha}) , \\ \sigma^2{}_{\mathrm{i}}(\vec{\alpha}) &= M_2^{\mathrm{i}}(\vec{\alpha}) - [M_1(\vec{\alpha})]^2 , \\ \sigma^2{}_{\mathrm{e}}(\vec{\alpha}) &= M_2^{\mathrm{e}}(\vec{\alpha}) . \end{aligned} \tag{10}$$

Here we have written $P(\vec{\alpha})$ for the projection operator onto the set $\vec{\alpha}$.

The separation above is that of $\vec{m}$ into two parts, $\vec{m} = \vec{\alpha} + (\vec{m} - \vec{\alpha})$. If we have a group in mind, for which $\vec{m}$ supplies a reducible representation, a more complete decomposition into irreducible representations may be relevant; thus $\vec{m} = \vec{\alpha} + \vec{\beta} + \vec{\gamma} + \ldots$. In terms of this, or any other decomposition into disjoint parts, we then define *partial widths* by

$$\begin{aligned} \sigma^2(\vec{\alpha}; \vec{\beta}) &= M_2(\vec{\alpha}; \vec{\beta}) - \delta_{\alpha\beta} [M_1(\vec{\alpha})]^2 , \\ M_2(\vec{\alpha}; \vec{\beta}) &= d^{-1}(\vec{\alpha}) \sum_{\substack{\alpha \in \vec{\alpha} \\ \beta \in \vec{\beta}}} \langle \alpha | H | \beta \rangle \langle \beta | H | \alpha \rangle . \end{aligned} \tag{11}$$

Then it follows easily that

$$\begin{aligned} \sigma^2(\vec{\alpha}) &= \sum_{\vec{\beta}} \sigma^2(\vec{\alpha}; \vec{\beta}) , \\ \sigma^2{}_{\mathrm{i}}(\vec{\alpha}) &= \sigma^2(\vec{\alpha}; \vec{\alpha}) , \\ d(\vec{\alpha}) \sigma^2(\vec{\alpha}; \vec{\beta}) &= d(\vec{\beta}) \sigma^2(\vec{\beta}; \vec{\alpha}) . \end{aligned} \tag{12}$$

We see that $\sigma^2(\vec{\alpha}; \vec{\beta})$ is the average over the states of $\vec{\alpha}$ of the squares of the matrix elements connecting the two subspaces. A corresponding (but more complicated) decomposition can be made also for higher moments.

What now can one learn from spectral distributions? The simplest thing is about the general eigenfunction structure, for example, about which representations contribute significantly in a particular range of energy. This would follow in an obvious way from the centroids and widths of the representations, which we could regard for this purpose as defining the distributions, assumed, for example, to be normal.

If the distribution for a particular representation is then found to span an energy region which contains the centroid of another representation, we would naturally suspect (unless that is forbidden by the exact symmetries) that the two representations would be strongly admixed. But we might be misled in assuming this; for it could happen that the width of the representation is predominantly a self-width, in which case the admixings would be small, we would have a peaceful coexistence of the states of the two representations, and separate transformations in the separate subspaces would produce good eigenfunctions; and of course similarly if many representations are involved. Thus we see the need for the partial-width decomposition of the total widths.

For more detailed results, if, for example, our interest is with the low-lying states, we may use the partial widths and centroids to supply a basis for a truncation of the original space to produce a resultant space small enough for detailed calculation. These quantities might, for example, indicate a small number of representations strongly coupled to a low-lying $\vec{\alpha}$ (or set of such $\vec{\alpha}$'s) in which case we might discard the other representations and make a more detailed calculation in the reduced space. The criterion for weak coupling of two representations is, of course, that the partial width[8] be small compared with the difference between their centroid energies.

We have, of course, been assuming here that the distributions are decently represented via moments of order $p \leq 2$. We shall, in fact, go beyond that and assume that they are close to "normal" or Gaussian, an assumption which has been found valid in the various examples, calculable in detail, which we have studied; we hope, in a later paper, to discuss the theoretical basis for it. It is, of course, possible to consider higher moments, which would measure departures from normality and would enable us to judge the validity of the entire moment procedure for studying the eigenfunction structures; at present, in the cases which we consider, it is feasible to go as far as the fourth moment. Besides the sequence of higher and higher moments for a given decomposition of $\vec{\mathfrak{N}}$, there is also the sequence in which we restrict the order (say to $p = 2$) but take a nested set of finer and finer subspace decompositions. This technique is successful if, as we proceed along the sequence, the widths, or better the partial-width matrix, indicate a structure of weakly interacting subspaces. When it works, the second technique is superior; on the one hand, the low moments of the finer distribution do carry information about the high moments of the coarser, while, on the other hand, since the finer decomposition will always correspond to some group structure, we learn at the same time about the group symmetries. It is important in both cases to study some examples which we can calculate in detail.

Going in the opposite direction we can study what happens when we enlarge a space by including more single-particle states both above and below the Fermi surface, i.e., proceeding to an enlarged orbital space $\{N'\} \supset \{N\}$. Not only do new distributions enter but the original distributions will spread. Thus, one can study the stability of the makeup of low-excitation spectra to changes in the size of the vector space. One may also introduce variational methods to study, for example, the way distributions involving deformed orbitals vary with the deformation parameters.

It is also clear that such distributions can provide information about the localization of interesting wave-function components at excitation energies appropriate to nuclear reactions exploring single-particle resonances and intermediate structure. They supply also, in an obvious way, a foundation for the microscopic theory of level densities, and in a less obvious way, perhaps also a basis for a statistical theory of reactions.[9] And as mentioned above it should be possible, by averaging over states corresponding to exactly good quantum numbers, to make contact with the statistical theory of spectra.

The considerations above give a method for studying group symmetries, involving as they do the question of the admixing of different representations. As an alternative method one may use mean-squared energies as an orthonormalization measure in terms of which one can decompose H^2 into orthogonal parts, some of which preserve the symmetry while the others violate it. Then, roughly speaking, the relative weights of these parts give us a criterion for the goodness of the symmetry.[10]

Going beyond the spectral distributions, we can study also the strength distributions for various excitation operators,[11] making thereby much more immediate contact with experiment. This would, for example, supply a way of studying cascade processes in highly excited systems, a subject also of direct experimental interest.

We mention also the possible extension to the case where the orbital states form a continuous, rather than discrete, set and to the case where the basis states have intrinsic correlations going beyond those supplied by any restricted orbital space.

These general remarks should indicate the utility of spectral distributions and show that partial widths of representations (and their extension to higher-order quantities) are the really essential ingredients of the subject. We now must proceed to methods for calculating such quantities. Let it be said at the outset that there does not as yet exist a general theory which can be applied to an arbitrary group structure. However, several important structures are well understood, as are also some features of the general case. It will turn out to be convenient to deal at first with total widths of representations, leaving until a little later the decomposition into partial widths (and similarly for the higher moments).

In the next section we consider, by the most elementary methods, a very simple case: that of averaging over all the states of m particles. In Sec. 4 we give a more general discussion introducing various concepts which will be of importance to us in later papers. The later sections of this paper and all of the next (and these will not draw heavily on the results of Sec. 4) will be devoted to configuration distributions in the spherical shell model.

3. SCALAR DISTRIBUTION AS A SIMPLE EXAMPLE

By the m-particle *scalar* distribution we mean the spectral distribution averaged over all m-particle states. The appropriate decomposition of $\vec{\mathfrak{N}}$ is then $\vec{\mathfrak{N}} = \sum \vec{m}$. Let us first derive an elementary result for the *average* over $\vec{m}$ of the expectation value of a k-particle operator. This average we write as $\langle O(\vec{k})\rangle^{\vec{m}}$, and the corresponding *trace* as $\langle\langle O(\vec{k})\rangle\rangle^{\vec{m}}$. The particular k-particle operator which gives unity on states in which the single-particle states Nos. $1 \ldots k$ are filled and zero on states in which any of these are empty, obviously has a trace $\binom{N-k}{m-k}$, that number being the number of ways in which the $(m-k)$ "free" particles can be distributed over the $(N-k)$ "free" single-particle states. Other operators which differ only in specifying a different k subset have the same trace. The set of all such operators, supplemented by the set of nondiagonal k-particle operators [which, for example, convert $(1 \ldots m)$ into $(1 \ldots m-1,\ m+1)$] gives a basis for the general k-particle operator. These latter operators being traceless, we see then that, for the trace and the related average,

$$\langle\langle O(\vec{k})\rangle\rangle^{\vec{m}} = \binom{N-k}{N-m}\langle\langle O(\vec{k})\rangle\rangle^{\vec{k}} ,$$
$$\langle O(\vec{k})\rangle^{\vec{m}} = \binom{m}{k}\langle O(\vec{k})\rangle^{\vec{k}} . \tag{13}$$

The operators which we enounter will not in general have a specified particle rank; e.g., $O^2(\vec{k})$ has ranks $k, k+1, \ldots, 2k$, as long as $k \leqslant \frac{1}{2}N$. Rather than carry out the explicit decomposition by particle rank we can recognize that, since $\binom{m}{k}$ is a kth order polynomial in m, $\langle O\rangle^{\vec{m}}$ for an operator of maximum rank u will be a uth order polynomial in m. Writing such a polynomial as a linear function of its own values at $m = 0, 1, 2, \ldots, u$, we have then, for an operator of maximum rank u, that[4,12]

$$\begin{aligned}\langle O\rangle^{\vec{m}} &= \sum_{t=0}^{u}\langle O\rangle^{\vec{t}}\prod_{\substack{y=0\\(\neq t)}}^{u}\frac{m-y}{t-y}\\ &= \sum_{t=0}^{u}(-1)^{t-u}\frac{m-u}{m-t}\binom{m}{u}\binom{u}{t}\langle O\rangle^{\vec{t}}\\ &= \sum_{t=0}^{u}(-1)^{t-u}\binom{m-t-1}{u-t}\binom{m}{t}\langle O\rangle^{\vec{t}}\\ &= \sum_{t=0}^{u}\binom{u-m}{u-t}\binom{m}{t}\langle O\rangle^{\vec{t}} ,\end{aligned} \tag{14}$$

where the first form involves the usual Lagrange polynomials, and the other forms follow immediately from the first.[13] The last form demonstrates that for $m \leqslant u$ the result is simply an identity.

As a trivial application of Eq. (14), we have for the angular momentum,

$$\langle J^2\rangle^{\vec{m}} = -m(m-2)\langle J^2\rangle^{\vec{1}} + \binom{m}{2}\langle J^2\rangle^{\vec{2}} , \tag{15}$$

the first term then correcting for the fact that J^2 is not a pure two-body operator. For the usual case where the orbital space supplies a representation of the angular momentum group (defining therefore a set of spherical orbits), we have $\langle J^2\rangle^{\vec{N}} = 0$; then eliminating $\langle J^2\rangle^{\vec{2}}$ from Eq. (15) we have

$$\langle J^2\rangle^{\vec{m}} = \frac{m(N-m)}{(N-1)}\langle J^2\rangle^{\vec{1}} . \tag{16}$$

The result of Eq. (16) could have been written down immediately by using the fact that $\langle J^2\rangle^{\vec{m}}$ is a second-order polynomial in m which vanishes at $m = 0, N$. More generally, Eq. (14) is, as we have said, the equation which expresses the value for arbitrary m of the uth order polynomial in m as a linear function of its values at $m = 0, 1, \ldots u$; but from the linear independence of the m polynomials $\binom{m}{t}$ (or of the corresponding trace operators) it is obvious that we could choose any other $(u+1)$-dimensional set of m values (which we shall speak of as a "net") with which to define the polynomials. The solution for the new net is simple. If the net points are $t_1, t_2, \ldots, t_u$, then the new Lagrange polynomial for the net point t is

$$\prod_{\substack{i=1\\t_i\neq t}}^{u}\frac{m-t_i}{t-t_i} , \tag{17}$$

and with these we can express the trace of any u operator in terms of its traces for any $(u+1)$ particle numbers. We may say, incidentally, that the simplification in going from Eq. (15) to Eq. (16) has resulted from our choice of a net $(0, 1, N)$ simpler than the "elementary" net $(0, 1, 2)$. In general there is special interest in a net which is most economical with respect to the number of holes or particles involved; this is because the calculations required for the input quantities become increasingly complicated as the number of particles or holes increases. We achieve this economy by taking a two-segment net, one part proceeding upward from the origin, the other downward from the "plenum" state.[14] Taking these segments to be, respectively, $(t \leqslant u_1)$ and $(N-u_2 \leqslant t \leqslant N)$ where $u_1 + u_2 + 1 = u$, we find as the extension of the third form of Eq. (14) that[12]

$$\langle O\rangle^{\vec{n}} = \binom{N-m}{u_2+1}\sum_{t=0}^{u_1}(-1)^{t-u_1}\binom{N-t}{u_2+1}^{-1}\binom{m-t-1}{u_1-t}\binom{m}{t}\langle O\rangle^{\vec{t}} + \binom{m}{u_1+1}\sum_{t=N-u_2}^{N}(-1)^{t-u_2+N}\binom{t}{u_1+1}^{-1}\binom{t-1-m}{u_2-N+t}\binom{N-m}{N-t}\langle O\rangle^{\vec{t}}, \tag{18}$$

which gives $\langle O\rangle^{\vec{m}}$ as a linear combination of traces with $(t \leqslant u_1)$ particles and $(t \leqslant u_2)$ holes. If u is odd, then for maximum economy we would take $u_1 = u_2 = \frac{1}{2}(u-1)$, while if u is even we would take one of the limits to be $\frac{1}{2}u$, the other $\frac{1}{2}(u-2)$.

We have stressed that the scalar averaging results follow in a simple way from the properties of polynomials in particle number. Since the product of polynomials is itself a polynomial, it follows that the product of averages propagates by the same rule as does the average of a product, it being only necessary to take $u \geqslant$ (sum of particle ranks of factors). Thus we can deal with *central moments* or with *cumulants* in the same way as with the corresponding moments. We have then, for example, for the scalar centroid energy $\mathcal{E}(\vec{m})$ and the dispersion $\sigma^2(\vec{m})$ the results[12]

$$\mathcal{E}(\vec{m}) = \frac{m(m-1)}{N(N-1)}\mathcal{E}(\vec{N}) - \frac{(N-m)(m-1)}{N}\mathcal{E}(\vec{0}) + \frac{(N-m)m}{(N-1)}\mathcal{E}(\vec{1}), \tag{19}$$

$$\sigma^2(\vec{m}) = \frac{m(m-1)(m-2)(N-m)(N-m-1)}{(N-1)(N-2)(N-3)}\left[\frac{\sigma^2(\vec{N}-\vec{1})}{(N-m-1)} - \frac{(N-3)}{(m-1)}\sigma^2(\vec{1}) + \frac{(N-1)}{2(m-2)}\sigma^2(\vec{2})\right]. \tag{20}$$

In the first of these we have taken $u_1 = 1$, $u_2 = 0$, while in the second we have used $u_1 = 2$, $u_2 = 1$, along with the fact that the vacuum and plenum states are dispersion free. Both results could have been written down at sight.

As a simple example, consider the centroid energies in the (ds) shell, $(N = 24)$. Reasonable interpretations of the low-lying level structure for $A = 16$, 17, 39, 40 enable us to evaluate the centroid energies for these cases $(m = 0, 1, 23, 24)$ and, on the assumption[15] that H is of maximum rank 2, the question is whether the four centroids reasonably well satisfy Eq. (19). With the exception of the $A = 39$ $d_{5/2}$-hole energies (K^{39}, Ca^{39}) the assignments are all unambiguous. In each of the $A = 39$ nuclei three $\frac{5}{2}^+$ states are found[16] within a spread of 1 MeV centered at about 6-MeV excitation; we make the assumption that the $d_{5/2}$-hole strength is equally distributed among these. Then we find $\mathcal{E}(23) = -326.8$ MeV to be compared with the value -324.8 MeV found, via Eq. (19), from the other centroids. We would interpret this as satisfactory agreement (even though no convincing formal criterion for the goodness of agreement suggests itself). It is clear that this kind of simple analysis should precede any attempt to calculate, for example, effective interaction matrix elements by fits to empirical energy levels.

4. OPERATOR AVERAGING

We have used a simple combinatorial method to produce the results of the last section. While this kind of procedure would be adequate to deal with configuration distributions, that would not be so for more complicated cases. We therefore devote this section to a more formal (though still elementary) discussion in terms of operators, introducing thereby a number of important concepts.

We can interpret Eq. (13) as giving, for the trace of a k-body operator, a rule for propagation through the various subspaces $\vec{m}$ of $\mathfrak{N}$. It tells us first that linear traces of k-body operators do propagate in a simple fashion and that, if we are interested only in such traces, only one significant piece of information distinguishes one k-body operator from another. Indeed we have, where n is the number operator,

$$O(\vec{k}) = \langle O(\vec{k})\rangle^{\vec{k}}\binom{n}{k} + \text{traceless parts}. \tag{21}$$

We then identify $\binom{m}{k}$ as the $(\vec{k} \to \vec{m})$ *propagation coefficient* for the average of a k-body operator, and $\binom{n}{k}$ as a *propagation operator*. For a quadratic trace we have seen that we need $(k+1)$ pieces of information to distinguish the squares of k-particle operators, and similarly for other cases. It will be up to us, then, to calculate the traces which define, for a given moment, a complete set of pieces of "input" information.

The group structure involved in the trivial example of Eq. (14) is that of $U(N)$, the set of unitary transformations in the orbital space, and the m spaces are those which supply irreducible representations of $U(N)$. What is implied by the remark that $O(\vec{k})$ is a *k-body operator*? For one thing $O(\vec{k})$ vanishes in m spaces with $m < k$ [consistent with this, so does $\binom{m}{k}$]; beyond this the structure of $O(\vec{k})$ in spaces with $m > k$ is determined completely by its structure in the $(m = k)$ space. Recalling now that an algebraist thinks of an operator as a transformation on a space, we see that the notion of a

k-body operator involves more than this. It implies specifying an algebraist's operator in the ($m = k$) subspace together with a rule for propagating it into the other subspaces of $\mathfrak{N}$. This general idea of *propagation* along a line (or more generally on a lattice) of representations is also of major importance to us.

For $O(\mathfrak{k})$ the representation given by the subspace $\mathfrak{k}$ is special and we might call it the *defining representation* of $O(\mathfrak{k})$, this going along with the idea that the information about O is given to us in the $\mathfrak{k}$ space. This would clearly be inadequate for an operator which is not number conserving. But in general we can think of any operator O as having a *quadratic structure*, since it may be expanded in terms of operators which connect one particular subspace with another. Thus we have, for the $U(N)$ case, $O = \sum_{\mathfrak{k}\mathfrak{k}'} O(\mathfrak{k}, \mathfrak{k}')$, where $O(\mathfrak{k}, \mathfrak{k}')$ produces a transformation $\mathfrak{k}' \to \mathfrak{k}$ and is defined in the subspace pair $(\mathfrak{k}, \mathfrak{k}')$. Its behavior, when acting on a state of $\mathfrak{m} \neq \mathfrak{k}'$, is given by a propagation rule. Both the quadratic structure and the nature of the propagation rule become particularly clear in second quantization. For a general decomposition of $\mathfrak{N} = \sum \vec{\alpha}$, corresponding for example to some given chain of groups, we have the corresponding operator expansion

$$O = \sum_{\vec{\alpha}, \vec{\alpha}'} O(\vec{\alpha}, \vec{\alpha}') \ , \tag{22}$$

and we would say that $O(\vec{\alpha}, \vec{\alpha}')$ has the *defining representation pair* $(\vec{\alpha}, \vec{\alpha}')$. For the diagonal case we shall continue to speak of the *defining representation* $\vec{\alpha}$.

A separate decomposition of O is by its tensorial behavior with respect to the groups under consideration, $O = \sum_{\vec{\gamma}} O^{\vec{\gamma}}$. We combine the two decompositions to get

$$O = \sum_{\vec{\alpha}, \vec{\alpha}', \vec{\gamma}} O^{\vec{\gamma}}(\vec{\alpha}, \vec{\alpha}') \ . \tag{23}$$

Since $O(\vec{\alpha}, \vec{\alpha}')$ has the direct-product structure $\vec{\alpha} \times \vec{\alpha}'$, the allowed $\vec{\gamma}$'s are determined by the combination rule for $\vec{\alpha}, \vec{\alpha}'$. For example, if we are dealing with $U(N)$, a k-body operator has $\vec{\alpha} = \vec{\alpha}' = \mathfrak{k}$ and then it is well known that there are $(k+1)$ different representations $\vec{\gamma}$, those in fact with one- and two-columned Young shapes, $[N-\lambda, \lambda]$, with $\lambda = 0, 1, \ldots, k$. As another example, for the Coulomb interaction (defined in the isospin-one two-particle space) $\vec{\gamma}$ could define the isospin ranks 0, 1, 2; Eq. (23) then giving the usual scalar-vector-tensor separation.

Observe next that in the $U(N)$ case there is a natural ordering for the representations, defined by the fact that a k-body operator vanishes in m with $m < k$. The ordering is the trivial one $m = 0, 1, 2, \ldots, N$ but the general idea of ordering is not trivial. For the $U(N)$ case the representations form a *fully ordered set* (or a *chain*), the ordering operation being that $\vec{\alpha} \geq \vec{\alpha}'$ if every operator defined in $\vec{\alpha}$ vanishes in $\vec{\alpha}'$. We see now that propagation proceeds in order along a chain. When we come to more complicated group structures we shall encounter the more subtle notion of a partially-ordered set, with a correspondingly more complicated propagation.

Equation (13) may be rewritten as

$$\langle O(\mathfrak{k})\rangle^{\vec{m}} = \binom{m}{k}\binom{N}{k}^{-1} \langle O(\mathfrak{k})\rangle^{\vec{N}} \ . \tag{24}$$

Thus, while for $k > 0$ we cannot propagate from the vacuum representation (since the operator vanishes there), we *can* propagate from the plenum representation, which is just as simple and is indeed the hole-vacuum representation. The significant fact is that the plenum representation, giving only a highly nonfaithful representation for $O(\mathfrak{k})$, is still useful, since it does carry the information about the linear trace of $O(\mathfrak{k})$. We shall see later how a set of nonfaithful representations may be used as a *sieve* to decompose the moments of an operator.

Via a trivial example, we have introduced above the nontrivial notions of propagation, propagation coefficients and operators, defining-representation pairs, the linear and quadratic structure of operators, ordered sets of representations, and sieves. All of these are going, eventually, to be of importance to us.

We proceed now by introducing second-quantization forms for the operators and states of the system. This is not strictly necessary but it is highly advantageous. Corresponding to the N single-particle states which define the orbital space $\{N\}$ we introduce creation operators A_r and destruction operators $B_r = A_r^\dagger$, where $r = 1, 2, \ldots, N$. These satisfy the usual anticommutation relations,

$$\begin{aligned} [A_r, B_s]_+ &= \delta_{rs} \ , \\ [A_r, A_s]_+ &= [B_r, B_s]_+ = 0 \ . \end{aligned} \tag{25}$$

The normalized operator for m particles in state α is represented as $Z_\alpha(m)$. In a product (or Hartree-Fock) representation, a basis state $Z_\alpha(m)$ is a product of m A operators; more generally, it is a linear combination of such products. The operator $Z_\alpha(k) Z_{\alpha'}^\dagger(k)$ is a k-body operator which is defined by the single nonvanishing k-body matrix element $\langle k\beta | Z_\alpha(k) Z_{\alpha'}^\dagger(k) | k\beta'\rangle = \delta_{\alpha\beta}\delta_{\alpha'\beta'}$. Moreover, as α, α' run through the states of $\mathfrak{k}$, the set of such quadratic objects gives a basis for the k-body operators. It follows then that the general expansion of a k-body operator $O(\mathfrak{k})$ is

$$O(\vec{k}) = \sum_{\beta,\beta'} \langle k\beta|O(\vec{k})|k\beta'\rangle Z_\beta(k) Z_{\beta'}^\dagger(k) \,. \tag{26}$$

A general matrix element of $O(\vec{k})$ in m-particle states is

$$\langle m\alpha|O(\vec{k})|m\alpha'\rangle = \sum_{\beta\beta'} \langle k\beta|O(\vec{k})|k\beta'\rangle \times \langle Z_\alpha^\dagger(m) Z_\beta(k) Z_{\beta'}^\dagger(k) Z_{\alpha'}(m)\rangle_0 \,, \tag{27}$$

where $\langle\rangle_0$ denotes a vacuum expectation value (v.e.v.) which, of course, is entirely independent of the operator $O(k)$. Such a second-quantized representation makes explicit the defining properties of a k-body operator and its propagation, which we have discussed above; and similarly for more general operators.

The commutation rules (25) are invariant under the involution

$$\begin{aligned} A_r &\to B_r \equiv A_r(\mathrm{h}) \,, \\ B_r &\to A_r \equiv B_r(\mathrm{h}) \,, \end{aligned} \tag{28}$$

which, as indicated, defines holes to be fermions. This operation (applied in every element of the orbital space) we denote as $(\tilde{\ })$. Thus we have

$$\begin{aligned} \tilde{A}_r &= B_r = A_r^\dagger = A_r(\mathrm{h}) \,, \\ (A_r A_s)^T &= B_r B_s = -(A_r A_s)^\dagger = A_r(\mathrm{h}) A_s(\mathrm{h}) \,, \end{aligned} \tag{29}$$

and for general state operators

$$\begin{aligned} \tilde{Z}_\alpha(m) &= (-1)^{\frac{1}{2}m(m-1)} Z_\alpha^\dagger(m) \equiv Z_\alpha(m_{\mathrm{h}}) \,, \\ \tilde{Z}_\alpha^\dagger(m) &= (-1)^{\frac{1}{2}m(m-1)} Z_\alpha(m) \equiv Z_\alpha^\dagger(m_{\mathrm{h}}) \,, \end{aligned} \tag{30}$$

these equations then defining for us a correspondence between hole and particle state operators. The essential equivalence between the operations $\tilde{\ }$ $({}^T)$ and † when applied to state operators with a definite number of particles is not maintained with more general operators and we have

$$\begin{aligned} Z_\alpha(k) Z_\beta^\dagger(k') &\dot{=} (-1)^{\frac{1}{2}(k+k'-1)(k-k')} Z_\alpha^\dagger(k) Z_\beta(k') \\ &= Z_\alpha(k_{\mathrm{h}}) Z_\beta^\dagger(k'_{\mathrm{h}}) \,. \end{aligned} \tag{31}$$

We see that in general $\tilde{O}$ may be said to be an operator with the same representation in the hole states as O has in particles. If for example O is a k-body operator [as it would be if $k=k'$ in (31)] then we shall call $\tilde{O}$ the corresponding k-hole operator. Then, reexpressing $\tilde{O}$ in normal form with respect to particles, we get a k-body operator which, to within a phase $(-1)^k$, is simply the adjoint[17] of O, along with terms of lower particle rank arising from commutation. In other words a k-hole operator is the sum (from $l=0\ldots k$) of l-body operators. The $l=k$ term does *not* vanish and is trivially related to the original operator. The $l<k$ terms may vanish; we shall see later the significance of such vanishings.

Since two state operators essentially commute,[18] as do their adjoints also, we have the result that

$$\langle Z_\alpha^\dagger(m) Z_\beta^\dagger(k) Z_{\beta'}(k) Z_{\alpha'}(m)\rangle_0 = \langle Z_\beta^\dagger(k) Z_\alpha^\dagger(m) Z_{\alpha'}(m) Z_{\beta'}(k)\rangle_0 \,, \tag{32}$$

which tells us that to an m-particle matrix element of a k-hole operator, there corresponds an equal k-particle matrix element of an m-hole operator (and similarly for non-number-conserving operators). Writing now, in correspondence with Eq. (26), the expansion of a k-hole operator, $\tilde{O}(\vec{k}) = O(\vec{k}_{\vec{h}})$, viz.,

$$O(\vec{k}_{\vec{h}}) = \sum_{\beta\beta'} \langle k_{\mathrm{h}}\beta|O(k_{\mathrm{h}})|k_{\mathrm{h}}\beta'\rangle Z_\beta^\dagger(k) Z_{\beta'}(k) \,, \tag{33}$$

we have for the $(\vec{m}, \vec{\alpha})$ trace

$$\langle\langle O(\vec{k}_{\vec{h}})\rangle\rangle^{\vec{m},\vec{\alpha}} = \sum_{\substack{\alpha\in\vec{\alpha} \\ \beta\beta'}} \langle k_{\mathrm{h}}\beta|O(\vec{k}_{\vec{h}})|k_{\mathrm{h}}\beta'\rangle \times \langle Z_\alpha^\dagger(m) Z_\beta^\dagger(k) Z_{\beta'}(k) Z_\alpha(m)\rangle_0 \,. \tag{34}$$

Let us now introduce the trace operator for the set $(\vec{m}, \vec{\alpha})$

$$\rho(\vec{m}, \vec{\alpha}) = \sum_{\alpha\in\vec{\alpha}} Z_\alpha(m) Z_\alpha^\dagger(m) \,. \tag{35}$$

This self-adjoint operator is the operator extension of the $(\vec{m}, \vec{\alpha})$ trace of the density matrix. It has the defining properties:

(i) $\rho(\vec{m}, \vec{\alpha})$ is an m-particle operator and therefore gives zero on all k-particle states with $k<m$.

(ii) $\rho(\vec{m}, \vec{\alpha})$ gives unity on all m-particle states in $(\vec{m}, \vec{\alpha})$ and zero on all states in $(\vec{m}, \vec{\alpha}')$ with $\vec{\alpha}' \neq \vec{\alpha}$.

In terms of these operators we have from Eqs. (32), (34) that[19]

$$\langle\langle O(\vec{k}_{\vec{h}})\rangle\rangle^{\vec{m},\vec{\alpha}} = \sum_{\beta\beta'} \langle k_{\mathrm{h}}\beta|O(\vec{k}_{\vec{h}})|k_{\mathrm{h}}\beta'\rangle \times \langle Z_\beta^\dagger(k) \tilde{\rho}(\vec{m}, \vec{\alpha}) Z_{\beta'}(k)\rangle_0 \,, \tag{36}$$

which gives the trace for an operator which is in *antinormal form*.

A simple and well-known device now enables us to do the same for a normal form operator, which is more interesting. It is clear that Z_α applied to the particle vacuum and $\tilde{Z}_\alpha$ applied to the hole vacuum (the plenum state for particles) give related states; these we define as complementary. Then

$$\begin{aligned} Z_{\alpha_c}(N-m)|0\rangle &= \tilde{Z}_\alpha(m)|N\rangle = \tilde{Z}_\alpha(m) Z(N)|0\rangle \,, \\ \langle 0|Z_{\alpha_c}^\dagger(N-m) &= \langle N|\tilde{Z}_\alpha^\dagger(m) = \langle 0|Z^\dagger(N) \tilde{Z}_\alpha^\dagger(m) \,, \end{aligned} \tag{37}$$

where we have chosen $\vec{\alpha}_{\vec{c}}$, the $(N-m)$-particle basis set, to coincide with the complement of $\vec{\alpha}$, the m-particle basis set.[20]

The relationship between hole and particle v.e.v.'s

$$\langle Q\rangle_0 = \langle \tilde{Q}\rangle_{0_h} , \tag{38}$$

which follows from Eq. (28), when applied to the integrand operator of a matrix element, then leads to the result

$$\langle Z_\alpha{}^\dagger(m) K Z_{\alpha'}(m')\rangle_0 = \langle Z_{\alpha_c}{}^\dagger(N-m)\tilde{K} Z_{\alpha'_c}(N-m')\rangle_0 . \tag{39}$$

Thus, for a k-particle operator, we have

$$\begin{aligned}\langle\langle O(\vec{k})\rangle\rangle^{\vec{m},\vec{\alpha}} &= \sum_{\substack{\alpha\in\vec{\alpha}\\ \beta\beta'}} \langle k\beta|O(\vec{k})|k\beta'\rangle \\ &\quad\times\langle Z_\alpha{}^\dagger(m) Z_\beta(k) Z_{\beta'}{}^\dagger(k) Z_\alpha(m)\rangle_0 \\ &= \sum_{\beta\beta'} \langle k\beta|O(\vec{k})|k\beta'\rangle \\ &\quad\times\langle Z_\beta{}^\dagger(k)\tilde{\rho}(\vec{N}-\vec{m},\vec{\alpha}) Z_{\beta'}(k)\rangle_0 \\ &= \langle\langle\tilde{\rho}(\vec{N}-\vec{m},\vec{\alpha}) O(\vec{k})\rangle\rangle^{\vec{k}} ,\end{aligned} \tag{40}$$

where in the first step we have used the operator expansion of Eq. (26), in the second have used Eqs. (32), (35), and, in the last, either the reality of the matrix elements of O or the Hermiticity of ρ. Note that this result is derivable also from Eq. (36) via Eq. (39). For the corresponding compact form of Eq. (36) we use Eq. (39) and then

$$\langle\langle O(\vec{k}_{\vec{n}})\rangle\rangle^{\vec{m},\vec{\alpha}} = \langle\langle \rho(\vec{m},\vec{\alpha}) O(\vec{k}_{\vec{n}})\rangle\rangle^{\vec{N}-\vec{k}} . \tag{41}$$

We can think of the subspace $(\vec{m},\vec{\alpha})$ as one member in a subspace decomposition of $\mathfrak{N}$. There are important decompositions of $\mathfrak{N}$ for which it turns out that the trace operator for each subspace is diagonal with respect to subspaces and behaves like a multiple of unity in each one. If $\mathfrak{N} = \sum(\vec{m},\vec{\alpha})$ is such a decomposition, we have then from Eq. (40) that

$$\begin{aligned}\langle\langle O(\vec{k})\rangle\rangle^{\vec{m},\vec{\alpha}} &= \sum_{\vec{\alpha}'} \langle k\alpha'|\tilde{\rho}(\vec{N}-\vec{m},\vec{\alpha})|k\alpha'\rangle\langle\langle O(\vec{k})\rangle\rangle^{\vec{k},\vec{\alpha}'} \\ &= \sum_{\vec{\alpha}'} \langle\tilde{\rho}(\vec{N}-\vec{m},\vec{\alpha})\rangle^{\vec{k},\vec{\alpha}'}\langle\langle O(\vec{k})\rangle\rangle^{\vec{k},\vec{\alpha}'},\end{aligned} \tag{42}$$

where, in the first form, α' is any state in $\vec{\alpha}'$. We see that in this case a subspace trace of $O(\vec{k})$ propagates from its defining representation throughout the entire lattice of subspaces with a propagating coefficient given by the matrix element indicated. An operator, then, which is traceless (or a multiple of unity) in each of the defining subspaces is traceless (or a multiple of unity) in all the subspaces. We shall describe this general situation, which is characteristic of the decomposition, as one of *simple propagation*. For the scalar distribution of Sec. 3, it should by this time be obvious that[21]

$$\rho(\vec{k}) = \binom{n}{k}, \quad \tilde{\rho}(\vec{k}) = \binom{N-n}{k} , \tag{43}$$

so that simple propagation does pertain in this case (as it will also in the case of configurations to be discussed in Sec. 5). The general reason for this should also be clear, namely that all m-particle states have equal standing, no preferred axes being specified in $\vec{m}$ by the operations considered. In these first papers our interest will be solely in cases which admit simple propagation, and in the discussion which follows here we make that assumption also.

The propagation coefficients have a symmetry which follows directly from Eq. (32), viz.,

$$\begin{aligned} d(\vec{k},\vec{\alpha}')\langle\tilde{\rho}(\vec{N}-\vec{m},\vec{\alpha})\rangle^{\vec{k},\vec{\alpha}'} &= \langle\langle\tilde{\rho}(\vec{N}-\vec{m},\vec{\alpha})\rangle\rangle^{\vec{k},\vec{\alpha}'} \\ &= \langle\langle\rho(\vec{k},\vec{\alpha}')\rangle\rangle^{\vec{m},\vec{\alpha}} \\ &= d(\vec{m},\vec{\alpha})\langle\rho(\vec{k},\vec{\alpha}')\rangle^{\vec{m},\vec{\alpha}} ,\end{aligned} \tag{44}$$

where, as always, $d(\vec{\gamma})$ is the dimensionality of $\vec{\gamma}$. Using this we have[22] (for the *average* instead of the trace)

$$\langle O(\vec{k})\rangle^{\vec{m},\vec{\alpha}} = \sum_{\vec{\alpha}'} \langle m\alpha|\rho(\vec{k},\vec{\alpha}')|m\alpha\rangle\langle O(\vec{k})\rangle^{\vec{k},\vec{\alpha}'} , \tag{45}$$

$$O(\vec{k}) \equiv \sum_{\vec{\alpha}'} \langle O(\vec{k})\rangle^{\vec{k},\vec{\alpha}'}\rho(\vec{k},\vec{\alpha}') , \tag{46}$$

where Eq. (46) displays a linear-trace equivalence. The operator $\rho(\vec{k},\vec{\alpha}')$ is then, in the sense of the discussion following Eq. (21), a propagation operator. It functions like a net-point Green's function in propagating, throughout the lattice, trace information given in the defining subspace $(\vec{k},\vec{\alpha}')$. Alternatively it may be regarded, *for fixed k* as a $(\vec{k},\vec{\alpha}')$ projection operator, since it satisfies the condition

$$\begin{aligned}\langle k\alpha|\rho(\vec{k},\vec{\alpha}')|k\alpha\rangle &= 1 \quad \text{if } \alpha\in\vec{\alpha}' \\ &= 0 \quad \text{if } \alpha\notin\vec{\alpha}' .\end{aligned} \tag{47}$$

If the operator with which we deal is not of definite particle rank [e.g., $O^2(\vec{k})$ which we would need for the second moment], we could decompose it into operators of definite rank, each part of which would then propagate via Eq. (45). Though we shall sometimes use this procedure, it is basically uneconomical to carry out a microscopic transformation of an operator solely for the purpose of evaluating a few traces. We may alternatively proceed as follows, which will be recognized as a natural extension of what we have already done in Sec. 3.

Suppose that the operator has maximum particle rank which is not greater than u; e.g., we would take $u=4$ for the square of a $(0+1+2)$-particle Hamiltonian. Then those terms in the defining expansion of the operator which are nondiagonal in representations cannot contribute to traces, because we are dealing with cases of simple propagation. Therefore the operator is defined on a *net of representations* $(\vec{k}, \vec{\alpha})$ with $k=0, 1, \ldots, u$. Now, while a density operator $\rho(\vec{k}, \vec{\alpha}')$ does give $\delta(m, k)\delta(\vec{\alpha}, \vec{\alpha}')$ in any representation $(\vec{m}, \vec{\alpha})$ with $m \leqslant k$, it cannot vanish in all those representations of the defining net which have $m > k$. This follows from the fact that $\rho(\vec{k}, \vec{\alpha}')$, being a k-body operator whose defining trace does not vanish, must necessarily have a nonvanishing trace in states with $m > k$. Thus $\rho(\vec{k}, \vec{\alpha}')$ does not behave on the defining net like a Green's function. We therefore introduce a new set of density operators $\rho_u(\vec{k}, \vec{\alpha}')$,

$$\rho_u(\vec{k}, \vec{\alpha}') = \binom{u-n}{u-k}\rho(\vec{k}, \vec{\alpha}') = (-1)^{u-k}\binom{n-k-1}{u-k}\rho(\vec{k}, \vec{\alpha}'), \tag{48}$$

where, as always, n is the number operator. The n factor, being then a product of linear factors which give zero on all representations with $k < m \leqslant u$, forbids propagation from the "source point" $(\vec{k}, \vec{\alpha}')$ to any representation of the defining net. The $\rho_u(\vec{k}, \vec{\alpha}')$ are operators of maximum rank $\leqslant u$ satisfying

$$\rho_u(\vec{k}, \vec{\alpha}') \to \delta(m, k)\delta(\vec{\alpha}, \vec{\alpha}') \text{ on } (\vec{m}, \vec{\alpha}) \text{ if } m, k \leqslant u \tag{49}$$

and act collectively as a Green's function on the defining net.[23]

In terms of these operators we have, as an extension of Eq. (46),

$$O \equiv \sum_{(\vec{k}, \vec{\alpha}')} \langle O \rangle^{\vec{k}, \vec{\alpha}'} \rho_u(\vec{k}, \vec{\alpha}') , \tag{50}$$

the sum being over the defining net, and the operator being of maximum rank $\leqslant u$. The equation is valid for linear traces and follows from the fact that the right-hand side is of the correct maximum particle rank and, via Eq. (49), reproduces the defining traces of O. Then, corresponding to Eq. (45), we have

$$\langle O \rangle^{\vec{m}, \vec{\alpha}} = \sum_{(\vec{k}, \vec{\alpha}')} \langle m\alpha | \rho_u(\vec{k}, \vec{\alpha}') | m\alpha \rangle \langle O \rangle^{\vec{k}, \vec{\alpha}'} . \tag{51}$$

We may use Eq. (50) to expand the density operators themselves. Then for $t \leqslant u$ we have

$$\rho(\vec{t}, \vec{\gamma}) = \sum_{k=t}^{u} \sum_{\vec{\alpha}} \langle \rho(\vec{t}, \vec{\gamma}) \rangle^{\vec{k}, \vec{\alpha}} \rho_u(\vec{k}, \vec{\alpha}) \tag{52}$$

and since $\langle \rho(\vec{t}, \vec{\gamma}) \rangle^{\vec{t}, \vec{\gamma}'} = \delta_{\vec{\gamma}\vec{\gamma}'}$, we may rewrite this as[24]

$$\rho(\vec{t}, \vec{\gamma})\left[1 - \binom{u-n}{u-t}\right] = \sum_{k=t+1}^{u} \sum_{\vec{\alpha}} \langle \rho(\vec{t}, \vec{\gamma}) \rangle^{\vec{k}, \vec{\alpha}} \rho_u(\vec{k}, \vec{\alpha}) . \tag{53}$$

Similarly we find

$$\tilde{\rho}(\vec{t}, \vec{\gamma}) = \sum_{k=0}^{t} \sum_{\vec{\alpha}} \langle \tilde{\rho}(\vec{t}, \vec{\gamma}) \rangle^{\vec{k}, \vec{\alpha}} \rho_u(\vec{k}, \vec{\alpha}) , \tag{54}$$

and we have also the expansions ($\tilde{\rho}$ versus $\tilde{\rho}$ and ρ versus $\tilde{\rho}_u$) which come from applying the operation $(\tilde{\ })$ to these equations.

It is worthwhile to have the inverse of Eq. (52), which would express a trace operator which acts as a Green's function on the net, in terms of trace operators for single representations. We can invert the equation by a process which will be recognized as that of an elementary sieve. Remember that, by construction, $\rho_r(\vec{k}, \vec{\alpha})$ and $\rho_s(\vec{k}, \vec{\alpha})$ $(k \leqslant r \leqslant s \leqslant u)$ coincide in the $(m \leqslant r)$-particle representations; $\rho_s(\vec{k}, \vec{\alpha})$, however, has higher-rank terms $(r < \text{rank} \leqslant s)$ whose function is to cancel the $\rho_r(\vec{k}, \vec{\alpha})$ traces which propagate into the $(m > r,\ m \leqslant s)$-particle domain. Then we have

$$\begin{aligned}\rho_u(\vec{k}, \vec{\alpha}) &= \rho(\vec{k}, \vec{\alpha}) + \sum_{r=k+1}^{u} O(r) \\ &= \rho(\vec{k}, \vec{\alpha}) - \sum_{\vec{\alpha}_1} \langle \rho_k(\vec{k}, \vec{\alpha}) \rangle^{\vec{k}+\vec{1}, \vec{\alpha}_1} \rho(\vec{k}+\vec{1}, \vec{\alpha}_1) \\ &\quad + \sum_{r=k+2}^{u} O(r) ,\end{aligned} \tag{55}$$

where the last step comes from the vanishing of $\rho_u(\vec{k}, \vec{\alpha})$ in $(k+1)$-particle states (if $k+1 \leqslant u$). The first two terms on the right-hand side of this equation give $\rho_{k+1}(\vec{k}, \vec{\alpha})$; proceeding in order we find the desired inversion,

$$\rho_u(\vec{k}, \vec{\alpha}) = \rho(\vec{k}, \vec{\alpha}) - \sum_{t=k+1}^{u} \sum_{\vec{\beta}} \langle \rho_{t-1}(\vec{k}, \vec{\alpha}) \rangle^{\vec{t}, \vec{\beta}} \rho(\vec{t}, \vec{\beta}) . \tag{56}$$

Combining Eqs. (50) and (56) yields a trace expansion of an operator O, of maximum rank $\leqslant u$, in terms of t-particle operators; with the aid of this we can make the trace decomposition of an operator whose traces are given on the elementary net into a set of operators each defined in a single subspace of that net. The result is

$$\begin{aligned} O &\equiv \sum_{k=0}^{u} \sum_{\vec{\alpha}} \Lambda(\vec{k}, \vec{\alpha}) \rho(\vec{k}, \vec{\alpha}) , \\ \Lambda(\vec{k}, \vec{\alpha}) &= \langle O \rangle^{\vec{k}, \vec{\alpha}} - \sum_{t=0}^{k-1} \sum_{\vec{\beta}} \langle \rho_{k-1}(\vec{t}, \vec{\beta}) \rangle^{\vec{k}, \vec{\alpha}} \langle O \rangle^{\vec{t}, \vec{\beta}} ,\end{aligned} \tag{57}$$

which we can recognize as the natural extension (for the more general lattice) of the usual trace separation of an operator according to its particle ranks. For that case, Eq. (57) gives immediately the result (easily derivable also by other methods)

$$\begin{aligned}\Lambda(k) &= \sum_{t=0}^{k} (-1)^{t-k} \binom{k}{t} \langle O \rangle^{t} , \\ O &\equiv \sum_{k=0}^{u} \sum_{t=0}^{k} (-1)^{k-t} \binom{k}{t} \langle O \rangle^{t} \binom{n}{k} .\end{aligned} \tag{58}$$

As a minor application we find the trace expansion of an l-hole operator to be

$$\bar{O}(\overleftarrow{1}) \equiv \langle \bar{O}(\overleftarrow{1})\rangle^{\overleftarrow{N}-\overleftarrow{1}} \sum_{k=0}^{l} \binom{n}{k} \sum_{t=0}^{k} (-1)^{t-k} \binom{k}{t}\binom{N-t}{l}$$

$$= \langle \bar{O}(\overleftarrow{1})\rangle^{\overleftarrow{N}-\overleftarrow{1}} \sum_{k=0}^{l} (-1)^{k} \binom{N-k}{N-l}\binom{n}{k}, \qquad (59)$$

where we have used a combinatorial identity. A different derivation of this result is worthwhile. Since $\langle \rho(\overleftarrow{k})\rangle^{\overleftarrow{N}} = \binom{N}{k}$, it follows that $\rho(\overleftarrow{k}) - \binom{N}{k}\bar{\rho}(\vec{0})$ vanishes in $(\vec{N})$; it gives $\binom{N-1}{k} - \binom{N}{k} = -\binom{N-1}{k-1}$ in $(\vec{N}-\vec{1})$; thus, $\rho(\overleftarrow{k}) - \binom{N}{k}\bar{\rho}(\vec{0}) + \binom{N-1}{k-1}\bar{\rho}(\vec{1})$, which $\equiv \binom{N-2}{k-2}$ in $(\vec{N} - \vec{2})$, and vanishes in $(\vec{N})$ and $(\vec{N}-\vec{1})$. Continuing in this way, we have that

$$\rho(\overleftarrow{k}) = \sum_{t=0}^{k} (-1)^{t} \binom{N-t}{k-t} \bar{\rho}(\vec{t}), \qquad (60)$$

since the difference between the operators on both sides, being an operator of maximum hole rank k which vanishes in $(l \leqslant k)$-hole states, must necessarily vanish. The hole-particle adjoint operation ($\bar{\ }$) applied to Eq. (60) will give the density-operator equivalent of Eq. (59). These combinatorial results are elementary and, in fact, are not of very great interest; but their extensions for more complicated lattices are not trivial at all (a point which, in a more general way, has been made by Rota[25]) and, in some cases, will be of considerable interest.

When there is simple propagation, we are now able to propagate the trace of an operator of mixed particle rank without recourse to the normal-form decomposition. To do this via Eq. (51) we need to calculate the input traces over the entire defining net, but for high h in particular this may be very difficult. We can, however, proceed to a new net for which the required input calculations are simpler; in particular, one which discards the high-number members of the defining net in favor of representations involving only a few holes. An elementary way to do this is to use Eq. (51) to eliminate elementary-net points on the right-hand side in favor of simple representations on the left, as we have already done for the simple scalar case in Sec. 3. We shall see later how this works out in practice for more complicated cases. We shall give also more elegant ways of proceeding which get around difficulties arising from the fact that not every net of the correct dimensionality carries all the necessary input information, this because of a failure of linear independence.[26] It might be noted that the defining net gives the simplest faithful array of representations for the operator being considered and that the simpler net does not then define a faithful array. But for traces of limited order, such faithfulness is not required and, as we shall see, the use of a nonfaithful array makes things often very much easier.

Finally, we mention a different way of looking at some of the things discussed above. We have in Eq. (27) written the matrix elements of an operator in terms of the vacuum expectation value of a quartic object. We can consider this v.e.v. as measuring the overlap of two quadratic objects. For example, the nonnegative quantity $\langle Z_{\alpha}^{\dagger}(m)Z_{\beta}^{\dagger}(k)Z_{\beta}(k)Z_{\alpha}(m)\rangle_0$ determines the normalization of the state produced by combining an m-particle and a k-particle system as indicated, while $\langle Z_{\alpha}^{\dagger}(m)Z_{\beta}(k)Z_{\beta}^{\dagger}(k)Z_{\alpha}(m)\rangle_0$ measures the amplitude with which the state (k, β) can be found in the state (m, α). It is easy to see that in each case we have $\langle\ \rangle \leqslant 1$ (consider the evaluation in a Hartree-Fock basis).

We are really interested in the combination of *subspaces* instead of states. A natural measure then of the "strength" with which two subspaces combine is

$$\mathfrak{a}(\vec{m}, \vec{\alpha}; \overleftarrow{k}, \vec{\beta}) = d^{-1}(\vec{m}, \vec{\alpha})d^{-1}(\overleftarrow{k}, \vec{\beta}) \times \sum_{\substack{\alpha \in \vec{\alpha} \\ \beta \in \vec{\beta}}} \langle Z_{\alpha}^{\dagger}(m)Z_{\beta}^{\dagger}(k)Z_{\beta}(k)Z_{\alpha}(m)\rangle_0$$

$$= d^{-1}(\overleftarrow{k}, \vec{\beta})\langle \bar{\rho}(\overleftarrow{k}, \vec{\beta})\rangle^{\vec{m}, \vec{\alpha}}, \qquad (61)$$

which incidentally gives us an upper limit on the propagation coefficients (since $\mathfrak{a} \leqslant 1$). The departure of $\mathfrak{a}$ from unity represents what is commonly called a "blocking effect." For the simple $U(N)$ case, Eq. (43) gives, for combining an m particle and k particle subspaces, that

$$\mathfrak{a}(\vec{m}; \overleftarrow{k}) = \binom{N}{k}^{-1}\binom{N-m}{k}$$

$$= \left(1 - \frac{m_>}{N}\right)\left(1 - \frac{m_> - 1}{N-1}\right) \cdots \left(1 - \frac{m_> - m_<}{N - m_<}\right), \qquad (62)$$

where $m_>$ is the larger of (m, k) and $m_<$ the smaller. The "blocking effect" measured by $(1 - \mathfrak{a})$ is seen to be small, as expected, if we combine two subspaces each small compared with the total space available.

In the same way a measure for the probability of finding a subspace $(\overleftarrow{k}, \vec{\beta})$ in $(\vec{m}, \vec{\alpha})$ is

$$d(\vec{m}, \vec{\alpha})\mathfrak{B}(\vec{m}, \vec{\alpha}; \overleftarrow{k}, \vec{\beta}) = d^{-1}(\overleftarrow{k}, \vec{\beta}) \sum_{\substack{\alpha \in \vec{\alpha} \\ \beta \in \vec{\beta}}} \langle Z_{\alpha}^{\dagger}(m)Z_{\beta}(k)Z_{\beta}^{\dagger}(k)Z_{\alpha}(m)\rangle$$

$$= d(\vec{m}, \vec{\alpha})d^{-1}(\overleftarrow{k}, \vec{\beta})\langle \rho(\overleftarrow{k}, \vec{\beta})\rangle^{\vec{m}, \vec{\alpha}}. \qquad (63)$$

The measure adopted here ensures a unit value for the measure of a space within itself and leads to the simple result of Eq. (65) below. The normalization of $\mathcal{B}$ is such as to give the simplest form for the symmetry properties which follow from Eq. (44); they are

$$\mathcal{B}(\vec{m}, \vec{\alpha}; \vec{k}, \vec{\beta}) = \mathcal{A}(\vec{N} - \vec{m}, \vec{\alpha}; \vec{k}, \vec{\beta}) = \mathcal{A}(\vec{k}, \vec{\beta}; \vec{N} - \vec{m}, \vec{\alpha})$$
$$= \mathcal{B}(\vec{N} - \vec{k}, \vec{\beta}; \vec{N} - \vec{m}, \vec{\alpha}) \ . \qquad (64)$$

$\mathcal{B}$ of course vanishes if $k > m$, and $\mathcal{A}$ if $m + k > N$, and both quantities are bounded by unity. Indeed, they satisfy elementary sum rules arising from the the fact that $\sum_{\vec{\gamma}} \rho(\vec{t}, \vec{\gamma}) = \rho(\vec{t})$.

If we have simple propagation, then, on combining two spaces (i.e., taking the product of an orthonormal basis set in one space with a set in another), the states of an orthonormal basis for the resultant space appear with equal intensity; and, similarly, for the operation measuring one space in another. In this case, also, we have a very simple picture of the trace propagation. For an operator defined in a subspace $(\vec{k}, \vec{\beta})$, we have

$$\text{Trace of } O \text{ in } (\vec{m}, \vec{\alpha}) = [\text{Trace of } O \text{ in } (\vec{k}, \vec{\beta})] \times [\text{Measure of } (\vec{k}, \vec{\beta}) \text{ in } (\vec{m}, \vec{\alpha})] \ . \qquad (65)$$

The $\mathcal{A}$, $\mathcal{B}$ quantities correspond to looking at the quartic object, whose vacuum expectation value is of interest to us, according to its [(12)(34)] structure rather than [1(23)4] which is natural when we think of density operators. When later we replace the *scalar* multiplication of the four objects by a *tensor* coupling, with subspaces and tensors defined by the group structure being considered, we shall be led naturally to decompose $\mathcal{A}$ and $\mathcal{B}$ according to the intermediate subspaces, viz.,

$$\mathcal{A}(\vec{m}, \vec{\alpha}; \vec{k}, \vec{\beta}) = \sum_{\vec{\gamma}} A(\vec{m}, \vec{\alpha}; \vec{k}, \vec{\beta}; \vec{m} + \vec{k}, \vec{\gamma}) ,$$
$$\mathcal{B}(\vec{m}, \vec{\alpha}; \vec{k}, \vec{\beta}) = \sum_{\vec{\gamma}} B(\vec{m}, \vec{\alpha}; \vec{k}, \vec{\beta}; \vec{m} + \vec{k}, \vec{\gamma}) . \qquad (66)$$

The resultant quantities, as we shall see later, are vital for the decomposition of widths into partial widths. They have a considerably extended set of symmetry properties and sum rules. The relationship between the two kinds of decompositions of the quartic structure will be that of a Racah transformation involving, of course, the Racah algebra of the relevant group or set of groups.

5. CONFIGURATION DISTRIBUTIONS

The scalar distribution of Sec. 3 is always of interest, since it has to do with the distribution of the eigenstates themselves, but it tells us essentially nothing about the structure of the eigenstates. We thus turn to a more detailed decomposition of the $\mathfrak{M}$ space, that corresponding to configurations.

In the configuration or fixed-partition distribution we imagine a partitioning of the orbital set $\{N\}$ into l subsets, $\{N\} \to \{N_1, N_2, \ldots, N_l\} = \{[N]\}$, say,[27] in which $\sum N_i = N$, N_k being then the number of single-particle states in the kth subset. Such partitioning is naturally suggested by the orbitals of spherical or deformed shell models, but we may in fact have no shell model in mind at all. Corresponding to this partition we may distribute particles according to $m \to (m_1, m_2, \ldots, m_l) = [m]$, with $\sum m_i = m$, this arrangement defining a configuration and $[\vec{m}]$ then standing for a basis set for that configuration.

We observe that even a number-conserving operator may not conserve the separate numbers m_i. Operators which change configurations are of course traceless in m; among the m-conserving operators the simplest are those of the type $O([\vec{k}])$ which have definite particle ranks with respect to each subset of $\{N\}$. For these operators we have an immediate extension of Eqs. (13) and (43), viz.,

$$\langle O([\vec{k}])\rangle^{[\vec{m}]} = \langle O([\vec{k}])\rangle^{[\vec{k}]} \prod_i \binom{m_i}{k_i} , \qquad (67)$$

$$\rho([\vec{m}]) = \prod_i \binom{n_i}{m_i} , \qquad (68)$$

which will give the averages and trace equivalents for a general operator as long as its decomposition $O = \sum O([\vec{k}; \vec{k}'])$ is available.

What now is the extension which, without recourse to normal-form expansions, will enable us to deal with operators of indefinite (vectorial) particle rank? Just as we did before, in the transition from Eq. (13) to Eq. (14) or (46)→(50), we may introduce a set of Lagrange polynomials or trace operators orthogonal on a defining net. If we wish to deal with an operator O whose vectorial rank $\leqslant [u]$ (i.e., rank in the ith orbit $\leqslant u_i$), then it follows from the direct-product nature of the system that the density operator on the u-net factors according to the orbits, and we have

$$\rho_{[u]}([\vec{k}]) = \prod_i \rho_{u_i}(k_i)$$
$$= \prod_i (-1)^{u_i - k_i} \binom{n_i - k_i - 1}{u_i - k_i} \binom{n_i}{k_i} , \qquad (69)$$

the average of O being then given by the natural extension of Eq. (14).

This in fact is not very useful, for usually we shall deal with operators O which have a specified maximum *scalar* rank u. Then the maximum ranks in the separate orbits are also restricted to $\leqslant u$ but there is a correlation between them. With a (0 + 1 + 2)-body Hamiltonian in a two-orbit configuration, we have for example $[\vec{k}] = [\vec{2}, \vec{0}]$, $[\vec{1}, \vec{1}]$, $[\vec{0}, \vec{2}]$, but $[\vec{2}, \vec{1}]$ and $[\vec{2}, \vec{2}]$ would be ruled out. In the $l=2$ case, then, we should take the $\binom{u+2}{2}$ dimensional *triangular* net $(t_1+t_2 \le u)$ rather than the square net $(t_i \le u)$ appropriate to Eq. (69); this constitutes a major simplification, since the new net has fewer and simpler elements; and similarly for arbitrary l. We shall indicate the new density operators as $\rho_u([\vec{t}])$.

Consider as an example $l=2$, $u=5$, $[\vec{t}]=[\vec{1},\vec{2}]$ as shown in Fig. 2. We need an operator of (mixed) rank ≤ 5 which gives zero for all points on the net except for unity on $(\vec{1},\vec{2})$. Then noting that there are three varieties of linear factors [(0 + 1)-particle operators] which annihilate along a line we see that

$$\rho_5([\vec{1},\vec{2}]) = \tfrac{1}{4} n_1 n_2 (n_2-1)(n-4)(n-5) \; . \qquad (70)$$

A little consideration will show that the same kind of linear factoring is valid for all points of the net, for arbitrary u, and finally for an arbitrary dimensionality l. The result then, as may easily be verified, is

$$\rho_u([\vec{t}]) = \binom{u-n}{u-t}\prod_i \binom{n_i}{t_i} = (-1)^{t-u}\binom{n-t-1}{u-t}\prod_i \binom{n_i}{t_i} \; , \qquad (71)$$

which gives for the average

$$\langle O\rangle^{[\vec{m}]} = \sum_{t_1=0}^{u} \cdots \sum_{t_l=0}^{u} (-1)^{t-u}\binom{m-t-1}{u-t} \times\left[\binom{m_1}{t_1}\cdots\binom{m_l}{t_l}\right]\langle O\rangle^{[\vec{t}]} , \quad (t \le u), \qquad (72)$$

where $t=\sum t_i$, $m=\sum m_i$. The $[\vec{m}]$ dependence of $\langle O\rangle^{[\vec{m}]}$ is, of course, that of a uth-rank polynomial in the l variables $m_1 \ldots m_l$.

Once again, just as in the scalar case, the elementary net (the triangular net about the origin which constitutes the defining space for O) will usually not be convenient to deal with, because of the complexity of the necessary input traces. In the scalar case a "most economical" net and the corresponding density operators were easily found but, because of a failure of linear independence, this is not at all true in the present case. It has been shown by Kim[28] that a "most economical" lattice in the l-dimensional case derives from the elementary net by a kind of reflection-contraction process in which $[\vec{t}] \to [\vec{t}^*]$ where $t_i^* = (N_i - \frac12 t_i + \frac12)$ if t_i is odd, $t_i^* = \frac12 t_i$ if t_i is even. This replaces the triangular elementary net by a set of triangular nets, one at each corner, the largest of which corresponds to $\frac12 u$ or $\frac12(u-1)$ according to whether u is even or odd. The general solution for ρ_u for this net is very complicated but in any case of interest we may numerically invert Eq. (72) to express the old input traces in terms of the new ones.

Instead of using this method we shall in the next sections use a normal-form decomposition of H^2 [propagating the separate parts via Eq. (67)] to express the configuration widths in terms of the Hamiltonian matrix elements, and shall give also a decomposition into partial widths. In a later paper a quite different solution for the width problem will be derived by use of a unitary-group decomposition of the Hamiltonian and expressed in terms of physically significant quantities. The entire structure of the width expressions will then become transparent.

6. SCALAR AND CONFIGURATION MOMENTS IN THE SPHERICAL SHELL MODEL

In this section we derive and discuss expressions, in terms of the Hamiltonian matrix elements, for the centroids and widths of the scalar and configuration spectral distributions. We are particularly interested in the case where the set of single-particle states corresponds to a number of spherical orbits and we work, therefore, in the usual shell-model or spherical-tensor representation in which both states and operators carry definite angular momentum (and isospin where that is relevant). Actually, since we do not now consider distributions for fixed angular momentum, this choice of representation is not at all essential and it is, in fact, a slightly restrictive one since, at first sight, it precludes a consideration of nonspherical orbits. This restriction however is more apparent than real, because it is clear from the discussions above that the centroids and widths must be expressible in terms of linear and quadratic traces in the matrix-element space, and the translation of such quantities from one representation to another is not difficult. We choose the explicit

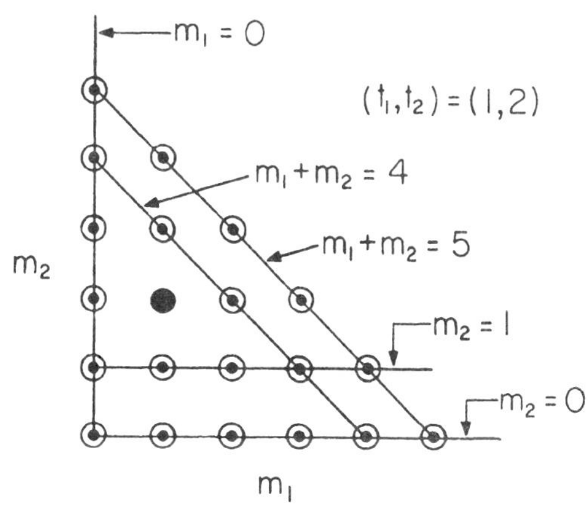

FIG. 2. Representation of the density operator in Eq. (70).

spherical basis because that will be particularly convenient for future work. In dealing with the propagation of the quadratic traces we shall use the normal-form method, leaving until later the application of other methods. The essential formal problem is then that of decomposing H and H^2 in an appropriate way and evaluating the N-state expectation value of the various parts.

We adopt a notation[29] which emphasizes the tensorial properties of the various quantities arising in the spherical shell model. Single-particle orbitals are characterized by principal quantum number, parity, angular momentum, and, where relevant, the isospin (whose value is $t=\frac{1}{2}$). We can for most purposes describe all these by a composite label r which, among other things, then defines the tensorial ranks. In the isospin formalism, the $2(2j_r+1)$ single-particle states corresponding to such an orbital are distinguished by their projection quantum numbers $(\mu_j, \mu_t) \equiv \mu$. The basic creation operator is A^r_μ which creates a normalized single-particle state

$$A^r_\mu|0\rangle \equiv \psi^r_\mu \ . \tag{73}$$

The destruction operator B^r_μ is related to the Hermitian conjugate of A^r_μ as

$$B^r_\mu = (-1)^{r+\mu}(A^r_{-\mu})^\dagger \ , \tag{74}$$

the phase being chosen to cause B^r_μ to transform as the μ component of a spherical tensor of rank r. The notation is a direct-product one so that in a (j,t) formalism we have, for example,

$$(-1)^r = (-1)^{j_r+t_r} = (-1)^{j_r+1/2} \ . \tag{75}$$

But we can also interpret the equations via a j formalism in which isospin does not occur (at least not as a tensorial rank) or even via a t formalism. In some cases it will be found advantageous to introduce more explicitly some of the single-particle state labels.

The normalized two-particle state is

$$-\zeta_{rs}(A^r \times A^s)^\Gamma_\mu|0\rangle \equiv |r, s; \Gamma, \mu\rangle \ , \tag{76}$$
$$\zeta_{rs} = 1, \quad r \neq s \ ,$$
$$\zeta_{rs} = 1/\sqrt{2} \ , \quad r = s \ ,$$

where $(\Gamma, \mu) \equiv (JT, \mu_J\mu_T)$ in an isospin formalism and the indicated $(\times)$ product defines the usual angular momentum coupling. The coupled commutation rules for the basic operators are then

$$\begin{aligned}
&(A^r \times A^s)^\Gamma + (-1)^{r+s-\Gamma}(A^s \times A^r)^\Gamma = 0 \ , \\
&(B^r \times B^s)^\Gamma + (-1)^{r+s-\Gamma}(B^s \times B^r)^\Gamma = 0 \ , \\
&(A^r \times B^s)^\Gamma + (-1)^{r+s-\Gamma}(B^s \times A^r)^\Gamma = [r]^{1/2}\delta_{rs}\delta_{\Gamma 0} \ ,
\end{aligned} \tag{77}$$

where $[r]$ denotes the statistical weight of orbital r,

$$[r] = N_r = (2j_r+1)(2t_r+1) = 2(2j_r+1) \tag{78}$$

in the (j,t) formalism. $\delta_{rs} = 0$ unless r, s define the same orbital. It will be found convenient, allowing for the fact that there may be active orbits differing only in radial quantum number, to introduce the symbol $\hat{\delta}_{rs}$ which distinguishes between orbits only according to their tensorial rank

$$\hat{\delta}_{rs} = \delta(j_r, j_s) \qquad [\text{in } j \text{ or } (jt) \text{ formalisms}] \ . \tag{79}$$

For the Hamiltonian we take a $(0+1+2)$-particle scalar operator, the separate parts of which via Eq. (26) have the form

$$\begin{aligned}
&H(\vec{0}) = H_0 \ , \\
&H(\vec{1}) = \sum_{r,s} \epsilon_{rs}[r]^{1/2}(A^r \times B^s)^0 \ , \\
&H(\vec{2}) = -\tfrac{1}{4}\sum_{rstu\Gamma}[\Gamma]^{1/2}\zeta_{rs}^{-1}\zeta_{tu}^{-1}W^\Gamma_{rstu}[(A^r \times A^s)^\Gamma \times (B^t \times B^u)^\Gamma]^0 \ .
\end{aligned} \tag{80}$$

The one- and two-body matrix elements of H, denoted by ϵ and W, respectively, have the following symmetry properties:

$$\epsilon_{rs} \equiv \langle r|H(1)|s\rangle = \epsilon_{sr} = \hat{\delta}_{rs}\epsilon_{rs} \ , \tag{81}$$

$$W^\Gamma_{rstu} \equiv \langle rs, \Gamma|H(2)|tu, \Gamma\rangle = (-1)^{r+s-\Gamma}W^\Gamma_{srtu} = (-1)^{t+u-\Gamma}W^\Gamma_{rsut} = (-1)^{r+s-t-u}W^\Gamma_{srut} = W^\Gamma_{turs} \ . \tag{82}$$

We shall sometimes write the diagonal single-particle energies ϵ_{rr} as ϵ_r.

Similarly, $H^2 \equiv K$ is a $(0-4)$-particle operator[30]

$$H^2 = \sum_{t=0}^{4} K(t) \tag{83}$$

and on carrying out the commutations required to put K into normal form, we find

$$K(\vec{0}) = (H_0)^2 , \tag{84}$$

$$K(\vec{1}) = \sum_{rs} \{2H_0[r]^{1/2}\epsilon_{rs} + \sum_t [r]^{1/2}\epsilon_{rt}\epsilon_{ts}\} (A^r \times B^s)^0 , \tag{85}$$

$$\begin{aligned} K(\vec{2}) = \sum_{rstu\Gamma} [\Gamma]^{1/2}(&-\tfrac{1}{2}H_0\zeta_{rs}^{-1}\zeta_{tu}^{-1}W^{\Gamma}_{rstu} - \epsilon_{rt}\epsilon_{su} - \tfrac{1}{2}\sum_e \zeta_{rs}^{-1}\zeta_{te}^{-1}W^{\Gamma}_{rste}\epsilon_{eu} \\ &- \tfrac{1}{2}\sum_e \zeta_{es}^{-1}\zeta_{tu}^{-1}\epsilon_{re}W^{\Gamma}_{estu} - \tfrac{1}{8}\sum_{ef} \zeta_{rs}^{-1}\zeta_{ef}^{-2}\zeta_{tu}^{-1}W^{\Gamma}_{rsef}W^{\Gamma}_{eftu})[(A^r \times A^s)^{\Gamma} \times (B^t \times B^u)^{\Gamma}]^0 , \end{aligned} \tag{86}$$

$$\begin{aligned} K(\vec{3}) = \sum_{\substack{rstuef \\ \Gamma\Gamma'\Delta}} [\Delta]^{1/2}\{&- \tfrac{1}{2}\zeta_{rs}^{-1}\zeta_{tu}^{-1}\epsilon_{ef}W^{\Gamma}_{rstu}\delta_{\Gamma\Gamma'}(-1)^{e+\Gamma-\Delta} + \tfrac{1}{4}\sum_g \zeta_{rs}^{-1}\zeta_{fg}^{-1}\zeta_{eg}^{-1}\zeta_{tu}^{-1}[\Gamma]^{1/2} \\ &\times W^{\Gamma}_{rsfg}W^{\Gamma'}_{egtu}(-1)^{\Delta+\Gamma-e}W(\Gamma' ef\Gamma{:}g\Delta)\}\{[(A^r \times A^s)^{\Gamma} \times A^e]^{\Delta} \times B^f \times (B^t \times B^u)^{\Gamma'}]^{\Delta}\}^0 , \end{aligned} \tag{87}$$

$$K(\vec{4}) = \sum_{\substack{rstuefgh \\ \Gamma\Gamma'\Delta}} \tfrac{1}{16}[\Delta]^{1/2}\zeta_{rs}^{-1}\zeta_{tu}^{-1}\zeta_{ef}^{-1}\zeta_{gh}^{-1}W^{\Gamma}_{rstu}W^{\Gamma'}_{efgh}\{[(A^r \times A^s)^{\Gamma} \times (A^e \times A^f)^{\Gamma'}]^{\Delta} \times [(B^t \times B^u)^{\Gamma} \times (B^g \times B^h)^{\Gamma'}]^{\Delta}\}^0 . \tag{88}$$

All orbital summations are unrestricted. The sumbol $W(\Gamma' ef\Gamma{:}g\Delta)$ in Eq. (87) is the usual Racah coefficient. It is clear *a priori*, and demonstrated in these equations, that the product $H(\vec{k})H(\vec{k}')$ (for $k \geqslant k'$) contributes to $K(\vec{k})$, $K(\vec{k}+\vec{1}), \ldots, K(\vec{k}+\vec{k}')$. Note also that the matrix elements of the k-particle part of $H(\vec{k})H(\vec{k})$ are those which arise from the usual matrix product in the k spacc, this coming about because the k space supplies a representation of $K(\vec{k})$ as well as of $H(\vec{k})$. Examples are the ϵ^2 terms in $K(\vec{1})$ and the W^2 terms in $K(\vec{2})$. The matrix elements of $K(\vec{k}+\vec{k}')$ arising from $H(\vec{k})H(\vec{k}')$ are those of a kind of direct-product matrix, examples being the ϵW terms in $K(\vec{3})$ and W^2 in $K(\vec{4})$. These terms, of course, arise from transforming the product to normal form but ignoring all the commutator contractions.

For the scalar centroid and width we use Eq. (24) for which the necessary input is the set of N-state values of $H(\vec{t})$ and $K(\vec{t})$. These we evaluate by the standard methods and we find

$$\langle H(\vec{0})\rangle^N = H_0 , \tag{89}$$

$$\langle H(\vec{1})\rangle^N = \sum_r N_r\epsilon_{rr} , \tag{90}$$

$$\langle H(\vec{2})\rangle^N = \tfrac{1}{2}\sum_{rs} A_{rsrs} , \tag{91}$$

$$\langle K(\vec{0})\rangle^N = H_0^2 , \tag{92}$$

$$\langle K(\vec{1})\rangle^N = \sum_r (2H_0N_r\epsilon_{rr} + \sum_k N_r\epsilon_{rk}\epsilon_{kr}) , \tag{93}$$

$$\langle K(\vec{2})\rangle^N = \sum_{rs}[H_0A_{rsrs} + N_rN_s\epsilon_{rr}\epsilon_{ss} - N_r\epsilon_{rs}\epsilon_{sr} + \sum_k (A_{rsrk}\epsilon_{ks} + \epsilon_{rk}A_{ksrs}) + \tfrac{1}{4}\sum_{kl} B_{rskl}] , \tag{94}$$

$$\langle K(\vec{3})\rangle^N = \sum_{rst}[N_t\epsilon_{tt}A_{rsrs} + \sum_k N_s^{-1}A_{rsrk}A_{tkts}\hat{\delta}_{sk} - A_{srst}\epsilon_{tr} - \epsilon_{tr}A_{rsts} - \tfrac{1}{2}\sum_k B_{rstk}] , \tag{95}$$

$$\langle K(\vec{4})\rangle^N = \sum_{rstu} (\tfrac{1}{4}A_{rsrs}A_{tutu} + \tfrac{1}{4}B_{rstu} - N_s^{-1}A_{rsrt}A_{tusu}\hat{\delta}_{st}) , \tag{96}$$

where the orbital summations are all unrestricted.

We have introduced here the quantities

$$\begin{aligned} A_{rstu} &= \zeta_{rs}^{-1}\zeta_{tu}^{-1}\sum_{\Gamma}[\Gamma]W^{\Gamma}_{rstu} , \\ B_{rstu} &= \zeta_{rs}^{-2}\zeta_{tu}^{-2}\sum_{\Gamma}[\Gamma](W^{\Gamma}_{rstu})^2 , \end{aligned} \tag{97}$$

which to within normalization may be regarded as deriving from a Γ averaging of the matrix elements of $H(\vec{2})$ and $[H(\vec{2})]^2$ between the two-particle states (r, s) and (t, u). In terms of these it is clear that

$$W_{rs} = (N_rN_s)^{-1}\sum_{J,T}[JT]W^{JT}_{rsrs} = (N_rN_s)^{-1}A_{rsrs} , \quad r \neq s \tag{98}$$

is the average two-body interaction between two nucleons in different orbits, r, s. When they are in the same orbit the average is

$$W_{rr} \equiv W_r = \binom{N_r}{2}^{-1}\sum_{JT}[JT]W^{JT}_{rrrr} = [N_r(N_r - 1)]^{-1}A_{rrrr} . \tag{99}$$

Then, considering all the two-nucleon states, we have (for the trace instead of the average), writing $\mathcal{W}$ for the two-particle *matrix* of $H(\vec{2})$ and W for the over-all average interaction energy

$$\langle\langle H(\vec{2})\rangle\rangle^{\vec{2}} = \mathrm{Tr}(\mathcal{W}) = \binom{N}{2} W = \sum_{\substack{r,s \\ r<s}} N_r N_s W_{rs} + \sum_r \binom{N_r}{2} W_r = \tfrac{1}{2}\sum_{r,s} A_{rsrs} \;, \tag{100}$$

$$\langle\langle [H(\vec{2})]^2\rangle\rangle^{\vec{2}} = \mathrm{Tr}(\mathcal{W}\times\mathcal{W}) = \tfrac{1}{4}\sum_{rstu} B_{rstu} \;. \tag{101}$$

Finally, the following symmetry properties for the A's and B's follow immediately from Eq. (82).

$$A_{rsrs} = A_{srsr} \;, \tag{102}$$

$$A_{rstu} = A_{turs} = (-1)^{r+s-t-u} A_{srut} \;, \tag{103}$$

$$B_{rstu} = B_{srtu} = B_{rsut} = B_{srut} = B_{turs} \;. \tag{104}$$

The centroid of the complete scalar distribution now follows via Eq. (24) and is

$$\mathcal{E}(\vec{m}) = \langle H\rangle^{\vec{m}} = H_0 + \frac{m}{N}\sum_r N_r \epsilon_{rr} + \frac{m(m-1)}{N(N-1)}\,\tfrac{1}{2}\sum_{rs} A_{rsrs} \;. \tag{105}$$

For the dispersion we may do the same for $\langle H^2\rangle^{\vec{m}}$ and then subtract the square of the centroid polynomial. Or equivalently we can recognize that $\sigma^2(\vec{m})$ is the second moment (and similarly for the pth central moment) of

$$H - \sum_{k=0}^{2} \langle H(k)\rangle^{\vec{k}} \binom{n}{k} = H - H_0 - \epsilon n - W\binom{n}{2} \;, \tag{106}$$

the subtracted term being simply the linear trace equivalent of H. The result is

$$\begin{aligned}\sigma^2(\vec{m}) &= \langle H^2\rangle^{\vec{m}} - (\langle H\rangle^{\vec{m}})^2 \\ &= \frac{m}{N} Q + \frac{m(m-1)}{N(N-1)}(X+Y-Q) + \frac{m(m-1)(m-2)}{N(N-1)(N-2)}(Z-X-2Y) + \frac{m(m-1)(m-2)(m-3)}{N(N-1)(N-2)(N-3)}(Y-Z)\;,\end{aligned} \tag{107}$$

which, as will be recognized from the m polynomials, exhibits separately the dispersions from the $(1-4)$-particle parts of H^2. Here

$$\begin{aligned}Q &= \sum_{rs} N_r(\epsilon_{rs})^2 - \frac{1}{N}\Big(\sum_r N_r \epsilon_{rr}\Big)^2 \\ &= \mathrm{Tr}(\epsilon\times\epsilon) - \frac{1}{N}\Big[\mathrm{Tr}(\epsilon)\Big]^2\end{aligned} \tag{108}$$

$$X = 2\sum_{rst} A_{rsrt}\,\epsilon_{ts} - \frac{2}{N}\Big(\sum_r N_r\epsilon_{rr}\Big)\Big(\sum_{st} A_{stst}\Big)\;, \tag{109}$$

$$\begin{aligned}Y &= \frac{1}{4}\sum_{rstu} B_{rstu} - \frac{1}{2N(N-1)}\Big(\sum_{rs} A_{rsrs}\Big)^2 \\ &= \mathrm{Tr}(\mathcal{W}\times\mathcal{W}) - \binom{N}{2}^{-1}(\mathrm{Tr}\mathcal{W})^2\;,\end{aligned} \tag{110}$$

$$Z = \sum_{rstu} N_s^{-1} A_{rsru} A_{tuts}\,\delta_{su} - \frac{1}{N}\Big(\sum_{rs} A_{rsrs}\Big)^2 \;, \tag{111}$$

where, in correspondence with $\mathcal{W}$ as used in Eqs. (101) and (110), ϵ in Eq. (108) is the one-particle matrix of $H(\vec{1})$. The terms Q and Y then have a simple significance, being the one- and two-particle matrix dispersions, i.e., the dispersions of the one- and two-particle Hamiltonians, each considered as a matrix operator in its defining space. The parameters X, Z involve less familiar "partial" matrix products. Observe that each of these four separate terms propagates in its own characteristic way, the Q and Y polynomials, for example, being proportional to $\binom{m}{k}\binom{N-m}{k}$ with $k=0,\ 1$, respectively. The significance of the new forms and of the partial products will become clear when we consider the unitary-group decompositions of the widths.

We turn now to configurations. In this case we need a further decomposition of H and H^2 in order to isolate those parts which are number conserving and have a definite rank in each orbit separately. The N-state expectation values of these operators are then the input [see Eqs. (24) and (67)]. The decomposition follows by inspection in Eqs. (80) and (85)–(88). Using $H(r^a s^b t^c \ldots)$ to indicate that part of H which has particle ranks $a, b, c, \ldots$ in orbits $r, s, t, \ldots$, respectively, we see for example that

$$H(rs) = -\sum_{\Gamma} [\Gamma]^{1/2} W^{\Gamma}_{rsrs} [(A^r \times A^s)^{\Gamma} \times (B^r \times B^s)^{\Gamma}]^0 \ , \tag{112}$$

it being understood of course that $r \neq s$. The relevant N-state expectation values are then

$$\langle H(\vec{0})\rangle^N = H_0 \ , \tag{113}$$

$$\langle H(r)\rangle^N = N_r \epsilon_{rr} \ , \tag{114}$$

$$\langle H(r^2)\rangle^N = \tfrac{1}{2} A_{rrrr} \ , \tag{115}$$

$$\langle H(rs)\rangle^N = A_{rsrs} \ , \tag{116}$$

$$\langle K(\vec{0})\rangle^N = (H_0)^2 \ , \tag{117}$$

$$\langle K(r)\rangle^N = 2H_0 N_r \epsilon_{rr} + N_r \sum_k (\epsilon_{rk})^2 \ , \tag{118}$$

$$\langle K(r^2)\rangle^N = N_r (N_r - 1)(\epsilon_{rr})^2 + H_0 A_{rrrr} + 2\sum_k \epsilon_{rk} A_{rrrk} + \tfrac{1}{4}\sum_{kl} B_{rrkl} \ , \tag{119}$$

$$\langle K(rs)\rangle^N = 2N_r N_s \epsilon_{rr} \epsilon_{ss} - 2N_r (\epsilon_{rs})^2 + 2H_0 A_{rsrs} + 2\sum_k (\epsilon_{rk} A_{srsk} + \epsilon_{sk} A_{rsrk} + \tfrac{1}{2}\sum_{kl} B_{rskl} \ , \tag{120}$$

$$\langle K(r^3)\rangle^N = (N_r - 2)\epsilon_{rr} A_{rrrr} + N_r^{-1} \sum_k (A_{rrrk})^2 \hat{\delta}_{rk} - \tfrac{1}{2}\sum_k B_{rrrk} \ , \tag{121}$$

$$\begin{aligned}\langle K(r^2 s)\rangle^N = {} & 2(N_r - 1)\epsilon_{rr} A_{rsrs} + N_s \epsilon_{ss} A_{rrrr} - 4\epsilon_{rs} A_{rrrs} + 2N_r^{-1}\sum_k A_{rrrk} A_{ksrs} \hat{\delta}_{kr} \\ & + N_s^{-1}\sum_k (A_{rsrk})^2 \hat{\delta}_{sk} - \sum_k B_{rsrk} - \tfrac{1}{2}\sum_k B_{rrsk} \ , \end{aligned} \tag{122}$$

$$\begin{aligned}\langle K(rst)\rangle^N = {} & 2N_r \epsilon_{rr} A_{stst} + 2N_s \epsilon_{ss} A_{rtrt} + 2N_t \epsilon_{tt} A_{rsrs} - 4\epsilon_{rt} A_{rsts} - 4\epsilon_{rs} A_{rtst} - 4\epsilon_{st} A_{srtr} \\ & - \sum_k (B_{rstk} + B_{rtsk} + B_{strk}) + 2N_r^{-1}\sum_k A_{ksrs} A_{ktrt} \hat{\delta}_{kr} + 2N_s^{-1}\sum_k A_{krsr} A_{ktst} \hat{\delta}_{ks} + 2N_t^{-1}\sum_k A_{krtr} A_{ksts} \hat{\delta}_{kt} \ , \end{aligned} \tag{123}$$

$$\langle K(r^4)\rangle^N = \tfrac{1}{4} B_{rrrr} + (\tfrac{1}{4} - N_r^{-1})(A_{rrrr})^2 \ , \tag{124}$$

$$\langle K(r^3 s)\rangle^N = B_{rrrs} + (1 - 2N_r^{-1}) A_{rrrr} A_{rsrs} - 2N_r^{-1} (A_{rrrs})^2 \hat{\delta}_{rs} \ , \tag{125}$$

$$\langle K(r^2 s^2)\rangle^N = \tfrac{1}{2} B_{rrss} + B_{rsrs} + \tfrac{1}{2} A_{rrrr} A_{ssss} + (1 - N_r^{-1} - N_s^{-1})(A_{rsrs})^2 - 4N_r^{-1} A_{rrrs} A_{rsss} \hat{\delta}_{rs} \ , \tag{126}$$

$$\begin{aligned}\langle K(r^2 st)\rangle^N = {} & B_{rrst} + 2B_{rsrt} + A_{rrrr} A_{stst} + 2(1 - N_r^{-1}) A_{rsrs} A_{rtrt} - 2N_s^{-1} (A_{rsrt})^2 \hat{\delta}_{st} \\ & - 4N_r^{-1} A_{rrrs} A_{rtst} \hat{\delta}_{rs} - 4N_r^{-1} A_{rrrt} A_{rsts} \hat{\delta}_{rt} \ , \end{aligned} \tag{127}$$

$$\begin{aligned}\langle K(rstu)\rangle^N = {} & 2(B_{rstu} + B_{rtsu} + B_{rust}) + 2(A_{rsrs} A_{tutu} + A_{rtrt} A_{susu} + A_{ruru} A_{stst}) - 4N_r^{-1} A_{srst} A_{rutu} \hat{\delta}_{rt} \\ & - 4N_r^{-1} A_{srsu} A_{rtut} \hat{\delta}_{ru} - 4N_r^{-1} A_{trts} A_{rusu} \hat{\delta}_{rs} - 4N_s^{-1} A_{rsrt} A_{sutu} \hat{\delta}_{st} \\ & - 4N_s^{-1} A_{rsru} A_{stut} \hat{\delta}_{su} - 4N_t^{-1} A_{rtru} A_{tsus} \hat{\delta}_{tu} \ . \end{aligned} \tag{128}$$

The centroid and dispersion polynomials are now formed just as in the scalar case. The terms in Eqs. (113)–(116) propagate via Eq. (65) to give the energy polynomial for the configuration $[\vec{m}] \equiv [\vec{m}_1, \vec{m}_2, \ldots \vec{m}_l]$;

$$\mathcal{E}([\vec{m}]) = \langle H\rangle^{[\vec{m}]} = H_0 + \sum_r M_r(1) \times N_r \epsilon_{rr} + \sum_r M_r(2) \times \tfrac{1}{2} A_{rrrr} + \sum_{rs}{}' M_r(1) M_s(1) \times \tfrac{1}{2} A_{rsrs} \ , \tag{129}$$

where

$$M_r(k) = \binom{m_r}{k}\binom{N_r}{k}^{-1} = \frac{m_r(m_r - 1)\cdots(m_r - k + 1)}{N_r(N_r - 1)\cdots(N_r - k + 1)} \ , \tag{130}$$

with these polynomial quantities varying in value from 0 to 1 as the orbit fills. Equation (129) would follow immediately from our definitions of the average energies. For the dispersion we may similarly propagate the terms in Eqs. (117)–(128) and then subtract from the resultant polynomial the square of the centroid polynomial [Eq. (129)]. Alternatively, we may subtract from H its linear trace equivalent

$$H \to \mathcal{H} = H - H_0 - \sum_r \epsilon_{rr} n_r - \sum_r W_r \binom{n_r}{2} - \sum_{r<s} W_{rs} n_r n_s \ , \tag{131}$$

which is equivalent in matrix elements to

$$\begin{aligned} &H_0 \to 0, \ \epsilon_{rr} \to 0, \quad W^{\Gamma}_{rrrr} \to W^{\Gamma}_{rrrr} - W_r \ , \\ &W^{\Gamma}_{rsrs} \to W^{\Gamma}_{rsrs} - W_{rs} \ . \end{aligned} \tag{132}$$

The forms here, if not obvious, are derived from Eqs. (97) and (129). The second moment of the new operator is then simply the variance. The result is

$$\begin{aligned} \sigma^2([\vec{m}]) = {} & \sum_r M_r(1)[N_r \sum_k (\epsilon_{rk})^2(1-\delta_{rk})] + \sum_r M_r(2)\{\tfrac{1}{4}\sum_{kl} B_{rrkl} - \tfrac{1}{2}[N_r(N_r-1)]^{-1}(A_{rrrr})^2 + 2\sum_k \epsilon_{rk} A_{rrrk}(1-\delta_{rk})\} \\ & + \sum_{rs}{}' M_r(1)M_s(1)\{\tfrac{1}{4}\sum_{kl} B_{rskl} - \tfrac{1}{2}[N_r N_s]^{-1}(A_{rsrs})^2 + 2\sum_k \epsilon_{rk} A_{srsk}(1-\delta_{rk}) - N_r(\epsilon_{rs})^2\} \\ & + \sum_r M_r(3)\{[N_r(N_r-1)]^{-1}(A_{rrrr})^2 - \tfrac{1}{2}\sum_k B_{rrrk} + N_r^{-1}\sum_k (A_{rrrk})^2 \hat{\delta}_{rk}(1-\delta_{rk})\} \\ & + \sum_{rs}{}' M_r(2)M_s(1)\{[N_r N_s]^{-1}(A_{rsrs})^2 - \sum_k B_{rsrk} - \tfrac{1}{2}\sum_k B_{rrsk} - 4\epsilon_{rs}A_{rrrs} + 2N_r^{-1}\sum_k A_{rrrk}A_{srsk}\hat{\delta}_{rk}(1-\delta_{rk}) \\ & + N_s^{-1}\sum_k (A_{rsrk})^2 \hat{\delta}_{sk}(1-\delta_{sk})\} + \sum_{rst}{}' M_r(1)M_s(1)M_t(1)[-\tfrac{1}{2}\sum_k B_{rstk} - 2\epsilon_{rt}A_{srst} \\ & + N_t^{-1}\sum_k A_{rtrk}A_{stsk}\hat{\delta}_{tk}(1-\delta_{tk})] + \sum_r M_r(4)\{\tfrac{1}{4}B_{rrrr} - \tfrac{1}{2}[N_r(N_r-1)]^{-1}(A_{rrrr})^2\} \\ & + \sum_{rs}{}' M_r(3)M_s(1)[B_{rrrs} - 2N_r^{-1}(A_{rrrs})^2\hat{\delta}_{rs}] + \sum_{rs}{}' M_r(2)M_s(2)\{\tfrac{1}{4}B_{rrss} + \tfrac{1}{2}B_{rsrs} \\ & - \tfrac{1}{2}[N_r N_s]^{-1}(A_{rsrs})^2 - 2N_r^{-1}A_{rrrs}A_{rsss}\hat{\delta}_{rs}\} + \sum_{rst}{}' M_r(2)M_s(1)M_t(1)[\tfrac{1}{2}B_{rrst} + B_{rsrt} - N_s^{-1}(A_{rsrt})^2\hat{\delta}_{st} \\ & - 4N_r^{-1}A_{rrrs}A_{trts}\hat{\delta}_{rs}] + \sum_{rstu}{}' M_r(1)M_s(1)M_t(1)M_u(1)(\tfrac{1}{4}B_{rstu} - N_t^{-1}A_{rtru}A_{stsu}\hat{\delta}_{tu}) \ . \end{aligned} \tag{133}$$

The summation $\sum{}'$ indicates that terms with any two summation variables equal must be omitted; otherwise the summations are unrestricted. The product $\hat{\delta}_{rs}(1-\delta_{rs})$ designates a term which survives only when orbitals r and s are distinct from each other but differ only in their principal quantum number, such as the 1s and 2s orbitals. The origin of each term in Eq. (133) is clearly indicated by the polynomial factors. Thus terms $M_r(2)M_s(1)M_t(1)$ arise from that four-body part of H^2 which is a two-body operator in orbit r and a one-body operator in orbits s and t. The quantities $B_{rrrr} - \binom{N_r}{2}^{-1}(A_{rrrr})^2$ and $B_{rsrs} - [N_r N_s]^{-1}(A_{rsrs})^2$ which appear in Eq. (133) are proportional to the variance of the two-body interaction between a pair of nucleons in the same and different orbits, respectively. The configuration widths do not depend on the diagonal single-particle energies $\epsilon_{rr} \equiv \epsilon_r$, as is clear formally from Eqs. (132) and (133) and otherwise from the fact that an operator which is a multiple of unity in a subspace does contribute to the energy but cannot contribute to the dispersion. The same is clearly true of the partial width. There is an incipient paradox here; a partial width connecting two subspaces is unchanged if we move one of them so high up in energy that its effects on the other subspace become negligible. But the paradox disappears when we remember that the admixing intensities depend in the limit inversely as the square of the centroid differences.

A decomposition of the width into partial widths, whose importance we have already stressed, is achieved via an intermediate-state expansion

$$\sigma^2([\vec{m}]) = \sum_{\vec{m}'} \sigma^2([\vec{m}; \vec{m}']) = \sum_{\substack{[\vec{m}'] \\ \alpha \in [\vec{m}] \\ \beta \in [\vec{m}']}} |\langle [\vec{m}]\alpha | \mathcal{H} | [\vec{m}']\beta \rangle|^2 \ . \tag{134}$$

We may think of $[\vec{m}']$ as generated from $[\vec{m}]$ by the excitation of a particular number and type of particle-hole pairs; $[\vec{m}_1 + \vec{1}, \ \vec{m}_2 - \vec{1}, \ \vec{m}_3, \dots, \vec{m}_l]$ is a particular 1p-1h excitation, and so on. We see easily then that, in Eq. (133), the terms B_{rrrr}, B_{rsrs}, $(A_{rrrr})^2$, and $(A_{rsrs})^2$ are due to the internal mixing of 0p-0h states; the terms $(\epsilon_{rs})^2$, $\epsilon_{rs}A_{rrrs}$, $\epsilon_{rs}A_{rtst}$, B_{rsss}, B_{rsts}, $(A_{rrrs})^2$, $(A_{rsrt})^2$, $A_{rrrt}A_{srst}$, and $A_{rtru}A_{stsu}$ are due to the admixing of 0p-0h states with 1p-1h states; the terms B_{rrss}, B_{rrst}, and B_{rstu} are due to the admixing of 0p-0h and 2p-2h states.

We may thus decompose the variance into terms each of which involves intermediate states of a unique configuration. We now list these partial variances using the notation that, for example, $\sigma^2([\vec{m}]; r^{-2}s^{+1}t^{+1})$ arises from the process

$$[\vec{m}] \to [\vec{m}_1, \vec{m}_2, \ldots, \vec{m}_r - \vec{2}, \ \vec{m}_s + \vec{1}, \ \vec{m}_t + \vec{1}, \ldots, \vec{m}_l] \to [\vec{m}] \ ,$$

where, of course, the orbits r, s, t are distinct. We find

$$\sigma^2([\vec{m}]:0) = \sum_r [M_r(2) - 2M_r(3) + M_r(4)]\left[\frac{1}{4}B_{rrrr} - \frac{1}{4}\binom{N_r}{2}^{-1}(A_{rrrr})^2\right]$$
$$+ \sum_{rs}{}'[M_r(1)M_s(1) - 2M_r(2)M_s(1) + M_r(2)M_s(2)]\left[\frac{1}{2}B_{rsrs} - \frac{1}{2N_rN_s}(A_{rsrs})^2\right] \ , \tag{135}$$

$$\begin{aligned}
\sigma^2([\vec{m}]:r^{-1}s^{+1}) = {} & M_r(1)[1 - M_s(1)] \times N_r(\epsilon_{rs})^2 + M_r(2)[1 - M_s(1)] \times 2\epsilon_{rs}A_{rrrs} \\
& + M_r(1)[M_s(1) - M_s(2)] \times 2\epsilon_{rs}A_{rsss} + [M_r(2) - M_r(3)][1 - M_s(1)] \times \tfrac{1}{2}B_{rrrs} \\
& + M_r(1)[M_s(1) - 2M_s(2) + M_s(3)] \times \tfrac{1}{2}B_{rsss} + M_r(2)[M_s(1) - M_s(2)] \times 2N_r^{-1}A_{rrrs}A_{rsss}\hat{\delta}_{rs} \\
& + M_r(3)[1 - M_s(1)] \times N_r^{-1}(A_{rrrs})^2\hat{\delta}_{rs} + M_r(1)[M_s(2) - M_s(3)] \times N_r^{-1}(A_{rsss})^2\hat{\delta}_{rs} \\
& + M_r(1)[1 - M_s(1)]\sum_t{}'[M_t(1) - M_t(2)] \times B_{rtst} + M_r(1)[1 - M_s(1)]\sum_t{}'M_t(1) \times 2\epsilon_{rs}A_{rtst} \\
& + M_r(1)[M_s(1) - M_s(2)]\sum_t{}'M_t(1) \times 2N_r^{-1}A_{rsss}A_{rtst}\hat{\delta}_{rs} + M_r(2)[1 - M_s(1)]\sum_t{}'M_t(1) \\
& \times 2N_r^{-1}A_{rrrs}A_{rtst}\hat{\delta}_{rs} + M_r(1)[1 - M_s(1)]\sum_t{}'M_t(2) \times N_r^{-1}(A_{rtst})^2\hat{\delta}_{rs} \\
& + M_r(1)[1 - M_s(1)]\sum_{tu}{}'M_t(1)M_u(1) \times N_r^{-1}A_{rtst}A_{rusu}\hat{\delta}_{rs} \ ,
\end{aligned} \tag{136}$$

$$\sigma^2([\vec{m}]:r^{-2}s^{+2}) = M_r(2)[1 - 2M_s(1) + M_s(2)] \times \tfrac{1}{4}B_{rrss} \ , \tag{137}$$

$$\sigma^2([\vec{m}]:r^{-2}s^{+1}t^{+1}) = M_r(2)[1 - M_s(1)][1 - M_t(1)] \times \tfrac{1}{2}B_{rrst} \ , \tag{138}$$

$$\sigma^2([\vec{m}]:r^{-1}s^{-1}t^{+2}) = M_r(1)M_s(1)[1 - 2M_t(1) + M_t(2)] \times \tfrac{1}{2}B_{rstt} \ , \tag{139}$$

$$\sigma^2([\vec{m}]:r^{-1}s^{-1}t^{+1}u^{+1}) = M_r(1)M_s(1)[1 - M_t(1)][1 - M_u(1)] \times B_{rstu} \ . \tag{140}$$

The summation symbol $\sum'$ in Eq. (136) indicates that the summation variables are distinct from r and s. If we add together the terms from Eqs. (135)–(140) we reproduce the results of Eq. (133). The separate terms have certain obvious vanishings; for example, any r^{-2} term vanishes for a configuration with less than two particles in orbit r. It should be clear that the widths are unaffected by the $(H \to \mathcal{H})$ transformation of Eq. (131). The $(r^{-1}s^{+1})$ and $(r^{+1}s^{-1})$ terms involve the same matrix elements of H but are not equal, because of an asymmetry arising from different occupancies; a way will be described in a later paper in which this can be used to give an inductive method for partial widths when we deal with more complicated group symmetries.

A familiar result emerges if one asks for the change in the centroid energy due to the addition of a nucleon in the rth orbit. Calculating the difference $\Delta\mathcal{E}_r$ between $\mathcal{E}([\vec{m}_1, \vec{m}_2, \ldots, \vec{m}_r + \vec{1}, \ldots, \vec{m}_l])$ and $\mathcal{E}([\vec{m}_1, \vec{m}_2, \ldots, \vec{m}_r, \ldots, \vec{m}_l])$, one finds via Eqs. (98), (99), and (129) that

$$\Delta\mathcal{E}_r = \epsilon_{rr} + \sum_t m_t W_{tr} \ , \tag{141}$$

a result entirely to be expected. Similarly, on adding a pair of particles in the same or different orbits we find, in an obvious notation,

$$\begin{aligned}
\Delta\mathcal{E}_{rr} &= 2\Delta\mathcal{E}_r + W_r \ , \\
\Delta\mathcal{E}_{rs} &= \Delta\mathcal{E}_r + \Delta\mathcal{E}_s + W_{rs} \ .
\end{aligned} \tag{142}$$

7. MOMENT RENORMALIZATIONS

Shell-model calculations being feasible only in rather small spaces, we are forced to restrict ourselves, usually, to dealing with a small set of configurations, a particular class of particle-hole states, or to a set of representation subspaces of a group which defines an approximate symmetry of H (this latter being an extension of the notion of configurations). In this section we cast the configuration polynomials into forms relevant to the first two types of restriction.[31] The idea is to consider the subspace embedded in a larger one, whose effects on the widths we then take account of by permitting transitions to intermediate states which lie outside the smaller subspace, this then generating a width external to the original subspace.

As a minor example, if we consider $(d_{5/2})^m$, firstly as a subspace with variance $\sigma^2(\vec{m})$, and secondly as a complete space with variance $\sigma^2([\vec{m}, \vec{0}])$ [the zero occupancy being for the $(d_{3/2}, s_{1/2})$ orbits], we have

$$\sigma^2([\vec{m}, \vec{0}]) = \sigma^2(\vec{m}) + P^3(\vec{m}) \; . \tag{143}$$

Since the external-width polynomial must involve at least one particle in $(d_{3/2}, s_{1/2})$, it can have no quartic terms in m, being then a third-order polynomial as indicated. In fact, since $\sigma^2(\vec{m})$ and $\sigma^2([\vec{m}, \vec{0}])$ are both known, the explicit form of the external-width polynomial can be easily written down.

For the general case we should take account also of orbits which are inert because they are filled. Then dividing our l orbits into l_A active orbits, l_E inert empty orbits, and l_F inert full orbits, we may rewrite the polynomials of Eqs. (129) and (133) for configuration centroids and dispersions. Depending on whether orbit r belongs to class A, E, or F, the single-orbit polynomial $M_r(m_r)$ is unchanged, zero, or unity; the dependence in Eqs. (129) and (133) on the occupancies of the inert orbits is thereby eliminated and the polynomials originally in l variables become polynomials in l_A variables. The algebraic form of the resulting renormalized width polynomial is complicated and we do not write it here. The energy polynomial becomes

$$\mathcal{E}([\vec{N}_F|\vec{m}_A|\vec{0}_E]) = H_0 + \sum_\alpha N_\alpha \tilde{\epsilon}_{\alpha\alpha} - \tfrac{1}{2}\sum_{\alpha\beta} A_{\alpha\beta\alpha\beta} + \sum_r M_r(1) \times N_r \tilde{\epsilon}_{rr} + \sum_r M_r(2) \times \tfrac{1}{2} A_{rrrr} + \sum_{rs}{}' M_r(1) M_s(1) \times \tfrac{1}{2} A_{rsrs} \; . \tag{144}$$

Here the symbols r, s, range over the active (A) orbits, and α, β over the inert filled (F) orbits. The $\tilde{\epsilon}$ quantities are given by

$$\tilde{\epsilon}_{ij} = (\epsilon_{ij} + N_i^{-1} \sum_\beta A_{\beta i \beta j}) \hat{\delta}_{ij} \; , \tag{145}$$

where i, j refer to any of the orbits.

The energies defined by Eq. (144) may be regarded as "renormalized" in the sense of the above discussion, and similarly for the variances. If, in them, we put to zero all the coefficients which make reference to the inert orbits (F, E) we are left with the unrenormalized quantities which take no account at all of the inert orbits. The renormalization effects on the centroid energies are, of course, trivial but this is not at all so for the variances.

The single-particle ϵ parameters enter in the renormalized quantities only in the $\tilde{\epsilon}$ quantities of Eq. (145), which are obviously the Hartree-Fock energies (as defined by the original orbits, no variation having been made to produce the optimal results). The condition that a closed-core state $[\vec{N}_F|\vec{0}_A|\vec{0}_E]$ can be an eigenstate of H is that

$$\sigma^2([\vec{N}_F|\vec{0}_A|\vec{0}_E]) = 0 = \sum_{\alpha k} (\tilde{\epsilon}_{\alpha k})^2 \delta_{\alpha k} + \tfrac{1}{4} \sum_{\alpha\beta k l} B_{\alpha\beta k l} \; , \tag{146}$$

where the explicit form comes via the operation defined above. Since each term here is positive definite [see Eq. (97) for B], it must vanish separately. The vanishing of the $\tilde{\epsilon}_{\alpha k}$ implies zero single-particle nondiagonal energies involving an occupied and an unoccupied orbit (the usual Hartree-Fock condition) while the vanishing of $B_{\alpha\beta k l}$ would imply the absence of two-particle–two-hole excitations in this representation. The energy of the HF state is then the centroid energy

$$\mathcal{E}([N_F|0_A|0_E]) = \sum_\alpha N_\alpha \tilde{\epsilon}_{\alpha\alpha} - \tfrac{1}{2}\sum_{\alpha\beta} A_{\alpha\beta\alpha\beta} \; , \tag{147}$$

the usual result. Thus, if a HF calculation has been performed the quantity $(\frac{1}{4}\sum_{\alpha\beta k l} B_{\alpha\beta k l})^{1/2}$ computed in the HF basis is the width of the distribution of the HF ground state over the actual eigenstates of the system, and thus represents a measure of how close the HF ground state is to an eigenstate of the complete Hamiltonian. The HF energies also have the expected interpretation, as single-particle and single-hole energies, in terms of the differences between centroids. We have

$$\bar{\epsilon}_{rr} = \mathcal{E}([N_F|r^{+1}|0_B]) - \mathcal{E}([N_F|0_A|0_B]) , \tag{148}$$

$$\bar{\epsilon}_{\alpha\alpha} = \mathcal{E}([N_F|0_A|0_B]) - \mathcal{E}([\alpha^{-1}|0_A|0_B]) \tag{149}$$

in an obvious notation.

Finally we turn to the particle-hole truncation scheme in which one treats a subset of normally filled orbits as a vacuum whose excitations give the various particle-hole states. We could go back to the original Hamiltonian, make p⇌h transformations for the orbits filled in the vacuum, and then proceed just as above. It is often simpler, however, to introduce polynomials for holes instead of particles. The basic transformation (for the density operators) is given by Eq. (60) and this in turn gives for the polynomials of Eq. (130) the result

$$M_\alpha(k) = \sum_{t=0}^{k} (-1)^t \binom{k}{t} \tilde{M}_\alpha(t) , \tag{150}$$

where

$$\tilde{M}_\alpha(t) = \binom{h_\alpha}{t}\binom{N_\alpha}{t}^{-1} , \tag{151}$$

$h_\alpha = (N_\alpha - m_\alpha)$ being the number of holes in orbit α. We then find for a configuration $[\vec{h}|\vec{m}]$, with h holes distributed over the hole orbits and m particles over the particle orbits, that

$$\begin{aligned}\mathcal{E}([\vec{h}|\vec{m}]) = \mathcal{E}([\vec{0}|\vec{0}]) + \sum_r M_r(1) \times N_r \bar{\epsilon}_{rr} - \sum_\alpha \tilde{M}_\alpha(1) \times N_\alpha \bar{\epsilon}_{\alpha\alpha} + \sum_r M_r(2) \times \tfrac{1}{2} A_{rrrr} \\ + \sum_\alpha \tilde{M}_\alpha(2) \times \tfrac{1}{2} A_{\alpha\alpha\alpha\alpha} + \sum_{rs}{}' M_r(1) M_s(1) \times \tfrac{1}{2} A_{rsrs} + \sum_{\alpha\beta}{}' \tilde{M}_\alpha(1) \tilde{M}_\beta(1) \times \tfrac{1}{2} A_{\alpha\beta\alpha\beta} - \sum_{\alpha r} M_r(1) \tilde{M}_\alpha(1) \times A_{r\alpha r\alpha} ,\end{aligned} \tag{152}$$

where the particle-hole vacuum energy $\mathcal{E}([\vec{0}|\vec{0}])$ is given in terms of previously defined particle matrix elements, by Eq. (148). In particular, we find for the (1p-1h) and (2p-2h) configurations, written in terms of the average interaction energies of Eqs. (98) and (99),

$$\begin{aligned}\mathcal{E}([\alpha^{-1}|r^{+1}]) - \mathcal{E}([0|0]) &= \bar{\epsilon}_{rr} - \bar{\epsilon}_{\alpha\alpha} - W_{r\alpha} , \\ \mathcal{E}([\alpha^{-2}|r^{+2}]) - \mathcal{E}([0|0]) &= 2\bar{\epsilon}_{rr} - 2\bar{\epsilon}_{\alpha\alpha} + W_{rr} + W_{\alpha\alpha} - 4W_{r\alpha} .\end{aligned} \tag{153}$$

These intuitively obvious results, when coupled with the corresponding results for widths (which derive in the same way but whose explicit forms we do not give) will be of consequence in studying the structure of nuclei near closed shells. Renormalizations, as discussed in the first part of this section, will also be significant here.

8. SUMMARY AND FINAL REMARKS

We have discussed the essential ideas involved in the application of spectral-distribution methods to nuclear structure (and, indeed, to other many-particle systems). The distributions are characterized by their moments, expectation values of powers of the Hamiltonian averaged over subsets of states, and these become the central elements of the theory. A new technique, that of propagation throughout the lattice of subsets, has been introduced for the evaluation of moments in a certain class of cases, and in particular to write the configuration moments as polynomials in occupation numbers. These have been rearranged in ways corresponding to common vector-space truncations used in spectroscopic calculations, this focussing attention upon the difference between truncations of the operator whose moments are being considered and truncation of powers of the operator; the first of these is equivalent to a restriction of the space in a detailed calculation, while the second goes beyond that. Having explicit polynomials whose coefficients are Hamiltonian integrals, we shall be able, for example, to see which parts of the interaction are dominant in various parts of the total space. We have indicated a number of domains in which moment methods may be usefully applied; the following paper and others in the future will be concerned with that. It may turn out that a valuable use of these methods will be in the development of a kind of perturbation theory, based on moments, which will permit the extension of finite-space calculations.

Since we cannot look forward to dealing with high-order moments, it will be a requirement on the subspaces and operators with which we deal that the distributions be well described by a few low-order moments, this in turn requiring that they be close to normal. The experience has been that this is liable to be an excellent assumption, although a satisfactory theoretical justification for it is not presently available. We rely instead on detailed study of examples, some of which have already been given[32] and others of which are given in the next paper. A very important aspect of the moment methods is that they display the explicit dependence of the quantities of interest on the Hamiltonian parameters and structure, at the same time reducing a highly nonlinear matrix problem to one whose nonlinearities are of low order; basic to this procedure is the assumption of near normality which in turn implies that the higher cumulants (which, for example, determine fluctuations in the distributions) carry only little of the interesting physics.

We mention finally that besides giving applications of the methods above, future papers will describe new methods which are particularly relevant for dealing with group symmetries and more complex representation subspaces, and will also discuss strength distributions for various processes.

---

*Work supported in part by the U. S. Atomic Energy Commission under Contract Nos. AT(30-1)-875 and AT(30-1)-2098. Work in México was done while the first-named author was Senior Postdoctoral Fellow of the National Science Foundation.

†Now at Department of Physics, State University of New York, Albany, New York.

[1]I. Talmi and I. Unna, Ann. Rev. Nucl. Sci. 10, 353 (1960); A. de-Shalit, in *Nuclear Physics, Proceedings of the International School of Physics, "Enrico Fermi," Course XXIII,* edited by V. F. Weisskopf (Academic Press Inc., New York, 1963).

[2]G. E. Brown, *Unified Theory of Nuclear Models and Nucleon Forces* (North-Holland Publishing Company, Amsterdam, The Netherlands, 1967), 2nd ed.; M. Baranger, in *Nuclear Structure and Nuclear Reactions, Proceedings of the International School of Physics, "Enrico Fermi," Course XL,* edited by M. Jean and R. A. Ricci (Academic Press Inc., New York, 1969).

[3]I. Kelson and C. A. Levinson, Phys. Rev. 134, B269 (1964); G. Ripka, in *Advances in Nuclear Physics,* edited by M. Baranger and E. Vogt (Plenum Press, Inc., New York, 1968), Vol. 1; D. Rowe, Rev. Mod. Phys. 40, 153 (1968).

[4]J. B. French, in *Nuclear Structure,* edited by A. Hossain (North-Holland Publishing Company, Amsterdam, The Netherlands, 1967), and earlier references given therein. See also P. Löwdin, Rev. Mod. Phys. 39, 259 (1967).

[5]States regarded as remaining inert, either filled or empty, would not usually be counted in $\{N\}$. In some cases some of the subsets of $\{N\}$ would be better describable in terms of holes, in which case we could use m to indicate the number of active fermions, noting, of course, that this number does not necessarily remain constant and that H is not necessarily fermion-number conserving. Modifications to look after these things will be simple.

[6]The point of this is that we could instead define H to be the projection into $\vec{\mathfrak{N}}$ of an operator defined in the general many-fermion-Hilbert space. Then we would have a notational ambiguity when writing H^p. As things stand now we proceed as if single-particle states outside the orbital space $\{N\}$ did not "exist." Of course, the relationship between projections of powers of H and powers of projections of H is the heart of the matter as far as general many-body problems are concerned, and we shall approach this later by considering $\{N\}$ to be a subspace of a larger orbital space $\{N'\}$.

[7]C. E. Porter, *Statistical Theories of Spectra: Fluctuations* (Academic Press Inc., New York, 1965).

[8]For representations of different dimensionalities, the partial-width matrix is not symmetrical, as shown by Eq. (12). But this simply corresponds to the fact that, if the dimensionalities are very different, the smaller representation could be completely fragmented, thereby losing its identity, without this having any appreciable effect on the individual states of the larger representation.

[9]J. Griffin, Phys. Rev. Letters 17, 478 (1966).

[10]J. B. French, Phys. Letters 26B, 75 (1967).

[11]J. B. French and L. S. Hsu, Phys. Letters 25B, 75 (1967).

[12]J. B. French, Phys. Letters 23, 248 (1966).

[13]Remember that $\binom{a}{b}$ is defined for a negative as well as positive, and

$$\binom{-a}{b} = (1)^b \binom{a+b-1}{b} = (-1)^{a+b} \binom{-b-1}{-b-a}.$$

[14]I.e., the N-particle state (in which all single-particle states are filled). We borrow the term from L. C. Biedenharn.

[15]In such light nuclei, and when we span a large domain of A we must expect variations in the effective interactions due, among other things, to nuclear size effects. Taking input data from both ends of the domain tends to minimize the effects which these variations have on the calculated centroid energies; whether this is a good thing or bad is another matter.

[16]S. Hinds and R. Middleton, Nucl. Phys. 84, 651 (1966).

[17]We assume that we deal with real vector spaces of operators so that the matrix elements in Eq. (26) are real.

[18]I.e., commute or anticommute depending upon the particle numbers involved.

[19]Note carefully that the m-hole trace operator $\tilde{\rho}(\vec{m}, \vec{\alpha}) = \sum Z_\alpha^\dagger(m) Z_\alpha(m)$ is not at all the same as the $(N-m)$-particle operator.

[20]The subscript c is then redundant and we shall usually omit it. Note, as a point of notation, that $Z_{\alpha_c}(k)$ is a state operator for k particles, *not* for $(N-k)$. Note also that $Z_{\alpha_c}(N-m) \neq \tilde{Z}_\alpha(m) Z(N)$; we do have though that $Z_{\alpha_c}(N-m)$ is the lowest-rank part [the $(A)^k$ part] in the normal expansion of $\tilde{Z}_\alpha(m) Z(N)$.

[21]Equations (42) and (43) will reproduce the elementary result of Eq. (13). A combinatorial derivation of Eq. (43)

will be found both simple and instructive.

[22]If the subspace decomposition allows simple propagation, Eqs. (45) and (46) are valid for all operators $O(\vec{k})$. It is easily seen that if this should not be so, we would still have propagation of traces for those operators whose form is that of Eq. (46), behaving as a multiple of unity in every subspace. This should be remembered for those equations ahead in this section which are written for operators O. On the other hand, equations written explicitly for density operators [e.g., Eq. (52)] are valid for arbitrary (disjoint) decompositions.

[23]While the density operator ρ is analogous to the elementary Green's function of electrostatics which gives potentials in terms of charges, ρ_u corresponds to the more sophisticated function which gives potentials in terms of surface potentials.

[24]The right-hand side of Eq. (53) involves "higher" subspaces (with respect to particle number) than $(\vec{t}, \vec{\gamma})$ itself, but if we take $t \geq N/2$, these higher subspaces are, of course, simpler. Even so, Eq. (53) does not yield an inductive method for constructing the ρ operators, since $\rho(\vec{t}, \vec{\gamma})$ appears on both sides.

[25]G. C. Rota, Z. Wahrscheinlichtkeitstheorie 2, 340 (1964).

[26]The criterion for linear independence of a net is that the determinant of the propagating coefficients should not vanish. It follows from Eq. (49) that the defining net is linearly independent.

[27]Square brackets define a "vector" in l dimensions. Note that $\{N\} \equiv \vec{1}$ and $\{[N]\} \equiv [\vec{1}]$. For a $[\vec{k}'] \rightarrow [\vec{k}]$ operator, as in Eq. (22), we shall often write $O([\vec{k}; \vec{k}'])$, and shall as well often write $O([\vec{k}; \vec{k}])$ as $O([\vec{k}])$.

[28]J. Kim, private communication.

[29]J. B. French, in *Many-Body Description of Nuclear Structure and Reactions, Proceedings of the International School of Physics, "Enrico Fermi," Course XXXVI, 1966*, edited by C. Bloch (Academic Press Inc., New York, 1966).

[30]We write $K \equiv H^2$ to avoid an ambiguity. According to our notation $H^2(t)$ should stand for $[H(t)]^2$, the squares of the t-particle part of H, while $K(t)$ is the t-particle part of the square of H.

[31]More complex group structures will be considered in later papers.

[32]F. S. Chang, J. B. French, and K. F. Ratcliff, Phys. Letters 23, 251 (1966).

PHYSICAL REVIEW C VOLUME 3, NUMBER 1 JANUARY 1971

Applications of Spectral Distributions in Nuclear Spectroscopy*

K. F. Ratcliff†
Laboratory for Nuclear Science and Department of Physics, Massachusetts Institute of Technology, Cambridge, Massachusetts 02139

(Received 8 May 1970)

The energy moments of spectral distributions are used to investigate the structure of spectroscopic calculations. Eigenvalue distributions are used to predict energy spectra, which are compared with the results of matrix diagonalization. The nature of the corresponding eigenvectors is similarly analyzed. The propagation of moments for scalar and configuration distributions is illustrated throughout the *sd* shell. These scalar moments are then used to estimate the trend in theoretical binding energies for these nuclei and in turn are compared with empirical data. Finally, we investigate the dependence of the energy and wave function of the ground state of O^{16} upon the vector space underlying a theoretical calculation. The unrenormalized Kuo-Brown matrix elements are employed in this analysis, and the role of multiple particle-hole excitations is discussed.

I. INTRODUCTION

The concepts of spectral distributions, of operator averaging, and the propagation of such averages through a sequence of subspaces have been the subject of recent investigations.[1-6] In this paper we apply these ideas to a number of examples that are closely related to conventional nuclear spectroscopy. The techniques which we develop below are founded on the moments of spectral distributions, i.e., the averages of powers of the Hamiltonian over various subspaces of the complete many-particle space. One may adopt various attitudes towards these moments. They may be regarded as measures of the average behavior of the Hamiltonian in various subspaces and then be used to study various formal problems. Alternatively they may be used in a fashion that is complementary to conventional spectroscopy in which one sets up and diagonalizes the Hamiltonian matrix in a representation generated by a finite number of single-nucleon states. This latter attitude is adopted in this paper. The spectral moments are used to make predictions of spectroscopic details for spaces that are too large for conventional matrix analysis. The general quality of these predictions may be assessed by working in smaller spaces in which comparison with exact diagonalization is possible. When it is necessary to have details beyond that provided by such a moment-generated spectroscopy, then these predictions may be used as the basis for making those truncations of either the vector space or operator space which are required by the conventional methods.

In Sec. II we use the spectral moments of H for the invariant subspaces, those defined by the exact symmetries of the Hamiltonian, to approximate the actual eigenvalue spectrum of H. The moments of H over smaller subspaces, in which a shell-model configuration is also specified, are used to predict the relative contribution of each such subspace in the structure of the low-lying eigenvectors. These moment-generated results are compared with the exact results of matrix diagonalization for four nucleons in the $2s$-$1d$ shell.

In Sec. III we deal with the scalar distribution and use the corresponding moments to discuss the variation of ground-state energies throughout the $2s$-$1d$ shell. In Sec. IV we investigate configuration centroids, widths, and partial variances, and their use as the basis for making vector-space truncations.

In Sec. V we consider the problem of the dependence of a theoretically determined ground-state energy and wave function upon the size of the vector space underlying the calculation. We focus our attention on the filled $1s$ and $1p$ shells and consider the change in the moments of this 16-nucleon state as it is embedded in an ever enlarging vector space. From this analysis we infer the corresponding changes to be found in the calculated energy and eigenvector of the ground state of O^{16}.

II. MOMENTS OF SPECTRAL DISTRIBUTIONS, ENERGY SPECTRA, AND EIGENSTATES

In this section we describe a method by which knowledge of low-order energy moments may be used to approximate the eigenvalue spectrum of a Hamiltonian, which operates in a finite-dimensional vector space. The method consists of constructing a moment-equivalent frequency function for eigenvalue distributions. An approximate spectrum is then constructed from the energy dependence of the integrated frequency function. The eigenvalue spectrum thus produced is compared with the results of exact diagonalization of the Hamiltonian matrix for the case of the shell-model space $(2s$-$1d)$.[4] Finally, we use a similar method to predict the relative importance of various shell-model components in the low-excitation eigenfunctions, and again compare the results with those of exact matrix diagonalization.

A. Eigenvalue Distributions and Spectra

We denote by $(N;\vec{m})$ the $\binom{N}{m}$-dimensional vector space formed by distributing m nucleons among N single-particle states.[7] The space $(N;\vec{m})$ may be decomposed into subspaces $(N;\vec{m},\vec{\Gamma})$ corresponding to the exact symmetries of the nuclear Hamiltonian, i.e., spin, isospin, and parity, denoted collectively by $\Gamma=(J,T,\pi)$. The submatrix of the Hamiltonian in this space is denoted by $H(\vec{m},\vec{\Gamma})$. The eigenvalue distribution of $H(\vec{m},\vec{\Gamma})$ is represented by the discrete normalized frequency function $f(\vec{\Gamma},E)$, where

$$f(\vec{\Gamma},E)=d^{-1}(\vec{m},\vec{\Gamma})\sum_i\delta(E-E_{i\Gamma})\,, \tag{1}$$

$$1=\int_{-\infty}^{\infty}f(\vec{\Gamma},E)dE\,, \tag{2}$$

and where the sum runs over the eigenvalues of $H(\vec{m},\vec{\Gamma})$. The dimensionality of $(N;\vec{m},\vec{\Gamma})$ is denoted by $d(\vec{m},\vec{\Gamma})$. Each eigenvalue distribution may be characterized by its moments, e.g. the centroid $\mathcal{E}(\vec{m},\vec{\Gamma})$ and pth central moment $\mu_p(\vec{m},\vec{\Gamma})$ which are related to the frequency function in the usual way[8]

$$\mathcal{E}(\vec{m},\vec{\Gamma})=\int_{-\infty}^{\infty}Ef(\vec{\Gamma},E)dE\,, \tag{3}$$

$$\mu_p(\vec{m},\vec{\Gamma})=\int_{-\infty}^{\infty}[E-\mathcal{E}(\vec{m},\vec{\Gamma})]^p f(\vec{\Gamma},E)dE\,. \tag{4}$$

These moments are also related to traces of powers of the submatrix $H(\vec{m},\vec{\Gamma})$

$$\mathcal{E}(\vec{m},\vec{\Gamma})=d^{-1}(\vec{m},\vec{\Gamma})\,\mathrm{Tr}[H(\vec{m},\vec{\Gamma})]\,, \tag{5}$$

$$\mu_p(\vec{m},\vec{\Gamma})=d^{-1}(\vec{m},\vec{\Gamma})\,\mathrm{Tr}\{H(\vec{m},\vec{\Gamma})-\mathcal{E}(\vec{m},\vec{\Gamma})\underline{1}]^p\}\,, \tag{6}$$

and where $\underline{1}$ is the unit matrix in this space. The normalized distribution function $\mathfrak{F}$ is defined as usual by integrating the frequency function

$$\mathfrak{F}(\vec{\Gamma},E)=\int_{-\infty}^{E}f(\vec{\Gamma},E')dE'\,, \tag{7}$$

$$\mathfrak{F}(\vec{\Gamma},\infty)=1\,. \tag{8}$$

It is convenient to define a function F, which is proportional to $\mathfrak{F}$

$$F(\vec{\Gamma},E)=\frac{d(\vec{m},\vec{\Gamma})}{(2J+1)(2T+1)}\mathfrak{F}(\vec{\Gamma},E)\,, \tag{9}$$

$$F(\vec{\Gamma},\infty)=\frac{d(\vec{m},\vec{\Gamma})}{(2J+1)(2T+1)}\,. \tag{10}$$

We shall also refer to F as a distribution function. We refer to the asymptotic value of F, i.e., $d/(2J+1)(2T+1)$ as the strength of the distribution. It is also equal to the number of distinct eigenvalues of $H(\vec{m},\vec{\Gamma})$. The distribution function F is a monotonic function which increases by unity at each distinct eigenvalue of $H(\vec{m}, \vec{\Gamma})$. Thus $F(\vec{\Gamma}, E)$ has a value equal to the number of distinct eigenvalues such that $E_{i\Gamma} < E$.

Our procedure will be to replace the discrete frequency function f by a continuous frequency function $\bar{f}$, so chosen as to share a common set of low-order moments with f. The eigenvalue spectrum is then obtained from the energy dependence of the continuous frequency function $\bar{F}$ that is produced using $\bar{f}$ in Eq. (7).

The functional form of standard one-dimensional continuous frequency functions of statistics[8,9] involves parameters which are fixed by the value of the centroid $\mathcal{E}$ and the width $\sigma(=\mu_2^{1/2})$ of the distribution. The choice of one of these functions to represent the eigenvalue distribution is made on the basis of the values of two standard measures, the skewness γ_1 and excess γ_2. These two quantities are dimensionless combinations of the second, third, and fourth central moments which measure the asymmetry and spread of the distribution about its centroid

$$\gamma_1 = \mu_3 \mu_2^{-3/2}, \tag{11}$$

$$\gamma_2 = \mu_4 \mu_2^{-2} - 3. \tag{12}$$

As examples of symmetrical ($\gamma_1 = 0$) frequency functions we cite[9] the normal or Gaussian distribution ($\gamma_2 = 0$), the Laplace or double exponential distribution ($\gamma_2 = 3.0$), the rectangular distribution ($\gamma_2 = -1.2$), and the semicircular distribution ($\gamma_2 = -1.0$) which arises in the statistical analysis of spectra.[10]

In Fig. 1 we display four frequency functions which illustrate the meaning of the skewness and excess parameters. Negative values of γ_2 indicate a broad central peak with a more rapid decay in the tail relative to the Gaussian ($\gamma_1 = \gamma_2 = 0$). Positive values of γ_2 indicate a sharper central peak and more gradual decay in the tail region. A positive skewness ($\gamma_1 > 0$) is associated with a rapid fall-off for energies below the centroid and a gradual decrease for energies above the centroid.

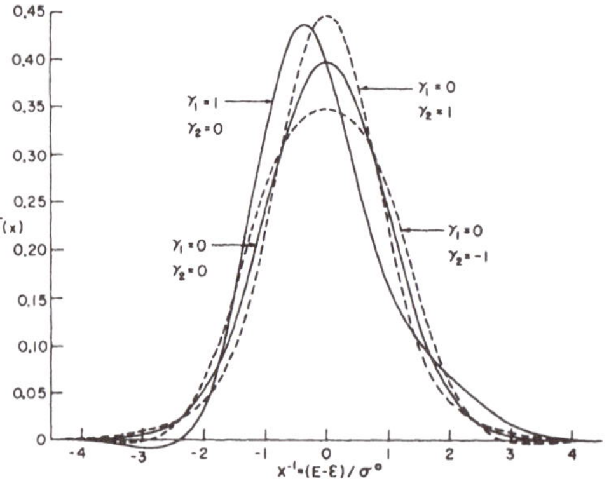

FIG. 1. Four-moment frequency functions of Eq. (15) for various large values of the skewness (γ_1) and excess (γ_2) parameters. The pure Gaussian has $\gamma_1 = \gamma_2 = 0$. Distance from the centroid is measured in units of the width σ.

We next illustrate the use of these ideas for the vector space in which four nucleons are distributed among the 24 single-particle states of the $2s$-$1d$ shell. We employ a two-body Rosenfeld interaction[11] and consider the eigenvalue distributions of the 640 distinct levels in this space. In Table I we list the moments and parameters for 19 JT groups. These groups have the largest strengths $[9 \leq d/(2J+1)(2T+1) \leq 69)]$ and may be employed to illustrate the use of continuous frequency functions. These data were obtained by computing the trace of the appropriate power of $H(\vec{4}; \vec{J}\vec{T})$. It is found that the eigenvalues in each group tend to be distributed symmetrically about their centroid, i.e., γ_1 is small. It is also observed that the excess γ_2 is small and has a tendency to decrease in magnitude with increasing strength of the distribution. From the value of these parameters, we infer that the eigenvalues of these JT groups are more closely represented by a Gaussian distribution than by any of the other standard frequency functions of statistics.

We therefore adopt the Gaussian function as a starting point and proceed to modify it to take into account our knowledge of the third and fourth moments of the distribution. It is convenient to construct the continuous frequency function $\bar{f}$ in terms of the dimensionless variable $x = (E - \mathcal{E})/\sigma$ where σ is the width $(=\mu_2^{1/2})$. The frequency function may then be expanded in terms of the series of orthogonal Hermite) polynomials which involve the Gaussian as a weighting function[8]

$$\bar{f}(\vec{\Gamma}, x) = (2\pi)^{-1/2} e^{-\frac{1}{2}x^2} \sum_j C_{j\Gamma} H_j(x), \tag{13}$$

The number of terms in Eq. (13) will be determined by the number of known moments of the discrete frequency function. The expression for the coefficients $C_{j\Gamma}$ may be obtained by minimizing the weighted, least-squares difference Δ between the true and approximate frequency functions,[12] where

$$\Delta = \int_{-\infty}^{\infty} [f(\vec{\Gamma}, E) - e^{-\frac{1}{2}x^2} \sum_{j=0} C_{j\Gamma} H_j(x)]^2 e^{\frac{1}{2}x^2} dE. \tag{14}$$

In the least-squares fit represented by Eq. (14) we use a finite series which reflects our knowledge of only the first k moments of the actual distribution. We also note that the weighting factor in Eq. (14) emphasizes the tail of the distribution in the determination of the coefficients. The result of this analysis is a continuous frequency function $\bar{f}(\vec{\Gamma},x)$ given by

$$f(\vec{\Gamma},x)=(2\pi)^{-1/2}e^{-\frac{1}{2}x^2}\times[1+\tfrac{1}{6}\gamma_{1\Gamma}(x^3-3x)+\tfrac{1}{24}\gamma_{2\Gamma}(x^4-6x^2+3)+\cdots]. \quad (15)$$

In Eq. (15) the dimensionless parameters $x=(E-\mathcal{E})/\sigma, \gamma_{1\Gamma}, \gamma_{2\Gamma},\ldots$ are those of the discrete frequency function f, and guarantee that

$$\int_{-\infty}^{\infty}(E-\mathcal{E}_\Gamma)^p[f(\vec{\Gamma},E)-\bar{f}(\vec{\Gamma},E)]dE=0 \quad (16)$$

is satisfied for $p=0,1,\ldots,k$ if the series in Eq. (14) is terminated after the kth term. In such a case we refer to the continuous function $\bar{f}$ as being the k-moment equivalent of f. The frequency functions which are displayed in Fig. 1 represent the four-moment functions of Eq. (15) for various combinations of γ_1 and γ_2.

The k-moment continuous distribution function $\bar{F}$ is defined by the insertion of the k-moment equivalent of f, i.e., $\bar{f}$, into Eq. (7). In Fig. 2 we compare the discrete two-, and four-moment continuous distribution functions for the case $J=2$, $T=0$. This case represents the poorest over-all agreement between the discrete and two-moment distribution functions of all JT groups. This difference is due to the large width and skewness resulting from the two most highly excited (and physically uninteresting) states, which are found to be separated by 6.6 MeV. The value of including the third and fourth moments is evident in the success of the four-moment distribution function. In Fig. 3 we compare the discrete and two-moment distribution functions for the case $J=2$, $T=0$. This case represents the best two-moment fit, and inclusion of the third and fourth moments has a negligible effect on the distribution function.

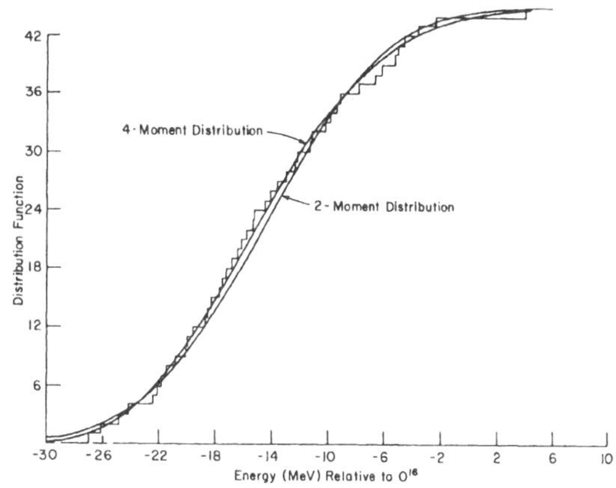

FIG. 2. The discrete $J=3$, $T=0$ eigenvalue distribution function compared with the two- and four-moment distribution functions. The energy zero coincides with the O^{16} ground state. The moments and parameters used in Eq. (15) for this case are $\mathcal{E}=-14.54$ MeV, $\sigma=6.85$ MeV, $\gamma_1=0.40$, and $\gamma_2=-0.24$.

The predictive value of a good continuous approx-

TABLE I. Moments and parameters of JT eigenvalue distributions for the space $(2s\text{-}1d)^4$.

J	T	Strength $\frac{d(4,JT)}{(2J+1)(2T+1)}$	Centroid (MeV) $\mathcal{E}(4,JT)$ [a]	Width (MeV) $\sigma(4,JT)$	Skewness $\gamma_1=\mu_3\mu_2^{-3/2}$	Excess $\gamma_2=(\mu_4\mu_2^{-2})-3$
0	0	21	−15.76	9.37	−0.40	−0.50
1	0	31	−13.36	7.08	0.56	−0.22
2	0	56	−14.56	8.05	0.04	−0.0002
3	0	45	−14.54	6.85	0.40	−0.24
4	0	44	−15.76	7.01	−0.05	−0.36
5	0	24	−16.40	5.57	0.24	−0.68
6	0	17	−18.32	5.69	0.11	−0.79
0	1	16	−11.89	7.18	0.29	−0.41
1	1	54	−12.44	7.06	0.21	−0.29
2	1	66	−12.77	6.97	0.22	−0.34
3	1	69	−13.32	6.63	0.15	−0.40
4	1	50	−13.96	5.88	0.11	−0.48
5	1	34	−15.04	5.32	0.14	−0.57
6	1	16	−16.22	4.46	0.15	−0.58
0	2	9	−10.21	7.01	0.22	−0.48
1	2	12	−8.61	5.46	0.36	−0.89
2	2	21	−10.28	5.38	0.08	−0.80
3	2	15	−9.64	4.29	0.12	−0.86
4	2	15	−11.14	4.70	−0.03	−0.99

[a] Energy zero is the ground state of O^{16}. The ground-state energy of $(2s\text{-}1d)^4$ is −36.01 MeV.

imation to the discrete distribution function is immediately evident from Figs. 2 and 3. If the unit increase in the discrete distribution function, at the ith eigenvalue, were smeared locally about E_i, the resulting continuous distribution would assume the value $i-\frac{1}{2}$ at E_i. Conversely, if one has a moment-equivalent continuous distribution function, then an approximation to the discrete spectrum may be generated by identifying the ground state with that energy at which $F(\Gamma, E_g)=\frac{1}{2}$ and the jth excited state with $\overline{F}(\vec{\Gamma}, E_j)=j+\frac{1}{2}$. When used with the four-moment distribution function of Fig. 2, this criterion gives rise to an approximate spectrum which differs from the true one by as much as 1 MeV for only 4 of 45 eigenvalues. In the case of Fig. 3 and the two-moment distribution, only the 2 extreme eigenvalues out of 56 would be in error by 1 MeV. The other JT groups give rise to approximate spectra of comparable quality.

The physically important question for spectroscopy is the quality of the prediction of the low-excitation spectrum. This tests the adequacy of the asymptotic behavior of the continuous frequency function. Here the importance of third and fourth moments is emphasized because of the relatively greater contribution of the extremes of the spectrum in determining the value of higher moments.

In Fig. 4 we compare the shell-model spectra resulting from matrix diagonalization with the relative spectra generated using two-, three-, and four-moment distribution functions. This figure is restricted to the lowest five eigenvalues for 12 JT groups consisting of 16 to 69 nondegenerate eigenstates. In spite of this restriction to the most physically interesting part of the spectrum, it should be remembered that Figs. 2 and 3 show that the entire spectrum may be predicted by this procedure. For the 19 JT groups of Table I, the average error in the absolute energy of the predicted ground state is 1.57, 0.92, 0.80 MeV, respectively. The absolute energy of these ground states, considered of course as $(2s\text{-}1d)^4$, and their four-moment predictions is displayed in Fig. 5. The corresponding errors in the absolute energy of the first excited states is 0.85, 0.59, 0.55 MeV. The central moments that we use are sensitive to the absolute position of eigenvalues relative to the centroid. For example, it is obvious that if the two lowest eigenvalues of a JT group are shifted by an amount ΔE in the same direction, the central moments will be changed more than if identical shifts of ΔE are made in opposite directions. Thus the low-order moments are less directly sensitive to the relative spacing of eigenvalues. This fact appears in the average errors of 1.33, 1.21, 1.06 MeV that are found in the excitation energy of the first excited state of the three approximate spectra. The discrepancy between levels of the exact and approximate spectra will, on the average, decrease with increasing excitation energy in the energy range below the centroid. The worst prediction of a low-excitation spectrum occurs for the case $J=1$, $T=1$, where only 98 keV separates the first two shell-model states. Clearly such a situation can never be identified using just low-order moments. Nevertheless the general features of the shell-model calculation seem to be reasonably approximated by the use of a few low-order moments of spectral distributions.

Our general success in reproducing the details of a spectrum from a small set of moments may seem surprising. This success would appear to give some support to the basic assumptions made in the statistical theories of level spacings in nu-

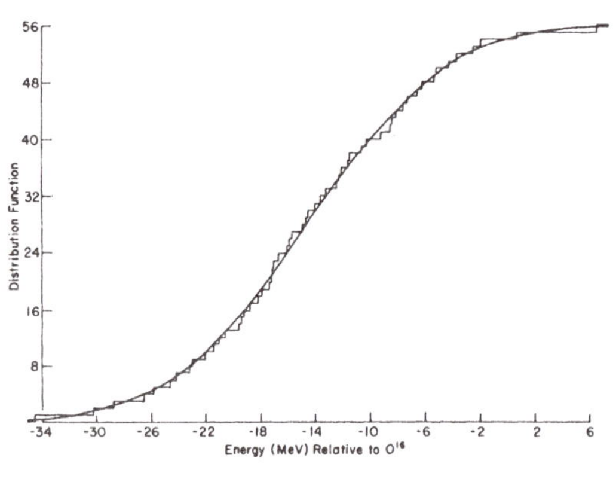

FIG. 3. The discrete $J=2$, $T=0$ eigenvalue distribution function compared with the two-moment distribution function. The moments and parameters for this case are $\mathcal{E}=-14.58$ MeV, $\sigma=8.06$ MeV, $\gamma_1=0.04$, and $\gamma_2=-0.002$.

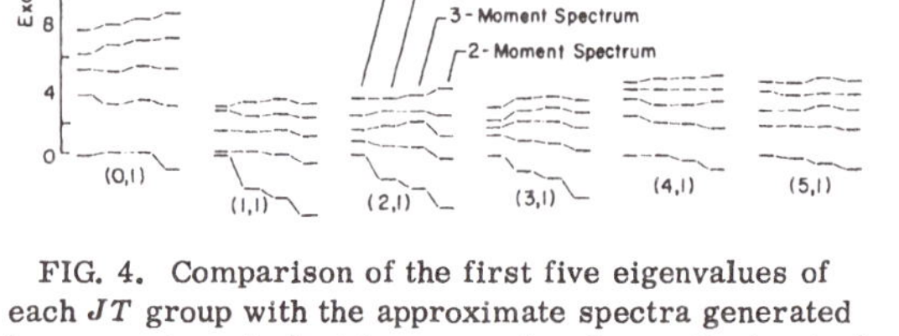

FIG. 4. Comparison of the first five eigenvalues of each JT group with the approximate spectra generated by moment-equivalent frequency functions. Each set of four spectra is ordered from left to right as indicated. All energies are excitation energies relative to the true ground state of each JT group. The moments and parameters for each case may be found in Table I.

clei.[13-15] There it is found that levels of a group, corresponding to exact symmetries of the Hamiltonian, are rarely found to be clustered into sets characterized internally by small level spacings. The absence of such a clustering is known as level repulsion and has a decided smoothing effect on eigenvalue distributions. This underlying fact would then support the idea that exact and moment-generated spectra should have an over-all agreement when they share a common set of low-order moments. Still unexplained, however, is the Gaussian behavior of these eigenvalue distributions.

One obvious weakness of our procedure is evident in Fig. 1. The continuous function $\bar{f}$ that we define in Eq. (13) does not satisfy the requirement that a frequency function be positive definite. Both positive skewness and negative excess, the conditions which dominate Table I, will cause the frequency function to become negative sufficiently far below the centroid. The usefulness of Eq. (13) in approximating spectra is then limited by the maximum negative value assumed by the distribution function. This in turn is determined by the strength, skewness, and excess of the distribution. For the parameters typical of Table I ($\gamma_1 < 0.2, \gamma_2 > -0.5$) the moment-equivalent function of Eq. (13) is found to be useful for spaces $(N; \vec{m}, \vec{\Gamma})$ with strengths $d/(2J+1)(2T+1)$ up to a few hundred. The negative asymptotic tail of $\bar{f}$ is therefore not a problem in the applications of this section. For larger spaces an alternative to Eq. (13) should be developed in which greater attention is paid to the positive definite nature of the frequency function. Indeed, for larger spaces one must be very confident of the asymptotic behavior of $\bar{f}$ before it may be used to predict detailed spectra as was done above.

Finally we use our method to generate a graph which may be used to quickly estimate the position of the lowest eigenvalue of a spectrum relative to its centroid. Assuming that the eigenstates are essentially distributed according to a Gaussian frequency function, the corresponding distribution is given by

$$\bar{F}(E) = \frac{d}{(2J+1)(2T+1)\sigma(2\pi)^{1/2}} \int_{-\infty}^{E} e^{-\frac{1}{2}[(E-\mathcal{E})/\sigma]^2} dE = \frac{d}{2(2J+1)(2T+1)} \left[1 - \operatorname{erf}\left(2^{-1/2}\frac{E-\mathcal{E}}{\sigma}\right)\right]. \tag{17}$$

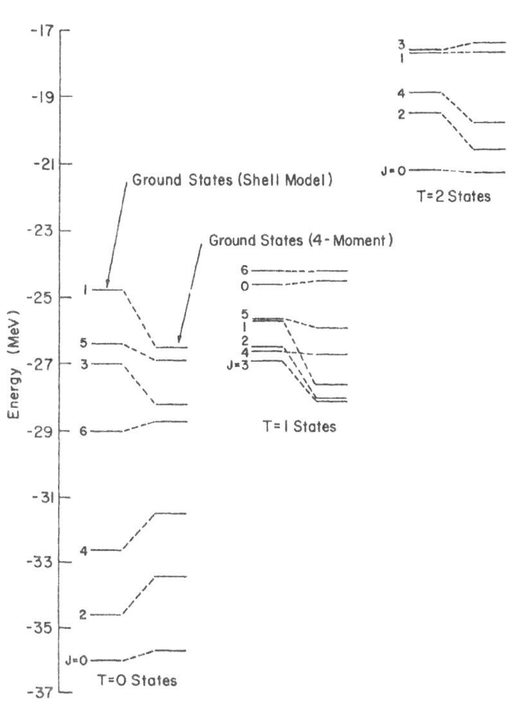

FIG. 5. Comparison of the absolute energy of the lowest eigenvalue of each JT group and the estimated eigenvalue from the four-moment distribution function. The energy zero is provided by the ground state of O^{16}.

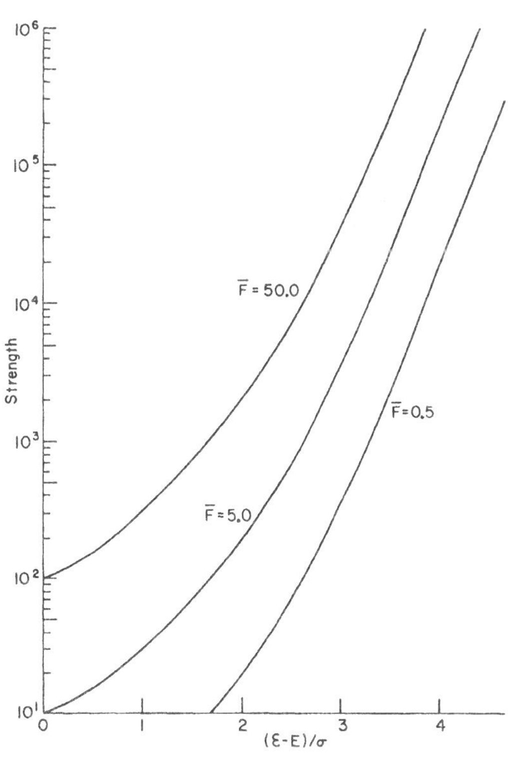

FIG. 6. Three values of the distribution function of Eq. (17) as a function of the strength of the distribution. These values are located relative to the centroid and are measured in units of the width.

In Eq. (17) the distribution function has been expressed in terms of erf(x), the usual error function. In Fig. 6 we show the location of three values of the distribution function for a large range of strengths. Each such value occurs at an energy relative to the centroid that is conveniently expressed in energy units of the width or standard deviation σ. As an example, we see that the lowest eigenvalue ($\bar{F} = 0.5$) of a space $(N; \vec{m}, \vec{\Gamma})$ with a strength of 400 would be found about 3σ below the centroid, while five eigenvalues would be found below 2.2σ.

B. Spectral Moments and Structure of Eigenvectors

We next consider the use of spectral distributions in the determination of the structure of low-lying eigenvectors of JT groups. The vector space $(N; \vec{m}, \vec{\Gamma})$ may be decomposed into subspaces $(N; [\vec{m}], \vec{\Gamma})$, where by $[\vec{m}]$ we indicate the set of orbital occupation numbers m_r $(\sum_r m_r = m)$, which specifies an m-nucleon shell-model configuration. The dimensionality of the subspace $(N; [\vec{m}], \vec{\Gamma})$ is denoted by $d([\vec{m}], \vec{\Gamma})$ and is related to that of $(N; \vec{m}, \vec{\Gamma})$ by

$$\sum_{[\vec{m}]} d([\vec{m}], \vec{\Gamma}) = d(\vec{m}, \vec{\Gamma}) . \tag{18}$$

We will be interested in the distribution of squared amplitudes of the states of the subspace $(N; [\vec{m}], \vec{\Gamma})$ over the eigenstates of $(N; \vec{m}, \vec{\Gamma})$. These distributions will be used to indicate the relative importance of the various subspaces $(N; [\vec{m}], \vec{\Gamma})$ in the low-lying eigenstates of physical interest. One use of this analysis is to act as a guide in making any vector-space truncation prior to a detailed spectroscopic calculation. Again our application will be for the case of four nucleons in the $2s$-$1d$ shell employing the Rosenfeld interaction[11] used above.

We denote by $|[m], \Gamma, \alpha\rangle$ a set of normalized vectors which span the subspace $(N; [\vec{m}], \vec{\Gamma})$. The label α includes all quantum numbers other than $[m]$ and Γ necessary to specify an orthogonal labeling in this subspace. The eigenvectors of $(N; \vec{m}, \vec{\Gamma})$ will be denoted by $|m, \Gamma, E_{i\Gamma}, i'\rangle$, where we specify a label i' to distinguish among the $(2T+1)(2J+1)$ degenerate eigenvectors with eigenvalue $E_{i\Gamma}$. For clarity of notation we list the usual completeness relations

$$\sum_{[\vec{m}], \alpha} |\langle [m], \Gamma, \alpha | m, \Gamma, E_{i\Gamma}, i'\rangle|^2 = 1 , \tag{19}$$

$$\sum_{E_{i\Gamma}, i'} |\langle [m], \Gamma, \alpha | m, \Gamma, E_{i\Gamma}, i'\rangle|^2 = 1 . \tag{20}$$

The normalized frequency function of the distribution of the space $(N; [\vec{m}], \vec{\Gamma})$ over the eigenstates of $(N; \vec{m}, \vec{\Gamma})$ is given by

$$f([\vec{m}], \vec{\Gamma}, E) = d^{-1}([\vec{m}], \vec{\Gamma}) \sum_{\alpha, E_{i\Gamma}, i'} |\langle [m], \Gamma, \alpha | m, \Gamma, E_{i\Gamma}, i'\rangle|^2 \delta(E - E_{i\Gamma}) , \tag{21}$$

$$\int_{-\infty}^{\infty} f([\vec{m}], \vec{\Gamma}, E)\, dE = 1 . \tag{22}$$

We introduce the distribution function of $(N; [\vec{m}], \vec{\Gamma})$ in analogy with Eq. (9)

$$F([\vec{m}], \vec{\Gamma}, E) = \frac{d([\vec{m}], \vec{\Gamma})}{(2J+1)(2T+1)} \int_{-\infty}^{E} f([\vec{m}], \vec{\Gamma}, E')\, dE' , \tag{23}$$

$$F([\vec{m}], \vec{\Gamma}, \infty) = \frac{d([\vec{m}], \vec{\Gamma})}{(2J+1)(2T+1)} . \tag{24}$$

As before Eq. (24) identifies $d([\vec{m}], \vec{\Gamma})/(2J+1)(2T+1)$ as the strength of the distribution. The sum of the subspace distribution functions then reproduces that of $(N; \vec{m}, \vec{\Gamma})$ which was studied above:

$$F(\vec{m}, \vec{\Gamma}, E) = \sum_{[\vec{m}]} F([\vec{m}], \vec{\Gamma}, E) . \tag{25}$$

The centroid and central moments of the distribution are given by

$$\begin{aligned} \mathcal{E}([\vec{m}], \vec{\Gamma}) &= \int_{-\infty}^{\infty} E f([\vec{m}], \vec{\Gamma}, E)\, dE , \\ &= d^{-1}([\vec{m}], \vec{\Gamma}) \sum_{\alpha} \langle [m], \Gamma, \alpha | H | [m], \Gamma, \alpha \rangle , \end{aligned} \tag{26}$$

$$\begin{aligned} \mu_p([\vec{m}], \vec{\Gamma}) &= \int_{-\infty}^{\infty} [E - \mathcal{E}([\vec{m}], \vec{\Gamma})]^p f([\vec{m}], \vec{\Gamma}, E)\, dE , \\ &= d^{-1}([\vec{m}], \vec{\Gamma}) \sum_{\alpha} \langle [m], \Gamma, \alpha | [H - \mathcal{E}([\vec{m}], \vec{\Gamma}) \underline{1}]^p | [m], \Gamma, \alpha \rangle , \end{aligned} \tag{27}$$

where $\underline{1}$ is the unit operator. In the second forms, the moments have been expressed in terms of partial traces of powers of the Hamiltonian matrix $H(\vec{m}, \vec{\Gamma})$.

The distribution function in Eq. (23) may also be written as

$$F([\vec{m}], \vec{\Gamma}, E) = [(2J+1)(2T+1)]^{-1} \sum_{\alpha, E_i, i'}' |\langle [m], \Gamma, \alpha | m, \Gamma, E_{i\Gamma}, i' \rangle|^2 , \tag{28}$$

where $\sum'$ indicates that the sum over eigenstates is restricted to those with $E_{i\Gamma} < E$. In this form we see that $F([\vec{m}], \vec{\Gamma}, E)$ may be identified with the sum of the squared amplitudes of the states $|[m], \Gamma, \alpha\rangle$ with those eigenstates of $(N; \vec{m}, \vec{\Gamma})$ which satisfy $E_{i\Gamma} < E$. Thus it is obvious that a reasonable approximation to $F([\vec{m}], \vec{\Gamma}, E)$ is pertinent in the determination of the relative importance of the subspaces $(N; [\vec{m}], \vec{\Gamma})$ in the low-lying eigenstates of $(N; \vec{m}, \vec{\Gamma})$. In analogy with the treatment of the eigenvalue distribution above, we may approximate f by a moment-equivalent continuous frequency function $\bar{f}$. Then replacement of f by $\bar{f}$ in Eq. (23) will generate a continuous approximate distribution function $\bar{F}([\vec{m}], \vec{\Gamma}, E)$.

It is obvious, however, that we have no license to use moment-equivalent distribution functions $\bar{F}([\vec{m}], \vec{\Gamma}, E)$ in the detailed fashion that eigenvalue distributions were used. The distribution function $F([\vec{m}], \vec{\Gamma}, E)$ increases its value at $E_{i\Gamma}$ by

$$\sum_{\alpha, i} \frac{|\langle [m], \Gamma, \alpha | m, \Gamma, E_{i\Gamma}, i' \rangle|^2}{(2J+1)(2T+1)} .$$

This quantity varies greatly from eigenvalue to eigenvalue unlike the constant unit increment of the eigenvalue distribution function $F(\vec{m}, \vec{\Gamma}, E)$. However, the average of these increases over a few neighboring eigenvalues will have a smoother behavior as a function of energy. In terms of distribution functions such an average suggests the approximation

$$\bar{F}([\vec{m}], \vec{\Gamma}, E + \Delta E) - \bar{F}([\vec{m}], \vec{\Gamma}, E) \approx F([\vec{m}], \vec{\Gamma}, E + \Delta E) - F([\vec{m}], \vec{\Gamma}, E) , \tag{29}$$

where ΔE includes a few eigenvalues.

To illustrate this idea, we again consider the shell-model space $(2s\text{-}1d)^4$ with the same Rosenfeld interaction used above. The sum of the distribution functions $F([\vec{m}], \vec{\Gamma}, E)$ reproduces the eigenvalue distribution $F(\vec{m}, \vec{\Gamma}, E)$ which was successfully approximated above by use of a moment-equivalent Gaussian frequency function. We therefore introduce continuous distribution functions $\bar{F}([m], \Gamma, E)$ generated by Gaussian frequency functions

$$\bar{F}([\vec{m}], \vec{\Gamma}, E) = \frac{d([\vec{m}], \vec{\Gamma})}{(2J+1)(2T+1)} \int_{-\infty}^{E} \frac{1}{\sigma([\vec{m}], \vec{\Gamma})(2\pi)^{1/2}} \exp\left[-\frac{1}{2}\left(\frac{E - \mathcal{E}([\vec{m}], \vec{\Gamma})}{\sigma([\vec{m}], \vec{\Gamma})}\right)^2\right] dE \tag{30}$$

with centroid and width identical to that of $f([\vec{m}], \vec{\Gamma}, E)$. We choose an energy interval in Eq. (29) which includes the first three eigenstates of a JT group. Recalling that F vanishes below the ground state, our application of Eq. (29) then tests the approximation $\bar{F}([\vec{m}], \vec{\Gamma}, E_3) \approx F([\vec{m}], \vec{\Gamma}, E_3)$. Specifically we find an energy E_3 which satisfies

$$\sum_{[\vec{m}]} \bar{F}([\vec{m}], \Gamma, E_3) = 3 , \tag{31}$$

and then compare each $\bar{F}([\vec{m}], \vec{\Gamma}, E_3)$ with the actual sum of squared amplitudes $F([\vec{m}], \vec{\Gamma}, E_3)$ obtained by matrix diagonalization.

Two typical examples of this comparison are shown in Fig. 7 for $J=0$, $T=0$ and in Fig. 8 for $J=2$, $T=1$. In both cases the configurations subspaces $(N; [\vec{m}], \vec{\Gamma})$ have been ordered from left to right in order of increasing centroid $\mathcal{E}([\vec{m}], \vec{\Gamma})$. Such predictions of the results of a detailed spectroscopic calculation may be improved by the use of higher moments or by the decomposition of $(N; \vec{m}, \vec{\Gamma})$ into even smaller subspaces. A trivial improvement would also follow from increasing the energy in Eq. (31) to include additional eigenstates. Nevertheless, the inclusion of only two moments and three eigenstates is found to yield a reasonably good ordering of the various subspaces in terms of their importance in the determination of the physically interesting part of the shell-model spectrum.

The choice of the examples presented in this section was determined by the most complex shell-model calculation available to us. This permitted the evaluation of various ideas on the use of spectral distributions by comparison with detailed spectroscopy. All spectral moments were computed from partial traces of powers of the Hamiltonain matrix in $(2s\text{-}1d)^4$. French and Mugambi[16] are analyzing methods in which the moments of $(N; [\vec{m}], \vec{\Gamma})$ may be expressed in terms of those moments in which only a few holes or particles are present in each orbital. Such methods will then permit the propagation of these moments into spaces too large for conventional matrix methods. Thus our success in the prediction of the eigenvalues and the nature of low-lying eigenfunctions of $(2s\text{-}1d)^4$ is encouraging. In the event that detailed spectroscopy is needed in the larger spaces, these moment techniques may be used as a guide in making the necessary vector-space truncation.

III. SCALAR DISTRIBUTIONS AND GROUND-STATE ENERGIES

In this section we use moment techniques to discuss the m particle scalar distribution, i.e., the distribution of all eigenvalues of H in the space $(N; \vec{m})$ giving full weight to all degeneracies. We use the scalar propagation laws to generate scalar moments for each $(N; \vec{m})$. It is then shown how these moments may be used to estimate the lowest eigenvalue of H in $(N; \vec{m})$. Numerical examples are again drawn from the $(2s\text{-}1d)$ shell using the Rosenfeld interaction. We investigate the ground-state energy of nuclei in the range $16 \leq A \leq 40$ and show how these energies may be compared with empirical binding energies.

Moments of the scalar distribution in $(N; \vec{m})$ are proportional to traces of powers of the Hamiltonian matrix $H(N; \vec{m})$ in that space

$$\mathcal{E}(\vec{m}) = \binom{N}{m}^{-1} \mathrm{Tr}[H(N; \vec{m})], \tag{32}$$

$$\mu_k(\vec{m}) = \binom{N}{m}^{-1} \mathrm{Tr}[H(N; \vec{m}) - \mathcal{E}(\vec{m})\underline{1}]^k. \tag{33}$$

This relation to traces results in simple propagation formulas[1-4] which relate the scalar moments

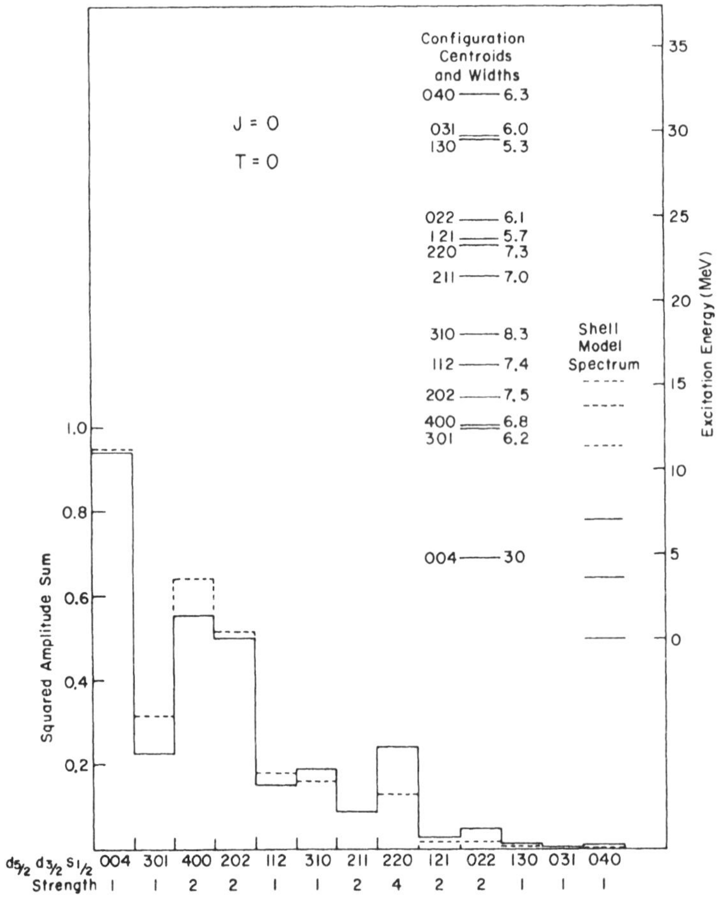

FIG. 7. Sum of squared amplitudes for each four-nucleon configuration for the first three eigenstates of the space $J=0$, $T=0$ as derived by detailed spectroscopy (solid line) and by the two-moment distribution function (dashed line). Each configuration is identified by the number of nucleons in each of the three orbitals in the order $1d_{5/2}$, $1d_{3/2}$, $2s_{1/2}$. The strength of each configuration is listed. Also shown are the centroids of each configuration and the width in MeV. The shell-model spectrum consists of 21 eigenstates and we have shown the positions of the first three (solid lines), for which the comparison is being made, as well as the position of the next three levels (dashed lines). The squared-amplitude sums have been arranged from left to right in order of increasing energy of the configuration centroid.

in spaces containing different numbers of nucleons. Since H is a two-body operator, it follows that there are at most $2k+1$ independent values of $\mu_k(\vec{m})$. Any particular $\mu_k(\vec{m})$ can, therefore, be expressed as a linear combination of $\mu_k(\vec{t})$,

$$\mu_k(\vec{m}) = \sum_t \mu_k(\vec{t}) \rho_t(k) , \tag{34}$$

where

$$\rho_t(k) = \prod_{t' \neq m} (m - t') / \prod_{t' \neq t} (t - t') , \tag{35}$$

and where t and t' run over any set of $2k+1$ integers between 0 and N. Examples of these propagation formulas for the centroid and variance are given in the preceding paper[1] and earlier by French.[4]

In the example which follows we choose the set of spaces $(N; \vec{t})$, $t = \{0, 1, \ldots, k, N-k+1, \ldots, N\}$, to evaluate μ_k in any space $(N; \vec{m})$. This choice of input data presents two clear advantages. This set involves the minimum number of particles and holes and therefore is of maximum economy in calculating the input traces of powers of $H(N; \vec{t})$ which are needed for propagation. The demands on the numerical accuracy of the input data are also minimized by the above choice. This latter point follows from the nature of the polynomial coefficients in Eq. (35). If the input is taken just from the few-particle spaces, then the coefficients are found to increase monotonically and rapidly as we propagate through the sequence of spaces to the plenum state ($m = N$). With the defining spaces chosen with maximum symmetry between particles and holes, as indicated above, the corresponding coefficients are not monotonic and remain relatively small. As an example in the $2s$-$1d$ shell ($N = 24$), we find that the variance defined by the set $\{0, 1, 2, 3, 4\}$ requires some three additional significant figures in the input variances over that required when using the set $\{0, 1, 2, 23, 24\}$. The comparison is more striking for higher moments.

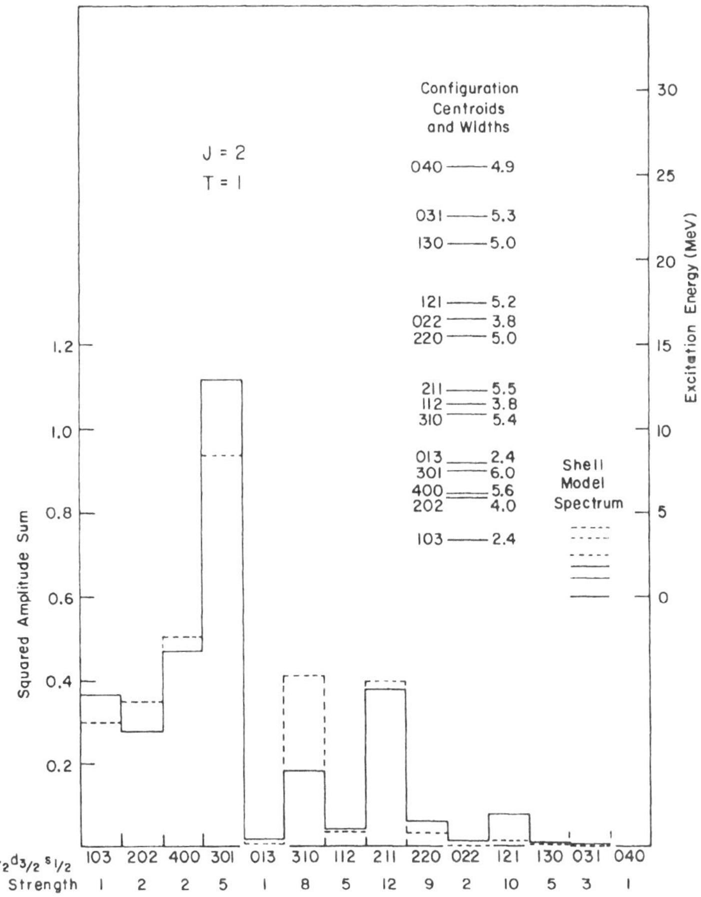

FIG. 8. Sum of squared amplitudes for each four-nucleon configuration in the space $J=2$, $T=1$. The notation is identical with that of Fig. 7. The total number of eigenstates in this space is 66.

The scalar propagation formulas of Eq. (34) may be used to evaluate moments throughout the $2s$-$1d$ shell. We again employ the Rosenfeld interaction described in the preceding section. The single-particle energies are chosen to reproduce the three single-neutron states of O^{17}. The set of input moments to be used in Eq. (34) are listed in Table II. The results of this propagation are displayed in Table III. In Table III the third and fourth moments have been combined with the variance to form the dimensionless skewness ($\gamma_1 = \mu_3/\sigma^3$) and excess ($\gamma_2 = \mu_4 \sigma^{-4} - 3$) parameters which were discussed in the preceding section. As in the case of JT distributions, the value of these parameters ($\gamma_1 \approx 0, \gamma_2 \approx 0$) is interpreted as a signature of the Gaussian frequency function.

At an energy corresponding to the eigenvalue E_i, the discrete scalar distribution function will rise by an amount equal to the degeneracy $(2J_i+1) \times (2T_i+1)$ associated with that eigenvalue. If the maximum degeneracy in $(N; \vec{m})$ is small compared with the number of distinct eigenvalues, then we may expect that the discrete distribution function may be reasonably approximated by a continuous distribution function. An inspection of Table III reveals that this criterion for the introduction of a continuous distribution function is satisfied for four or more particles or holes in the $2s$-$1d$ shell. To each such scalar distribution we then assign a continuous distribution function obtained by integrating the Gaussian frequency function

$$\bar{F}(\vec{m}, E) = \binom{N}{m} \frac{1}{\sigma(\vec{m})(2\pi)^{1/2}} \int_{-\infty}^{\infty} \exp\left[-\frac{1}{2}\left(\frac{E - \mathcal{E}(\vec{m})}{\sigma(\vec{m})}\right)^2\right] dE \tag{36}$$

and by using the moments listed in Table III. From the form of Eq. (36) we see that the dimensionality of $(N; \vec{m})$, i.e., $\binom{N}{m} = \bar{F}(\vec{m}, \infty)$, is identified as the strength of the scalar distribution. The present case is an example of where Eq. (13) cannot be profitably used to modify the Gaussian frequency function by exploiting our knowledge of third and fourth moments. Inspection of Fig. 6 reveals that for scalar strengths $\binom{24}{m}$, typical of the middle of the $2s$-$1d$ shell, the ground state may be expected to occur some four or five widths below the centroid. In this energy range the modified Gaussian of Eq. (13) is generally not positive. Thus we restrict ourselves to the two-moment approximation represented by Eq. (36).

A good approximate distribution function may be used to estimate the ground-state energy relative to the scalar centroid. By a chain of arguments identical to that used above with JT distributions,

TABLE II. Scalar moments used for propagation according to Eq. (34). In each case, the moments are chosen with maximum symmetry between particles and holes. These input data for the propagation formulas were specified with the accuracy indicated below, which was required to insure numerical stability in the use of Eq. (34).

m	$\mathcal{E}$ (MeV)	μ_2 (MeV^2)	μ_3 (MeV^3)	μ_4 (MeV^4)
0	0.0	0.0	0.0	0.0
1	−2.3117	5.3488	8.0971	42.6653
2		14.4845	11.1157	468.8420
3			16.7654	1819.7322
4				4388.0937
21				1864.2319
22			−17.9032	486.9856
23		5.4384	−8.6307	44.2769
24	−256.7161	0.0	0.0	0.0

we might expect to identify the m-nucleon ground state by satisfying

$$\bar{F}(\vec{m}, E_g) = \tfrac{1}{2}(2J_g+1)(2T_g+1) , \tag{37}$$

where J_g and T_g are the spin and isospin of the ground state. These quantum numbers are uncertain prior to actual matrix diagonalization. However, the tensorial rank (J_g, T_g) of the ground states in the $2s$-$1d$ shell might be expected to fall within an energy range identified from Eq. (37) to be $0.5 \leq \bar{F}(\vec{m}, E_g) \leq 5.0$. An inspection of Fig. 6 reveals that, for strengths $\binom{24}{m}$ characteristic of the $2s$-$1d$ shell, this energy range is at most $\frac{1}{2}\sigma(\vec{m})$ in width. Furthermore, we may refer to the results for JT distributions in Fig. 4, where it is found that the solution of $\bar{F}(m, \Gamma, E) = \frac{1}{2}$ is shifted in energy by an average of some $\frac{1}{4}\sigma$ because of the inclusion of γ_1 and γ_2 in $\bar{F}$. Thus, were we to construct and use a four-moment frequency function in Eq. (36) employing the small values of γ_1 and γ_2 typical of Table III, we would expect that our estimate of the value of E_g would differ by a small fraction of σ from that obtained using the two-moment frequency function.

This discussion then suggests that we use Eq. (36) to locate an energy $E_{1/2}$ defined by

$$\bar{F}(\vec{m}, E_{1/2}) = \tfrac{1}{2}. \tag{38}$$

The ground-state eigenvalue of $(N; \vec{m})$ should be found within a fraction of $\sigma(\vec{m})$ of $E_{1/2}$. In the case of four nucleons in the $2s$-$1d$ shell, the actual ground-state energy obtained by matrix diagonalization is found to lie 2.2 MeV above $E_{1/2}$, the difference being about $\frac{1}{3}\sigma$. The difference in energy between the scalar centroid and the ground state may then be estimated by Δ, where

$$\Delta(m)=\mathcal{E}(m)-E_{1/2}\ . \tag{39}$$

This difference has been evaluated throughout the $2s$-$1d$ shell and the results are listed in the last column of Table III.

Having located the ground-state energy relative to the scalar centroid, we next consider how these results may be compared with empirical binding energies of nuclei in the range $16\le A\le 40$. Such a comparison would then assess the usefulness of the Rosenfeld interaction as an effective interaction for $2s$-$1d$-shell nuclei. Before making any direct comparison of theoretical and empirical energies, we should consider a number of possible additions to the theoretical Hamiltonian. Already included in H is the nuclear interaction between $2s$-$1d$-shell nucleons given by the Rosenfeld interaction. We have also included the kinetic energy and the average nuclear interaction between $2s$-$1d$-shell nucleons and the inert O^{16} core by using empirical single-particle energies as determined by O^{17}. Necessary additions to H should include the energy of the inert core given by the binding energy of O^{16}. Any size effects over the range $16\le A\le 40$ could be approximated by a function of the number operator for $2s$-$1d$-shell nucleons. The Coulomb interaction of $2s$-$1d$-shell protons with the inert core may be represented by the product of the active proton number operator n_p with a constant taken empirically from F^{17}-O^{17} data. Finally, the Coulomb interaction of $2s$-$1d$-shell protons with each other may be represented by the product of the number of active proton pairs $\frac{1}{2}n_p(n_p-1)$ and a constant taken from $A=18$ data or Coulomb energy differences.

We now demonstrate an approximation that will facilitate easy comparison of theoretical and empirical energies. This approximation has the further advantage of being independent of specific assumptions about the magnitude of Coulomb and size effects as we move across the mass range $16\le A\le 40$. In first-order approximation the Coulomb operators may be averaged over $(N;\vec{m})$ as in Eq. (34), with the result

$$\langle n_p\rangle_m=\tfrac{1}{2}m\ , \tag{40}$$

$$\langle \tfrac{1}{2}n_p(n_p-1)\rangle_m=\frac{m(m-1)(N-2)}{8(N-1)}\ ; \tag{41}$$

The average represented by Eq. (41) follows from the fact that $\frac{1}{2}n_p(n_p-1)$ is a two-body operator with scalar averages 0, 0, and $\frac{1}{2}(\frac{1}{2}N)(\frac{1}{2}N-1)$ for $m=0$, 1, and N, respectively. In this order of approximation the entire set of additions to H has been

TABLE III. Scalar moments throughout the $2s$-$1d$ shell, and the energy difference (Δ) between the centroid and the estimated position of the ground state.

m	Strength	Number of distinct eigenvalues	Maximum degeneracy	$\mathcal{E}$ (MeV)	σ (MeV)	γ_1	γ_2	Δ (MeV)
0	1	1	1	0.0	0.0	0.0	0.0	
1	24	3	12	−2.31	2.31	0.65	−1.51	
2	276	28	27	−5.35	3.80	0.26	−0.76	
3	2024	128	48	−9.12	5.12	0.12	−0.36	17.9
4	10 626	640	65	−13.62	6.29	0.11	−0.19	24.5
5	42 504	1935	84	−18.85	7.32	0.12	−0.13	30.7
6	134 596	6116	105	−24.81	8.21	0.13	−0.11	36.1
7	346 104	13 029	120	−31.49	8.96	0.14	−0.10	41.2
8	735 471	28 503	133	−38.91	9.57	0.15	−0.11	45.9
9	1 307 504	44 133	144	−47.05	10.05	0.15	−0.11	49.2
10	1 961 256	69 784	153	−54.93	10.39	0.16	−0.12	50.9
11	2 496 144	80 115	160	−65.53	10.60	0.16	−0.12	53.0
12	2 704 156	93 710	161	−75.86	10.67	0.15	−0.12	53.4
13	2 496 144	80 115	160	−86.92	10.60	0.15	−0.12	53.0
14	1 961 256	69 784	153	−98.71	10.40	0.14	−0.12	51.0
15	1 307 504	44 133	144	−111.23	10.06	0.13	−0.11	49.3
16	735 471	28 503	133	−124.48	9.58	0.11	−0.10	46.0
17	346 104	13 029	120	−138.46	8.97	0.09	−0.10	41.3
18	134 596	6116	105	−153.16	8.22	0.06	−0.10	36.2
19	42 504	1935	84	−168.60	7.34	0.01	−0.12	30.8
20	10 626	640	65	−184.76	6.31	−0.05	−0.18	24.6
21	2024	128	48	−201.64	5.14	−0.16	−0.33	18.0
22	276	28	27	−219.28	3.83	−0.32	−0.73	
23	24	3	12	−237.63	2.33	−0.68	−1.50	
24	1	1	1	−256.72	0.0	0.0	0.0	

cast in the form of a function of the number operator for $2s$-$1d$-shell nucleons. Such additions to H will shift the centroid but will leave all central moments unchanged. Therefore, these modifications of H will leave invarient the energy difference $\Delta = \mathcal{E} - E_{1/2}$ obtained above, the approximation then being that Δ is determined solely by the effective nuclear interaction.

A corresponding energy difference may also be obtained from empirical data with the assumption that the ground states of O^{16} and Ca^{40} be regarded as the vacuum and plenum states, respectively, of the sequence of spaces $(24; \vec{m})$. Empirical centroids are generated by using the binding energy of O^{16} and Ca^{40} for $\mathcal{E}'(0)$ and $\mathcal{E}'(24)$. The centroid of the single-particle levels in O^{17} and F^{17} yields $\mathcal{E}'(1)$. The remaining empirical centroids follow by propagation

$$\mathcal{E}'(\vec{m}) = \frac{m(m-1)}{552}\mathcal{E}'(24) - \frac{(24-m)(m-1)}{24}\mathcal{E}'(0) + \frac{(24-m)m}{23}\mathcal{E}'(1) . \qquad (42)$$

The ground-state energies E_g are taken from the mass tables.[17] These data and the energy differences $\Delta' = \mathcal{E}'(\vec{m}) - E_g$ are displayed in Table IV. The empirical (Δ') and theoretical (Δ) energy differences are found to differ by a small fraction of the scalar width for $m < 18$. This small difference $\Delta' - \Delta$ would then represent the difference between the theoretical and empirical ground-state energies if the additions to H, mentioned above, were to bring into coincidence the theoretical and empirical scalar centroids. This suggests that a Hamiltonian, consisting of the Rosenfeld interaction and an appropriate function of the nucleon number operator, may be expected to yield ground-state energies in reasonable agreement with experiment for about three quarters of the way through the $2s$-$1d$ shell. Beyond this point the discrepancy may be attributed either to the inadequacy of our effective interaction or to an insufficiently large vector space. Perhaps the major error made in our approximation is in ignoring the contribution to the scalar width which arises from the differences in Coulomb energies of states in $(N; \vec{m})$ having different numbers of protons. The increase in $\sigma(\vec{m})$ which would result from a more careful treat-

TABLE IV. Comparison of empirical (Δ') and theoretical (Δ) estimates of the energy separation of the scalar centroid and ground state. The difference $\Delta' - \Delta$ is a small fraction of the scalar width for $m < 18$. If the empirical and theoretical centroids were to coincidence, then $\Delta' - \Delta$ would represent the difference between experimental and theoretical ground-state energies.

m	Scalar width σ (MeV)	Empirical centroid (MeV)	Empirical ground state (MeV)	Δ' (MeV)	Δ (MeV)	$\Delta' - \Delta$
0	0.0	−127.6	−127.6	0.0		
1	2.3	−128.2	−131.8	3.6		
2	3.8	−129.5	−139.8	10.3		
3	5.1	−131.5	−147.8	16.3	17.9	−1.6
4	6.3	−134.3	−160.6	26.3	24.5	1.8
5	7.3	−137.8	−167.4	29.6	30.7	−1.1
6	8.2	−142.0	−177.8	35.8	36.1	−0.3
7	9.0	−146.9	−186.6	39.7	41.2	−1.5
8	9.6	−152.6	−198.3	45.7	45.9	−0.2
9	10.0	−159.0	−205.6	46.6	49.2	−2.6
10	10.4	−166.1	−216.7	50.6	50.9	−0.3
11	10.6	−174.0	−225.0	51.0	53.0	−2.0
12	10.7	−182.5	−236.5	54.0	53.4	0.6
13	10.6	−191.8	−245.0	53.2	53.0	0.2
14	10.4	−201.8	−255.6	53.8	51.0	2.8
15	10.1	−212.6	−262.9	50.3	49.3	1.0
16	9.6	−224.1	−271.8	47.7	46.0	1.7
17	9.0	−236.3	−281.0	44.7	41.3	3.4
18	8.2	−249.2	−291.8	42.6	36.2	6.4
19	7.3	−262.9	−298.8	35.9	30.8	5.1
20	6.3	−277.3	−308.7	31.4	24.6	6.8
21	5.1	−292.4	−317.1	24.7	18.0	6.7
22	3.8	−308.2	−327.4	19.2		
23	2.3	−324.8	−333.7	8.9		
24	0.0	−342.1	−342.1	0.0		

ment of Coulomb energies would increase with m and would shift Δ in the direction of Δ' towards the end of the shell.

Clearly ground-state energies may be estimated with greater confidence, and additional detail may be achieved using the moments which arise from the decomposition of $(N; \vec{m})$ into the subspaces $(N; \vec{m}, \vec{J}\vec{T}\vec{\pi})$. Indeed a fixed-isospin estimate of these energies is quite feasible. Nevertheless, the treatment of this section has the virtue of simplicity, can handle interactions of any complexity, and does not require additional assumptions concerning the nature of the ground state, such as the assignment of a definite configuration or seniority.

IV. CONFIGURATION DISTRIBUTIONS AND SHELL MODEL

In this section we analyze the spectral distributions of shell-model configurations throughout the $2s$-$1d$ shell by exploiting the polynomial expressions[1-5] for the centroid and variance of such distributions. We also consider the partial variances which measure the mixing of the states of a pair of configurations. One objective is to learn what such an analysis may suggest as the basis for vector-space truncations prior to a detailed spectroscopic calculation.

A. Configuration Moments and Vector-Space Truncation

The vector space $(N; \vec{m})$, consisting of m nucleons distributed among N single-nucleon states, may be decomposed into subspaces $(N; [\vec{m}])$ in which we specify the nucleon configuration $[\vec{m}]$; i.e., the partition of the m nucleons among λ single-nucleon orbitals with $\sum_{r=1}^{\lambda} m_r = m$, m_r denoting the occupancy of the rth orbital which consists of N_r single-nucleon states. The dimensionality of $(N; [\vec{m}])$ and, in this case, the strength of the configuration distribution are denoted by $d([\vec{m}])$ and are given by

$$d([\vec{m}]) = \prod_{r=1}^{\lambda} \binom{N_r}{m_r}. \tag{43}$$

The moments of the spectral distributions over

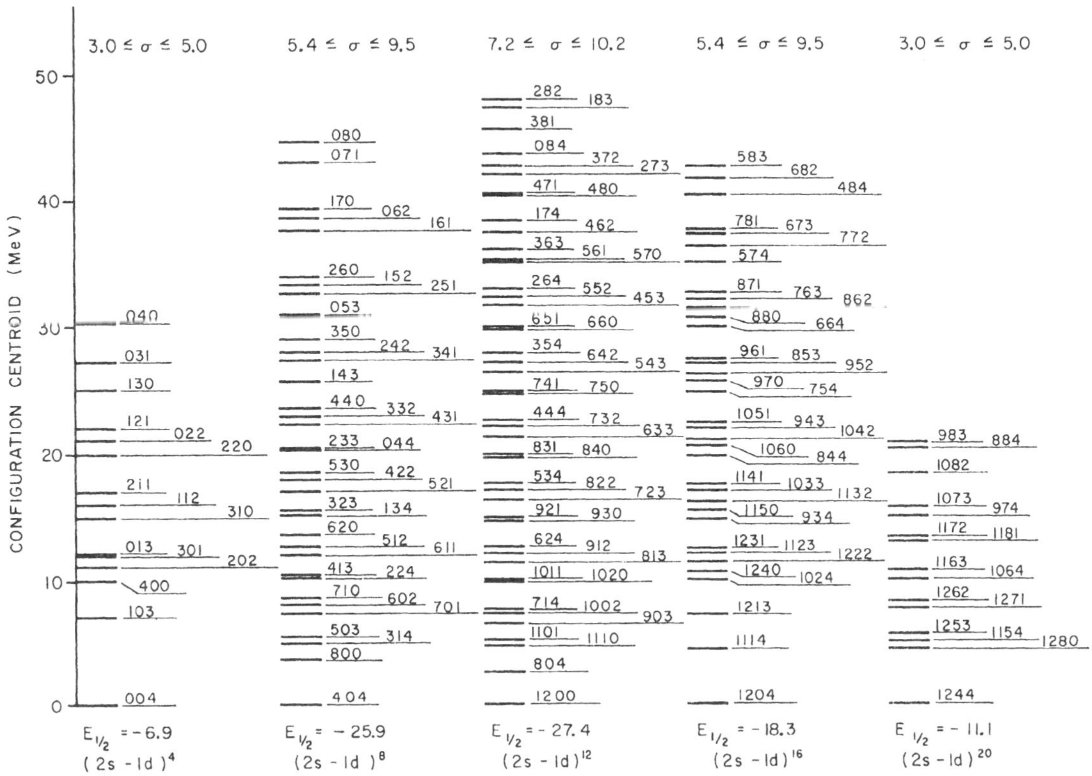

FIG. 9. Configuration centroids for 4, 8, 12, 16, and 20 nucleons in the $2s$-$1d$ shell. Each configuration is identified by a set of three numbers representing the number of nucleons in orbitals $1d_{5/2}$, $1d_{3/2}$, and $2s_{1/2}$, respectively. Configuration widths (in MeV) are found to fall in a narrow range of energy which is exhibited above each group. Below each group is listed the value of $E_{1/2}$ (MeV) defined in Eq. (38). This represents an estimate, based on scalar distributions, of the location of the theoretical ground state of each space relative to the lowest configuration centroid.

$(N;[\vec{m}])$ are given in terms of traces of powers of H over $(N;[\vec{m}])$, the centroid and central moments being

$$\mathcal{E}([\vec{m}])=d^{-1}([\vec{m}])\sum_{\alpha}\langle[m]\alpha|H|[m]\alpha\rangle\,, \tag{44}$$

$$\mu_p([\vec{m}])=d^{-1}([\vec{m}])\sum_{\alpha}\langle[m]\alpha|[H-\mathcal{E}([\vec{m}])]^p|[m]\alpha\rangle\,, \tag{45}$$

and where α denotes the set of labels other than the configuration necessary to specify the states of $(N;[\vec{m}])$.

The trace operators responsible for propagating the configuration moments may be found in Ref. 1 and are simple polynomials in the number operators of each orbital. From this fact it follows that these moments can be expressed as polynomials in m_r. The polynomial coefficients are given in terms of the N-nucleon expectation value of those parts of H^p which have definite particle rank in each orbital, i.e., which do not break the symmetry $U(N_1)\times U(N_2)\times\cdots\times U(N_\lambda)$. In this procedure the configuration moments are expressed directly in terms of the one- and two-body matrix elements of H. An alternative procedure exists[2,4] in which all configuration moments may be expressed in terms of an independent subset or elementary net of these moments. This latter procedure is analogous to that used for the propagation of scalar moments in the preceding section.

The centroid polynomial is second degree, and for λ orbitals consists of $\binom{\lambda+2}{2}$ independent terms,

$$\mathcal{E}([\vec{m}])=H_0+\sum_r m_r\epsilon_r+\sum_r \tfrac{1}{2}m_r(m_r-1)\overline{W}_{rr}+\sum_{r\neq s}{}' \tfrac{1}{2}m_r m_s\overline{W}_{rs}\,. \tag{46}$$

The constants H_0 and ϵ_r are the zero-body part of H and single-particle energy of the rth orbital, respectively. The average interaction of a pair of nucleons in the same or different orbitals has been denoted by $\overline{W}$ where

$$\overline{W}_{rr}=\binom{2(2j_r+1)}{2}^{-1}\sum_{JT}(2J+1)(2T+1)W^{JT}_{rrrr}\,, \tag{47}$$

$$\overline{W}_{rs}=[4(2j_r+1)(2j_s+1)]^{-1}\times\sum_{JT}(2J+1)(2T+1)W^{JT}_{rsrs}\,,\quad r\neq s\,; \tag{48}$$

and where W^{JT}_{rstu} is the coupled, antisymmetrized,

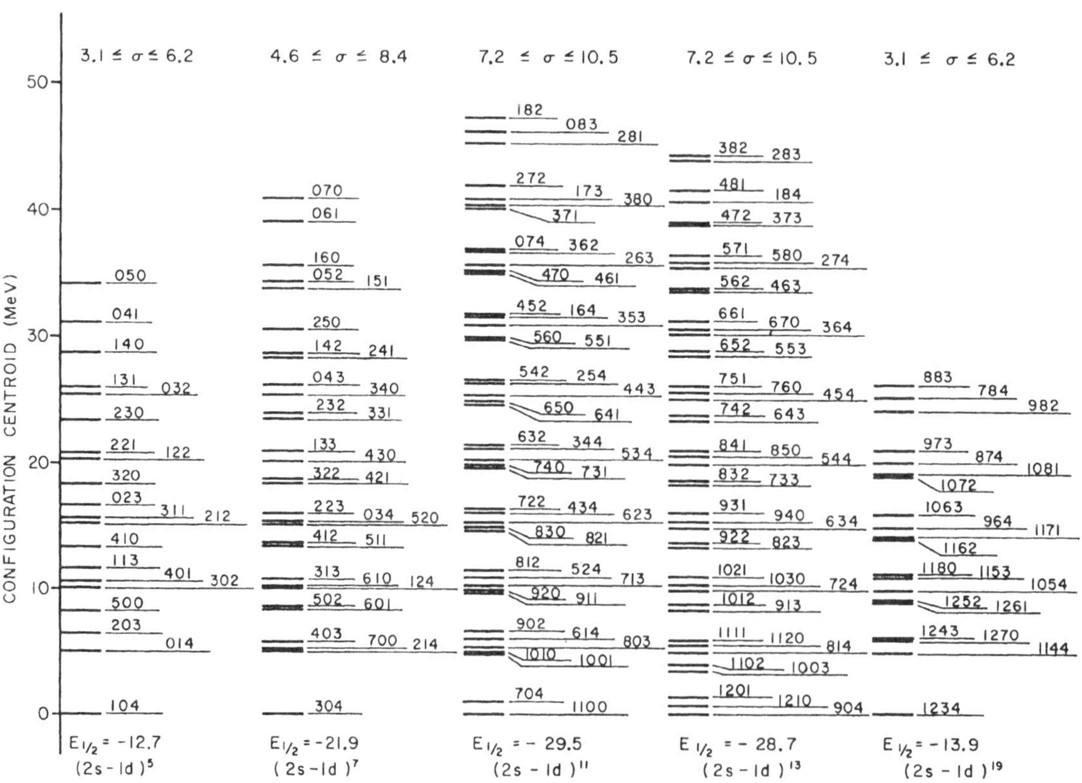

FIG. 10. Configuration centroids for 5, 7, 11, 13, and 19 nucleons in the $2s$-$1d$ shell. The notation here is identical to that used in Fig. 9.

two-body matrix element $\langle rs;JT|V|tu;JT\rangle$. The dispersion polynomial is fourth degree and for λ orbitals consists of $\binom{\lambda+4}{4}-1$ independent terms, the -1 arising from the absence of a zero-body term in $(H-\mathcal{E})^2$. The explicit expression of the dispersion polynomial in terms of the one- and two-body matrix elements of H may be found in Ref. 1.

We next turn to the evaluation of these configuration moments throughout the $2s$-$1d$ shell. We employ the Rosenfeld interaction defined and used in the preceding sections. The three orbitals of this shell are denoted by the subscripts 5 ($1d_{5/2}$), 3 ($1d_{3/2}$), and 1 ($2s_{1/2}$). Configurations are specified by the set of three occupation numbers in the order $[m_5, m_3, m_1]$. The centroid polynomial consists of $\binom{5}{2}=10$ independent terms and the variance polynomial of $\binom{7}{4}-1=34$ independent terms, some of which vanish under the requirement of angular momentum conservation. The total number of configurations in the $2s$-$1d$ shell is 585. Inserting the single-particle energies and average interactions in Eq. (46) yields the following expression for the configuration centroid[5]

$$\begin{aligned}\mathcal{E}([\vec{m}]) &= m_5(-4.15)+m_3(0.93)+m_1(-3.28)\\ &+\tfrac{1}{2}m_5(m_5-1)(-0.79)+\tfrac{1}{2}m_3(m_3-1)(-0.78)\\ &+\tfrac{1}{2}m_1(m_1-1)(-3.01)+m_5m_3(-0.81)\\ &+m_5m_1(-0.40)+m_3m_1(-0.40)\ ,\end{aligned} \tag{49}$$

all energies being in MeV. Some results of the evaluation of configuration centroids and variances are given in Figs. 9 and 10.

We turn next to the use of this information about configuration distributions as a guide to making vector-space truncations. The idea is simply that knowledge of the strength and width of a distribution tells us the range of energy, about the centroid, over which that part of the vector space is important in the makeup of the eigenfunctions of H. The sum of squared overlaps of the components of $(N;[\vec{m}])$ with the eigenstates of H lying between E and E' is then approximated by $\bar{F}([\vec{m}];E')-\bar{F}([\vec{m}];E)$. We check this idea with an example $(2s\text{-}1d)^4$ for which the detailed spectroscopy is known. For this space the Rosenfeld interaction yields 27 distinct eigenvalues and 301 eigenvectors below $E=-25$ MeV, i.e., in the first 11 MeV of excitation energy. We evaluate each $\bar{F}([\vec{4}];-25)$ using a Gaussian frequency function in the manner of Eq. (17). In Table V we compare $\bar{F}([\vec{4}];-25)$ with the discrete distribution function $F([\vec{4}];-25)$ obtained by matrix diagonalization. It is found that the moment-generated frequency functions provide a reasonable measure of the relative importance of each subspace $(N;[\vec{m}])$ in the makeup of the low-lying eigenstates of H. The most serious disagreement in Table V involves the configuration [220]. The probable source of this error is indicated by the partial variances considered below. In summary, the use of a few low-order moments of spectral distributions would appear to serve as a useful guide for making vector-space truncations prior to a detailed spectroscopic calculation. Our use of a Gaussian frequency function has not been supported by any evaluation of third or fourth moments.

TABLE V. Comparison of the moment-generated and discrete configuration distribution functions at 11-MeV excitation energy for the space $(2s\text{-}1d)^4$.

Configuration[a]	Centroid[b] (MeV)	Width (MeV)	Strength	Moment-generated $\bar{F}(-25)$	Matrix-diagonalization F
004	−31.21	2.99	1	1.0	0.9
103	−24.22	2.96	48	18.9	18.0
400	−21.33	4.73	495	107.7	89.6
202	−20.25	4.05	396	47.2	47.2
301	−19.28	4.74	880	99.5	60.6
013	−19.14	2.96	32	0.8	1.0
310	−16.31	4.46	1760	45.0	37.0
112	−15.19	3.95	576	3.7	8.3
211	−14.24	4.56	2112	19.3	18.0
220	−11.24	4.42	1848	1.7	12.9
022	−10.08	4.11	168	0.0	0.7
121	−9.15	4.61	1344	0.4	5.0
130	−6.11	4.60	672	0.0	1.5
031	−4.01	4.90	224	0.0	0.2
040	−0.94	5.06	70	0.0	0.0
Total			10 626	345.2	301.

[a]The three occupation numbers are in order $d_{5/2}$, $d_{3/2}$, $s_{1/2}$.
[b]The ground-state energy is −36.01 MeV.

The present example should then be regarded as a lowest-order estimate to be improved upon if warranted by one's requirement of greater spectroscopic detail. It should be emphasized that for $(N;[\vec{m}])$ the information necessary for this lowest-order estimate is trivially obtained from the one- and two-body matrix elements.

B. Partial Variances and Symmetry Breaking

We have suggested various uses of the low-order moments of spectral distributions. To obtain more detailed spectroscopic information one may turn to higher-order moments or to the moments of even smaller subspaces of the original vector space. Another alternative is to consider the partial moments which exist among a set of subspaces. In this section we consider the partial variance which measures the mixing of the states of $(N;[\vec{m}])$ and $(N;[\vec{m}'])$. The contribution to the variance of the states of configuration $[m]$ due to the presence of configuration $[\vec{m}']$ is denoted by $\sigma^2([\vec{m}];[\vec{m}'])$ and is given by

$$\sigma^2([\vec{m}];[\vec{m}'])=d^{-1}([\vec{m}])\sum_{\alpha\beta}\langle[m]\alpha|H-\mathcal{E}([\vec{m}])|[m']\beta\rangle^2\ , \tag{50}$$

$$d([\vec{m}])\sigma^2([\vec{m}];[\vec{m}'])=d([\vec{m}'])\sigma^2([\vec{m}'];[\vec{m}])\,. \tag{51}$$

Summing the partial variances of $[\vec{m}]$ over all m-nucleon configurations included in the space reproduces the configuration variance

$$\sigma^2([\vec{m}])=\sum_{[\vec{m}']}\sigma^2([\vec{m}];[\vec{m}'])\ . \tag{52}$$

The partial variances may also be expressed as fourth-order polynomials in the occupation numbers of the various orbitals. A complete set of these partial variance polynomials is given in the preceding paper.[1]

As an example, we may consider the mixing of the states of $[\vec{m}]$ with the states of a configuration which differs from $[\vec{m}]$ by the addition of two $1d_{3/2}$ nucleons and two $2s_{1/2}$ holes. The contribution to the variance due to such a mixing of configurations is given by

$$\begin{aligned}\sigma^2([m_5,m_3,m_1];[m_5,m_3+2,m_1-2])\\ =\tfrac{1}{12}m_1(m_1-1)[1-\tfrac{1}{4}m_3+\tfrac{1}{56}m_3(m_3-1)]\\ \times[\sum_{JT}(2J+1)(2T+1)(W^{JT}_{1313})^2]\,.\end{aligned} \tag{53}$$

For the Rosenfeld interaction, the value of $\sum_{JT}(2J+1)(2T+1)(W^{JT}_{1313})^2$ is 1.5 MeV^2. The polynomial coefficients are evaluated in Table VI and display the effect on this partial variance of the availability of $2s_{1/2}$ particles and $1d_{3/2}$ holes in $[\vec{m}]$.

TABLE VI. Partial dispersion coefficients in Eq. (53).

m_1	$\frac{1}{12}m_1(m_1-1)$	m_3	$1-\frac{1}{4}m_3+\frac{1}{56}m_3(m_3-1)$
0	0.0	0	1.0
1	0.0	1	0.75
2	0.167	2	0.536
3	0.5	3	0.357
4	1.0	4	0.204
		5	0.107
		6	0.036
		7	0.0
		8	0.0

The use of operator averaging to measure the goodness of various group symmetries in $(N;\vec{m})$ has been explored by French.[18] In the present case of configurations in the $2s$-$1d$ shell, the partial variances $\sigma^2([\vec{m}];[\vec{m}'])$, with $[\vec{m}]\neq[\vec{m}']$, may be used to calculate the norm of the symmetry-breaking part of H for the group $U(12)\times U(8)\times U(4)$. Similarly, the partial variances $\sigma^2([\vec{m}];[\vec{m}])$ express the norm of the symmetry-preserving part of H after removal of that trivial part of H proportional to the Casimir operators of $U(12)\times U(8)\times U(4)$. Such measures, however, pay no attention to the difference in the centroids of representations $[\vec{m}]$ and $[\vec{m}']$. The importance to spectroscopy of the strong coupling of two configurations should be greatly reduced if their centroids are sufficiently separated in energy.

An approximate measure of this effect is suggested by perturbation theory. In lowest order the change in the energy of a state $|[m]\alpha\rangle$ due to the inclusion in the vector space of the states $|[m']\beta\rangle$ of a different configuration may be expressed as

$$\Delta E_\alpha([\vec{m}];[\vec{m}'])=\sum_\beta\frac{\langle[m]\alpha|H-\mathcal{E}([\vec{m}])|[m']\beta\rangle^2}{\mathcal{E}([\vec{m}'])-\mathcal{E}([\vec{m}])}\ , \tag{54}$$

where the usual energy denominator in Eq. (54) has been replaced by its average expressed as the difference between centroids. The average of $\Delta E_\alpha([\vec{m}];[\vec{m}'])$ over the states of $[\vec{m}]$ is denoted by $\Delta E([\vec{m}];[\vec{m}'])$ where

$$\Delta E([\vec{m}];[\vec{m}'])=\frac{\sigma^2([\vec{m}];[\vec{m}'])}{\mathcal{E}([\vec{m}']-\mathcal{E}([\vec{m}])}\,. \tag{55}$$

This expresses the average change in energy of a state $|[m]\alpha\rangle$ in the presence of the states of configuration $[\vec{m}']$. The ratio represented by $\Delta E([\vec{m}];[\vec{m}'])$ may be easily calculated for any pair of configurations via the polynomial expressions for partial variances and centroids.

Fifteen configurations arise from the distribution of four nucleons in the $2s$-$1d$ shell. In Table

TABLE VII. Four-nucleon partial dispersion measured in MeV^2 for configurations $[d_{5/2}d_{3/2}\,s_{1/2}]$. The configurations are in order of increasing centroid energy.

$[m']\backslash[m]$	004	103	400	202	301	013	310	112	211	220	022	121	130	031	040
004		0.00		0.01		0.00		0.01			0.01				
103	0.00			0.08	0.07	2.34		0.05	0.04		0.03	0.03			
400				0.37	0.43		2.16		0.06	2.10					
202	3.25	0.66	0.30		0.33	0.25	0.13	3.96	0.14	0.05	3.08	0.04			
301		1.35	0.77	0.72			0.25	0.14	3.75	0.06		2.56			
013	0.00	1.00		0.02				0.04	0.02		0.07	0.04		0.08	
310			7.67	0.58	0.49			0.45	0.57	4.36		0.14	7.69		
112	4.20	0.60		5.76	0.09	0.74	0.15		0.24	0.18	7.17	0.26	0.16	0.15	
211		1.92	0.25	0.73	9.00	1.62	0.68	0.89		0.51	0.34	7.34	0.18	9.22	
220			7.83	0.24	0.11		4.57	0.56	0.45		0.54	0.61	5.42	0.26	18.46
022	1.47	0.12		1.31		0.39		2.09	0.03	0.05		0.10	0.13	0.28	0.31
121		0.73		0.16	3.91	1.84	0.11	0.61	4.68	0.44	0.84		0.45	9.34	0.40
130							2.93	0.18	0.06	1.97	0.52	0.34		0.53	3.70
031						0.55		0.06	0.98	0.03	0.37	1.56	0.18		0.59
040										0.61	0.13	0.02	0.38	0.18	
$\sigma_p^2([m])$	0.00	1.81	5.56	6.44	8.03	1.03	8.97	6.54	9.81	9.20	3.78	8.21	6.44	3.95	2.08
$\sigma_b^2([m])$	8.92	6.95	16.81	9.99	14.44	7.74	10.98	9.05	11.02	10.36	13.11	13.04	14.59	20.04	23.46
$-\mathcal{E}([m])$	31.21	24.22	21.33	20.25	19.28	19.14	16.31	15.19	14.24	11.24	10.08	9.15	6.11	4.01	0.94

VII we display the matrix of partial variances for these configurations. Also displayed are two collections of partial variances representing the symmetry-preserving (σ^2_p) and symmetry-breaking (σ^2_b) contributions to the total variance

$$\sigma^2_p([\vec{m}]) = \sigma^2([\vec{m}];[\vec{m}]) \,, \tag{56}$$

$$\sigma^2_b([\vec{m}]) = \sum_{[\vec{m}'] \neq [\vec{m}]} \sigma^2([\vec{m}];[\vec{m}']) \,, \tag{57}$$

$$\sigma^2([\vec{m}]) = \sigma^2_p([\vec{m}]) + \sigma^2_b([\vec{m}]) \,. \tag{58}$$

We observe that the total variance $\sigma^2([\vec{m}])$ is generally dominated by the symmetry-breaking component σ^2_b. The configuration widths $\sigma([\vec{m}])$ may thus be regarded as essentially pure mixing widths. Two classes of partial variances have been underlined in Table VII. These classes identify pairs of configurations which differ by the transfer of one (singly underlined) or two (doubly underlined) nucleons between the $1d_{5/2}$ and $1d_{3/2}$ orbitals. These two classes are observed to include most of the large contributions to σ^2_b, thus indicating the part of the Rosenfeld interaction mainly responsible for strong coupling of different configurations in the case of four nucleons. The importance of these two classes of coupled configurations is further enhanced when one considers the energy shifts $\Delta E([\vec{m}];[\vec{m}'])$ for pairs of configurations which are sufficiently separated in energy as to make Eq. (55) meaningful. In particular, the strong coupling $\sigma^2([400];[220]) = 7.83\ \text{MeV}^2$ and energy shift $\Delta E([400];[220]) = 0.77$ MeV help to explain our underestimate in Table V of the importance of the [220] configuration in the low-lying spectrum.

We have clearly chosen the most trivial example of a group for discussing ideas like symmetry breaking and the relative importance of various irreducible representations. Nevertheless, a general approach to the discussion of group symmetries in spectroscopy has been constructed along these lines by French[18] and is now being applied to several cases.

V. GROUND STATE OF O^{16}

In this section we apply moment techniques to an investigation of the ground state of O^{16}. The importance of the filled $1s$ and $1p$ orbitals (hereafter referred to as the core state) in the 16-particle

TABLE VIII. Diagonal and off-diagonal single-particle energies using the Kuo-Brown matrix elements.

Single nucleon orbitals	Kinetic energy (MeV)	Potential energy (MeV)	$\bar{\epsilon}$ (MeV)	Experiment[a]
$1s_{1/2}$	10.50	−56.28	−45.78	−47.7
$1p_{3/2}$	17.50	−39.13	−21.63	−21.82
$1p_{1/2}$	17.50	−35.42	−17.92	−15.67
$1d_{5/2}$	24.50	−25.51	−1.01	−4.15
$2s_{1/2}$	24.50	−25.21	−0.71	−3.28
$1d_{3/2}$	24.50	−20.47	4.03	0.93
$2p_{3/2}$	31.50	−19.95	11.55	3.57
$2p_{1/2}$	31.50	−18.64	12.86	5.58
$1f_{7/2}$	31.50	−16.50	15.00	1.55
$1f_{5/2}$	31.50	−11.65	19.85	7.74
$1s_{1/2}\,2s_{1/2}$	8.57	−14.36	−5.79	
$1p_{3/2}\,2p_{3/2}$	11.07	−15.55	−4.48	
$1p_{1/2}\,2p_{1/2}$	11.07	−14.67	−3.60	

[a]Empirical single-particle energies taken from Spicer and Eisenberg. See Ref. 22.

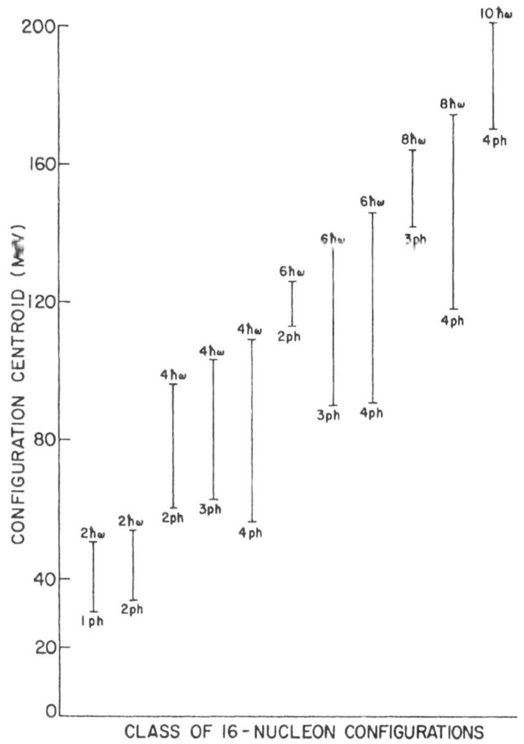

FIG. 11. Range of centroids of classes of 16-nucleon positive-parity configurations. All centroids are measured relative to the core consisting of filled 1s and 1p shells. Each class of configurations is identified by the number of excited particle-hole pairs and the shell-model excitation energy in units of $\hbar\omega$. Thus the 4p-4h $4\hbar\omega$ configuration centroids are found between 55 and and 110 MeV above the s^4p^{12} centroid.

ground state is strongly suggested by the large depression of the core centroid below that of other 16-particle configurations. It is then argued that the ground-state energy and the squared amplitude of the core component ($\equiv s^4p^{12}$) in the ground-state wave function may be inferred from the spectral distribution of the core state. Configuration centroids, widths, and partial variances are used in conjunction with a model of the core frequency function to obtain estimates of these two ground-state parameters.

Our primary interest is to understand how the ground state depends upon our choice of the finite vector space that underlies a spectroscopic calculation. To this end we adopt an intentionally naive attitude toward spectroscopy, and neglect any modification of the effective interaction generated by the act of vector-space truncation. Thus, given a vector space consisting of the core state and a class of excited configurations, we ignore any piece of the Hamiltonian which takes us outside this vector space and simply ask what ground state would result from diagonalizing H in this finite vector space. We then investigate what happens to the ground state as we enlarge the class of configurations that defines our vector space. Spectral distributions are particularly suited for handling such questions.

The interaction used in this section is derived from the nucleon-nucleon potential of Hamada and Johnson.[19] Specifically, we use the reaction matrix elements of this potential as calculated by Kuo and Brown[20] which have been suggested as being realistic two-body matrix elements for use in spectroscopic calculations. These matrix elements are calculated in a harmonic-oscillator representation with $\hbar\omega = 14$ MeV. We consider vector spaces containing as many as 10 active orbitals (80 single-particle states), and therefore use the unrenormalized version of these matrix elements.[21] The total dynamical information of our system is contained in the 2304 two-body matrix elements relating the 1s, 1p, 2s-1d, and 2p-1f major shells.

A. Core Centroid and Partial Dispersions

The distribution of 16 nucleons among the 80 available single-particle states of the first 10 shell-model orbitals generates a $\binom{80}{16}$-dimensional vector space. This space may be decomposed into subspaces corresponding to shell-model configurations. It is convenient to make the usual transformation to particle-hole notation relative to the core (s^4p^{12}) as a reference state. Particle-hole configurations are labeled by $[\vec{h}|\vec{m}]$ denoting the collection of occupation numbers h_α and m_r for hole orbital α and particle orbital r, respectively. The core state itself then carries the label $[\vec{0}|\vec{0}]$.

In the 10-orbital vector space, the configuration-centroid polynomial consists of $\binom{12}{2} = 66$ independent terms,

$$\mathcal{E}([\vec{h}|\vec{m}]) = \mathcal{E}([\vec{0}|\vec{0}]) + \sum_r m_r \bar{\epsilon}_{rr} - \sum_\alpha h_\alpha \bar{\epsilon}_{\alpha\alpha} + \sum_r \tfrac{1}{2} m_r (m_r - 1)\overline{W}_{rr} + \sum_\alpha \tfrac{1}{2} h_\alpha (h_\alpha - 1)\overline{W}_{\alpha\alpha}$$
$$+ \sum_{r\neq s} \tfrac{1}{2} m_r m_s \overline{W}_{rs} + \sum_{\alpha\neq\beta} \tfrac{1}{2} h_\alpha h_\beta \overline{W}_{\alpha\beta} - \sum_{\alpha r} h_\alpha m_r \overline{W}_{\alpha r}\,. \tag{59}$$

The average interaction $\overline{W}$ has been previously defined in Eq. (48). The single-particle energy $\bar{\epsilon}$ is given by

$$\bar{\epsilon}_{rs} = \langle r|\mathcal{T}|s\rangle + \frac{1}{2(2j_r+1)} \sum_{\beta J T} \xi_{r\beta}^{-1}\xi_{s\beta}^{-1}(2J+1)(2T+1)\langle r\beta, JT|V|S\beta, JT\rangle\hat{\delta}_{rs} , \tag{60}$$

where

$$\begin{aligned} \xi_{rs}^{-1} &= 1 \quad (r \neq s) \\ &= 2^{1/2} \quad (r = s) , \end{aligned} \tag{61}$$

and $\hat{\delta}$ distinguishes between orbits only according to their tensorial rank, i.e.,

$$\hat{\delta}_{rs} = \delta(j_r, j_s) . \tag{62}$$

The kinetic energy and potential energy operators are denoted by $\mathcal{T}$ and V, respectively. The single-particle energies in this 10-orbital space are tabulated in Table VIII along with a separate listing of their contributions from $\mathcal{T}$ and V. The experimental single-particle energies in Table VIII are taken from the work of Spicer and Eisenberg.[22] These experimental numbers represent excitation energies relative to the true O^{16} ground state and are derived mostly from stripping and pickup reactions. Strictly speaking, the theoretical energies $\bar{\epsilon}$ are centroid energies of 15- and 17-particle configurations relative to the core state as shown in Sec. IX of the preceding paper. These two sets of single-particle energies show good correlation within a major shell. The theoretical energies, however, show a greater spacing between major shells.

The three off-diagonal single-particle energies of Table VIII are each less than 0.15 of the difference between the corresponding diagonal energies. This suggests that the admixture of the core (s^4p^{12}) in the theoretical ground state should approximate that of the Hartree-Fock state in this space.

In Fig. 11 we summarize the calculated centroids of various classes of 16-particle positive-parity configurations. The centroids in Fig. 11 are expressed relative to the core centroid as the energy zero. The absolute energy of the core centroid is given by

$$\mathcal{E}([\vec{0}|\vec{0}]) = \sum_{\alpha} 2(2j_\alpha+1)\bar{\epsilon}_{\alpha\alpha} - \tfrac{1}{2}\sum_{\alpha\beta} \xi_{\alpha\beta}^{-2} \sum_{JT}(2J+1)(2T+1)W^{JT}_{\alpha\beta\alpha\beta} , \tag{63}$$

where W is the coupled, antisymmetrized, two-body matrix element of the interaction V. The core centroid is found to have the value -87.9 MeV. The empirical ground-state energy of O^{16} is -127.6 MeV. The Coulomb interaction contributes about 18 MeV to the empirical energy. Thus the core centroid is perhaps 58 MeV above the true ground-state energy of O^{16} after Coulomb subtraction.

In Fig. 11 we observe that the core centroid lies some 30 MeV below the nearest centroids of configurations with which it may mix. This large energy gap strongly argues for the importance of the core component in the 16-particle ground state of the Kuo-Brown Hamiltonian. We denote by I_c the squared amplitude of the core component in the theoretical ground state. The remaining intensity of the core state, $1-I_c$, is distributed over the many excited 0^+ states. The pth central moment of the core distribution,

$$\langle \text{core}|[H - \mathcal{E}([\vec{0}|\vec{0}])]^p|\text{core}\rangle ,$$

includes a contribution $I_c[E_g - \mathcal{E}([\vec{0}|\vec{0}])]^p$ from the core component I_c at the ground-state energy E_g. This contribution should dominate the pth moment by reason of the presumed size of I_c and the magnitude of the energy difference $E_g - \mathcal{E}([\vec{0}|\vec{0}])$. Thus, from the moments of the core-state spectral distribution, one may hopefully learn about the ground-state energy E_g and its core component I_c.

To proceed with this program we next consider the variance of the core configuration. The configuration dispersion polynomial is given in analytic form in the preceding paper.[1] It is a fourth-order polynomial in the occupation numbers of the various single-particle orbitals. For our 10-orbital space the dispersion polynomial would have $\binom{14}{4} - 1 = 1000$ independent terms. However, greater insight into the spectroscopic problem is obtained by concentrating on the more detailed partial dispersion polynomials which are also listed in the preceding paper. These polynomials become trivial in the case of a configuration consisting of filled orbitals and are given by

$$\sigma^2(\text{core:}0) = 0 , \tag{64}$$

$$\sigma^2(\text{core:}[\vec{\alpha}|\vec{r}]) = 2(2j_\alpha+1)(\bar{\epsilon}_{\alpha r})^2\delta_{\alpha r} , \tag{65}$$

$$\sigma^2(\text{core:}[\vec{\alpha}^2|\vec{r}^2]) = \tfrac{1}{2}\sum_{JT}(2J+1)(2T+1)(W^{JT}_{\alpha\alpha rr})^2 , \tag{66}$$

$$\sigma^2(\text{core}:[\vec{\alpha}^2|\vec{r}\vec{s}]) = 2^{-1/2}\sum_{JT}(2J+1)(2T+1)(W^{JT}_{\alpha\alpha rs})^2 , \tag{67}$$

$$\sigma^2(\text{core}:[\vec{\alpha}\vec{\beta}|\vec{r}^2]) = 2^{-1/2}\sum_{JT}(2J+1)(2T+1)(W^{JT}_{\alpha\beta rr})^2 , \tag{68}$$

$$\sigma^2(\text{core}:[\vec{\alpha}\vec{\beta}|\vec{r}\vec{s}]) = \sum_{JT}(2J+1)(2T+1)(W^{JT}_{\alpha\beta rs})^2 . \tag{69}$$

In Eqs. (64)–(69) the notation $\sigma^2(\text{core}:[\vec{\alpha}^2|\vec{r}\vec{s}])$ denotes the partial variance arising from the coupling of the core to the states of a configuration with two holes in orbital α and one particle each in orbitals r and s. The partial variances of the core configuration are tabulated in Table IX.

If the Kuo-Brown Hamiltonian were diagonalized in a vector space consisting of the core state and any set of the excited configurations of Table IX, then the total variance of the core state in that vector space would be the sum of the partial variances to that set of configurations. Thus, the variance of the core state when the entire 10 single-particle orbitals are active is 2295.3 MeV^2, representing the combined influence of 3 1p-1h and 74 2p-2h configurations. If the space is truncated such that holes are not excited in the $1s$ shell, then the core variance is reduced to 1187.6 MeV^2. A vector-space truncation which keeps only the 2p-2h $2\hbar\omega$ configurations reduces the core variance to 539.6 MeV^2. The 1p-1h $2\hbar\omega$ configurations that would be eliminated in a proper Hartree-Fock representation are responsible for 346.4 MeV^2 of the total variance of the core state in the 10-orbital space.

The contribution of excited configurations to the ground-state energy may be estimated by perturbation theory. In second order a set of $T=0$, $J=0^+$ states $|i\rangle$ lowers the ground-state energy by an amount given by

$$E^{(2)} = \sum_i \frac{|\langle \text{core}|H - \mathscr{E}([\vec{0}|\vec{0}])|i\rangle|^2}{E_i - \mathscr{E}([\vec{0}|\vec{0}])} , \tag{70}$$

where E_i is the centroid of state $|i\rangle$, i.e., $\langle i|H|i\rangle$. If the energy denominators are replaced by an average, i.e., the configuration centroid of these states, then Eq. (70) reduces to the ratio of the partial dispersion to the centroid difference. In this approximation the contribution $\Delta E_g([\vec{h}|\vec{m}])$ of states of the configuration $[\vec{h}|\vec{m}]$ to the ground-state energy is given by

$$\Delta E_g([\vec{h}|\vec{m}]) = \frac{\sigma^2(\text{core}:[\vec{h}|\vec{m}])}{\mathscr{E}([\vec{h}|\vec{m}]) - \mathscr{E}([\vec{0}|\vec{0}])} . \tag{71}$$

These contributions are exhibited in Table IX. The difference between the theoretical ground-state energy and the core centroid is then obtained by summing the contributions of Eq. (71) over all configurations included in the vector space

$$\mathscr{E}([\vec{0}|\vec{0}]) - E_g = \sum_{[\vec{h}|\vec{m}]} \Delta E_g([\vec{h}|\vec{m}]) . \tag{72}$$

For the entire 10-orbital space the ground state is then estimated to lie 38.2 MeV below the core centroid. The three 1p-1h configurations contribute 9.2 MeV of this total. The remaining 29.0 MeV arises from five classes of 2p-2h configurations. A total of 12.0 MeV arises from the seventeen $2\hbar\omega$ configurations involving the $1p$ and $2s$-$1d$ shells. The five $4\hbar\omega$ configurations involving holes in the 1s shell and particles in the $2s$-$1d$ shell gives 1.8 MeV. Another 6.0 MeV comes from the twenty-six $4\hbar\omega$ configurations having $1p$-shell holes and $2p$-$1f$-shell particles. The nineteen $4\hbar\omega$ configurations in which the $1s$ and $1p$ shells each have one hole and the $2s$-$1d$ and $2p$-$1f$ shells each have one particle contribute 8.0 MeV to the ground-state binding energy. Finally the seven $6\hbar\omega$ configurations supply the remaining 1.2 MeV.

This additional energy of 38.2 MeV combined with the core centroid at −87.9 MeV would yield a predicted ground-state binding energy of 126.1 MeV for the 16-nucleon system in the 10-orbital vector space. This is to be compared with the empirical value of 143.6 MeV (after Coulomb subtraction). These estimated contributions to the ground-state binding energy are probably too low. In replacing the energy denominators in Eq. (70) by the configuration centroid, we have used an average energy that heavily weights $T \neq 0$ states of the configuration that lie at a higher excitation energy. Our energy denominators are then presumably too large and thus our estimate of the theoretical ground-state binding energy should probably be regarded as a lower bound.

In Fig. 12 we summarize the contribution of configurations to the ground-state energy. Here we plot the energy difference $\mathscr{E}([\vec{0}|\vec{0}]) - E_g$ as a function of the centroid of configurations in our 10-orbital space. The major contributions to the ground-state energy are found in two groups centered about 40 and 80 MeV (centroid excitation energies relative to the core). These correspond to $2\hbar\omega$ and $4\hbar\omega$ configurations, respectively. The contribution of the few $6\hbar\omega$ configurations is evident beyond 110 MeV. If the vector space were enlarged beyond the first 10 orbitals, then further contributions to E_g would be made by the additional

TABLE IX. Partial dispersions of core due to excited configurations which are identified by their excitation (units of $\hbar\omega$), the particle orbitals involved, the hole orbitals involved, and the configuration centroid relative to the core centroid. The first three entries involve 1p-1h configurations. As an example, a contribution of 34 MeV^2 to the core variance is made by the configuration $(1p_{3/2})^{-2}(1d_{5/2})^2$ whose centroid lies 43 MeV above that of the core. The last two columns record ratios of partial variance to centroid spacing which are used in the evaluation of Eq. (71) and Eq. (72), respectively. Active particle orbitals are labeled by 1 $(1d_{5/2})$, 2 $(2s_{1/2})$, 3 $(1d_{3/2})$, 4 $(2p_{3/2})$, 5 $(2p_{1/2})$, 6 $(1f_{7/2})$, 7 $(1f_{5/2})$. Active hole orbitals are labeled by 1 $(1s_{1/2})$, 2 $(1p_{3/2})$, 3 $(1p_{1/2})$.

Excitation	Particle orbitals	Hole orbitals	σ^2 (MeV^2)	$\Delta\mathcal{E}$ (MeV)	$\frac{\sigma^2}{\Delta\mathcal{E}}$ (MeV)	$\left(\frac{\sigma}{\Delta\mathcal{E}}\right)^2 \times 10^3$
2	2	1	134.0	47.6	2.8	59.1
2	4	2	160.6	34.2	4.7	137.3
2	5	3	51.7	31.4	1.6	52.4
2	11	22	34.1	43.3	0.8	18.1
2	11	33	22.0	38.9	0.6	14.5
2	33	22	47.5	54.4	0.9	16.1
2	33	33	0.3	45.2	0.0	0.2
2	22	22	1.0	42.0	0.0	0.6
2	22	33	7.0	33.7	0.2	6.2
2	11	23	20.6	40.3	0.5	12.6
2	33	23	16.4	49.0	0.3	6.8
2	22	23	10.6	37.0	0.3	7.7
2	13	22	103.9	48.2	2.2	44.6
2	13	33	3.1	41.4	0.1	1.8
2	12	22	14.6	43.8	0.3	7.6
2	23	22	4.2	49.2	0.1	1.7
2	23	33	4.6	40.5	0.1	2.8
2	13	23	216.1	44.0	4.9	111.4
2	12	23	0.8	39.8	0.0	0.5
2	23	23	32.7	44.1	0.7	16.8
4	11	11	4.7	87.3	0.0	0.6
4	33	11	6.4	95.7	0.1	0.7
4	22	11	40.4	89.6	0.4	5.0
4	13	11	41.8	90.9	0.5	5.1
4	23	11	71.9	93.7	0.8	8.2
4	66	22	6.9	74.0	0.1	1.2
4	66	33	6.7	68.7	0.1	1.4
4	77	22	11.5	83.6	0.1	1.6
4	77	33	1.1	76.0	0.0	0.2
4	44	22	24.7	67.0	0.4	5.5
4	44	33	3.0	60.7	0.0	0.8
4	55	22	3.0	71.5	0.0	0.6
4	55	33	2.0	62.0	0.0	0.5
4	66	23	3.2	70.6	0.0	0.6
4	77	23	7.3	79.0	0.1	1.2
4	44	23	7.1	63.0	0.1	1.8
4	55	23	6.8	65.9	0.1	1.5
4	67	22	47.4	78.3	0.6	7.7
4	67	33	8.5	71.9	0.1	1.6
4	46	22	1.2	70.8	0.0	0.2
4	56	22	1.1	72.7	0.0	0.2
4	47	22	46.8	75.5	0.6	8.2
4	47	33	0.3	68.5	0.0	0.0
4	57	22	20.1	77.8	0.3	3.3
4	45	22	7.2	68.7	0.1	1.5
4	45	33	6.6	60.8	0.1	1.8
4	67	23	99.1	74.3	1.3	17.9
4	46	23	1.4	67.2	0.0	0.3
4	47	23	69.8	71.2	1.0	13.8
4	57	23	4.0	72.7	0.1	0.8
4	45	23	38.9	64.0	0.6	9.5
4	16	12	10.4	82.5	0.1	1.5
4	16	13	12.0	80.3	0.2	1.9

TABLE IX *(Continued)*

Excitation	Particle orbitals	Hole orbitals	σ^2 (MeV^2)	$\Delta\mathcal{E}$ (MeV)	$\frac{\sigma^2}{\Delta\mathcal{E}}$ (MeV)	$\left(\frac{\sigma}{\Delta\mathcal{E}}\right)^2 \times 10^3$
4	36	12	84.4	86.7	1.0	11.2
4	17	12	50.1	86.7	0.6	6.7
4	17	13	88.9	83.9	1.1	12.6
[illegible]	[illegible]	[illegible]	[illegible]	[illegible]	[illegible]	[illegible]
4	37	13	5.2	88.0	0.1	0.7
4	27	12	80.0	88.2	0.9	10.3
4	14	12	6.7	80.2	0.1	1.0
4	14	13	15.4	77.7	0.2	2.5
4	34	12	56.4	84.9	0.7	7.8
4	34	13	92.2	81.2	1.1	14.0
4	24	12	41.8	80.8	0.5	6.4
4	24	13	21.4	77.2	0.3	3.6
4	15	12	33.0	81.1	0.4	5.0
4	35	12	12.1	86.1	0.1	1.6
4	35	13	5.8	81.7	0.1	0.9
4	25	12	21.4	82.6	0.2	3.1
4	25	13	12.6	78.3	0.2	2.0
6	66	11	0.6	116.9	0.0	0.0
6	77	11	1.6	125.4	0.0	0.1
6	44	11	4.8	113.2	0.0	0.4
6	55	11	29.4	114.4	0.3	2.2
6	67	11	10.3	120.7	0.1	0.7
6	47	11	44.8	119.6	0.4	3.1
6	45	11	49.1	113.4	0.4	3.8

2p-2h configurations with $4\hbar\omega$, $6\hbar\omega$, etc. excitation energy. It is not clear just how much Fig. 12 would be changed by opening up the vector space to include additional orbitals. For that reason one cannot conclude from Fig. 12 that one is close to the asymptotic ground-state energy of the Brown-Kuo Hamiltonian in the absence of truncation. One may also question the use of such effective interaction matrix elements as the vector space is indefinitely enlarged.

An additional feature of Table IX is that the binding-energy contributions from the 2p-2h configurations are found to be widely distributed among many configurations. A sum of the contributions of the 9 configurations with $\Delta E_g > 0.8$ MeV yields only 14.3 MeV of the total 29.0 MeV contributed by all 2p-2h configurations. A total of 33 of the 74 configurations with $\Delta E_g > \frac{1}{4}$ MeV yields 25.8 MeV of the binding energy. This cooperative effect of many configurations in the determination of the binding energy may be said to be evidence of the collective nature of the ground state, but this is more correctly a statement about the representation which is being used. It is also found that those configurations that make the biggest contribution do not share a common simple relationship, e.g. via pairing excitations, to the core state. There is no indication of a class of configurations that would be expected to be strongly coupled to the core if additional major shells were included in our vector space.

B. Model for Core Distribution

We next turn our attention to estimating the squared amplitude (I_c) of the core state in the ground-state wave function. As in the case of the binding energy, our interest is in the dependence of I_c on the vector space in which H is diagonalized. We proceed by adopting a model for the frequency function of the core distribution. The parameters of this frequency function are then fixed by calculated moments of the distribution. Our model is based on the idea that the Hartree-Fock state is a good first approximation to the ground state and that the core state is closely related to the Hartree-Fock state for closed-shell nuclei. This idea is reflected in our model frequency function shown in Fig. 13. The energy zero is chosen to occur at the core centroid. The ground state occurs at an energy E_0 below the centroid and has a squared amplitude I_c with the core state. The remaining core intensity, $1 - I_c$, is distributed with constant density among the excited 0^+ states up to a cutoff energy E_c above the core centroid. The assumption of constant density in Fig. 13 reflects the combined effects with increasing excitation energy of decreasing core-state amplitudes and an increasing density of 0^+ states. The abrupt cutoff of the frequency function reflects the finite size of the vector space and should occur in the region of energy where the rapid increase in the density of 0^+ states terminates. We would estimate that the cutoff should occur roughly in the vicinity of the centroids of those configurations with the greatest statistical weight. Our model frequency function, $f(E)$, has the form

$$f(E) = I_c\delta(E+E_0) + \frac{1-I_c}{E_c+E_0}\,, \quad -E_0 \le E \le E_c\,, \qquad (73)$$
$$f(E) = 0 \quad \text{otherwise}\,.$$

The moments of the distribution are related to $f(E)$ by

$$1 = \int_{-\infty}^{\infty} f(E)dE\,, \qquad (74)$$

$$0 = \int_{-\infty}^{\infty} Ef(E)dE\,, \qquad (75)$$

$$\sigma^2 = \int_{-\infty}^{\infty} E^2 f(E)dE\,, \qquad (76)$$

$$\mu_k = \int_{-\infty}^{\infty} E^k f(E)dE\,. \qquad (77)$$

Substituting the model Eq. (73) into Eqs. (75)–(77) we generate a family of parameter values (E_0, E_c, I_c) consistent with a given value of the width σ. This parameter space is displayed in Fig. 14.

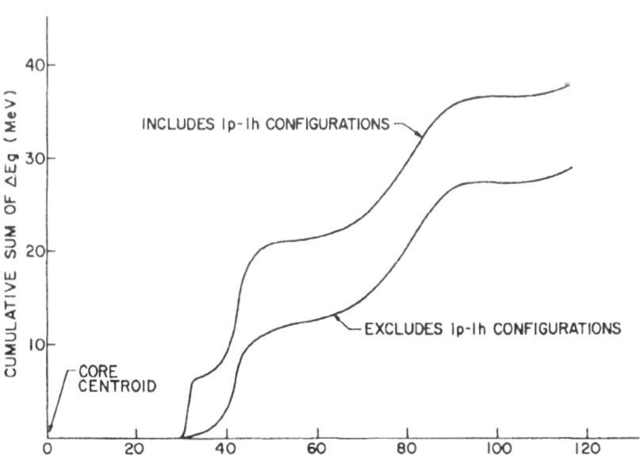

FIG. 12. Cumulative sum of the contributions of excited configurations to the 16-nucleon ground-state energy as evaluated in Eq. (71). The two curves differ by the inclusion of the three 1p-1h configurations which would be eliminated in a Hartree-Fock representation. As an example, a vector space consisting of all 2p-2h configurations whose centroids lie within 80 MeV of the core centroid would result in a theoretical ground state lying 20 MeV below the core centroid. This ground state would be lowered another 10 MeV by the inclusion of the 1p-1h configurations.

As the vector space is enlarged by the addition of 2p-2h configurations at high excitation energy, the core variance increases monotonically. In terms of our model, this increased variance implies that core intensity will be found at higher energies, i.e., E_c increases. However this must be accompanied in general by an increase in E_0 and decrease in I_c sufficient to leave the position of the centroid invariant. This interplay of the model parameters with increasing width is evident in Fig. 14. Given the core variance consistent with a particular vector space we may obtain limits on the ground-state parameters E_0 and I_c. As an example, the six-orbital vector space in which the $2p$-$1f$ shell is empty has a core width of 29.0 MeV. For this case a lower limit ($E_c > 50$ MeV) is established on the cutoff energy by our model. From the centroids of configurations included in this space and listed in Fig. 11, one may expect an upper limit on E_c of about 100 MeV. Interpreting such a limit in terms of Fig. 14 then requires that the ground state must lie at least 14 MeV below the core centroid and the squared overlap I_c must be less than 0.75. When the vector space is expanded to 10 orbitals the core width grows to 47.9 MeV. A similar analysis for this space in terms of Fig. 14 yields $E_c < 130$ MeV, $E_0 > 30$ MeV, and $I_c < 0.62$.

It is clear that knowledge of any higher moments of the core distribution would eliminate ambiguity in the values of the model parameters. It would perhaps also indicate necessary modifications in our assumed model for the frequency function. In the absence of knowledge of moments of greater than second order we use our perturbation estimate from Eq. (72) of $\mathcal{E}([\vec{0}|\vec{0}]) - E_g$ to determine E_0, and then use our model to determine the value of I_c and E_c. In Table X we record the results of this analysis for various vector spaces which both include and omit the off-diagonal single-particle energies $\bar{\epsilon}$. The value of E_c in each case is consistent with the centroids of the highest configurations included in that space. These results indi-

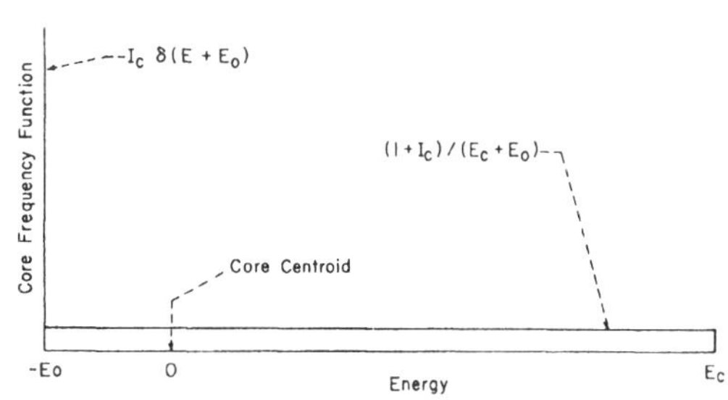

FIG. 13. The model for the core frequency function defined in Eq. (73).

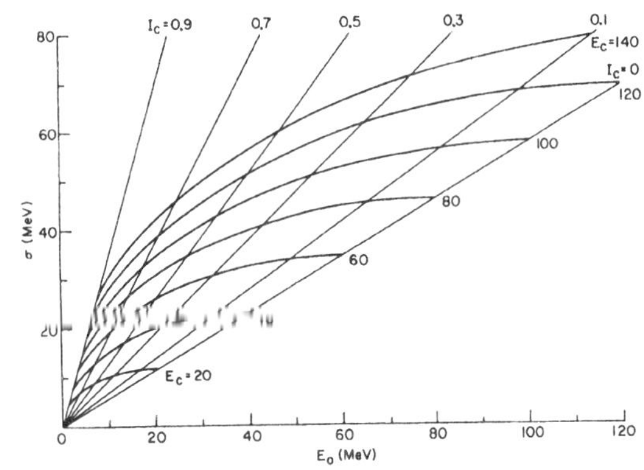

FIG. 14. Parameter space of the model frequency function of Eq. (73) indicating possible solutions of Eqs. (75)–(77). The two grids correspond to constant values of I_c, the squared amplitude of the core component in the ground state, and of E_c (MeV), the cutoff parameter.

cate the value of the Hartree-Fock analysis as a first approximation, as well as the necessity for going beyond such an analysis to obtain a reasonable approximation to the ground state.

Also shown in Table X are the perturbation estimates for I_c obtained from the last column of Table IX and denoted here by I_c'

$$I_c' = \exp\left(-\sum \frac{\sigma^2(\mathrm{core}:[\bar{h}|\bar{m}])}{\{\mathcal{E}([\vec{0}|\vec{0}]) - \mathcal{E}([\vec{h}|\vec{m}])\}^2}\right) . \tag{78}$$

By using centroid differences for the energy denominators as in Eq. (71), we have averaged the terms in a linked cluster perturbation expansion for the squared amplitude of the core component in the ground state. The good correlation between I_c and I_c' as reported in Table X is an argument for the reasonableness of the model we have used for the core distribution. Alternative models for the core frequency function were examined which increased asymptotically in a smooth way (e.g.) a Lorentzian tail at high excitation energies). All such models required that the core frequency function extend to excitation energies that were judged to be incompatible with the finite vector space being used.

We have investigated the dependence of moments of spectral distributions upon the size of the vector space underlying a spectroscopic calculation. With the assumption that the ground state of the 16 nucleon system is dominated by the closed 1*s* and 1*p* shells, we infer, from the lowest moments of this core configuration, the dependence of the ground state upon the vector space in which the Hamiltonian is diagonalized. All the examples presented are for large but finite vector spaces. The infinite vector space has not been considered. The asymptotic values or rates of divergence of a few low-order moments of the core (or better yet of the Hartree-Fock state) should be investigated with the idea of establishing limits on the ground-state energy.

C. Role of Complex Configurations

We next consider the role that *n*p-*n*h configurations ($n > 2$) might play in the determination of the 16-particle ground state. The influence of such configurations upon the moments of the core distribution is regulated by the two-body nature of the interaction. In the evaluation of the *k*th central moment $\langle \mathrm{core}|(H-\mathcal{E})^k|\mathrm{core}\rangle$ the inclusion of 3p-3h and 4p-4h configurations will only affect moments of order four and higher, 5p-5h and 6p-6h configurations contribute beginning with the sixth central moment, etc. Thus if H is diagonalized in two vector spaces which differ only by the inclusion in one of them of an arbitrary number of 3p-3h, 4p-4h, etc. configurations, the two distributions of

TABLE X. Parameters of the model frequency function of Eq. (73) representing solutions of Eqs. (75)–(77) for vector spaces of various size, with and without the inclusion of off-diagonal single-nucleon energies.

Active particle shells in space	Active hole shells in space	σ (MeV)	E_0 (MeV)	I_c	I_c' [a]	E_c (MeV)
2*s*-1*d*	1*p*	23.2	12.0	0.72	0.76	73.3
2*s*-1*d*	1*s*, 1*p*	29.0	16.6	0.67	0.70	84.3
2*s*-1*d*, 2*p*-1*f*	1*p*	34.5	24.3	0.56	0.58	85.6
2*s*-1*d*, 2*p*-1*f*	1*s*, 1*p*	47.9	38.2	0.48	0.48	110.0
All $\bar{\epsilon}$(off diagonal) = 0						
2*s*-1*d*	1*p*	23.2	12.0	0.72	0.76	73.3
2*s*-1*d*	1*s*, 1*p*	26.6	13.8	0.72	0.75	83.8
2*s*-1*d*, 2*p*-1*f*	1*p*	31.2	18.0	0.67	0.70	90.1
2*s*-1*d*, 2*p*-1*f*	1*s*, 1*p*	44.1	29.0	0.60	0.62	115.1

[a] Linked cluster perturbation estimate of squared amplitude of the core in the ground state. See Eq. (78).

the core state must be identical in their first three moments. The contribution of the ground state to the third moment of the core distribution is $I_c[E_g - \mathcal{E}([\vec{0}|\vec{0}])]^3$. Thus, if the inclusion of, say, 4p-4h configurations were to lower the ground-state energy {i.e., increase $\mathcal{E}([\vec{0}|\vec{0}]) - E_g$}, then the core amplitude in the ground state would be expected to decrease even more rapidly in order to leave the third moment invariant. These considerations suggest that, in the determination of the theoretical ground state, it is more important to include high-lying 2p-2h configurations than lower-lying configurations of a more complicated nature. Detailed treatment of the more complex configurations must await the computation of higher moments of the core distribution. It may be that the value of third and fourth moments will prove incompatible with our model frequency function even in the finite space being considered. More optimistically, we note that such ambiguities may be less important as the complexity of the configuration increases. Presumably two frequency functions with a common set of moments up to order four should be very similar in spite of differences in higher-order moments. Thus, the consideration of higher and higher moments is not only uninteresting but probably also unnecessary.

D. Width of Excited Configurations

Inserting the Kuo-Brown two-body matrix elements into the configuration dispersion polynomial, one can quickly generate the width of any configuration in the 10-orbital space being considered. It is found that these configuration widths are very uniform. In the 10-orbital space some 5000 configuration widths were observed to be distributed about an average of some 42 MeV, this distribution itself having a width of about 2 MeV. The dimensionalities of these configurations varied from 1 to 10^8 without correlation between width and dimensionality. This fact suggests that these widths are dominated by mixing rather than by the distribution of centroids of the states of each configuration. It should be emphasized that configuration variances are independent of the single-particle energies $\bar{\epsilon}_{rr}$. These diagonal matrix elements reappear, of course, in the third- and higher-order moments. Thus the average mixing width of perhaps 40 MeV which is found for configurations in the 10-orbital space should be compared with the 14-MeV major shell spacing of this shell-model representation. Such a comparison would argue against the usual shell-model truncation to one major shell as being a sufficient representation of the many-body problem for states of low excitation energy. Of course, this well-known fact merely requires that one must construct the appropriate effective interaction[23-25] as the penalty for keeping only a small vector space. The result of such investigations may merely emphasize the inadequacy of a representation based on uncorrelated single-particle states for approaching the true many-body problem. Indeed, Bethe[26] has emphasized that one may expect the theoretical ground state to have vanishingly small overlap with the true ground state. The extension of moment techniques to spaces with intrinsic correlations built into it should be interesting but must be deferred to the future.

Configuration centroids and widths may also be used with frequency functions in the manner of preceding sections to approximate the scalar density of states in the vector space. The state densities usually considered would require moments of JT spectral distributions. However, we recognize that the large ratio of configuration width to major shell spacing suggests that the state density retains little memory of the underlying shell structure with increasing excitation energy.

VI. CONCLUSIONS

The usefulness of spectral distributions and moment techniques has been demonstrated. The success of an approximate spectroscopy based on a few low-order moments is at first surprising. This success is probably due to the relative infrequency with which the eigenvalues of an invariant subspace are found to be bunched into clusters which are themselves widely separated in energy. This fact has been shown to follow from the assumptions made in the theory of eigenvalue spacing distributions[13-15] and lends support to the use of moment-equivalent continuous frequency and distribution functions. But we stress again that though we make considerable use of the ideas and language of statistics, our method of using these ideas is not the conventional one used in statistical theories.

We emphasize again that the study of moments of spectral distributions does not obviate the necessity for doing conventional spectroscopy. These two approaches are in a sense complementary, and each often provides answers to questions for which the other is ill suited. The power of moment techniques lies in the facts that neither matrix construction nor diagonalization is required, and that the propagation of moments is inherently simpler than the propagation of individual matrix elements (fundamentally because of the group-theoretical significance of the moments with which we deal). Moment techniques measure changes in either the Hamiltonian or the vector space underlying a calculation by the corresponding changes in moments of spectral distributions, and enable one to estimate the influence of such changes upon the spectroscopy of a single nucleus or an entire class of neighboring nuclei.

ACKNOWLEDGMENTS

I am deeply grateful to J. B. French for his interest and encouragement in the development of these ideas. I am also appreciative of the computing assistance rendered by F. S. Chang and S. S. M. Wong and of matrix elements supplied by T. T. S. Kuo. Much of this work was performed at the Center for Theoretical Physics, and the hospitality of H. Feshbach is most gratefully acknowledged.

---

*Work supported in part through funds provided by the U. S. Atomic Energy Commission under Contract No. AT(30-1)-2098.

†Permanent address: Department of Physics, State University of New York at Albany, Albany, New York.

[1]J. B. French and K. F. Ratcliff, preceeding paper, Phys. Rev. C, 3, 94 (1970).

[2]J. B. French, in *Nuclear Structure,* edited by A. Hossain *et al.* (North-Holland Publishing Company, Amsterdam, The Netherlands, 1967), p. 85.

[3]H. Banerjee and J. B. French, Phys. Letters 23, 245 (1966).

[4]J. B. French, Phys. Letters 23, 248 (1966).

[5]F. S. Chang, J. B. French, and K. F. Ratcliff, Phys. Letters 23, 251 (1966).

[6]J. B. French and L. S. Hsu, Phys. Letters 25B, 75 (1967).

[7]As in Ref. 1, an arrow notation is used to label a set of states or subspace.

[8]M. G. Kendall, *The Advanced Theory of Statistics* (Charles Griffin and Company Ltd., London, England, 1943), Vol. I.

[9]*Handbook of Mathematical Functions with Formulas, Graphs, and Mathematical Tables,* edited by M. Abramowitz and I. A. Stegun (U. S. Government Printing Office, Washington, D. C., 1964), p. 930.

[10]C. E. Porter, *Statistical Theories of Spectra: Fluctuations* (Academic Press Inc., New York, 1965), pp. 188, 235, 342, 379.

[11]The shell-model Hamiltonian is a Rosenfeld interaction of Yukawa shape with isospin-singlet spin-triplet strength = 40 MeV and range parameter $r_0 = 1.37$ fm. The single-particle (harmonic-oscillator) radial functions are proportional to $e^{-\frac{1}{2}\nu r^2}$, and $\lambda = r_0(\frac{1}{2}\nu)^{\frac{1}{2}} = 0.589$. The Hamiltonian parameters are adjusted to give O^{16} as an energy zero and to reproduce the single-particle levels of O^{17}.

[12]A. Fisher, *The Mathematical Theory of Probabilities* (The Macmillan Company, New York, 1928), 2nd ed.

[13]Porter, *Statistical Theories of Spectra: Fluctuations* (Academic Press Inc., New York, 1965).

[14]N. Rosenzweig, in *Statistical Physics, Brandeis Summer Institute,* edited by K. W. Ford (W. A. Benjamin, Inc., New York, 1963), Vol. 3.

[15]M. L. Mehta, *Random Matrices* (Academic Press Inc., New York, 1967).

[16]J. B. French, private communication.

[17]J. H. E. Mattauch *et al.*, Nucl. Phys. 67, 1 (1965).

[18]J. B. French, Phys. Letters 26B, 75 (1967).

[19]T. Hamada and I. D. Johnson, Nucl. Phys. 34, 382 (1962).

[20]T. T. S. Kuo and G. E. Brown, Nucl. Phys. 85, 40 (1966).

[21]T. T. S. Kuo, private communication.

[22]B. M. Spicer and J. M. Eisenberg, Nucl. Phys. 63, 520 (1965).

[23]R. J. Eden and N. C. Francis, Phys. Rev. 97, 1366 (1955).

[24]C. Bloch and J. Horowitz, Nucl. Phys. 8, 91 (1958).

[25]M. H. Macfarlane, in *Nuclear Structure and Nuclear Reaction, Proceedings of the International School of Physics "Enrico Fermi," Course XL,* edited by M. Jean (Academic Press Inc., New York, 1969).

[26]H. A. Bethe, Phys. Rev. 103, 1353 (1965).

ANNALS OF PHYSICS: 66, 137–188 (1971)

Distribution Methods for Nuclear Energies, Level Densities, and Excitation Strengths*

F. S. CHANG, J. B. FRENCH, AND T. H. THIO†

Department of Physics and Astronomy, University of Rochester, Rochester, N.Y. 14627

Received September 15, 1970

The centroid energies and widths for configurations, both with and without a specification of isospin, are given in compact forms in which the various terms have a clear physical significance. The forms are derived by use of elementary unitary-group methods which lead to an orthogonal symmetry decomposition of the Hamiltonian, in which Hartree–Fock–like terms representing the induced single-particle energies, and other related quantities, enter in a natural way. Appropriate measures for the various terms are given and discussed. The same method is applicable to the more complicated operators encountered when dealing with excitation strengths of various kinds. Some applications are given to determining ground-state energies, orbit occupancies, and low-lying spectra for complicated systems, and comparisons made with shell-model calculations and with experiment. An account, with some elementary applications, is given of a distribution theory of level densities which takes into consideration the shell-model structures and residual interactions; to specify the densities for fixed angular momentum, the configuration distributions are used to derive the energy variation of the J_z^2 and J_z^4 averages, the first of these fixing the "spin cut-off factor" $\sigma(J)$, and the second verifying the accuracy of the Gaussian form commonly used. Finally, an introductory account is given of the distributions of single-particle transfer and electromagnetic-transition strengths in large shell-model spaces.

1. INTRODUCTION

The purpose of spectral distributions [1–4] is to study general properties of nuclei. We assume, as in almost all spectroscopy, that the states of interest may be defined in terms of particles distributed over some finite set of single-particle states, and that we have available effective interactions and effective operators which act in the shell-model space so formed. But the spaces which we allow may be

* Supported in part by the U. S. Atomic Energy Commission. Part of this work was done while the second-named author was N.S.F. Senior Postdoctoral Fellow at Instituto de Fisica, Universidad de México, México, D. F.

† Present Address: Physics Department, Nanyang University, Singapore.

very large and we do not require that the effective interactions be oversimplified so as to lead to solvable matrix problems. Instead of seeking the expansion of the Hamiltonian eigenfunctions in terms of a natural basis, we consider the inverse problem, that of expanding a basis state in terms of the eigenfunctions; the corresponding distribution is examined in terms of its energy moments which are expressible as expectation values in the selected basis state of powers of H, being in principle therefore calculable without any knowledge of the eigenfunctions.

If our testing state should be close to an eigenstate that fact would be immediately revealed by the moments, the width in particular then being small. But, there being of course a multiparameter continuous set of states, it is clear that our chances of coming upon such a state by accident are negligible, and of course any direct procedure for finding one would be equivalent to a procedure for solving the original problem; besides that the evaluation of matrix elements of high powers of H is an impossibly complicated problem in any large space. So we leave the distributions of single states and turn instead to distributions averaged over some subset of states. If the subset is selected with symmetry considerations in mind, being a representation space for some group, then we can expect that new methods which rely on the invariance properties of traces will become available for the evaluation of the moments; thus we might expect to escape the long-standing limitations imposed on spectroscopy by the dimensionality problem. Broadly speaking we are then attempting to find approximate eigenfunctions in the subspaces defined by the chosen symmetries. If such are not to be found it would turn out that the distributions are broad (perhaps even spanning the entire spectral domain) and are then of no interest; we should turn to subspaces defined by other symmetries. If they are found we learn incidentally about the goodness of the symmetries. We shall see actually that there is far more flexibility in the procedure that this might suggest and, indeed, even quite badly broken symmetries might be very effective for our purpose. This being so, we may then hope to escape two other difficulties, often associated with applications of symmetry methods to spectroscopy; on the one hand we shall not have to assume that the symmetry is good or nearly good (if it is hopelessly bad we shall learn that quickly), and we shall not have to truncate our spaces to those described by one or two representations (for when we consider the distribution for one subspace we take account of its interactions with the others). From a more technical standpoint it will be the general structures and the Racah algebra which will concern us and not the great mass of detail and the tedious and less interesting Clebsch–Gordan algebra.

Symmetries then are dominant, the other key word being "statistics". If the distributions should turn out to be very complicated, we would not in fact be able to evaluate enough moments to describe them. The absolutely essential fact here is that the spectroscopic spaces are (antisymmetrized) direct products of the single-particle space. In such a direct-product space there operates an extended central limit theorem which tells us that for additive quantities defined in a factor subspace (the usual Hamiltonian is defined in the $(0+1+2)$-factor subspace) the distribution in a many-particle space will be asymptotically normal. This is at best vaguely understood at present, considerable extensions of the usual theorem being required, but there is very little doubt that the result is essentially true. It seems indeed that normality describes much of the general structure of complex nuclei. As examples of the behavior we show in Fig. 1 the fixed-angular-momentum distribution function (or cumulative spectrum), $\int_{-\infty}^{E} \rho(E')\,dE'$, where ρ is the density function, for a particular large spectroscopic calculation [5], (that for ^{63}Cu with $J = 5/2$, $T = 5/2$; the dimensionality is 273) compared with the normal distribution result; the agreement is excellent. In Fig. 2 we show results [6] of diagonalizing two ensembles of 50-dimensional random matrices. The conventional ensemble (Fig. 2a) gives the Wigner "semicircular" spectrum (no central limit theorem being operative because H is not defined in a small factor subspace); when the interaction is restricted to be two-body an excellent normal spectrum (Fig. 2b) is produced.

The direct-product nature of the spectroscopic spaces is of course responsible for the occurrence of nuclei in families ((*d*, *s*)-shell etc.). In conventional calculations, fractional parentage procedures make an inductive use of the direct-product

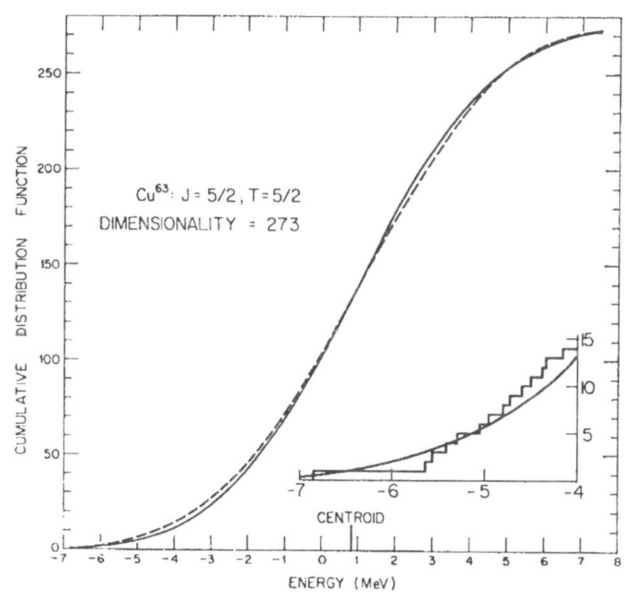

FIG. 1. Distribution functions $F(E) = \int_{-\infty}^{E} \rho(E')\,dE'$ for ^{63}Cu, $(f_{5/2}p_{3/2}p_{1/2})^7$ with $J = 5/2$, $T = 5/2$. The function derived by exact diagonalization of the 273-dimensional matrix [5] is compared with its continuous normal approximation (solid line). The exact function is a series of unit steps as shown, for the ground-state domain, by the expanded figure in the inset.

nature, but in a trivial way, since for example they usually regard a 10-particle system as (9 + 1) or (8 + 2) and ignore the structure of the larger subspace, that being of course an advantage when one is involved in detailed calculations. We shall see as we proceed that direct-products are important in distribution methods, not only because they give rise to simple distributions, but also because they yield methods for evaluating the moments which describe the distributions.

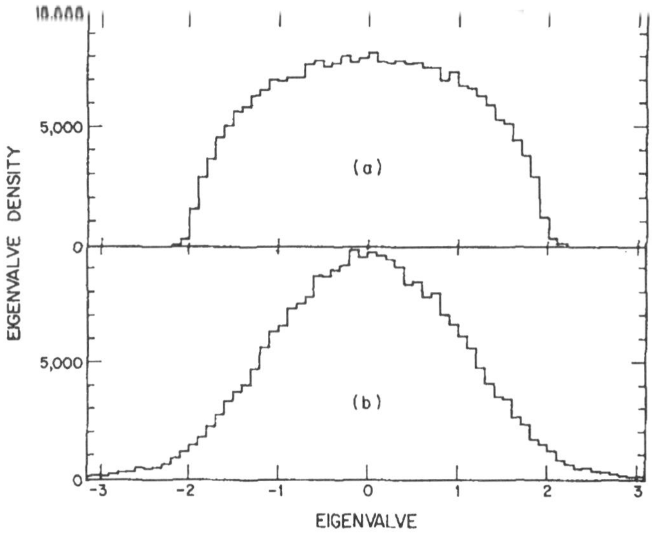

FIG. 2. Eigenvalue distributions from diagonalizing random ensembles consisting of 500 50-dimensional matrices arising from $(f_{7/2} + f_{5/2})^7$ with $T = 7/2$ (atomic f^7) and $J = 7/2$. In (a) the matrix elements are selected independently taking values between ± 1 with equal probability. In (b) the matrices are for 2-body interactions, the selection being in the space of the 21 two-body matrix elements. The results are due to S. S. M. Wong.

In this paper we shall deal with distributions defined by configurations with or without a specification of isospin. We first consider some elementary properties of the moments, and then, by introducing simple unitary-group considerations, give a separation of the Hamiltonian according to its unitary symmetries. This will lead to very compact and physically transparent forms for the moments. Some attention is given to determining the energies and orbit occupancies of ground states and low-lying excited states, following here procedures which have used particularly by Ratcliff [4]. We discuss at some length a theory for level densities, including densities for fixed angular momentum, which takes account of orbital structures and residual interactions. We give finally a preliminary discussion of strength distributions for various excitations.

It should be remarked that, in anticipation of future needs, we give the unitary group results in more generality than is presently needed. And, in order to make this paper self-contained, we give throughout, but particularly in the next section, results and methods which have already been given in the published papers [1–4].

2. Moment Polynomials

The centroid energies and widths of a configuration can be expressed as polynomials in the particle numbers which define the configuration. Consider first the case where the testing set consists of all the m-particle states (the "scalar" or 1-orbit case). It is easy to see that the average over this set of the expectation value of an operator O is a polynomial in m of order u, where u is the maximum particle rank of the operator (2 for the conventional H, 4 for H^2, etc). This average, $\langle O \rangle^m$, may be written in the alternative forms (where in the second $u_1 + u_2 = u - 1$)

$$\langle O \rangle^m = \sum_{s=0}^{u} \binom{u-m}{u-s}\binom{m}{s} \langle O \rangle^s = \sum_{s=0}^{u} (-1)^{s-u} \binom{m-s-1}{u-s}\binom{m}{s} \langle O \rangle^s, \tag{2.1}$$

$$\begin{aligned} \langle O \rangle^m = {} & \binom{N-m}{u_2+1} \sum_{s=0}^{u_1} (-1)^{s-u_1} \binom{N-s}{u_2+1}^{-1} \binom{m-s-1}{u_1-s}\binom{m}{s} \langle O \rangle^s \\ & + \binom{m}{u_1+1} \sum_{s=0}^{u_2} (-1)^{s-u_2} \binom{N-s}{u_1+1}^{-1} \binom{N-m-s-1}{u_2-s}\binom{N-m}{s} \langle O \rangle^{N-s}. \end{aligned} \tag{2.2}$$

These forms are simply ways of expressing the coefficients of a polynomial in terms of its values at a set of values of its variable; they may be derived via elementary Lagrange polynomials and verified by inspection. The second form represents a great simplification over the first, since, by using it, we can, for the usual H, express $\langle H^2 \rangle^m$ in terms of (0, 1 and 2)-fermion traces and need not calculate (3, 4)-fermion ones as needed for the first form. We would take $u_1 = 2$, $u_2 = 1$, or vice versa, and would need both the particle and the hole representation of H, since, for $F = H^2$, we would write $\langle F \rangle^{N-t} = \langle \tilde{F} \rangle^t$, where $F \to \tilde{F}$ under a hole-particle transformation (every $A_i \rightleftharpoons A_i^+ = B_i$, where A_i is a creation operator). For a two-body H we could take advantage also of the fact that H^2 has no $(0 + 1)$-body parts; with $u_1 = 1$, $u_2 = 2$ the result for this case is expressible in terms of $(0 + 1 + 2)$-hole averages of H^2; and similarly for other products of operators of specified particle ranks.

As indicated here, our main problem will be with squares or products of operators: It will not be adequate however to deal only with operators which are number conserving. So, extending the notion of a k-body operator $F(k)$, we introduce a (k, μ) operator belonging to the set $\{\psi_\alpha(k+\mu) \times \psi_\beta^+(k-\mu)\}$; then $F(k,\mu)^+$ is a $(k, -\mu)$ operator, while $F(k, 0)$ is a k-body operator. We shall continue to call k the "particle rank" of F. $F(k, \mu)$, which creates 2μ particles (or destroys -2μ, depending on the sign of μ), acts between pairs of fermion subspaces as $(t - \mu) \to (t + \mu)$, where $k \leqslant t \leqslant N - |\mu|$, and the labels indicate particle numbers. Obviously any of these pairs will give a faithful representation of the operator, in the sense that the matrix elements connecting the two spaces, together with the specified value of k, define the operator. For values of t outside the range specified, the operator gives zero. Then for an operator product $F^{+}(k, \mu)\, G(k', \mu)$, for which $u = k + k'$, we take $u_1 = k_{>} - \mu - 1$, $u_2 = k_{<} + \mu$, are once again left only with the hole contributions in (2.2), and have

$$\begin{aligned}
&\langle F^{+}(k, \mu)\, G(k', \mu)\rangle^{m} \\
&\quad = \binom{m}{k_{>} - \mu} \sum_{s=0}^{k_{<}+\mu} \binom{N-s}{k_{>} - \mu}^{-1} (-1)^{s-k_{<}-\mu} \binom{N-m-s-1}{k_{<} + \mu - s} \binom{N-m}{s} \\
&\qquad \times \langle F^{+}(k, \mu)\, G(k', \mu)\rangle^{N-s}.
\end{aligned} \tag{2.3}$$

We have been dealing so far with the simple linear lattice defined by fermion numbers $m = 0, 1, \ldots, N$. The notion of the propagation of trace information along this chain is obvious from the equations given. Let us proceed now to divide up the space according to configurations, by partitioning the single-particle space into l "orbits", $N = \sum_{\alpha=1}^{l} N_\alpha$, and making a similar partition of the particles, $m = \sum m_\alpha$. Clearly now we expect that the average of O over a configuration $\mathbf{m} = [m_1, m_2, \ldots, m_l]$ will be expressible as a polynomial, again of maximum order u but now in the l variables m_α. One finds easily

$$\langle O\rangle^{\mathbf{m}} = \sum_{\substack{s_1=0 \\ (s \leqslant u)}}^{u} \cdots \sum_{s_l=0}^{u} (-1)^{s-u} \binom{m-s-1}{u-s} \left\{ \binom{m_1}{s_1} \cdots \binom{m_l}{s_l} \right\} \langle O\rangle^{\mathbf{s}}, \tag{2.4}$$

where $s = \sum s_\alpha$, $m = \sum m_\alpha$. This is the analog of (2.1); but no analog of (2.2) will be found for an operator which is general except for a restriction on the maximum particle rank. This difference, which arises because we now have an l-dimensional lattice of subspaces instead of the simple linear chain of the one-orbit case, has in earlier treatments made a great deal of trouble and has led to exceeding complicated forms for the moments. If, on the other hand, we consider first a decomposition of F, G according to $\mathbf{k}$, $\boldsymbol{\mu}$ (l-dimensional vectors with components k_α, μ_α in the α orbit), we have an immediate vectorial extension[1] of (2.3) and the problem is solved.

As an illustration take a two-orbit case and consider the contribution to the

[1] In which m, k, k', μ, s, and N are all replaced by vectors, each binomial coefficient being then interpreted as a product of single-orbit coefficients; and similarly for the summation. We shall use bold-faced symbols $\mathbf{m}$, $\mathbf{k}$, etc., for these vectors, as we have done in (2.4) above.

width from the pairing interaction $H(P)$ which acts between them. We label the pair states as Λ and Ω in orbits 1, 2, respectively, and write $W(\Lambda, \Omega)$ for the pairing matrix elements. The interaction being traceless, the variance σ^2 (the square of the width) is simply the second moment. Then $k_1 = k_2 = 1$, and $(\mu_1, \mu_2) = (-1, 1)$ or $(1, -1)$, the first describing a process $\Lambda \to \Omega$ and the second the inverse. The first term gives, as its contribution to $\sigma^2(m_1, m_2)$,

$$\binom{m_1}{2}\binom{N_1}{2}^{-1} \sum_{s_2=0}^{2} (-1)^{s_2} \binom{N_2 - m_2 - s_2 - 1}{2 - s_2}\binom{N_2 - m_2}{s_2} \langle \tilde{H}^+(P)\, \tilde{H}(P)\rangle^{0,s_2}. \quad (2.5)$$

Since the contributing interaction has the form $\sum_{\Lambda,\Omega} W(\Lambda, \Omega)\, \psi_\Omega \psi_\Lambda{}^+$, we have $\tilde{H} = H^+$, the only contribution comes from $s_2 = 2$, and

$$\langle \tilde{H}^+(P)\, H(P)\rangle^{0,2} = \langle H(P)\, H^+(P)\rangle^{0,2} = \binom{N_2}{2}^{-1} \sum_{\Lambda,\Omega} W^2(\Lambda, \Omega),$$

where the binomial factor is the $(0, 2)$ dimensionality. Adding in the adjoint $(\Omega \to \Lambda)$ term we have

$$\begin{aligned} \langle (H(P))^2 \rangle^{m_1,m_2} &= \sigma_P{}^2(m_1, m_2) \\ &= \binom{N_1}{2}^{-1} \binom{N_2}{2}^{-1} \left\{ \binom{m_1}{2}\binom{N_2 - m_2}{2} + \binom{N_1 - m_1}{2}\binom{m_2}{2} \right\} \sum_{\Lambda\Omega} W^2(\Lambda, \Omega). \end{aligned} \quad (2.6)$$

With a little thought the result here could have been written at sight and its extension to the many-orbit case (still involving two-orbit pairing) is immediate. The two terms of (2.6) differ in the intermediate states whose excitation gives rise to the width, in the first term, $[m_1 - 2, m_2 + 2]$, in the second, $[m_1 + 2, m_2 - 2]$; this separation gives the standard "partial-width" decomposition of the width. Observe that the configuration variances for a given pair of orbits are all multiples of the basic variance which we encounter for the two-particle system, this, we shall see, coming about because the pairing interaction is irreducible with respect to a relevant group; the number dependence is quite strong and, since we can expect it to be different for different parts of the interaction, we see that under varying circumstances different parts of the interaction might predominate; we return to this point later.

The pairing interaction is defined by the orbital structure of its matrix elements, these having the general form $W(\Lambda, \Omega) = W_{\alpha\alpha\beta\beta}$, where α and β correspond to orbits. Moreover the orbital structure defines an orthogonal decomposition of H in the sense that no crossterms, e.g., of the form $W_{\alpha\alpha\beta\beta} W_{\alpha\beta\alpha\beta}$, (pairing $\times$ multipole) can enter into σ^2. However, the simplicity encountered above that, for given orbits, the variances are all proportional to a basic one, will be found, on applying (2.3), not to occur for all orbital structures, one-exception being the two-orbit multipole interaction $W_{\alpha\beta\alpha\beta}$. This, in fact, occurs because these interactions are reducible with respect to the unitary transformations which define orbits; or equivalently we can generate from them, by a contraction process involving summation over an orbit, a Hartree–Fock-like one -body effective interaction. In the next section, therefore, we consider the unitary decomposition of operators; although the treatment given will be quite simple, it will in fact go further than we need for our immediate purposes, and much of it may be ignored by the reader more immediately interested in applications. Indeed if his only interest is in spherical orbits, and particularly, if the set of such orbits contains none which differ only in radial quantum number, the general result given in (4.18) and (4.19) ahead, follows by only a slight extension of the argument here.

3. Unitary Group Decompositions

The classification of operators by the number of particles which they create has a group-theoretical significance[2]; in the one-orbit case the group is $U(N)$ (or $SU(N)$), of unitary transformations on the single-particle basis, while in the multiorbit case it is the "direct-sum" subgroup of separate transformations in the various orbits, which we write as $\sum_{\alpha=1}^{l} U(N_\alpha)$. The averaging (or trace) operation being scalar, the vanishing of the averages for the product $F^+ \times G$, when F and G have different μ's, informs us that operators of different μ, while not necessarily irreducible, have no representations in common. The decomposition by μ is therefore a partial symmetry decomposition. The complete decomposition will give a finer separation of the widths and will introduce some new notions which are physically significant.

We first review some elementary facts about the transformation properties of operators. With respect to $U(N)$, a creation operator A_i transforms as [1], and a k-body state operator $\psi(k)$ as $[k]$, these symbols defining the *column* structure of the Young shape. A destruction operator $\sim [N-1]$ and $\psi^+(k) \sim [N-k]$. The coupling of two ψ or ψ^+ spaces is trivial, the resultant spaces being irreducible. The product of a $\psi(k)$ and $\psi^+(k)$ space can however support two-columned representations. For $m = k = 1$ we have $A \times B \sim [1] \times [N-1] = [N] + [N-1, 1]$, the explicit reduction being

$$\begin{aligned} \{A_iB_j\} &\to \left\{\sum A_iB_i\right\} + \{A_iB_j - \delta_{ij}N^{-1}n\}, \\ \{F_i\} &\to \left\{\sum F_i\right\} + \left\{F_i - \delta_{ij}N^{-1}\sum F_j\right\}, \end{aligned} \tag{3.1}$$

[2] See for example Refs. [7, 8]. A review of unitary group operators, which goes far beyond our present needs, is given by Louck [9], who also gives extensive references. For hole-particle structures in particular see Ref. [10]. Note that in several instances below (e.g., in the statement that an $F(k, \mu)$ operator transforms as $(N + 2\mu)$-block representations) we should really be thinking about $SU(N)$, in which case complete columns may be ignored.

where we have also written the configuration-space form (i being a particle number), and, in the first form, n is as always the number operator. Note that the $[N-1, 1]$ operator goes simply to the negative of its adjoint under the hole-particle involution $A_i \rightleftharpoons B_i$; the significance of this result in general form, will become clear shortly.

More generally we see that an operator $F(k, \mu)$ transforms as representations (always one or two columned) with $(N+2\mu)$ blocks. We label these as $[N-\nu+\mu, \nu+\mu] = D^{\nu}(\mu)$, thereby introducing the new symmetry label ν. The dimensionalities are

$$d(\nu, \mu) = \frac{N-2\nu+1}{N+1}\binom{N+1}{\nu+\mu}\binom{N+1}{\nu-\mu}. \tag{3.2}$$

It is clear from the column structure that $D^{\nu}(\mu)$ and $D^{\nu}(-\mu)$ form a contragredient pair, the $D^{\nu}(0)$ representations encountered with number-conserving operators being self-contragredient. Representations exist for the following values of (ν, μ):

$$\begin{aligned} -N \leqslant 2\mu \leqslant N \qquad & 2\mu = \text{integer}; \\ |\mu| \leqslant \nu \leqslant 1/2N, \qquad & \nu+\mu = \text{integer}. \end{aligned} \tag{3.3}$$

Thus for example with $N = 4$, we have $D^0(0)$, $D^{1/2}(\pm 1/2)$, $D^1(0, \pm 1)$ $D^{3/2}(\pm 1/2, \pm 3/2)$, and $D^2(0, \pm 1, \pm 2)$, fifteen in all.

Since state operators $\psi(k)$ exist for $0 \leqslant k \leqslant N$ we see that there are (k, μ) operator spaces as long as $(k \pm \mu)$ are nonnegative integers $\leqslant N$. For these allowed (k, μ) we find, by the usual combination of one-columned shapes, that the space of operators $F(k, \mu)$ contains representations $D^{\nu}(\mu)$ once and once only as long as

$$\begin{aligned} &|\mu| \leqslant k \leqslant N - |\mu|, \\ &|\mu| \leqslant \nu \leqslant (k, N-k)_< , \end{aligned} \tag{3.4}$$

where the indicated upper limit on ν is the lesser of k and $(N-k)$, and of course ν is integral or 1/2-integral according as k is integral or 1/2-integral. Note that in every case, since $\nu \leqslant N/2$, we have $k \geqslant \nu$. As a dimensionality check we have, for given k and μ,

$$\sum_{\nu=|\mu|}^{\nu_{\max}} d(\nu, \mu) = \binom{N}{k+\mu}\binom{N}{k-\mu}, \tag{3.5}$$

which may easily be verified. Observe too that if an operator belongs to $D^{\nu}(\mu)$, its adjoint belongs to $D^{\nu}(-\mu)$.

Since $D^{\nu}(\mu)$ occurs once and only once for each particle rank satisfying $\nu \leqslant k \leqslant N-\nu$ and not at all for other values, we have a chain of $D^{\nu}(\mu)$ operator spaces which moreover are related to each other by a contraction process. To derive this, consider the relationship between the operators $F^\nu(k, \mu)$ and $\binom{n-k-\mu}{p-k} F^\nu(k, \mu)$. Since the binomial factor is a unitary scalar with maximum particle rank $(p - k)$, the product operator is also of symmetry (ν, μ) but of maximum rank p; in fact the rank is exactly p. Consider the $\mu = 0$ case for example; then the product vanishes in $(m < k)$ subspaces by the F factor and in $k \leqslant m < p$ by the binomial factor, so that the minimum and maximum ranks are both p (and similarly for $\mu \neq 0$). The binomial factor is then a unitary scalar $(k \rightarrow p)$ shift operator, shifting the pair of particle spaces between which the operator is defined; it establishes therefore a correspondence between the different $D^\nu(\mu)$ operator spaces. Looking at this inversely we see that any (ν, μ) operator of particle rank k must contain $\binom{n-\nu-\mu}{k-\nu}$ as a left factor; dividing it out is an operation with respect to (A, B) pairs closely analogous to the (A, A) seniority contraction of state operators.

It now follows that, for a general operator $F(k, \mu)$,

$$F(k, \mu) = \sum_\nu F^\nu(k, \mu) = \sum_\nu \binom{n - \nu - \mu}{k - \nu} \mathscr{F}(\nu, \mu) = \sum_\nu \mathscr{F}(\nu, \mu) \binom{n - \nu + \mu}{k - \nu}, \tag{3.6}$$

where the contracted operator $\mathscr{F}$ might also be written as $\mathscr{F}^\nu(\nu, \mu)$, if we wish to emphasize the symmetry, or $\mathscr{F}_k(\nu, \mu)$ if we wish to stress the particle rank of the operator form which it derives. $\mathscr{F}(\nu, \mu)$ is irreducible with respect to $U(N)$, cannot be further reduced in particle rank by (A, B) contractions, and has a very important hole $\rightleftharpoons$ particle symmetry[3]

$$\mathscr{F}^+(\nu, \mu) = (-1)^{\nu(2\nu-1)} \tilde{\mathscr{F}}(\nu, \mu), \tag{3.7}$$

a special case of which we have seen following (3.1). This general symmetry arises from the fact that the commutator terms which would normally distinguish between the two operations (note that to within phases $\psi_\alpha \psi_\beta{}^+ \xrightarrow{\sim} \psi_\alpha{}^+ \psi_\beta$ and $\xrightarrow{(+)} \psi_\beta \psi_\alpha{}^+$) are automatically of lower particle rank but the same unitary symmetry; hence, since $\mathscr{F}$ is not further contractible, these terms must vanish. The phase in (3.7) comes on carrying out the operations indicated.

An inverse result which follows immediately is that if an operator has both a definite particle rank k and hole rank k', then $k = k'$, and the operator transforms as some set of representations $D^k(\mu)$, is irreducible under contractions, and satisfies (3.7). It should be clear that these results can be extended to give a basis for the discussion of the commutation of many-particle state operators and adjoints.

A number conserving k-body operator $F(k)$ vanishes in $(t < k)$-particle

[3] It is satisfactory to assume that, with respect to a basis given by the various products and powers of A and B operators, the operators which we consider form a *real* vector space, their expansion coefficients being real. The general extension would be immediate. Note that $\mathscr{F}^+(\nu, \mu)$ is the adjoint of $\mathscr{F}(\nu, \mu)$ and transforms as $D^\nu(-\mu)$; similarly for $\tilde{\mathscr{F}}(\nu, \mu)$.

subspaces, but in general the subspaces with $t \geqslant k$ all give representations of the operator. The t-hole representations however with $t < k$ cannot be faithful[4] since they cannot support any irreducible component of F with symmetry $\nu > t$, as follows easily from (3.7). An elementary extension gives the result for more general operators. A fixed-μ operator, which we write as $F(\cdot, \mu)$ operates between two subspaces, $(t - \mu) \to (t + \mu)$; obviously then $F(\cdot, \mu)$ vanishes outside the domain given by $|\mu| \leqslant t \leqslant N - |\mu|$. Further specifications on the operator narrow the domain; we find easily

$$\begin{aligned} |\mu| \leqslant t \leqslant N - |\mu|, &\quad \text{for } F(\cdot, \mu), \\ k \leqslant t \leqslant N - |\mu|, &\quad \text{for } F(k, \mu), \\ k \leqslant t \leqslant N - \nu, &\quad \text{for } F^{\nu}(k, \mu), \\ \nu \leqslant t \leqslant N - \nu, &\quad \text{for } \mathscr{F}_k(\nu, \mu) \end{aligned} \tag{3.8}$$

as shown schematically in Fig. 3.

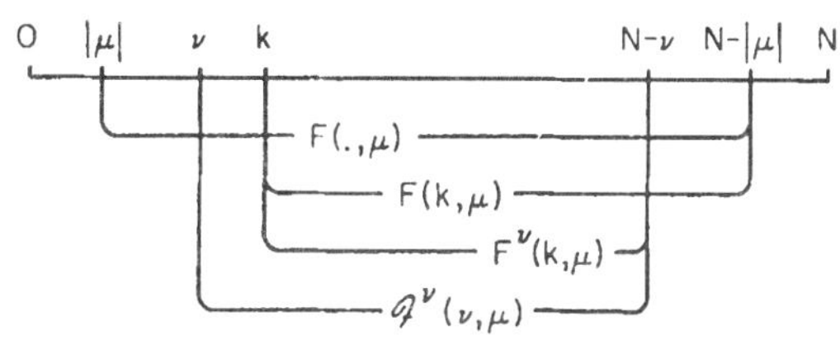

FIG. 3. The domain of operators with fixed μ as labeled by the intermediate-subspace particle number. The operators act between two subspaces defined by (t, μ) as $(t - \mu) \to (t + \mu)$.

Only the $\nu = \mu = 0$ part of an operator survives under scalar averaging, and then

$$\langle F(k, 0)\rangle^{m} = \binom{m}{k} \mathscr{F}(0, 0). \tag{3.9}$$

But for widths we need $\langle F^{+}F\rangle^{m}$; we then get contributions from each term in the unitary decomposition of F, but without cross terms in different symmetries since these cannot contribute to the scalar part of $F^{+}F$. We can therefore decompose the quadratic trace according to the decomposition of the linear factors, a much more effective procedure than one requiring normal forms for $F^{+}F$.

[4] In the sense that a specification of the rank and of the matrix elements in the subspace cannot define the operator in all subspaces. It is easy to see however that the t-particle representations with $k \leqslant t \leqslant N - k$ are all faithful.

Introducing a product of two operators instead of a square (and writing an adjoint of one of them for consistency with later results) we have from (3.6)

$$\langle F^+(k,\mu)\, G(k',\mu)\rangle^m = \sum_\nu \binom{m-\nu+\mu}{k-\nu}\binom{m-\nu+\mu}{k'-\nu} \langle \mathscr{F}^+(\nu,\mu)\, \mathscr{G}(\nu,\mu)\rangle^m. \quad (3.10)$$

Evaluating the m-average via (2.3) we note that for a given ν only a single term, that for $(\nu+\mu)$ holes, survives, the others defining subspaces which are too simple to support operators with (ν, μ) symmetry. Thus the quadratic averages for operators with given symmetry are all multiples of a single defining average and carry with them only a single piece of trace information. We may say that such averages propagate multiplicatively.

From (2.3) and (3.10) we have now

$$\begin{aligned} &\langle F^+(k,\mu)\, G(k',\mu)\rangle^m \\ &\quad = \sum_\nu \binom{m-\nu+\mu}{k-\nu}\binom{m-\nu+\mu}{k'-\nu}\binom{N-m}{\nu+\mu}\binom{N-\nu-\mu}{\nu-\mu}^{-1}\binom{m}{\nu-\mu} \\ &\quad\quad \times \langle \mathscr{F}^+(\nu,\mu)\, \mathscr{G}(\nu,\mu)\rangle^{N-\nu-\mu} \\ &\quad = \sum_\nu \binom{m-\nu+\mu}{k-\nu}\binom{m-\nu+\mu}{k'-\nu}\binom{N-m}{\nu+\mu}\binom{N-\nu+\mu}{\nu+\mu}^{-1}\binom{m}{\nu-\mu} \\ &\quad\quad \times \langle \mathscr{F}^+(\nu,\mu)\, \mathscr{G}(\nu,\mu)\rangle^{\nu-\mu}. \end{aligned} \quad (3.11)$$

The second form here, which would have emerged automatically had we written (2.3) with $u_1 = k_> - \mu$, $u_2 = k_< + \mu - 1$, comes from the first by an integral identity. There are a set of these which derive from (a) the ordinary $h \rightleftharpoons p$ involution; (b) the result that if operators S and T act between two subspaces $\boldsymbol{\alpha}$, $\boldsymbol{\beta}$ in the sense $\boldsymbol{\alpha} \xrightarrow{(S)} \boldsymbol{\beta}$, $\boldsymbol{\beta} \xrightarrow{(T)} \boldsymbol{\alpha}$ then, for *traces*[5] in the separate spaces, $\langle\!\langle TS \rangle\!\rangle^{\boldsymbol{\alpha}} = \langle\!\langle ST \rangle\!\rangle^{\boldsymbol{\beta}}$; (c) $\langle\!\langle O \rangle\!\rangle^m = \langle\!\langle O^+ \rangle\!\rangle^m$ since only the (self-adjoint) $\mu = 0$ component survives in a trace. Combining these we have

$$\begin{aligned} \langle\!\langle \mathscr{F}^+(\nu,\mu)\, \mathscr{G}(\nu,\mu)\rangle\!\rangle^{\nu-\mu} &= \langle\!\langle \mathscr{G}^+(\nu,\mu)\, \mathscr{F}(\nu,\mu)\rangle\!\rangle^{\nu-\mu} = \langle\!\langle \mathscr{F}(\nu,\mu)\, \mathscr{G}^+(\nu,\mu)\rangle\!\rangle^{N-\nu+\mu} \\ &= \langle\!\langle \mathscr{G}(\nu,\mu)\, \mathscr{F}^+(\nu,\mu)\rangle\!\rangle^{N-\nu+\mu} = \langle\!\langle \mathscr{G}(\nu,\mu)\, \mathscr{F}^+(\nu,\mu)\rangle\!\rangle^{\nu+\mu} \\ &= \langle\!\langle \mathscr{F}(\nu,\mu)\, \mathscr{G}^+(\nu,\mu)\rangle\!\rangle^{\nu+\mu} = \langle\!\langle \mathscr{G}^+(\nu,\mu)\, \mathscr{F}(\nu,\mu)\rangle\!\rangle^{N-\nu-\mu} \\ &= \langle\!\langle \mathscr{F}^+(\nu,\mu)\, \mathscr{G}(\nu,\mu)\rangle\!\rangle^{N-\nu-\mu} \end{aligned} \quad (3.12)$$

and it is one of these, combined with

$$\binom{N-\nu-\mu}{\nu-\mu} d(\nu+\mu) = \binom{N-\nu+\mu}{\nu+\mu} d(\nu-\mu),$$

which we have used for the second form of (3.11).

[5] We write $\langle\!\langle\ \rangle\!\rangle^m$ for the trace, related to the average by $\langle\!\langle\ \rangle\!\rangle^m = d(m)\, \langle\ \rangle^m$.

We have discussed all this in terms of $U(N)$ averages which are in any case very simple. But the extension to the configuration case is now immediate. Just as before we introduce vectors (with components in the l segments of the single-particle set) for various quantities including now the representation symbols; a binomial coefficient becomes a product of such, a summation becomes a set of independent summations, and $|\mu|$ is replaced by a set $|\mu_i|$. We encounter then direct sums of the structures which appear in the one-orbit case, so that the configuration widths are formally as simple as the scalar widths. Besides that the quadratic traces of irreducible operators propagate multiplicatively and without cross terms. We saw this in the pairing example (2.6), for which the interactions are indeed irreducible with $\boldsymbol{\nu} = (1_\alpha, 1_\beta)$ and $\pm\boldsymbol{\mu} = (1_\alpha, -1_\beta)$, but it is clear from (3.11) that the result is general.

We have seen in the $U(N)$ case that behavior under the $h \rightleftharpoons p$ involution gives a partial classification of operators (with respect to ν), besides which knowledge of such behavior makes for much easier evaluation of needed averages. With configurations the same is true for the classification, and the technical simplifications which come in are much more significant. In the configuration case we have l independent "partial" hole-particle transformations; we may introduce as well a "partial" adjointing in which elementary A, B operators not referring to the orbit in question are left undisturbed. Then a $\boldsymbol{\mu}$-operator $\mathscr{F}(\boldsymbol{\nu}, \boldsymbol{\mu})$, irreducible with respect to the configuration group and fully contracted, is one for which the partial hole-particle and adjointing operations in each orbit α are related to each other by the basic (3.7) with $\nu \to \nu_\alpha$.

We need the input integrals in the vector equivalent of (3.11) and, exactly as in the pairing example, we can write them in terms of the defining matrix elements of the reduced operators. We must not forget, taking $F = G = H$, that H has parts of various particle ranks (0, 1, 2 for $U(N)$, further possibilities with respect to configurations), and while there is orthogonality in μ, there is none in k, so that, for example, we get cross terms for certain symmetries between the one-particle and two-particle Hamiltonians. Then writing $H = \sum H(\mathbf{k}, \boldsymbol{\mu})$, we have

$$\langle H^2\rangle^{\mathbf{m}} = \sum_{\boldsymbol{\mu}} \langle H^+(\cdot, \boldsymbol{\mu})\, H(\cdot, \boldsymbol{\mu})\rangle^{\mathbf{m}} = \sum_{\boldsymbol{\mu}} \left\{ \sum_{\mathbf{k},\mathbf{k}'} \langle H^+(\mathbf{k}, \boldsymbol{\mu})\, H(\mathbf{k}', \boldsymbol{\mu})\rangle^{\mathbf{m}} \right\} \qquad (3.13)$$

and in using (3.11) we would then further label each reduced operator by the rank of the operator from which it derives. Then finally for the basic averages

$$\langle \mathscr{H}_{\mathbf{k}}^+(\boldsymbol{\nu}, \boldsymbol{\mu})\, \mathscr{H}_{\mathbf{k}'}(\boldsymbol{\nu}, \boldsymbol{\mu})\rangle^{\boldsymbol{\nu}-\boldsymbol{\mu}}$$
$$= d^{-1}(\boldsymbol{\nu} - \boldsymbol{\mu}) \sum \langle \boldsymbol{\nu} - \boldsymbol{\mu} \,|\, \mathscr{H}_{\mathbf{k}}^+(\boldsymbol{\nu}, \boldsymbol{\mu}) \,|\, \boldsymbol{\nu} + \boldsymbol{\mu}\rangle \langle \boldsymbol{\nu} + \boldsymbol{\mu} \,|\, \mathscr{H}_{\mathbf{k}'}(\boldsymbol{\nu}, \boldsymbol{\mu}) \,|\, \boldsymbol{\nu} - \boldsymbol{\mu}\rangle, \qquad (3.14)$$

where the sum is over all the matrix elements of the form indicated, just as in (2.6), and $d(\mathbf{v} - \boldsymbol{\mu})$ is the dimensionality of the $(\mathbf{v} - \boldsymbol{\mu})$ states, which may, for the conventional Hamiltonian, be 0, 1 or 2-particle states according as $\sum \nu_i = 0, 1,$ or 2.

Equation (3.14) gives the "basic" second moments of the irreducible parts of H, which, if we exclude the $\mathbf{v} = (000 \cdots)$ case (for which $\boldsymbol{\mu} = (000 \cdots)$ also), are also the basic variances σ^2. The many-particle variances come from adding the basic ones, each multiplied by its own polynomial as in (3.11). The resulting expression will be of course equivalent to that given earlier, though it is given here in "Cartesian" rather than in "spherical" form (i.e., in terms of single-particle states rather than in "spherical" form); we shall later see how to go easily from one to the other. But the present forms are valid for general operators and are not restricted to $(k = 2)$ number-conserving ones, they give the widths decomposed according to the unitary symmetries of the Hamiltonian, and the explicit separation into configuration partial widths (since fixed $\mathbf{m}$ and $\boldsymbol{\mu}$ vectors uniquely fix the intermediate states).

We need now the explicit decomposition of H with respect to the configuration subgroup. We consider first the general one-orbit decomposition problem using a procedure of applying unitary scalar contractions to the operator[6]; these reduce the particle rank by various amounts, and in each case therefore eliminate symmetries which are incompatable with the smaller rank.

It is advantageous to distinguish between bosonlike operators, for which μ and k are integral and fermionlike operators, for which μ and k are 1/2-integral; we shall take for granted that every operator with which we deal belongs to one of these classes (not being of mixed nature). We introduce unitary-scalar contractions of an operator F by

$$D_{\pm}F = \sum_i \, [A_i[B_i\,, F]_{\mp}]_{\pm}\,, \tag{3.15}$$

where D_+ will be used only for bosonlike operators and D_- for fermionlike. This associate being understood, we shall be able to drop the $(\pm)$ subscript on D. The operations are "contractions" because DF is of maximum particle rank one less than that of F itself, this coming about because of the double commutation, and are unitary scalar because so is $\sum A_i B_i = n$.

It is obvious now that $D\mathscr{F}(\nu, \mu) = 0$, since $\mathscr{F}$ cannot be further contracted (this equation in fact defines an operator as belonging to one of the ν representa-

[6] In an alternative method, one projects the operator successively into a sequence of t-hole subspace pairs (t labeling the intermediate subspaces in the sense of Fig. 3, this generating a sieve action based on the fact that the smaller the value of t the fewer the symmetries of the operator which can be supported. In a quite different procedure one can make a unitary decomposition of the width itself without decomposing the operator at all. This latter procedure is more economical and thus might be favored; but the method which we give has the advantage of dealing with quantities whose physical significance is more apparent.

tions). Just as easily we have that $DF^{\nu}(k,\mu) \sim \binom{n-\nu-\mu}{k-\nu-1}\mathscr{F}(\nu,\mu)$, valid to within a multiplying constant which we must evaluate. To derive the constant we use the elementary results

$$\begin{aligned}\sum_i A_i\mathscr{F}(\nu,\mu)\,B_i &= (-1)^{2\nu}\,(n-\nu-\mu)\mathscr{F},\\ \sum_i B_i\mathscr{F}(\nu,\mu)\,A_i &= (-1)^{2\nu}\,(N-n-\nu+\mu)\mathscr{F},\\ A_i f(n) = f(n-1)\,A_i\,; &\qquad B_i f(n) = f(n+1)\,B_i\,,\end{aligned} \tag{3.16}$$

where $f(n)$ is a function only of the number operator n, and, as always, $\mathscr{F}$ is irreducible and fully contracted. The first result follows from $\sum_i A_i\psi_\beta(p)\,\psi_\gamma{}^+(p')\,B_i = \psi_\beta(p)\,n\psi_\gamma{}^+(p') = (n-p)\,\psi_\beta(p)\,\psi_\gamma{}^+(p')$, while for the second we use also the fact that the factors of an irreducible operator essentially commute. On writing out the double commutators of (3.15), we find that

$$\begin{aligned}&D\{f(n)\,\mathscr{F}(\nu,\mu)\} \qquad\qquad (3.17)\\ &= \{f(n+1)(N-n-\nu+\mu) - f(n-1)(n-\nu-\mu) - f(n)(N-2n+2\mu)\}\,\mathscr{F}\end{aligned}$$

and on applying this find that

$$\begin{aligned}DF^{\nu}(k,\mu) = D\left\{\binom{n-\nu-\mu}{k-\nu}\mathscr{F}(\nu,\mu)\right\} &= (N+1-k-\nu)\binom{n-\nu-\mu}{k-1-\nu}\mathscr{F}(\nu,\mu)\\ &\equiv (N+1-k-\nu)\,F^{\nu}(k-1,\mu).\end{aligned} \tag{3.18}$$

In the last equivalence of (3.18) we have associated with an operator $F^{\nu}(k,\mu)$ a related operator with particle rank one less. More generally we have a propagation of the operator throughout the whole set of $(t\pm\mu)$-subspace pairs which can support symmetry ν, any two operators in such a chain carrying the same information, and an attempt to go outside this chain resulting in an annihilation.

On iterating the contraction (3.18) we then have, for a general (k,μ) operator

$$D^r F(k,\mu) = \sum_\nu r!\binom{N+r-k-\nu}{r}F^{\nu}(k-r,\mu), \tag{3.19}$$

the inverse of which, namely

$$\begin{aligned}\mathscr{F}(\nu,\mu) = \frac{1}{(k-\nu)!}\binom{N-2\nu}{k-\nu}^{-1}\sum_{t=k-\nu}^{k-\mu}\frac{(-1)^{t-k+\nu}}{(t-k+\nu)!}\binom{N-\nu-k+t+1}{t-k+\nu}^{-1}\\ \times\binom{n-k+t-\mu}{t-k+\nu}D^tF(k,\mu)\end{aligned} \tag{3.20}$$

expresses the contracted irreducible part of $F(k, \mu)$ in terms of a series of commutator contractions carried out on $\tilde{F}$.

Because of the simple-direct-product nature of the configuration spaces, the extension of (3.20) to the configuration case is immediate, just as before.

4. The Decomposition and Structure of the Hamiltonian

We take the usual (0 + 1 + 2)-particle Hamiltonian

$$H = H_0 + \epsilon(i, j)\, A_i B_j - \tfrac{1}{4} W_{ijkl} A_i A_j B_k B_l \,. \tag{4.1}$$

We regard the single-particle states, which, as indicated, we take to be ordered, as forming l orbits, $\alpha = 1 \cdots l$, the state ordering being such that orbit #1 has states $i = 1 \cdots N_1$, etc, where the N_α are the orbit degeneracies ($\sum N_\alpha = N$). Our problem is to decompose H according to the unitary symmetry defined by the orbits, so that the variance becomes a sum of variances, one for each irreducible part of H; each of these will then be, for a many-particle system, a multiple of a basic variance which we can write in terms of the defining matrix elements.

In a single orbit, terms in the Hamiltonian may behave as $1(0, 0)$, $A(\frac{1}{2}, \frac{1}{2})$, $A^2(1, 1)$, $AB(1, 0; 0, 0)$, $A^2B(\frac{3}{2}, \frac{1}{2}; \frac{1}{2}, \frac{1}{2})$, $A^2B^2(0, 0; 1, 0; 2, 0)$ or their adjoints, the numbers in brackets giving (ν, μ) values for the orbit in question. Classing together terms which differ only in orbit labeling, we see that the one-particle Hamiltonian has two (k, μ) structures and three representation types as follows: $(A \mid B)$ which transforms as $D^{\frac{1}{2}\frac{1}{2}}(\frac{1}{2}, -\frac{1}{2})$ and $(AB \mid\ )$ which transforms as $D^1(0)$ and $D^0(0)$. Similarly the two-particle H involves nine structures and twelve representation types which we list ahead.

Certain of the interactions are easily seen to be irreducible. For the others the procedures of the last section may be used to carry out the decomposition, or alternatively, since the H operator is not very complicated, we may use the behavior under hole-particle conjugations (3.7) carried out in separate orbits; in any case, the process of reduction brings in various Hartree–Fock-type single-particle operators and some elementary averages, which we introduce now. We use i, j, k, l as labels for the states, a summation convention being always understood. For the orbits we use α, β, γ, and δ, with no summation unless specified; moreover, since a main concern is with partitionings according to orbits, we shall understand that two orbit symbols α and β appearing in an equation must not be taken equal; that case we would write separately. We now let

$$W_\alpha = \binom{N_\alpha}{2}^{-1} W_{ijij}\,, \qquad i < j; \quad i, j \text{ in } \alpha, \tag{4.2}$$

$$W_{\alpha\beta} = (N_\alpha N_\beta)^{-1}\, W_{ijij}\,, \qquad i \text{ in } \alpha; \quad j \text{ in } \beta, \tag{4.3}$$

$$\epsilon(i, j; \alpha) = W_{ikjk}\,, \qquad k \text{ in } \alpha; \quad i, j \text{ arbitrary}. \tag{4.4}$$

We shall be interested in partial traces over orbits, and introduce therefore the notation that $\epsilon(\beta; i, j) = \epsilon(i, j)$ when i, j are restricted to orbit β, $\epsilon(\beta, \delta; i, j) = \epsilon(i, j)$, when i is restricted to β and j to δ (it being agreed that $\beta \neq \delta$), and similarly for $\epsilon(\beta; i, j; \alpha)$, etc. The new quantities are understood to vanish if the indicated restrictions are not satisfied. When we write out operators themselves, and not simply their matrix elements, we shall also often write, for example, $g_i A_i(\alpha)$ which indicates that the i summation is restricted to range over the states of orbit α. We have now a certain (helpful) redundancy of notation; for example, $\epsilon(\alpha; i, j)\, A_i(\alpha)\, B_j(\alpha) = \epsilon(\alpha; i, j)\, A_i B_j = \epsilon(i, j)\, A_i(\alpha)\, B_j(\alpha)$.

The ϵ quantities define one-body operators; we shall need the corresponding traceless operators and the "scalar" operators which carry the trace. We therefore introduce averages $\epsilon(\beta)$, $\epsilon(\beta : \alpha)$, and traceless energy parameters ξ:

$$\epsilon(\beta) = N_\beta^{-1}\epsilon(\beta; i, i), \tag{4.5}$$

$$\epsilon(\beta : \alpha) = N_\beta^{-1}\epsilon(\beta; i, i; \alpha), \tag{4.6}$$

$$\epsilon(\alpha : \alpha) = N_\alpha^{-1}\epsilon(\alpha; i, i; \alpha), \tag{4.7}$$

$$\xi(\beta; i, j) = \epsilon(\beta; i, j) - \delta_{ij}\epsilon(\beta), \tag{4.8}$$

$$\xi(\beta; i, j; \alpha) = \epsilon(\beta; i, j; \alpha) - \delta_{ij}\epsilon(\beta : \alpha). \tag{4.9}$$

Since operators acting between two orbits have zero partial traces, we can write $\epsilon(\alpha, \beta; i, j) = \xi(\alpha, \beta; i, j)$, etc.

Of the quantities defined in (4.2)–(4.8), the first two give the average interaction energy between two particles in the same or different orbits; the third, $\epsilon(i, j; \alpha)$, and its refinements $\epsilon(\beta; i, j; \alpha)$ and $\epsilon(\beta, \delta; i, j; \alpha)$, define the effective one-body interactions[7] induced by two-body interactions summed over orbit α. The $\epsilon(i, j)$, the "primary" one-body parameters of (4.1), may be thought of as arising in part from the kinetic energy, and in part from the $\epsilon(i, j; \omega)$ summed over all orbits ω which are excluded from the space which we consider. $\epsilon(\beta)$ is its one-orbit average, while $\epsilon(\alpha : \alpha)$ and $\epsilon(\beta : \alpha)$ define the one-orbit average of the interaction induced into α and β from orbit α. Similarly, the ξ quantities define traceless one-body operators; for any ξ, $\xi(i, i) = 0$ and $\xi(i, j)\, A_i B_j$ is unitarily irreducible[8] with $\nu = 1$.

[7] These parameters define two-body interactions which, for a given configuration, behave as one-body operators and may therefore be regarded as number-dependent or "effective" one-body interactions. See (4.11)–(4.17) ahead. It will be appreciated that it is not really the particle rank but rather the unitary rank which is significant.

[8] For the multi-orbit case we shall often write $\sum \nu_\alpha = \nu$.

The parameters are of course not all independent and they have some obvious symmetries; in particular,

$$\begin{aligned} W_{\alpha\beta} &= W_{\beta\alpha}\,; \qquad & N_\beta \epsilon(\beta : \alpha) &= N_\alpha \epsilon(\alpha : \beta), \\ \epsilon(\beta : \alpha) &= N_\alpha W_{\alpha\beta}\,; \qquad & \epsilon(\alpha : \alpha) &= (N_\alpha - 1)\, W_\alpha\,. \end{aligned} \tag{4.10}$$

We can now make a partial decomposition of H by separating according to orbital structure, and complete the decomposition by introducing induced one-body and zero-body terms. Below, for each interaction, we give; a description (pairing, etc.); the operator and matrix element structure (for example, $W_{\alpha\alpha\beta\beta}$ describes $\beta \to \alpha$ pairing, the equal matrix element $W_{\beta\beta\alpha\alpha}$ describing $\alpha \to \beta$); the allowed representations for a particular set of orbits involved (the interaction will be scalar with respect to those not mentioned); the irreducible Hamiltonians (remember that, in the sums over orbits, two orbit symbols, α, β, etc, must not be taken equal; otherwise sums are unrestricted unless otherwise specified; and of course $\sum_{\alpha<\beta}$ is a double sum over α, β, restricted as shown):

(1) *Self-interaction.* $A^2B^2 \mid;\ W_{\alpha\alpha\alpha\alpha} \cdot AB \mid;\ \epsilon(\alpha, \alpha) \cdot 1;\ H_0:$

$$\begin{aligned} D^0(0) &: H_0 + \sum_\alpha \left\{ \epsilon_\alpha n_\alpha + W_\alpha \binom{n_\alpha}{2} \right\}, \\ D^1(0) &: \sum_\alpha \{\xi(i, k) + (N_\alpha - 2)^{-1} (n_\alpha - 1)\, \xi(i, k; \alpha)\}\, A_i(\alpha)\, B_k(\alpha), \\ D^2(0) &: -\frac{1}{4} \sum_\alpha \{W_{ijkl} - (N_\alpha - 2)^{-1} \{\delta_{ik} \xi(j, l; \alpha) + \delta_{jl} \xi(i, k; \alpha) \\ &\qquad - \delta_{il} \xi(j, k; \alpha) - \delta_{jk} \xi(i, l; \alpha)\} \\ &\qquad - \{\delta_{ik}\delta_{jl} - \delta_{il}\delta_{jk}\}\, W_\alpha\}\, A_i(\alpha)\, A_j(\alpha)\, B_k(\alpha)\, B_l(\alpha); \end{aligned} \tag{4.11}$$

(2) *Two-Orbit Pairing.* $A^2B^2 \mid;\ W_{\alpha\alpha\beta\beta}:$

$$D^{11}(1, -1) : -\frac{1}{4} \sum_{\alpha,\beta} W_{ijkl} A_i(\alpha)\, A_j(\alpha)\, B_k(\beta)\, B_l(\beta); \tag{4.12}$$

(3) *Two-Orbit Multipole.* $AB \mid AB;\ W_{\alpha\beta\alpha\beta}:$

$$\begin{aligned} D^0(0) &: \sum_{\alpha<\beta} W_{\alpha\beta} n_\alpha n_\beta\,, \\ D^1(0) &: \sum_{\alpha,\beta} N_\beta^{-1} n_\beta \xi(i, k; \beta)\, A_i(\alpha)\, B_k(\alpha), \\ D^{11}(0, 0) &: -\sum_{\alpha<\beta} \{W_{ijkl} - N_\beta^{-1} \delta_{jl} \xi(i, k; \beta) \\ &\qquad - N_\alpha^{-1} \delta_{ik} \xi(j, l; \alpha) - \delta_{ik}\delta_{jl} W_{\alpha\beta}\}\, A_i(\alpha)\, A_j(\beta)\, B_k(\alpha)\, B_l(\beta); \end{aligned} \tag{4.13}$$

(4) *Two-Orbit Hole-Particle.* $A^2B \mid B;\ W_{\alpha\alpha\alpha\beta} \cdot AB^2 \mid A;\ W_{\beta\alpha\alpha\alpha} \cdot A \mid B;\ \epsilon(\alpha, \beta)$:

$$D^{1/2,1/2}(1/2, -1/2) : \sum_{\alpha,\beta} \{\xi(i, l) + (N_\alpha - 1)^{-1} (n_\alpha - 1)\, \xi(i, l; \alpha)\}\, A_i(\alpha)\, B_l(\beta),$$

$$\begin{aligned} D^{3/2,1/2}(1/2, -1/2) : -\tfrac{1}{2} \sum_{\alpha,\beta} \{W_{ijkl} &- (N_\alpha - 1)^{-1}\, \delta_{ik}\xi(j, l; \alpha) \\ &+ (N_\alpha - 1)^{-1}\, \delta_{jk}\xi(i, l : \alpha)\}\, A_i(\alpha)\, A_j(\alpha)\, B_k(\alpha)\, B_l(\beta), \end{aligned} \tag{4.14}$$

$$D^{3/2,1/2}(-1/2, 1/2) : \{H^{3/2,1/2}(1/2, -1/2)\}^+;$$

(5) *Three-Orbit Pairing.* $A^2 \mid B \mid B;\ W_{\alpha\alpha\beta\gamma}\ ;\ A \mid A \mid B^2 : W_{\beta\gamma\alpha\alpha}$:

$$\begin{aligned} &D^{1,1/2,1/2}(1, -1/2, -1/2) : -\tfrac{1}{4} \sum_{\alpha\beta\gamma} W_{ijkl} A_i(\alpha)\, A_j(\alpha)\, B_k(\beta)\, B_l(\gamma), \\ &D^{1,1/2,1/2}(-1, 1/2, 1/2) : \{H^{1,1/2,1/2}(1, -1/2, -1/2)\}^+; \end{aligned} \tag{4.15}$$

(6) *Three-Orbit Multipole.* $AB \mid A \mid B;\ W_{\alpha\beta\alpha\gamma}$;

$$\begin{aligned} &D^{0,1/2,1/2}(0, 1/2, -1/2) : \tfrac{1}{2} \sum_{\alpha\beta\gamma} N_\alpha^{-1} n_\alpha \xi(j, l; \alpha)\, A_j(\beta)\, B_l(\gamma), \\ &D^{1,1/2,1/2}(0, 1/2, -1/2) : -\tfrac{1}{2} \sum_{\alpha\beta\gamma} \{W_{ijkl} - N_\alpha^{-1}\delta_{ik}\xi(j, l : \alpha)\}\, A_i(\alpha)\, A_j(\beta)\, B_k(\alpha)\, B_l(\gamma); \end{aligned} \tag{4.16}$$

(7) *Four-Orbit.* $A \mid A \mid B \mid B;\ W_{\alpha\beta\gamma\delta}$:

$$D^{1/2,1/2,1/2,1/2}(1/2, 1/2, -1/2, -1/2) : - \sum_{\substack{\alpha<\beta \\ \gamma<\delta}} W_{ijkl} A_i(\alpha)\, A_j(\beta)\, B_k(\gamma)\, B_l(\delta). \tag{4.17}$$

There are l different self-interaction terms of each allowed rank, $\binom{l}{2}$ two-orbit pairing terms, and so forth; though we have combined these above, the resultant further orthogonalities should not be forgotten. The operator of highest symmetry in each case is defined, for a given orbital structure, by subtracting the irreducible operators of lower symmetry from the total H with the given structure; but for convenience we have written them also in terms of their two-body matrix elements. Some of the terms for a given reduced H are equal (for example the $\delta_{ik}\delta_{jl}$ and $\delta_{il}\delta_{jk}$ terms in the self-interaction $D^2(0)$), but have been written separately because it is convenient to preserve the standard matrix-element symmetries. The scalar terms ($\nu = 0$) arising in (1) and (3) combine to give the scalar Hamiltonian which fixes the configuration centroids. Note that this term is not a trivial constant, this because of the number dependence (the induced parts are of course one- and two-body operators, whose unitary symmetry is lower than the maximum allowed by their orbital structure). Similarly, we must combine the (1, 0) terms arising from

(3) and (1) and the $(\frac{1}{2}, \frac{1}{2})$ terms from (6) and (4). Inspection of the results will make more or less clear the rules for writing the irreducible parts of H.[9]

Recalling that each term has a basic variance given (except for the scalar interaction) by Eq. (3.14), and a polynomial propagator given by (3.11) and (3.13), we see that we have the general results for configuration widths, given in terms of a physically meaningful orthogonal symmetry decomposition, and expressed directly in terms of the defining matrix elements. The partial widths, as we have indicated earlier, are given automatically. And, since each interaction term behaves simply under hole-particle transformations in any orbit, all the hole-particle symmetries of widths and centroid are made explicit.

Spherical orbits are of course of particular importance, and the angular-momentum-scalar property of H gives rise to special simplicities when our configurations involve only such orbits. The most important feature is that all the $(\nu = 1)$ ξ-interactions, the primary and induced "splitting" interactions, either vanish or can be transformed away; for the single-particle potential produced by averaging an angular-momentum-scalar interaction over a spherical orbit is itself spherical and cannot therefore split the states of a single particle. Then if the orbits are all distinguishable by parity and angular momentum, which is the common case, the only $(\nu < 2)$ interactions for a given configuration are unitary scalar ones, which, while they do move centroids, do not spread the states. As far as widths are concerned we may separate them or not as we wish; it is convenient to do so for then the variances become simply second moments.

We label spherical orbits by r, s, t, and u, where, in contrast with the general unitary orbit case, it is convenient now to permit orbital labelings to become equal; we assume at first no "radial degeneracy" (the orbital set does not include for example $1p_{\frac{1}{2}}$ and $2p_{\frac{1}{2}}$); matrix elements are $W^{\Gamma}_{rstu} = \langle rs \mid H \mid tu \rangle_{\Gamma}$, where Γ denotes J or (J, T) as appropriate. We eliminate the unitary scalar interactions by putting to zero the primary $(0 + 1)$-body energies and proceeding to[10] the Hamiltonian defined by

$$\mathscr{W}^{\Gamma}_{rstu} = W^{\Gamma}_{rstu} - \delta_{rt}\delta_{su} d^{-1}(2; r, s) \sum_{\Gamma} [\Gamma]\, W^{\Gamma}_{rsrs}, \tag{4.18}$$

where $d(2; r, s) = (1 + \delta_{rs})^{-1} N_r(N_s - \delta_{rs})$, is the two-particle dimensionality, the subtracted term in (4.18) being then simply W_{rs}.

[9] We hope to make available a report containing programs for computing configuration and configuration-isospin centroids and widths, as well as heuristic derivations of the rules for these cases.

[10] Since we write the results below in terms of a restricted orbital summation $(r \leqslant s,\ t \leqslant u)$ it is not worthwhile here, contrary to the discussion following (4.17), to add in the $\delta_{ru}\delta_{st}$ terms; and similarly in (4.20) below.

The orbital structure now uniquely defines the unitary symmetry (and vice versa), all seven interaction types being irreducible. Thus, the exercise of Section 2 for the two-orbit pairing can be extended to give the complete variance; for a given orbital structure the basic variance is $d^{-1}(2; r, s) \sum_\Gamma [\Gamma](\mathscr{W}^\Gamma_{rstu})^2$, where $[\Gamma] = (2J+1)$ or $(2J+1)(2T+1)$ looks after the counting of z-components. For the propagating polynomial corresponding to W^Γ_{rstu}, any or all of whose orbital labels may be equal, we have the conditions that it is a fourth-order polynomial in numbers which goes to zero in states which do not have at least one particle in each of orbits t, u (two in t if $t = u$), and similarly for holes in orbits r, s, and goes to unity when the fermion numbers have precisely these limiting values. This polynomial is constructible on inspection, and putting it together with the basic variance, we find

$$\sigma^2(\mathbf{m}) = \sum_{\substack{r\leqslant s\\ t\leqslant u}} \frac{(N_r - m_r)(N_s - m_s - \delta_{rs})\, m_t(m_u - \delta_{tu})}{(N_r - \delta_{rt} - \delta_{ru})(N_s - \delta_{st} - \delta_{su} - \delta_{rs})\, N_t(N_u - \delta_{tu})} \cdot \sum_\Gamma [\Gamma]\{\mathscr{W}^\Gamma_{rstu}\}^2, \tag{4.19}$$

a remarkably simple expression, which, like its unitary generalization, also yields the decomposition into partial widths.

If there is a radial degeneracy, not every orbit being then distinguishable by angular momentum and parity, there are effective one-body ($\nu = 1$) interactions which may produce an admixing of the orbits with different radial quantum numbers. These interactions are of the two-orbit and three-orbit multipole type and can only produce a rotation in the radial quantum number space, thereby generating new radial functions. Formally this comes about because, in the natural representation, which orders in the same way the states of two spherical orbits (r, t), which differ only in radial properties, the induced-energy parameter $\xi(r, t; i, j; s)$, as follows easily from (4.13) and (4.16) can connect only corresponding states, and these with the same amplitude, being then a multiple of the "spin-angular delta function" which we write as $\hat{\delta}_{ij}$. It is the essence of the Hartree–Fock "balancing" that the potential and kinetic radial admixings should cancel each other, and one commonly assumes, even without calculation, that this has been arranged; we see however that, because of the induced interactions, the balancing and hence the appropriate definition of the radial functions, will depend on the configurations.

Taking account of this effect we find easily, from the unitary-orbit results, that for the variance in the most general spherical-orbit case, we must modify (4.18) in that

$$\delta_{rt}\delta_{su} \to \tfrac{1}{2}\{\delta_{rt}\hat{\delta}_{su} + \hat{\delta}_{rt}\delta_{su}\} \tag{4.20}$$

and add an additional term in (4.19), the ($\nu = 1$) variance, namely

$$\sum_{r \neq s} m_r(N_s - m_s)(N_r N_s)^{-1} \left\{ \epsilon_{rs} + \sum_t (m_t - \delta_{rt})(N_t - \delta_{rt})^{-1} \epsilon_{rs}(t) \right\}^2. \quad (4.21)$$

Here ϵ_{rs}, which necessarily vanishes unless r, s are radially degenerate, is the conventional parameter which comes into the primary one-body Hamiltonian[11] [11], and $\epsilon_{rs}(t)$, which defines similar induced interactions is given by

$$\epsilon_{rs}(t) = \sum_{\Gamma} [\Gamma] \, W_{rtst}^{\Gamma}. \quad (4.22)$$

The centroid energies are, as always, given simply by

$$E(\mathbf{m}) = H_0 + \sum_r m_r \epsilon_{rr} + \sum_{r \leqslant s} m_r(m_s - \delta_{rs})(1 + \delta_{rs})^{-1} W_{rs}. \quad (4.23)$$

Realizing next that isospin is essential if we wish to deal with nuclei with a neutron excess, and is of great interest in other cases as well, we ask how we may calculate moments for configurations with fixed isospin. A formal basis for this lies in the connection of isospin with a direct-product $U(N)$ subgroup $U(\frac{1}{2}N) \times U(2)$, and starting with this, we could proceed along the same lines as before, introducing induced isospin splittings and similar related extensions of the quantities introduced above. We can however avoid all these formal problems by using a (p, n) basis in which protons and neutrons are treated separately, and we may recover the isospin by making the usual elementary subtraction of distributions with nucleon numbers (m_p, m_n) and $(m_p - 1, m_n + 1)$. The interaction must of course be defined in terms of the (p, n) matrix elements, the relationship between the two sets of matrix elements being well known [11]. This is what we have chosen to do and the results given ahead are derived in this way.[9]

Finally, as far as the Hamiltonian is concerned, we consider very briefly the magnitudes of the various kinds of interaction terms which we have given above. The width produced by an operator (or more precisely its second moment) is a natural measure of its magnitude, being, in the usual mathematical sense, a norm (the "Euclidean" norm), and one moreover which is properly compatible with the unitary norm for states. So, not surprisingly, asking about the width produced by an operator is equivalent to asking for its measure, and a decomposition, orthogo-

[11] For orbits r and s, which are identical or radially degenerate, ϵ_{rs} is given in terms of $\epsilon(i, j)$ of (4.1) by $\epsilon_{rs} = \epsilon(i, j)$, when states i (in orbit r) and j (in orbit s) are "corresponding" (identical with respect to spin and orbital angular momentum). Formally $\epsilon_{rs} = N_r^{-1} \epsilon(r, s; i, j)\mathcal{S}_{ij}$ and similarly $\epsilon_{rs}(t) = N_r^{-1}\epsilon(r, s; i, j; t)\mathcal{S}_{ij}$, where $\mathcal{S}$ is a delta function in the angular-momentum variables. Do not confuse ϵ_{rs} with $\epsilon(r{:}s) \sim \epsilon(r; i, i,; s)$ defined in (4.6). In (4.20) we have assumed that the orbit orderings are independent of radial quantum numbers (if $1p$ comes before $1f$, then $2p$ comes before $2f$); else further terms would be needed.

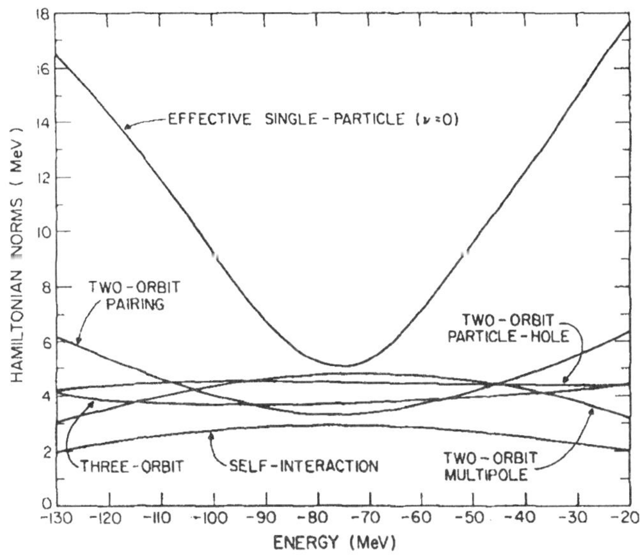

FIG. 4. The variation with excitation energy of various orthogonal parts of a Rosenfeld interaction in $(d_{5/2}d_{3/2}s_{1/2})^{12}$ calculated by the technique of Section 5. Since the orbits are spherical, all interactions, except the effective single-particle, have $\nu = 2$ and for these the norm (the square root of the second moment) is also a width. For the ($\nu = 0$) interaction, the $U(N)$ scalar, which gives the overall centroid, has been subtracted out. Interactions which differ only in orbital labeling (e.g., the six different two-orbit pairings) have been combined, as have also the three-orbit pairing and multipole. The ground-state is at approximately -130 MeV.

nal for widths, describes one for which we can meaningfully discuss the sizes of the separate parts without concern about cross terms. The natural spaces then in which to measure our Hamiltonian are those for single configurations, but in the (d, s) shell, for example, this produces far more information than we can use for a simple description of relative magnitudes. Proceeding to the space of all states for a fixed particle number is unsatisfactory because the states in common cases span too large an energy domain. We introduce therefore a procedure which we discuss in detail ahead and use there for other purposes, of using configurations as an indicator of the excitation energy in order to produce the variation with excitation energy of the various parts of H. Results for a Rosenfeld interaction acting in the large space $(d, s)^{12}$ are given in Fig. 4. The configuration-scalar part of the interaction (with the $U(N)$-scalar part, which simply moves the entire spectrum, subtracted out) is dominant for low energies, verifying that configurations do have some significance there, while two-orbit pairing is the strongest configuration-admixing interaction in the same energy domain.

We hope to return, in a later paper, to a more systematic discussion of Hamiltonian norms discussing among other things, the special simplicities which may enter when we deal with very large spaces, in which case there is an *a priori* favoring (arising in the propagation) of operators of low symmetry.

5. Ground-State and Low-Lying Energies and Structures

The low-moment-approximation distribution functions may be expected to give good information near the center of the spectrum where the level density is high, but one hardly expects at the outset that they will describe things well in the low-lying region where the density is low (although, of course, the exact distributions are valid here as well). Though the limitations on this kind of application are as yet by no means clear, considerable evidence has been given [4] that the low-moment functions often do very well indeed in the low density domain. In the course of applying them to nuclei in the (d, s) and (f, p) shells we shall give much more evidence to this effect.

We start with some general remarks. If we have a distribution for which all the exact symmetries (including J_z and, if relevant, T_z) are specified, then it refers to a set of nondegenerate states and the ground-state energy is that value at which the (cumulative) distribution function has a $(0 \rightarrow 1)$ jump; we may define it therefore by $F(E_1) = 1/2$ and similarly the j-th state comes at E_j with $F(E_j) = (j - 1/2)$, these definitions then being usable with the approximate functions. In this way we get a natural continuous $\rightarrow$ discrete transcription, first stressed and used by Ratcliff [4]; it is, in a way, the inverse of converting a discrete spectrum to a continuous one by use of a resolution function (or a histogram), and of greater use for low-lying spectra.

If J is specified but not J_z, the modification is trivial; $F(E_1) = 1/2(2J + 1)$ (and similarly with isospin). But if J is not specified, then different levels have different degeneracies; the discontinuities in the exact function measure the J values for us, but with the approximate functions we lose that feature and cannot even locate the level energies without more information. With even–even nuclei we may assume that any reasonable interaction gives a 0^+ ground state, and perhaps a $0^+ - 2^+ - 4^+$ sequence (the isospins if relevant also being obvious), and we are then able to locate these states. In other cases we may simply proceed on the assumption that the interaction will give the *observed* spectrum, and thereby produce the distribution spectrum. If the fixed-J distribution were available (we *shall* use fixed isospin) this kind of assumption would be unnecessary, but in the meantime it will have to be borne in mind when assessing the comparison between the distribution spectrum and either an exact shell-model spectrum or the experimental one.

First, we consider the low-lying distribution spectra for ^{63,65}Cu treated as $(f_{5/2}p_{3/2}p_{1/2})^m$, with the Kuo–Brown interaction already used by Wong [5] in his shell-model calculations. The distributions are fixed-isospin, to look after the neutron excess; the total dimensionalities for fixed m, T, and T_z are approximately 10 000 and 6000, while the largest matrix dimensionalities (J, J_z, also fixed) are 273 and 167. We give in Figs. 5 and 6, both the (m, T) and the finer configuration-isospin distributions, as well as the exact shell-model spectra and the experimental spectra (though of course comparison with shell model is of greater interest in the

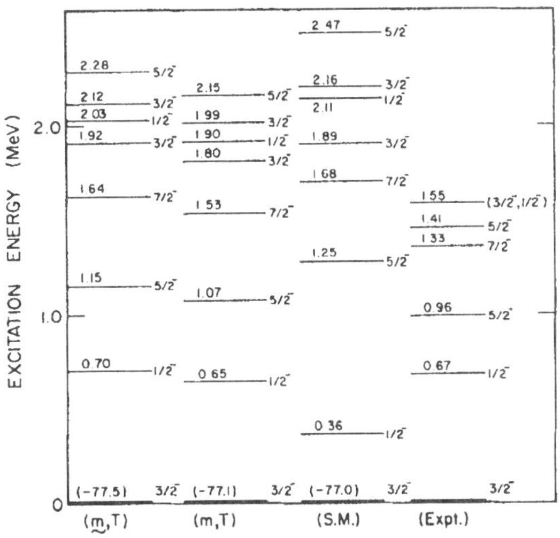

FIG. 5. The low-lying configuration-isospin and isospin distribution spectra for ^{63}Cu compared with shell-model results [5] using the same B–K interaction, and with experiment. The absolute energies of the ground states with respect to ^{56}Ni are also given; these include the $\binom{n}{2}$ and T^2 contributions described in the text ahead.

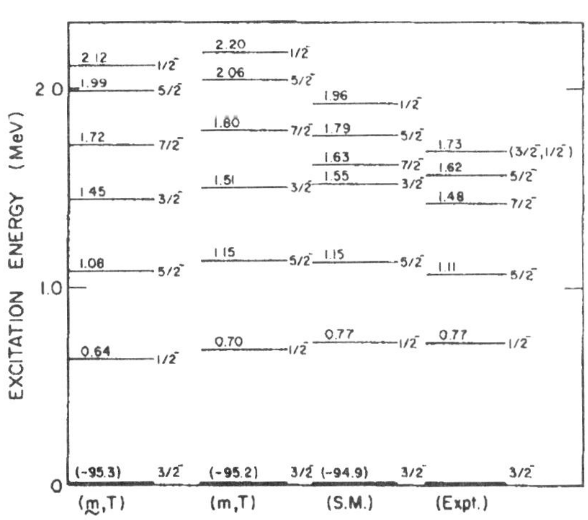

FIG. 6. Spectra for ^{65}Cu.

present context, a major concern being with the accuracy of the method). The two distribution spectra agree well with each other, and well also with the exact calculations (quite remarkably so for ^{65}Cu). We remark that the ground states in these cases lie [(3.6)–(3.8)] σ below the overall centroids (see Fig. 6 of Ratcliff [4]) and the total spectrum spans for states with the ground-state isospin are about 18 MeV. We emphasize again that, fixed-J distributions being as yet unavailable, the J-orderings given are not derived from, but are imposed upon, the results. The computations involved in these distribution spectra are quite minor, the total IBM-360-65 computing time for 35 complete configuration-isospin (f, p) spectra being about one min.

If the distribution subspaces are defined in part by symmetries which are not exact, configurations for example, then a sum of these distributions will define the spectrum, and at the same time the relative intensities of the different distributions, evaluated at the derived spectrum energies, will in principle yield the intensities of the various components in the eigenfunctions. Because we know from example that the component intensities fluctuate strongly even from one state to the next, we should in fact regard the derived intensities as representing an average over several states; Ratcliff [4], for ^{20}Ne with fixed (J, T), takes configuration intensities summed over the first three states (see his Figs. 7 and 8) and finds excellent agreement with detailed calculation. In the Cu isotopes also, the shell-model calculations find strong intensity fluctuations from state-to-state and corresponding disagreements with single-state distribution intensities. But instead of averaging the intensities over states, we can use them to produce relative orbit occupancies; since these already represent an average, we may with more confidence apply them to single states, deriving thereby quantities directly comparable with stripping and pickup results. Table 1 gives results for a number of (f, p) nuclei, agreement with shell-model being once again excellent. The individual proton and neutron orbit occupancies, also measureable, would be derivable from the finer distribution in which the individual orbit isospins are specified, but we have not yet made use of these distributions.

With Ni isotopes the same kind of agreements with shell model are found as for Cu. For other nuclei in the upper (f, p) shell ($f_{7/2}$ still closed), we have calculated distribution spectra and have made comparisons with experiment. For the even–even nuclei we find the first $(0^+ - 2^+)$ spacing, for the six cases available, to agree reasonably, being, on the average 150 keV larger than the 1.0–1.3 MeV observed. In fact there is for these nuclei generally good agreement for all the states identified, just as with the odd–even ^{63,65}Cu, and the same is true for many of the other odd–even spectra as well.

However, we can only expect detailed agreement with low-moment distribution results when there is a certain smoothness in the observed spectrum, for the distribution will not predict close-lying states. The "repulsion" between levels of the

TABLE I

Fractional occupancy (%) of the single-particle orbits for ground states of (f, p) nuclei, as predicted by the distribution method and detailed spectroscopic calculations

Nucleus	Distribution			Detailed Spectroscopy		
	$f_{5/2}$	$p_{3/2}$	$p_{1/2}$	$f_{5/2}$	$p_{3/2}$	$p_{1/2}$
^{60}Ni	2.0	40	13	5.5	38	8.3
^{61}Cu	3.5	48	18	6.0	47	13
^{61}Ni	6.2	44	18	10	48	17
^{62}Cu	4.8	55	26	5.2	55	24
^{62}Ni	8.3	49	27	9.3	46	30
^{63}Cu	8.9	58	32	8.7	59	32
^{64}Zn	3.1	71	48	—	—	—
^{64}Cu	13	60	42			
^{64}Ni	17	50	48	22	48	40
^{65}Zn	13	69	46	—	—	—
^{65}Cu	19	61	45	20	60	43
^{66}Zn	19	72	50	—	—	—
^{67}Ga	20	81	53	—	—	—
^{67}Zn	27	72	50	—	—	—
^{68}Zn	34	74	50			

same exact symmetries makes for such a smooth behavior since it discourages anomolously small spacings [12, 4]. With our distributions however, the J's are mixed and thus we must expect failures when, as happens especially with odd–odd nuclei (but also with some odd–even, three excited levels for example being within 200 keV of the ground state for ^{65}Zn), levels of different J are close. For these cases we should have distributions for fixed J in order to make detailed comparisons which are meaningful for the very lowest states. There is no way at present to guess how well the fixed-J distributions will do for these nuclei.

Instead of trying so hard with the lowest-lying states in unfavorable nuclei, we can choose a point higher up in the spectrum as the reference point, leaving aside then comparisons for the states below. This is helpful in two ways; we escape from the immediate domain of the ground state where things may be uncertain; but also we move closer to the centroid and make then less stringent demands on our distributions. There is a lot of evidence that even moving up a few states, one MeV or less in (f, p) nuclei, yields a marked improvement.

The B–K interaction used by Wong [5] for ^{63,65}Cu slightly underestimates the excitation energies of the isobaric analogs of the Ni ground states, and overestimates the binding energies relative to ^{56}Ni. Adding a term $\{0.11T^2 + 0.07\binom{n}{2}\}$ to correct these deficiencies, we have calculated the ground-state energies, shown, compared with experiment, and with detailed calculations where available, in Table 2. For Coulomb energy corrections we have used the measured Coulomb displacement energies to correct the observed energies, and, except where indicated, have used a low excited reference state, measuring down from that via the observed excitation energy to produce the energy of the ground state. For the 34 nuclei given in the table, the average error in locating the ground state is about 700 keV, the largest errors arising in general for odd–odd nuclei where insufficient information is available to use an excited reference state.

TABLE II

Ground-state energies of (f, p) nuclei, as predicted by configuration-isospin distributions, by empirical binding energies, and by shell-model calculations. Energies are in MeV with respect to ^{56}Ni. Coulomb-corrections have been applied to empirical energies. The interaction is B–K supplemented by $\{0.07\binom{n}{2} + 0.11T^2\}$.

	Distribution	*B.E.*	*S.M.*		Distribution	*B.E.*
^{60}Ni	−42.8	−42.8	−42.9	^{65}Ni	−84.8	−83.9
^{60}Cu	−45.9	−45.4	—	^{65}Zn	−102.5	−101.9
^{61}Ni	−50.9	−50.7	−51.1	^{65}Ga	−110.1[b]	−107.9
^{61}Cu	−57.2	−57.0	−56.9	^{65}Ge	−113.4[b]	−111.0
^{61}Zn	−60.3[a]	−60.5	—	^{66}Ni	−92.1	−92.8
^{62}Ni	−61.8	−61.3	−62.1	^{66}Cu	−102.2[a]	−101.4
^{62}Cu	−66.3	−65.9	−66.6	^{66}Zn	−113.3	−112.9
^{62}Zn	−73.2	−73.2	—	^{66}Ga	−119.3[a]	−116.9
^{63}Ni	−68.8	−68.1	—	^{66}Ge	−126.5[a]	−123.3
^{63}Cu	−77.5	−76.7	−77.4	^{67}Cu	−109.6[a]	−110.4
^{63}Zn	−83.7[a]	−82.3	—	^{67}Zn	−119.5	−119.8
^{63}Ga	−86.9[b]	−86.1	—	^{67}Ga	−129.5[a]	−127.9
^{64}Ni	−78.6	−77.8	−78.6	^{67}Ge	−135.6[b]	−133.0
^{64}Cu	−86.0[a]	−84.5	—	^{68}Cu	−116.7	−116.6
^{64}Zn	−94.8	−94.1	—	^{68}Zn	−129.7	−129.9
^{64}Ga	−98.9[a]	−96.2	—	^{68}Ga	−138.0[a]	−136.0
^{65}Cu	−95.3	−94.4	−94.9	^{68}Ge	−146.9[a]	−144.9

[a] No "excited-state" correction, due to lack of data.
[b] Ground-state J-value is assumed to be 3/2.

There is, then, excellent agreement between the distribution and shell-model energies in the smaller spaces where shell-model calculations are feasible, it being understood that it is often necessary and always advisable to use an excited reference energy. But with the larger spaces encountered, where our comparisons are with experimental energies, we meet simultaneously the two problems of the accuracy and limitations of the methods, and the adequacy of the interaction. A significant feature of the (f, p) results is that we see no systematic disagreements dependent on the sizes of the distribution spaces, even though these range from 80 to 35 000 (the number of states with given m, T, T_z), the range of matrix sizes being up to 6000. In particular, the binding energy differences for isobaric nuclei (for which the $\binom{n}{2}$ correction to the interaction does not enter, while the T^2 correction is in any case small) are given very well despite the large variation of dimensionality with isospin. Altogether then we have considerable confidence that the (f, p) distribution results and interaction are both good.

It is clear however that, as we go to larger and larger spaces, the accuracy of the energy determination for any low-lying state must decrease, for, measured in terms of the natural unit σ, a low-lying state is further out from the center of the distribution the larger the dimensionality. We have seen that in the immediate neighborhood of the ground state there are "local" sources of inaccuracy, especially for odd–odd nuclei, which can be minimized by use of an excited reference state. This at the same time brings us closer to the distribution center, thereby alleviating the dimensionality problem. There are however practical limitations on the excitation energy which we may use, and we are left with the feeling that dimensionality imposes the most significant limitation on the entire procedure (fixed-J distributions, which involved much smaller dimensionalities, may then be of greater value because of that).

We propose later to study the limitations by means of soluble models, one for noninteracting particles (which we briefly refer to in the next section), and one involving Hamiltonians made from Casimir operators for compatible groups.[12] We believe at present that two-moment distributions may be unreliable beyond (3.5–4.0) σ from the center, which corresponds to dimensionalities say 5000–100 000 depending on the reference energy and the degeneracies. A failure often shows as a marked increase in the low-lying spacings (an example is given in Fig. 7 ahead, and we have earlier seen a trace of the effect in the $0^+ - 2^+$ (f, p) spacings). This indicates, as one would expect, that the approximate distribution has a smaller energy derivative near the ground state then is compatible with the exact distribution. It may well be in fact that this gives a good indication whether or not our procedures are accurate; if so, a distribution is trustworthy if it gives a good level density in the neighborhood of the reference energy.

We turn briefly to the (d, s) shell which, though it involves the same orbits as our (f, p), has larger spectrum spans and a correspondingly larger energy scale, the 12-particle configuration widths being characteristically $\sim$8 MeV compared

[12] Another method is to study how well the distribution spectra reproduce the effects of adding a term αJ^2 to the Hamiltonian. Some preliminary calculations suggest that, in the large even–even (d, s) cases, the entire spectrum moves properly except for the ground state which, as discussed ahead, is known in these cases to be badly given.

with the (f, p) value $\sim$2.5 MeV. The (d, s) cases have been discussed by Ratcliff [4], largely via scalar (fixed-m) distributions with moments as high as fourth order, and using a Rosenfeld interaction. For ^{20}Ne however, fixed (m, J, T) distributions, with moments calculated via detailed spectroscopy, have been used and the results discussed in detail. The average error, when compared with exact calculation, in the energy of the lowest state of given (J, T) is (1.6, 0.9, 0.8) MeV for (2, 3, 4)-moment spectra, the errors for the second state being smaller, (0.9, 0.6, 0.6) MeV. For $A = 22$, in which the matrix sizes range up to 500, the spectrum span being $\sim$60 MeV, there are also exact results [13], in fact, for a variety of reasonable interactions; the configuration distributions for this odd–odd case give ground-states too low for each interaction by 2.7–4.0 MeV, but once again, when we use the third (or higher) excited state as the reference, we get excellent agreement, the distribution ground states then lying 0.3–1.0 MeV too low. Figure 7 shows the experimental ^{28}Si spectrum compared with spectra calculated via $\mathbf{m}$ and $(\mathbf{m} - T)$ distributions, the interaction in this case B–K (12.5) of Ref. [14,]; except for the lowest spacing, remarked on earlier, the agreement is satisfactory.

For binding energies we can proceed as in the (f, p) shell, but to make a coherent picture and to separate the effects of inadequacy of the interaction and errors in the method is much harder. This is in part perhaps due to the fact that the relative span in A (16–40) is larger than in the (f, p) case so that, as pointed out by many

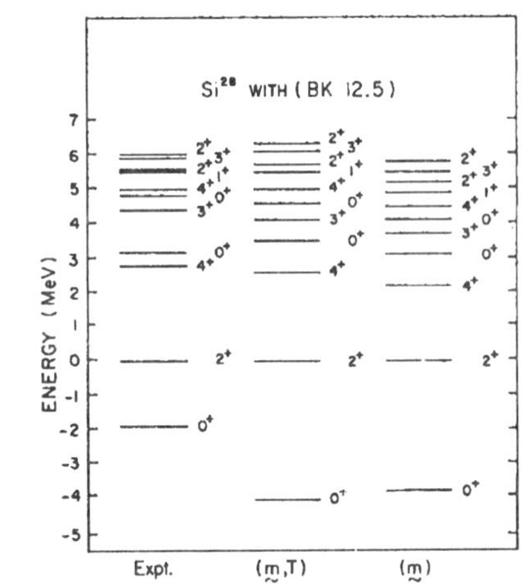

FIG. 7. Distribution spectra, configuration-isospin and configuration, for ^{28}Si compared with experiment, drawn with the first 2^+ states aligned. The $(\mathbf{m} - T)\,2^+$ state is calculated to be 0.2 MeV above the corresponding $\mathbf{m}$ state. For absolute comparison with experiment see Table III, noting that, because of the use of an excited reference, the discrepancies given there are really relevant for the low-lying excited states. The depression of the ground states is discussed in the text.

authors, one does not expect any $(0 + 1 + 2)$-particle Hamiltonian to work well over the entire domain of A. In any case, many interactions have been considered and none is adequate throughout the entire shell. This raises a problem about the "secular" ($\binom{n}{2}$ and T^2) terms which we think it quite reasonable to add in for binding energies (we have done so in (f, p)). If, as we shall do, we use one interaction for the lower-half shell and one for the upper half, we no longer have a pair of calculable energies at each end of the range for use in fixing the secular terms. We shall therefore use the $A = 28$ distribution ground-state energy for this purpose, thereby, of course making an assumption (about whose validity we feel fairly confident) concerning the accuracy of that theoretical energy. For the Coulomb-energy correction to the empirical energies we use $3.45\,Z' + 0.365\binom{Z'}{2} + 0.19[\frac{1}{2}Z']$, where $Z' = (Z - 8)$ and $[\frac{1}{2}Z']$ is $\frac{1}{2}Z'$ or $\frac{1}{2}(Z' - 1)$ according as Z' is even or odd. This correction is very slightly different (to better accomodate things at the upper end) than that used in Ref. [13]. For the distribution ground states we use configuration-isospin distributions with an excited reference energy ($\leqslant 8$ MeV in the lower half shell, $\leqslant 4$ MeV in the upper half). As we have indicated above, the theoretical results are stable with respect to this reference once we get above a few MeV, but of course, if an interaction does not reproduce the empirical spectrum well, we get a fluctuation (which we find may be (1–2) MeV in the (d, s) shell) depending on which state we use. We note also that, the energy unit being larger in (d, s), agreement with the same quality as in (f, p) will imply discrepancies which may be as large as 2 MeV.

For the upper half-shell we use the B–K ($\hbar\omega = 12.5$ MeV) interaction [14] with a (small) secular correction $0.6T^2$. The agreement is excellent. For the lower half-shell, seven Hamiltonians are given by Halbert *et al* [13] and we have considered in detail the two most favored by these authors, $(K + 12FP)$ and $(KB + {}^{17}\mathrm{O})$. Up to $A = 22$, agreements are for the most part good without secular correction, which is not surprising since these interactions are regarded as optimal for this domain and since we have already indicated good agreement between shell-model and distribution energies. Beyond $A = 22$, however, we get an overbinding, increasing with particle number to $\sim$10 MeV at $A = 28$ (it would give $\sim$50 MeV at $A = 40$). We use then, for both interactions, a secular correction $0.175\binom{n}{2}$ and then the agreements are good. The $(K + 12FP)$ results are given in Table 3, together with B-K (12.5) results for the upper half.

We conclude this section with a remark about comparisons between different distributions. We would naturally expect a scalar (fixed-m) to be inferior to a configuration distribution, and similarly for the isospin cases. Surprisingly enough however, while this appears to be true for the "realistic" interactions used in the (d, s) shell, the simpler distributions give essentially the same results as the more complicated ones when a Rosenfeld interaction is used. There is no present understanding of this curious fact.

TABLE III

Ground-state energies in (d, s) shell as predicted by configuration-isospin distributions, and by binding energies. Energies are with respect to ^{16}O. In the lower half shell the interaction is $(K + 12FP)$ with $0.175\binom{n}{2}$ correction; in upper half the interaction is B–K (12.5) with $0.6T^2$ correction

A	T	Distribution	$B.E.$	A	T	Distribution	$B.E.$
20	0	−39.2	−40.7	28	0	−134.7	−136.2
	1	−31.0	−30.3		1	−126.2	−126.8
	2	−23.3	−23.8		2	−120.0	−120.8
21	1/2	−46.0	−47.4	29	1/2	−146.5	−146.2
	3/2	−38.3	−38.4		3/2	−145.0	−144.7
22	0	−57.5	−58.4	30	0	−155.9	−156.0
	1	−57.3	−57.8		1	−155.8	−155.3
23	1/2	−70.4	−70.9	31	1/2	−169.2	−168.3
	3/2	−64.7	−63.0		3/2	−161.1	−161.9
24	0	−86.1	−87.4	32	0	−184.2	−183.5
	1	−78.9	−77.8		1	−177.1	−176.3
25	1/2	−94.2	−94.7	33	1/2	−193.1	−192.1
	3/2	−89.8[a]	−86.9		3/2	−188.4[a]	−186.4
26	0	−106.8	−106.0	34	0	−207.4[a]	−203.8
	1	−106.6	−105.8		1	−204.3	−203.6
27	1/2	−119.0	−119.1	35	1/2	−218.1	−216.4
	3/2	−113.2	−112.2		3/2	−211.5	−210.5
28	0	−136.2	−136.2	36	0	−231.8	−231.9
	1	−127.8	−126.8		1	−224.9	−225.0

[a] No "excited-state" correction, due to lack of data.

6. Level Densities

We outline now a theory of nuclear level densities which, in strong contrast to the conventional theories, takes detailed account of shell-model structures and internucleon interactions. It is based on a quite straightforward application of spectral distributions, and its general nature should be quite obvious at the outset since, after all, spectral distributions are nothing more than partial level densities. This theory undertakes to give *a priori* evaluations of level density parameters, in terms of the parameters of the effective interaction, and to make, in other ways as well, closer contact with modern nuclear structure procedures and ideas.

There are two domains of level density, according as we are or are not dealing with separate countable levels. The first domain includes the low-lying spectra, which we have just considered, and the spectra of unbound levels which begins at the lowest separation energy, while the second involves higher overlapping or not easily separated levels. In each domain the densities for fixed angular momentum and parity have been always of great interest and we might consider fixed isospin as well. In the second domain, where one is really concerned more with strength functions than with level densities *per se*, there is interest also in the decomposition of the densities according to the number of hole-particle-pairs which are present, or more generally according to configurations. Our methods will handle all these cases easily.

The conventional theories (reviewed by Ericsson [15], Lynn [16], Bloch [17], and Huizenga *et al* [18]) are based largely on thermodynamic notions whose applications to nuclei are, as discussed by Bloch, of doubtful validity. In their simplest form they pay attention neither to shell structure nor to residual interactions, and consequently even if a good theoretical foundation were available for them, they would largely fail to make contact with much of the phenomena of greatest current interest. A fair amount of work has been done on shell-model effects insofar as they be included by largely ad hoc modifications of the single-particle spectrum, but residual interaction effects, with very few exceptions (the use of quasiparticles for pairing, the use of Casimir-type Hamiltonians which lead to soluble shell model problems) have been ignored. We shall see that there is no proper basis for any such theory built on spherical orbits. It is not ruled out, though it seems improbable, that there is, in some cases, a single-particle spectrum involving nonspherical orbits which does take account of the major interaction effects[13], but the ad hoc theories do not investigate that question. And even for such a case, as we demonstrate below, the distribution method is far simpler than and superior to the methods generally used.

Random matrix theories (reviewed by Porter [12], Wigner [19] and Mehta [20]) also produce level densities, but they make even less contact with the standard theories of nuclear structure, and, besides that, the densities which they produce (usually described by the Wigner "semicircular" distribution) are regarded as being unsatisfactory.[14] The random matrix theories are then mainly used for more sophisticated quantities, level spacings and so forth, though whether there is a fundamental justification for this application is a difficult open question. We believe incidentally, and this is supported by the recent calculations of Wong [6] that the

[13] In *very large* spaces with a large number of particles and holes ($m \gg 1$, $N - m \gg 1$) there is an *a priori* tendency for the low-symmetry ($\nu = 1$) part of the high-particle-rank part of an arbitrary operator to dominate, this arising from the $m^{2k}\{(N-m)/mN\}^{\nu}$ asymptotic dependence of the $U(N)$ polynomials. Since ($\nu = 1$) is effectively one-body, we can therefore see a possible source of behavior determined by a single-particle spectrum.

[14] We agree with this (see Fig. 2). However most authors who make this comment ignore the relationship between "bounded" and "unbounded" level densities, which we discuss ahead, and leave unclear then the basis for their conclusion.

weakness of random matrix theories is not in the use of an ensemble of Hamiltonians to describe a situation which, after all, does have a definite Hamiltonian. Instead the flaws are in the use of an improper ensemble, one which ignores the direct-product nature of the fermion space and the fact that the conventional Hamiltonian is at most a two-body operator, defined therefore in a factor subspace of the entire space of the system.[15] Ensembles then, if they should be used at all, should be taken in this subspace; the results will be to produce distributions which in finite many-particle spaces are close to normal and depend on only a few known parameters of H. But it should be clear that an ensemble average is not really required at all for level densities.

The two most striking features of the results of the conventional theories are the exponential increase of level density with increasing energy, and the Gaussian form which gives the variation at a given energy with angular momentum (and isospin). We shall verify the Gaussian forms and will in fact calculate the angular-momentum and isospin widths (or "spin cut off factors") in terms of the orbit and interaction parameters. With regard to the energy variation, since our procedure allows only a finite number N of single-particle states, our spectrum then has some upper limit, and our level densities must first increase to a maximum somewhere near the centroid energy and then fall off to zero at the upper limit. This is contrary both to experience and to the results of conventional theories, but the difficulty is a trivial one. As the excitation energy increases, more and more single-particle states become involved and thus, if we wish to consider level densities up to a certain excitation energy, we must be careful to take a wide-enough single-fermion spectrum[16] (holes and particles). We shall see that, for total level densities, we must in fact stop in energy far short of the centroid. Besides that a finite single-particle spectrum is essential anyway when we deal with configuration structures as in statistical reaction studies (doorway states, the Griffin model for the transition between direct and compound nuclear reactions [23] and so on).

Level densities are conventionally, and for good reason, considered as a function of the excitation energy (with respect to the ground state). The theories which neglect residual interactions necessarily ignore all the subtleties which come into play in fixing the ground state energy for a complex system. We, on the other hand, shall take for granted that the considerations of the preceding section give an adequate solution to the ground-state problem, though we should keep in mind that higher moments or finer subspaces may in some cases be necessary. Finally, for the higher-lying level densities there are going to be many uncertainties about the interactions to be used. But this is simply a sign that problems of defining and

[15] See in this connection a very perceptive remark near the end of p. 210 in the book by Grenander [21].

[16] For noninteracting fermions, the relationship between the bounded and unbounded cases has been studied analytically by F. C. Williams [22].

constructing effective interactions automatically intrude into level density studies as soon as one treats the densities in a more realistic way; it is not ruled out that in the future we may learn something about the interactions by studying the densities.

We have mentioned in the last chapter that the spectrum for noninteracting particles is useful for testing the accuracy of distribution methods. Since it is also the basis of most of the level density studies to date, we comment briefly about it. H is now a $(0 + 1)$-body operator; ignoring H_0 and choosing a natural basis which diagonalizes H, we have $H = \sum_{i=1}^{N} \epsilon_i n_i$ (n_i being the number operator for state i); the centroid energy is $mN^{-1} \sum \epsilon_i$ and the variance

$$\sigma^2(m) = m(N - m)\, N^{-1}(N - 1)^{-1} \left\{\sum \epsilon_i^2 - N^{-1} \left(\sum \epsilon_i\right)^2\right\},$$

these of course being valid for arbitrary spacings and for arbitrary degeneracies. Thus we can study, almost without effort, such things as the effects of gaps or high degeneracies in the single-particle spectrum [24]. Expressions for higher moments are easy to give in terms of standard symmetric polynomials so that we can study departures from normality. If we wish to use spherical orbits, and study angular-momentum distributions, we may proceed as we do later in this paper for the interacting-particle case.[17] We shall not pursue these things but instead consider the example of Fig. 8, due to F. C. Williams [22], in which the single-particle

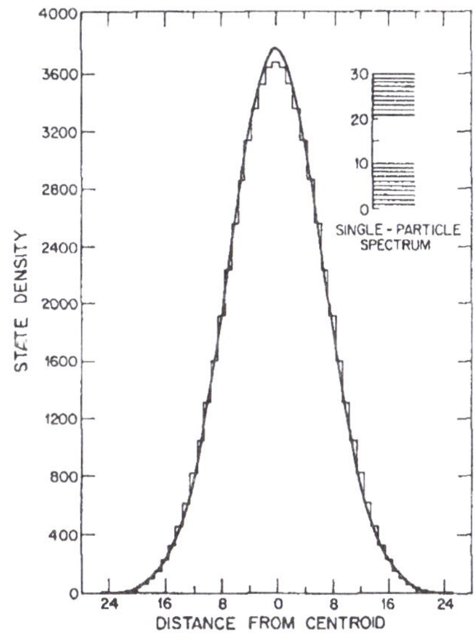

FIG. 8. The exact $(5p - 5h)$ state density, and the two-moment approximation, for 10 non-interacting particles with the single-particle spectrum shown in the inset. The single-particle degeneracies are unity, the total number of states is 63 504, $\sigma = 6.770$, and the exact ground state comes at 25 units $(= 3.7\sigma)$ below the centroid. The energies have integral values, and the exact spectrum and level degeneracies follow from inspection of the exact function.

states are divided into two sets (two "orbits") and we consider the configuration distributions, which we may characterize by the number of particle-hole pairs. The exact spectrum is calculated by counting, and the distribution moments via (2.4). For variety we show the density (frequency) function rather than its integral, the distribution function. The agreement is excellent. For the ground state determination it turns out that, if we can tolerate a maximum error of 2% of the spectrum span, we must use a reference energy at least 5 units above the exact ground state, where $F(E) \sim 50$; on the other hand, the densities match within that error as low as 2 units above, where $F(E) \sim 5$. This is consistent with our discussion in Section 5.

Let us now consider level densities for interacting particles, at first ignoring angular momentum but allowing for configuration and isospin structure. Imagine that we start without interactions and turn on the residual interaction, focussing our attention on states of single configurations. Two things happen: The centroids of the configurations are shifted by the scalar interaction, and the configurations are broadened[18] both by self-interactions and by the interactions between particles in different orbits (the $\mu \neq (0, 0, \ldots)$ terms in the mutual interaction H also produce configuration *admixings* but these are of no present concern).

TABLE IV

The average interactions W_{rs}, and the range of the differences Δ_{rs} in the effective single-particle energies induced by the interaction, for the (d, s) shell with Rosenfeld and B–K (12.5) interactions. Values are in MeV. The orbit ordering is (5/2, 3/2, 1/2)

	Rosenfeld		B–K (12.5)	
r.s.	W_{rs}	Δ_{rs}	W_{rs}	Δ_{rs}
1.1	−0.78	—	−0.88	—
1.2	−0.81	$-0.09 \leqslant \Delta \leqslant 0.18$	−1.02	$-2.51 \leqslant \Delta \leqslant 0.77$
1.3	−0.40	$-3.77 \leqslant \Delta \leqslant 3.93$	−0.70	$-3.51 \leqslant \Delta \leqslant 3.00$
2.2	−0.78	—	−0.48	—
2.3	−0.40	$-3.85 \leqslant \Delta \leqslant 3.93$	−0.39	$-1.66 \leqslant \Delta \leqslant 3.46$
3.3	−1.51	—	−1.32	—

[17] Alternatively, a theory for fixed-J distributions, in which configurations and isospin may also be specified, has been given in Ref. [24], and this could be easily applied to this case, although there are still problems connected with more general applications.

[18] They could in principle be narrowed.

The shifting of the centroids may be taken care of by a modification of the orbit energies

$$\epsilon(\alpha) \rightarrow \epsilon(\alpha) + \tfrac{1}{2}(m_\alpha - 1)\, W_\alpha + \tfrac{1}{2} \sum_\beta m_\beta W_{\alpha\beta}$$

$$= \epsilon(\alpha) + \tfrac{1}{2}(m_\alpha - 1)(N_\alpha - 1)^{-1} \epsilon(\alpha : \alpha) + \tfrac{1}{2} \sum_\beta m_\beta N_\beta^{-1} \epsilon(\alpha : \beta), \qquad (6.1)$$

the added terms being of course the induced energies discussed in Section 4. Table 4 gives values for B–K and Rosenfeld parameters for the (d, s) shell considered as three spherical orbits. It is seen that for Rosenfeld the effective spin-orbit splitting[19] varies little, and, while the $s_{1/2}$ orbit does move with respect to the other two, this has little effect on level densities since the orbit is small. For B–K, things are different; the induced splittings are large and strongly dependent on configuration. Figure 9 shows the $(d, s)^{12}$ configuration centroids for these cases as compared with the noninteracting-particle case. Figure 10 shows the state

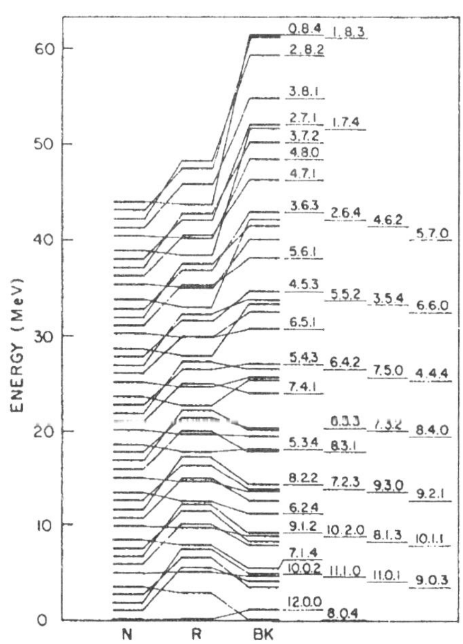

FIG. 9. Configuration centroid energies for $(d, s)^{12}$, given with respect to the lowest centroid. The single-particle energies are taken from ^{17}O. The three cases are for noninteracting particles, and for particles interacting via Rosenfeld and B–K (12.5) interactions.

[19] A spin-orbit splitting is possible with jj configurations even for a central interaction.

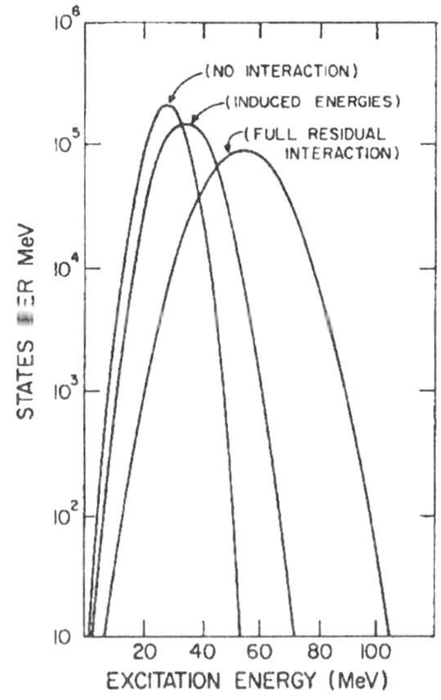

FIG. 10. State densities (all T, T_z) for $(d, s)^{12}$: (i) As derived from configuration centroids calculated with $A = 17$ single-particle energies, converted to a continuous curve via a 2-MeV Gaussian window for each configuration; (ii) same, but including the effects of induced B–K (12.5) single-particle energies; (iii) taking account of the complete interaction by centroids and widths.

densities for the cases where the residual interactions are ignored, where the induced B–K energies only are taken into account, and where the complete residual interaction is considered. We see that including the effects of the induced energies results in lower state densities at a given excitation, as we would expect from the fact (Fig. 9) that they produce an extended centroid spectrum. If now the effects of the nonunitary-scalar interactions were small, we would have a new noninteracting particle model (with orbit energies somewhat different than the primary ones), and a general justification of the conventional counting methods for treating the densities. But this is not at all so[20], as follows directly from the fact that the widths, and indeed the mixing widths, are much larger than the centroid spacings. The effects of the nonscalar interactions are in fact remarkably large as we see in Fig. 10.

The procedures which produce these densities automatically give the partial densities for separate configurations, and one envisions that these will be of value

[20] We point out again that a different nonspherical basis might reduce these effects, and we have in fact seen an appreciable reduction with a Hartree–Fock basis supplied by J. C. Parikh [25]. This basis is, of course, designed to minimize an energy rather than a width. To find the proper optimum basis is a difficult problem and interesting (though not relevant to the present theory). We shall return to it later.

in statistical reaction and cascade theory. We may specify isospin also if we wish; Fig. 11 shows the isospin decomposition for the $(d, s)^{12}$ density. But the procedures do not specify the angular momentum for, as we have said earlier, fixed-J-spectral distributions involve difficult technical problems not yet fully solved. If these were available they would have many uses, but in the meantime we can find a good solution to the fixed-J level density problem by evaluating moments of J_z for fixed energy. We produce these by using configurations as an indicator of the excitation energy, as we have already done to produce Fig. 4.

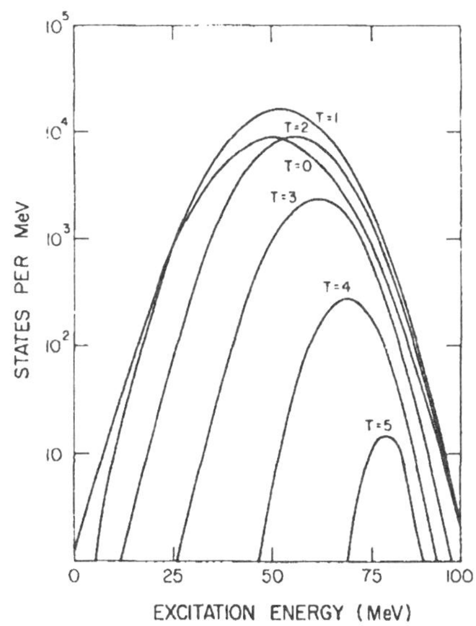

FIG. 11. The isospin decomposition of the $(d, s)^{12}$ state density for a Rosenfeld interaction, as calculated via configuration-isospin distributions. The densities plotted are for fixed (m, T, T_z).

The elementary procedure uses the fact that J_z defines an *additive* quantum number, and thus, by an elementary application of the central limit theorem, we expect that in the limit of large particle number, the distribution of values of the total J_z should be normal and with variance $\sigma^2(m)$ equal to $m\sigma^2(1)$; thus [15]

$$\begin{aligned} \rho(E, M) &= (2\pi\sigma^2)^{-1/2} \exp(-M^2/2\sigma^2)\, \rho(E), \\ \sigma^2 &= \overline{M^2} = m\langle J_z^2\rangle^1, \end{aligned} \tag{6.2}$$

where $\rho(E, M)$ is the density of states with fixed J_z $(= M)$ and $\rho(E)$ is the total density of states.

The two ingredients of this expression are the incoherent addition, whereby the mean squares add linearly, and the normality of the resulting distribution, at least for large m. However we can not really expect the variance to be properly given by (6.2) for the angular momenta for different particles do not combine independently, for one reason because of Pauli principle "blocking effects" (or more formally because we have used a limit theorem not valid for the algebraic structure involved). In fact, we see by a simple application of (2.1) that if we should average σ^2 over all the states formed with m particles in N single-particle states, then

$$\overline{\sigma^2} = m(N-m)(N-1)^{-1}\langle J_z{}^2\rangle^1, \tag{6.3}$$

the average blocking correction being then $(N-m)(N-1)^{-1}$. It will not usually be adequate to take $N \gg m$ (remember that we in any case have $m \gg 1$) since we would then be averaging over a very wide energy domain, and our interest is in the angular-momentum dependence at a fixed energy. The difficulty with the σ^2 values has been well appreciated, and people analyzing level densities have demanded freedom to maintain the normal form for the distribution while ignoring the simple value given above for σ^2. They have often chosen a value in terms of a "moment of inertia"; we wish instead to derive the σ^2 as a function of the excitation energy and the nuclear model and interaction parameters. We shall maintain the assumption of normality, but by evaluating the fourth moment of J_z will verify that it is adequate, and will produce a fourth-moment correction which could easily be applied (note that the odd powers of J_z average to zero so that we have only the even moments).

The procedure begins by using the configuration distributions (or configuration-isospin for finer results), along with the ground state energy determination as discussed in Section 5, to give the intensities of the various configurations as a function of the excitation energy. Writing the intensities as $I_{\mathbf{m}}(E)$ we have

$$\langle J_z{}^k\rangle^{\mathbf{m}}_{(E)} = \sum_{\mathbf{m}} I_{\mathbf{m}}(E)\langle J_z{}^k\rangle^{\mathbf{m}} \Big/ \sum_{\mathbf{m}} I_{\mathbf{m}}(E). \tag{6.4}$$

The configuration intensities come, as always, from the configuration distributions. The $(J_z)^k$ averages come by the usual propagation from the values for a few particles or holes and are easy to calculate as far as $k = 4$ since J_z is a one-body operator.

In order to get some genèral idea about the validity of the Gaussian form we consider a single shell of identical particles for which the moment evaluation is particularly simple; we find for the variance, and the "excess" (defined below in terms of the second and fourth (central) moments), that

$$\begin{aligned} \sigma^2 &= \frac{1}{12} m(N-m)(N+1), \\ \gamma_2 &\equiv (\mu_2)^{-2}\{\mu_4 - 3\mu_2{}^2\} = -\frac{6}{5}\frac{\{(N(N+1) - m(N-m)\}}{m(N-m)(N+1)} \xrightarrow[m \text{ fixed}]{N\to\infty} -\frac{6}{5m} \end{aligned} \tag{6.5}$$

The magnitude of γ_2 decreases as particle number increases, in line with the fact that the central limit theorem implies an approach to normality for large m (the departures due to blocking effects, which increase with m, then not invalidating this general feature). γ_2 is always negative (implying a favoring of higher-J values), but its value is so small for even a few particles that, on the one hand, this favoring is negligible (Ratcliff's [4] Fig. 1 shows that a major departure from normality requires $|\gamma_2| \sim$ unity or larger), and, on the other hand, the central limit theorem becomes operative very quickly as m increases.

The single-configuration (or configuration-isospin) results can be worked out just as easily. For the configuration case we may decompose J_z by orbits, $J_z = \sum_{r=1}^{l} J_z(r)$ and then, ignoring odd powers in the results,

$$J_z^2 \equiv \sum_r J_z^2(r),$$
$$J_z^4 \equiv \sum_r J_z^4(r) + 6 \sum_{r<s} J_z^2(r)\, J_z^2(s), \tag{6.6}$$

to which we can apply single-shell propagation (2.2). Table 5 gives values for the $(d, s)^{12}$ configurations. The spin cut-off factors are largest for the more complicated configurations which allow higher angular momenta. The excess γ_2 is always negative and small in magnitude except for some of the relatively small configurations. The variation of these quantities with excitation energy, derived via (6.4) is given in Fig. 12. Since we are dealing with (d, s) configurations only, the parameters

TABLE V

Spin cut-off factor, σ, and the excess, γ_2, for the $(d, s)^{12}$ configurations. Configurations are labeled by particle numbers in 5/2, 3/2, 1/2, orbits, respectively. Hole-particle-conjugate pairs, (7, 3, 2) and (5, 5, 2) for example, have the same values of σ and γ_2

Configuration	σ	γ_2	Configuration	σ	γ_2
0, 8, 4	0	—	4, 6, 2	3.30	−0.23
1, 8, 3	1.79	−1.10	4, 5, 3	3.38	−0.21
1, 7, 4	2.05	−0.75	4, 4, 4	3.36	−0.23
2, 8, 2	2.37	−0.56	5, 7, 0	3.24	−0.19
2, 7, 3	2.61	−0.44	5, 6, 1	3.42	−0.18
2, 6, 4	2.72	−0.37	5, 5, 2	3.51	−0.18
3, 8, 1	2.72	−0.41	5, 4, 3	3.52	−0.18
3, 7, 2	2.90	−0.32	5, 3, 4	3.45	−0.21
3, 6, 3	3.10	−0.28	6, 6, 0	3.42	−0.22
3, 5, 4	3.13	−0.27	6, 5, 1	3.54	−0.18
4, 8, 0	2.92	−0.35	6, 4, 2	3.56	−0.18
4, 7, 1	3.15	−0.27			

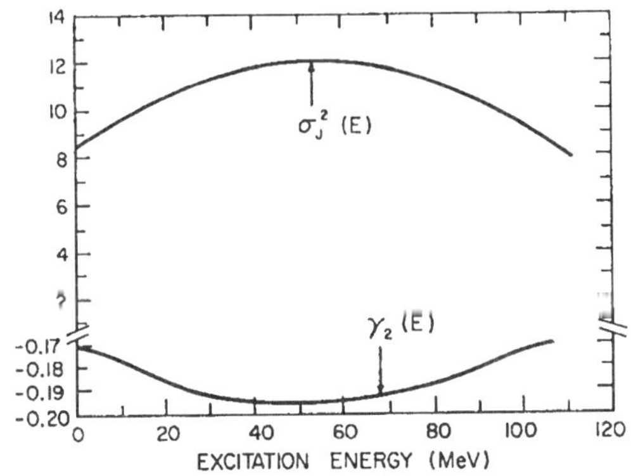

FIG. 12. Energy variation of the spin cut-off factor $\sigma_J(E)$ and the distribution excess $\gamma_2(E)$ for $(d, s)^{12}$ with Rosenfeld interaction. Note the change of scale.

refer to these partial densities; if for example we allowed excitations involving other active orbits the energy variation above, say, 20 MeV excitation would be different.

The conversion from M- to J-distributions is trivial and works the same way here as in the elementary theory. We have for the density of *levels*, by subtracting the state densities for successive M's, that

$$\begin{aligned} \rho_l(E, J) &= \rho(E, M = J) - \rho(E, M = J + 1) \\ &= \frac{\rho(E)}{(2\pi\sigma^2)^{1/2}} \{\exp(-J^2/2\sigma^2) - \exp(-(J + 1)^2/2\sigma^2\} \\ &\simeq \rho(E) \frac{(2J + 1)}{2\sigma^2} \exp\{-(J + 1/2)^2/2\sigma^2\}, \end{aligned} \tag{6.7}$$

where now our values of both $\rho(E)$ and σ^2 take account of the orbital structures and the residual interactions involved, and, as always, we have available the decomposition of the level density according to configurations. It should be remarked that, if we use an isospin formalism and take configuration averages, our state density above sums over all T_z values and so does the resultant level density; in order to produce densities which refer to single isotopes we should either use configuration-isospin averages, or apply exactly the same procedure as used above to isospin as well, making the standard normality assumption (and calculating the correction) exactly as for angular momentum. The situation here is even simpler since, unlike the j case, every particles has $t = \frac{1}{2}$, and moreover we have both the detailed fixed-isospin distributions and the central limit theorem result. For brevity however, we shall not give the isospin decomposition results. As a useful curiosity we quote however an approximate result, perhaps known to others, for the matrix or level dimensionality $d_l(m, J, T)$ for m particles with a set of spherical orbits. We have

$$\sigma^2 = \tfrac{1}{3}m(N-m)\,N^{-1}(N-1)^{-1}\sum_r N_r j_r(j_r+1),$$

$$d_l(m, J, T) = 2(2J+1)(2T+1)(N+2)^{-1}\,(2\pi\sigma^2)^{-1/2}\binom{\tfrac{1}{2}N+1}{\tfrac{1}{2}m-T}\binom{\tfrac{1}{2}N+1}{\tfrac{1}{2}m+T+1} \times \exp\{-(J+\tfrac{1}{2})^2/2\sigma^2\} \qquad (6.8)$$

The first of these is valid in (p, n) or isospin formalisms (or for identical particles). In the second we could replace the exact isospin dimensionality by another approximate Gaussian factor, or ignore it if we are not using isospin. For $(d, s)^{12}$ the formulas give 5–10 % errors except for the very smallest matrices.

In Fig. 13 we compare the prediction of our level density theory with Wong's

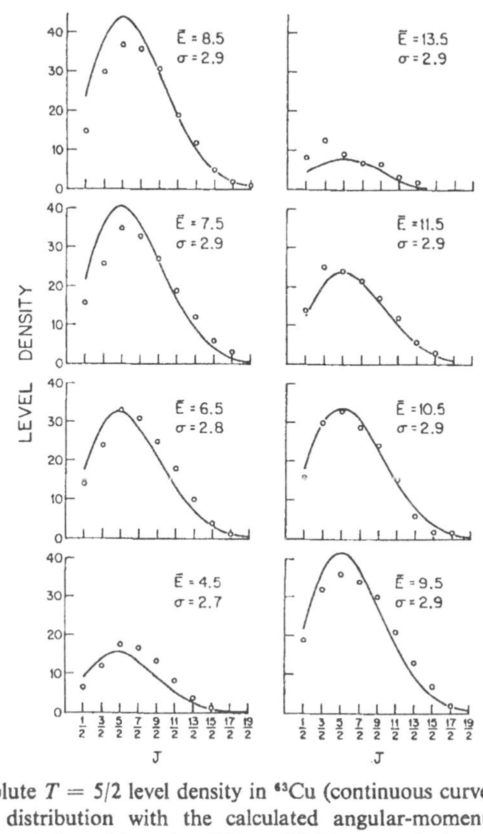

FIG. 13. The absolute $T = 5/2$ level density in ^{63}Cu (continuous curves) as derived from the configuration-isospin distribution with the calculated angular-momentum cut-off factor σ, compared with the exact shell-model results of Wong [5]. Except at the lowest and highest energies, where 3-MeV bands are used, the exact counting is done in 1-MeV intervals.

shell-model results [5] for the $T = 5/2$ states in ^{63}Cu. We stress that the comparison is an absolute one, the distribution density being calculated (very simply) from a specification of the orbits, single-particle energies, and interaction matrix elements, there being no free parameters whatever. Agreement is excellent.

We turn briefly to comparisons with experiment. At low excitation energies we have shown generally good agreements for various spectra and thus for low-lying state and level densities. For $(d, s)^{12}$, for example, we predict $\sim$50 states below 8 MeV, and $\sim$50 even parity states are found. Beyond that, for ^{28}Si [26] there are 35 observed levels between 12 and 13 MeV, 26 of these with known (J, π). For positive-parity levels the average J-degeneracy is found to be about 6.6 and, since $(2J + 1)_{av} = (2\pi\sigma^2)^{1/2}$, this yields a value $\sigma \sim 2.7$, while our $\sigma(E)$ is 3.0 in this region. If we assume that the other 9 levels have approximately the same average J-degeneracy as the identified 26, then experiment gives 140 even-parity states (and 90 odd-parity states), while our distribution gives 75 even-parity states. Thus we successfully predict σ, but our even-parity state density is somewhat lower than that obtained by experiment. This clearly is due to the fact that we are working in a too small model space (opening up the space however will have a negligible effect on σ so that the σ agreement will not be destroyed).

For densities at effectively higher excitation, consider the resonances found by Hibdon [27] in the reaction ^{27}Al(n, γ) ^{28}Al. There are 66 levels found in (7.73–8.16) MeV above the ^{28}Al ground state, (corresponding to 17 MeV above the ^{28}Si ground state), to which Hibdon has made angular-momentum and parity assignments (33 levels of each parity). These yield an extremely small value of the spin cut-off factor ($\sigma \sim 1.7$), as has already been discussed by Kanestrom [28]; we find the value $\sigma \sim 3.3$ and we can think of no mechanism whatever which would give a value much different from this. This then strongly suggests that many high-J states have been missed by experiment or have been assigned a low-J value. This uncertainty of course affects the state density also, the true density being presumably considerably higher than the values found in this experiment. These values are about 270 even-parity (and about 320 odd-parity) states per MeV, while our $(d, s)^{12}$ calculations give about 200 even-parity states distributed over 27 levels.

We are of course at too high an excitation to deal only with (d, s) orbits, as is emphasized by the relatively high density of negative-parity levels. When we increase our orbital space, to take account of $f_{7/2}$ for example, we need a newly renormalized interaction, and, besides the negative-parity states, we will encounter double-excitation even-parity ones. We have made some preliminary studies along this line using a B–K (10-shell) interaction [29]. We allow only the $f_{7/2}$ extension of the orbital space, and choose the $f_{7/2}$ orbital energy so that the lowest negative-parity state, as calculated by distributions, comes at the observed value. Various resultant densities (derived via configuration averages so that there is a $(2T + 1)$ weighting which could be extracted by using the isospin cut-off factor) are given in Fig. 14.

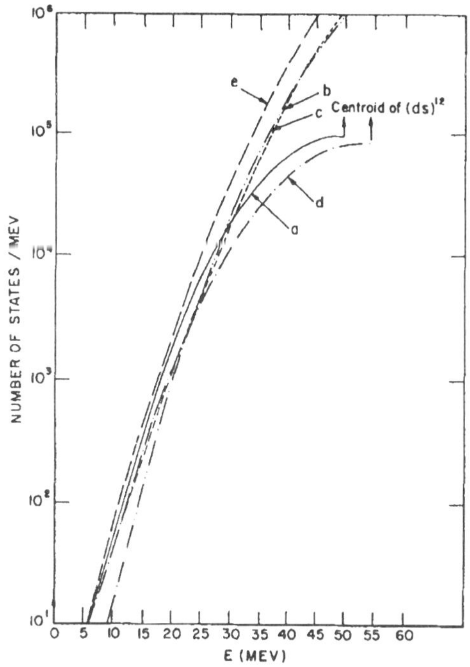

FIG. 14. State densities for ^{28}Si in various model spaces, calculated by configuration distributions with a B–K 10-shell renormalized interaction. Distribution (a) is for $(d, s)^{12}$; (b) for the even-parity states of $(d, s + f_{7/2})^{12}$; (c) for the odd-parity states of $(d, s + f_{7/2})^{12}$; (d) for $(d, s)^{12}$ $(f_{7/2})^0$ (in which excitation into $f_{7/2}$ is allowed for intermediate states, in contrast to (a) in which no such excitation occurs); (e) the total density for $(d, s + f_{7/2})^{12}$. Note that for (b, c, e) an arbitrary number of particles are allowed into the $f_{7/2}$ orbit.

We find that up to 10 MeV excitation (in ^{28}Si) the effects on the densities are negligible. At 17 MeV we also get an almost unchanged even-parity density but now find also about 50 negative-parity states per MeV. It appears then that simply allowing particle excitations into $f_{7/2}$ is not enough, but there is no good reason why it should be, since we should be considering the other (f, p) orbits, as well as excitations from the first p orbit. The major problems involved in making this further extension involve the accuracy of the distribution methods in large spaces, and the choice of the Hamiltonian, the single-particle energies in particular being quite critical. Work along this line is now in progress.

7. Strength Distributions

In the preceding two sections our attention has focussed on nuclear energies regarded as eigenvalues of a Hamiltonian operator defined in a restricted spectroscopic space. We have been concerned, for one thing, with actual values of the low-lying energies and, for another, with the density of such eigenvalues at higher energies. But, besides energies, it is important that we discuss excitations, asking for example about the distribution of the excitation strength when a particle is added to a nucleus, or taken away, or when a nuclear state decays via electromagnetic emission.

In the usual procedure for dealing with such processes we need wavefunctions rather than densities. We might in some cases use the spectral intensities as a guide to truncating the space and reduce it thereby to one which can be handled in detail; but more often than not this will be unsatisfactory, and it is moreover contrary to the spirit of the distribution methods. Instead we should construct *strength* distributions [30] and use them to tell us whether, for example, the transition strength is mainly concentrated between close-lying states or far separated ones, and whether the strength is centered about a high or low excitation energy. The importance of this kind of question in the statistical theory of reactions is obvious. For more detailed spectroscopy we could consider to what extent the general nature of the strength distribution is determined by the nature of the interaction (the role of the quadrupole force for example), making thereby some contact with the theory of collective excitations. Beyond that, if the distributions were available to sufficient accuracy, we could use them in detailed studies of low-lying transitions, in analogy with our discussion of low-lying energies. Just as in that case, we have no *a priori* right to expect that low-moment distributions will be adequate for this purpose, and whether they will be or not is quite unknown. As a final general remark, we stress that, just as level densities are used as a rough theoretical device to parametrize strength distributions relevant to reaction studies, so also theoretical spectra and wavefunctions may be regarded as playing the same role with respect to the observed states. An example of this is given by Zuker *et al* [31] who demonstrate that it is profitable (the strength widths being small) to regard $A = 18$ states as two-particle excitations of $A = 16$.

Just as with spectra, it is essential that the strength distributions be representable in terms of a few low-order moments. We have investigated a few cases for which exact results are available, one such being for $M1$ strengths in the $T = 5/2$ $(f_{5/2}p_{3/2}p_{1/2})^7$ space, for which we can once again use Wong's results [5]. Figure 15 shows a typical comparison of a detailed transition-energy distribution (defined ahead) with a normal distribution. Agreement is generally satisfactory though it would be improved significantly with a third (and perhaps fourth) moment correction.

We give now a very brief introductory account of strength distributions, leaving until later papers the more detailed theory and significant applications. We introduce the distributions by considering *all* the transitions of a given variety (electromagnetic $E2$ for example) which are possible inside a given model space, or

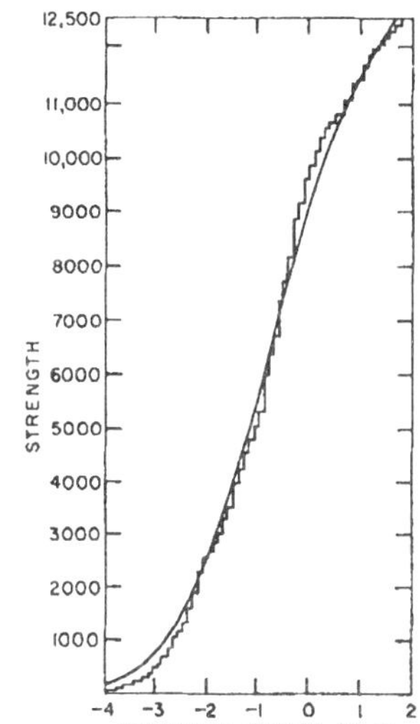

FIG. 15. The $M1$ ($J = 13/2 \to 15/2$) transition-energy distribution, and its normal approximation, in the $T = 5/2$ states of $(f_{5/2}p_{3/2}p_{1/2})^7$. The total strength is 12 500, the centroid energy is -0.68 MeV, and the width $\sigma = 1.44$ MeV.

between two model spaces (m- and $(m+1)$-particle spaces, for example, for single-particle transfer). We assume as before that these are "spectroscopic" spaces (definite numbers of particles and single-particle states) and that there are defined, in the sum of these spaces, a Hamiltonian H, and an operator O which produces the excitation in question. Then if i and j label the initial and final Hamiltonian eigenstates (which may be the same set), the $(i \to j)$ strength is defined as $S_{ji} = |\langle j \mid O \mid i\rangle|^2$, and the collection of these for all i and j tells the whole story. For a given i, we may regard S_{ji} as a function of the final-state eigenenergy E_j, and therefore as a distribution, whose moment of order f is $\sum_j (E_j)^f S_{ji} = \langle i \mid O^+H^fO \mid i\rangle$. This quantity is simply the f-th energy-weighted sum rule for state i, and we thus have a simple connection between sum rules and distributions. Rather than pursue this well-known subject we point out that this moment, considered now as a function of the initial energy E_i is itself a distribution, whose g-th moment is easily seen to be given by the initial-state trace, $\langle\langle O^+H^fO\,H^g\rangle\rangle$. We see then that the strengths form a bivariate distribution corresponding to the fact that a transition involves two eigen-energies. Instead of the two end-point energies involved in the transition, it is convenient to use the difference energy and the mean; then, for example, the transition-energy distribution is univariate, produced by integration (taking the zero-th moment) with respect to the other variable.

Anticommutators involving H, O, O^+, linear in the last two, come in naturally in the mean-energy moments, and similar commutators in the transition-energy moments. For example, the second moment for the latter distribution (which is also the strength-weighted mean-square transition energy) is

$$d^{-1} \sum_{i,j} |\langle j \mid O \mid i \rangle|^2 (E_j - E_i)^2 = \langle O^+[H[HO]]\rangle, \tag{7.1}$$

where d is the dimensionality of the initial-state space and d^{-1} has been inserted so that our distribution now corresponds to an *average* over the initial states, just as with the spectral distributions. An expression for the general (p, q) moment is given in [30]. Since H is bosonlike, commutators are of lower-particle rank than corresponding anticommutators and consequently the transition-energy distribution is much easier to deal with than the mean-energy one. We restrict ourselves now to this simpler distribution, whose q-th moment is given by an average involving a q-fold commutator, $\langle O^+[H[H \cdots [H, O] \cdots]]\rangle$. For the conventional H, the operator is of maximum particle rank $(q + 2k)$ where k is the maximum rank of the excitation (1/2 for single-nucleon transfer, 1 for electromagnetic transitions, and so on).

A technical difficulty arises when the initial and final states span the same space, for then, to every $(i \rightarrow j)$ transition, another one enters with the states interchanged so that, in the electromagnetic case for example, our distributions would combine emission and absorption.[21] If our interest is with transitions which change an exact symmetry, isospin for example, this problem does not enter. Otherwise however we should, for accuracy divide the space into subspaces, as by configurations, so that most of the transitions are between states in disjoint subspaces. The new distributions of course give us at the same time more detailed information about the transitions and may be used also for localization in excitation energy, as discussed before.

Formally now we wish to make a symmetry decomposition of both the initial and final spaces, the latter showing up as an intermediate space in the initial-state averages which define the moments. To do this we can expand the operators H, O, O^+ as tensors according to the symmetry in question and the multiple commutators then as coupled tensors. When we have done this, an elementary recoupling of the variety standard in sum-rule studies [11] exhibits the intermediate state and

[21] These correspond respectively to the sides of the distribution with negative and positive $(E_j - E_i)$. When initial and final states span the same space, the odd moments vanish. More formally, if the averaging space supplies a representation of the self-adjoint excitation O, the odd-order multiple commutators vanish because of the invariance of a trace under cyclic permutations of the factors. For (non-self-adjoint) operators acting between two spaces, permutations are also allowed but, with any permutation which interchanges the order of O, O^+, we must change from one averaging space to the other. Permutations can also be achieved by explicit commutation, and, putting the two procedures together, we get an inductive method for evaluating moments which applies also to tensor products of operators with specified symmetry. It is this method which has been applied by Mugambi [24] to calculate fixed-(J, T) moments.

decomposes the moments. All this then is quite straightforward provided only that the appropriate explicit Racah algebra is available. For configurations, there is no problem at all, while for isospin [32] and angular momentum (and identical-particle-pairing quasispin) everything proceeds with the usual R_3 recoupling, so that even the intermediate-state J problem becomes in principle straightforward, in contrast with the initial-state J averaging.[22]

By carrying out these expansions and transformations we reduce the problem to one entirely similar to the spectral problem, and, just as there, we use a propagation method to express the many-particle moments in terms of few-fermion moments. It is not worthwhile here to go further into technicalities. There are some preliminary results. Consider single-nucleon transfer distributions, which tell us, among other things, about a correlation between m-particle and $(m + 1)$-particle structures. There are two extremes; if the two structures are completely uncorrelated the resulting strength distribution will be wider than the corresponding spectral distributions, whereas, if the two are completely correlated (each target state being connected to only one final state), it will become narrower, and perhaps much narrower depending on the level spacings of the two spectra. In characteristic cases, both in (d, s) and (f, p), we find the fixed-isospin strength widths to be about one-half the corresponding spectral widths, indicating a strong correlation. These strength distributions, when refined to specify configurations, will be of special interest for statistical reaction theory [23] in which we think of the transfer of a nucleon to a definite configuration and the subsequent approach to equilibrium via a succession of hole-particle excitations.

We have similarly made studies of $M1$ and $E2$ fixed-isospin distributions, which reproduce some of the well-known results (weakness of $M1$ $\Delta T = 0$ transitions in self-conjugate nuclei and other things of that sort) and give some indications about more interesting questions; but until they are refined it is not worthwhile reporting them in any detail.

8. Conclusion

The spectral distribution methods which we have described and applied are of a general nature, opposed in spirit to the highly detailed calculations which have been conventional in spectroscopy, and equally opposed to the use of strong approximations concerning the model space and the interaction which have been often used to render problems soluble by conventional procedures. In their present form they do function however in restricted, even though large, model spaces and con-

[22] It should be feasible moreover to extend, to the initial-state averaging problem, the method used for fixed-J level densities in Section 6.

sequently do assume that effective Hamiltonian and excitation operators are available for use in these spaces. The principal notions are that we should exploit certain elementary mathematical concepts, norms, etc., which are a natural part of vector-space theory, and exploit as well the simplicities in the description of many-particle systems which are introduced by simultaneous consideration of symmetry and statistics.

Because of the generality of the methods there are many extensions possible of the work described in this and the earlier papers. Most straightforward perhaps (though a very large project) would be to give a systematic treatment of nuclear level densities in various parts of the periodic table, making comparisons with data and considering to what extent level densities are influenced by, and may be used to study features of the effective interaction. Besides this, one should be able to establish contact with the earlier level-density theories (which essentially ignore interactions not describable by a single-particle or single-quasiparticle spectrum) either by a suitable parametrization of the residual-interaction theory, or by the unitary-symmetry decomposition followed by a maximization of the effective one-body interaction.

Since level densities are in most cases quantities used in a parametrization of strength distributions, level-density studies will lead naturally to further developments of strength-distribution theory. Though more work has been done with this than appears in Section 7, the theory is still in a rudimentary state. With a properly extended theory, one should be in a good position to study statistical cascades and reactions, and in particular the processes, described by Griffin, which are intermediate between direct and compound-nucleus reactions. Ground-state energies being the natural reference energy for level densities, we shall have to learn much more about the limitations and the accuracy of our methods of calculating such energies, and the low-lying spectra more generally. It should be clear of course that easy methods for calculating ground-state energies (and low-lying strength distributions as well) are of interest far beyond their application to level densities.[23]

We have given methods for making significant comparisons between different effective interactions. Systematic comparisons should be made in different parts of the periodic table. But, beyond that, there are a large number of questions connected with the effects, on the effective interaction, of space truncations and extensions. Since much of the essentials of the interaction are contained in low-order moments of various kinds it would seem reasonable, instead of calculating the effective interaction in detail, to attempt a direct calculation of these moments and their renor-

[23] As an example of its use in testing the validity of a nuclear model we remark (Parikh [25]) that, in the (d, s) shell, the distribution ground state comes considerably below that given by the projected Hartree–Fock theory. For most of the interactions of Ref. [13] the amount is $\sim$5 MeV for the $T = 0$ ground state in $A = 22$ (in which case we know that the distribution ground state is accurate) and $\sim$15 MeV for $(d, s)^{12}$.

malizations, by an appropriately simplified version of effective interaction theory (in which moreover the intermediate state energies could presumably be calculated by distributions). Nothing has been done along this line. We have however made some calculations in which relatively mild space extensions are made by a method of stochastic approximations (unpublished, but for similar ideas and an application of them see Refs. [33, 34]).

Finally we stress that the work which we have described may be regarded as part of a study of the goodness and the consequences of certain unitary symmetries in nuclei. The problem of finding a single-particle basis which accomodates itself best to the residual interaction is then one involving the goodness of a more general unitary symmetry. There are of course other symmetries known to be important, and, for some of these, preliminary distribution studies have been made. It seems certain moreover that a systematic search for further relevant symmetries will be fruitful. The methods described seem well suited for the search, but we shall need to know much more about Racah algebras, and should have also an understanding of the limit theorems appropriate to symmetry subspaces, and more generally of the extremely interesting connections between group symmetries and statistics.

Acknowledgments

For supplying us with unpublished results, and permitting us to quote them, we are indebted to E. C. Halbert, J. C. Parikh, F. C. Williams, and S. S. M. Wong. We thank M. Blann, J. R. Huizenga and F. C. Williams, for general education and much specific information about level densities, P. E. Mugambi, L. S. Hsu and K. F. Ratcliff for insights about general features of spectral distributions, and J. G. Flores, P. A. Mello, M. Moshinsky, and J. C. Parikh for discussions about nuclear symmetries.

References

1. J. B. French, *in* "Nuclear Structure" (A. Hossain, Harun-ar-Raschid, and M. Islam, Eds.), North-Holland, Amsterdam, 1967.
2. J. B. French, *in* "Proceedings of Eastern Theoretical Conference" (F. Rohrlich, Ed.), Syracuse University Report (1969).
3. J. B. French and K. F. Ratcliff, *Phys. Rev.* (in press).
4. K. F. Ratcliff, *Phys. Rev.* (in press).
5. S. S. M. Wong, *Nucl. Phys.* (in press), and private communication.
6. S. S. M. Wong, private communication.
7. M. Moshinsky, *in* "Physics of Many-Particle Systems" (E. Meeron, Ed.), Gordon and Breach, New York, 1965.
8. C. M. Vincent, *Phys. Rev.* **163** (1967), 1044.
9. J. D. Louck, *Amer. J. Phys.* **38** (1970), 3.
10. J. G. Flores and M. Moshinsky, *Nucl. Phys. A* **93** (1967), 81.

11. J. B. FRENCH, *in* "Proceedings of the International School of Physics, Enrico Fermi, Course 36" (C. Bloch, Ed.), Academic Press, New York, 1966.
12. "Statistical Theories of Spectra: Fluctuations" (C. E. Porter, Ed.), Academic Press, New York, 1965.
13. E. C. HALBERT, J. B. MCGRORY, B. H. WILDENTHAL, AND S. P. PANDYA, *in* "Advances in Nuclear Physics" (M. Baranger and E. Vogt, Eds.), Plenum, New York, to be published.
14. T. T. S. KUO AND G. E. BROWN, *Phys. Letters* **18** (1965), 241.
15. T. ERICSON, *Advances Phys.* **9** (1960), 425.
16. J. E. LYNN, "The Theory of Neutron Resonance Reactions," Oxford Univ. Press, Oxford, 1968.
17. C. BLOCH, in "Physique Nucleaire" (C. Dewitt and V. Gillet, Eds.), Gordon and Breach, New York, 1969.
18. J. R. HUIZENGA, L. VAZ, F. C. WILLIAMS, AND M. BLANN, University of Rochester Report UR-NSRL-28, 1970.
19. E. P. WIGNER, *SIAM Rev.* **9** (1967), 1.
20. M. L. MEHTA, "Random Matrices and the Statistical Theory of Energy Levels," Academic Press, New York, 1967.
21. U. GRENANDER, "Probabilities on Algebraic Structures," Wiley, New York, 1963.
22. F. C. WILLIAMS, private communication.
23. J. GRIFFIN, *Phys. Rev. Letters* **17** (1966), 478.
24. P. E. MUGAMBI, Ph.D. Thesis, University of Rochester, 1970.
25. J. C. PARIKH, private communication.
26. P. M. ENDT AND C. VAN DER LEUN, *Nucl. Phys. A* **105** (1967), 1.
27. C. T. HIBDON, *Phys. Rev.* **114** (1959), 179.
28. I. KANESTROM, *Nucl. Phys.* **83** (1966), 380.
29. Private communication from T. T. S. KUO TO K. F. RATCLIFF.
30. J. B. FRENCH AND L. S. HSU, *Phys. Letters* **25B** (1967), 75.
31. A. P. ZUKER, B. BUCK AND J. B. MCGRORY, *Phys. Rev. Letters* **21** (1969), 39.
32. J. B. FRENCH, *in* "Isospin in Nuclear Physics" (D. H. Wilkinson, Ed.), North-Holland, Amsterdam, 1969.
33. N. ULLAH, private communication, and to be published in *Canad. J. Phys.*
34. N. ULLAH, S. S. M. WONG, AND L. E. H. TRAINOR, *Nucl. Phys.* (in press).

Volume 34B, number 4 PHYSICS LETTERS 1 March 1971

TWO-BODY RANDOM HAMILTONIAN AND LEVEL DENSITY

O. BOHIGAS and J. FLORES ‡

Institut de Physique Nucléaire, Division de Physique Théorique ‡, *91 - Orsay - France*

Received 22 December 1970

It is shown numerically that strong deviations from Wigner's semi-circle law for the level probability density of a random matrix are obtained if the two-body nature of the hamiltonian is taken into account. Instead of Wigner's law the probability density approaches closely a normal (gaussian) distribution.

In the statistical theory of spectra the hamiltonian is represented by a hermitian random matrix; level densities, level spacings and other average properties of complex nuclear and atomic systems are then predicted. However, and in order to avoid an enormous amount of complications, these results are obtained under the fundamental assumption of independence: the elements of the random matrix are assumed to be distributed independently of each other [1]. Using this basic assumption, the distribution of eigenvalues can be obtained analytically for an $N \times N$ hermitian matrix [2]. In the limit of large N, Wigner's semi-circle law is obtained [3]:

$$P(\epsilon) = \begin{cases} \frac{2}{\pi}\sqrt{1-\epsilon^2} & \text{for } |\epsilon| < 1 \\ 0 & \text{for } |\epsilon| > 1 \end{cases} \qquad (1)$$

where $\epsilon = E/2m\sqrt{N}$ is the energy E measured in units $2m\sqrt{N}$, m being the standard deviation of the off-diagonal matrix elements and where $P(\epsilon)$ is normalized to unity. As a matter of fact, the asymptotic law (1) is obtained rather closely for $N \geqslant 10$ [4]. (see figs. 1 and 2).

However, the assumption of independance is not justified from a dynamical point of view. In fact, as Ratcliff has pointed out recently [5] ↓, the level density obtained after diagonalizing a realistic shell-model matrix, does not follow eq. (1) at all, but rather a normal (gaussian) distribution †.

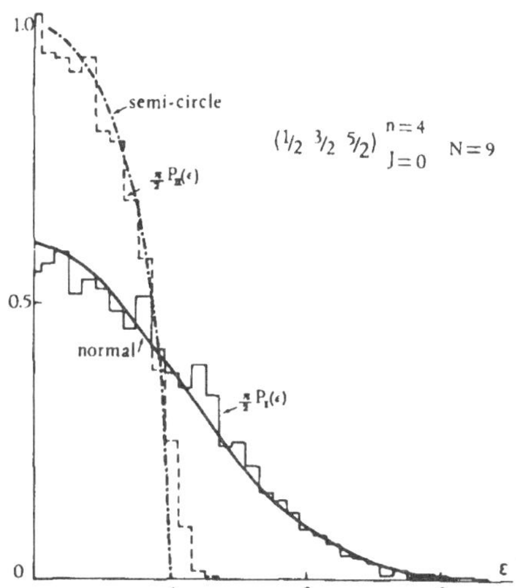

Fig. 1. Histograms corresponding to the distributions P_I(solid) and P_{II}(dotted) obtained with 10 000 counts for the case $n = 4$, $J = 0$ and $N = 9$. The normal (solid) curve has the same standard deviation as $P_I(\epsilon)$. The semi-circle (dotted) curve appears distorted because the vertical scale is twice that on the abscissa.

It is the purpose of this note to analyze, for some simple cases, the probability density $P(\epsilon)$ calculated using a random shell-model residual interaction and antisymmetric wave functions, instead of the usual random matrix. We shall see that strong departures from the law (1) are then obtained.

In order to proceed, we compute the level

‡ Work supported by the Comisión Nacional de Energía Nuclear, Mexico. On leave of absence from Instituto de Física, Universidad de México, México, D.F.

‡ Laboratoire Associé au C.N.R.S.

↓ In fact, the idea that distributions are close to normal has formed the basis of the work done on spectral distributions by J. B. French and his collaborators since 1966 [e.g. 6].

† Dubner [7] has introduced correlations among random matrix elements using a gaussian ensemble of cyclic matrices and has obtained a normal level density. However, from the physical point of view, it is not clear which correlations are thereby introduced. We would like to thank Drs. A. Gervois and M. L. Mehta for pointing out to us the existence of this work.

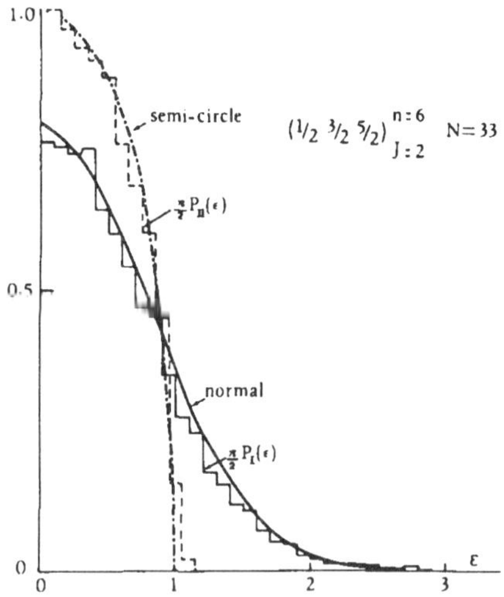

Fig.2. The same as in fig.1, but for the case $n=6$, $J=2$ and $N=33$.

density in two different ways:

I. - We consider a system of n identical particles and a set of degenerate single-particle states a, b, c,... (which we take as $p_{1/2}$, $p_{3/2}$ and $f_{5/2}$ in our calculation), and define a residual interaction by its independent two-body matrix elements

$$\langle ab|V|cd\rangle_\Gamma, \text{ with } a \leqslant b,\ c \leqslant d \text{ and if } \begin{matrix} a \neq c, & a < c \\ a = c, & b \leqslant d, \end{matrix} \quad (2)$$

where Γ is the two-particle total angular momentum. We then take the matrix elements (2) as independent random numbers, according to the prescriptions of the gaussian orthogonal ensemble (1). Using them we compute, by standard shell-model techniques, the $N\times N$ hamiltonian matrix corresponding to n particles and a given total angular momentum J. After repeating this procedure a sufficiently large number of times, in order to have good statistics, we obtain a level probability density which we shall call $P_{\rm I}(\epsilon)$.

II. - We follow Porter and Rosenzweig [8] and diagonalize a set of $N\times N$ random matrices, obtaining a probability density $P_{\rm II}(\epsilon)$. In the limit of large N, $P_{\rm II}(\epsilon)$ should obey the semi-circle law. The set of random $N\times N$ matrices was constructed so that they form a gaussian orthogonal ensemble [1].

The result of the comparison between I and II is shown in fig. 1 for the case $n=4$, $J=0$, $N=9$, and in fig. 2 for the case $n=6$, $J=2$ and $N=33$. As is clear from the figures, the function $P_{\rm I}(\epsilon)$, obtained with a two-body random hamiltonian, differs considerably from the semi-circle law. In these, as well as in all other cases that we have calculated, the function $P_{\rm I}(\epsilon)$ approaches closely a normal distribution. This can be seen from table 1, where the mean value $\bar{\epsilon}$, the standard deviation σ and the cumulants in standard measure κ_i, $i=4$, 6 and 8 are shown [9], for several values of both n and N. In the table we also give the values for the normal and semi-circle distributions.

Table 1
Comparison of distributions $P_{\rm I}(\epsilon)$ and $P_{\rm II}(\epsilon)$

	n	J	N	ϵ	σ	κ_4	κ_6	κ_8
I	3	3/2	10	-0.001	0.62	-0.39	1.61	-11.6
II				-0.001	0.53	-0.90	4.27	-45.4
I	4	0	9	0.005	1.04	-0.26	0.99	- 6.9
II				-0.002	0.53	-0.89	4.16	-43.9
I	5	11/2	8	-0.031	1.12	0.01	0.56	0.69
II				0.001	0.53	-0.85	3.99	-41.5
I	6	6	8	0.067	1.48	-0.02	-0.32	0.20
II				0.001	0.53	-0.85	3.99	-41.5
I	4	2	21	-0.003	0.59	-0.38	1.18	- 7.0
II				0.000	0.52	-0.95	4.64	-50.7
I	5	7/2	23	0.002	0.68	-0.10	0.41	- 0.64
II				-0.001	0.51	-0.95	4.58	-49.7
I	6	1	19	0.015	0.94	0.23	0.80	- 0.41
II				0.000	0.51	-0.93	4.43	-47.5
I	6	2	33	-0.037	0.79	0.09	0.39	- 4.7
II				0.002	0.51	-0.97	4.75	-52.3
normal				0	σ	0	0	0
semi-circle				0	0.5	-1.0	5.0	-56.0

Several systematic features emerge from the analysis of table 1:

i) A normal distribution is obtained for $P_{\rm I}(\epsilon)$ of about the same quality for all cases considered This is in contrast with the semi-circle law, to which $P_{\rm II}(\epsilon)$ approaches as N gets larger.

ii) The standard deviation σ, for matrices of approximately the same dimensionality grows with the number n of particles.

As a result of this numerical analysis, we could say that the correlations among matrix elements induced by the two-body character of the hamiltonian and by the Pauli principle, produce a normal level density, in agreement with shell-model calculations. It seems to us an interesting problem to justify this result from a theoretical point of view.

Toward the final stages of this work we learned from Prof. J. B. French that he, in collaboration with S.S.M. Wong, had carried out a similar work [10]. We would like to acknowledge interesting discussions with Profs. J. B. French and R. Arvieu. We are grateful to Centre de

Volume 34B, number 4 PHYSICS LETTERS 1 March 1971

Calcul de la Faculté des Sciences d'Orsay for computing facilities. One of the authors (J. F.) would like to thank the director of the Theoretical Division Prof. M. Vénéroni for his kind hospitality.

References

[1] C.E.Porter, Statistical theories of spectra, (Academic Press 1965, New York); M.L.Mehta, Random matrices, (Academic Press 1967, New York)
[2] M.L.Mehta and M.Gaudin, Nuclear Physics 18 (1960) 420
[3] E.P.Wigner, Ann. of Math. 67 (1958) 325
[4] C.E.Porter, Journ. Math. Phys. 4 (1963) 1039
[5] K.F.Ratcliff, to be published.
[6] J.B.French in Proceedings of the 8th Eastern Theoretical Physics Conference, Syracuse University, 1969, and references quoted therein.
[7] V.M.Dubner, Theor. Prob. Appl. 14 (1969) 342
[8] C.E.Porter and N.Rosenzweig, Ann.Acad.Scient. Fennicae A VI 44 (1960) 235
[9] M.G.Kendall and A.Stuart, The advanced theory of statistics, (Charles Griffin and Co. 1969, London)
[10] J.B.French and S.S.M.Wong, private communication.

* * * * *

Volume 35B, number 1 PHYSICS LETTERS 26 April 1971

SPIN-ISOSPIN SU(4) SYMMETRY IN THE (ds) SHELL †

J. B. FRENCH and J. C. PARIKH ‡

Department of Physics and Astronomy, University of Rochester, Rochester, New York. 14627, USA

Received 16 March 1971

A spectral distribution study shows that a conventional central interaction gives a fairly good SU(4) symmetry, in the groundstate energy domain, throughout the shell, but more realistic interactions give rise to large symmetry admixings.

Much of the general behavior of light nuclei is determined by the strong energy favoring of space-symmetric states which is produced by the effective interaction, and by the spin-orbit admixing of the corresponding spin-isopsin SU(4) symmetries. In this paper (see also ref. [1]) we discuss the goodness of the symmetry in low-lying states throughout the (ds) shell as deduced from the SU(4) spectral distributions, whose form and construction we first describe.

In the distribution method [2] one considers the energy distribution of the intensities of the various representations. Then the relative intensity with which representations contribute in a given energy domain is given by the relative magnitudes of the distributions. From these we can in some cases ⟂ learn about the goodness of the symmetry in low-lying states, and, as described below, can determine approximately the ground-state energy and the low-lying spectrum. The extension to higher excitation energy gives a theory of level density [3] which takes account of orbital structures and residual interactions; the level density emerges as a sum of partial densities interesting in their own right.

Let $\rho(\alpha) = \sum_{\alpha\epsilon\alpha} \psi_\alpha(m)\, \psi_\alpha^+(m)$, the trace in second quantization of the density operator over the states α of the m-particle representation α. Then the m-particle trace over α of an operator O whose maximum particle rank is not greater than u (a p-body operator has rank p and the conventional H^2 has $u = 4$) is found to be given by a trace over the set $[u]$ of all k-particle states with $k = 0, 1, \ldots u$. Explicitly we have, with n the number operator, that ‡‡

$$\langle\langle 0 \rangle\rangle^{m,\,\alpha} = \langle\langle \tilde{\rho}(N-m, \alpha_c) \binom{u-m}{u-n} 0 \rangle\rangle^{[u]} \tag{1}$$

For $U(N/r) \times U(r)$, a direct-product subgroup of $U(N)$, it may be shown that $\tilde{\rho}$ for any represenation is a group scalar expressible as a polynomial in n and the various Casimir invariants, $G_i (i = 1 \ldots r\text{-}1)$, of $U(r)$. It is easily constructed as long as the eigenvalues of the G_i are known for the various representations. In the spin-isospin SU(4) case, $r = 4$ and we encounter three invariant operators. Their eigenvalues have been found by a standard procedure of operating on maximum-weight states.

Let us now, in the trace above, decompose $[u]$ into subspaces with definite symmetry, $[u] = \sum_{k=0}^{u} \sum_\beta (k, \beta)$ where β is the SU(4) label; we see then that, because of the scalar nature of the ρ operators, the trace 'propagates' so that the m-particle traces are given as a linear combination of k-particle ones; viz,

$$\langle\langle 0 \rangle\rangle^{m,\,\alpha} = \sum_{k=0}^{u} \sum_{\beta} \binom{u-m}{u-k} \langle \hat{\rho}(N-m, \alpha_c)\rangle^{k,\,\beta} \langle\langle 0 \rangle\rangle^{k,\,\beta} \tag{2}$$

† Supported in part by the U.S. Atomic Energy Commission.

‡ Present Address: Physical Research Laboratory, Ahmedabad, India.

⟂ We could do this more generally by means of the partial widths but these are not yet available for SU(4).

‡‡ As described in ref. [2], $\tilde{\rho}(\alpha_c)$ derives from $\rho(\alpha)$ by a double hole $\rightleftharpoons$ particle transformation, the $(\sim)$ indicating that we use $\psi^*\psi$ instead of $\psi\psi^*$ and α_c denoting the $(N-m)$-particle representation which is complementary to α. For example if α denotes all m-particle states, then $\rho(\alpha) = \binom{n}{m}$, $\tilde{\rho}(\alpha) = \binom{N-n}{m}$, $\tilde{\rho}(\alpha_c) = \binom{N-n}{N-m}$. We have also that $\binom{u-m}{u-n} = (-1)^{n-u} \binom{m-n-1}{u-n}$. Eq. (1), which is entirely general, represents an extension of the results given in the references.

where the *average* denoted by $\langle\tilde{\rho}\rangle^{k,\beta}$ is a 'propagation coefficient' whose value follows immediately from the explicit form for ρ. For (0 + 1 + 2)-body H's the centroids and widths of the distributions involve operators with $k \leq 4$ (H and H^2) and this gives us an 'input net' of 12 SU(4) representations. It is convenient in fact to eliminate some of the $k = 4$ input representations in favor of representations with $m \sim N$, an operation which can be carried out by using (2) to solve for the unwanted representations, and which results in forms for the traces which are less sensitive to inaccuracies in the input values.

The input traces may be evaluated by means of explicit shell-model calculations for $(ds)^k$ with $k = 0...4$ (the hole representations giving the same results with different single-particle energies). These calculations are made with the ORNL-Rochester computing code [4] and the results emerge in a jj representation. The transformation to SU(4) comes by diagonalizing, also in jj, the SU(4) two-particle Casimir operator. On transforming the H and H^2 matrices and evaluating the partial traces we have the necessary input to the propagating equations. These then produce the centroids and widths for the multi-particle representations.

Since the (k, β) trace operation occuring in (2) is SU(4) scalar we see that the traces are determined by the scalar part of O. Thus for determining the relative centroids the Hamiltonian is equivalent to a multiple of the SU(4) 2-particle Casimir operator. The coefficient and the constant depend of course on H but, except for that, we have a unique pattern for the centroid ordering and spacing (changing the sign of H would invert the centroid spectrum, but that possibility we can ignore). The spacing between the two lowest centroids then defines a natural unit D, whose value is characteristically 3-8 MeV in the (ds) shell, in terms of which all (0 + 1 + 2)-body H's give the same relative centroid spectrum. For $(ds)^{12}$, as an example, the lowest representation is the one with maximum space symmetry [444] and the next lowest is its one-particle one-hole excitation [4431] (we quote always the space or SU(6) symmetry; the SU(4) symmetry has the associated shape, [3333] and [4332] for the above cases). The total numbers of states with these symmetries are 4 116 and 170 000 respectively, these being the SU(6) × SU(4) dimensionalities. The $(ds)^{12}$ centroid spectrum span is 15 D/2, about 50 MeV for the usual interactions, compared with 100 MeV for the full eigenvalue span.

The pattern for the widths is not unique since these are determined by a (0 - 4)-body polynomial in the three Casimir operators. The variations from representation to representation are however limited and, for each of the interactions which we have used, the widths σ; $(\sigma^2 = \overline{H^2} - (\overline{H})^2)$, are quite close; for example all but the highest two of the $(ds)^{12}$ Rosenfeld widths [1] lie between 6.9 and 7.9 MeV (the spacing unit in this case is 7.1 MeV). Since it will turn out that our interest will be in the lowest few representations, and especially the lowest two, we may reasonably characterize an interaction in terms of σ/D, the average width of the two lowest representations measured in terms of the spacing unit which the interaction produces. The reciprocal of this quantity may be regarded as a natural measure of the distace between representaions, and the closer the representations the larger the admixings to be expected.

We can in fact learn something from the σ/D quantities alone about the relative symmetry breakings produced by different interactions. Table 1 gives σ/D in the lower half (ds) shell for a Rosenfeld interaction, and for the 'realistic' interaction (K + 12 FP) defined by Halbert et al. in a recent review article [5]; values in the upper half shell are given for the Brown-Kuo interaction K.B. - 12.5 [6]. Also given are values which result when the single-particle energies (single-hole energies in the upper half shell) are put to zero. It is clear that the representations are effectively much closer with the more realistic interactions, the admixings being then expected to be correspondingly larger. If perturbation

Table 1
Ratio of width to spacing (σ/D) for various interactions. σ is the average width of the two lowest-lying representations and D is their centroid separation. m = particle number for Rosenfeld and (K + 12 FP) interactions; = hole number for (KB - 12.5). The widths increase with fermion number and are 5 - 7.5 MeV and 7 - 10 MeV for the central and non-central interactions respectively. In two of the cases, as indicated, the single particle hole energies have been put to zero, the widths then ranging from 2.5 - 5 MeV and 4.5 - 6.5 MeV.

m	R	R s.p.e.=0	K + 12 FP	KB - 12.5	KB - 12.5 s.h.e.=0
4	0.7	0.4	1.1	1.2	0.7
5	1.0	0.6	1.5	1.7	1.0
6	1.7	0.9	2.6	2.9	1.7
7	1.2	0.7	1.9	2.0	1.2
8	1.0	0.6	1.5	1.6	1.0
9	1.3	0.8	2.0	2.1	1.3
10	2.0	1.2	3.2	3.2	2.0
11	1.4	0.8	2.1	2.1	1.3
12	1.1	0.7	1.6	1.6	1.0

Volume 35B, number 1 PHYSICS LETTERS 26 April 1971

Table 2

Approximate intensities in percent for the space symmetries dominant in the ground-state domain. m = particle number for Rosenfeld and (K+12 FP) interactions:= hole number for (KB - 12.5). Representations are ordered according to the eigenvalue of the two-body Casimir (Majorana) operator, which for space symmetry $[\lambda_1, \lambda_2, \ldots \lambda_k]$ is $\frac{1}{2}\{\sum \lambda_i^2 - \bar{\lambda}_i^2\}$ where $\bar{\lambda}_i$ refers to the shape with rows and columns interchanged; for example, for $m = 9$, the ordering is [441], [432], [4311], [333] and [4221]. For m=6, 10 there is a degeneracy, between $[4^k11]$ and $[4^{k-1}11]$; we list them in that order. For two of the cases, as indicated, the single-particle or hole energies have been put to zero.

m	R	R s.p.e.=0	K+12 FP	KB - 12.5	KB - 12.5 s.h.e.=0
4	92-8	100	70-27-3	51-44-4-1	100
5	90-9-1	100	75-20-4-1	70-24-5-1	99-1
6	87-9-4-1	99	73-17-5-5	68-17-8-6	97-3
7	80-19-1	100	50-45-3-1-1	55-39-4-1-1	93-7
8	79-19-1	100	36-49-11-1-2	39-52-6-1-2	96-4
9	82-16-1	100	52-35-9-0-4	61-31-6-0-1	96-4
10	80-10-8-2	98-1	57-15-10-16-0-1	65-14-12-9	94-4-2
11	79-19-1	99-1	43-44-7-4-0-1-1	59-34-5-1	94-6
12	78-20-1	100	28-52-14-2-0-3	39-52-6-1-0-1	95-5

theory were trustworthy one would expect admixings about three times higher than in the central case. Since in the central-force calculations of Akiyama et al. [7] an intensity admixing ~20% is found for the ^{24}Mg ground state, it is clear that, for non-central interactions, perturbation theory is inadequate but that approximately 50% admixing might be expected. Going in the opposite direction, we would expect from the results in table 1 that 'turning off' the single-particle energies would, for the central interaction, result in good symmetry throughout the shell.

The centroid and width calculations are exact. However, instead of considering only a comparison of the σ/D ratios for different interactions, we may use the dimensionalities, centroids and widths, to construct normal ‡ (Gaussian) approximations to the intensity distributions of the individual representations. We can draw much more definite conclusions about the symmetry breaking from the values of these intensities near the ground state. Since our distributions do not specify angular momentum and isospin, we take the ground state (Ratcliff [2]) to be the energy at which the total distribution function has the value one half of the ground-state degeneracy, and similarly for the low-lying excited states. We take the degeneracies (defined by J,T) from experiment. The resultant energies may be compared with those derived via configuration-isospin distributions. For the realistic interactions [3] they are found to be somewhat lower, by up to 2 MeV in the lower half shell and up to 4 MeV in the upper. We might expect the differences to be smaller, at least in the upper half, if we had available the SU(4) distributions with fixed isospin, but these values are quite satisfactory †.

Table 2 gives the intensities near the ground state for $(ds)^{4 \ldots 20}$. Some comparisons with

‡ When there are many particles, and the interactions are of low particle rank, the distribution over *all* states is approximately normal because of the operation of a central limit theorem. This is more or less understood, but normality for distributions over fixed-symmetry subspaces is farr less so (though presumably there is an extended theorem for this case also). The almost unique pattern of the centroids and widths (in which the large representations are centrally located in energy, with the small ones at the ends) gives a clue in this case, for the summed distributions will be approximately normal, at least as the individual distributions are not too wide compared with the spacing. An example of this, for configurations, is given in fig. 4 in the first paper of [2]. We could not for example generate a markedly bimodal distribution except by moving far apart some of the large representations and, this requiring a high-order polynomial in the Casimir invariants, can be achieved only with high-rank interactions. To use this argument for normality in SU(4) we should of course consider the pattern for the representations of an appropriate subgroup.

† In determining the ground-state energy for reasonable interactions, not of course for those with zero single-fermion energies, one actually focusses [3] on a low-lying excited state which will be more accurately given since it is determined by a larger value of the distribution function; one then uses the experimental results to measure down in energy from this 'excited reference state'. Because of the energy uncertainties one must be cautious if the relative intensities have a strong energy dependence; in the present case this happens only near the ends of the shell.

shell-model calculations are possible, though most of these unfortunately are for $(ds)^4$ for which the distribution intensities are found to vary more rapidly with energy than for the more complicated cases, being then less reliable. With a Rosenfeld interaction the [4] and [31] ground-state percentages in ^{20}Ne are found to be (92,8) compared with shell-model values (95,5), while with the (K+12 FP) interaction we find (75,23) compared with (84,13). For ^{24}Mg the Rosenfeld [44] and [431] intensities are (79,19); in this case the corresponding shell-model calculation is not available but (80,20) ground-state percentages are found [7] for a somewhat different central interaction.

For the Rosenfeld interaction (given only for the lower half shell; the results however are little different for the upper half) the symmetry is seen to be fairly good, the maximum admixings being ~20%. That these may be ascribed, as they usually are, to the spin-orbit interaction represented by the $A = 17$ single-particle energies is seen from the R(s.p.e.=0) values which result when these energies are put to zero; we find 98-100% intensity throughout for the maximum space symmetry.

In line with the predictions made from the σ/D values and in contrast with the central-interaction results, the 'realistic' interactions are seen to give quite large symmetry admixings, predominantly involving the three lowest-lying representations. There are many uncertainties about the proper interaction to use ‡, and the distribution intensities by their very nature must not be expected to give highly accurate values of the admixings; still, whether one considers either the simple σ/D argument or the more general distribution results, the conclusion seems inescapable that the SU(4) symmetries are badly admixed throughout the (ds) shell.

The calculations reported here represent an early application of distribution methods to a study of symmetries, and a first tentative one to SU(4) in particular. We have mentioned technical extensions which are needed, one to distributions with fixed isospin (and fixed J would be very much harder), and another to the decomposition of the widths into partial widths. With the first of these we would have more precise intensities and, more important than that, would be able to extend SU(4) studies to nuclei with neutron excess. With the second we would learn more about the nature of the admixings and would for example be able to distinguish between true admixings of symmetries and interweaving of them (in which nearby states have good but different symmetries). There seems little chance that the second alternative applies for SU(4) near the ground state, but it could be important for other symmetries and would be for SU(4) at higher energies. We remark also that the individual distributions, being partial level densities, could be of great interest in studying such reactions as α transfer, in which SU(4) would be expected to be significant.

Even the large admixings indicated would not destroy the significance of the SU(4) symmetry. On the one hand SU(4) is 'strong' in that large energy changes are involved in a change of representation, the effects of this being observable even with large admixings. On the other hand it is a fundamental symmetry, not confined to restricted shell-model spaces and possible even better preserved in extended ones. See the recent discussions [8] of some of these things by Radicati and by Wigner.

‡ The (KB-12.5) interaction for example generates too large a spin-orbit interaction, and this presumably is responsible for the unusually large admixing seen for $A = 36$ with this interaction.

We are indebted to S. S. M. Wong for discussions about SU(4) and for advice about the calculations, and to D. S. Koltun for comments about the manuscript.

References

[1] J.C.Parikh, Intern.Conf. on Properties of nuclear states (University of Montreal Press, Montreal, 1969).
[2] J.B.French, in: Nuclear structure, eds.A.Hossain, Harum-ar-Raschid, and M. Islam, (North-Holland, Amsterdam, 1967);
J.B.French and K.F.Ratcliff, Phys.Rev.C3 (1971) 94;
K.F.Ratcliff, Phys.Rev. C3 (1971) 117.
[3] F.S.Chang, J.B.French and T.H.Thio, Annals of Physics, to be published.
[4] J.B.French, E.C.Halbert, J.B.McGrory and S.S. M.Wong, Advances in Nuclear Physics, 3 eds.M. Baranger and E.Vogt (Plenum, New York, 1969).
[5] E.C.Halbert, J.B.McGrory, B.H.Wildenthal and S.P.Pandya, Advances in nuclear physics, eds.M. Baranger and E.Vogt (Plenum, New York), to be published.
[6] T.T.S.Kuo and G.E.Brown, Phys.Letters 18 (1965) 241.
[7] Y.Akiyama, A.Arima and T.Sebe, Nucl.Phys. A 138 (1969) 273.
[8] L.Radicati, E.P.Wigner, 1970 Solvay Conf. on Nuclear symmetries.

* * * * *

VOLUME 31, NUMBER 20 PHYSICAL REVIEW LETTERS 12 NOVEMBER 1973

Angular-Momentum Dependence of the Density of States*

J. N. Ginocchio
Wright Nuclear Structure Laboratory, Yale University, New Haven, Connecticut 06520
(Received 13 August 1973)

An approximate angular-momentum dependence of the energy centroid and width of the distribution of shell-model states derived from the theory of continuous distributions is compared with the exact results obtained from the diagonalization of shell-model states. The agreement is found to be very good. A density of shell-model states as a function of energy and angular momentum is presented which should be valid for nuclei with a large number of valence particles.

For nuclei with a large number of valence particles (or holes) in a large but finite set of shell-model orbits, the shell-model states with a definite angular momentum are distributed almost normally with respect to the energy,[1,2] that is, the energy dependence is Gaussian with some possible correction to the third and fourth moment. Hence one need only calculate the first few energy moments of the distribution, but these are difficult to calculate for a fixed angular momentum. However, for these nuclei the states are distributed almost normally with respect to the angular-momentum projection along a fixed axis as well.[1] Hence the theory of continuous distributions[3] can be used to generate the needed energy moments for a fixed angular-momentum projection. Then using the fact that the Hamiltonian is rotationally invariant the moments for a fixed angular momentum can be obtained.

Let $\mathcal{N}(M)$ be the number of states per unit spacing of angular-momentum projection that exist for the angular-momentum projection along the Z axis, M, which is the eigenvalue of the angular-momentum operator along the Z direction, J_Z. The states with a given angular-momentum projection $M=J$ will contain all those with angular momentum $\geqslant J$. Hence, to obtain the number of states with angular momentum J, $n(J)$, we need to subtract from the number of states with $M=J$ those with $M=J+1$:

$$n(J)=\mathcal{N}(J)-\mathcal{N}(J+1)\simeq-(d/dJ)\mathcal{N}(J+\tfrac{1}{2}). \quad (1)$$

The last step is an approximation which improves as J increases but is always within 5% of the exact result.

The density $\mathcal{N}(M)$ is given formally by the trace of the operator which projects out all states with angular-momentum projection M:

$$\mathcal{N}(M)\equiv(2\pi)^{-1/2}\int_{-\infty}^{\infty}d\varphi\,\mathrm{Tr}\exp[i\varphi(J_Z-M)]. \quad (2)$$

The density weighted by the expectation value of some operator K is

$$\mathcal{N}_K(M)\equiv(2\pi)^{-1/2}\int_{-\infty}^{\infty}d\varphi\,\mathrm{Tr}\{\exp[i\varphi(J_Z-M)]K\}. \quad (3)$$

If K is rotationally invariant, then, following arguments similar to those in (1), the mean value of K for a fixed angular momentum will be

$$\overline{K}(J)=\frac{-(d/dJ)\mathcal{N}_K(J+\frac{1}{2})}{n(J)}. \quad (4)$$

The next important point is that $\mathcal{N}_K(M)$ can be derived from a knowledge of $\mathcal{N}(M)$.[3] If we replace J_Z by J_Z+yK in (2), denoting the new distribution by $\mathcal{N}(M,y)$ with $\mathcal{N}(M,0)=\mathcal{N}(M)$, the change in $\mathcal{N}(M,y)$ from $\mathcal{N}(M)$ as y vanishes will be related to the derivative of $\mathcal{N}_K(M)$ with respect to M. That is, given this more general distribution

$$\mathcal{N}(M,y)=(2\pi)^{-1/2}\int_{-\infty}^{\infty}d\varphi\,\mathrm{Tr}\exp[i\varphi(J_Z+yK-M)], \quad (5)$$

then it follows from simple algebra and from (4) that

$$\bar{K}(J)=[(d/dy)\mathfrak{N}(J+\tfrac{1}{2},y)]_{y=0}/n(J). \tag{6}$$

The distribution $\mathfrak{N}(M)$ is approximately given by a Gaussian corrected for its third and fourth moments:

$$\mathfrak{N}(M)=\frac{n_0}{(2\pi)^{1/2}\lambda}\exp\left(\frac{-(M-\bar{M})^2}{2\lambda^2}\right)\left\{1+\frac{\mu}{6}\left[\left(\frac{M-\bar{M}}{\lambda}\right)^3-3\frac{M-\bar{M}}{\lambda}\right]+\frac{\gamma}{24}\left[\left(\frac{M-\bar{M}}{\lambda}\right)^4-6\left(\frac{M-\bar{M}}{\lambda}\right)^2+3\right]+\ldots\right\}. \tag{7}$$

In the above n_0 is the total number of states for all M, $\bar{M}$ is the mean angular-momentum projection ($=\langle J_z\rangle$), λ the width of the distribution $\{=[\langle(J_z-\bar{M})^2\rangle]^{1/2}\}$, μ the skewness $[=\langle(J_z-\bar{M})^3\rangle/\lambda^3]$, and γ the excess $[=\langle(J_z-\bar{M})^4\rangle/\lambda^4-3]$. For a system in which all the angular-momentum projections of a given spherical orbit are included in the valence orbits, the mean $\bar{M}$, the skewness μ, and all the odd moments will be zero. However, we include these explicitly because in general the odd moments of J_z+yK will not vanish even though the odd moments of J_z do.

Replacing J_z by J_z+yK in (7) we have, using (6),

$$\bar{K}(J)\simeq\langle K\rangle\left[1+\frac{\alpha D_1(x)+\beta D_2(x)}{1+\frac{1}{24}\gamma D_2(x)}\right], \tag{8a}$$

where we have introduced the variable $x=(J+\frac{1}{2})/\lambda$ for convenience. The polynomials are

$$D_1(x)=x^2-3, \tag{9a}$$

$$D_2(x)=x^4-10x^2+15, \tag{9b}$$

and the parameters α and β are

$$\alpha=\tfrac{1}{2}[\langle J_z^2K\rangle/\lambda^2\langle K\rangle-1], \tag{10a}$$

$$\beta=\tfrac{1}{24}[\langle J_z^4K\rangle/\lambda^4\langle K\rangle-\langle J_z^4\rangle/\lambda^4]-\tfrac{1}{2}\alpha. \tag{10b}$$

Furthermore, the number of states for a given angular momentum is obtained by putting (7) into (1):

$$n(J)=\frac{n_0}{(2\pi)^{1/2}\lambda^2}x\exp\left(\frac{-x^2}{2}\right)\left[1+\frac{\gamma}{24}D_2(x)\right], \tag{8b}$$

We can use this general result to calculate the angular-momentum dependence of the moments of H, the Hamiltonian, by letting K be the various powers of H. For example, if we put $K=H$, we will get the angular-momentum dependence of the energy centroid, $\epsilon(J)=\bar{H}(J)$, and if we put $K=[H-\epsilon(J)]^2$, we will get the energy width, $\sigma(J)=[H^2(J)-\epsilon^2(J)]^{1/2}$.

The expansion in (7) and (8) will be valid only if $|\gamma|$, $|\alpha|$, and $|\beta|$ are small. For N, the total number of single-particle states, and m, the number of valence particles, large, but m/N small, then it can be shown, using techniques developed to evaluate the m and N dependence of operator averages,[4] that $|\gamma|$ is small and $O(m^{-1}, m/N)$. The parameters α and β depend on the operator K. For the centroid α is $O(m^{-1},m/N)$ and for the width it falls off faster, like $O(m^{-2}, m/N)$. In general β falls off faster than α, which makes the expansion (8) in powers of J reasonable. For the centroid, β falls off as $O(m^{-2},m/N)$ and for the width as $O(m^{-3},m/N)$.

An exact shell-model calculation was performed for ^{20}Ne with ^{16}O an inert core, using the Kuo interaction[5] and the single-particle energies $\epsilon(0d_{5/2})=-4.15$ MeV, $\epsilon(1s_{1/2})=-3.28$ MeV, $\epsilon(0d_{3/2})=0.93$ MeV, and for ^{62}Ni with ^{56}Ni inert core, using an effective interaction[6] and the single-particle energies $\epsilon(1p_{3/2})=0.00$ MeV, $\epsilon(0f_{5/2})=0.78$ MeV, and $\epsilon(1p_{1/2})=1.08$ MeV. With the exact eigenvalues generated from this shell-model calculation the exact energy centroid and width for each angular momentum J can be calculated. Using (7) and (8) the approximate results are also calculated with all the parameters determined from the exact eigenvalues. The parameters are tabulated in Table I. The mean centroid over all angular momentum is denoted by ϵ_0 ($=\langle H\rangle$) and α

TABLE I. Parameters calculated from the exact shell-model calculations which describe the angular-momentum dependence of the number of states, the centroid, and the width.

	m	N	n_0	λ	γ	ϵ_0 (MeV)	α	β	σ_0 (MeV)	$\bar{\alpha}$	$\bar{\beta}$
^{20}Ne	4	24	1716	2.7	−0.25	−18.9	0.03	−0.002	6.9	−0.11	0.01
^{62}Ni	6	12	924	2.5	−0.40	4.7	0.02	−0.003	2.2	−0.10	0.01

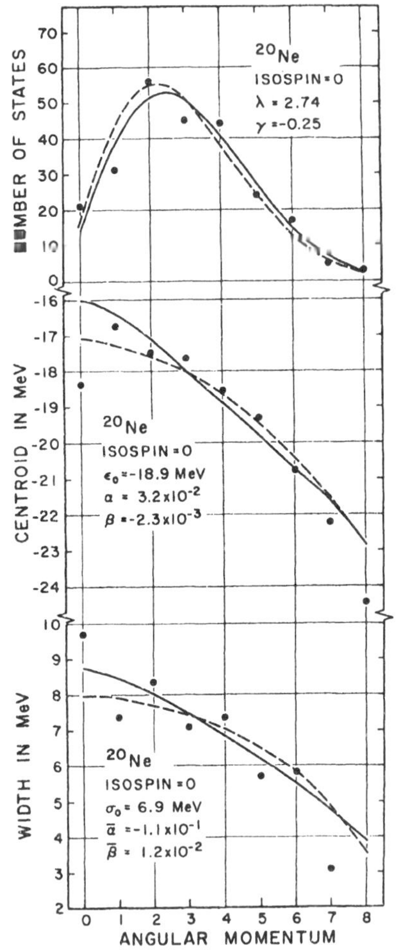

FIG. 1. The number of states, the centroid, and the width of the isospin-0 states of ^{20}Ne. The dots are the exact values calculated from the shell model, whereas the dashed line is the approximate values given by Eq. (12) in the text and the solid line is the approximate values given by Eqs. (8) in the text.

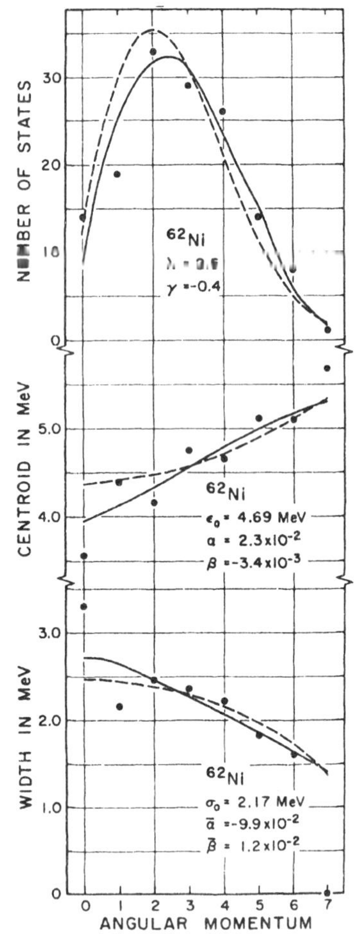

FIG. 2. The number of states, the centroid, and the width of the states of ^{62}Ni. For an explanation of the dots and dashed and solid lines see caption to Fig. 1.

and β are given by (10) with $K \to H$. The mean variance is denoted by σ_0^2 $\{=\langle[H-\epsilon(J)]^2\rangle\}$ but, to distinguish from the centroid, we denote the quantities in (10a) and (10b) with $K=[H-\epsilon(J)]^2$ as $\bar{\alpha}$ and $\bar{\beta}$, respectively. We see from Table I that

$$|\beta| \ll |\alpha| \ll 1 \text{ and } |\bar{\beta}| \ll |\bar{\alpha}| \ll 1 \tag{11}$$

which gives us faith in the approximation. In Figs. 1 and 2 the exact results are compared to the approximate results, and the agreement is good except for the lowest angular momenta in some cases. In fact good agreement is also obtained if one sets γ, β, and $\bar{\beta}$ equal to zero obtaining even a simpler approximation:

$$n(J) \simeq [n_0 x/(2\pi)^{1/2}\lambda^2]\exp(-x^2/2), \tag{12a}$$

$$\epsilon(J) \simeq \epsilon_0[1+\alpha D_1(x)], \tag{12b}$$

$$\sigma^2(J) \simeq \sigma_0^2[1+\bar{\alpha} D_1(x)]. \tag{12c}$$

The nuclei we examined are relatively small systems and yet the conditions (11) are satisfied. For nuclei with m and N larger, we can then certainly expect the approximations (7) and (8) and even (12) to be valid. For such nuclei the density of states of such systems as a function of both energy E and angular momentum will then be

$$\rho(E,J) = \frac{n(J)}{(2\pi)^{1/2}\sigma(J)}\exp\left(\frac{-[E-\epsilon(J)]^2}{2\sigma^2(J)}\right) \tag{13}$$

with $n(J)$, $\epsilon(J)$, and $\sigma(J)$ approximated by (12).

VOLUME 31, NUMBER 20 PHYSICAL REVIEW LETTERS 12 NOVEMBER 1973

The author would like to thank S. Ayik, J. B. French, S. S. M. Wong, and M. M. Yen for discussions and Edith Halbert for ^{20}Ne eigenvalues which were used to check the shell-model code.

*Work supported by U.S. Atomic Energy Commission Contract No. AT(11-1)-3074.

[1]F. S. Chang, J. B. French, and T. H. Thio, Ann. Phys. (New York) 66, 137 (1971).

[2]F. S. Chang and A. Zuker, Nucl. Phys. A198, 417 (1972).

[3]F. S. Chang and J. B. French, Phys. Lett. 44B, 131 (1973). In the above paper the average of an operator K for a fixed energy is derived. In this paper we use the same formalism but average the operator K with respect to the angular momentum projection along the Z axis.

[4]J. N. Ginocchio, Phys. Rev. C 8, 135 (1973).

[5]T. T. S. Kuo, Nucl. Phys. A103, 71 (1967).

[6]R. D. Lawson, M. H. MacFarlane, and T. T. S. Kuo, Phys. Lett. 22, 168 (1966).

OCCUPATION PROBABILITIES AND SPIN CUT-OFF FACTORS☆

F.S. CHANG

Department of Physics, University of Pennsylvania, Philadelphia, Pennsylvania 19104, USA

and

J.B. FRENCH[*]

Department of Physics, Princeton University, Princeton, New Jersey 08540, USA

Received 1 March 1973

Examples are given, for both non-interacting and interacting particles, of the evaluation of orbit occupancies and spin cut-off factors by parametric differentiation on the state density

We illustrate the theory of the preceding paper [1] by considering single-state occupancies and spin cut-off factors. For the occupancy of the sth single-particle state we take $K = n_s$, the number operator, and observe that the parametric derivative is really with respect to ϵ_s, the single-particle energy. Consider to begin with the linear theory which derives from a Gaussian density via eq. (7) of ref. [1] or the first two terms of eq. (10) of ref. [1]. We have $\langle n_s \rangle^m = m/N$; for the variance term we recall that, when there are no interactions, the m-particle variance must be a second-order polynomial in m. Then, since the $m = 0$, N states are dispersion-free, the form must be $m(N-m)(N-1)^{-1}\langle \xi^2(1) \rangle$ where $\xi_i(1) = \epsilon_i(1) - N^{-1}\Sigma\epsilon_j(1)$ is the traceless single-particle energy ($\Sigma\xi_i^2(1) = 0$) and $\langle \xi^2(1) \rangle = N^{-1}\Sigma\xi_i^2(1)$, the mean square, is the 1-particle variance. The interaction, defined by the two-body matrix elements W_{abcd}, generates a Hartree-Fock-like shift in the single-particle energies

$$\epsilon_i(1) \to \epsilon_i = \epsilon_i(1) + (m-1)(N-1)^{-1} \sum_j W_{ijij}$$

$$\xi_i(1) \to \xi_i = \xi_i(1) + (m-1)(N-2)^{-1} \left\{ \sum_j W_{ijij} - N^{-1} \sum_{jk} W_{jkjk} \right\} \tag{1}$$

and this, through changing $\langle \xi^2 \rangle$, gives an ϵ_s-dependent modification of σ^2. The "residual" part of the interaction generates an ϵ_s-independent addition to σ^2, and there are no cross-terms because the separation of H is an orthogonal one, the three parts of H transforming according to different irreducible representations under unitary transformations in the single-particle space. Using

$$N \frac{\partial}{\partial \epsilon_s} \langle \xi^r \rangle = \frac{\partial}{\partial \epsilon_s} \sum_i (\epsilon_i - \langle \epsilon \rangle)^r = r\{\xi_s^{r-1} - \langle \xi^{r-1} \rangle\} \tag{2}$$

we have the linear occupancy equation

$$n_s(E) = \frac{m}{N} \left\{ 1 + \frac{(N-m)}{(N-1)} \xi_s \frac{(E-\xi)}{\sigma^2} \right\} \tag{3}$$

in which of course the total variance σ^2 must be used. The equation is valid, with or without interactions, as long as the density is Gaussian and remains so to first order when ϵ_s is varied (or more generally as long as the shape of

☆Supported in part by the US Atomic Energy Commission.

[*]On leave from the University of Rochester.

the density function is unchanged). The occupancy curves ($n_s(E)$ versus E) are straight lines, all intersecting at the spectrum centroid (at which point all the states have the same occupancy) and with slopes proportional to the traceless single-particle energy. At a given excitation energy the occupancies also vary linearly with ξ_s. The linearities aside, these features are qualitatively what we would expect. We shall see that they are well maintained even when we allow for non-Gaussian corrections to the density.

Now, going beyond the linear theory, we include the $\nu = 3, 4$ terms of eq. (10) of ref. [1], the result being to add to the right-hand side of eq. (3)

$$\frac{\sigma}{6}\left\{1 + \frac{1}{6}(x^3 - 3x) S_3\right\}^{-1} \left\{(x^2-1)\frac{\partial S_3}{\partial \epsilon_s} + \frac{1}{4}(x^3-3x)\frac{\partial S_4}{\partial \epsilon_s}\right\} \tag{4}$$

where $x = (E-\xi)/\sigma$. In writing the corrections it appears better to proceed in each series to the same maximum order of Hermite polynomial; thus the terms as written give the cubic theory, while for the quadratic we would drop the S_3 and $\partial S_4/\partial\epsilon_s$ terms. If, as in the case of the occupancy examples below, we are dealing not with a single state but with a degenerate set of such states (as with a spherical orbit), it is easy to see that the extended equation gives the fractional occupancy as long as we differentiate with respect to the *single-state* energy (or, equivalently with respect to the single-orbit energy and then divide by the orbit degeneracy). Moreover, $\Sigma_s \xi_s = 0 = \Sigma_s \partial S_\nu/\partial\epsilon_s$, the latter (valid for $\nu \geqslant 2$) via translation invariance, and with these it follows that the constraint $\Sigma_s n_s(E) = m$ is automatically satisfied.

As a first illustration of the extended theory, consider the case of non-interacting particles, for which the further constraint $\Sigma_s \epsilon_s n_s(E) = E$ is also satisfied ($\Sigma_s \epsilon_s \partial S_\nu/\partial\epsilon_s = 0$ for $\nu \geqslant 3$ when we take account of scale invariance). One finds easily for the m-particle system that

$$\begin{aligned}
&\xi = m\langle\epsilon\rangle\;;\;(N-1)\,\sigma^2 = m(N-m)\langle\xi^2\rangle,\\
&(N-1)(N-2)\,S_3 = m(N-m)(N-2m)\,\sigma^{-3}\langle\xi^3\rangle\\
&(N-1)^2(N-2)(N-3)\,S_4 = m(N-m)(N-1)[N(N+1) - 6m(N-1)]\,\sigma^{-4}\langle\xi^4\rangle + 3m(N-m)[2m(N-m)(2N-3) - N(N-1)^2]
\end{aligned} \tag{5}$$

which give for the linear theory $n_s(E) = N^{-1}\{m + \langle\xi^2\rangle^{-1}\xi_s(E-\xi)\}$, a special case of (3), and when combined with (2) give the quadratic and cubic corrections.

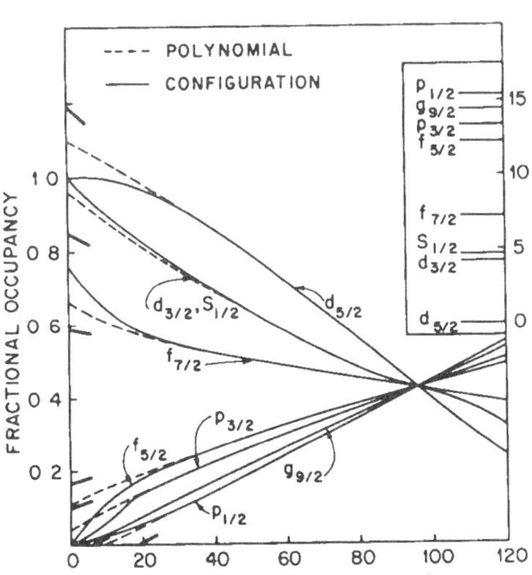

Fig. 1. Occupancies for 18 non-interacting protons distributed over the (N = 42) single-particle spectrum indicated. Calculations are made via the cubic polynomial theory and by using 1074 four-moment configuration distributions. The intercepts of the linear theory are indicated by short heavy lines. The dimensionality is 3.5×10^{11}, too large for the polynomial theory to give high accuracy at low energies. The centroid is at 96.2 MeV and the spectrum extends to 184 MeV

Fig. 1 gives occupancies for 18 non-interacting protons distributed over the $N = 42$ set of 8 spherical orbits indicated. The skewness and excess parameters, S_3 and S_4, are small, −0.023 and −0.091 respectively (a departure from Gaussian is recognizable by eye only if $|S_3|$ or $|S_4| \gtrsim 0.25$). Moreover their single-state parametric derivatives are also small, $|\sigma \partial S_\nu / \partial \epsilon_s| \leqslant 0.01$, 0.02 for $\nu = 3, 4$ respectively, so that under first-order changes in the single-particle energies the almost-Gaussian shape is well maintained. For comparison we show the essentially exact occupancies calculated via a partitioning of the single-particle space. The dimensionality (3.5×10^{11}) is larger by several orders of magnitude than the limit suggested in [1] for good accuracy (the ground-state is 5.6 σ below the centroid) and deviations are seen at low energies. But these are surprisingly small (and for dimensionalities $\leqslant 10^7$ would have in fact not been apparent at all); the errors seem to increase only very slowly with dimensionality. To extend this example to take account of neutron excitations one would simply convolute the proton and neutron densities and carry out parametric differentiation on the resultant.

As an example of occupancies for interacting particles we consider $(ds)^{12}$ with the ^{17}O single-particle energies and the KLS interaction [2]. The dimensionality is 2.7×10^6. Formal expressions for $\langle H \rangle^m$ and $\langle H^2 \rangle^m$ are well known, the derivative of $\langle H^2 \rangle^m$ is given above and, since the derivative of $\langle H^3 \rangle^m$ is also simple, we have an elementary quadratic theory. The cubic correction needs $\langle H^3 \rangle$ itself and the derivative of $\langle H^4 \rangle^m$. Fortunately the third- and fourth-order spectral moments have been worked out by Li and Koltun [3] and Li has supplied us with a computing program for their numerical evaluation. To calculate a derivative with the program one may simply calculate S_ν for the interaction of interest and for one in which the single-particle energy is slightly changed. A more instructive procedure recognizes that the cumulant, K_ν, is a νth-order polynomial in the various single-particle energies so that from a finite set of K_ν values the form analytic in the ϵ_s can be written. As a minor example we give K_4 along the $d_{5/2}$ "ray" in the three-orbit space

$$S_4(\epsilon_1 + \alpha; \epsilon_2; \epsilon_3) = -0.165\{1 - 0.108\alpha + 0.0099\alpha^2\}^{-2}\{1 - 0.198\alpha + 0.0152\alpha^2 - 0.0041\alpha^3 - 0.0001{\cdot}1\alpha^4\} \qquad (6)$$

where α is measured in MeV. With the analytic forms we can of course easily find out what single-particle spectra are, for the given interaction, compatible with normality.

The cubic corrections are found to be negligible and we can thus have confidence in the quadratic theory; indeed the quadratic corrections are only 10% or so, as is clear from the near linearity of the occupancy lines given in fig. 2. We give for comparison the results of a configuration calculation (the space being divided into 41 fixed-isospin configurations, the subspace densities being taken as Gaussian). The agreement is excellent. For this case we are fortunate also to have for the $J = T = 0$ states (839-dimensional) the occupancies derived from a matrix di-

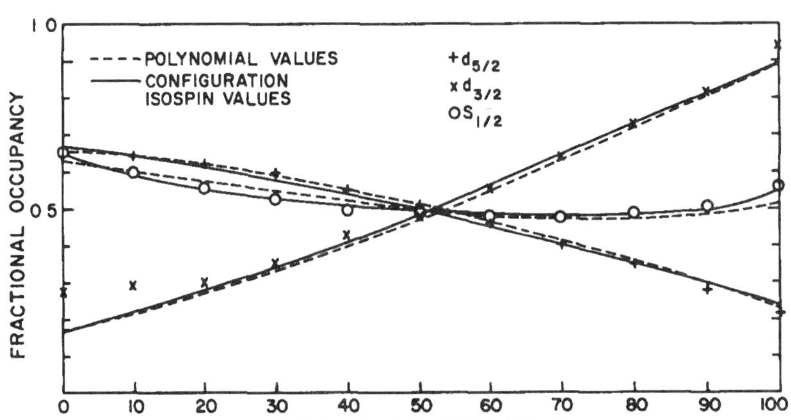

Fig. 2. Quadratic theory occupancies for $(ds)^{12}$ with a KLS interaction. Cubic corrections are negligible. The discrete values are for $J = 0$, $T = 0$ and come by smoothing the matrix results of Soyeur and Zuker.

Volume 44B, number 2 PHYSICS LETTERS 16 April 1973

agonalization done by Soyeur and Zuker [4] and these have been compared with ones derived via fixed-(J, T) configurations whose parameters also come from that calculation. Once again agreement is excellent except for the lowest range of the weakly occupied $d_{3/2}$ orbit. Strictly speaking, the latter sets of results, being for states of fixed J, T, are not comparable with the others in which an averaging over J, T is implicit; the agreement between the two sets, displayed in fig. 2, is evidence that fractional occupancies at a fixed energy may not depend strongly on angular momentum and isospin, which, at least for J, would not be surprising.

Finally consider the spin cut-off factor, given for the m-particle system by $\sigma_J^2 = \langle J_z^2 \rangle^m$, which we evaluate as a function of the energy by considering $H \to H + \alpha J_z^2$. For the constant term in $\sigma_J^2(E)$ we have

$$3(N-1)\langle J_z^2\rangle^m = m(N-m)\langle J^2\rangle^1 = m(N-m)N^{-1}\sum_r N_r j_r(j_r+1) \tag{7}$$

where we have used the polynomial nature of $\langle J_z^2\rangle^m$, as discussed above, and have assumed that the single-particle space is spanned by a set of spherical orbits j_r whose degeneracy $N_r = 2(2j_r+1)$ or half that according as we use or do not use isospin. The linear term requires $\langle J_z^2(H-\xi)\rangle^m$ which is straightforward also, while the shape corrections are evaluated via Li's program. Once again for good measure the cumulants may be given as polynomials in α.

As an example we consider 7 interacting particles in the (fp) shell outside the ^{56}Ni core, i.e., $(f_{5/2}p_{3/2}p_{1/2})^7$ for which $N = 24$, $d = 3.5 \times 10^5$. The single-particle energies are from ^{57}Ni and the interaction is Brown-Kuo [5]. The spectrum span is about 20 MeV and, since $\sigma = 3.5$ MeV, $|x| = |(E-\xi)/\sigma| \leqslant 2.8$. We find

$$\sigma_J^2 = 9.99 - 0.073x - 0.72(x^2-1) \tag{8}$$

the cubic term being negligible It is clear from (8) that this is true also for the linear term, this arising from the essential orthogonality of J^2 and $(H-\xi)$ which is found for other interactions as well and explains why we encounter only a mild variation in $J^2(E)$ in contrast for example to the variation of $T^2(E)$ or $Q\cdot Q(E)$. (8) gives $\sigma_J = 2.8$, 3.2 and 2.6 at $E = 2$, 9.5 and 17 MeV, respectively, compared with 3.2, 3.2, and 3.1 from a configuration-distribution calculation using 30 Gaussian subdistributions. The latter results are not necessarily superior to those of the very simple quadratic theory because, in contrast to the occupancy case where it is rigorously true, we *assume* in the configuration calculations that the operator has a constant average over a configuration In any case the differences are very small.

Finally we emphasize that the very simple method used here can be applied to other operators to study for example the contributions which different parts of H make at various energies, and to compare different H's (such as H's at different stages of renormalization). We may be interested, on the one hand, in accurate calculations, or, on the other, in displaying general features in which case the extremely simple linear theory may well be adequate. We shall present later an extended account which will give the technical details, will describe ways of dealing with larger spaces and spaces with specified symmetries, and will include a variety of applications.

We are indebted to S.Y. Li for making available to us the results and programs for the higher-moment corrections, and to A.P. Zuker for the detailed (ds)-shell results.

References

[1] F S Chang and J.B. French, Phys Lett 44B (1973) 131.
[2] S Kahana, H.C Lee and C.K Scott, Phys Rev. 185 (1969) 1378
[3] S Y Li and D S. Koltun, to be published;
S.Y Li, private communication.
[4] M Soyeur and A P Zuker, Phys. Lett. 41B (1972) 135;
A.P Zuker, private communication
[5] T.T.S. Kuo and G E. Brown, Phys. Lett. 18 (1965) 241.

PHYSICAL REVIEW C VOLUME 10, NUMBER 4 OCTOBER 1974

Recent results on energy-level fluctuations*

O. Bohigas†
Facultad de Ciencias, Universidad Autónoma de Madrid, Madrid, España

J. Flores‡
Instituto de Física, Universidad de México, México 20, D.F., Mexico

J. B. French
Department of Physics and Astronomy, University of Rochester, Rochester, New York 14627

M. J. Giannoni
Institut de Physique Nucléaire, Division Physique Théorique, 91406-Orsay, France

P. A. Mello‡§
Escuela Superior de Física y Matemáticas, Instituto Politécnico Nacional, México 14, D.F., México

S. S. M. Wong
Department of Physics, University of Toronto, Toronto, Canada
(Received 26 March 1974)

We comment on the present theoretical status of the distant-neighbor spacing distributions in view of recent experimental results. In particular, we stress that both the Gaussian orthogonal and the two-body random Hamiltonian ensembles predict essentially identical spacing distributions for the highly excited part of the spectrum; this is true also for the other available measures. Thus the statistical analysis of spectrum fluctuations does not inform us about the two-body or multibody nature of the interaction. We point out also that, when due attention is paid to the variation of level density with excitation, the fluctuation pattern observed in the slow-neutron domain is found to extend throughout the entire spectrum.

In a recent paper Camarda *et al.*[1] study the spacing distributions of slow-neutron resonances in ^{182}W and ^{184}W. The energy level fluctuations are analyzed using, among other measures, the width $\sigma(k)$ of the kth neighbor spacing distribution $p(k; x)$; here x is the kth neighbor spacing ($k=0$ corresponds to nearest neighbors) measured in units of the nearest neighbor spacing average D. Camarda *et al.* compare the experimental values[2] for $\sigma(k)$ with those obtained for the Gaussian orthogonal ensemble (GOE)[3] as well as with the quite different values for the two-body random Hamiltonian ensemble (TBRE)[4] which were presented in a preliminary report[5] at the International Conference on Statistical Properties of Nuclei held at Albany, New York, in 1971. Their conclusion, from this work and from recent studies on other nuclei,[6] is that experiment favors the GOE prediction, which, as discussed at the conference, would suggest a multibody interaction.

We wish to draw attention however to numerical calculations published after the Albany meeting,[7] and to theoretical results,[8] which modify the preliminary values for the TBRE spacing widths to be expected. These lead to TBRE results which are not distinguishable from GOE and of course render invalid the conclusion that the latter results are favored by the experiments. Because of its importance we think it worthwhile to summarize briefly the present status of the problem, mentioning also a remarkable extension of the slow-neutron fluctuation pattern right down into the ground-state domain.[9-11]

The preliminary results, both GOE and TBRE, emerged from Monte Carlo calculations using rather small matrices (dimensionality ≤ 61) and were subject therefore to possible errors from finite-size effects of two varieties. One arose from the "curvature" in the individual spectra, giving rise to a *secular* variation in D over the subspectrum chosen for each matrix; the other arose from *fluctuations*, from member to member of the ensemble, in the average spacing D. In the first results reported the errors of the first kind were eliminated by choosing only a single centrally located spacing from each matrix (it turns out that they could have equally well have been eliminated by a spectrum "unfolding," a procedure which, as we indicate below, would permit the use of spacings from a large part of the spectrum). The D-fluctuation effect mentioned above was however not considered; it should be clear on the one hand that its importance increases with the spacing order [for such a spacing is the sum of $(k+1)$ nearest neighbor spacings], and it is moreover of greater importance with the TBRE ensembles than with GOE for the same matrix size (since the number of matrix-element choices is smaller, and often very much smaller, with TBRE). It follows then that the original method of analysis runs into trouble when the spacing order increases and does so more quickly for TBRE than for GOE. In fact the preliminary GOE results are quite satisfactory while the TBRE results must be revised as in Ref. 7.

The D-fluctuation error can be eliminated in either of two ways. While still selecting a single spacing from each matrix we may renormalize the spectra so that they all have constant variance,[12] or alternatively we may analyze separately a run of levels from each matrix using then a D value computed from that matrix. Both of these procedures give the same results, and give rise, in the cases which have been considered, to a major change in the preliminary TBRE predictions and to a negligible change for GOE.

One can in fact also calculate analytically the correction factor which must be applied to the original $\sigma(k)$ results in order to eliminate the D-fluctuation effect. Suppose that each ensemble is decomposed into "fixed-strength" subensembles characterized by fixed values of the spectrum variance; let the weight with which each subensemble enters into the total be defined by a frequency function g. Then one finds that the apparent spacing width $\hat{\sigma}(k)$, as originally given, is related to the true one $\sigma(k)$ by

$$\hat{\sigma}^2(k)=\sigma^2(k)+[\sigma^2(k)+(k+1)^2]\frac{\Sigma^2}{\overline{D}^2}, \tag{1}$$

where $\overline{D}$, Σ are, respectively, the centroid and variance of g. Values for these quantities come directly from the Monte Carlo calculations. Note that the deviation between $\hat{\sigma}(k)$ and $\sigma(k)$ increases with k as predicted above.

In deriving (1) the assumption has been made that the subensembles differ only in the energy scale; that this is so may in fact be proved,[8] if one takes for granted that the specification of fixed angular momentum (and isospin when relevant) has no effect on the spectrum fluctuations, as seems to be well borne out by calculation. In fact, for k-body ensembles in which J,T constraints are specifically ignored, the quantity $\Sigma^2/\overline{D}^2$ of (1) may be calculated so that we have a complete analytic form for the modifications needed for the original $\hat{\sigma}(k)$ results.

The same conclusion emerges from these various procedures, namely, that the TBRE and GOE spacing widths cannot be distinguished from each other.[13] The same thing is found by considering other statistics, for example[14] Δ_3. The experimental results thus do not favor GOE over TBRE, and by the same token they yield, with the present methods of analysis, no information about the two-body or multibody nature of the interaction.

An important general result derived for the GOE of asymptotic dimensionality is that there is a long-range correlation between the energy levels, indeed a strong correlation of arbitrarily long range, which led Dyson and Mehta[14] to characterize these spectra as "essentially crystalline." We may ask whether this correlation also is to be found with TBRE: further, whether it survives under the secular changes in the density which arise when the spectra are finite; and still further whether the correlation extends throughout the whole spectral domain. Information about those questions arise from three sources. Very large matrix calculations using "realistic" interactions,[9] and TBRE ensemble results,[15] indicate quite clearly that, if the spectrum is "unfolded" (i.e., mapped into one which is essentially of constant density to within fluctuations) by using the (slowly varying) density function, the resultant spectrum has quite closely the same fluctuation pattern[16] as observed in the central regions of the spectra and in the slow-neutron resonance experiments. Perhaps even more striking is the realization of both the Poisson[10] and Wigner[11] nearest neighbor spacing distributions derived by using single spacings in the ground-state region taken throughout the Periodic Table, the first distribution arising when spacings are taken independently of angular momentum and the second when each member of a spacing pair is taken to have the same J value.

It seems quite remarkable that the fluctuation patterns which show up in their purest form in the slow-neutron resonance experiments are apparently those which obtain much more generally in nuclear spectra. However, there is no theoretical understanding of this general result.

---

*Work supported by the Consejo Nacional de Ciencia y Tecnología, México, by the U.S. Atomic Energy Commission, and by the National Research Council of Canada.

†Present address: Institut de Physique Nucléaire, 91406-Orsay, France.

‡Technical Consultant, Instituto Nacional de Energía Nuclear, México.

§On sabbatical leave from Instituto de Física, Universidad de México.

[1]H. S. Camarda *et al.*, Phys. Rev. C 8, 1813 (1973).

[2]One should perhaps mention here that all the values of $\sigma(k)$ for ^{182}W are somewhat smaller than the corresponding GOE values, and that all those for ^{184}W are larger. However, the $\sigma(k)$ for different k values are correlated among themselves, as follows from their definition, so that if the widths for the first few values of k, say $k=0$ and $k=1$, are larger than the corresponding widths for GOE, all tend to be larger and conversely. This is borne out by the shell-model calculations.

[3]M. L. Mehta, *Random Matrices* (Academic, New York, 1967).

[4]J. B. French and S. S. M. Wong, Phys. Lett. 33B, 447 (1970); 35B, 5 (1971); O. Bohigas and J. Flores, *ibid.* 34B, 261 (1971); 35B, 383 (1971).

[5]O. Bohigas and J. Flores, in *Statistical Properties of Nuclei*, edited by J. B. Garg (Plenum, New York, 1972), p. 195.

[6]H. I. Liou *et al.*, Phys. Rev. C 5, 974 (1972); F. Rahn *et al.*, *ibid.* 6, 251 (1972); H. I. Liou *et al.*, *ibid.* 7, 823 (1972); and earlier papers cited therein.

[7]S. S. M. Wong and J. B. French, Nucl. Phys. A198, 188 (1972).

[8]J. B. French, Rev. Mex. Fís. 22, 221 (1973).

[9]F. S. Chang and A. Zuker, Nucl. Phys. A198, 417 (1972).

[10]J. Flores and P. A. Mello, Rev. Mex. Fís. 22, 185 (1973).

[11]E. Cota, J. Flores, P. A. Mello, and E. Yépez, unpublished.

[12]O. Bohigas and M. J. Giannoni, unpublished.

[13]It is known (see Mehta in Ref. 5, p. 182) that for infinite dimensionality the GOE and the corresponding semicircular ensemble of Dyson give rise to the same fluctuations and, on the evidence supplied by Monte Carlo calculations, one takes for granted that this applies also for finite dimensionality (as long as the spectrum curvature is properly looked after) and to similar ensembles in which the individual matrix elements are independently distributed but in non-Gaussian fashion. However, since the results of Refs. 7 and 8 involve Gaussian, we refer in this paper to GOE rather than OE.

[14]F. J. Dyson and M. L. Mehta, J. Math. Phys. 4, 701 (1963).

[15]E. Yépez, Ph.D. thesis, University of Mexico, 1974 (unpublished).

[16]There is some evidence for small deviations: The ω statistic [T. A. Brody, Lett. Nuovo Cimento 7, 482 (1973)] which has a value of 0.95 for the entire GOE spectrum, decreases to 0.80 ± 0.03 for the ground-state region of TBRE and seems therefore to show a small but definite difference between the two ensembles; the nuclear table data yield an ω of 0.69 ± 0.21, so that these may also deviate from the GOE model.

ANNALS OF PHYSICS 95, 90–111 (1975)

Statistical Properties of Many-Particle Spectra*

K. K. MON† AND J. B. FRENCH‡

Physics Department, Princeton University, Princeton, New Jersey 08540

Received June 25, 1975

Many-particle eigenvalue densities, generated by k-body interactions in asymptotically large spectroscopic spaces, are studied in terms of the density moments of a Gaussian orthogonal ensemble of such interactions. It is shown that the densities approach Gaussian as particle number increases; the rate of approach is determined, as is the rate of departure from the semicircular form which obtains in the few-particle limit, as well as the low-moment representation for intermediate cases. By considering the variances of the moments it is shown that a convergence in probability of the individual densities to the ensemble-averaged density, first demonstrated by Grenander for the few-particle limit, is valid in general. In part as a preliminary to a study of the density and spectrum fluctuations, explicit forms are derived for the covariances of low-order moments for arbitrary particle number and of moments of general order in the few-particle and many-particle limits. In the limiting cases the ensemble eigenvalue densities are exhibited in terms of a set of statistically uncorrelated Chebyshev or Hermite-polynomial excitations of the ensemble-averaged densities. Some explicit results for GOE fluctuations are given.

1. INTRODUCTION

It is by now clear that much of many-particle spectroscopy is dominated by two fundamental principles, the first of which fixes the essential form of the (smoothed) eigenvalue density, and the second of which asserts that the energy-level fluctuations are small so that the actual eigenvalues found in typical calculations depart little from those that would be implied by the fluctuation-free density. Moreover, by considering the relationship between the density and the partition function, we see that these principles should simplify not only the spectra but other features of the many-particle system, such as strength distributions.

An enormous amount of attention has been devoted [1] to the Gaussian Orthogonal Ensemble (GOE) of infinite-dimensional real symmetric matrices, which has

* Supported in part by the U.S. Atomic Energy Commission and the National Science Foundation, GP32884. Based on a Princeton University A. B. dissertation by K. K. Mon, May 1973.

† Present Address: Physics Department, Cornell University, Ithaca, New York 14850.

‡ On leave during 1972–1973 from University of Rochester, Rochester, New York 14627.

been introduced in order to study the statistical properties, while for finite dimensionality there has been a variety of Monte-Carlo calculations. However, the GOE, when used as a model for the behavior of the many-particle system, describes a situation in which all the particles interact simultaneously; this unsatisfactory feature obtains because of the large number of Hamiltonian degrees-of-freedom implied by the statistical independence of the GOE matrix elements. A few years ago, operator ensembles of two-body interactions (TBRE), which act in the m-particle spaces, were introduced as a better model [2] and extensive numerical calculations have been made with them; moreover, some formal results for the eigenvalue density have been given by Gervois [3] for the cases of one and two-body Hamiltonians.

Our purpose in this series of papers is to give considerably extended density results for k-body ensembles for both finite and infinite systems (finite and infinite number N of single-particle states), to consider the "ergodic" property of the ensembles (how close to the ensemble-averaged spectrum should the spectrum for an arbitrarily chosen member of the ensemble be?), and to introduce and give results for further ensembles in which the strengths of various "parts" of the Hamiltonian are specified. We also propose to study the fluctuation properties, hoping thereby to make clear the origin of two remarkable features [4]: first that the fluctuations appear to be identical for the k-body ensembles and the GOE, and second, that the fluctuation patterns as calculated for the center of these ensembles (and measured experimentally in the slow neutron resonances) are in fact valid throughout the spectrum, even in the groundstate domain.

Apart from the interest that has been attached to ensemble calculations for a long time, the present results and methods, and natural extensions of them, would appear on several counts to be of physical interest. For one thing ensemble averaging, in the $N \to \infty$ limit, defines characteristic forms for the density. These may be expected to be generally valid in many-particle spectroscopy, and therefore relevant to studies by shell-model methods, by the combinatorial methods used predominantly for noninteracting particles and by spectral distribution methods. With regard to the latter, it should be clear that the present papers will be dealing with ensemble-averaged spectral distributions (Wigner's [5] calculation of the semicircular GOE density being perhaps the first application of such a method). Moreover, we envision a further and quite different application to ensemble-averaged perturbation theory for many-particle systems. It is known by this time [6] that the statistical methods and the more conventional perturbation methods are complementary in that the simplest versions of the first are valid only when the interactions are strong enough to generated a kind of "chaos," a requirement that is quite opposite to the conditions needed for the validity of perturbation methods. On the other hand, we will see that the same partial widths (mean-squared matrix elements connecting one set of states with another), and higher partial moments as well, enter, though in different ways, into both theories so that moments required for the averaged distributions may be adapted also to averaged perturbation theories. We would hope in particular that the procedure implied here will be of value in the theory of effective interactions.

All of the TBRE calculations have been made with Hamiltonians that preserve angular momentum (and isospin where that is relevant). This has been done not because of the belief that the statistical properties are critically dependent on this feature of H, but instead in order to make use of the available shell-model computing programs that incorporate this feature. However, if one is interested in analytic results the (J, T) specifications represent an enormous complication. Therefore, we drop them and, at the same time extending from two-body to k-body interactions, we take the H's to form a Gaussian orthogonal ensemble (GOE) of operators defined in the $\binom{N}{k}$-dimensional space $\mathbf{k}$, formed by distributing k particles over N single particle states. Thus, $H = \sum W_{\alpha\beta}\psi_\alpha(k)\, \psi_\beta^+(k)$, where the ψ_α are the operators for the k-particle states, α denoting an orthogonal labeling. The operators act in the m-particle spaces $\mathbf{m}$, k, and $m \geqslant k$ being arbitrary, as is $N \geqslant m$, which may be finite or infinite. The $W_{\alpha\beta}$ are real, statistically independent (except that $W_{\alpha\beta} = W_{\beta\alpha}$), and have Gaussian distributions with $\overline{W_{\alpha\beta}} = 0$, $\overline{W_{\alpha\beta}^2} = (1 + \delta_{\alpha\beta})\, v^2$; we fix v^2, which supplies the scale, by $\binom{N}{k} v^2 = 1$, so that the ensemble average of the k-particle trace of H^2 will have the value $\binom{N}{k} + 1$, the ensemble average of the average expectation value of H^2 being then $1 + \binom{N}{k}^{-1} \xrightarrow{N\to\infty} 1$. The ensemble, which we shall speak of as an "embedded Gaussian ensemble" (EGOE) is now defined by a triple of quantities $(k, m \geqslant k, N \geqslant m)$; our interest in this paper is with large "dilute" systems $(N \to \infty, m/N \to 0)$. In this limit, it is intuitively clear that the diagonal matrix elements $W_{\alpha\alpha}$ may be ignored almost at will (an exception will be seen in Eq. (21)) since their number is down by a factor $\binom{N}{k}$ from the number of off-diagonal elements. Similarly, we may rewrite H as a k-hole operator, $H \to (-1)^k \sum W_{\alpha\beta}\psi_\beta^+\psi_\alpha$, ignoring the operators of lower hole rank that come from the commutation (or anticommutation). For the logic of this suppose that α, β are direct-product states (ψ_α and ψ_β being then simple products of k creation operators); then the (α, β) commutator is nonzero only if α, β have at least one filled state in common. However, for given α, the number of β's that have t-filled states in common with α is $\binom{k}{t}\binom{N-k}{N-t} \sim N^{k-t}$, the relative weight for $t > 0$ then going to zero in our limit.

We consider the ensemble-averaged m-particle densities in terms of their moments $M_p(m)$, a procedure that was used by Wigner for $m = k$ (i.e., the GOE) in the first evaluation of an ensemble density [5]. We have easily that $M_p(m) = \overline{\langle H^p\rangle^{\mathbf{m}}}$ where $\langle O\rangle^{\mathbf{m}}$ is the average expectation value of O in the m-particle states $(= \binom{N}{m}^{-1} \mathrm{Tr}(O))$, and the bar denotes an ensemble averaging. It is clear that in evaluating M_p we need consider only even-order correlations between the H factors (e.g., $H \times H \to \sum (W_{\alpha\beta})^2\psi_\alpha(k)\psi_\beta^+(k)\, \psi_\beta(k)\, \psi_\alpha^+(k) = HH$ say) since an uncorrelated term will vanish on ensemble averaging. Moreover, for $N \to \infty$, only pairwise correlations will be of consequence, once again by a dimensionality argument. Then, since the eigenvalue distributions depend only on the second moment of the $W_{\alpha\beta}$ distribution, the Gaussian form for the distribution of the matrix elements is irrelevant and, subject only to very mild restrictions, all distributions that are symmetrical about zero lead to the same eigenvalue densities. This is a trivial extension of a result given by Wigner [7] for the GOE case, $m = k$. We have, of course, that $M_p = 0$ for odd p with arbitrary N; and since HH sums uniformly over the k-particle states it must be invariant under unitary transformations in the single-particle space, and hence, a function only of the number operator n, commuting therefore with every H.

In evaluating M_p we must take into account all possible patterns of pairwise correlations. Thus for $p = 4$ we have $AABB$, $ABBA$, $ABAB$, three in all; for arbitrary even $p = 2\nu$ the number is $(2\nu)!/2^{\nu}\nu! = (2\nu - 1)!!$. (From now on we shall for typographical convenience use a notation in which correlated H's are assigned the same letter, and uncorrelated ones different letters.) In an obvious sense the first of the $p = 4$ terms is *unlinked*; but by the cyclic invariance of a trace so is the second; and so for example, since CC commutes with B, is the $p = 6$ term $(AABCCB)$. The *unlinked* terms of order $p = 2\nu$ are then all of those that reduce to $\{AABB...\} \equiv \{AA\}^{\nu}$ when we use cyclic invariance and the fact that $[AA, B] = 0$; each of these makes a contribution $\{M_2(m)\}^{\nu}$ to $M_{2\nu}(m)$. All other terms are linked; a linked term may be factorable, for example $ABCCAB = (CC)(ABAB)$, the contribution to the moment being then similarly factorable. A term that is not factorable we shall describe as *irreducible*.

2. The Gaussian and Semicircular Limits

When the dilute system has a indefinitely large number of particles (i.e., in the limit $N \to \infty$, $m \to \infty$, $m/N \to 0$) every term, linked or unlinked, makes the same contribution to $M_{2\nu}$ for any finite ν. For, in that limit, different correlated pairs of H's will be able to act on different particles (i.e., on particles in different single-particle states), the relative weight of the contributions involving the same particles going to zero (this time as an inverse power of m). Then all H's effectively commute, every term is effectively unlinked, and, since the total number of terms is $(2\nu - 1)!!$ we have that

$$M_{2\nu}(m \gg k) \xrightarrow{N \to \infty} (2\nu - 1)!! \{M_2(m)\}^{\nu} \tag{1}$$

which defines a Gaussian distribution,

$$\rho_G(x) = (2\pi\sigma^2(m))^{-1/2} \exp\{-x^2/2\sigma^2(m)\} \tag{2}$$

the reduced moment $M_{2\nu}/\sigma^{2\nu}$ being the number of pairwise correlation patterns. We shall see ahead that $M_2(m) \equiv \sigma^2(m) = \binom{m}{k}$.

At the GOE limit, in which we have the smallest number of particles, $m = k$, the linked terms do not contribute at all. For in this limit every H acts on the same particles, and, for example, the linked term $ABAB$ could contribute only via a quadruple correlation whose relative weight goes to zero in the $N \to \infty$ limit. At this limit the moment has been given by Wigner [5] as

$$M_{2\nu}(m = k) \xrightarrow{N\to\infty} t_\nu\{M_2(m = k)\}^\nu = t_\nu = (\nu + 1)^{-1} \binom{2\nu}{\nu}$$
$$= (2\nu + 1)^{-1} \binom{2\nu + 1}{\nu} \tag{3}$$

where we have used $M_2(m = k) \to 1$. The *Catalan numbers* t_ν are the moments for the semicircular distribution

$$\rho_0(x) = (2\pi)^{-1}(4 - x^2)^{1/2} \tag{4}$$

and have as a generating function [5, 8] $t(x) = (2x)^{-1}\{1 - (1 - 4x)^{1/2}\} = \sum t_\nu x^\nu$. We observe now that the reduced moment is given as the number of unlinked pairwise-correlation patterns. Wigner's calculation, which we do not reproduce though we give later an extension of it, was expressed in matrix terms. But in the $m = k$ space the matrix for the k-body part of H^p, the only part which can contribute, is simply the pth power of the H matrix and thus the connection is immediate.

We see then that as we increase the number of particles, starting with $m = k$, more and more of the linked patterns contribute to the moments and the distribution changes gradually from semicircle to Gaussian. We wish to understand more about this transition.

3. Noninteracting Particles

The $k = 1$ case, noninteracting particles, is very simple. In the diagonal representation $H = \sum \epsilon_i n_i$ where, for consistency with the $k > 1$ cases, we would normalize things by $N^{-1} \sum \epsilon_i^2 = (N + 1)/N \to 1$. The energies for different particles being additive and statistically independent (since there is no effective Pauli principle when $N \to \infty$) the m-particle density $\rho_m(E)$ is simply the m-fold convolution of the single-particle density ρ_1 with itself. Then, by the most elementary version of the central limit theorem, $\rho_m \to$ Gaussian for large m.

It should be noted carefully that ensemble averaging is not needed for this asymptotic result, which is valid for all reasonable single-particle densities. (As

$N \to \infty$ the single-particle spectrum must be redefined continually. The CLT requires that the variance of the density remain finite during this process.) This is in contrast with the interacting-particle cases ($k > 1$) for which many quite reasonable operators have non-Gaussian spectra; for example, the square of a one-body operator will, in the large-N limit, be distributed as $\chi^2(1)$. For $k > 1$, a Gaussian spectrum is assured only on averaging over an ensemble that assigns a negligible relative weight to these special interactions. In order that the ensemble-averaged results should be relevant to physics we must assume that the Hamiltonians actually encountered are very different from these special ones. But beyond that, in order to be confident that the ensemble-averaged distributions do represent the characteristic densities to be expected for a given H, we will need to consider the variations to be found over the ensemble, which we shall do in terms of the variances of the moments.

Before proceeding to interacting particles we derive from the $k = 1$ example a few other results, more or less well-known, which, while not essential, will nonetheless be of value to us. The characteristic function $\sum M_p(it)^p(p!)^{-1}$, which is the Fourier transform of the density, multiplies[1] under convolution, and then writing $M_r(1) = \langle H^r \rangle^1 = N^{-1} \sum (\epsilon_i)^r$ as $\langle \epsilon^r \rangle$, we have

$$\left\{\sum_r \langle \epsilon^r \rangle\, x^r (r!)^{-1}\right\}^m = \sum M_p(m)\, x^p (p!)^{-1} \tag{5}$$

so that the left-hand side is the generating function for the quantities $M_p(m)(p!)^{-1}$. On the other hand, we have from the theory of partition polynomials [8] that the coefficient of x^p in $(1 + a_1 x + a_2 x^2 + \cdots)^m$ is

$$\sum_{\omega(p)} \binom{m}{l} \frac{l!}{\pi(l_s)!} \prod_r (a_r)^{l_r} \tag{6}$$

where the sum is over all the partitions of p; l is the total number of parts in a partition and l_s is the number of times that s is found in it; thus, $l = \sum l_s$, $p = \sum s l_s$. We then find for the m-particle moment the well-known result that

$$M_p(m) = p! \sum_{\omega(p)} \prod_r \left\{\frac{\langle \epsilon^r \rangle}{r!}\right\}^{l_r} \frac{l!}{\pi(l_s!)} \binom{m}{l}. \tag{7}$$

The extension of this result to the finite-N case has been given recently by Ginocchio [9]. This equation with $\langle \epsilon^{2\nu} \rangle \to t_\nu$, $\langle \epsilon^{2\nu+1} \rangle \to 0$, then gives the m-particle ensemble-averaged moments for noninteracting particles (i.e., for the $k = 1$ EGOE) while $\{\sum_\nu t_\nu x^{2\nu}/(2\nu)!\}^m$ is a generating function for those moments, as in (5). The

[1] Its logarithm is then additive and this leads to the additivity of the cumulants $K_p(m) = mK_p(1)$. Eq. (7) then simply reflects the relationship between moments and cumulants.

m-particle density is of course the m-fold convolution of the semicircle. Fig. 1 shows the semicircular → Gaussian transition for this case.

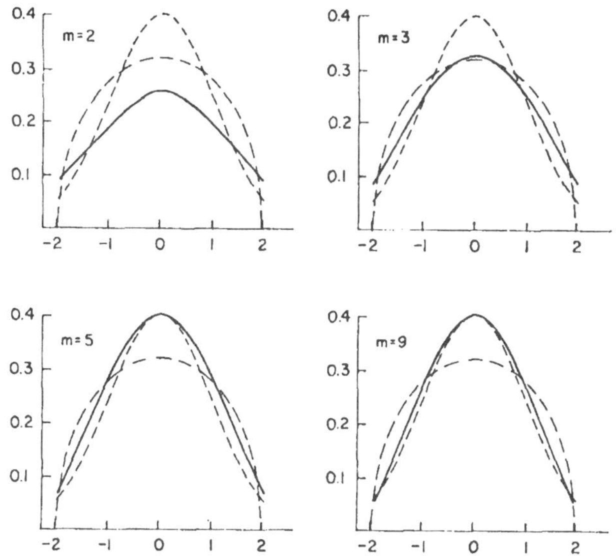

FIG. 1. The density for m noninteracting particles (solid line) showing the transition from semicircular, which obtains when $m = 1$, to Gaussian.

We shall have use for a counting theorem [8] related to these results. Suppose that we have f slots and that we associate with μ objects in any one of them a factor g_μ. Agree moreover that for any arrangement of q *identical* objects among the slots, the g-factors multiply so that $[\mu_1\mu_2\mu_3 \cdots]$ associates with $g_{\mu_1} g_{\mu_2} g_{\mu_3}$; agree also that these products add for different arrangements. Then on enumerating the partitions we see from (6) that for all possible arrangements of the q objects, the factor is the coefficient of x^q in $\{\sum g_\mu x^\mu\}^f$. Moreover for $g_\mu = t_\mu$ we have the result [10] for powers of the Catalan generator that

$$\left\{\sum_\mu t_\mu x^\mu\right\}^f = \sum_q \frac{f}{2q+f}\binom{2q+f}{q} x^q \tag{8}$$

so that in this case the counting gives a factor $f/(2q+f)\binom{2q+f}{q}$.

This result, with $f = 2\nu' - 1$, may be used in counting the number $R(\nu, \nu')$ of equivalent correlation patterns that derive from inserting $q = (\nu - \nu')$ correlated pairs of H's into one irreducible pattern of order $2\nu' > 0$, generating thereby a corresponding contribution to the moment of order 2ν. For example, two pairs can be inserted into each of the three "internal" slots in $ABAB$ in t_2 ways, giving a factor $3t_2$, and we can insert one pair in each giving $3t_1^2$; but $3t_2 + 3t_1^2 = 9 = 3/7\binom{7}{2}$ in agreement with the result given. Here, we have carried out our counting on the line, but because of the cyclic invariance of the trace we really should count on the circle. To do this we must allow for one internal and one external pair and no internal and two external pairs; in the first case, the external pair gives patterns of type XXO, XOX, OXX and a factor $3t_1$, while in the second we get a factor $5t_2$. In the general case, taking account of both kinds of terms in the promotion from $2\nu' \to 2\nu$ and the various internal external partitions of the pairs $(\nu - \nu' \to (\nu - \nu' - y) + y)$, and noting that $(2\nu + 1)t_\nu = \binom{2\nu+1}{\nu}$, we have if $\nu \neq \nu'$

$$\begin{aligned} R(\nu, \nu') &= (2\nu' - 1) \sum_{y=0}^{\nu-\nu'} (2\nu - 2y - 1)^{-1} \binom{2y+1}{y} \binom{2\nu - 2y - 1}{\nu - \nu' - y} \\ &= \binom{2\nu}{\nu - \nu'} \end{aligned} \tag{9}$$

the last step by a slight extension of the Vandermonde identity. The internal–external separation is not valid for $\nu' = 0$ (because of the circular symmetry) but in that case we already have $R(\nu, 0) = t_\nu = (\nu + 1)^{-1}\binom{2\nu}{\nu}$.

4. Densities for Interacting Particles

We turn to the case of particles interacting with k-body forces (which with $k = 1$ will include the noninteracting case). Now when a particle is added to a system, the energies, except for $k = 1$, are not additive because of the effect of the interactions, and thus, the straightforward convolution argument fails. We therefore consider the ensemble-averaged distributions in terms of their moments. Only the even-order moments are nonvanishing so that they coincide with the central moments, $M_p = \mu_p$. For the second moment we need AA which, as indicated above, is a polynomial in n. This is a special case of the operator $AO(t)A$ which we may think of as an ensemble-averaged contraction about the t-body operator $O(t)$. Since $N \to \infty$ we may regard $O(t)$ also as a t-hole operator, $O(t) = (-1)^t\tilde{O}(t)$ and we use this in the evaluation. We adopt a short-hand notation in which $\alpha \equiv \alpha(k) = \psi_\alpha(k)$, and a summation convention is understood; $\tilde{n} = (N - n)$ is the hole number operator. We have the results that

$$\alpha\alpha^+ = \binom{n}{k}, \qquad \beta^+\beta = \binom{\tilde{n}}{k} \to \binom{N}{k} \tag{10}$$

the first of which comes from the fact that both $\alpha\alpha^+$ and $\binom{n}{k}$ are k-body operators ($\binom{n}{k}$ vanishes on m-particle states with $m < k$) with unit diagonal defining ($m = k$)

matrix elements), while the second follows from the first by hole $\rightleftharpoons$ particle conjugation. Then we have, writing O_{rs} for the t-particle matrix elements of O, that

$$\begin{aligned} AOA &= v^2\alpha\beta^+O\beta\alpha^+ = (-1)^t\, v^2 O_{rs}\alpha\beta^+\psi_s{}^+(t)\, \psi_r(t)\, \beta\alpha^+ \\ &= (-1)^t\, v^2 O_{rs}\alpha\psi_s{}^+\beta^+\beta\psi_r\alpha^+ = O_{rs}\alpha\psi_r\psi_s{}^+\alpha^+ = O_{rs}\psi_r\alpha\alpha^+\psi_s{}^+ \\ &= O_{rs}\psi_r \binom{n}{k} \psi_s{}^+ = \binom{n-t}{k} O(t) \\ AA &= \binom{n}{k} \end{aligned} \tag{11}$$

Note that $AO(t)A$ is then a $(t+k)$-body operator (vanishing in m with $m < (t+k)$). It might be remembered also that the AOA form (11) is derived for the large-N limit and, as will be made clear in the next paper, relies on the property that in this limit $O(t)$ is unitarily irreducible with the highest unitary symmetry (i.e., of Young-column structure $[N+t, N-t]$). If on the other hand only parts of O of lower unitary rank should contribute in a given calculation then (11) would have to be modified; for example $\langle F(t')\, AO(t)A\rangle^{m} = \langle AF(t')\, AO(t)\rangle^{m}$ and using (11) as it stands would lead to a paradox if $t' \neq t$. However, no cross-terms in unitary symmetry can contribute to a trace so that if $t' < t$, the highest symmetry parts of O would be ineffective and the trace would have the value $\binom{n-t'}{k}\langle FO\rangle^{m}$.

We are now ready to evaluate the low-order moments. We write operator forms in which, for the m-particle moment, we would simply write $n \to m$.

$p = 2$:

$$M_2 = AA = \binom{n}{k} \tag{12}$$

Although H^2 has mixed particle ranks $k, k+1, \ldots, 2k$, only the lowest survives, the higher-rank terms being describable as "surface" terms that vanish for $N \to \infty$. A similar result applies for $H^{2\nu}$ where parts with rank $> \nu k$ vanish in the limit.

$p = 4$:

As above, we have three correlation patterns, two of which yield $\{M_2\}^2$. For the irreducible pattern we have, via (11),

$$ABAB = \binom{n-k}{k} BB = \binom{n-k}{k}\binom{n}{k} = \binom{2k}{k}\binom{n}{2k}. \tag{13}$$

Then for the fourth moment M_4 and the "excess", defined in terms of the central moments by $\gamma_2 = \mu_2^{-2}(\mu_4 - 3\mu_2^2)$, we have

$$
\begin{aligned}
M_4 &= \binom{n}{k}\left\{2\binom{n}{k} + \binom{n-k}{k}\right\} \\
\gamma_2 &= \binom{n-k}{k}\binom{n}{k}^{-1} - 1 \xrightarrow{m \gg k} \frac{-k^2}{n}.
\end{aligned} \tag{14}
$$

The excess $\gamma_2(m)$ increases with m from the value -1 at $m = k$ (semicircle) to zero for large m (Gaussian), the approach to Gaussian being slower when the k value is larger. A departure from Gaussian defined by $|\gamma_2| > 0.3$ say, which is recognizable by eye, is reached for $k = 1, 2, 3, 4, 5$ with $m \simeq 4, 12, 26, 47, 74$, respectively. Similarly near the small-m limit the departure away from the semicircle is slower when the k-value is higher. The negative excess is unity when $m < 2k$ and at $m = 2k$ decreases to

$$
1 - \binom{2k}{k}^{-1} = 1/2,\ 5/6,\ 19/20,\ 69/70\ldots \qquad \text{for} \quad k = 1, 2, 3, 4.
$$

As a technical point observe that $ABAB$ ($\sim \binom{n}{2k}$) is a $2k$-body operator and thus cannot contribute until we have at least $2k$ particles, and similarly for any higher-order term that contains this as a factor or even as a *sequence* (other H's standing between the four $ABAB$ members, as for example in $ACDBCADB$). Moreover by repeated use of (11) or by simply recognizing that uncorrelated H's act on different particles we see that the sixth-order term $ABCABC$ is $3k$-body $= (3k)!/(k!)^3\binom{n}{3k}$ and the 2νth order term $ABC\ldots XABC\ldots X$ is νk-body $= (\nu k)!/(k!)^\nu\binom{n}{\nu k}$, where the constant is a multinomial coefficient.

For most purposes moments of order higher than fourth are unnecessary but we nonetheless give those of order 6, 8 and indicate how one could proceed somewhat further.

$p = 6$:

We have $5!! = 15$ correlation patterns, for each of which the contribution to the moment can be written at sight by use of (11); for example,

$$
ABCBAC = \binom{n-k}{k} ACAC = \binom{n}{k}\binom{n-k}{k}^2 = \binom{2k}{k}\binom{n-k}{k}\binom{n}{2k}.
$$

On writing all the terms we find:

$$
\begin{aligned}
M_6 &= 5\binom{n}{k}^3 + 3\binom{2k}{k}\binom{n}{2k}\left\{2\binom{n}{k} + \binom{n-k}{k}\right\} + \frac{(3k)!}{(k!)^3}\binom{n}{3k} \\
&= 5\{M_2\}^3 + 3\binom{n-k}{k} M_4 + \frac{(3k)!}{(k!)^3}\binom{n}{3k}.
\end{aligned} \tag{15}
$$

We see that, in the first way of writing, the terms divide into four classes containing 5, 6, 3, 1 members, the first class generating (for arbitrary ν) the semicircular moment, the last the $\binom{n}{\nu k}$ term discussed above. The terms of the second class are factorable 4×2, and derive from adding a pair to *ABAB*; by the counting theorem (9), with $2\nu = 6$, $\nu - \nu' = 1$, this set has $R(6, 1) = 6$ terms. Finally we have the single class of irreducible operators, *ABCBAC* and its two relatives.

$p = 8$:

There $7!! = 105$ terms; we could easily write these down, and all but two of them we could evaluate at sight. Let us instead proceed more formally. $t_4 = 14$ of the terms are unlinked and give the semicircular moment. We have $R(4; 2) = 28$ terms deriving from two correlated pairs on *ABAB*, which by (13) give 28[31]; there are $3R(4; 3) = 24$ terms arising from 1 pair on the three-dimensional irreducible class *ABCBAC*, and these by (15) give 24[22]; $R(4; 3) = 8$ terms arise from 1 pair on *ABCABC* giving 8[211]; 4 terms are 4×4 factorable generated from *ABABCDCD* giving then 4[22]. (We temporarily use a notation in which $[abc...] = \binom{n}{k}^a\binom{n-k}{k}^b\binom{n-2k}{k}^c \ldots$.) Thus, we have 78 reducible terms; we now need the irreducible ones, 27 in all.

Let us consider the arrangements on a circle of four pairs of objects *ABCD* starting with those that contain an *ABCD* subset of adjacent members. Let us catalogue the 14 arrangements according to the set of unordered distance between members of each pair, e.g., $ABCDACDB \equiv 4233$, $ABCDBADC \equiv 3333$. We find seven distinct classes of irreducible patterns 4233, 4242, 4433, 4444, 2332, 3333, 2222, the last of which, being factorable 4×4, we disregard; members of each of the classes are respectively *ABCDACDB*, *ABCDADCB*, *ABCDABDC*, *ABCDABCD*, *ABCDBDAC*, *ABCDBADC*. Under cyclic operations and order inversions these patterns generate respectively 8, 4, 4, 1, 8, 2 distinct members (no member of a class being convertible to another by relabeling of A, B, C, D). Since the numbers add to 27 it follows that we have identified all the irreducible operators. The contributions from the first five classes can be written at sight giving respectively 8[121], 4[13], 4[112], [1111], 8[13]. We have by now evaluated 103 of the 105 contributions to M_8.

In *ABCDBADC*, a member of the missing class 3333, we cannot escape considering a contraction over a product of two operators which are oppositely ordered to their pairing partners (in contrast to the situation, e.g., in *ABCABC*, in which, as indicated above, *BC* is effectively a $2k$-body operator). We evaluate by a procedure which, with its obvious variants, can be used as well with higher-order products encountered with higher p. We have, using a combinatorial identity [8], that the particle-rank decomposition of *CDDC* is

$$CDDC = \binom{n}{k}^2 = \sum_{\nu=k}^{2k} \binom{k}{2k-\nu}\binom{\nu}{k}\binom{n}{\nu} \tag{16}$$

and since $\binom{n}{\nu} = \alpha(\nu)\,\alpha^{+}(\nu)$ we encounter $AB\alpha(\nu)\,AB\alpha^{+}(\nu)$, which is seen to give $\binom{n-\nu}{k}^2\binom{n}{\nu}$. Taking account of the two terms we have then that the missing contribution to the $p = 8$ moment is

$$2\sum_{\nu=k}^{2k}\binom{k}{2k-\nu}\binom{\nu}{k}\binom{n-\nu}{k}^2\binom{n}{\nu} \tag{17}$$

and thus

$$\begin{aligned} M_8 = {} & 14(M_2)^4 + 28[31] + 28[22] + 8[211] \\ & + 8[121] + 12[13] + 4[112] + [1111] \\ & + 2\sum_{\nu=k}^{2k}\binom{k}{2k-\nu}\binom{\nu}{k}\binom{n-\nu}{k}^2\binom{n}{\nu} \end{aligned} \tag{18}$$

which, of course, can be written in a variety of forms.

For noninteracting particles ($k = 1$) it is easily verified that the values given here reduce to those of (7) with the appropriate values of $\langle\epsilon^r\rangle$ as specified above; (7) of course solves the $k = 1$ problem generally. $k = 1$ results up to sixth order (and $k = 2$ results to fourth order) also have been derived by Gervois [3] using different methods. With regard to the general moment problem for ($k > 1$)-body interactions we stress that the procedures given and illustrated above are adequate for any term in any moment M_p. Nonetheless, except in the two limiting cases $m = k$ and $m \gg k$, the combinatorial problem of the classification of terms is such that, for the moment of arbitrary order, a new procedure, probably an inductive one, would be required.

5. Variance, Ergodicity, and Fluctuations

For ensemble averaging to be relevant to an individual system, the properties of interest must vary little from member to member of the ensemble. Therefore, we should consider the variation over the ensemble of the density at fixed energy, which we may do in terms of the covariances $\sum^2_{p,q}(m)$ (and the variances $\sum^2_{p,p}(m) \equiv \sum^2_p(m)$) of the spectral moments,

$$\sum\nolimits^2_{p,q}(m) = \overline{\langle H^p\rangle^{\mathrm{m}}\,\langle H^q\rangle^{\mathrm{m}}} - \overline{\langle H^p\rangle^{\mathrm{m}}}\;\overline{\langle H^q\rangle^{\mathrm{m}}}. \tag{19}$$

It has already been shown by Grenander [11] that, in a normalization in which the moments are finite in the $N \to \infty$ limit (which is the case with our choice, $\binom{N}{k}v^2 = 1$), the GOE variances ($m = k$) vanish in this limit. This result can be interpreted as convergence in probability of the individual densities to the ensemble-averaged density; results that are more refined from the mathematical standpoint are available also [12]. We shall give, *inter alia*, the extension of Grenander's results to the EGOE Gaussian limit, as well as explicit large-N forms for moment variances and covariances that are of consequence for the fluctuation properties of the spectra and in deciding the circumstances under which we have an ergodic behavior [13].

As with the moments we need consider only binary correlations among the $W_{\alpha\beta}$, but we have the difference that $H_{\alpha\beta}$ and its correlated partner may both act in the same space, contributing then to one of the $\langle H^p\rangle^m$ factors, or they may be "cross-linked," one operator of the pair then acting in each of the factors. With the cross-linked operators we must admit correlations of both types, $H_{\alpha\beta}H_{\beta\alpha}$ and $H_{\alpha\beta}H_{\alpha\beta}$, only the first of which contributes for large N when the operators act in the same space. We have now a decomposition of $\sum_p^2$ according to ζ the number of cross-linked operator pairs; $\zeta = 0, 1, 2,..., p$, but with $(p - \zeta)$ necessarily even (observe that p itself may be even or odd, for while the odd moments vanish their variances do not). When there is no cross-linkage ($\zeta = 0$) the separate factors are statistically independent and the two terms of $\sum_p^2$ then cancel; the variance is thus the sum of the $\zeta > 0$ terms in $\{\langle H^p\rangle^m\}^2$.

Consider first the completely cross-linked terms, $\zeta = p$, with $p > 2$, a particular one of which is the ensemble average of

$$\langle ABC \cdots X\rangle^m \langle ABC \cdots X\rangle^m.$$

In the direct-product representation (Hartree–Fock like, in which a state operator is a simple product of single-particle operators) we may by an argument given in Section 1, ignore operators $H_{\alpha\beta} = W_{\alpha\beta}\psi_\alpha(k)\,\psi_\beta^+(k)$ in which (α, β) have any single-particle states in common. Thus, every $H_{\alpha\beta}$ may be regarded as transferring k particles to different states. Consider now the action of H^p, in the first space, on a particular m-particle state which we may think of as the "vacuum;" the first operation generates k hole-particle pairs, the number of possible transformations being $\binom{m}{k}\binom{N-m}{k} \to \binom{m}{k}\binom{N}{k}$; after this the dominant sequence (dimensionality of highest order in N) is a succession of $(p-2)$ scatterings into "particle" states of the k particles generated in the first step, each of these involving a selection among $\binom{N}{k}$ choices, and recurrences during a sequence being ignorable; the final transformation, with no free choice, simply destroys the k hole-particle pairs. The result is that the completely cross-linked part of the H^p matrix element is represented by k-body matrix rings, $W_{\alpha\pi}W_{\pi\sigma} \cdots W_{\delta\gamma}W_{\gamma\beta}W_{\beta\alpha}$, the number of terms being $\binom{m}{k}\binom{N}{k}^{p-1}$. (So also is the GOE *moment*, but in that case the number of contributing rings is strongly reduced by the need for binary correlations among the $W_{\alpha\beta}$.) The statistical correlation then essentially determines the sequence of operators in the other $\langle H^p\rangle^m$ factor, the only further choices with the operators arising from invariance under cyclic permutations (e.g., $W_{\gamma\beta}W_{\beta\alpha}W_{\alpha\pi} \cdots W_{\delta\gamma}$ is allowable) which gives a factor p, and under inversions of the order $(W_{\alpha\beta}W_{\beta\gamma} \cdots W_{\pi\alpha})$ which gives a factor 2; e.g., with $\langle ABC\rangle^m$ for the first factor, we have ABC, BCA, CAB

and their inverses for the second. The $\binom{m}{k}$ factor allows for the fact that there are $\binom{m}{k}$ m-particle states which contain the k-particle state α as a "core." There is also a choice of "starting" state in the second factor; a given ring requires that a specified set of k single-particle states be filled ($\equiv \alpha(k)$). The number of corresponding m-particle states is then $\binom{N}{m-k}$ and averaging yields the factor $\binom{N}{m-k}\binom{N}{m}^{-1} \to \binom{m}{k}\binom{N}{k}^{-1}$. All states in the first factor give the same results and no further averaging is needed. Collecting things together, and using $\overline{W_{\alpha\beta}^2} \to v^2 = \binom{N}{k}^{-1}$, we have then, for the fully linked ($\zeta = p$) contribution to the variance, that for fixed p, arbitrary m,

$$\sum^2 (m: \zeta = p) \to 2\zeta \binom{m}{k}^2 \binom{N}{k}^{-2} \tag{20}$$

in the large-N limit.

The result is valid for arbitrary p though the derivation is so only for $p > 2$. For $p = 1$, we must consider the diagonal transformations $H_{\alpha\alpha}$, the only ones which contribute to a linear trace. For $p = 2$, the diagonal elements may indeed be ignored but a complication arises because H^2 must be diagonal, this imposing a binary correlation in each factor, $(H^2)_{\alpha\alpha} = \sum H_{\alpha\beta}H_{\beta\alpha}$; thus, with the cross-linkage, a fourth-order correlation must be considered for the variance. However, we have for $p = 1$, that $\langle H\rangle^m$ is a pth order polynomial in m, and thus

$$\langle H\rangle^m = \binom{m}{k} \langle H\rangle^k = \binom{m}{k}\binom{N}{k}^{-1} \sum W_{\alpha\alpha} \tag{21}$$

being then a sum of $\binom{N}{k}$ independent identically distributed Gaussian random variables with centroid 0 and width $\binom{m}{k}\binom{N}{k}^{-1}2^{1/2}(v) = 2^{1/2}\binom{m}{k}\binom{N}{k}^{-3/2}$ (where the $2^{1/2}$ factor comes because the diagonal variance is $2v^2$ rather than v^2). This sum is itself Gaussian $(0, 2^{1/2}\binom{m}{k}\binom{N}{k}^{-1})$, the variance then being given properly by (20). For $p = 2$, we could use a similar polynomial argument (observing that $\langle H^2\rangle^m$ is a fourth-order polynomial in m, symmetric for large N under $m \to (N - m)$, or we could proceed from (21) noting that $\langle H^2\rangle^m$ is quadratic and uniform in the W's; we have then that

$$\langle H^2\rangle^m \xrightarrow{N\to\infty} 2\binom{m}{k}\binom{N}{k}^{-1} \sum_{\alpha<\beta} (W_{\alpha\beta})^2 \tag{22}$$

being a sum of $\frac{1}{2}\binom{N}{k}^2$ (non-Gaussian) random variables and therefore distributed as χ^2 in $\frac{1}{2}\binom{N}{k}^2$ variables, which is Gaussian for large N. Since $(W_{\alpha\beta})^2$ has centroid v^2 and variance $(\overline{W_{\alpha\beta}^4} - (\overline{W_{\alpha\beta}^2})^2) = 2v^2$ (because $W_{\alpha\beta}$ is Gaussian) we see that $\langle H^2\rangle$ is $(\binom{m}{k}, 2\binom{m}{k}\binom{N}{k}^{-1})$ so that its variance once again is properly given by (20).

Since with our normalization the pth moment becomes constant when $N \to \infty$ for fixed m, we see from (20) that in this limit the rms deviation in the fully-linked contribution to the moment goes to zero as $\binom{N}{k}^{-1}$. This conclusion is unaffected by inclusion of the $\zeta < p$ terms; a correlated pair of operators standing together in

$\langle H^p \rangle^m$ gives by (12) a factor $\binom{m}{k}$, which goes to $\binom{m-\tau}{k}$ if the correlated operators are separated by an operator of rank τ. Since these values are independent of N, as is the number of such correlated pairs for any fixed p, we see that the convergence in probability demonstrated by Grenander [11] for the GOE applies also for the embedded k-body ensembles (EGOE); the resultant density, defined by (m, k), is of course semicircle for $m = k$, Gaussian for $m \gg k$, and more complicated in between.

We may now assert that, in the limit of large N, all members of the ensemble, except for a negligible fraction which becomes a zero-measure set as $N \to \infty$, give the same density. This behavior may be reasonably described by the statement that, as $N \to \infty$, the EGOE ensembles become ergodic for the density. (We are dealing now not with a single ensemble but with a chain of ensembles specified by N. For fixed N we can also consider the chain generated by varying m though there is no $m \to \infty$ limit and we have already restricted ourselves to $m \ll N$.) However, this does not imply ergodicity for the spectrum, nor for measures of spectrum fluctuations (especially measures that span a prescribed number of levels as opposed to a fixed energy domain) since the level spacing D goes very rapidly to zero (as N^{-m}) as $N \to \infty$.

To make things clear and to consider fluctuations we need more complete evaluations of the covariances. We begin by giving large-N variances for $p = 3-6$ and arbitrary m, which are derived easily by the methods of Section 4. The corresponding $p = 1, 2$ results are given by (20). For $p = 3$ we have nine equal terms with structure $\langle XXA \rangle \langle YYA \rangle$ which, added to the $\zeta = 3$ result (20) give

$$\Sigma_3^2(m) = 6\binom{m}{k}^2\binom{N}{k}^{-2}\left\{1 + 3\binom{m}{k}^2\right\}. \tag{23}$$

Similarly, we find

$$\begin{aligned}
\Sigma_4^2(m) &= 8\binom{m}{k}^2\binom{N}{k}^{-2}\left\{1 + 2\left[2\binom{m}{k} + \binom{m-k}{k}\right]^2\right\} \\
\Sigma_5^2(m) &= 10\binom{m}{k}^2\binom{N}{k}^{-2} \\
&\quad \times \left\{1 + 15\left[\binom{m}{k} + \binom{m-k}{k}\right]^2 + 5\binom{m}{k}^2\left[2\binom{m}{k} + \binom{m-k}{k}\right]^2\right\} \\
\Sigma_6^2(m) &= 12\binom{m}{k}^2\binom{N}{k}^{-2}\left\{1 + 6\left[2\binom{m}{k} + 3\binom{m-k}{k}\right]^2\right. \\
&\quad \left. + 3\left[5\binom{m}{k}^2 + 6\binom{m}{k}\binom{m-k}{k} + 3\binom{m-k}{k}^2 + \binom{m-k}{k}\binom{m-2k}{k}\right]^2\right\}
\end{aligned} \tag{24}$$

in each of which the separate terms are for definite ζ, the terms being arranged in descending order; thus, in the last one we have $\zeta = 6, 4, 2$, respectively. As described later, the corresponding contributions to the covariances, being factorable, may now be written at sight for $(p, q) \leqslant 6$; and of course, for nonvanishing terms, we must require that (p, q, ζ) all have the same parity.

For covariances of arbitrary order we shall deal with the two extreme cases, GOE $(m = k)$ and Gaussian $(m \gg k)$, beginning with GOE. For cross-linked terms of order ζ in the covariance $\Sigma^2_{p,q}$, (19), the ζ uncorrelated operators in each factor define $(\zeta + 1)$ slots, $(\zeta - 1)$ internal ones and two external ones in the sense of the counting leading to (9). Into these we must insert $(p - \zeta)/2$ correlated pairs with the condition that the members of a pair cannot be in different internal slots; nor can one member be in an internal slot and its companion in an external one, for in these cases the correlated pair contracts about the k-body operator (or the product of such) which separates them and, as in (11), gives an operator of higher rank than k, which vanishes then for $m = k$. Thus, we have the same counting that gave rise to (9) with $2\nu' \to \zeta$ (which now may be even or odd) and $(\nu - \nu') \to (p - \zeta)/2$. In the $\langle H^p \rangle^m$ factor, the correlated pairs now yield a factor $R(\nu, \nu') = \binom{p}{(p-\zeta/2)}$ and similarly in $\langle H^q \rangle^m$ so that the large-N GOE covariances are

$$\Sigma^2_{p,q}(m = k) = 2\binom{N}{k}^{-2} \sum_{\zeta=1}^{p} \zeta \binom{p}{(p-\zeta)/2}\binom{q}{(q-\zeta)/2} = \sum_{\zeta=1}^{p} \Lambda_p^{(\zeta)} \Lambda_q^{(\zeta)}, \tag{25}$$

the sum being restricted by parity. Carrying out the summation for $p = q$ we have for the GOE variances

$$\begin{aligned} \Sigma^2_p(m = k) &= 2p \binom{N}{k}^{-2} \binom{p-1}{\mu}^2 \\ 2\mu &= p - 1 \quad \text{if} \quad p = \text{odd} \\ &= p - 2 \quad \text{if} \quad p = \text{even}. \end{aligned} \tag{26}$$

We observe that all allowed orders of cross-linkage contribute to the GOE variances, the fractional contribution of the lowest order being in fact $8p/(p + 2)^2$ or $4p/(p + 1)^2$ according as p is even or odd. Similarly for the covariances.

Going beyond the question of the convergence of the densities to the ensemble value, eq. (25), in particular the factoring of the partial convariances that it displays, tells us a considerable amount about the energy-level fluctuations. For corresponding to the decomposition of the covariance into a sum of products there is a decomposition of the density $\rho(x)$ and the distribution function $F(x) = \int_{-\infty}^{x} \rho(x')dx'$ which is adequate for calculating two-point density correlations and corresponding fluctuation measures. There already exists of course a different, elegant, and difficult fluctuation theory for GOE, developed by Wigner, Dyson, Mehta, and many others [1]. But this theory is quite intractable for TBRE and EGOE ensembles, so that a formalism that can give any results for these as well is of considerable interest, both theoretically and because of experiment.

Writing the density in terms of excitations of the semicircle (4) by

$$\rho(x) = \rho_0(x) \left\{1 + \sum_{\zeta \geqslant 1} R_\zeta \mathscr{G}_\zeta(x)\right\} \tag{27}$$

we define $\mathscr{G}_\zeta(x)$ by its moments according to

$$\int \mathscr{G}_\zeta(x)\, \rho_0(x)\, x^p\, dx = (2\zeta)^{-1/2} \binom{N}{k} \Lambda_p^{(\zeta)} = \binom{p}{(p-\zeta)/2}. \tag{28}$$

Then inspection shows that we properly reproduce the average density and the covariances if we regard the amplitudes R_ζ as random variables satisfying

$$\begin{aligned} \overline{R}_\zeta &= 0 \\ \overline{R_\zeta R_{\zeta'}} &= 2\zeta \binom{N}{k}^{-2} \delta_{\zeta\zeta'} \end{aligned} \tag{29}$$

the variables being then uncorrelated (though not necessarily statistically independent). Thus, covariances define the two-point density correlations; for higher-order fluctuations we would need to evaluate higher-order ensemble averages.

To construct the functions $\mathscr{G}_\zeta(x)$ we first evaluate the Fourier transform of the partial density:

$$g_\zeta(t) = \int e^{itx} \rho_0(x)\, \mathscr{G}_\zeta(x)\, dx = (2\zeta)^{-1/2} \binom{N}{k} \sum_p (it)^p\, (p!)^{-1}\, \Lambda_p^{(\zeta)} = i^\zeta J_\zeta(2t) \tag{30}$$

where J_ζ is the Bessel function. The inverse Fourier transform gives [14]:

$$\pi\rho_0(x)\, \mathscr{G}_{2\nu}(x) = (-1)^\nu (4 - x^2)^{-1/2} T_{2\nu}(\cos\phi) - (-1)^\nu (4 - x^2)^{-1/2} \cos 2\nu\phi$$
$$\pi\rho_0(x)\, \mathscr{G}_{2\nu+1}(x) = (-1)^\nu 2^{-1} (4 - x^2)^{-1/2} x U_{2\nu}(\cos\phi) = (-1)^\nu (4 - x^2)^{-1/2} \sin(2\nu+1)\phi$$

where

$$\phi \equiv \phi(x) = \sin^{-1}(x/2); \qquad -\pi/2 \leqslant \phi \leqslant \pi/2; \qquad \pi\rho_0(x) = \cos\phi. \tag{31}$$

Here $T_{2\nu}$, $U_{2\nu}$ are Chebyshev polynomials of the first and second kind and ϕ is identifiable on the semicircle as the angle between the radial coordinate and the normal through the center.

We have now, via (27), (29), and (31), a pleasing picture of the ensemble density in terms of a series of uncorrelated excitations of the semicircle, whose wavelength, which varies inversely with increasing cross-linkage order, is constant when measured along the semicircle ($\lambda = 2\pi/\zeta$). In the central region, observing that

$2\phi(E)d/\pi$, with $d = \binom{N}{k}$ = total number of levels, is the average number of levels in the energy interval (0–E), we see that for order $\zeta \gg 1$ a fraction $4/\zeta$ of the levels is spanned by a single wavelength. It should be noted that we have, quite naturally, made use of the polynomials appropriate to the continuous semicircle rather than the lattice polynomials appropriate to the discrete spectrum. In order to make a (partial) correction for this we must restrict the ζ summation to $\zeta_{max} = d$, that being the number of lattice polynomials.

The distribution function follows immediately.

$$F(x) = F_0(x) + \sum_{\nu \geqslant 1} (-1)^\nu (2\pi\nu)^{-1} R_{2\nu} \sin 2\nu\phi + \sum_{\nu \geqslant 0} (-1)^{\nu+1} [(2\nu+1)\pi]^{-1} R_{2\nu+1} \cos(2\nu+1)\phi \tag{32}$$

where $F_0(x)$, the ensemble-averaged (semicircular) distribution function is given by

$$2\pi F_0(x) = \sin 2\phi + 2\phi + \pi = \{4 - x^2\}^{1/2} x/2 + 2\sin^{-1}(x/2) + \pi. \tag{33}$$

As a check on these results we rederive an important result of Dyson and Mehta [15], which demonstrates the rigidity of the GOE spectrum; more general applications for GOE (and EGOE) we leave for a later paper. Recalling that $d \times F(E)$ counts the number of levels with energy $< E$, we see that the number of levels in the interval (E_1, E_2), with of course $|E_i| \leqslant 2$, is represented by the random variable

$$n = n_0 + \frac{d}{2\pi} \sum_{\nu \geqslant 1} (-1)^\nu \nu^{-1} R_{2\nu} \{\sin 2\nu\phi_2 - \sin 2\nu\phi_1\} - \frac{d}{2\pi} \sum_{\nu \geqslant 0} (-1)^\nu (\nu + \tfrac{1}{2})^{-1} R_{2\nu+1} \{\cos(2\nu+1)\phi_2 - \cos(2\nu+1)\phi_1\} \tag{34}$$

where n_0, the expected number, follows simply from (33). The variance of n is then

$$\begin{aligned} \Sigma^2(n) &= \frac{4}{\pi^2} \sum_{\nu \geqslant 1} \nu^{-1} \sin^2 \nu(\phi_2 - \phi_1) \cos^2 \nu(\phi_2 + \phi_1) \\ &\quad + \frac{4}{\pi^2} \sum_{\nu \geqslant 0} (\nu + \tfrac{1}{2})^{-1} \sin^2(\nu + \tfrac{1}{2})(\phi_2 - \phi_1) \sin^2(\nu + \tfrac{1}{2})(\phi_2 + \phi_1) \\ &\to \frac{4}{\pi^2} \int_0^{d/2} \frac{d\nu}{\nu} \sin^2 \nu(\phi_2 - \phi_1) \simeq \frac{2}{\pi^2} \ln 2d(\phi_2 - \phi_1) \\ &\to \frac{2}{\pi^2} \ln(\pi n_0) \end{aligned} \tag{35}$$

where, in the second last step, we have dropped "small" terms, so that the result is valid for the case $\ln n_0 \gg 1$, and, in the last step, have used (33) to give the variance for the central region of the spectrum. This final result agrees with that of Dyson and Mehta. On the other hand, a more general result that

$$\Sigma^2(n) \to \frac{2}{\pi^2} \ln(\pi n_0/\cos^2 \phi) \tag{36}$$

follows also for an angular interval centered at ϕ, and small enough so that the variation of the density over the interval may be ignored (and of course large enough to contain many levels). The belief, deriving from Monte-Carlo calculations, that measures of the fluctuation, when expressed in terms of local spacing units, vary only slowly over the spectrum is seen to be justified in this case.

We turn now to the Gaussian case, EGOE with $m \gg k$, which as explained above should give a better model for studying observed fluctuations. Using the same procedure as for GOE we see that, for cross-linked terms of order ζ in the co-variance $\Sigma^2_{p,q}(m)$, we have $\binom{p}{\zeta}\binom{q}{\zeta}$ ways of selecting the operators to be cross-linked, and, for $\zeta > 0$, each contributes a factor (20). The $\zeta = 0$ term gives $\overline{M}_p\overline{M}_q$ and then once again, as in (25), we find a factored covariance

$$\Sigma^2_{p,q} = 2\binom{m}{k}^2\binom{N}{k}^{-2}\sum_\zeta{}' \zeta \binom{p}{\zeta}\binom{q}{\zeta} T_p^{(\zeta)}T_q^{(\zeta)} \equiv \sum_\zeta{}' \Lambda_p^{(\zeta)}\Lambda_q^{(\zeta)} \tag{37}$$

where the prime indicates a sum restricted to even $(p-\zeta)$, $(q-\zeta)$, and the $T_p^{(\zeta)}$ takes account of the $(p-\zeta)$ operators in $\langle H^p\rangle^m$ which are *not* cross-linked. If $m \gg (p-\zeta)k$ we may regard these operators as acting on different particles (as we have done in deriving (1)); they give then the Gaussian moment of order $(p-\zeta)$ so that

$$T_p^{(\zeta)} \to (p-\zeta-1)!!\binom{m}{k}^{(p-\zeta)/2} + O(m^{-1}), \qquad m \gg (p-\zeta)\,k \tag{38}$$

where, as indicated and may be verified from the exact forms of (24), the terms ignored are of relative magnitude m^{-1} (and smaller). (For the long-range excitations (small ζ) the implied limit on p is of no consequence, but, as shown ahead, things are very different with the short-range excitations for which the form given for $T_p^{(\zeta)}$ does not apply.)

The largest terms in (37) correspond to the lowest-order linkages, $\zeta = 1, 2$ for odd and even moments, respectively. For these we have

$$\begin{aligned} \Sigma^2_p(m) &\to 2(p!!)^2\binom{m}{k}^{p+1}\binom{N}{k}^{-2} && p \text{ odd} \\ &\to p^2(p-1)!!^2\binom{m}{k}^{p}\binom{N}{k}^{-2} && p \text{ even} \\ \Sigma^2_{p,q}(m) &\to \Sigma_p\Sigma_q. \end{aligned} \tag{39}$$

The ensemble density defined by (39), which results when all the $\zeta > 2$ excitations are quiescent, is very simple, being Gaussian with a centroid $\mathscr{E}$ and variance σ^2 which are uncorrelated random variables with ensemble centroids and variances $(0, \Sigma_1^2)$ for $\mathscr{E}$ and $(\sigma_0^2 = \binom{m}{k}, \Sigma_2^2)$ for σ^2. We already know that these variables are Gaussian. An elementary calculation gives for the pth moment of such a distribution that $M_p = (-i\sigma)^p He_p(i\mathscr{E}/\sigma)$ which, to lowest order in $\binom{N}{k}^{-1}$, gives properly back the results of (39).

These long-range position and scale fluctuations are large. We have for the centroid fluctuation ($\zeta = 1$), with D the central level spacing, that the rth eigenvalue fluctuates in position by an rms amount $\Sigma_1 = \pi^{-1/2}\binom{N}{m}\binom{N}{k}^{-1}\binom{m}{k}^{1/2}D$, thereby spanning a number of levels which (in contrast to GOE for which $\Sigma_1 = 2^{1/2}D/\pi$) grows rapidly with N (and also with m which we have however restricted, $m \ll N$). The ($\zeta = 2$) variance describes a fractional fluctuation in the energy scale, whose rms value is $\binom{N}{k}^{-1}$; under its influence the fluctuation of the given level, located at E say, spans a number of levels $\binom{N}{k}^{-1}\binom{N}{m}\, E\rho(E)$ which also increases indefinitely with N (and m). We could say that the ($m \gg k$) EGOE ensembles are "not ergodic" for the position of a given (numbered) eigenvalue. The $\zeta = 1$ fluctuation has no effect on level-spacing distributions; $\zeta = 2$ on the other hand induces a "spurious" addition to the variance of the tth-nearest-neighbor spacing distributions whose value increases with t, is independent of m for fixed (k, t, N), and goes rapidly to zero as N increases. This effect has given rise to serious ambiguities in Monte-Carlo analyses. For a better statistical model of many-particle systems we should use centered-renormalized ensembles in which these fluctuations are put to zero, and for which there are no $m \to \infty$ divergences.

For the intermediate-order excitations we observe that for even $(p - \zeta)$ and with $\rho_0 \equiv G(0, \sigma_0)$

$$\int \rho_0(E)\, E^p He_\zeta(E/\sigma_0)\, dE = \sigma_0{}^p \zeta!\, (p - \zeta - 1)!! \binom{p}{\zeta} \tag{40}$$

and then recognize from (37, 40) that for $\zeta \geq 3$, we have Hermite-polynomial *shape* fluctuations in the density, representable, by

$$\begin{aligned} \rho(x) &= \rho_G(x) \left\{ 1 + \sum_{\zeta \geq 3} \zeta!^{-1} S_\zeta He_\zeta((x - \mathscr{E})/\sigma) \right\} \\ F(x) &= F_G(x) - \sigma\rho_G(x) \sum_{\zeta \geq 3} \zeta!^{-1} S_\zeta He_{\zeta-1}((x - \mathscr{E})/\sigma) \end{aligned} \tag{41}$$

in which $\rho_G(x)$ is Gaussian ($\mathscr{E}$, σ) with the $\mathscr{E}$ and σ^2 fluctuations built in. The S_ζ, which for a fixed Hamiltonian would be the Hermite-polynomial moments, are now uncorrelated random variables with zero centroids and with variances which follow from (37), (38), and (41). We find easily that

$$\overline{S_\zeta S_{\zeta'}} = 2\zeta\delta_{\zeta\zeta'} \binom{m}{k}^{2-\zeta} \binom{N}{k}^{-2} \tag{42}$$

a result which extends to $\zeta = 1, 2$, as in (39), if we define $S_1 = \mathscr{E}/\sigma_0$ and $S_2 = (\sigma^2/\sigma_0^2 - 1)$ which are the Hermite-polynomial moments of the orders indicated; but we prefer to put to zero the $\zeta = 1, 2$ variances. (Since $S_\zeta \sim \binom{N}{k}^{-1}$ we can ignore the $\zeta = 1, 2$ fluctuations in dealing with the S_ζ contributions to the moments; i.e., we can put $\mathscr{E} \to 0$, $\sigma^2 \to \sigma_0^2$ in (41), the shape fluctuations being therefore unaltered by centering and scale renormalization. We can also put $S_\zeta \to 0$ in (41) in dealing with the centroid and variance fluctuations as we have done in deriving (39).)

The most striking feature of these excitations is the very rapid decrease in their amplitude as the order increases; this behavior, which is quite different from that for GOE, arises because the $\binom{m}{k}$ "counting" factor generated by a correlated operator pair acting in the same space is lost when the pair is cross-linked; for given ζ it is of course these "self-correlated" pairs which determine the moments that describe the excitation (37), (38), and (40).

However, as we proceed to excitations of still higher order a "saturation" effect begins to set in when the system has too few particles to permit the self-correlated pairs to act independently. In this case (38) is no longer valid and a transition occurs to a system of which the GOE (in which all the self-correlated pairs act on the same particles) is the limiting case. It seems clear that this must be the origin of the agreements which have been found [16] for Monte-Carlo calculations of various EGOE fluctuations and their GOE counterparts, and coupled with the GOE results, it also should explain why the EGOE fluctuation patterns also extend over the complete spectrum. It would appear that when we embed the k-body GOE in the m-particle space, the "medium" has little effect on the shortest-range fluctuations. We hope later to derive the details of the transition, to make clearer thereby the origin and the limitations of the agreements described, and to discuss whether or not significant information about nuclear structure can be deduced, for example from energy-level fluctuations as determined in slow-neutron experiments.

6. Conclusion

The transition from a semicircular shape to Gaussian for the ensemble-averaged eigenvalue density in a large many-particle model space has been shown to be generated by the action of a central limit theorem. The rate of the transition has been calculated for k-body interactions. That the ensemble-averaged result is relevant for characteristic Hamiltonians has been shown by considering the mean-squared deviations in the density as determined by the variances of its moments. Some preliminary results about the energy-level fluctuations, derived from the moment covariances, are also given. Extensions of these results in several directions and applications to slow neutron and related fluctuations will be given later.

Acknowledgments

We are indebted to G. T. Garvey for discussions and general support, to G. V. Chester and D. F. Holcomb for making funds available for computing, to P. A. Mello for an important remark about two-point correlations, and to S. S. M. Wong for making available to us a large array of Monte-Carlo calculations that have been invaluable in checking results and in indicating the structures involved.

References

1. Statistical theories of spectra: C. E. Porter (Ed.), "Fluctuations," Academic Press, New York, 1965; M. L. Mehta, "Random Matrices and the Statistical Theory of Energy Levels," Academic Press, New York, 1967; E. P. Wigner, *SIAM Rev.* **9** (1967), 1; M. L. Mehta, *in* "Statistical Properties of Nuclei," (J. B. Garg, Ed.), Plenum Press, New York, 1972.
2. J. B. French and S. S. M. Wong, *Phys. Letters* **33B** (1970), 449; **35B** (1971), 5; O. Bohigas and J. Flores, *Phys. Letters* **34B** (1971), 261; **35B** (1971), 383; S. S. M. Wong and J. B. French, *Nucl. Phys.* **A198** (1972), 188.
3. A. Gervois, *Nucl. Phys.* **A184** (1972), 507.
4. O. Bohigas, J. Flores, J. B. French, M. J. Giannoni, P. A. Mello, and S. S. M. Wong, *Phys. Rev.* **C10** (1974), 1551.
5. E. P. Wigner, *Ann. Math.* **62** (1955), 548.
6. J. B. French, *Rev. Mex. Fis.* **23** (1974), 189.
7. E. P. Wigner, *Ann. Math.* **67** (1958), 325.
8. J. Riordan, "Combinatorial Identities," John Wiley and Sons, New York, 1968.
9. J. N. Ginocchio, *Phys. Rev.* **C8** (1973), 135.
10. J. Riordan, Ref. [8], p. 154.
11. U. Grenander, "Probabilities on Algebraic Structures, John Wiley and Sons, New York, 1963.
12. L. Arnold, *J. Math. Anal. Applic.* **20** (1967), 262; W. H. Olson and V. R. R. Uppuluri, *in* "Proceedings of the Sixth Berkeley Symposium on Probability and Statistics," University of California Press, Berkeley, California, 1970.
13. T. Brody and P. A. Mello, *Phys. Letters* **37A** (1971), 429; O. Bohigas and M. J. Giannoni, *Ann. Phys.* (*N.Y.*) **89** (1975), 393.
14. A. Erdelyi *et al.*, "Higher Trancendental Functions," Vol. 2, p. 92, McGraw-Hill, New York, 1953.
15. F. J. Dyson and M. L. Mehta, *J. Math. Phys.* **4** (1963), 701.
16. S. S. M. Wong and J. B. French, Ref. [2]; O. Bohigas *et al.*, *Phys. Rev.* **C10** (1974), 1551.

Group theory and propagation of operator averages

C. Quesne*

Physique Théorique et Mathématique, Université Libre de Bruxelles, Brussels, Belgium
(Received 17 June 1975)

The propagation of operator averages, which is the basis of French's spectral distribution method, is reformulated in the framework of group theory. The concept of complementary groups is extensively used. It is shown that the possibility of propagating averages is intimately connected with the absence of state labeling problem. The construction of the propagation operators is examined, and for those cases where it is not trivial, a new way of approach is suggested by establishing a link with recent group theoretical advance in the construction of subgroup invariants in the universal covering algebra of a group. Finally the discussion is illustrated by some examples taken, or not, from current literature.

1. INTRODUCTION

During the last few years, the spectral distribution method mainly developed by French and coworkers[1-3] has proved quite useful in nuclear spectroscopy. Its use has been based on the possibility of propagating the operator averages, defined in given subspaces of the fixed number of particles spaces, from low to high values of the number of particles (or holes). Although it has been obvious from the beginning that this procedure was intimately connected with group theory, the precise relationship has not been clearly understood up to now.

In this paper we shall reformulate the average propagation in the framework of group theory. For this purpose, the concept of complementary groups, introduced previously by Moshinsky and Quesne,[4] turns out to be useful and for this reason is briefly reviewed in Sec. 2.

Section 3 is devoted to the discussion of the average propagation in the light of the concepts just introduced. We show the existence of close connections between some difficulties encountered in the propagation process and classical problems of group theory, such as those coming from the use of noncanonical chains of subgroups. In particular we stress the usefulness of the construction of an integrity basis for the subgroup invariants in the enveloping algebra of a group when there is a state labeling problem in the group to subgroup reduction. This establishes a link between an interesting group theoretical problem, whose solution was formulated recently in general terms,[5] and some physical applications.

Finally, in Sec. 4, we illustrate the general discussion by some examples taken from the current physical literature or corresponding to new averaging processes.

2. NONINVARIANCE GROUPS AND COMPLEMENTARY SUBGROUPS

In the second quantized picture, fermions are represented by creation and annihilation operators, b_α^+ and b_α respectively, where α denotes all the quantum numbers specifying the one-particle states. In the case of nucleons in a central field, α stands for $nljm$ or nlm_lm_s for identical nucleons, and $nljmm_t$ or $nlm_lm_sm_t$ for neutrons and protons in the isospin formalism. All n-particle states can be written as linear combinations of the states

$$|\alpha_1\alpha_2\cdots\alpha_n\rangle = b_{\alpha_1}^+ b_{\alpha_2}^+ \cdots b_{\alpha_n}^+ |0\rangle, \quad \alpha_1 < \alpha_2 < \cdots < \alpha_n, \tag{2.1}$$

where $|0\rangle$ is the vacuum state.

When the number of one-particle states is finite and equal to $\mathcal{N}$ (so that $0 \leqslant n \leqslant \mathcal{N}$), it is well known[4,6] that the largest noninvariance group that can be built is the group $U(2^{\mathcal{N}})$, the generators of whose Lie algebra $\mathrm{C}^{\alpha_1'\alpha_2'\cdots\alpha_{n'}'}_{\alpha_1\alpha_2\cdots\alpha_n}$ can transform any state (2.1) into any other one,

$$\mathrm{C}^{\alpha_1'\alpha_2'\cdots\alpha_{n'}'}_{\alpha_1\alpha_2\cdots\alpha_n} |\alpha_1''\alpha_2''\cdots\alpha_{n''}''\rangle$$

$$= \delta_{n'n''}\delta_{\alpha_1'\alpha_1''}\cdots\delta_{\alpha_{n'}'\alpha_{n'}''}|\alpha_1\alpha_2\cdots\alpha_n\rangle. \tag{2.2}$$

This group was called supergroup by Judd.[6] It is clear from Eq. (2.2) that all n-particle states, $0 \leqslant n \leqslant \mathcal{N}$, belong to the irreducible representation (IR) [1] of $U(2^{\mathcal{N}})$.

The supergroup contains a smaller noninvariance group $O^+(2\mathcal{N}+1)$, the generators of whose Lie algebra are the one-particle creation and annihilation operators, and their commutators.[7] All n-particle states, $0 \leqslant n \leqslant \mathcal{N}$, belong to the single IR $[\frac{1}{2}\mathcal{N}]$ of this group. The group $O^+(2\mathcal{N}+1)$ contains all the symmetry groups, the unitary group $U(\mathcal{N})$, and its subgroups.

In general we are concerned with a chain of groups

$$\begin{array}{cccc} U(2^{\mathcal{N}}) \supset & O^+(2\mathcal{N}+1) \supset & U(\mathcal{N}) \supset & G, \\ [1] & [\frac{1}{2}\mathcal{N}] & [1^n] & \lambda \end{array} \tag{2.3}$$

below each one of which we give the corresponding IR to which the states belong. The n-particle states can be written as

$$|n\varphi\lambda\mu\rangle = P^+_{n\varphi\lambda\mu}|0\rangle, \tag{2.4}$$

where μ characterizes the row of the IR λ of G, φ distinguishes between the various equivalent IR's λ of G contained in the IR $[1^n]$ of $U(\mathcal{N})$, and $P^+_{n\varphi\lambda\mu}$ is an appropriate linear combination of products of n creation operators b_α^+. The generators of $U(2^{\mathcal{N}})$ can be rewritten in the basis (2.4) as $\mathrm{C}^{n'\varphi'\lambda'\mu'}_{n\varphi\lambda\mu}$ and are characterized by the property

$$\mathrm{C}^{n'\varphi'\lambda'\mu'}_{n\varphi\lambda\mu}|n''\varphi''\lambda''\mu''\rangle = \delta_{n'n''}\delta_{\varphi'\varphi''}\delta_{\lambda'\lambda''}\delta_{\mu'\mu''}|n\varphi\lambda\mu\rangle. \tag{2.5}$$

They are rather complicated linear combinations of the operators

$$\mathrm{D}^{n'\varphi'\lambda'\mu'}_{n\varphi\lambda\mu} = P^+_{n\varphi\lambda\mu}P^{n'\varphi'\lambda'\mu'}, \tag{2.6}$$

which can be inverted to give the latter in terms of the former.[4] Here

$$P^{n\varphi\lambda\mu}=[P^{\dagger}_{n\varphi\lambda\mu}]^{\dagger}. \tag{2.7}$$

The noninvariance groups $U(2^{\mathcal{N}})$ and $O^{+}(2\mathcal{N}+1)$ are especially useful when discussing the concept of complementary groups.[4] Two groups G and G^{C}, whose direct product is a subgroup of a larger group H, are referred to as "complementary" within a definite IR of H, if there is a one-to-one correspondence between all the IR's of G and G^{C} contained in this IR of H. All Casimir operators of one of the subgroups depend linearly on the Casimir operators of the other one in this IR of H.[4] In the following, we shall take for H either noninvariance group, whose IR is unique, so that we shall not mention it anymore and shall speak of "complementary groups within a group H."

In Ref. 4, it is shown that any subgroup G of $U(\mathcal{N})$, including itself, does have a complementary group within the supergroup and that its structure can be obtained easily. We shall call it the trivial complementary group of G and denote it by G^{TC}:

$$U(2^{\mathcal{N}})\supset G\times G^{TC}. \tag{2.8}$$

Its generators are obtained from the generators of $U(2^{\mathcal{N}})$ which are scalar with respect to G, and are thus

$$C^{n'\varphi'\lambda}_{n\varphi\lambda}=\sum_{\mu}\mathbf{C}^{n'\varphi'\lambda\mu}_{n\varphi\lambda\mu}. \tag{2.9}$$

G^{TC} is a direct sum of unitary groups:

$$G^{TC}=\sum_{\lambda}\oplus U(d(\lambda)), \tag{2.10}$$

each of which is associated with a given IR λ of G and has a dimension equal to the number of times $d(\lambda)$ that the IR λ of G is contained in the IR $[1]$ of the supergroup. All the states belonging to the various equivalent IR's λ of G belong to the IR $[1]$ of the group $U(d(\lambda))$ and to the IR's $[0]$ of the groups $U(d(\lambda'))$ with $\lambda'\neq\lambda$.

On the contrary all the subgroups G of $U(\mathcal{N})$ do not have a complementary group within the noninvariance group $O^{+}(2\mathcal{N}+1)$. When a complementary group does exist, it is not trivial at all, and will be denoted by G^{C}:

$$O^{+}(2\mathcal{N}+1)\supset G\times G^{C}. \tag{2.11}$$

The generators of its Lie algebra are obtained from those of $O^{+}(2\mathcal{N}+1)$, which are scalar with respect to G.

Known examples of complementary groups are those associated with unitary,[9] orthogonal,[10] or symplectic[11] subgroups of $U(\mathcal{N})$. The corresponding chains of groups are

$$\begin{array}{lll} O^{+}(2\mathcal{N}+1)\supset & U(q)\times & U(r), \quad \mathcal{N}=qr, \\ \quad [\tfrac{1}{2}\mathcal{N}] & \quad\lambda & \quad\tilde{\lambda} \end{array} \tag{2.12}$$

$$\begin{array}{lll} O^{+}(2\mathcal{N}+1)\supset & O^{+}(2q+1)\times & O(2r), \quad \mathcal{N}=(2q+1)r, \\ \quad [\tfrac{1}{2}\mathcal{N}] & \quad\lambda & \quad\lambda'' \end{array} \tag{2.13}$$

and

$$\begin{array}{lll} O^{+}(2\mathcal{N}+1)\supset & Sp(2q)\times & Sp(2r), \quad \mathcal{N}=2qr. \\ \quad [\tfrac{1}{2}\mathcal{N}] & \quad\lambda & \quad\lambda' \end{array} \tag{2.14}$$

Below each group we give the IR's to which the states belong. For the unitary groups, λ denotes a partition of n $(0\leq n\leq\mathcal{N})$ into q integers not exceeding r, and $\tilde{\lambda}$ its conjugate partition. For the orthogonal and symplectic groups, λ denotes a partition into q integers not exceeding r, λ' is the partition into r integers not exceeding q, whose conjugate partition is such that $\tilde{\lambda}'_i=r-\lambda_{q+1-i}$, and λ'' is the set of r half-integers defined by $\lambda''_i=\lambda'_i+\frac{1}{2}$.

Whenever the complementary group G^{C} does exist, the generators of the Lie algebra of G^{TC} can be expressed as polynomials in the generators of the Lie algebra of G^{C}, i.e., the algebra of G^{TC} is a (finite-dimensional) representation of the universal enveloping algebra[12] of G^{C}, built in terms of fermion creation and annihilation operators. Indeed the generators of G^{TC} carry any state transforming according to a definite IR of G into zero or a state transforming irreducibly in the same way. But all these states belong to the same IR of G^{C}, so that the transformation of any one of them into any other one can always be produced by successive applications of generators of G^{C}. There is no such simple prescription for the generators of G^{TC} when G^{C} does not exist, and in this case their explicit building is generally quite complicated.

A special type of trivial complementary group will play an important role in the following section. It is associated with a subgroup $\mathcal{G}$ of $U(\mathcal{N})$ defined as follows: If G is a subgroup of $SU(\mathcal{N})$, then

$$\mathcal{G}=U(1)\times G, \tag{2.15a}$$

the single generator of $U(1)$ being the number operator N; if G is not a subgroup of $SU(\mathcal{N})$, i.e., if N belongs to its Lie algebra, then

$$\mathcal{G}=G. \tag{2.15b}$$

The IR's of $\mathcal{G}$ are characterized by $\Lambda=(n,\lambda)$. The algebra of

$$\mathcal{G}^{TC}=\sum_{\Lambda}\oplus U(d(\Lambda)) \tag{2.16}$$

is generated by the operators

$$C^{\varphi'\Lambda}_{\varphi\Lambda}=\sum_{\mu}\mathbf{C}^{n\varphi'\lambda\mu}_{n\varphi\lambda\mu}, \tag{2.17}$$

which are linear combinations of the operators

$$D^{\varphi'\Lambda}_{\varphi\Lambda}=\sum_{\mu}\mathbf{D}^{n\varphi'\lambda\mu}_{n\varphi\lambda\mu}. \tag{2.18}$$

It is made of all the scalars with respect to G which leave the number of particles invariant, i.e., the polynomials in the generators of $U(\mathcal{N})$ which are scalar with respect to G. Therefore, the algebra of $\mathcal{G}^{TC}$ is nothing less than a (finite-dimensional) representation of the subalgebra of scalars with respect to G of the universal enveloping algebra of $U(\mathcal{N})$, which representation is built in terms of fermion creation and annihilation operators. It contains itself as a subalgebra a representation of the set of all polynomials in the Casimir operators of $\mathcal{G}$, i.e., polynomials in N [Casimir operator of $U(\mathcal{N})$], and in the Casimir operators of G.

3. PROPAGATION OF OPERATOR AVERAGES

Let us consider the average of a k-body operator,

$$\mathcal{O}(k)=\sum_{\substack{\varphi\lambda\mu\\ \varphi'\lambda'\mu'}} \langle k\varphi\lambda\mu|\mathcal{O}(k)|k\varphi'\lambda'\mu'\rangle \mathbf{D}^{k\varphi'\lambda'\mu'}_{k\varphi\lambda\mu}, \qquad (3.1)$$

in a subspace of dimension $\dim(\lambda)$ of the n-particle space, defined as the representation space of an IR λ of a subgroup G of $U(\mathcal{N})$, and specified by additional quantum numbers φ if necessary,

$$\langle\mathcal{O}(k)\rangle^{n\varphi\lambda}=[\dim(\lambda)]^{-1}\sum_{\mu}\langle n\varphi\lambda\mu|\mathcal{O}(k)|n\varphi\lambda\mu\rangle. \qquad (3.2a)$$

Instead of characterizing the subspace in which the average is calculated by some IR's of the chain (2.3) of subgroups of $O^{+}(2\mathcal{N}+1)$, it will be more useful in the following discussion to specify it by some IR Λ of the subgroup $\mathcal{G}$ of $O^{+}(2\mathcal{N}+1)$, defined in relation (2.15). As $\dim(\Lambda)=\dim(\lambda)$, relation (3.2a) becomes

$$\langle\mathcal{O}(k)\rangle^{\varphi\Lambda}=[\dim(\Lambda)]^{-1}\sum_{\mu}\langle\varphi\Lambda\mu|\mathcal{O}(k)|\varphi\Lambda\mu\rangle, \qquad (3.2b)$$

where φ distinguishes between the various equivalent IR's Λ of $\mathcal{G}$ when necessary. As it will appear in the examples of Sec. 4, this way of writing the average gives a unified description of all the cases treated up to now.

Introducing expansion (3.1) into relation (3.2b), we get

$$\langle\mathcal{O}(k)\rangle^{\varphi\Lambda}=\sum_{\substack{\varphi'\Lambda'\mu'\\ \varphi''\Lambda''\mu''}}\{\langle\varphi'\Lambda'\mu'|\mathcal{O}(k)|\varphi''\Lambda''\mu''\rangle$$
$$\times[\dim(\Lambda)]^{-1}\sum_{\mu}\langle\varphi\Lambda\mu|\mathbf{D}^{\varphi''\Lambda''\mu''}_{\varphi'\Lambda'\mu'}|\varphi\Lambda\mu\rangle\}, \qquad (3.3)$$

where the summation over Λ' and Λ'' is restricted to $\Lambda'=(k,\lambda')$, and $\Lambda''=(k,\lambda'')$. Only the part of the operator $\mathbf{D}^{\varphi''\Lambda''\mu''}_{\varphi'\Lambda'\mu'}$ which is scalar with respect to $\mathcal{G}$ can contribute to the sum over μ and its matrix elements are independent of μ, so that

$$[\dim(\Lambda)]^{-1}\sum_{\mu}\langle\varphi\Lambda\mu|\mathbf{D}^{\varphi''\Lambda''\mu''}_{\varphi'\Lambda'\mu'}|\varphi\Lambda\mu\rangle$$
$$=\delta_{\Lambda'\Lambda''}\delta_{\mu'\mu''}\langle\varphi\Lambda|[\dim(\Lambda')]^{-1}\sum_{\mu'}\mathbf{D}^{\varphi''\Lambda'\mu'}_{\varphi'\Lambda'\mu'}|\varphi\Lambda\rangle. \qquad (3.4)$$

Relation (3.3) becomes

$$\langle\mathcal{O}(k)\rangle^{\varphi\Lambda}=\sum_{\varphi'\varphi''\Lambda'}\{\langle\varphi\Lambda|D^{\varphi''\Lambda'}_{\varphi'\Lambda'}|\varphi\Lambda\rangle$$
$$\times[\dim(\Lambda')]^{-1}\sum_{\mu'}\langle\varphi'\Lambda'\mu'|\mathcal{O}(k)|\varphi''\Lambda'\mu'\rangle\}, \qquad (3.5)$$

where the operator $D^{\varphi''\Lambda'}_{\varphi'\Lambda'}$ is defined by Eq. (2.18).

Following French,[2] we say that the average of $\mathcal{O}(k)$ can be propagated from its defining subspaces $\varphi'\Lambda'$, $\Lambda'=(k,\lambda')$, if it can be expressed, for any φ and Λ, as a linear combination of the averages of $\mathcal{O}(k)$ in its defining subspaces. From relation (3.5), it is clear that a necessary and sufficient condition for this to happen whatever $\mathcal{O}(k)$ may be is that

$$\langle\varphi\Lambda|D^{\varphi''\Lambda'}_{\varphi'\Lambda'}|\varphi\Lambda\rangle$$
$$=\delta_{\varphi'\varphi''}\langle\varphi\Lambda|D^{\varphi'\Lambda'}_{\varphi'\Lambda'}|\varphi\Lambda\rangle, \qquad (3.6)$$

for any $\varphi,\varphi',\varphi'',\Lambda$ and Λ' such that $\Lambda'=(k,\lambda')$, because then

$$\langle\mathcal{O}(k)\rangle^{\varphi\Lambda}=\sum_{\varphi'\Lambda'}\langle\varphi\Lambda|D^{\varphi'\Lambda'}_{\varphi'\Lambda'}|\varphi\Lambda\rangle\langle\mathcal{O}(k)\rangle^{\varphi'\Lambda'}. \qquad (3.7)$$

In this case, it is possible to define an operator

$$\hat{\mathcal{O}}(k)=\sum_{\varphi'\Lambda'}\langle\mathcal{O}(k)\rangle^{\varphi'\Lambda'}D^{\varphi'\Lambda'}_{\varphi'\Lambda'}, \qquad (3.8)$$

which is trace equivalent to $\mathcal{O}(k)$, i.e., such that

$$\langle\hat{\mathcal{O}}(k)\rangle^{\varphi\Lambda}=\langle\mathcal{O}(k)\rangle^{\varphi\Lambda} \qquad (3.9)$$

for any φ and Λ. The operators $D^{\varphi'\Lambda'}_{\varphi'\Lambda'}$, of which it is a linear combination, are called propagation operators. The calculation of average (3.9) in any subspace $\varphi\Lambda$ then reduces to the construction of the propagation operators, and the determination of their diagonal matrix elements.

At this point there arise two main questions. The first one is to know in which cases the propagation is possible, i.e., conditions (3.6) are satisfied. The second one is to give prescriptions for building the propagation operators when these do exist. We now proceed to examine both problems successively.

Conditions (3.6) are trivially satisfied if there is no need for additional quantum numbers φ. It is thus obvious that a sufficient condition for the propagation to be possible for any k is that any IR Λ' of $\mathcal{G}$ be contained at most once in the IR $[\frac{1}{2}\mathcal{N}]$ of $O^{+}(2\mathcal{N}+1)$, or that any IR λ' of G be contained at most once in the IR $[1^n]$ of $U(\mathcal{N})$. If this is not the case, and that additional quantum numbers are necessary for some value of n, the propagation is still possible for a given value of k if any IR $\Lambda'=(k,\lambda')$ of $\mathcal{G}$ is contained at most once in $[\frac{1}{2}\mathcal{N}]$. However, if for a given k and a given λ', the IR $\Lambda'=(k,\lambda')$ is contained at least twice in $[\frac{1}{2}\mathcal{N}]$, conditions (3.6) will generally not be fulfilled because a state $|n\varphi\lambda\mu\rangle$ may have simultaneously parents of the type $|k\varphi'\lambda'\mu'\rangle$ and $|k\varphi''\lambda'\mu'\rangle$ (some simple examples of this property can be easily constructed). We thus see coming out for the first time connections between the propagation procedure and the state labeling problem. Other relations will appear later on.

When the propagation is possible, how can the propagation operators be constructed? The operators $D^{\varphi''\Lambda'}_{\varphi'\Lambda'}$ are linear combinations of the generators $C^{\varphi''\lambda'}_{\varphi'\lambda'}$ of the trivial complementary group $\mathcal{G}^{TC}$ of $\mathcal{G}$. It is thus clear that the study of the structure of this group is intimately connected with the construction of the propagation operators. This implies that a procedure has to be found to build the generators of $\mathcal{G}^{TC}$ in a direct way, without constructing explicitly the n-particle states.

When the algebra of $\mathcal{G}^{TC}$ coincides with its subalgebra of polynomials in the Casimir operators of $\mathcal{G}$, the problem is easily solved because the Casimir operators are well known for all classical groups, as well as their eigenvalues in all IR's. Sufficient conditions for this to happen can be found:

(i) $\mathcal{G}$ is the first step of a canonical chain of subgroups of $U(\mathcal{N})$ [therefore $\mathcal{G}=U(\mathcal{N}-1)$], because then the

Casimir operators of the groups of the chain form a complete set of commuting operators[13];

(ii) $\mathcal{G}$ is the direct product of two complementary groups, $\mathcal{G}=\bar{G}\times\bar{G}^{C}$, because then the generators of $\bar{G}^{TC}$ are polynomials in the generators of $\bar{G}^{C}$, and $\mathcal{G}^{TC}$ is the subgroup of $\bar{G}^{TC}$ which is left invariant by the transformations of $\bar{G}^{C}$, so that its generators are polynomials in the Casimir operators of $\bar{G}^{C}$ (or of $\bar{G}$ as the latter can be expressed in terms of the former);

(iii) $\mathcal{G}$ is the direct product of a group $\bar{G}$ and a canonical subgroup $\bar{G}'$ of its complementary group $\bar{G}^{C}$, $\mathcal{G}=\bar{G}\times\bar{G}'$, $\bar{G}'\subset\bar{G}^{C}$, because then the generators of $\mathcal{G}^{TC}$ are those polynomials in the generators of $\bar{G}^{C}$ which are left invariant by the transformations of $\bar{G}'$, and are therefore polynomials in the Casimir operators of $\bar{G}^{C}$ (or $\bar{G}$) and $\bar{G}'$.

The construction of the generators of $\mathcal{G}^{TC}$ is not as trivial when the algebra of polynomials in the Casimir operators of $\mathcal{G}$ is a proper subalgebra of the algebra of $\mathcal{G}^{TC}$. In this case the propagation operators can be easily built only for those k values for which the number of linearly independent polynomials in the Casimir operators of $\mathcal{G}$ is the same as that of defining subspaces (k,λ').

Such a situation may happen even when there is no state labeling problem in the reduction $U(\mathcal{N})\supset G$ for the IR's $[1^n]$ of $U(\mathcal{N})$. In Sec. 4, we give an example of such a case, corresponding to the reduction $U(6)\supset O^{+}(3)$, which occurs for identical fermions in a single shell of angular momentum 5/2.

When there is a state labeling problem in the reduction $U(\mathcal{N})\supset G$, the above-mentioned subalgebra is always proper because the algebra of $\mathcal{G}^{TC}$ contains at least one generator of the type $C^{\varphi''\Lambda'}_{\varphi'\Lambda'}$ $(\varphi'\neq\varphi'')$, which cannot be expressed in terms of Casimir operators of $\mathcal{G}$ as it connects two different IR's of $\mathcal{G}$. In fact it is well known[4] that in this case the algebra of $\mathcal{G}^{TC}$ contains the supplementary operators whose eigenvalues define the additional quantum numbers which complete the classification of n-particle states.

In both cases, the problem of building the generators of $\mathcal{G}^{TC}$ will be solved if we can construct an integrity basis, i.e., a minimal set of operators of $\mathcal{G}^{TC}$ in terms of which all the generators of $\mathcal{G}^{TC}$ can be written as polynomial expressions. This is a difficult problem to tackle, but some definite progress was made recently in its general formulation and its explicit solution in some particular cases.[5] Therefore, there is some hope that it will be possible in a near future to build the propagation operators in their full generality at least for some of those chains of groups for which we are restricted now to low k values.

It is quite easy to generalize the results of the discussion to the case of an operator $\mathcal{O}$ of mixed particle rank, not greater than u,

$$\mathcal{O}=\sum_{k=0}^{u}\mathcal{O}(k). \qquad (3.10)$$

We may distinguish two cases:

(i) When any IR Λ' of $\mathcal{G}$ is contained at most once in the IR $[\frac{1}{2}\mathcal{N}]$ of $O^{+}(2\mathcal{N}+1)$, the average of any operator $\mathcal{O}$ can be propagated from its defining subspaces $\Lambda'=(k,\lambda')$, $0\leq k\leq u$, whatever u may be:

(ii) When $\bar{u}$ is the first value of n for which an IR Λ' of $\mathcal{G}$ is contained more than once in the IR $[\frac{1}{2}\mathcal{N}]$ of $O^{+}(2\mathcal{N}+1)$, the average of any operator $\mathcal{O}$ of maximum particle rank $u<\bar{u}$ can be propagated from its defining subspaces $\Lambda'=(k,\lambda')$, $0\leq k\leq u$; however, the average of an operator $\mathcal{O}$ for which $u\geq\bar{u}$ cannot be propagated in general. The problems encountered in the explicit construction of the propagation operators remain of course the same as before.

Finally, let us mention that a modification of definition (3.2) is used by several authors when dealing with subgroups G of $U(\mathcal{N})$ for which there is a state labeling problem.[14,15] It consists in taking the average of the operator $\mathcal{O}(k)$ over all the equivalent IR's Λ of $\mathcal{G}$ contained in the IR $[\frac{1}{2}\mathcal{N}]$ of $O^{+}(2\mathcal{N}+1)$,

$$\langle\mathcal{O}(k)\rangle^{\Lambda}=[d(\Lambda)\dim(\Lambda)]^{-1}\sum_{\varphi u}\langle\varphi\Lambda\mu|\mathcal{O}(k)|\varphi\Lambda\mu\rangle. \qquad (3.11)$$

Here $d(\Lambda)$ is the number of times the IR Λ of $\mathcal{G}$ is contained in the IR $[\frac{1}{2}\mathcal{N}]$ of $O^{+}(2\mathcal{N}+1)$ or $[1]$ of $U(2^{\mathcal{N}})$, and is clearly a member of the set of numbers which define the dimensions of the unitary subgroups of $\mathcal{G}^{TC}$. The average (3.11) is thus taken in the representation space of the IR $\Lambda\times[1]$ of the subgroup $\mathcal{G}\times U(d(\Lambda))$ of $U(2^{\mathcal{N}})$. The operators $D^{\varphi''\Lambda'}_{\varphi'\Lambda'}$ which contribute to such an average are therefore linear combinations of the generators of the subgroup $\sum_{\Lambda'\neq\Lambda}\oplus U(d(\Lambda'))$ of $\mathcal{G}^{TC}$. Trivial changes have to be applied to the relations derived above, and all the main conclusions remain valid. Moreover, it should be clear that whenever other scalar operators enter the construction of the propagation operators in addition to the Casimir operators, average (3.11) is more natural and much easier to calculate than (3.2) because in general those operators cannot be simultaneously diagonalized so that their diagonal matrix elements in the subspaces characterized by φ and Λ are difficult to determine.

We proceed now to illustrate the discussion of this section by same examples mostly taken from current applications of the spectral distribution method.

4. DISCUSSION OF SOME PARTICULAR CASES

(i) There is no state labeling problem and the algebra of $\mathcal{G}^{TC}$ is generated by the polynomials in the Casimir operators of $\mathcal{G}$.

Let us consider first the case where $\mathcal{G}=\bar{G}\times\bar{G}^{C}$. Then $\mathcal{G}$ is one of the three direct product groups of relations (2.12), (2.13), and (2.14). Only examples of the first type have been considered up to now because the other two imply averaging over states with different particle numbers. They correspond to r values equal to 1 (scalar averaging[1,2]), 2 (fixed isospin averaging[1]), or 4 (fixed supermultiplet averaging[16]). Any operator can be propagated, and the propagation operators are polynomials in the Casimir operators of $U(r)$, $G_1=N$, G_2, G_3, . . . , G_r. For scalar averaging, they are thus polynomials in the number operator, for fixed isospin averaging polynomials in N and $\mathbf{T}^2$, and for fixed supermultiplet averaging polynomials in N, G_2, G_3, and G_4.

Let us consider next a case where $\mathcal{G}=\bar{G}\times\bar{G}'$, and $\bar{G}'$ is a canonical subgroup of $\bar{G}^C$: fixed seniority averaging for identical nucleons, the seniority being either that in a single subshell, or the multi-shell generalized seniority.[1,17] We have indeed $\mathcal{G}=\mathrm{Sp}(2q)\times\mathrm{U}(1)$, where $U(1)$ is a canonical subgroup of the quasispin group Sp(2), generated by the operator $S_0=\frac{1}{2}(N-q)$. Any operator can be propagated, and the propagation operators are polynomials in N and $\mathbf{S}^2$, the square of the quasispin operator.

(ii) There is no state labeling problem, but the algebra of $\mathcal{G}^{TC}$ is not generated by the polynomials in the Casimir operators of $\mathcal{G}$ only.

An example of such a case is given by the quite trivial fixed angular momentum averaging for identical fermions in a single shell of angular momentum 5/2, corresponding to the chain of groups

$$\mathrm{U}(2^6)\supset\mathrm{O}^+(13)\supset\mathcal{G}=\mathrm{U}(1)\times\mathrm{O}^+(3). \quad (4.1)$$

The IR's of the last group are (0,0), (1,5/2), (2,0), (2,2), (2,4), (3,3/2), (3,5/2), (3,9/2), (4,0), (4,2), (4,4), (5,5/2), and (6,0), and are contained only once in the IR $[\frac{1}{2}^6]$ of $\mathrm{O}^+(13)$. Moreover, it can be easily verified that the generators of $\mathcal{G}^{TC}$ cannot be built in terms of polynomials in the Casimir operators N and $\mathbf{J}^2$ only. However, it is sufficient to add to them the square of the quasispin operator $\mathbf{S}^2$ to get an integrity basis of $\mathcal{G}^{TC}$ in terms of which all the propagation operators can be explicitly constructed.

(iii) There is a state labeling problem. Many examples of such a situation are known. Let us mention the fixed supermultiplet and SU(3) symmetry averaging,[18] the fixed supermultiplet, spin and isospin averaging,[19] and the fixed seniority, isospin, and reduced isospin averaging for both types of nucleons. The corresponding groups are $\mathcal{G}=\mathrm{SU}(3)\times\mathrm{U}(4)$, $\mathcal{G}=\mathrm{U}(q)\times\mathrm{SU}(2)\times\mathrm{SU}(2)$, and $\mathcal{G}=\mathrm{Sp}(2q)\times\mathrm{U}(2)$, and the state labeling problem occurs in the reduction $\mathrm{U}(q)\supset\mathrm{SU}(3)$, $\mathrm{U}(4)\supset\mathrm{SU}(2)\times\mathrm{SU}(2)$, and $\mathrm{Sp}(4)\supset\mathrm{U}(2)$, respectively. Another important case corresponds to the fixed angular momentum averaging for high enough N values.

The usefulness of the construction of an integrity basis of $\mathcal{G}^{TC}$ can be illustrated by the following example. Let us consider the fixed supermultiplet and orbital angular momentum averaging for the p-shell nuclei. The corresponding group is $\mathcal{G}=\mathrm{O}^+(3)\times\mathrm{U}(4)$, where the rotation group in orbital space is a subgroup of the complementary group U(3) of U(4). The propagation operators for $u=2$ operator averages can be built in terms of the Casimir operators 1, N, N^2, G_2, and $\mathbf{L}^2$. However, for $u=4$ operator averages, the full integrity basis is needed. It was shown recently[5] that it includes the operators N, G_2, G_3, $\mathbf{L}^2$, as well as a third-order operator $X^{(3)}$, and a fourth-order operator $X^{(4)}$. In terms of these, the propagation operators for $u=4$ operator averages can be written as linear combinations of 1, N, N^2, N^3, N^4, G_2, NG_2, N^2G_2, G_3, NG_3, G_2^2, $\mathbf{L}^2$, $N\mathbf{L}^2$, $N^2\mathbf{L}^2$, $\mathbf{L}^4$, $G_2\mathbf{L}^2$, $X^{(3)}$, $NX^{(3)}$, and $X^{(4)}$.

*Maître de recherches F.N.R.S.

[1]J. B. French, in *Nuclear Structure*, edited by A. Hossain, Harun-ar-Rashid, and M. Islam (North-Holland, Amsterdam, 1967).
[2]J. B. French and K. F. Ratcliff, Phys. Rev. C 3, 94 (1971).
[3]F. S. Chang, J. B. French, and T. H. Thio, Ann. Phys. (N.Y.) 66, 137 (1971).
[4]M. Moshinsky and C. Quesne, J. Math. Phys. 11, 1631 (1970).
[5]B. R. Judd, W. Miller Jr., J. Patera, and P. Winternitz, J. Math. Phys. 15, 1787 (1974).
[6]B. R. Judd and J. P. Elliott, *Topics in Atomic and Nuclear Theory* (University of Canterbury Press, New Zealand, 1970), p. 21.
[7]B. R. Judd, in *Group Theory and Applications*, edited by E. M. Loebl (Academic, New York, 1968), Vol. 1.
[8]M. C. K. Aguilera-Navarro and V. C. Aguilera-Navarro, J. Math. Phys. 16, 72 (1975).
[9]M. Moshinsky, *Group Theory and the Many Body Problem* (Gordon and Breach, New York, 1968).
[10]K. Helmers, Nucl. Phys. 69, 593 (1965).
[11]K. Helmers, Nucl. Phys. 23, 594 (1961).
[12]N. Jacobson, *Lie Algebras* (Wiley, New York, 1962).
[13]I. M. Gel'fand and M. L. Tseitlin, Dokl. Akad. Nauk SSSR 71, 825 (1950).
[14]K. F. Ratcliff, Phys. Rev. C 3, 117 (1971).
[15]C. Jacquemin, Phys. Lett. 43B, 253 (1973).
[16]J. C. Parikh, Ann. Phys. (N.Y.) 76, 202 (1973).
[17]C. Quesne and S. Spitz, Ann. Phys. (N.Y.) 85, 115 (1974).
[18]J. C. Parikh, Phys. Lett. 41B, 468 (1972).
[19]R. U. Haq and J.C. Parikh, Nucl. Phys. A 220, 349 (1974).

ANNALS OF PHYSICS **106**, 472–502 (1977)

Strength Distributions and Statistical Spectroscopy. I. General Theory*

J. P. DRAAYER

Department of Physics and Astronomy, Louisiana State University, Baton Rouge, Louisiana 70803

J. B. FRENCH

Department of Physics and Astronomy, University of Rochester, Rochester, New York 14627

AND

S. S. M. WONG

Department of Physics, University of Toronto, Toronto, Ontario, Canada

Received December 1, 1976

The strength distribution for an arbitrary excitation is given in terms of a double expansion, and its sum rules by single expansions, in polynomials defined by the initial and final energy spectra. In model spaces which are not too large, a rapid convergence, to within fluctuations, is assured by the action of a central limit theorem, as is shown in particular by considering the response of the system under infinitesimal deformations of the Hamiltonian. When larger spaces are decomposed into subspaces defined by a partitioning of the single-particle space a similar convergence results. At the same time, close contact is made with, and important corrections are found to, intuitive procedures which are often used for approximating strength distributions. The general features of the distribution are often easily understood in terms of a simple geometry made effective in the model space by the central limit theorem, and further features by exploiting the connection of this geometry to the unitary group of transformations in the single-particle space. Extensions are given for multipole strengths and sum rules, appropriate when the angular momenta (and isospins) are specified for the states involved in the transitions. Measures for the RMS fluctuations in the sum-rule quantities, and correlations between them, are given by combining the low-order-polynomial (statistically smoothed) strengths with an assumed Porter–Thomas distribution for the (high-order) strength fluctuations.

* Supported in part by the U.S. Energy Research and Development Administration, the U.S. National Science Foundation, and the National Research Council of Canada.

ISSN 0003–4916

1. Introduction

We give in this paper a theory of strength distributions for a wide class of nuclear excitations, applicable for example to β- and γ-transitions and to excitation via particle transfer. As always there is an excitation operator O which induces transitions, $| W\rangle \rightarrow^{O} | W'\rangle$, between the starting states $| W\rangle$ and final states $| W'\rangle$. In the usual way we define the microscopic strength as

$$R(W', W) = |\langle W' | O | W\rangle|^2 \tag{1}$$

the square of the expansion coefficient of the state $O | W\rangle$ in terms of the final Hamiltonian eigenstates $| W'\rangle$, which may (as in $2^+ \rightarrow^{E2} 2^+$) or may not (as in particle transfer) belong to the same model subspace. The pth-order energy-weighted sum rule is

$$\begin{aligned} \mathscr{M}_p(W) &= \sum_{W'} R(W', W)(W')^p \\ &= \int I'(W')(W')^p R(W', W)\, dW' = \langle W | O^{+} H^p O | W\rangle \end{aligned} \tag{2}$$

where $I'(z)$ is the eigenstate density in the final-state space whose dimensionality is then $\int I'(z)\, dz = d'$. We see that $\mathscr{M}_p(W)$ is, on the one hand, a starting-state expectation value and, on the other, the p'th moment of the strength distribution which originates with $| W\rangle$. Other types of strength moments, closely related to the $\mathscr{M}_p$, are the polynomial moments which we encounter ahead (13), and moments in which the strength weighting is $(W' - W)^p$ which give rise to expectation values involving multiple commutators. For finite-dimensional spaces, the only ones we encounter, the strength is determined by the set $\mathscr{M}_p(W)$ for all $p \geqslant 0$ or by the corresponding sets of the other moments.

The fact that sum rules determine everything is, however, not of *direct* practical consequence because of a number of intervening difficulties. The first difficulty is that, because of the undue weighting which they give to small unobservable strengths at high excitations, high-order $\mathscr{M}_p(W)$, say $p > 2$, and consequently the strength distributions themselves, are not measurable. This is not to say, however, that these quantities are devoid of interest for there are many cases in which measurable quantities depend on the strengths; an example would be radiation emitted during a cascade reaction. It will often happen moreover, that even for low-order moments one can experimentally determine only the endothermic ($W' > W$) or exothermic ($W' < W$) strengths, and these are adequate only to give lower limits to the low-order $\mathscr{M}_p$: in order to decompose the starting-state expectation values into exothermic and endothermic parts we also need the strength $R(W', W)$ itself.

In some cases when the sums *are* measurable the analysis will yield quantities which are of immediate significance (as is often the case with occupancies), but, particularly with complicated sum rules, the significance may not be clear at all. In this second case, and very often in the first as well, one will wish to proceed further with the analysis and ask, for example, what the measurements imply about the nature of the interaction which is effective in the target state. In the conventional procedure one would construct the starting state, for example by a shell-model diagonalization, evaluate the required expectation values, and perhaps attempt thereby to determine the sensitivity of the expectation values to the input parameters. In that procedure the starting state is treated differently from the final states for which no explicit construction is needed. But beyond that the construction of a state will often be hopelessly complicated; and if to achieve it we make radical simplifications in the Hamiltonian and the model space considered, we must compound that error by using the simplified H in the operators $O^{+}H^{p}O$ which come into the energy weighted sum rules (2) if we want self-consistency, or else give up self-consistency which in itself involves errors beyond our control. Very often then the further analysis is not carried out at all and the significance of the measurements is not explored. And of course the same argument applies if one wishes to make sum-rule *predictions*, and *a forteriori* if one's interest is with the strength itself rather than in its low-order moments (the sum-rule quantities).

The purpose of the present paper is to demonstrate that there are major simplicities associated with strength distributions and sum rules in many-particle spaces, and to exploit them in such a manner as to dispose of these difficulties. The theory which we give now, which has been briefly discussed in earlier papers [1–3], makes use of spectral distribution methods. These in turn derive from a recognition of the role of the central limit theorem (CLT) in the usual many-particle model spaces constructed by distributing nucleons over some finite set of single-particle states. In an asymptotic limit, that of large particle number, the *smoothed* eigenvalue density for "almost all" Hamiltonians which could act in the model space will, by the action of the CLT, be close to Gaussian [4, 5]. Since the state density determines the partition function it is then not surprising that the strength distribution will, to within fluctuations which we shall have to consider separately, also have a characteristic simple asymptotic form. In favorable cases it is in fact bilinear in the energies (linear in both W' and W); the sum rules for the strength originating at W will then determine starting-state expectation values $\langle W \mid O^{+}H^{p}O \mid W\rangle$, which are similarly (again to within fluctuations) linear in W. The sum rules and strengths are thus determined by a small number of quantities, which in fact are traces over the model space, and some of its subspaces, of operators $O^{+}H^{p}OH^{q}$ with small p, q, being calculable therefore by methods which do not involve the construction of any Hamiltonian (or other) eigenfunctions. Because of this feature the methods can be applied in model spaces of arbitrarily large dimensionality. They have a further advantage that they give the results as more or less explicit functions (involving low-order powers) of the Hamiltonian matrix elements so that one can see what features of the Hamiltonian are implied by particular properties of the strength distribution; with standard methods in which explicit starting-state wavefunctions are required this is feasible only in most exceptional circumstances. The problem discussed above of the endothermic–exothermic separation is solved by the construction of the strength itself, which of course solves also the problem of the high-order moments of the smoothed distribution. The implication of the latter remark is that moments, of order high enough to fix the *secular* variations of the strength (not its fluctuations), *are* in fact calculable because of the simplicities generated by the central limit theorem.

These preliminary remarks raise a number of questions. The first is about the way in which the CLT originates and acts in the many-particle model space; the important thing to keep in mind here is that the CLT is a very "strong" principle whose operation dominates much of the behavior and gives rise to quite different approximations than one would come to if its operation were ignored. We shall briefly review these matters ahead.

As a second general problem we must face the fact that only a small fraction of the model-space states can have any physical significance; this arises from the fact that the many-particle spectrum span is far greater than the corresponding single-particle span (100 MeV compared with 5 MeV, for example, in $(ds)^{12}$) so that all but the lowest states have an inadequate single-particle basis. This feature is of course shared by shell-model and other calculations which make use of *spectroscopic model spaces* (those with a finite single-particle basis) but they are put more plainly into view in the spectral methods in which many things are "tied to" the spectrum centroid which lies far above the states which are adequately treated. The questions here are whether the CLT is adequate to extend things into the physical domain and whether we can deal with model spaces large enough to treat things adequately when our interest is in physical domains which themselves span a considerable energy range (say $\gtrsim$ 10 MeV); these matters also are by now well understood, and the answers are positive.

A third general question concerns the nature and magnitude of the fluctuations which are ignored and the consequences of ignoring them. Ignoring the fluctuations will, of course, induce an error in the predictions for the strengths and so forth involving individual levels, and for that reason, if no other, we should understand their general magnitude. Concerning the nature of the fluctuations we remark first that the eigenvalue density rapidly smooths out and converges to Gaussian as the particle number increases, so that the *secular* deviations (measured over a fixed energy interval) from the asymptotic Gaussian rapidly disappear. The natural unit for the spectrum fluctuations is, however, the local level spacing which itself would rapidly decrease as we increase the particle number in order to take advantage of the CLT smoothing. The consequence (not yet completely understood) is that the CLT smoothing has essentially no effect on the spectrum fluctuations which may be regarded as having an existence of their own, undisturbed by the changes in the eigenvalue density generated by changing the particle number. To the extent that the Porter–Thomas distribution [6] describes the strength fluctuations they also may be regarded as having an independent existence; we shall see, moreover, how the secular variation in the strength will combine with the Porter–Thomas fluctuations to give measures for the fluctuations in, and correlations between, the sum-rule quantities.

There is still the general question whether level-to-level fluctuations carry information, or equivalently, whether the errors involved in ignoring them are of physical consequence. It is known [7] that energy-level fluctuations at high excitation are determined by parameter-free statistical laws and therefore carry little or no information; the same seems to be true for the strengths. It would not be safe to assume that this behavior holds at all excitations, and indeed it seems quite clear, and is understood, that fluctuations in the ground-state domain are considerably modified by collective effects. On the other hand, it is also known that when collectivity is not dominant, those aspects of the fluctuations which are measurable both at high and low excitations do indeed extend into the ground-state domain [8]. This is incompatible with the point of view, encountered for example in discussions of effective-interaction theory [9], that the major physical interest involves only the first few levels and that high accuracy is essential, this implying that fluctuation [illegible] results are without interest. These arguments do not seem to be valid even for phenomena involving ground states and of course high accuracy is pointless for almost all phenomena occurring at high excitations.

In summary then, while questions may be raised concerning statistical methods for strength distributions, it appears that their simplicity does not involve a high-handed neglect of much of the physics involved but rather derives from the recognition that the phenomena involved are fairly simple. We shall demonstrate this in the examples of the following paper.

In Section 2 of this paper we derive orthogonal-polynomial expansions for the smoothed strength function $R(W', W)$ and the associated sum-rule quantities. The polynomials are those defined by taking the eigenvalue densities in the W and W' subspaces as weight functions. The whole success of our procedure derives from the rapid convergence of the expansions (Eqs. (11), (16) ahead) which is guaranteed by the CLT, whose origin and method of operation we describe in Section 3. We show there moreover, that there is, for the operators of the system, a geometry which determines much of the behavior. In Section 4 we show the connection of the strength and sum-rule results with a theory of the linear response of the system under a perturbation of the Hamiltonian. The details of this are not as important as the general connections which are established between Hamiltonian "deformations" and the associated strength distributions and expectation values.

The basic energy dependences which are generated by the CLT, bilinear in the energy for the strengths and linear for the sum rules, give at least a semiquantitative description for an extremely wide range of cases. They derive from the first two terms in the polynomial expansions, the only terms which are uninhibited by the action of the CLT. We can take a few further terms in the expansions to generate the corrections to the linear forms; alternatively we can partition the single-particle space into "orbits" and the model space thereby by configurations, in order to produce a simpler theory which should be accurate in arbitrarily large spaces. The results of this are given in Section 5, which starts with a detailed derivation of the simplest of all theories for the strength distribution; it then goes on to demonstrate the remarkable way in which the polynomial expansions take account of the correlations between Hamiltonian and excitation operator in order to convert the simple schematic theory into an exact one. Section 6 gives, for future reference, the basic equations required to deal with the angular-momentum decomposition of the strengths and sum rules.

In Section 7 we give an introduction to the relationships between the geometry of the model-space operators and the symmetry generated by unitary transformations in the single-particle space. As an example we use these results to show, in a particular case, how information is "propagated" over a wide energy domain in the spectrum. In Section 8 we combine the Porter–Thomas distribution for the elementary transition strengths with the polynomial theory for the secular energy variation in order to calculate the mean-squared fluctuations in the sum-rule quantities and the correlations between them.

The purpose of this paper then is to give the general theory. In the following paper we assess its accuracy via a comparison with detailed shell-model results for a variety of electromagnetic transitions. In later papers we shall give further methods for evaluating the necessary input traces and shall apply the theory to particle transfer, to β and γ giant resonances, and to other excitations. We should stress that "excitations" which are not physically realizable, and expectation values which do not derive from sum rules, are often of great interest also, in some cases for example because of their connection with symmetries. We shall discuss these also.

2. Polynomial Expansions

As in (1), (2) the Hamiltonian is H and the excitation operator O. We are interested in transitions $|W\rangle \to^{O} |W'\rangle$ where we label the states by their energies and where the $|W\rangle$ states span one model subspace (an eigenspace for H) and the $|W'\rangle$ another (the two subspaces may coincide, and then so do the corresponding states); in case of degeneracies a more complete labeling will be adopted as needed. Besides the state density $I(z)$ it is convenient to introduce the corresponding probability density

$$\rho(z) = d^{-1}I(z); \qquad \int \rho(z)\,dz = 1. \tag{3}$$

In the case where the subspaces are distinct we can of course, consider the inverse transitions by a simple relabeling ($W \leftrightarrow W'$, $O \leftrightarrow O^{+}$) but that will not be convenient if we are interested in the relationships between the two distributions $|W\rangle \to^{O} |W'\rangle$ and $|W'\rangle \to^{O^+} |W\rangle$.

The total strength originating with $|W\rangle$, its centroid, and its variance (square of the width), are given respectively in terms of the moments $\mathcal{M}_p(W)$ (2) by

$$\mathcal{M}_0(W);\ \mathscr{E}(W) = \mathcal{M}_1(W)/\mathcal{M}_0(W); \qquad \hat{\sigma}^2(W) = \mathcal{M}_2(W)/\mathcal{M}_0(W) - (\mathscr{E}(W))^2 \tag{4}$$

and similarly for higher-order quantities. Since finite-space distributions are determined by their moments the set of $\mathcal{M}_p(W)$ for all p determines the distribution of the strength which originates in W; its Fourier transform (characteristic function) is

$$\int R(z, W)\,e^{-itz}I'(z)\,dz = \langle W \mid O^{+}e^{-itH}O \mid W\rangle = \sum \frac{(-it)^p}{p!}\mathcal{M}_p(W) \tag{5}$$

where the prime on $I'(z)$ indicates that we are dealing with the state density for the W' space, and inversion gives

$$I'(W')R(W', W) = \langle W \mid O^{+}\delta(H - W')\,O \mid W\rangle \tag{6}$$

a result which we can verify by making an intermediate-eigenstate expansion of the expectation value and using $\sum_{W''}$ [ ] $\to \int I'(W'')\, dW''$ [ ]. We could of course have started with (6) but we have derived it instead in order to stress the relationship between sum rules and strength moments.

As they stand these moment expansions are not directly applicable. For one thing, there is no convergence in the moment order and, for another, as described in Section 1, it is impossible to measure high moments; and to calculate any moments at all we need the model states $|\, W\rangle$ which, in complicated nuclei, we are unable to construct. Even in "simple" cases most model wavefunctions promise more than they can really deliver, and give many-particle correlations which are inadequate for dealing with operators of high "particle rank" (a k-body operator has particle rank k).

To circumvent the last difficulty we sum (6) over the starting states as well; letting $|\, z\rangle$ run over the $|\, W\rangle$ states we have

$$\begin{aligned} I'(W')\, R(W', W)\, I(W) &= \langle W \,|\, O^{+}\delta(H - W')O \,|\, W\rangle\, I(W) \\ &= \int \langle z \,|\, O^{+}\delta(H - W')O \,|\, z\rangle\, I(z)\, \delta(z - W)\, dz \\ &= \sum_{z} \langle z \,|\, O^{+}\delta(H - W')\, O\delta(H - W) \,|\, z\rangle \\ &= \langle\!\langle O^{+}\delta(H - W')\, O\delta(H - W)\rangle\!\rangle^{m} \\ &= d \times \langle O^{+}\delta(H - W')\, O\delta(H - W)\rangle^{m} \end{aligned} \tag{7}$$

where we have introduced the notation that $\langle\!\langle G\rangle\!\rangle^{m} = d \times \langle G\rangle^{m}$ is the trace of G over the $|\, W\rangle$ model space so that $\langle G\rangle^{m}$ is the average expectation value.

To deal with the convergence problem we replace the moment (*Taylor series*) expansion by a polynomial (*normal mode*) expansion. The appropriate polynomials are $P_{\mu}(x)$, the orthonormal polynomials defined by the (nonnegative-definite) $|\, W\rangle$ density, $\rho(x)$, as weight function (there is of course a corresponding set P_{μ}' for the final-state space). Then

$$\begin{aligned} \int P_{\mu}(x)\, P_{\nu}(x)\, \rho(x)\, dx &= \delta_{\mu\nu}\,, \\ \delta(x - y) &= \rho(x) \sum_{\mu} P_{\mu}(x)\, P_{\mu}(y). \end{aligned} \tag{8}$$

The polynomials can be constructed directly from (8) and appear as functions of the density moments

$$M_{p} = \int \rho(z)\, z^{p}\, dz = \langle H^{p}\rangle^{m}. \tag{9}$$

For the first two we have $P_0(z) = 1$, $P_1(z) = (z - \mathscr{E})/\sigma$ where $\mathscr{E} = M_1$ and $\sigma^2 = M_2 - (M_1)^2$ are the density centroid and variance. The unnormalized polynomial of order ν requires density moments of orders up to $(2\nu - 1)$, while the normalization requires also the moment of order 2ν. We have in general that [10]

$$[D_{\nu-1}D_\nu]^{1/2}\, P_\nu(z) = \begin{vmatrix} 1 & M_1 & M_2 & \cdots & M_\nu \\ M_1 & M_2 & \cdots & & M_{\nu+1} \\ \vdots & & & & \\ M_{\nu-1} & \cdots & & & M_{2\nu-1} \\ 1 & z & z^2 & \cdots & z^\nu \end{vmatrix} \tag{10}$$

where D_ν is the determinant with the last row replaced by $[M_\nu, M_{\nu+1} \cdots M_{2\nu}]$. As a practical matter we shall often be able to calculate, analytically, moments of maximum order 3 or 4 but may for accuracy need polynomials of somewhat higher order than this limit would permit. But the density will commonly be very close to Gaussian, the "skewness," $\gamma_1 = \mu_3/\sigma^3$, and "excess," $\gamma_2 = \mu_4/\sigma^4 - 3$ (where the μ_p are central moments) being less, say, than 0.3 in magnitude. When that is so, we can make the necessary extension of our moment calculations by equating to zero the higher-order cumulants (which vanish for Gaussian), these relationships then defining the necessary additional moments. We remark also that the number of polynomials is finite, there being the same number of polynomials as eigenvalues; but if we approximate the densities as continuous the number of polynomials is of course infinite. The distinction in fact will be of little concern to us since, for the secular variations of the strengths, only very low polynomial orders will be required, while, for the fluctuations, we proceed quite differently as we shall see ahead.

From (7), (8) we have for the strength

$$R(W', W) = R(W'm', Wm) = \sum_{\mu\nu} G_{\mu\nu}(m', m)\, P_\mu'(W')\, P_\nu(W), \tag{11}$$

$$\begin{aligned} G_{\mu\nu}(m', m) &= (dd')^{-1} \langle\!\langle O^+ P_\mu'(H)\, O P_\nu(H) \rangle\!\rangle^m \\ &= (dd')^{-1} \langle\!\langle O P_\nu(H)\, O^+ P_\mu'(H) \rangle\!\rangle^{m'} = G_{\nu\mu}^{(+)}(m, m'), \end{aligned} \tag{12}$$

where the second form in (11) emphasizes that O acting on a state $|\, W\rangle$ in the m-model-space generates an m'-state. The second equality in (12) has used the elementary result that, for operators, F, G, of that type, $\langle\!\langle FG \rangle\!\rangle^m = \langle\!\langle GF \rangle\!\rangle^{m'}$, and $G_{\nu\mu}^{(+)}$ denotes a coefficient associated with the inverse transitions, $|\, W'\rangle \rightarrow^{O^+} |\, W\rangle$. For the inverse strength we have of course $R^+(Wm, W'm') = R(W'm', Wm)$.

For the sum rules it is often better to use, not the "power moments" $\mathcal{M}_p(W)$ of (2) but the "polynomial moments"

$$\begin{aligned} \Lambda_\alpha(W) &= \sum_{W'} R(W', W)\, P_\alpha'(W') = \int R(z, W)\, P_\alpha'(z)\, I'(z)\, dz \\ &= \langle W \,|\, O^+ P_\alpha'(H) O \,|\, W\rangle = d' \times \sum_\nu G_{\alpha\nu}(m', m)\, P_\nu(W) \end{aligned} \tag{13}$$

where the last form derives from (8), (11) and expresses the polynomial sum rules (in which the strengths are weighted, as in the second form, by polynomials in the energy rather than by simple powers). Observe that the Λ_α for $\alpha \leqslant t$ are linear combinations of the $\mathcal{M}_p$ for $p \leqslant t$, and inversely, so that the sets carry the same information; specifically if $x^p = \sum c_{p\mu} P_\mu(x)$ then $\mathcal{M}_p(W) = \sum c_{p\mu} \Lambda_\mu(W)$. From (11), (13) we have

$$d' \times G_{\alpha\nu}(m', m) = \int \Lambda_\alpha(z)\, P_\nu(z)\, \rho(z)\, dz, \tag{14}$$

$$\begin{aligned} d' \times R(W', W) &= \sum_{\alpha,\nu} P_\alpha'(W')\, P_\nu(W) \int \Lambda_\alpha(z)\, P_\nu(z)\, \rho(z)\, dz \\ &= \sum_\nu \Lambda_\nu'(W')\, P_\nu(W) = \sum_\mu \Lambda_\mu(W)\, P_\mu'(W'), \end{aligned} \tag{15}$$

which gives alternative expressions for the strength in terms of the (polynomial) moments.

We observe also that for the expectation value of an arbitrary operator K, not necessarily one which derives from a sum rule, we have

$$\langle W \mid K \mid W \rangle \equiv K(W) = \rho^{-1}(W) \langle K \delta(H - W) \rangle^m = \sum_\nu \langle K P_\nu(H) \rangle^m P_\nu(W) \tag{16}$$

With $K = O^+ H^p O$ this becomes a polynomial expansion for the strength moments $\mathcal{M}_p(W)$. Finally the density itself is expressible as a trace,

$$\rho(W) = \langle \delta(H - W) \rangle^m. \tag{17}$$

3. Convergence and the Central Limit Theorem; Correlation Coefficients and Simple Geometry

Simply writing the strength and its sum rules in the forms (11), (13) does not of course solve the problems which we have described above (following (6)); the new feature which does solve them derives from the many-particle nature of the model spaces of interest, more specifically from the *direct-product* structure of these spaces, which ensures a rapid convergence (to within fluctuations) in the polynomial expansions (11), (13). Everything indeed will derive from the fact, which has been known for some years and is by now well understood, that as we increase the number of particles the model-space eigenvalue density goes rapidly to Gaussian, this feature deriving from the central limit theorem. We shall briefly describe the origin and function of the CLT and show then how it leads to convergence of our expansions.

For the simplest possible case, that of noninteracting particles, the density convolutes as we add particles; thus

$$\rho_m(x) = \int \rho_{m-1}(x - y)\, \rho_1(y)\, dy \equiv \rho_{m-1} \otimes \rho_1[x] = \rho_1 \otimes \rho_1 \otimes \rho_1 \cdots \otimes \rho_1[x] \tag{18}$$

in which m now stands for the particle number. Now, by the simplest version of the

CLT, $\rho_m(x) \to$ Gaussian for large enough particle number (for characteristic single-particle spectra a half-dozen particles is enough for a good Gaussian). It must not be forgotten however that, as described in Section 1, convergence of the density to Gaussian does not imply convergence of the spectrum to a characteristic form. We shall therefore always have in mind convergence of "smoothed" functions or "convergence to within fluctuations."

Equation (18) represents an approximation, even for noninteracting particles, since in deriving it we have ignored a Pauli "blocking effect" (which would be of no consequence however for a "dilute" system, one with $m \ll N$ where N is the number of single-particle states). If we construct the m-particle states by adding one particle at a time we may think of the energy of the single-particle state into which a particle is placed as a random variable, and the convolution defines the density for a sum of random variables. The CLT requirements are that the variables be additive (which is satisfied for noninteracting particles) and that they be independent (which is disturbed by the blocking effect since a state filled by one particle is not accessible to another). An extremely nonuniform single-particle spectrum might in fact require such a high order of convolution (large number of particles) that a Gaussian spectrum is not achievable. In practical cases, however, this does not occur and we can say simply that noninteracting-particle systems generate Gaussian spectra.

For interacting particles on the other hand the energies are not additive and there is no *necessary* convergence to Gaussian even if the usually weak blocking effects are ignored (suppose for example that H is a type of separable interaction, represented by the square of a one-body operator whose centroid is zero and whose spectrum *is* Gaussian; then H has a $\chi^2(1)$ distribution, $\rho(x) \sim x^{-1/2} \exp(-x/2)$, which is highly non-Gaussian). Despite this, it is found that all nuclear Hamiltonians which give reasonable agreements with experimental data have model-space spectra which are quite close to Gaussian. To understand this one introduces an ensemble of Hamiltonians, for example a Gaussian Orthogonal Ensemble [7] (GOE) of two-body interactions (in which the two-body matrix elements are chosen independently according to similar zero-centroid Gaussian laws) which contains members giving rise to every possible spectrum of the specified dimensionality. One can now show [5] that the ensemble-averaged density becomes Gaussian for large enough particle number, and that moreover, the relative weight of those H's in the ensemble which give deviant densities becomes negligible for large systems (an aspect of ergodicity). For "most" Hamiltonians then (the measure being defined by the GOE) we have effectively a CLT for interacting particles as well as for noninteracting. Drawing now on our experience we assert that successful model-space Hamiltonians are characteristic members of the ensemble and it follows then that, as we increase the particle number, the eigenvalue density converges to Gaussian even when the particles are interacting.

In case the sudden appearance of an ensemble is mysterious we remark that its real function is to reproduce, for interacting particles, the dominance [5, 11] of the binary correlations between matrix elements which is found automatically for noninteracting particles when $1 \ll m \ll N$ (the corresponding physical assumption, in accord with experience, is that such dominance is indeed found with reasonable model Hamiltonians). To show this consider a one-body Hamiltonian, $H = \sum \epsilon_\alpha n_\alpha$ with n_α the number operator and $\sum \epsilon_\alpha = 0$ (so that $\langle H^p\rangle^m$ is the *central* moment). Then $\langle H^p\rangle^m$ decomposes into a number of parts according to the partitions of p (5 for $p = 4$) a characteristic [2, 2] term for example being $\epsilon_\alpha^2\epsilon_\gamma^2 \langle n_\alpha^2 n_\gamma^2\rangle^m$ with $\alpha \neq \gamma$. Since $n_\alpha^2 = n_\alpha$ and $\langle n_\alpha n_\beta \cdots n_\sigma\rangle^m \propto \binom{m}{l}$ where l is the number of different operators in the product, it is almost obvious, and easy to verify, that the partition with the largest number of parts will be dominant, with the proviso however that partitions containing 1's ([31]; [211] for example) will contain a power of $(\sum \epsilon_\alpha) = 0$ as a factor and are therefore ruled out. Thus binary correlations are dominant, corresponding to counting of particle pairs. Doing the same calculation in a general (nondiagonal) basis, with $H = \sum \epsilon_{\alpha\beta} a_\alpha^\dagger a_\beta$, we find correlations between the Hamiltonian matrix elements $\epsilon_{\alpha\beta}$ and $\epsilon_{\beta\alpha}$. The argument leading to this simple result is obviously invalid for interacting particles, but on the other hand, the binary correlations are precisely those which are dominant for the GOE in our $(N \gg m \gg 1)$ limit. Their dominance arises in this case because the "1"-partitions do not contribute (nor do others involving an odd factor) because of the vanishing of the ensemble-averaged matrix element, $\overline{W}_{\alpha\beta}$, while, compared with the number of binary-correlation terms, the number involving quaternary and higher-order correlations goes to zero in our limit.[1]

The ensemble methods used in exploring these questions are closely analogous to the methods of statistical mechanics (involving ensembles of Hamiltonians rather than states however). They are useful for many general purposes, for example to derive the rate of convergence and other properties of the transition to Gaussian, and they give the only available method for studying the fluctuations. It must be emphasized however, that ensembles are used in this paper only for *general* arguments. When dealing with a specific system, as we shall be from now on, we work with the given Hamiltonian; we shall make no use of ensembles nor shall we ignore blocking effects nor the nonadditivity of the energy.

Returning now to the strength distributions, we consider first the sum rules, or more generally the expectation value of an operator K. The polynomial expansion of $K(W)$ is given by (16). The convergence of the densities to Gaussian implies an asymptotic linearity in $K(W)$. This is demonstrated in Fig. 1, the corresponding argument being that if H has eigenvalues W_i then $(H + \alpha K)$ with small α has eigenvalues $W_i + \alpha K(W_i)$; if both spectra are Gaussian, being identical therefore to within a scale change, it follows that $K(W_i)$ is linear in W_i, and is given therefore by the first two terms of (16);

$$K(W) \xrightarrow{\text{CLT}} \langle K\rangle^m + \langle K(H - \mathscr{E})\rangle^m((W - \mathscr{E})/\sigma^2). \tag{19}$$

This result is, of course, valid to within level-to-level fluctuations. It relies on the fact that, starting in the operator space at the point H we may, when there are enough particles, move an infinitesimal distance in "almost" any direction without modifying the smoothed spectrum shape, at least to first order in α. With $K = H$ (19) is an identity but, on the other hand, $K = H^2$ (whose expectation value is quadratic in W and

[1] This does *not* prove that their relative contribution vanishes, for that depends on the matrix-element distribution and with a "singular" distribution they could dominate.

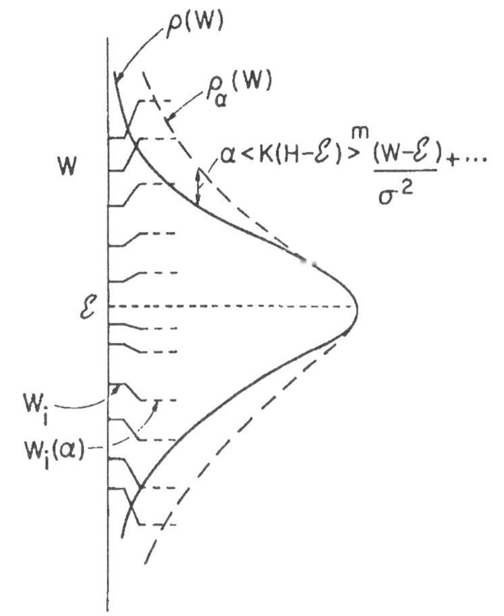

FIG. 1. The expectation values $\langle W | K | W \rangle \equiv K(W) = (\partial W(\alpha)/\partial\alpha)_{\alpha=0}$, as determined by the response of the system under $H \rightarrow H_\alpha = H + \alpha K$ for infinitesimal α. The figure is for the CLT limit, assuming that $\langle K \rangle^m = 0$, so that the only response (to within fluctuations) is a linear scale variation, $\sigma^2 = \langle (H - \mathscr{E})^2 \rangle^m \rightarrow \langle (H - \mathscr{E} + \alpha K)^2 \rangle^m$, and then the level shifts are linear in the energy as shown. Allowing for a nonzero value of $\langle K \rangle^m$ gives a (constant) centroid shift as well.

which by itself would generate a non-Gaussian spectrum) defines an exceptional or singular direction in the operator space, as would other operators closely correlated with a power of H, which would similarly require further terms in (19).

If linearity obtains for the moments of the strengths generated by the excitation operator O and its adjoint O^+ we see from (12), (13) that the polynomial strength moments $\Lambda_\alpha(W)$ vanish when $\alpha > 1$; thus in the CLT limit the strength itself (always to within fluctuations) is bilinear in the energies

$$\begin{aligned} R(W', W) \xrightarrow{\text{CLT}} & \langle O^+O \rangle^m + \langle O^+(H - \mathscr{E}')O \rangle^m \frac{(W' - \mathscr{E}')}{(\sigma')^2} \\ & + \langle O^+O(H - \mathscr{E}) \rangle^m \frac{(W - \mathscr{E})}{\sigma^2} \\ & + \langle O^+(H - \mathscr{E}')\, O(H - \mathscr{E}) \rangle^m \frac{(W' - \mathscr{E}')(W - \mathscr{E})}{(\sigma'\sigma)^2} \end{aligned} \tag{20}$$

a special case of (11), important in giving for many cases the general trend of the energy variation of the strength.

Although the low-order shape corrections, giving nonlinearities in $K(W)$ and $R(W', W)$, are not necessarily small, it is worthwhile paying special attention to the asymptotic case because there we can see very easily an important geometrical structure which is essentially maintained when we proceed to higher order. In particular, the linear form (19) for $K(W)$ has a simple geometrical interpretation. For an operator F the bilinear trace $\langle F^+F\rangle$ (we shall often drop the m-superscript which labels the model space; note too that $\langle F^+F\rangle$ is the trace of $d^{-1}F^+F$) defines a proper *norm* and then, for Hermitian operators, $\langle FG\rangle$ defines an inner product in terms of which magnitudes of operators, $\| F \| = \{\langle F^2\rangle\}^{1/2}$, and angles between operators, $\cos \theta_{FG} = \langle FG\rangle / \| F \| \times \| G \|$, may be defined, the operators then residing in a (real) Euclidean space. Subtracting out the centroid for an operator ($K \to K - \langle K\rangle^m = \mathscr{K}$; $H \to H - \mathscr{E} = \mathscr{H}$) has the geometrical significance of projecting in a plane perpendicular to the "one" operator 1. For these centered operators we write the *unit operators* or *unit vectors* as

$$k = \mathscr{K}/\| \mathscr{K} \|; \qquad h = \mathscr{H}/\| \mathscr{H} \| \tag{21}$$

and then (19) takes on the forms

$$\begin{aligned} \mathscr{K}(W) &= \langle \mathscr{K} \cdot \mathscr{H}\rangle((W - \mathscr{E})/\sigma^2), \\ k(W) &= (k \cdot h)((W - \mathscr{E})/\sigma) = \zeta_{K-H}((W - \mathscr{E})/\sigma), \end{aligned} \tag{22}$$

in which $\zeta_{K-H} = (k \cdot h)$, whose value lies in the interval $(-1, 1)$ may be regarded, in geometrical terms, as the cosine of the angle between the two centered operators or, in statistical terms, as the correlation coefficient between them.

A strong correlation between $\mathscr{K}$ and $\mathscr{H}$ (i.e., alignment, either parallel or antiparallel) implies a strong energy dependence of $K(W)$, the states of maximum expectation value lying low if the operators are antiparallel and high if parallel. A good example occurs with the isoscalar-quadrupole transition strength, for which $K = Q \cdot Q$ while $(H - \mathscr{E})$ "contains" a large part of $Q \cdot Q$ with negative coefficient; the "strong" states are then low-lying. On the other hand, if $(H - \mathscr{E})$ should contain only a small part of K (i.e., if the operators are essentially orthogonal) then the smoothed expectation value $K(W)$ will be essentially constant over the entire spectrum. Of course the notion that one operator contains a multiple of another is given a precise formulation by the geometry which we have introduced, the appropriate operation being that of projecting one operator along the other, or, more completely, of decomposing one of the operators into a part directed along the other and a part orthogonal. Thus the projection of $(H - \mathscr{E})$ along $(G - \langle G\rangle)$ is the operator $\| G - \langle G\rangle\|^{-2}\langle G(H - \mathscr{E})\rangle(G - \langle G\rangle)$, in which for the inner product we could just as well have written $\langle (G - \langle G\rangle)(H - \mathscr{E})\rangle$ or $\langle (G - \langle G\rangle)H\rangle$ since $\langle (H - \mathscr{E})\rangle = \langle (G - \langle G\rangle)\rangle = 0$; more generally still we could write as an operator equivalent of (16)

$$K = \sum_\nu \langle KP_\nu(H)\rangle\, P_\nu(H) \tag{23}$$

valid of course only for linear expectation values of K.

As a minor illustration we have that the correlation coefficient in $(ds)^{12}$ between the isoscalar $Q \cdot Q$ operator and any reasonable interaction is negative and large in magnitude; for the Kuo interaction [12] $\zeta = -0.56$ and comparable values are found for other "realistic" interactions. With a normalization such that $Q \cdot Q = 4\mathscr{G} - 3L^2$, where $\mathscr{G}$, the $SU(3)$ bilinear Casimir operator, has eigenvalue $\lambda^2 + \mu\lambda + \mu^2 + 3(\lambda + \mu)$

in the (λ, μ) representation, we have in linear approximation then that $Q \cdot Q(W) \rightarrow 156.5 - 52.3x$ where $x = (W - \mathscr{E})/\sigma$ has the value -5.1 at the ground state (for a nondegenerate ground state and a Gaussian density x is given in terms of an error function by $x = \text{erfc}^{-1}(1/d)$ which is very roughly $(2\ln 2d)^{1/2}$; excellent approximations are available [13]). This would give a $Q \cdot Q$ ground-state expectation value $\simeq 420$ quite compatible with the fact that the largest $Q \cdot Q$ eigenvalue in $(ds)^{12}$ is 720 deriving from the representations (12, 0) or (0, 12).

This example, in fact, is quite instructive, for at the upper end of the energy spectrum (where we would find the small-eigenvalue $SU(3)$ representations) the linear form for $Q \cdot Q(W)$ goes negative even though $Q \cdot Q$ is a positive definite operator. This comes about because the circumstances in this example are [14] unfavorable for a Gaussian density and a corresponding linear $K(W)$. The space is too large ($d = 2.7 \times 10^6$) for these approximations to extend over the entire spectrum span (the shape deformations of course show up first near the extremes); $Q \cdot Q$ is strongly correlated with H so that $Q \cdot Q(W)$ has a large slope, and, $Q \cdot Q$ being the sum of five squares, is somewhat related to the square of a one-body operator which we have mentioned above as a prototype of an operator with a non-Gaussian spectrum. In fact the necessary cut-off in the $Q \cdot Q$ spectrum at small eigenvalues generates quite a strong deviation from Gaussian at the large-eigenvalue end, which would correspond to low-lying states in the energy spectrum; the deviations in density are also large but the expectation value (which has a polynomial rather than an exponential form) is given reasonably. Even so for good accuracy we should take into account the shape deformations either by including some more terms in the expansions given above or by using the partitioning procedure (to be discussed in the next section) which is more efficient and more instructive. It will be seen, however, from the examples in the next paper that the geometry will survive, giving quantitative results in favorable circumstances and at least a qualitative description in unfavorable ones.

A Euclidean (or Unitary) geometry can be introduced into any model space. However, the physical significance of the geometry as we have introduced it above derives from the CLT which for centered operators, and always to within fluctuations, assures us that the norms of operators and the scalar products which derive from the norms, determine everything about the asymptotic behavior. In other words the significance of the geometry derives from the direct-product nature of the model spaces. But this same structure determines also symmetries, defined by the group $U(N)$ of unitary transformations among the N single-particle states with which we build the model spaces, and its various subgroups. Quite new consequences follow from this as we shall discuss briefly in Section 7.

4. Linear Response Theory

The asymptotic results (19), (20) may be thought of as deriving from the linear response of the system under the deformation $H \rightarrow H_\alpha = H + \alpha K$. In the general (nonasymptotic) case there will be a deformation of the spectrum *shape*, as well as the variations in the centroid and width which lead to (19), (20). We might expect then to relate the higher terms in the polynomial expansions to the shape deformation, and thereby the rate of convergence of the expansions to the rate determined by the CLT for the density, which is well understood.

Let us first derive a compact form for $K(W)$ in terms of the density deformation. Under $H \to H_\alpha = H + \alpha K$, with small α, we have $W_i \to W_i + \alpha K(W_i)$, as in Fig. 1. Let $W_i < W < W_{i+1}$ and agree that $K(W_i), K(W_{i+1}) \geqslant 0$ (the other cases lead to the same result). Since $d \times F(W)$ counts the number of levels below W, the deformation will either produce no change in $F(W)$, or a change in $F(W)$ of value $(-d^{-1})$, the probability of the latter being the ratio of the shift $\alpha K(W_i)$ to the local level spacing $(W_{i+1} - W_i)$. For smoothed distributions, adequate for our purpose, this ratio is $\alpha d \times \rho(W) K(W)$ and then, taking the small-α limit, we have the result [2]

$$\rho(W)\,K(W) = -\left(\frac{\partial F_\alpha(W)}{\partial \alpha}\right)_{\alpha=0} = -\left(\frac{\partial}{\partial \alpha}\int_{-\infty}^{W} \rho_\alpha(z)\,dz\right)_{\alpha=0}. \tag{24}$$

Recognizing that $F(W)$ is a staircase function and that the change in $F(W)$ takes place only for W in the neighborhood of an eigenvalue W_i, we see that (24) is valid in general.

A density $\rho_\alpha(x)$ may be defined in terms of its *location*, *extent*, and *shape*, respectively by its centroid $\mathscr{E}_\alpha \equiv S_1(\alpha) = \langle H_\alpha \rangle^m$, its variance $\sigma_\alpha^2 \equiv S_2(\alpha) = \langle (H_\alpha - \mathscr{E}_\alpha)^2 \rangle^m$, and a set of shape parameters $S_{\nu\geqslant 3}(\alpha)$ which must then be translation and scale invariant. It follows by linear independence of the moments that we can take the $S_{\nu\geqslant 3}$ to be νth order polynomial moments of the form

$$S_\nu(\alpha) = \int [\rho_\alpha(z) - \rho(z)] s_\nu(z)\,dz = \langle [s_\nu(H + \alpha K) - s_\nu(H)] \rangle^m = \alpha \langle K s_\nu'(H) \rangle^m + \cdots \tag{25}$$

where $s_\nu(z)$ is a νth order polynomial (in which the coefficient of z^ν does not vanish) and $s_\nu'(z)$ is its z-derivative. The precise form of the $S_{\nu\geqslant 3}$ we will fix later; it is clear however that, in order to satisfy the invariance requirements of the shape parameters, the $S_{\nu\geqslant 3}(z)$ must be functions only of $(z - \mathscr{E})/\sigma$. Under the Hamiltonian deformation the density $\rho(x)$ becomes $\rho_\alpha(x)$ and the variation of each parameter S_ν gives a contribution to $K(W)$; specifically, from (24)

$$\rho(W)\,K(W) = -\sum_{\nu\geqslant 1} \left(\frac{\partial S_\nu(\alpha)}{\partial \alpha}\right)_{\alpha=0} \frac{\partial F(W)}{\partial S_\nu}. \tag{26}$$

The $(\nu = 1, 2)$ terms, exactly as we would expect, give back the CLT result (19); formally $\rho(x)$ has the functional form $\sigma^{-1} f((x - \mathscr{E})/\sigma)$ and then (19) follows from the results that [2]

$$\frac{\partial F(W)}{\partial S_1} = -\rho(W); \qquad \frac{\partial F(W)}{\partial S_2} = -\frac{(W - \mathscr{E})}{2\sigma^2}\,\rho(W). \tag{27}$$

We now have for the deformed density and distribution function

$$\begin{aligned} \rho_\alpha(z) &= \rho(z) \Big\{1 + \sum_{\nu \geq 3} S_\nu(\alpha)\, T_\nu(z)\Big\}, \\ F_\alpha(z) &= \int_{-\infty}^{z} \rho_\alpha(x)\, dx. \end{aligned} \tag{28}$$

The centroid and variance deformations are included in $\rho(z)$ but since their contributions to $K(W)$ are already known (26), (27) we may as well ignore them here. In (28) the $S_\nu(\alpha)$ are given by (25) and the functions $T_\nu(z)$ will follow from the fact that the S_ν are polynomial moments. Since from (25), (28)

$$\begin{aligned} (\partial S_\nu/\partial\alpha)_{\alpha=0} &= \langle K s_\nu{}'(H)\rangle^m, \\ (\partial F(W)/\partial S_\nu) &= \int_{-\infty}^{W} \rho(z)\, T_\nu(z)\, dz, \end{aligned} \tag{29}$$

we have, on writing $K(W) = K_{\mathrm{CLT}}(W) + \Delta K(W)$, that $\Delta K(W)$, the shape-deformation correction to $K(W)$, is given (26), (29) by

$$\rho(W)\, \Delta K(W) = -\sum_{\nu \geq 3} \langle K s_\nu{}'(H)\rangle^m \int_{-\infty}^{W} \rho(z)\, T_\nu(z)\, dz. \tag{30}$$

We fix the s_ν polynomials and the T_ν functions in terms of the orthogonal polynomials P_ν by

$$\begin{aligned} s_\nu{}'(z) &= P_{\nu-1}(z), \\ \rho(z)\, T_\nu(z) &= -(d/dz)(\rho(z)\, P_{\nu-1}(z)), \end{aligned} \tag{31}$$

(the integration constant being of no consequence in the first of these may be taken zero) and then (30) reproduces the $K(W)$ expansion (16).

We have now a term-by-term association between the polynomial expansions of the strength and that of the expectation values (and therefore of the sum-rule quantities and the strengths themselves) which assures us that the CLT convergence of the density implies a similar convergence of these quantities as well. In particular as we add particles the shorter-wavelength (higher polynomial order) deviations from linearity are attenuated more rapidly so that the most significant corrections to the CLT forms for $K(W)$ and $R(W', W)$ will be those of low order.

For completeness we mention two technical points. (1) There is a derivative relationship (as in (31)) between the density and strength deformations which, for a given order, are thus not strictly related to each other on a point-by-point basis. However for Gaussian $\rho(x)$ we have $\rho^{-1}(x)(d/dx)(\rho(x)\, P_{\nu-1}(x)) \sim P_\nu(x)$ so that in this case the correspondence is precise; a slightly more complicated correspondence exists for a small class of other density forms, namely those for which there is a generalized Rodriguez formula [13]. (2) Even for Gaussian the convergence is not best exhibited by an orthogonal polynomial expansion. In the standard Gram–Charlier expansion about a Gaussian density the ($\nu \geqslant 6$) coefficients are not strictly Hermite-polynomial moments, and the differences are more pronounced with the Edgeworth series which exhibits the CLT convergence in a more uniform fashion [10]. Neither of these points, however, will have any real effect on our arguments and results.

The results above have followed from the *linear* response of the system under the ($H \rightarrow H + \alpha K$) "deformation" of the Hamiltonian. The quadratic response, in which we take account of the $(\partial^2 \rho_\alpha / \partial \alpha^2)_{\alpha=0}$ terms, and the response of higher order, can be used to give a theory for the inverse-energy-weighted sum rules, a subject which will be discussed and applied in a later paper.

5. Partitioning for Strengths, Sum Rules, and Expectation Values

Simple notions about the distribution of shell-model configurations throughout the spectrum lead naturally to procedures for calculating strength distributions which have been often used though they are approximate and in many cases quite inaccurate. We start with a sequence of such procedures and then give an exact polynomial expansion which in form is closely related to the simple ones but which, unlike them, takes full account of the interactions and in particular of the correlations between the Hamiltonian and the excitation operator.

We consider the single-particle space as composed of orbits (perhaps spherical as in the *ds* shell with $d_{5/2}$, $d_{3/2}$, $s_{1/2}$) and the m-particle space as decomposed thereby into configurations (3 for $m = 1$, 45 for $m = 12$ in our *ds* example); schematically

$$N = \sum N_i\,, \qquad m = \sum [m_1\,, m_2\,, \ldots, m_l] = \sum \mathbf{m}, \tag{32}$$

where l is the number of orbits; for simplicity we ignore the angular-momentum and isospin labels so that we consider together the transitions involving the various (J, T) values.

We can now give an approximate strength in the m-particle space in terms of the dimensionalities of the configurations $d(\mathbf{m})$, the centroids, $\mathscr{E}(\mathbf{m})$ $(= \langle H \rangle^{\mathbf{m}} \rightarrow \sum \epsilon_\alpha m_\alpha$ for noninteracting particles) and the mean-squared excitation matrix element $M^2(\mathbf{m}', \mathbf{m})$ which connects two configurations. For the (average) strength connecting two states we have, with $\mathscr{E}(\mathbf{m})$, $\mathscr{E}(\mathbf{m}')$ the configuration centroids, that

$$R(W', W) \rightarrow \sum_{\mathbf{m}', \mathbf{m}} M^2(\mathbf{m}', \mathbf{m})\, \delta(W' - \mathscr{E}(\mathbf{m}'))\, \delta(W - \mathscr{E}(\mathbf{m})) \tag{33}$$

so that the strength appears as a series of "spikes"; for the total strength in a spike we must of course include the configuration-dimensionality factors.

To give a form for M^2 we observe that, for F, G defined in m,

$$\langle\langle FG\rangle\rangle^{\mathbf{m}} = \sum_{\mathbf{m}'} \langle\langle FG(\mathbf{m}' - \mathbf{m})\rangle\rangle^{\mathbf{m}} = \sum_{\mathbf{m}'} \langle\langle G(\mathbf{m}' - \mathbf{m})F\rangle\rangle^{\mathbf{m}'}. \tag{34}$$

Here we have written $G = \sum_{\mathbf{q}} G(\mathbf{q})$ where $G(\mathbf{q})$ acting on any vector in $\mathbf{m}$ gives a vector in $\mathbf{m} + \mathbf{q}$ (i.e., with partition numbers $(m_i + q_i)$). This decomposition is quite immediate; for example with a one body operator, $G = \sum g_{\alpha\beta} A_\alpha{}^{\dagger} A_\beta$, a restriction of β to the orbit N_i and of α to orbit N_r, gives an operator which generates only transformations $[m_1, m_2, \ldots, m_l] \to [m_1, m_2, \ldots, m_i - 1, \ldots, m_r + 1, \ldots, m_l]$. F could similarly be decomposed but the simultaneous decomposition of F and G is obviously unnecessary for quadratic traces. As a point of notation we shall write $F^{+}(\mathbf{q}) = (F(\mathbf{q}))^{+}$ and is therefore an operator of type $(-\mathbf{q})$. Observe that $\langle\langle F^{+}(\mathbf{q})F(\mathbf{q})\rangle\rangle^{\mathbf{m}}$ is a partial variance of F, the contribution to the variance of the term $\mathbf{m} \to \mathbf{m} + \mathbf{q} \to \mathbf{m}$.

Using (34) we have that $d(\mathbf{m}')\, M^2(\mathbf{m}', \mathbf{m}) = \langle\langle O^{+}(\mathbf{m}' - \mathbf{m})O\rangle\rangle^{\mathbf{m}}$ so that (33) gives for the simplest version of the strength,

$$R(W', W) \to \sum_{\mathbf{m}',\mathbf{m}} d(\mathbf{m}')^{-1} \langle\langle O^{+}(\mathbf{m}' - \mathbf{m})O\rangle\rangle^{\mathbf{m}}\, \delta(W' - \mathscr{E}(\mathbf{m}'))\, \delta(W - \mathscr{E}(\mathbf{m})). \tag{35}$$

This very crude "schematic" theory already accomplishes something. In the ds shell for example, there are considerable energy differences between the three j-orbits (the relative primary single-particle energies are $\epsilon_j = 0$, 0.88, and 5.08 MeV for $d_{5/2}$, $s_{1/2}$, $d_{3/2}$, respectively) and as a consequence the centroids of the 45 $(ds)^{12}$ configurations themselves span a sizeable part of the entire spectrum (with the primary energies the centroid span is $(8\epsilon_{3/2} + 4\epsilon_{1/2}) \sim 44$ MeV and renormalization of the single-particle energies (which takes some account of the interaction) will increase this somewhat. Thus (35) spreads the strength over a considerable energy domain, not adequate however, because the spectrum span will be 100–125 MeV depending on the interaction and we would expect the strength to spread over the entire spectrum.

We can improve things by taking account of the spread of the configurations and making the simple assumption that the strength "follows" the densities. Then, since the $\mathbf{m}$-states form, at W, a fraction $I_{\mathbf{m}}(W)/I_m(W)$ of the total states, we have, as the extension of the simplest model, the approximation

$$R(W', W) \to \sum_{\mathbf{m}',\mathbf{m}} \frac{I_{\mathbf{m}'}(W')}{I_{m'}(W')} \frac{I_{\mathbf{m}}(W)}{I_m(W)} \cdot d(\mathbf{m}')^{-1} \langle\langle O^{+}(\mathbf{m}' - \mathbf{m})O\rangle\rangle^{\mathbf{m}} \tag{36}$$

which now does spread the strength over the entire spectrum. We have introduced here the decomposition of the density into partial (configuration) densities,

$$\begin{aligned} I_m(z) &= \sum I_{\mathbf{m}}(z) = \sum d(\mathbf{m})\, \rho_{\mathbf{m}}(z) = \sum \langle\langle \delta(H - z)\rangle\rangle^{\mathbf{m}}, \\ d(m)\, M_p(m) &= \langle\langle H^p\rangle\rangle^m = \sum_{\mathbf{m}} d(\mathbf{m})\, M_p(\mathbf{m}) = \sum_{\mathbf{m}} \langle\langle H^p\rangle\rangle^{\mathbf{m}}. \end{aligned} \tag{37}$$

In contrast to the total density, the partial densites are not densities of eigenvalues except in the special case where the configuration subspaces are Hamiltonian eigenspaces; instead $I_{\mathbf{m}}(z)\,dz$ measures the total **m**-intensity in the eigenstates located in the energy interval dz at z.

Observe very carefully that this linear decomposition involving, as it does, quantities nonlinear in the energy, does *not* imply that the configuration distributions are noninteracting; in evaluating $\langle H^p \rangle^{\mathbf{m}}$ for $p > 1$ we would logically take account of intermediate excitations to other configurations so that, taking $p = 2$ for example, (which along with the centroids would often suffice) the variance for one configuration has both an "internal" part ($\mathbf{m} \to \mathbf{m} \to \mathbf{m}$) and a set of "external" parts ($\mathbf{m} \to \mathbf{m}' \neq \mathbf{m} \to \mathbf{m}$). Then (36) does take account of some of the correlations induced by the interaction.

With the strength as given by (36) we find, for the energy-weighted sum rules, the expected result that

$$\mathscr{M}_p(W) \to \sum_{\mathbf{m},\mathbf{m}'} \frac{I_{\mathbf{m}}(W)}{I_m(W)} M_p(\mathbf{m}') \langle O^+(\mathbf{m}' - \mathbf{m}) O \rangle^{\mathbf{m}} \tag{38}$$

with $M_p(\mathbf{m}')$ a final-state density moment. The expectation value result follows from $\mathscr{M}_0(W)$.

$$K(W) \to \sum_{\mathbf{m}} \frac{I_{\mathbf{m}}(W)}{I_m(W)} \langle K \rangle^{\mathbf{m}}. \tag{39}$$

But this intuitively simplest form, which has been used, for example [15], in evaluating spin cutoff factors (with $K = J_z^2$), does not even take account of the first-order K–H correlation which, depending on its sign, raises or lowers the states with larger-than-average K expectation values, and which *is* taken account of in the CLT limit. We see then that the approximate results (38), (39), and *a forteriori* (36) have no proper basis at all except for operators very weakly correlated with H (as J_z^2 usually is, so that the earlier spin-cutoff evaluations are still more or less satisfactory). The major difficulty with these approximations is with the assumption that the strength follows the densities, which ignores the correlation between the Hamiltonian and the excitation and which is in general not at all correct. A related difficulty is the partial neglect of interferences between configurations, which there is no natural way to take account of in the schematic theory. To give an exact theory which will include the schematic theory as a very poor approximation, and the CLT-limit theory as a much better one, it is simplest to start all over again.

For very large spaces the CLT approximations (19), (20), will not be sufficiently accurate in the ground-state domain and consequently correction terms are required. It is quite satisfactory in some cases to include a few higher terms in the general expansions (11), (16). The traces required for these corrections, however, become rapidly harder and harder to evaluate as the order increases. Just as in the schematic models just described, it pays us to partition by orbits and configurations, which we do exactly as above. Let us now rewrite the polynomial expansions (11), (16) in accord with the new forms (37) for the density.

For the expectation values and hence the sum-rule quantities we define

$$K(W:\mathbf{m}) = \sum_{\mu} \langle KP_{\mu}^{(\mathbf{m})}(H)\rangle^{\mathbf{m}} P_{\mu}^{(\mathbf{m})}(W) \tag{40}$$

and then from (16)

$$\begin{aligned} K(W) &= I_m^{-1}(W)\langle\langle K\delta(H-W)\rangle\rangle^m = I_m^{-1}(W)\sum_{\mathbf{m}} \langle\langle K\delta(H-W)\rangle\rangle^{\mathbf{m}} \\ &= \sum_{\mathbf{m}} \frac{I_{\mathbf{m}}(W)}{I_m(W)} \sum_{\mu} \langle KP_{\mu}^{(\mathbf{m})}(H)\rangle^{\mathbf{m}} P_{\mu}^{(\mathbf{m})}(W) \\ &= \sum_{\mathbf{m}} \frac{I_{\mathbf{m}}(W)}{I_m(W)} K(W:\mathbf{m}) \end{aligned} \tag{41}$$

which is quite different from (39). We have here first decomposed the trace according to $m = \sum \mathbf{m}$ and have then taken the *physically motivated* step of introducing separate polynomial sets for the separate subspaces, namely for a given $\mathbf{m}$ the polynomials defined by the partial density $\rho_{\mathbf{m}}(W)$. Since a configuration density is to some extent localized in the spectrum a given term is not now required to give results over the entire spectrum and much lower polynomial orders are appropriate. It should be remarked also that, while there will be more of them, configuration traces are no harder to evaluate than the m-space ones (whereas the complexity increases extremely rapidly with polynomial order). The net result is that our partitioning gives a major simplification. If, as we usually do, we take the partial densities $\rho_{\mathbf{m}}(W)$ to be Gaussian the $P_{\mu}^{(\mathbf{m})}(W)$ are of course Hermite polynomials, $(\mu!)^{-(1/2)}\, He_{\mu}(x)$ where $\sigma(\mathbf{m})x = W - \mathscr{E}(\mathbf{m})$. In some cases, particularly for low-lying configurations, it would be worthwhile to include skewness and excess corrections.

In (41), just as in (36), $I_{\mathbf{m}}(W)/I_m(W)$ is a natural intensity ratio while $K(W:\mathbf{m})$, as defined in (40) is given by the standard $K(W)$ form (16) but restricted to the $\mathbf{m}$ subspace. Superficially then (41) has exactly the form to be expected, $K(W)$ appearing as a properly weighted fixed-energy average of its subspace values. This expectation is quite logical if the subspaces are Hamiltonian eigenspaces (which in practice does not really happen with configurations) for then $K(W:\mathbf{m})$ is indeed a K expectation value. But in the general case it is not *a priori* so because of the $(\mathbf{m} \rightarrow \mathbf{m}')$ excitations induced by H. It is a remarkable property of (41), that, without changing its form, it properly takes account of correlations and interferences of all orders, and gives then a formally exact expansion for $K(W)$. Note incidentally that, in the CLT limit in which we take only polynomials of order (0, 1), $K(W:\mathbf{m})$ is a true expectation value as long as $\mathbf{m}$ is an eigenspace for *either H or K*, so that with configurations the CLT limit of (41) does give for occupancies an uncorrelated superposition. The $\mu = 0$ terms of (41) give back the schematic-theory result (39) but, as indicated above, there is in general no logic in ignoring the $\mu = 1$ terms which are similarly uninhibited by the CLT.

For the strength we proceed similarly

$$
\begin{aligned}
I_{m'}(W')\, R(W', m'; W, m)\, I_m(W) \\
&= \langle\langle O^{+}\delta(H - W')\, O\delta(H - W)\rangle\rangle^{m} \\
&= \sum_{\mathbf{m}',\mathbf{m}} \langle\langle O^{+}(\mathbf{m}' - \mathbf{m})\, \delta(H - W')\, O\delta(H - W)\rangle\rangle^{\mathbf{m}} \\
&= \sum_{\mathbf{m}',\mathbf{m}} I_{\mathbf{m}'}(W')\, I_{\mathbf{m}}(W)\, d^{-1}(\mathbf{m}') \\
&\quad \times \sum_{\mu\nu} \langle\langle O^{+}(\mathbf{m}' - \mathbf{m})\, P_{\mu}^{(\mathbf{m}')}(H)\, O P_{\nu}^{(\mathbf{m})}(H)\rangle\rangle^{\mathbf{m}}\, P_{\mu}^{(\mathbf{m}')}(W')\, P_{\nu}^{(\mathbf{m})}(W) \\
&= \sum_{\mathbf{m}',\mathbf{m}} I_{\mathbf{m}'}(W')\, I_{\mathbf{m}}(W)\, R(W', \mathbf{m}'; W, \mathbf{m})
\end{aligned}
\tag{42}
$$

and thus

$$
R(W', m'; W, m) = \sum_{\mathbf{m}',\mathbf{m}} \frac{I_{\mathbf{m}'}(W')\, I_{\mathbf{m}}(W)}{I_{m'}(W')\, I_{m}(W)}\, R(W', \mathbf{m}'; W, \mathbf{m}) \tag{43}
$$

with $R(W', \mathbf{m}'; W, \mathbf{m})$ defined by the fourth form occurring in (42). We have introduced here, for each initial and each final suspace $(\mathbf{m}, \mathbf{m}')$, a set of polynomials generated by the subspace densities, the argument for doing this being identical with that given for expectation value (41). In (43) we find the intuitively obvious form for the strength decomposition as a superposition, with the natural initial- and final-state branching ratios, of the strengths connecting the subspaces. But once again all correlations and interferences are properly included and the result is formally exact. The CLT limit takes account of all polynomials of order $\leqslant 1$, thus $(\mu, \nu) = (0, 0)$, $(0, 1)$, $(1, 0)$, and $(1, 1)$, just as in (20) for the spaces not partitioned. The $(0, 0)$ terms gives for R the mean-squared matrix elements connecting two configurations which leads to the schematic theory derived above. It is easily verified that an integration (2) over the strength (43) reproduces the sum rules (41), just as, in the nonpartitioned case, an integration over (11) gives back (16). The same is true for the schematic version of (43) and the corresponding sum rules (39). We stress also that, in all these cases, integration over a restricted energy domain will produce the exothermic or endothermic sums disposing then of a difficulty discussed in the Introduction.

6. Multipole Strengths and Sum Rules

If we consider $J = 0^{+} \to 2^{+}$ quadrupole excitations we may start with a $J = 0$ model space and then the selection rule automatically yields $J = 2$. For a case such as this no formal extension of the procedures described above is needed (although of course the technical problem of evaluating the traces for fixed particle number and angular momentum will be harder than for fixed number only). By symmetry (12) the same is true for the inverse transitions $J = 2^{+} \to 0^{+}$ even though final states $J = 0$–4

are accessible, but things are quite different for example for $J = 2^+ \to 1^+$ and more generally for nonscalar excitations connecting nonscalar states. We need thus to decompose the strengths according to angular momentum.

In principle this can be done by introducing a projection operator for the final states (intermediate ones in the sum-rule expectation values!). Such an operator is easily written as a polynomial in J^2 but, in general calculations, as opposed, for example, to matrix ones in which they are automatically taken care of, it is hardly feasible to deal *directly* with operators of such high particle rank. (The fact though that there is an alternative procedure for solving the problem implies that there is an *indirect* way of dealing with them but we shall not explore that.)

We follow instead the methods [16] used in generating the multipole sum rules from the conventional "scalar" ones. We assume that the model spaces "accommodate" the angular momentum in the sense that if a *state* with given J is found in it, so will the complete *level* (i.e., all $(2J + 1)\ J_z$ values); this simply implies that the model spaces supply representations (in general reducible) of the rotation group. The basic rules are now very simple; every quantity should be displayed as a spherical tensor (or a linear combination of such), ordinary multiplication is replaced by tensor (Clebsch–Gordan) coupling, ordinary matrix elements by zero-coupled ones (zero because only scalars can survive integration over the coordinates of the system; the zero-coupled matrix elements may also be written as vacuum-expectation values or as Racah's double-barred matrix elements (dbme)). Then elementary recoupling algebra will solve for us all the problems connected with tensorial behavior. The same operations work also for isospin and in fact we shall use a direct-product notation which will work for angular-momentum or isospin or both together (e.g., $\Gamma = J$ or T or J, T). In this connection we remark parenthetically that, although the J and T coupling operations are identical, things are very different for the trace evaluation. This comes about because all the states of given T belong to an *irreducible* representation of a (direct-product) group, but this is not true at all for J.

Specifically we use the notation of Ref. [16]. We also need a definition of "double-barred" traces. Let us label the states by (m, Γ, M, x) where M is the z-component of Γ so that the $[\Gamma]$-dimensional set with fixed (m, Γ, x) constitutes a *level* and the non-angular-momentum label x counts the levels. Let $d(m, \Gamma)$ be the state dimensionality for fixed (m, Γ), and $d_l(m, \Gamma)$ the level dimensionality, and similarly for $I(m, \Gamma; W)$ and $I_l(m, \Gamma; W)$. Then

$$
\begin{aligned}
d(m, \Gamma) &= [\Gamma]\, d_l(m, \Gamma), \\
I(m, \Gamma; W) &= [\Gamma]\, I_l(m, \Gamma; W) = d(m, \Gamma)\, \rho(m, \Gamma; W), \\
\sum_W [\ ] &\to \int I_l(m, \Gamma; W)[\ ]\, dW = d_l(m, \Gamma) \int \rho(m, \Gamma; W)[\ ]\, dW.
\end{aligned}
\tag{44}
$$

We define

$$
\begin{aligned}
\langle\!\langle \| O^\nu \| \rangle\!\rangle^{m\Gamma} &= d_l(m, \Gamma) \langle \| O^\nu \| \rangle^{m\Gamma} = \sum_x \langle m\Gamma x \| O^\nu \| m\Gamma x \rangle \\
&= \int I_l(m, \Gamma; W) \langle m\Gamma W \| O^\nu \| m\Gamma W \rangle\, dW
\end{aligned}
\tag{45}
$$

in the last form of which we have used the Hamiltonian basis, taking for granted that we have a scalar Hamiltonian.

Consider first the (tensorial) expectation values of K^λ in the (m, Γ) eigenfunctions.

$$\begin{aligned}\langle\| K^\lambda \delta(H - W)\|\rangle^{m\Gamma} &= \rho(m, \Gamma; W) \sum_\mu \langle\| K^\lambda P_\mu^{(m,\Gamma)}(H)\|\rangle^{m\Gamma} P_\mu^{(m,\Gamma)}(W) \\ &= \int \rho(m, \Gamma; W)\, \delta(W' - W) \langle m\Gamma W' \| K^\lambda \| m\Gamma W'\rangle\, dW' \\ &= \rho(m, \Gamma; W) \langle m\Gamma W \| K^\lambda \| m\Gamma W\rangle \end{aligned} \tag{46}$$

and hence

$$\langle m\Gamma W \| K^\lambda \| m\Gamma W\rangle \equiv K^\lambda(W, \Gamma) = \sum_\mu \langle\| K^\lambda P_\mu^{(m,\Gamma)}(H)\|\rangle^{m\Gamma} P_\mu^{(m,\Gamma)}(W) \tag{47}$$

which gives the tensorial extension of (16); taking $\mu = 0,1$ only we have the CLT limit, the corresponding extension of (19). The polynomials introduced here are of course, those defined by the (m, Γ) partial density.

We may derive the sum rules from the strength itself, which we consider next. We have then *for a pure multipole transition*, $(m\Gamma M W) \to^{O^\lambda} (m'\ \Gamma'\ M'\ W')$, in which only the $(M' - M)$ component of O^λ contributes to the transition,

$$\begin{aligned} R^\lambda(m'\Gamma'M'W'; m\Gamma M W) &\equiv |\langle m'\Gamma'M'W' \mid O^\lambda_{M'-M} \mid m\Gamma M W\rangle|^2 \\ &= [\Gamma']^{-1} (C^{\Gamma\,\lambda\,\Gamma'}_{M,M'-M})^2 |\langle \Gamma' W' \| O^\lambda \| \Gamma W\rangle|^2 \\ &= [\Gamma']^{-1} (C^{\Gamma\,\lambda\,\Gamma'}_{M,M'-M})^2 R^\lambda(\Gamma' W'; \Gamma W), \end{aligned} \tag{48}$$

$$R^\lambda(\Gamma' W'; \Gamma W) = |\langle \Gamma' W' \| O^\lambda \| \Gamma W\rangle|^2 = \sum_{M,M'} R^\lambda(\Gamma' M' W'; \Gamma M W), \tag{49}$$

where we have used the Wigner–Eckart theorem and, for convenience, have suppressed the (m, m') labels. The polarized strength has been defined precisely as in Section 2 while the *unpolarized strength* $R^\lambda(\Gamma' W'; \Gamma W)$ $(= R^\lambda(\Gamma W; \Gamma'\ W')$ for a self-adjoint excitation) is, for electromagnetic and other hole-particle transitions, essentially the $BE\lambda$ or $BM\lambda$ value, differing by a statistical factor since we have *summed* over initial states instead of averaging. For angular momentum $(\Gamma \equiv J)$ our interest is usually in the unpolarized strength, for isospin $(\Gamma \equiv T)$ usually in the polarized strength. Then for $(\Gamma \equiv J, T)$ we are left with a Clebsch–Gordan isospin factor.

In the polarized cases there is, of course, interference between multipoles. We therefore need the more general quantity

$$\begin{aligned} &R^{\sigma\lambda}(m'\Gamma'M'W'; m\Gamma M W) \\ &\quad= \langle m\Gamma M W \mid (O^\sigma_{M'-M})^+ \mid m'\Gamma'M'W'\rangle\langle m'\Gamma'M'W' \mid O^\lambda_{M'-M} \mid m\Gamma M W\rangle \\ &\quad= [\Gamma']^{-1}\, C^{\Gamma\,\sigma\,\Gamma'}_{M,M'-M} C^{\Gamma\,\lambda\,\Gamma'}_{M,M'-M} R^{\sigma\lambda}(\Gamma' W'; \Gamma W), \end{aligned} \tag{50}$$

$$R^{\sigma\lambda}(\Gamma' W'; \Gamma W) = \langle \Gamma' W' \| O^\sigma \| \Gamma W\rangle\langle \Gamma' W' \| O^\lambda \| \Gamma W\rangle. \tag{51}$$

We have dropped a complex conjugate since all dbme can be taken real. Observe that $R^{\lambda} \equiv R^{\lambda\lambda}$, so that a polynomial expansion for $R^{\sigma\lambda}$ will be adequate for all cases, polarized or unpolarized, pure multipole or mixed.

We have now, using (3.17) of Ref. [16] which we label as (V 3.17)

$$\rho(\Gamma'; W')\, R^{\sigma\lambda}(\Gamma' W'; \Gamma W)\, \rho(\Gamma; W) = (-1)^{\Gamma'+\sigma-\Gamma} \int \rho(\Gamma'; z')\, dz' \int \rho(\Gamma; z)\, dz \times \langle \Gamma z \| \bar{O}^{\sigma}\delta(H - W') \| \Gamma' z' \rangle \langle \Gamma' z' \| O^{\lambda}\delta(H - W) \| \Gamma z \rangle \tag{52}$$

and then from (44), (45), (V 3:23)

$$I_l(m'\Gamma'; W')\, R^{\sigma\lambda}(m'\Gamma' W'; m\Gamma W)\, I_l(m\Gamma; W) = (-1)^{\Gamma'-\lambda-\Gamma}[\Gamma']^{1/2} \sum_{\omega} (-1)^{\omega}\, U(\Gamma\sigma\Gamma\lambda; \Gamma'\omega) \langle\!\langle \| (\bar{O}^{\sigma}\delta(H - W') \times O^{\lambda}\delta(H - W))^{\omega} \| \rangle\!\rangle^{m,\Gamma} \tag{53}$$

It is seen from the LHS that the intermediate states in the RHS trace must all be Γ' states and thus our expansion is completed by expanding $\delta(H - W')$ and $\delta(H - W)$ in $\rho^{(\Gamma')}$ and $\rho^{(\Gamma)}$ polynomials respectively, whence

$$d_l(m'\Gamma')\, R^{\sigma\lambda}(m'\Gamma' W'; m\Gamma W)\, d_l(m\Gamma) = (-1)^{\Gamma'-\lambda-\Gamma}[\Gamma']^{1/2} \sum_{\omega} (-1)^{\omega}\, U(\Gamma\sigma\Gamma\lambda : \Gamma'\nabla) \cdot \sum_{\mu,\nu} \langle\!\langle \| (\bar{O}^{\sigma} P_{\mu}^{(\Gamma')}(H) \times O^{\lambda} P_{\nu}^{(\Gamma)}(H))^{\omega} \| \rangle\!\rangle^{m\Gamma} P_{\mu}^{(\Gamma')}(W')\, P_{\nu}^{(\Gamma)}(W) \tag{54}$$

which extends (11). It should be clear of course that the ω-sum in (53), (54) is responsible for the Γ' projection. In fact, for the diagonal case ($\sigma = \lambda$), $U(\Gamma\lambda\Gamma\lambda{:}\Gamma'\omega) \sim P_{\omega}(\cos\theta)$ where θ is the ($\Gamma + \lambda \rightarrow \Gamma'$) coupling angle and the symbol ($\sim$) denotes the Racah Legendre correspondence (V 3.29); then $\sum_{\omega}$ generates the usual spherical harmonic transform. We have, with the subscript zero defining a vacuum-expectation value, that

$$\sum_{\omega} (-1)^{\omega}\, U(\Gamma\sigma\Gamma\lambda : \Gamma'\omega) \langle \Gamma z \| G^{\omega} \| \Gamma z \rangle = \langle (\bar{\psi}^{\Gamma}(z) \times G^{\omega} \times \psi^{\Gamma}(z))^{0} \rangle_0 \tag{55}$$

and then (53), (54) become

$$I_l(m'\Gamma'; W')\, R^{\sigma\lambda}(m'\Gamma' W'; m\Gamma W)\, I_l(m\Gamma; W) = (-1)^{\Gamma'-\lambda-\Gamma}[\Gamma']^{1/2} \sum_{z} \langle ((\bar{\psi}^{\Gamma}(z) \times \bar{O}^{\sigma}\delta(H - W'))^{\Gamma'} \times (O^{\lambda}\delta(H - W) \times \psi^{\Gamma}(z))^{\Gamma'})^{0} \rangle_0 , \tag{56}$$

$$d_l(m'\Gamma')\, R^{\sigma\lambda}(m'\Gamma' W'; m\Gamma W)\, d_l(m\Gamma) = (-1)^{\Gamma'-\lambda-\Gamma}[\Gamma']^{1/2} \sum_{z,\mu,\nu} \langle ((\bar{\psi}^{\Gamma}(z) \times \bar{O}^{\sigma} P_{\mu}^{(\Gamma')}(H))^{\Gamma'} \times (O^{\lambda} P_{\nu}^{(\Gamma)}(H) \times \psi^{\Gamma}(z))^{\Gamma'})^{0} \rangle_0 \times P_{\mu}^{(\Gamma')}(W')\, P_{\nu}^{(\Gamma)}(W), \tag{57}$$

in which the z sums define a new variety of coupled trace (a spherical transform of the original given by (45)). The technical problem of trace evaluation is of course essentially the same for the two varieties.

The multipole sum rules follow immediately from these basic equations, as does also the decomposition by configurations (which is somewhat complicated however by the interorbit couplings). We leave these matters until we come to applications in later papers.

7. Unitary Geometry

In Section 3 we have introduced a geometry for the (centered) operators which operate in a model space by taking the square of the norm of an operator to be defined by its variance. It has been stressed that the effectiveness of this geometry (the fact that we can understand phenomena and even calculate parameters by making use of it) derives from the CLT which, in the many-particle limit, assigns great significance to the widths (norms). The CLT in turn derives from the direct-product nature of the model spaces, which is best described in group theoretical terms; specifically the $\binom{N}{m}$ states formed by distributing m fermions over N single-particle states form an *irreducible* representation of the group $U(N)$ of unitary transformations in the single-particle space. It is not surprising then that symmetry considerations (1) will give a deeper understanding of the results derived above; (2) will make explicit the particle number dependence (see for example (59) ahead) so that we can explore how a given phenomenon changes as we consider a set of neighboring nuclei; (3) will give group theoretical methods for evaluating the necessary traces; (4) will lead to major extensions of the geometry when we consider subgroups of $U(N)$.

In this section we give an introduction to these matters, considering $U(N)$ only and not its subgroups. We leave the extensions and applications for later papers. We shall, however, answer, in a particular case, a general question discussed in the Introduction, namely: How is information propagated over a very wide energy range in the model spectrum so that, although most of the model states are highly "unphysical," the space nonetheless does supply real information about the low-lying states? Remember that the Hamiltonian norm is determined by the sum of squares of all its model-space eigenvalues, only a negligible fraction of which will be "physical" so that this question is very important for us.

The m-particle states generate the $U(N)$ irrep defined by a one-columned Young shape with m blocks. Since all the representations we encounter will be predominantly antisymmetric (many rows, few columns) we shall label them by their *columnar* structure rather than by rows, thus $[m]$ for the m-particle antisymmetric states. A general k-body operator does not transform irreducibly but instead [15]

$$F(k) = \sum_{\nu=0}^{k} F^{\nu}(k) = \sum_{\nu=0}^{k} \binom{n-\nu}{k-\nu} \mathscr{F}^{\nu}(\nu) \tag{58}$$

where ν, which describes the unitary symmetry, and which we shall define as the *unitary rank*, indicates that $F^\nu(k)$ transforms as one- or two-columned representations $[N-\nu, \nu]$. n is the number operator and the last equality expresses the fact that any k-body operator of definite unitary rank contains (if $\nu < k$) an n-polynomial factor which is the result of a $(k-\nu)$-fold contraction of $F(k)$. We may describe the $\mathscr{F}^\nu(\nu)$ (or more simply $\mathscr{F}(\nu)$) as fully reduced or contracted tensors. They vanish in $(s<\nu)$-hole or particle spaces and behave simply under the hole $\leftrightarrow$ particle transformation $(\mathscr{F}(\nu) \leftrightarrow (-1)^{\nu(2\nu-1)}\mathscr{F}^+(\nu))$. Their unitary and particle ranks are equal.

We have thus that the space of the k-body operators (whose dimensionality is $\binom{N}{k}^2$, that being the number of independent matrix elements) decomposes into a set of $(k+1)$ subspaces each defining an irreducible tensor; the subspace dimensionalities are $D(\nu) = \binom{N}{\nu}^2 - \binom{N}{\nu-1}^2 = (N+1)^{-1}(N-2\nu+1)\binom{N+1}{\nu}^2$ (in spaces generated by distributing fermions over N single-particle states, every contracted tensor of rank ν is expandible in terms of $D(\nu)$ tensors). A general k-body operator has a component in each subspace. In particular a two-body interaction has a $(\nu=1)$ part which behaves like a number-dependent (because of the $(n-1)$ factor) irreducible (traceless) one-body operator and which renormalizes the single-particle energies; it contains also a "number-dependent zero-body operator," a multiple of $\binom{n}{2}$, which for $m=N$ gives the closed-shell energy and for arbitrary m the average energy, or eigenvalue centroid; this is subtracted out of course, when the operator is centered.

The square of the norm of $F(k)$ now decomposes as a sum of squares, no cross-terms occurring because only the scalar $(\nu=0)$ part of the product contributes to the trace

$$\begin{aligned}\langle F^+(k)\,F(k)\rangle^m &= \sum_{\nu=0}^{k}\binom{m-\nu}{k-\nu}^2\langle\mathscr{F}^+(\nu)\,\mathscr{F}(\nu)\rangle^m\\ &= \sum_{\nu=0}^{k}\binom{m-\nu}{k-\nu}^2\binom{m}{\nu}\binom{N-m}{\nu}\binom{N-\nu}{\nu}^{-1}\langle\mathscr{F}^+(\nu)\,\mathscr{F}(\nu)\rangle^\nu \qquad (59)\\ \langle\mathscr{F}^+(\nu)\,\mathscr{F}(\nu)\rangle^\nu &= \binom{N}{\nu}^{-1}\sum_{\alpha,\beta}|\mathscr{F}_{\alpha\beta}|^2\end{aligned}$$

where $\mathscr{F}_{\alpha\beta}$ is the matrix element of $\mathscr{F}(\nu)$ between states α, β which belong to an orthonormal basis for the ν-particle space. The first equality here derives from (58) and the second from the fact that, since $\mathscr{F}^+(\nu)\mathscr{F}(\nu)$ is an operator of various particle ranks $\leqslant 2\nu$, $\langle\mathscr{F}^+(\nu)\mathscr{F}(\nu)\rangle^m$ is a polynomial in m of order 2ν, which must be invariant under $m \leftrightarrow (N-m)$, must vanish for $m<\nu$, and must reduce to unity for $m=\nu$, these conditions then defining the polynomial to have the form indicated. $\langle\mathscr{F}^+(\nu)\mathscr{F}(\nu)\rangle^\nu$, defining the *basic norm* of $F^\nu(k)$, is simply the mean square defining matrix element of $\mathscr{F}$. Observe that the different tensorial parts of $F(k)$ grow in magnitude (norm) at different rates as the particle number increases, the simpler tensors (smaller ν) growing more rapidly than the more complex ones. A completely similar decomposition follows immediately for the scalar product $\langle F^+(k)G(k')\rangle^m$. Note that if $k \neq k'$ (say $k<k'$) then the $(\nu>k)$ parts of $G(k')$ contribute nothing to the scalar product, an obvious result but, as we shall see, an important one. (This property can indeed be used to define the irreducible tensors and to carry out the explicit reduction of $F(k)$: the key statement here is that any k-body operator with highest unitary rank ($\nu = k$) is orthogonal in all m-particle spaces to every operator of lower particle rank.)

In the Introduction we commented on the fact that, with the methods we are using, everything is "tied to" the spectrum centroid which may lie very high (this will change somewhat when we partition the space but its essential truth will remain). The natural question is how is the significant information transmitted over the wide energy span extending from the centroid to the ground state. We are now in a position to deal with a much more specific version of the same question, namely, how is "single-particle information" transmitted, and above all how does it "survive" in the presence of strong two-body interactions. From the fact that level densities can be adequately calculated in terms of single-particle energies, and from the success of the Strutinsky method for shell corrections, we know that single-particle information does survive; we shall deal here with a different aspect of the problem, involving ground-state occupancies which can be deduced from single-nucleon-transfer experiments. The result which we draw on is the fact that the number operators for the single-particle states, having $\nu = (0 + 1)$, are not correlated with the unitary-rank-two part of H, so that the scalar product $\langle n_s(H - \mathscr{E})\rangle^m$ which enters into the CLT result (19) for $n_s(W)$ (where n_s is the operator for the single-particle state numbered s) depends only on the (traceless) one-body interaction including the interaction operator $(n-1)\mathscr{H}(1)$ which renormalizes it. Since $\langle n_s\rangle^m = m/N$ and σ^2 is the sum of two parts ($\nu = 1, 2$) we have, with $H(1)$ the unrenormalized one-body H and $\mathscr{E}(1) = \langle H(1)\rangle^1 n$ its centroid operator

$$n_s(W) \xrightarrow{\text{CLT}} \frac{m}{N} + \frac{\langle n_s[(H(1) - \mathscr{E}(1) + (n-1)\,\mathscr{H}(1)]\rangle^m}{\sigma^2(\nu = 1) + \sigma^2(\nu = 2)}(W - \mathscr{E}) \tag{60}$$

whose slope $n_s'(W)$ is proportional then to the correlation coefficient with the (renormalized) one-body H. The only role played by the "intrinsic" ($\nu = 2$) interaction is to damp the correlation coefficient and only its norm is of consequence for that. If the two parts are comparable in magnitude so that the damping is not too large, and provided that the CLT result is adequate (and both of these things are often true), the one-body information $n_s(W)$ extends to the ground state. If, moreover, the renormalization of $H(1)$ is appreciable $n_s(W)$ carries with it information about the two-body interaction (through its ($\nu = 1$) part) and the experimental occupancy results can be used to test the interaction. This has been recently done for ds-shell nuclei [3], the occupancies being considered throughout the entire shell. It is of course a major advantage of the methods being used that the number dependences of the norms and scalar products (as well as the dependence on the matrix elements) are explicitly exhibited [15].

In the asymptotic limit for a dilute system ($N \to \infty, m \to \infty, m/N \ll 1$) the "induced" one-body interaction ($(n-1)\,\mathscr{H}(1)$), assumed not to vanish identically, will dominate both the primary $\mathscr{H}(1)$ and the ($\nu = 2$) interaction (in each case because of its explicit number dependence $\sim(n-1)$; from (59) we have $\langle n_s H^{\nu=1}\rangle^m \to^{m\to\infty} m^2\langle n_s\mathscr{H}(1)\rangle^1$;

$\sigma^2(\nu = 1) \to m^3\langle(\mathscr{H}(1)^2\rangle^1$; $\sigma^2(\nu = 2) \to \frac{1}{2}m^2\langle(\mathscr{H}(2)^2\rangle^2)$). Then the correlation coefficient $\sim m^{-1}$ so that in the limit all occupancies become constant over the spectrum. When there is no induced one-body interaction the ($\nu = 2$) damping dominates everything and once again the occupancies are asymptotically constant. Although these cases are not of great interest in themselves, they give us an insight into the role played by the unitary gometry. Major extensions of the geometrical arguments are possible based in part on refinements arising from the existence of $U(N)$ subgroups.

8. Fluctuations and Correlations

It will be useful in order to calculate a measure for the probable errors (deviations between statistically smoothed and exact values), and instructive for other reasons as well, to consider the level-to-level fluctuations in the moments $\mathscr{M}_p(W)$ and the correlations between moments of different order. In principle of course the complete (exact) polynomial expansion (11) for $R(W', W)$ determines everything but, as stressed in Section 1, we can deal explicitly only with the polynomial terms of low order which determine the *secular* variations, and we need a different method for handling those of high order which determine the fluctuations.

Specifically we shall combine the Porter–Thomas distribution for the local behavior of the individual matrix elements $\langle W' \mid O \mid W\rangle$, with the secular variations of $\overline{\langle W' \mid O \mid W\rangle^2}$, as given above, to derive the mean-squared deviations in $\mathscr{M}_p(W)$ and, more generally, the covariance of $\mathscr{M}_p$ and $\mathscr{M}_q$. Leaving some general remarks about the procedure until the end of this section we take for granted then that, for one energy fixed and the other varying locally (over a domain in which the secular energy variations can be ignored), the matrix elements $\langle W' \mid O \mid W\rangle$ have a zero-centered Gaussian distribution. We assume, moreover, that with respect to *local* energy averaging (defined then by low-order polynomials) different strengths are linearly uncorrelated. Thus

$$\langle W' \mid O \mid W\rangle \text{ distributes as Gaussian } (0, \sigma). \tag{61}$$

$$\overline{R(W_2', W_2)\, R(W_1', W_1)} = \overline{R(W_2', W_2)}\;\overline{R(W_1', W_1)}, \tag{62}$$

where, in (61) σ^2 is the variance and in (62) the strengths are distinct. We now have

$$\overline{\mathscr{M}_p(W)} = \sum_{W'} \overline{R(W', W)}(W')^p, \tag{63}$$

$$\begin{aligned}
\overline{\mathscr{M}_p(W)\,\mathscr{M}_q(W)} &= \sum_{W'W''} (W')^p(W'')^q\, \overline{R(W', W)\, R(W'', W)} \\
&= \Big(\sum_{W' \neq W''} + \sum_{W' = W''}\Big)(W')^p(W'')^q\, \overline{R(W', W)\, R(W'', W)} \\
&= \sum_{W'} (W')^p\, \overline{R(W', W)} \sum_{W''} (W'')^q\, \overline{R(W'', W)} \\
&\quad + \sum_{W'} (W')^{p+q}(\overline{R^2(W', W)} - \{\overline{R(W', W)}\}^2),
\end{aligned} \tag{64}$$

so that

$$\begin{aligned}\hat{\Sigma}^2_{p,q}(W) &= \overline{\mathscr{M}_p(W)\mathscr{M}_q(W)} - \overline{\mathscr{M}_p(W)}\,\overline{\mathscr{M}_q(W)} \\ &= d' \times \int \rho(z)\, z^{p+q}\{\overline{R^2(z, W)} - \{\overline{R(z, W)}\}^2\} \\ &= 2d' \times \int \rho(z)\, z^{p+q}\{\overline{R(z, W)}\}^2\, dz\end{aligned} \tag{65}$$

where, in the second to the last step, we have converted the sum to an integral and, in the last, have used the fourth-moment result from (61) that $\overline{\langle z \mid O \mid W \rangle^4} = 3\{\overline{\langle z \mid W \mid O \rangle^2}\}^2$. (Note that $\overline{R(z, W)}$ is by (61) distributed as $\chi^2(1)$ but we do not need the entire distribution, only the relationship between the centroid and variance.) In (65), $\overline{R(z, W)}$ is simply the low-order polynomial approximation to $R(z, W)$ given therefore by a restricted summation in (11).

Consider first the total strength originating with $\mid W\rangle$. Our interest in particular is in the relative mean-squared fluctuation of the non-energy-weighted sum. We write $\hat{\Sigma}^2_{pp} \equiv \hat{\Sigma}_p{}^2 \equiv (\hat{\Sigma}_p)^2$. Then

$$\begin{aligned}\hat{\Sigma}_0{}^2(W)/(\overline{\mathscr{M}_0(W)})^2 &= \frac{2}{d'}\, \frac{\int \rho(z)(\overline{R(z, W)})^2\, dz}{\{\int \rho(z)\, \overline{R(z, W)}\, dz\}^2} \\ &= \frac{2}{d'}\, [\overline{\Lambda_0(W)}]^{-2} \sum_{\mu}^{\mu_{\max}} [\overline{\Lambda_\mu(W)}]^2 \equiv \frac{2}{\tilde{d}'(W)}\,.\end{aligned} \tag{66}$$

The second form here follows immediately from (15), (65) and the third, in which the μ-sum is restricted to those values used in defining the smoothed strength, by using the last form of (15). In the expansion (66) we have incoherent contributions to the mean-squared deviation, one for each value of μ, indicating a normal-mode decomposition. But, by (15), $\Lambda_\mu(W)$ is the coefficient in the final-state-polynomial decomposition of $d' \times R(W', W)$ for fixed W, and this defines then the normal modes involved. Observe that, for a real separation of secular and fluctuation effects, the results of (66) should be independent of the value of $\mu_{\max}$ provided only that it is large enough so that $\overline{R(z, W)}$ gives the secular behavior properly but small enough so that it takes no account of the level-to-level fluctuations. In the corresponding case of the energy-level fluctuations (for particle number much larger than the particle rank of the interaction) this is in fact well understood; we have at first a very rapid convergence as μ increases and a result then essentially independent of $\mu_{\max}$ until the fluctuations set in. From the results of Section 4 we can expect that the same will apply here.

If the smoothed strength for fixed $\mid W\rangle$ is constant over the final spectrum we can see, from either the second or third form in (66) (in the third only $\mu = 0$ contributes), that the relative mean-squared fluctuation $\to 2/d'$, exactly as expected for then $R(z, W)$ distributes as $\chi^2(d')$, being the sum of d' squares of similar zero-centered Gaussian random variables (even more simply the ratio would be 2 for a single random variable and therefore $\to 2/d'$ for d' of them). Conversely we can use (66), as indicated by its last form, to *define* $\hat{d}'$, the effective number of final states which participate in the excitation (the analog of the number of "open channels" in reaction theory).

For the CLT-limit result we recall that $P_0'(z) = 1$, $P_1'(z) = (z' - \mathscr{E}')/\sigma'$ and then, by (13), $\Lambda_0(W) = \mathscr{M}_0(W)$ and $\Lambda_1(W) = (\hat{\mathscr{E}}(W) - \mathscr{E}')\mathscr{M}_0(W)/\sigma'$ where as always $\mathscr{E}'$, σ' are the centroid and width of the final *state* density (independent therefore of W) while $\hat{\mathscr{E}}(W)$ is the centroid of the *strength* which originates with W. Then

$$\hat{d}'(W) = 2(\overline{\mathscr{M}_0(W)})^2/\hat{\Sigma}_0^2(W) \xrightarrow{\text{CLT}} d'\left\{1 + \frac{(\overline{\hat{\mathscr{E}}(W)} - \mathscr{E}')^2}{(\sigma')^2}\right\}^{-1}. \tag{67}$$

Thus, in the CLT limit, when W is such that the strength and final-state-density centroids coincide, we have $\hat{d}' = d'$, the strength should divide evenly through the final states and the fluctuations are minimal. When the CLT is inadequate (as would happen in huge vector spaces, or even in smaller ones if the excitation strength is localized in energy) we must of course use higher-order polynomials in (66) or have resort to configuration decomposition.

Analogous results of course follow from (65) for the mean-squared fluctuation $\hat{\Sigma}_p^2(W)$ in the power-weighted sums, and similarly for the covariances and the polynomial-weighted sums. For the correlation coefficients we have (with $\hat{\Sigma}_p = \{\hat{\Sigma}_{pp}^2\}^{1/2}$ as above)

$$\zeta_{p,q}(W) = \hat{\Sigma}_{p,q}^2(W)/\hat{\Sigma}_p(W)\,\hat{\Sigma}_q(W) \tag{68}$$

which, as always, lies in the interval $(-1, 1)$. It should be clear from (65) that, if the strength spreads through many states (large $\hat{d}'$), then the fluctuations in the various strength sums are always small. They are closely correlated if the strengths lie in a narrow energy band ($\hat{\sigma} \ll \sigma'$) and essentially uncorrelated in the opposite case. On the other hand, if $\hat{d}'$ is very small the fluctuations are large and correlated.

Concerning the Porter–Thomas distribution we remark finally that it is very well supported by experimental evidence and follows easily from the CLT for sums of random variables if we are willing to *assume* them uncorrelated. More formally we can introduce a Hamiltonian ensemble, such as GOE, and make certain plausible assumptions about the variation, over the ensemble, of a scalar product of the "giant-resonant state" and the states of a general basis. But both of these procedures are rather unsatisfactory, the first because it requires an assumption about matrix-element correlations and the second mainly because the ergodicity question, the equivalence of ensemble and spectral averaging, is ignored. No entirely satisfactory derivation of this general law appears to have been given. It might be possible to produce one by ensemble averaging the polynomial expansion for the strength; we shall look into that later.

9. Conclusion

We have outlined a theory of considerable generality and have indicated several directions in which it can be extended, to inverse energy weighting for example and to make use of geometries defined by $U(N)$ subgroups. In order to assess the accuracy which can be achieved, we give in the next paper a comparison of the predictions with shell-model results for several electromagnetic excitations. After that of course the important thing is to proceed to an analysis of the experimental data; a little of this has already been done for single-nucleon transfer but we have in mind considerable more systematic and extensive studies and for a wide variety of excitations. At the same time we shall extend the theory and give methods for evaluating the traces which it makes use of.

REFERENCES

1. J. P. DRAAYER, J. B. FRENCH, V. POTBHARE, AND S. S. M. WONG, *Phys. Lett. B* **55** (1975), 263, 349; J. P. DRAAYER, J. B. FRENCH, V. POTHBARE, M. PRASAD, AND S. S. M. WONG, *Phys. Lett. B* **57** (1975), 130.
2. F. S. CHANG AND J. B. FRENCH, *Phys. Lett. B* **44** (1973), 131,135.
3. V. POTBHARE AND S. P. PANDYA, *Nucl. Phys. A* **256** (1976), 253.
4. K. F. RATCLIFF, *Phys. Rev. C* **3** (1971), 117; A. GERVOIS, *Nucl. Phys. A* **184** (1972), 507.
5. K. K. MON AND J. B. FRENCH, *Ann. Physics* **95** (1975), 90.
6. C. E. PORTER AND R. G. THOMAS, *Phys. Rev.* **104** (1956), 483; J. M. C. SCOTT, *Phil. Mag.* **45** (1954), 1322; J. E. JYNN, "The Theory of Neutron Resonance Reactions," Oxford Univ. Press (Clarendon), London, 1968.
7. C. E. PORTER (Ed.), "Statistical Theories of Spectra: Fluctuations," Academic Press, New York, 1965; M. L. MEHTA, "Random Matrices and the Statistical Theory of Energy Levels," Academic Press, New York, 1967; E. P. WIGNER, *SIAM Rev.* **9** (1967), 1.
8. J. FLORES AND P. A. MELLO, *Rev. Mexicana Fis.* **22** (1973), 185; T. A. BRODY, E. COTA, J. FLORES, AND P. A. MELLO, *Nucl. Phys. A* **259** (1976), 87.
9. M. W. KIRSON, *in* "Effective Interactions and Operators in Nuclei" (B. R. Barrett, Ed.), Springer-Verlag, Berlin, 1975.
10. H. CRAMER, "Mathematical Methods of Statistics," Princeton, 1946; G. SZEGO, "Orthogonal Polynomials," American Mathematical Society Colloquium Publications, Vol. XXIII, 1939.
11. E. P. WIGNER, *Ann. of Math.* **62** (1955), 548.
12. T. T. S. KUO, *Nucl. Phys. A* **103** (1967), 71.
13. M. ABRAMOWITZ AND I. A. STEGUN (Eds.), "Handbook of Mathematical Functions," NBS Applied Mathematics Series, 55, U.S. Government Printing Office, Washington, 1964.
14. M. NOMURA, *Prog. Theoret. Phys.* **48** (1972), 442.
15. F. S. CHANG, J. B. FRENCH, AND T. H. THIO, *Ann. Physics* **66** (1971), 137.
16. J. B. FRENCH, *in* "Proceedings of the International School of Physics, Enrico Fermi, Course 36" (C. Bloch, Ed.), Academic Press, New York, 1966.

SYMMETRIES AND STATISTICAL BEHAVIOR IN FERMION SYSTEMS[+]

J. B. French[†]

Department of Physics and Astronomy, University of Rochester
Rochester, New York 14627

and

J. P. Draayer

Department of Physics and Astronomy, Louisiana State University
Baton Rouge, Louisiana 70803

1. Introduction

The subject is the interplay between statistical behavior and symmetries in nuclei, as revealed for example by spectra and by distributions for various kinds of excitations. Much of the argument and general procedure may be applied also to other many-fermion spaces which have (as do nuclear shell-model spaces) a *direct-product* structure, being representable by distributing a certain number, m, of fermions over N single-particle states ($0\leq m\leq N$). I shall talk about methods and general results because that seems appropriate here, rather than about specific applications, of which however there have been many.

Most people think of statistical behavior as being a relevant way to describe things only at high excitation, whereas the natural domain of symmetries, with the obvious exception of those which are exactly (or almost exactly) conserved[*], is near the nuclear ground state. But even rather badly broken symmetries are useful in roughly partitioning an energy spectrum and play thereby an important role in determining locally averaged strengths, etc; thus symmetry effects extend *upward* in energy away from the ground-state region. On the other hand, if we introduce Hamiltonian ensembles to describe the statistical behavior (which is necessary for fluctuations, though not for *locally averaged* eigenvalue densities, etc.), it turns out that appropriate ensembles gives results which are *ergodic* and, when

[+]Supported in part by the U.S. Department of Energy and National Science Foundation.

[†]John Simon Guggenheim Memorial Fellow, 1977-78

[*]We speak of a symmetry as conserved if each wave function belong to a single irreducible representation

expressed in an energy scale supplied by the local eigenvalue density, are *stationary* in energy.[1] Thus statistical behavior extends *downward* into the ground-state domain so that the interplay which we discuss exists in all parts of the spectrum.

As I have indicated, statistical behavior can be discussed in terms of secular averages and of fluctuations about the average.* A finer decomposition is in terms of the sequence of ℓ-point correlation functions ($\ell = 1,2,\ldots$), of which $\ell = 1$ gives the density, while $\ell = 2$ dominates all of the measures which have been introduced (and perhaps all which can be experimentally measured). There are interesting connections between fluctuations and symmetries, for, while fluctuations are "small" and carry little information, the little which they do carry is closely connected with the existence or non-existence of a conserved or almost conserved symmetry; this is because the existence of such a symmetry moderates the "level repulsion" which dominates the fluctuation behavior. Methods for studying these connections have only recently been developed. I shall therefore say nothing more about fluctuations but focus instead on the (secular) behavior of locally averaged quantities.

2. Elementary Principles

The m-fermion wave functions for fixed m (=0,1,...,N) form an irreducible (anti-symmetric) tensor, of dimensionality $d(m) = \binom{N}{m}$, with respect to the group U(N) of unitary transformations in the single-fermion space. To each state $\Psi_\alpha(m)$ there exists a *complementary* one $\Psi_{\alpha_c}(N-m)$ which forms a contragradient tensor. The system has two vacuum states $|0\rangle$ and $|N\rangle$ and any state $\Psi_\alpha(m)$ can be described in terms of either:

$$\Psi_\alpha(m) = \psi_\alpha(m)|0\rangle = \tilde{\psi}_{\alpha_c}(N-m)|N\rangle \tag{1}$$

in which the operation (~) replaces every creation and destruction operator by its adjoint so that $\tilde{\psi}_\alpha(m) = \psi_\alpha(m: a_i^+ \rightleftarrows a_i)$. Since holes are also fermions (the fundamental commutation rules being invariant under (~)) we have that for any real

*In nuclear model systems, at least, there is a remarkable (well-understood) separation between the two phenomena which allows us to discuss them separately

operator, F

$$[\text{Trace }(F)\text{ in m-particle space}] \equiv \langle\langle F\rangle\rangle^m \equiv d(m)\langle F\rangle^m = \langle\langle \tilde{F}\rangle\rangle^{N-m} \qquad (2)$$

where, as indicated, $\langle\langle F\rangle\rangle^m$ is the trace and $\langle F\rangle^m$ the average eigenvalue (or expectation value).

A k-body operator, $F(k) = \Sigma\langle k\beta'|F(k)|k\beta\rangle\psi_{\beta'}(k)\psi^+_\beta(k)$, transforms according to (k+1) irreps of U(N); for example, with n the number operator,

$$F(2) \sim \binom{n}{2}\left([1^N]\right) + \binom{n-1}{1}\left([21^{N-2}]\right) + \left([2^2 1^{N-4}]\right) = \binom{n}{2}F(0) + \binom{n-1}{1}F(1) + F(2) = \sum_{\nu=0}^{2} F^{(\nu)} \qquad (3)$$

where $\binom{n}{2}F(0)$ measures the spectrally-averaged eigenvalue (which then "propagates" with particle number as $\binom{m}{2}$), $\binom{n-1}{1}F(1)$ (in which F(1) is an SU(N) generator) gives the modification of the single-particle energies, and F(2) may be regarded, in the context of the U(N) decomposition, as the "true" residual interaction.

There is, for each particle number m, a natural geometry for traceless operators in which

$$(\text{norm } F)^2 = d^{-1}(m)\langle\langle F^+F\rangle\rangle^m = \langle F^+F\rangle^m \qquad (4)$$

The unitary decomposition of F is orthogonal in this geometry (true also for operators which do not conserve particle number). With this geometry we can consider such questions as "how much Q·Q is contained in a given H, acting in an m-particle space?" (answer: a large amount; the scalar product between the unit vectors for $H-\langle H\rangle^m$ and Q·Q (in statistical terms the "correlation coefficient" between H and Q·Q) is negative and, for light nuclei at least, ~ -0.5). Note that Q·Q is a sum of squares of traceless 1-body operators; it appears that realistic H's are rather well representable in this way.

There are central-limit theorems (CLT's) which ensure that, if N is large enough the smoothed spectrum converges, as we increase the particle number m, to a characteristic density describable by a few parameters. For example with non-interacting particles (H=H(k=1)) the single-particle spectrum convolutes and rapidly becomes Gaussian as we increase the particle number. The real function of the CLT is to make the unitary geometry an *effective* one. There are two aspects of this; the unitary norm for an operator might well be criticized on physical grounds because it treats all states on the same footing and therefore concerns itself mostly with states which lie high in energy (where the density is highest), even though most of these are not of direct physical significance and our interest might well be in the lowest ones. This difficulty disappears when the energy variation of the eigenvalue density is closely specified by a CLT, for what happens then at all energies is determined by the trace of H and the norms of low powers of H. At the same time we avoid the hopeless technical problem of evaluating high-order moments.

3. A Simple Example: Single-state Occupancies

The Strutinsky explanation of shell effects in fission may be taken as demonstrating the "survival" of single-particle information in the presence of strong interactions. As a simpler example of this let us ask whether the ground-state sum rule for the single-nucleon-transfer strength (which measures single-particle information) can tell us anything about the effective interaction.

The sum rule for transfer in an orbit j determines the ground-state occupancy $\langle W_g|n_j|W_g\rangle$. In general, for any H-eigenstate $|W\rangle$ of eigenenergy W, we have, with $\rho(W)$ the eigenvalue density function, $P_\nu(W)$ the orthonormal polynomials defined with $\rho(W)$ as weight function, and $\langle\rangle^m$ the m-particle (or model-space) average with $E\equiv E(m) = \langle H\rangle^m$ the eigenvalue centroid, that

$$\langle W|K|W\rangle = \rho^{-1}(W)\langle\delta(H-W)\rangle^m = \sum_\nu \langle K\ P_\nu(H)\rangle^m P_\nu(W) = \langle K\rangle^m + \frac{\langle K(H-E)\rangle^m}{\langle(H-E)^2\rangle^m}(W-E) + \ldots \qquad (5)$$

It is easy to see that the ν-th order term would arise from the ν'th order deformations of $\rho(W)$ under the infinitesimal change of the Hamiltonian, $H \to H + \delta\alpha\times K$. But then, in the CLT limit, the terms of order larger than 1 must vanish (except for very high order, involving "wavelengths" comparable with the level spacing, for which the CLT smoothing is ineffective). Thus we have a linear form for the smoothed $\langle W|K|W\rangle$, the slope of which is proportional to the scalar product $\langle K(H-E)\rangle^m$. For any one-body operator, $K = n_j$ for example, only the $[2,1^{N-2}]$ part of H contributes (by orthogonality); the part of this deriving from the two-body H measures the Hartree-Fock-like contribution to the effective single-particle energy, proportional to $\{\sum_s W_{sjsj} - N^{-1}\sum_{sj} W_{sjsj}\}$. Observe how the CLT in (5) and the unitary geometry here have combined to modify a single-particle property of a complicated system. The modification has been measured in the ds shell, and used in determining which model interactions are acceptable.[2] Occupancies can of course be used in many other ways also.

4. Subgroup Structures, Spectra and Characteristic Forms

If, in the example above, we focus our attention on an N_1-dimensional subset of the single-particle states we are led naturally to think about the $U(N_1)$ subgroup of U(N), and more generally about "additive" decompositions of the U(N) algebra. If we consider neutron and proton transfer separately we are led to $U(N/2) \otimes U(2)$ and a direct-product decomposition. A great variety of other subgroups are encountered in other circumstances and so we must consider the subgroup problem more generally.

A given representation of a subgroup G, as with $SU_2(T)$ or $O_3(J)$, may occur very many times in a large space. We then naturally ask what representations of G are found and how often. This question is answered to some extent by considering the spectrum, in the m-particle space, of the bilinear Casimir operator C_2, which is, of course discrete with degeneracies equal to the number of times that the appropriate* representation occurs, multiplied by its dimensionality.

For any U(N) subgroup, C_2 may be written as a sum of squares of (non-commuting) traceless one-body operators (SU(N) generators) which have the same norm in all m-particle spaces, and which may be taken as orthogonal in the 1-particle space (and hence via (3) in m-particle spaces). Thus $C_2 = \sum_{\alpha=1}^{\ell} h_\alpha^2$ with $\langle h_\alpha\rangle^1 = 0$ and $\langle h_\alpha h_\beta\rangle^1 = \sigma^2\delta_{\alpha\beta}$. Let us consider the moments of C_2 in the limit of a large dilute multi-particle space ($N\to\infty$, $m\to\infty$, $\frac{m}{N}\to 0$). For a traceless h we can decompose $\langle h^{2\nu}\rangle^m$ in terms of correlation patterns defined by the partition of 2ν into integers ≥ 2; then for example the partition $[2^\nu]$ gives a term proportional to $\{\langle h^2\rangle^m\}^\nu = \{m(N-m)(N-1)^{-1}\sigma^2\}^\nu$. From the m-dependence it is easy to see that only binary correlations survive in the asymptotic limit.[3] Extending this to the sum of squares and observing that no cross-terms contribute we see that, in the present limit, C_2 is equivalent to a sum

*C_2 does not by itself distinguish all irreducible representations. We may say that it classes together all "C_2-equivalent" irreps.

of *commuting* squares so that its distribution is an ℓ-fold convolution; in other words only the dimensionality, ℓ, of the space of generators matters, nothing further about the nature of the group or about the way in which it is realized in the space. Thus the asymptotic spectrum of C_2 has a χ^2_ℓ form,

$$\rho_{C_2}(x) = \{2^{\ell/2}\sigma^{\ell-2}\Gamma(\ell/2)\}^{-1}\; x^{\ell/2-1}\; \exp(-x/2\sigma^2) \qquad (6)$$

in which, as above, σ^2 is the one-particle variance of a generator. An SU(3) example (ℓ=8) is shown in Fig. 1. For ℓ=3 (and thus for angular momentum) the distribution is Maxwellian, a result given in 1936 by Bethe[4] for non-interacting particles and used ever since, with minor corrections, for the J-dependence of the level density. For a quadrupole interaction (ℓ=5) Nomura[5] evaluated the moments to order four, and stressed that the asymptotic spectrum is non-Gaussian. From the argument above we see that the distribution is χ^2_5.

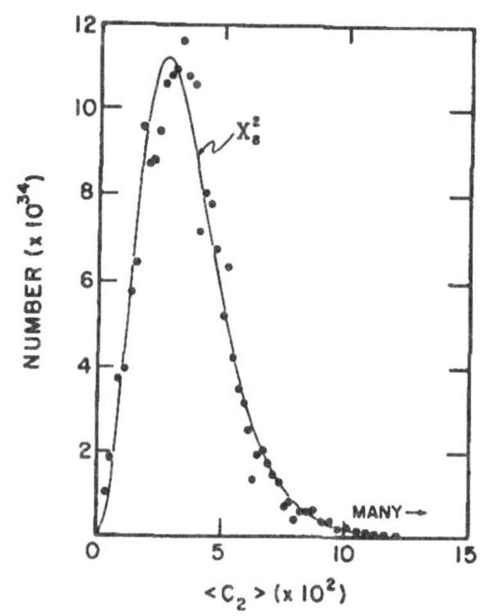

Fig. 1 The spectrum of C_2 for SU(3) realized as a direct-product factor of U(N=300). The particle number is 25. A running 5-point average has been used to reduce fluctuations. The smooth curve is χ^2_8.

This distribution (6) can be extended to finite particle number by explicit evaluation of the low-order moments (and, much less easily, to finite N as well). In general terms we can expect that an approximate χ^2_ℓ distribution will obtain as long as we have a half-dozen particles or more. For large ℓ, by the standard CLT, the χ^2_ℓ distribution is close to Gaussian. Since, as we have said, the Hamiltonian itself is often reasonably representable as a sum of squares we see that *its* asymptotic spectrum likewise should be approximately χ^2_ℓ (ℓ being now the number of squares of comparable magnitude) and in many cases then close to Gaussian, as is observed in detailed calculations.

Spectral distributions, as used in practice, involve the distribution in energy of states of a given symmetry, the distribution being a Hamiltonian-eigenvalue distribution if the symmetry is a good one, but otherwise an "intensity" distribution (giving for each Hamiltonian eigenvalue, the contribution to $<\Psi^+\Psi> = 1$ which arises from basis states of the specified symmetry). The moments of such a distribution follow from the simplest moments, $<H^p>^m$, by the model-space decomposition,* $\underset{\sim}{m} \to \sum_{\Gamma}(\underset{\sim}{m},\Gamma)$ where Γ defines the symmetry irrep; we come to that in Section 5. For now observe that if the C_2-H correlation coefficient, ζ, is small it follows from (5) that a given G irrep will distribute more or less uniformly over the whole energy spectrum, the decomposition (6) being then locally valid (this goes far towards justifying the use of Bethe's J-decomposition in level density studies). At the other extreme ($|\zeta|\sim 1$), the H spectrum will be close to that of C_2. In many cases the Hamiltonian centroids are determined by a single group invariant, C_2; in these cases, if we start with $H=C_2$ (to within a sign) and allow it to move away from C_2 ($|\zeta|$ going gradually to 0) the overall χ^2_ℓ distribution will be maintained for at least part of the range of ζ) even while the degenerate C_2 "spikes" are broadened by the movement of the individual levels and their admixing with levels of other symmetries. The line shape (which one would guess should be Gaussian, except perhaps near the extremes of the spectrum) and more generally the energy distribution of the fixed-symmetry states can be studied by the polynomial methods referred to in Section 3.

5. Propagation of Information

The asymptotic distribution (in energy) for a fixed symmetry is defined by three quantities only, the dimensionality, centroid and variance. For distributions which are not asymptotic the same characteristic form will obtain, but with m^{-1} and m/N corrections for which we need a few more pieces of information. Since the Hamiltonian and the other interesting operators of the system are *defined* in few-particle subspaces "embedded" in the m-particle space the general problem of understanding what goes on in a many-particle subspace may be thought of in terms of propagation of information from the simple subspaces to the more complicated ones. Alternatively if we measure things in a many-particle space we are "probing" the

*We use $\underset{\sim}{m}$ to denote the m-particle space.

structure of the physical few-particle spaces. Because of the filtering action of the CLT much of the information is lost* and conversely the many-particle probe is sensitive to only a few properties of the system.

Intuitively we might guess for a k-body operator, averaged over m, that

$$\langle F(k)\rangle^{m} = \langle F(k)\rangle^{k} \ \{\text{weight with which k is found in m}\} \tag{7}$$

in which the natural definition of the weight is

$$\langle \sum_{\alpha} \psi_{\alpha}(k)\psi_{\alpha}^{+}(k)\rangle^{m} = \langle \rho(k)\rangle^{m} = \langle \binom{n}{k}\rangle^{m} = \binom{m}{k} \tag{8}$$

This result is easily seen to be correct; schematically (with particle number increasing upward so that $\psi_{\beta'}(k) \equiv \beta'\uparrow$ and $\psi_{\beta'}(k)\psi_{\beta}^{+}(k) \equiv \beta'\vee\beta$) $F(k) = \Sigma\, F_{\beta'\beta}(\beta'\vee\beta)$ and then

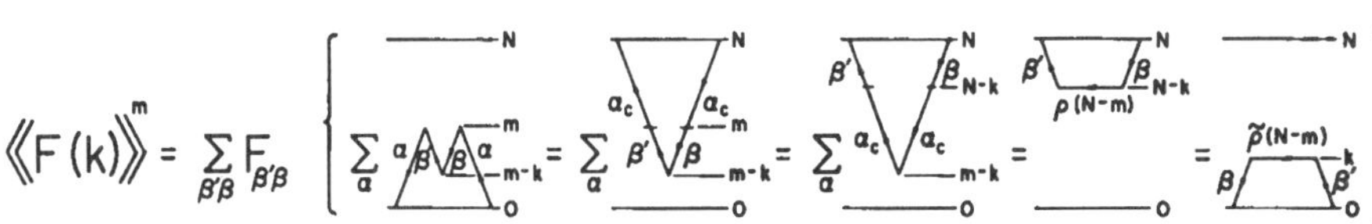

(9)

with ρ the eigenvalue density operator defined in (8). In going from the first to the second forms we have used the contragradient representation of the m-particle states, this followed by a commutation of state operators (which would not be valid for the first form) and a use of (2). Note that $\rho_{c}(N-m) = \rho(N-m)$; ρ is of course diagonal in β in the present case but not always so for its symmetry extensions.

Carrying out the sums we have

$$\langle\langle F(k)\rangle\rangle^{m} = \langle\langle \tilde{\rho}(N-m)F(k)\rangle\rangle^{k} = \langle\langle \rho(N-m)\tilde{F}(k)\rangle\rangle^{N-k} \tag{10}$$

which, with (8), reproduces the elementary result (7). The intuitive interpretation of (10) follows from the fact that $\tilde{\rho}(N-m)$ measures the number of holes in the complementary space $(\underset{\sim}{N}-\underset{\sim}{m})$. It is easy enough to see that (10) extends to any subspace Γ of $\underset{\sim}{m}$ (even to single states) so that, with $\langle\langle F\rangle\rangle^{m\Gamma}$ the trace over all m-

*The CLT does not smooth out the fluctuations but the "lost information is not to be found there. Fluctuations, as we have said, carry almost no information.

particle states belonging to all irreducible representations Γ of a group G, a subgroup of U(N), (involving therefore a sum over *all* irreps Γ)

$$\langle\langle F(k)\rangle\rangle^{m\Gamma} = \langle\langle \tilde{\rho}_{\Gamma_c}(N-m)F(k)\rangle\rangle^{k} = \sum_{\Gamma'} \langle\langle \tilde{\rho}_{\Gamma_c}(N-m)F(k)\rangle\rangle^{k\Gamma'}$$

$$\xrightarrow{(s)} \sum_{\Gamma'} \langle\langle \tilde{\rho}_{\Gamma_c}(N-m)\rangle\rangle^{k\Gamma'}\langle F(k)\rangle^{k\Gamma'} = \sum_{\Gamma'} \langle\langle \rho_{\Gamma'}(k)\rangle\rangle^{m\Gamma}\langle F(k)\rangle^{k\Gamma'} \qquad (11)$$

the last two forms of which (the second of these following from the first via the same operations used in (9)) are valid when $\rho_{\Gamma_c}(N-m)$ behaves as a multiple of unity in each subspace, in which case we have simple (s) propagation, from the space in which F is defined, to the more complicated spaces of interest to us. This form displays beautifully how the trace information propagates throughout all the subspaces, the propagation coefficient being dependent only on the pair of representations involved, in fact on the weight with which one representation space is found in the other. The nature of simple propagation (in this case for "non-scalar" information) will become clearer on inspection of Fig. 2, for which the density operators are also available.[6]

By construction the operators $\rho_\Gamma(k)$ are k-body G scalars which satisfy $\rho_\Gamma(k)\psi_\alpha^{\Gamma'}(k) = \delta_{\Gamma\Gamma'}\psi_\alpha^{\Gamma'}(k)$. These properties completely define the operators and show that the set of them for fixed k forms a Green's Function, on the m=k subspace, by means of which we can calculate traces away from the defining "surface", i.e. for m>k. We can obviously construct one operator for each set of G irreps Γ. But there are also off-diagonal G scalars that can be constructed by coupling states from two equivalent irreps. Suppose that G is a subgroup of a factor U(k) where U(N) ⊃ U(N/k) ⊗ U(k) and we specify also the

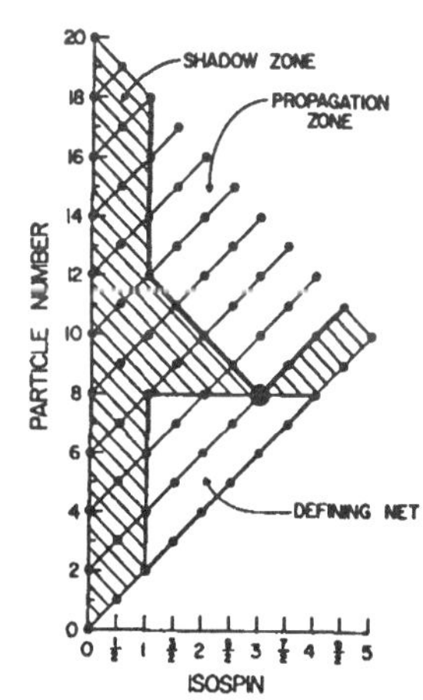

Fig. 2 Propagation on the isospin lattice for an operator with isospin-rank 2 and maximum particle rank 8. The triangle shows the effective defining lattice and the cross-hatching the "shadow zone" into which no propagation occurs for the part of the operator defined in the (m=8,T=3) subspace.

(single) irrep Λ of U(k); if now in the reduction $\Lambda\to\Gamma$ there is a multiplicity κ we have κ diagonal G scalars and $\frac{1}{2}\kappa(\kappa-1)$ off-diagonal ones, the total being $\frac{1}{2}\kappa(\kappa+1)$. Similarly if G is the last member of a chain, as in $U(N)\supset K\supset G$ and we also specify a single irrep Λ of κ. In these circumstances for each (Λ,Γ) pair there is a matrix of density operators which should enter into the propagation of information; we shall see that they do.

For U(N) itself (G=U(N)), Γ is of course unique and $\rho_\Gamma(k)=\binom{n}{k}$ as we have seen. For G=U(k) a factor of $U(N)\supset U(N/k)\otimes U(k)$ (and similarly for an additive decomposition) Γ is not unique but one sees easily that ρ_Γ is a function only of the generators of G and hence of its Casimir invariants. The total number of G irreps for $k\leq m$ particles then equals the number of terms in a polynomial, of maximum particle rank m, involving the U(N) invariant, n, and the G Casimir operators, this being true as long as N is large enough so that all the irreps can in fact be generated. In this case, which includes the nuclear physicist's isospin and spin-isospin SU(4), the propagation is simple in the sense of (11). We should mention here that the first systematic discussion in group theoretical terms of simple propagation and some of its extensions is given in an important paper by Quesne.[7]

Going beyond simple propagation, consider[8,9] $U(3)\supset O(3)$. The density operators are O(3) scalars constructed from the generators of U(3). By a simple counting of polynomial terms and representations one finds immediately that the polynomials are not constructible from the Casimir invariants of U(3) and O(3) alone; one new invariant is needed* for m=3 and one for m=4. We make contact here with the cataloging of invariants by means of the "integrity basis", that basis in the present case consisting of n, $C_2(O_3)$, $C_2(U(3))$, $C_3(U(3))$, X_3, X_4. As a demonstration that there are no higher ones we have for m=50 (and N large enough to generate the irreps) a polynomial of 35,993 terms and the same number of density operators (including the $(\kappa(\kappa-1)/2)$ off-diagonal ones which enter when there is a κ-fold U(3)-O(3) multiplicity). It should be noted that X_4 is redundant for the state-

*The first four operators give 5 polynomial terms for $m\leq 2$ and 9 for $m\leq 3$; there are 5 irreps $(\lambda\mu;L)$ for $m\leq 2$ and 10 for $m\leq 3$. In constructing the polynomials all of the scalars including the new ones can be taken as effectively commuting since commutation reduces the particle rank by unity giving scalars which have already been counted.

labeling problem, and thus the propagation of information, for which X_4 is necessary, makes more demands on the group structure than does the simpler problem.

For the decomposition[10] $U(N) \supset U(N/4) \otimes U(4)$ with $U(4) \supset SU(2) \otimes SU(2)$, which is of considerable interest to nuclear physicists, we find as the integrity basis (by counting the $U(4) \supset SU(2) \otimes SU(2)$ irreps) n, G_2, G_3, G_4, S^2, T^2 and non-Casimir invariants X_3, X_4, Y_4, Z_4, X_5, X_6, Y_6, explicit forms for which have in fact been given by Miller.[11] But here the integrity-basis counting gives numbers which grow slightly more quickly than the number of density operators, for m=7,9,25,50 the numbers are (73,201,57125,11766281) for the former, and (73,200,48653,6541506) for the latter. The source of the difference cannot be that we have too many basic invariants; it must therefore be that not all products and powers of the basic invariants are independent. The redundancies here could be resolved once again by counting irreps. Alternatively these redundancies can be anticipated from generating-function forms[9,12] for the number of scalars of given polynomial order. In this case we find that Z_4 ($=S \cdot E^2 \cdot T$ where E is the Gamow-Teller operator) and X_5 ($=S \cdot E^3 \cdot T$) can only occur linearly while no products of X_6 ($=(S \times S)^2 \cdot (E^4)^2$) and Y_6($=(E^4)^2 \cdot (T \times T)^2$) can occur at all. With these restrictions the polynomial counting precisely reproduces the irrep counting above.

As another example we mention Elliott's SU(3) which appears in the decomposition $U(N) \supset U(4) \otimes U(N/4)$; $U(N/4) \supset SU(3)$. Here one finds easily that when N=12,24 the invariants up to m=2 have a Casimir basis so that SU(3) centroids propagate simply in the p and ds-shell spaces; not so however for N=40 (fp shell), for which another invariant of particle rank 2 is needed.

We shall not discuss the explicit construction of the density operators for the non-Casimir case. We remark however that, in general the final forms of the propagation equation (11) will not apply, while, for the other forms we might have to resort to an explicit evaluation for each irrep (not just for the sum over all equivalent ones).

Finally we remark that we should be interested also in "non-scalar" information, as in Fig. 2. As a trivial example the quadrupole moment operator Q_0 has a zero average expectation value in the substates of a J-level but nonetheless carries information (about the intrinsic quadrupole moment) via the "double-barred" or tensor-coupled matrix element. Then there enter naturally the non-scalar density operators $\rho_\Gamma^\nu(m) \sim \sum_{\alpha\varepsilon\Gamma} (\psi_\alpha^\Gamma(m) \times \bar{\psi}_\alpha^\Gamma(m))^\nu$ (in which $(\times)$ denotes tensor coupling and $\bar{\psi}^\Gamma$ the tensor adjoint) and the corresponding extension of the integrity basis. This is particularly important for strength distributions.[12]

6. Information Propagation in Dilute Spaces

Looking at things from the standpoint of many-body physics we might well ask for the source of the great complications which show up in constructing the integrity basis. We recall that one type of unpleasantness arises in many-body-theory from the Pauli Principle which forbids two fermions to be in the same state. The (blocking) corrections involved disappear in dilute systems, those in which the single-state occupancies are much less than one. Thus $m \ll N$; the more detailed requirement arises from the need to avoid "condensations" in subsets of single-particle states whose occupancies then would be comparable with unity. There's no trouble calculating the occupancies (just go back to section 3) so that whether or not a system is dilute can in fact be determined.

If we first use completeness to write $\langle\langle \psi_\alpha(k)\psi_\beta^+(k)\rangle\rangle^m = \langle\langle \psi_\beta^+(k)\psi_\alpha(k)\rangle\rangle^{m-k}$, and then use (10) we find different propagation forms,

$$\begin{aligned}\langle\langle F(k)\rangle\rangle^m = \langle\langle \tilde{F}(k)\rangle\rangle^{m-k} &= \langle\langle F(k)\rangle\rangle^{N-m+k} = \langle\langle \tilde{\rho}(m-k)F(k)\rangle\rangle^k \\ &\downarrow \\ \sum_{p=0}^{k} \langle\langle (\tilde{F}(k))_p\rangle\rangle^{m-k} &= \sum_{p=0}^{k} \langle\langle \tilde{\rho}(m-k-p)(\tilde{F}(k))_p\rangle\rangle^p \end{aligned} \tag{12}$$

In the second branch of this equation $(\tilde{F}(k))_p$ is the p-body part of the k-hole operator $\tilde{F}(k)$ which of course has particle ranks $0,1,\ldots k$. The p-decomposition given here is important for dilute systems ($N\to\infty$, $m/N\to 0$) because in such systems $\langle\langle \tilde{F}(k)\rangle\rangle^{m-k}$ is dominated by its $p=0$ part; this comes about because the system has a great many holes, which are essentially counted by $\tilde{F}(k)$, but very few particles to inhibit the counting. The p'th term in fact is proportional to $(m/N)^p$, as one sees by noting that for the prototype operator $\binom{n}{k}$ we have* $\widetilde{\binom{n}{k}}_p = (-1)^p \binom{N-p}{k-p}\binom{n}{p}$.

Equation (12) is trivial as it stands; we need its decomposition according to a group G, producing thereby the equivalent of (11). But things are quite different now because any decomposition $\underset{\sim}{m} \to \sum_{\Gamma}(\underset{\sim}{m},\Gamma)$ will, for most operators, destroy the completeness used in the first step of (12); we get around this**, just as one does with multipole sum rules, by making use of the non-scalar density operators $\rho_{\Gamma}^{\nu}(m)$ introduced above and using recoupling algebra to isolate single intermediate Γ spaces. We end up then with a tensorial extension of (12) which we can use for propagating scalar or non-scalar information, and which is particularly appropriate for dilute systems.

In p=0 approximation we find for a k-body operator, $F^0(k{:}\lambda,\lambda)$, whose tensorial structure is $(\psi^{\lambda}(k)\times\bar{\psi}^{\lambda}(k))^0$, and which therefore is defined in the (k,λ) subspace, that

$$\langle\langle F^0\rangle\rangle^{m\Gamma} = d(\Gamma)d^{-1}(\lambda)\ \langle\langle F^0\rangle\rangle^{k} \sum_{\Gamma_1} d_{\ell}(m-k,\Gamma_1)\delta_{\Gamma\lambda\Gamma_1} \tag{13}$$

where d_{ℓ} is a multiplicity, $d(\Gamma)$ a dimensionality, and $\delta_{\Gamma\lambda\Gamma_1}$ the triangle function. For the coupled trace of a non-scalar operator we would find not a sum over multiplicities but instead a Racah transform of the multiplicity function. The structure revealed here is as simple as could be; F^0 is measured in terms of its defining trace and the contribution to any $(\underset{\sim}{m},\Gamma)$ irrep is simply proportional to the number of parents which $(\underset{\sim}{m},\Gamma)$ has in the space $(\underset{\sim}{m}-\underset{\sim}{k})$. By taking F^0 as the (k,λ) density operator we evaluate the propagation coefficient precisely in terms of the weight of one space in the other. The complexities of propagation then disappear in the dilute-system limit. They gradually re-enter of course as we take into account terms of higher p.

*For scalar traces $F(k)\equiv\langle F(k)\rangle^k\binom{n}{k}$ so that the result is general. Note that the higher-p terms in (12) can also be expanded similarly, so that we have an expansion in (m/N). The error made in truncating at any stage is of course calculable in this simple case, but is significant because it can be applied to group decompositions as well.

**In defining ρ^{ν} above we have assumed that the tensor coupling is multiplicity free, but this restriction is in fact not necessary.

7. Conclusion

Our subject has been spectral distribution and group symmetries though we have been able to refer to only a few papers in each domain. We have stressed that group theory constructions enter in a particularly elegant way when many-body problems are treated via spectral distributions; and by the same token, many-body results shed some new light on problems in group theory.

References

1) A. Pandey, Ann. Phys. (N.Y.), in press.
2) V. Potbbare and S. P. Pandya, Nucl. Phys. A256 (1976) 253.
3) K. K. Mon and J. B. French, Ann. Phys. (N.Y.) 95 (1975) 90.
4) H. A. Bethe, Phys. Rev. 50 (1936) 332; J. G. Cleary and B. G. Wybourne, J. Math. Phys. 12 (1971) 45; T. R. Halemane, unpublished.
5) M. Nomura, Progr. Theoret. Phys. 48 (1972) 442.
6) J. B. French, in Isospin in Nuclear Physics, D. H. Wilkinson editor, (North-Holland, Amsterdam, 1969).
7) C. Quesne, J. Math. Phys. 16 (1975) 2427.
8) J. P. Elliott, Proc. Roy. Soc. A245 (1958) 128, 562; V. Bargmann and M. Moshinsky, Nucl. Phys. 18 (1960) 697, 23 (1961) 177; G. Racah, in Group Theoretical Concepts and Methods in Elementary Particle Physics, F. Gürsey editor (Gordon and Breach, New York, 1964).
9) B. R. Judd, W. Miller, Jr., J. Patera and P. Winternitz, J. Math. Phys. 15, (1974) 1787.
10) R. U. Haq and J. C. Parikh, Nucl. Phys. A220 (1974) 349; C. Quesne, J. Math. Phys. 17 (1976) 1452.
11) W. Miller, Jr., as quoted in Ref. 11.
12) R. Gaskell, A. Peccia and R. T. Sharp, J. Math. Phys. 19 (1978) 727.
13) J. P. Draayer, J. B. French and S. S. M. Wong, Ann. Phys. (N.Y.) 106 (1977) 472, 503.

SPECIAL TOPICS IN SPECTRAL DISTRIBUTIONS

J. B. French

Department of Physics and Astronomy
University of Rochester
Rochester, N.Y. 14627

ABSTRACT

We discuss two problems which relate to the foundations of the subject, and a third about asymptotic properties of spectral distributions. We give also a brief list of topics which should be further explored.

I. FLUCTUATIONS

In general, events distinguished by a great disparity in size have little influence on one another; they do not communicate and so the phenomena associated with each scale can be treated independently (K. G. Wilson[1]).

Let us agree that by fluctuations we mean phenomena which are ignored when we consider only the smoothed eigenvalue density, $\rho_{smoothed}(E)$ of* (I-1). A basic question is whether $\rho_{sm}(E)$ really converges rapidly to the true density as we take account of higher and higher moment corrections. Experience, supported by the CLT, might appear to guarantee that it does, but in fact, as Wilson's principle would suggest, it does not. We must understand something about this, and about the magnitude and nature of the fluctuations, in order to estimate the errors implicit in spectral distribution calculations, and to learn about the information content of the fluctuations. If they should have no content then the errors made in ignoring them would be irrelevant, and higher accuracy than that yielded by $\rho_{sm}(E)$ beside the point.

*We refer to the first talk at this Conference as I.

The subject of eigenvalue fluctuations is large and complicated. The first theoretical papers, by Wigner[2] and Landau and Smorodinski[3], made use of the level-repulsion arguments of von Neumann and Wigner[4], and found agreement with the early experimental results. A large theoretical development, including a definitive series of papers by Dyson and Mehta[5], was matched by high quality slow-neutron-resonance data[6] involving heavy targets, and proton data involving lighter ones[7]. The essential point of the theory[8] is that one gives up the idea of *spectral* (or *energy*) averaging for a specific Hamiltonian, replacing it by averaging over an ensemble of H's, this being justified by invoking an aspect of ergodic behavior[9].

There are three "canonical" ensembles, introduced to take account of differing behaviors of the system under rotations and time reversal, while assigning equal *a priori* probabilities to H's which are compatible with the specified behavior. Invariance of the canonical ensembles under the group of unitary transformations, or an orthogonal or symplectic subgroup, makes it possible to exhibit explicitly the joint probability distribution of the eigenvalues A very difficult integration over all but k of the eigenvalues produces then, with k=1, the *ensemble-averaged* density $\overline{\rho}(x)$; the two point correlation function

$$S^{\rho}(x,y) = \overline{\rho(x)\rho(y)} - \overline{\rho}(x)\,\overline{\rho}(y) \qquad (1)$$

which determines almost all the observable fluctuation measures derives from K = 2.

Our present interest is with only one of the canonical ensembles, the Gaussian orthogonal ensemble (GOE) of large real-symmetric matrices with distinct elements chosen independently in accordance with a Gaussian law. It is not obvious, though it turns out to be true, that the GOE should supply a good model for the fluctuations; for choosing the matrix elements independently has the inevitable consequence that the Hamiltonians generated involve only many-body interactions - between all the particles of the system. For a better *a priori* model we can choose an ensemble of k-body interactions (k = 2 is an interesting case) by generating a GOE in the k-particle space and using it in the space of m particles[10]. For most purposes the resulting *embedded* GOE (or EGOE) is very difficult to deal with but, by good luck, we can use it to study the questions we have posed and the answers are different from, and much more enlightening than, those which would come from GOE.

From the viewpoint of this conference it is interesting that, with moment methods, we can give an alternative, approximate but surprisingly accurate, evaluation of the two-point function for the canonical ensembles and for various modifications and extensions of them. We invoke the same assumption used in deriving the density

(I, section 3), that binary correlations between H's are dominant in the moments. Then[11]

$$\int S^{\rho}(x,y)x^p y^q dxdy = \overline{\langle H^p\rangle\langle H^q\rangle} = \sum_{\zeta>0} \overline{\langle H^p\rangle\langle H^q\rangle}_{\zeta}$$

$$= \sum_{\zeta>0} \mu^p_{\zeta}\mu^q_{\zeta}\overline{\langle H^{\zeta}\rangle\langle H^{\zeta}\rangle}_{\zeta} = \sum_{\zeta>0} \mu^p_{\zeta}\mu^q_{\zeta}T_{\zeta} \tag{2}$$

where ζ, the number* of contributing binary correlations between H's in different traces ("cross correlations") indicates the complexity of the coupling between them. $\mu^p\zeta$, which for GOE is an extension of the Catalan numbers encountered with the density, is the number of allowed ways in which $(p-\zeta)/2$ pairs of H's can be correlated in a trace which already contains ζ uncorrelated ones. T_ζ is the completely cross-correlated trace product. Let us now expand the density, considered as a random function, in terms of a series of polynomial "excitations" of its ensemble average (semicircular for the canonical ensembles, Gaussian for EGOE with m >> k and intermediate for $m \neq k$ but not asymptotic). Thus

$$\rho(x) = \bar{\rho}(x)\left\{1 + \sum_{\zeta>0} R_{\zeta}P_{\zeta}(x)\right\} \tag{3}$$

where the P_ζ are closely related to the polynomials v_ζ which are orthonormal with $\bar{\rho}(x)$ as weight function, and the R_ζ are then random variables.

It is easy to see that, to within a normalization constant, R_ζ coincides with the T_ζ defined by (2), so that the completely cross-correlated product determines the magnitude of the coupled excitations. μ^p_ζ is the p'th moment of the "excitation density" $\bar{\rho}(x)P_\zeta(x)$; since μ^p_ζ vanishes for $p<\zeta$ we see that the higher the ζ value the shorter is the wavelength of the excitation.

Once we determine the first and second-order statistical properties of the random variables R_ζ we can determine everything about the two-point fluctuations. For the GOE, normalized as in (I), with d the matrix dimensionality and $P_\zeta(x) = (\zeta\bar{\rho}(x))^{-1}\frac{d}{dx}(\bar{\rho}v_{\zeta-1}(x))$,

---

*Just as with the density only a subset of the binary correlation patterns are contributing ("allowed") when m=k, the GOE case, for which the density is semicircular. The other patterns are "inhibited". In the limit of a Gaussian density (m>>k) all binary correlations are allowed. For both limits (but not in intermediate cases) all allowed patterns contribute equally, so that combinatorial counting is the relevant operation.

where now v_ζ is the Chebyshev polynomial of the second kind, defined for the range (-2,2), we find

$$\mu_\zeta^p = \binom{p}{\frac{p-\zeta}{2}} ; \quad \bar{R}_\zeta = 0 ; \quad \overline{R_\zeta R_{\zeta'}} = \frac{2\zeta}{d^2}\,\delta_{\zeta\zeta'} \tag{4}$$

and then the density is represented as a series of uncorrelated Chebyshev-polynomial excitations of the semicircle. Much more interesting to us is the EGOE case, for which the basic density is Gaussian, while the excitations are of Hermite-polynomial type. Specifically[11], with $\sigma_o^2 = \binom{m}{k}$, d the dimensionality of the GOE which is embedded (i.e. of the m=k space), and $P_\zeta(x) = \zeta!^{-\frac{1}{2}} v_\zeta(x)$ $= \zeta!^{-1} He_\zeta(x/\sigma_o)$,

$$\mu_\zeta^p = \sigma_o^p\,(p-\zeta-1)!!\binom{p}{\zeta} ; \quad \bar{R}_\zeta = 0 ; \quad \overline{R_\zeta R_{\zeta'}} = \frac{2\zeta}{d^2}\binom{m}{k}^{2-\zeta}\delta_{\zeta\zeta'} \tag{5}$$

which once again expresses the density in terms of uncorrelated excitations.

The remarkable difference between the two cases is that while the GOE amplitudes are essentially constant for all excitations, the EGOE amplitudes are extremely rapidly damped, by the factor $\binom{m}{k}^{2-\zeta}$ as the wave length decreases. This rapid damping arises from the smoothing action of the CLT which does not of course operate in the GOE. We have now a general and explicit demonstration of the CLT convergence for interacting as well as non-interacting particles.

If this damping should extend to all orders ζ all but the longest-range deviations from the averaged density would be eliminated and the spectra would be locally uniform. But it does not extend to the very high ζ values which describe fluctuations of wavelength comparable with the spacing. Suppose that we normalize things so that the spectral variance is independent of particle number m. Then the density becomes smoother and smoother *on a fixed energy scale* as m increases. But at the same time the spacing D rapidly becomes smaller and smaller so that, if our interest is in level-to-level fluctuations, we are continually examining the density on a smaller and smaller scale. The short-range fluctuations in fact escape the actions of the CLT* and may be said to have an existence (and a character?) independent of the long-range variations of the density, the whole phenomenon being in accord with Wilson[1].

EGOE has then, for large particle number, a remarkable excitation spectrum. There are a few long-range (low ζ) excitations of appreciable amplitude, which represent the secular behavior, and no intermediate-range excitations at all; this behavior is generated by the CLT. But at high ζ the excitations reappear and are responsible for the level-to-level fluctuations. The separation implied by (I-1) is then a very sharply defined one. For GOE on the other hand there is no such separation at all. The difference represents a violation of the Wilson principle. It arises here because the GOE involves physically unreasonable multibody interactions; whether such violations occur in nature, and for a similar reason, is an interesting question.

To make all this more explicit we consider in particular the deviation of a level (the "level motion") from its ensemble-averaged position. The mean-squared motion for a numbered level in the neighborhood of x, which we write as $\overline{(\delta x)^2}$, is essentially $S^F(x,x)$ where $S^F(x,y)$ is the two-point *distribution* function, the double integral of $S^\rho(x',y')$ which is given by (1). Then from (3,5) we find for the mode decomposition of $\overline{(\delta x)^2}$, measuring δx in units of D, the local average level spacing, and taking x to be at the center of the spectrum

$$\frac{\overline{(\delta x)^2}}{D^2} = \frac{1}{\pi}\binom{N}{m}^2\binom{N}{k}^{-2}\binom{m}{k}\left\{1 + \frac{1}{12}\binom{m}{k}^{-2} + \frac{1}{320}\binom{m}{k}^{-4} + \ldots\right\} \tag{6}$$

the terms, for $\zeta=1,3,5,\ldots$, displaying the rapid damping. Here N is the number of underlying single-particle states, so that $\binom{N}{k}$ is the GOE dimensionality and $\binom{N}{m}$ is that of the model space. The expansion (6) is of course not valid for ζ in the fluctuation domain.

For GOE the corresponding decomposition is

$$\frac{\overline{(\delta x)^2}}{D^2} \simeq \frac{2}{\pi^2}\sum_{\substack{\zeta=1\\(\zeta\ \text{odd})}} \zeta^{-1} = \frac{1}{\pi^2}\left\{\ln 2s + \gamma\right\} \tag{7}$$

---

*Formally this comes about because, with asymptotic (m>>k) EGOE, but not with GOE, the evaluation of fluctuation moments by counting the uninhibited binary correlations breaks down for high ζ. The argument is satisfactory as long as different H's in the moments act on different particles, but for high ζ there are simply too many H's. Incidentally, the "action of the CLT" is one of convolution in the case of non-interacting particles and gives rise[12] to a Poisson process for the fluctuations. With interactions however the eigenvalue densities do not convolute; it is not surprising in this case that the fluctuations are similar to or identical with those of GOE as suggested by experimental data and Monte-Carlo calculations.

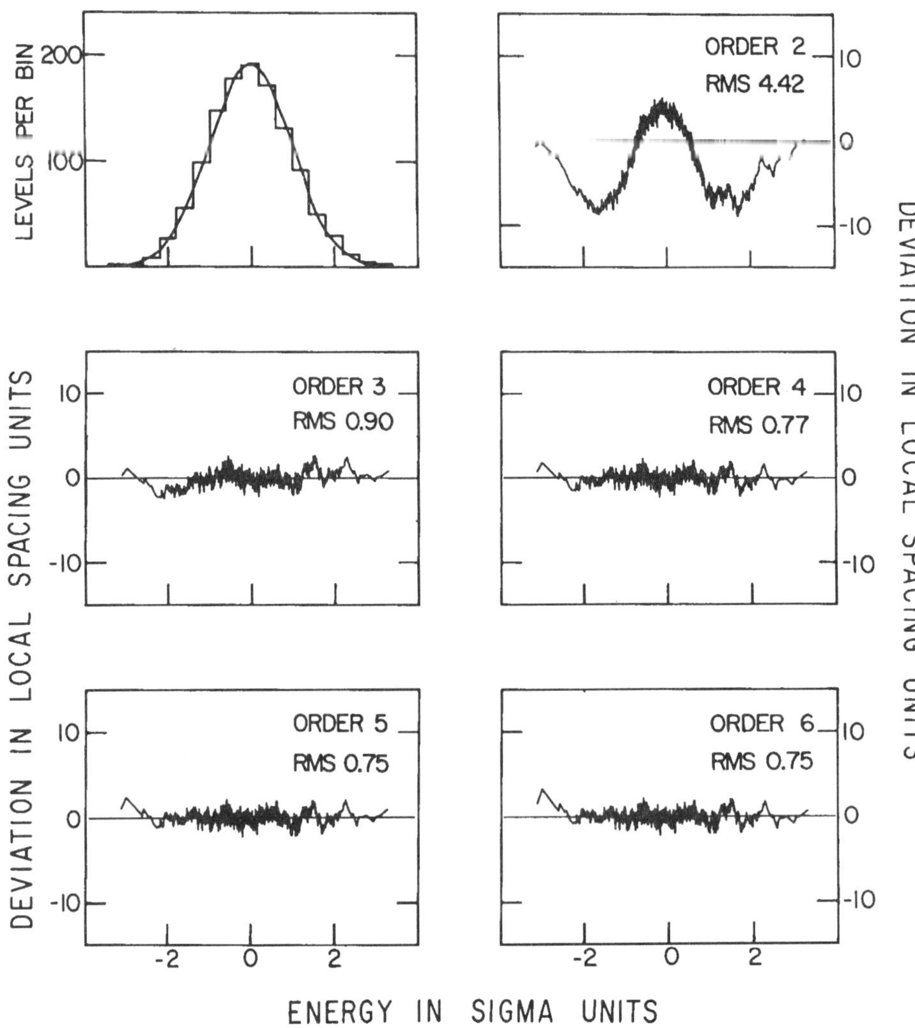

Figure 1. A partial normal-mode decomposition of a 1206-dimensional $(ds)^8$ J=T=0 spectrum. The spectral histogram in the upper left shows that the smoothed distribution is essentially Gaussian. See the text for discussion of the analysis and results. The original spectrum is due to M. Soyeur and A. Zuker and the numerical analysis to S. S. M. Wong (private communication).

where s=t+1 for odd t and s=t for even t. Since the total number of modes is $\binom{N}{k}$ = d we take t=d to find the total variance of the level motion*. It is the intermediate form here which is analogous to the expansion in (6); the final form gives the partial sum, not individual terms as in (6).

It is an essential feature of ergodicity that we should be able to transcribe these ensemble results to their spectral equivalent, and deal then with fluctuations in a single spectrum. Figure 1 shows an example[13] for a 1206-dimensional $(ds)^8$ J=T=0 model space, appropriate to ^{24}Mg. The exact spectrum is compared with a series of p'th-moment Gram-Charlier spectra generated from the exact spectrum by calculating the moments, constructing thereby the Gram-Charlier distribution functions and using Ratcliff's procedure[15] to produce the spectra**. The deviations δx_i found for each level are shown for p=2 (Gaussian), 3, ...,6. The skewness and excess corrections have large effects, but after that very little happens and probably wouldn't happen for orders less than a few hundred (for p=6 there are about 200 sign changes in δx_i). The limiting mean-squared deviation found (0.75) is very close to the theoretical GOE value (more precisely the value for its circular equivalent) which is 0.74 as derived from the Dyson result in the preceding footnote. Since however the numerical example is not strictly EGOE, involving, as it does, a mixture of k=1 and k=2 interactions and with specified J, T, a numerical comparison for the long-wavelength part of the excitation spectrum is not appropriate. For the same reason we cannot give good estimates yet of the errors in spectral-distribution calculations.

---

*The asymptotic result given by Dyson[14], for the closely related circular orthogonal ensemble, is $\pi^{-2}\{\ell n\ 2d+\gamma+1-5\pi^2/24\}$, which shows a characteristic small constant difference from the binary-correlation result. This is a good place to stress that $(\delta x)^2$, which grows indefinitely but very slowly with d, is not an observable fluctuation measure, which would for large d be independent of d. For some measures, or combinations of standard measures, their binary-correlation approximations are identical with the exact results.

**The precise procedure indicated would not give convergence in the sense of a $\langle(\delta x)^2/D^2\rangle_{av}$ non-increasing as p increases. The lack of convergence can be seen in the results of Chang and Zuker[16] who made the first numerical studies of this kind; it arises from the fact that the G.C. expansion is continuous while the x_i spectrum is discrete. Instead of using polynomial orthogonality one can fix the expansion coefficients by minimizing $\langle(\delta x^2)/D^2\rangle_{av}$ for each p value. This much more complicated procedure is used in the present example.[13]

In summary then, a very sharp separation between secular behavior and fluctuations gives a complete justification for our Equation (I-1). The operation of the CLT produces convergence *only to within fluctuations* so that there are irreducible errors in spectral calculations; attempts to eliminate them by proceeding to higher and higher moments must necessarily fail*, and for preciesly the reason given by Wilson[1]. For energies above the ground-state domain there is much evidence that the fluctuations carry little or no information, so that the errors are not harmful - quite to the contrary since they are a reminder of the economy of spectral methods. For the ground-state region however, it is difficult to assess the significance of agreements or disagreements between exact and smoothed level spectra. It is known that statistical behavior extends to the ground-state region, but the ensembles used for treating the fluctuations there should specify the exact symmetries and take account of collectivities and broken symmetries. Many technical problems involved here are quite unsolved.

II. MODEL SPACES

Spectral moments $\langle H^p \rangle^m$ involve no energy denominators. Thus the shift of a ground-state energy, insofar as it is represented by a partial variance, taking account of excitations between a low-lying configuration and a high-lying one, is independent *ceteris paribus* of how high the second configuration lies. On the other hand a second-order perturbation calculation will give a ground-state energy shift inversely proportional to the separation of the configurations. One (or both) of these calculations must be in error.

For an unrealistic example consider m particles in two orbits, separated by an energy Δ, with a pairing interaction $H_p + H_p^+$ connecting them. H_p promotes two particles from orbit #1 up to orbit #2. Then if Δ is large compared with the widths of the configurations, and if the ground state belongs to configuration (m,0) we find

$$\Delta E^{\text{pert.}}_{\text{g.s.}} = -\frac{1}{2\Delta}\langle H_p^+ H_p\rangle^{m,0} = -\frac{1}{2\Delta}\sigma_e^2(m,0) \quad (8)$$

where $\sigma_e^2(m,0)$ (an external partial variance) is the amount which pairing excitations contribute to the variance of the ground-state

---

*To produce methods for evaluating higher order traces is still however, a most important technical problem. On the one hand we would need them to reach the CLT limit (or to verify that we have reached it) and, on the other hand, if we could evaluate them we could use projection operators for dealing with symmetries.

configuration. This same quantity enters into the statistical calculation, which gives however a different result, independent of Δ. Since the g.s. energy is fixed by $F_g \equiv F(E_g) = 1/2d$ it follows that, when F(E) changes because of a change in the spectral variance σ^2,

$$\delta F_g = 0 = \left(\frac{\partial F}{\partial \sigma^2}\right)_g \delta\sigma^2 + \left(\frac{\partial F}{\partial E}\right)_g \Delta E_g^{stat.} = \left(\frac{\partial F}{\partial \sigma^2}\right)_g \delta\sigma^2 + \rho(E_g)\Delta E_g^{stat.} \tag{9}$$

If we assume that the *shape* of the density is unchanged by including the $1 \leftrightarrow 2$ excitations we have easily that, with $\mathcal{E}$ the (m,0) spectral centroid,

$$\left(\frac{\partial F}{\partial \sigma^2}\right)_g = \frac{(\mathcal{E}-E_g)}{2\sigma^2}\rho(E_g) \tag{10}$$

and then

$$\Delta E_g^{stat.} = -\frac{(\mathcal{E}-E_g)}{2\sigma^2}\delta\sigma^2 = -\frac{(\mathcal{E}-E_g)}{2\sigma^2}\sigma_e^2(m,0) \tag{11}$$

which should be compared with (8).

The statistical result is obviously wrong in the large-Δ case. The $1 \leftrightarrow 2$ excitations in this case have forced upward, by an energy 2Δ, a small part of the (m,0) density, generating thereby a bimodal distribution*. It is by our ignoring this Δ-dependent deformation of the original density that the error has arisen. We could have guarded against this kind of error by calculating the skewness, the dominant contribution to which involves a pairwise excitation and de-excitation with a measurement in between of the intermediate energy; this gives a value $\gamma_1 \sim \sigma_e^2(m,0)\Delta/\sigma^3$ which, unlike the variance, increases linearly with Δ. More generally then a large value of the partial skewness gives an indication that the effect being considered should be treated, instead, by perturbation theory or simply ignored.

Our example is unrealistic but it does remind us that our spectral methods are really appropriate when strong interactions generate a statistical "chaos", and that caution must be used in "huge" spaces, separate parts of which are not strongly interacting.

* It belongs in fact to the class of singular distributions mentioned in Section I-2. When Δ is large enough the contributions from the small high-lying spike dominate all the central moments of order >2. This kind of thing occurs in many calculations, as it did for example in early attempts to calculate the spectral variance generated by the optical potential.

It appears that calculations of third partial moments, and perhaps fourth for centrally located configurations, might be needed as a matter of routine.

This elementary notion seems of consequence because, if spectral methods are going to be useful for studying fundamental aspects of nuclear structure, and not simply as a competitor of shell-model methods, they must be usable in indefinitely large spaces. It would be of great interest to investigate the convergence of low lying energies and other quantities as we enlarge the model space while at the same time "undoing" renormalizations of the interaction which have been carried out in order to adapt it to the original space. This of course is just one aspect of the study of effective interactions by spectral methods. Another aspect is of consequence for the information content of the spectral fluctuations in the ground-state region, for which, as we have said, adequate direct spectral methods are not now available. Perhaps the low-lying states should be treated by matrix methods, the connection between the corresponding subspace and the rest of the model space being established by the methods to be discussed by Dr. Chang.

III. INFORMATION PROPAGATION

We introduce a density-operator method for writing many-particle traces, and use it: (1) to make clearer the situations in which we have simple propagation: (2) to discuss an expansion appropriate for dilute systems: (3) to say something about propagation of information which is not scalar with respect to the group involved.

In (I-13) we have seen that the elementary equation $\langle F(k)\rangle^m = \binom{m}{k}\langle F(k)\rangle^k$, or equivalently, for the proper trace,

$$\langle\langle F(k)\rangle\rangle^m = \binom{N-k}{N-m}\binom{N}{k}\langle F(k)\rangle^k = \langle\langle\rho(k)\rangle\rangle^m\langle F(k)\rangle^k \tag{12}$$

gives the ratio of the m- and k-particle averages as the weight $\langle\rho(k)\rangle^m = \binom{N}{m}^{-1}\langle\langle\rho(k)\rangle\rangle^m$ with which the space $\underline{k}$ is contained in $\underline{m}$. Here $\rho(k)$, the k-particle trace operator, is given in terms of the k-particle state operators $\psi_\beta(k)$ by

$$\rho(k) = \sum_\beta \psi_\beta(k)\psi_\beta^+(k) = \binom{n}{k} . \tag{13}$$

Note that $\Psi_\beta(k) = \psi_\beta(k)|0\rangle$ and that, for k=1, $\psi_\beta = a_\beta^+$. The $\binom{n}{k}$ form, with n the number operator, follows from the fact that both $\rho(k)$ and $\binom{n}{k}$ are k-body operators which give unity in the k-particle space.

For the partitioning* $\underline{m} \to \sum (m,\Gamma)$, the analog of (12) would be

$$\langle\langle F(k)\rangle\rangle^{m\Gamma} = \sum_{\Lambda} \langle\langle \rho_{\Lambda}(k)\rangle\rangle^{m\Gamma} \langle F(k)\rangle^{k\Lambda} = \sum_{\Lambda} \langle\langle \rho_{\Lambda}(k)\rangle\rangle^{m\Gamma} \langle F(k:\Lambda)\rangle^{k} \tag{14}$$

where we mean by $F(k,\Lambda)$ the projection of $F(k) = \sum_{\Lambda} F(k,\Lambda)$ into the (k,Λ) subspace. This equation however, is correct (for arbitrary $F(k)$) only for certain subgroup decompositions, namely for those in which $\rho_{\Lambda}(k)$ is Casimir representable, in which case the propagating coefficients can be easily evaluated and the propagation is simple and elegant. For particular $F(k)$ operators, for example those of low particle rank, (14) may apply for some other decompositions as well.

A general result which reduces to (14) in these special cases follows by noting that, since holes are also Fermions, we have for an arbitrary operator K

$$\langle\langle K\rangle\rangle^{m} = \langle\langle \tilde{K}\rangle\rangle^{N-m} \,. \tag{15}$$

*Notation: As in (I), Γ and Λ are labels for irreducible representations of a U(N) subgroup G, and (m,Γ) is the m-particle subspace spanned by the set of equivalent irreps with that label. $[\Gamma]$ will stand for the dimensionality of a single irrep, and $[\Gamma\Lambda/\Gamma']$ for $[\Gamma][\Lambda][\Gamma']^{-1}$ etc; $d_{\ell}(m,\Gamma) \equiv d_{\ell}(\Gamma)$ will denote the number of equivalent irreps (in the case of angular momentum, d_{ℓ} counts the number of levels) and $d(\Gamma)=[\Gamma]d_{\ell}(\Gamma)$, the total number of states labeled by Γ. Thus $\sum_{\Gamma} d(\Gamma)=d(m)=\binom{N}{m}$. The corresponding density operator $\rho(m,\Gamma) = \sum_{\alpha\epsilon\Gamma} \psi_{\alpha}^{\Gamma}(m)\psi_{\alpha}^{\Gamma+}(m)$, which is an m-body operator scalar with respect to G, is then a sum of $d(\Gamma)$ terms. The notation will accomodate various extensions, e.g. to the case of several commuting groups (J,T for example). The multiplicity d_{ℓ} may be enumerated in terms of a larger group, or chain of groups, containing G as a subgroup but we always sum over all the equivalent irreps of G. P^{Λ} will denote a tensor with respect to G, $\bar{P}^{\Lambda}$, the adjoint tensor, and $(P^{\Lambda} \times Q^{\Lambda'})^{\nu}$ a coupled tensor. To an operator F there corresponds a "hole-particle adjoint", $\tilde{F}$, generated from F by every $a_i \leftrightarrow a_i^{+}$. Thus $\tilde{\rho}(k) = \binom{N-n}{k}$ is a k-hole operator. For angular momentum $(\psi^{\Gamma} \times \bar{\psi}^{\Gamma'})_{\nu} \xrightarrow{\sim} (-1)^{\Lambda-\nu} (\bar{\psi}^{\Gamma} \times \psi^{\Gamma'})^{\Lambda}_{-\nu}$. To an m-particle irrep Γ there corresponds a complementary, or contragredient, (N-m)-particle irrep Γ_c, generated by replacing every $\psi_{\alpha}^{i}(m)$ by its complementary state operator $\psi_{\alpha}^{\Gamma_c}(N-m)$, the single-particle states filled in one being empty for the other. Finally $\langle G\rangle_0$ will be a vacuum expectation value.

Applying this to F(k), noting that a characteristic term of $\tilde{F}(k)$ is $F_{\beta'\beta}\psi^{+}_{\beta}(k)\psi_{\beta'}(k)$, and using $\sum_{\alpha\epsilon\Gamma} \langle\psi^{+}(N-m)\psi^{+}_{\beta}(k)\psi_{\beta'}(k)\psi_{\alpha}(n-m)\rangle_{o}$

$$= \langle\psi^{+}_{\beta}(k)\tilde{\rho}_{\Gamma_c}(N-m)\psi_{\beta'}(k)\rangle_{o}$$

we have

$$\langle\langle F(k)\rangle\rangle^{m\Gamma} = \langle\langle\tilde{\rho}_{\Gamma_c}(N-m)F(k)\rangle\rangle^{k} = \sum_{\Lambda} \langle\langle\tilde{\rho}_{\Gamma_c}(N-m)F(k)\rangle\rangle^{k\Lambda} \quad (16)$$

For an operator F of mixed particle ranks $\leq u$, the corresponding result is

$$\langle\langle F\rangle\rangle^{m\Gamma} = \langle\langle\rho_{\Gamma_c}(N-m)\binom{u-m}{u-n}F\rangle\rangle^{\{u\}} \quad (17)$$

where {u} is the space of all states with $k \leq u$ particles; however the fixed rank result (16) is adequate for our present purpose.

When $\rho_{\Gamma_c}(N-m)$ behaves as a multiple of unity in each subspace, the (k,Λ) traces factor and using (15) for the propagating coefficient gives us the intuitive result (14). As we have mentioned in (I) there is both a simple test for, and a general understanding of, this case.

For other cases we could argue that, since $\sum_{\Lambda}\rho_{\Lambda}(k) = \binom{n}{k}$, we have that, in some average way, $\rho_{\Lambda}(k)\to d(k,\Lambda)d^{-1}(k)\binom{m}{k}$, which then gives the crude approximation that $\langle F(k)\rangle^{m\Gamma} = \langle F(k)\rangle^{m}$. But this trivial form could be quite misleading since it pays no attention to the tensorial character of F(k) with respect to the group G. This information is contained in the decomposition of F(k) according to the coupling scheme $(\psi^{\Lambda}_{\beta}(k) \times \tilde{\psi}^{\Lambda'}_{\beta'}(k))^{\Omega}$. For traces, only $\Omega=0$ contributes* and therefore $\Lambda' = \Lambda$. The action of a characteristic term of F(k) on a state (m,Γ) is represented by a sequence $(m,\Gamma) \xrightarrow{\Lambda'} (m-k,\Gamma') \xrightarrow{\Lambda} (m,\Gamma''=\Gamma)$ with two Clebsch-Gordan couplings as indicated, and it is here that the G-structure of F comes into play. Inverting the order of the two transformations generates an inductive method for trace evaluation and, perhaps more important,

*By $\Omega=0$ we mean of course the scalar representation of G. Though the results which follow are valid much more generally we write them for the case G=0(3), e.g. angular momentum, isospin or quasispin, or any combination of them. For other groups the coupling is not usually multiplicity free and then an extra label is needed to describe it.[17]

a series which should be very useful for dilute systems ($m<<N$, or, better, with particle or hole occupancies $<<1$ for all single-particle states).

For the trivial case, $G=U(N)$, in which we average over all m-particle states, we first use (15) for $F(k)$, then expand the resulting k-hole operator as a sum of p-*particle* operators to which we apply (16). Then

$$<<F(k)>>^m = <<\tilde{F}(k)>>^{m-k} = \sum_{p=0}^{k} <<(\tilde{F}(k))_p>>^{m-k}$$

$$= \sum_{p=0}^{k} <<\tilde{\rho}(m-k-p)(\tilde{F}(k))_p>>^p \simeq (F(k))_0 \binom{N}{m-k}$$

$$= \binom{N}{m-k} <<F(k)>>^m \tag{18}$$

a "p=0" approximation which compares well, for $m<<N$, with the exact result (12). The p-decomposition used here is useful because for dilute systems $F(k)$ essentially counts the number of holes; there are few particles to inhibit the counting, and, by ignoring them, we see that $<<F(k)>>^m$ is dominated by its p=0 part. For the last step we have used the fact that, for U(N) averaging, $F(k) \equiv <F(k)>^k \binom{n}{k}$, its U(N)-scalar part, and then the constant term $(\tilde{F}(k))_{p=0}$ is given by $<F(k)>^k \binom{N-n}{k}_{p=0} = \binom{N}{k}<F(k)>^k = <<F(k)>>^k$.

For the analog of this result for a non-trivial group G we need the tensorial extension of the first equality in (17). Let

$$<<\| F^{\Omega} \|>>^{m\Gamma} = \sum_{x} <m\Gamma x \| G^{\Omega} \| m\Gamma x> \xrightarrow{\Omega=0} [\Gamma]^{-\frac{1}{2}} <<F^0>>^{m\Gamma} \tag{19}$$

where x labels the $d_{\Omega}(m,\Gamma)$ different equivalent irreps (m,Γ) and $<m\Gamma x\|G^{\Omega}\|m\Gamma x>$ is the (diagonal) Racah double-barred matrix element. Then with U a unitary Racah coefficient, $Q^{\Lambda'}$ and P^{Λ} any operators which respectively destroy and create k particles,

$$<<\| (P^{\Lambda} \times Q^{\Lambda'})^{\Omega} \|>>^{m\Gamma} \equiv \sum_{x} <m\Gamma x \| (P^{\Lambda} \times Q^{\Lambda'})^{\Omega} \| m\Gamma x>$$

$$= \sum_{\Omega_1 \Gamma_1} (-1)^{\Omega-\Omega_1} [\Gamma/\Gamma_1]^{\frac{1}{2}} U(\Gamma\Lambda\Gamma\Lambda' : \Gamma_1\Omega) U(\Gamma_1\Lambda\Gamma_1\Lambda' : \Gamma\Omega_1) \cdot$$

$$<<\| (Q^{\Lambda'} \times P^{\Lambda})^{\Omega_1} \|>>^{m-k,\Gamma_1}$$

$$\xrightarrow[\Lambda=\Lambda']{\Omega=0} \sum_{\Omega_1 \Gamma_1} (-1)^{\Omega_1+\Gamma_1-\Lambda-\Gamma} [\Lambda]^{-\frac{1}{2}} U(\Gamma_1\Lambda\Gamma_1\Lambda : \Gamma\Omega_1) <<\| (Q^{\Lambda} \times P^{\Lambda})^{\Omega_1} \|>>^{m-k,\Gamma_1}$$

$$\xrightarrow[\Lambda=\Lambda']{\Omega=0=\Omega_1} [\Lambda]^{-1} \sum_{\Gamma_1} (-1)^{2\Lambda} [\Gamma/\Gamma_1]^{\frac{1}{2}} <<\| (Q^{\Lambda} \times P^{\Lambda})^{0} \|>>^{m-k,\Gamma_1} \tag{20}$$

The general result here, which emerges easily if we express the trace as a coupled vacuum expectation value and carry out the hole $\leftrightarrow$ particle transformation and recoupling, is applicable, with a slight extension of notation, to other groups. It is of course not simply a geometrical result since its derivation required both recoupling and trace results. p=0 gives immediately for the trace of a k-body operator

$$\langle\langle F(k)\rangle\rangle^{m\Gamma} = [\Gamma] \sum_{\Lambda} [\Lambda]^{-1} \langle\langle F(k)\rangle\rangle^{k\Lambda} \sum_{\Gamma_1} d_{\ell}(m-k,\Gamma_1)\delta_{\Gamma_1\Lambda\Gamma} \qquad (21)$$

where δ_{abc} is the triangle function and $\langle\langle F(k)\rangle\rangle^{k\Lambda} = \langle\langle F(k,\Lambda)\rangle\rangle^{k}$ is given by the defining matrix elements. The structure of this equation is transparent, and indeed it would be a nice exercise to derive it without any of the formal operations; for given Λ the Γ_1 sum runs over all the (m-k)-particle representations which couple by $\psi^{\Lambda}(k)$ and the (m,Γ) representation of interest, while $F(k,\Lambda)$ the corresponding part of F(k) is measured by its k-particle trace.

The p=0 approximation which leads to the final form of (17), and to (21), can be improved by treating the p=1 part as a new operator to be transformed and truncated in the same way; this gives a "(0) + (1,0)" approximation. More generally let:

$$\begin{aligned} F(k)_{p_1:} &= p_1\text{-body part of } \tilde{F}(k) \\ F(k)_{p_1:p_2:} &= p_2\text{-body part of } \tilde{F}(k)_{p_1:} \qquad (22) \\ &\;\;\vdots \\ F(k)_{p_1:\dots:p_\ell:} &= p_\ell\text{-body part of } \tilde{F}(k)_{p_1:p_2:\dots:p_{\ell-1}:} \end{aligned}$$

Then we find

$$\langle\langle F(k)\rangle\rangle^{m} = \sum_{[p]_\ell} \langle\langle F(k)_{p_1:p_2:\dots:p_\ell:}\rangle\rangle^{m-k-p+p_\ell} \qquad (23)$$

where $[p]_\ell \equiv [p_1 \geq p_2 \dots \geq p_\ell]$, so that the sum, a restricted one over $p_1, \dots, p_\ell$ may be regarded as a sum over all the ℓ-dimensional "path vectors" $[p]_\ell$ for which $p_1 \leq k$. Thus for example, with k=1 and $\ell=2$, we have $[p]_2 = [0],[10]$ and $[11]$; note that the result, though not the form of (23), is independent of ℓ. We have written $p = \sum p_i$ so that $m-k-\sum_{i\leq\ell-1} p_i = m-k-p+p_\ell$.

For path vectors with the last component p = 0 ("zero vectors") the corresponding operator is a constant which, multiplied by the final dimensionality $\binom{N}{m-k-p}$, gives its contribution to the original trace. For a non-zero vector on the other hand, e.g. [211] with $\ell=3$ we have $[211] \equiv [2110] + [2111]$ the [2111] term being smaller than [2110] by a factor $\sim N^{-1}$. We thus see a natural expansion in terms of zero-vector contributions, with the terms grouping naturally according to p. The explicit decomposition for the U(N) case follows easily by using $F(k) \equiv \langle F(k)\rangle^{k}\binom{n}{k}$ and

$$\binom{\tilde{n}}{k} = \binom{N-m}{k} = \sum_{p} (-1)^{p}\binom{N-p}{N-k}\binom{n}{p} \tag{24}$$

which both displays the decrease of the operator magnitude with p and gives the correction needed to convert a zero-vector result into an exact one; this same correction we would apply also for averaging over a non-trivial group G.

All this of course applies immediately to averaging over configurations, and may well give real simplifications in the calculation of higher-order configuration traces. But, beyond that, it transcribes directly to the more general case. At every step of the expansion we encounter a further Racah recoupling, as in (20), so that the Racah algebra for G must be available to us; and even then, as with O(3), the interest will probably be in a few terms rather than the complete expansion. The p=(0) + (1,0) approximation is quite straightforward and useful; but so far only simple examples have been dealt with.

It should be noted that in general we encounter non-scalar $(\Omega_1 \neq 0)$ traces in (20) even when the original operator is scalar $(\Omega=0)$. Thus some information about the "action" of a scalar operator is carried by coupled traces of the corresponding non-scalar ones, $(\psi^{\Lambda} \times \bar{\psi}^{\Lambda})^{0} \longrightarrow (\bar{\psi}^{\Lambda} \times \psi^{\Lambda})^{\Omega_1}$; similarly operators of tensorial rank Ω generate transformed operators of general rank. In the particular case of an O(3)-scalar operator the recoupling coefficient in (20) is the analog of a Legendre polynomial $P_{\Omega_1}(\cos\theta)$, with θ the angle between Γ,Λ as defined by the $(\Gamma\ \Lambda\ \Gamma_1)$ coupling; the Γ_1 sum reduces then to an P_{Ω_1}-weighted integration over the multiplicity function $d_{\ell}(m-k,\Gamma_1)$, i.e. to a Legendre transform of it.

This extension to non-scalar operators should remind us that not all of the significant information is scalar in nature. With multipole moments and multipole sum rules for example, we encounter[18] tensor-coupled traces (19) which propagate, in "simple" cases, by non-scalar density operators constructed from the generators of G. Similarly for the more complicated strength distribution, as recently studied by Kar[19] for Gamow-Teller β-decay. The expansion given above, which is not restricted to "simple" cases, and other procedures related to it[20], will be useful for such problems. They may also give new insight into the way group representations couple in many-particle spaces.

IV. FURTHER PROBLEMS

Among the problems mentioned above are: spectroscopy in indefinitely large spaces, and its connection with more general many-body theory: level fluctuations and correlations, and the way they are modified by symmetries, an understanding of which is necessary to deal with low-energy spectra: trace evaluation and information propagation in spaces defined by "complicated" symmetries, which is of interest in its own right and for its connection with many other topics. We end with some comments about further problems, questions and applications, not all of them by any means of the same difficulty or significance.

1. Spectral Distributions in the Continuum: Is there anything to be gained from a spectral-distribution approach to nuclear reactions, perhaps analogous to the theory of Mahaux and Weidenmuller[21]?

2. Quasiparticles: We can make good calculations of single-particle occupancies, and probably also of correlations between them. Can this replace the BCS minimization which leads to a quasiparticle theory, and can it, for example, generate such a theory, valid at higher excitations?

3. Symmetries: Correlation coefficients and norms can be calculated as explicit functions of the Hamiltonian matrix elements. Does this lead to a systematic way of searching for relevant symmetries? Are spectral methods useful for understanding or applying the interacting-boson model?[22]

4. Collectivities: It has been verified by shell-model examples that one can predict collectivity generated by a given transition operator, by combining spectral theory for the secular behavior of the strength sum (as we vary the starting state) with the Porter-Thomas theory for its fluctuations.[23] Extensions can be made in which the final J,T values are specified. Are these procedures of any general use? New studies of schematic interactions might be profitable also since it is now easy, by the "geometrical" methods described in (I) to calculate the parameters for such models. The schematic Hamiltonian could be supplemented by a random ensemble of calculable strength, in order to take account of the damping of the collective states.

5. Level Densities, Partition Functions and Spin Cut-off Factors: Because the conventional theory has been for non-interacting quasiparticles there has been little connection between this subject and nuclear physics at low excitations. Do the

experimental results show deviations from the conventional theory, which can be identified as interaction effects by comparison with the predictions of spectral theory which does make such a connection?

6. Electromagnetic and β-Decay Sum Rules and Giant Resonances: Recent work[19,24] on these subjects makes clear that spectral methods are ideally suited for them. A thorough survey, especially of electromagnetic sum rules, would seem to be called for.

7. Statistical Mechanics in Spectroscopic Model Spaces: This is another way of describing the theory of spectral distributions and fluctuations. A more conscious exploitation of the methods of statistical mechanics should be profitable.

This work was supported in part by the U.S. Department of Energy.

REFERENCES

1. K. G. Wilson, Scientific American 241, 158 (August, 1979).
2. E. P. Wigner, in "Gatlinberg Conference on Neutron Physics", ORNL Report #2309 (1957). This article is reprinted in: C.E. Porter (ed.), "Fluctuations", Academic Press, N.Y., 1965.
3. L. D. Landau and Ya. Smorodinski, "Lectures on the Theory of the Atomic Nucleus", (Consultants Bureau) N.Y., 1958; the relevant short section is reprinted in Porter[2].
4. J. von Neumann and E. P. Wigner, Phys. Zeitschr. 30 (1929) 467.
5. F. J. Dyson, J. Math. Phys. 3 (1962) 140; 3 (1962) 157: 3 (1962) 166; 3 (1962) 1191; 3 (1962) 1199; F. J. Dyson and M. L. Mehta, J. Math. Phys. 4 (1963) 701; M. L. Mehta and F. J. Dyson, J. Math Phys. 4 (1963) 713. These papers are reprinted in Porter[2].
6. H. Camarda, H. I. Liou, F. Rahn, G. Hacken, M. Slagowitz, W. W. Havens, Jr., and J. Rainwater, in "Statistical Properties of Nuclei", J. B. Garg, editor, Plenum Press, N.Y., 1972. See also later papers in Physical Review by this Columbia group.
7. E. G. Bilpuch, A. M. Lane, G. E. Mitchell and J. D. Moses, Phys. Reports 28 (1976) 145.
8. M. L. Mehta, "Random Matrices and the Statistical Theory of Energy Levels", Academic Press, N.Y., 1967; C. E. Porter in Porter[2]; E.P. Wigner, SIAM Review 9 (1967) 1;M.L. Mehta in Garg[6]; M. Carmeli, J. Stat. Phys. 10 (1974) 259; T. A. Brody, J. Flores, J. B. French, P. A. Mello, A. Pandey and S. S. M. Wong, to be published; J. B. French in Proceedings of the 1979 Michigan Summer School, to be published.
9. A. Pandey, Ann. Phys. (N.Y.) 119 (1979) 170.

10. J. B. French, Rev. Mex. Fis. 22 (1973) 221. Closely related "two-body random ensembles" (TBRE), which are intractable analytically, were introduced earlier and studied via Monte-Carlo calculations, by J. B. French and S. S. M. Wong, Phys. Lett. 33B (1970) 449, 35B (1971) 5 and by O. Bohigas and J. Flores, Phys. Lett. 34B (1971) 261, 35B (1971) 383.
11. K. K. Mon and J. B. French, Ann. Phys. (N.Y.) 95 (1975) 90.
12. J. B. French and A. Pandey, to be published.
13. J. B. French and S. S. M. Wong, unpublished.
14. F. J. Dyson, the third paper of the series[5], Eq. (75).
15. K. F. Ratcliff, Phys. Rev. C3 (1971) 117.
16. F. S. Chang and A. Zuker, Nucl. Phys. A198 (1972) 417.
17. J. P. Draayer, private communication.
18. J. B. French, in "Isospin in Nuclear Physics," D. H. Wilkinson, editor, North-Holland, Amsterdam, 1969.
19. K. Kar, Ph.D. Thesis, University of Rochester, 1979.
20. P. Mugambi, Ph.D. Thesis, University of Rochester, 1970; G. D. Lougheed and S. S. M. Wong, Nucl. Phys. A243 (1975) 215, A246 (1975) 29.
21. C. Mahaux and H. A. Weidenmuller, Shell-Model Approach to Nuclear Reactions, North-Holland, Amsterdam, 1969.
22. A. Arima and F. Iachello, Ann. Phys. (N.Y.) 99 (1976) 253, 111 (1978) 201, 123 (1979) 468. Some results on spectral distributions for bosons are due to M. Nomura, Ann. Phys. (N.Y.) 91 (1975) 83; P. F. Bortignon, V. R. Manfredi and S. Spitz, Lett. Nuovo Cimento 23 (1978) 210; V. K. B. Kota, Jour. de Physique 40 (1979) L-579; V. K. B. Kota and V. Potbhare, in press.
23. J. P. Draayer, J. B. French and S. S. M. Wong, Ann. Phys. (N. Y.) 106 (1977) 472, 503.
24. T. R. Halemane, Ph.D. Thesis, University of Rochester, 1979.

PHYSICAL REVIEW C VOLUME 21, NUMBER 6 JUNE 1980

Shape of eigenvalue distribution for bosons in scalar space

V. K. B. Kota
Physical Research Laboratory, Navrangpura, Ahmedabad 380 009, India

V. Potbhare
Department of Physics, Faculty of Science, M. S. University, Baroda 390 002, India

(Received 7 February 1979; revised manuscript received 7 November 1979)

Analytic expressions for obtaining first four moments of noninteracting as well as interacting bosons have been given in terms of matrix elements. These expressions have been utilized to obtain general conditions under which the eigenvalue distributions for bosons tend towards the Gaussian distribution. Results for pairing and $Q \cdot Q$ operators in an s-d boson shell have also been given.

NUCLEAR STRUCTURE Analytic expressions, first four moments, noninteracting and interacting bosons, eigenvalue distribution, conditions leading to Gaussian distribution, pairing and Q-Q operators, s-d boson space.

The assumption of applicability of a strong central limit theorem (CLT) to two-body interactions has brought forth a tremendous simplification in the nuclear shell model spectroscopy.[1] The conditions on traces of matrix elements, which give rise to normal eigenvalue distributions (EVD) for fermion systems (noninteracting and interacting), have been clearly formulated.[2,3] The question of normality of EVD for boson systems was recently raised by Bortignon *et al.*[4,5] They have given analytical expressions for the first four moments of noninteracting boson EVD. These expressions, however, do not reveal the conditions for normality of EVD; also, the more interesting and more realistic case of interacting bosons was not treated. In this paper, we present simple analytic expressions for the first four moments of EVD for noninteracting and interacting bosons. We shall also discuss the conditions under which EVD for bosons tend to Gaussian distribution.

The m-particle scalar average $\langle H(k)\rangle^m$ of a k-body operator $H(k)$ in the space defined by N single-particle states, corresponds to the average of expectation values of $H(k)$ over all the states of m particles, and can be expressed in terms of $\langle H(k)\rangle^k$:

$$\langle H(k)\rangle^m = \binom{m}{k} \langle H(k)\rangle^k , \tag{1}$$

where $\binom{m}{k}$ is the usual binomial factor. Equation 1 is valid for both the fermions and the bosons. The number of states of m bosons in the space of N single-particle states is given by $\binom{N+m-1}{m}$; hence, in terms of the trace of $H(k)$ in k-particle space $\langle\langle H(k)\rangle\rangle^k$, we obtain

$$\langle H(k)\rangle^m = \binom{m}{k}\binom{N+k-1}{k}^{-1} \langle\langle H(k)\rangle\rangle^k , \tag{2}$$

where the double bracket $\langle\langle\ \rangle\rangle$ indicates the trace. For noninteracting bosons, the Hamiltonian H is a one-body operator $H=\sum_i \epsilon_i n_i$; here ϵ_i is single-particle energy and n_i is the number operator for the ith single-particle state. Except for the centroid which defines the location of a distribution, the shape of the EVD is determined by its central moments. Therefore, we shall deal only with traceless single-particle energies $\tilde{\epsilon}_i = \epsilon_i - N^{-1}\sum \epsilon_i$, and shall evaluate moments of $\tilde{H} = H - nN^{-1}\sum \epsilon_i$, where $n=\sum_i n_i$, for noninteracting bosons. We evaluate the second moment of H by separating $(\tilde{H})^2$ into its one-body and two-body parts:

$$(\tilde{H})^2 = \sum_i \tilde{\epsilon}_i^2 n_i + \sum_i \tilde{\epsilon}_i^2 n_i(n_i-1) + \sum_{i\neq j} \tilde{\epsilon}_i \tilde{\epsilon}_j n_i n_j . \tag{3}$$

The first term above is the one-body part of $(\tilde{H})^2$ and its scalar average in m-particle space is given by $m\sum \epsilon_i^2/N$. The remaining terms form the two-body part of $(\tilde{H})^2$ and their scalar average equals $[m(m-1)/N(N+1)]\sum_i \tilde{\epsilon}_i^2$. Hence the average value of $(\tilde{H})^2$ in m-particle space can be expressed as

$$\langle \tilde{H}^2\rangle^m = \frac{m(N+m)}{N(N+1)} \sum \tilde{\epsilon}_i^2 . \tag{4}$$

Similarly the scalar averages of $(\tilde{H})^3$ and $(\tilde{H})^4$

can be obtained by separating $\tilde{H}^3$ into (1+2+3)-body parts and $\tilde{H}^4$ into (1+ 2+3+4)-body parts. Adding the contributions from different k-body parts gives us the following expressions:

$$\langle\tilde{H}^3\rangle^m = \frac{m(N+m)(N+2m)}{N(N+1)(N+2)} \sum_i \tilde{\epsilon}_i^{\,3},$$

$$\langle\tilde{H}^4\rangle^m = \frac{m(N+m)}{N(N+1)} \sum_i \tilde{\epsilon}_i^{\,4} + \frac{m(m-1)(N+m)(N+m+1)}{N(N+1)(N+2)(N+3)} \times\left[3\left(\sum_i \tilde{\epsilon}_i^{\,2}\right)^2 + 6\sum_i \tilde{\epsilon}_i^{\,4}\right]. \tag{5}$$

It is quite interesting to note that, by substituting $-N$ for N in the corresponding expressions for fermions, and then by taking the absolute value of each term, one obtains the above expressions for the boson systems. For example,

$$\langle\tilde{H}^3\rangle^m_{\text{fermions}} = \frac{m(N-m)(N-2m)}{N(N-1)(N-2)} \sum_i \tilde{\epsilon}_i^{\,3} \tag{6}$$

goes to its boson counterpart by substituting $-N$ for N and taking modulo of the final result (however, one cannot obtain moments of EVD for fermions from those of bosons as we will be left with phase ambiguities). This was previously conjectured by Nomura[6]; using this principle one can obtain several new results. For example, a class of Wigner coefficients of the O(N) group can be obtained from those of the Sp(N) group via moments in quasispin subspaces.[7]

Another interesting one-body Hamiltonian is $H=\sum \epsilon_{ij}A_iB_j$, where ϵ_{ij} is an off diagonal single-particle matrix element when $i\neq j$, ϵ_{ii} is the usual single-particle energy, and A_i, B_j are single-particle creation and destruction operators, respectively. Such a Hamiltonian is widely used in solid-state physics, where ϵ_{ij} $(i\neq j)$ can be identified with the exciton hopping integral between sites i and j. The traceless single-particle matrix elements $\tilde{\epsilon}_{ij}$, in this case, are given by $\tilde{\epsilon}_{ij}=\epsilon_{ij}-\delta_{ij}N^{-1}\sum_k \epsilon_{kk}$. The expressions for second, third, and fourth central moments of this Hamiltonian can be obtained by substituting $\sum\tilde{\epsilon}_i^2$, $\sum\tilde{\epsilon}_i^3$, and $\sum\tilde{\epsilon}_i^4$ by $\sum\tilde{\epsilon}_{ij}\tilde{\epsilon}_{ji}$, $\sum\tilde{\epsilon}_{ij}\tilde{\epsilon}_{jk}\tilde{\epsilon}_{ki}$, and $\sum\tilde{\epsilon}_{ij}\tilde{\epsilon}_{jk}\tilde{\epsilon}_{kl}\tilde{\epsilon}_{li}$, respectively, in Eqs. 4 and 5 given above.

The preceding treatment of a one-body Hamiltonian was simple and elementary. We shall start with some general results before going to the case of moments of EVD for bosons interacting via two-body interactions. A number conserving, k-body operator for bosons in the space defined by N single-particle states can be decomposed according to the irreducible (IR) representations of the U(N) group; $F(k)=\sum_{\nu=0}^{k} F^\nu(k)$, where $F^\nu(k)$ is the rank-ν part of $F(k)$ and transforms as a $[2\nu, \nu^{N-2}]$ partition under a U(N) transformation. $F^\nu(k)$ can be written as a product of a unitary scalar operator and a completely IR ν-body and rank-ν operator $\mathcal{F}(\nu)$:

$$F^\nu(k)=\binom{n-\nu}{k-\nu}\mathcal{F}(\nu). \tag{7}$$

In order to find out the different unitary IR parts systematically, we introduce a contraction operator D_-, defined by the following double commutator:

$$D_-F=\sum_{i=1}^{N}\left[[B_i,F]_-,A_i\right]_-. \tag{8}$$

The operator D_- reduces the particle rank (or the body rank) of F by one, without altering its unitary rank. Since $\mathcal{F}(\nu)$ is completely irreducible, any reduction of its particle rank by D_- destroys it; $D_-\mathcal{F}(\nu)=0$. It should be noted that this contraction operator is different from similar contraction operators for fermions.[1,8] Contracting $F(k)$ r times by D_- gives us

$$D_-^rF(k)=\sum_\nu r!\binom{N+k+\nu-1}{r}F^\nu(k-r). \tag{9}$$

By inverting this expression we obtain

$$\mathcal{F}(\nu)=\frac{1}{(k-\nu)!}\binom{N+\nu+k-1}{k-\nu}^{-1} \times\sum_{t=k-\nu}^{k}\frac{(-1)^{t-k+\nu}}{(t-k+\nu)}\binom{N+2\nu-2}{t-k+\nu}^{-1} \times\binom{n-k+t}{t-k+\nu}D_-^tF(k). \tag{10}$$

Application of this to a one-body operator $H=\sum\epsilon_in_i$ gives us the IR $\nu=0$ part as $n\sum\epsilon_i/N$, and the IR $\nu=1$ part as $\sum\tilde{\epsilon}_in_i$ in terms of the traceless single-particle energies $\tilde{\epsilon}_i$. Similar application of Eq. (10) to the two-body interaction given by $\frac{1}{4}\sum V_{ijkl}A_iA_jB_kB_l$, expressed in terms of the two-body matrix elements V_{ijkl}, gives us the following three parts: (i) the IR $\nu=0$ part

$$n(n-1)\sum V_{ijij}/[2N(N+1)],$$

(ii) the IR $\nu=1$ part

$$\sum_{i,k}(n-1)(N+2)^{-1}\left[\sum_j\left(V_{ijkj}-\delta_{ik}\sum_{ij}V_{ijij}/N\right)A_iB_k\right],$$

and (iii) the IR $\nu=2$ part

$$\tfrac{1}{4}\sum\tilde{V}_{ijkl}A_iA_jB_kB_l.$$

Here $\tilde{V}_{ijkl}$ is the IR $\nu=2$ matrix element, obtained by subtracting $\nu=0$ and $\nu=1$ contributions from V_{ijkl}.

We have already given expressions for the first four central moments of EVD generated by the one-body (which is essentially IR $\nu=1$) interaction. Now we give the final expressions for second, third, and fourth central moments of EVD generated by a two-body, IR rank $\nu=2$ Hamiltonian H. The procedure leading to these final expressions is straightforward, and consists of (a) decomposing H^n into parts according to the body (particle) rank, and (b) using Eq. (2) separately for each part. In order to take care of the bookkeeping of various terms, which is the only difficult part in the whole procedure, we use the Feynman diagram type approach.[8,9] The following is the expression for the ith central moment μ_i [$\mu_i^m = \langle(\tilde{H})^i\rangle^m$, $\tilde{H}$ is the traceless part of H in m-particle scalar space] of EVD for m bosons interacting via two-body, IR rank $\nu=2$ interaction $\tilde{V}$:

$$\langle(\tilde{V})^i\rangle^m=\sum_{s=2}^{i}\binom{m}{s}\binom{N+s+m-1}{s}\times\binom{2s}{s}^{-1}\binom{N+2s-1}{2s}C_i^s. \tag{11}$$

Here i goes from 2 to 4. All the necessary information about the two-body, $\nu=2$ interaction goes in the evaluation of coefficients C_i^s. The expressions for these in terms of the two-body matrix elements are given below:

$$\begin{aligned}
&C_2^{\,2}=\tfrac{1}{4}\sum \tilde{V}_{ijkl}\tilde{V}_{klij},\quad C_3^{\,2}=\tfrac{1}{8}\sum \tilde{V}_{ijkl}\tilde{V}_{klmn}\tilde{V}_{mnij},\\
&C_3^{\,3}=2C_3^{\,2}+\sum \tilde{V}_{iijk}\tilde{V}_{knlm}\tilde{V}_{mjni},\\
&C_4^{\,2}=\tfrac{1}{16}(AA1),\\
&C_4^{\,3}=\tfrac{1}{4}(AA1)+CC1+\tfrac{1}{2}BA1+2(CA1),\\
&C_4^{\,4}=\tfrac{3}{8}(AA1)+6(CC1)+3(BA1)+6(CA1)\\
&\qquad +3(AB1)+3(C_2^{\,2})^2,
\end{aligned} \tag{12}$$

where

$$\begin{aligned}
&AA1=\sum \tilde{V}_{ijkl}\tilde{V}_{klmn}\tilde{V}_{mnop}\tilde{V}_{opij},\\
&AB1=\sum \tilde{V}_{ijkl}\tilde{V}_{lmjn}\tilde{V}_{nomp}\tilde{V}_{pkoi},\\
&BA1=\sum \tilde{V}_{ijkl}\tilde{V}_{klin}\tilde{V}_{mnop}\tilde{V}_{opmj},\\
&CA1=\sum \tilde{V}_{ijkl}\tilde{V}_{kmno}\tilde{V}_{olmp}\tilde{V}_{npij},\\
&CC1=\sum \tilde{V}_{ijkl}\tilde{V}_{mnjo}\tilde{V}_{olnp}\tilde{V}_{pkmi}.
\end{aligned} \tag{13}$$

Here the summation is over all labels. These results can be very easily extended to evaluation of the moments of EVD generated by a general (1+2)-body interaction H; the corresponding expressions are given in the Appendix.

The necessary information C_i^s for the evaluation of the ith moment of EVD in the m-boson space, generated by IR $\nu=2$ interaction, can also be obtained from the ith moment in 2 to the i-boson spaces only. One, however, would expect, due to the two-body character of the interaction, that the ith moments in 2 to $2i$ particle spaces would have been necessary[10] for evaluation of the ith moment of EVD in m-boson space. The validity of this fact in fermion spaces[3,10] can be attributed to the particle-hole symmetry of the IR $\nu=2$ interaction. There is no concept of hole for bosons, and hence one can see that the IR $\nu=2$ nature of interaction is responsible for this simplicity, and in fermion spaces, the particle-hole symmetry is a consequence of the IR $\nu=2$ nature of the interaction.

We shall now discuss the conditions for the normality of the EVD. The first two moments of the EVD tell us about the location (centroid) and the spread (width) of the EVD. The information about the shape of the spectrum is given by the third and higher central moments of the EVD. In principle, all the higher moments are necessary for determining the shape of EVD; but here we shall deal only with the moments up to fourth order. The corresponding shape parameters are the skewness $\gamma_1=\mu_3/(\mu_2)^{3/2}$, and the excess $\gamma_2=\mu_4/(\mu_2)^2-3$. For the Gaussian or the normal distribution, both γ_1 and γ_2 are equal to zero.

Let us start with a pure one-body interaction; the central moments for such an interaction are given by Eqs. (4) and (5). In the dilute limit, defined by $m\to\infty$, $N\to\infty$, but $m/N\ll 1$, we find that the shape parameters $\gamma_1(m)$ and $\gamma_2(m)$ for m-boson EVD approach asymptotically to values given by the expressions

$$\begin{aligned}
&\gamma_1(m)=m^{-1/2}\gamma_1(1),\\
&\gamma_2(m)=[\gamma_2(1)+3](m^{-1}+6N^{-1}).
\end{aligned} \tag{14}$$

Thus, in the dilute limit, both $\gamma_1(m)$ and $\gamma_2(m)$ approach zero if the shape parameters $\gamma_1(1)$ and $\gamma_2(1)$ for one-boson EVD are finite. The same result was obtained for the fermion EVD in the dilute limit.[2] Thus in the dilute system, one gets the same result irrespective of the fermion or the boson nature of particles, since the Pauli's exclusion principle has negligibly small effect when $N\gg m$. However, for bosons, since there is no limit to the number of particles in any single-particle state, unlike fermions, we can consider the limit $m \to \infty$ with finite N. This is called the dense limit; the second central moment and the shape parameters in this limit are given by

$$
\begin{aligned}
\langle(\bar{H})^2\rangle^m &= \sigma^2(m) = m^2\sigma^2(1)/(N+1)\,, \\
\gamma_1(m) &= 2\gamma_1(1)\frac{(N+1)^{1/2}}{(N+2)}\,, \\
\gamma_2(m) &= 6\frac{\{[\gamma_2(1)+3](N+1)-(2N+3)\}}{(N+2)(N+3)}
\end{aligned}
\tag{15}
$$

The width of the m-boson EVD $\sigma(m)$ increases linearly with m. The shape parameters, however, approach values which are independent of m, but which very much depend on the shape parameters for one-boson EVD and the size of the single-boson space N. For large values of N again, the EVD can be considered Gaussian if $\gamma_1(1)$ and $\gamma_2(1)$ are reasonably small. For example, in the case of noninteracting particles, with N equidistant single-particle states, we have $\gamma_1(1)=0$ and $\gamma_2(1)=-1.2\ [(N^2+1)/(N^2-1)]$. We can see that for $N=5$, the absolute value of excess in the dense limit is less than 0.24; for $N=10$ it is less than 0.12. Thus, the EVD in this case approaches Gaussian in the dense limit if $N>5$.

We now consider the dilute and the dense limits of the shape of EVD generated by a purely two-body, IR rank-2 interaction V, by introducing dimensionless quantities $P(m,n)$, defined by

$$
P(m,n) = \frac{C_m^n}{(C_2^{\,2})^{m/2}}\,. \tag{16}
$$

These are the ratios of traces of various combinations of the matrix elements, and are characteristic of the interaction used to generate the EVD. In the dilute limit, $\gamma_1(m)$ approaches $NP(3,2)/m + P(3,3)$ and $[\gamma_2(m)+3]$ tends to $N^2P(4,2)/m^2 + NP(4,3)/m + P(4,4)$. Thus the EVD approach Gaussian distribution if $P(3,2) \ll m/N$, $P(3,3) \ll 1$, $P(4,2) \ll m^2/N^2$, $P(4,3) \ll m/N$, and $(P(4,4)-3) \ll 1$. One can translate these conditions to conditions on the shape parameters of EVD for 2, 3, and 4 bosons. The skewness and excess in dilute limit approach constant values $-\sqrt{2}\,\gamma_1(2)+\sqrt{6}\,\gamma_1(3)$, and $\gamma_2(2)-6\gamma_2(3)+6\gamma_2(4)$, respectively. Thus the approach to Gaussian distribution in the dilute limit is decided by these shape parameters of EVD for 2, 3, and 4 bosons. Since in the dilute limit, bosons and fermions behave similarly, the same conditions will apply for fermions. This clearly shows that for purely two-body, IR rank-2 interaction, the assumption of applicability of a strong central limit theorem is not valid. The shape of the EVD in the dilute limit does not necessarily approach the Gaussian distribution with increasing m, as the limiting values of $\gamma_1(m)$ and $\gamma_2(m)$ reach constant values depending on the shape parameters of the EVD for $m=2$, 3, and 4. Of course, the addition of the one-body part of the interaction and the IR rank-1 part of the two-body interaction to the two-body IR rank-2 interaction is likely to alter the whole picture, and the conditions for normality of the EVD will have complicated structures. It is known that the IR rank-1 part of the interaction is quite important for the nuclear structure studies.[11]

In the dense limit ($m \to \infty$, N finite) valid only for bosons, the values of $P(3,3)$ and $P(4,4)$ completely determine the shape of the EVD. For,

$$
\begin{aligned}
\gamma_1(m) &\to \frac{P(3,3)[N(N+1)(N+2)(N+3)]^{1/2}}{(N+4)(N+5)}\,, \\
\gamma_2(m)+3 &\to \frac{P(4,4)N(N+1)(N+2)(N+3)}{(N+4)(N+5)(N+6)(N+7)}\,.
\end{aligned}
\tag{17}
$$

For large N, we get $\gamma_1(m) = P(3,3)$ and $\gamma_2(m)+3 = P(4,4)$ in the dense limit. Again one can express these conditions in terms of the shape parameters of EVD for 2, 3, and 4 bosons, using the following expressions for $P(3,3)$ and $P(4,4)$:

$$
\begin{aligned}
P(3,3) &= (N+4)\left(\frac{2}{N(N+1)}\right)^{1/2} \times\left[\left(\frac{3(N+4)}{(N+2)}\right)^{1/2}\gamma_1(3)-\gamma_1(2)\right], \\
P(4,4) &= \frac{(N+4)}{N(N+1)}\left(6[\gamma_2(4)+3]\frac{(N+4)(N+5)^2}{(N+2)(N+3)} - 6[\gamma_2(3)+3]\frac{(N+6)(N+4)}{(N+2)} + [\gamma_2(2)+3](N+7)\right).
\end{aligned}
\tag{18}
$$

So in the dense limit also, the skewness and the excess of EVD for m bosons approach constant values depending on the values of $\gamma_1(i)$ for $i=2,3$ and $\gamma_2(j)$ for $j=2,3,4$. The approach to Gaussian limit depends only on these numbers.

Two sets of matrix elements have been given by Bortignon *et al.*[4] for the d bosons of the interacting boson approximation model.[12] Using these two sets of matrix elements, we obtain $P(3,3) = 1.78$ and $P(4,4) = 23.12$ for one interaction, while the second one gives $P(3,3) = 2.09$ and $P(4,4) = 25.55$. The size of the single-particle space is given by $N=5$. So, in the dense limit, the limiting values of the shape parameters are

$\gamma_1(m)=0.81, 0.95$ and $\gamma_2(m)=0.27, 0.61$, for the two interactions, respectively; hence the shape of the EVD is nowhere near Gaussian. A pairing interaction in the space of s-d bosons with $N=6$ gives $P(3,3)=1.46$ and $P(4,4)=15.64$; hence the limiting values of γ_1 and γ_2 in the dense limit are 0.73 and -0.24, respectively. Similarly, a $Q \cdot Q$ interaction in the s-d boson space (having both IR rank-1 and IR rank-2 parts) gives the following limiting values for $\gamma_1(m)=-0.76$ and $\gamma_2(m)=0.86$ in the dense limit. Incidentally, for fermions and the $Q \cdot Q$ interaction in a large j orbit, the values of γ_1 and γ_2 in the middle of the shell have been obtained by Nomura,[13] and these are 1.26 and -0.6, respectively.

One interesting feature of the boson systems is the complete domination of the two-body part of the interaction, over the one-body part for large m. This can be seen from the expressions given in the Appendix for moments of the $(1+2)$-body interaction; each factor containing the external single-particle energy term (ϵ) gets multiplied by m, while each factor containing a two-body matrix element (V) has a coefficient proportional to m^2 in the large m limit. For example, the one-body contribution to μ_4 has a coefficient which goes as m^4, and the corresponding coefficient for the two-body contribution to μ_4 goes as m^8 in the dense limit. One also sees that in this limit the shape parameters approach constant values, which are not necessarily those of a Gaussian distribution. For the dilute limit, one can get the result of noninteracting particles only if the one-body part (external single-particle energies) is dominant over the two-body part.

Finally, let us consider the shape of ensemble-averaged EVD, using an ensemble of random interactions. The ensemble averaged skewness will be zero for any ensemble (with matrix element mean $=0$) of random interactions, because each necessary matrix element trace for the evaluation of the ensemble averaged third moment will have an odd number of matrix elements. We therefore need to consider only the ensemble averaged excess in this case. For the dilute limit one can show that the ensemble averaged eigenvalue distribution is Gaussian. For the dense limit, one gets the Gaussian distribution if N is sufficiently large; the shape parameters for finite N in the dense limit have complicated structure and shall be discussed elsewhere.

In conclusion, we have obtained analytic expressions for the first four moments of the EVD of interacting (via two-body interactions) and noninteracting boson systems. The conditions leading to normality of the EVD have been discussed. Extension of the present work for configuration subspaces is straightforward.

APPENDIX

Here we give expressions for the first four moments of the EVD of a general $(1+2)$-body Hamiltonian (H):

$$\begin{aligned} H &= \sum \epsilon_{ij} A_i B_j + \tfrac{1}{4} \sum V_{ijkl} A_i A_j B_k B_l \\ &= H^{\nu=0} + H^{\nu=1} + H^{\nu=2} \\ &= H_0 + \tilde{H} , \end{aligned} \tag{A1}$$

$$H^{\nu=0} = H_0 = n \sum \epsilon_{ii}/N + \frac{n(n-1)}{N(N+1)} \sum_{i\leq j} V_{ijij} ,$$

$$H^{\nu=1} = h = \sum \xi_{ij} A_i B_j ,$$

$$\xi_{ij} = \epsilon_{ij} - \frac{1}{N} \sum \epsilon_{ii} \delta_{ij} + \frac{n-1}{N+2} \times \left[\sum_k \left(V_{ikjk} - \frac{\delta_{ij}}{N} \sum V_{mnmn} \right) \right] ,$$

$$H^{\nu=2} = \tilde{V} = \tfrac{1}{4} \sum \tilde{V}_{ijkl} A_i A_j B_k B_l , \tag{A2}$$

$$\langle H \rangle^m = \frac{m}{N} \sum \epsilon_{ii} + \frac{m(m-1)}{N(N+1)} \sum_{i\leq j} V_{ijij} ,$$

$$\langle (H)^2 \rangle^m = \frac{m(N+m)}{N(N+1)} \sum \xi_{ij} \xi_{ji} + \langle (\tilde{V})^2 \rangle^m ,$$

$$\langle (H)^3 \rangle^m = \frac{m(N+m)(N+2m)}{N(N+1)(N+2)} \sum \xi_{ij} \xi_{jk} \xi_{ki} + \frac{\binom{m}{2}\binom{N+m+1}{2}}{\binom{4}{2}\binom{N+3}{4}} (3D_1 + \tfrac{3}{2} E_1) + \frac{\binom{m}{3}\binom{N+m+1}{2}}{\binom{5}{2}\binom{N+4}{5}} (3E_1) + \langle (\tilde{V})^3 \rangle^m ,$$

where

$$D_1 = \sum \tilde{V}_{ijkl} \xi_{ik} \xi_{jl} ,$$

$$E_1 = \sum \tilde{V}_{ijkl} \tilde{V}_{kmij} \xi_{lm} , \tag{A3}$$

$$
\langle\langle (H)^4 \rangle\rangle^m = \frac{m(N+m)}{N(N+1)} M_2 + \frac{\binom{m}{2}\binom{N+m+1}{2}}{\binom{4}{2}\binom{N+3}{4}} [12K_1 + 2G_1 + 2G_2 + 2G_3 + F_1 + 3(M_1)^2 + 6M_2]
$$

$$
+ \frac{\binom{m}{3}\binom{N+m+1}{2}}{\binom{5}{2}\binom{N+4}{5}} [24K_1 + 2F_1] + \frac{\binom{m}{3}\binom{N+m+2}{3}}{\binom{6}{3}\binom{N+5}{6}} [12G_1 + 6G_2 + 12G_3 + 12G_4 + \tfrac{3}{2} G_5 + 2F_1 + 12F_2 + 6F_3]
$$

$$
+ \frac{\binom{m}{4}\binom{N+m+2}{3}}{\binom{7}{3}\binom{N+6}{7}} [4F_1 + 24F_2 + 12F_3] + \langle\langle (\tilde{V})^4 \rangle\rangle^m ,
$$

where

$$
\begin{aligned}
M_1 &= \sum \xi_{ij}\xi_{ji}, \quad M_2 = \sum \xi_{ij}\xi_{jk}\xi_{kl}\xi_{li} , \\
K_1 &= \sum \tilde{V}_{ijkl}\xi_{ik}\xi_{jm}\xi_{ml} , \\
G_1 &= \sum \tilde{V}_{ijkl}\tilde{V}_{knij}\xi_{lm}\xi_{mn} , \\
G_2 &= \sum \tilde{V}_{ijkl}\tilde{V}_{mnij}\xi_{km}\xi_{ln} , \\
G_3 &= \sum \tilde{V}_{ijkl}\tilde{V}_{kmin}\xi_{lm}\xi_{nj} , \\
G_4 &= \sum \tilde{V}_{ijkl}\tilde{V}_{kmin}\xi_{jl}\xi_{mn} , \\
G_5 &= \sum \xi_{ij}\xi_{ji} \sum \tilde{V}_{klmn}\tilde{V}_{mnkl} , \\
F_1 &= \sum \tilde{V}_{ijkl}\tilde{V}_{kpmn}\tilde{V}_{mnij}\xi_{lp} , \\
F_2 &= \sum \tilde{V}_{ijkl}\tilde{V}_{lmjn}\tilde{V}_{npmi}\xi_{kp} , \\
F_3 &= \sum \tilde{V}_{ijkl}\tilde{V}_{klmj}\tilde{V}_{mpin}\xi_{np} .
\end{aligned}
\tag{A4}
$$

Note that the expressions for $\langle\langle (\tilde{V})^i \rangle\rangle^m$ are given in Eq. (11) of the text.

[1]F. S. Chang, J. B. French, and T. H. Thio, Ann. Phys. (N.Y.) 66, 137 (1971).

[2]K. K. Mon and J. B. French, Ann. Phys. (N.Y.) 95, 90 (1975).

[3]M. Nomura, Prog. Theor. Phys. 59, 1199 (1978).

[4]P. F. Bortignon, M. Dalla Francesca, and V. R. Manfredi, Lett. Nuovo Cimento 19, 15 (1977).

[5]P. F. Bortignon, V. R. Manfredi, and S. Spitz, Lett. Nuovo Cimento 23, 210 (1978).

[6]M. Nomura, Ann. Phys. (N.Y.) 91, 83 (1975).

[7]V. K. B. Kota, J. Phys. Lett. (Paris) (to be published).

[8]J. N. Ginocchio, Phys. Rev. C 8, 135 (1973); S. Ayik and J. N. Ginocchio, Nucl. Phys. A221, 285 (1974).

[9]V. Potbhare, Ph.D. thesis, University of Rochester, 1975.

[10]J. B. French and K. F. Ratcliff, Phys. Rev. C 3, 94 (1971).

[11]V. Potbhare and S. P. Pandya, Nucl. Phys. A256, 253 (1976); V. K. B. Kota and V. Potbhare, *ibid.* (to be published).

[12]A. Arima and F. Iachello, Ann. Phys. (N.Y.) 99, 253 (1976).

[13]M. Nomura, Prog. Theor. Phys. 48, 442 (1972).

VOLUME 48, NUMBER 16 PHYSICAL REVIEW LETTERS 19 APRIL 1982

Fluctuation Properties of Nuclear Energy Levels: Do Theory and Experiment Agree?

R. U. Haq
Physique Théorique et Mathématique, Université Libre de Bruxelles, B-1050 Brussels, Belgium, and Department of Physics, University of Toronto, Toronto M5S 1A7, Canada

and

A. Pandey
Division de Physique Théorique, Institut de Physique Nucléaire, F-91406 Orsay, Cédex, France, and Department of Physics and Astronomy, University of Rochester, Rochester, New York 14627

and

O. Bohigas
Division de Physique Théorique, Institut de Physique Nucléaire, F-91406 Orsay, Cédex, France

(Received 11 December 1981)

The fluctuation properties of nuclear energy levels are analyzed with new spectrally averaged measures. A remarkably close agreement between the predictions of random-matrix theories and experiment is found.

PACS numbers: 21.10.Ma, 24.60.+m

Since the pioneering work of Wigner and its systematic development by many authors, it has become clear that random-matrix theories[1,2] provide a natural theoretical framework for nuclear energy-level fluctuations, i.e., departures from spectral uniformity. The Gaussian orthogonal ensemble (GOE) of asymptotically large real symmetric random matrices is the model that has been most extensively investigated. Almost twenty years ago Dyson and Mehta[3] emphasized one of the main questions at issue: "We would be very happy if we could report that our theoretical model had been strikingly confirmed by the statistical analysis of neutron capture levels. We would be even happier if we could report that our theoretical model had been decisively contradicted.... Unfortunately, our model is as yet neither proved nor disproved." A large amount of effort has been devoted since then to providing accurate data on slow-neutron resonances of heavy nuclei as well as on proton resonances of light nuclei, and to their analysis. Despite this, only relatively vague, although significant, "overall good agreement" between theory and experiment can be quoted from subsequent work on the subject. It is the purpose of this Letter to settle once and for all this question for energy-level fluctuations.

Our analysis will differ from the previous ones in two main respects: (i) The *whole body of high-quality data* is analyzed simultaneously, and (ii) several *new spectral measures* are introduced. In particular, special care regarding sample-size effects can then be taken when comparing data with predictions. Use will be made of what is known analytically, supplemented with new Monte Carlo results when analytic expressions are not available.

In the conventional analysis one treats the data from each nucleus, containing about 100 levels, separately and consequently the statistical significance is poor. On the other hand, one expects that the resonance energies of every compound nucleus will share the same fluctuation properties and it seems therefore natural, and in the spirit of random-matrix theory, to treat the combined set of nuclear resonance-energy data of different nuclei—in short, the nuclear data ensemble (NDE)—as a sampling of eigenvalues of GOE matrices; see Rosenzweig and Porter[4] for a similar notion for complex atomic spectra. The data analyzed here consist of 1407 resonance energies corresponding to 30 sequences of 27 different nuclei: (i) slow-neutron resonance data[5] on ^{110,112,114}Cd, ^{152,154}Sm, ^{154,156,158,160}Gd, ^{160,162,164}Dy, ^{166,168,170}Er, ^{172,174,176}Yb, ^{182,184,186}W, ^{232}Th, and ^{238}U (1146 levels); (ii) proton resonance data[6] on ^{44}Ca ($J=\frac{1}{2}^-$), ^{44}Ca ($J=\frac{1}{2}^+$), and ^{48}Ti ($J=\frac{1}{2}^+$) (157 levels); and (iii) (n,γ) reaction data[7] on ^{177}Hf ($J=3$), ^{177}Hf ($J=4$), ^{179}Hf ($J=4$), and ^{179}Hf ($J=5$) (104 levels). The relative scarcity of data useful for our purpose is due to the very stringent conditions that are imposed. For each nucleus the sequence should ideally be *complete* (no missing levels) and *pure* (the same angular momentum and parity).

Fluctuation properties are in general characterized by the set of ν-level cluster functions[8] Y_ν ($\nu \geq 2$). Although all cluster functions have been evaluated[9] for the GOE, mostly measures deriving from Y_2 only have been used so far. There are several reasons for this: One should follow a hierarchy and start with $\nu = 2$; the higher the value of ν, the more complicated the structure of Y_ν, and therefore the harder it is to produce meaningful measures that can be derived analytically; moreover, detection of higher-order effects requires larger samples. For properties related to Y_ν with $\nu > 2$, we shall mostly rely on Monte Carlo results that we have obtained from a sample, adequate for our purpose, of 100 GOE matrices of dimensionality 300.

To begin with, we consider the function Y_2. For the GOE,[8]

$$Y_2(r) = \left(\frac{\sin\pi r}{\pi r}\right)^2 + \left(\int_r^\infty \frac{\sin\pi r'}{\pi r'}dr'\right)\frac{d}{dr}\left(\frac{\sin\pi r}{\pi r}\right), \tag{1}$$

which has the value unity at $r=0$ and goes to zero as $(\pi r)^{-2}$ for large r ($\gtrsim 1$). However, since the large-r form cannot be tested accurately from the data, one considers measures which involve integrations over Y_2. One such measure[2,3] is $\Sigma^2(\bar{n})$, the variance of the number n of levels in a given interval of length[10] $\bar{n}$:

$$\begin{aligned}\Sigma^2(\bar{n}) &= \bar{n} - 2\int_0^{\bar{n}}(\bar{n}-r)Y_2(r)dr \\ &= (2/\pi^2)\{\ln(2\pi\bar{n}) + \gamma + 1 + \tfrac{1}{2}[\mathrm{Si}(\pi\bar{n})]^2 - \tfrac{1}{2}\pi\,\mathrm{Si}(\pi\bar{n}) - \cos(2\pi\bar{n}) - \mathrm{Ci}(2\pi\bar{n}) + \pi^2\bar{n}[1-(2/\pi)\mathrm{Si}(2\pi\bar{n})]\}.\end{aligned} \tag{2}$$

Another is the Δ_3 statistic of Dyson and Mehta[3] (higher moments of Δ_3 involve functions with $\nu > 2$ as well). It measures, for a fixed interval $[x, x+\bar{n}]$, the least-squares deviation of the staircase $N(E)$, the number of levels with energy less than or equal to E, from the best straight line fitting it:

$$\Delta_3(\bar{n};x) = (1/\bar{n})\min_{A,B}\int_x^{x+\bar{n}}[N(E) - AE - B]^2 dE. \tag{3}$$

Its ensemble average is related[11] to Σ^2, and hence to Y_2, by

$$\bar{\Delta}_3(\bar{n}) = (2/\bar{n}^4)\int_0^{\bar{n}}(\bar{n}^3 - 2\bar{n}^2 r + r^3)\Sigma^2(r)dr, \tag{4}$$

which may be integrated numerically; see Fig. 1(a). Note that the ensemble properties, due to stationarity,[11] are independent of the position x of the interval $[x, x+\bar{n}]$ and that the asymptotic-$\bar{n}$ results[3] [e.g., $\bar{\Delta}_3(\bar{n}) \simeq \pi^{-2}\ln\bar{n} - 0.007$, $\mathrm{Var}\Delta_3 \simeq 0.012$], the only ones used so far to analyze data, may not be of much interest for small $\bar{n}$; see Fig. 1(b).

We have made an extensive test of the theory by studying the aforementioned functions $Y_2(r)$, $\Sigma^2(r)$, and $\Delta_3(\bar{n})$ as well as others (spacing distributions, etc.). A complete account will be given elsewhere and from now on we shall restrict ourselves to Δ_3. This choice is sensible because Δ_3 is adapted to describe one of the most characteristic features of the GOE (the spectral rigidity) and also because it provides among the sharpest comparisons between theory and experiment.

In order to make optimal use of the data, we introduce *spectral-averaged quantities*. Consider a sequence containing p levels. As an example consider the spectral average

$$\langle\Delta_3(\bar{n})\rangle_p = (p-\bar{n})^{-1}\int_x^{x+(p-\bar{n})}\Delta_3(\bar{n},y)dy \tag{5}$$

which estimates $\bar{\Delta}_3(\bar{n})$. Such an estimate will be meaningful only if sample errors due to finite p are also given. The variance of $\langle\Delta_3\rangle_p$, which is the square of the sample error, is given by

$$\mathrm{Var}\langle\Delta_3(\bar{n})\rangle_p \equiv \overline{\langle\Delta_3(\bar{n})\rangle_p^2} - \overline{\langle\Delta_3(\bar{n})\rangle_p}^2 = \frac{2\mathrm{Var}\Delta_3(\bar{n})}{(p-\bar{n})^2}\int_0^{p-\bar{n}}(p-\bar{n}-r)C(r;\bar{n})dr \simeq \frac{2\mathrm{Var}\Delta_3(\bar{n})}{p}\int_0^1 C(\lambda\bar{n};\bar{n})d\lambda. \tag{6}$$

In (6) $C(r;\bar{n})$ is the autocorrelation function of Δ_3:

$$C(r;\bar{n}) = [\mathrm{Var}\Delta_3(\bar{n})]^{-1}\{\overline{\Delta_3(\bar{n};x)\Delta_3(\bar{n};x+r)} - [\bar{\Delta}_3(\bar{n};x)]^2\}. \tag{7}$$

The last step of Eq. (6), which has been checked by Monte Carlo calculations, is valid for $p \gg \bar{n} \gtrsim 2$. We emphasize that it is the theoretical model itself, with which one compares data, that provides the sample errors. Notice moreover that as $p \to \infty$, $\mathrm{Var}\langle\Delta_3(\bar{n})\rangle_p$ goes to zero (ergodic property[11]). In practice, instead of (5) we will use a less smooth average in which the integral is replaced by a summation.

We now turn to the data (NDE). We calculate for each of the thirty sequences the spectral-averaged quantity $\langle\Delta_3(\bar{n})\rangle_p$ and then take their average, weighted according to the size of each se-

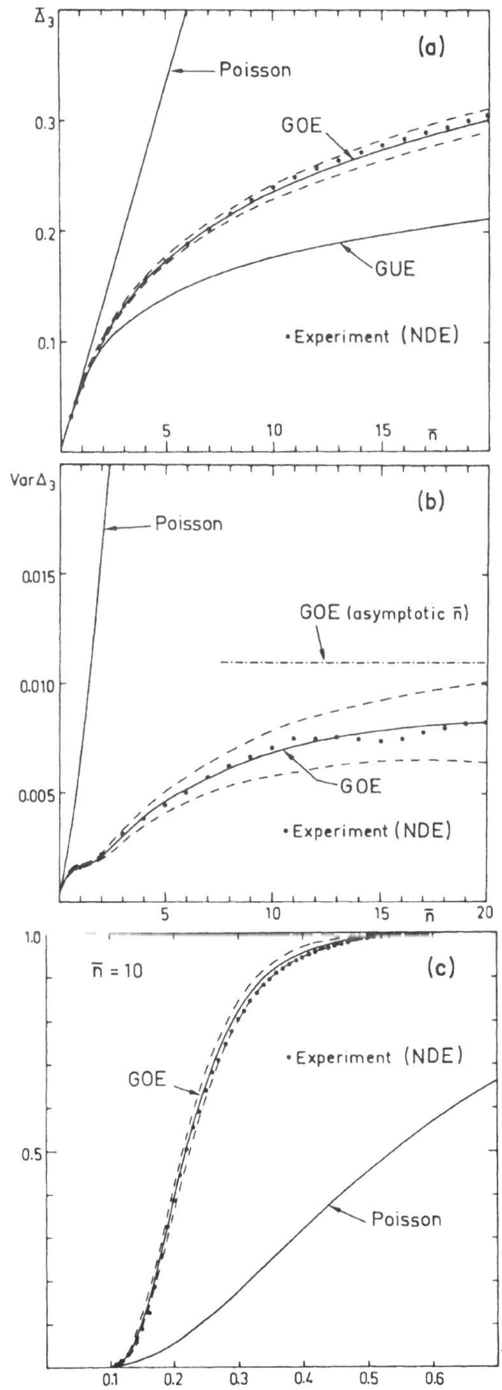

FIG. 1. (a) $\bar{\Delta}_3$ and (b) VarΔ_3 as functions of $\bar{n}$; (c) distribution function of Δ_3 for $\bar{n} = 10$. Dashed lines correspond, for GOE, to one standard deviation from the average.

quence. We have considered the range $0 \leq \bar{n} \leq 100$. The results for $\bar{n} \leq 20$ are shown in Fig. 1(a). They are in a remarkably good agreement with the GOE. We emphasize that typically the "figure of merit" $(\text{Var}w)^{1/2}/\bar{w}$ for a measure w is of the order of a few percent in the present analysis, whereas in the conventional one it is ~20% or more. That the data points for various values of $\bar{n}$ all lie within one sample error is partly due to the fact that they are not totally independent. Furthermore, since $\langle\Delta_3(\bar{n})\rangle$ is a smooth function of $\bar{n}$, there is no need to insert more points in the range considered.

To explore effects of $\nu > 2$ we have also considered the ensemble variance of $\Delta_3(\bar{n})$ and the distribution of Δ_3 for various values of $\bar{n}$. We proceed in the same way as before except that now the GOE predictions come from Monte Carlo calculations. The results for VarΔ_3 are shown in Fig. 1(b) and one example ($\bar{n} = 10$) of the distribution function of Δ_3 is shown in Fig. 1(c). The agreement between GOE predictions and data (NDE) is again very good. The values corresponding to the Poisson case (a spectrum which unlike the GOE has no level repulsion nor any correlation between spacings) are also given for comparison.

We have thus established an astonishingly good agreement between a parameter-free theory (GOE) and the data. We emphasize that, apart from *rotation and time-reversal invariance* resulting in the real symmetric nature of the matrices, the GOE takes no account of the specific properties of the nuclear Hamiltonian, e.g., its (1 + 2)-body nature, its large pairing and quadrupole components, etc. We also recall that eigenvalues of realistic (nonrandom) nuclear shell-model matrices show the same fluctuation patterns.[2,12] How can this be understood theoretically? Monte Carlo calculations[12] have indicated that ensembles of two-body operators acting in many-particle (shell-model) spaces also yield fluctuations characteristic of the GOE. More recently[13] it has been shown that adding a GOE matrix H to any real symmetric matrix K $(K + \alpha H)$ leads very quickly, as α increases, to the same fluctuations. Intermediate fluctuation patterns are to be expected only when the random-matrix elements are not much larger than the local average spacing of the given (nonrandom) matrix K. The good agreement with experiment, coupled with the theoretical understanding which is slowly emerging, reinforces the belief that the GOE fluctuations are to be found in nature under very general conditions.

We mention finally that one can use the close agreement to impose restrictions on mechanisms which would change the fluctuations. In particular if time reversal is not an exact symmetry, the appropriate model would be an ensemble of complex Hermitian matrices, in which the real and imaginary parts are sampled independently. {The Gaussian unitary ensemble [GUE, see Fig. 1(a)] in which the two parts have the same norm, is one such example.} But even a small magnitude of the imaginary part induces major changes in the fluctuation properties. This notion is being pursued[14] to derive an upper bound on the time-reversal–noninvariant part of the nuclear Hamiltonian.

We are grateful to J. B. French for valuable suggestions and a critical reading of the manuscript. We also acknowledge helpful discussions with S. S. M. Wong. One of us (A.P.) was a Fellow of the Fondation Joliot-Curie, 1980–1981. The Division de Physique Théorique del' Institut de Physique Nucléaire is a Laboratoire associé au Centre National de la Recherche Scientifique.

[1] *Statistical Theories of Spectra: Fluctuations*, edited by C. E. Porter (Academic, New York, 1965); M. L. Mehta, *Random Matrices and the Statistical Theory of Energy Levels* (Academic, New York, 1967).

[2] T. A. Brody, J. Flores, J. B. French, P. A. Mello, A. Pandey, and S. S. M. Wong, Rev. Mod. Phys. 53, 385 (1981); this review covers recent developments and contains up-to-date references in the field.

[3] F. J. Dyson and M. L. Mehta, J. Math. Phys. 4, 701 (1963).

[4] N. Rosenzweig and C. E. Porter, Phys. Rev. 120, 1698 (1960).

[5] H. I. Liou, G. Hacken, J. Rainwater, and U. N. Singh, Phys. Rev. C 11, 462 (1975), and earlier references therein; for ^{155}Gd, see C. Coceva and M. Stefanon, Nucl. Phys. A315, 1 (1979), and private communication.

[6] W. M. Wilson, E. G. Bilpuch, and G. E. Mitchell, Nucl. Phys. A245, 285 (1975).

[7] C. Coceva, F. Corvi, P. Giacobbe, and M. Stefanon, in *Statistical Properties of Nuclei*, edited by J. B. Garg (Plenum, New York, 1972).

[8] F. J. Dyson, J. Math. Phys. 3, 166 (1962).

[9] F. J. Dyson, Commun. Math. Phys. 19, 235 (1970); M. L. Mehta, Commun. Math. Phys. 20, 245 (1971).

[10] For a random variable w, we denote its average by $\overline{w}$ and its ensemble variance $\overline{(w-\overline{w})^2}$ by Var$w$. Note, moreover, that we are considering spectra normalized to unit local average spacing.

[11] A. Pandey, Ann. Phys. (N.Y.) 119, 170 (1979).

[12] S. S. M. Wong and J. B. French, Nucl. Phys. A198, 188 (1972); O. Bohigas and M. J. Giannoni, Ann. Phys. (N.Y.) 89, 393 (1975).

[13] A. Pandey, Ann. Phys. (N.Y.) 134, 110 (1981).

[14] J. B. French, V. K. B. Kota, and A. Pandey, to be published.

Volume 114B, number 2,3 PHYSICS LETTERS 22 July 1982

SHELL MODEL AND SPECTRAL DISTRIBUTION STUDY OF SU(4) SYMMETRY MIXING IN THE S–D SHELL

R.U. HAQ [1] and C. QUESNE [2]

Physique Théorique et Mathématique CP 229, Université Libre de Bruxelles, Boulevard du Triomphe, B 1050 Brussels, Belgium

Received 11 February 1982
Revised manuscript received 21 April 1982

Ground state intensities of SU(4) irreducible representations are evaluated by shell model diagonalization in various nuclei of the first half of the s–d shell. They are well reproduced by using SU(4) spin–isospin distributions with exactly calculated widths. Such distributions can then serve to predict the intensities in the other nuclei. The SU(4) symmetry is most often badly broken.

During the last few years, Wigner spin–isospin SU(4) symmetry [1] has been studied in light nuclei using the spectral distribution method of French [2]. Most studies have been based upon SU(4) distributions, i.e., distributions over the m-nucleon spectrum of the average intensity of the m-nucleon states belonging to given irreducible representations (IR) of SU(4). From a knowledge of the centroid energies and widths of such distributions, Parikh [3] evaluated the ground-state intensities of the various SU(4) IR's in s–d shell nuclei using realistic interactions. These relative intensities may be interpreted as mixing intensities [4]. Further information on SU(4) symmetry mixing was obtained by Hecht and Draayer [5], who evaluated the partial widths of SU(4) distributions by taking advantage of the full power of group theory and then used them to calculate a simple measure of SU(4) symmetry mixing. Recently, Chakraborty et al. [6] adopted a similar approach based upon a new measure of symmetry breaking that is theoretically more sound than the one used in ref. [5].

SU(4) distributions do not take into account the existence of exact symmetries in the problem, namely those connected with the conservation of angular momentum and isospin (when neglecting the Coulomb interaction in the latter case). The isospin effect should be important when considering nuclei with neutron excess. In view of this, Haq and Parikh [7] considered distributions of states having fixed SU(4) symmetry, spin S, and isospin T. However, due to technical reasons connected with the group theoretical structure of the problem, it was not possible to calculate in a straightforward way the widths of the distributions. They introduced therefore the concept of an equivalent width and proposed two different methods to calculate it. Later it was shown by Quesne [8] that the widths of SU(4) ST distributions could be calculated exactly through a group theoretical approach. In a recent publication [9], we carried out a numerical application to the s–d shell nuclei of the method proposed in ref. [8], and used the results to analyse the SU(4) symmetry mixing near the ground state again. However, since there has been no comparison between the mixing intensities deduced from SU(4) ST distributions and those obtained from shell model (SM) calculations, the reliability of the former remained uncertain.

To fill in this gap, we have performed shell model calculations and computed SU(4) mixing intensities near the ground state with the Preedom and Wildenthal (PW) interaction [10] for some nuclei in the first

[1] Present address: Department of Physics, University of Toronto, Toronto, Canada.
[2] Maître de recherches FNRS.

0 031-9163/82/0000–0000/$02.75 © 1982 North-Holland

Volume 114B, number 2,3 PHYSICS LETTERS 22 July 1982

half of the s–d shell. In the present note we report on the results of such calculations and confront them with the corresponding ones deduced from SU(4)ST distributions. Once the reliability of the latter has been established, we may use them to study SU(4) symmetry mixing in neighbouring nuclei wherein no SM calculations have been performed.

Let us first briefly review how the intensities of the various SU(4) IR's near the ground state can be derived from SU(4)ST distributions [7]. For every m-particle subspace $S(\boldsymbol{\alpha})$ characterized by $\boldsymbol{\alpha} = m[f]ST$, where $[f]$ labels the SU(4) IR, the centroid energy $E_c(\boldsymbol{\alpha})$ and the width $\sigma(\boldsymbol{\alpha})$ are, respectively, calculated using the methods of refs. [7] and [8] based upon propagation techniques: $E_c(\boldsymbol{\alpha})[\sigma^2(\boldsymbol{\alpha})]$ is expressed as a linear combination of the centroid energies (variances) of the subspaces with $m \leqslant 2$ ($m \leqslant 4$), with coefficients (the so-called propagation coefficients) independent of the interaction. From $E_c(\boldsymbol{\alpha})$, $\sigma(\boldsymbol{\alpha})$, and the dimensionality $d(\boldsymbol{\alpha})$ of $S(\boldsymbol{\alpha})$, it is possible to build a gaussian frequency function

$$\rho(\boldsymbol{\alpha},E) = d(\boldsymbol{\alpha})[2\pi\sigma^2(\boldsymbol{\alpha})]^{-1/2} \times \exp\{-[E - E_c(\boldsymbol{\alpha})]^2/2\sigma^2(\boldsymbol{\alpha})\}, \tag{1}$$

which is considered as a continuous approximation to the discrete distribution of the states belonging to $S(\boldsymbol{\alpha})$. By summing the partial densities $\rho(\boldsymbol{\alpha},E)$ over all the subspaces characterized by given values of m and T, the total density of isospin T states is obtained in the form

$$\rho(mT,E) = \sum_{[f]S} \rho(m[f]ST,E). \tag{2}$$

As was suggested by Ratcliff [11], the ground-state energy $E_g = E_0$ is obtained by solving the equation

$$F(mT_0,E_0) = \tfrac{1}{2}(2J_0 + 1)(2T_0 + 1), \tag{3}$$

where J_0 and T_0 are the ground-state angular momentum and isospin, respectively, and $F(mT,E)$ is the distribution function defined by

$$F(mT,E) = \int_{-\infty}^{E} \rho(mT,E')\,\mathrm{d}E'. \tag{4}$$

Similarly, the energy of the ith excited isospin T_0 state is determined by the following equation

$$F(mT_0,E_i) = \left(\sum_{j=0}^{i-1} (2J_j + 1) + \tfrac{1}{2}(2J_i + 1)\right)(2T_0 + 1), \tag{5}$$

if one knows the JT sequence up to this state. Since the ground-state energy, as determined by eq. (3), is sensitive to inaccuracies in the tail of the distribution, it is advantageous to first determine an excited state and then subtract the observed excitation energy to get the binding energy, i.e.,

$$E_g = E_i - [E_i(\text{exp.}) - E_g(\text{exp.})].$$

Once the ground-state energy E_g has been determined by either procedure, it is possible to separate the relative contributions of the various symmetries $[f]$ to the distribution function (4) at the energy E_g. These relative intensities $I(m[f]T_0,E_g)$, defined by

$$I(m[f]T_0,E_g) = [F(mT_0,E_g)]^{-1} \times \sum_{S_0} \int_{-\infty}^{E_g} \rho(m[f]S_0T_0,E)\,\mathrm{d}E, \tag{6}$$

provide a measure of symmetry breaking in the ground-state region.

In table 1, we compare some results obtained for the binding energies and the ground-state mixing intensities with SU(4)ST distributions with those coming from an exact SM diagonalization. The s–d shell nuclei considered correspond to m values ranging from $m = 4$ up to $m = 10$. In the SM case, only the intensity of the lowest SU(4) IR was extracted for the heaviest nuclei. For the SU(4)ST distributions, an excited reference state was used to determine the binding energy, except for three nuclei for which no experimental data were available. In order to study the influence of the ground state position on the mixing intensities, we also give the values obtained for the latter when taking the ground-state energy equal to E_0, as determined by eq. (3), or equal to the shell model binding energy E_g(SM).

Most results for the ground state mixing intensities do not depend much upon the approximation used to determine the ground-state energy. Ten intensities obtained using an excited reference state always lie around the middle of the variation range, e.g.,

Volume 114B, number 2,3 PHYSICS LETTERS 22 July 1982

Table 1
Comparison of SM and SU(4)ST distribution results for the ground-state energies (in MeV) and mixing intensities (in %) of the lowest SU(4) IR's calculated with PW interaction in some s–d shell nuclei. For each nucleus the ground-state angular momentum and isospin are given in column 2, and the dimensionality of the corresponding SM matrix in column 3.

Nucleus	J_0	T_0	Dimension of matrix	E_g(SM)	E_g	I(SM)	$I(E_g)$	$I(E_0)$	$I[E_g$(SM)]
^{20}Ne	0	0	21	−40.6	−41.1	88-11	88-10	95-4	87-10-2
^{20}F, ^{20}Na	2	1	66	−30.6	−32.6	89-6-5	93-4-2	96-3-1	91-5-4
^{20}O	0	2	9	−23.8	−25.2	65-34	66-34	70-30	63-37
^{21}Ne, ^{21}Na	3/2	1/2	188	−47.4	−50.2	85-10-5	82-16-1	87-12	77-20-2
^{21}F, ^{21}Mg	5/2	3/2	148	−38.4	−41.8	64-27-8	61-37-2	63-34-1	55-41-2
^{22}Ne, ^{22}Mg	0	1	148	−57.8	−62.8	60	51-30	57-28	43-31
^{22}F	4	2	195	−43.4	–	76	–	92-7	88-10
^{23}F	5/2	5/2	273	−51.1	–	45	–	55-44	50-49
^{24}Mg	0	0	325	−87.8	−93.7	57	61-27	70-22	49-33
^{24}Ne	0	2	287	−72.2	−79.4	39	41-32	44-32	33-33
^{26}Ne	0	3	191	−82.2	–	32	–	39-30	30-31

$I(E_g(\mathrm{SM})) < I(E_g) < I(E_0)$ with $I(E_0) - I(E_g(\mathrm{SM})) \approx 10\%$ on the average for the lowest IR. In view of this, we may estimate $I(E_g)$ to be around 90, 52 and 35% for the lowest IR in ^{22}F, ^{23}F, and ^{26}Ne, respectively.

When concentrating now on the results obtained with an excited reference state, we find out that the ground-state energies predicted by the spectral distribution method, although in reasonable agreement with the SM ones around $A = 20$, become too negative by as much as 7 MeV when going to $A = 24$. For the determination of binding energies, SU(4)ST distributions are therefore not so accurate as configuration-isospin ones, which predict them within 1 MeV for $A = 22$ [12]. Yet the mixing intensities are reproduced very well: the difference is at most 10% (except perhaps in ^{22}F) for the lowest SU(4) IR and often much less than that. The same kind of agreement is obtained for the distribution of the intensities over the two or three lowest IR's.

The overall agreement obtained for the ground-state mixing intensities between SU(4)ST distributions and SM is very striking especially as we have interpreted the spectral distribution values as pertaining to a single SM state and not to a few neighbouring ones as we should have done in such a statistical method. It allows us to use with confidence SU(4)ST distributions to get information on SU(4) symmetry mixing in those nuclei wherein no SM calculations can be performed.

Table 2
Ground-state energies (in MeV) and mixing intensities (in %) of the lowest SU(4) IR's calculated with PW interaction using SU(4)ST distributions and an excited reference state. Experimental ground-state energies (corrected for the Coulomb energy) are given in column 4.

Nucleus	$I(E_g)$	E_g	E_g(exp.)
^{20}Ne	88-10-2	−41.1	−40.7
^{20}F	93-4-3	−32.6	−30.3
^{20}O	66-34	−25.2	−23.9
^{21}Ne	82-16-1	−50.2	−47.6
^{21}F	61-37-2	−41.8	−38.4
^{22}Na	93-3	−63.6	−58.4
^{22}Ne	51-30-13-5	−62.8	−58.0
^{23}Na	62-34-1	−76.9	−70.8
^{23}Ne	69-21-5	−69.8	−63.2
^{24}Mg	61-27-11	−93.7	−87.4
^{24}Na	82-7-3	−85.7	−77.9
^{24}Ne	41-32-15-9	−79.4	−72.0
^{25}Mg	59-33-5-2	−103.6	−94.7
^{25}Na	51-38-2-5-2	−95.5	−86.9
^{26}Al	81-4-3-11	−115.0	−106.0
^{26}Mg	36-25-20-15	−117.9	−105.9
^{27}Al	49-40-6-3	−130.0	−119.1
^{27}Mg	52-33-5-5	−124.9	−112.3
^{28}Si	43-35-19	−146.6	−136.2
^{28}Al	70-9-8-8	−138.4	−126.9

For completeness we collect in table 2 all the SU(4)ST distribution results for E_g and $I(E_g)$ in the first half of the s–d shell, together with the experimental binding energies (corrected for the Coulomb energy). The general trends of the results are the same as those previously obtained [9] with other realistic interactions, namely the Kuo–Brown KB (12.5) [13], and the K + 12fp [14] interactions. The amount of mixing is, however, different in the three cases, being the largest for pW and the smallest for K + 12fp: in ^{28}Si, for instance, the percentage intensities of the lowest three IR's are 43-35-19, 56-30-13, and 77-18-5 for PW, KB (12.5) and K+12fp interactions, respectively. As the effective interaction in the first half of the s–d shell is expected to be closer to the PW interaction than to the other two, we therefore conclude that SU(4) symmetry is badly broken in this region of the periodic table.

In the present note, we have tested the accuracy with which SU(4)*ST* distributions reproduce the SU(4) mixing intensities near the ground state by comparing the results with those from shell model calculations corresponding to large dimensionalities. The agreement is good, which indicates that matters as subtle as symmetry mixing can be well looked after by the distribution theory provided that the spectroscopic space is partitioned in an appropriate way. As the computational effort needed to calculate SU(4)*ST* distributions is rather small, we have at our disposal a quite efficient method to study SU(4) symmetry mixing over an extended range of nuclei. The results obtained along these lines in the nuclei of the first half of the s–d shell confirm the large SU(4) symmetry breaking which is known to prevail in such nuclei.

We are grateful to A. Zuker for the shell model code and computer facility at Strasbourg.

References

[1] E.P. Wigner, Phys. Rev. 51 (1937) 106.
[2] J.B. French, in: Nuclear structure, eds. A. Hossain, Harun-ar-Raschid and M. Islam (North-Holland, Amsterdam, 1967).
[3] J.C. Parikh, Ann. Phys. (NY) 76 (1973) 202.
[4] J.B. French and J.C. Parikh, Phys. Lett. 35B (1971) 1.
[5] K.T. Hecht and J.P. Draayer, Nucl. Phys. A223 (1974) 285.
[6] M. Chakraborty, V.K.B. Kota and J.C. Parikh, Ann. Phys. (NY) 127 (1980) 413.
[7] R.U. Haq and J.C. Parikh, Nucl. Phys. A220 (1974) 349; A273 (1976) 410.
[8] C. Quesne, J. Math. Phys. 17 (1976) 1452.
[9] R.U. Haq and C. Quesne, Physica 114A (1982) 327.
[10] B.M. Preedom and B.H. Wildenthal, Phys. Rev. C6 (1972) 1633.
[11] K.F. Ratcliff, Phys. Rev. C3 (1971) 117.
[12] F.S. Chang, J.B. French and T.H. Thio, Ann. Phys. (NY) 66 (1971) 137.
[13] T.T.S. Kuo and G.E. Brown, Phys. Lett. 18 (1965) 241.
[14] E.C. Halbert, J.B. McGrory, B.H. Wildenthal and S.P. Pandya, in: Advances in nuclear physics, Vol. 4, eds. M. Baranger and E. Vogt (Plenum, New York, 1969).

PHYSICAL REVIEW C VOLUME 25, NUMBER 4 APRIL 1982

Electromagnetic sum rules by spectral distribution methods

T. R. Halemane*
Department of Physics and Astronomy, University of Rochester, Rochester, New York 14627
and Department of Physics and Astronomy, Rutgers University, Piscataway, New Jersey 08854

J. B. French
Department of Physics and Astronomy, University of Rochester, Rochester, New York 14627
(Received 6 July 1981)

The spectral distribution methods are applied to calculate non-energy-weighted and linear-energy-weighted sum rules for electric and magnetic multipole excitations in the *ds*-shell nuclei ^{20}Ne, ^{24}Mg, ^{28}Si, ^{32}S, and ^{36}Ar. We see the inadequacy of *ds*-shell model space to explain the observed isoscalar quadrupole transition strengths. Isospin admixing of 1^+ levels in ^{24}Mg and ^{28}Si are also confirmed. For isovector $M1$, a detailed study of the Kurath sum rule is made. The strength sums for the higher order multipoles where experimental data are not accurate enough are also evaluated.

[NUCLEAR STRUCTURE ^{20}Ne, ^{24}Mg, ^{28}Si, ^{32}S, ^{36}Ar; sum rules for $E2$, $E4$, $M1$, $M3$, $M5$ transitions; Brown-Kuo Hamiltonian; spectral distribution methods used for calculation; Kurath sum rule extended.]

I. INTRODUCTION

A statistical approach to nuclear structure calculations has been recently developed,[1–5] in part, to overcome a major practical limitation in the conventional shell-model approach. By dispensing with the need to know the nuclear wave functions, these statistical methods make it possible to handle calculations in large model spaces. The theory is based on the recognition of certain simplifying features arising out of the many-particle nature of the model spaces, by virtue of which, strengths and expectation values exhibit, in most cases, a smooth behavior (with small fluctuations) over the model space spectrum. Recently, the method was successfully applied to the study of beta-decay giant resonances[6] and electromagnetic sum rules.[7] We report here on the latter application.

A brief review of the spectral distribution method is given in Sec. II. In Sec. III we discuss the non-energy-weighted sum rules (NEWSR). The NEWSR is evaluated for electric and magnetic multipole excitations from the ground state in the *ds*-shell nuclei ^{20}Ne, ^{24}Mg, ^{28}Si, ^{32}S, and ^{36}Ar. A generally applicable procedure for evaluating the eigenvalue bound to the NEWSR is presented and numerical results obtained for the said excitations and nuclei. Comparisons are made with experimental data and shell-model results. By this sum-rule analysis we are able to prove the known inadequacy of the *ds*-shell model space to explain the observed isoscalar quadrupole strengths in these nuclei. In addition, isospin admixing of 1^+ levels in ^{24}Mg and ^{28}Si are confirmed. The strength sums for the higher order multipoles, where experimental data is not accurate enough, are also predicted.

The linear-energy-weighted sum-rules (LEWSR) are discussed in Sec. IV. When the Hamiltonian is one body, this has a very simple form (expressible in terms of occupancies) and amounts to an extension of the Kurath sum rule to other types of excitations and to arbitrary one-body Hamiltonians. This is then extended to a unitary sum rule which takes account of a part of the two-body contribution to LEWSR, but can still be expressed in terms of occupancies. Further, the contribution to the LEWSR from the two-body interaction is fully evaluated. Comparisons are made with experimental data and with other theories. For isovector $M1$, a detailed study is made of the Kurath sum rule in the *ds* shell and the importance of the two-body contribution is established.

II. REVIEW OF SPECTRAL DISTRIBUTION METHODS

In a conventional nuclear shell-model[8] calculation, one first generates the eigenvalues and eigenvectors of the Hamiltonian in a many-particle model space consisting of m particles distributed among N single-particle states with two-body interactions between the particles. Then the expectation values and strength functions of other operators of interest are calculated in the Hamiltonian eigenstates. All properties of the nuclear system in the chosen model space, and with the chosen interaction, are then defined, and one expects to get better and better results (as compared with experiment) by enlarging the model space and improving the model interaction. But, in practice, this soon becomes impossible since, even for relatively small values of m and N (and even if there are simplifying symmetries), the dimensionalities of the matrices to be constructed and diagonalized become too large to handle even for the most sophisticated computers.

The purpose of statistical nuclear spectroscopy[1–5] is to overcome this limitation by essentially starting at the other end of the problem. The idea is to look at the global properties of the Hamiltonian and other operators in the model space, and at the correlations between them. Some general properties of the system can be obtained most readily this way. Level densities and spectra can also be constructed from the moments. In fact, at least in finite-dimensional model spaces, one could in principle produce all the microscopic details by evaluating moments and covariances up to the order determined by the dimensionality of the model space. But this, by itself, would not be of much practical significance, if it were not for the existence of two helpful factors that contribute towards making this approach more successful than would be presumed otherwise.

The first simplifying factor derives from the recognition of the role played by a central limit theorem[2–4] (CLT) in many-particle model spaces constructed by distributing nucleons over some finite set of single-particle states. Then, by virtue of this CLT, in the limit of large particle number, the smoothed eigenvalue distributions for most Hamiltonian operators in the model space become close to Gaussian. This, in turn, implies[5] closely related characteristic asymptotic forms, for expectation values and strengths of other operators, defined by only a small number of traces over the model space of operators and their products. Such operator traces can be calculated[3,9] by methods which do not involve construction of any Hamiltonian eigenfunctions. The fact that such traces can be obtained[4] in a many-particle space by "propagation" from lower particle spaces is the second simplifying factor. This requires one to evaluate those traces only in a minimum defining set of spaces of low particle or hole number. These two features make the statistical approach attractive, especially in model spaces of large dimensionality. Moreover, by its very nature, it gives results as more or less explicit functions of the Hamiltonian matrix elements so that one can easily connect the features of the Hamiltonian with the corresponding properties of the strengths and expectation values.

A. Polynomial expansions

Apart from the spectra and level densities, most calculations in statistical nuclear spectroscopy are based on two (usually rapidly convergent) polynomial expansions[5]; one for the expectation values and another for strengths.

Let H be the Hamiltonian for the nuclear system and O be an excitation operator which induces transitions $|W\rangle \xrightarrow{O} |W'\rangle$ between starting states $|W\rangle$ and final states $|W'\rangle$. Then the microscopic transition strength is defined as

$$R(W',W)=|\langle W'|O|W\rangle|^2 . \qquad (2.1)$$

The states $|W\rangle$ and $|W'\rangle$ may be in different eigensubspaces of H or the eigensubspaces may coincide. Let d and d' be their dimensionalities. It is also convenient to consider the eigenvalue densities $I(z)$ and $I'(z)$ in the initial and final subspaces. Also define

$$\rho(z)=d^{-1}I(z), \quad \rho'(z)=(d')^{-1}I'(z) . \qquad (2.2)$$

Associated with these densities are complete sets of orthonormal polynomials P_μ and P'_μ such that

$$\int P_\mu(x)P_\nu(x)\rho(x)dx=\delta_{\mu\nu} , \qquad (2.3)$$

$$\delta(x-y)=\rho(x)\sum_\mu P_\mu(x)P_\mu(y) , \qquad (2.4)$$

with similar relations between ρ' and P'_μ. These polynomials[10] can be constructed explicitly in terms of the density moments

$$M_p=\int \rho(z)z^p dz=\langle H^p\rangle^m = \frac{1}{d}\langle\langle H^p\rangle\rangle^m , \qquad (2.5)$$

where we use the notation that, for any operator G, $\langle\langle G\rangle\rangle^m$ denotes the trace over the $|W\rangle$ model space and $\langle G\rangle^m$ denotes the average expectation value. The first two polynomials are

$$P_0(z)=1,\quad P_1(z)=(z-\mathscr{E})/\sigma\ , \tag{2.6}$$

where $\mathscr{E}=M_1$ and $\sigma^2=M_2-M_1^2$ are the centroid and variance, respectively. The polynomial of order ν requires density moments of orders up to 2ν and is given by

$$[D_{\nu-1}D_\nu]^{1/2}P_\nu(z)=\begin{vmatrix} 1 & M_1 & M_2 & \cdots & M_\nu \\ M_1 & M_2 & & \cdots & M_{\nu+1} \\ \vdots & \vdots & & & \\ M_{\nu-1} & M_\nu & & \cdots & M_{2\nu-1} \\ 1 & z & z^2 & \cdots & z^\nu \end{vmatrix}\ , \tag{2.7}$$

where D_ν is the determinant in Eq. (2.7) with the last row replaced by $[M_\nu, M_{\nu+1},\ldots M_{2\nu}]$.

When the density is Gaussian, i.e.,

$$\rho(W)=\frac{1}{\sqrt{2\pi\sigma^2}}\exp\left[-\frac{(W-\mathscr{E})^2}{2\sigma^2}\right]\ , \tag{2.8}$$

the polynomials P_μ are related to the Hermite polynomials H_μ by

$$P_\mu(W)=(\mu!)^{-1/2}H_{e\mu}[(W-\mathscr{E})/\sigma]\ , \tag{2.9}$$

where

$$H_{e\mu}(z)=2^{-\mu/2}H_\mu(z/\sqrt{2})\ .$$

When the density is of chi-squared type, the P_μ are related to the Laguerre polynomials.

In terms of the orthogonal polynomials $P_\mu(x)$, one derives

$$\begin{aligned} R(W',W)&=(I'(W)I(W))^{-1}\langle\langle O^\dagger\delta(H-W')O\delta(H-W)\rangle\rangle^m \\ &=(dd')^{-1}\sum_{\mu,\nu}\langle\langle O^\dagger P'_\mu(H)OP_\nu(H)\rangle\rangle^m P'_\mu(W')P_\nu(W) \end{aligned} \tag{2.10}$$

and similarly for the expectation value of an operator K,

$$\begin{aligned} K(W)&=(I(W))^{-1}\langle\langle K\delta(H-W)\rangle\rangle^m \\ &=\sum_\nu\langle KP_\mu(H)\rangle^m P_\mu(W)\ . \end{aligned} \tag{2.11}$$

B. Central limit theorem

The above polynomial expansions would not be of much practical value if it were not for their rapid convergence. This results from the fact that, as we increase the number of particles, the model space eigenvalue density for "almost all" Hamiltonians goes[2] rapidly to Gaussian (by CLT). In the case of noninteracting particles this comes about because the density convolutes as we add particles. Thus, if we construct an m-particle system by adding one particle at a time and consider the energy of the state into which a particle is placed as a random variable, then the total energy is the sum of random variables and its density is given by the convolution of m single particle energy densities. The CLT then assures us that the convoluted density approaches Gaussian as the number of particles increases.

The convolution argument requires the variables to be additive and independent. These conditions are not met for a system of interacting nucleons. The Pauli blocking effect (which can be ignored for dilute systems, i.e., $m<<N$) violates statistical independence, and additivity is violated if interactions have to be considered. Despite all this, it is found that all nuclear Hamiltonians which give reasonable agreements with experimental data have model space spectra[3,4] which are close to Gaussian. This is better understood by studying an ensemble of Hamiltonians of two-body interactions.[11,12] It is found that the ensemble-averaged density is Gaussian and that for large systems ($1<<m<<N$), only a negligible fraction of the members of the ensemble give deviant densities. However, in actual calculations, corrections to Gaussian are often necessary and will be incorporated.

In the CLT limit, Eqs. (2.10) and (2.11) can be approximated by[5]

$$K(W)=\langle K\rangle^m+\langle K(H-\mathscr{E})\rangle^m\frac{(W-\mathscr{E})}{\sigma^2}\ , \tag{2.12}$$

$$R(W',W)=(d')^{-1}\left[\langle O^{\dagger}O\rangle^{m}+\langle O^{\dagger}(H-\mathscr{E}')O\rangle^{m}\frac{(W'-\mathscr{E}')}{(\sigma')^2}+\langle O^{\dagger}O(H-\mathscr{E})\rangle^{m}\frac{(W-\mathscr{E})}{\sigma^2}\right.$$
$$\left.+\langle O^{\dagger}(H-\mathscr{E}')O(H-\mathscr{E})\rangle^{m}\frac{(W'-\mathscr{E}')(W-\mathscr{E})}{\sigma'^2\sigma^2}\right]. \quad (2.13)$$

Equation (2.12) and, in fact, (2.11) also can be derived by a linear-response theory.[5,13] It can also be interpreted by a geometrical picture.[5]

Finally, it should be noted that there are operators, like J^2, $Q\cdot Q$, and pairing, which do not have asymptotic Gaussian spectra.[14] $Q\cdot Q$, in fact, has an asymptotic density which is chi-squared in five variables, and in general, the spectrum of an operator of the multipole form $T^{\lambda}\cdot T^{\lambda}$ is chi-squared in $(2\lambda+1)$ variables, which becomes Gaussian when λ is large enough.

C. Configuration partitioning

Partitioning the model space into subspaces is often necessary to treat properly the exact symmetries conserved by a Hamiltonian. On the other hand, partitioning by other properly chosen nonexact symmetries can provide, as discussed below, a simple way to get better accuracy, especially when dealing with huge model spaces. For that purpose we could add high-order terms to the CLT results [Eqs. (2.12) and (2.13)] but these terms involve more complicated traces that rapidly become harder to evaluate. Configuration partitioning provides an easier solution.[5] In this, the single-particle space is divided into orbits, say l in number. Then $\sum_{i=1}^{l}N_i=N$, where N_i is the number of states in the ith orbit. Correspondingly, the m-particle space is divided into configurations defined, as usual, by assigning numbers of particles to the various orbits—m_i particles in the ith orbit—such that $\sum_{i=1}^{l}m_i=m$. Each configuration is then defined by a partition $\vec{m}\equiv[m_1,m_2,,\ldots,m_l]$ and additional attributes (like isospin T or angular momentum J) if any. For a properly-chosen partitioning, the configuration densities are to some extent localized in the spectrum and it is natural to introduce orthonormal polynomials for each configuration and produce the overall result by combining the contributions of all the configurations. Each configuration term looks after a part of the total spectrum and when many such configurations are present, one gets high accuracy even with low order polynomials. It is thus equivalent to going to considerably higher orders in the polynomial expansion over the entire model space. With configuration partitioning, Eqs. (2.10) and (2.11) can be replaced by[5]

$$K(W)=\sum_{\vec{m}}\frac{I_{\vec{m}}(W)}{I_m(W)}K(W:\vec{m}), \quad (2.14)$$

where

$$K(W:\vec{m})=\sum_{\mu}\langle KP_{\mu}^{(\vec{m})}(H)\rangle^{\vec{m}}P_{\mu}^{(\vec{m})}(W) \quad (2.15)$$

and

$$R(W',m';W,m)=\sum_{\vec{m}',\vec{m}}\frac{I_{\vec{m}'}(W')I_{\vec{m}}(W)}{I_{m'}(W')I_m(W)}R(W',\vec{m}';W,\vec{m}), \quad (2.16)$$

where

$$R(W',\vec{m}';W,\vec{m})=(d_{\vec{m}'})^{-1}\sum_{\mu,\nu}\langle O^{\dagger}(\vec{m}'-\vec{m})P_{\mu}^{(\vec{m}')}(H)OP_{\nu}^{(\vec{m})}(H)\rangle^{\vec{m}}P_{\mu}^{(\vec{m}')}(W')P_{\nu}^{(\vec{m})}(W). \quad (2.17)$$

The superscripts and suffixes $\vec{m}$ or $\vec{m}'$ indicate that the corresponding quantities refer to or are evaluated in the configuration space $\vec{m}$ or $\vec{m}'$, respectively. As the configuration traces are no harder to evaluate[3,9] than the scalar ones, better accuracy is obtained in a simple way. It should also be remarked that in the configuration traces above, there is no restriction on the intermediate states. All the correlations and interferences are thus properly included and the result is formally exact.

D. Unitary geometry and trace propagation

The $\binom{N}{m}$ states formed by distributing m fermions over N single-particle states form an irreducible representation of the group U(N) of unitary transformations in the single-particle space. This symmetry plays an important role in calculations because it provides simple methods for the evaluation of traces of operators.[3] The trace operators in the m-particle space are scalars under this U(N) group. Operators can be decomposed according to their unitary ranks $\nu=0,1,\ldots,\frac{1}{2}N$ (where each ν corresponds, in the language of Young shapes, to a column structure $[N-\nu,\nu]$, or as is more commonly used, to a row structure $[2^\nu,1^{N-2\nu}]$). Then, a k-body operator $F(k)$ can be written as

$$F(k)=\sum_{\nu=0}^{k} F^\nu(k)=\sum_{\nu=0}^{k} \begin{bmatrix} n-\nu \\ k-\nu \end{bmatrix} F_k^\nu(\nu)\,, \tag{2.18}$$

where $F^\nu(k)$ is the unitary[15] rank-ν part of $F(k)$ and it has been further factored into its irreducible[15] part $F_k^\nu(\nu)$ which is a ν-body operator derived from a k-body operator, as the subscript indicates. The binomial factor $\binom{n-\nu}{k-\nu}$ converts $F_k^\nu(\nu)$ into a k-body operator. Then, for a $(0+1+\cdots+u)$-body operator, we have

$$F=\sum_{\nu=0}^{u}\sum_{k=\nu}^{u} \begin{bmatrix} n-\nu \\ k-\nu \end{bmatrix} F_k^\nu(\nu)\,. \tag{2.19}$$

The $\nu=0$ part gives the centroid of F. The scalar product of two irreducible operators F^ν and $G^{\nu'}$ is then

$$\langle F^\nu G^{\nu'}\rangle^m=\begin{bmatrix} m \\ \nu \end{bmatrix}\begin{bmatrix} N-m \\ \nu \end{bmatrix}\begin{bmatrix} N-\nu \\ \nu \end{bmatrix}^{-1}\langle F^\nu G^\nu\rangle^\nu \delta_{\nu\nu'}\,, \tag{2.20}$$

where

$$\langle F^\nu G^\nu\rangle^\nu=\begin{bmatrix} N \\ \nu \end{bmatrix}^{-1}\sum_{\alpha,\beta} F_{\alpha\beta}^\nu G_{\alpha\beta}^\nu\,. \tag{2.21}$$

The defining ν-particle matrix elements $F_{\alpha\beta}^\nu$ and $G_{\alpha\beta}^\nu$ are related to those of F and G by contraction operators.[3] (Here we have assumed F and G to have real matrix elements. Then the Hermitian property makes them symmetric too.) Then, we have for the scalar product of two operators F and G, of maximum particle ranks u and v, respectively,

$$\langle FG\rangle^m=\sum_{\nu=0}^{(u,v)_<}\sum_{k=\nu}^{u}\sum_{k'=\nu}^{v}\begin{bmatrix} m-\nu \\ k-\nu \end{bmatrix}\begin{bmatrix} m-\nu \\ k'-\nu \end{bmatrix}\begin{bmatrix} m \\ \nu \end{bmatrix}\begin{bmatrix} N-m \\ \nu \end{bmatrix}\begin{bmatrix} N-\nu \\ \nu \end{bmatrix}^{-1}\langle F_k^\nu G_{k'}^\nu\rangle^\nu\,, \tag{2.22}$$

where $(u,v)_<$ is the minimum of the pair. Without the $\nu=0$ term, Eq. (2.22) would give the covariance of F and G, viz., $\langle (F-\langle F\rangle)(G-\langle G\rangle)\rangle^m$.

These methods can be easily extended for configuration traces. Related methods for more complicated trace evaluation have been developed[9,16,17] recently, and are expected to be important in the computation of higher-order traces.

E. Spectra and occupancies

From the smoothed eigenvalue density $\rho(W)$, it is not possible to extract the actual discrete spectra exactly, but the following prescription (Ratcliff[18] procedure) gives the most probable positions of the energy levels. The energy E_i of the ith level is obtained by requiring that

$$d\int_{-\infty}^{E_i}\rho(W)dW=\sum_{j=1}^{i-1}d_j+\frac{d_i}{2}\,, \tag{2.23}$$

where d_j denotes the degeneracy of jth energy level and d is the total dimensionality. In view of the CLT, good results are obtainable, by assuming a four moment Gram-Charlier expansion for $\rho(W)$, viz.,

$$\rho(W)=\frac{\exp(-x^2/2)}{\sqrt{2\pi\sigma^2}}\left[1+\frac{\gamma_3}{6}(x^3-3x)+\frac{\gamma_4}{24}(x^4-6x^2+3)\right], \tag{2.24}$$

where $x=(W-\mathscr{E})/\sigma$ and $\mathscr{E}$, σ, γ_3, and γ_4 are the centroid, width, skewness, and excess, respectively, of the Hamiltonian H. When the model space is partitioned by configurations, $\rho(W)$ is taken to be a sum of configuration densities, each configuration density usually assumed to be a Gaussian.

The position of energy levels determined by these methods are found[4] to deviate from their exact positions by within half a local level spacing.

The occupancy $n_r(W)$ of the rth orbit can be[19] calculated by putting $K=n_r$ in Eq. (2.12). Better accuracy is obtained if the model space is partitioned into these orbits, because then in Eq. (2.15) only the $\mu=0$ term survives, and we get

$$n_r(W)=\frac{\sum_{\vec{m}} m_r I_{\vec{m}}(W)}{I_m(W)}\ , \tag{2.25}$$

where m_r denotes number of particles in the rth orbit corresponding to the configuration $\vec{m}$.

III. NON-ENERGY-WEIGHTED SUM RULES

The non-energy-weighted sum rule[20] (NEWSR) for the excitations generated by an excitation operator O acting on a state $|\alpha\rangle$ measures the expectation value $\langle\alpha|O^{\dagger}O|\alpha\rangle$. As pointed out by Hsu,[21] it is bounded by the largest eigenvalue of the (positive definite) operator $O^{\dagger}O$ acting in the model space of the system. Such sum-rule results can be of use in assessing whether a given model space and a given interaction Hamiltonian would be adequate to explain experimentally measured transition strengths, the point being that the individual strength is dominated by the sum-rule quantity which in turn is dominated by the maximum $O^{\dagger}O$ eigenvalue. If the experimental measurement is larger than the maximum eigenvalue then one has to enlarge the model space.

In small model spaces the eigenvalues can be found by constructing and diagonalizing the $O^{\dagger}O$ matrices; in certain other cases (as with isoscalar $E2$ transitions within a single major shell) $O^{\dagger}O$ is equivalent to a Casimir operator whose eigenvalues may be taken to be known. However, we shall present a generally applicable procedure for evaluating the maximum eigenvalue using spectral distribution methods.

These eigenvalue limits, being independent of the Hamiltonian, are useful only in cases where the individual strength, or at least the total strength originating in a given state, is large (as happens sometimes with $E2$ excitations). But in other cases this limit may not be very useful. In such situations we need to know the NEWSR (i.e., the $O^{\dagger}O$ expectation value) itself, which takes account of the fact that the strength distribution depends on the Hamiltonian. The spectral distribution method can be used to do this too.

We shall present below, in Secs. III B and III C, procedures for evaluating both the NEWSR and its eigenvalue bound. Then in Sec. III D we apply these methods to electromagnetic excitations in even-even self-conjugate ds-shell nuclei. Before we do all this, we give, in Sec. III A, a brief account of the NEWSR in the spherical tensor formalism. Finally, in Sec. III E, we discuss average strengths and units.

A. Sum rule analysis

Consider a nuclear transition from an initial state $(x\Gamma_i\mu_i)$ to a final state $(y\Gamma_f\mu_f)$. Let $\sum_{t=0}^{1}O^{\lambda}$ be the excitation operator. We use a direct-product notation (of Ref. 20) here wherein $\Gamma\equiv(J,T)$, $\lambda\equiv(k,t)$, $\mu\equiv(M,\mathscr{T})$, $[\Gamma]\equiv(2J+1)(2T+1)$, and

$$U(\Gamma_1\Gamma_2\Gamma_3\Gamma_4{:}\Gamma_5\Gamma_6)\equiv U(J_1J_2J_3J_4{:}J_5J_6)U(T_1T_2T_3T_4{:}T_5T_6)\ ,$$

where x and y denote quantum numbers other than angular momentum and isospin. M and $\mathscr{T}$ are the z-components of angular momentum J and isospin T, respectively. $\mathscr{T}_i$ and $\mathscr{T}_f$ would be the same for transitions in a given nucleus (as with examples in the next section) but we keep the generality for the present. We shall make use of the techniques of spherical tensors and Racah algebra.

Nuclear levels are degenerate with respect to the z-component M of angular momentum J and hence it is natural to define the transition strength (also called the reduced transition probability) $B(O^{\lambda},x\Gamma_i\rightarrow y\Gamma_f)$ by averaging over the initial z components M_i and summing over the final ones. Thus,

$$\begin{aligned}B(O^{\lambda},x\Gamma_i\rightarrow y\Gamma_f)&=[J_i]^{-1}\sum_{M_iM_f}|\langle\psi^{\Gamma_f}_{y\mu_f}|\sum_{t=0}^{1}O^{k,t}|\psi^{\Gamma_i}_{x\mu_i}\rangle|^2\\&=[J_i]^{-1}[T_f]^{-1}\left[\sum_{t=0}^{1}C^{T_itT_f}_{\mathscr{T}_i\mathscr{T}_f}\langle y\Gamma_f||O^{k,t}||x\Gamma_i\rangle\right]^2\ ,\end{aligned}\tag{3.1}$$

where the second equality obtained by using the Wigner-Eckart theorem involves double barred matrix elements (dbme) of O^{λ}. Here $t=0$ for isoscalar and $t=1$ for isovector excitations. Now let

$$B(O^{\lambda},x\Gamma_i\rightarrow\Gamma_f)=\sum_{y}B(O^{\lambda},x\Gamma_i\rightarrow y\Gamma_f)\ . \tag{3.2}$$

This gives the total reduced transition probability from the level $x\Gamma_i$ in the initial nucleus to all the levels of angular momentum Γ_f in the final nucleus. Further simplification can be obtained by using the spherical tensor formulas[20]

$$\langle y\Gamma_f||O^\lambda||x\Gamma_i\rangle=(-1)^{\Gamma_f+\lambda-\Gamma_i}\langle x\Gamma_i||\bar{O}^\lambda||y\Gamma_f\rangle\ , \tag{3.3}$$

$$\sum_y \langle x\Gamma_i||\bar{O}^\lambda||y\Gamma_f\rangle\langle y\Gamma_f||O^{\lambda'}||x'\Gamma_i\rangle =[\Gamma_f]^{1/2}\sum_\Delta(-1)^{\lambda+\lambda'-\Delta}U(\Gamma_i\lambda\Gamma_i\lambda':\Gamma_f\Delta)\langle x\Gamma_i||(\bar{O}^\lambda\times O^{\lambda'})^\Delta||x'\Gamma_i\rangle\ . \tag{3.4}$$

Then,

$$B(O^\lambda,x\Gamma_i\to\Gamma_f)=[J_iT_f]^{-1}\sum_{t=0}^{1}\sum_{t'=0}^{1}C^{T_itT_f}_{\mathcal{T}_i\mathcal{T}_f}C^{T_it'T_f}_{\mathcal{T}_i\mathcal{T}_f}\times\sum_\Delta[\Gamma_f]^{1/2}(-1)^{\Gamma_f+\lambda'-\Gamma_i-\Delta}U(\Gamma_i\lambda\Gamma_i\lambda':\Gamma_f\Delta)\langle x\Gamma_i||(\bar{O}^\lambda\times O^{\lambda'})^\Delta||x\Gamma_i\rangle, \tag{3.5}$$

where $\lambda=(k,t)$ and $\lambda'=(k,t')$. For transitions from $\Gamma_i=(0,0)$ this expression reduces to

$$B(O^\lambda,0\to\Gamma_f)=[t]^{-1}\langle x,0\,|\,O^\lambda\cdot O^\lambda\,|\,x,0\rangle\delta_{\lambda\Gamma_f}\ , \tag{3.6}$$

where we have used the Hermiticity property

$$\bar{O}^\lambda=(-1)^\lambda O^\lambda \tag{3.7}$$

and the relation

$$O^\lambda\cdot O^\lambda=(-1)^\lambda[\lambda]^{1/2}(O^\lambda\times O^\lambda)^0\ . \tag{3.8}$$

Then, Eq. (3.6) gives the NEWSR for transitions induced by the excitation operator O^λ from the initial state $x,\Gamma_i=(0,0)$. We shall therefore be interested in the eigenvalues and expectation values of $O^\lambda\cdot O^\lambda$. The explicit construction of the excitation operators O^λ for electric and magnetic multipoles is given in Appendix A and the evaluation of $O^\lambda\cdot O^\lambda$ in the standard form,[20] from O^λ, is given in Appendix B.

B. Eigenvalue bounds

The eigenvalue densities of $O^\lambda\cdot O^\lambda$ are highly non-Gaussian.[14] The skewness and excess of these operators are given in Table I. For example, $E2\ (T=0)\cdot E2\ (T=0)$ has a skewness of 0.96 and excess of 0.88 in the $(ds)^8$ space. By partitioning we approximate this density by a sum of configuration densities, each assumed to be a Gaussian. Thus, in the $(ds)^{m,T}$ model space, the eigenvalue density of $O^\lambda\cdot O^\lambda$ is assumed to be

$$I^{O\lambda}_{m,T}(x)=\sum_{\vec{m}}I^{O\lambda}_{\vec{m},T}(x)\ , \tag{3.9}$$

$$I^{O\lambda}_{\vec{m},T}(x)=\frac{d(\vec{m},T)}{\sqrt{2\pi\sigma_{O\lambda}{}^2(\vec{m},T)}}\times\exp\left\{-\frac{[x-\mathscr{E}_{O\lambda}(\vec{m},T)]^2}{2\sigma_{O\lambda}{}^2(\vec{m},T)}\right\}, \tag{3.10}$$

where $\mathscr{E}_{O\lambda}(\vec{m},T)$ and $\sigma_{O\lambda}{}^2(\vec{m},T)$ are the centroid and variance of the eigenvalues of $O^\lambda\cdot O^\lambda$ in the configuration space $(\vec{m},T)$, and $d(\vec{m},T)$ is the dimensionality of this $(\vec{m},T)$ space. The highest eigenvalue $\chi_{O\lambda}$ of $I^{O\lambda}_{m,T}(x)$ is then evaluated by requiring that

$$\int_{\chi_{O\lambda}}^{\infty}I^{O\lambda}_{m,T}(x)dx=\tfrac{1}{2}\ . \tag{3.11}$$

The validity of a configuration-Gaussian approximation to a non-Gaussian density lies in the fact that each configuration density looks after a more or less localized region of the spectrum, and when the whole space is divided into many configurations, these many pieces of information combine rather well to reproduce the original density. Crudely speaking, one could compare this procedure to the process of approximating an irregular patch of area (in two dimensions) by a large number of tiny regular squares.

In Fig. 1, we plot the distribution function for $E2\ (T=0)\cdot E2\ (T=0)$ in the $(ds)^{4,T=0}$ space, along with the configuration-Gaussian approximation to it. The agreement is good except at the lower end

TABLE I. Skewness and excess parameters for $E\lambda \cdot E\lambda$ and $M\lambda \cdot M\lambda$ operators in $(ds)^m$ spaces [for isoscalar $M1$, the $\vec{J}$ term has been left out—see Eq. (3.14)]. Owing to the hole ⇄ particle symmetry of these operators, the values for $(ds)^{16}$ and $(ds)^{20}$ are the same as those for $(ds)^8$ and $(ds)^4$, respectively, and hence are not tabulated.

	Skewness									
	Isoscalar					Isovector				
m	$E2$	$E4$	$M1$	$M3$	$M5$	$E2$	$E4$	$M1$	$M3$	$M5$
4	0.993	0.835	0.760	1.27	0.647	0.520	0.130	1.09	0.438	−0.137
8	0.957	0.846	1.37	1.31	0.568	0.664	0.241	0.908	0.386	−0.239
12	0.950	0.839	1.47	1.31	0.488	0.694	0.232	0.873	0.326	−0.277
	Excess									
	Isoscalar					Isovector				
m	$E2$	$E4$	$M1$	$M3$	$M5$	$E2$	$E4$	$M1$	$M3$	$M5$
4	1.18	0.864	−0.604	1.55	0.824	−0.118	0.231	1.85	0.071	−0.242
8	0.877	0.646	2.35	1.72	0.255	0.173	0.036	0.860	−0.125	−0.079
12	0.794	0.583	3.01	1.73	0.078	0.273	0.017	0.615	−0.215	−0.105

where the configuration-Gaussian approximation has a tail extending to the negative side. Our interest, however, is only at the upper end. In Table II we give the largest eigenvalues calculated by this method for $E\lambda \cdot E\lambda$ and $M\lambda \cdot M\lambda$ in the ds shell. In the $(ds)^{4,T=0}$ cases we also calculate exact values by shell model and, for isoscalar $E2$, by SU(3). Good agreement is found in all cases [except for isoscalar $M1$, for which the larger discrepancy is due to the fact that $M1\ (T=0) \cdot M1\ (T=0)$ being proportional to S^2 (see ahead) has only three distinct eigenvalues in the $(ds)^{4,T=0}$ space].

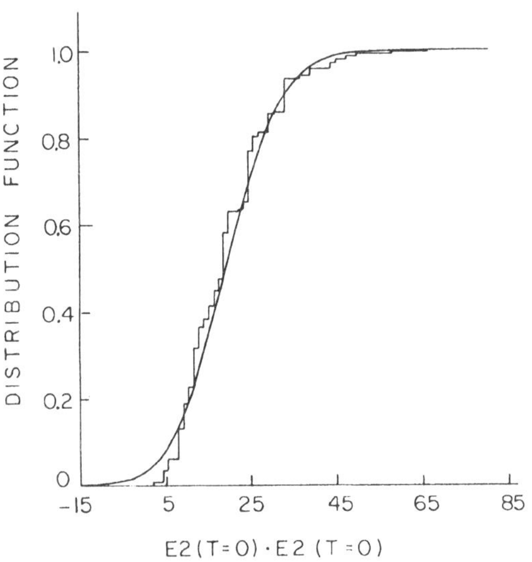

FIG. 1. The eigenvalue distribution function for $E2\ (T=0) \cdot E2\ (T=0)$ in the $(ds)^{4,T=0}$ space. The staircase curve is the exact distribution function obtained by a shell-model calculation and the other curve is the configuration-Gaussian approximation to it.

C. Expectation values

The expectation value of $O^\lambda \cdot O^\lambda$ can be obtained by the polynomial expansion (2.11) or its CLT approximation (2.12) with $K = O^\lambda \cdot O^\lambda$. In a $(ds)^{m,T}$ model space, the first term of the expansion is $\langle O^\lambda \cdot O^\lambda \rangle^{m,T}$, which is the average total strength from a single state and gives the 0th approximation to the NEWSR. These values are tabulated in Table III, for electric and magnetic multipoles in the ds shell. The second term in the expectation value expansion,

$$\langle O^\lambda \cdot O^\lambda (H - \langle H \rangle^{m,T}) \rangle^{m,T} \frac{(W - \langle H \rangle^{m,T})}{\sigma_H{}^2(m,T)} ,$$

takes account of the fact that the strength distribution depends on the Hamiltonian. It depends on the $O^\lambda \cdot O^\lambda$ component of H and gives a contribution which is proportional to the correlation coefficient between $O^\lambda \cdot O^\lambda$ and H. These correlation coefficients are given in Table IV. The ground state (see Table V) expectation values thus determined are given in Tables VI and VII.

We shall get even better accuracy for the expectation values by using the configuration partitioning technique described in Sec. II, with $(\vec{m}, T)$ configurations and CLT approximation in each configuration. These are also given in Tables VI and VII.

D. Examples

We consider NEWSR for electric and magnetic multipole[8] transitions from the ground state in

TABLE II. Upper bounds for the NEWSR of electric and magnetic transitions from the ground state in even-even self-conjugate nuclei. These are the largest eigenvalues of $O^\lambda \cdot O^\lambda$ when the excitation O^λ is isoscalar and of $\frac{1}{3} O^\lambda \cdot O^\lambda$ when the excitation is isovector [see Eq. (3.6)], and were obtained by the configuration-distribution method in $(ds)^{m,T=0}$ spaces. Also given (in parenthesis) are the bounds from the exact largest eigenvalues, calculated via SU(3) in the $E2$ case and via shell model in other cases. Units are $e^2\text{fm}^{2L}$ for the electric (EL) cases and $\mu_N{}^2\text{fm}^{2L-2}$ for the magnetic (ML) cases. All values should be multiplied by the corresponding scale factors given in the last row.

$m=A-16$	Isoscalar					Isovector				
^{A}Nucleus	$E2$	$E4$	$M1$	$M3$	$M5$	$E2$	$E4$	$M1$	$M3$	$M5$
^{20}Ne	62	73	35.6	20.3	4.9	50	54	46.5	29.0	9.5
	(70)	(83)	(20.6)	(17.9)	(4.4)	(43)	(50)	(47.7)	(27.1)	(7.7)
^{24}Mg	125	174	76.1	38.0	8.3	95	121	80.8	52.8	16.1
	(133)									
^{28}Si	160	241	92.3	45.6	10.8	116	169	91.5	64.5	20.8
	(179)									
^{32}S	152	255	76.1	46.0	12.2	111	178	80.8	64.0	23.6
	(161)									
^{36}Ar	92	160	35.6	30.1	10.6	74	119	46.5	43.0	20.8
	(103)									
Scale factor	1	10^2	10^{-2}	10^2	10^5	1	10^2	1	10^3	10^6

even-even self-conjugate ds-shell nuclei. The Hamiltonian used is the Brown-Kuo (BK) interaction[22] with experimental ^{17}O single-particle energies. The ground state energies are determined by the Ratcliff procedure[18] and are given in Table V. The construction of the excitation operators O^λ and their scalar product $O^\lambda \cdot O^\lambda$ are discussed in an appendix. The upper bounds to NEWSR, obtained from the highest eigenvalues of $O^\lambda \cdot O^\lambda$ calculated in $(ds)^{m,T=0}$ spaces using the statistical method of Sec. III B, are given in Table II. In some cases we also give, for comparison, the bounds obtained from the actual largest eigenvalues. The 0th approximation to the NEWSR are given in Table III. Given in Tables VI and VII are the NEWSR calculated using the CLT approximations in the scalar $(m,T=0)$ and also the configuration $(\vec{m},T=0)$ spaces.

The experimental values of NEWSR, also given

TABLE III. Zeroth approximation to the NEWSR for electric and magnetic multipole excitations from the ground state in even-even self-conjugate ds-shell nuclei. These are the centroids of $[t]^{-1}\, O^\lambda \cdot O^\lambda$ in $(ds)^{m,T=0}$ spaces, where $t=0$ for isoscalar and $t=1$ for isovector excitations. The units are $e^2\text{fm}^{2L}$ for the electric (EL) cases and $\mu_N{}^2\text{fm}^{2L-2}$ for the magnetic (ML) cases. All values should be multiplied by the corresponding scale factors in the last row.

$m=A-16$	Isoscalar					Isovector				
^{A}Nucleus	$E2$	$E4$	$M1$	$M3$	$M5$	$E2$	$E4$	$M1$	$M3$	$M5$
^{20}Ne	19.5	26.5	10.1	5.57	1.38	16.7	22.7	14.3	11.0	3.37
^{24}Mg	35.2	54.0	16.2	10.1	2.81	30.1	46.3	22.9	19.8	6.90
^{28}Si	43.9	74.6	18.2	12.6	3.88	37.7	64.0	25.8	24.7	9.53
^{32}S	42.6	79.3	16.2	12.2	4.12	36.7	68.0	22.9	24.0	10.1
^{36}Ag	28.8	58.0	10.1	8.25	3.01	24.7	49.7	14.3	16.2	7.4
Scale factor	1	10^2	10^{-2}	10^2	10^5	1	10^2	1	10^3	10^6

TABLE IV. Correlation coefficients in the $(ds)^{m,T=0}$ spaces between the BK Hamiltonian and $O^{\lambda}\cdot O^{\lambda}$ for electric and magnetic excitation operators O^{λ} [for isoscalar $M1$, the $\vec{J}$ term has been left out see Eq. (3.14)].

$m=A-16$	Isoscalar					Isovector				
m	$E2$	$E4$	$M1$	$M3$	$M5$	$E2$	$E4$	$M1$	$M3$	$M5$
4	−0.523	−0.244	+0.012	−0.331	−0.378	+0.290	+0.208	+0.376	−0.273	−0.512
8	−0.557	−0.323	+0.172	−0.059	−0.112	+0.231	+0.137	+0.512	−0.015	−0.326
12	−0.552	−0.333	+0.195	+0.080	+0.134	+0.215	+0.116	+0.563	+0.145	0.052
16	−0.519	−0.308	+0.161	+0.216	+0.382	+0.219	+0.121	+0.585	+0.289	+0.274
20	−0.431	−0.220	+0.010	+0.458	+0.644	+0.250	+0.159	+0.594	+0.471	+0.562

in Tables VI and VII, have been deduced from the data compiled in the literature. These data are from Refs. 23–27 for isovector $M1$ and from Ref. 28 for all others. We now turn to a more detailed discussion.

1. Electric quadrupole

In the isoscalar quadrupole case, along with the eigenvalue bound (to the NEWSR) calculated by the configuration-distribution method, we have also given (in Table II) the upper bounds from the exact largest eigenvalues calculated by the SU(3) model.[29] These are the values given in parenthesis under the isoscalar $E2$ column. They were obtained using the expression

$$e^2b^4\frac{5}{16\pi}[\lambda^2+\mu^2+\lambda\mu+3\lambda+3\mu-\tfrac{3}{4}L(L+1)]$$

for the eigenvalues of $E2\ (T=0)\cdot E2\ (T=0)$ when the model space is restricted to a single major shell. Here b is a scale parameter for the oscillator wave functions; we have assumed the standard value $b^2=1.04\ A^{1/3}\,\text{fm}^2$ in all of our calculations. For a given number of particles, the leading (λ,μ) representation defines the highest eigenvalue. For example, in $(ds)^{8,T=0}$ corresponding to ^{24}Mg, the exact largest eigenvalue of $E2\ (T=0)\cdot E2\ (T=0)$ from SU(3) is 133 $e^2\,\text{fm}^4$, whereas the spectral distribution method gives 125 $e^2\,\text{fm}^4$. In all cases, the latter method is seen to give values within 10% of the exact values. SU(3) and similar simple and exact methods are not usually available, in other kinds of model spaces and with other types of excitations, but the configuration-distribution method is no harder to apply to other cases.

We find that the ground state expectation values for $E2\ (T=0)\cdot E2\ (T=0)$ (see Table VI) are a large fraction of the largest eigenvalues (see Table II), a result of the high negative correlation (see Table IV) of this operator with the Hamiltonian. For example, the correlation coefficient in $(ds)^{12,T=0}$ is −0.56 and it takes the ground state expectation value to 102 $e^2\,\text{fm}^4$ from an average expectation value of 44 $e^2\,\text{fm}^4$. The corresponding maximum eigenvalue is 160 $e^2\,\text{fm}^4$ by spectral distribution method and 179 $e^2\,\text{fm}^4$ exact.

For purposes of comparison, we quote here the results of some exact shell-model calculations. In ^{20}Ne, Countee *et al.*,[30] get (with BK Hamiltonian) a value of 49 $e^2\,\text{fm}^4$ for the isoscalar quadrupole transition strength $B(E2,0^+\rightarrow 2_1^+)$ from the ground state to the first excited 2^+ state and very small (<0.3) contributions to the higher lying 2^+ states. We get 43 $e^2\,\text{fm}^4$ for the NEWSR. The experimental value of $B(E2,0^+\rightarrow 2_1^+)$ is 292 $e^2\,\text{fm}^4$. McGrory and Wildenthal,[31] using a suitably modified Kuo interaction and a truncated basis (with no more than four particles outside $d_{5/2}$ and no more than two in $d_{3/2}$) get a value of 79 $e^2\,\text{fm}^4$ for $B(E2,0^+\rightarrow 2_1^+)$ in ^{24}Mg and 74 $e^2\,\text{fm}^4$ in ^{28}Si. Our values for NEWSR using the BK Hamiltonian and

TABLE V. Ground state parameters for even-even self-conjugate *ds*-shell nuclei. The ground state energy was determined by the configuration-distribution method (Ratcliff procedure) with $(\vec{m},T=0)$ configurations. The Hamiltonian assumed is that of Brown-Kuo with ^{17}O single particle energies.

Nucleus	Active nucleons in *ds* shell m	Ground state J	T	Energy
^{20}Ne	4	0	0	−40.1
^{24}Mg	8	0	0	−97.0
^{28}Si	12	0	0	−158.
^{32}S	16	0	0	−222.
^{36}Ar	20	0	0	−287.

TABLE VI. NEWSR for isoscalar electric and magnetic multipole excitations from the ground state in even-even self-conjugate *ds*-shell nuclei. Values calculated by the scalar theory as well as by the more accurate configuration partitioning method [with $(\vec{m}, T=0)$ configurations] are given. The Hamiltonian assumed is that of Brown-Kuo with ^{17}O single-particle energies. Experimental values are also given where available, and include the strengths to levels below ~ 15 MeV excitation only. The units are $e^2\,\text{fm}^{2L}$ for the electric (EL) cases and $\mu_N{}^2\,\text{fm}^{2L-2}$ for the magnetic (ML) cases. All values should be multiplied by the corresponding scale factors given in the last column.

	Nucleus →	^{20}Ne	^{24}Mg	^{28}Si	^{32}S	^{36}Ar	Scale factor
	Scalar	36	79	102	92	50	1
$E2$	Config.	43	89	111	94	49	1
	Expt.	292±36	518±38	358±22	387±43	364±42	1
	Scalar	35	85	122	123	76	10^2
$E4$	Config.	42	105	159	174	108	10^2
	Expt.						
	Scalar	9.8	7.3	5.8	7.8	9.9	10^{-2}
$M1$	Config.	9.3	4.1	1.3	3.8	8.7	10^{-2}
	Expt.		29.7±11.5	2.41±0.08	0.23±0.07	1.73±0.08	10^{-2}
	Scalar	8.7	11.4	10.2	6.3	1.6	10^2
$M3$	Config.	6.6	8.5	8.7	6.7	2.1	10^2
	Expt.						
	Scalar	2.3	3.3	3.1	1.6		10^5
$M5$	Config.	1.5	2.3	2.6	2.0	0.6	10^5
	Expt.						

TABLE VII. NEWSR for isovector electric and magnetic multipole excitations from the ground state in even-even self-conjugate *ds*-shell nuclei. Values calculated by the scalar theory as well as by the configuration partitioning method are given (except for isovector $M1$, where we give the 0th approximation to NEWSR—see text). The Hamiltonian assumed is that of Brown-Kuo with ^{17}O single-particle energies. Experimental values are also given where available, but include the strengths to levels below ~ 15 MeV excitation only. The units are $e^2\,\text{fm}^{2L}$ for the electric (EL) cases and $\mu_N{}^2\,\text{fm}^{2L-2}$ for the magnetic (ML) cases. All values should be multiplied by the corresponding scale factors given in the last column.

	Nucleus	^{20}Ne	^{24}Mg	^{28}Si	^{32}S	^{36}Ar	Scale factor
	Scalar	9.6	18	22	22	15	1
$E2$	Config.	6.5	15	20	21	15	1
	Expt.			0.34	1.79	2.11	1
	Scalar	18	38	55	58	41	10^2
$E4$	Config.	19	43	65	75	57	10^2
	Expt.						
	0th approx.	14	23	26	23	14	1
$M1$	Expt.	2.04	5.81	6.68	6.50	3.11	1
	Scalar	15	20	20	15	6.7	10^3
$M3$	Config.	10	15	17	15	6.6	10^3
	Expt.						
	Scalar	5.9	9.5	10	6.8	1.2	10^6
$M5$	Config.	5.0	8.2	8.9	6.2	1.6	10^6
	Expt.						

the full ds-space are 89 $e^2\,\mathrm{fm}^4$ for ^{24}Mg and 111 $e^2\,\mathrm{fm}^4$ for ^{28}Si. The experimental values of $B(E2,0^+\to 2_1^+)$ are 486 $e^2\,\mathrm{fm}^4$ in ^{24}Mg and 337 $e^2\,\mathrm{fm}^4$ in ^{28}Si. And for ^{32}S, with an MSDI interaction and a truncated basis (not less than ten particles in $d_{5/2}$) Wildenthal *et al.*[32] get a $B(E2,0^+\to 2_1^+)$ value of 54 $e^2\,\mathrm{fm}^4$. We get 94 $e^2\,\mathrm{fm}^4$ for the NEWSR and the experimental value is 107 $e^2\,\mathrm{fm}^4$. In ^{36}Ar, Wildenthal *et al.*[33] get a $B(E2,0^+\to 2_1^+)$ value of 73 $e^2\,\mathrm{fm}^4$ using a realistic interaction in the full ds-shell model space. Our value is 49 $e^2\,\mathrm{fm}^4$ for the NEWSR and the experimental value is 364 $e^2\,\mathrm{fm}^4$.

Experimental transition strengths for isoscalar quadrupole are seen to be much larger than the predicted NEWSR and their eigenvalue bounds. For example, the $0^+\to 2^+$ transition connecting the two lowest states of ^{28}Si is experimentally determined to have a strength of (337 ± 15) $e^2\,\mathrm{fm}^4$ while the upper bound for the NEWSR, calculated in $(ds)^{12,T=0}$, is 179 $e^2\,\mathrm{fm}^4$. This proves that the ds-orbital space is inadequate. It has become customary to account for the large experimental $E2$ strength by the assumption of an effective charge, the justification being that the same effective charge works over a region of several nuclei. A proper treatment would, however, require enlarging the model space to include excitations across major shells. Such extended treatment would seem to be out of the question by any method except our own, and we propose to do such calculations in the future. We have not assumed here any effective charges in our calculations, and the values of other calculations quoted here have been corrected to discount any effective charges assumed by the original authors.

The isovector quadrupole strengths are known to be extremely fragmented and to lie at high excitation energies. The experimental values of NEWSR quoted in Table VII include the contribution due to the lowest 2^+, $T=1$ states only. Thus, in ^{28}Si the experimental value of transition strength from the ground state to the first 2^+, $T=1$ level at 9.38 MeV is only 0.34 $e^2\,\mathrm{fm}^4$, whereas the calculated value of the NEWSR is 20 $e^2\,\mathrm{fm}^4$ and its eigenvalue bound is 116 $e^2\,\mathrm{fm}^4$.

2. *Magnetic dipole*

The isoscalar magnetic dipole operator $M1$ $(T=0)$ is $\sqrt{3/16\pi}[\vec{J}+(g_n+g_p-1)\vec{S}]$, but the $\vec{J}$ term does not contribute to the transitions that we consider. The $\vec{J}$ term is much larger in magnitude than the $\vec{S}$ term, and if it is not excluded, the fixed-T traces needed in the calculation are mostly determined by the diagonal contribution from the $\vec{J}$ term. This amounts to adding nonexistent transition strength. This difficulty would not arise in fixed-JT model spaces (except for $\Delta J=0$, $J\neq 0$ transitions, in which case, the diagonal contribution can be eliminated by adding a term $\beta\vec{J}$ to the excitation operator so as to minimize the NEWSR quantity with respect to β). However, in the present case, dropping the $\vec{J}$ term essentially removes the said difficulty, because, then the correlation coefficient between $M1\ (T=0)\cdot M1\ (T=0)$ and J^2 drops from very large to extremely small values. For example, in $(ds)^{12,T=0}$, this correlation coefficient is 0.94 when the $\vec{J}$ term is included and 0.003 when it is left out.

Hence, we take

$$M1\ (T=0)\approx\sqrt{3/16\pi}(g_n+g_p-1)\vec{S}\ . \tag{3.12}$$

The isovector $M1$ has the form

$$M1\ (T=1)=\sqrt{3/16\pi}\sum_i[-\vec{j}_i\vec{\tau}_i+(g_n-g_p+1)\vec{s}_i\vec{\tau}_i]\ . \tag{3.13}$$

Since $(g_n+g_p-1)^2/(g_n-g_p+1)^2=0.008$, the isovector $M1$ transitions are estimated to be about 100 times stronger than their isoscalar counterparts. We shall discuss more about this in Sec. III E.

For isoscalar $M1$, it is easy to calculate the eigenvalue bounds exactly. By Eq. (3.12)

$$M1\ (T=0)\cdot M1\ (T=0)=\frac{3}{16\pi}(g_n+g_p-1)^2S^2\ . \tag{3.14}$$

Then the maximum value of S in the model space defines the upper bound for the NEWSR. In the examples, we consider S takes the maximum value of $w/2$, where w is the number of nucleons or holes (in the ds shell), whichever is smaller. This implies an upper bound of

$$\frac{3}{16\pi}(g_n+g_p-1)^2\frac{w(w+2)}{4}$$

for the NEWSR. These are tabulated below (in units of $10^{-2}\,\mu_N{}^2$, where μ_N is the nuclear magneton).

Nucleus	^{20}Ne	^{24}Mg	^{28}Si	^{32}S	^{36}Ar
upper bound (in $10^{-2}\mu_N{}^2$)	20.6	68.9	145	68.9	20.6

Thus, for ^{28}Si we get an upper bound of 1.45 $\mu_N{}^2$ by this method and 0.92 $\mu_N{}^2$ by the spectral distribution method (see Table II). The larger discrepancy seen here is, however, expected because S^2 has a spectrum that is highly degenerate (taking only three distinct eigenvalues in the ^{20}Ne and ^{36}Ar cases and going up to seven in the case of ^{28}Si), but the fact that even in this case the spectral distribution method gives bounds well within a factor of 2 compared to the exact ones is very encouraging (in many cases, a factor of 2 would be tolerable for such upper bounds).

These eigenvalue bounds are very large compared to the experimental values. For example, in ^{32}S the one observed $M1$ ($T=0$) transition to the ground state has a $B(M1,0^+\rightarrow 1^+)$ strength of 0.23×10^{-2} $\mu_N{}^2$ only. The eigenvalue bounds for the NEWSR in this case are 76×10^{-2} $\mu_N{}^2$ by the spectral distribution method and 69×10^{-2} $\mu_N{}^2$ exactly. It is then necessary to look at the NEWSR itself in order to get a better understanding. These are given in Table VI.

In ^{24}Mg, the $M1$ ($T=0$)$\cdot M1$ ($T=0$) centroid, i.e., the zeroth approximation to NEWSR, is 16×10^{-2} $\mu_N{}^2$. As the correlation coefficient with H is positive ($+0.17$), the value of NEWSR at the ground state is even lower and we obtain 7.3×10^{-2} $\mu_N{}^2$ by scalar and 4.1×10^{-2} $\mu_N{}^2$ by configuration calculations. The eigenvalue bound is 69×10^{-2} $\mu_N{}^2$. The experimental transition strengths are[28] 0.8×10^{-2} $\mu_N{}^2$ to the 7.75 MeV level and 29×10^{-2} $\mu_N{}^2$ to the 9.83 MeV level, thus exceeding the calculated NEWSR value, and also the larger zeroth approximation to it, but well within the eigenvalue bound. The large strength from the 9.83 MeV level suggests that[34] there is some isospin mixing due to the neighboring $J=1^+$, $T=1$ level at 9.97 MeV. (The 9.83 and 9.97 MeV levels in ^{24}Mg cannot both have $T=1$ because there is only one 1^+ analog state in the corresponding region of excitation energy in ^{24}Na.) Since the transition strength[24] to the 9.97 MeV level is 193×10^{-2} $\mu_N{}^2$, the mixing amounts to a Coulomb matrix element of 53 keV. If the experimental errors in the measurements corresponding to the 9.83 and 9.97 MeV levels are considered, this value could go down to 38 keV.

Isospin admixture has also been suspected[28] in ^{28}Si. The experimental values of $M1$ strengths are 0.12×10^{-2} $\mu_N{}^2$ and 2.3×10^{-2} $\mu_N{}^2$ to the $J=1^+,T=0$ levels at 8.33 and 10.73 MeV, respectively. The large $M1$ strength of the latter level suggests isospin admixing due to the neighboring $J=1^+,T=1$ level at 10.90 MeV which has an $M1$ strength of 113×10^{-2} $\mu_N{}^2$ from the ground state. This would amount to a Coulomb matrix element of 24 keV. The calculated values of NEWSR are 18×10^{-2} $\mu_N{}^2$ by zeroth approximation, 5.8×10^{-2} $\mu_N{}^2$ by scalar, and 1.3×10^{-2} $\mu_N{}^2$ by configuration calculations (the correlation coefficient with H is $+0.195$). This is quite compatible with isospin admixing—even if we allow for large inaccuracies that may arise in this case because the NEWSR value is small and far away (towards zero) from its value at the spectral centroid.

In ^{32}S the theoretical NEWSR is 3.8×10^{-2} $\mu_N{}^2$. Experimentally, for the one observed transition, $B(M1,0^+\rightarrow 1_1^+)$ is 0.23×10^{-2} $\mu_N{}^2$. Wildenthal *et al.*[32] find, by a shell-model calculation, a $B(M1,0^+\rightarrow 1_1^+)$ strength of 0.27×10^{-2} $\mu_N{}^2$ in ^{32}S using an MSDI interaction. They did not calculate the NEWSR. The 1^+ levels at 7.00 MeV ($T=1$) and 7.19 MeV ($T=0$) are plausible candidates for isospin admixing, but there are no experimental measurements of transition rates corresponding to these.

In ^{36}Ar, the spin assignments are ambiguous. The 4.95 MeV level may be $J=1^+$ or $J=2^+$. Assuming $J=1^+$, the experimental $B(M1,0^+\rightarrow 1_1^+)$ strength is 1.73×10^{-2} $\mu_N{}^2$ compared to the calculated NEWSR value of 8.7×10^{-2} $\mu_N{}^2$. (The 2^+ assignment would mean a $E2$ strength of 16.8 $e^2\,\mathrm{fm}^4$.)

For isovector magnetic dipole, the 0th approximation to the NEWSR are given in Table III. This value for ^{24}Mg is 23 $\mu_N{}^2$, compared to the experimental NEWSR of 5.6 $\mu_N{}^2$ (for two transitions). The $M1$ ($T=1$)$\cdot M1$ ($T=1$) operator is strongly and positively correlated with H with a correlation coefficient of $+0.51$ in $(ds)^{8,T=0}$. Because of this high positive correlation, the NEWSR at the ground state is expected to be considerably less than the 0th approximation and hence closer to the experimental value. But the CLT approximations to the ground state NEWSR gives negative values ($-2.2\ \mu_N{}^2$ in a scalar calculation and $-14.6\ \mu_N{}^2$ in a configuration calculation), thus indicating that in this case it is necessary to consider the higher order terms of the polynomial expansions (2.11) and (2.15). This we shall investigate in the future.

3. Other multipoles

We also calculate the NEWSR and their eigenvalue bounds for $E4$, $M3$, and $M5$ transitions, both isoscalar and isovector. These are also given in Tables II, III, VI, and VII. The correlation coefficients with H are given in Table IV. $E4$ $(T=0)\cdot E4$ $(T=0)$ has a correlation coefficient of -0.33 with H in $(ds)^{12,T=0}$, and a largest eigenvalue of $2.4\times 10^4\ e^2\,\mathrm{fm}^8$. Its ground state expectation value is $1.6\times 10^3\ e^2\,\mathrm{fm}^8$. For isovector $E4$ the correlation coefficient is small and positive [0.12 in $(ds)^{12,T=0}$]. For $M3$ and $M5$, both isoscalar and isovector, the correlation coefficients with H are strongly dependent on the number of active nucleons, changing from negative values in the beginning of the shell to positive values at the upper end. For example, $M5$ $(T=0)\cdot M5$ $(T=0)$ has a correlation coefficient of -0.38 in $(ds)^{4,T=0}$ and $+0.64$ in $(ds)^{20,T=0}$.

The experimental strengths for $E4$, $M3$, and $M5$ are not known well enough for comparsion with the calculated values. The branching ratios for them are indicated only by an upper limit, in the experimental literature; they cannot be measured accurately because of competing lower order electromagnetic transitions which are much stronger. Our calculations indicate that the actual branching ratios are expected to be several orders of magnitude smaller than the upper limits indicated in the experimental literature. For example, in ^{28}Si, the γ-ray branching ratio from the $4^+,T=0$ level at 4.62 MeV to the ground state is[28] $<0.5\%$ (and $\sim 100\%$ to the 2^+ state at 1.78 MeV) and since the mean lifetime $\tau_m=59$ fs the transition strength $B(E4,0^+\rightarrow 4_1^+)$ is $<47\times 10^8\ e^2\,\mathrm{fm}^8$. The theoretical value is only $1.6\times 10^4\ e^2\,\mathrm{fm}^8$ for the NEWSR and $2.4\times 10^4\ e^2\,\mathrm{fm}^8$ for its upper bound. The corresponding Weisskopf unit is $454\ e^2\,\mathrm{fm}^8$. Similar situations are found in other nuclei and with $M3$ and $M5$ transitions.

E. Average strengths and units

The Weisskopf single particle unit[35] for the electric and magnetic multipole transitions· is often found to be too crude an estimate of the typical transition strengths in nuclei. For example, the Weisskopf unit for magnetic dipole transition has the value $1.79\ \mu_N{}^2$, whereas the average experimental strengths are usually weaker by one or two orders of magnitude. A more realistic estimate of the typical magnitudes would be given by the average transition strength in the model space one is working in. This would also enable the shell effects to be taken into account in such a strength unit.

The average strength can be calculated by dividing the the NEWSR by the number of transitions. Using the NEWSR given in Tables VI and VII and taking the number of transitions to be the dimensionality[36] of the $\Gamma=\lambda$ subspace of $(ds)^m$, we tabulate, in Table VIII, the average strengths in the five ds-shell nuclei under consideration. For comparison purposes, the Weisskopf units are tabulated in Table IX. For example, in ^{24}Mg the average strength for isoscalar $E2$ is $0.74\ e^2\,\mathrm{fm}^4$, whereas the Weisskopf unit is $4.11\ e^2\,\mathrm{fm}^4$. For a given multipole the average strength is smaller in the middle of the shell than at either end.

The relative magnitudes of the different multipoles can also be inferred from the average strengths. Thus, the isovector $M1$ is seen to be stronger than isoscalar $M1$, by a factor of 100 in ^{20}Ne, 300 in ^{24}Mg, and 1000 in ^{28}Si. Isoscalar $E2$ is about ten times stronger than isovector $E2$ and isoscalar $E4$ is about five times stronger than isovector $E4$, in the midshell region. On the other hand, for $M3$ and $M5$, the isovector parts are stronger than the isoscalar ones, by a factor of 10 for $M3$ and 20 for $M5$.

The ratio of an individual transition strength to the average strength can be considered as a measure of collectivity for the transition. In Table X we give these values for the strongest of the observed transitions from the ground state for each multipole. Thus, the experimental $B(E2,0^+\rightarrow 2_1^+)$ of 486 $e^2\,\mathrm{fm}^4$ in ^{24}Mg and 337 $e^2\,\mathrm{fm}^4$ in ^{28}Si correspond to collectivities of about 6500 and 9900, respectively. Comparing different multipoles, isoscalar $E2$ is seen to be the most collective, isovector $M1$ coming next. The collectivities of the strongest nonadmixed isoscalar and isovector $M1$ transitions in ^{28}Si are 190 and 550, respectively.

IV. LINEAR-ENERGY-WEIGHTED SUM RULES

The linear-energy-weighted sum rules (LEWSR) are often used in the analysis of experimental data and theoretical models. But a generally applicable

TABLE VIII. Average strengths for electric and magnetic transitions in the *ds* shell. The units are $e^2\,\mathrm{fm}^{2L}$ for the electric (EL) and $\mu_N{}^2\,\mathrm{fm}^{2L-2}$ for the magnetic (ML) cases. All values should be multiplied by the corresponding scale factor in the last row.

$m=A-16$	Isoscalar					Isovector				
^{A}Nucleus	$E2$	$E4$	$M1$	$M3$	$M5$	$E2$	$E4$	$M1$	$M3$	$M5$
^{20}Ne	77	96	30	15	6.3	9.9	38	26	14	15
^{24}Mg	7.4	8.0	0.53	0.65	0.22	0.75	2.0	1.6	0.66	0.46
^{28}Si	3.4	4.2	0.061	0.23	0.079	0.35	0.99	0.65	0.25	0.16
^{32}S	7.8	13	0.49	0.51	0.19	1.1	3.5	1.6	0.66	0.35
^{36}Ar	88	246	28	4.7	2.5	2.3	110	26	9.6	4.7
Scale factor	10^{-2}	1	10^{-4}	1	10^{3}	10^{-2}	1	10^{-2}	10	10^{4}

method to calculate the LEWSR has been lacking. The shell-model approach cannot be used (due to practical limitations) in large model spaces. Other approaches, like the classical sum rule[37] for electric multipoles and the Kurath[38] sum rule for isovector $M1$, have involved crude approximations.

We shall give here a quite general theory of LEWSR and then calculate it for electric and magnetic multipole transitions from the ground state in even-even self-conjugate *ds*-shell nuclei. When the Hamiltonian is one body, this has a very simple form, expressible in terms of ground-state occupancies, single-particle energies, and the double-barred matrix elements of the excitation operator. This amounts to an extension of the Kurath sum rule (which is for isovector $M1$ and for a one-body spin-orbit Hamiltonian) to other types of excitations and to arbitrary one-body Hamiltonians. We shall also discuss a unitary sum rule which takes account of a part of the two-body contribution, but can still be expressed in terms of occupancies. Further, we shall also evaluate the contributions to the LEWSR from the two-body interactions. Comparisons are also made with experimental data and with some shell-model results. For isovector $M1$, the relative importance of spin-flip and orbital contributions is examined explicitly.

The LEWSR operator is discussed in Sec. IV A and the applications to isovector magnetic dipole, isoscalar magnetic dipole, and electric quadrupole excitations are given in Secs. IV B, C, and D. Other multipoles are considered in Sec. IV E. In Sec. IV F, the strength centroids are evaluated using the NEWSR quantities calculated earlier in Sec. III.

A. The LEWSR operator

The linear-energy-weighted sum for the excitations $x\Gamma_i \xrightarrow{O\lambda} \Gamma_f$ (using the notation of Sec. III and Ref. 20) generated by a one-body excitation operator O^{λ} is defined by [where $\lambda=(k,t)$, $\Gamma=(J,T)$]

TABLE IX. Weisskopf units in $e^2\,\mathrm{fm}^{2L}$ for electric (EL) and $\mu_N{}^2\,\mathrm{fm}^{2L-2}$ for magnetic (ML) transitions. All values should be multiplied by the corresponding scale factors in the last row.

Nucleus	$E2$	$E4$	$M1$	$M3$	$M5$
^{20}Ne	3.22	185	1.79	89.6	5.67
^{24}Mg	4.11	301	1.79	114	9.22
^{28}Si	5.05	454	1.79	140	13.9
^{32}S	6.04	649	1.79	168	19.9
^{36}Ar	7.06	888	1.79	196	27.2
Scale factor	1	1	1	1	10^{3}

TABLE X. Collectivities for the strongest observed transition from the ground state for each nucleus and multipole.

$m=A-16$	Isoscalar		Isovector	
^{A}Nucleus	$E2$	$M1$	$E2$	$M1$
^{20}Ne	380			7.9
^{24}Mg	6600	160		220
^{28}Si	9900	190	98	550
^{32}S	4200	46	170	220
^{36}Ar	380	6.1	9.3	6.0

$$S_{\rm EW}=[J_i]^{-1}\sum_{y}\sum_{M_i}\sum_{M_f}\left|\langle\psi^{\Gamma_f}_{y\mu_f}|\sum_{t=0}^{1}O^{k,t}|\psi^{\Gamma_i}_{x\mu_i}\rangle\right|^2(E_{y\Gamma_f}-E_{x\Gamma_i})\ . \tag{4.1}$$

Using the methods[20] of spherical tensors and Racah algebra, Eq. (4.1) can be rewritten as

$$S_{\rm EW}=[J_iT_f]^{-1}\sum_{t=0}^{1}\sum_{t'=0}^{1}C^{T_itT_f}_{\mathscr{T}_i\mathscr{T}_f}C^{T_it'T_f}_{\mathscr{T}_i\mathscr{T}_f}(-1)^{\Gamma_f+\lambda-\Gamma_i}[\Gamma_f]^{1/2}$$

$$\times\sum_{\nu}(-1)^{\lambda+\lambda'-\nu}U(\Gamma_i\lambda\Gamma_i\lambda':\Gamma_f\nu)\langle x\Gamma_i||(\bar{O}^{\lambda}\times[H,O^{\lambda'}]^{\lambda'}_{-})^{\nu}||x\Gamma_i\rangle\ , \tag{4.2}$$

where for any two spherical tensors T^{λ} and U^{σ} the commutator is defined by

$$[T^{\lambda},U^{\sigma}]^{\nu}_{-}\equiv(T^{\lambda}\times U^{\sigma})^{\nu}-(-1)^{\lambda+\sigma-\nu}(U^{\sigma}\times T^{\lambda})^{\nu}\ . \tag{4.3}$$

The electric and magnetic multipole excitations are Hermitian [satisfying the condition $\bar{O}^{\lambda}=(-1)^{\lambda}O^{\lambda}$], and then the double barred expectation value in Eq. (4.2) can be replaced by

$$\tfrac{1}{2}\langle x\Gamma_i||(\bar{O}^{\lambda}\times[H,O^{\lambda'}]^{\lambda'}_{-})^{\nu}-([H,O^{\lambda'}]^{\lambda'}_{-}\times\bar{O}^{\lambda})^{\nu}||x\Gamma_i\rangle\ ,$$

and if in addition $\Gamma_i=(0,0)$ we get

$$S_{\rm EW}=\tfrac{1}{2}(-1)^{\lambda}[\lambda]^{1/2}[t]^{-1}\langle x,0|[O^{\lambda},[H,O^{\lambda}]^{\lambda}_{-}]^{0}_{-}|x,0\rangle\ . \tag{4.4}$$

Here $t=0$ for isoscalar and $t=1$ for isovector excitations. Since O^{λ} is one-body, the operator $[O^{\lambda},[H,O^{\lambda}]_{-}]^{0}_{-}$ has the same particle rank as H itself. If H is of mixed particle rank, the p-body part of $[O^{\lambda},[H,O^{\lambda}]_{-}]^{0}_{-}$ is determined by the p-body part of H only. Explicit construction of this double commutator operator when H is a $(1+2)$-body operator is given in Appendix C. The contribution to $S_{\rm EW}$ in Eq. (4.4) from the one-body part of H has the simple form

$$S^{(1)}_{\rm EW}=\sum_{r}e_r\langle x,0|n_r|x,0\rangle\ , \tag{4.5}$$

where $\langle x,0|n_r|x,0\rangle$ is the ground state occupancy of the rth orbit, and

$$e_r=[t]^{-1}\sum_{b}\alpha_{rb}\alpha_{br}(-1)^{r-b}[r]^{-1}(\epsilon_{bb}-\epsilon_{rr})\ . \tag{4.6}$$

Here for any r and b, ϵ_{bb} is the one-body energy in orbit b and α_{rb} is the double barred matrix element $\langle r||O^{\lambda}||b\rangle$. Moreover, $\sum_r[r]e_r=0$.

A part of the two-body contribution to the expectation value in Eq. (4.4) can also be expressed in terms of occupancies by making a unitary decomposition of the operator $[O^{\lambda},[H,O^{\lambda}]^{\lambda}_{-}]^{0}_{-}$. Noting that this operator is traceless in all m-particle spaces (and hence does not have a unitary rank zero part), we define the unitary sum by

$$S^{\nu=1}_{\rm EW}=\sum_{r}\left[e_r+\frac{m-1}{N-2}\xi_r\right]\langle x,0|n_r|x,0\rangle\ . \tag{4.7}$$

When there is no radial degeneracy, this is all of the unitary rank-one contribution. Here ξ_r is the traceless induced one-body matrix element of orbit r, defined (see Ref. 3) in terms of the standard two-body matrix elements V^{Γ}_{abcd} (see Appendix C) of $[O^{\lambda},[H,O^{\lambda}]^{\lambda}_{-}]^{0}_{-}$ by (since the average induced one-body part vanishes)

$$\xi_r=\tfrac{1}{2}(-1)^{\lambda}[\lambda]^{1/2}[t]^{-1}\left[[r]^{-1}\sum_{b}[\Gamma]V^{\Gamma}_{rbrb}\right]\ . \tag{4.8}$$

We evaluate the ground state expectation value and occupancies appearing in Eqs. (4.4), (4.5), and (4.7) by the spectral distribution method using the configuration decomposition with fixed isospin ($T=0$). The Hamiltonian used is (BK) that of Brown-Kuo[22] with ^{17}O single-particle energies. For electric and magnetic multipole excitations in the ds shell, the parameters e_r and ξ_r are given in Table XI. The ground state occupancies are given in Table XII. We also tabulate the skewness, excess, and correlation coefficients with H of the LEWSR operator appearing in Eq. (4.4). These are given in Tables XIII and XIV.

B. Isovector magnetic dipole

The excitation operator is

$$M1\ (T=1)=\left[\frac{3}{16\pi}\right]^{1/2}\sum_{i}[-\vec{l}_i\vec{\tau}_i+(g_n-g_p)\vec{s}_i\vec{\tau}_i]. \tag{4.9}$$

TABLE XI. One-body and induced one-body parts of the LEWSR operator $\frac{1}{2}(-1)^{\lambda}[\lambda]^{1/2}[t]^{-1}[O^{\lambda},[H,O^{\lambda}]_{-}^{\lambda}]_{-}^{0}$ of Eq. (4.4) for isoscalar ($t=0$), isovector ($t=1$), and electric and magnetic multipole excitations (O^{λ}) in the ds shell. The values given are for ^{20}Ne and are to be multiplied by the corresponding scale factors in the last row. For the other even-even self-conjugate ds-shell nuclei of mass number A, multiply these ^{20}Ne values by $(A/20)^{L/3}$ for electric multipole (EL) and by $(A/20)^{(L-1)/3}$ for magnetic multipole (ML).

	$E2$	$E4$	$M1$	$M3$	$M5$	$E2$	$E4$	$M1$	$M3$	$M5$
$e_{5/2}$	4.21	27.5	6.99	0.546	0.0	4.21	27.5	8.58	3.48	0.0
$e_{3/2}$	−10.94	−41.2	−10.49	−0.032	0.0	−10.94	−41.2	−12.87	−2.97	0.0
$e_{1/2}$	9.24	0.0	0.0	−1.574	0.0	9.24	0.0	0.0	−4.50	0.0
$\xi_{5/2}$	6.26	14.3	3.62	1.828	0.0	6.26	14.3	4.46	6.23	0.0
$\xi_{3/2}$	−1.62	−21.4	−5.43	−0.016	0.0	−1.62	−21.4	−6.69	−1.55	0.0
$\xi_{1/2}$	−15.54	0.0	0.0	−5.446	0.0	−15.54	0.0	0.0	−15.59	0.0
Scale factor	1	10^2	10^{-2}	10^2	1	1	10^2	1	10^3	1

Excitations of nuclei using the 180° electron scattering technique[39] have been found to be very convenient for studying these transitions. The strengths of these transitions are usually discussed[40] in terms of the Kurath sum rule.

Kurath[38] made detailed studies of $1p$-shell nuclei using the Hamiltonian

$$\mathscr{H}=\sum_i \mathscr{H}_0(i)+a\sum_i \vec{l}_i\cdot\vec{s}_i+\sum_{i>j} V(i,j) \ , \tag{4.10}$$

which is a harmonic oscillator shell-model with spin-orbit coupling and two-body interactions of a central force nature. The relative strength a/K (where $K\sim 1$ MeV is a representative integral of the two-body interaction) of spin-orbit coupling was varied to study the effects and determine the best fits. A value of $a/K\approx 3$ in ^{8}Be and $a/K\approx 4.5-6$ in ^{12}C were needed for the theoretical estimates to fit with experimental strengths.

The LEWSR for this model was then calculated by forming the double commutator [see Eq. (4.4)] between $\mathscr{H}$ of Eq. (4.10) and $M1$ $(T=1)$ of Eq. (4.9). The $\mathscr{H}_0(i)$ term commutes with the $M1$ $(T=1)$ operator. The contribution from the spin-orbit term is

$$S_{\mathrm{EW}}^{\mathrm{Ku}}=\frac{3}{16\pi}a(g_p-g_n-1)^2\langle 0|\sum_i \vec{l}_i\cdot\vec{s}_i|0\rangle \ . \tag{4.11}$$

Kurath found that for $4N$ nuclei the total calculated LEWSR was dominated by this spin-orbit contribution. Moreover, his calculations agreed well with the experimental data in p-shell nuclei. These conclusions were then extrapolated to ds-shell nuclei, although no corresponding shell-model studies

TABLE XII. The ground state occupancies of even-even self-conjugate ds-shell nuclei in the extreme single particle model and also those obtained by the configuration-distribution method [calculated in $(ds)^{m,T=0}$ model spaces] using the BK Hamiltonian.

$m=A-16$		Extreme single particle model			Distribution method	
^{A}Nucleus	$d_{5/2}$	$d_{3/2}$	$s_{1/2}$	$d_{5/2}$	$d_{3/2}$	$s_{1/2}$
^{20}Ne	4	0	0	3.09	0.26	0.65
^{24}Mg	8	0	0	5.76	0.93	1.31
^{28}Si	12	0	0	8.14	1.77	2.09
^{32}S	12	0	4	10.21	2.84	2.95
^{36}Ar	12	4	4	11.63	4.65	3.73

TABLE XIII. Skewness and excess parameters in $(ds)^m$ spaces of the LEWSR operators [see Eq. (4.4)] for isoscalar and isovector, and electric and magnetic multipole from the ground states of even-even self-conjugate ds-shell nuclei.

					Skewness					
$m=A-16$			Isoscalar					Isovector		
^{A}Nucleus	$E2$	$E4$	$M1$	$M3$	$M5$	$E2$	$E4$	$M1$	$M3$	$M5$
^{20}Ne	0.380	−0.164	−1.06	−0.568	−0.763	−0.090	−0.221	−0.234	+0.031	+0.098
^{24}Mg	0.223	−0.141	−0.819	−0.481	−0.667	−0.284	−0.255	−0.063	−0.207	−0.143
^{28}Si	0.201	−0.127	−0.669	−0.428	−0.651	−0.314	−0.243	−0.027	−0.219	−0.184
^{32}S	0.266	−0.094	−0.556	−0.350	−0.667	−0.266	−0.188	−0.051	−0.118	−0.143
^{36}Ar	0.461	+0.014	−0.356	−0.132	−0.763	−0.044	−0.030	−0.081	+0.193	+0.098
					Excess					
$m=A-16$			Isoscalar					Isovector		
^{A}Nucleus	$E2$	$E4$	$M1$	$M3$	$M5$	$E2$	$E4$	$M1$	$M3$	$M5$
^{20}Ne	+0.094	+0.232	+1.23	0.832	1.65	+0.055	−0.099	−0.430	−0.419	−0.043
^{24}Mg	−0.200	+0.050	+1.05	0.208	0.462	−0.070	−0.196	−0.084	−0.106	−0.029
^{28}Si	−0.192	−0.007	+0.755	0.112	0.248	−0.113	−0.249	−0.002	+0.013	−0.027
^{32}S	−0.163	−0.068	+0.359	0.292	0.462	−0.107	−0.253	−0.112	0.079	−0.029
^{36}Ar	+0.038	−0.250	−0.360	0.906	1.63	−0.060	−0.254	−0.428	+0.347	−0.043

were made.

In the ds shell, the Kurath sum $S_{\rm EW}^{\rm Ku}$ of Eq. (4.11) can be written as

$$S_{\rm EW}^{\rm Ku}=\frac{3}{32\pi}a(g_p-g_n-1)^2(2n_{5/2}-3n_{3/2})$$

$$=4.286(2n_{5/2}-3n_{3/2})\mu_N{}^2\ \text{MeV}\ . \qquad (4.12)$$

Here $n_{5/2}$ and $n_{3/2}$ are the ground state occupancies of the orbits $d_{5/2}$ and $d_{3/2}$, respectively, and we have used the commonly accepted value $a=2.03$ MeV corresponding to the splitting of $d_{5/2}$ and $d_{3/2}$ levels as experimentally observed in ^{17}O spectra. This is then equivalent to the contribution $S_{\rm EW}^{(1)}$ of Eq. (4.5) to the LEWSR from the one-body part of the BK Hamiltonian, because the $s_{1/2}$ orbit does not affect single-particle excitations of magnetic dipole character (or, simply, $\langle f||M1||s_{1/2}\rangle=0$, unless $f=s_{1/2}$ and $\langle s_{1/2}||M1||i\rangle=0$, unless $i=s_{1/2}$).

Here we make a thorough study of the LEWSR for isovector magnetic dipole transitions from the ground state in even-even self-conjugate ds-shell nuclei. First, we calculate the Kurath sum rule using three sets of occupancies: (i) from the extreme single-particle model (j-j coupling), (ii) from the spectral distribution method using the BK Hamiltonian, and (iii) experimental occupancies[41,42] (obtained from single-nucleon transfer reactions); the values are given in Table XV. $S_{\rm EW}^{\rm Ku}$ is very sensitive to the occupancies. Thus, in ^{28}Si, the Kurath sum rule is 103 $\mu_N{}^2$ MeV in the extreme single particle limit, but goes down to 47 $\mu_N{}^2$ MeV when the ground state occupancies obtained by the spectral distribution method are used. This compares favor-

TABLE XIV. Correlation coefficients in $(ds)^{m,T=0}$ spaces, between the BK Hamiltonian and the LEWSR operators [see Eq. (4.4)] for isoscalar and isovector, electric and magnetic multipole transitions from the ground states of even-even self-conjugate ds-shell nuclei.

$m=A-16$			Isoscalar					Isovector		
^{A}Nucleus	$E2$	$E4$	$M1$	$M3$	$M5$	$E2$	$E4$	$M1$	$M3$	$M5$
^{20}Ne	−0.585	−0.745	−0.632	−0.622	−0.546	−0.680	−0.872	−0.754	−0.760	−0.762
^{24}Mg	−0.552	−0.704	−0.578	−0.663	−0.584	−0.707	−0.881	−0.640	−0.729	−0.742
^{28}Si	−0.556	−0.710	−0.587	−0.665	−0.579	−0.713	−0.887	−0.635	−0.708	−0.689
^{32}S	−0.585	−0.745	−0.637	−0.643	−0.546	−0.711	−0.895	−0.691	−0.681	−0.586
^{36}Ar	−0.663	−0.826	−0.749	−0.584	−0.459	−0.702	−0.909	−0.836	−0.631	−0.343

TABLE XV. Linear-energy-weighted sum rules for isovector magnetic dipole transitions from the ground state in some *ds*-shell nuclei. The values in parenthesis are from the (less accurate) scalar theory and are given for comparison purposes. The spin-flip contribution is the value of LEWSR obtained by excluding the $\vec{l}\,\vec{\tau}$ term in Eq. (4.9) for the isovector $M1$ operator. The units are μ_N^2 MeV.

	Kurath sum rule				Total LEWSR		
Nucleus	Extreme single particle	Distribution method	Experimental occupancies	Unitary sum rule	Distribution method	Experiment[a]	Spin-flip contribution
^{20}Ne	34	23		25	67	23[d]	38
					(75)		(50)
^{24}Mg	69	37	38[b]	44	106	58[e]	57
					(126)		(80)
^{28}Si	103	47	49[c]	59	127	77[f]	70
					(152)		(96)
^{32}S	103	51		69	125	69[g]	75
					(148)		(96)
^{36}Ar	51	40		58	91	33[h]	61
					(103)		(70)

[a]See Ref. 39.
[b]See Ref. 41.
[c]See Ref. 42.
[d]See Ref. 23.
[e]See Ref. 24.
[f]See Ref. 25.
[g]See Ref. 26.
[h]See Ref. 27.

ably with the value of 49 μ_N^2 MeV obtained by using the occupancies determined experimentally by Gove *et al.*[42] from the ^{28}Si(d,^{3}He)^{27}Al reaction. However, the total LEWSR is experimentally determined (by the known isovector $M1$ strengths) to be at least 77 μ_N^2 MeV, indicating substantial contributions from the two-body part of the Hamiltonian.

A part of the contribution from the two-body interaction to the LEWSR is included in the unitary sum rule $S_{\rm EW}^{\nu=1}$ [see Eq. (4.7)] which have also been given in Table XV. The value is 59 μ_N^2 MeV for ^{28}Si and is also below the experimental value.

Thus, except in the extreme single-particle model (which is too crude, anyway), neither the Kurath sum rule nor its unitary extension is capable of explaining satisfactorily the experimental situation. We therefore calculate the total LEWSR, using the full $(1+2)$-body BK Hamiltonian. The calculations were made by configuration partitioning in fixed-$(m,T=0)$ spaces and the values are given in Table XV. Also given, for comparison purposes, are the values obtained by the less accurate scalar theory. Thus, for ^{28}Si, the total calculated LEWSR is 127 μ_N^2 MeV (152 by scalar calculation) compared to the experimental value of at least 77 μ_N^2 MeV. The experimental measurements include transitions to a few low lying states only, and hence are expected to be somewhat lower than the theoretical result.

Finally, we estimate the relative importance of spin-flip and orbital contributions to the isovector $M1$ strengths. The spin-flip part is obtained by neglecting the $\vec{l}\,\vec{\tau}$ term in Eq. (4.9) for the $M1$ $(T=1)$ operator. The values are tabulated in Table XV. Indeed, we find that the spin-flip contribution predominates. Thus, in ^{28}Si on neglecting the $\vec{l}\,\vec{\tau}$ term in Eq. (4.9) we get an LEWSR value of 70 μ_N^2 MeV compared to the value 127 μ_N^2 MeV obtained with the full operator. This means, assuming the spin-flip and orbital-flip contributions to be in phase (for otherwise, the orbital part would have to be unreasonably large; it could also be easily verified by a direct calculation), the spin-flip amplitude is about 0.74 of the total amplitude.

C. Isoscalar magnetic dipole

These transitions are generally weak because the neutron and proton spin contributions tend to cancel.[37] In this case also, the contribution $S_{\rm EW}^{(1)}$ from the one-body part of H to the LEWSR can be expressed in a form similar to the Kurath sum rule. Since the $\vec{J}$ term in the isoscalar $M1$ operator

$$\sqrt{3/16\pi}[\vec{J}+(g_n+g_p-1)\vec{S}]$$

commutes with H, it is only the $\vec{S}$ term that contributes, and with the parameter a chosen to represent the splitting of the single-particle levels $d_{5/2}$ and $d_{3/2}$, the one-body contribution can be written as

$$S_{\rm EW}^{(1)}=\frac{3}{16\pi}a(g_n+g_p-1)^2\langle 0|\sum_i \vec{l}_i\cdot\vec{s}_i|0\rangle$$

$$=0.0350(2n_{5/2}-3n_{3/2})\mu_N{}^2\ \text{MeV}\ . \quad (4.13)$$

We calculate $S_{\rm EW}^{(1)}$ using the occupancies presented in Table XII and give the values in Table XVI. Also given there are the unitary sum $S_{\rm EW}^{\nu=1}$ and the total LEWSR for isoscalar $M1$.

For example, in ^{28}Si, the one-body contribution to LEWSR is 0.33 $\mu_N{}^2$ MeV (0.84 $\mu_N{}^2$ MeV in j-j coupling) and the unitary rank-one part of the two-body contribution increases this to the unitary sum $S_{\rm EW}^{\nu=1}$ value of 0.48 $\mu_N{}^2$ MeV. The total LEWSR is 0.50 $\mu_N{}^2$ MeV. The one-body contribution is seen to dominate the total LEWSR in all cases.

The experimental values given in Table XVI include the one or two observed transitions only and hence are not complete. However, the following conclusions can be made.

In ^{24}Mg, the total calculated value for the total LEWSR is 0.37 $\mu_N{}^2$ MeV whereas the experimental value is 2.9 $\mu_N{}^2$ MeV, for which the 7.75 MeV level contributes 0.064 $\mu_N{}^2$ MeV and the 9.83 MeV level contributes 2.83 $\mu_N{}^2$ MeV. As discussed in the NEWSR case (Sec. III), the 9.83 MeV level is suspected to have admixtures from the $T=1$ level at 9.97 MeV, thus accounting for its larger transition strength and hence larger energy-weighted strength.

In ^{28}Si, the 10.73 MeV level contributes[28] 0.247 $\mu_N{}^2$ MeV and the 8.33 MeV level contributes 0.01 $\mu_N{}^2$ MeV to LEWSR, the theoretical value of total LEWSR being 0.5 $\mu_N{}^2$ MeV. This is compatible with the isospin admixing of the 10.73 MeV ($T=0$) and 10.90 MeV ($T=1$) levels suggested by the NEWSR discussion.

TABLE XVI. Linear-energy-weighted sum rules for isoscalar magnetic dipole transitions from the ground state in some ds-shell nuclei. Values given in parenthesis are from a scalar calculation. The experimental values include the one or two observed transitions only and hence are not complete. The units are $\mu_N{}^2$ MeV. The experimental data are from Ref. 28.

Nucleus	$S_{\rm EW}^{(1)}$ Extreme single particle	$S_{\rm EW}^{(1)}$ Distribution method	$S_{\rm EW}^{\nu=1}$	Total LEWSR Distribution method	Total LEWSR Expt.
^{20}Ne	0.28	0.19	0.20	0.19 (0.42)	
^{24}Mg	0.56	0.30	0.36	0.37 (0.76)	2.9 $\pm$1.1[b]
^{28}Si	0.84	0.38	0.48	0.50 (0.93)	0.26 $\pm$0.01[b]
^{32}S	0.84	0.41	0.56	0.53 (0.91)	0.011$\pm$0.003[a]
^{36}Ar	0.42	0.33	0.47	0.41 (0.63)	0.009$\pm$0.004[a]

[a]For one transition.
[b]For two transitions.

D. Electric quadrupole

The LEWSR for isoscalar and isovector electric quadrupole transitions are given in Table XVII. For the isoscalar quadrupole, the one-body contribution is a significant fraction of the total LEWSR. For example, in ^{28}Si, the one-body contribution $S_{\rm EW}^{(1)}$ is 43 $e^2\,{\rm fm}^4$ MeV and the unitary rank-one part of the two-body contribution increases this to $S_{\rm EW}^{\nu=1}$ value of 53 $e^2\,{\rm fm}^4$ MeV. The total LEWSR is 164 $e^2\,{\rm fm}^4$ MeV, calculated by the configuration partitioning method in $(ds)^{12,T=0}$. The scalar theory, which gives 165, is seen to be equally good. From the shell-model calculation of McGrory and Wildenthal[31] (who used a suitably modified Kuo interaction and a truncated basis), the $0^+\rightarrow 2_1^+$ isoscalar quadrupole transition contributes a value 131 $e^2\,{\rm fm}^4$ MeV to the LEWSR. As seen in Sec. III the experimental transition strengths are much larger than can be explained by any *ds*-shell calculation (thus proving the need to enlarge the model space), and hence give large values of LEWSR too.

For isovector quadrupole, the one-body and the unitary rank-one parts of the LEWSR are the same as for the isoscalar case. However, the total LEWSR is much larger. This is because $T=1$ levels, in general, lie at higher excitation energies compared to $T=0$ levels. For ^{28}Si, the total LEWSR is 449 $e^2\,{\rm fm}^4$ MeV, of which 43 $e^2\,{\rm fm}^4$ MeV is from the one-body part.

A commonly used linear-energy-weighted sum rule for electric multipole (EL) transitions is the classical quantity[37]

$$S_{\rm EW}^{\rm cl}({\rm EL},T)=\frac{A}{4}\frac{e^2\hbar^2}{2m_p}\frac{L(2L+1)^2}{4\pi}\langle 0\,|\,r^{2L-2}\,|\,0\rangle\,, \tag{4.14}$$

where $\langle 0\,|\,r^{2L-2}\,|\,0\rangle$ is the ground state expectation value of r^{2L-2}. [The result is the same for both isoscalar and isovector electric multipole (EL) transitions from the ground state in even-even self-conjugate nuclei.] This result is obtained by assuming the two-body interaction $V(i,j)$ in the Hamiltonian

$$\sum_i \frac{-\hbar^2}{2m_p}\nabla_i^{\,2}+\sum_{i<j}V(i,j)$$

to be local and neglecting velocity dependent and exchange forces. Then the only nonvanishing contribution to LEWSR is from the kinetic energy term

$$\frac{-\hbar^2}{2m_p}\sum_i \nabla_i^{\,2}$$

TABLE XVII. Linear-energy-weighted sum rules for isoscalar and isovector electric quadrupole transitions from the ground states of some *ds*-shell nuclei. The values in parenthesis are those obtained by a scalar calculation. $S_{\rm EW}^{(1)}$, $S_{\rm EW}^{\nu=1}$, and $S_{\rm EW}^{\rm cl}$ are common to both the isoscalar and isovector transitions. The experimental and shell-model values take into account transitions to the first few levels only. Except for ^{20}Ne, the Hamiltonians used in the shell-model calculations are different from the BK Hamiltonian we have used (see Sec. III for a brief account of the interactions and model spaces they used). No effective charges are assumed. The units are $e^2\,{\rm fm}^4$ MeV. The experimental data are from Ref. 28.

	$S_{\rm EW}^{(1)}$		$S_{\rm EW}^{\nu=1}$	Isoscalar			Isovector	$S_{\rm EW}^{\rm cl}$
Nucleus	Extreme single particle	Distribution method		Total LEWSR	Shell model	Expt.	Total LEWSR	
^{20}Ne	17	17	18	48 (57)	72[a]	477± 60	152 (135)	2630
^{24}Mg	38	30	35	115 (123)	108[b]	835± 63	337 (308)	3560
^{28}Si	63	43	53	164 (165)	131[b]	756± 82	449 (408)	4600
^{32}S	120	54	66	174 (164)	106[c]	987±126	430 (382)	5750
^{36}Ar	65	48	58	101 (106)	136[d]	724± 93	252 (217)	6990

[a]See Ref. 30.
[b]See Ref. 31.
[c]See Ref. 32.
[d]See Ref. 33.

and it yields Eq. (4.14). In a simple liquid drop picture, all of the LEWSR of Eq. (4.14) lies in one collective state. For a uniform mass distribution of radius R, $\langle r^l \rangle = 3R^l/(l+3)$, and using the commonly accepted value of $R = 1.2A^{1/3}$ fm, we get, for $L=2$,

$$S_{\mathrm{EW}}^{\mathrm{cl}}(E2,T) = 17.82A^{5/3}e^2\,\mathrm{fm}^4\ \mathrm{MeV}\ . \qquad (4.15)$$

These are given in the last column of Table XVII. Experimentally, one finds that no more than 10% of this sum rule is exhausted by a single low-lying collective state. Theoretically, this kinetic energy part of the sum rule is found to be much larger than the total LEWSR calculated in the ds shell. For example, in ^{24}Mg, the total LEWSR calculated by the spectral distribution method for isoscalar quadrupole excitations is 115 $e^2\,\mathrm{fm}^4$ MeV compared to 3560 $e^2\,\mathrm{fm}^4$ MeV obtained from Eq. (4.15). The connection between the classical result (4.14) and calculations in restricted model spaces like the ds shell is not yet fully understood.

E. Other multipoles

The linear-energy-weighted sum rules for isoscalar and isovector $E4$, $M3$, and $M5$ transitions from the ground state in even-even self-conjugate ds-shell nuclei are given in Tables XVIII, XIX, and XX.

For isoscalar $E4$, the one-body contribution to LEWSR is as much as 50% of the total LEWSR. The values of $S_{\mathrm{EW}}^{(1)}$, $S_{\mathrm{EW}}^{\nu=1}$, and S_{EW} in ^{28}Si are 236×10^2, 298×10^2, and 572×10^2, respectively, in units of $e^2\,\mathrm{fm}^8$ MeV. The classical sum rule [see Eq. (4.14)], on the other hand, is much larger (as in the $E2$ case), its value for ^{28}Si being 2.92×10^6 $e^2\,\mathrm{fm}^8$ MeV.

For isovector $E4$, the one-body part and also the unitary rank-one part of the LEWSR are a small, but still an appreciable fraction of the total LEWSR, the values in ^{24}Mg being 153×10^2, 178×10^2, and 856×10^2 $e^2\,\mathrm{fm}^8$ MeV, respectively.

For both isoscalar and isovector $M3$, the one-body part as well as the full unitary rank-one part of the LEWSR is only a small part of the total LEWSR. These three quantities, in ^{28}Si, are 1.37×10^2, 3.53×10^2, and 89.2×10^2 $\mu_N{}^2\,\mathrm{fm}^4$ MeV for isoscalar $M3$, and 17.1×10^3, 26.7×10^3, and 216×10^3 $\mu_N{}^2\,\mathrm{fm}^4$ MeV for isovector $M3$.

The one-body as well as the unitary rank-one contributions vanish for both isoscalar and isovector $M5$. The LEWSR in ^{28}Si is calculated to be 45.2×10^5 $\mu_N{}^2\,\mathrm{fm}^8$ MeV for isoscalar $M5$ and 116×10^6 $\mu_N{}^2\,\mathrm{fm}^8$ MeV for isovector $M5$.

TABLE XVIII. Linear-energy-weighted sum rules for isoscalar and isovector $E4$ transitions from the ground state for some ds-shell nuclei. The values in parenthesis are those obtained by a scalar calculation. $S_{\mathrm{EW}}^{(1)}$, $S_{\mathrm{EW}}^{\nu=1}$, and $S_{\mathrm{EW}}^{\mathrm{cl}}$ are common to both the isoscalar and the isovector transitions. The units are $e^2\,\mathrm{fm}^8$ MeV. No effective charges are assumed. All values should be multiplied by the corresponding scale factors in the last row.

	$S_{\mathrm{EW}}^{(1)}$		$S_{\mathrm{EW}}^{\nu=1}$	Isoscalar	Isovector	$S_{\mathrm{EW}}^{\mathrm{cl}}$
Nucleus	Extreme single particle	Distribution method		Total LEWSR	Total LEWSR	
^{20}Ne	110	74	80	148 (198)	348 (337)	1.06
^{24}Mg	281	153	178	370 (466)	856 (840)	1.84
^{28}Si	517	236	298	572 (701)	1270 (1250)	2.92
^{32}S	618	306	415	666 (793)	1380 (1320)	4.36
^{36}Ar	362	281	407	491 (604)	983 (886)	6.21
Scale factor	10^2	10^2	10^2	10^2	10^2	10^6

TABLE XIX. Linear-energy-weighted sum rules for isoscalar and isovector $M3$ transitions from the ground state for some ds-shell nuclei. The values in parenthesis are those obtained by a scalar calculation. The units are $\mu_N{}^2\text{fm}^4$ MeV. All values should be multiplied by the corresponding scale factors in the last row.

	Isoscalar				Isovector			
	$S_{\text{EW}}^{(1)}$		$S_{\text{EW}}^{\nu=1}$		$S_{\text{EW}}^{(1)}$		$S_{\text{EW}}^{\nu=1}$	
Nucleus	Extreme single particle	Distribution method		Total LEWSR	Extreme single particle	Distribution method		Total LEWSR
^{20}Ne	2.2	0.65	0.94	35.4 (22.9)	14	7	8	82 (75)
^{24}Mg	4.9	1.2	2.4	74.7 (57.9)	31	13	18	170 (164)
^{28}Si	8.2	1.4	3.5	89.2 (78.0)	52	17	27	216 (214)
^{32}S	0.35	1.1	3.5	68.7 (70.7)	33	19	31	191 (200)
^{36}Ar	0.19	0.49	1.6	26.9 (35.7)	18	15	24	94 (115)
Scale factor	10^2	10^2	10^2	10^2	10^3	10^3	10^3	10^3

TABLE XX. Linear-energy-weighted sum rules for isoscalar and isovector $M5$ transitions from the ground state for some ds-shell nuclei. The values in parenthesis are those obtained by a scalar calculation. The units are $\mu_N{}^2\text{fm}^8$ MeV. All values should be multiplied by the corresponding scale factors in the last row. $S_{\text{EW}}^{(1)}$ and $S_{\text{EW}}^{\nu=1}$ are not given because they vanish.

Nucleus	Isoscalar total LEWSR	Isovector total LEWSR
^{20}Ne	16.5 (8.9)	36.8 (28.9)
^{24}Mg	36.7 (25.2)	87.5 (71.5)
^{28}Si	45.2 (36.9)	116 (98.3)
^{32}S	34.5 (35.0)	99.0 (88.4)
^{36}Ar	13.0 (16.8)	39.7 (38.3)
Scale factor	10^5	10^6

F. Strength centroids

The strength centroids are defined by the ratio LEWSR/NEWSR. In general, because of the nature of their use, larger inaccuracies are tolerable for sum rules than for strength centroids. Hence, although with the present accuracy of our calculations we have been able to do meaningful comparison with experiment as far as sum rules are concerned (some other existing sum rules are much more crude than ours), strength centroids are a different matter and require better accuracy. However, for completeness, we give a brief discussion of the strength centroids calculated with our sum rules, but exclude the magnetic dipole case where larger inaccuracies can be expected in the NEWSR.

When $O^\lambda \cdot O^\lambda$ is positively correlated with H, its ground state expectation value is below the $O^\lambda \cdot O^\lambda$ centroid (both the expectation value and the centroid are positive since $O^\lambda \cdot O^\lambda$ is a positive definite operator). The spectral distribution method actually estimates how many widths away from the centroid does this expectation value lie, and if it is far enough below the centroid to be a low value (as with isoscalar $M1$), even a small inaccuracy in the separation of the expectation value from the centroid could lead to a large inaccuracy in the expectation value thus estimated. And in some extreme cases (like isovector $M1$) we may even get a negative value (in the CLT limit). It would then be necessary to include the higher order terms in the polynomial expansions. Detailed shell-model comparisons made by Draayer, French, and Wong[5] indicate that terms up to fourth order would be sufficient to get good accuracy.

Another source of inaccuracy is that our calculations are with fixed (m,T) traces and not with fixed (m,J,T) ones. This amounts to an averaging over the various J values, and J being an exact symmetry, could sometimes lead to significant inaccuracies. But detailed studies made by Kar[6] (for β decay) have shown that reasonably good results should be obtainable even without fixed J averaging. Although evaluation of fixed (m,J,T) traces are difficult at present, the theory has been worked out by French and Mugambi[43] and a computer code has been written by Lougheed and Wong.[44] We hope to be able to use it in the future. A related problem is that for isoscalar excitations, with fixed (m,T) spaces, the contributions from the diagonal terms (which do not cause transitions) to the NEWSR cannot be properly eliminated (for isovector excitations they are automatically excluded because of fixed T).

The strength centroids we calculated are given in Table XXI. For isoscalar $E2$, if the assumption of effective charges is justified, one expects to get strength centroids in agreement with experiment even though the individual sums are not. We find a discrepancy of about 20% between theory and experiment [we expect even better accuracy when fixed (m,J,T) calculations are made]. The low values of strength centroids predict collectivity for these transitions and this, of course, is what is seen experimentally, with the lowest 2^+, $T=0$ state carrying most of the transition strength.

V. CONCLUSION

The spectral distribution method is opposed in spirit to the conventional shell-model approach where the calculations are of a highly detailed nature, but strong approximations about the model

TABLE XXI. Strength centroids for the different multipoles. Given for each nucleus are (a) strength centroid by scalar theory, (b) strength centroid by configuration theory, (c) strength centroid by experiment, and (d) energy of the lowest experimentally seen level of the corresponding multipolarity. The units are MeV.

		Isoscalar				Isovector			
Nucleus		$E2$	$E4$	$M3$	$M5$	$E2$	$E4$	$M3$	$M5$
	a	1.58	5.63	2.62	3.9	14.1	19.1	5.16	4.9
^{20}Ne	b	1.12	3.52	5.38	11.1	23.4	18.6	7.95	7.3
	c								
	d	1.63	4.25			10.27			
	a	1.56	5.49	5.08	7.61	17.4	21.9	8.12	7.5
^{24}Mg	b	1.29	3.52	8.84	16.2	23.0	19.9	11.2	10.7
	c	1.61							
	d	1.37	4.12	5.24		10.06	9.52		
	a	1.62	5.75	7.66	12.0	18.2	22.9	10.8	9.8
^{28}Si	b	1.48	3.60	10.2	17.5	22.8	19.6	12.6	13.1
	c	2.11							
	d	1.78	4.62	6.28		9.38		9.32	
	a	1.77	6.45	11.2	22.3	17.3	22.9	13.7	13.0
^{32}S	b	1.84	3.83	10.3	17.7	20.7	18.3	12.9	16.1
	c	2.55							
	d	2.23	4.46	5.41		7.12			
	a	2.14	7.99	22.3		14.0	21.6	17.2	
^{26}Ar	b	2.05	4.55	13.0		16.8	17.3	14.2	
	c	1.99							
	d	1.97	4.41			6.61		7.34	

space and the model interaction are often made for practical reasons; instead, by disregarding some of the details, it seeks to extend the calculations to less restricted model spaces, at the same time keeping the essential physical content intact. This is achieved by making use of certain simplicities that exist in many-particle model spaces.

We have seen here some applications of this method to electromagnetic transition data in ^{20}Ne, ^{24}Mg, ^{28}Si, ^{32}S, and ^{36}Ar. The model space for these calculations was the full ds shell. So far there have not been any exact shell-model strength calculations in the full ds-shell model space for ^{24}Mg, ^{28}Si, and ^{32}S.

By calculations of the NEWSR and its upper bound, we were able to demonstrate the inadequacy of the ds-shell space to explain low-lying isoscalar $E2$ excitations in these nuclei. We could also predict isospin admixings of 1^+ levels in ^{24}Mg and in ^{28}Si. Moreover, we predicted the strength sums for the higher order multipoles where the experimental data are not yet accurate enough.

We made a detailed study of the Kurath sum rule in these ds-shell nuclei, and evaluated the corrections to it from the two-body interaction. This two-body contribution to LEWSR has not been evaluated before for ds-shell nuclei. We have also given "Kurath-type" sum rules for other multipoles. These were then extended to a unitary sum rule, which takes into account a part of the two-body interaction and then we evaluated the total LEWSR also.

Among the extensions and improvements of this work that we propose to do in the future are (1) extending the model space to allow for excitations across major shells—this cannot be done by the conventional approach—such extended calculations are necessary to treat properly the giant isoscalar quadrupole strengths; (2) including in the calculation, one or two higher order terms of the polynomial expansions—this would enable us to calculate the ground state expectation values with better accuracy, especially when $O^\lambda \cdot O^\lambda$ is positively correlated with H; and (3) making the calculations with fixed $(\vec{m}, J, T)$ traces—this would give better accuracy and also enable us to extend our calculations to initial states with nonzero angular momentum.

This work was supported in part by grants from the Department of Energy and the National Science Foundation.

APPENDIX A: ELECTRIC AND MAGNETIC MULTIPOLE OPERATORS

For a nucleus consisiting of A active nucleons, the electric multipole operator of order L is defined by[35]

$$E^L = \sum_{i=1}^{A} e_i r_i^L Y^L(\Theta_i \phi_i) ,$$

where e_i is the charge on the ith nucleon (or its "effective charge"), $(r_i,\ \Theta_i,\ \phi_i)$ are its spherical coordinates, and Y^L is the spherical harmonic of order L. In the isospin notation ($\tau_i = +1$ for neutron, -1 for proton), E^L can be broken up into isoscalar and isovector parts such that

$$E^L = E^{L,T=0} + E^{L,T=1} ,$$

where

$$E^{L,T=0} = \frac{e_p + e_n}{2} \sum_{i=1}^{A} r_i^L Y^L(\Theta_i \phi_i) 1_{\text{isospin}} ,$$

$$E^{L,T=1} = -\frac{e_p - e_n}{2} \sum_{i=1}^{A} r_i^L Y^L(\Theta_i \phi_i) \tau_i ,$$

where e_p and e_n are the effective charges of proton and neutron, respectively.

The double barred matrix elements (dbme) of the one-body electric and magnetic multipole operators are obtained as follows. Each single particle state can be denoted by $| nlsjmtt_z \rangle$ (where $s = \frac{1}{2}$ and $t = \frac{1}{2}$ for all nucleons). It can be separated into radial, angular, and isospin parts,

$$| nlsjmtt_z \rangle = | nl \rangle \, | lsjm \rangle \, | tt_z \rangle .$$

[The notation used is the standard one with n, l, s, j, m, t, and t_z representing, respectively, the radial, orbital angular momentum, spin, total angular momentum ($j = l + s$), z component of total angular momentum, isospin, and z component of isospin quantum numbers.] Then, using properties of spherical tensors and Racah algebra one obtains[35]

$$\langle n_f l_f s j_f t \,||\, E^{L,T} \,||\, n_i l_i s j_i t \rangle = \langle n_f l_f | r^L | n_i l_i \rangle (-1)^{j_f + l_i - 1/2} C_{0\,0\,0}^{l_f L l_i} \times \left[\frac{(2L+1)(2l_f+1)(2j_f+1)(2j_i+1)}{16\pi} \right]^{1/2} W(l_f j_f l_i j_i ; \tfrac{1}{2} L) \times \begin{cases} (e_p + e_n)\sqrt{2}, & \text{for } T=0 \text{ (isoscalar)} , \\ -(e_p - e_n)\sqrt{6}, & \text{for } T=1 \text{ (isovector)} . \end{cases}$$

Similarly, the magnetic multipole operators are defined by

$$M^L=\mu_N \sum_i [\text{grad}_i r_i^L Y^L(\Theta_i\phi_i)] \cdot \left[e_i \frac{2l_i}{L+1} + g_i s_i \right] ,$$

which is then divided into isoscalar and isovector parts,

$$M^{L,T=0}=\mu_N \sum_{i=1}^{A} [\text{grad}_i r_i^L Y^L(\Theta_i\phi_i)] \cdot \left[(e_p+e_n)\frac{l_i}{L+1} + \frac{(g_p+g_n)}{2} s_i \right] 1_{\text{isospin}} ,$$

$$M^{L,T=1}=-\mu_N \sum_{i=1}^{A} [\text{grad}_i r_i Y^L(\Theta_i\phi_i)] \cdot \left[(e_p-e_n)\frac{l_i}{L+1} + \frac{(g_p-g_n)}{2} s_i \right] \tau_i ,$$

so that

$$M^L=M^{L,T=0}+M^{L,T=1} .$$

Here $\mu_N=e\hbar/2m_pc$ is the nuclear magneton, l_i and s_i are the orbital and spin angular momenta of the ith particle, and g_p and g_n are the spin-gyromagnetic factors for proton and neutron, respectively (we assume the standard values $g_p=5.5855$ and $g_n=-3.8256$). The dbme are then given by (see also Ref. 35),

$$\langle n_f l_f s j_f t \,||\, M^{L,T} \,||\, n_i l_i s j_i t \rangle = \langle n_f l_f | r^{L-1} | n_i l_i \rangle (-1)^{L-1}(2L+1) \left[\frac{(2j_f+1)(2j_i+1)(2l_f+1)(2L-1)}{L\, 16\pi} \right]^{1/2} F_T ,$$

where

$$F_T = \left\{ \frac{L}{L+1} \sqrt{j_i(j_i+1)(2j_i+1)}\, W(l_f \tfrac{1}{2} L-1 j_i; j_f l_i) W(j_f j_i L-1 1; L j_i)(e_p+e_n) + \sqrt{\tfrac{3}{2}} \left[L\frac{(g_p+g_n)}{2} - (e_p+e_n)\frac{L}{L+1} \right] \begin{Bmatrix} l_f & \tfrac{1}{2} & j_f \\ l_i & \tfrac{1}{2} & j_i \\ L-1 & 1 & L \end{Bmatrix} \right\} \sqrt{2}, \text{ for } T=0 \text{ (isoscalar)} ,$$

and

$$F_T = \left\{ -\frac{L}{L+1} \sqrt{j_i(j_i+1)(2j_i+1)}\, W(l_f \tfrac{1}{2} L-1 j_i; j_f l_i) W(j_f j_i L-1 1; L j_i)(e_p-e_n) - \sqrt{\tfrac{3}{2}} \left[L\frac{(g_p-g_n)}{2} - (e_p-e_n)\frac{L}{L+1} \right] \times \begin{Bmatrix} l_f & \tfrac{1}{2} & j_f \\ l_i & \tfrac{1}{2} & j_i \\ L-1 & 1 & L \end{Bmatrix} \right\} \sqrt{6}, \text{ for } T=1 \text{ (isovector)} .$$

The radial matrix elements $\langle n_f l_f | r^k | n_i l_i \rangle$ ($k=L$ for EL, $k=L-1$ for ML) are to be evaluated in the wave functions of a harmonic oscillator potential since they form the basis states in the shell model. The wave functions for a harmonic oscillator potential $U(r)=\frac{1}{2}M\omega^2 r^2$ are given by

$$| nlsjmtt_z \rangle = R_{nl}(r) | lsjm \rangle | tt_z \rangle ,$$

where the radial part $R_{nl}(r)$ is

$$R_{nl}(r)=N_{nl} e^{-\nu r^2/2} r^l \mathscr{V}_{nl}(\nu r^2) ,$$

with

$$\nu \equiv \frac{1}{b^2} \equiv \frac{M\omega}{\hbar}$$

(the standard value is $\nu = 0.96A^{-1/3}\,\mathrm{fm}^{-2}$). The eigenvalue corresponding to the above wave function is

$$E_{nl} = (2n + l + \tfrac{3}{2})\hbar\omega,\ n = 0, 1, \ldots$$

[for example, $n=0$ for $1d$, $n=1$ for $2s$ in the (ds) shell]. We also have

$$N_{nl}{}^2 = \frac{(2l+2n+1)!!(2\nu)^{l+3/2}}{2^n n![(2l+1)!!]^2}\sqrt{2/\pi}$$

and

$$\mathscr{V}_{nl}(\rho) = \sum_{\mu} \begin{bmatrix} n \\ \mu \end{bmatrix} \frac{(2l+1)!!}{(2l+2+1)!!}(-2\rho)^{\mu} .$$

Using the integrals

$$\int_0^{\infty} e^{-\nu r^2} r^{2l+2} dr = \begin{cases} \dfrac{(2l+1)!!}{(2\nu)^{l+3/2}}\sqrt{\pi/2}, & \text{if } l \text{ is an integer ,} \\[2ex] \dfrac{(2l+1)!!}{(2\nu)^{l+3/2}}, & \text{if } l \text{ is half integer ,} \end{cases}$$

we can readily obtain the following expression

$$\langle n_f l_f \,|\, r^k \,|\, n_i l_i \rangle = \int R_{n_f l_f}(r) r^k R_{n_i l_i}(r) r^2 dr$$

$$= \sum_{\mu_f=0}^{n_f} \sum_{\mu_i=0}^{n_i} (-1)^{\mu_f+\mu_i} \left[\frac{1}{2\nu}\right]^{k/2} \begin{bmatrix} n_f \\ \mu_f \end{bmatrix} \begin{bmatrix} n_i \\ \mu_i \end{bmatrix} \left[\frac{(2l_f+2n_f+1)!!(2l_i+2n_i+1)!!}{n_f! n_i! 2^{n_i+n_f}}\right]^{1/2}$$

$$\times \frac{(l_f+l_i+2\mu_f+2\mu_i+k+1)!!}{(2l_f+2\mu_f+1)!!(2l_i+2\mu_i+1)!!} \begin{cases} 1, & \text{if } l_1+l_2+k \text{ is even ,} \\ \sqrt{2/\pi}, & \text{if } l_1+l_2+k \text{ is odd .} \end{cases}$$

APPENDIX B: SCALAR PRODUCT $O^{\lambda}\cdot O^{\lambda}$ OF TENSORIAL OPERATOR O^{λ}

We use the notation of Ref. 20. Let

$$O^{\lambda} = \sum_{r,s} \alpha_{rs} \mathscr{U}_{rs}^{\lambda}$$

be a one-body Hermitian $[\bar{O}^{\lambda} = (-1)^{\lambda} O^{\lambda}$, and hence $\alpha_{rs} = (-1)^{r-s}\alpha_{rs}]$ operator of tensorial rank λ, where

$$\mathscr{U}_{rs}^{\lambda} = [\lambda]^{-1/2}(A^r \times B^s)^{\lambda} .$$

Here A^r and B^s are the creation and destruction tensor operators for orbits r and s, respectively. We use a spin-isospin direct product notation[20] where $\Gamma \equiv (J,T)$, $(-1)^{\Gamma} = (-1)^{J+T}$, $[\Gamma] = (2J+1) \times (2T+1)$, and for the Racah W coefficients, $W(\Gamma_1\Gamma_2\Gamma_3\Gamma_4{:}\Gamma_5\Gamma_6) = W(J_1J_2J_3J_4{:}J_5J_6)\,W(T_1T_2T_3T_4{:}T_5T_6)$, etc. Then, by the methods[20] of spherical tensors and Racah algebra, we can write $O^{\lambda}\cdot O^{\lambda}$ in the standard form

$$O^{\lambda}\cdot O^{\lambda} = \sum_{r,s} \epsilon_{rs} [r]^{1/2} (A^r \times B^2)^0$$

$$+ \sum_{\substack{r \le s \\ t \le u \\ \Gamma}} [\Gamma]^{1/2} W_{rstu}^{\Gamma} (\Psi^{\Gamma}(r,s) \times \bar{\Psi}^{\Gamma}(t,u))^0 .$$

Here r, s, t, and u denote orbit labels and

$$\Psi^{\Gamma}(r,s)=-\zeta_{rs}(A^r\times A^s)^{\Gamma},$$

$$\zeta_{rs}=\begin{cases}1, & \text{if } r\neq s,\\ 1/\sqrt{2}, & \text{if } r=s.\end{cases}$$

We then get

$$\epsilon_{rs}=\frac{\sum_{p}\alpha_{pr}{}^2}{[r]}\delta_{rs},$$

$$W^{\Gamma}_{rstu}=\zeta_{rs}\zeta_{tu}[\beta^{\Gamma}_{rstu}-(-1)^{r+s-\Gamma}\beta^{\Gamma}_{srtu} -(-1)^{t+u-\Gamma}\beta^{\Gamma}_{rsut}+(-1)^{r+s-t-u}\beta^{\Gamma}_{srut}],$$

where $\beta^{\Gamma}_{rstu}=(-1)^{r+u-\Gamma}\alpha_{rt}\alpha_{su}W(rtsu:\lambda\Gamma)$. Because of symmetry relations among β^{γ}_{rstu}, a further reduction is possible, giving finally

$$W^{\Gamma}_{rstu}=2\zeta_{rs}\zeta_{tu}[\beta^{\Gamma}_{rstu}-(-1)^{r+s-\Gamma}\beta^{\Gamma}_{srtu}].$$

APPENDIX C: EVALUATION OF $[O^{\lambda},[H,O^{\lambda}]^{\lambda}_{-}]^{0}_{-}$

When O^{λ} is a one-body operator and H is a (1 + 2)-body operator, $[O^{\lambda},[H,O^{\lambda}]^{\lambda}_{-}]^{0}_{-}$ is also a (1 + 2)-body operator, with its one-body and two-body parts determined solely by the one-body and two-body parts, respectively, of H. We now express this double commutator operator in the standard form for (1 + 2)-body operators. The notation used is that of Ref. 20. Let

$$O^{\lambda}=\sum_{a,b}\alpha_{ab}[\lambda]^{-1/2}(A^a\times B^b)^{\lambda}$$

and

$$H=\sum_{r,s}\epsilon_{rs}[r]^{1/2}(A^r\times B^s)^0 +\sum_{\substack{rstu:\Gamma\\ r\leq s\\ t\leq s}}[\Gamma]^{1/2}W^{\Gamma}_{rstu} \times(\Psi^{\Gamma}(r,s)\times\overline{\Psi}^{\Gamma}(t,u))^0 .$$

Here a, b, r, s, t, and u denote the orbit labels,

$$\Psi^{\Gamma}(r,s)=-\zeta_{r,s}(A^r\times A^s)^{\Gamma},$$

$$\zeta_{rs}=\begin{cases}1, & \text{if } r\neq s,\\ 1/\sqrt{2}, & \text{if } r=s,\end{cases}$$

and a direct product spin-isospin notation is used, with

$$[r]\equiv 2(2j_r+1),$$

$$(-1)^r\equiv(-1)^{j_r+1/2},$$

etc., as in Appendix B. We also set

$$W^{\Gamma}_{rstu}=-(-1)^{r+s-\Gamma}W^{\Gamma}_{srtu} =-(-1)^{t+u-\Gamma}W^{\Gamma}_{rsut} =(-1)^{r+s-t-u}W^{\Gamma}_{srut}.$$

We then get, using the methods of spherical tensors and Racah algebra [see also Eqs. (3.10), (4.4), and (A5.16) in Ref. 20]:

$$[O^{\lambda},[H,O^{\lambda}]^{\lambda}_{-}]^{0}_{-}=\sum_{r,s}\eta_{rs}[r]^{1/2}(A^r\times B^s)^0 +\sum_{\substack{rstu:\Gamma\\ r\leq s\\ t\leq u}}[\Gamma]^{1/2}V^{\Gamma}_{rstu} \times(\Psi^{\Gamma}(r,s)\times\overline{\Psi}^{\Gamma}(t,u))^0 ,$$

where (assuming no radial degeneracy),

$$\eta_{rs}=\delta_{rs}2\sum_{b}\alpha_{rb}\alpha_{br}(-1)^{r-b+\lambda}[\lambda]^{-1/2}[r]^{-1}(\epsilon_{bb}-\epsilon_{rr}),$$

$$V^{\Gamma}_{rstu}=\zeta_{rs}{}^2\zeta_{tu}{}^2[X^{\Gamma}_{rstu}-(-1)^{r+s-\Gamma}X^{\Gamma}_{srtu}-(-1)^{t+u-\Gamma}X^{\Gamma}_{rsut}+(-1)^{r+s-t-u}X^{\Gamma}_{srut}].$$

This step is for antisymmetrization, and the X^{Γ}_{rstu} appearing here are given by

$$X^{\Gamma}_{rstu}=[\lambda]^{-1/2}\{T_1+T_2+T_3+T_4+T_5+T_6\},$$

where

$$T_1=\sum_{i,j,\Omega}\alpha_{ri}\alpha_{ju}(-1)^{r+i+j+u+\Omega+\Gamma+\lambda}\zeta_{is}{}^{-1}\zeta_{tj}{}^{-1}W^{\Omega}_{istj}W(\lambda i\Gamma s:r\Omega)[\Omega]W(\Gamma t\lambda j:u\Omega)\zeta_{rs}{}^{-1}\zeta_{tu}{}^{-1},$$

$$T_2=\sum_{i,j,\Omega}\alpha_{it}\alpha_{sj}(-1)^{\lambda+\Omega+\Gamma}[\Omega]\zeta_{iu}{}^{-1}\zeta_{rj}{}^{-1}W^{\Omega}_{iurj}W(\lambda i\Gamma u:t\Omega)W(\Gamma r\lambda j:s\Omega)\zeta_{rs}{}^{-1}\zeta_{tu}{}^{-1},$$

$$T_3=-\sum_{i,j}\alpha_{ri}\alpha_{sj}(-1)^{2r}\zeta_{rs}{}^{-1}\zeta_{ij}{}^{-1}W(\Gamma si\lambda{:}rj)W^{\Gamma}_{ijtu}f(t,u)\ ,$$

$$T_4=-\sum_{i,j}\alpha_{it}\alpha_{ju}(-1)^{i+t+j-u}\zeta_{ij}{}^{-1}\zeta_{tu}{}^{-1}W^{\Gamma}_{ijrs}W(\Gamma ui\lambda{:}tj)f(r,s)\ ,$$

$$T_5=-\sum_{i}\alpha_{ri}\alpha_{ir}(-1)^{r-i+\lambda}[r]^{-1}W^{\Gamma}_{rstu}\zeta_{rs}{}^{-2}f(t,u)\ ,$$

$$T_6=-\sum_{i}\alpha_{it}\alpha_{ti}(-1)^{t-i+\lambda}[t]^{-1}W^{\Gamma}_{turs}\zeta_{tu}{}^{-2}f(r,s)\ ,$$

and

$$f(r,s)=\begin{cases}1, & \text{if } r\le s\ ,\\ 0, & \text{otherwise}\ .\end{cases}$$

---

*Present address: Department of Physics, State University College, Fredonia, New York 14063.

[1]J. B. French, in *Nuclear Structure,* edited by A. Hossain *et al.* (North-Holland, Amsterdam, 1967).

[2]J. B. French and K. F. Ratcliff, Phys. Rev. C 3, 94 (1971); K. F. Ratcliff, *ibid.* 3, 117 (1971); A. Gervois, Nucl. Phys. A184, 507 (1972).

[3]F. S. Chang, J. B. French, and T. H. Thio, Ann. Phys. (N.Y.) 66, 137 (1971); F. S. Chang, University of Rochester Report UR-425, 1973 (unpublished).

[4]J. B. French, in *Dynamic Structure of Nuclear States,* edited by D. J. Rowe, L. E. H. Trainor, S. S. M. Wong, and T. W. Donnelly (University of Toronto, Ontario, Canada, 1972).

[5]J. P. Draayer, J. B. French, V. Potbhare, and S. S. M. Wong, Phys. Lett. B55, 263 (1975); B55, 349 (1975); J. P. Draayer, J. B. French, and S. S. M. Wong, Ann. Phys. (N.Y.) 106, 472 (1977); 106, 503 (1977).

[6]K. Kar, Ph.D. thesis, University of Rochester, 1979.

[7]T. R. Halemane, Ph.D. thesis, University of Rochester, 1979.

[8]A. de-Shalit and I. Talmi, *Nuclear Shell Theory* (Academic, New York, 1963); E. C. Halbert, J. B. McGrory, B. H. Wildenthal, and S. P. Pandya, in *Advances in Nuclear Physics,* edited by M. Baranger and E. Vogt (Plenum, New York, 1971), Vol. 4.

[9]S. Y. Li and D. S. Koltun (unpublished); V. Potbhare, Ph.D. thesis, University of Rochester, 1975; S. Ayik and J. N. Ginocchio, Nucl. Phys. A221, 285 (1974).

[10]H. Cramer, *Mathematical Methods of Statistics* (Princeton University, Princeton, New Jersey, 1946); G. Szego, *Orthogonal Polynomials* (American Mathematical Society Colloquium Publications, Providence, Rhode Island, 1939), Vol. XXIII.

[11]J. B. French and S. S. M. Wong, Phys. Lett. 33B, 449 (1970); 35B, 5 (1971); O. Bohigas and J. Flores, Phys. Lett. 34B, 261 (1971); 35, 383 (1971); S. S. M. Wong and J. B. French, Nucl. Phys. A198, 188 (1972).

[12]K. K. Mon and J. B. French, Ann. Phys. (N.Y.) 95, 90 (1975).

[13]F. S. Chang and J. B. French, Phys. Lett. B44, 131 (1973); B44, 135 (1973).

[14]M. Nomura, Prog. Theor. Phys. 48, 442 (1972); J. B. French and J. P. Draayer, in *Group Theoretical Methods in Physics, Proceedings of the Seventh Internal Colloquium and Integrative Conference on Group Theory and Mathematical Physics,* edited by W. Beiglböck, A. Bohm, and E. Takasugi (Springer, Berlin, 1979); T. R. Halemane (unpublished).

[15]J. D. Louck, Am. J. Phys. 38, 3 (1970).

[16]J. N. Ginocchio, Phys. Rev. C 8, 135 (1973).

[17]B. D. Chang and S. S. M. Wong, Nucl. Phys. A294, 19 (1978).

[18]K. F. Ratcliff, Phys. Rev. C 3, 117 (1971).

[19]V. Potbhare and S. P. Pandya, Nucl. Phys. A256, 253 (1976); J. P. Draayer, J. B. French, M. Prasad, V. Potbhare, and S. S. M. Wong, Phys. Lett. 57B, 130 (1975).

[20]J. B. French, in *Proceedings of the International School of Physics,* Enrico Fermi, Course XXXVI, edited by C. Bloch (Academic, New York, 1967).

[21]L. S. Hsu, Phys. Lett. 25B, 588 (1967).

[22]T. T. S. Kuo, Nucl. Phys. A103, 71 (1967).

[23]W. L. Bendel, L. W. Fagg, S. K. Numrich, E. C. Jones, Jr., and H. F. Kaiser, Phys. Rev. C 3, 1821 (1971).

[24]L. W. Fagg, E. C. Jones, and W. L. Bendel, Nucl. Instrum. Methods 77, 136 (1970).

[25]L. W. Fagg, W. L. Bendel, E. C. Jones, Jr., and S. Numrich, Phys. Rev. 187, 1378 (1969).

[26]L. W. Fagg, W. L. Bendel, L. Cohen, E. C. Jones, Jr., H. F. Kaiser, and H. Uberall, Phys. Rev. C 4, 2089 (1971).

[27]L. W. Fagg, W. L. Bendel, E. C. Jones, Jr., L. Cohen, and H. F. Kaiser, Phys. Rev. C 5, 120 (1972).

[28]F. Ajzenberg-Selove, Nucl. Phys. A190, 1 (1972); P. M.

Endt and C. Van der Leun, *ibid.* A214, 1 (1973).

[29]J. P. Elliott, Proc. R. Soc. London A245, 128 (1958); A245, 562 (1958); M. Harvey, in *Advances in Nuclear Physcis,* edited by M. Baranger and E. Vogt (Plenum, New York, 1968), Vol. I.

[30]C. R. Countee, J. P. Draayer, T. R. Halemane, and K. Kar, Nucl. Phys. A356, 1 (1981).

[31]J. B. McGrory and B. H. Wildenthal, Phys. Lett. 34B, 373 (1971).

[32]B. H. Wildenthal, E. C. Halbert, J. B. McGrory, and H. D. Graber, Phys. Rev. C 4, 1708 (1971).

[33]B. H. Wildenthal, E. C. Halbert, J. B. McGrory, and T. T. S. Kuo, Phys. Rev. C 4, 1266 (1971).

[34]O. Titze, Z. Phys. 220, 66 (1969).

[35]P. J. Brussard and P. W. M. Glaudemans, *Shell-Model Applications in Nuclear Spectroscopy* (North-Holland, Amsterdam, 1977).

[36]T. Sebe and M. Harvey, Atomic Energy of Canada Report AECL-3007, 1968.

[37]E. K. Warburton and J. Wenser, in *Isospin in Nuclear Physics,* edited by D. H. Wilkinson (North-Holland, Amsterdam, 1969).

[38]D. Kurath, Phys. Rev. 130, 1525 (1963).

[39]L. W. Fagg, Rev. Mod. Phys. 47, 683 (1975).

[40]S. S. Hanna, in *Isospin in Nuclear Physics,* edited by D. H. Wilkinson (North-Holland, Amsterdam, 1969); S. S. Hanna, in, *Photonuclear Reactions,* edited by J. Ehlers *et al.* (Springer, Berlin, 1977); H. Uberall, R. A. Lindgren, and L. W. Fagg, Acta Phys. Austr. 41, 341 (1975).

[41]R. O. Nelson and N. R. Roberson, Phys. Rev. C 6, 2153 (1972).

[42]H. E. Gove, K. H. Purser, J. J. Schwartz, W. P. Alford, and D. Cline, Nucl. Phys. A116, 369 (1968).

[43]J. B. French and P. E. Mugambi (unpublished).

[44]G. D. Lougheed and S. S. M. Wong, Nucl. Phys. A246, 29 (1975).

VOLUME 51, NUMBER 24 PHYSICAL REVIEW LETTERS 12 DECEMBER 1983

Nuclear Level Densities and Partition Functions with Interactions

J. B. French and V. K. B. Kota[(a)]
Department of Physics and Astronomy, University of Rochester, Rochester, New York 14627
(Received 29 August 1983)

A general theory, based on central limit theorems and unitary-group decompositions of the microscopic H, is given for the nuclear level density. The density appears in terms of convolutions of noninteracting-particle densities with easily calculable interaction functions given explicitly in terms of the Hamiltonian matrix elements.

PACS numbers: 21.10.Ma, 21.60.Cs

The nuclear level density, and its decomposition by conserved symmetries and by configurations (partitionings of the active particles among orbits), are important on practical grounds—they are sometimes measurable, and they enter into the calculation of reaction cross sections—and, formally, because, when given as functions of the Hamiltonian parameters, they determine the smoothed partition function and hence much of the statistical behavior of the nucleus. The purpose of this paper is to outline a general theory for interacting nucleons which derives, directly and without free parameters, from the microscopic Hamiltonian. Among other things it will explain why the conventional noninteracting-particle (NIP) level-density theories[1,2] are often successful, and will give forms and measures for the various kinds of orbital admixings generated by the interaction which the NIP theories do not deal with. From the formal standpoint it should lead to a more fundamental statistical theory of nuclear reactions and should make contributions also to the theory of the effective interaction.

The basis of the theory is an elementary unitary-group decomposition of the interaction Hamiltonian, coupled with the use of spectral- and ensemble-averaging methods. It has been argued before,[3] and to a considerable extent demonstrated,[3,4] that spectral averaging should lead to a proper level-density theory. A problem of form, however, arises, with spectral averaging, from the initial restriction of the single-particle (sp) spectrum to a finite set of sp states. This leads to the level density as a superposition of configuration densities valid only for a restricted part of the spectrum. The consequence is that one loses contact with the essentially combinatorial NIP form. Moreover, as we extend the sp spectrum in order to proceed to higher energies we encounter a problem of substance, namely that the variances of all the configuration distributions increase indefinitely; this, on the face of it, generates ever larger orbital admixings and spreadings, destroying the shell-model stability. We shall dispose of both of these problems.

Let the single-particle model space be partitioned as usual according to major shells, and further via spherical (j) orbits. Then the m-particle states are described via configurations which we may catalog by s, the major-shell excitation; for example, if the $1s$ and $1p$ shells are regarded as filled, $(1d, 2s)^m$ with $0 < m \leq 24$ has $s=0$ while $(1d, 2s)^{m-2}(1f)^2$ has $s=2$. Writing $G(k)$ for a k-body operator we have $H=H(0)+h(1)+V(2)$ where $h=\sum \epsilon_i(1)n_i$ is determined by the single-particle energies (spe), and the interaction V by the matrix elements V_{ijkl}. V also has a major-shell decomposition, $V=\sum V_t$ with $t=0, 2, \ldots,$ where V_t generates $\pm t\hbar\omega$ excitations. When $V=0$ we have the NIP density, $I_0(E)$ say, which consists of a number of degenerate sets ("spikes") labeled by configurations (c) with energies $\mathcal{E}_c$ and degeneracies d_c. The smoothed form of $I_0(E)$, which is essentially $\exp(bE)^{1/2}$, can be calculated by either conventional or spectral methods. For brevity, we do not deal with that here, nor with the (J, T) decomposition which, even with interactions, can be carried out spectrally; the parity decomposition comes automatically with the configurations.

Unitary transformations come into play because of the antisymmetry of the fermion states. With that in mind we first extract from V that part ($V^{[0]}$ with $[0]=[0, 0, \ldots]$) which is invariant with respect to the group of simultaneous unitary transformations in every orbit α. Thus $V=V^{[0]} + V^{[\Lambda]}$ where $[\Lambda]$ then stands for all the nonscalar irreps. Since it is necessarily a quadratic polynomial in the orbital number operators $n_\alpha = \sum_{i\in\alpha} a_i^\dagger a_i$, $V^{[0]}$ preserves the configuration degeneracies, has no matrix elements connecting configurations, and receives contributions only from V_0. It does, however, shift the centroids, $\mathcal{E}_c \to \mathcal{E}_c + \Delta\mathcal{E}_c$. Moreover, because of its J-scalar two-body nature, $V^{[\Lambda]}$ cannot transform as a nontrivial ($\neq 1$) scalar in *any* orbit since any n_α operator appearing in V must be multiplied by some $n_\beta - \delta_{\alpha\beta}$ and must therefore belong to $V^{[0]}$. Thus all the centroids are fixed by $V^{[0]}$. Our first problem now is to investigate the centroid shifts $\Delta\mathcal{E}_c$, asking in particular whether they are compatible with a renormalization of the spe.

Anticipating that the abnormally large orbital admixings will be eliminated, we see that at energy E there will be some finite subset of active orbits, containing, say, $N = N(E)$ sp states. We can now decompose that part of $V^{[0]}$ which involves the active orbits into U(N) tensors, $V^{[0]} = \sum V^{\nu,[0]}$ where V^ν, with "unitary rank" $\nu = 0, 1, 2$, belongs to the Young columnar shape $[N-\nu, \nu]$ and contains $\binom{n-\nu}{2-\nu}$ as a factor[3,5]; here $n = \sum n_\alpha$, summed over the active orbits. For fixed particle number m, V^0 behaves as a constant, V^1 as a one-body operator, and V^2 as a "true" two-body operator.

Now we "turn on" V as follows:

(1) $V^{0,[0]}$ which locally shifts the NIP spectrum. This adds to $H(0)$ and may for present purposes be ignored.

(2) $V^{1,[0]}$ which contains $(n-1)$ as a factor and therefore gives only a number-dependent renormalization of the spe. Thus $I_0(E) \to I_1(E)$ where $I_1(E)$ is locally of NIP form.

(3) $V^{2,[0]}$ which preserves the NIP degeneracies but moves the $\mathcal{E}_c$ in a way *not* given by changes in the spe. We shall see that $V^{2,[0]}$ is negligible.

(4) $V_0^{[\Lambda]}$, the part of the residual H which generates excitations only within a major shell, preserving the oscillator symmetry. This term produces the $0\hbar\omega$ admixings and spreadings (variances $\hat{\sigma}_c^2$) of the configurations. In conjunction with $I_1(E)$ it gives rise to the level density $I_2(E)$ which is valid for $E \lesssim \hbar\omega$.

(5) $V_{t>0}^{[\Lambda]} = V_{t>0}$. If treated like $V_0^{[\Lambda]}$ this term would generate the anomalously large orbital admixings. But by connecting configurations which are too far apart to be strongly interacting, it generates multimodal distributions which require a special treatment.

$V^{[0]}$ is obviously defined by the quantities $V_{\alpha\beta}$, the average interaction between two particles, one in orbit α and one in β (which may coincide). With N_α the orbital degeneracy, the dimensionality of the corresponding two-particle space is $N_{\alpha\beta} = N_\alpha(N_\beta - \delta_{\alpha\beta})(1 - \frac{1}{2}\delta_{\alpha\beta})$. Recognizing that $n_{\alpha\beta}$, defined similarly, is the projection operator for these states, we have, with $i<j$ when $\alpha = \beta$,

$$V^{[0]} = \sum_{\alpha\leq\beta} V_{\alpha\beta} n_{\alpha\beta}; \qquad V_{\alpha\beta} = N_{\alpha\beta}^{-1} \sum_{i\in\alpha} \sum_{j\in\beta} V_{ijij}. \tag{1}$$

To carry out the local decomposition referred to above we restrict the orbits to those active in the region of interest; then $N = \sum N_\alpha$. By simple intuitive arguments, or more formally,[3] we find

$$V^{0,[0]} = \overline{V}\binom{n}{2}; \quad \overline{V} = \binom{N}{2}^{-1} \sum_{\alpha\leq\beta} N_{\alpha\beta} V_{\alpha\beta},$$
$$V^{1,[0]} = (n-1)\sum \epsilon_\alpha(2) n_\alpha, \tag{2}$$
$$\epsilon_\alpha(2) = (N-2)^{-1} \sum (N_\gamma - \delta_{\alpha\gamma})(V_{\alpha\gamma} - \overline{V}).$$

$V^{0,[0]}$ is of no interest to us. The form of $V^{1,[0]}$ displays explicitly the renormalization of the sp Hamiltonian, $h \to \hat{h} = h + V^{1,[0]}$, while

$$\epsilon_\alpha(1) \to \epsilon_\alpha = \epsilon_\alpha(1) + (m-1)\epsilon_\alpha(2).$$

The unitary decomposition of $\hat{h}$ is

$$\hat{h} = \hat{h}^0 + \hat{h}^1 = \overline{\epsilon} n + \sum(\epsilon_\alpha - \overline{\epsilon}) n_\alpha$$

with $\overline{\epsilon} = N^{-1}\sum N_\alpha \epsilon_\alpha$; similarly for h with $\epsilon_\alpha(1)$ instead of ϵ_α. The results that $\binom{n}{2} = \sum_{\alpha\leq\beta} n_{\alpha\beta}$ and $(n-1)n_\alpha = \sum_{\alpha\leq\beta}(1+\delta_{\alpha\beta})n_{\alpha\beta}$, while $V^{2,[0]} = V^{[0]} - V^{0,[0]} - V^{1,[0]}$, enables us to write "two-body" forms $\sum V_{\alpha\beta}{}^\nu n_{\alpha\beta}$ for each V^ν.

To understand the action of $V^{[0]}$ in the m-particle space of the active orbits we need, for the traceless operators h^1, $V^{1,[0]}$, and $V^{2,[0]}$, their relative magnitudes and their correlation coefficients. These are defined (for traceless operators F) in terms of the unitary norm $|F|^m$ whose square is $\langle F^\dagger F\rangle^m \equiv \binom{N}{m}^{-1}\mathrm{Tr}(F^\dagger F)$, the average eigenvalue of the Hermitian square, which is easily evaluated.[3] Three general results are that operators of different unitary rank are uncorrelated, that the correlation coefficient for operators of the same ν are independent of m, and that the norm of $F^\nu(k)$ grows asymptotically with m as $m^{k-\nu/2}$ so that $|h^1|^m \sim m^{1/2}$, $|V^{1,[0]}|^m \sim m^{3/2}$, and $|V^{2,[0]}|^m \sim m$. With typical interactions[6] we find that, in the two-particle space, $V^{1,[0]}$ and $V^{2,[0]}$ are of comparable magnitude, but very much smaller than h^1; moreover h^1 and $V^{1,[0]}$ have a strong positive correlation [≥ 0.9 in (ds), 0.99 in the space generated by the first fifteen orbits, the primary one-body H in the latter case being taken as that of a harmonic oscillator]. Then, because the primary and induced one-body Hamiltonians add coherently, and because the latter grows more rapidly than $V^{2,[0]}$, the $V^{2,[0]}$ effects are negligible, giving a norm correction

~2% in (ds) and ~0.2% in the larger case. Taking for granted that this result extends to larger nuclei, and leaving for the future on explanation of its origin in terms of the spatial and spin properties of V, we conclude that the entire centroid-generating Hamiltonian $V^{[0]}$ is well represented by an effective one-body operator so that, under its inclusion, the NIP level-density form is still maintained.

As we have said, the $t=0$ part of the remaining interaction $V^{[\Lambda]} = V^{2,[\Lambda]}$ generates the $0\hbar\omega$ admixings and configuration spreadings. Though there will be fluctuations in the shapes of individual configuration distributions it is adequate for us to take them as Gaussian,[7] defined therefore by $(d_c, \mathcal{E}_c, \hat{\sigma}_c^2)$ with d_c the dimensionality, and $\hat{\sigma}_c^2 = \sigma_c^2(t=0)$. It has been observed in many numerical calculations that m-particle configuration variances, calculated for realistic interactions, fluctuate very little about the average which, properly weighted, is

$$\sigma_A^2 = d^{-1}\sum d_c \sigma_c^2 = \langle (V^{[\Lambda]})^2\rangle^m , \tag{3}$$

where $\langle\cdots\rangle^m$, essentially a trace, is defined above. To calculate the variance of the σ_c^2 (similarly for $\hat{\sigma}_c^2$) we first construct a variance-generating operator K (analogous to the centroid generator above, and also expressed in terms of the n_α operators) and then evaluate *its* variance in the m-particle space. We find, as a compact operator version of an earlier result,[3] that

$$K = \sum_{\alpha\leq\beta,\, \gamma\leq\delta} \{\langle n_{\alpha\beta} n_{\gamma\delta}^{\times}\rangle^2\}^{-1} K_{\alpha\beta\gamma\delta}\, n_{\alpha\beta} n_{\gamma\delta}^{\times} . \tag{4}$$

Here $N_{\alpha\beta}K_{\alpha\beta\gamma\delta}$ is the sum of squares of the matrix elements of $V^{[\Lambda]}$ connecting the two-particle configurations $(1_\alpha, 1_\beta)$ and $(1_\gamma, 1_\delta)$, so that $K_{\alpha\beta\gamma\delta}$ is a partial $(1_\alpha, 1_\beta)$ variance, while $F^\times$ is the hole-particle adjoint of F, $n_\alpha^\times$ then being $N_\alpha - n_\alpha$, and of course $\langle K\rangle^m = \sigma_A^2$. The fractional rms deviation of the σ_c is then $\delta(m) = (|K|^m)^2/2\sigma_A^4(m)$; schematically $\sigma_c(m) = \sigma_A(m)[1\pm\delta(m)]$. The evaluation of the quadratic and quartic traces, in terms of the V_{ijkl}, follows easily[3] from their values for two, three, and four particles.

For ds-shell interactions ($N=24$) we find $\sigma_c(m) \sim 8(1\pm 0.03)$ MeV. Spaces whose sp spectra span several major shells are immense, and then simple averaging may be misleading. The appropriate averages, also easily calculable, are over the subspaces with fixed $s\hbar\omega$ excitation, which so far are noninteracting since $t=0$. For example, a ten-orbit interaction ($N=80$) with $m=24$ has $s_{\max}=44$ and gives $\hat{\sigma}_c = 30\pm 1.5$ MeV on the gross average. But for $s=0$, 2, and 4 we find $\hat{\sigma}_c = 7.5(1\pm 0.03)$, $11.9(1\pm 0.11)$, and $15.2(1\pm 0.10)$ MeV. We note incidentally a secular increase of the spreading widths with energy, $\hat{\sigma}_A^2(m) \to \hat{\sigma}_A^2(m, s)$.

Ignoring the small variance fluctuations we see now that for $E \lesssim \hbar\omega$, for which only $s=0$ is relevant, the $t=0$ level density is a simple convolution[7] of the renormalized NIP density $I_1(E)$ with an "interaction Gaussian" of unit integral and variance $\hat{\sigma}_A^2(s=0)$, $I_2(E) = I_1 \otimes \rho_G^{(s=0)}[E]$. For the general case we have a sum of convolutions,

$$I_3(E) = \sum I_3^{(s)}(E) = \sum I_1^{(s)} \otimes \rho_G^{(s)}[E] , \tag{5}$$

which displays the simple connection between the interacting- and noninteracting-particle densities. The s decomposition of the NIP level density follows directly from the configuration decomposition[3] and is very simply done by spectral methods. It is noteworthy that, in their phenomenological introduction of interactions into an NIP theory, Haq and Wong[9] have *assumed* a somewhat similar convolution form, as well as an increase of the effective configuration variances with energy.

The $t\geq 2$ excitations make such large contributions to all the configuration variances that the corresponding Gaussian distributions could not be reconciled with the low-energy shell structure; with $(ds)^8$ configurations in our ten-orbit case $\sigma_c^2(t) = \sum_{c'}\sigma_{cc'}^2(t) \simeq 64$, 280, and 550 MeV^2 for $t=0$, 2, and 4, giving widths ~8, 19, and 30 MeV. But these variances are still small compared with $(t\hbar\omega)^2$ where $t\hbar\omega$ is the separation of spherical configurations connected by V_t. Hence V_t generates admixings too weak to produce a unimodal (Gaussian) distribution. Formally the admixing parameter, which must be $\gtrsim 1$ for Gaussian, is $\tau(t) = \sigma^2(t)/(t\hbar\omega)^2 = 0.18$, 0.09, and 0.03 for $t=2$, 4, and 6 excitations of $s=0$ configurations in our example. When $\tau(t) \ll 1$ the configuration distributions become multimodal,[10] each of the well separated modes having a dominant s value. The $t\geq 2$ variances then mainly describe not configuration spreading but rather the transport of small configuration intensities over a wide energy range. This effect, of no present interest, is measured by the anomalous variances. But the interesting quantities are the modal variances.

For small τ we can ignore the t admixings or treat them perturbatively. A procedure valid for arbitrary τ represents the intershell interactions by a random matrix. We illustrate this for $t=2$ excitations of an $s=0$ configuration (adequate for $E \lesssim 30$ MeV), ignoring the internal variance σ_{00}^2 to begin with, as well as the fine structure associated with the $t=2$ excitations, both being much smaller than Δ^2. We thus have two spikes separated by $\Delta \simeq 2\hbar\omega$ with a two-body interaction whose elements we draw from an off-diagonal Gaussian orthogonal ensemble defined in the two-particle space but acting in the $(m \gg 2)$-particle space. Solving this we find that the $s=0$ distribution, ρ_G in (5), is replaced by a bimodal form which vanishes in $0 < W < \Delta$, where $W = E - \mathscr{E}_c$, and otherwise is

$$\rho(W; s=0) = \sigma_{02}^{-2} |W - \Delta| \exp\{-W(W-\Delta)/\sigma_{02}^2\}. \tag{6}$$

In the weak-interaction limit ($\tau \ll 1$) (6) gives the perturbation result, a Gaussian modal distribution with variance $\sigma_{00}^2 + \tau\sigma_{02}^2$. At the other limit ($\tau \gg 1$) we get $(1-\pi/4)\sigma_{02}^2$ for the modal variance; the model in this limit is artificial because of the $t \leqslant 2$ restriction, without which the configuration distributions would of course be Gaussian with anomalous variances. For the physically relevant intermediate strengths we have a true bimodal distribution, the "lost" $s=0$ intensity being replaced by $s=2$ and the modal variance taking on part of the intershell variance. For $\tau = 0.18$, found for $(ds)^8$ in the ten-orbit case, the modal variance is $0.08\sigma_{02}^2 \simeq 20$ MeV^2 which will increase the effective $s=0$ configuration widths from 8 to 9 MeV rather than to 19 MeV as in the naive treatment. We thus have disposed of the problem of the anomalous variances. It is the procedures used here which should have special relevance to effective-interaction theory.

A more complete account, with an extended treatment of the multi-$\hbar\omega$ excitations, and applications to level densities and other statistical quantities, will be given later. We acknowledge useful conversations with J. P. Draayer and C. Jacquemin. This work has been supported in part by the U. S. Department of Energy.

[(a)]On leave from Physical Research Laboratory, Ahmedabad, India.

[1]H. A. Bethe, Phys. Rev. 50, 338 (1936).

[2]*Theory and Applications of Moment Methods in Many-Fermion Spaces*, edited by B. J. Dalton *et al.* (Plenum, New York, 1980).

[3]F. S. Chang, J. B. French, and T. H. Thio, Ann. Phys. (N.Y.) 66, 137 (1971).

[4]S. Ayik and J. N. Ginocchio, Nucl. Phys. A234, 13 (1974); S. M. Grimes *et al.*, Phys. Rev. C 27, 2893 (1983).

[5]C. M. Vincent, Phys. Rev. 163, 1044 (1967).

[6]T. T. S. Kuo, private communication.

[7]A better treatment of the configurations which contribute strongly in the ground state is easy, and would better locate the ground-state energy.

[8]The sum of Gaussians with the same variance is the convolution of the centroid distribution with a zero-centered Gaussian of that variance.

[9]R. U. Haq and S. S. M. Wong, Can. J. Phys. 60, 1502 (1982).

[10]J. B. French, in Ref. 2, p. 98.

FLUCTUATION PROPERTIES OF NUCLEAR ENERGY LEVELS AND WIDTHS : COMPARISON OF THEORY WITH EXPERIMENT

O. Bohigas
Division de Physique Théorique*, Institut de Physique Nucléaire
91406 Orsay Cedex, France

R.U. Haq
Department of Physics, University of Toronto, Toronto
Ontario M5S 1A7, Canada

A. Pandey
Service de Physique Théorique, CEN-Saclay
91191 Gif-sur-Yvette Cedex, France

*Laboratoire associé au CNRS.

We analyze the fluctuation properties of nuclear energy levels and widths with new spectrally averaged measures. A remarkably close agreement between the predictions of random-matrix theories and experiment is found.

[Fluctuation properties of energy levels, width fluctuations, level repulsion, random matrices, neutron resonances]

1. Introduction

This paper deals with the current status of the random matrix theories with regard mainly to the nuclear energy level fluctuations i.e. departures from spectral uniformity, and also to width fluctuations. Specifically the question is how well the ensemble results agree with experiment. And if the agreement is good, then : what is the source of this agreement and what limits can one impose on the mechanisms which affect the fluctuations. We are mainly concerned here with the first question for which we rely for the most part on our recent work[1]. We refer only briefly to the latter two questions for which also the answers are emerging.

The nuclear data relevant for fluctuation studies consists of slow-neutron resonances of medium and heavy nuclei and also of proton resonances of some light nuclei. Most of the neutron data come from systematic measurements of total neutron cross-sections performed at Columbia during the sixties and early seventies. There are also measurements in which the spin of the levels are measured directly. These include mainly (n,γ) measurements performed at Geel, Livermore and Oak Ridge. Besides the neutron measurements, detailed nuclear resonance parameters have been extracted from high resolution inelastic proton scattering experiments performed at Duke.

The principal random matrix model, valid for space rotation and time reversal invariant systems, is Wigner's Gaussian Orthogonal Ensemble (GOE), consisting of asymptotically large real symmetric matrices, in which the distinct matrix elements are distributed independently with essentially the same zero-centered Gaussian law. For energy levels the main predictions are due to Dyson and Mehta[2]. The close-lying levels behave as if they are repelling each other and the distant ones display a long range order, the resultant spectral rigidity being then akin to a crystalline structure. For widths the model predicts a χ_1^2-distribution known after the names of Porter and Thomas[3]. The early theoretical work in the subject is reviewed in Refs.[4-6]. For later developments as well as uptodate references see Ref.[7].

The GOE predictions are very sensitive to spurious and missing levels, especially for energy level fluctuations, imposing stringent conditions on the quality of data. Ideally the observed level sequence should be pure (all levels should have the same spin and parity) and complete (no missing level). Because of this, while the main GOE predictions have been known for almost two decades, good quality data required to test the model have come relatively recently. Earlier analyses of individual level sequences indicate "overall good agreement" between theory and experiment ; see Ch.VI of Ref.[7] for a review.

The present analysis makes much sharper comparisons between theory and experiment than the previous ones. The main reason for this improvement is not due to the additional data of the last few years, but rather to making better use of the theory when compared with experiment. Indeed, there are two disconcerting features in the previous analyses : the predictions of the theory have not been tested in a systematic way nor much attention has been paid to sample size effects. We resolve these difficulties here. We ask : what features of the theory are observed in the data and with what accuracy. We attempt to test directly the (fundamental) k-level correlation functions which are essentially the joint probability functions and of which the two level properties are the most significant. We go on to other fluctuation measures (which in fact derive from the correlation functions) so as to improve the statistical significance of the testing procedure. Moreover, since the theory is parameter-free, we combine all the data into a nuclear data ensemble (NDE) of approximately 1700 levels, this improving the statistics even more dramatically. Our main conclusion is that the agreement between theory and experiment is remarkably good.

Some comparison between theory and experiment is also made for widths, although in this case conclusions are less significant because the theoretical distribution itself has been used to a large extent to make spin and parity assignment of the levels.

In the next section we review the GOE predictions for the energy levels and in section 3 we present the results of comparison with the NDE. Section 4 deals with width fluctuations. The concluding section discusses the implications of the agreement.

2. GOE Predictions for the Energy Levels

Dyson's[8] k-level cluster functions* $Y_k(k\geq 1; Y_1\equiv 1)$ provide a natural hierarchy for energy level fluctuation studies. Essentially these are the k-level correlation functions, viz the joint-probability densities of observing k levels at given positions, from which the lower order correlation effects have been subtracted out. In particular $(1-Y_2(r))dr$ gives the probability of observing a level in an infinitesimal interval dr

---

*For a random variable W, we denote its ensemble average by $\overline{W}$ and its ensemble variance $\overline{(W-\overline{W})^2}$ by Var W. Note, moreover, that we are considering spectra normalized to unit local average spacing.

K. H. Böckhoff (ed.), Nuclear Data for Science and Technology, 809-813.

at a distance r from a given level. These functions are designed in such a way that for the Poisson ensemble, in which the energy levels are chosen randomly in a given interval, the Y_k are all zero for $k>1$.

For GOE[5,8]

$$Y_2(r) = \left(\frac{\sin\pi r}{\pi r}\right)^2 + \left(\int_r^\infty \frac{\sin\pi r'}{\pi r'}\, dr'\right)\frac{d}{dr}\left(\frac{\sin\pi r}{\pi r}\right), \qquad (1)$$

which is normalized to a total unit integral. It has the value unity at r=0 implying level repulsion and goes to zero as $(\pi r)^{-2}$ for large $r(\gtrsim 1)$ implying spectral rigidity. The latter aspect is more apparent from the number variance[2,7] $\Sigma^2(\bar{n})$, namely the variance of the number statistic n, the number of levels in a given interval of length $\bar{n}$:

$$\Sigma^2(\bar{n}) = \bar{n} - 2\int_0^{\bar{n}} (\bar{n}-r)Y_2(r)dr$$

$$= \frac{2}{\pi^2}\left\{\ln(2\pi\bar{n})+\gamma+1+\frac{1}{2}[Si(\pi\bar{n})]^2 - \frac{\pi}{2}\,Si(\pi\bar{n}) - \cos(2\pi\bar{n}) - Ci(2\pi\bar{n}) + \pi^2\bar{n}\left[1-\frac{2}{\pi}\,Si(2\pi\bar{n})\right]\right\}$$

$$\xrightarrow{\bar{n}\gtrsim 1} \frac{2}{\pi^2}\left\{\ln(2\pi\bar{n}) + \gamma + 1 - \frac{\pi^2}{8}\right\}. \qquad (2)$$

Note that its value is $\sim$ unity for $\bar{n}=100$ and even for $\bar{n} \sim 10^6$ the fluctuation is not more than a couple of levels. This should be compared with $\Sigma^2(\bar{n})=\bar{n}$ for the Poisson ensemble.

Another measure of the spectral rigidity is the Δ_3-statistic of Dyson and Mehta[2]. It measures, for a fixed interval $[x,x+\bar{n}]$, the least square deviation of the staircase N(E), the number of levels with energy less than or equal to E, from the best straight line fitting it :

$$\Delta_3(\bar{n};x) = (1/\bar{n}) \min_{A,B} \int_x^{x+\bar{n}} (N(E)-AE-B)^2 dE \quad . \qquad (3)$$

Its ensemble average is related to Σ^2, and hence to Y_2, by[9]

$$\bar{\Delta}_3(\bar{n}) = (2/\bar{n}^4) \int_0^{\bar{n}} (\bar{n}^3 - 2\bar{n}^2 r + r^3)\,\Sigma^2(r)\, dr \quad , \qquad (4)$$

which may be integrated numerically, or for $\bar{n}\gtrsim 15$ one may use[2] $\bar{\Delta}_3 \simeq \pi^{-2} \ln \bar{n} - 0.007$ with good accuracy. Again, in comparison with the Poisson value $\bar{\Delta}_3=\bar{n}/15$, the spectrum is seen to exhibit a long range order. (Note that because of stationarity[10] the ensemble properties are independent of x, the position of the interval).

Usually the most one can hope for is to test the theory for the two level functions and its integrals such as Σ^2 and $\bar{\Delta}_3$. However from a very long sequence of levels or from a collection of many short sequences as is the case for nuclear spectra, it may be possible to observe some higher-than-two level effects as well. Moreover the higher order functions are needed for the calculation of the sample errors of the testing procedure ; thus for example for testing two-point measures one needs to know the k-level functions for $k\lesssim 4$. Although all cluster functions have been evaluated for GOE[11], very few analytic results are available for higher-than-two-point measures. For properties related to Y_k with $k>2$, we rely mostly on Monte Carlo results that we have obtained from a sample, adequate for our purpose, of 100 GOE matrices of dimensionality 300.

Some examples of higher-than-two-level measures are provided by the distributions and correlation properties of the Δ_3 as well as the number (n) statistics. In this paper some results for Δ_3 are considered (see Fig.2c for Var Δ_3 and Figs.3a,3b for the Δ_3 distribution for $\bar{n}$=10,20). For very large $\bar{n}$, Var Δ_3 approaches[2] the value .011 which should be compared with the Poisson result Var $\Delta_3 = (1+11\bar{n}/30)\bar{n}/210$. Once again it is seen that the fluctuations are small for GOE.

In order to make optimal use of an observed spectrum, we introduce spectral averaged quantities. Consider a sequence containing (p+1) levels. For a statistic $W(\bar{n};x)$ defined in an interval of length $\bar{n}$ at x, the spectral average

$$\langle W(\bar{n})\rangle_p = (p-\bar{n})^{-1} \int_x^{x+(p-\bar{n})} W(\bar{n},y)\, dy \qquad (5)$$

is an estimator of $\overline{W}(\bar{n})$. The variance of $\langle W\rangle_p$, which is the square of the sample error, is also given by the theory :

$$\text{Var}\,\langle W(\bar{n})\rangle_p \equiv \overline{\langle W(\bar{n})\rangle_p^2} - \overline{\langle W(\bar{n})\rangle_p}^{\,2}$$

$$= \frac{2\,\text{Var}\,W(\bar{n})}{(p-\bar{n})^2} \int_0^{(p-\bar{n})} (p-\bar{n}-r)\,C(r;\bar{n})\,dr$$

$$\simeq \frac{2\,\text{Var}\,W(\bar{n})}{(p/\bar{n})} \int_0^1 C(\lambda\bar{n};\bar{n})\,d\lambda \;, \qquad (6)$$

where $C(r;\bar{n})$ is the autocorrelation function of W :

$$C(r;\bar{n}) = (\text{Var}\,W(\bar{n}))^{-1}\left\{\overline{W(\bar{n};x)W(\bar{n};x+r)} - (\overline{W}(\bar{n}))^2\right\}. \qquad (7)$$

The last step of Eq.(6), valid for $p\gg\bar{n}$, follows from the fact that the autocorrelation functions for measures discussed in this paper fall off rapidly, being close to zero for $r\gtrsim\bar{n}$ for $\bar{n}\gtrsim 1$. Notice that as $p\to\infty$, Var $\langle W\rangle_p$ goes to zero, which is nothing but the ergodic property[10] : the spectral average coincides with the ensemble average.

It is seen from (6) that for the sample errors we need Var W as well as the integral $\int C(\lambda\bar{n};\bar{n})d\lambda$ both of which we estimate from our Monte Carlo sample. The details will be published elsewhere. In practice instead of (5) we use a less smooth average in which the integral is replaced by a summation.

3. Comparison with Experiment for Energy Levels

In the conventional analysis one treats the data from each nucleus, containing about 50-100 levels, separately and consequently the statistical significance is poor. As seen from (5) the testing procedure improves as the sample size increases. Moreover since the correlation functions, Eq.(7), fall off rapidly it is not necessary to have one long sequence but a collection of many relatively short sequences could also be used.

Since one expects that the resonance energies of every compound nucleus will share the same fluctuation properties, it seems natural, and in the spirit of random matrix theory, to treat the set (NDE) of nuclear resonance energy data of different nuclei as a sampling of eigenvalues of GOE matrices ; see Ref.[12] for a similar notion for complex atomic spectra. As an example of this procedure, the nearest neighbor spacing distribution is shown in Fig.1. It is clear that a histogram containing $\sim$ 1700 spacings corresponding to the NDE (Fig.1b) is statistically much more significant, when compared to a theoretical prediction, than a histogram containing only $\sim$ 100 spacings corresponding to a single nucleus (Fig.1a).

Our NDE consists of 1762 resonance energies corresponding to 36 sequences of 32 different nuclei :

i) Slow-neutron resonance data on ^{64,66,68}Zn (103,65,45 levels[13]), ^{114}Cd (17 levels[14]), ^{152,154}Sm (70, 27 levels[15]), ^{154,156,158,160}Gd (19,47,21,54 levels[16]), ^{160,162,164}Dy (18,46,20 levels[17]), ^{166,168,170}Er (109,48,31 levels[18]), ^{172,174,176}Yb (55,19,23 levels[19]), ^{182,184,186}W(41,30,14 levels[20]), ^{186,190}Os (56,17 levels[21]), ^{232}Th (178 levels[22]) and ^{238}U (146 levels[22]),

ii) (n,γ) reaction data on ^{177}Hf (34'J=3' levels, 23'J=4' levels[23]), ^{179}Hf(25'J=4' levels, 22'J=5' levels[23]) and ^{235}U(58'J=3' levels, 68'J=4' levels[24]),

iii) proton resonance data on ^{44}Ca(52'$1/2^-$' levels, 39'$1/2^+$' levels[25]), ^{48}Ti(66'$1/2^+$' levels[25]) and ^{56}Fe (56'$1/2^+$'levels[26]).

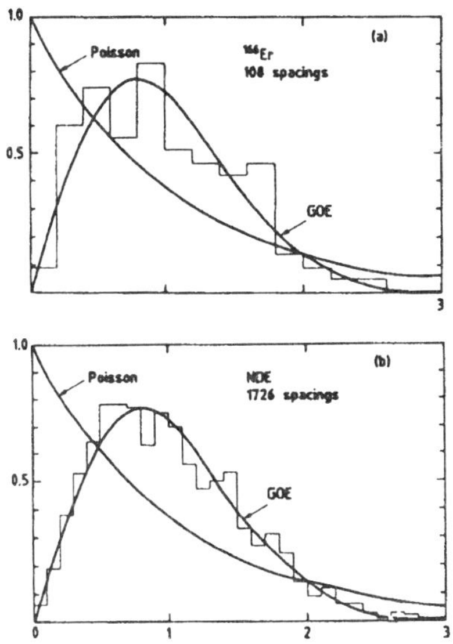

Fig.1. Nearest neighbor spacing histogram for : (a) ^{166}Er, (b) Nuclear Data Ensemble (NDE).

The criterion* for inclusion in the NDE is that the individual sequences be in general agreement with GOE.

We have made an extensive test of the theory by studying the aforementioned functions $Y_2(r)$, $\Sigma^2(\bar{n})$, $\Delta_3(\bar{n})$ as well as others. A complete account will be given elsewhere. Here we restrict ourselves to a few typical examples for the NDE.

Consider first the two-level measures $\Sigma^2(\bar{n})$ and $\bar{\Delta}_3(\bar{n})$. The procedure for calculations is to evaluate for each of the 36 sequences the spectral-averaged measure, say $\langle\Delta_3(\bar{n})\rangle$ for $\bar{\Delta}_3(\bar{n})$, and then take their average, weighted according to the size of each sequence. We have considered the range $0<\bar{n}<100$. The results for $\bar{n}<25$ are shown in Figs.2a, 2b. They are in a remarkably good agreement with GOE. We emphasize that typically the "figure of merit", e.g. $(\text{Var}\ \langle\Delta_3\rangle)^{1/2}/\bar{\Delta}_3$ for $\bar{\Delta}_3$, is of the order of a few percent in the present analysis, whereas in the conventional analysis it is $\sim$ 20% or more. The values corresponding to the Poisson case as well as the Gaussian Unitarity Ensemble (GUE ; see section 5 below) are also shown for comparison.

To explore higher-than-two level effects we consider the ensemble variance of $\Delta_3(\bar{n})$, which is sensitive to Y_k upto k=4, and the distribution function of Δ_3 for various values of $\bar{n}$ (which involves in principle all order effects). The results are presented in Fig.2c for Var Δ_3 for $\bar{n}\leq 25$ and two examples ($\bar{n}$=10,20) of the distribution function of Δ_3 are shown in Fig.3. The agreement between the GOE predictions and data is again very good.

----

*In the previous paper[1] our NDE consisted of 1407 resonance energies corresponding to 30 sequences of 28 different nuclei. In the present NDE we have added 8 new sequences. Moreover, we have not included the modifications suggested by the authors of the Columbia data to account for some missing levels. Instead, we have considered shorter sequences in a few cases and have dropped from the list the ^{110,112}Cd-spectra which involved large modifications. Our conclusions, as exemplified by Figs.2,3, are still the same as in Ref.[1].

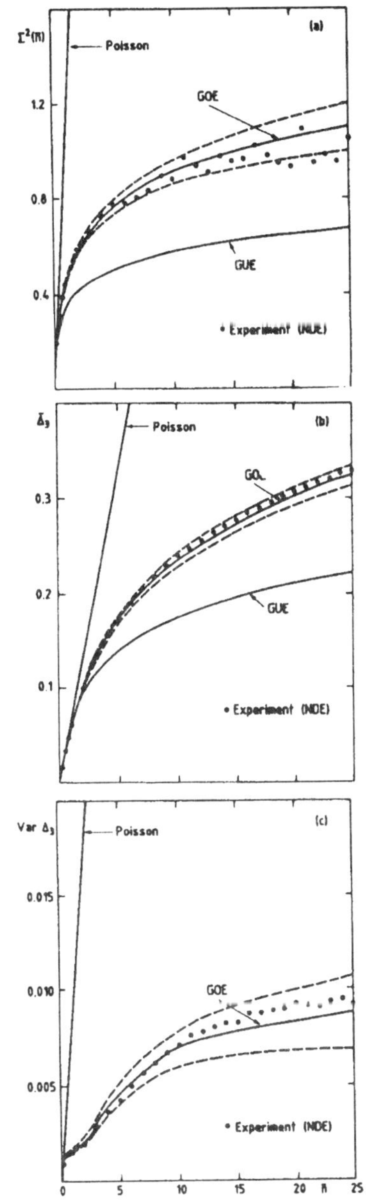

Fig.2. (a) Σ^2,(b) $\bar{\Delta}_3$,(c) Var Δ_3 as functions of $\bar{n}$. Dashed lines correspond, for GOE, to one standard deviation from the average.

4. Width Fluctuations

The main question is whether the transition widths (via which the resonance levels are detected and identified) have a χ_1^2 (Porter-Thomas[3]) distribution :

$$\rho(x) = (2\pi x)^{-1/2} \exp(-x/2) \ ; \ x = \Gamma_{ic}|\bar{\Gamma}_{ic} , \qquad (8)$$

where the width Γ_{ic} measures the rate of transition from an initial (statistical) state i to a channel state c. Or, equivalently the question is whether the transition amplitudes, whose squares (to within a penetrability factor) are the widths, are Gaussian random variables. This prediction follows directly from the orthogonal invariance of the random matrix ensembles as is the case with GOE ; see Ch.VII of Ref.[7].

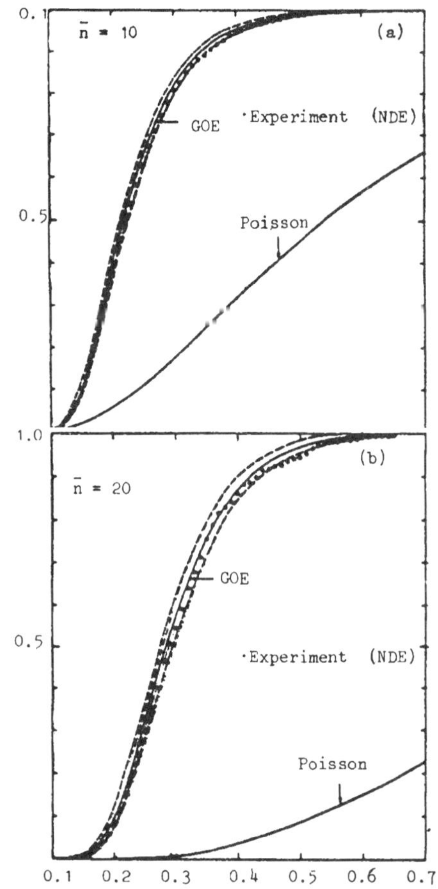

Fig.3. Distribution function of Δ_3 : (a) $\bar{n}$=10, (b) $\bar{n}$=20. Dashed lines correspond, for GOE, to one standard deviation from the average.

The amount of data that can be used to test the theory is much larger than in the case of energy levels. This is because even if some of the resonances are not detected due to very small widths, Eq.(8) can be tested for $x \geq \alpha$, an experimental threshold. In this section we present the results for some of the neutron-widths. A more detailed analysis will be taken up later. For a similar analysis of much older data, see Ref.[27].

We consider an NDE containing 1182 widths corresponding to 21 level sequences* of the previous section : ^{64,66,68}Zn, ^{114}Cd, ^{152,154}Sm, ^{154,156,158,160}Gd, ^{162}Dy, ^{166,168,170}Er, ^{172,174,176}Yb, ^{182,184}W, ^{232}Th and ^{238}U. The transition amplitude distributions are shown in Fig.4 for an individual sequence as well as the NDE. The agreement between theory and data is seen to be very good.

For further tests we calculate ν, the number of degrees of freedom, assuming that the width distribution itself is χ^2_ν with ν not necessarily an integer. A statistically significant departure from ν=1 would then be an indication of a failure of the Porter-Thomas model. A commonly-used procedure is to construct the likelihood function which is the product of the χ^2_ν-densities of the observed (renormalized) widths and then to maximize this function with respect to ν. The sample errors associated with this estimation procedure are obtained by studying the deviations from the maximum. The result would however show departure if some small widths are missing. For a better procedure we may assume a (relative) cut-off α below which the widths are likely to be missed and then do the above analysis for all widths of values $\geq\alpha$ with a truncated

----

* The widths in each sequence are renormalized to unit average.

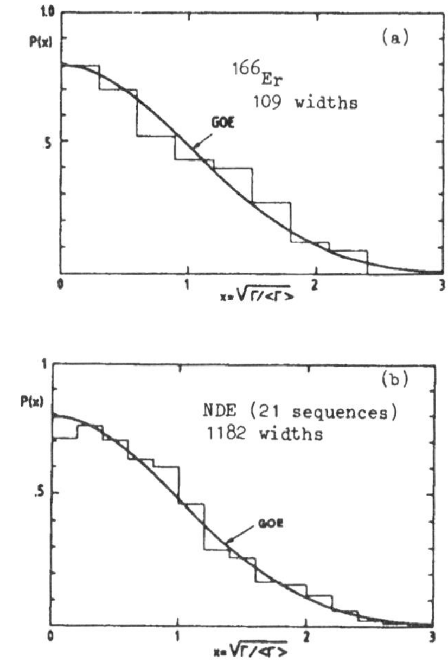

Fig.4. Histogram for the absolute values of transition amplitudes : (a) ^{166}Er, (b) NDE. The theory predicts $P(x)=(2/\pi)^{1/2}\exp(-x^2/2)$.

χ^2_ν-distribution;for details see Refs.[3,27]. We have studied ν as a function of α and a plot is given in Fig.5. The departures from theory for $\alpha \lesssim 10^{-3}$ may be due to the missing small widths. For larger values of α the agreement is again very good.

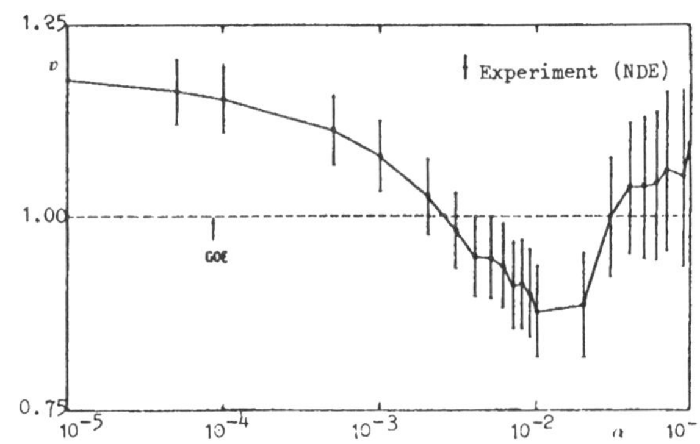

Fig.5. The degree of freedom ν as a function of cut-off α. The error bars correspond to one standard deviation from the estimated ν.

Finally we turn to the question of correlations between the width and energy level fluctuations. The theory predicts the two to be statistically independent. To test this we calculate the correlation coefficient

$$r = \frac{1}{N}\sum_{i=1}^{N}\left(\frac{\Gamma_{ic}-\langle\Gamma\rangle}{\sigma_\Gamma}\right)\left(\frac{E_i-\langle E_i\rangle}{\sigma_E}\right) , \qquad (9)$$

for each sequence. Here N is the number of levels in the sequence and Γ_{ic} the width associated with the level E_i. $\langle\Gamma\rangle$ and σ_Γ are the mean and standard deviation of the Γ_{ic}, and

$$\langle E_i\rangle = Ai+B,\ \sigma_E^2 = N^{-1}\sum_{i=1}^{N}(E_i-\langle E_i\rangle)^2 , \qquad (10)$$

where A and B are fixed by minimizing* the expression for σ_E^2 . For the NDE we take the average of the correlation coefficients for each sequence, weighted according to the size of the sequence. We obtain r(NDE)= 0.017 as compared to the theoretical value 0±0.029 confirming again the predictions.

5. Conclusion

We have established an astonishingly good agreement between a parameter-free theory (GOE) and the data. We emphasize that, apart from rotation and time-reversal invariance resulting in the real symmetric nature of the matrices, GOE takes no account of the specific properties of the nuclear Hamiltonian, e.g. its (1+2)-body nature, its large pairing and quadrupole components, etc . We also recall that the eigenvalues and eigenvector components of realistic (non-random) nuclear shell model matrices show the same fluctuation patterns[7,28]. How can this be understood theoretically ?

To answer this we recall first that there are other classical ensembles in which, like GOE, the orthogonal invariance plays a key role and which also give the same fluctuations[11,29]. But more significantly for the nuclear Hamiltonian, there are Monte Carlo results[28,30] which indicate that ensembles of two-body operators acting in many-particle (shell model) spaces also yield fluctuations characteristic of GOE. More recently[31] it has been demonstrated that adding a GOE matrix H to any real symmetric matrix K,(K+αH), leads very quickly, as α increases, to the same fluctuations. Intermediate fluctuation patterns are to be expected only when the random matrix elements are not much larger than the local average spacing of the given (non-random) matrix K. The good agreement with experiment, coupled with the theoretical understanding which is now emerging, reinforces the belief that GOE fluctuations are to be found in nature under very general conditions.

We mention finally that one can use the close agreement to impose restrictions on mechanisms which would change the fluctuations. In particular if time reversal is not an exact symmetry, the appropriate model would be an ensemble of complex Hermitian matrices in which the real and imaginary parts are sampled independently. The Gaussian Unitary Ensemble (GUE) in which the two parts have the same norm is one such example. It gives very different fluctuation properties than GOE[4-7]. For example, $Y_2(r)=(\sin\pi r/\pi r)^2$ implying stronger level repulsion than GOE (see Figs. 2a,2b for Σ^2 and $\overline{\Delta}_3$) and widths follow a χ_2^2-distribution. Analytic results for ensembles in which the real and imaginary parts have unequal weights have recently been given[31-33]. It is shown for energy level fluctuations (and is expected to be valid for width fluctuations as well) that even a small magnitude of the imaginary part induces major changes in fluctuation properties, leading very rapidly from GOE to GUE type fluctuations. The results are being used[32] to derive an upper bound on the time reversal non-invariant part of the nuclear Hamiltonian.

References

1. R.U. Haq, A. Pandey and O. Bohigas, Phys. Rev. Lett. 48, 1086 (1982)
2. F.J. Dyson and M.L. Mehta, J. Math. Phys. 4, 701 (1963),reprinted in Ref.[4]
3. C.E. Porter and R.G. Thomas, Phys. Rev. 104, 483 (1956), reprinted in Ref.[4]
4. C.E. Porter (Ed.), Statistical Theories of Spectra: Fluctuations, Academic Press, New York, 1965

---

* We remark parenthetically that the expression for A thus obtained is a good estimator of the average spacing ; see section X.B of Ref.[7]. The spectra used in section 3 were normalized by this procedure.

5. M.L. Mehta, Random Matrices and the Statistical Theory of Energy Levels, Academic Press, New York 1967
6. E.P. Wigner, SIAM Rev. 9, 1 (1967)
7. T.A. Brody, J. Flores, J.B. French, P.A. Mello, A. Pandey and S.S.M. Wong, Rev. Mod. Phys. 53, 385 (1981)
8. F.J. Dyson, J. Math. Phys. 3, 166 (1962), reprinted in Ref.[4]
9. A. Pandey, Ann. Phys. (N.Y.) 119, 170 (1979), Appendix II
10. A. Pandey, Ref.[9]
11. F.J. Dyson, Commun. Math. Phys. 19, 235 (1970) ; M.L. Mehta, Commun. Math. Phys. 20, 245 (1971)
12. N. Rosenzweig and C.E. Porter, Phys. Rev. 120, 1698 (1960), reprinted in Ref.[4]
13. J.B. Garg, V.K. Tikku and J.A. Harvey, Phys. Rev. C23, 671 (1981); J.B. Garg, V.K. Tikku, J.A. Harvey, R.L. Macklin and J. Halperine, Phys. Rev. C24, 1922 (1981) ; J.B. Garg, private communication
14. H.I. Liou, G. Hacken, F. Rahn, J. Rainwater, M. Slagowitz and M. Makofske, Phys. Rev. C10, 709 (1974)
15. F. Rahn, H.S. Camarda, G. Hacken, W.W. Havens,Jr., H.I. Liou, J. Rainwater, M. Slagowitz and S. Wynchank, Phys. Rev. C6, 251 (1972)
16. F. Rahn, H.S. Camarda, G. Hacken, W.W. Havens, Jr., H.I. Liou and J. Rainwater, Phys. Rev. C10, 1904 (1974) ; for ^{156}Gd, C. Coceva and M. Stefanon, Nucl. Phys. A315, 1 (1979) and private communication
17. H.I. Liou, G. Hacken, J. Rainwater and U.N. Singh, Phys. Rev. C11, 462 (1975)
18. H.I. Liou, H.S. Camarda, S. Wynshank, M. Slagowitz, G. Hacken, F. Rahn and J. Rainwater, Phys. Rev. C5, 974 (1972)
19. H.I. Liou, H.S. Camarda, G. Hacken, F. Rahn, J. Rainwater, M. Slagowitz and S. Wynchank, Phys. Rev. C7, 823 (1973)
20. H.S. Camarda, H.I. Liou, G. Hacken, F. Rahn, W. Makofske, M. Slagowitz, S. Wynchank and J. Rainwater, Phys. Rev. C8, 1813 (1973)
21. J.C. Browne and B.L. Berman, Phys. Rev. C23, 1434 (1981) and private communication
22. F. Rahn, H.S. Camarda, G. Hacken, W.W. Havens,Jr., H.I. Liou, J. Rainwater, M. Slagowitz and S. Wynchank, Phys. Rev. C6, 1854 (1972)
23. C. Coceva, F. Corvi, P. Giacobbe and M. Stefanon, in Statistical Properties of Nuclei, J.B. Garg (Ed.), Plenum Press, New York, 1972
24. M.S. Moore, J.D. Moses, G.A. Keyworth, J.W.T. Dabbs and N.W. Hill, Phys. Rev. C18, 1328 (1978)
25. W.M. Wilson, E.G. Bilpuch and G.E. Mitchell, Nucl. Phys. A245, 285 (1975)
26. W.A. Watson, E.G. Bilpuch and G.E. Mitchell, Z. Phys. A300, 89 (1981)
27. J.D. Garrison, Ann. Phys. (N.Y.) 30, 269 (1964)
28. S.S.M. Wong and J.B. French, Nucl. Phys. A198, 188 (1972)
29. M.L. Mehta, J. Math. Phys. 17, 2198 (1976)
30. O. Bohigas and M.J. Giannoni, Ann. Phys. (N.Y.) 89, 393 (1975)
31. A. Pandey, Ann. Phys. (N.Y.) 134, 110 (1981)
32. J.B. French, V.K.B. Kota and A. Pandey, to be published
33. A. Pandey and M.L. Mehta, Commun. Math. Phys., to be published.

PHYSICAL REVIEW LETTERS

VOLUME 52 | 2 JANUARY 1984 | NUMBER 1

Characterization of Chaotic Quantum Spectra and Universality of Level Fluctuation Laws

O. Bohigas, M. J. Giannoni, and C. Schmit
Division de Physique Théorique, Institut de Physique Nucléaire, F-91406 Orsay Cedex, France

(Received 2 August 1983)

It is found that the level fluctuations of the quantum Sinai's billiard are consistent with the predictions of the Gaussian orthogonal ensemble of random matrices. This reinforces the belief that level fluctuation laws are universal.

PACS numbers: 05.40.+j, 03.65.Bz

The statistical study of spectra of quantum systems is almost as old as quantum mechanics itself. One distinguishes two types of properties: *global* ones and *local* ones. An example of the former is provided by the density of levels as a function of excitation energy. In this Letter we shall discuss local properties, or more precisely, fluctuations (departures of the energy-level distribution from uniformity). We shall deal with time-independent systems and energies of stationary states.

There exists a well established theory to describe fluctuation properties of quantal spectra, namely the random matrix theory (RMT) initiated by Wigner, developed mainly by Dyson and Mehta,[1] and later extended by several authors.[2] Recently,[3] the predictions of RMT [specifically, the predictions of the Gaussian orthogonal ensemble (GOE)] have been compared in great detail with the whole body of available nuclear data coming mainly from *compound-nucleus resonances*. No discrepancy between theory and experiment has been detected. In particular the data have been shown to exhibit two of the salient phenomena predicted by the theory—the *level repulsion* (tendency of the levels to avoid clustering) and especially the *spectral rigidity* (very small fluctuation around its average of the number of levels found in an interval of given length), which is a property due to correlations between level spacings. We should also mention that recently a comparison between data of *atomic levels* and GOE has been performed.[4] Although the significance of the comparison is lower than in the nuclear case (due to the method used in Ref. 4 as well as to the relatively small number of available data), a good agreement between GOE predictions and experiment was found.

Once the ability of the theory (GOE) to predict the fluctuation properties exhibited by data is established, one could think that no major question in this field is still open. It is one purpose of this Letter to show that this is not the case. Indeed, as it will be discussed in what follows, the origin of the success of the theory as well as its domain of validity remain to be clarified.

In connection with the study of regular and irregular motion some very interesting results have been obtained in recent years. Strong arguments have been given which indicate that for integrable systems with more than one degree of freedom,[5] the nearest-neighbor spacing distribution $p(x)$ of the quantum energy levels should be Poisson-like [$p(x)=\exp(-x)$] and the spacings should not be correlated.[6,7] In contrast, evidence of level repulsion has been put forward by studying numerically systems having two degrees of freedom and known to be chaotic[8] in the classical case (stadium,[9] Sinai's billiard[10]). Although this feature is clearly appealing, it is up to now only

VOLUME 52, NUMBER 1 PHYSICAL REVIEW LETTERS 2 JANUARY 1984

qualitative [one would need a spectrum with very many levels to get a precise evaluation of the behavior of $p(x)$ at the origin]. On the other hand, the information carried by $p(x\sim 0)$ is rather limited. In particular, it does not give any indication about the *correlations* between spacings which are responsible for the degree of regularity of the spectrum. The purpose of this Letter is to use some of the systematic tools developed in RMT to make a detailed comparison of the level fluctuations of the quantum Sinai's billiard (SB) with GOE predictions. The choice of a two-dimensional billiard is convenient for our aim for several reasons: (i) Billiards have the lowest possible number of degrees of freedom allowing for chaotic motion; (ii) for billiards, it is possible to make a precise separation between global and local properties [cf. the Weyl formula, Eq. (1)]; (iii) billiards have a discrete spectrum with an infinite number of eigenvalues and by computing a large number of them one can reach a high statistical significance of the results. Finally, SB is known to be strongly chaotic (K system) and there exists an efficient method to compute its eigenvalues.

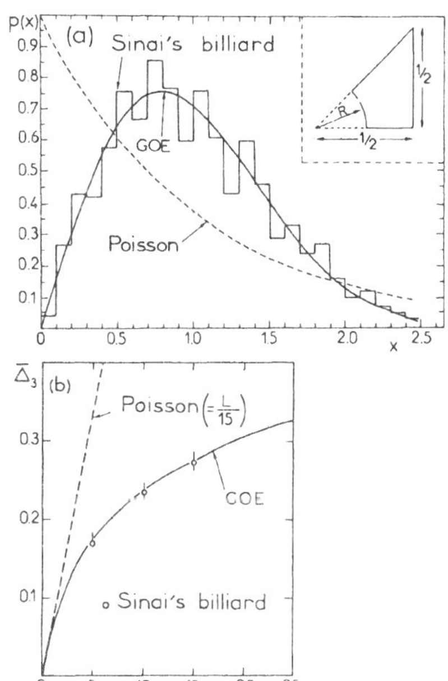

FIG. 1. Results of energy-level fluctuations for desymmetrized Sinai's billiards as specified in the upper right-hand corner of (a). 740 levels have been included in the analysis corresponding to the 51st to 268th level for $R=0.1$, 21st to 241st level for $R=0.2$, 16th to 194th level for $R=0.3$, 11th to 132nd level for $R=0.4$. (a) Results for the nearest-neighbor spacing distribution. (b) Results for the average value of the $\Delta_3(L)$ statistic of Dyson and Mehta for $L=5$, 10, and 15. Curves corresponding to the Poisson case (stretch of uncorrelated levels) and to the random-matrix-theory predictions (GOE) are drawn for comparison. The error bars in (b) (one standard deviation) correspond to finite-sampling effects as predicted by GOE.

We proceed as follows. We determine the eigenvalues $E_n = k_n^2/2m$ of the Schrödinger equation $(\Delta + k_n^2)\psi_n = 0$ for the "desymmetrized" SB [see upper right-hand corner of Fig. 1(a)] with Dirichlet boundary conditions by using the Korringa-Kohn-Rostoker method as described in Ref. 10. We compute several sets of eigenvalues $\{E_i(R)\}$ for different values of the parameter R (see caption of Fig. 1). By using the Weyl-type formula,[11] which gives the average number of levels up to energy E,

$$\bar{N}(E) = (\tfrac{1}{4}\pi)(SE - L\sqrt{E} + K), \qquad (1)$$

where S and L are, respectively, the surface and the perimeter of the billiard and K is a constant of the order of unity, we can map the spectrum $\{E_i(R)\}$ onto the spectrum $\{\epsilon_i(R)\}$ through $\epsilon_i(R) = \bar{N}(E_i(R))$. Each spectrum $\{\epsilon_i(R)\}$ has on the average a constant mean spacing $D(R)$ which will be taken as the energy unit. The cumulative density $n(\epsilon)$ of levels ϵ_i will therefore have a staircase shape fluctuating around a straight line of slope equal to unity. In order to investigate the fluctuations we study the nearest-neighbor spacing distribution $p(x)$ and the Dyson-Mehta statistic Δ_3. Δ_3 is defined, for a fixed interval $[x, x+L]$, as the least-squares deviation of the staircase function $n(\epsilon)$ from the best straight line fitting it:

$$\Delta_3(L,x) = (1/L)\mathrm{Min}_{A,B}\int_x^{x+L}[n(\epsilon) - A\epsilon - B]^2\, d\epsilon. \qquad (2)$$

It provides a measure of the degree of rigidity of the spectrum: For a given L, the smaller Δ_3 is, the stronger is the rigidity, signifying the long-range correlations between levels. We proceed as described in Ref. 3: Given a stretch of levels on the ϵ axis, we compute $\Delta_3(L)$, for instance, for the intervals $[a, a+L]$, $[a+L/2, a+3L/2]$, $[a+L, a+2L]$, $[a+3L/2, a+5L/2]$, ... until the stretch $[a,b]$ has been covered. If the spectrum fluctuations are translationally invariant on the ϵ axis, then the average value $\bar{\Delta}_3$ of Δ_3 will be independent of the chosen interval $[a,b]$ [equivalently, the average of Δ_3 computed with Eq. (2) will not depend on the initial value x of the interval]. We have checked numerically that this is the case for the SB spectra, provided that the first lowest levels of each spectrum $\{\epsilon_i(R)\}$ are omitted (see caption of Fig. 1). This procedure seems appropriate if one observes the spectra corresponding to different values of R: Indeed, the smaller the value of R the larger the number of levels, starting from the ground state, that can be attained by perturbation theory from the spectrum $\{\epsilon_i(R=0)\}$ (triangular billiard). We emphasize that the fluctuation properties we are looking for will appear to be essentially different from the ones corresponding to the triangular billiard[12] and cannot be attained from it by perturbation theory. To increase the statistical significance of the results, four spectra $\{\epsilon_i(R)\}$ corresponding to different values of R will be analyzed as corresponding to a single stretch of 740 levels (see caption of Fig. 1). Care has been taken that one is dealing with "independent information": The different values of R should not be chosen too close to one another. Otherwise, two different spectra corresponding to R and $R+\delta R$ would be almost deducible one from the other and one would just be analyzing redundant information.

Let us now discuss the results. In Fig. 1(a) is shown the nearest-neighbor spacing distribution $p(x)$ which is compared to the GOE and Poisson predictions. As can be seen, the SB results follow very closely GOE not only for small spacings (level repulsion) but over the whole range of spacings. The variance of $p(x)$ for SB is 0.273 which is close to the GOE value 0.286 ± 0.015 (the error bar takes into account the finite sampling effects) and far from the Poisson value 1.0. We next consider quantities related with the spacing correlations. In Fig. 1(b) are shown the average values of $\Delta_3(L)$ for $L=5$, 10, and 15 for SB; they are close to the corresponding GOE values. We have also computed the correlation factor between two adjacent spacings. For SB we obtain -0.30, to be compared to -0.27 ± 0.04 (GOE) and 0.0 (Poisson). We can summarize the numerical results as follows: All fluctuation properties of SB investigated so far are fully consistent with GOE predictions.

Is this a surprising result? With a few inconclusive exceptions (see a discussion on small metallic particles, for instance in Ref. 2), the basic hypotheses leading to RMT have always been put forward by invoking the complexity of the system. In other words, it has been taken as essential that one is dealing with a many-particle system (system with many degrees of freedom). Our results indicate that this is by no means a necessary condition. Indeed, the quantum chaotic system with two degrees of freedom studied here (a one-particle system in two dimensions) shows also GOE fluctuations. The present work should have further developments [for instance, when time-reversal invariance does not hold, the adequate model in RMT is the Gaussian unitary ensemble (GUE) and one should look for "simple" chaotic systems having GUE fluctuations]. It is an attempt to put in close contact two areas—random matrix physics and the study of chaotic motion—that have remained disconnected so far. It indicates that the methods developed in RMT to study fluctuations provide the adequate tools to characterize chaotic spectra and that, conversely, the generality of GOE fluctuations is to be found in properties of chaotic systems. In summary, the question at issue is to prove or disprove the following conjecture: Spectra of time-reversal-invariant systems whose classical analogs are K systems show the same fluctuation properties as predicted by GOE (alternative stronger conjectures that cannot be excluded would apply to less chaotic systems, provided that they are ergodic). If the conjecture happens to be true, it will then have been established the *universality of the laws of level fluctuations* in quantal spectra already found in nuclei and to a lesser extent in atoms. Then, they should also be found in other quantal systems, such as molecules, hadrons, etc.

We are grateful to C. Jacquemin and A. Pandey for valuable suggestions and discussions. The Division de Physique Théorique is a Laboratoire associé au Centre National de la Recherche Scientifique.

[1] *Statistical Theories of Spectra: Fluctuations*, edited by C. E. Porter (Academic, New York, 1965); M. L. Mehta, *Random Matrices and the Statistical Theory of Energy Levels* (Academic, New York, 1967).

[2] T. A. Brody, J. Flores, J. B. French, P. A. Mello, A. Pandey, and S. S. M. Wong, Rev. Mod. Phys. 53, 385 (1981).

[3] R. U. Haq, A. Pandey, and O. Bohigas, Phys. Rev. Lett. 48, 1086 (1982); O. Bohigas, R. U. Haq, and A. Pandey, in *Nuclear Data for Science and Technology*, edited by K. H. Böckhoff (Reidel, Dordrecht, Netherlands, 1983), pp. 809–813.

[4]H. S. Camarda and P. D. Georgopulos, Phys. Rev. Lett. 50, 492 (1983).

[5]The harmonic-oscillator case is an exception and must be studied separately; see M. V. Berry and M. Tabor, Proc. Roy. Soc. London 356, 375 (1977); O. Bohigas, M. J. Giannoni, and A. Pandey, to be published.

[6]Berry and Tabor, Ref. 5.

[7]M. V. Berry, in Lectures of the Les Houches Summer School, 1981 (North-Holland, Amsterdam, to be published).

[8]We use here the term chaotic in a qualitative way. In the classical case it designates a large class of systems whose trajectories are unstable with respect to the initial conditions. In fact, the specific systems we call chaotic in the Letter are K systems. Via the correspondence principle we cover the quantum case, for which the asymptotic behavior of the spectrum at high energies is of most relevance.

[9]S. W. McDonald and A. N. Kaufman, Phys. Rev. Lett. 42, 1189 (1979); G. Casati, F. Valz-Gris, and I. Guarneri, Lett. Nuovo Cimento 28, 279 (1980).

[10]M. V. Berry, Ann. Phys. (N.Y.) 131, 163 (1981).

[11]H. P. Baltes and E. R. Hilf, *Spectra of Finite Systems* (Wissenschaftsverlag, Mannheim, 1976).

[12]For the triangular billiard ($R=0$) the system is integrable and the spectrum is asymptotically dominated by degeneracies (Ref. 10).

Z. Phys. A - Atoms and Nuclei 315, 91-98 (1984)

Zeitschrift für Physik A Atoms and Nuclei

Bivariate Distributions in Statistical Spectroscopy Studies: Fixed-J Level Densities, Fixed-J Averages and Spin Cut-Off Factors

V.K.B. Kota*

Department of Physics and Astronomy, University of Rochester, Rochester, New York, USA

Received July 19, 1983; revised version September 29, 1983

Analytical studies with the TBRE ensemble, supplemented by numerical calculations with realistic interactions show that for large particle numbers, the bivariate cumulants (k_{rs}) defined for the Hamiltonian operator (H) and the J_Z operator, are very small; $|k_{rs}| \lesssim 0.3$ for $3 \leqq r+s \leqq 6$. As a result the expansions around a bivariate normal density are meaningful for the fixed-M densities ($\rho(E,M)$). We adopt a bivariate Edgeworth expansion for $\rho(E,M)$ and give a compact form for the same. Finally using this $\rho(E,M)$ (which also define fixed-J densities uniquely), new series expansions are given for fixed-M (and hence for fixed-J) averages of the powers of H and also for the spin cut-off factors.

1. Introduction

In the statistical spectroscopy studies [1-3], ignoring fluctuations, one deals with the smoothed part of the level density function $\rho(E)$. The action of the central limit theorem (CLT) in model spaces of interest allow us to construct this $\rho(E)$, in terms of a few lower order moments of the Hamiltonian (H). The smoothened density[1] then defines the ground state and one can also read off the smoothened spectrum from it. The parametric differentiation of the density [4, 5] or the polynominals defined by it [6], yield results for other observables in terms of trace quantities. To learn about the structure of the wavefunction at a given energy (E) and also at the same time to attain better accuracy with a few lower order moments, one partitions the many particle space and the partial level densities defined over these subspaces then determine all the spectroscopic quantities of interest. Partitioning by the irreducible representations (irreps) of groups play an important role, as they yield propagation laws [7, 8] for the traces of interest.

The partial level densities $\rho(E,[x_1 x_2 \ldots])$ involve more than one variable, viz. E and the irrep labels $x_1, x_2, \ldots$. As a result, the question of using bivariate and multivariate distributions and the related ingredients (for example the bivariate cumulants), in statistical spectroscopy studies will naturally arise. However, so far all the studies avoided them by using the conditional probability densities $\rho(E/[X])$; $\rho(E,[X]) = \rho([X])^* \rho(E/[X])$ and $[X] = [x_1, x_2, \ldots]$. The conditional moments $\langle H^p \rangle^{[X]}$, which are the averages of H^p over all states belonging to the irrep $[X]$, define $\rho(E/[X])$. With $d(m)$ and $d([X])$ being the dimensionalities of the m-particle space and the irrep $[X]$ respectively, $\rho([X]) = d([X])/d(m)$. The second question one can ask is, will the bivariate cumulants etc., yield any new insights into the structure of $\rho(E,[X])$? The present paper provides answers in the affirmative to the above two questions. We will workout the example of fixed-M level densities $\rho(E,M)$; fixed-J level densities $\rho(E,J)$, follow from (15) given ahead. Note that M is the eigenvalue of the J_Z operator. Throughout this article, we in fact deal directly with $\rho(E,M)$, rather than $\rho(E,J)$. The reasons should be clear; as $J^2 = J_x^2 + J_y^2 + J_z^2$, its spectrum is of χ_3^2 shape [8] and hence the limiting shape of $\rho(E,J)$ cannot be that of a bivariate normal distribution. As a result one can not use, for $\rho(E,J)$,

---

* Work supported in part by the U.S. Department of Energy

[1] As the present article does not consider energy level fluctuations, we make no distinction between the true level density and its smoothened counterpart. Therefore, one should understand that all the convergences we mention in the text, are all only "to within fluctuations"

the series expansions of the type given in (10). It should be pointed out that, as our H are rotationally invariant, H and J_Z operator commute and hence $\rho(E,M)$ is a joint probability density. However, for other groups of interest, the group generators in general do not commute with H and hence the interpretation of $\rho(E,[X])$ here, in probability terms is difficult and it is deferred till the end.

Analytical and numerical studies on bivariate cumulants (k_{rs}) for the two operators H and J_Z are presented in Sect. 2. These studies show that $|k_{rs}|$ are quite small ($\lesssim 0.3$) for $3 \leqq k_{rs} \leqq 6$ and they also make it plausible that $k_{rs} \alpha 1/m^{(r+s-2)/2}$ for $r+s \geqq 3$. As a consequence, expansions around a bivariate normal probability density are meaningful and we adopt a bivariate Edgeworth expansion (as suggested by the behavior of the k_{rs}) for $\rho(E,M)$ and give a compact form for the same in Sect. 3. Here we will also discuss its relationship with other methods of constructing fixed-M level densities. In Sect. 4, we derive series expansions for fixed-M traces of H^p which involve only the bivariate cumulants. In Sect. 5 we derive a similar expansion for the spin cut-off factors. The series expansions given in Sects. 3-5, are all formally exact and their truncation to lower order terms is made meaningful by the results of Sect. 2. Numerical comparisons with the results of other series expansions, is made in Sects. 4 and 5. Some concluding remarks are made in Sect. 6.

2. Bivariate Cumulants for the Operators H and J_Z

The general references here are Kendall [9] and Cramers [10]. The bivariate central moments for any two operators X_1 and X_2 (which commute with each other) are defined by,

$$\mathcal{M}_{rs} = \langle \tilde{X}_1^r \tilde{X}_2^s \rangle^m \tag{1}$$

where $\tilde{X} = X - \langle X \rangle^m$ and $\langle X \rangle^m = \langle\langle X^m \rangle\rangle^m / d(m)$ with $d(m)$, the dimensionality of the m-particle space. The m-particle trace $\langle\langle X \rangle\rangle^m$ is,

$$\langle\langle X \rangle\rangle^m = \sum_{\alpha \in m} \langle m\alpha | X | m\alpha \rangle. \tag{2}$$

Now, the dimensionless reduced central moments are $\mu_{rs} = \mathcal{M}_{rs} / [\mathcal{M}_{20}^r \mathcal{M}_{02}^s]^{\frac{1}{2}}$. Note that $\mathcal{M}_{k0}$ are the central moments for the distribution of the eigenvalues (x_1) of X_1; the variance $\sigma^2(m, X_1) = \mathcal{M}_{20}$. Similarly $\mathcal{M}_{0k}$ are for the operator X_2. The joint probability density $\rho(x_1, x_2)$ generates $\mathcal{M}_{rs}$ and alternatively,

$$\mathcal{M}_{rs} = \int_{-\infty}^{+\infty} \int_{-\infty}^{+\infty} (x_1 - \langle X_1 \rangle^m)^r \cdot (x_2 - \langle X_2 \rangle^m)^s \rho(x_1 x_2) dx_1 dx_2. \tag{3}$$

Now, in terms of the standardized variable $\hat{x} = (x - \langle X \rangle)/\sigma(m, X)$, the joint probability density $\eta(\hat{x}_1, \hat{x}_2)$ generates the reduced moments and its fourier transform generates the reduced bivariate cumulants k_{rs}. The relationship between k_{rs} and μ_{rs} can be easily worked out [9, p. 84]. The operators of interest for us are H and J_Z with eigenvalues denoted by E and M. Throughout this article we adopt the convention that $X_1 = H$ and $X_2 = J_Z$, then for example $\mathcal{M}_{31} = \langle \tilde{H}^3 \tilde{J}_Z \rangle^m$ and $\mathcal{M}_{24} = \langle \tilde{H}^2 \tilde{J}_Z^4 \rangle^m$ etc. Because of the symmetry of the J_Z operator,

$$\langle H^p J_Z^q \rangle = 0 \quad \text{for } q \text{ odd}. \tag{4}$$

As a result $\tilde{J}_Z = J_Z$ and only the cumulants k_{pq} (q even) will be non-zero. So the correlation coefficient, given by k_{11}, for H and J_Z will vanish. However H and J_Z, though uncorrelated, are not independent as $\langle H^p J_Z^{2q} \rangle \neq 0$ in general. A few lower order cumulants of interest are k_{30}, k_{12}, k_{40}, k_{22}, k_{04}, k_{50}, k_{14} etc. and they are related to μ_{rs} by

$$\begin{aligned} &k_{30} = \mu_{30} && k_{12} = \mu_{12} \\ &k_{40} = \mu_{40} - 3 && k_{22} = \mu_{22} - 1 \quad k_{04} = \mu_{04} - 3 \\ &k_{50} = \mu_{50} - 10\mu_{30} && k_{32} = \mu_{32} - \mu_{30} - 3\mu_{12} \\ &k_{14} = \mu_{14} - 6\mu_{12}. \end{aligned} \tag{5}$$

For realistic interactions in $2s-1d$ shell, we evaluated the bivariate cumulants for $r+s \leq 6$ and they are given in Table 1. All these results are generated by using the programmes that calculate average for product of three and four two body operators only. As a result, all k_{rs} for $r+s \leqq 6$ could not be obtained. The results of Table 1 show that the cumulants k_{rs} are rather small; generally $|k_{rs}| \lesssim 0.3$. This, then suggests that a series expansion around a bivariate normal density $\hat{\eta}(\hat{E}, \hat{M})$,

$$\begin{aligned} \hat{\eta}(\hat{E}, \hat{M}) &= \hat{\eta}_1(\hat{E}) \hat{\eta}_2(\hat{M}) \\ &= \left[\frac{1}{\sqrt{2\pi}} \exp \frac{-\hat{E}^2}{2}\right] \left[\frac{1}{\sqrt{2\pi}} \exp \frac{-\hat{M}^2}{2}\right] \\ &= \frac{1}{2\pi} \exp -\tfrac{1}{2}(\hat{E}^2 + \hat{M}^2), \end{aligned} \tag{6}$$

is sensible, in that a few terms in the series *suffice*. One can in general use a bivariate Edgeworth or Gram-Charlier expansion for $\eta(\hat{E}, \hat{M})$. If the cumulants $k_{rs} \alpha 1/p^{(r+s-2)/2}$, with p being a parameter of the system, then under fairly general conditions, the Edgeworth expansion is really an asymptotic expan-

Table 1. Bivariate Cumulants k_{rs} for $2s-1d$ Shell Interactions

m	k_{12}	k_{30}	k_{04}	k_{22}	k_{40}	k_{32}	k_{14}	k_{24}	k_{06}
2	−0.219[a]	0.254	−0.438	−0.119	−0.589	0.087	0.198	0.953	0.278
	−0.222	0.069	−0.438	−0.079	−0.731	0.106	0.181	0.247	0.278
	−0.217	0.238	−0.438	−0.141	−0.501	0.112	0.202	0.348	0.278
4	−0.162	0.120	−0.279	−0.123	−0.181	0.071	0.109	0.400	0.159
	−0.168	−0.034	−0.279	−0.090	−0.253	0.100	0.107	0.111	0.159
	−0.159	0.117	−0.279	−0.139	−0.083	0.072	0.115	0.188	0.159
8	−0.120	0.099	−0.206	−0.122	−0.145	0.061	0.086	0.217	0.087
	[illegible]	0.019	0.206	−0.103	−0.197	0.083	0.094	0.057	0.087
	−0.116	0.094	−0.206	−0.124	−0.108	0.063	0.089	0.087	0.087
12	−0.088	0.085	−0.193	−0.120	−0.171	0.041	0.066	0.188	0.090
	−0.091	−0.027	−0.193	−0.110	−0.206	0.061	0.071	0.078	0.090
	−0.082	0.084	−0.193	−0.111	−0.156	0.042	0.063	0.080	0.090
16	−0.042	0.048	−0.206	−0.117	−0.182	0.010	0.020	0.217	0.117
	−0.041	−0.072	−0.206	−0.115	−0.214	0.036	0.018	0.118	0.117
	−0.036	0.055	−0.206	−0.097	−0.198	0.012	0.014	0.097	0.117
20	0.071	−0.033	−0.277	−0.105	−0.294	−0.028	0.133	0.400	0.127
	0.066	−0.177	−0.277	−0.110	−0.354	−0.011	−0.129	0.127	0.127
	0.062	−0.057	−0.277	−0.074	−0.396	−0.029	−0.114	0.080	0.127

[a] For each particle number there are three rows, the first row is for the $k+12fp$ interaction [15], the second row is for the CW interaction given in [16], and the third row is for the KLS interaction [17]. The single particle energies for $k+12fp$ interaction are $\varepsilon_{d_{5/2}} = -4.24$ MeV, $\varepsilon_{d_{3/2}} = 1.51$ MeV and $\varepsilon_{s_{1/2}} = -3.34$ MeV. For the remaining two interactions, they are taken to be the ^{17}O energies

sion of $\eta(\hat{E},\hat{M})$ in powers of $p^{-\frac{1}{2}}$. This is not true of the Gram-Charlier series; see Cramer [10; p. 221-231]. Now we will see below that $p=m$, the particle number.

We know that in general, for a two body H, $k_{40}=-4/m$ [12] and $k_{60}\alpha 1/m^2$ [French, private cummunication]. Moreover, it is easy to see that for the J_Z operator $k_{0,2r}\alpha 1/m^{(2r-2)/2}$. As a result, it is clear that, for $\eta(\hat{E})$ and $\eta(\hat{M})$, Edgeworth expansion is appropriate. See Ref. [13] for a good discussion on this point for $\eta(\hat{M})$. To learn about the behavior of k_{rs}, for $r,s\neq 0$ we proceed as follows.

As in [12], we represent H by a two body random ensemble (TBRE), with the two body matrix elements $\overline{V_{ijkl}}=0$ and $\overline{V_{ijkl}^2}=q^2$, where the bar denotes ensemble averaging. All μ_{pq} (and k_{pq}) with p,q odd, on ensemble averaging, will vanish. So the lowest cumulant of interest is k_{22}. Let us assume that H acts in a space of N single particle states and $m\to\infty$, $N\to\infty$ and $m/N\to 0$. Now H behaves essentially as a unitary rank-2 operator (with respect to the unitary group $U(N)$) and using Eq. (A4) of [14], we obtain

$$\langle\tilde{H}^2 J_Z^2\rangle^m = \frac{m(m-1)(N-m)(N-m-1)}{N(N-1)(N-2)(N-3)} \left[\frac{1}{3}(G_1+G_2+G_3)\right] + \frac{m(m-1)(m-2)(N-m)(N-m-1)(N-m-2)}{N(N-1)(N-2)(N-3)(N-4)(N-5)} \cdot\left[2G_1+G_2+2G_3+2G_4+\frac{1}{4}G_5\right] \tag{7}$$

and also,

$$\langle J_Z^2\rangle^m = \frac{m(N-m)}{(N-1)}\langle J_Z^2\rangle^1$$

$$\langle\tilde{H}^2\rangle^m = \frac{m(m-1)(N-m)(N-m-1)}{2(N-2)(N-3)}\langle\tilde{H}^2\rangle^2. \tag{8}$$

In (7), the G's are some functions of the two body Matrix elements. It can be shown that, on ensemble averaging, the contribution of the diagrams G_2, G_3 and G_4 will vanish, while $G_5\to N^5q^2\langle J_Z^2\rangle^1$ and $G_1\to N^4q^2\langle J_Z^2\rangle^1$. Using these results and collecting terms of the order of 1, $1/m$, $1/m^2$ etc. in k_{22}, we obtain

$$k_{22}=\frac{\overline{\langle\tilde{H}^2J_Z^2\rangle^m}}{\overline{\langle\tilde{H}^2\rangle^m}\,\langle J_Z^2\rangle^m}-1 \xrightarrow[m/N\to 0]{m\to\infty,\ N\to\infty} \frac{-2}{3m}+O\left(\frac{1}{m^2}\right). \tag{9}$$

This is the result we are looking for and establishes that $k_{rs}\alpha\frac{1}{m}$ for $r+s=4$. As we do not have expressions similar to (7) for $r+s>4$, we could not proceed further. The equation (9), together with the fact that k_{60}, $k_{06}\alpha 1/m^2$, makes it plausible that $k_{rs}\alpha 1/m^{(r+s-2)/2}$ in large spaces.

The results of Table 1 do not show the m-dependence of (9) for k_{22} nor k_{12} is zero as it should be for TBRE. However it is *gratifying* that k_{12} and k_{22} are very small. The reasons for this discrepancy lie in the fact that there is no ensemble averaging involved in generating the numbers of Table 1, the H

of TBRE do not in general preserve J^2 and also the significant number N/k (k is the body rank of the operator and is equals to six for $\langle \tilde{H}^2 J_Z^2 \rangle^m$) is quite small. In fact for a single H, using equations (7) and (8), it can be shown that $k_{22}(m) = [3k_{22}(3) - 2k_{22}(2)] + 6/m * [k_{22}(2) - k_{22}(3)] + O\left(\frac{1}{m^2}\right)$.

Our results do not supply a complete proof that $k_{rs} \alpha 1/m^{(r+s-2)/2}$ for TBRE, but they make it meaningful and from now on we assume that for realistic-H this is the case and adopt the bivariate Edgeworth expansion for $\eta(\hat{E}, \hat{M})$. We shall test some of its consequences in Sects. 4 and 5.

3. Bivariate Edgeworth Expansion for $\eta(\hat{E}, \hat{M})$

The bivariate Edgeworth expansion [11] for $\eta(\hat{E}, \hat{M})$ is given by

$$\eta(\hat{E}, \hat{M}) = \exp\left\{ \sum_{r+s \geq 3} (-1)^{r+s} \frac{k_{rs}}{r!s!} D^r_{\hat{E}} D^s_{\hat{M}} \right\} \hat{\eta}_1(\hat{E}) \hat{\eta}_2(\hat{M}) \tag{10}$$

where the D are differential operators. Now we can expand the RHS and obtain

$$\eta(\hat{E}, \hat{M}) = \left\{ 1 + \sum_{r=1}^{\infty} \sum_{P} \sum_{[(p_1q_1)^{\pi_1}(p_2q_2)^{\pi_2}\dots(p_mq_m)^{\pi_m}]} \left\{ \prod_{i=1}^{m} [k'_{p_iq_i}]^{\pi_i}/\pi_i! \right\} * He_{\Sigma p_i\pi_i}(\hat{E})\, He_{\Sigma q_i\pi_i}(\hat{M}) \right\} \hat{\eta}_1(\hat{E}) \hat{\eta}_2(\hat{M}) \tag{11}$$

where the second summation is over all the partitions P of the integer r. For a partition $P = [P_1, P_2, \dots P_l]$ or r, we have $P_1 \geq P_2 \geq \dots \geq P_l$, $\sum P_i = r$ and $P_l > 0$. The third summation is over all bipartitions $[(p_1q_1)^{\pi_1}(p_2q_2)^{\pi_2}\dots(p_mq_m)^{\pi_m}]$ corresponding to the partition P and they are defined as follows. First generate all possible (but distinct) sets $[(p_1q_1)(p_2q_2)\dots(p_lq_l)]$ such that $p_i + q_i = P_i + 2$, p_i and q_i are non-negative integers (but they can be zero). Now, if (p_iq_i) is repeated π_i times in the bipartition, then it is rewritten as $[(p_aq_a)^{\pi_a}(p_bq_b)^{\pi_b}\dots(p_mq_m)^{\pi_m}]$ and $\sum_{i=a}^{m} \pi_i = l$. As $k_{pq} \alpha 1/m^{(p+q-2)/2}$, all terms belonging to a given r in (11) behave as $1/m^{r/2}$. In equation (11), the $He_v(\hat{O})$ are the Hermite polynomials and $k'_{pq} = k_{pq}/p!q!$. It is important to point out that the RHS of (11) can also be expanded using the partitions of bi-partite numbers [18]. However, as we need to collect, for a given integer r, all the terms that behave as $1/m^r$, we felt the need to introduce the new kind of partitions, the bipartitions. Table 2 gives the bipartitions for the integers 3 and 4. Now, retaining terms only up to $1/m$ we obtain explicitly,

Table 2. Bipartitions $(p_aq_a)^{\pi_a}(p_bq_b)^{\pi_b}\dots(p_mq_m)^{\pi_m}$, with q_i even, for Integers 3 and 4

Integer	P	Bipartitions
3	[3]	(50), (32), (14)
	[21]	(40)(30), (22)(30), (04)(30), (40)(12), (22)(12), (04)(12)
	[111]	$(30)^3$, $(30)^2(12)$, $(30)(12)^2$, $(12)^3$
4	[4]	(60), (42), (24), (06)
	[31]	(50)(30), (50)(12), (32)(30), (32)(12), (14)(30), (14)(12)
	[22]	$(40)^2$, (40)(22), (40)(04), $(22)^2$, (22)(04), $(04)^2$
	[211]	$(40)(30)^2$, $(40)(12)^2$, (40)(30)(12), $(22)(30)^2$, $(22)(12)^2$, (22)(30)(12), $(04)(30)^2$, $(04)(12)^2$, (04)(30)(12)
	[1111]	$(30)^4$, $(30)^3(12)$, $(30)^2(12)^2$, $(30)(12)^3$, $(12)^4$

$$\eta(\hat{E}, \hat{M}) = \left[1 + \left\{ \frac{k_{30}}{3!} He_3(\hat{E}) + \frac{k_{12}}{2!} He_1(\hat{E}) He_2(\hat{M}) \right\} + \left\{ \frac{k_{40}}{4!} He_4(\hat{E}) + \frac{k_{22}}{2!2!} He_2(\hat{E}) He_2(\hat{M}) + \frac{k_{04}}{4!} He_4(\hat{M}) + \frac{k_{30}^2}{2!3!3!} He_6(\hat{E}) + \frac{k_{12}^2}{(2!)^3} He_2(\hat{E}) He_4(\hat{M}) + \frac{k_{30}k_{12}}{3!2!} \cdot He_4(\hat{E}) He_2(\hat{M}) \right\} \right] \hat{\eta}_1(\hat{E}) \hat{\eta}_2(\hat{M}) \tag{12}$$

Instead of using (11, 12), one can use the conditional probability densities and these give two different expansions for $\rho(E, M)$ or $\eta(\hat{E}, \hat{M})$. Using the Edgeworth expansion for the conditional probability densities, we obtain

$$\begin{aligned} \rho(E, M) &= \rho_1(E)\, \rho_2(M/E) \\ &= \left[\exp\left\{ \sum_{v \geq 3} (-1)^v \frac{\langle k_v(H) \rangle^m}{v!} \frac{\partial^v}{\partial E^v} \right\} \cdot \frac{1}{\sqrt{2\pi \langle \tilde{H}^2 \rangle^m}} \exp - \tfrac{1}{2} \frac{(E - \langle H \rangle^m)^2}{\langle \tilde{H}^2 \rangle^m} \right] \\ &\quad * \left[\exp\left\{ \sum_{v \geq 3} (-1)^v \frac{\langle k_v(J_Z) \rangle^E}{v!} \frac{\partial^v}{\partial M^v} \right\} \cdot \frac{1}{\sqrt{2\pi \langle J_Z^2 \rangle^E}} \exp - \tfrac{1}{2} \frac{(M)^2}{\langle J_Z^2 \rangle^E} \right] \end{aligned} \tag{13}$$

or

$$\begin{aligned} \rho(E, M) &= \rho_a(M)\, \rho_b(E/M) \\ &= \left[\exp\left\{ \sum_{v \geq 3} (-1)^v \frac{\langle k_v(J_Z) \rangle^m}{v!} \frac{\partial^v}{\partial M^v} \right\} \right. \end{aligned}$$

[2] Note that, as the TBRE H do not in general preserve J, $\langle \tilde{H}^2 J_Z^2 \rangle$ in (7) is calculated as the average of the three basic correlated diagrams $\langle \tilde{H}\tilde{H} J_Z J_Z \rangle$, $\langle \tilde{H} J_Z \tilde{H} J_Z \rangle$ and $\langle \tilde{H} J_Z J_Z \tilde{H} \rangle$

$$\cdot \frac{1}{\sqrt{2\pi\langle J_Z^2\rangle^m}} \exp - \frac{M^2}{2\langle J_Z^2\rangle^m}\Bigg]$$

$$* \left[\exp\left\{\sum_{\nu \geq 3} (-1)^\nu \frac{\langle k_\nu(H)\rangle^M}{\nu!} \frac{\partial^\nu}{\partial E^\nu}\right\}\right.$$

$$\left.\cdot \frac{1}{\sqrt{2\pi\langle \tilde{H}^2\rangle^M}} \exp - \frac{(E - \langle H\rangle^M)^2}{2\langle \tilde{H}^2\rangle^M}\right]. \tag{14}$$

In the above equations $\langle k_\nu(H)\rangle^m$, $\langle k_\nu(J_Z)\rangle^m$ are the reduced cumulants for the operators H and J_Z respectively, while $\langle k_\nu(H)\rangle^M$ and $\langle k_\nu(J_Z)\rangle^E$ are conditional cumulants and the corresponding conditional moments are expectation values. The conditional moments are evaluated using the polynomial expansion method or the parametric derivative method which yields series expansions for them. As a result, in addition to the two series expansions in (13) and (14), each of the conditional moments again have series expansions, and one truncates all these series to lower order terms and obtain $\rho(E, M)$. Because of the several truncations one does not very well know the convergence properties of (13) and (14). Unlike this, equation (11) involves only one series, it is symmetrical in the two variables E and M, and its convergence is tied up with the convergence of $\eta(\hat{E}, \hat{M})$ to bivariate normal density. We can derive $\rho_1(E)$ and $\rho_a(M)$ from $\eta(\hat{E}, \hat{M})$, by integrating over M and E respectively. Then, equations (13) and (14) will allow us to derive the conditional densities, which now involve only the bivariate cumulants. Chang et al. [19] used (13), while Ginocchio and Yen [20] used (14).

Finally, let us mention that one can derive the fixed-J densities from fixed-M densities by using the well known difference equation,

$$d(J)\,\rho(E, J) = d(M)\,\rho(E, M = J) - d(M = J+1)\,\rho(E, M = J+1) \tag{15}$$

where $d(M)$ are fixed-M dimensionalities and $d(J)$ are fixed-J dimensionalities.

4. Fixed-J Averages of Powers of H

Using the bivariate probability density $\eta(\hat{E}, \hat{M})$, we can derive expressions for fixed-M traces of H^p and thereby obtain fixed-J averages using,

$$\langle\!\langle H^p \rangle\!\rangle^{m,J} = \langle\!\langle H^p \rangle\!\rangle^{m,M=J} - \langle\!\langle H^p \rangle\!\rangle^{m,M=J+1}. \tag{16}$$

It is easy to see that

$$\langle \tilde{H}^p \rangle^M = \sigma^p(m, H) \int_{-\infty}^{+\infty} \hat{E}^p \eta(\hat{E}, \hat{M})\, d\hat{E} \Big/ \int_{-\infty}^{+\infty} \eta(\hat{E}, \hat{M})\, d\hat{E}. \tag{17}$$

Now using (11) we obtain, for $\langle \tilde{H}^p \rangle^M$, expressions which involve ratio of polynomial series in $\hat{M}$. Note that the integrations in (17) are quite straight forward, as $\int_{-\infty}^{+\infty} He_\nu(x) He_{\nu'}(x) \eta(x)\, dx = \delta_{\nu\nu'}(\nu!)$. Using the proposition that $k_{rs} \alpha 1/m^{(r+s-2)/2}$, we can make an expansion in terms of the small parameter $1/m$ and obtain for example,

$$\langle \hat{H} \rangle^M = \left[\frac{k_{12} He_2(\hat{M})}{2}\right] + \left[\frac{k_{14} He_4(\hat{M})}{24} + \frac{k_{04} k_{12}}{48} He_6(\hat{M}) - \frac{k_{12} k_{04}}{48} He_2(\hat{M}) He_4(\hat{M})\right] + O(1/m^{5/2}) \tag{18}$$

and

$$\begin{aligned}\langle \tilde{H}^2 \rangle^M = \sigma^2(m, H) \Bigg\{ 1 &+ \left[\frac{k_{22}}{2} He_2(\hat{M}) + \frac{k_{12}^2}{4} He_4(\hat{M})\right] \\ &+ \left[\frac{k_{24}}{24} He_4(\hat{M}) + \frac{k_{22} k_{04}}{48} He_6(\hat{M}) + \frac{k_{12} k_{14}}{24} He_6(\hat{M}) \right. \\ &+ \frac{k_{12}^2 k_{04}}{96} He_8(\hat{M}) - \frac{k_{04}}{24} He_4(\hat{M}) \left\{ \frac{k_{12} He_2(\hat{M})}{2} \right. \\ &\left.\left. + \frac{k_{12}^2}{4} He_4(\hat{M}) \right\}\right] + O(1/m^3) \Bigg\}. \end{aligned} \tag{19}$$

The parametric derivative method of [4, 20] or the polynomials defined by $\rho(M)$ [21] or $\rho(J)$ [22] can also be used to derive series expansions similar to (18) and (19). Although exact methods are available [23–25] for evaluating $\langle H^p \rangle^M$, in our judgement, they are useful only in small spaces. Using the polynomials (p_μ) defined by the density $\rho(M)$, we obtain for example,

$$\begin{aligned}\langle K \rangle^M &= \sum_\mu \langle K p_\mu(J_Z) \rangle^m p_\mu(M) \\ &= \langle K \rangle^M + \langle \tilde{K} \tilde{J}_Z^2 \rangle^m \frac{\hat{M}^2 - 1}{(k_{04} + 2)} \\ &\quad + \left[\langle \tilde{K} \tilde{J}_Z^4 \rangle^m + \frac{B}{A} \langle \tilde{K} \tilde{J}_Z^2 \rangle^m\right] \\ &\quad \cdot \frac{\left[\hat{M}^4 + \frac{B}{A} \hat{M}^2 + \frac{C}{A}\right]}{\left[\mu_{08} + \mu_{06} \frac{B}{A} + \mu_{04} \frac{C}{A}\right]} \\ &\quad + \ldots \end{aligned} \tag{20}$$

where $A = \mu_{04}\mu_{06} - \mu_{04}^3 - \mu_{06} + \mu_{04}^2$, $B = \mu_{04}\mu_{06} - \mu_{04}^3 - \mu_{06}^2 + \mu_{04}^2 \mu_{06}$ and $C = (\mu_{06} - \mu_{04})^2$. The successive terms in (20) arise from $p_0(J_Z)$, $p_2(J_Z)$ and $p_4(J_Z)$ respectively.

We compare in Table 3 the predictions of the bivariate Edgeworth expansion for $\eta(\hat{E}, \hat{M})$ and the polynomial expansion method, for fixed-J centroids

Table 3. Fixed-J Dimensions (d), Centroids (E_C) and Variances (σ^2) for 12 Particles in $2s-1d$ Shell

J	$d(J)$	$E_C(J)$ (MeV)	$\sigma(J)$ (MeV)
0	11,434	−92.38[a]	13.05
		−92.25	13.28
		−92.35	13.09
1	31,332	−92.45	12.99
		−92.38	12.99
		−92.34	13.09
2	45,640	−92.66	12.83
		−92.61	12.89
		−92.63	12.86
3	51,267	−92.94	12.61
		−92.92	12.61
		−92.93	12.63
4	49,441	−93.34	12.31
		−93.35	12.32
		−93.33	12.31
5	41,456	−93.83	11.93
		−93.85	11.89
		−93.83	11.91
6	31,064	−94.43	11.45
		−94.46	11.44
		−94.44	11.42
7	20,461	−95.12	10.89
		−95.11	10.85
		−95.14	10.86
8	11,975	−95.93	10.22
		−95.94	10.22
		−95.96	10.19
9	6,027	−96.83	9.42
		−96.79	9.44
		−96.86	9.44
10	2,597	−97.86	8.48
		97.89	8.58
		−97.88	8.58
11	887	−98.98	7.35
		−99.02	7.49
		−98.99	7.61
12	237	−100.25	5.88
		−100.47	6.45
		−100.23	6.48
13	35	−101.61	3.84
		−101.88	4.75
		−101.56	5.11
14	1	−102.97	2.08
		−103.28	0.0
		−102.88	3.68

[a] All the results are for the $k+12fp$ interaction [15]. The centroids and variances given in the first row for each J value, are obtained by using (18) and (19) respectively. The numbers in the third row are obtained by using (20), while the numbers in the second row are from exact calculations [26]

($E_C(J)$) and variances ($\sigma^2(J)$), as given by (18, 19) and (20) respectively, with the exact results [26]. We note that the results of the present theory are in good agreement with both the exact results and also the polynomial expansion method. We see that there are noticeable deviations for $J=12$–14 (this is similar to the results of Ref. [20]). One expects that $\eta(\hat{E}, \hat{M})$ with a few terms in (11) works well for M within $\pm 3-4\sigma$ and this corresponds to $M \leq 12$, as $\sigma(m=12, J_z) \simeq 3.4$. Now, it is easy to see that the use of (16) yields errors in fixed-J averages for $J=12$–14. Inaccuracies, in the averages for very high J is not a real problem; if necessary one can always perform shell model calculations just for these high angular momenta since they have small dimensions. Alternatively one can partition the space, as discussed ahead, and improve the accuracy.

5. Spin Cut-Off Factors

The spin cut-off factors, as given by $\langle J_Z^2 \rangle^E$, are of particular interest in level densities. As in equations (18, 19), we can derive a series expansion for $\langle J_Z^2 \rangle^E$ using,

$$\langle J_Z^2 \rangle^E = \langle J_Z^2 \rangle^m \cdot \left\{ 1 + \int_{-\infty}^{+\infty} He_2(\hat{M}) \eta(\hat{E}, \hat{M}) d\hat{M} \Big/ \int_{-\infty}^{+\infty} \eta(\hat{E}, \hat{M}) d\hat{M} \right\}. \tag{21}$$

Once again the expansion of (21) in powers of $(1/m)$ yields,

$$\langle J_Z^2 \rangle^E = \langle J_Z^2 \rangle^m \left\{ 1 + [k_{12} He_1(\hat{E})] + \left[\frac{k_{22}}{2} He_2(\hat{E}) + \frac{k_{12} k_{30}}{6} He_4(\hat{E}) - \frac{k_{12} k_{30}}{6} He_1(\hat{E}) He_3(\hat{E}) \right] + O(1/m^{3/2}) \right\}. \tag{22}$$

We compare in Table 4 the predictions of (22) with that of the polynomial expansion (defined by $\rho(E)$),

$$\langle J_Z^2 \rangle^E = \langle J_Z^2 \rangle^m + \langle J_Z^2 \hat{H} \rangle^m \hat{E} + \langle J_Z^2 \hat{H}^2 - k_{30} J_Z^2 \hat{H} - J_Z^2 \rangle^m * \frac{\hat{E}^2 - k_{30} \hat{E} - 1}{k_{40} + 2 - k_{30}^2}, \tag{23}$$

and we see good agreement. Unfortunately we did not have the exact results here. However the previous comparisons (see Grimmes in [1], p, 273) show that the results of (23) agree rather well with exact results, and hence from Table 4 we can conclude that the expansion (22) for $\langle J_Z^2 \rangle^E$ is meaningful. As with the traces of H^p, to calculate $\langle J_Z^2 \rangle^E$ beyond the 3σ limit, $|\hat{E}| = |(E - \langle H \rangle^m)/\sigma(m, H)| > 3$, one should partition the space. Partitioning by some symmetry group which supplies the irreps $[f]$, we have,

$$\eta(\hat{E}, \hat{M}) = [\sum d([f]) \eta^{[f]}(\hat{E}([f]), \hat{M}([f]))]/d(m) \tag{24}$$

Table 4. Spin Cut-Off Factors

m	$\hat{E}$						
	−3	−2	−1	0	1	2	3
4	2.67[a]	2.79	2.78	2.66	2.37	1.86	0.81
	2.59	2.76	2.78	2.65	2.36	1.86	0.70
9	3.17	3.45	3.54	3.45	3.16	2.62	1.58
	3.07	3.41	3.53	3.45	3.17	2.62	1.50
12	3.13	3.48	3.61	3.56	3.31	2.81	1.86
	3.01	3.43	3.60	3.57	3.32	2.79	1.76

[a] For each particle number, the spin cut-off factors in the first row are obtained by using (22) and the numbers in the second row are obtained by using (23). Note that $\hat{E}=(E-\langle H\rangle^m)/\sigma(m,H)$, where $\langle H\rangle^m$ and $\sigma(m,H)$ are the m-particle space centroids and widths

where the $\eta^{[f]}(\ldots)$ is defined by the fixed symmetry bivariate moments,

$$\mu_{p,\nu}([f])=\langle \tilde{H}^p \tilde{J}_Z^\nu\rangle^{[f]}/\sigma^p([f],H)\,\sigma^\nu([f],J_Z) \tag{25}$$

and $\tilde{O}=O-\langle O\rangle^{[f]}$.

The convergence of the series in (18), (19) and (22) is tied up with the convergence of $\eta(\hat{E},\hat{M})$ to bivariate normal density. The structure of the correction terms (they are the bivariate cumulants) and their behavior with respect to particle number, is fairly well understood. This is not the case with the parametric derivative method or the polynomial expansion method; though their convergence criterion is known, the nature and behavior of the correction terms is not very clear. Numerically these two methods (especially the latter one) are shown to work very well. We feel that these two methods and our new method are all complementary and they reveal different structures of fixed-J averages and fixed-J level densities.

Finally let us point out that the agreement shown in Table 3 and 4 show that the bivariate Edgeworth expansion with a few terms provides an adequate representation for fixed-M densities.

6. Conclusions

In this article we made a beginning in applying bivariate distributions and the related quantities in statistical spectroscopy studies. The fixed-J densities are worked out as the first example. Using the bivariate Edgeworth expansion for fixed-M densities, new series expansions for fixed-J traces of H^p and spin cut-off factors are derived. Numerical calculations show that these series expansions are quite meaningful. There remain, however several unanswered questions in this paper.

To strengthen our understanding of the applicability of bivariate Edgeworth expansion for $\rho(E,M)$, the analytic studies of Sect. 2 need to be extended for k_{rs} with $r+s=5$ and 6 also. We point out that Pandey [27] studied bivariate moments for H represented by a Gaussian orthogonal ensemble of matrices and what we need is its extension to TBRE. Moreover, the behavior of the cumulants for TBRE with J-preserving H need to be understood. Once these are known, equation (17) will enable us to learn about the origin of the Gaussian shape for fixed-J densities, as observed in shell model calculations.

Although we employed the bivariate Edgeworth expansion for $\rho(E,M)$, it might be worthwhile making a comparative study of various bivariate expansions for $\rho(E,M)$. The series expansions they yield for the traces of H^p and also for the spin cut-off factors, need to be compared with each other. These studies, together with partitioning as discussed in Sect. 5, will be presented in a subsequent publication.

The extension of the present work for fixed-JT level densities (T being the isospin) is quite straight-forward. To this end, one has to replace all fixed-m averages by fixed mT averages. Alternatively one can recover fixed-JT densities from fixed-$(m_p m_n M)$ densities using a difference equation similar to (15). Here one can use trivariate and multivariate Edgeworth expansions (see Ref. [28]).

As pointed out in the beginning, it is not clear whether one can define a bivariate expansion (as in Eq. (10)) for the eigenvalues of non-commuting operators (one can certainly define an expansion similar to (14)). Their exploration will enable us to learn about partial level densities and their convergence to Gaussian form etc.

The author thanks Prof. J.B. French for provoking his interest on the subject of bivariate distributions and also for many useful discussions.

References

1. Dalton, B.J., Grimes, S.M., Vary, J.P., Williams, S.A. (eds.): Moment methods in many fermion systems. New York, Plenum 1980
2. French, J.B., Kota, V.K.B.: Annu. Rev. Nucl. Part. Sci. **32**, 35 (1982)
3. French, J.B.: Nucl. Phys. A**396**, 87C (1983)
4. Chang, F.S., French, J.B.: Phys. Lett. **44**B, 131 (1973)
5. Halemane, T.R.: J. Math. Phys. **22**, 2961 (1981)
6. Draayer, J.P., French, J.B., Potbhare, V., Wong, S.S.M.: Phys. Lett. **55**B, 263 (1975)
   Draayer, J.P., French, J.B., Wong, S.S.M.: Ann. Phys. **106**, 483 (1977)
7. Parikh, J.C.: Group symmetries in nuclear structure. New York: Plenum Press 1978
8. French, J.B., Draayer, J.P.: In: Group theoretical methods in physics Beiglböck, W., Bohm, A., Takasugi, E. (eds.). Berlin, Heidelberg, New York: Springer 1979

9. Kendall, M.G., Stuart, A.: Advanced theory of statistics. Vol. 1, 3rd Edn. New York: Hafner Publishing Company 1969
10. Cramer, H.: Mathematical methods in statistics. Princeton, NJ: Princeton University Press 1946
11. Edgeworth, F.Y.: Trans. Camb. Philos. Soc. **20** (Part I), 36 (1905)
Edgeworth, F.Y.: Trans. Camb. Philos. Soc. **20** (Part I), 113 (1905)
12. Mon, K.K., French, J.B.: Ann. Phys. (NY) **95**, 90 (1975)
13. Zirnbauer, M.R., Brink, D.M.: Z. Phys. A - Atoms and Nuclei **301**, 237 (1981)
14. Kota, V.K.B., Potbhare, V.: Phys. Rev. C**21**, 2637 (1980)
15. Halbert, E.C., McGrory, J.B., Wildenthal, B.H., Pandya, S.P.: Adv. Nucl. Phys. **4**, 315 (1971)
16. Vary, J.P., Yang, S.N.: Phys. Rev. C**15**, 1545 (1977)
17. Kahana, S., Lee, H.C., Scott, C.K.: Phys. Rev. **185**, 1378 (1969)
18. Macmahon, M.P.A.: Combinatory analysis. New York: Chelsea Publishing Company 1960
19. Chang, F.S., French, J.B., Thio, T.H.: Ann. Phys. **66**, 137 (1971)
20. Ginocchio, J.N., Yen, M.M.: Nucl. Phys. A**239**, 365 (1975)
21. Haq, R.U., Wong, S.S.M.: Nucl. Phys. A**327**, 314 (1979)
22. Halemane, T.R.: Z. Phys. A - Atoms and Nuclei **301**, 177 (1981)
23. Jacquemin, C., Spitz, S.: Z. Phys. A - Atoms and Nuclei **290**, 251 (1979)
24. Mugambi, P.: Ph.D. Thesis, University of Rochester, Rochester, NY (1970)
25. Verbaarschot, J.J.M., Brussaard, P.J.: Phys. Lett. **102**B, 201 (1981)
26. Haq, R.U.: Ph.D. Thesis, Gujarat University, Ahmedabad, India 1975
27. Pandey, A.: Ph.D. Thesis, University of Rochester, Rochester, NY 1978
Brody, T.A., Flores, J., French, J.B., Mello, P.A., Pandey, A., Wong, S.S.M.: Rev. Mod. Phys. **53**, 472 (1981)
28. Chambers, J.M.: Biometrika **54**, 367 (1967)

V.K.B. Kota
Theory Group
Physical Research Laboratory
Navarangpura
Ahmedabad 380009
Gujarat
India

VOLUME 54, NUMBER 15 PHYSICAL REVIEW LETTERS 15 APRIL 1985

Higher-Order Correlations in Spectra of Complex Systems

Oriol Bohigas
Division de Physique Théorique, Institut de Physique Nucléaire, 91406 Orsay Cedex, France

and

Rizwan U. Haq
Department of Physics and Astronomy, Laurentian University, Sudbury, Ontario P3E 2C6, Canada

and

Akhilesh Pandey
Department of Physics and Astronomy, University of Rochester, Rochester, New York 14627
(Received 21 September 1984)

The presence and significance of three- and four-level correlation effects in nuclear spectra are demonstrated for the first time. New ways to study two-level properties are also presented. Agreement with random-matrix predictions is found.

PACS numbers: 24.60.Ky, 05.40.+j

The two-level correlation function for the spectra of complicated systems[1] determines essentially all of the fluctuation measures which have so far been seriously studied. Moreover, it determines two general properties, namely the Wigner–von Neumann level repulsion (tendency of levels to avoid clustering) and the Dyson-Mehta *long-range order* (large correlations between distant levels). In an earlier analysis,[2] by combining all the high quality nuclear data into a *nuclear data ensemble* (NDE) and by introducing new spectrally averaged measures, we found an extremely close agreement with the *Gaussian orthogonal ensemble* (GOE) predictions for two-level correlations. This goes a considerable distance toward confirming Wigner's suggestion[3] that "the Hamiltonian which governs the behavior of a complicated system is a random symmetric matrix, with no particular properties except for its symmetric nature." Because of that, and because of recent interest in problems of quantum extensions of classically chaotic motion,[4] it is important to ask whether the GOE, which follows naturally from Wigner's proposition for time-reversal–invariant systems, gives proper predictions for fluctuations of higher order, say three- and four-level correlations. These of course are not determined by the two-level function, even though a lower-order function does impose certain complicated constraints which are largely unknown on the higher-order ones.

Consider an infinite stationary spectrum with unit average spacing. The k-level correlation function[5] $R_k(r_1, \ldots, r_k)$ is the probability density of observing a level at each of the k points $r_1, \ldots, r_k$, irrespective of the location of the other levels. The R_k's for $k > 1$ ($R_1 = 1$) depend only on the relative variables $r_i - r_j$, e.g., $R_2(r_1,r_2) = R_2(r_1 - r_2)$. The probability $E_k(r)$ that the number statistic $n(r)$, the number of levels in an interval of length r, takes the value k is given by[6]

$$E_k(r) = \frac{1}{k!}\sum_{l=0}^{\infty}(-1)^l \frac{\hat{R}_{k+l}(r)}{l!}, \tag{1}$$

$$\hat{R}_k(r) = \int_0^r \cdots \int_0^r R_k(r_1, \ldots, r_k)\, dr_1 \cdots dr_k. \tag{2}$$

For small r, $\hat{R}_k/k!$ gives the probability that in an interval of length r there are k levels. From (1) we have[1,6]

$$\bar{n} = r, \tag{3}$$

$$\Sigma^2(r) = \hat{R}_2(r) - r(r-1), \tag{4}$$

$$\kappa_3(r) = \gamma_1(r)\Sigma^3(r) = \hat{R}_3(r) - 3(r-1)\Sigma^2(r) - r(r-1)(r-2), \tag{5}$$

$$\kappa_4(r) = \gamma_2(r)\Sigma^4(r) = \hat{R}_4(r) - (4r-6)\kappa_3(r) - 3\Sigma^4(r) - (6r^2 - 18r + 11)\Sigma^2(r) - r(r-1)(r-2)(r-3), \tag{6}$$

for the average, variance, and third and fourth cumulants of n, respectively; γ_1, γ_2 defined in (5), (6) are the skewness and excess, respectively. Note that in general the kth cumulant depends on all $\hat{R}_\nu(r)$ with $\nu \le k$ so that Σ^2, κ_3, and κ_4 derive from the 2-, (2+3)-, and (2+3+4)-level functions, respectively.

To understand the distinct role played by level repulsion and long-range order on the k-level fluctuation measures we shall, apart from GOE, also consider the following: (i) *Poisson ensemble*: This is an ensemble characterized by an exponential $[\exp(-x)]$ nearest-neighbor spacing distribution $p(x)$ and no correlations between spacings. One has $R_k = 1$ for all

TABLE I. Small-r behavior of $\hat{R}_2$, $\hat{R}_3$, and $\hat{R}_4$.

	GOE	Poisson	UW	GUE
$\hat{R}_2(r)$	$\pi^2r^3/18$	r^2	$\pi r^3/6$	$\pi^2r^4/18$
$\hat{R}_3(r)$	$\pi^4r^6/1350$	r^3	$\pi^2r^5/80$	$\pi^6r^9/48\,600$
$\hat{R}_4(r)$	$\dfrac{\pi^8r^{10}}{6\,615\,000}$	r^4	$\dfrac{\pi^3r^7}{1680}$	$\dfrac{\pi^{12}r^{16}}{53\,581\,500\,000}$

k, so that $\hat{R}_k = r^k$, and $\Sigma^2(r) = \kappa_3(r) = \kappa_4(r) = r$. This ensemble exhibits neither level repulsion nor long-range order. (ii) *Uncorrelated Wigner (UW) ensemble*: There are, as in (i), no correlations between spacings but the spacing distribution $p(x)$ is that of GOE, for which the Wigner surmise $[(\pi x/2)\exp(-\pi x^2/4)]$, used here, is an excellent approximation. This ensemble exhibits no long-range order but it does show level repulsion and in fact has the same small-r behavior as that of GOE for $R_2(r)$. (iii) *Gaussian unitary ensemble* (GUE): This is an ensemble of complex Hermitian matrices, valid, unlike GOE, for systems in which time-reversal invariance is completely broken. This ensemble exhibits both features more strongly than GOE. Poisson and UW, unlike GOE and GUE, are not matrix ensembles.

GOE and GUE predictions are known in closed form[1] for $k = 2$. For $k > 2$ we have obtained the results from the tabulated values of the E_k functions.[7] GOE-sample errors needed for comparison with data of limited size (such as the NDE) have been estimated from Monte Carlo calculations. The errors for other ensembles are in general larger than in GOE for Poisson and UW, and smaller for GUE. For the experimental data analyzed here, an NDE of 1762 levels, see Ref. 2.

We discuss first a new analysis of the R_2 function providing efficient measures of level repulsion and long-range order. For level repulsion it is necessary that $R_2(r)$ be zero at $r = 0$ [implying that $p(x) = 0$ at the origin] and the slope is then a measure of level repulsion. Thus by considering $R_2(r) = \alpha + \beta r$ for small r, so that $\hat{R}_2(r) = \alpha r^2 + (\beta/3)r^3$, we can estimate α and β from data [see Tables I and II and Fig. 1(a)]. Using $r = 0.25$ and 0.5 we obtain $\alpha = -0.02$ (consistent with zero) and $\beta = 1.56$ (consistent with the GOE value $\pi^2/6$ to within $\sim 10\%$). To measure long-range order, consider

$$\Sigma^2(r) = ar + 2b\ln r + c \qquad (7)$$

valid for the above ensembles for $r \gtrsim 1$. In (7), a is given by

$$a = 1 - \int_{-\infty}^{+\infty} Y_2(s)\,ds, \qquad (8)$$

where $R_2 = 1 - Y_2$; for long-range order we must have $a = 0$, *a condition on the total integral of the two-level cluster function* Y_2 which expresses that the spectrum is "incompressible"[8]; b may be nonzero if Y_2 falls off as b/s^2 (as it does for GOE and GUE) and provides a measure of the long-range order (the smaller b, the more rigid the spectrum is, provided that $a = 0$); c becomes independent of r for large r. NDE values of a, b, and c obtained by a least-squares fit of $\Sigma^2(r)$ with (7) in the range $1 \leqslant r \leqslant 25$ are -0.007, 0.11, and 0.45, respectively. They agree with GOE predictions (0 ± 0.005, 0.10 ± 0.02, 0.44 ± 0.02) but disagree with Poisson (1,0,0), UW (0.273,0,0.17), and GUE (0,0.05,0.34) values.

We now turn to the *higher-order correlations*. In Figs. 1(b) and 1(c) are displayed γ_1 and γ_2 as functions of r and in Table II are given values of $\hat{R}_3$ and $\hat{R}_4$. We see that only GOE agrees with the data for the whole range of r and for all quantities.

Let us focus first on the small-r behavior ($r \lesssim 1$). UW and GOE give identical results for $\hat{R}_2$ but differ

TABLE II. Values of $\hat{R}_2$, $\hat{R}_3$, and $\hat{R}_4$ for NDE, GOE ($\pm$ one sample error), UW, and GUE.

	$\hat{R}_2(\times 10^2)$				$\hat{R}_3(\times 10^2)$				$\hat{R}_4(\times 10^2)$			
r	NDE	GOE	UW	GUE	NDE	GOE	UW	GUE	NDE	GOE	UW	GUE
0.25	0.7	0.8±0.1	0.8	0.2								
0.50	5.9	6.4±0.4	6.3	3.0	0.1	0.1±0.1	0.3	0.005				
0.75	20.3	20.3±0.7	20.1	13.2	1.2	1.1±0.3	2.3	0.1				
1.0	44.8	44.6±0.9	45.2	34.4	5.8	5.4±0.8	10.2	1.1	0.1	0.1±0.1	1.3	∼0.001
1.5	128.5	127.6±1.3	133.0	113.8	48.2	45.3±2.3	68.2	21.9	3.2	4.2±1.4	22.0	∼0.3
2.0	259.0	258.4±1.5	211.2	241.6	180.7	178.0±4.8	239.4	125.5	44.7	46.8±5.0	144.3	∼12

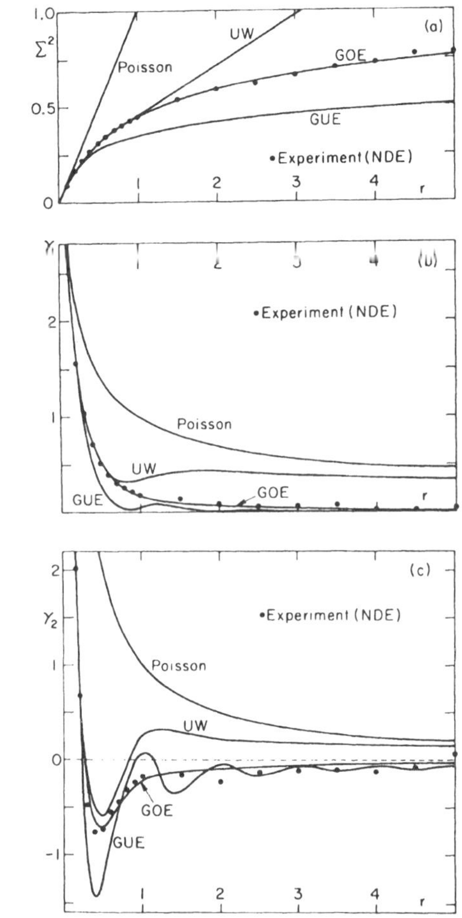

FIG. 1. (a) $\Sigma^2(r)$, (b) $\gamma_1(r)$, and (c) $\gamma_2(r)$ for $r \leqslant 5$. The sample errors corresponding to GOE theory are as follows. For Σ^2, they rise monotonically from 0.001 for $r = 0.25$, to 0.009 for $r = 1$, and 0.03 for $r = 5$. For γ_1, they are of the order of 0.02 for $0.25 \leqslant r \leqslant 1.5$ and then rise to 0.05 for $r = 5$. For γ_2, they decrease from 0.08 for $r = 0.25$ to 0.05 for $r = 1.5$ and then rise again to 0.08 for $r = 5$. Although not shown in the figure, the results for $5 \leqslant r \leqslant 25$ have also been calculated, and NDE and GOE values are consistent.

for $\hat{R}_3$ and $\hat{R}_4$. In fact (cf. Table I) as $r \to 0$, $\hat{R}_2$ goes to zero in the same way but $\hat{R}_3$ and $\hat{R}_4$ go to zero more rapidly for GOE than UW. The difference can be attributed to intrinsic higher-order correlation effects. Experimental values (cf. Fig. 1 and Table II) do distinguish between UW and GOE for γ_1, γ_2, $\hat{R}_3$, and $\hat{R}_4$, showing therefore three- and four-level effects of the GOE type in the data.

Consider finally the large-r ($r \gtrsim 1$) results. γ_1 and γ_2 go to zero for all ensembles, indicating that the distribution of $n(r)$ becomes Gaussian. The approach to zero, however, is faster for GOE and data (and even more for GUE) than for Poisson and UW. In fact κ_3 and κ_4 become small for GOE (and GUE), not ruling out the possibility that they go to zero. One can prove that for Poisson and UW the leading term in the expansion of κ_3 and κ_4 is linear in r, whereas for GOE and GUE it will be at most logarithmic. This is a consequence of the relations[8] satisfied by the usual integrals of the k-level cluster functions and may be regarded as an aspect of long-range order in the three- and four-level correlations.

We have thus presented a new analysis of the two-level correlation properties and introduced precise measures of level repulsion and long-range order. In particular, the total integral of the two-level cluster function, essential for determining long-range order, is deduced from the data. We have also introduced methods for analyzing higher-order correlations. We have shown for the first time that the three- and four-level effects predicted by the GOE are present in the nuclear resonance data.

Significant progress in understanding Wigner's prescription for complicated systems has been made recently in the form of the following conjecture[4] concerning the relationship between classical and quantum chaos: Generically an integrable quantum system should exhibit Poisson-type energy-level fluctuations; on the other hand, quantization of nonintegrable time-reversal–invariant systems should yield fluctuations intermediate between Poisson and GOE, the latter applying for systems for which almost all classical trajectories are chaotic. The conjecture is well supported by numerical calculations for two-dimensional model Hamiltonians.[4] The best experimental evidence for level fluctuations comes from nuclear spectra,[2] which we have used, but there is also evidence from atomic[9] and molecular[10] spectra and from the low-temperature properties of small metallic particles.[1] All of these tests have centered on the two-level correlation properties. Present results confirm the validity of the GOE even for higher-order correlations and go a considerable distance toward showing the "universality" of the level fluctuations, which is implied by Wigner's prescription and the above conjecture.

We are indebted to J. B. French for many valuable discussions and for critical comments on the manuscript. This work has been partly carried out while two of the authors (O.B. and R.U.H.) were visiting the Department of Physics and Astronomy, University of Rochester; they acknowledge the hospitality extended to them. This work has been supported in part by the U. S. Department of Energy and the Natural Sciences and Engineering Research Council of Canada. Division de Physique Théorique is a laboratoire associé au Centre National de la Recherche Scientifique.

[1]T. A. Brody, J. Flores, J. B. French, P. A. Mello, A. Pandey, and S. S. M. Wong, Rev. Mod. Phys. **53**, 385 (1981), and references therein.

[2]R. U. Haq, A. Pandey, and O. Bohigas, Phys. Rev. Lett. **48**, 1086 (1982); O. Bohigas, R. U. Haq, and A. Pandey, in *Nuclear Data for Science and Technology*, edited by K. H. Böckhoff (Reidel, Dordrecht, 1983), p. 809; the data (NDE) used in the present analysis are the same as in the second paper which contains a few more sequences than the first one.

[3]E. P. Wigner, in *Symmetries and Reflections*, edited by E. P. Wigner (Ox Bow, Woodbridge, Connecticut, 1979), p. 203.

[4]I. C. Percival, J. Phys. B **6**, L229 (1973); M. V. Berry and M. Tabor, Proc. Roy. Soc. London, Ser. A **356**, 375 (1977); O. Bohigas, M. J. Giannoni, and C. Schmit, Phys. Rev. Lett. **52**, 1 (1984); T. H. Seligman, J. J. Verbaarschot, and M. R. Zirnbauer, Phys. Rev. Lett. **53**, 215 (1984); M. V. Berry, in *Chaotic Behavior of Deterministic Systems*, edited by G. Iooss *et al.* (North-Holland, Amsterdam, 1983); O. Bohigas and M. J. Giannoni, in *Mathematical and Computational Methods in Nuclear Physics*, edited by J. S. Dehesa *et al.*, Lecture Notes in Physics Vol. 209 (Springer-Verlag, New York, 1984), and references therein.

[5]F. J. Dyson, Commun. Math. Phys. **19**, 235 (1970); M. L. Mehta, Commun. Math. Phys. **20**, 245 (1971). These papers contain expressions for R_k for GOE and GUE.

[6]A. Pandey, Ph.D. thesis, University of Rochester, 1978 (unpublished).

[7]M. L. Mehta and J. des Cloizeaux, Indian J. Pure Appl. Math. **3**, 329 (1972).

[8]M. L. Mehta and F. J. Dyson, J. Math. Phys. **4**, 713 (1963).

[9]N. Rosenzweig and C. E. Porter, Phys. Rev. **120**, 1698 (1960); H. S. Camarda and P. D. Georgopulos, Phys. Rev. Lett. **50**, 492 (1983).

[10]E. Haller, H. Köppel, and L. S. Cederbaum, Chem. Phys. Lett. **101**, 215 (1983); E. Abramson, R. W. Field, D. Imre, K. K. Innes, and J. L. Kinsey, J. Chem. Phys. **80**, 2298 (1984); S. Mukamel, J. Sue, and A. Pandey, Chem. Phys. Lett. **105**, 134 (1984).

Propagation coefficients for fixed-isospin (T,T_z) average and related spectroscopic sum rules

Masao Nomura
Institute of Physics, College of Arts and Sciences, University of Tokyo, Tokyo, 153, Japan

(Received 3 June 1985; accepted for publication 30 September 1985)

The reduction relation for the fixed-isospin (T,T_z) average of a general operator in the model space of many fermions is described in two forms with and without recourse to factorization of isospin z components. Algebraic treatment is developed to deduce various types of expressions for each propagation coefficient that plays the role of the Green's function in each form of the reduction relation. Propagation coefficients are described also in relation to sum rules as to fixed-isospin spectroscopic factors. These results lead to novel identities among n-j symbols and factorials.

I. INTRODUCTION

A many-fermion system is often described by the model space generated from several interacting particles being distributed over a definite number of single-particle orbits or lattice sites.[1–9] Typical cases are seen in the study of atomic and nuclear structures[1–8] and statistical mechanical treatment of various phenomena.[9] A striking feature of the model space is the existence of the reduction relation for the operator average,[1–8] implying that a type of operator average in a many-body space could be expressed in terms of the same average in a few-body space. It, called propagation of operator average, underlies the statistical treatment of nuclear spectroscopy.[1–8]

The present work is devoted to the operator average in the model space with the total isospin T and its z component Tz being fixed, i.e., free from the average. The system of isospin-$\frac{1}{2}$ fermions is treated. The discussion can be extended to the average with intrinsic spin being fixed, if the terms (T,T_z) are only replaced by (S,S_z), respectively. The orbit or the site in the model space need not be specified explicitly. We can keep in mind the Fermi-gas model, or the spherical or deformed shell model with the Pauli principle being rigorously taken into account.

The fixed-isospin average of a general k-body operator $O(k)$ in the n-body space is defined by

$$\langle O(k)\rangle^{nTT_z} \equiv \frac{\sum_\mu \langle nTT_z\,\mu|O(k)|nTT_z\,\mu\rangle}{d(nTT_z)}, \qquad (1)$$

where the index μ stands for a set of quantum numbers that specify, together with T and T_z, the complete orthonormalized states of the n-body space. The dimensionality of the n-body states with a given T and T_z is denoted as $d(nTT_z)$, a number independent of T_z. The propagation of the average (1) is expressed as[1,7,8]

$$\langle O(k)\rangle^{nTT_z} = \sum_{tt_z} Z(nTT_z,ktt_z)\langle O(k)\rangle^{ktt_z}, \qquad (2)$$

where Z stands for the proportional coefficient, called the propagation coefficient.[1–7] The k-body state is specified by $|ktt_z\omega\rangle$ as the n-body state is by $|nTT_z\,\mu\rangle$. Some other types of averages, such as the fixed-seniority average,[1,3,4] lead to the same type of reduction relations. The propagation coefficient Z fulfills an analog of the Chapman–Kolmogorov equation and plays the role akin to the Green's function.[8] It is possible to express Z as

$$Z(nTT_z,ktt_z) = \sum_{t'} S'(nTT_z,ktt_z + k't't'_z), \qquad (3)$$

with $k' = n - k$, $t'_z = t_z - T_z$, and

$$S'(nTT_z,ktt_z + k't't'_z) = \binom{n}{k}(tt't_zt'_z|TT_z)^2 \sum_{\omega\omega'} \langle kt\omega + k't'\omega'|nT\mu\rangle^2 \qquad (4)$$

$$\equiv (tt't_zt'_z|TT_z)^2 S(nT,kt + k't'), \qquad (5)$$

where the symbol $\langle\ +\ |\ \rangle$ stands for the coefficient of fractional parentage (cfp).[10] The quantity S' or S, in which the z components of isospins are reduced, represents the sum of spectroscopic factors with all the isospins being fixed.

Previous works[2,4,6] concerned, in place of (2), the alternative form

$$\frac{\sum_\mu \langle nT\mu\|O^{(r)}(k)\|nT\mu\rangle}{d(nTT)} = \sum_t \frac{R(nT,kt,r)\,\sum_\omega \langle kt\omega\|O^{(r)}(k)\|kt\omega\rangle}{d(ktt)}, \qquad (6)$$

where the double-barred matrix element is reduced with respect to isospins, and $O^{(r)}(k)$ is the k-body irreducible isotensor of rank r. The coefficient R relies on r, while Z in (2) relies on T_z and t_z. The relations (2) and (6) are associated with each other by virtue of the Wigner–Eckart theorem.[10]

French[2] deduced the explicit form of R. Another formalism[4] led to the same result. However, the result cited later [see Eq. (57)], is much involved and hardly manifests inherent properties.

The purpose of the present work is to deduce various types of expressions separately for Z and R that characterize (2) and (6), respectively. Each of the new expressions manifests itself in a transparent form and has the practical advantage of ease of manipulation. The quantity S, defined by (5), is shown to be expressed in terms of dimensionality of representations described by the same Young tableau,[11] which gives a sum rule as to fixed-isospin spectroscopic factors. Another spectroscopic sum rule is given to the sum of Z's. Apparently different expressions for the same quantity

$Z(R)$, as well as the sum rules, lead to types of novel identities among n-j symbols and factorials.

The search for the possibly simplest form of Z is intended for the use of (2) to the average of a product of several operators. Using (2) in place of (6), we can escape from complexity to decompose the operator product as a sum of irreducible isotensors.

We interpret Z and S (or S') as spectroscopic factors for the nuclear fragmentation in the high energy nuclear reaction. Simple expressions of Z and S will promote analyses of mass and charge distributions of the target or projectile residues.

A survey on quantities S, Z, and R is given in Sec. II as a preparation for later discussions. In particular, these quantities are shown to be equivalent to each other. Sections III, IV, and V are devoted to getting new expressions for S, Z, and R, respectively. In Sec. III, a recurrence relation for S is presented. Solving it, we get the compact expression of S so that Z and R can be expressed in terms of a Clebsch–Gordan (CG) coefficient squared and a Racah coefficient, respectively. In Sec. IV, we deduce a remarkably simple form of Z in the case of $|T_z| = T$ or $|t_z| = t$, the most familiar case in nuclear physics. In Sec. V, this result is translated into an expression of R that involves a single CG coefficient. Subsequently, another form of R is deduced, which is particularly suitable in the case where the isotensor rank of the operator is small. The result is elucidated in terms of information propagation that obeys the difference equation for R. Section VI concerns the prescription to deal with the average of a product of several operators. In Sec. VII, the quantities Z and S are described as spectroscopic factors for the target (or projectile) fragmentation.

New expressions for Z are given in (36) and (50). Those for R are in (37), (55), and (59), each of which is to be algebraically identical to the previous result (57). Spectroscopic sum rules are presented in (34) and (69).

II. A SURVEY ON PROPAGATION COEFFICIENTS

A general k-body operator $O(k)$ is written in the form

$$O(k) = \sum_{\gamma\gamma'} \langle k\gamma|O(k)|k\gamma'\rangle A^+(k\gamma)A(k\gamma'), \quad (7)$$

where $\gamma = (tt_z\omega)$, i.e., all the quantum numbers specifying the k-body state. The symbol $A^+(k\gamma)$ indicates the state operator[12,13] that creates the state $|k\gamma\rangle$ in case it acts on vacuum. The conjugate of $A^+(k\gamma)$ is denoted as $A(k\gamma)$. It is postulated that $A(k=0) = 1$. For the tensor analysis in the isospin space, it is convenient to define the operator $\widetilde{A}$ by

$$\widetilde{A}(ktt_z\omega) = (-1)^{t-t_z} A(kt(-t_z)\omega). \quad (8)$$

The expression (2) in the case of $O(k) = A^+A$ reduces to

$$\sum_\mu \langle n\lambda\mu|A^+(kv\omega)A(kv'\omega')|n\lambda\mu\rangle = \delta(v,v')\delta(\omega,\omega')Z(n\lambda,kv)\frac{d(n\lambda)}{d(kv)}, \quad (9)$$

where $\lambda = (TT_z)$ and $v = (tt_z)$. An inherent feature of (9) is ω independence of Z. The dimensionality $d(ktt_z)$ is given by[1]

$$d(nTT_z) = (2T+1)(N/2+1)^{-1} \times \binom{N/2+1}{n/2+T+1}\binom{N/2+1}{n/2-T}\Delta(t-|t_z|), \quad (10)$$

where N indicates the number of the single-particle states and

$$\Delta(a) = 1, \quad \text{if } a>0, \quad \text{and 0 otherwise.} \quad (11)$$

In the fermion system, there exists a particle–hole symmetry of the matrix element[1]

$$\langle n\lambda\mu|A^+(kv\omega)A(k'v'\omega')|n'\lambda'\mu'\rangle = \langle N-n'\lambda'\mu'|A(k'v'\omega')A^+(kv\omega)|N-n\lambda\mu\rangle. \quad (12)$$

By virtue of (9) and (12) we get

$$Z(n\lambda,kv) = Z(N-kv,N-n\lambda)\frac{d(kv)}{d(n\lambda)}. \quad (13)$$

The expression (9) is rewritten by virtue of (12) as

$$\left\langle n\lambda\mu\left|\sum_\omega A^+(kv\omega)A(kv\omega)\right|n\lambda'\mu'\right\rangle = \delta(\lambda,\lambda')\delta(\mu,\mu')Z(n\lambda,kv). \quad (14)$$

The matrix element of the state operator is proportional to cfp as[12,13]

$$\langle n\lambda\mu|A^+(kv\omega)|n'\lambda'\mu'\rangle = \sqrt{\binom{n}{k}}\langle kv\omega+n'\lambda'\mu'|n\lambda\mu\rangle, \quad (15)$$

where $k+n'=n$. The relation (12) with $k'=0$ reduces to a particle–hole symmetry of cfp:

$$\sqrt{\binom{n}{k}}\langle kv\omega+n'\lambda'\mu'|n\lambda\mu\rangle = \sqrt{\binom{N-n'}{k}}\langle kv\omega+(N-n)\lambda\mu|(N-n')\lambda'\mu'\rangle. \quad (16)$$

Expressing the left-hand side (lhs) (14) in terms of cfp yields

$$\binom{n}{k}\sum_{\omega\omega'v'}\langle kv\omega+k'v'\omega'|n\lambda\mu\rangle\langle kv\omega+k'v'\omega'|n\lambda'\mu'\rangle = \delta(\lambda,\lambda')\delta(\mu,\mu')Z(n\lambda,kv), \quad (17)$$

which leads to (3) followed by (4). The factorization of cfp[14] gives

$$\binom{n}{k}\sum_{\omega\omega'}\langle k\nu\omega+k'\nu'\omega'|n\lambda\mu\rangle\langle k\nu\omega+k'\nu'\omega'|n\lambda\mu'\rangle$$
$$=\delta(\mu,\mu')\times \text{a term independent of } \mu$$
$$=\delta(\mu,\mu')S'(n\lambda,k\nu+k'\nu'), \tag{18}$$

where (4) is used in the last step. We see that the quantity S' and, therefore, S are independent of μ. Substituting $O^{(r)}(k)=[A^{+}\tilde{A}]^{(r)}$ into (6) yields

$$\sum_{\mu}\langle nT\mu\|[A^{+}(kt\omega)A(kt'\omega')]^{(r)}\|nT\mu\rangle$$
$$=\delta(t,t')\delta(\omega,\omega')(-1)^{2t}\sqrt{2r+1}\,R(nT,kt,r)\frac{d(nTT)}{d(ktt)}. \tag{19}$$

Applying the Wigner–Eckart theorem to the lhs of (9) to link (9) with (19), we get

$$Z(nTT_z,ktt_z)=\frac{\Sigma_r(trt_z0|tt_z)(TrT_z0|TT_z)(2r+1)R(nT,kt,r)}{\sqrt{(2t+1)(2T+1)}}. \tag{20}$$

The inverse is given by

$$R(nT,kt,r)=\sqrt{\frac{(2T+1)}{(2t+1)}}\,\frac{\Sigma_{t_z}(trt_z0|tt_z)Z(nTT,ktt_z)}{(TrT0|TT)}. \tag{21}$$

Combining (21) with (20) yields

$$Z(nTT_z,ktt_z)$$
$$=\sum_r(2r+1)(trt_z0|tt_z)(TrT_z0|TT_z)$$
$$\times\frac{\Sigma_{t_z'}(trt_z'0|tt_z')Z(nTT,ktt_z')}{(2t+1)(TrT0|TT)}. \tag{22}$$

It indicates that the set $\{Z(nTT_z,ktt_z);\,T_z=T\}$ is sufficient to determine all the set of Z's. Expanding the lhs of (19) in terms of cfp's, we get

$$R(nT,kt,r)=(2T+1)\sum_{t'}W(Trt't;Tt)S(nT,kt+k't'), \tag{23}$$

which is inverted as

$$S(nT,kt+k't')=(2t'+1)$$
$$\times\frac{\Sigma_r(2r+1)W(Trt't;Tt)R(nT,kt,r)}{(2T+1)}. \tag{24}$$

We see from (20)–(24) that the quantities S, R, and Z with $T_z=T$ are transformed into each other.

The propagation coefficient Z fulfills[8]

$$\sum_{\lambda'}Z(n\lambda,n'\lambda')Z(n'\lambda',k\nu)=\binom{n-k}{n'-k}Z(n\lambda,k\nu), \tag{25}$$

a relation akin to the Chapman–Kolmogorov equation. The arguments n and λ play the roles of time and space, respectively. The relation (25) with $n'=n-1$ reads

$$\sum_{\lambda'}Z(n\lambda,n-1\,\lambda')Z(n-1\,\lambda',k\nu)=(n-k)Z(n\lambda,k\nu). \tag{26}$$

The quantity $Z(n\lambda,n-1\,\lambda')$ here is written as

$$Z(n\lambda,n-1\,\lambda')$$
$$=(T'1/2T_z't_z'|TT_z)^2S(nT,(n-1)T'+1), \tag{27}$$

with

$$S(nT,(n-1)T'+1)$$
$$=\frac{2T'+1}{2T+1}\times\begin{cases}n/2+T+1, & \text{for } T'=T-\frac{1}{2},\\ n/2-T, & \text{for } T'=T+\frac{1}{2},\end{cases} \tag{28}$$

where in (27) we have used (3) and (5), and in (28) the fixed-isospin sum rule on one-particle cfp's.[4] Substituting the explicit form of $Z(n\lambda,n-1\,\lambda')$ into (26), we get the recurrence relation,[8] i.e., the difference equation for Z to be solved under the condition

$$Z(k\nu,k\nu')=\delta(\nu,\nu'), \tag{29}$$

a property of Z seen from (9). The difference equation for Z is translated into that for R as

$$(n-k)(2T+1)R(nT,kt,r)$$
$$=(n/2+T+1)\sqrt{(2T-r)(2T+r+1)}$$
$$\times R(n-1\,T-\tfrac{1}{2},kt,r)$$
$$+(n/2-T)\sqrt{(2T-r+1)(2T+r+2)}$$
$$\times R(n-1\,T+\tfrac{1}{2},kt,r). \tag{30}$$

III. THE SUM OF FIXED-ISOSPIN SPECTROSCOPIC FACTORS

Here we express S, defined by (5), in a compact form that leads to a new expression for each of Z and R.

Let us show the recurrence relation for S

$$S(nT,kt+(n-k)t')$$
$$=S((n-k)t',(n-k-1)t''+1)^{-1}\sum_{T'}(2t'+1)$$
$$\times(2T'+1)\,W(tTt''1/2;t'T')^2$$
$$\times S(nT,(n-1)T'+1)$$
$$\times S((n-1)T',kt+(n-k-1)t''). \tag{31}$$

Proof: We start from the identity

$$\langle i|C_\gamma A(ktt_z\omega)|j\rangle=(-1)^k\langle i|A(ktt_z\omega)C_\gamma|j\rangle, \tag{32}$$

where $\langle i|=\langle(n-k-1)t''t_z''\omega''|$ and $|j\rangle=|nTT_z\mu\rangle$. Expressing both sides of (32) in terms of cfp's, we factorize isospin z components in cfp's. Then, we get

$$\sum_{\omega'}\langle kt\omega+(n-k)t'\omega'|nT\mu\rangle\langle (n-k-1)t''\omega''+1\omega_0|(n-k)t'\omega'\rangle$$
$$=\sum_{T'}(-1)^{t-t''+T'}\sqrt{(2t'+1)(2T'+1)}\,W(tTt''\tfrac{1}{2};t'T')$$
$$\times\sum_{\mu'}\langle kt\omega+(n-k-1)t''\omega'|(n-1)T'\mu'\rangle\langle (n-1)T'\mu'+1\omega_0|nT\mu\rangle\,. \tag{33}$$

Let us square the expression on each side and subsequently impose the sum over ω, ω'', and ω_0 on each side. After carrying out the sum by virtue of (18), we obtain (31).

The recurrence relation (31) leads to

$$S\,(nT,kt+k't')\equiv\binom{n}{k}\sum_{\omega\omega'}\langle kt\omega+k't'\omega'|nT\mu\rangle^2$$
$$=\binom{n}{k}\frac{f\,(k,t)f\,(k',t')}{f\,(n,T)}\,\Delta'(tt'T),$$

with (34)

$$f(n,T)=(2T+1)\binom{n+1}{n/2-T}(n+1)^{-1}, \tag{35}$$

where Δ' is the triangle condition among T, t, and t'.

Proof: We use induction on n with k, t, and ω being fixed. The expression (34) in the case of $n=k$ properly reduces to unity. In the case of $n=k+1$, it agrees with (28). Supposing (34) for $n=p$, let us deduce it for $n=p+1$. We substitute (34) for $n=p$ into the right-hand side (rhs) of (31) for $n=p+1$. After applying the orthogonality for the Racah coefficient to the sum over T' in (31), we obtain (34) for $n=p+1$.

The quantity $f(n,T)$ is just the dimensionality of the representations described by the same Young tableau[11] $[h_1,h_2]$, where $h_1-h_2=2T$ and $h_1+h_2=n$. In the branching diagram,[4] it is illustrated as the number of allowed paths between (n,T) and $(1,\frac{1}{2})$. We can elucidate (34) by the branching rule of the Young tableau.

The relation (34) is viewed as the sum rule on spectroscopic factors with the involving isospin quantum numbers being fixed. Rewriting it in terms of intrinsic spins in place of isospins, we get the fixed-intrinsic-spin sum rule, which gives much fine information in comparison with the sum rule discussed in Ref. 15.

The relation (34) incorporated into (3) and (23), respectively, yields

$$Z\,(nTT_z,ktt_z)=\sum_{t'}(tt't_zt_z'|TT_z)^2\binom{n}{k}\frac{f(k,t)f(k',t')}{f\,(n,T)} \tag{36}$$

and

$$R\,(nT,kt,r)$$
$$=(2T+1)\sum_{t'}W\,(Trt't;Tt)\binom{n}{k}\frac{f(k,t)f(k',t')}{f(n,T)}. \tag{37}$$

We see from (34), (36), and (37) that all of S, Z, and R are independent of N, the number of the single-particle states. The relation (36) shows that the quantity

$$Z\,(nTT_z,ktt_z)f(n,T)/\{(2T+1)f(k,t)\} \tag{38}$$

is symmetric with respect to the simultaneous interchanges of T and t and of T_z and t_z. The similar symmetry property of R is found from (37). The expression of R given by (37) should be a solution of (30), though it is far from obvious.

IV. THE PROPAGATION COEFFICIENT Z

In this section, the propagation coefficient Z is shown to be a very simple form in the case of $|T_z|=T$, regardless of the magnitude of T relative to n.

In order to deduce another type of expression (equation) for Z other than (36) [(26)], we utilize the p-n formalism that deals with protons and neutrons separately. The state operator of creating k_1 neutrons and k_2 protons is written as

$$A^+\,(k_1k_2\omega_1\omega_2)=A^+\,(k_1\omega_1)A^+\,(k_2\omega_2), \tag{39}$$

where ω_1 stands for all the quantum numbers specifying the k_1-body state with isospin $t_1=t_{1z}=k_1/2$, and ω_2 for the quantum numbers specifying the k_2-body state with $t_2=-t_{2z}=k_2/2$. The operator defined by (39) produces the state $|k_1k_2\omega_1\omega_2\rangle$ of k_1 neutrons and k_2 protons. Notice that this state does not have a definite isospin.

The operator specified in the framework of isospin formalism is expanded as

$$A^+\,(ktt_z\omega)A\,(kt't_z'\omega')$$
$$=\sum A^+\,(k_1k_2\omega_1\omega_2)A\,(k_1'k_2'\omega_1'\omega_2')$$
$$\times\langle k_1k_2\omega_1\omega_2|\,A^+\,(ktt_z\omega)A\,(kt't_z'\omega')|k_1'k_2'\omega_1'\omega_2'\rangle, \tag{40}$$

where

$$k_1+k_2=k\quad\text{and}\quad(k_1-k_2)/2=t_z, \tag{41}$$

and the same condition is applied to k_1' and k_2'. The sum on the rhs of (40) is taken over ω_1, ω_2, ω_1', and ω_2'.

There exists the completeness relation

$$\sum_{\mu T}|nTT_z\mu\rangle\langle nTT_z\mu|$$
$$=\sum_{\mu_1\mu_2}|n_1n_2\mu_1\mu_2\rangle\langle n_1n_2\mu_1\mu_2|, \tag{42}$$

where T on the lhs runs over $|T_z|$, $|T_z|+1$, ..., and $n/2$. The condition (41) is applied to n_1 and n_2. By virtue of (42), the transformation coefficient on the rhs of (40) fulfills

$$\sum_{\omega_1\omega_2}\langle k_1k_2\omega_1\omega_2|\,A^+\,(ktt_z\omega)A\,(kt't_z'\omega')|k_1k_2\omega_1\omega_2\rangle$$
$$=\delta(\gamma,\gamma'), \tag{43}$$

where $\gamma = (tt_z\omega)$, etc.

By virtue of (40) and (42), the T_z-fixed trace in the framework of the isospin formalism is transformed as

$$\sum_{\mu T}\langle nTT_z\mu|A^+(ktt_z\omega)A(kt't'_z\omega')|nTT_z\mu\rangle$$
$$=\sum\langle n_1n_2\mu_1\mu_2|A^+(k_1k_2\omega_1\omega_2)$$
$$\times A(k'_1k'_2\omega'_1\omega'_2)|n_1n_2\mu_1\mu_2\rangle$$
$$\times\langle k_1k_2\omega_1\omega_2|A^+(ktt_z\omega)A(kt't'_z\omega)|k'_1k'_2\omega'_1\omega'_2\rangle\,, \tag{44}$$

where the sum on the rhs is taken over $\mu_1,\mu_2,\omega_1,\omega_2,\omega'_1$, and ω'_2. The sum over μ_1 (μ_2) is done by virtue of

$$\sum_{\Gamma}\langle n\Gamma|A^+(k\gamma)A(k\gamma')|n\Gamma\rangle=\delta(\gamma,\gamma')\binom{N/2-k}{n-k}, \tag{45}$$

the well-known relation[1] that describes the propagation of the operator average with all the quantum numbers being summed. The subsequent sum over ω_1 and ω_2 on the rhs of (44) is done by virtue of (43). Then, we get

$$\sum_{\mu T}\langle nTT_z\mu|A^+(ktt_z\omega)A(kt't'_z\omega')|nTT_z\mu\rangle$$
$$=\delta(\gamma,\gamma')\binom{N/2-k_1}{n_1-k_1}\binom{N/2-k_2}{n_2-k_2}. \tag{46}$$

Applying (9) to the lhs and using (41) to rewrite the rhs in the framework of the isospin formalism, we get the basic equation for Z, in a form different from (26),

$$\sum_T Z(nTT_z,ktt_z)d(nTT_z)$$
$$=d(ktt_z)\binom{(N-k)/2-t_z}{(n-k)/2+T_z-t_z}$$
$$\times\binom{(N-k)/2+t_z}{(n-k)/2-T_z+t_z}\equiv F(N). \tag{47}$$

The isospin T in any of (44), (46), and (47) runs over $|T_z|$, $|T_z|+1,\ldots$, and $t+(n-k)/2$.

It is a little too involved to solve (47) in a direct way. A device to circumvent the difficulty is to substitute $N=n+2T_0$ into (47), where T_0 is a parameter fulfilling

$$|T_z|<T_0<t+(n-k)/2,$$

and

$$T_0+n/2=\text{an integer}. \tag{48}$$

This device yields

$$\sum_T Z(nTT_z,ktt_z)d(nTT_z)_{N=n+2T_0}=F(n+2T_0), \tag{49}$$

where T runs only over $|T_z|$, $|T_z|+1,\ldots$, and T_0. Derivation of (49) relies first on N independence of Z, and second on vanishing of $d(nTT_z)$ with $N=n+2T_0$ in the case of $T_0<T$: These are distinctive features seen in (36) and (10), respectively.

Let us put $T_0=T_z$ (>0) in (49) so that T can be fixed uniquely to T_z. Then, it follows that

$$Z(nTT,ktt_z)$$
$$=(2t+1)\binom{(n-k)/2+T+t_z}{2T}\binom{n/2+T+1}{k/2+t+1}$$
$$\times\binom{n/2+T+1}{k/2-t}\left\{(2T+1)\binom{n/2+T+1}{2T+1}\right\}^{-1}$$
$$\times\Delta(t-|t_z|). \tag{50}$$

The simple form on the rhs is worth notice. The rhs should be identical to that of (36) with $T_z=T$, though apparently very different. Combining (36) and (50) yields a novel identity involving a CG coefficient squared. The rhs of (50) with $k=1$ reduces to $n/2\pm T$ for $t_z=\pm\frac{1}{2}$, respectively, as is expected from the numbers of neutrons and protons of the state with $T_z=T$. Using (13) and (50), we get

$$Z(nTT_z,ktt)=\binom{(n-k)/2+t+T_z}{2t}\binom{n/2-T}{k/2-t}$$
$$\times\binom{n/2+T+1}{k/2-t}\binom{k/2+t+1}{2t+1}^{-1}$$
$$\times\Delta(T-|T_z|)\,. \tag{51}$$

In this derivation, the N independence of the rhs is associated with the property that the rhs of (13), comprised of N-dependent factors, does not depend on N. The symmetry property of the quantity (38) can be checked from (50) and (51).

The expression (50) together with the symmetry

$$Z(nT(-T_z),kt(-t_z))=Z(nTT_z,ktt_z), \tag{52}$$

seen from (3) or (36), will facilitate the statistical treatment of nuclear spectroscopy, since low-lying states in the nucleus are characterized by $T=|T_z|$, i.e., the lowest isospin.

The propagation coefficient $Z(nTT_z,ktt_z)$ in case T_z is close to T is easily deduced from (50) being incorporated with the recurrence

$$(T+T_z)(T-T_z+1)Z(nT(T_z-1),ktt_z)$$
$$=2\{T(T+1)-T_z^2-t(t+1)+t_z^2\}Z(nTT_z,ktt_z)$$
$$-(T-T_z)(T+T_z+1)Z(nT(T_z+1),ktt_z)$$
$$+(t-t_z)(t+t_z+1)Z(nTT_z,kt(t_z+1))$$
$$+(t+t_z)(t-t_z+1)Z(nTT_z,kt(t_z-1)), \tag{53}$$

which results from the n-body average of the identity

$$T_+A^+AT_-$$
$$=T_-T_+A^+A-T_-A^+AT_++A^+AT_+T_-$$
$$-[T_-[T_+,A^+]]A-A^+[T_-[T_+,A]]$$
$$-[T_+,A^+][T_-,A]-[T_-,A^+][T_+,A]. \tag{54}$$

V. THE PROPAGATION COEFFICIENT R

In this section, types of expressions for R are shown, which are apparently different from (37).

Substituting (50) into (21) yields

$$R(nT,kt,r) = \sqrt{(2t+1)(2T-r)!(2T+r+1)!} \times\left(\frac{n}{2}-T\right)!\left\{\left(\frac{k}{2}-t\right)!\left(\frac{n-k}{2}+T+t+1\right)!\right\}^{-1} \times\binom{n/2+T+1}{k/2+t+1}\sum_{t_z}(trt_z0|tt_z) \times\binom{(n-k)/2+T+t_z}{2T}, \tag{55}$$

where the CG coefficient is explicitly written as[16]

$$(trt_z0|tt_z) = \sqrt{(2t+1)}\{(2t-r)!(2t+r+1)!\}^{-1/2} \times\sum_u(-1)^u\frac{(r+u)!(2t-u)!}{u!(r-u)!}\binom{t-t_z}{u}\Delta(t+t_z). \tag{56}$$

The explicit form of $(TrT0|TT)$ is used to get (55). We notice that the CG coefficient of the form (56) appears also in the Gram expansion of the finite rotation operator.[17]

French[2] got the following result [(7.40) of Ref. 2 being multiplied by $\sqrt{2t+1}$]:

$$R(nT,kt,r) = \frac{(2t+1)\,(n/2-T)!(n/2+T+1)!}{(n-k)!(k/2-t)!(k/2+t+1)!}\sqrt{\frac{(2t-r)!(2T+r+1)!}{(2T-r)!(2t+r+1)!}} \times\sum_p\frac{(-1)^p(2t-p)!(n+2t-r-k-2p)!}{p!(2t-r-2p)!((n-k)/2+t-T-p)!((n-k)/2+T+t+1-p)!}. \tag{57}$$

Comparing (55) with (57), we find that the involved sum in the latter is rewritten in the former as the sum of products comprised of a CG coefficient and a binomial coefficient.

Let us carry out the sum over t_z on the rhs of (55). We discuss first the case when the value of r is only a few. Expanding the CG coefficient as a polynomial in t_z, we make use of the identity

$$\sum_{k=-t}^{t}\binom{n+k}{r} = \binom{n+1+t}{r+1} - \binom{n-t}{r+1}, \tag{58}$$

etc. to sum over t_z. The result for $r<2$ is summarized as

$$R(nT,kt,r) = (2t+1)(2T+1)\sqrt{\frac{(2t-r)!(2T-r)!}{(2t+r+1)!(2T+r+1)!}} \times R'(nT,kt,r), \tag{59a}$$

$$R'(nT,kt,r) = \left\{\binom{n/2+T+1}{k/2+t+1}\binom{n/2-T}{k/2-t}B_+ - \binom{n/2+T+1}{k/2-t}\binom{n/2-T}{k/2+t+1}B_-\right\}(2T+1)^{-1}, \tag{59b}$$

$$B_\pm = \begin{cases}1, & \text{for } r=0,\\ -(n-k+1)\pm(2t+1)(2T+1), & \text{for } r=1,\\ 3(n-k+1)^2\mp 3(n-k+1)(2t+1)(2T+1) + 16t(t+1)T(T+1), & \text{for } r=2.\end{cases} \tag{59c}$$

The expression for the case of $r=0$ is particularly simple. It was deduced previously in Ref. 4 as the propagation coefficient for the average with T (but not T_z) being fixed. We see from (59c) that the coefficient B_- is related to B_+ by

$$B_- = (B_+)_{T\to -T-1}. \tag{60}$$

From this we find that two terms in the bracket on the rhs of (59b) are interchanged under the replacement of T by $-T-1$ and that the quantity R' is a polynomial in $T(T+1)$.

Next, let us extend (59c) to a general r, retaining (59a) and (59b) as they are. In order to get the form of B_+, we consider the case of $(n-k)/2<T+t$, under which the term containing B_- vanishes. Substituting (56) into (55), let us sum over t_z as

$$\sum_{t_z}\binom{t-k}{u}\binom{(n-k)/2+T+t_z}{2T}\Delta(t+t_z) = \binom{(n-k)/2+T+t+1}{2T+u+1}, \tag{61}$$

valid only for $(n-k)/2<T+t$ because of existence of $\Delta(t+t_z)$ on the lhs. We then get

$$B_+ = \sum_{u=0}^{r}(-1)^u\frac{(2t-u)!(r+u)!}{(2t-r)!} \times\binom{(n-k)/2-T+t}{u}\binom{2T+r+1}{r-u}. \tag{62}$$

We easily deduce the expression of B_-, applying (60) to (62).

The implication of (59a)–(59c) as well as (57) is elucidated in terms of solutions of the difference equation (30). As discussed in Ref. 4, every term with the definite p on the rhs of (57) fulfills (30). The linear combination of terms with various p's is to reproduce the condition (29). Each term with the definite p describes information propagation between the points (n,T) and $(k+r-2t+2p,r/2)$ in the branching diagram.

The expressions (59a)–(59c) are split into two parts, one containing only B_+ and the other containing B_-. We find that each part fulfills (30). The part containing B_+ can be interpreted as information propagation between the points (n,T) and (k,t) in the branching diagram, as is seen from the vanishing of the expression in the case of $n=k$ and $T\neq t$. Similarly, the part containing B_- describes the connection between (n,T) and $(k,-t-1)$. Notice that the point $(k,-t-1)$ is unphysical, since the isospin is negative. It is a mirror image of (k,t) with respect to the isospin $= -\frac{1}{2}$ axis. The mirror image takes the place of the condition that propagation should be confined to the region of isospin$>r/2$. The symmetry (60) implies that the propagation between (n,T) and $(k,-t-1)$ is the same as that between $(n,-T-1)$ and (k,t).

New expressions (37), (55), and (59) as well as the previous result (57), which are given to the same quantity R, are to be algebraically identical to each other, though we hardly see it except in the case of $r=0$, the isoscalar operator. Combining two of them leads to a novel identity among n-j symbols and factorials. From the first two, for example, we get, for any integer p,

$$\sum_{t'} \frac{(2t'+1)\,W([\text{illegible}]\,T)}{(p-t')!(p+t'+1)!} = \sqrt{\frac{(2T-r)!(2T+r+1)!}{(p+T+t+1)!(p+T-t)!}} \sum_{t_z} (trt_z0|tt_z) \times \binom{p+T+t_z}{2T} (2t+1)^{-1/2}, \qquad (63)$$

a relation somewhat akin to the novel identity (1) of Ref. 18.

VI. THE FIXED-ISOSPIN AVERAGE OF THE OPERATOR PRODUCT

The operator subject to the average is often a power of an operator or a product of several operators so that the direct use of (2) would not be available. Here, we present a prescription to deal with the n-body average of the operator

$$O \equiv A^+(n_1\gamma_1)\,A(n_1'\,\gamma_1')\,A^+(n_2\gamma_2)\cdots A(n_f'\,\gamma_f'), \qquad (64)$$

where $\gamma_1 = T_1 T_{1Z}\,\mu_1$, etc. and any of n_1, n_2, etc. can be zero. Usually, it is easy to express a given operator as a sum of O's.

Averaging the operator O in the n-body space requires three steps. First, the extended Wick's theorem[13] is applied to O to rewrite it as a sum of normal products. The resultant expression is summarized in (38) of Ref. 13. The normal product that is relevant to the operator average is of the form (7) being subdivided as

$$O(k) = O_+(k)O_-(k) \qquad (65)$$

with

$$O_+(k) = A^+(k_1\alpha_1)\,A^+(k_2\alpha_2)\cdots A^+(k_f\,\alpha_f) \qquad (66a)$$

and

$$O_-(k) = A(k_1'\,\alpha_1')\,A(k_2'\,\alpha_2')\cdots A(k_f'\alpha_f'), \qquad (66b)$$

where $\alpha_1 = t_1 t_{1z}\omega_1$, etc. The sum of $k_1, k_2, \cdots$, and k_f is equal to k, the sum of k_1', k_2',$\cdots$, and k_f'. Next, the reduction relation (2) is applied to the n-body average of $O(k)$ so as to express it as a sum of k-body averages. The last step concerns modification of the resultant k-body average, which is rearranged as

$$\langle O(k)\rangle^{ktt_z} = \langle 0|O_-(k)P(tt_z)O_+(k)|0\rangle / d(ktt_z), \qquad (67)$$

where $P(tt_z)$ stands for projection operator onto the isospin-tt_z space. Operating $P(tt_z)$ to the right, we state it in terms of CG coefficients so that Wick's theorem can be applied to the antinormal ordering form O_-O_+ in the matrix element. The n-body average of O is finally given as a sum of vacuum expectation values.

Notice that rearranging any operator as a sum of isotensors is not required in the first two steps. This advantage becomes remarkable when the number of A^+'s and A's consisting of O increases.

VII. USE OF *Z* AND *S* AS SPECTROSCOPIC FACTORS FOR NUCLEAR FRAGMENTATION

Here, we point out that it is available to use Z and S' as formation factors of target (or projectile) residues in high energy nuclear reaction.

Let us consider the case when the [illegible] target nucleus in any state with isospin (T,T_z) is fragmented into two pieces (residues), one being a cluster of k nucleons coupled to isospin (t,t_z) and the other a cluster of k' $(=n-k)$ nucleons coupled to (t',t_z'). We assume that specification of the target residue by angular momentum, etc. other than isospin is not a matter of concern, which will be meaningful in analyses of mass and charge distributions of target (projectile) residues.[19]

We set up three types of spectroscopic factors for the fragmentation. The first, used in the prevalent treatment,[19] is of the hypergeometric expression

$$\binom{n_1}{k_1}\binom{n_2}{k_2}. \qquad (68)$$

The condition (41) is applied to n_1, k_1, etc. That is, the argument n_1 (n_2) implies the number of neutrons (protons) of the target nucleus and k_1 (k_2) the number of neutrons (protons) of one of residues. The second and the third types of spectroscopic factors are $Z(nTT_z, ktt_z)$ and $S'(nTT_z, ktt_z + k't't_z')$. These are to be used for inclusive and exclusive analyses, respectively, regarding isospin t.

None of the isospins t and t' are taken into account in (68) contrary to Z or S'. From this, we infer the spectroscopic sum rule

$$\sum_t Z(nTT_z, ktt_z) = \binom{n_1}{k_1}\binom{n_2}{k_2}. \qquad (69)$$

To prove it, we have only to substitute (13) into the lhs of (47) after replacing n and k in (47) by $N-n$ and $N-k$, respectively. The relation (69) linked with (36) leads to a novel identity containing a CG coefficient squared.

It has little been discussed how the fragmentation process is influenced by the magnitude of t relative to that of t_z. It is, however, known that the nucleus with $t>|t_z|$ has fairly large excitation energy and is unstable as to particle decays. Considering this, we present an extreme model that target residues are formed mainly under the condition $|t_z|=t$. The relevant spectroscopic factor is expressed as

$$Z(nTT,kt\,|t_z|=t) = \frac{(n_1+1)(k'-k''+1)}{(k'+1)(n_1-k''+1)}\binom{n_1}{k_1}\binom{n_2}{k_2}, \qquad (70)$$

where k' (k'') indicates the larger (smaller) number between k_1 and k_2. The relation (50) is used to get (70). Comparing (70) with (68), we find that the former leads to a slight enhancement of production of β-unstable nuclei. Existing experimental data is not sufficient to support further discussion on this point.

It is easy to extend (34) so as to get spectroscopic factors for a nucleus being fragmented into several clusters. We present here only the expression

$$
\begin{aligned}
S(nTT_z, & k_1t_1t_1 + k_2t_2t_2 + \cdots + k'_1t'_1(-t'_1) \\
& + k'_2t'_2(-t'_2) + \cdots) \\
& = (tt't - t'|TT_z)^2 \\
& \times \frac{n!f(k_1,t_1)f(k_2,t_2)\cdots f(k'_1,t'_1)f(k'_2,t'_2)\cdots}{k_1!k_2!\cdots k'_1!k'_2!\cdots f(n,T)}, \quad (71)
\end{aligned}
$$

where t (t') stands for the sum of $t_1, t_2, \cdots$ ($t'_1, t'_2, \cdots$).

[1]J. B. French, in *Nuclear Structure*, edited by A. Hossain, Harum-ar-Raschid, and M. Islam (North-Holland, Amsterdam, 1967), p. 85.

[2]J. B. French, in *Isospin in Nuclear Physics*, edited by D. H. Wilkinson (North-Holland, Amsterdam, 1969), p. 259.

[3]M. Nomura, Prog. Theor. Phys. **47**, 1858 (1972).

[4]M. Nomura, Prog. Theor. Phys. **51**, 489 (1974).

[5]K. T. Hecht and J. P. Draayer, Nucl. Phys. A **223**, 285 (1974).

[6]J. P. Draayer, in *Theory and Applications of Moment Methods in Many-Fermion Systems*, edited by B. J. Dalton, S. M. Grimes, J. P. Vary, and S. A. Williams (Plenum, New York, 1980), p. 213.

[7]C. Quesne, in Ref. 6, p. 257.

[8]M. Nomura, J. Math. Phys. **26**, 965 (1985).

[9]D. Walikainen and R. B. McQuistan, J. Math. Phys. **26**, 815 (1985).

[10]de-Shalit and I. Talmi, *Nuclear Shell Theory* (Academic, New York, 1963).

[11]I. G. Kaplan, *Symmetry of Many-Electron Systems*, translated by J. Gerratt (Academic, New York, 1975).

[12]M. Nomura, Prog. Theor. Phys. **59**, 451 (1978).

[13]M. Nomura, J. Math. Phys. **26**, 732 (1985).

[14]G. Racah, Phys. Rev. **76**, 1352 (1949).

[15]E. McGuire, Phys. Rev. A **10**, 32 (1974).

[16]A. R. Edmonds, *Angular Momentum in Quantum Mechanics* (Princeton U. P., Princeton, NJ, 1957), p. 44.

[17]W. Happer, Ann. Phys. NY **48**, 579 (1968).

[18]B. I. Dunlap and B. R. Judd, J. Math. Phys. **16**, 318 (1975).

[19]D. J. Morrissey, W. R. Marsh, R. J. Otto, W. Loveland, and G. T. Seaborg, Phys. Rev. C **18**, 1267 (1978).

VOLUME 58, NUMBER 23 PHYSICAL REVIEW LETTERS 8 JUNE 1987

Symmetries and Quantum Chaos: Time-Reversal Invariance in the Nucleon-Nucleon Interaction

J. B. French, V. K. B. Kota,[a] A. Pandey, and S. Tomsovic[b]

Department of Physics and Astronomy, University of Rochester, Rochester, New York 14627

(Received 9 February 1987)

Let a be the relative norm of a symmetry-breaking term in the Hamiltonian of a many-particle system, and Λ the energy-dependent transition parameter which characterizes the quantum chaos via spectral and strength fluctuations. Combining a compact theory for Λ/a^2 with fluctuation theories by which Λ can be deduced from (neutron-resonance) data gives, for the time-reversal nonvariant nucleon-nucleon interaction, $a \leq (1-2)\times 10^{-3}$, which would improve with better small-strength data in nuclei with dense spectra. Diffusion equations involving Λ as the "time" variable are also discussed.

PACS numbers: 24.60.Ky, 05.45.+b, 11.30.Er, 21.30.+y

Our principal purpose is to understand the way in which the nature of quantum chaos, as characterized by spectral (energy level) and strength fluctuations, varies when a good symmetry is gradually broken, i.e., as $H=h(1)+V(k)\rightarrow H_a=H+aU(p)$, where $V(k)$ is a k-body operator; we shall often have $k=p=2$ and will then define the symmetry-breaking operator U to have, in the two-particle space, the same norm as the symmetry-preserving V. The choice of U to be representable by an imaginary antisymmetric matrix will accommodate our first application, that of determining from observed nuclear quantum chaos values for or bounds upon $a\geq 0$, the relative time-reversal–nonvariant (TRNI) component of the nucleon-nucleon interaction.[1] The problem breaks into two parts.[2] The $a\leftrightarrow\Lambda$ part is to calculate, from H_a, the energy-dependent transition parameter $\Lambda=a^2|U_{ij}|^2/D^2$, the mean squared admixing matrix element between neighboring states measured in units of the local spacing D; our solution of this problem will apply, or extend immediately, to symmetries other than time-reversal invariance (TRI) and to the more general problem of calculating excitation strengths in complicated systems. The "Λ problem" is that of deducing, from the data, values or bounds for Λ. This is simple when, as, e.g., with parity, the symmetry breaking is detected by the occurrence of an otherwise forbidden process, but is complicated when, as with TRI, Λ derives from fluctuations[2] or via detailed-balance experiments.[3] Though the mere existence of a nonzero Λ could be of interest, indicating the occurrence of a symmetry breaking, the significance, *if any*, of an upper bound on Λ will usually not be obvious, so that the further $\Lambda\rightarrow a$ analysis is essential.

From its definition we recognize that Λ/a^2 is the locally smoothed (sm) value, for $x\simeq y$ in the energy region of interest, of the bivariate strength density $S(y,x)$, defined with U as the excitation operator; thus, with m the number of "active" particles (those in single-particle states neither essentially filled nor empty), Γ the conserved quantum numbers ($=J$, J_z, T, T_z, π when one deals with nuclear TRI, as we shall mostly take for granted in what follows),

$$\begin{aligned} S^{H,m,\Gamma}(y,x) &= I^{H,m,\Gamma}(y)\,|\langle y|U|x\rangle|^2 I^{H,m,\Gamma}(x) \\ &= \langle\langle U^\dagger\delta(H-y)U\delta(H-x)\rangle\rangle^{m,\Gamma} \\ &\underset{\text{sm}}{\longrightarrow} \Lambda/a^2. \end{aligned} \tag{1}$$

Here $\langle\langle\ \rangle\rangle$ indicates the trace, $|x\rangle$ and $|y\rangle$ the H eigenstates, and $I^{H,m,\Gamma}(x)=\langle\langle\delta(H-x)\rangle\rangle^{m,\Gamma}$ gives, when smoothed, the level density $[D^{m,\Gamma}(x)]^{-1}$. It is good to start with noninteracting particles (NIP), $H\rightarrow h$; then, in h-diagonal representation, when $m\gg p$ the matrix elements of $U(p)$ necessarily lie in a narrow diagonal band [since $U(p)$ simultaneously transfers only a fraction $\leq p/m$ of the particles]. This implies that the x,y variables in the bivariate density $S^{h,m,\Gamma}$ are strongly correlated, the density peaking along the line $x\simeq y$. Since the h eigenstates are m-fold products of single-particle eigenstates it is not difficult, even when good symmetries are specified, to give the *exact* strength density, in terms of eigenvalue densities, for an arbitrary excitation operator. But it will be adequate for us to deal with large spaces for which we can disregard those matrix elements of U, e.g., U_{ijik}, which transfer fewer than p particles. [Ignoring Γ to begin with, we find a propagation equation for the m-particle strength density in the form of a bivariate convolution,

$$\begin{aligned} S^{h,m} &= T^{h,m-p}\otimes S^{h,p}, \\ T^{h,m}(u,v) &= \delta(u-v)I^{h,m}(u). \end{aligned} \tag{2}$$

To appreciate the nature of this result take, for example, U to have p-particle matrix elements essentially equal in magnitude, the single-particle spectrum to be finite and nonsingular, and m large enough so that, by the central limit theorem,[4] $I^{h,m}$ is Gaussian. Then using additivity of bivariate cumulants under convolution, we find $S^{h,m}$ to be close to an "almost singular" bivariate Gaussian, almost singular because the x,y correlation coefficient is $\zeta_h=1-p/m$ so that the distribution for large m becomes extremely sharp in the variable $x-y$, as predicted above.

The strong positive correlation between the variables x,y survives when V is turned on because V admixes and spreads the h eigenstates over a narrow energy range, typically ± 2 MeV for heavy nuclei in the resonance region. This gives an enhancement of the admixing matrix element and an increase in the sensitivity of deducing α from Λ. A separate enhancement is sensitivity[2] arises from the $d^{-1/2}$ compression of the level spacing D due to the effect of distant levels (d here being the "effective" dimensionality, roughly the number of interacting states in the spreading interval above). Without these enhancements no interesting α bound would emerge from TRI fluctuation studies.

It is known that the spreading is Gaussian and moreover[5] that the spreading width is closely constant for excitation energies of interest to us (at higher energies it increases slowly with energy). The initial and final state densities have therefore the form $I^{H,m}=I^{h,m}\otimes\rho_G^{(1)}$ where $\rho_G^{(1)}$ is (univariate) Gaussian with unit integral, zero centroid, and m-particle variance $\sigma^2(m)$, related[4] to the defining k-particle variance by $\sigma^2(m)=\binom{m}{k}\sigma^2(k)$. For a heuristic derivation of S we assume now that the strength "follows" the densities so that the interacting-particle strength density becomes a convolution of the NIP density with a (unit integral) bivariate Gaussian spreading function $\rho_G^{(2)}$; thus

$$S^{H,m}=S^{h,m}\otimes\rho_G^{(2)}$$
$$=\{\langle\langle U^\dagger U\rangle\rangle^m\}^{-1}S^{h,p}\otimes T^{h,m-p}\otimes S^{V,m}, \qquad (3)$$

where the final form comes by consideration of $h=0$. With $h\rightarrow V$ in (1) we have, as the exact moments of $S^{V,m}$, $M_{rs}=\langle\langle V^rU^\dagger V^sU\rangle\rangle^m$ which, with $r+s\le 2$, determines the six Gaussian parameters. (Note incidentally that for small spaces $S^{H,m}$ will itself be a bivariate Gaussian.[6]) Since for TRI we have no form for U it is natural to regard it as a random operator, specifically, one defined in the p-particle space by an imaginary antisymmetric random matrix, and to perform then the ensemble average. Thus, using (1), we have

$$S^{h,p}(u,v)=d^{-2}(p)\langle\langle U^\dagger U\rangle\rangle^p I^{h,p}(u)I^{h,p}(v).$$

Moreover, for $r=0$ or $s=0$ the moment traces factor and the marginal densities of $S^{V,m}$ coincide with $d(m)^{-1}\langle\langle U^\dagger U\rangle\rangle^m I^{V,m}$ in line with the heuristic argument. For the correlation coefficient ζ_V we need[4]

$$\langle\langle U^\dagger VUV\rangle\rangle^m\rightarrow\langle\langle U^\dagger U\rangle\rangle^m\sigma^2(m-p),$$

and then

$$\zeta_V\rightarrow\sigma^2(m-p)/\sigma^2(m)\sim 0.80\pm 0.05$$

for the spaces considered below. The intuitive argument involved here is that, since U_{ij} must couple with $U_{ij}^\dagger$, the number of particles "available" for the operation of V is reduced by p; a formal derivation gives also a Pauli-blocking correction. We mention too that a derivation of (3) follows by our taking V to be a k-body embedded[4] Gaussian orthogonal ensemble (GOE) (i.e., defined in the k-particle space but operating in an $m>k$-particle space) and then evaluating the low-order moments for large m; shell-model and partitioning tests have shown that this ensemble is adequate for our spaces. The variances need to evaluate $S^{H,m}$ follow directly[5] from the matrix elements of V, and the NIP eigenvalue densities by spectral averaging involving low-order partitioned h moments.[7]

The parity and T_z quantum numbers, dropped above, are restored by a simple partitioning. For the angular momentum the same J decomposition via a spin-cutoff factor which is encountered in the theory of level densities is appropriate also for the strength density; the cutoff factor itself is easily calculable. The spreading widths and the correlation coefficient ζ_V are essentially J independent, this arising from the fact that the correlation coefficient between V and J_z^2 (whose expectation value gives the spin-cutoff factor) is small, typically $\lesssim 0.03$. For heavy nuclei, which give the best fluctuation bounds for TRI, isospin (T) is irrelevant.

The calculations are simple, once the model space, the single-particle energies, and the interaction V have been decided upon and tested. For heavy nuclei the first two come by inspection of the zero-deformation Nilsson spectrum which, e.g., makes plausible that for ^{232}Th eight active protons should be distributed over $h_{9/2}$, $i_{13/2}$, and $f_{7/2}$, and 17 neutrons over $g_{9/2}$, $k_{15/2}$, $i_{11/2}$, and $d_{5/2}$, the gross dimensionality being $\binom{32}{8}\binom{44}{17}\sim 7\times 10^{18}$. We choose for V a surface-delta interaction, perhaps the one best studied for heavy nuclei, and verify that the calculated level density fits experiment for a reasonable interaction strength G (which turns out to be 0.17 MeV for ^{167}Er and 0.13 MeV for ^{233}Th), and is stable under small extensions of the model space. In the slow-neutron resonance region, with $J=\frac{1}{2}^+$, we find, using an early version of (3), that $\Lambda/\alpha^2=3900$, 730, 11600, and 6000, respectively, for ^{167,169}Er, ^{233}Th, ^{239}U, and, with $J=3^-$, $\Lambda/\alpha^2=1.08\times 10^5$ for ^{236}U, these being the nuclei needed below the TRI. The spacings for these nuclei are 38, 100, 17, 21, and 1.1 eV so that, at the resonance energies, $\Lambda D/\alpha^2$ is essentially constant, this implying that, other things being equal, the best α bounds will come from dense spectra (small D) found in heavy nuclei. Moreover, since D decreases with angular momentum J, odd-target ($J_t\neq 0$) nuclei, which give $J=J_t\pm\frac{1}{2}$ resonances, will do better than even targets ($J_t=0$) if the two J values can be separated, as in a polarized experiment.

We turn now to the Λ problem for the TRI, the theory for which derives from the ensemble[2] $\{H_\alpha\}$ which interpolates between the Gaussian orthogonal ($\alpha=0$) and unitary ($\alpha=1$) ensembles. Once the transition theory is known, values for or bounds on Λ can be determined directly from experimental data. This is so because the theory applies not only to ensembles but equally well to the individual realistic systems described above. For the spectral Λ problem see Ref. 2. For the strengths the main TRNI effect is that the locally renormalized strength density makes a transition from a χ_1^2 (Porter-Thomas) form to χ_2^2. Let $|\lambda\rangle$ be the renormalized projection of the final state of the excitation into the d-dimensional statistical space in which $\{H_\alpha\}$ operates.[8] For each member of the fixed-α ensemble we consider the strength to the jth H_α eigenstate, the ordering being by energy. For a simple notation we write the corresponding energy $E_{\alpha:j}$, a mildly fluctuating random variable (variance $\sim\pi^{-2}D^2\ln d$), as E, and the complex eigenvector as $|E\rangle=t|R\rangle+i(1-t^2)^{1/2}|I\rangle$ with the phase chosen so that the real and imaginary parts of $|E\rangle$ are orthogonal, $\langle R|I\rangle=0$. Renormalized to unit mean, the strength, the absolute square of the overlap between $|\lambda\rangle$ and the eigenvector, is now

$$K(E)=d|\langle\lambda|E\rangle|^2$$
$$=d\{t^2|\langle\lambda|R\rangle|^2+(1-t^2)|\langle\lambda|I\rangle|^2\}.$$

The invariance of $\{H_\alpha\}$ under orthogonal transformations insures that, for fixed t and asymptotic d, the overlaps $\langle\lambda|R\rangle$ and $\langle\lambda|I\rangle$ vary across the ensemble as independent Gaussian random variables. Introducing $\rho_K(z)$ and $\rho_\theta(u)$ as the ensemble probability densities for $K(E)$ and $\theta=4t^2(1-t^2)$, and $\rho_K(z|u)$ as the $\theta=u$ conditional strength density, we have $\rho_K(z)=\int\rho_K(z|u)\rho_\theta(u)du$. We find then, with I_0 a modified Bessel function, that

$$\rho_K(z|u)=u^{-1/2}e^{-(z/u)}I_0((1-u)^{1/2}z/u). \qquad (4)$$

Monte Carlo calculations have shown that the first-moment approximation (exact for $\Lambda=0,\ \infty$), $\rho_K(z)\rightarrow\rho_K(z|\bar{\theta})$, is more than adequate for the data analysis. To evaluate $\bar{\theta}$, the first moment of ρ_θ, we use first-order (renormalized) perturbation theory, including degenerate theory for nearest neighbors, to expand the H_α eigenstate in the H_0-diagonal basis. $\bar{\theta}$ then appears in terms of an inverse-square energy-weighted sum with a damping term to take account of near degeneracies. The final result, exact to order Λ, turns out to be $3\bar{\theta}=2\pi^2\Lambda[2-\gamma-\ln(2\pi^2\Lambda)]$, good for $\Lambda^{1/2}\lesssim0.2$.

For the strength bounds the data come from a major subset of the neutron-resonance sequences used for the spectral case,[2,9] essentially groups 1–3 of Table I in Ref. 2. We combine sequences in the same group after renormalizing the strengths to unit means. Since the χ_1^2 and χ_2^2 forms differ most significantly for small arguments, our main interest is in the small strengths. It is good therefore to consider as a measure the distribution function $F(x;\epsilon)$, the density integrated from ϵ to x, for which the sample error from p strengths is $\{F(x;\epsilon)[1-F(x;\epsilon)]/p\}^{1/2}$. Our analysis shows that the data for groups 2,3 agree with GOE for all $x\geq\epsilon$ if we choose $\epsilon=0.004$; this choice of ϵ as a cutoff is dictated by the fact that very few levels with strengths $\leq\epsilon$ have been observed. The optimal measure (which depends on the number of strengths) turns out to lie between $F(0.02;0.004)$ and $F(0.07;0.004)$ for all groups. Proceeding as in the spectral analysis with a 3σ confidence criterion we find that the $\Lambda^{1/2}$ bounds for groups 2 and 3 are essentially identical (0.17, 0.07) with the spectral bounds; uncertainties about the data prevent us at present from giving a bound for the first group. If an $\epsilon=0$ cutoff had been appropriate the strength bound would have been smaller by a factor of 4.

The GOE ($\Lambda=0$) strength analysis would suggest that a subset of the levels, randomly ordered in energy, is missing, the expected fraction being 5%, whereas the GOE spectral analysis would give essentially zero. A detailed analysis shows that consistency of these fractions has a low ($\lesssim10^{-3}$) probability. They can, however, be reconciled by the assumption of a nonzero Λ, the optimal value, with expected fraction $\simeq1.5\%$, being $\Lambda^{1/2}\simeq0.03$ which is properly less than the derived bounds. Three possible explanations for the nonzero effective Λ are errors in the data, small-strength departures from the statistical model, and the actual occurrence of TRI breaking. It would not be prudent at this time to argue that the symmetry breaking has been observed.

For the $\Lambda\rightarrow\alpha$ reduction we observe that, while the Λ/α^2 values above are for individual nuclei, the Λ bounds[2] come from groups of nuclei, the grouping being by their complexity as displayed by D and being necessary to obtain adequate sample size. The resultant spectral α bounds are $\alpha_{\rm sp}\leq(2.9,\ 3.3,\ 1.5,\ 2.6,\ 0.6)\times10^{-3}$, the ordering being as given for Λ/α^2 above. The strength bounds are $\alpha_{\rm st}\leq(2.8,\ 2.6,\ldots,\ldots,\ldots)\times10^{-3}$. It seems proper to quote, as the final bound for the relative norm of the TRNI nucleon-nucleon interaction, $\alpha\leq(1-2)\times10^{-3}$. If we were to take seriously, as a measurement of TRNI, the value (for the combined data) derived from reconciliation of the data, the α *value*, e.g., for ^{233}Th, would be 3×10^{-4}. It should be clear that to improve the α bounds (or even the values) we need more information about small strengths in complex nuclei, i.e., nuclei with small average spacings such as ^{236}U. For a theoretical review of CP violation see Wolfenstein.[10]

We have stressed the generality of our solution to the $\Lambda\leftrightarrow\alpha$ problem. The generality of the Λ theory follows from its appearance as the "time" variable in a set of hierarchic equations[11] for the n-level correlation functions. This derives in turn from Dyson's diffusion equation[12] for the joint probability distribution of the eigenvalues, with α^2 as time and β^{-1} as temperature, by use of a renormalization process to eliminate the effects of distant levels. Dyson's equation applies specifically to the ensembles $\{H_\alpha=H+\alpha U\}$ where $\{U\}$ in H-diagonal basis is a classical ensemble G_β, respectively GOE, Gaussian unitary ensemble (GUE), Gaussian sympletic ensemble (GSE) for $\beta=1,2,4$. However the renormalization process is of such generality that the hierarchic set should apply to any $\{H_\alpha\}$ in which $\{U\}$, in H-diagonal basis, is *locally* G_β. For finite Λ, approximate[2] and complete[13] solutions are available for GOE $\to$ GUE as well as GSE $\to$ GUE. For the important Poisson $\to G_\beta$ transitions (relevant,[14] e.g., with $\beta=1$, for atomic L,S symmetries) the small-Λ solution gives, for the number variance,[7]

$$\Sigma^2(r;\Lambda) \to \Sigma^2(r;0) - 2\beta\Lambda\{\ln(r^2/2\Lambda) - \psi(\beta/2+1)+1\}$$

with ψ the digamma function.

Since the classical ensembles involve only multibody interactions, it is important to ask whether the equations extend to the embedded ensembles defined above. There are strong theoretical arguments and experimental indications that the local-G_β assumption is valid for these ensembles. Then the above $\{H_\alpha\}$, involving a realistic Hamiltonian, when embedded[4] in m-particle spaces, should yield the same results for all $m \geq p > 1$ even though the global invariance of $\{U\}$ is lost for $m > p$. The NIP case, $m > p = 1$, is exceptional (physically because level crossing is not forbidden); for example, with H also NIP, it is described by the same equations with $\beta \to 0$, the large-Λ solution being then Poisson. This many-particle behavior is consistent with recent numerical and semiclassical theoretical results[15] for systems with few degrees of freedom, not surprising since NIP relates to an integrable classical system and interacting particles almost always, to a chaotic one.

A full account will be ready shortly. We acknowledge computing help from the University of Rochester High Energy Group and information about the data from R. U. Haq and P. Ribon. The work has been supported in part by the U. S. Department of Energy.

[a]On leave from Physical Research Laboratory, Ahmedabad, India.

[b]Presently at Institut de Physique Nucléaire, Orsay, France.

[1]E. P. Wigner, SIAM Rev. **9**, 1 (1967).

[2]J. B. French, V. K. B. Kota, A. Pandey, and S. Tomsovic, Phys. Rev. Lett. **54**, 2313 (1985). This paper deals with the spectral Λ theory and data analysis for nuclear time-reversal invariance (TRI); it does not deal with strengths and uses a crude theory for the $\Lambda \leftrightarrow \alpha$ reduction, giving an α bound much larger than the present value.

[3]E. Blanke, H. Driller, W. Glöckle, H. Genz, A. Richter, and G. Schrieder, Phys. Rev. Lett. **51**, 355 (1983); D. Boosé, H. L. Harney, and H. A. Weidnmüller, Phys. Rev. **56**, 2012 (1986). The latter paper deals only with a the Λ problem; we shall later derive the α bound.

[4]K. K. Mon and J. B. French, Ann. Phys. (N.Y.) **95**, 90 (1975).

[5]J. B. French and V. K. B. Kota, Phys. Rev. Lett. **51**, 2183 (1983).

[6]J. J. M. Verbaarschot and P. J. Brussard, Z. Phys. A **321**, 125 (1985). See also J. P. Draayer, J. B. French, and S. S. M. Wong, Ann. Phys. (N.Y.) **106**, 472, 503 (1977).

[7]J. Smith, private communication.

[8]T. A. Brody, J. Flores, J. B. French, P. A. Mello, A. Pandey, and S. S. M. Wong, Rev. Mod. Phys. **53**, 385 (1981).

[9]R. U. Haq, A. Pandey, and O. Bohigas, Phys. Rev. Lett. **48**, 1086 (1982).

[10]L. Wolfenstein, Ann. Rev. Nucl. Part. Sci. **36**, 137 (1986).

[11]A. Pandey, in *Quantum Chaos and Statistical Nuclear Physics*, edited by T. H. Seligman and H. Nishioka (Springer-Verlag, Berlin, 1986).

[12]F. J. Dyson, J. Math. Phys. (N.Y.) **3**, 1191 (1962).

[13]M. L. Mehta and A. Pandey, J. Phys. A **16**, 2655 (1983).

[14]N. Rosenzweig and C. E. Porter, Phys. Rev. **120**, 1698 (1960).

[15]O. Bohigas, M. J. Giannoni, and C. Schmit, Phys. Rev. Lett. **52**, 1 (1984); M. V. Berry, Proc. Roy. Soc. London, Ser. A **400**, 229 (1985).

80

COMPOUND-NUCLEAR TESTS OF TIME REVERSAL INVARIANCE IN THE NUCLEON-NUCLEON INTERACTION*†

J. B. French, A. Pandey and J. Smith
Department of Physics & Astronomy, University of Rochester
Rochester, NY 14627

Abstract

The theory for the effects of time-reversal noninvariance (TRNI) in complex systems is reviewed. Applied to the compound-nuclear data for energy-level, width and cross-section fluctuations (the latter for detailed-balance pairs of reactions proceeding through the compound nucleus) this gives bounds on multiparticle TRNI Hamiltonian matrix elements. Using a fluctuation-free form of statistical spectroscopy the results are reduced to bounds on α, the relative magnitude of the TRNI nucleon-nucleon interaction. The level and width analyses for heavy nuclei give $\alpha \lesssim 2\times10^{-3}$ at high (~99%) statistical confidence; preliminary calculations for detailed balance with $^{24}Mg(\alpha,p)^{27}Al$ and its inverse gives $\alpha \lesssim 4\times10^{-3}$ at the same high confidence, but $\lesssim 0.2\times10^{-3}$ at 80% confidence. Suggestions are made about experiments which should yield sharper bounds.

1. Introduction

We shall discuss three methods which have yielded bounds on time-reversal noninvariance (TRNI) in Hamiltonians of complex nuclei. All three methods require an understanding of fluctuations[1,2] in complex systems, respectively (in the order that we shall consider them) fluctuations in energy levels, transition strengths (particle widths) and cross-sections. Specifically the first method[3,4] involves the analysis of energy-level fluctuations which are found mostly in the neutron-resonance spectra of heavy nuclei and the proton-resonance spectra of lighter ones. The

---

* Presented by the second-named author.

† Supported in part by the U.S. Department of Energy.

second method[4,5] involves a corresponding analysis of the particle widths; since the levels are detected via their widths these two methods are closely related, and indeed a simultaneous analysis of levels and widths is sometimes appropriate. The third method[6-9] makes use of detailed-balance pairs of reactions which proceed through a compound nucleus, the major example to date being $^{24}Mg + \alpha \leftrightarrow {}^{27}Al + p$, with intermediate ^{28}Si.

The first two methods are based on statistical theories of spectra and strengths for TRI systems, these generating what we shall call the "statistical model". It is essential that accuracy of the model be first confirmed by an analysis of the available data; only then can we have any confidence that the absence of statistically significant departures from the statistical-model TRI predictions may be used to give TRNI bounds (or of course the presence of such to give TRNI measures). The detailed-balance method on the other hand starts from the simple statement that $\sigma_{AB} = \sigma_{BA}$, and, since this is certain for TRI systems, any true deviations are necessarily due to TRNI. The function of the compound nucleus in this method is, by a mechanism discussed below, to enhance any interference terms between TRI and TRNI reaction amplitudes which may exist; it is this interference which will give rise to the deviations. Though its primary origin is very different it will turn out that the third method (detailed balance in compound-nucleus reactions) is formally related to the first two. The main difference in fact is that the first two theories operate in a region of discrete or quasi-discrete (resonance) spectra while that associated with detailed balance operates in the continuum domain of strongly-overlapping resonances.

We remark next that the TRNI bounds will emerge from the data analysis in the form of bounds on a quantity[3,4,10] Λ, the mean-squared matrix element of the TRNI (symmetry breaking) interaction, measured in units of the local average level spacing and evaluated between nearby states in the chaotic region of the spectrum from which the data is taken. For weak symmetry breaking (which of course is the most we can expect) this follows from the simplest perturbation argument, but in fact Λ is the significant parameter over the whole range of symmetry breaking (and for symmetries other than TRI). On the other hand, while a measure for Λ would be significant in many cases, TRI in particular, indicating the occurrence of symmetry breaking, bounds on Λ are of little or no immediate interest. One must, for example, resist the temptation to argue that if a bound or value for a forbidden process in a complex system is smaller by a certain factor than the value for a corresponding allowed one, then that factor automatically applies to the relative magnitude of the symmetry-breaking fundamental Hamiltonian (i.e. that of the TRNI nucleon-nucleon interaction in the case of present interest). Instead, a proper (and complicated) study[4] of the way in which the fundamental H operates in the complex system must be carried out in order to deduce the significant bound (or value), and to decide whether or not it is of interest.

After a brief discussion of the statistical model we shall turn to the "Λ bounds" as given by the three methods, and then outline the procedure (which applies to other symmetries as well) for extracting from them the information about the interaction. We shall then be in a good position to discuss present TRI results and future prospects for the three methods.

2. Wigner Universality and the Statistical Model

> "The assumption is that the Hamiltonian which governs the behavior of a complicated system is a random symmetric matrix, with no particular properties except for its symmetric nature."

The complicated system referred to in Wigner's 1961 statement[11] may be a nucleus, atom, molecule, or a system much more complicated than any of these; Wigner's original application was to slow-neutron fluctuations in nuclei but the explicit statement above referred to living systems. Observe that, since it specifies the Hamiltonian, it has implications not only for spectra but for other things as well, but not (as we know now) for <u>global</u> phenomena such as the secular behavior of the state density. It applies instead to <u>local</u> properties, specifically to fluctuations in energy levels, transition strengths and cross-sections which will be our direct concern.

The natural questions are about the accuracy with which Wigner's very general proposition deals with the data, and the origin of the universality which it implies. A rough general agreement of statistical-model predictions with nuclear data has been known for many years; quite recently new methods of analysis have made clear that the agreement for spectral fluctuations is very good,[12] so good in fact that the values of properly chosen statistical measures supply rather sensitive tests for TRNI (at the same time extending the validity of Wigner's assumption); the same is true for transition strengths, specifically for particle widths.[5,13] For the third method we need only repeat that any departure from $\sigma_{AB} = \sigma_{BA}$ is necessarily due to TRNI.

With respect to the origin of Wigner's universality it has now become clear that when classically-chaotic (non-integrable) TRI systems are quantized the resultant discrete spectra are locally describable by a Hamiltonian ensemble of real symmetric matrices, precisely as called for in Wigner's theorem. Thus the origin is connected with a subject, "quantum chaos", of immense current interest.[2,14]

Though Wigner's universality is fundamental for the first two processes which concern us and, in a sense, secondary for the third we can nonetheless assert that all three processes derive from the same statistical model of the compound nucleus. We introduce first the Hamiltonian

$$H_\alpha = H + i\alpha U = h + V + i\alpha U \tag{1}$$

where h, V, $i\alpha U$ are respectively the single-particle, TRI interaction and TRNI interaction parts of H_α. It is a well-known result, again due to Wigner,[15] that every (J-conserving) TRI H may be given simultaneously a real symmetric representation and then U is real antisymmetric. We agree that V and U have the same magnitude (norm) in the two-particle space, and then α, which measures the relative TRNI magnitude, is the quantity of basic interest. We shall return to (1) when we deal ahead with the $\Lambda \to \alpha$ reduction, but for the fluctuations we invoke Wigner's theorem and represent H locally as a Gaussian orthogonal ensemble (GOE) whose distinct matrix elements are chosen independently from a zero-centered Gaussian distribution with (off-diagonal) variance v^2 whose value is immaterial since all our fluctuation measures will be scale-free. (The fact that matrix-element correlations are put to zero need not be a matter of concern; if this ensemble is defined in a two-particle space and then embedded in a many-particle one, very large matrix element correlations are generated but the results for observable statistical measures are unchanged.[1,3]) The anti-symmetric U is independently generated in precisely the same way, H_α then interpolating between the orthogonal (GOE) and unitary (GUE) cases as α increases from $0 \to 1$. This is the natural way to implement Wigner's theorem.

For the levels the TRI ensemble (GOE) generates the von-Neumann-Wigner linear repulsion between nearest neighbors, this becoming quadratic for GUE. This indicates that, as TRI is gradually broken, the spectra become increasingly rigid. A simple demonstration with a 2-dimensional matrix shows why; the discriminant, which must vanish for a level crossing, is the sum of two squares when the matrix is real and of three squares when complex. The nearest-neighbor level repulsion is a special case of a much more general phenomenon, the Dyson-Mehta[16] spectral rigidity, which dominates the fluctuations in complex systems.

For the strengths the GOE gives the Porter-Thomas[17] law that distinct transition amplitudes, when locally normalized, are real independent zero-centered Gaussian random variables; the strength then has a χ^2_1 distribution which changes to χ^2_2 when TRI is completely broken, the amplitudes then becoming complex Gaussian variables with equal variances for the independent real and imaginary parts. It is of course the changes[18] in the spectral rigidity and the strength distribution which are used to give the TRNI bounds in the first two methods. The relevance of the statistical model to the detailed-balance experiments will be made clear ahead.

3. Spectral and Width Fluctuations

At the outset we ask whether it is at all possible to derive an interesting α-bound ($\sim 10^{-3}$) from the statistical analysis of a few hundred levels in the chaotic domain of the nuclear spectrum. The answer is "yes" because there are two major enhancements of the TRNI effects on fluctuations. The first is that the TRI (GOE) $\rightarrow$ TRNI (GUE) transition is governed not by α^2, but by $\sim\alpha^2 d$ where d is the matrix dimensionality. This comes about because, for fixed v, the GOE spectral width $[d^{-1}\mathrm{tr}(H^2)]^{1/2}$ is proportional to $d^{1/2}$ so that the local average spacing $D \propto d^{-1/2}$; the effect of the distant states is thereby to enhance by a factor $d^{1/2}$ the imaginary amplitudes $i\alpha U_{ij}/(E_j-E_i)$ which for small α are admixed into the real H eigenstates. More generally the transition parameter is, with i,j denoting close-lying states and the bar denoting either ensemble averaging or local spectral averaging,

$$\Lambda = \alpha^2 \overline{|U_{ij}|^2}/D^2 \xrightarrow[\text{center}]{\text{GOE}} \alpha^2 d/\pi^2 \tag{2}$$

where, by virtue of the local GOE result referred to above, the first form may be applied directly to the data. The second form in (2) derives from the GOE state density at the center and displays the enhancement. The second enhancement - in the evaluation of Λ/α^2 for many-body systems - will be discussed in Section 5 ahead.

To derive Λ-bounds from data, we must first calculate the TRI $\rightarrow$ TRNI "transition curves", i.e. the fluctuation measures as functions of Λ. This has been recently solved, for spectral fluctuations, at three different levels which we illustrate briefly by considering $\Sigma^2(r;\Lambda)$, the variance of the number of levels in intervals of length rD. A second-order perturbation calculation[3] yields, to order Λ,

$$\Sigma^2(r;\Lambda) = \Sigma^2(r;0) - 4[1-Y_2(r)]\Lambda \tag{3}$$

where $Y_2(r)$ ($\simeq 1/\pi^2 r^2$ for $r \gtrsim 1$) is Dyson's[19] two-level GOE cluster function. This gives the exact slope at $\Lambda=0$ and is adequate for data analysis. The "binary correlation" approximation[3] yields Σ^2 to all orders in Λ:

$$\Sigma^2(r;\Lambda) = \Sigma^2(r;\infty) + \frac{1}{2\pi^2} \ln \left\{1 + \frac{\pi^2 r^2}{4(\tau+2\pi^2\Lambda)^2}\right\} \tag{4}$$

in which $\tau = 0.615$ is fixed from the known GOE(Λ=0) - GUE($\Lambda=\infty$) difference. Finally the level correlation functions of all orders and all Λ have been obtained[10]; these give, inter alia, the exact result for Σ^2 and confirm the accuracy of (4) for $r \gtrsim 1$.

The data used in the spectral analysis consist of 1762 levels from 32 nuclei. To derive the Λ-bound, we first determine the measure which maximizes the GOE-GUE difference in units of the corresponding sample error (σ). The $\Sigma^2(1)$ measure turns out to be optimal. The condition for agreement is defined by a 3σ criterion (99.7% confidence). The sample error is $\sigma(p;\Lambda) \simeq 0.35\ p^{-1/2}$ where p is the number of levels in the sample. For a run of 100 levels $3\sigma \simeq 0.11$ so that the confidence band may span the entire GOE-GUE difference for $\Sigma^2(1)$ (= 0.446 - 0.344). To obtain reasonable bounds, we group the data into sets of ~400 levels, the grouping being according to the complexity of the nuclei as measured by D. We refer to Fig. 1 and Table 1 of Ref. 3 for the details of the analysis. The outcome is that, for all the groups, $\Lambda^{1/2} \lesssim 0.1 - 0.2$. Bounds on Λ for heavy nuclei, each multiplied by the corresponding experimental D, are given in Table I ahead.

We now turn briefly to the theory and data analysis for width fluctuations.[4,5] In this case the transition curves have been obtained only for small Λ. The probability density for the width (normalized to unit mean) is given by

$$\rho(x) = u^{-1/2} \exp(-x/u)\ I_0\left((1-u)^{1/2}x/u\right) \tag{5}$$

valid for $\Lambda^{1/2} \lesssim 0.2$ and therefore adequate for data; here, with γ the Euler constant, $3u = 2\pi^2\Lambda[2-\gamma-\ln(2\pi^2\Lambda)]$ and I_0 is a modified Bessel function. For GOE, (5) properly gives back the χ^2_1 density. The data come from a major subset of the neutron-resonance sequences used in the spectral case. We combine the renormalized widths in the same groups as before. Since the χ^2_1 and χ^2_2 densities differ most significantly for small arguments the best Λ bounds would emerge from the distribution of small widths. It turns out that, due to uncertainties in the data, it is not possible to derive bounds from every group. Of the heavy nuclei listed in Table I ahead, only ^{167,169}Er give proper Λ-bounds; their magnitudes are essentially the same as those given by the spectra.

A major source of uncertainty in the width data is that, without doubt, there are many missing very small widths. This[4,5] on the one hand increases significantly the width bounds (which in principle should be superior to the spectral bounds) and, on the other, introduces a small inconsistency between the GOE analyses of the spectral and strength data. For the latter, it is possible to reconcile the two analyses by assuming a small non-zero value of $\Lambda^{1/2}$ ($\simeq$ 0.03, appropriately smaller than the bounds); but for obvious reasons it would be imprudent at present to argue that TRNI has been observed.

4. Ericson Fluctuations and Detailed Balance

The above results apply to sharp resonances in compound nuclei ($\Gamma \ll D$, Γ being the average decay width). At higher energies the resonances begin to overlap. The spectral and width fluctuations still follow the same statistical rules as above, but they are no longer directly observable. We must then deal with the cross-section fluctuations, which, for $2\pi\Gamma \gg D$, a large number of channels, and no direct reactions, have a particularly simple pattern ("Ericson fluctuations").

We consider the differential cross sections $\sigma_{AB}(A \neq B)$ in reactions where only one channel spin contributes, this providing the most favorable conditions for testing the statistical theories, especially for TRNI. In this case[20] the scattering matrix element S_{AB} is a zero-centered complex Gaussian random variable with independent real and imaginary parts, each having the same variance. Then $\sigma_{AB} = |S_{AB}|^2$ has an exponential distribution, i.e. $\rho_{\sigma/\bar{\sigma}}(x) = \exp(-x)$. This is an ensemble distribution which, by the ergodic theorem[21], can be realized in individual systems as a distribution over the energy; typically the cross sections at energies separated by $\gtrsim \Gamma$ can be taken as independent.[20]

These results are valid for TRNI systems ($\sigma_{AB} = \sigma_{BA}$) as well as TRNI ($\sigma_{AB} \neq \sigma_{BA}$). However, in the latter case, the symmetric $S^{(s)}$ and antisymmetric $S^{(a)}$ parts of the S-matrix are also independently distributed; we encounter then a joint-probability density involving three independent variables,

$$\rho_{\mu^2,\nu^{-1},\phi}(x,y,z) = (2\pi)^{-1} \exp(-x-y) \; ; \; \mu^2 = |S_{AB}^{(a)}|^2 \; / \; \overline{|S_{AB}^{(a)}|^2}$$

$$\nu^{-1} = |S_{AB}^{(s)}|^2 \; / \; \overline{|S_{AB}^{(s)}|^2} \quad ; \; \phi = \text{Phase} \left\{ S_{AB}^{(a)} / S_{AB}^{(s)} \right\} \tag{6}$$

Thus[6] the distribution of

$$\Delta = \frac{2(\sigma_{AB}-\sigma_{BA})}{\sigma_{AB}+\sigma_{BA}} = 4\xi\sqrt{\nu}\mu \cos\phi + 0(\xi^2) \tag{7}$$

depends on the magnitude of the "local" TRNI Hamiltonian via the parameter

$$\xi^2 = \overline{|S_{AB}^{(a)}|^2} \; / \; \overline{|S_{AB}^{(s)}|^2}. \tag{8}$$

In particular for small ξ, $\Delta/4\xi\sqrt{\nu}$ is a zero-centered Gaussian variable with variance 1/2. Note that Δ is experimentally observable, as also is $\nu(= \overline{\sigma}_{AB}/\sigma_{AB} + 0(\xi^2))$. Let us suppose that the experiment gives the Δ-bounds, $|\Delta| \leq \delta_j$, at n independent energy points E_j. Then, for small ξ, a statistical bound ξ_0 on ξ can be deduced with a confidence given by[9]

$$P(\xi \leq \xi_0) = 1 - \prod_{j=1}^{n} \text{erf}\left(\frac{\delta_j}{4\xi_0\sqrt{\nu}_j}\right) \tag{9}$$

where ν_j is $\nu(E_j)$. Since, for n=1, $\xi_0 \propto \nu^{-1/2}$ for given confidence, TRNI effects would be most pronounced in "deep" cross-section minima[6]; as a corollary it follows also that the maxima could be used to normalize the forward and backward reactions.

For small values of $\delta/\sqrt{\nu}$, even a single observation (n=1) can give a sharp bound on ξ. Indeed a recent detailed-balance experiment[7], $^{24}Mg+\alpha \leftrightarrow {}^{27}Al+p$, involving the compound nucleus ^{28}Si

at ~19 MeV above the ground state with the differential cross section measured at $\theta \simeq 180°$, gives $\delta = 5\times10^{-3}$ and $\xi_0 = 5\times10^{-4}$ (80% confidence). At higher confidence the bounds[9,22] are $\xi_0 \simeq 10^{-3}$ (90%), 2×10^{-3} (95%), 10^{-2} (99%). Note the unusual sensitivity to the confidence which arises directly from (9) but is not found in the spectral and width analyses.

To relate the bound on ξ to that on the local TRNI interaction (i.e., on Λ), we recall[8] first that, at energy E, the S-matrix is given by $S = I - i\gamma G\gamma^+$ where $G = (E - H_\alpha - i\gamma^+\gamma/2)^{-1}$, $H_\alpha = H + i\alpha U$ is the compound-nuclear Hamiltonian as in (1), and γ is the reduced width matrix which couples the channels to the H_α-eigenstates. As above, we take H_α to be the interpolating ensemble. Since there are no direct reactions ($\overline{S}=0$), we take $\gamma\gamma^+$ to be diagonal. Let $\Gamma_A = d^{-1}(\gamma\gamma^+)_{AA}$, $\Gamma = \Sigma\Gamma_A$, and d_c = number of open channels. We consider the case $D >> \Gamma_A$ (small transmission coefficients) and $d >> d_c >> 1$. Then orthogonal invariance of the ensembles implies that $\gamma^+\gamma$ in G can be effectively replaced by $d^{-1}\mathrm{tr}\gamma^+\gamma = \Gamma$. Thus, for example, for $A\neq B$, the average cross section is given by

$$\overline{\sigma}_{AB} = \Gamma_A\Gamma_B\, \mathrm{tr}\, \overline{GG^+} = \Gamma_A\Gamma_B\, \Sigma\, \overline{g}_i(\alpha) = \frac{2\pi\Gamma_A\Gamma_B}{\Gamma D} \tag{10}$$

consistent with the Hauser-Feshbach formula; here $g_i(\alpha) = \left((E-E_i(\alpha))^2 + (\Gamma/2)^2\right)^{-1}$, the $E_i(\alpha)$ being the H_α eigenvalues. For ξ^2 we have[9]

$$\xi^2 = \frac{\mathrm{tr}\overline{GG^+} - \mathrm{tr}\overline{GG^*}}{\mathrm{tr}\overline{GG^+} + \mathrm{tr}\overline{GG^*}} \xrightarrow{\text{small } \alpha^2} \alpha^2 v^2 \sum_{i\neq j} \overline{g_i(0) g_j(0) / \Sigma_k\, \overline{g}_k(0)}$$

$$= \frac{2\pi\Lambda D}{\Gamma}\, w\left(\frac{\Gamma}{D}\right) \xrightarrow{\Gamma >> D} \frac{2\pi\Lambda D}{\Gamma} \tag{11}$$

Here the first form (yielding $\xi^2 = 0,1$ for $\alpha = 0,1$, as expected) follows from orthogonal invariance and the second by expansion to order α^2 followed by averaging over the U-matrix elements. The third form, expressed in terms of $\Lambda = \alpha^2 v^2/D^2$, completes the averaging over the GOE eigenvalues $E_i(0)$; w, which lies in (0,1), is related to the GOE Y_2 by

$$w(\epsilon) = 1 - \frac{1}{\pi}\int_{-\infty}^{\infty} \frac{Y_2(r\epsilon)}{1 + r^2}\, dr \xrightarrow{\epsilon \gg 1} 1 - \frac{1}{\pi\epsilon} \longrightarrow 1 \qquad (12)$$

The final result in (11), first given by Ericson,[23] has been recently rederived (with a $\pi/2$ instead of 2π factor) by Boosé et al.[8]

The above derivation of (11) ignores the fact that H_α conserves (J,π). It appears however that, under very general conditions, ξ^2 should be a weighted average of the result in (11) taken as valid for fixed (J,π), the weighting factor being the corresponding contribution to $\overline{\sigma}_{AB}$. It is reasonable to assume that ΛD and Γ are essentially (J,π)-independent. Calculations described in the next section show that the density of <u>states</u> (not levels) is ~2000 per MeV; then, using the fact that the spin cutoff factor is about 3.7, we find that D_{level}, the average spacing between levels of all (J,π), is about 5 keV, and that $D_{J,\pi}$, the average spacing between levels of the same (J,π), is typically 60-250 keV. The experimental value[24] of Γ is about 45 keV. Which of these D values must now be used in checking the condition $(2\pi\Gamma/D \gg 1)$ for Ericson fluctuations, and in evaluating $w(\Gamma/D)$, and thereby ξ^2, from (11)? A little consideration shows that, in the first, $D \to D_{level}$ while, in the second, $D \to D_{J,\pi}$. We see then that the condition is satisfied and that the argument of w is in the range 0.2 - 0.8, which gives rise to $w \simeq 0.2 - 0.7$. We take as a typical value $w \simeq 0.5$. Then[9], from the bound on ξ^2, we obtain the bound on ΛD as $\simeq$ 1.5 eV (99% confidence) and 0.004 eV (80% confidence); the second of these agrees with the result of Boosé et al.[8] to within a factor of 2 (they give $4\pi\Lambda D \lesssim 0.1$ eV). See Table I for comparison of ΛD bounds by the three methods.

5. <u>Fluctuation-Free Spectroscopy and the $\Lambda \to \alpha$ Reduction</u>

In order to deal with the fluctuation analysis we replaced the Hamiltonian H_α of our basic equation (1) by an ensemble, and implicitly agreed to take an ensemble average over the predictions of the resultant Hamiltonian. This high-handed procedure ignores everything about the nuclear physics of the problem except for one simple TRNI/TRI ratio which determines the variation of the fluctuation measures as the symmetry is gradually broken, i.e. as α increases. That it is a sensible thing to do follows directly from Wigner's universality for $\alpha=0$ and its natural extension for arbitrary α. A direct application to nuclear spectral and strength fluctuations and to detailed-balance then resulted in bounds on Λ, the multiparticle transition parameter, or on ΛD.

To learn now what Λ tells us about the fundamental α in the original version of (1) we return to that nuclear Hamiltonian and ask for the Λ value, proportional of course to α^2, which it would generate at a given excitation in a particular nucleus. Our interest here will be in the nuclei which give the experimental Λ bounds, though in fact the general behavior of Λ/α^2, Eq. (2), is also of interest in quite different contexts. To avoid confusion we repeat that the significance of α will be fixed by taking U to have the same norm (square root of the Hermitian square) as V in the two-particle space.

The calculation of Λ/α^2 for a given nucleus is a non-trivial matter in the huge model spaces encountered here and we approach it in stages. The first thing is to decide upon the model space and the model Hamiltonian, i.e. upon the number of active nucleons to be considered, the single-particle (s.p.) orbits which they will inhabit, the single-particle spectrum represented by h, and the interaction V; U then will look after itself. Since this part of the problem is standard we need say nothing more about it except that we should test the adequacy of the space and the Hamiltonian by calculating the state density and comparing with experimental results (which are *a fortiori* available when dealing with the first two methods), and ideally also by verifying the stability of the density under an enlargement of the model space. The final stage requires the calculation of the strength (conventionally defined as the absolute matrix element squared of a transition operator between H-eigenstates) induced by U. Thus our principal needs are for theories of interacting-particle (IP) state densities and strengths; it will turn out that the second will follow in a fairly straightforward way from the first.

Hundreds of papers have been written about state densities, all but a few of them ignoring the interaction or using ad hoc devices (quasiparticles for pairing, "back-shifting", etc.) to take some account of it. To do that properly it is good to recognize that at any given energy the spherical orbits of the system divide into those which are essentially filled or essentially empty (and can therefore be dealt with in a trivial way) and the others which are "active", defining, as they do, the Fermi surface. If the number of active single-particle states is $N = \Sigma N_i$, where $N_i = (2j_i+1)$ refers to a single spherical orbit, there are two significant symmetries, defined by the group U(N) of unitary transformations in the N-dimensional s.p. space, and by a U(N) subgroup $\Pi U(N_i)$, the direct product of single-orbit unitary groups. On decomposing V with respect to tensors defined by these groups we find:[25]

(i) A harmless (number-dependent) constant.

(ii) A term which renormalizes the s.p. energies. We shall take for granted that this renormalization has been carried out in (1).

(iii) A non-renormalizing effective one-body term which is negligible in all many-particle cases of interest.

(iv) An effective two-body term which spreads and admixes multi-particle states of the same total harmonic-oscillator excitation $\underline{s}$ (the definition of $\underline{s}$ would be modified appropriately in heavy nuclei).

(v) A term which admixes states only of different $\underline{s}$; this can be ignored (as we shall do) or treated perturbatively because admixing states, for which by parity $|\underline{s}'-\underline{s}| = 2,4,\ldots$, are far apart.

It follows then that V gives a renormalization, by (ii), of the s.p. energies, and a spreading (and admixing), by (iv), of states which would otherwise be degenerate; moreover a close study of effective interactions shows that the spreading is Gaussian and that the width is essentially constant (±5%) for all states of given $\underline{s}$, increasing slowly with $\underline{s}$ as is to be expected. The outcome of all this is that the state density for given $\underline{s}$ is a convolution[25] of the

NIP density with a zero-centered spreading Gaussian, ρ_G, with variance σ_s^2,

$$I_{IP} = \sum_s I_{NIP}^{(s)} \otimes \rho_G\left(\epsilon=0;\sigma_s^2\right) \tag{13}$$

The non-interacting-particle density I_{NIP}, which is numerically representable by the Bethe form, $\sim \exp\{bE\}^{1/2}$, is, along with its s decomposition, easily calculated as a function of the s.p. energies by elementary spectral averaging; we need not go into details of that. The fixed-s NIP densities are approximately Gaussian, often very closely so, and then the IP density is approximately a sum of Gaussians with NIP centroids and with variances which are increased by the interaction spreading. Since $\delta s = 2$ corresponds to a $2\hbar\omega$ change in excitation it is clear that we need very few terms in (13); a single one, $s = 0$, is adequate for the resonance region of heavy nuclei, while for ^{28}Si at 19 MeV (which we shall discuss for the detailed-balance analysis) it is necessary to consider also $s=2$.

Agreeing now that we can calculate interacting-particle densities, not to mention spin-cutoff factors, occupancies and so forth, let us turn to strengths. A little consideration will show that the (bivariate) strength density for an m-particle system

$$S^{H,m}(y,x) = I(y)\,|\langle y|U|x\rangle|^2 I(x) \xrightarrow[y\sim x]{sm} \Lambda/\alpha^2 \tag{14}$$

whose near-diagonal form, when smoothed (sm), becomes our Λ/α^2 as indicated, can, for the NIP case (V=0, H=h), be calculated exactly in terms of the state density I(E) and, of course, the two-particle matrix elements of U. The reason for this is that the h-eigenstates are simple products of s.p. states (which we label by i, j, k, l,...), and in that basis the operator U is a sum of operators each of which simply transfers 0, 1 or 2 particles to different unoccupied single-particle states, thereby generating one eigenstate from another. This calculation of $S^{h,m}$, which would be pointless in small model spaces, is extremely useful in large ones for which h plays a major or even dominant role, as we have seen with the level density. Let us now think of one component of U, say $U_{ij,kl}a_j^+a_i^+a_ka_l$, as acting in a 2-particle subspace with 4 s.p. states, the remaining (m-2) particles distributed over (N-4) s.p.

states then forming a "medium" in which the "action", defined in the subspace, propagates. It is an easy exercise to show that (if we ignore certain Pauli "blocking" corrections which in practice we would not) the NIP strength density is given by a bivariate convolution involving the 2-particle strength and the (m-2)-particle density. Thus[4]

$$S^{h,m} = S^{h,2} \otimes T^{h,m-2} \quad ; \quad T^{h,m-2}(x,y) = \delta(x-y) I^{h,m-2}(x) \qquad (15)$$

which is the NIP result promised. An important property of $S^{h,m}(x,y)$ is that the (x,y) variables are very highly correlated in large spaces (correlation coefficient = $\zeta \simeq 1 - 2/m$), the distribution then becoming extremely sharp in the (x-y) direction. This result derives physically from the fact that a component of U can transfer at most two particles, generating thereby only a small fractional change (~ 1/m) in the multiparticle energy; moreover we can expect this strong positive correlation to survive in the IP system since turning on V introduces only a narrow spreading (σ ~ 2 MeV for heavy nuclei). Since for the admixing matrix element which defines Λ, we have x ~ y we see that this correlation effect strongly enhances the admixing and thereby the sensitivity for deducing α from Λ. Without this and the separate Λ enhancement mentioned above no interesting α bound would emerge from fluctuation studies (including those associated with detailed balance). Stated otherwise we now understand why fluctuation studies involving only a few thousand levels in total can yield interesting bounds on the fundamental α.

For the final stage of the argument, the transition from NIP → IP, or equivalently $S^{h,m} \to S^{H,m}$, we make the natural assumption (more or less inescapable if U, V are independent as is the case in the present problem where, having no model for U, we naturally represent it via an ensemble) that, when the states admix under the action of V, each fragment of an NIP state carries with it its share of the original V = 0 strength. If the initial and final state admixings were independent this would immediately give, for states of a single s̲, separate monovariate convolutions of $S^{h,m}$ with the

Gaussian spreading functions, or equivalently the bivariate convolution with a $\zeta = 0$ bivariate Gaussian. Now to take account of the (x,y) correlation we need only change to a bivariate Gaussian $\rho_G^{(2)}$ <u>with non-zero correlation</u>; thus[4]

$$\underset{\sim}{S}^{H,m} = S^{h,m} \otimes \rho_C^{(2)} = \left[tr^{(m)}(U^+U)\right]^{-1} S^{h,2} \otimes T^{h,m-2} \otimes S^{V,m} \quad (16)$$

where the final form, which comes by consideration of h = 0, fixes the parameters of $\rho_G^{(2)}$ in terms of low-order moments of $S^{V,m}$. The quantities appearing in (16) are all easily calculated[4,9,25] once the model space, the s.p. energies and the interaction V have been decided upon and tested. This includes the correlation coefficient in $S^{V,m}$ which for large spaces is shown by an ensemble argument to be $\binom{m-2}{2}\binom{m}{2}^{-1}$, close to unity as predicted. The parity, T_z, and $\underline{s}$ quantum numbers dropped above can be restored by simple partitioning of the space. For the angular momentum the spin cutoff factor which decomposes the level density does the same for the strengths Λ/α^2.

For the heavy nuclei of Table I the calculations have been carried out[4,5] for a surface-delta interaction with a zero-deformation Nilsson s.p. spectrum. The model-space dimensionalities are 10^{15} to 10^{20} even though only $\underline{s} = 0$ is relevant. Instead of giving Λ/α^2 we give in Table I the calculated values of $\Lambda D/\alpha^2$, which, though D itself varies by a factor 100, are seen to be essentially the same for these nuclei; this is in line with a prediction[8] based on experimental results for isospin symmetry[26]. A general derivation, based on (16), of the variation of $\Lambda D/\alpha^2$ with energy and mass number will be given later.

The more recent of the two detailed-balanced experiments[6,7] involves ^{28}Si at 19 MeV (as opposed to resonances at ~ 6 MeV in the heavy nuclei), but the same equations and formal procedures are applicable. For the first calculations we have taken $(ds)^{12}$ + $(ds)^{10}(fp)^2$ for the model space. Also (unlike the heavy-nuclear cases) we have now an essentially parameter-free Hamiltonian, V being a 10-orbit Kuo-Brown[27] interaction, and the s.p. spectrum more or less deriving from the fine structures near A = 16, 40. The one uncertain parameter is the (fp)-(ds) separation; it turns out however that the bounds are relatively insensitive to this parameter. The calculated value of $\Lambda D/\alpha^2$ (see Table I) is essentially identical with the values for heavy nuclei.

The resultant bounds on α (the relative TRNI nucleon-nucleon interaction) are given in Table I.

6. Summary and Outlook

By a process which amounts to extracting information from quantum chaos Wigner's proposal[18] that nuclear spectral and width fluctuations should give information about the time-reversal noninvariant part of the nuclear Hamiltonian has been implemented. Specifically bounds on Λ, a complex-system transition parameter (or equivalently on ΛD where D is the known spacing) have been determined for classes of heavy nuclei. The same thing has been done (the relevant parameter in this case being ΛD) for detailed-balance pairs of reactions proceeding through a region of a compound nucleus where the levels are strongly overlapping; the general considerations for this were first given by Ericson[23] and by Witsch et al.,[6] and a theory has been recently outlined by Boose et al.[8] and used in the analysis of a later experiment.[7] Our own very similar theory pays more attention to the statistical confidence limits (which turn out to be of consequence); for all three methods it goes much beyond the above-mentioned theories in using fluctuation-free statistical spectroscopy to give explicit procedures and forms for deriving, from the bound on the multiparticle parameter Λ, a bound for the much more fundamental quantity, α, the ratio of the magnitudes (norms) of the TRNI and TRI nucleon-nucleon interactions. At high confidence (~ 99%) we find that all three methods give about the same α bounds[4,9] (a few parts per thousand); but at "low confidence" (80%) the detailed-balance experiment gives a better bound, smaller (at the present state of our calculations) by a factor ~10. Looking to the future, it would be good on theoretical grounds[28] if the α-bound could be brought down to ~ 10^{-4}. It is clear from the constancy of $\Lambda D/\alpha^2$ that to improve the spectral bounds we need more data from nuclei with small spacings. For example with ^{236}U (for which the present data are inadequate) a bound of $\Lambda \sim 0.04$ would lead to $\alpha \lesssim 5\times10^{-4}$, and the strength bound would be better than this, even by a factor 4, if very small strengths could be measured and the levels identified. With detailed balance the outlook is better; observations of quality comparable to that of Ref. 7 made at three independent energy points could, according to (9), improve the detailed-balance bound by an order of magnitude.

Acknowledgement

We are indebted to V.K.B. Kota and S. Tomsovic with whom a large part of the this work has been done.

Table I

Parameters and high-confidence bounds for heavy nuclei via spectral and width fluctuations, and for ^{28}Si via detailed balance. At 80% confidence the α-bounds would improve in the spectral case by less than a factor 2, but in the detailed balance case by a factor 20.

	D in eV (exper.)	ΛD bound in eV (~99% conf.)	$\Lambda D/\alpha^2$ in eV (calculated)	α-bound (~99% conf.)
^{167}Er	38	1.2	1.5×10^5	3×10^{-3}
^{169}Er	100	0.8	0.7×10^5	3×10^{-3}
^{233}Th	17	0.4	2.0×10^5	1.5×10^{-3}
^{239}U	21	0.8	1.3×10^5	2.5×10^{-3}
^{236}U	1.1	-	1.2×10^5	-
^{28}Si	-	1.5	1.3×10^5	3.5×10^{-3}

References

1. T. A. Brody, J. Flores, J. B. French, P. A. Mello, A. Pandey, and S. S. M. Wong, Rev. Mod. Phys. 53, 385 (1981).
2. T. H. Seligman and H. Nishioka (eds), "Quantum Chaos and Statistical Nuclear Physics" (Springer-Verlag, Berlin, 1986).

3. J. B. French, V. K. B. Kota, A. Pandey, and S. Tomsovic, Phys. Rev. Lett. 54, 2313 (1985).
4. J. B. French, V. K. B. Kota, A. Pandey, and S. Tomsovic, Phys. Rev. Lett. 58, 2400 (1987), and to be published.
5. S. Tomsovic, Ph.D. Thesis (University of Rochester, unpublished, 1986).
6. W. von Witsch, A. Richter, and P. von Brentano, Phys. Rev. 169, 923 (1968).
7. E. Blanke, M. Driller, W. Glöckle, H. Genz, A. Richter, and G. Schrieder, Phys. Rev. Lett. 51, 355 (1983).
8. D. Boosé, H. L. Harney, and H. A. Weidenmüller, Phys. Rev. Lett. 56, 2012 (1986).
9. J. B. French, A. Pandey, and J. Smith, to be published.
10. A. Pandey and M. L. Mehta, Commun. Math. Phys. 87, 449 (1983); M. L. Mehta and A. Pandey, J. Phys. A16, 2655 (1983), L601 (1983).
11. E. P. Wigner (1961), reprinted in "Symmetries and Reflections" (Ox Bow Press, Woodbridge, 1979), p. 203.
12. R. U. Haq, A. Pandey, and O. Bohigas, Phys. Rev. Lett. 48, 1086 (1982); O. Bohigas, R. U. Haq, and A. Pandey, Phys. Rev. Lett. 54, 1645 (1985).
13. O. Bohigas, R. U. Haq, and A. Pandey, in "Nuclear Data for Science and Technology", edited by K. H. Böckhoff (Reidel, Dordrecht, 1983), p. 809.
14. O. Bohigas, M. J. Giannoni, and C. Schmit, Phys. Rev. Lett. 52, 1 (1984); M. V. Berry, Proc. Roy. Soc. London A400, 229 (1985).
15. C. E. Porter, in "Statistical Theories of Spectra: Fluctuations", edited by C. E. Porter (Academic Press, New York, 1965), p. 3; M. L. Mehta, "Random Matrices" (Academic Press, New York, 1967).
16. F. J. Dyson and M. L. Mehta, J. Math. Phys. 4, 701 (1963).
17. C. E. Porter and R. G. Thomas, Phys. Rev. 104, 483 (1956).
18. E. P. Wigner, SIAM Rev. 9. 1 (1967).
19. F. J. Dyson, J. Math. Phys. 3, 1191 (1962).
20. T. Ericson and T. Mayer-Kuckuk, Ann. Rev. Nuc. Sci. 16, 183 (1966).

21. J. B. French, P. A. Mello, and A. Pandey, Phys. Lett. B80, 17 (1978). A similar ergodic question is also encountered for the spectral and width fluctuations; see Ref. 1.
22. The ν-value is not reported in Ref. 7, but can be estimated from the 80% bound via Eq. (9).
23. T. E. O. Ericson, Phys. Lett. 23, 97 (1966).
24. B. W. Allardyce, W. R. Graham, and I. Hall, Nucl. Phys. 52, 239 (1964).
25. J. B. French and V. K. B. Kota, Phys. Rev. Lett. 51, 2183 (1983).
26. E. Kuhlmann, Z. Phys. A322, 527 (1985); H. L. Harney, A. Richter, and H. A. Weidenmüller, Rev. Mod. Phys. 58, 607 (1986).
27. T. T. S. Kuo, private communication.
28. L. Wolfenstein, Ann. Rev. Nucl. Part. Sci. 36, 137 (1986).

ANNALS OF PHYSICS 181, 235–260 (1988)

Statistical Properties of Many-Particle Spectra VI. Fluctuation Bounds on *N–N* *T*-Noninvariance*

J. B. FRENCH, V. K. B. KOTA,† A. PANDEY, AND S. TOMSOVIC‡

Department of Physics and Astronomy,
University of Rochester, Rochester, New York 14627

Received October 13, 1987

The preceding paper (V) shows that fluctuation analysis applied to experimental data in a chaotic region of the nuclear spectrum, determines values or bounds for a parameter Λ, proportional to the mean-squared near-diagonal matrix element of a symmetry-breaking interaction. The immediate interest is with time-reversal invariance (TRI) for which bounds, not values, are found. We show here how this information can be reduced to a bound on α, the relative magnitude of the time-reversal noninvariant (TRNI) part of the nucleon–nucleon interaction. This is done in terms of the level and transition-strength densities, both given as explicit functions of the Hamiltonian parameters. It is shown that, just as univariate Gaussian densities are essential ingredients of the state density, so also bivariate Gaussians enter naturally into a theory for the transition strengths. Calculations made for a subset, namely ^{167,169}Er, ^{233}Th, and ^{239}U, of the nuclei considered in (V), give $\alpha \lesssim 2 \times 10^{-3}$, a value which by some arguments is at the boundary of fundamental interest. It is found that $\Lambda D/\alpha^2$ is roughly constant over the periodic table and varies only slowly with energy. This implies that, for α-bounds better by perhaps a factor 4 (or α-values), we need more small strengths in nuclei with small spacings such as ^{236}U (which presently yields no Λ-bound). The methods introduced here, which admit a number of simplifying approximations, are applicable to TRI studies via detailed-balance experiments proceeding through a compound nucleus, to other symmetries, and to a wide class of other problems.

1. INTRODUCTION

Determining Λ-values or Λ-bounds, as we have done in the preceding paper [1] (which we shall refer to as (V)), amounts to extracting information from fluctuations, specifically values for or bounds upon a local RMS multiparticle matrix element of a symmetry-breaking Hamiltonian. Similar information can be derived in other ways, for example, by measuring cross sections or lifetimes for highly forbidden processes or, in the case of time-reversal invariance (TRI), from detailed-balance pairs of reactions [2]. We have stressed that, while a nonzero *value* of Λ may be of interest, a *bound* on Λ tells us nothing of consequence until it has been reduced to an "α-bound," i.e., a bound on the relative norm α of the elementary

* Supported in part by the U.S. Department of Energy.

† Permanent address: Physical Research Laboratory, Ahmedabad 380009, India.

‡ Present address: Institut de Physique Nucléaire, 91406 Orsay Cedex, France.

0003-4916/88 $7.50

symmetry-breaking interaction (as measured in the nucleon–nucleon system in our TRI case). This point, clearly stated by Simonius [3] ("manifestations of symmetries and symmetry breaking in complex systems ... in most cases ... are not of great help ... due to uncertainty in the complex nuclear physics part of their analysis"), has been well appreciated by some authors but not considered at all by others. Solving the "complex nuclear physics" reduction problem referred to here could be dealt with in some cases, though with very large uncertainties, by comparison with data and analysis for a different well-understood symmetry, isospin in particular [4]. Instead we shall in this paper deal directly with the general $\Lambda \to \alpha$ problem. Although we shall write things specifically for the TR symmetry it will be clear that our procedures will apply to other symmetries and indeed to a wide class of other problems not connected at all with symmetry breaking.

As explained in (V) the definition of Λ as the squared ratio of the admixing matrix element to the locally averaged level spacing is valid for both the ensemble theory of fluctuations and the analysis of experimental data. The fact that the ensemble Hamiltonian had little resemblance to a realistic one was irrelevant at the Λ stage of the analysis. But this is not true at all for the reduction problem. Most important is the fact that, unlike the H's of the classical ensembles which are always multibody (all particles interacting simultaneously when the matrix dimensionality is large), the physical Hamiltonian is of low particle rank (a k-body operator has particle rank k), the single-particle ($k=1$) part generating very large shell effects. We write

$$H_\alpha = h + V + i\alpha U = H + i\alpha U, \tag{1}$$

an equation which we may consider at three levels (we destinguish them via subscripts). In the first it is the general Schrödinger equation with h_1 the one-body kinetic-energy operator while V_1 and $i\alpha U_1$ are respectively the two-body TRI and TRNI (TR noninvariant) interactions. On proceeding to the shell-model picture the model space becomes finite-dimensional and h_1 is renormalized to h_2 which will be still TRI; the resultant single-particle energies (spe) are then to be regarded as the eigenvalues for a particle in a well deriving from the averaged interaction of the particle with those filled shells which are explicitly left out of the calculation; V_2 would then be the usual TRI residual interaction and $i\alpha U_2$ the noninvariant (TRNI) one, these two-body operators being representable respectively by symmetric and antisymmetric matrices. Our calculation really begins here, dealing, therefore, as is usual in the physics of complex nuclei, not with the scattering nucleon–nucleon interaction, but with an effective interaction in which the very-short-range effects of the fermion sea have been eliminated. All of our TRNI values or bounds will therefore refer to such an interaction, as would results derived otherwise (e.g., via detailed balance) by other studies of complex nuclei, as opposed to results from high-energy elementary nucleon–nucleon scattering.

For the third stage there is a further renormalization [5], according to representations of the group $U(N)$ of unitary transformations in the N-dimensional space spanned by the "active" single-particle (sp) states (those neither essentially filled nor essentially empty; the intuitive initial assignments can be verified a posteriori by calculating single-state occupancies) which constitute the Fermi surface. For a two-body operator V_2 there are three representations, with Young columnar shape $[N-\nu, \nu]$ where $\nu = 0, 1, 2$. Thus $V_2 = \sum V_2^\nu$ where for fixed particle number V_2^0 is a harmless constant, V_2^1 is an effective one-body operator which adds to h_2, and V_2^2 is the "true" two-body operator which we shall call simply V. The antisymmetric operator U makes no contribution to this renormalization of h since it is of type $\nu = 2$ only. We are thus left with the original equation (1) in which

$$h = h_2 + V_2^0 + V_2^1; \qquad V = V_2 - V_2^0 - V_2^1 = V_2^2; \qquad U = U_2 = U_2^2 \tag{2}$$

and it is this form of the equation that we shall deal with from now on. At this point we arrange things so that, in a sense to be made clear ahead, V and U will have the same magnitude (norm), and then α will be the relative strength of the $\nu = 2$ parts of the TRNI and TRI effective interaction. After we have produced the α-bound we may multiply it by the norm ratio $|V_2^2|/|V_2|$, evaluated in the two-particle space, in order to take account of that part of V_2 which has gone into the renormalization of h_2. We remark that α plays the same role here as in the H_α-ensemble equation ($V-2$) for which $h = 0$. But confusion will arise if the two α's, one connected with the fluctuations and one the density, are regarded as identical.

The norms of U and V (i.e., of U_2^2 and V_2^2) increase with particle number in exactly the same way [5]. Thus we can measure α in the many-particle space and then, with the superscript bars indicating a local average, (i, j) referring to close-lying H-eigenstates in the resonance region, and D the local average spacing,

$$\alpha^2 = \overline{|H_{ij}(\text{TRNI})|^2}/\overline{(U_{ij})^2}; \qquad \Lambda/\alpha^2 = \overline{U_{ij}|^2}/D^2. \tag{3}$$

The analysis of the spectral and strength data in Section 5 of V has given a bound on the first numerator; we must now calculate the denominator.

The one-body h generates the large orbital splittings characteristic of the nuclear shell model, and thereby, in a complicated way, imposes on the system an effective dimensionality d_{eff} which rapidly increases with the energy; this is determined by the number of active particles and single-particle states which can contribute significantly to the multiparticle states in the region of interest. Unlike the situation with fluctuations in TRI systems, in which the dimensionality d is a "harmless" parameter (since it disappears when measures are expressed in terms of local spacing units), d_{eff} is physically significant when TRI is broken; see the discussion surrounding (V Sect. 5). For example if one takes for granted that d is irrelevant as long as it is large we would conclude from the subspectra of (V, Fig. 1) that the smallest detectable α would, even with low confidence, be a disappointing few percent (say 3%). However, the same subspectra could well be found, say, with matrices of $d = 21{,}000$ or 2.1×10^6 rather than $d = 210$ as is the case; then the corresponding α's would be 0.3% or 0.03%, the difference resulting from the $d^{-1/2}$ spectral compression which accelerates the GOE → GUE transition. Our problem now could be described (V) as that of calculating the effective d though we shall not proceed directly along that path.

It follows from this discussion that at least the essential shell structure, represented by h, must be taken into account in the reduction problem. But the norms of V and U in (1) grow [5] with increasing particle number faster than the norm of h; thus any *ensemble* satisfactory for this problem would have to deal explicitly with the particle ranks of H (i.e., must be of the embedded type, Section 8 of V), must contain a nonrandom component to represent h, and should ideally be adaptable to the treatment of the exact symmetries, J, J_z, π, and, where relevant, isospin T, T_z. For low excitation energies, where h is relatively unimportant, we shall in fact use ensemble arguments (Section 3.5) for a more general derivation of results which we shall derive otherwise for arbitrary h. For the latter derivation we shall use a different procedure, first describing in conventional shell-model terms the problem to be solved and then turning to spectral averaging methods [5, 6] to solve it.

To deal with the shell-model problem we would in principle take, for each nucleus, an appropriate form for V, an adequate model space, defined in our case by a set of spherical orbits, and a one-body h, derived for example from a zero-deformation Nilsson-model spectrum (zero deformation because we specifically deal with the V which generates the deformations); note that this spectrum would itself give a good indication of the orbits to be included in the model space. Then constructing and diagonalizing the H-matrices would give the TRI spectrum and eigenstates. If necessary to give the correct level spacing $D = I^{-1}$ in the resonance region, adjustments of the Hamiltonian parameters (for example, the magnitude parameter G if, as is usual with heavy nuclei, a functional form of V is assumed) would be made and new calculations carried out. We would demand moreover that this density be essentially stable under small enlargements of the original single-particle space. By making the very strong requirement that we produce a correct and stable level density for the resonance domain we can be reasonably certain that we adequately understand the nuclear structure there, important because of the role played by the effective dimensionality d_{eff}. For the same reason it is good to test against experiment the level densities I for neighboring nuclei. These tests being satisfied, we would construct in the H-diagonal representation the matrix of U, read off its near-diagonal elements, and thereby evaluate the first denominator in (3) and hence the value of α. If, as happens with TRNI, there is no appropriate functional form for U we could represent it in terms of an (embedded) ensemble or, equivalently, by specifying certain relationships between traces of low powers and products of U and V_2^2.

This straightforward procedure cannot be carried out as described. The model spaces involved are impossibly large, the dimensionality for the relatively simple ^{169}Er being $\sim 10^{15}$ for an adequate set of single-particle orbits and active particles, and, for nuclei with small average resonance spacings, much larger than that. But since we have already dealt with the fluctuations we need now only the locally smoothed version of the microscopically detailed shell-model results, specifically theories for the interacting-particle (IP) smoothed level and strength densities. With the first we can test the adequacy of the TRI Hamiltonian and model space as described above. For the second we recognize that the squared matrix element of U which enters into the form (3) for α^2 may be interpreted as a locally averaged strength (square of a transition-operator matrix element between H-eigenstates) generated by U, so that Λ/α^2 is a strength density. Both of these theories will follow by spectral-averaging methods, which will give also appropriate definitions of the operator norms and decompositions called for above, and methods for evaluating them. These methods are not new but recent work [7, 8] has greatly simplified their application in very large model spaces. Broadly speaking the (IP) level and strength densities are given in terms of convolutions of noninteracting-particle (NIP) densities with spreading univariate and bivariate Gaussians, all the parameters involved being directly calculable from the given Hamiltonian. In the next two sections we give a brief account of the whole subject in order to make clear what goes on in the $\Lambda \rightarrow \alpha$ reduction.

2. Spectral Averaging and Level Densities

Spectral averaging [6] produces a fluctuation-free form of theoretical spectroscopy which can be applied in (m, N) spectroscopic model spaces (i.e., spaces formed by distributing m fermions over N single-particle states) where the dimensionality $\binom{N}{m}$ may be indefinitely large, and gives smoothed quantities of interest as explicit functions of the defining parameters of the system (Hamiltonian-matrix elements, etc.). The level density, given in such functional form, then plays the same role as the smoothed partition function, the two being related by Laplace transformation, and carries therefore much of the information about the system (which can be extracted for example by a parametric differentiation).

For such a theory to be satisfactory it is desirable or even essential that there be a sharp separation between fluctuations and local averages, that the fluctuations be small, and that some principle determine the general form of the smoothed eigenvalue density. The separation, needed in order that the smoothing be well defined, is in fact found [9] in the power spectrum of interacting-particle EGOE systems which display fluctuations only of very long and very short wavelengths, comparable, respectively, for finite model spaces, with the spectrum span and the level spacing. The smallness of the fluctuations is guaranteed by the Dyson–Mehta spectral-rigidity theorem [10]; both of these matters have been discussed in (V). Central limit theorems (CLTs) supply the general functional form of the density, its symmetry decompositions (which we shall need for J), smoothed expectation values, strength distributions, and other quantities of interest.

For finite d-dimensional spaces it has usually been standard to write $I(x) = d \times \rho(x)$ and $\int_{-\infty}^{x} I(z)\, dz = d \times F(x)$ where then $\int \rho(x)\, dx = 1 = F(\infty)$. The smoothed ground state x is then defined [11] by $F(x) = g/2d$ where g is the ground-state (gs) degeneracy. If all the degeneracies are known (they would be unity for example if we are dealing with a "pure" spectrum, i.e., with states of the same exact symmetries) the same procedure produces the complete smoothed spectrum, only a low-lying part of which could of course be directly related to physical states. If, as in the present case, our interest is in high lying states the ground state is a poor reference energy since it is known [12] that, except for nuclei with simple low-lying structures, the fluctuations ($\sim D$) in the pure-spectrum spacings of nearest-neighbor states extend to the ground state. Ignoring them, as the smoothing procedure does, propagates an error in $I(E)$ which, measured in the natural spacing units, becomes large at high energy. Besides that the smoothed density itself fails first in its tail region. We should therefore choose [13] our reference energy as the highest level up to which all the degeneracies are known, the pure spectrum D being smaller there, often very much so.

The eigenvalue density $I(x)$, usually close to Gaussian in small spectroscopic spaces, is specified by very few parameters, say the centroid ε and variance σ^2, which define a Gaussian, and perhaps the third and fourth moments which, via an Edgeworth or Gram–Charlier expansion, give corrections to it. For the huge spaces which we encounter a simple Gaussian density, even with an elevated reference energy, is usually inadequate, for CLT convergence always fails first in the tail of the distribution which is a domain of interest. Typically this begins to happen at $\sim(3\text{–}4)\,\sigma$ below the centroid, which occurs at the ground state for $d \sim 10^4$; on the other hand for the resonance region of heavy nuclei we need $d \sim 10^{14}\text{–}10^{20}$ which gives a ground state (8–9) σ below the centroid (very roughly, $\varepsilon - E_{gs} \sim (2 \ln 2d)^{1/2}\, \sigma$ for a Gaussian density). Therefore we divide the set of N active single-particle states into subsets according to $N = \sum N_i$, the density then becoming a sum of configuration densities where the configurations are defined by the corresponding partition of the particle number; thus $\mathbf{m} = [m_1, m_2, \ldots]$ where $\sum m_i = m$. We have then for the state density, its decomposition, and its moments which, in contrast to the bivariate strength moments in (8) ahead, we give for the normalized density $\rho(x)$ instead of $I(x)$ (we do this for consistency with earlier papers)

$$\begin{aligned} I^m(x) &= d(m)\, \rho^m(x) = \langle\!\langle \delta(H-x) \rangle\!\rangle^m = d(m) \langle \delta(H-x) \rangle^m \\ &= \sum_{\mathbf{m}} \langle\!\langle \delta(H-x) \rangle\!\rangle^{\mathbf{m}} = \sum_{\mathbf{m}} I^{\mathbf{m}}(x) \\ M_p &= \int \rho^m(x)\, x^p\, dx = \langle H^p \rangle^m; \\ d(m)\, M_p &= \int I^m(x)\, x^p\, dx = \langle\!\langle H^p \rangle\!\rangle^m = \sum_{\mathbf{m}} \langle\!\langle H^p \rangle\!\rangle^{\mathbf{m}}, \end{aligned} \tag{4}$$

where, as indicated, $\langle\langle G\rangle\rangle_N^m$ is, for an arbitrary operator G, its trace in the spectroscopic space; $d(m)=\binom{N}{m}$ is the dimensionality of the space so that $\langle G\rangle^m$ is the average eigenvalue. We have written the density here in an operator form (easily verified by writing the trace in H-diagonal representation as an integral) which will be useful later. We shall occasionally add a subscript N as in I_N^m. Observe carefully that the $\mathbf{m}$-decomposition is formally exact. The configuration densities are interacting since, for example, with the variance in $\mathbf{m}$ we encounter intermediate-state excitations, $\mathbf{m}\to\mathbf{m}'\to\mathbf{m}$, the exceptional case is where $\mathbf{m}$ (or the corresponding label for other decompositions of the model space) defines a good symmetry. In that case only $I^{\mathbf{m}}$ will be, like I^m, an eigenvalue density. Otherwise, with the ϕ_j forming an orthonormal basis for the $\mathbf{m}$ subspace, we have

$$I^{\mathbf{m}}(x)=\sum_i \delta(x-E_i)\left\{\sum_j |\langle\phi_j \Psi_i^m\rangle|^2\right\}; \qquad H\Psi_i=E_i^m\Psi_i; \qquad I^m(x)=\sum_i \delta(x-E_i) \quad (5)$$

so that, comparing $I^{\mathbf{m}}$ with I^m, we have at each eigenvalue E_i an extra factor equal to the total "intensity" $\leqslant 1$ of the E_i eigenvector in the $\mathbf{m}$ subspace.

If one defines the partitioning by single spherical orbits, or for example by sets of such orbits whose single-particle energies are not too widely separated, it will turn out that the individual configuration densities are adequately represented as Gaussian (with the exception perhaps of very low-lying ones for which higher-moment corrections may be significant in determining the reference state). The formal problem reduces then to a calculation of a large number of centroids and variances (which are given easily as functions of the defining Hamiltonian matrix elements), the resultant densities being then added together.

This procedure is adequate for quite large spaces. But it is inelegant and it makes no contact with the conventional Bethe NIP theory [14], $I(E)\propto\exp(b\sqrt{E})$, which with proper choice of b gives, for E not too close to the gs, a good account of the density as roughly determined by experiment. Moreover the line of argument involved above does not lead easily to an adequate theory for the strengths in very large spaces, and in particular does not give a compact form for Λ/α^2, the value for which solves completely our "reduction" problem. All of these deficiencies have been recently eliminated [7, 13] by considering the significance of an empirical observation, namely that all spherical-orbit configurations of the same harmonic-oscillator excitation, which when $V=0$ would form "spikes" of degenerate states located at the configuration centroids ε_c, have quite closely ($\pm 5\%$) the same variance σ_A^2. Their sum, the IP density which we are seeking, is then given by a convolution $I_{\text{IP}}=I_{\text{NIP}}\otimes\rho_G$ where ρ_G, a zero-centered Gaussian with unit integral and variance σ_A^2, then defines the broadening of the individual spikes, while I_{NIP} is the NIP density as modified by the centroid shifts $\langle V\rangle^{\mathbf{m}}$ also generated by V. This simple form is adequate for dealing with the resonance region of heavy nuclei (excitation energy ~ 6 MeV) which is our immediate interest. At higher excitation energy, involving states of different oscillator excitation $\underset{\sim}{s}$, we have, since σ_A^2 increases with $\underset{\sim}{s}$, a sum of convolutions. Thus

$$I_{\text{IP}} = \sum_{\underset{\sim}{s}} I_{\text{NIP}}^{(\underset{\sim}{s})} \otimes \rho_G^{(\underset{\sim}{s})} \tag{6}$$

in which only a few $\underset{\sim}{s}$ values will be needed for the density up to 30 MeV. Note then that the effective spreading variance increases with energy as would be given by any complex optical model. Observe too that $\underset{\sim}{s}$ determines the parity so that we have automatically the parity decomposition of the state density. The calculation of the NIP density and its $\underset{\sim}{s}$ decomposition follows by an application [13] of spectral averaging methods which we need not describe here.

The formal basis which gives rise to the state-density equation (6) derives from the simultaneous decomposition of the Hamiltonian with respect to the $U(N)$ group described above and its direct-product subgroup $\prod U(N_i)$ whose factors are associated with the individual spherical orbits. In terms of the defining Hamiltonian parameters (spe and interaction matrix elements) this decomposition gives the configuration centroids and variances; the average variance σ_A^2, and its variance which shows the essential constancy of the configuration variances for physically reasonable interactions; the renormalizations (2) of the Hamiltonian; a proof of the result quoted above that the ratio of the norms of U_2^2 and V_2^2 is independent of particle number (the norm of a traceless operator O being defined as $\{\langle O^+O\rangle^m\}^{1/2}$, the usual unitary norm); and via parametric differentiation of the density, smoothed expectation values of operators in the H-eigenstates, and thereby single-state occupancies and (squares of) spin cut-off factors which are defined respectively by expectation values of n_i and J_z^2. See [7, 15] for the details.

For the huge spaces required to deal with the resonance regions of heavy nuclei we now have the level density as a convolution of a (renormalized) NIP density, defined by the configuration dimensionalities and centroids, with a spreading Gaussian whose width in the domain of present interest is usually 2–3 MeV. The resultant form is both simple and physically reasonable; it contains the Bethe level density as a convolution factor, and the smallness of the spreading width makes clear why the common neglect of the interactions (i.e., the assumption that the level density has the Bethe form) is, for some restricted purposes, not a bad approximation. On the other hand the parameters of the Bethe level density and its many extensions, not hitherto determined by the theory, also follow by spectral averaging [7, 13] which very easily replaces the combinatorial and statistical-mechanical techniques used in the conventional NIP theory and gives the final density completely in terms of Hamiltonian parameters.

Finally we mention [7] that spreadings due to interactions between configurations which are very far apart in energy must not be treated by statistical methods, these being appropriate only for strongly interacting subspaces. The proper procedure is to ignore such interactions or treat them in lowest-order perturbation theory (which itself supplies a criterion to distinguish strongly from weakly interacting configurations). This gives a straightforward extension of the theory sketched above, one which is essential in dealing with phenomena at considerably higher energies.

3. Strength Distributions

As indicated in the first section, a theory of strength distributions for systems of interacting particles is essential for evaluating the near-diagonal matrix element of U in the resonance region, and thereby Λ/α^2 (3) which, coupled with the experimental Λ-bound, will yield the required α-bound. A formal theory of strengths has already been given [16] but, as we shall see, a theory of strength *distributions*, which involves quite different expansions, will be more relevant. It is worthwhile therefore to describe it in some detail.

3.1. *Preliminaries*

Consider the strength density S for the transition $|m, x\rangle \xrightarrow{O} |m, y\rangle$, generated by the transition operator O and connecting an initial m-particle H-eigenstate with energy x and a final one with energy y; we take O to be number-conserving but the extension to more general O is immediate. With I the eigenvalue density as above, we have for S and its marginal densities [16]

$$S_N^{H,m}(x, y) = I_N^{H,m}(y)\, |\langle m, y|\, O\, |m, x\rangle|^2\, I_N^{H,m}(x) = \langle\!\langle O^+\delta(H-y)\, O\delta(H-x)\rangle\!\rangle_N^m$$

$$S_N^{H,m}(x) = \int S_N^{H,m}(x, y)\, dy = \langle\!\langle O^+ O\delta(H-x)\rangle\!\rangle_N^m; \tag{7}$$

$$S_N^{H,m}(y) = \int S_N^{H,m}(x, y)\, dx = \langle\!\langle O^+\delta(H-y)\, O\rangle\!\rangle_N^m,$$

where we have assumed that the states belong to an (m, N) spectroscopic space. Where no confusion will result we shall often drop one or more of the labels H, m, N.

The bivariate moments of S are

$$M_{pq} = \int S^m(x, y)\, x^p y^q\, dx\, dy = \langle\!\langle O^+ H^q O H^p\rangle\!\rangle^m. \tag{8}$$

The bivariate central moments $\mu_{p,q}$ follow from (8) by $x \to x - \varepsilon_X$, $y \to y - \varepsilon_Y$. Note that the strength density is not normalized to unit integral ($M_{00} \neq 1$ in general); it is $M_{00}^{-1} S$ which is analogous to the densities of probability theory, and it is in terms of this function that the marginal centroids $\varepsilon_X = M_{10}/M_{00}$, $\varepsilon_Y = M_{01}/M_{00}$, and variances $\sigma_X^2 = M_{20}/M_{00} - \varepsilon_X^2$, $\sigma_Y^2 = M_{02}/M_{00} - \varepsilon_Y^2$, are naturally defined. The normalization of the bivariate cumulants K_{rs} for $r+s \geq 2$ is irrelevant since we shall deal only with the reduced cumulants $k_{rs} = K_{rs}/(K_{20}K_{02})^{(r+s)/2}$; see [17, 18]. The correlation coefficient is $\zeta=\mu_{11}/(\mu_{20}\mu_{02})^{1/2}=\mu_{11}/M_{00}\sigma_X\sigma_Y=k_{11}$. If the reduced cumulants of third order and higher ($p+q\geqslant 3$) are negligible the centroids, variances, and correlation coefficient define the (normalized) bivariate Gaussian density

$$\rho_G(x,y)=(2\pi\sigma_X\sigma_Y\sqrt{1-\zeta^2})^{-1}\exp\left[-\frac{1}{2(1-\zeta^2)}\left\{\frac{(x-\varepsilon_x)^2}{\sigma_X^2}-\frac{2\zeta(x-\varepsilon_X)(y-\varepsilon_Y)}{\sigma_X\sigma_Y}+\frac{(y-\varepsilon_Y)^2}{\sigma_Y^2}\right\}\right]. \tag{9}$$

When $\zeta=0$ it is obvious that the Gaussian density factors $\rho_G(x,y)=\rho_G(x)\,\rho_G(y)$, each factor defined by the appropriate ε and σ. An expansion especially useful for small ζ is

$$\rho_G(x,y)=\rho_G(x)\,\rho_G(y)\sum_\mu \zeta^\mu P_\mu(x)\,P_\mu(y), \tag{10}$$

where the $P_\mu(x)$ are orthonormal (Hermite) polynomials defined by the corresponding density as weight function. The form of $\rho_G(x,y)$ is unchanged, but of course not the parameters, under linear transformations of the variables. Just as a monovariate density is reduced to standard form ($\varepsilon=0$, $\sigma^2=1$) by using $Z=(X-\varepsilon_X)/\sigma_X$ so also is the case with the bivariate form where $Z_1=(X-\varepsilon_X)/\sigma_X$, $Z_2=(Y-\varepsilon_Y)/\sigma_Y$, in which the ζ has its original value, or with $(Z_1+Z_2)/\sqrt{2(1+\zeta)}$ and $(Z_1-Z_2)/\sqrt{2(1-\zeta)}$ which gives the standard form with zero correlation.

3.2. *Strength Distributions in Noninteracting Particle Systems*

Instead of proceeding via a direct evaluation of the bivariate strength moments it is good, just as with the level density, to start with noninteracting particles, $h=\sum \varepsilon_i n_i$. Then when $O=O(k)$ and $m\gg k$, so that O simultaneously transfers only a small fraction $\leqslant k/m$ of the particles, the matrix elements of O in h-diagonal representation necessarily lie in a narrow diagonal band. This implies that the (X, Y) variables defining $S^{h,m}(x,y)$ are strongly correlated, the density peaking along the line $x=y$. In the same basis we have, as the natural expansion of O,

$$O=\sum_{\alpha,\beta} C_{\alpha\beta}O_{\alpha\beta}=\sum C_{\alpha\beta}\psi_\alpha(k)\,\psi_\beta^+(k); \qquad \psi_\beta(k)\,|0\rangle=\Psi_\beta(k), \tag{11}$$

where, as indicated, $\psi_\beta(k)$ is the state operator (a product of k creation operators) for the k-particle h-eigenstate $\Psi_\beta(k)$, whose energy is written as ε_β. Now, since $O_{\alpha\beta}$ acting on a given m-particle h-eigenstate (in which the β sp states are filled; it gives zero on other states) generates one particular eigenstate, there are no cross terms in the $(\alpha\beta)$ expansion. Because of this simple structure $S^{h,m}$ can be written down by inspection; but to illustrate certain useful techniques we give first a formal derivation. Then

$$S^{h,m} = \sum_{\alpha,\beta} |C_{\alpha\beta}|^2 \, S^{h,m}_{\alpha,\beta};$$

$$\begin{aligned} S^{h,m}_{\alpha,\beta}(x, y) &= \langle\!\langle O^+_{\alpha\beta}\delta(h-y)\, O_{\alpha\beta}\delta(h-x)\rangle\!\rangle^m_N = \langle\!\langle \delta(h-y+\varepsilon_\alpha-\varepsilon_\beta)\,\delta(h-x)\, O^+_{\alpha\beta}O_{\alpha\beta}\rangle\!\rangle^m_N \\ &= \delta(x-\varepsilon_\beta-y+\varepsilon_\alpha)\langle\!\langle \delta(h-x)\prod_{r\in\alpha} n_r \prod_{\substack{s\in\beta\\ s\notin\alpha}} (1-n_s)\rangle\!\rangle^m_N \\ &= \delta(x-\varepsilon_\beta-y+\varepsilon_\alpha)\, I^{h,m-k}_{N-N(\alpha\beta)}(x-\varepsilon_\beta) = T^{h,m-k}_{N-N(\alpha\beta)}(x-\varepsilon_\beta, y-\varepsilon_\alpha) \end{aligned}$$

$$S^{h,k}_{\alpha,\beta}(x, y) = \delta(x-\varepsilon_\beta)\,\delta(y-\varepsilon_\alpha); \qquad T^{h,m}_N(u, v) = \delta(u-v)\, I^{h,m}_N(u), \tag{12}$$

where we have used $O^+_{\alpha\beta}\delta(h-y) = \delta(h+y+\varepsilon_\alpha-\varepsilon_\beta)\, O^+_{\alpha\beta}$, have then evaluated $O^+_{\alpha\beta}O_{\alpha\beta}$ in terms of single-state number operators $n_i = a_i^+ a_i$, and, for the final form, have used the simple theorems that, for an arbitrary operator F written as a function of the n_i (which is always possible in a trace),

$$\langle\!\langle n_i F\rangle\!\rangle^m_N = \langle\!\langle F(n_i \to 1)\rangle\!\rangle^{m-1}_{N-1_i}; \qquad \langle\!\langle (1-n_j)\, F\rangle\!\rangle^m_N = \langle\!\langle F(n_j \to 0)\rangle\!\rangle^m_{N-1_j}, \tag{13}$$

where $(N-1_i)$ indicates that sp state i is to be omitted in the evaluation. The $m=k$ strength in (12) which follows from $I^{h,0}(z) = \delta(z)$, or directly from the definition of O, may be called the "elementary strength density." Our problem is to find how this strength, defined in the simplest space (in more general cases between two such spaces) propagates to more complex ones. The answer comes directly from (12), which may be rewritten in bivariate convolution form as

$$S^{h,m}_{\alpha\beta} = S^{h,k}_{\alpha\beta} \otimes T^{h,m-k}_{N-N(\alpha\beta)}; \qquad S^{h,m} \xrightarrow{N \gg N(\alpha\beta)} S^{h,k} \otimes T^{h,m-k}_N, \tag{14}$$

where the large-N form comes by summing over (α, β).

Before proceeding we indicate how, as mentioned above, (12) and hence (14) could be derived by considering the simple underlying structure. As one factor we need an energy-conserving δ function defined by $x-\varepsilon_\beta+\varepsilon_\alpha = y$; the other factor is simply the density of states at energy $(x-\varepsilon_\beta)$ generated by the "free" particles distributed over the "free"sp states. The number of free particles is $(m-k)$ since k particles are initially fixed in the sp states of β. If α and β are disjoint the number of free sp states is $N-2k$, those of α and β not being free since initially the k states of β must be filled and those of α must be empty; in the general case the set of free states is $N-N(\alpha, \beta)$ where $N(\alpha, \beta)$ counts the states in the union of α and β.

To understand the nature of the strength density (12) we should specify the $C_{\alpha\beta}$ coefficients which define O, or more precisely their absolute squares. For a reasonable example we can simply take $|C_{\alpha\beta}|^2 = 1$. Alternatively (and more appropriately for TRNI, for which we have no detailed model) we can represent O as a random operator defined by specifying the distinct $C_{\alpha\beta}$ to be independent random variables with zero centroids and unit variances, and then carry out the ensemble average. In our NIP case these two prescriptions give the same result. For the first we have from (12) that

$$S^{h,k}(x, y) = \sum_{\alpha\beta} \delta(y - \varepsilon_\alpha)\, \delta(x - \varepsilon_\beta) = I^{h,k}(x)\, I^{h,k}(y). \tag{15}$$

For the second simply take $H = h$ in (7) and carry out the ensemble averaging.

Let us now take the many-particle dilute limit, $N \gg m \gg 1$. Then in (14) the asymptotic form of S is appropriate. Its bivariate cumulants, which are additive under convolution, are then given in terms of the univariate cumulants K_r of the m-particle density by

$$K_{pq}(m) = K_{p+q}(m-k) + K_p(k)\, \delta_{q0} + K_q(k)\, \delta_{p0} \tag{16}$$

which follows from (14), (15) and the additivity of bivariate cumulants under convolution. Since in the dilute limit the m-particle density is an m-fold convolution of the one-particle density we have $K_r(m) = mK_r(1)$; then, the reduced cumulants k_{pq} and the correlation coefficient ζ are given by

$$\begin{gathered} k_{pq}(m) = \{(m-k)\, k_{p+q}(1) + kk_p(1)\, \delta_{q0} + kk_q(1)\, \delta_{p0}\}\, m^{-(p+q)/2} \\ \zeta = k_{11} = 1 - k/m. \end{gathered} \tag{17}$$

For large m, $\zeta \to 1$ as predicted above, while the cumulants with $p+q \geqslant 3$ go to zero, this suggesting that we have in the limit a (singular) bivariate Gaussian. However, in terms of new variables $(X+Y)/2$ and $(X-Y)$, the fourth cumulants k'_{pq}, arranged in order of increasing q, behave as m^{-1}, $m^{-3/2}$, m^{-1}, $m^{-1/2}$, m^0 ($k'_{04} = k_4(1)/(2k)$), indicating a departure from true Gaussian connected with the narrowly constrained $(X-Y)$ variable. This minor deviation, however, which is found even with interactions (Section 3.5) should be of little or no consequence in almost all cases.

It should be observed that the bivariate Gaussian result found here is by no means a general NIP result, that being given by (14) whose specific form depends on the parameters. A restriction arises in the above case from our agreement that the level density is of m-fold convolution type and therefore, for large m, of Gaussian (appropriate for finite N) rather than Bethe form.

3.3. *Strength Distributions in Interacting-Particle Systems*

Returning now to the general case where the $N(\alpha\beta)$ correction is not ignored and introducing the two-body interaction, we have $H = h + V$ where we will understand, just as with the level density, that V is of unitary type $\nu = 2$, the $\nu = 0, 1$ parts of V having gone into a renormalization of h. For a heuristic derivation of the strength distribution let us assume that "the spreading of the strength follows that of the density," i.e. that when a pair (initial and final) of h-eigenstates spreads among the H-eigenstates as V is turned on, the initial strength connecting them spreads in proportion to the product of the admixing intensities. This should be essentially exact when, as we shall do for TRI, we represent O by a featureless ensemble. In other cases there could be a difficulty if O is strongly correlated with H; we would in that case replace O by $O - \langle OH \rangle^m H/\langle H^2 \rangle^m$ which has the same (off-diagonal) strengths as O but is linearly uncorrelated with H. We could, at the expense of evaluating more complicated traces than those linear in H, also eliminate, say, the H^2 and H^3 components of O.

Since for TRNI in heavy nuclei we shall have only $s=0$ in (6), V generates, for the initial and final states, a Gaussian spreading represented by ρ_G with centroid zero, variance $\sigma_A^2(m)$, and unit integral. The NIP strength then takes on a convolution in each variable; if we could regard these as uncorrelated the IP strength would be represented by a convolution of the NIP strength with the product $\rho_G \rho_G$ of two Gaussians, one for each variable. Now when we take account of the fact that the initial and final spreadings are in fact correlated, the separable bivariate Gaussian becomes a bivariate Gaussian with zero centroid, variance $\sigma_A^2(m)$, and unit integral as before, but now with a correlation coefficient to be determined. This function, being generated by V, may be written as $\rho_G^{V,m}$ so that

$$S_{\alpha\beta}^{H,m} = S_{\alpha\beta}^{h+V,m} \simeq S_{\alpha\beta}^{h,m} \otimes \rho_G^{V,m} = S_{\alpha\beta}^{h,k} \otimes \rho_G^{V,m} \otimes T_{N-N(\alpha,\beta)}^{m-k}. \tag{18}$$

We have from (7) that $S_{\alpha\beta}^{0,m}(x,y) = \langle\!\langle O_{\alpha\beta}^{+} O_{\alpha\beta} \rangle\!\rangle^m \delta(x)\,\delta(y) = M_{00}(\alpha\beta)\,\delta(x)\,\delta(y)$ where $M_{00}(\alpha\beta)$ is the zeroth (normalization) moment (8) for $O_{\alpha\beta}$. Then taking $h=0$ in (18) we find, as an explicit form for the bivariate spreading function, that

$$\rho_G^{V,m}(x,y) \simeq M_{00}^{-1}(\alpha\beta)\, S_{\alpha\beta}^{V,m}(x,y) = M_{00}^{-1}(\alpha\beta) \langle\!\langle O_{\alpha\beta}^{+} \delta(V-y)\, O_{\alpha\beta} \delta(V-x) \rangle\!\rangle^m. \tag{19}$$

Then (18) may be written as

$$\begin{aligned} &\langle\!\langle O_{\alpha\beta}^{+} O_{\alpha\beta} \rangle\!\rangle^m \langle\!\langle O_{\alpha\beta}^{+} \delta(h+V-y)\, O_{\alpha\beta} \delta(h+V-x) \rangle\!\rangle^m \\ &\quad \simeq \int du\, dv \langle\!\langle O_{\alpha\beta}^{+} \delta(h-y-v)\, O_{\alpha\beta} \delta(h-x-u) \rangle\!\rangle^m \langle\!\langle O_{\alpha\beta}^{+} \delta(V+v)\, O_{\alpha\beta} \delta(V+u) \rangle\!\rangle^m \end{aligned} \tag{20}$$

which displays the structural content of our "heuristic derivation." Note also that (19) defines the needed Gaussian correlation coefficient, if we take for granted, as we shall (that assumption involving an incidental use of an ensemble notion coupled with ergodicity; see Section 3.5) that $(\alpha\beta)$ cross terms are negligible.

The evaluation of (18) is particularly simple, indeed trivial, in spaces large enough so that the Pauli blocking effects represented by the $N(\alpha, \beta)$ correction to N may be ignored. Then carrying out the (α, β) sum gives $S^{H,m}$ as a convolution of three well understood and easily calculable functions. The exact evaluation is also quite feasible and has been recently used [19] for studying TRI via detailed-balance experiments involving ^{28}Si as compound nucleus. An essential simplification arises from the fact that for J-scalar operators (and there is a corresponding result for nonscalar) all (α, β) pairs involving the same spherical orbits give rise to the same strength; thus the number of distinct terms in (18) is not $\sim \binom{N}{k}^2$, where the number of sp *states* $N \sim 100$ for heavy nuclei, but rather $\sim \binom{N'}{k}^2$ with the number of spherical *orbits* $N' \sim 5$–10. Finally we note that if we replace the renormalized NIP density and the IP spreading variance by the sets of spherical-configuration centroids (dimensionality-weighted) and variances, we recover the inelegant solution in terms of sums over very many configurations which we implicitly started with and which in fact first gave the numerical results ahead.

Finally we remark that in small enough IP spaces, $I^{H,m}$ and $S^{H,m}$ are representable by single Gaussians, formally because the (Gaussian) densities for h and V are reproduced in form under convolution, and physically because all shell-model basis states in such spaces, being strongly interacting, are susceptible to simple statistical (CLT) treatment. It is entirely essential to understand that one cannot extend, from these small spaces to the huge spaces necessary for our applications, simply by calculating higher-order Edgeworth corrections. That way is neither feasible nor effective. Thus the shell-model tests described in the next section are really directed toward confirming that strength densities are *locally* bivariate Gaussian. The general form would emerge in a huge space when we properly include the secular effects of the (NIP) h, which "carries" via (18) the bivariate Gaussian form to higher energies.

3.4. *Shell-Model Tests*

In order, among other things, to test our strength distribution we turn to shell-model results [16] for the 307-dimensional $(ds)^6$ space with $J=2$, $T=0$, which use spe taken from the ^{17}O spectrum along with a "realistic" interaction V. We take V itself to be the transition operator. From the shell-model matrix for H, and its eigenfunctions, we have constructed the diagonalizing matrix and thereby the V matrix (which would be diagonal in the shell-model basis if h were zero) in H-diagonal representation. We are dealing here not with a GOE but instead with an EGOE representative, so that, to begin with, we ask whether our V matrix is locally of GOE type, as proposed in V Sect. 8. Instead of dealing with the diagonal elements by the correlation procedure of Section 3.3 above, it is simpler to put them to zero, a procedure which though formally not correct for GOE will have very little effect on the analysis since only a small fraction of the elements are diagonal. Ignoring the highest eigenvector we partition the matrix into 9×9 34-dimensional blocks, and analyze the 45 matrix element distributions.

In every block the centroid is close to zero, values for the diagonal blocks lying in the range $(-0.02, +0.03)$ MeV. The skewness, k_1, and excess, k_2, are similarly small, typically $\sim(0.03$–$0.10)$ for $|k_1|$ and (0.08–0.30) for $|k_2|$. The only exceptions are the two extreme diagonal blocks for which $k_1 = (0.14, 0.23)$ and $k_2 = (0.17, 1.6)$. These results are entirely consistent with the idea that, in H-diagonal basis, the embedded V behaves locally like a GOE. Things are very different in the shell-model basis where the block centroids are small but the skewnesses are large and the excesses huge.

The RMS values of the strengths (the square roots of the strength variance)

$|\langle E'| V |E\rangle|^2$ in every block agree remarkably well with the predictions of the bivariate Gaussian, the level densities being divided out. For example, in the second row of the 9×9 array we find, in MeV, (0.35, 0.36, 0.36), (0.32, 0.34, 0.33), (0.28, 0.31, 0.30), (0.27, 0.28, 0.29), (0.28, 0.26, 0.27), (0.25, 0.23, 0.24), (0.21, 0.19, 0.21), (0.14, 0.16, 0.17), (0.08, 0.10, –) where the first number in each triplet derives from the shell model and the second from the bivariate Gaussian with $\zeta = 0.58$. The falling off in the width as we go away from the diagonal arises from the NIP mechanism discussed at the beginning of Section 3.2, moderated by the admixing generated by V. It still survives because the admixing energy range is small on the natural scale defined by h acting in $(ds)^6$, and it is responsible for the fact that V is only locally like a GOE. The quality of the agreement shown above between shell model and bivariate Gaussian extends over the entire matrix; the Edgeworth corrections [18], for which in each block $|k_{rs}| \lesssim 0.25$ for $3 \leqslant r+s \leqslant 6$, slightly improve the agreements for the end blocks.

The third number in each triplet follows from a bipolynomial expansion of the strength itself [16] rather than the strength density. Involving polynomial products of different orders, as it does, this is a different expansion from that given by (10) which of course also yields a strength expansion (for small ζ). The excellent agreement with bivariate Gaussian which is demonstrated (and found over the entire array) has however required traces which are quartic in H rather than quadratic as needed for the Gaussian. Agreement at the level indicated is *not* found at lower order in H; on the other hand proceeding to higher order in H (which is possible in the numerical example but not analytically feasible) does not improve the agreement. Thus we observe the clear superiority of the bivariate Gaussian form. We refer finally to [20] for what is probably the first discussion about bivariate Gaussian densities (in small spaces, as in the present example).

As a further test we consider the pth order ($p = 0, 1, 2$) energy-weighted strength sums,

$$M_p(E) = \sum_W W^p |\langle W| V |E\rangle|^2 = \int I(W)\, W^p |\langle W| V |E\rangle|^2 \, dW$$

$$= I(E)^{-1} \int W^p S(E, W)\, dW, \tag{21}$$

the final integral being the pth moment of the conditional strength density $S(W \mid E)$. In (21) the first form, for every individual eigenvalue E, will be used for the shell model, and the second, with smoothed functions, for the Gaussian. $M_0(E)$ is the total strength originating with E, $M_1(E)/M_0(E) = \varepsilon_S(E)$ is its centroid, and $M_2(E)/M_0(E) - \varepsilon_S^2(E) = \sigma_S^2(E)$ is its variance. Figures 1–3 show excellent agreements, to within fluctuations, between shell model and theory. The lowest order (m^{-1}) Edgeworth correction is helpful near the spectral extremes for the centroid and variance. $M_0(E)$ is proportional to the ratio of two Gaussian densities; its curvature, clear both in the shell model and the Gaussian theory, arises from the shifts in centroid and variance generated by the strength weighting of the level den-

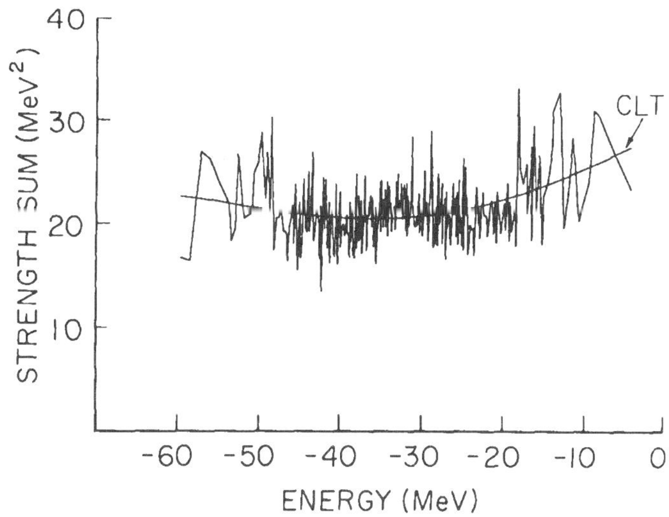

FIG. 1. The non-energy-weighted sum, i.e., the total strength, $M_0(E)$, originating at E. Exact shell-model results are plotted along with the bivariate Gaussian prediction. The Edgeworth corrections are negligible. This figure and the two which follow are patterned after figures in [16], given there for electromagnetic operators instead of V as here.

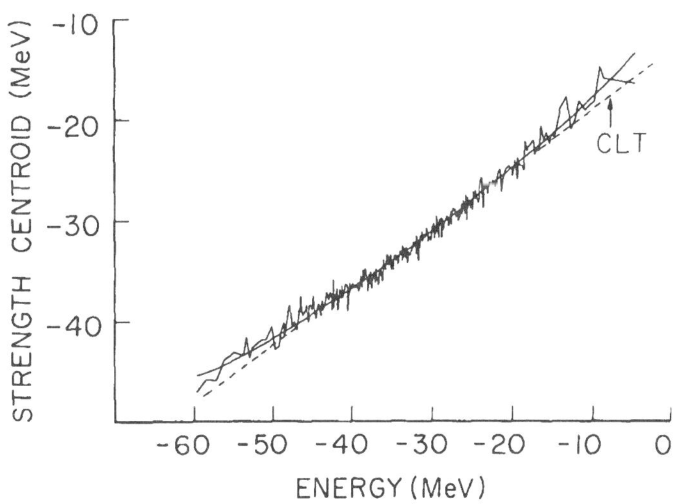

FIG. 2. The linear-energy-weighted sum, $M_1(E)/M_0(E)$. The exact strength centroid is compared to the Gaussian prediction with and without order m^{-1} Edgeworth corrections. The dashed line gives the linear energy variation characteristic of bivariate Gaussian (Eq. (22)).

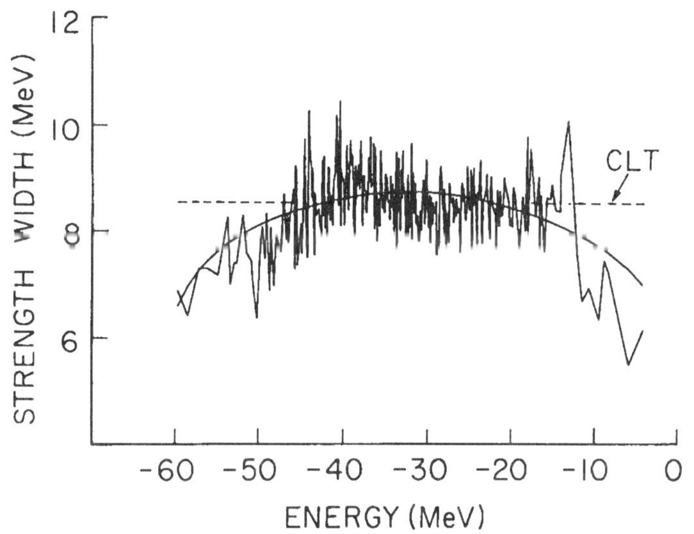

FIG. 3. The strength width, $[M_2(E)/M_0(E)-(M_1(E)/M_0(E))^2]^{1/2}$, with and without order m^{-1} Edgeworth corrections. The dashed line gives the constant Gaussian prediction (Eq. (22)).

sity in (21). The linearity of the CLT centroid curve and the constancy of the CLT variance

$$\varepsilon(E)=\varepsilon_S\zeta(E-\varepsilon_S);\ \sigma^2(E)=(1-\zeta^2)\,\sigma_S^2, \tag{22}$$

which are characteristic properties of a bivariate Gaussian density, are obvious in the figures, as are also the effects of the first Edgeworth corrections. Further positive evidence about the applicability of (22) is given in Refs. [16, 21].

3.5. *Ensemble Derivations*

We turn briefly to an ensemble derivation of the Gaussian densities. The logic of this is that because of the strong ergodic behavior of the embedded GOE (EGOE; see [9] and V, Sect. 8), averaging over such an ensemble for an operator V will give results valid for almost all members of the ensemble. One purpose is to understand the formal basis for our heuristic derivation of the strength density in Section 3.3, specifically to show that the correlations ignored in that derivation are those which vanish on ensemble averaging.

We take V to be a t-body EGOE ($t=2$ in the nuclear case). Because of the dominance of the maximum-ν $U(N)$ representations when N is large (the dimensionality [5] is $\sim N^{2\nu}/(\nu!)^2$) the ensemble operators will be automatically unitarily irreducible with $\nu=t$. We can, without cost, take h to be traceless and therefore irreducible with $\nu=1$, and similarly for O with $\nu=k$. There will be no point in deriving asymptotic Gaussian densities in spaces so large that, because of failure in the distribution tails, they become invalid in domains of physical interest. Therefore we restrict ourselves to excitations within the same oscillator shell (e.g., $s=0$ in (6)). To go beyond that would call for partitioned V ensembles.

It was shown in an earlier EGOE study [9] that for any operator T of fixed unitary rank v

$$\underbrace{VTV} = \overline{\langle V^2\rangle^t}\binom{n-v}{t}T, \tag{23}$$

where n is the number operator, the superscript bar denotes ensemble averaging, and the binary link $\sqcup$ denotes averaging over the pair of (correlated) V's connected by it. This result, valid for large N, can be used repeatedly (along with $\bar{V}=0$) to derive the low-order ensemble-averaged moments of $I^{H,m}$ and $S^{H,m}$. We consider below the dilute limit, $1 \ll m \ll N$.

We start with the moments of $d^{-1}(m)\, I^{V,m}$. The odd-order moments vanish. The variance is given by

$$\sigma^2(m) = \overline{\langle V^2\rangle^m} = \binom{m}{t}\sigma^2(t), \tag{24}$$

where we have used (23) with $T=1$. Similarly the fourth moment is given by

$$\begin{aligned}\overline{\langle V^4\rangle^m} &= \langle \underbrace{VV}\,\underbrace{VV}\rangle^m + \langle \underbrace{V\,\underbrace{VV}\,V}\rangle^m + \langle \underbrace{VVVV}\rangle^m \\ &= \sigma^4(m)\left\{2 + \binom{m-t}{t}\binom{m}{t}^{-1}\right\} \xrightarrow{m\to\infty} 3\sigma^4(m). \end{aligned}\tag{25}$$

It should be clear from (23)–(25) that for sufficiently large m we can regard each binary link as making the same contribution. Then, since the number of ways of pairing 2μ objects is $(2\mu-1)!!$, the moment of order 2μ is $(2\mu-1)!!\,\sigma^{2\mu}(m)$ which defines a zero-centered (univariate) Gaussian ρ_G so that [9] $I^{V,m} = d(m)\,\rho_G$. By the same argument $\overline{VhV} \simeq \overline{VVh}$ and then [22] $I^{H,m}$ is obtained as a convolution: $I^{H,m} = d^{-1}(m)\, I^{h,m} \otimes I^{V,m}$ as in (6).

Implicit in the above reasoning is the rule that in large-m spaces almost all operators of low-particle rank commute. Pursuing this for S^H gives the zeroth-order approximation, $S^H(x,y) = \delta(x-y)\, I^H(x)\, M_{00} d^{-1}(m)$, which corresponds to perfectly correlated variables and hence no transitions at all. It is essential to understand however that the commutation is not exact, that the operators are not simultaneously diagonalizable, and therefore that the small departures from exact commutativity must be taken into account when dealing, for example, with the off-diagonal matrix elements of O. We may then allow V to commute, as above, with h as well as with another V of a different binary link, but *not* with O. With this rule we find

$$\bar{M}_{pq}(H) = \overline{\langle\langle O^+(h+V)^q\, O(h+V)^p \rangle\rangle^m} \to \sum_{r,s} \binom{p}{r}\binom{q}{s} \langle\langle O^+ h^{q-s}\, \overline{V^s O V^r}\, h^{p-r} \rangle\rangle^m$$

$$= M_{00}^{-1} \sum_{r,s} \binom{p}{r}\binom{q}{s} \bar{M}_{p-r,q-s}(h)\, \bar{M}_{rs}(V), \tag{26}$$

where in the last step we have used, from (23), $\overline{V^r O V^s} = f(n)\, O$ where $f(m) = M_{00}^{-1} \bar{M}_{rs}(V)$. Moreover $\bar{M}_{pq}(V)$, which is zero for $p+q$ = odd, is given, for $p+q$ = even, by

$$\bar{M}_{pq}(V) = \overline{\langle\langle O^+ V^q O V^p \rangle\rangle^m} \to \sum_r \binom{p}{r}\binom{q}{r} r! \langle\langle O^+ \overline{V^{q-r}}\, \underbrace{V \cdots \underbrace{V O V} \cdots}\, \overline{V^{p-r}} \rangle\rangle^m$$

$$\to \sum_r \binom{p}{r}\binom{q}{r} r!(p-r-1)!!(q-r-1)!!\, \zeta^r \sigma^{p+q}(m)\, M_{00}. \tag{27}$$

Here in the second step we have made an expansion in r (the number of V's correlated across O) and then allowed the V's of different binary links to cummute; the factor $r!$ is included so that in the multiply linked expression centered around O only one linkage pattern is to be counted. In the last step

$$\zeta = \frac{\overline{\langle\langle O^+ VOV \rangle\rangle^m}}{\langle\langle O^+ O V^2 \rangle\rangle^m} = \frac{\sigma^2(m-k)}{\sigma^2(m)} = \binom{m-k}{t}\binom{m}{t}^{-1} = 1 - \frac{kt}{m} + O\left(\frac{1}{m^2}\right) \tag{28}$$

is the correlation coefficient. The final form of (26) yields the convolution result $S^{H,m} = M_{00}^{-1} S^{h,m} \otimes S^{V,m}$ and that of (27) yields, via the (bivariate) Gaussian $\rho_G^{[2]}$ (with zero centroids, $\sigma^2(m)$ variances, and ζ correlation coefficient) of (9), (10), $S^{V,m} = M_{00}\rho_G^{[2]}$, thereby completing a formal derivation of the dilute limit IP strength density of Section 3.3. Note carefully that it is the *very small* departure from the exact commutativity of V and O (as well as of h and O) which gives rise to an *almost* singular bivariate strength density.

In this derivation we have ignored the $O(m^{-1})$ corrections to all reduced cumulants except to the correlation coefficient. For the low-order cumulants these corrections are also easily calculable and could be incorporated into an appropriate Edgeworth expansion. We have calculated them exactly ($N \to \infty$, finite m) for $p+q \leqslant 6$ for S^V. On expansion we find that $k_{pq} = -(t^2/m) + O(1/m^2)$, $(t^3(6t-1)/m^2) + O(1/m^3)$ for $p+q = 4, 6$, respectively. The leading terms do not depend on k nor on p, q separately. The next terms are needed for verifying whether the uncorrelated variables $(X+Y)/\sqrt{2(1+\zeta)}$ and $(X-Y)/\sqrt{2(1-\zeta)}$ are independent Gaussians. We find that, as $m \to \infty$, the two variables are indeed independent. However, the difference variable is not a Gaussian, its fourth and sixth cumulants being respectively $(2k)^{-1}(2t-3)$ and $(4k^2)^{-1}(16t^2 - 46t + 35)$ which for our application ($k = t = 2$) are small (4^{-1} and 7/64, respectively). Details will be given elsewhere. Finally it should be noted that, by taking $t = 1$, the results here agree with the small-space (NIP) results of Section 3.2 for the correlation coefficient and for the anomalous behavior of the difference variable; the latter is now seen to extend to the IP case also. All results would agree had we taken h as a one-body EGOE or, equivalently, had given it a semicircular $m = 1$ spectrum.

4. The Spectroscopy of Heavy Nuclei; α-Bounds on TRNI

We focus our attention on heavy nuclei since it will become clear that, other things being equal, these will give the best fluctuation bounds (as opposed to detailed-balance bounds) on α. From the nuclei considered in (V) we select the five nuclei of Table I, chosen respectively from groups (3, 4, 2, 1, 2) of Table V-1), for which the spacing D varies from 1 to 100 eV. We take for granted that the A-bounds, given for the groups (except 1) containing each of these nuclei, apply to the individual nuclei. The fact that the spacings span two orders of magnitude will enable us to say something about the systematic behavior of certain parameters.

Broadly speaking a first choice of the model space (numbers of particles, and the contributing orbits) follows by inspection of the zero-deformation Nilsson spectrum. The results of this are given in the second and third columns of Table I; they yield model spaces of dimensionality 10^{15}–10^{20}. A fitting of the state density, calculated vs observed, gives the interaction-magnitude parameter G for the assumed "surface-delta" interaction [23], which, as expected, to within fluctuations slowly decreases with A. In making this calculation [24] we have chosen the reference states to be the ground states for the parities of interest; the exceptional case is ^{236}U, for which we have taken two reference states with widely different J values as a test and find G's and average configuration widths differing by about 15%. Since G is determined by fitting D (for fixed J) the calculations are not "absolute." With an error of about 5% on average we find $G \sim 28/A$ MeV, to be compared with $25/A$ which is proposed [23] for low-lying spectroscopy in light and intermediate nuclei and perhaps also in heavy nuclei. Had we chosen the latter value, our calculations, which would have produced the same α-bounds, could reasonably be described as absolute. Moreover it has been very recently shown [13] that absolute calculations of the resonant spacings D can be made to 20% accuracy over the whole periodic table by choosing the magnitude parameter to fit the number of *states* (not levels) in a (0.1–0.5) MeV band just below the second reference energy defined in the third paragraph of Section 2. This value of G, determined by considering a few hundred states, then extrapolates, with the accuracy quoted, to the resonance energy, below which there may be many millions of states. This result incidentally tells us something significant about the simplicity of secular behavior in complex systems, or at least in complex nuclei. Except for ^{169}Er we have not redone the calculations of Table I with the elevated reference energy because our experience tells us that, as long as the strength is calculated in the region where the level density is given correctly and is stable under model-space enlargement, and as long as the values of G do not differ markedly from expectations, the α-bounds do not depend on the precise procedures followed.

TABLE I

Nucleus	n_p	Orbits	E (ref.)	$\sigma(V)_{RMS}$	Λ/α^2	α-Bound
J^π (res.)	n_n	(spe/$\hbar\omega$)	(MeV)	ζ		
D (eV)				σ_J		$\Lambda D/\alpha^2$
E (res: MeV)			J^π (ref.)			(keV)
$\Lambda^{1/2}$						
^{167}Er	10	5/2+, 11/2−, 3/2+	gs	2.35	3,900	3.3×10^{-3}
1/2+		(5.68, 5.80, 6.00)	7/2+	0.76		
38.5	17	7/2−, 9/2−, 13/2+		10.2		150
6.44		(6.54, 6.61, 6.72)				
0.18		3/2−				
		(6.92)				
^{169}Er	10	5/2+, 11/2−, 3/2+	0.24	2.15	730	2.5×10^{-3}
1/2+		(5.68, 5.80, 6.00)	7/2+	0.74		
97.0		1/2+		10.1		
6.00		(6.04)				71
0.09	11	9/2−, 13/2+, 3/2−				
		(6.61, 6.72, 6.92)				
		5/2−, 1/2−				
		(6.97, 7.11)				
^{233}Th	8	9/2−, 13/2+, 7/2−	gs	2.25	11,600	2.0×10^{-3}
1/2+		(6.47, 6.56, 6.63)	1/2+	0.85		
16.8	17	9/2+, 15/2−, 11/2+		12.8		195
4.79		(7.39, 7.63, 7.64)				
0.16		5/2+				
		(7.81)				
^{236}U	10	See ^{233}Th	1.00	2.10	108,000	—
3−	18	"	7−	0.80		
1.08				13.3		117
6.55			0.69	2.50	145,000	—
—			1−	0.80		
				13.4		157
^{239}U	10	See ^{233}Th	gs	2.40	6,030	4×10^{-3}
1/2+	21	"	5/2+	0.80		125
20.8				13.8		
4.81						
0.20						

Note. Given for each nucleus are the resonance parameters including $\Lambda^{1/2}$ from (V); the number of active protons and neutrons; the sp orbits and the spe where $\hbar\omega \simeq 41/A^{-1/3}$ MeV; the spectroscopic reference state (two of these for ^{236}U); the RMS state-density spreading width, the correlation coefficient, and the calculated spin cut-off factor; and the calculated strength Λ/α^2. The final column gives the α-bounds calculated at high statistical confidence, and $\Lambda D/\alpha^2$ which is seen to be roughly constant.

As mentioned in Section 3.3 what is formally the most elementary (also the most accurate) procedure has been used for dealing with our general strength equation (18). It is described in detail in [24]. The construction of the level density via spherical-orbit partitioning [5] calls for a set of configuration dimensionalities, centroids, and variances; there will be 10,000 to 30,000 of them depending on the nucleus, these numbers reducing to 1000–2500 when configurations with very low intensities in the resonance region are dropped. The next step is to associate with each pair of configurations connected by the transition operator a bivariate Gaussian strength density. In the compact form of the strength density $S = \sum S_{\alpha\beta}$ with $S_{\alpha\beta}$ given by (18) this is automatically achieved by choosing the pairs of orbits associated with α, β, and this obviously applies to the more explicit form of the strength density which has been used.

There being no general agreed-upon form for the TRNI interaction operator U (but see [25]) we take it as an ensemble patterned after V. Ideally then, in the two-particle defining space the mean-square fixed-J nondiagonal elements of U should on the average equal those of V (which are roughly proportional to $(2J+1)^{-1}$). The diagonal elements of U in j-coupled representation must vanish by antisymmetry. In practice we can set the nonzero squared elements of U equal to those of V except where U, V correlations, which the ensemble model puts to zero, would be significant. Finally a renormalization factor is applied to U to restore the norm equality of U, V.

We are still in need of the correlation coefficients for the bivariate Gaussian partial strength densities connecting pairs of configurations. Exact calculations for a given transition operator (but not for fixed J) are feasible but difficult; fortunately we do not need them. To begin with we can assume that the exact ζ's are only mildly fluctuating for the configuration pairs dominant in the resonance region; therefore we take ζ to be the same for all pairs, relying on the fact that the α-bound is insensitive to small fluctuations in the range of interesting ζ-values given in Table I (changing ζ from 0.6 to 0.9 reduces the bound by less than a factor 2 for ^{169}Er). For the ζ-calculations we use EGOE techniques [9] as in Section 3.5, extensions to fixed proton and neutron numbers m_p, m_n, to finite N_p, N_n, and to combinations of symmetric and antisymmetric matrices being called for. Successful tests have been carried out [24], in the 6×10^{14}-dimensional ^{169}Er space, of the applicability of these techniques using the operators h, V, and J^2; the test is demanding since of these operators, only V is a natural EGOE candidate, h having high degeneracies and J^2 a non-Gaussian ($\sim \chi_3^2$) spectrum. As examples we find, giving, in order, (1) the exact value, (2) the ensemble (m, N) value, (3) the ensemble $(m_p, N_p; m_n, N_n)$ value, and (4) the ensemble dilute limit $(N \to \infty)$ value, 0.818, 0.863, 0.878, 0.905 for the trace ratio $\langle hVhV \rangle / \langle h^2 \rangle \langle V^2 \rangle$ and 1.012, 1, 1, 1 for $\langle h^2 V^2 \rangle / \langle h^2 \rangle \langle V^2 \rangle$. Similarly the exact value of $\langle h^2 J^2 \rangle / \langle h^2 \rangle \langle J^2 \rangle$ is found to be 0.981 which, being close to unity, implies that h^2 and J^2 are uncorrelated. Other examples are given in [24] which gives also the explicit trace formulae used. It is found that neither the partitioning by proton and neutron orbits nor the use of finite N changes significantly the correlation coefficient from the low-density value,

(28) with $t=k=2$, $\binom{m-2}{2}/\binom{m}{2} \simeq 1-4/m$ which gives 0.81–0.87 for the nuclei of Table I, and, for ^{169}Er, 0.81 to be compared with the exact EGOE value 0.74.

Before proceeding to the α-bound we must understand the J-dependence of the strength density. The fact that the H–J^2 correlation coefficient is small over the entire periodic table (the resonant-region values for our nuclei are ~ 0.01 for U and Th and ~ 0.03 for Er) indicates that the J effects are simple. The shell-model tests of Section 3.4, carried out for fixed J, suggest that if J has an effect at all on the shape of the strength density, it is to better the agreement with bivariate Gaussian. Our problem then is not with the form but with the J-dependence of dimensionalities, marginal centroids and variances, and correlation coefficients. Since $[J^2, H]=0=[J^2, U]$ the nonvanishing trivariate moments of $S(E, E', J)$ are of the form $\langle\langle UH^pUH^qJ^{2r}\rangle\rangle$. The trace tests [24] for h, V, J^2, or equivalently h, H, J^2, described above with examples, indicate that, when $E' \simeq E$ which is what we need, the strength density factors according to

$$
\begin{aligned}
&S(E, E', J) \simeq S(E, E')\, \rho_M(J \mid E); \qquad \int_0^{\infty} (2J+1)\, \rho_M(J \mid E)\, dJ = 1 \\
&\rho_M(J \mid E) = (2J+1)(8\pi\sigma_J^6(E))^{-1/2} \exp[-(2J+1)^2/8\sigma_J^2(E)],
\end{aligned}
\tag{29}
$$

where $\rho_M(J \mid E)$ is a Maxwellian density with $\sigma_J(E)$ the spin cut-off factor at energy E. This is the simplicity predicted above. The same ρ_M gives the J-decomposition of the level density, so that the parameter $\Lambda D/\alpha^2$ is J-independent.

We summarize now. In order to complete the $\Lambda \to \alpha$ reduction, and thereby determine the relative bound α on the nucleon–nucleon TRNI interaction, we need to construct and test the state density I; to use its partitioned form, or proceed otherwise from (18); to evaluate the near-diagonal strength S generated by U as transition operator; to calculate from the state density the spin cut-off factor $\sigma_J(E)$ and thereby the J-decomposition of the strength which, taken for the resonance J value, gives Λ/α^2 and thereby α itself. The final results are shown, along with certain other quantities, in Table I. We think it reasonable to quote as our final bound $\alpha \leqslant 2 \times 10^{-3}$, a value which by some particle-theoretic arguments [3, 26] is at the boundary defining values of fundamental interest. Because Λ was determined in (V) by a 3σ confidence criterion this α-value may be described formally as one with high ($\sim 99\%$) confidence. At 80% confidence the bound would be about 1.3×10^{-3}.

We return to the effective dimensionality $d_{\mathrm{eff}}(E)$ mentioned in Section 1. It enters as the dimensionality of a random matrix which describes at E the nature of the fluctuations, and an intuitive argument for its value derives from the bivariate Gaussian strength density. The large correlation coefficients which we have encountered, $\zeta = 0.75$–0.85, a consequence of the approximate commutation of low-rank operators in many particle spaces [9], tell us that the strength in the H-diagonal basis is concentrated in a region centered on the diagonal. The dropoff of the strength at a distance from the diagonal, the width being given by $(2\sigma^2(V)(1-\zeta))^{1/2}$, limits the number of levels (d_{eff}) which are mixed by U to those, say, within $\pm 2\sigma(V)$. Then by integrating over the level density we typically find $d_{\text{eff}} \sim 5 \times 10^7/D$ (eV), and we assume that the strength is spread more or less uniformly over these levels. Since $\sigma(V) \simeq 2.25$ MeV (Table I) and $S = \Lambda/\alpha^2$ we have, with D the resonant level spacing given in electron volts,

$$\alpha \simeq 3 \times 10^{-3} \Lambda^{1/2} D^{1/2}. \tag{30}$$

This, among other things, confirms that the most sensitive fluctuation tests for TRNI involve nuclei having the smallest resonance spacings (as already seen). It gives an α-bound correct within a factor of 2 of the final result. If we were to combine this effective-dimensionality argument with the simplest derivation of Λ, which follows from the slope at $\Lambda = 0$ of the statistical measure being considered, we would end up with a simple reasonable estimate of α.

5. Conclusion

We have given in (18) the basic outlines of an intuitively simple theory of strength distributions in interacting-particle systems which makes use of a corresponding theory [13, 15] for the state density. Although it is not designed for dealing with transitions in the ground-state domain, shell-model studies described above and in [16] strongly suggest that, when strength fluctuations are unimportant, the theory extends right to the ground state. For applications there it might be necessary in some cases to take specific account of the correlation between the transition operator and the Hamiltonian; it is indicated above how to do that. It is the recognition of the bivariate Gaussian form which greatly simplifies the earlier version [16] of a strength theory. On the other hand if one is willing to treat fluctuations statistically rather than microscopically (as in the shell model) the method given in the first paper of [16] for predicting collectivity via a treatment of the fluctuations is still applicable and in fact implies a formal definition of d_{eff}.

Our specific application has been to determine, from the experimental transition parameter Λ of (V), the fluctuation bound on the TRNI nucleon–nucleon interaction, the final value being $\alpha \simeq 2 \times 10^{-3}$. It is clear, from the essential constancy, under variation of E, A, J, of $\Lambda D/\alpha^2$ (displayed in part in Table I, in part by (29) and in part by calculations mentioned below) coupled with the fact that the level density at fixed E peaks at $J \simeq \sigma_J \gtrsim 10$, that in order to improve this bound we need more information about nuclei with small D such as ^{236}U, which do not presently give a Λ-bound. With more complete data spectral analysis could improve the bound by a factor 2–3, and a similar factor could result from the measurement of small strengths, so that the ultimate limit of resonance-fluctuation bounds (or values) might be $\simeq 0.4 \times 10^{-3}$ at the high statistical confidence appropriate to the present analysis ($\sim 99\%$), and perhaps 0.3×10^{-3} at 80% confidence; these are to be compared with the present bounds $\simeq 2 \times 10^{-3}$ and $\simeq 1.3 \times 10^{-3}$. There would be many experimental difficulties in achieving these "ultimate" bounds, including the separation of two J values in the spectra of nuclei such as ^{236}U.

With regard to a comparison between our fluctuation bound and bounds derived from detailed-balance pairs of compound-nuclear reactions [2], the present situation is that at high (99%) statistical confidence, the detailed-balance bound is about 4×10^{-3}; but at low (80%) confidence this bound is much better, $\simeq 0.2\times10^{-3}$. These values have derived from the analysis of experimental results [2], coupled with a spectroscopic calculation [19] for ^{28}Si at about 19 MeV excitation, carried out directly from our Eq. (18). We shall comment elsewhere on experiments which could significantly improve this bound.

Calculations of $\Lambda D/\alpha^2 = \overline{|U_{ij}|^2}/D$ where i, j are close-lying show only a slow increase with energy. The ^{28}Si calculations give about the same value, this time with no variation between 5 and 30 MeV. The constancy over the periodic table was apparently first observed in isospin resonances; it was used (see the last paper of [4]) in a TRI detailed-balance study to extrapolate to ^{28}Si our fluctuation results for heavy nuclei. It should be straightforward to derive, directly from (18), the general variation of this parameter with A, E, and indeed to extend the result to the off-diagonal case; we shall do that as part of a study of complex nuclei. Finally after making obvious extensions to (18), for example to excitation operators which are not J-scalars, we intend to apply the above methods to other symmetries (π and isospin in particular) and other classes of problems.

ACKNOWLEDGMENTS

We acknowledge computing help from the UR High Energy Group. We thank Ms. Connie Gardner Jones for preparing the manuscript and for many courtesies.

REFERENCES

1. J. B. FRENCH, V. K. B. KOTA, A. PANDEY, AND S. TOMSOVIC, *Ann. Phys.* (*N.Y.*) **181** (1988), 198–234.
2. T. ERICSON, *Phys. Lett.* **23** (1966), 97; T. ERICSON AND T. MAYER-KUCKUK, *Annu. Rev. Nucl. Sci.* **16** (1966), 183; W. VON WITSCH, A. RICHTER, AND P. VON BRENTANO, *Phys. Rev.* **169** (1968), 923; E. BLANKE, M. DRILLER, W. GLÖCKLE, H. GENZ, A. RICHTER, AND G. SCHREIER, *Phys. Rev. Lett.* **51** (1983), 355; D. BOOSÉ, H. L. HARNEY, AND H. A. WEIDENMÜLLER, *Phys. Rev. Lett.* **56** (1986), 2012; see also Ref. [19].
3. M. SIMONIUS, *Nucl. Phys. A***396** (1983), 203c.
4. O. BOHIGAS, private communication (1984); E. KUHLMANN, *Z. Phys. A* **322** (1985), 527; H. L. HARNEY, A. RICHTER, AND H. A. WEIDENMÜLLER, *Rev. Mod. Phys.* **58** (1986), 607.
5. F. S. CHANG, J. B. FRENCH, AND T. H. THIO, *Ann. Phys.* (*N.Y.*) **66** (1971), 137.
6. J. B. FRENCH AND V. K. B. KOTA, *Annu. Rev. Nucl. Part. Sci.* **32** (1982), 35; J. B. FRENCH, *Nucl. Phys. A* **396** (1983), 87c.
7. J. B. FRENCH AND V. K. B. KOTA, *Phys. Rev. Lett.* **51** (1983), 2183.
8. J. B. FRENCH, V. K. B. KOTA, A. PANDEY, AND S. TOMSOVIC, *Phys. Rev. Lett.* **58** (1987), 2400.
9. K. K. MON AND J. B. FRENCH, *Ann. Phys.* (*N.Y.*) **95** (1975), 90.
10. F. J. DYSON AND M. L. MEHTA, *J. Math. Phys.* **4** (1963), 701.

11. K. F. Ratcliff, *Phys. Rev. C* **3** (1971), 117.
12. T. A. Brody, E. Cota, J. Flores, and P. A. Mello, *Nucl. Phys. A* **259** (1976), 87.
13. J. F. Smith, Ph. D. thesis, University of Rochester, 1987; UR Report 1023 (1987).
14. H. A. Bethe, *Phys. Rev.* **50** (1936), 332.
15. J. F. Smith and J. B. French, to be published.
16. J. P. Draayer, J. B. French, and S. S. M. Wong, *Ann. Phys.* (*N.Y.*) **106** (1977), 472, 503.
17. M. G. Kendall and A. Stuart, "The Advanced Theory of Statistics," Vol. 1, Hafner, New York, 1969.
18. V. K. B. Kota, *Z. Phys. A* **315** (1984), 91.
19. J. B. French, A. Pandey, and J. F. Smith, *in* "Proceedings of the International Conference on Time-Reversal Tests in Neutron Physics" (C. Gould and R. Roberson, Eds.), World, in press. A more detailed account will be available shortly.
20. J. J. M. Verbaarschot and P. J. Brussaard, *Z. Phys. A* **321** (1985), 125.
21. K. Kar, *Nucl. Phys. A* **368** (1981), 285.
22. A. Pandey and J. B. French, *J. Phys. A Math. Gen.* **12** (1979), L83.
23. P. W. M. Glaudemans and P. J. Brussaard, "Shell-Model Applications in Nuclear Spectroscopy," North-Holland, Amsterdam, 1977.
24. S. Tomsovic, Ph.D. thesis, University of Rochester, 1986; UR Report 974 (1986).
25. E. C. G. Sudarshan, *Proc. R. Soc. London Ser. A* **305** (1968), 319; M. Simonius, *Phys. Lett. B* **58** (1975), 147.
26. L. Wolfenstein, *Annu. Rev. Nucl. Part. Sci.* **36** (1986), 137.

"Universality" of Gaussian orthogonal ensemble fluctuations: the two-body random ensemble and shell model spectra

GAETAN J. H. LABERGE AND RIZWAN U. HAQ

Department of Physics, Laurentian University, Sudbury, Ont., Canada P3E 2C6

Received September 26, 1989

Starting from an appropriate decomposition of the level density into an average and fluctuating part, we studied the energy level fluctuations of an ensemble defined by two-body random Hamiltonians. A detailed analysis of several spectrally averaged fluctuation measures shows close agreement with the predictions of the Gaussian orthogonal ensemble (GOE). This confirms earlier indications that, except for noninteracting particles, fluctuation measures are insensitive to the rank of the interaction. Further, analysis of spectra obtained from realistic nuclear interactions agrees well with the GOE indicating that specific properties of the Hamiltonian have little or no influence on fluctuations. These results, therefore, strengthen our belief in the "universality" of GOE fluctuations.

Partant d'une décomposition appropriée de la densité de niveaux en une partie moyenne et une partie fluctuante, nous avons étudié les fluctuations des niveaux d'énergie dans un ensemble défini par des hamiltoniens aléatoires à deux corps. Une analyse détaillée de plusieurs mesures de la moyenne spectrale des fluctuations indique un accord étroit avec les prédictions de l'ensemble orthogonal gaussien (EOG). Cela confirme des indications antérieures à l'effet que, sauf dans le cas des particules sans interaction, les mesures de fluctuations sont insensibles au rang de l'interaction. De plus, l'analyse des spectres obtenu à partir d'interactions nucléaires réalistes est en bon accord avec EOG, ce qui indique que les propriétés spécifiques de l'hamiltonien ont peu ou pas d'influence sur les fluctuations. Ces résultats confirment donc notre croyance au caractère « universel » des fluctuations EOG.

[Traduit par la revue]

Can. J. Phys. **68**, 301 (1990)

1. Introduction

As a model for describing fluctuations (departures from uniformity), the Gaussian orthogonal ensemble (GOE) has been extremely successful (1, 2). A detailed analysis of the combined set of nuclear resonance-energy data of different nuclei, the nuclear data ensemble (NDE), has shown remarkable agreement with theory (3). Besides this detailed and extensive analysis of the fluctuations in the energy levels of the compound nucleus, the energy level spectra of some atoms (4, 5) and molecules (6–8) have also shown agreement with the GOE fluctuations. Further, it has been found that spectrum fluctuations of quantal time-reversal invariant systems whose classical analogues are strongly chaotic show GOE fluctuation patterns (9–15).

These results have gone a long way in confirming Wigner's suggestion (16) that "the Hamiltonian which governs the behavior of a complicated system is a random symmetric matrix, with no particular properties except for its symmetric nature". The GOE consists of asymptotically large real-symmetric matrices that apart from rotational and time-reversal invariance have no specific features. The GOE does not take account of the specific features of the Hamiltonians such as those used in conventional spectroscopy. One of the major differences is that the GOE matrix for a many-particle system describes simultaneous interaction between all particles because of the statistical independence of the matrix elements. Realistic Hamiltonians are usually limited to two-body interactions and therefore the many-particle states are linear combinations of the two-body matrix elements.

To keep the generality of GOE on the one hand and to conform to reasonable Hamiltonians, such as those used in shell model calculations, on the other hand, statistical extensions of the shell model have been proposed. A matrix ensemble of random two-body Hamiltonians called the two-body random ensemble (TBRE) was first introduced (17,18). This was later generalized to an ensemble of fixed rank interactions operating in many-particle spaces. Denoting the rank of the interaction by k, in an m-particle space, we have the GOE for $m = k$ and for $m \gg k$, the embedded Gaussian orthogonal ensemble (EGOE) (19). For $k = 2$, we obtain the TBRE provided $m \gg 2$.

In this work, we concentrate on two questions. Firstly, does the rank of the Hamiltonian influence the fluctuation properties of its energy levels? Since the GOE assumes many-body interactions, we study the other extreme case, that of two-body interactions. For a theory of fluctuations, the TBRE has proved to be mathematically difficult and at times intractable. While some progress has been made in obtaining analytical results (2), one has to rely mostly on numerical calculations (20, 21). We have made a detailed analysis using several spectrally averaged measures, developed for the study of the NDE (3). This allows a detailed comparison with the GOE and also takes into account statistical errors due to finite sample size. Secondly, we ask if specific properties, as opposed to the random nature of two-body interactions, like those pertaining to the realistic interactions used in nuclear spectroscopy change the fluctuations. To settle this question we have analyzed the eigenvalues obtained in large shell model calculations. One advantage in dealing with shell model eigenvalues is that, unlike experimental data, a long sequence of levels is available. This allows one to study higher order correlations (22) and thereby to provide a much more stringent test of the agreement with theory.

In Section 2, we make a few remarks about the TBRE and show how it is realized in a spectroscopic space. Section 3 deals with the decomposition of the level density into two distinct parts; one that describes the average behaviour and the other that consists of fluctuations about this average. The specific features of the Hamiltonian and the spectroscopic space are taken care of by the average density and the irreducible errors or fluctuations are left. This argument supposes that not only are the two parts of the density distinct but, furthermore, there exists a sharp separation. Analysis of the eigenvalues of large shell model (SM) calculations has indicated that the density separation is indeed distinct and sharp (23) and it is expected that the TBRE would show the same features. Section 4 deals with the analysis of fluctuations. We remark here that in previous calculations, only a small segment of the spectrum could be used. With the new method of density decomposition (24) on the one hand and use of spectrally averaged measures on the other, a much more detailed and finer comparison with theory can be made. Concluding remarks are presented in Sect. 5.

2. Two-body random ensemble

Let us consider m particles distributed over a set of orbits defined by N single particle states. For a two-body interaction between the particles, the Hamiltonian of the system is completely defined once its matrix elements between the two-particle states are known. The m-particle matrix elements are then linear combinations of these two-particle matrix elements and the coefficients depend on the single-particle states and angular-momentum coupling etc., but are independent of the interaction. By choosing the two-body matrix elements as random variables following a Gaussian distribution centered at zero, we obtain the TBRE. For the analysis we used the TBRE of ref. 20. We consider four orbits with $j = 5/2$, 3/2, 3/2, and 1/2. In this space we distribute six nucleons and look for states with total angular momentum $J = 7$ and isospin $T = 2$. This leads to a 294-dimensional matrix. The number of two-body matrix elements required is 190 and these are chosen randomly from a Gaussian distribution with centroid zero and variance 2 for diagonal elements and 1 for off-diagonal elements. The many-particle matrix is constructed using standard SM techniques and diagonalized to give the eigenvalues. A total of 50 such sets were obtained for the ensemble resulting in 14 700 eigenvalues.

The SM eigenvalues used later are due to Soyeur and Zuker (25). The first example is for 12 particles in the ds shell, $(ds)^{12}$ $J = 0$ $T = 0$, using the interaction of Kahana *et al.* (26). The total number of levels is 839. This will be referred to as SM(839). The second example corresponds to $(ds)^8$ $J = 2$ $T = 0$, with the same interaction as before. The dimensionality in this case is 1206 and will be called SM(1206).

3. Level density decomposition

Consider the density of levels $\rho(E)$, arising out of a Hamiltonian $\mathcal{H}$ acting in a spectroscopic space, obtained by distributing m particles in N single particle states. Throughout this work, we shall be dealing with levels that have the same quantum numbers corresponding to conserved quantities like angular momentum J and isospin T. Therefore, in principle, we should regard the density $\rho(E)$ as a partial density $\rho_{JT}(E)$. In the following we shall drop the subscript JT, it being understood that we are dealing with the eigenvalues of a SM matrix with specific values of J and T.

The rapidly increasing dimensionality of the space as a function of m and N precludes any possibility of determining the level density exactly. One then determines a smooth version via spectral distributions (27) and writes down the following decomposition:

$$[1]\qquad \rho(E) = \rho_s(E) + \rho_f(E)$$

where $\rho_s(E)$ represents the smooth or averaged part and $\rho_f(E)$ describes the fluctuations around this average.

The decomposition is useful if the smooth density has a characteristic form, describable in terms of low-order moments of the Hamiltonian. The density has been studied in detail, mathematically, by Mon and French (28). Denoting the rank of the Hamiltonian by k, they have shown that when the number of particles $m \gg k$, one obtains a Gaussian density whereas for $m = k$ one gets a semicircle. For the TBRE then, which corresponds to $k = 2$, the smooth density is well represented by a Gaussian distribution whereas for the GOE, corresponding to the latter case, one gets a semicircle (29). A beautiful numerical demonstration of the density transition from Gaussian to semicircle as the rank of the interaction is increased is given in ref. 20. The close to Gaussian approximation for shell model density has been numerically investigated by Chang and Zuker (30).

It is therefore clear that the average density is well described by a Gaussian plus corrections in which case only low-order moments of the Hamiltonian are required. This pleasing aspect forms the basis of spectral distribution methods (SDM) and for a recent review see French and Kota (31). The next question is whether $\rho_s(E)$ converges, apart from fluctuations, to the true density $\rho(E)$ as we take account of higher moments of the Hamiltonian. Ideally, one would hope that this convergence is rapid, a few moments being all that is required and the inclusion of higher ones making no substantial contributions.

Instead of the density $\rho(E)$, it is often more convenient to deal with its integrated version $F(E)$, which represents the total number of levels up to energy E. For the exact density $\rho(E)$ we have the staircase function $F(E)$ given by

$$[2]\qquad F(E) = \int_{-\infty}^{E} \rho(E')\,\mathrm{d}E'$$

and correspondingly for the smooth part $\rho_s(E)$ we define

$$[3]\qquad F_s(E) = \int_{-\infty}^{E} \rho_s(E')\,\mathrm{d}E'$$

The staircase function $F(E)$ is easily obtained from a sequence of levels $E_1, E_2, \ldots, E_d$, where d represents the dimensionality or total number of levels. We move along the spectrum and every time a level is encountered, the function rises by a step size of $1/d$. The step size is chosen as $1/d$ rather than 1 to keep $F(E)$ normalized.

Let us now define the deviation $\Delta(E)$ as

$$[4]\qquad \Delta(E) = d[F(E) - F_s(E)]$$

It is clear that $\Delta(E)$ carries information about the fluctuating part of the density $\rho_f(E)$. The nature and magnitude of $\Delta(E)$ relates directly to the decomposition of [1]. Its magnitude reflects how well $F_s(E)$ approximates $F(E)$ and its nature gives an idea of whether the desired separation has been achieved. Any secular variation in $\Delta(E)$ vs. E is an indication that the smooth part has not been properly taken care of. In the limit, a plot of $\Delta(E)$ vs. E should look like noise fluctuating rapidly around zero. As a measure of overall deviations one can calculate the root-mean-square deviation $\Delta_{\rm rms}$ given by

$$[5]\qquad \Delta_{\rm rms} = \sqrt{\frac{1}{d}\sum_{i=1}^{d} \Delta^2(E_i)}$$

We first consider the case where $\rho_s(E)$ is approximated by a Gaussian function, namely,

$$[6]\qquad \rho_s(E) = \frac{1}{\sigma\sqrt{2\pi}} \exp\left[\frac{(E - E_c)^2}{2\sigma^2}\right]$$

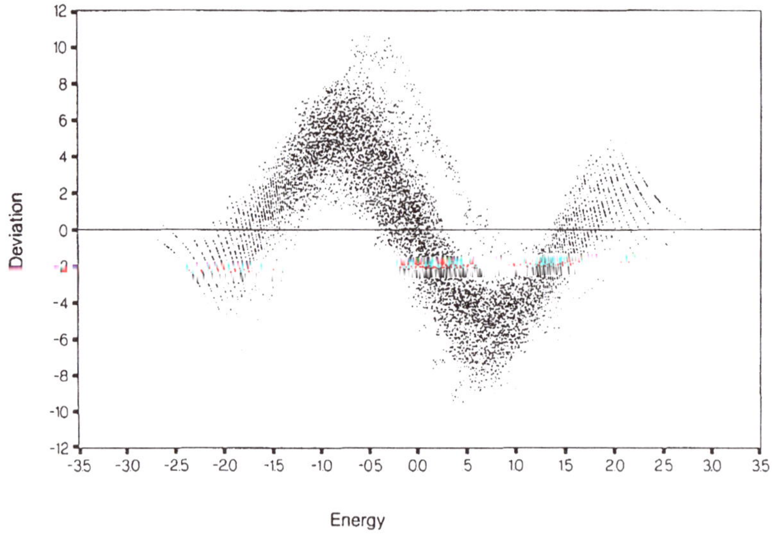

FIG. 1. Deviations $\Delta(E_i)$ vs. E_i for the TBRE. Total number of levels in the ensemble is 14 700 corresponding to 50 matrices of dimensionality 294. the smooth density $\rho_s(E)$ was taken as a Gaussian.

where E_c represents the centroid and σ^2 the variance of the distribution. In terms of the energies, we have

$$[7] \qquad E_c = \frac{1}{d}\sum_{i=1}^{d} E_i, \qquad \sigma^2 = \frac{1}{d}\sum_{i=1}^{d}(E_i - E_c)^2$$

In most cases E_c and σ^2, which represent the first- and second-central moment of H, can be easily calculated using the SDM. The cummulative frequency or distribution function $F_s(E)$ is obtained by integrating [6] and is given by

$$[8] \qquad F_s(E) = \frac{1}{2}\left[1 + \text{erf}\left(\frac{E - E_c}{\sigma\sqrt{2}}\right)\right]$$

where erf(x) is the error function (32). The staircase function $F(E)$ is given by

$$[9] \qquad F(E_i) = \frac{1}{d}\left(i - \frac{1}{2}\right)$$

where in accordance with the procedure adopted by Ratcliff (33), we have chosen the midpoints of the vertical step size.

Figure 1 shows the deviations $\Delta(E)$ vs. E for the TBRE. It is clear that the deviations show a distinct pattern varying slowly from level to level. This secular behavior is an indication that a Gaussian function [6] is not an adequate representation of $\rho_s(E)$ and corrections need to be made by including higher moments. Further, the value of Δ_{rms} is 3.95, which is rather large. Corrections or departures from a true Gaussian can be written in terms of the Gram–Charlier (GC) series expansion (34). The density in standardized variable is given by

$$[10] \qquad \rho_s(x) = \rho_G(x)\left[1 + \sum_{n\geq 3}\frac{C_n H_n(x)}{n!}\right], \qquad x = \frac{E - E_c}{\sigma}$$

where $\rho_G(x)$ is the Gaussian and $H_n(x)$ denotes the Hermite polynomials of degree n. The coefficients C_n can be expressed in terms of the moments of H. For example, C_3 represents the skewness of the distribution and C_4 the excess. From a practical point of view, the expansion [10] is useful only if a small number of terms are needed to describe the smooth density. Such an expansion has been considered by Chang and Zuker (30) for several SM examples. They found that the level to level deviations show a simple structure and although the overall deviations did get smaller by going to order 3 and 4, the secular component could not be eliminated.

The difficulty in the above calculation is due to the fact that the polynomials in the GC expansion are orthogonal with respect to a continuous Gaussian weight function. The essential problem is to define these with respect to a discrete spectra. We therefore carry out a least-squares minimization by treating the coefficients in [10] as free parameters. The least-squares equation can be written as

$$[11] \qquad \Delta_{\text{rms}}(C_3, \ldots, C_k) = \min\,(C_3 \ldots C_k)\left\{\sum_{i=1}^{d} d[F(x_i) - F_s(x_i)]^2\right\}^{1/2}$$

Equation [11] can then be solved giving the parameters C_n in closed form. Explicit forms for C_3 and C_4 are given in the Appendix. Further, because of the symmetry in the least-squares equation, one can deal with extremely high orders, if necessary, without increasing the complexity. For details we refer to ref. 24. It should be mentioned that a method that minimizes the level deviations rather than the differences in the distribution function had been used earlier by Wong (2). However, the method seems numerically complicated and no analytical least-squares solution seemed possible. Further, unlike the present method, the complexity increased with increasing order.

In Fig. 2 we show the deviations for order 3. The order is the value of the integer n in the expansion given in [10]. Order 3 means that only C_3 is obtained through minimization

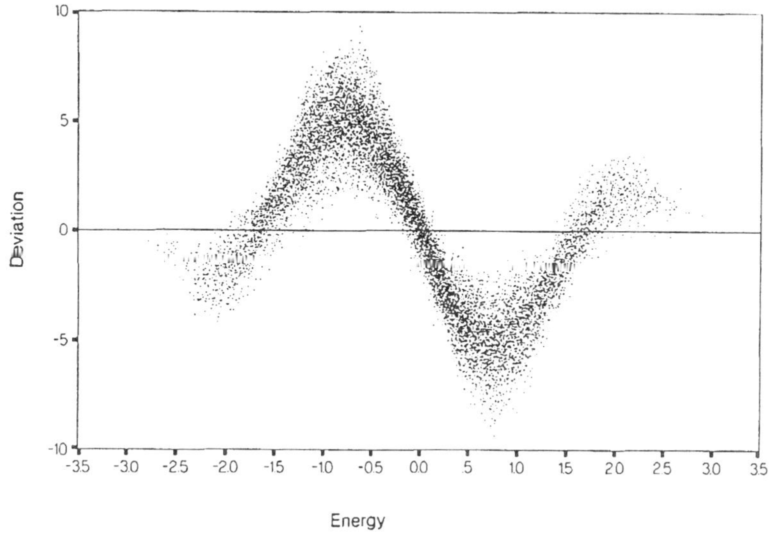

FIG. 2. The deviations for order 3 of the GC expansion with optimized constants for the TBRE of dimensionality 14 700.

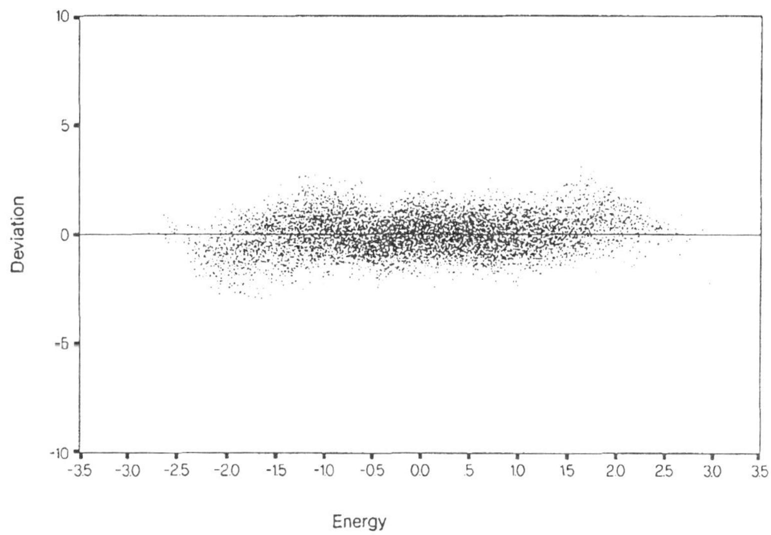

FIG. 3. The deviations for order 4 of the GC expansion with optimized constants for the TBRE of dimensionality 14 700.

and we shall refer to these as the optimized moments. Note that there is not much difference between Fig. 2 as compared with Fig. 1. The trend of the deviations is the same although the root-mean-square deviation Δ_{rms} has decreased slightly to 3.69. Inclusion of the next order makes a dramatic difference as seen in Fig. 3. The secular variation has disappeared and the deviations fluctuate rapidly around zero. This indicates that the deviations have now been reduced to fluctuations. A further test of this can be seen in Fig. 4 where we have gone up to order 10. The deviations are almost identical to those we had obtained for order 4. As a measure of the convergence we give, in Table 1, the values of Δ_{rms} for different orders. The values change rapidly up to order 4 and then show a slow monotonic decrease. This may be regarded as an indication of the fact that there exists a separation between the two parts of the density as shown in [1]. Essentially then, we have achieved an accurate description of the smooth density in terms of low-order moments provided we use optimized moments rather than those calculated from the Hamiltonian.

Similar conclusions are drawn from the study of SM examples (23). The Δ_{rms} values for the two SM examples are also given in Table 1. In both cases, Δ_{rms} decreases rapidly with increasing order and then converges to a fixed value. For SM(839) this critical value of the order is 6 whereas for SM(1206), this happens at order 4. The value of the critical order then shows a weak dependence on dimensionality and

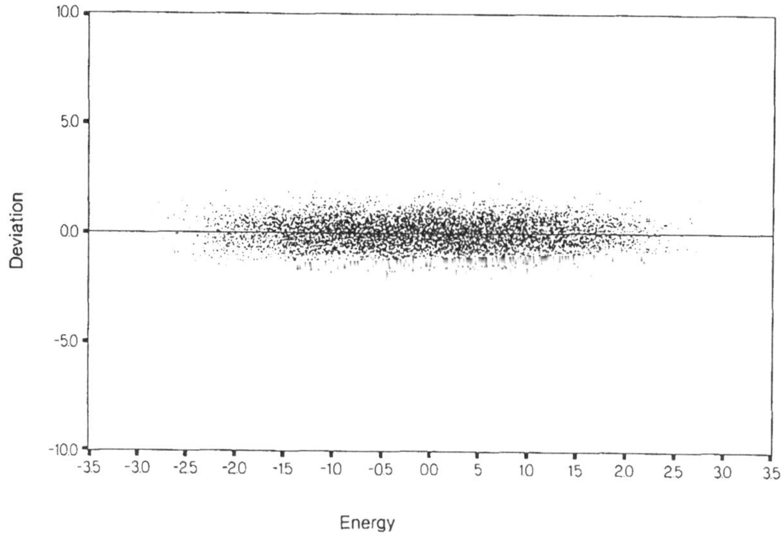

FIG. 4. The deviations for order 10 of the GC expansion with optimized constants for the TBRE of dimensionality 14 700.

TABLE 1. Root-mean-square deviation Δ_{arms} as a function of order n in the G-C expansion for the TBRE and SM spectra of dimensionality 839 and 1206

	Δ_{rms}		
n	TBRE	SM(839)	SM(1206)
2	3.95	9.20	7.7
3	3.69	1.80	1.4
4	0.77	1.10	0.77
5	0.74	1.10	0.75
6	0.71	0.77	0.75
7	0.68	0.77	0.75
8	0.63	0.77	0.75
9	0.62	0.77	0.74
10	0.60	0.76	0.74

we shall see later how this is reflected in the fluctuation statistics. Having established the separation, albeit numerically, in the next section we turn to the study of fluctuations.

4. Nature and magnitude of fluctuations

We shall define fluctuations as departures from uniformity. This means that the smooth part of the density should be a constant and therefore its integrated version should be a straight line. In this way of doing things, any features deriving from the particular form of the density are eliminated. This process called "unfolding" is easily achieved once $\rho_s(E)$ is known. Defining a new variable x, by conservation of probability we have

$$[12] \qquad \rho_s(E)\,\mathrm{d}E = \rho_s(x)\,\mathrm{d}x$$

Since we want $\rho_s(x)$ to be a constant, integrating [12] gives us the set of levels $x_i (i = 1, \ldots, d)$.

In the study of fluctuations, one defines certain measures or a certain statistic, which describe the essential features of the spectrum. Two such features for the GOE spectra are level repulsion and spectral rigidity. Level repulsion, which is the tendency of levels to avoid clustering, is measured by studying the nearest-neighbour spacing distribution. For random spacings S measured in terms of the mean spacing, the nearest-neighbour distribution of such a Poisson spectrum is given by

$$[13] \qquad P(S) = \exp(-S)$$

whereas for the GOE, $P(S)$ is well approximated by the Wigner distribution (35) written as

$$[14] \qquad P(S) \approx \frac{\pi S}{2} \exp\left(\frac{-\pi S^2}{4}\right)$$

The probability for zero spacing $P(0)$ is zero for the GOE but one for Poisson. The exact form of the GOE spacing distribution has been given by Mehta and des Cloizeaux (36) in terms of spheroidal functions. In Fig. 5, the spacing distribution for the TBRE is shown compared with the GOE distribution. The agreement is good. Figure 6 shows the spacing distribution for the SM example of dimensionality 839. Once again good agreement is obtained. The variance of the spacing distribution is 0.28 for the TBRE and 0.29 for the two SM examples. This compares well with the GOE value of 0.28.

The other aspect of the GOE spectra is spectral rigidity. Just as we defined the deviations Δ between the staircase and the distribution function of the smooth part of the density, the statistic Δ_3, due to Dyson and Mehta (37), determines the overall deviations between the staircase $F(x)$ and the best fit straight line. This is because we are dealing with a constant density and, therefore, the function $F_s(x)$ is either x or in general the straight line $Ax + B$. Hence we have, for a spectrum with unit mean spacing,

$$[15] \qquad \Delta_3(\bar{n}) = \min(A, B)\,\frac{1}{\bar{n}} \int_x^{x+\bar{n}} [F(x) - (Ax + B)]^2\,\mathrm{d}x$$

Here $\bar{n}$ represents the number of levels to be found on an

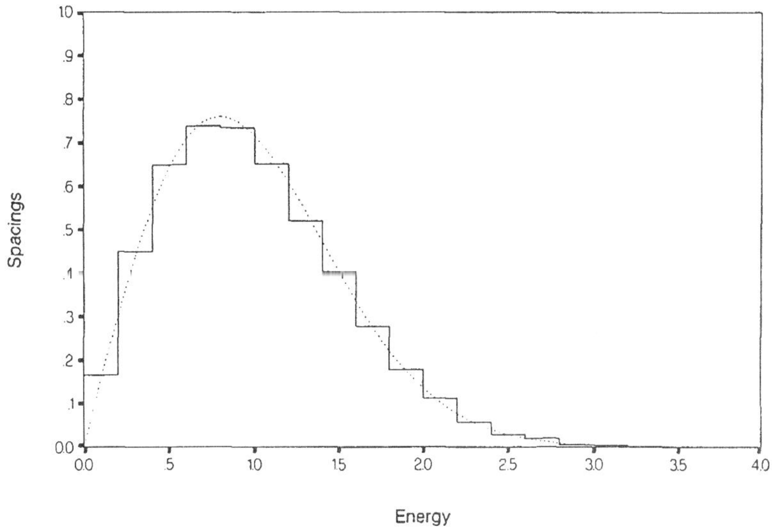

FIG. 5. Nearest-neighbour spacing distribution for the TBRE using 4th order unfolding.

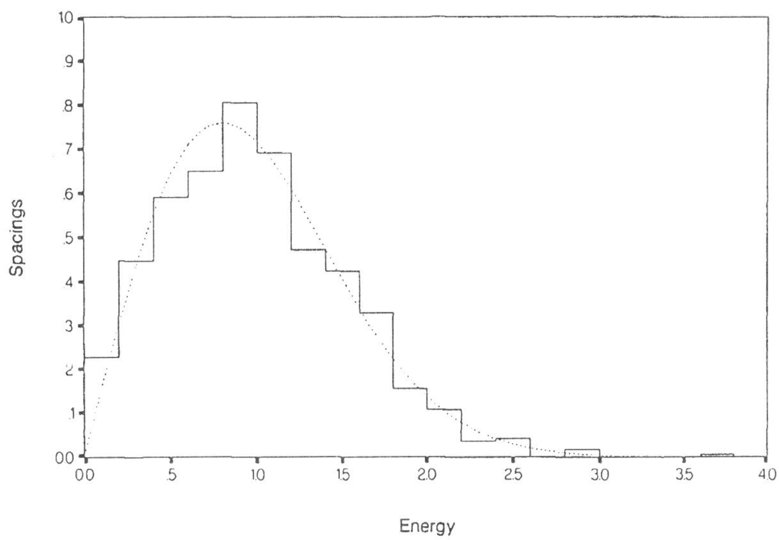

FIG. 6. Nearest-neighbour spacing distribution for SM of dimensionality 839.

average within the interval of the spectrum under consideration. The $\Delta_3(\bar{n})$ statistic measures the regularity or rigidity of a spectrum. To get an idea of how it differentiates among different kinds of spectra, we consider some examples. For a picket fence (P.F.), (equally spaced levels), it is given by

$$[16] \qquad \Delta_3(\bar{n}) = \frac{1}{12} \qquad \text{(P. F.)}$$

and for a Poisson spectrum, its ensemble averaged value is

$$[17] \qquad \bar{\Delta}_3(\bar{n}) = \frac{\bar{n}}{15}$$

For the GOE, its asymptotic form is

$$[18] \qquad \bar{\Delta}_3(\bar{n}) \approx \frac{1}{\pi^2}\,[\ln(\bar{n}) - 0.0687]$$

The mean-square deviations for the GOE are extremely small and hence the theory predicts that levels should be spaced with a high degree of regularity.

To make optimum use of this statistic we evaluate a spectrally averaged version of it (3). Given an energy interval $\bar{n}$, we calculate $\Delta_3(\bar{n})$ for overlapping or nonoverlapping segments of length $\bar{n}$ covering the entire spectrum. Finally we take the average. Results for the TBRE are given in Fig. 7. The errors due to finite sample size are also given and excellent agreement is obtained. In Table 2 we give the values calculated from the SM eigenvalues compared with those from the GOE.

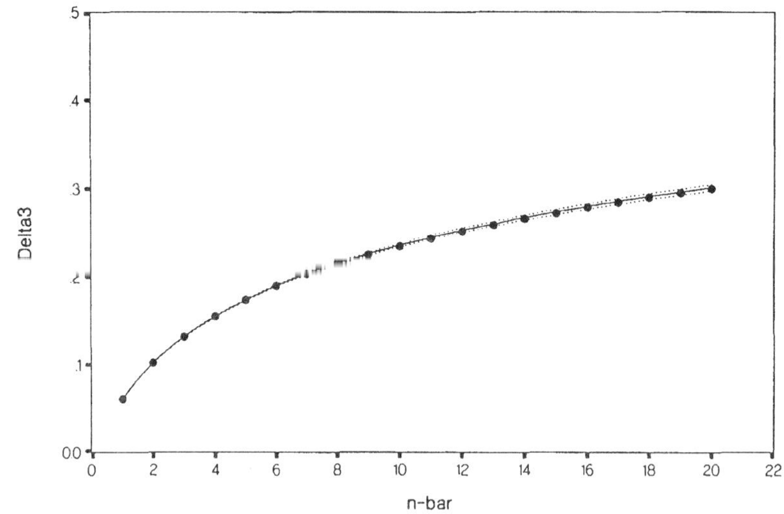

FIG. 7. Δ_3 statistic for TBRE.

TABLE 2. Δ_3 statistic for SM eigenvalues compared with the GOE. For SM(839) we have used 6th order unfolding and for SM(1206) a 4th order unfolding

$\bar{n}$	$\Delta_3(\bar{n})$ GOE	SM(839)	SM(1206)
1	0.06	0.056	0.056
2	0.102	0.097	0.095
3	0.132	0.128	0.126
4	0.155	0.148	0.155
5	0.173	0.165	0.169
6	0.189	0.184	0.191
7	0.203	0.197	0.204
8	0.215	0.206	0.223
9	0.225	0.219	0.221
10	0.236	0.232	0.238
15	0.274	0.279	0.276
20	0.301	0.318	0.330

TABLE 3. Δ_3 statistic for TBRE for large values of $\bar{n}$. To avoid the end effects, 25 eigenvalues from each end for each set were not used. This left a total of 12 200 eigenvalues in the ensemble

$\bar{n}$	$\Delta_3(\bar{n})$ GOE	TBRE
50	0.39	0.38
60	0.41	0.42
70	0.42	0.44
80	0.44	0.46
90	0.45	0.48
100	0.46	0.49
150	0.50	0.53
200	0.53	0.58

TABLE 4. Δ_3 statistic for the shell model example of 839 levels for large values of $\bar{n}$ and various orders in the G-C expansion

$\bar{n}$	$\Delta_3(\bar{n})$ GOE	SM(839) order 4	order 6	order 10
100	0.46	0.57	0.55	0.54
200	0.53	0.67	0.64	0.63
300	0.57	0.83	0.74	0.66
400	0.60	0.94	0.74	0.73
500	0.62	1.2	0.74	0.73
600	0.64	1.3	0.71	0.70
700	0.66	1.2	0.68	0.67

SM(839) uses an order 6 unfolding whereas SM(1206) uses order 4 only. Once again we have good agreement.

In the past, given a spectrum, the Δ_3 statistic was calculated for a short sequence in the center. This was due to the fact that levels in the center are more or less at constant density. If one wanted to extend further, one had to unfold using a GC or a similar expansion usually up to four moments where no optimization was involved. Even this way, only very few levels out of the whole sequence could be used. However, with the method of density separation suggested here, one can go to large values of $\bar{n}$ and the agreement is maintained. The results are shown in Table 3. In Table 4 we show the effect of the order of unfolding on the $\Delta_3(\bar{n})$ statistic. Whereas for small values of $\bar{n}$, ($\bar{n} \leq 20$), there is a very small difference, for large values of $\bar{n}$ we observe that agreement improves with increasing order as expected.

Another way of studying spectral rigidity is to calculate the moments of the number statistic. Given a spectrum with unit mean spacing $x_i (i = 1, 2, \ldots, d)$ we have seen that if we choose an energy interval of length $\bar{n}$, then the number of levels in each interval averaged over the spectrum give $\bar{n}$. We can now evaluate higher moments of the number statistic n. The number

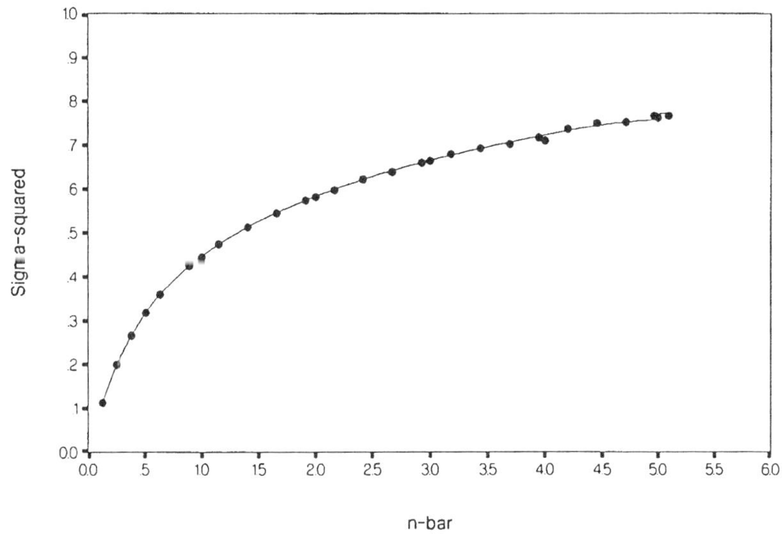

FIG. 8. Number variance $\sum^2(\bar{n})$ for TBRE.

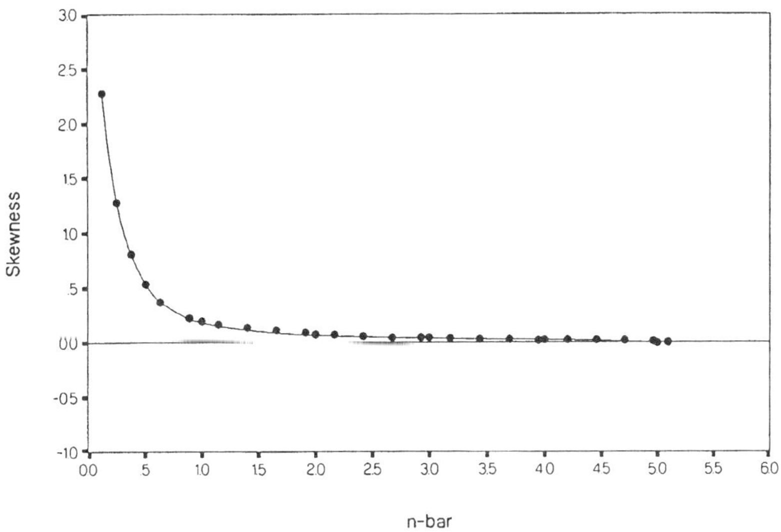

FIG. 9. Skewness $\gamma_1(\bar{n})$ for TBRE.

variance $\sum^2(\bar{n})$ is given by

$$\sum^2(\bar{n}) = \bar{n} \qquad \text{(Poisson)} \tag{19}$$

$$\sum^2(\bar{n}) = \frac{2}{\pi^2}\left[\ln(2\pi\bar{n}) + \gamma + 1 - \frac{\pi^2}{8}\right] \quad \text{large } \bar{n} \text{ (GOE)} \tag{20}$$

where γ is the Euler constant. For $\bar{n} = 100$, the Poisson spectrum value is 100 whereas for the GOE it is close to 1. This means that there are very small fluctuations about the average number of levels in an interval of given length.

The results for $\sum^2(\bar{n})$ for TBRE are shown in Fig. 8. For large values of $\bar{n}$, the results are shown in Table 5. In all cases agreement with the GOE is good. In Figs. 9–12 we compare the skewness $\gamma_1(\bar{n})$ and the excess $\gamma_2(\bar{n})$ of the number statistic n with the GOE predictions. Once again excellent agreement is obtained indicating that even higher order correlations (22) in the TBRE and SM eigenvalues are the same as in the GOE.

5. Conclusions

In this work we have looked at the density decomposition and fluctuations in the energy levels for the TBRE and some SM results using realistic nuclear interactions.

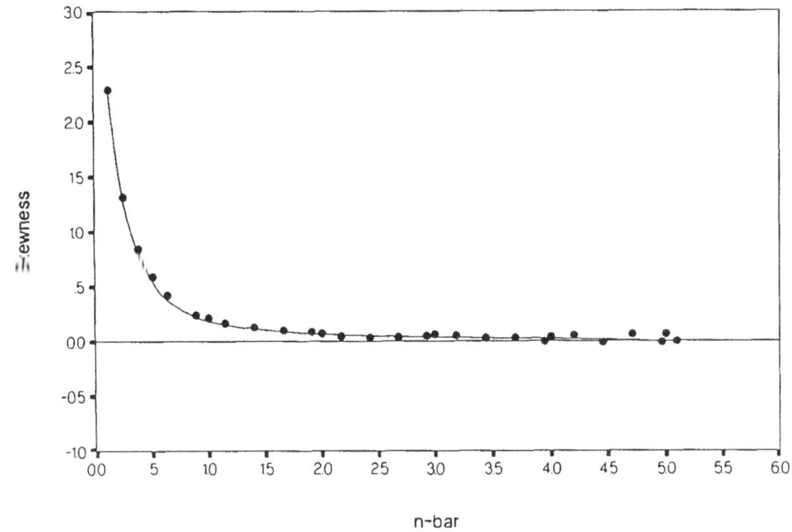

FIG. 10. Skewness $\gamma_1(\bar{n})$ for SM(1206).

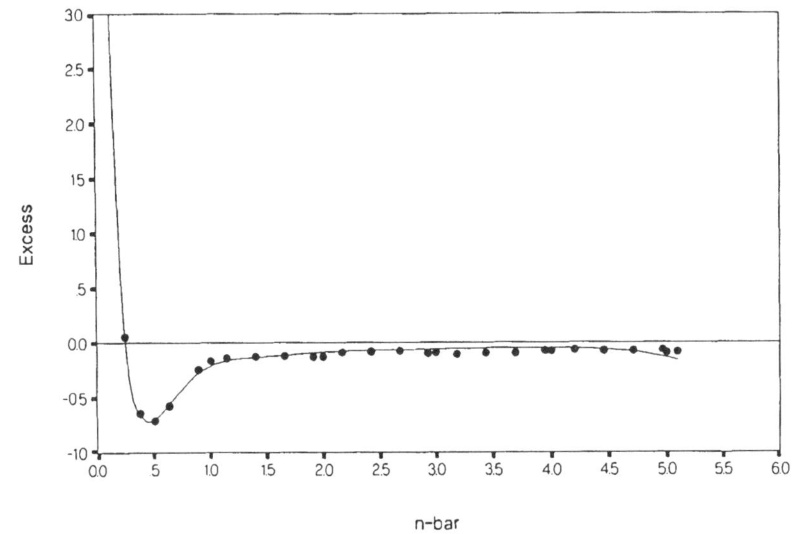

FIG. 11. Excess $\gamma_2(\bar{n})$ for TBRE.

Regarding the density decomposition, one expects that long wavelengths of the order of the spectrum span would be responsible for secular variations of the density, and short wavelengths of the order of the mean spacing would give rise to fluctuations. Using the GC expansion for the density, we have shown numerically that deviations between the exact density and its smooth approximation, in most cases, reduce to fluctuations by the time we get to order 4, provided we treat the expansion coefficients as free parameters to be determined through a least-squares fitting procedure. The usual expansion in which the coefficients are given in terms of the moments of the Hamiltonian does not show convergence.

This raises the important question of what coefficients to use when one is dealing with spectral distributions. So far these coefficients can be determined, as has been done here, if we know the eigenvalues. Ways to determine them directly from the Hamiltonian need to be studied. The pleasing aspect is that in most cases only low-order moments (optimized 3rd and 4th) are needed for a proper characterization of the smooth density.

Regarding the fluctuations, the excellent agreement between the TBRE and the SM results as compared with the GOE predictions confirms the belief that the GOE fluctuations are to be found universally and in this sense, carry very little information. Further it has been shown (38) that adding a GOE matrix $\mathcal{H}$ to any real symmetric matrix $K(K+\alpha H)$ leads very quickly, as α increases, to the same fluctuations. Intermediate

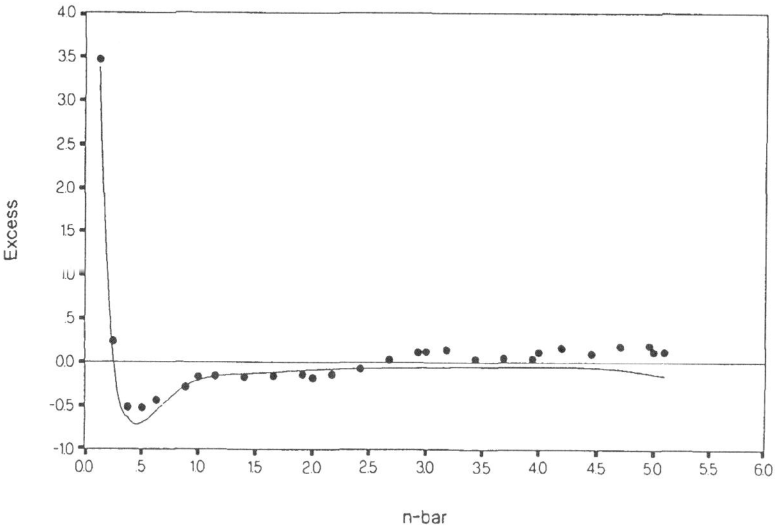

FIG. 12. Excess $\gamma_2(\bar{n})$ for SM(1206).

TABLE 5. Number variance $\Sigma^2(\bar{n})$ vs. $\bar{n}$ for the TBRE and the SM

	$\Sigma^2(\bar{n})$		
$\bar{n}$	GOE	TBRE	SM(839)
1	0.45	0.45	0.44
2	0.58	0.58	0.56
3	0.66	0.66	0.68
4	0.72	0.72	0.76
5	0.77	0.77	0.81
6	0.80	0.81	0.89
7	0.84	0.85	0.93
8	0.86	0.87	0.90
9	0.89	0.91	0.99
10	0.91	0.94	0.96
11	0.93	0.96	0.98
12	0.95	0.98	0.95
13	0.96	0.98	1.09
14	0.98	1.02	1.04
15	0.99	1.06	1.09
16	1.00	1.06	1.10
17	1.02	1.06	1.14
18	1.03	1.09	1.10
19	1.04	1.09	1.20
20	1.05	1.13	1.25

fluctuation patterns are to be expected only when the random-matrix elements are not much larger than the local average spacing of the given nonrandom matrix K. It is therefore no longer surprising that these fluctuations are encountered in a wide variety of systems.

It should be mentioned that the fact that fluctuations carry little information does not imply that they are devoid of any interest. Since we know that these agree with the GOE for a given symmetry class (conserved quantum numbers), it is advantageous to investigate what happens in the case of symmetry mixing. In such cases the fluctuation pattern changes. Hence the goodness of symmetries is intimately linked to the study of fluctuations (30, 40).

Acknowledgments

This work has been supported by the Natural Sciences and Engineering Research Council of Canada (NSERC). We are grateful to S. S. M. Wong for the TBRE eigenvalues and to A. Zuker for the SM spectra.

1. C. E. PORTER (*Editor*). Statistical theories of spectra: fluctuations. Academic Press Inc., New York. 1965.
2. T. A. BRODY, J FLORES, J. B. FRENCH, P. A. MELLO, A. PANDEY, and S. S. M. WONG. Rev. Mod. Phys. **53** 385 (1981).
3. R. U. HAQ, A. PANDEY, and O. BOHIGAS. Phys. Rev. Lett. **48**, 1086 (1982).
4. N. ROSENZWEIG and C. E. PORTER. Phys. Rev. **120**, 1698 (1960).
5. H. S. CAMARDA and P. D. GEORGOPULOS. Phys. Rev. Lett. **50**, 492 (1983).
6. E. HALLER, H. KOPPEL, and L. S. CEDERBAUM. Chem. Phys. Lett. **101**, 215 (1983).
7. E. ABRAMSON, R. W. FIELD, D IMRE, K. K. INNES, and J. L. KINSEY. J. Chem. Phys. **80**, 2298 (1984).
8. S. MUKAMEL, J. SUE, and A. PANDEY. Chem. Phys. Lett. **105**, 134 (1984).
9. I. C. PERCIVAL. J. PHYS. B, **6**, L229 (1973).
10. M. V. BERRY and M. TABOR. Proc. R. Soc. London. A, **356**, 375 (1977).
11. O. BOHIGAS, M. J. GIANNONI, and C. SCHMIT. Phys. Rev. Lett. **52** 1 (1984).
12. T. H. SELIGMAN, J. J. VERBAARSCHOT, and M. R. ZIRNBAUER. Phys. Rev. Lett. **52**, 215 (1984).
13. M. V. BERRY. In Chaotic behaviour of deterministic systems *Edited by* G. Iooss *et al.* North-Holland Publishing Co., Amsterdam. 1983.
14. O. BOHIGAS and M. J. GIANNONI. Lect. Notes Phys. **209**, 1 (1984).
15. O. BOHIGAS, M. J. GIANNONI, and C. SCHMIT. Lect. Notes Phys. **263**, 393 (1986).
16. E. P. WIGNER. *In* Symmetries and reflections. *Edited by* E. P. Wigner. OxBow, Woodbridge, CT. 1979 p. 203.
17. J. B. FRENCH and S. S. M. WONG. Phys. Lett. **33B**, 449 (1970).
18. O. BOHIGAS and J. FLORES. Phys. Lett. **34B**, 261 (1971).

19. J. B. French. Rev. Mex. Fis. **22**, 221 (1973).
20. S. S. M. Wong and J. B. French. Nucl. Phys. **A198**, 188 (19723).
21. O. Bohigas and M. J. Giannoni. Ann. Phys. (N.Y.) **89**, 393 (1975).
22. O. Bohigas, R. U. Haq, and A. Pandey. Phys. Rev. Lett. **54**, 1645 (1985).
23. R. U. Haq. Pramana, **32**, 475 (1989).
24. G. Laberge. M.Sc. Thesis. Laurentian University, Sudbury, Ont. 1989.
25. M. Soyeur and A. P. Zuker. Phys. Lett. **41B**, 135 (1972).
26. S. Kahana, H. C Lee, and C. K. Scott. Phys. Rev. **185**, 1378 (1969).
27. J. B. French. *In* Nuclear Structure. *Edited by* A. Hossain, Harun-ar-Raschid, and M. Islam. North Holland Publishing Co., Amsterdam. 1967.
28. K. K. Mon and J. B. French. Ann. Phys. (N.Y.) **95**, 90 (1975).
29. E. P. Wigner. Ann. Math. **67**, 325 (1958).
30. F. S. Chang and A. Zuker. Nucl. Phys. **A198**, 417 (1972).
31. J. B. French and V. K. B. Kota. Ann. Rev. Nucl. Sci. **32**, 35 (1982).
32. M. Abramowitz and I. A. Stegan. Handbook of mathematical functions. Dover Publications, New York. 1964.
33. K. F. Ratcliff. Phys. Rev. C, **3**, 117 (1971).
34. M. G. Kendall and A. Stuart. The advanced theory of statistics. Vol. 1. 3rd ed. Hafner Publishing Co. Inc., New York, 1969.
35. E. P. Wigner. *In* Statistical theories of spectra: fluctuations. *Edited by* C. E. Porter. Academic Press Inc., New York. 1965.
36. M. L. Mehta and J Des Cloizeaux. Indian. J. Pure Appl. Math. **3**, 329 (1972).
37. F. J. Dyson and M. L. Mehta. J. Math. Phys. **4**, 701 (1963).
38. A. Pandey. Ann. Phys. (N.Y.) **134**, 110 (1981).
39. J. B. French, V. K. B. Kota, A. Pandey and S. Tomsovic. Phys. Rev. Lett. **54**, 2313 (1985).
40. P. Bauer and R. Haq. Lect. Notes Phys. **263**, 157 (1986).

Appendix

To determine the coefficients of the fourth order GC expansion in closed form, we take $\Delta_{\rm rms}$ as in [11] and minimize it with respect to the coefficients C_3 and C_4 as follows:

$$\text{[A1]} \qquad \Delta_{\rm rms}(C_3, C_4) = \min\,(C_3, C_4) \left\{ \sum_{i=1}^{d} d[F(x_i) - F_s(x_i)]^2 \right\}^{1/2}$$

where

$$\text{[A2]} \qquad F_s(x) = F_G(x) - \rho_G(x) \left[\frac{C_3 H_2(x)}{3!} + \frac{C_4 H_3(x)}{4!} \right]$$

is the integrated version of the smooth density, ρ_s, given by

$$\text{[A3]} \qquad \rho_s(x) = \rho_G(x) \left[1 + \frac{C_3 H_3(x)}{3!} + \frac{C_4 H_4(x)}{4!} \right], \qquad x = \frac{E - E_c}{\sigma}$$

In terms of the standardized variable we have

$$\text{[A4]} \qquad F_G(x) = \frac{1}{2} \left[1 + \operatorname{erf} \left(\frac{x}{\sqrt{2}} \right) \right]$$

$$\text{[A5]} \qquad F(x_i) = \frac{1}{d} \left(i - \frac{1}{2} \right)$$

where erf is the error function [32]. We define

$$\text{[A6]} \qquad \xi = \sum_{i=1}^{d} [F(x_i) - F_s(x_i)]^2$$

and differentiate as follows

$$\text{[A7]} \qquad \frac{\partial \xi}{\partial C_3} = \frac{\partial \xi}{\partial C_4} = 0$$

to minimize ξ. The two equations are then solved for C_3 and C_4.

[A8]
$$C_3 = \frac{6\left[\sum_{i=1}^{d} PZH_2 - \sum_{i=1}^{d} FZH_2 - \frac{\sum_{i=1}^{d} PZH_3 \sum_{i=1}^{d} Z^2H_2H_3}{\sum_{i=1}^{d} Z^2H_3^2} + \frac{\sum_{i=1}^{d} FZH_3 \sum_{i=1}^{d} Z^2H_3H_2}{\sum_{i=1}^{d} Z^2H_3^2}\right]}{\left[\sum_{i=1}^{d} Z^2H_2^2 - \frac{\left(\sum_{i=1}^{d} Z^2H_2H_3\right)^2}{\sum_{i=1}^{d} Z^2H_3^2}\right]}$$

[A9]
$$C_4 = \frac{24\left[\sum_{i=1}^{d} PZH_2 - \sum_{i=1}^{d} FZH_2 - \frac{\sum_{i=1}^{d} PZH_3 \sum_{i=1}^{d} Z^2H_2^2}{\sum_{i=1}^{d} Z^2H_2H_3} + \frac{\sum_{i=1}^{d} FZH_3 \sum_{i=1}^{d} Z^2H_2^2}{\sum_{i=1}^{d} Z^2H_2H_3}\right]}{\left[\sum_{i=1}^{d} Z^2H_3H_2 - \frac{\sum_{i=1}^{d} Z^2H_3^2 \sum_{i=1}^{d} Z^2H_2^2}{\sum_{i=1}^{d} Z^2H_2H_3}\right]}$$

where F is [A5], P is identical to $F_G(x)$ of [A4], Z is [6] with zero centroid and unit variance, and H_2 and H_3 are the second and third order Hermite polynomials.

VOLUME 66, NUMBER 9 PHYSICAL REVIEW LETTERS 4 MARCH 1991

Influence of Saturation Properties on Shell-Model Calculations

A. Abzouzi, E. Caurier, and A. P. Zuker

Centre de Recherches Nucléaires, Institut National de Physique Nucléaire et de Physique des Particules, Université Louis Pasteur, B.P. 20, 67037 Strasbourg CEDEX, France

(Received 25 July 1990)

It is shown that the nuclear Hamiltonian $\mathcal{H}$ separates rigorously into a monopole field $\mathcal{H}_m$ and a multipole part $\mathcal{H}_M$. $\mathcal{H}_m$ is entirely responsible for saturation properties and can be treated phenomenologically with few parameters. When realistic interactions are used for $\mathcal{H}_M$ in regions from the p shell to the $N=82$ isotones, shell-model calculations yield excellent spectroscopy and demand nuclear radii very close to the observed ones.

PACS numbers: 21.60.Cs, 21.10.Ft, 21.30.+y

The use of realistic potentials (i.e., consistent with NN scattering) in shell-model calculations was pioneered by Kuo and Brown[1] (KB). Of the enormous body of work that followed we would like to extract two basic observations.

The first is that whatever the forces (hard or soft core, ancient or new) and the method of regularization (Brueckner G matrix,[1,2] Sussex direct extraction,[3] or Jastrow correlations[4]), the effective matrix elements are *extraordinarily similar.*[5,6] The most recent results[7] amount to a vindication of the original work of Kuo and Brown. We take this similarity to be the great strength of realistic interactions, since it confers on them a model-independent status as direct links to the phase shifts.

The second observation is that when used in shell-model calculations and compared with data these matrix elements lead to results that deteriorate rapidly as the number of particles increases.[8,9] It was found[5] that in the fp region a phenomenological cure, confirmed by exact diagonalizations[10] up to $A=48$, amounts to very simple modifications of the centroids of the KB interaction.[11] It may be obvious, but we shall demonstrate it explicitly, that the need for such a cure reflects the failure of nonrelativistic potentials to saturate properly.[12,13]

Saturation properties dictate the equilibrium radius and energy of nuclei through some form of Hartree-Fock (HF) self-consistency. Our program consists in showing that once these properties are correctly enforced phenomenologically, the realistic interactions give a very sound description of correlations and detailed structure. We proceed in three steps.

(I) The nuclear Hamiltonian is shown to separate rigorously in two terms $\mathcal{H}=\mathcal{H}_m+\mathcal{H}_M$. The monopole field $\mathcal{H}_m$ is entirely responsible for spherical HF self-consistency and its generalization to open shells. The multipole part $\mathcal{H}_M$ acts as residual interaction.

(II m) Good saturation properties in $\mathcal{H}_m$ are enforced by introducing directly as parameters observed closed-shell and single-particle energies.

(II M) The nuclear radii at which $\mathcal{H}_M$ is calculated are allowed to vary freely and adopt values consistently close to the observed ones.

(III) Shell-model calculations in the p and sd regions and for the $N=50,82$ isotones and $Z=28,50$ isotopes then establish the soundness of the realistic $\mathcal{H}_M$. Our choice of the Kahana, Lee, and Scott (KLS) interaction[2] is made owing to availability[14] but rests on the observation in the second paragraph above.

(I) Separation properties.— A properly regularized potential is density dependent, nonlocal, and of a rank higher than 2. However, since these complications are mostly related to saturation, which we shall treat phenomenologically, we assume that our Hamiltonian is a sum of kinetic (K) and two-body terms (V), and we perform on it a preliminary separation $\mathcal{H}=\mathcal{H}_m^d+\mathcal{H}_M^d$ (d for diagonal),

$$\mathcal{H}_m^d=\sum K_{rr}n_r+\sum_{r\le s}\left[V_{rs}n_{rs}+b_{rs}\left(T_r\cdot T_s-\frac{3n_r\bar{n}_s}{4(N_r-1)}\delta_{rs}\right)(1+\delta_{rs})^{-1}\right]. \tag{1}$$

Here n_r and T_r are the number and isospin operators for shell r which has degeneracy $N_r=2(2j_r+1)$. Also, $\bar{n}_r=N_r-n_r$, $n_{rs}=n_r(n_s-\delta_{rs})(1+\delta_{rs})^{-1}$ (same notation for N_{rs}). K_{rr} and V_{rstu}^{JT} are the one- and two-body matrix elements,

$$V_{rs}=\sum V_{rsrs}^{JT}(2J+1)(2T+1)\Big/\sum(2J+1)(2T+1)\,,\quad b_{rs}=V_{rs}^1-V_{rs}^0,\quad V_{rs}^T=\sum V_{rsrs}^{JT}(2J+1)\Big/\sum(2J+1)$$

(sums over Pauli-allowed values of J,T). $\mathcal{H}_M^d$ is the two-body part of the interaction with V_{rstu}^{JT} replaced by $W_{rstu}^{JT}=V_{rstu}^{JT}-V_{rs}^T\delta_{rt}\delta_{su}$.

The important property of $\mathcal{H}_m^d$ is that it reproduces the average energy of a configuration at fixed n_r,T_r for each shell.[15] In particular, it gives the expectation value of $\mathcal{H}$ for closed shells, which are configurations $(N_1N_2\cdots)$ with only one state. By unitary transformations of the underlying fermion fields $(a_r^\dagger = \sum U_{rs}b_s^\dagger)$, the n_r and T_r operators can only go to operators S_{rs} and T_{rs} coupled to $J=0$ and $T=0,1$, respectively (e.g., $n_r = a_r^\dagger \cdot a_r$ goes to $\sum U_{rs}U_{rt}b_s^\dagger \cdot b_t = \sum U_{rs}U_{rt}S_{st}$). Therefore the following must hold.[16]

Theorem.—There exists a decomposition $\mathcal{H} = \mathcal{H}_m + \mathcal{H}_M$ such that only $\mathcal{H}_m$ contributes to spherical HF calculations. $\mathcal{H}_m$ is written entirely in terms of unitary generators S_{rs} and T_{rs}.

The explicit form of the decomposition will be published elsewhere and will be made available on request.

Corollary.—By replacing expectation values by averages, HF variation can be extended to arbitrary configurations $(n_1n_2\cdots)$.[16]

Therefore we may assume that varying a good $\mathcal{H}_m$ produces an optimum field $\mathcal{H}_m(n)$ for each n and suggests a model space $\mathcal{M}$ spanned by configurations $(N_iN_jN_k\cdots n_rn_s\cdots)$ starting at $n=\sum n_r = 0$, the core and extending to $n=N=\sum N_r$, the closed shell. The Hamiltonian that enters a shell-model calculation $(H=H_m+H_M)$ is in first approximation the restriction of $\mathcal{H}$ to $\mathcal{M}$. We explain how its parameters "propagate" (a concept we borrow from the French), i.e., how they vary with n and how they are renormalized by states outside $\mathcal{M}$ (core polarization).

(II m) Propagation of H_m.—H_m has exactly the form of Eq. (1) with K_{rr} replaced by $\epsilon_r = K_{rr} + \sum N_{ir}V_{ir}$ (sum over occupied orbits i), but to incorporate observables as independent parameters we have to rewrite it[17,18] as

$$H_m = \mathcal{C} + \sum \Omega_r \eta_r + \sum_{r<s} \Omega_{rs} W_{rs} + \sum_{r\le s} \tau_{rs} b_{rs}\,,$$
$$\Omega_r = \left[\frac{n_r}{N_r} - \frac{n_1}{N_1}\right] N_r\,, \qquad (2)$$
$$\Omega_{rs} = \frac{N_{rs}}{2}\left[\frac{2n_{rs}}{N_{rs}} - \frac{n_{rr}}{N_{rr}} - \frac{n_{ss}}{N_{ss}}\right].$$

We have $\Omega_r = 0$ at $n=0,N$, $\Omega_{rs} = \tau_{rs} = 0$ [see Eq. (1)] at $n=0,1,N,N-1$. Furthermore, Ω_{rs} and τ_{rs} are invariant under $n_r \to \bar{n}_r$ and Ω_r changes sign. These properties explain the power of Eq. (2).

$\mathcal{C} = E + \epsilon n + Wn(n-1)/2$, with $E = \sum(K_{ii}N_i + V_{ij} \times N_{ij})$, $\epsilon = \sum N_r\epsilon_r/N$, $W = 2\sum N_{rs}V_{rs}/N(N-1)$, is the total centroid. All quantities depend on n, e.g., $\mathcal{C}(0) = E(0)$ is the energy of the core, but in general $E = E(n)$ is the energy of the occupied orbits. The minimal parametrization we choose is $\mathcal{C} = \mathcal{C}(n,4)$, a polynomial of third order in n that for $n=0,N,1,N-1$ reproduces the core, closed-shell energies, and the single-particle and single-hole centroids, respectively (all observables).

$\eta_r = \epsilon_r - \epsilon + (n-1)W_r$, with $W_r = \sum(N_r - \delta_{rs})(V_{rs} - W)/(N-2)$, are the central-field splittings (note $\sum N_r\eta_r = 0$). The minimal parametrization is $\eta_r(n,2)$, a linear form that reproduces the single-particle $(n=1)$ and single-hole $(n=N-1)$ splittings.

$W_{rs} = V_{rs} - W_r - W_s - W$ and b_{rs} are not directly observable. The minimal choice in this case follows hints from work in the fp shell[5] and it involves correcting the KLS values according to $W_{rs} = W_{rs}(\mathrm{KLS}) + W_{rs}^0(X)$ and $b_{rs} = b_{rs}(\mathrm{KLS}) + Y + 2(\delta_{rs} - 1)Z$, where W_{rs}^0 is the W_{rs}-type centroid of the monopole operator $X\sum(n_{rs} - n_{rr})$ $(r<s)$.

The minimal (or standard) parametrization we have described involves only three nonobservables (X,Y,Z). Physics may dictate more general forms as we shall see.

(II M) Propagation and core polarization in H_M.—The most conservative assumption about H_M is that matrix elements are calculated in an oscillator field common to all orbits (defined ideally by $\mathcal{H}_m$ variation). To relate its frequency $\hbar\omega$ (MeV) to the mean-square radius $\langle r^2\rangle^{1/2}$ (fm) we use[19]

$$\hbar\omega = 40(a+bn)/\langle r^2\rangle(c+n)\,, \qquad (3)$$

where $(a,b,c) = (6,2.5,4)$ in the p shell and $(36,3.5,16)$ in the sd shell. For heavier regions the estimate $\hbar\omega = 34.6A^{1/3}\langle r^2\rangle^{-1}$ (i.e., $40A^{-1/3}$ for $\langle r^2\rangle^{1/2} = 0.93A^{1/3}$) should be sufficient. The behavior of H_M as a function of ω (and hence $\langle r^2\rangle$ and n) follows from $W_{rstu}^{JT}(\omega) \cong W_{rstu}^{JT}(\omega_0)\omega/\omega_0$, which holds fairly well for KLS. To account for core polarization we argue that in the first approximation it should affect the width of the eigenvalue distribution in $\mathcal{M}$ and we incorporate it through two multiplicative parameters λ_T. This is, of course, no substitute for the sophisticated calculations now possible,[20] but it is consistent with *monopole-level corrections* that must precede core-polarization *improvements* of $\mathcal{H}$. Separating the $T=0$ and 1 parts of H_M and incorporating ω dependence we end up with

$$H_M = \frac{\omega(n)}{\omega_0}\lambda_0\left[H^0 + \frac{\lambda_1}{\lambda_0}H^1\right]. \qquad (4)$$

It should be stressed that $\omega(n)$ is an observable through Eq. (3) and, therefore, involves no free parameters.

(III) Results: p and sd shells.—We have selected the free parameters through a least-squares fit to the (Coulomb-corrected) absolute energies of 179 states in the sd shell and 41 states in the p shell. The evolution of the root-mean-square deviations (RMSD) can be found in Table I. For the Cohen and Kurath[21] (CK) and Wildenthal[9] (W) interactions we have extracted the $\mathcal{C}$ centroid and replaced it by a $\mathcal{C}(n,6)$ form. For KLS we have fixed $\lambda_0 = 1$, $\lambda_1 = 1.4$ (sd), 1.2 (p). The uncertainties on λ_T are less than 10%. NP (no parameters) incorporates $\mathcal{C}(n,4)$ and $\eta_r(n,2)$ in Eq. (1).

$S\omega(\nu)$ (standard with polynomials of order $\nu-1$ for ω) brings in only X,Y,Z as free parameters for b_{rs} and W_{rs} in Eq. (1). The drop in RMSD between the con-

TABLE I. Root-mean-square deviations in keV for different calculations in the *sd* and *p* shells. See text.

	sd	*p*
W-CK	235	−605
NP	2500	1800
Sω(1)	883	920
Sω(2)	486	913
Sω(3)	475	913
Sω(4)	445	739
Sω(5)	443	700
Sω(6)	443	692
Sω(7)	441	674
SVω(1)	832	703
SVω(7)	399	380
FP	285	290

stant $\omega(1)$ and linear $\omega(2)$ forms and the stabilization after $\omega(4)$ in the *sd* shell need no comment. The physical origin of the effect clearly shows in Fig. 1 for $\omega(4)$. In the *p* shell the calculated ω is more complicated and not as convincing.

For SV$\omega(v)$ [standard variations with $\omega(v)$], we incorporate as free parameters some of the observables. In this case we allow variation for η_r and $\bar{\eta}_r$. In the *sd* shell we do not learn much. In the *p* shell the gain is spectacular and Fig. 2 for $\omega(7)$ is now quite satisfactory: The improvement with respect to S$\omega(7)$ comes from a smaller η_r (the $p_{1/2}$-$p_{3/2}$ splitting, intermediate between CK and experiment). Although this amounts to *p*-shell idiosyncrasy, there is a general point to be made: With the (mild) exception of $\mathcal{C}(n,4)$, Sω and SVω amount to *constant* ω for the energies which is not very consistent if ω turns out to have the aspect of Figs. 1 and 2. It is nice to have very few parameters but if the physics commands greater freedom, we must take it.

For FP (full propagation), we come back to experimental values for the observables but allow for $\mathcal{C}(n,6)$ and $\eta_r(n,6)$ forms. Furthermore, all W_{rs} $(r\neq s)$, b_{rr}, and $\sum b_{rs}$ $(r<s)$ are left to vary freely and propagate linearly. At this stage we enter the realm of high-quality

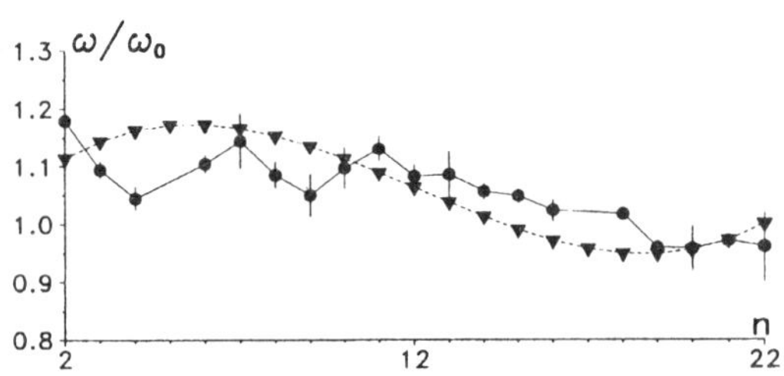

FIG. 1. Experimental (Ref. 26) Eq. (3) (circles and error bars) and calculated values of ω/ω_0 for S$\omega(4)$ in the *sd* shell (triangles) $\hbar\omega_0=11$.

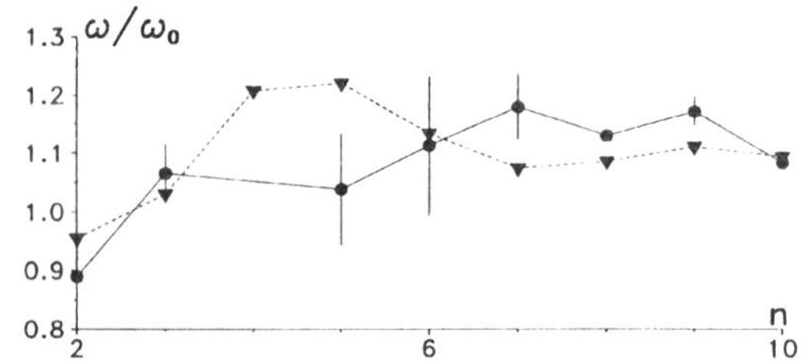

FIG. 2. Experimental (Ref. 26) Eq. (3) (circles and error bars) and calculated values of ω/ω_0 for SV$\omega(7)$ in the *p* shell (triangles) $\hbar\omega_0=13$.

spectroscopy and simultaneously explore saturation properties. Details will be given elsewhere.[22]

There are sources of uncertainty in the comparisons of Figs. 1 and 2: Eq. (3) may be naive, radii at fixed *n* may be *T* dependent (especially when deformation is present), and the λ_T parameters may be *n* and *T* dependent. However, these uncertainties are far from sufficient to cast doubts on a crucial result: *Spectra dictate choices of radii close to the experimental ones.* It should be noted that the *W* interaction yields RMSD=255, 235, and 215 keV for $\omega=\omega(2)$, $\omega=[18/(16+n)]^{0.3}$, and $\omega=\omega(6)$, respectively. The fact that a power law for ω is of no use in the *p* shell has led Van Hees, Booten, and Glaudemans[23] to introduce three-body forces that produce a drop in RMSD similar but smaller than those in Table I. It is an open question whether these results mock saturation effects or amount to genuine rank-3 forces.

(III) Results: N=50,82 isotones, Z=28,50 isotopes.—In these regions, SV$\omega(1)$ is sufficient to illustrate the luxuries we can afford with a realistic interaction.

(i) We select the ground states of even nuclei and the one-quasiparticle states of odd ones and do a seniority $v=0$ and 1 calculation (Table II). The values of η_r and $\bar{\eta}_r$ always come close to the experimental ones when these are known.

(ii) Having fixed the parameters, we increase v to 2,3,4,5, etc., and notice that v is a *very* good quantum number and that *many* states come quite naturally at observed positions to within the RMSD in Table II.

TABLE II. Root-mean-square deviations in keV for 0- and 1-quasiparticle fits in different isotone and isotope families. See text.

	λ_1	RMSD	$\hbar\omega_0$
$N=50$	1.35	160	9
$N=82$	1.35	128	7.9
$Z=28$	1.50	194	10.0
$Z=50$	1.40	129	8.4

Spectroscopy in the $N=82$ region is simply excellent.[22] For $N=50$, the results compare well with those of Ji and Wildenthal[24] except that our calculation misses a few levels they fit, brings the $g_{9/2}$ orbit at least 1 MeV lower, and conserves seniority better. The Ni and Sn isotopes are quite satisfactory in spite of a systematic discrepancy: For low n the first 2^+ state comes too high (up to 600 keV in Ni, 300 keV in Sn). Problems of this sort should be put to the *credit* of the interaction for signaling trouble in the calculation. In the case of Ni we know it must come from coupling to low-lying states in ^{56}Ni.[5] For Sn, the situation is bound to be similar: In general, $N=Z$ cores are easier to polarize than those with neutron excess and demand careful treatment.

Propagation plays a minor role in these regions, except through the linear behavior of the single-particle field, which is compatible with a constant two-body force.

In conclusion, the separation theorem provides the framework that dictates the necessary steps to enforce good monopole behavior. Some of them have been described previously[18,25] and now we propose a consistent implementation with bare forces and very few parameters, that stresses the influence of radii in light nuclei. Realistic forces are quite adequate for spectroscopic work provided saturation properties are taken care of.

[1]T. T. S. Kuo and G. E. Brown, Nucl. Phys. **A85**, 40 (1966).

[2]S. Kahana, H. C. Lee, and C. K. Scott, Phys. Rev. **180**, 956 (1969).

[3]J. P. Elliott, A. D. Jackson, H. A. Mavromantis, E. A. Sanderson, and B. Sing, Nucl. Phys. **A121**, 241 (1968); L. D. Skouras and J. C. Varvitsiotis, Nucl. Phys. **A513**, 239 (1990).

[4]J. Fiase, A. Hamoudi, J. M. Irvine, and F. Yazici, J. Phys. G **14**, 27 (1988).

[5]E. Pasquini and A. P. Zuker, in *Florence Conference on Medium and Light Nuclei,* edited by P. Blasi and R. Ricci (Editrice Compositrice, Bologna, 1978).

[6]M. L. Rustgi, H. W. Kung, R. Raj, R. A. Nisley, and H. H. Hull, Jr., Phys. Rev. C **4**, 854 (1971).

[7]M. F. Jiang, R. Machleidt, D. B. Stout, and T. T. S. Kuo, Phys. Rev. C **40**, 1857 (1989).

[8]E. C. Halbert, J. B. McGrory, B. H. Wildenthal, and S. P. Pandya, in *Advances in Nuclear Physics,* edited by M. Baranger and E. Vogt (Plenum, New York, 1971), Vol. 4, p. 315.

[9]B. A. Brown and B. H. Wildenthal, Annu. Rev. Nucl. Part. Sci. **38**, 29 (1988).

[10]E. Caurier, A. Poves, and A. P. Zuker, Université Louis Pasteur, Strasbourg, Report No. CRN/PN 89-22, 1989 (to be published).

[11]T. T. S. Kuo and G. E. Brown, Nucl. Phys. **A114**, 241 (1968).

[12]B. D. Day, in *From Nuclei to Particles,* International School of Physics "Enrico Fermi," Course LXXIX, edited by A. Molinari (North-Holland, Amsterdam, 1981).

[13]H. Kümmel, K. H. Lührmann, and J. G. Zabolitzky, Phys. Rep. **36C**, 1 (1978).

[14]H. C. Lee, computer code (private communication to A.Z.).

[15]J. B. French, in *Isospin in Nuclear Physics,* edited by D. H. Wilkinson (North-Holland, Amsterdam, 1969).

[16]A. P. Zuker, Czech. J. Phys. B **25**, 311 (1975).

[17]J. B. French and V. K. B. Kota, Phys. Rev. Lett. **51**, 2183 (1983).

[18]A. P. Zuker, in *Mathematical and Computational Methods in Nuclear Physics,* Lecture Notes in Physics Vol. 209, edited by J. S. Dehesa *et al.* (Springer, Berlin, 1984).

[19]A. Bohr and B. Mottelson, *Nuclear Structure I* (Benjamin, New York, 1964).

[20]P. J. Ellis and E. Osnes, Rev. Mod. Phys. **49**, 777 (1977); J. Shurpin, T. T. S. Kuo, and D. Strottman, Nucl. Phys. **A408**, 310 (1983).

[21]S. Cohen and D. Kurath, Nucl. Phys. **73**, 1 (1965).

[22]A. Abzouzi, E. Caurier, and A. P. Zuker (to be published).

[23]A. G. M. Van Hees, J. G. Booten, and P. W. M. Glaudemans, Phys. Rev. Lett. **62**, 2245 (1989).

[24]X. Ji and B. H. Wildenthal, Phys. Rev. C **37**, 1256 (1988); **40**, 389 (1989).

[25]A. Cortes and A. P. Zuker, Phys. Lett. **84B**, 25 (1979).

[26]H. De Vries, C. W. De Jager, and C. De Vries, At. Data Nucl. Data Tables **36**, 495 (1987).

Z. Phys. A 351, 377–383 (1995)

ZEITSCHRIFT FÜR PHYSIK A

Bivariate distributions in statistical spectroscopy studies: IV. Interacting particle Gamow Teller strength densities and β-decay rates of *fp*-shell nuclei for presupernova stars

V.K.B. Kota, D. Majumdar

Physical Research Laboratory, Ahmedabad 380 009, India

Received: 28 October 1994

Abstract. A method to calculate temperature dependent β-decay rates is developed by writing the expression for the rates explicitly in terms of bivariate GT strength densities ($\mathbf{I}^{H}_{\mathcal{O}(GT)}$) for a given hamiltonian $H = \mathbf{h} + \mathbf{V}$ and state densities of the parent nucleus besides having the usual phase space factors. The theory developed in the preceding paper (III) for constructing NIP strength densities is applied for generating $\mathbf{I}^{\mathbf{h}}_{\mathcal{O}(GT)}$ and then $\mathbf{I}^{H}_{\mathcal{O}(GT)}$ is constructed using the bivariate convolution form $\mathbf{I}^{H}_{\mathcal{O}(GT)} = \sum_S \mathbf{I}^{\mathbf{h},S}_{\mathcal{O}(GT)} \otimes \rho^{\mathbf{V},S}_{\mathcal{O}(GT);BIV-\mathcal{G}}$. The spreading bivariate Gaussian $\rho^{\mathbf{V}}_{\mathcal{O}(GT);BIV-\mathcal{G}}$, for *fp*-shell nuclei, is constructed by assuming that the marginal centroids are zero, the marginal variances are same as the corresponding state density variances and fixing the bivariate correlation coefficient $\bar{\zeta}$ using experimental β-decay half lifes. With the deduced values of ζ $\bar{\zeta} \sim 0.67$, β-decay rates for ^{61,62}Fe and $^{62-64}$Co isotopes are calculated at presupernova matter densities $\rho = 10^7$–10^9 gm/cc, temperatures $T = (3-5) \times 10^9$ °K and electron fractions $Y_e = 0.43-0.5$. The convolution form for $\mathbf{I}^{H}_{\mathcal{O}(GT)}$ led to a simple expression for calculating GT non-energy weighted sum rule strength and it describes (within 10%) the shell model results of *fp*-shell nuclei.

PACS: 21.10.Pc; 21.60.Cs; 23.40.Hc

1. Introduction

Given a one plus two-body hamiltonian (H) and a transition operator ($\mathcal{O}$), in statistical spectroscopy (SS), the smoothed form of the interacting particle (IP) transition strength densities ($\mathbf{I}^{H}_{\mathcal{O}}(E_i, E_f)$) in general takes a bivariate convolution form [1] with the non interacting particle (NIP) strength density being convoluted with a spreading bivariate Gaussian due to interactions. The construction of IP bivariate state densities allows one to address a wide variety of nuclear physics problems; earlier applications are given in [1, 2]. Applying it for the first time to problems of interest in nuclear astrophysics, this paper deals with the construction of β-decay Gamow–Teller (GT) strength distributions and in particular for $A > 60$ neutron excess *fp*-shell nuclei that are of interest in presupernova evolution calculations. The β-decay rates for these nuclei play a very important role [3, 4] in determining the structure of the core of massive presupernova stars and hence on their subsequent evolution towards gravitational collapse and supernova explosion phases.

The earlier calculations of β-decay rates at zero temperature mainly used the "gross theory" (which is indeed a statistical theory of β-decay) due to Takahashi and Yamada [5] and the microscopic quasiparticle random phase approximation (QRPA) [6]; shell model calculations are feasible and available for lighter ($A < 40$) nuclei [7]. Fuller et al. [8] (hereafter referred as FFN) for the first time dealt in detail with the problem of calculating β-decay rates (and also other steller weak interaction rates which include electron capture rates, neutrino energy loss rates etc.) at non-zero temperatures appropriate for presupernova evolution studies and tabulated the rates for 226 *ds* and *fp*-shell nuclei with $21 \leq A \leq 60$. They (FFN) use experimental data for energy levels and transition matrix elements (comparative half lifes $\log(ft)_{ij}$) where available and for the unmeasured matrix elements, zeroth order shell model calculations and also detailed shell model results (for GT strengths) wherever available are employed; recently Oda et al. [9] extended for *ds*-shell nuclei the FFN tabulations by using the results of complete shell model calculations with Wildenthal's interaction [10]. Steller evolution calculations including FFN rates showed that the FFN rates affect significantly the evolution of presupernova stars – they led to greater reduction of electron fraction throughout the core [11]. However more recently Aufderheide et al. [4] pointed out that the calculation of presupernova evolution generates neutron rich cores and therefore emphasized the need for reliable β-decay rates for neutron rich $A > 60$ *fp*-shell nuclei; they tentatively conclude that β-decay of neutron rich nuclei (particularly ^{63}Co; they in fact studied ^{60,62,63,64}Co isotopes) prevents the decrease of overall electron fraction and affects substantially the post-core-silicon-burning behavior. With this observation, the calculation of β-decay rates for $A > 60$ neutron excess nuclei has acquired new impetus. Calculations of the rates involve the construction of the bivariate GT strength densities. So far only [4, 12, 13] dealt with the neutron excess $A > 60$ fp-shell nuclei; in Ref. [4, 12] Aufderheide et al. used methods similar to that of FFN and in Ref. [13] Kar et al. (hereafter referred as KSR) introduced a method, combining the ideas of the "gross theory" with a zeroth order statistical spectroscopy result (see Sect. 6 ahead). In this paper a method is developed for β-decay rates calculations where GT strength densities are explicitly constructed, starting with an effective nucleon-nucleon intraction, using the bivariate convolution form (Eq. (2) ahead) for $\mathbf{I}^{H}_{\mathcal{O}(GT)}$. The theory for constructing NIP GT strength densities, which enter into IP theory as one of the convolution factors, is given in the preceding paper [14] (here after referred as III). We will now give a preview.

Section 2 gives the IP theory for constructing GT strength densities and Sect. 3 gives the formalism for calculating β-decay rates where $\mathbf{I}^{H}_{\mathcal{O}(GT)}$ and the state density $I^{H}(E)$ are explicitly used. In Sect. 4 the calculational procedure is described. Section 5 deals with GT non energy weighted sum rule (NEWSR) strength. Section 6 gives the results of β-decay rates at densities and temperatures typical of presupernova stars. Some concluding remarks are given in Sect. 7. Finally it should be mentioned that the present article follows paper (III).

2. IP strength densities for Gamow–Teller transition operator

Given $H = \mathbf{h} + \mathbf{V}$ the IP strength densities $\mathbf{I}^{H=\mathbf{h}+\mathbf{V}}_{\mathcal{O}}(E_i, E_f)$ produced by a transition operator $\mathcal{O}$, as derived in [1], will take a bivariate convolution form with the two convoluting functions being the NIP strength density $\mathbf{I}^{\mathbf{h}}_{\mathcal{O}}$ and a normalized spreading bivariate Gaussian ($\rho_{BIV-\mathcal{G}}$) due to $\mathbf{V}$ (interactions),

$$\mathbf{I}^{H=\mathbf{h}+\mathbf{V}}_{\mathcal{O}}(E_i, E_f) = \mathbf{I}^{\mathbf{h}}_{\mathcal{O}} \otimes \rho^{\mathbf{V}}_{\mathcal{O};BIV-\mathcal{G}}[E_i, E_f]. \tag{1}$$

In large spectroscopic spaces, just as in the case of state densities [15], one can use an extended version of (1) with some plausible approximations [1, 16] by decomposing the space into S-subspaces and unitary configurations. For the GT ($\beta^{\pm}$) operator, using pn unitary configurations and noting that GT operator do not connect two different S-subspaces [14], the IP strength density with partitioning is,

$$\begin{aligned}\mathbf{I}^{H=\mathbf{h}+\mathbf{V}}_{\mathcal{O}(GT)}(E_i, E_f) &= \sum_{S} \sum_{[\mathbf{m}^i_p, \mathbf{m}^i_n], [\mathbf{m}^f_p, \mathbf{m}^f_n] \in S} \mathbf{I}^{\mathbf{h};[\mathbf{m}^i_p, \mathbf{m}^i_n],[\mathbf{m}^f_p, \mathbf{m}^f_n]}_{\mathcal{O}(GT)} \\ &\quad \otimes \rho^{\mathbf{V};[\mathbf{m}^i_p, \mathbf{m}^i_n],[\mathbf{m}^f_p, \mathbf{m}^f_n]}_{\mathcal{O}(GT)}[E_i, E_f]; \\ \rho^{\mathbf{V};[\mathbf{m}^i_p, \mathbf{m}^i_n],[\mathbf{m}^f_p, \mathbf{m}^f_n]}_{\mathcal{O}(GT)}(x, y) &\\ = \rho^{\mathbf{V}}_{\mathcal{O}(GT);BIV-\mathcal{G}}(x, y; 0, 0, &\sigma_{\mathbf{V}}([\mathbf{m}^i_p, \mathbf{m}^i_n]), \sigma_{\mathbf{V}}([\mathbf{m}^f_p, \mathbf{m}^f_n]), \bar{\zeta})\end{aligned} \tag{2}$$

In (2) we are assuming that the marginal centroids M_{10}, M_{01} of $\rho^{\mathbf{V}}_{\mathcal{O}(GT)}$ to be zero and the marginal variances $\mathcal{M}_{20}, \mathcal{M}_{02}$ to be close to the corresponding state density variances. In $S = 0$ spaces, for the marginal centroids, with $(m_p, m_n) \to (m_p \pm 1, m_n \mp 1)$ for $\beta^{\mp}$ transitions, this is equivalent to assuming that

$$\begin{aligned} M_{10} &= \langle \mathcal{O}^{\dagger}\mathcal{O}\mathbf{V}\rangle^{m_p, m_n} / \langle \mathcal{O}^{\dagger}\mathcal{O}\rangle^{m_p, m_n} \simeq \langle \mathbf{V}\rangle^{m_p, m_n} = 0, \\ M_{01} &= \langle \mathcal{O}^{\dagger}\mathbf{V}\mathcal{O}\rangle^{m_p, m_n} / \langle \mathcal{O}^{\dagger}\mathcal{O}\rangle^{m_p, m_n} \\ &= \langle \mathcal{O}\mathcal{O}^{\dagger}\mathbf{V}\rangle^{m_p \pm 1, m_n \mp 1} / \langle \mathcal{O}\mathcal{O}^{\dagger}\rangle^{m_p \pm 1, m_n \mp 1} \\ &\simeq \langle \mathbf{V}\rangle^{m_p \pm 1, m_n \mp 1} = 0, \\ \mathcal{M}_{20} &= \langle \mathcal{O}^{\dagger}\mathcal{O}\mathbf{V}^2\rangle^{m_p, m_n} / \langle \mathcal{O}^{\dagger}\mathcal{O}\rangle^{m_p, m_n} - M^2_{10} \simeq \langle \mathbf{V}^2\rangle^{m_p, m_n} \end{aligned}$$

and

$$\begin{aligned} \mathcal{M}_{02} &= \langle \mathcal{O}^{\dagger}\mathbf{V}^2\mathcal{O}\rangle^{m_p, m_n} / \langle \mathcal{O}^{\dagger}\mathcal{O}\rangle^{m_p, m_n} - M^2_{01} \\ &= \langle \mathcal{O}\mathcal{O}^{\dagger}\mathbf{V}^2\rangle^{m_p \pm 1, m_n \mp 1} / \langle \mathcal{O}\mathcal{O}^{\dagger}\rangle^{m_p \pm 1, m_n \mp 1} - M^2_{01} \\ &\simeq \langle \mathbf{V}^2\rangle^{m_p \pm 1, m_n \mp 1} \end{aligned}$$

for $GT(\beta^{\mp})$ operators. Using the results of [14, 15], one can calculate $\langle \mathcal{O}^{\dagger}\mathcal{O}\mathbf{V}^2\rangle^{[m]}$ and $\langle \mathcal{O}^{\dagger}\mathcal{O}\rangle^{[m]}$ traces and hence test the above approximations. We verified explicitly that for fp-shell interactions (used are the MWH2 [17] and MFP13 [18] interactions), the above approximations are quite reasonable; centroid shifts are $\lesssim 0.1$ MeV and variances differ from $\langle \mathbf{V}^2\rangle$ by less than 3%. It should be clear that in general it is possible to take into account in (2) the centroid shifts and variance differences.

Besides the assumptions for the marginal centroids and variances, which are well justified for fp-shell nuclei, (2) also assumes that the bivariate correlation coefficient $\zeta = \bar{\zeta}$ is independent of the configurations involved. At present, because of the complexities in the proper definition of ζ as a function of the initial and final configurations, one has to make this assumption [1]. From NIP calculations in ds and fp-shells (carried out using results of [14]), one sees that $0.6 \leq \zeta \leq 0.9$. For example for $\mathcal{O}(GT(\beta^-))$ in ds-shell with ${}^{17}\mathcal{O}$ energies we have $\zeta = 0.78$ for ($m_p = 4$, $m_n = 4$) and $\zeta = 0.8$ for ($m_p = 6$, $m_n = 6$). Similarly in fp-shell with ^{41}Ca energies we have $\zeta = 0.84$ for ($m_p = 6$, $m_n = 16$), and $\zeta = 0.87$ for ($m_p = 6$, $m_n = 10$). In addition it is known from [1] that for various other operators, with realistic ds-shell interactions $0.6 \leq \zeta \leq 0.9$; some examples with GT operator are also given in [19]. Moreover in the two-body EGOE model [1], which extends the $\zeta \simeq 1 - 1/m$ result given in [14] for random one-body hamiltonians, one has the result $\zeta \simeq 1 - 2/m$ where m is the number of active nucleons. Thus in EGOE, $\zeta \simeq 0.8$ for $m = 10$ with a one-body transition operator and a two-body hamiltonian. Therefore in the present calculations we use (2) with $\bar{\zeta}$ a free parameter whose value will be determined by demanding that certain experimental data is represented by the theory but with the constraint $0.6 \leq \bar{\zeta} \leq 0.9$.

3. Formalism for β-decay rates at finite temperature

The β-decay rate $T_{\beta}(E_i \to E_f)$ is number of β-decays per second from a given initial state $|E_i\rangle$ of the parent nucleus to the final nuclear state $|E_f\rangle$ and $T_{\beta}(E_i \to E_f) \propto [g^2_V B_F(E_i \to E_f) + g^2_A B_{GT}(E_i \to E_f)]$, where g_V and g_A are respectively the vector and axial vector coupling constants and B_F and B_{GT} are the Fermi and Gamow–Teller transition strengths respectively. Including the phase space factor f that incorporates the dependence of the rate on nuclear charge Ze and the available β-decay energy (phase space available for the leptons), T_β takes the form $T_\beta(E_i \to E_f) = Cf(Z, Q_i - E_f)[g_V^2 B_F(E_i \to E_f) + g_A^2 B_{GT}(E_i \to E_f)]$ where C is a constant and with Q the Q-value for β-decay from ground state (GS), $Q_i = Q + E_i$. Fixing the value of C [20, 21] by using the $\log ft$ values ($ft = (\ln 2) f/T_\beta$) for pure Fermi transitions, one can write down the expressions for ground state half lifes and β-decay rates at finite temperature. Before going further it is important to note that in practice in β-decay rates calculations for fp-shell nuclei the B_F term can be neglected as the Fermi strength is concentrated in a narrow energy domain (width $\sim 0.157ZA^{-1/3}$ MeV) and high up in energy (centroid $\sim 1.44ZA^{-1/3}$ MeV) [5]. With this, by writing B_{GT} in terms of GT strength density $\mathbf{I}^h_{\mathcal{O}(GT)}$ and the state density $I^H(E)$, the expression for GS half lifes is,

$$t_{1/2}(GS) = \{6250(s)\} \left\{ \int_0^Q \left[\left(\frac{g_A}{g_V}\right)^2 3\mathcal{L} \right] \times \left[\frac{\mathbf{I}^H_{\mathcal{O}(GT)}(E_{GS}, E_f)}{I^H(E_{GS})} \right] f(Z, Q - E_f) dE_f \right\}^{-1}. \quad (3)$$

In (3), the factor 3 comes due to our definition of $\mathbf{I}^h$ in [14]★ and $\mathcal{L}$ is the so called quenching factor which is required because the calculated (shell model) GT sum rule strength is always found to be larger than the observed strength [22]. We adopt the usual value of 0.6 for $\mathcal{L}$. Similarly we use $(g_A/g_V)^2 = 1.4$ as given in [20]. Our primary concern in this paper is to calculate β-decay rates $\lambda(T)$ at finite temperature and they are defined by the thermal average (with weight factor $e^{-E/k_BT} I(E)$) of the rates from all parent nucleus states [8]. Then the expression for $\lambda(T)$ is,

$$\lambda(T) = \frac{\ln 2(s^{-1})}{6250} [\int e^{-E_i/k_BT} I^H(E_i) dE_i]^{-1} \times \left[\int dE_i e^{-E_i/k_BT} I^H(E_i) \left[\int_0^{Q_i} dE_f \left\{ \left(\frac{g_A}{g_V}\right)^2 3\mathcal{L} \right\} \times \left[\frac{\mathbf{I}^H_{\mathcal{O}(GT)}(E_i, E_f)}{I^H(E_i)} \right] f(Z, T, Q_i - E_f) \right] \right]. \quad (4)$$

The phase space factor $f(Z, T, Q_i - E_f)$ is given by the integral [4, 8, 21]

$$f(Z, T, E_0 = Q_i - E_f) = \int_1^{\varepsilon_0} \frac{F(Z, \varepsilon_e)\varepsilon_e(\varepsilon_e^2 - 1)^{1/2}(\varepsilon_0 - \varepsilon_e)^2}{\{1 + \exp[(\mu_e - \varepsilon_e)/(k_BT)_e]\}} d\varepsilon_e. \quad (5)$$

In (5) $\varepsilon_0 = E_0/m_e$, $\mu_e = \mu/m_e$ and $(k_BT)_e = k_BT/m_e$ where m_e is the rest mass of the electron in MeV. For the Coulomb factor $F(Z, \varepsilon)$, we use Schenter and Vogel [23] expression

$$F(Z, \varepsilon) = \frac{\varepsilon}{\sqrt{\varepsilon^2 - 1}} \exp[\alpha(Z) + \beta(Z)\sqrt{\varepsilon - 1}];$$

$$\alpha(Z) = -0.811 + 4.46(-2)Z + 1.08(-4)Z^2 \quad (\varepsilon - 1) < 1.2$$

$$= -8.46(-2) + 2.48(-2)Z + 2.37(-4)Z^2 \quad (\varepsilon - 1) \geq 1.2$$

$$\beta(Z) = 0.673 - 1.82(-2)Z + 6.38(-5)Z^2 \quad (\varepsilon - 1) < 1.2$$

$$= 1.15(-2) + 3.58(-4)Z - 6.17(-5)Z^2 \quad (\varepsilon - 1) \geq 1.2 \quad (6)$$

It should be mentioned that f in (3) follows from (5) by dropping the denominator in the integral in the R.H.S. of (5). The only unknown quantity left to be determined is the electron chemical potential μ. In the case of presupernova core where the density $\sim 10^8$ gms/cc, the electrons are close to being relativistic and hence μ is found by inverting the expression for the lepton number density. Given the density ρ_7 (ρ in units of 10^7 gms/cc), temperature $\bar{T}$ (T in MeV is denoted by $\bar{T}$) and the electron fraction Y_e in the presupernova star, there are hierarchy of approximate expressions that can be used for μ (see [4, 24]) and we adopt Eq. (14) of [4] in the present calculations and which is,

$$\mu = 1.11(\rho_7 Y_e)^{1/3} \left[1 + \left(\frac{\pi}{1.11}\right)^2 \frac{\bar{T}^2}{(\rho_7 Y_e)^{2/3}} \right]^{-1/3}. \quad (7)$$

Using (4–7) β-decay rates can be calculated by constructing state densities $I^H(E)$ and the strength densities $\mathbf{I}^H_{\mathcal{O}(GT)}(E_i, E_f)$. The procedure for calculating these densities is given in the following section and as an example the results for five neutron excess fp-shell nuclei ^{61,62}Fe, $^{62-64}$Co (^{61}Fe $\to$ ^{61}Co, ^{62}Fe $\to$ ^{62}Co, ^{62}Co $\to$ ^{62}Ni, ^{63}Co $\to$ ^{63}Ni, ^{64}Co $\to$ ^{64}Ni) are given in Sect. 6.

4. Calculational procedure for state and bivariate strength densities $I^H(E)$ and $\mathbf{I}^H_{\mathcal{O}(GT)}(E_i, E_f)$

Our primary concern in this paper is to demonstrate that the theory for β-decay rates calculation given by (3–5) combined with (2) and the methods of [14, 15] is applicable in practice. Because of this limited aim, we followed simpler methods for constructing state and strength densities unlike the more elaborate procedure used for example in [25] in the case of level densities.

4.1. State densities

For the calculation of state densities $I^H(E)$ the following procedure is adopted. The s.p. orbits are chosen to be the five orbits $1f_{7/2}$, $2p_{3/2}$, $1f_{5/2}$, $2p_{1/2}$ and $1g_{9/2}$ with $s = 0, 0, 0, 0$ and 1 respectively. Following the calculations in [25] for fp-shell level density data analysis, the traceless SPE for the four fp-orbits are chosen to be -2.664 MeV,

★In discrete version $B_{GT}(J_iE_i \to J_fE_f) = (2J_i + 1)^{-1} \sum_{M_i, M_f(\mu)} |\langle E_f J_f M_f | (\mathcal{O}(GT))^\dagger_\mu | E_i J_i M_i \rangle|^2$ which in the continuous version becomes $\{I(E_i)\}^{-1} \sum_{\alpha \in E_i, \beta \in E_f} |\langle E_f \alpha | (\mathcal{O}(GT))^\dagger_\mu | E_i \beta \rangle|^2$. Noting that $\sum_\alpha |E_i\alpha\rangle \Rightarrow |E_i\rangle I(E_i)$ and replacing the sum with $\mathbf{I}^h_{\mathcal{O}(GT)}$ using Eq. (5) of [14], gives rise to factor 3 in Eq. (3) above

-0.644 MeV, 3.526 MeV and 1.366 MeV respectively and the $(fp) - g_{9/2}$ separation $\Delta_{(fp)-g_{9/2}}$ is chosen to be 6 MeV; for ^{64}Co however it is taken to be 7 MeV. The calculations are performed in $S = 0 \oplus 1 \oplus 2$ spaces using pn unitary configurations with (fp) and $g_{9/2}$ orbits as unitary orbits; I^h is constructed using the above SPE (i.e. no renormalizations due to the interactions are included – their main effect is assumed to be in fixing the value of $\Delta_{(fp)-g_{9/2}}$) and spherical orbits. The density $\rho_{\mathscr{G}}^{V}$ is constructed by calculating the spreading variances using surface delta interaction (SDI) with strength $G = 20/A$ MeV (this value follows from [25]). The GS is fixed by demanding that the total level density $I_l(E)$ at 8 MeV excitation is same as the value given by the values of the level density parameters (a, Δ) that follow from Dilg et al. smoothed expression [26]. This choice is made because for the nuclei under consideration ($^{61-62}$Fe, $^{61-64}$Co, $^{62-64}$Ni), the low-energy spectrum is not known with certainty to be complete. For example, with Dilg et al. values for (a, Δ) together with the back shifted Lang-Le Couteur Fermi gas formula [26], for ^{62}Fe, ^{61}Co and ^{64}Co the $I_l(E)$ (in MeV^{-1}) values at (4, 8) MeV are (13, 455), (61, 1452) and (180, 3765) respectively. In general we find that with the calculated GS, the density at 4 MeV is quite close (within 30%) to the value predicted by Dilg et al (a, Δ) values. It is also verified that the calculated spin-cutoff factors $(\sigma_J(E))$ are quite close to the values given by the expression in [26] that involves (a, Δ) values. For example for ^{62}Fe, ^{61}Co, and ^{62}Ni nuclei the calculated and Dilg et al. values at 8 MeV are (3.9, 3.99), (3.72, 4.15), (4.24, 4.40) and (3.40, 3.99) respectively. Also for even-even and odd-odd nuclei for positive parity states and for odd-A nuclei negative parity states, the $S = 2$ intensities in the GS domain are $\leq 30\%$ of the $S = 0$ intensities (thus the $g_{9/2}$ orbit is seen to be important). In addition to these tests, we also verified that the GT NEWSR as predicted by the present calculations is quite close to shell model results. We will return to this in Sect. 5. The level density calculations here are carried out in the same spirit as in [1] and it is seen that as long as we construct the densities well upto 8 MeV (the β^- decay Q-values and (3, 4) show that densities only upto ~ 8 MeV are needed in the present exercise), the details of the SPE and interaction **V** do not alter the final results significantly. It should be stressed that, as in the studies of [15, 25], there are many other variants of the above procedure which can be used in practice.

4.2. *Strength densities*

The GT strength densities $\mathbf{I}^{H}_{\mathscr{O}(GT)}(E_i, E_f)$ are constructed for the $GT(\beta^-)$ operator using (2) with ζ as a free parameter. In our study the coulomb correction to GT centroids and the variances [5] are ignored and it is assumed that the parametrization of ζ takes care of the same (see Sect. 6 ahead); they can be easily put back. $\mathbf{I}^{h}_{\mathscr{O}(GT)}(E_i, E_f)$ is constructed using the results of [14]. In the first calculation ζ is determined for each nucleus by demanding the observed half lifes are reproduced within the constraint $0.6 \leq \zeta \leq 0.9$. The deduced values of ζ are 0.6, 0.7, 0.61, 0.68 and 0.72 and the corresponding half lifes (in seconds) are 244, 69.4, 88.8, 31 and 0.3 for ^{61}Fe, ^{62}Fe, ^{62}Co, ^{63}Co and ^{64}Co respectively. The half lifes are calculated using (3). These ζ values change quite substantially (also fluctuate) as we go from ^{61}Fe ($\zeta = 0.6$) to ^{64}Co($\zeta = 0.72$) while the (valence) particle number m changes only by 3 units. The EGOE form $\zeta \simeq a + b/m$ immediately shows that it is not possible to reproduce the above variation within the framework of SS. Therefore the above values of ζ are not acceptable in a SS calculation. The plausible alternative is to seek a best fit solution to the observed half lifes of the nuclei under consideration (in the present exercise they are $^{61-62}$Fe, $^{62-64}$Co isotopes) by assuming the form of ζ to be $\zeta_0 + \zeta_1/m$ where m is the number of valence nucleons/holes, (i.e.) by minimizing $\sum_{i=\text{nuclei}} (\log(\tau^i_{1/2})_{\text{cal}} - \log(\tau^i_{1/2})_{\text{expt}})^2$. The resulting ζ values and the corresponding calculated half lifes are given in Table 1. The ζ values are around 0.67 (changing from 0.68 to 0.668 as we go from ^{61}Fe to ^{64}Co). It should be obvious that the calculated half lifes, with these ζ values, are the predictions of SS. The differences between the calculated and experimental values are of the same order of magnitude as in other models [5, 6, 13]. Before going further it should be pointed out that in general for a given nucleus, as ζ decreases the calculated half life increases and sometimes the increase is fast. Using the ζ values given in Table 1, the IP strength densities $\mathbf{I}^{H}_{\mathscr{O}(GT)}$ are constructed using (2) and the β-decay rates are calculated. Before giving these results, we will briefly describe GT NEWSR.

Table 1. Calculated and experimental β^- decay half lifes. The method of determining the correlation coefficient ζ is given in the text. The calculated half lifes are given in the fourth column. The Q-values and experimental data for half lifes are taken from [27]. The KSR results are from [13, 24] and the QRPA result, which is available only for ^{64}Co in the set of nuclei considered in the table, is from the first reference of [6]

Nucleus	Q (MeV)	ζ	half life (s)			
			Calc.	Expt.	KSR	QRPA
^{64}Co	7.307	0.668	1.9	0.3	3.5	10.0
^{63}Co	3.662	0.671	52.7	27.5	52.1	—
^{62}Co	5.315	0.675	16.5	90	15.1	—
^{62}Fe	2.327	0.675	267.2	68	183.4	—
^{61}Fe	3.890	0.680	23.0	360	34.5	—

5. *GT* NEWSR strength

Using (2) and the explicit form the $\mathbf{I}^{h}_{\mathscr{O}(GT)}$ as given in [14], a simple expression for NEWSR strength $M_{\mathscr{O}(GT)}(E)$ for $\mathscr{O}(GT)$ can be written down,

$$M_{\mathscr{O}(GT)}(E) = \sum_{(\mathbf{m}_p, \mathbf{m}_n)} \frac{I^{H,(\mathbf{m}_p,\mathbf{m}_n)}(E)}{I^{H,(m_p,m_n)}(E)} \times \left\{ 3 \sum_{\alpha,\beta} [\varepsilon^2_{\alpha\beta}(\mathscr{O}(GT))] \frac{m_\beta(\mathbf{m}_p, \mathbf{m}_n)}{N_\beta} \frac{N_\alpha - m_\alpha(\mathbf{m}_p, \mathbf{m}_n)}{N_\alpha} \right\};$$

$$I^{H,(\mathbf{m}_p,\mathbf{m}_n)}(E) = d(\mathbf{m}_p, \mathbf{m}_n)\rho_{\mathscr{G}}(E; \varepsilon_{\mathrm{h}}(\mathbf{m}_p, \mathbf{m}_n), \sigma_{\mathrm{V}}(\mathbf{m}_p, \mathbf{m}_n))$$

$$I^{H,(m_p,m_n)}(E) = \sum_{(\mathbf{m}_p, \mathbf{m}_n) \in (m_p, m_n)} I^{H,(\mathbf{m}_p,\mathbf{m}_n)}(E). \tag{8}$$

Table 2. Calculated and shell model (SM) results for $GT(\beta^-)$ NEWSR

Nucleus	NEWSR	
	Calc.	SM
^{54}Fe	17.8	15.1^a
^{56}Fe	22.6	22.1^a
^{60}Fe	31.6	33.5^b
^{58}Ni	20.2	16.6^b
^{60}Ni	14.1	14.6^b

a) Ref. [30], b) combining $GT(\beta^+)$ NEWSR value 9.47 given in [29] with the sum rule result $S_{\beta^-} - S_{\beta^+} = 3(N - Z)$

In (8) $\varepsilon_{\alpha\beta}$ define $\mathcal{O}(GT)$ and α and β are proton and neutron spherical orbits respectively for $GT(\beta^-)$ and vice versa for $GT(\beta^+)$. The formula (8) whose origin lies in the NIP expression for $\mathbf{I}^{\mathbf{h}}_{\mathcal{O}(GT)}$, is essentially same as the sum rule formula used in earlier studies (though derived using different methods) [28, 13; see also Appendix A of [4]] of *fp*-shell nuclei. The corrections to (8) come from the centroid shifts and the variance corrections to $\rho_{\mathscr{G}}$ as discussed below (2) and they are small for *fp*-shell nuclei. In order to test the applicability of (8) for *fp*-shell nuclei, the $GT(\beta^-)$ NEWSR strengths are calculated for ^{54,56,60}Fe and ^{58,60}Ni isotopes by first carrying out level density calculations and determining the ground state, exactly as described in Sect. 4. These nuclei are chosen as there are corresponding shell model results available in literature [29, 30], though calculated using different interactions and including only *fp*-orbits (i.e. without the $g_{9/2}$ orbit). The result shown in Table 2 clearly demonstrates that (8) gives results that are in close agreement with shell model values. Thus (8), which is easy to apply, can be used in future studies of GT NEWSR and quenching of GT strengths. The calculated sum rule strengths (from GS) with quenching factor $\mathscr{L} = 0.6$ for ^{61}Fe, ^{62}Fe, ^{62}Co, ^{63}Co and ^{64}Co nuclei are 20, 18.9, 18.8, 19.7 and 20.8 respectively. The $^{61-62}$Fe sum rule strengths are compatible with the ^{60}Fe results given in Table 2. We conclude from these results that the various choices made in Sect. 4 are quite reasonable.

6. Results for β-decay rates at finite temperature

Using the formalism given in Sects. 2, 3 and the procedure outlined in Sect. 4, β-decay rates for the five nuclei ^{61}Fe, ^{62}Fe, ^{62}Co, ^{63}Co and ^{64}Co are calculated at densities $\rho = 10^9$ gm/cc, 10^8 gm/cc, 10^7 gm/cc, temperatures $T = 3\times10^9, 4\times10^9, 5\times10^9\,^\circ$K($T = 4\times10^9\,^\circ$K compared to 0.345 MeV) and electron fractions $Y_e = 0.5$, 0.47 and 0.43. The results are given in Table 3. It is observed that the excitation energy (of the parent nucleus) contribution to the decay rate are quite significant upto about $(6-8)\times k_BT$ and beyond that their importance ($\sim$ 10–15%) goes down rapidly; note that with $E_i = 5k_BT$, $e^{-E_i/k_BT} \simeq 1/150$. It is also observed that for a given nucleus the rates go down as the correlation coefficient ζ decreases. For example for ^{64}Co, with $\zeta = 0.668$, $\lambda(T) = 0.345$ s^{-1} while with $\zeta = 0.72$, $\lambda(T) = 1.33$ s^{-1} for $\rho = 10^8$ gm/cc, $T = 4\times10^9\,^\circ$K and $Y_e = 0.5$ and for the same set of (ρ, T, Y_e) values, for ^{62}Co, $\lambda(T) = 3.64\times10^{-2}$ s^{-1} for $\zeta = 0.675$ and $\lambda(T) = 0.46\times10^{-2}$ s^{-1} for $\zeta = 0.610$. By comparing the rates in Table 3 with the results given by Aufderheide et al. [4], we observe that the numbers for ^{63}Co are quite similar (in [4] the rates are tabulated for $\rho = 10^8$ gm/cc, 10^7 gm/cc, $T = 3\times10^9$, 4×10^9, $5\times10^9\,^\circ$K and $Y_e = 0.5$) while for ^{62}Co the numbers in Table 3 are smaller by a factor 5 and that for ^{64}Co by a factor 8. The numbers given in Table 3 are much closer to KSR results [13, 31] and this should not be surprising as the method employed by KSR* derives from (2). More importantly, the variation in $\lambda(T)$ with T as can be seen from our results in Table 3 is not to be found in the first KSR tabulation [13] but similar variation is found in their latest compilation of the rates [31]. For example for ^{63}Co nucleus with $\rho = 10^9$ gm/cc, $Y_e = 0.5$, for $T = 3\times10^9\,^\circ$K, $4\times10^9\,^\circ$K, $5\times10^9\,^\circ$K the rates $\lambda(T)$ from [31] and Table 3 are (in s^{-1}) (0.37×10^{-3}, 0.94×10^{-3}, 2.07×10^{-3}) and (0.12×10^{-3}, 0.65×10^{-3}, 2.26×10^{-3}). Similarly for ^{63}Co with $\rho = 10^8$ gm/cc, $Y_e = 0.5$ the numbers are (3.6×10^{-2}, 4.11×10^{-2}, 4.66×10^{-2}) and (3.06×10^{-2}, 3.64×10^{-2}, 4.41×10^{-2}) respectively. In the case of ^{62}Fe with $\rho = 10^8$ gm/cc, $Y_e = 0.47$ the numbers are (5.79×10^{-3}, 7.31×10^{-3}, 9.46×10^{-3}) and (2.57×10^{-3}, 6.86×10^{-3}, 16.7×10^{-3}) respectively. Thus it is plausible to conclude that the method proposed in this paper, where IP state and strength densities are constructed and used for the first time in β-decay rates calculations, is reliable. The goodness of the present method, which is based on the

*The KSR method is a simplified version of the method given in Sects. 2–5 and in parts it is an extended version of the "gross theory" given in [5]. The basic difference between the present method and KSR method is that the KSR method assumes single unitary orbit in applying (2) (without several unitary orbits the agreements found for level densities, as given in [15, 25], and Sect. 4, could not have been obtained and it is well known that the assumption of single unitary orbit is unsatisfactory [1, 15, 16, 32]) so that the summation in (2) disappears. Then the ratio $\mathbf{I}^H/I^H$ in (4) is the conditional density $\rho^H_{\mathcal{O}(GT)}(E_f\,|\,E_i)$ which will be a Gaussian (as $\mathbf{I}^H_{\mathcal{O}(GT)} = \mathbf{I}^{\mathbf{h}}_{\mathcal{O}(GT)} \otimes \rho^{\mathbf{V}}_{\mathcal{O}(GT);\mathscr{G}}$ is a bivariate Gaussian with CLT applied to $\mathbf{I}^{\mathbf{h}}$) with normalization given by GT NEWSR strength $M_{\mathcal{O}(GT)}(E)$; the NEWSR is calculated using an expression similar to (8). The conditional density $\rho^H_{\mathcal{O}(GT)}(E_f|E_i)$ is constructed by parametrizing its centroid and variance and in order to correct for the departures from a single Gaussian approximation for $\mathbf{I}^H_{\mathcal{O}(GT)}$, in some cases KSR [24] add Edgeworth corrections with (γ_1, γ_2) values chosen such that $|\gamma_1| \le 0.3$ and $|\gamma_2| \le 0.3$. As can be seen from [1, 16], the conditional centroid ε_c changes linearly with energy, $\varepsilon_c(E_i) = \varepsilon_0 + \zeta E_i$ and the conditional width is a constant, $\sigma_c = \sigma_0(1-\zeta^2)^{1/2}$ where $\sigma_0 \simeq \sigma_H = (\sigma_{\mathbf{h}}^2 + \sigma_{\mathbf{V}}^2)^{1/2}$. The centroid ε_c is parametrized (assuming $\zeta = 1$) and the value of σ_c is fixed, after adding Coulomb correction, by minimizing square of the difference in the logarithm of the calculated and experimental half lifes exactly as in the "gross theory". In addition, instead of carrying out integration over all E_i's in applying (4), KSR use the experimental energies and spins (wherever observed) so that the calculation of state densities is avoided. In the formalism presented in Sects. 2–4, the densities $\mathbf{I}^H$ and I^H are explicitly constructed and used alongwith (4) within a single framework unlike in the KSR method. It should be added that the s.p. spectrum and the $\mathbf{V}$ used in Sect. 4 give $\sigma_H^2 = \sigma_{\mathbf{h}}^2 + \sigma_{\mathbf{V}}^2 \simeq (8\text{ MeV})^2$ and the expression $\sigma_c = \sigma_H(1-\zeta^2)^{1/2}$ gives, with $\zeta \simeq 0.67$, $\sigma_c \simeq 6$ MeV which is compatible with the value of σ_c deduced by KSR

Table 3. β^--decay rates for ^{64}Co, ^{63}Co, ^{62}Co, ^{62}Fe and ^{61}Fe nuclei

Nucleus	ρ(gms/cc)	Y_e	Temperature (°K) 3 × 10⁹	4 × 10⁹	5 × 10⁹
			Rate (s^{-1})		
^{64}Co	10^9	0.50	0.85×10^{-1}	1.04×10^{-1}	1.30×10^{-1}
		0.47	0.91×10^{-1}	1.11×10^{-1}	1.38×10^{-1}
		0.43	1.00×10^{-1}	1.21×10^{-1}	1.49×10^{-1}
	10^8	0.50	3.10×10^{-1}	3.43×10^{-1}	3.84×10^{-1}
		0.47	3.14×10^{-1}	3.47×10^{-1}	3.87×10^{-1}
		0.43	3.19×10^{-1}	3.52×10^{-1}	3.93×10^{-1}
	10^7	0.50	3.83×10^{-1}	4.14×10^{-1}	4.54×10^{-1}
		0.47	3.83×10^{-1}	4.14×10^{-1}	4.54×10^{-1}
		0.43	3.84×10^{-1}	4.15×10^{-1}	4.55×10^{-1}
^{63}Co	10^9	0.50	0.12×10^{-3}	0.65×10^{-3}	2.26×10^{-3}
		0.47	0.15×10^{-3}	0.76×10^{-3}	2.54×10^{-3}
		0.43	0.21×10^{-3}	0.94×10^{-3}	2.98×10^{-3}
	10^8	0.50	0.85×10^{-2}	1.31×10^{-2}	2.09×10^{-2}
		0.47	0.88×10^{-2}	1.35×10^{-2}	2.14×10^{-2}
		0.43	0.93×10^{-2}	1.41×10^{-2}	2.21×10^{-2}
	10^7	0.50	1.63×10^{-2}	2.17×10^{-2}	3.06×10^{-2}
		0.47	1.64×10^{-2}	2.18×10^{-2}	3.07×10^{-2}
		0.43	1.65×10^{-2}	2.19×10^{-2}	3.08×10^{-2}
^{62}Co	10^9	0.50	0.29×10^{-2}	0.50×10^{-2}	0.84×10^{-2}
		0.47	0.34×10^{-2}	0.56×10^{-2}	0.92×10^{-2}
		0.43	0.41×10^{-2}	0.65×10^{-2}	1.04×10^{-2}
	10^8	0.50	3.06×10^{-2}	3.64×10^{-2}	4.41×10^{-2}
		0.47	3.12×10^{-2}	3.71×10^{-2}	4.48×10^{-2}
		0.43	3.22×10^{-2}	3.80×10^{-2}	4.58×10^{-2}
	10^7	0.50	4.41×10^{-2}	4.97×10^{-2}	5.74×10^{-2}
		0.47	4.42×10^{-2}	4.99×10^{-2}	5.76×10^{-2}
		0.43	4.44×10^{-2}	5.00×10^{-2}	5.77×10^{-2}
^{62}Fe	10^9	0.50	0.04×10^{-4}	1.16×10^{-4}	9.99×10^{-4}
		0.47	0.06×10^{-4}	1.42×10^{-4}	11.50×10^{-4}
		0.43	0.09×10^{-4}	1.88×10^{-4}	14.00×10^{-4}
	10^8	0.50	2.40×10^{-3}	6.56×10^{-3}	1.62×10^{-2}
		0.47	2.57×10^{-3}	6.86×10^{-3}	1.67×10^{-2}
		0.43	2.83×10^{-3}	7.29×10^{-3}	1.73×10^{-2}
	10^7	0.50	8.11×10^{-3}	1.43×10^{-2}	2.71×10^{-2}
		0.47	8.19×10^{-3}	1.44×10^{-2}	2.72×10^{-2}
		0.43	8.30×10^{-3}	1.46×10^{-2}	2.74×10^{-3}
^{61}Fe	10^9	0.50	0.36×10^{-3}	1.51×10^{-3}	4.52×10^{-3}
		0.47	0.45×10^{-3}	1.76×10^{-3}	5.07×10^{-3}
		0.43	0.62×10^{-3}	2.18×10^{-3}	5.93×10^{-3}
	10^8	0.50	1.89×10^{-2}	2.71×10^{-2}	3.98×10^{-2}
		0.47	1.96×10^{-2}	2.78×10^{-2}	4.07×10^{-2}
		0.43	2.06×10^{-2}	2.89×10^{-2}	4.19×10^{-2}
	10^7	0.50	3.44×10^{-2}	4.33×10^{-2}	5.72×10^{-2}
		0.47	3.46×10^{-2}	4.35×10^{-2}	5.74×10^{-2}
		0.43	3.48×10^{-2}	4.37×10^{-2}	5.76×10^{-2}

smoothed form for GT densities, derives partly from the fact that the β-decay rates involve an averaging (thermal average to be precise). In the present method, as the densities are constructed by superposing unitary orbit densities, it is easy to go beyond $0\hbar\omega$ spaces. Using the formalism given in Sects. 2–4, it is possible to investigate the sensitivity of the calculated β-decay rates to hamiltonian parameters, changes in ground states, ρ, T, Y_e etc. and these studies are defered for future.

7. Conclusions

A method based on smoothed forms, derived using statistical spectroscopy principles, for GT densities is developed for calculating β-decay rates which is simple as the earlier methods [4, 13] but at the same time it incorporates much more microscopic information as the desities $I^H_{O(GT)}$ and $I^H(E)$ are explicitly constructed using *interactions* and used. The method of constructing GT strength densities

$I^{H}_{c(GT)}$ given by (2) can be used in electron capture rates calculations and with this it is possible to generate detailed and complete tabulation of β-decay and electron capture rates for various (ρ, T, Y_e) values for all neutron excess *fp*-shell nuclei relevant for presupernova evolution calculations. Then they can be used, for example as carried out recently by Aufderheide et al. [12] using a tabulation they generated, in steller evolution codes and study the influence of the calculated rates on the evolution of presupernova stars. This study is being initiated and the results will be reported in a future publication. In addition, the GT strength densities which can be constructed using (2) have important applications in solar neutrino experiments [33]. The present exercise represents a first application of the bivariate convolution form for strength densities in nuclear astrophysics problems. This together with the earlier applications to the problem of time reversal non-invariance in nucleon–nucleon force where a random two-body transition operator enters [1] and the application to the single particle transfer strengths [2] show that the strength density formalism in terms of bivariate convolutions is a viable tool for studying wide variety of nuclear physics problems that involve strength densities. A different and new result that is derived from the β-decay rates calculations is that they determine a value for the bivariate correlation coefficient ζ; $\zeta \simeq 0.67$. Thus the level density calculations [15, 25] determine a magnitude parameter of the effective nucleon–nucleon interaction and β-decay rates calculations give a value for the bivariate correlation coefficient of the interaction with the GT operator. This subject of information propagation in complex systems in being studied [15, 34].

The work presented in this paper has grown out of many discussions and correspondence one of the authors (VKBK) has with J.B. French in the last five years and also out of an ongoing project one of the authors (VKBK) is carrying out with J.B. French and R.U. Haq. Thanks are due to K. Kar and S. Sarkar for supplying some of their results before publishing. The authors thank V. Potbhare for his interest in the work presented in this paper.

References

1. French, J.B., Kota, V.K.B., Pandey, A., Tomsovic, S.: Ann. Phys. (N.Y.) 181, 235 (1988); Phys. Rev. Lett. 58, 2400 (1987)
2. Potbhare, V., Tresseler, N.: Nucl. Phys. A 530, 171 (1991)
3. Bethe, H.A., Brown, G.E., Applegate, J., Lattimer, J.M.: Nucl. Phys. A324, 487 (1979)
4. Aufderhide, M.B., Brown, G.E., Kuo, T.T.S., Stout, D.B., Vogel, P.: Ap. J. 362, 241 (1990)
5. Takahashi, K., Yamada, M.: Prog. Theor. Phys. 41, 1479 (1969); Takahashi, K., Yamada, M., Kondoh, T.: At. Data Nucl. Data Tables 12, 101 (1973)
6. Klapdor, H.V., Metzinger, J., Oda, T.: At. Data Nucl. Data Tables 31, 81 (1984); Standt, A., Bender, E., Muto, K., Klapdor-Kleingrothaus, H.V.: At. Data Nucl. Data Tables 44, 79 (1990); Moller, P., Pandrup, J.: Nucl. Phys. A 514 1 (1990)
7. Brown, B.A., Wildenthal, B.H.: At. Data Nucl. Data Tables 33, 347 (1985)
8. Fuller, G.M., Flower, W.A., Newman, M.J.: Ap. J. Suppl. 42, 447 (1980); Ap. J. 252, 715 (1982); Ap. J. Suppl. 48, 279 (1982); Ap. J. 293, 1 (1985)
9. Oda, T., Hino, M., Muto, K., Takahara, M., Sato, K.: At. Data Nucl. Data Tables 56, 231 (1994)
10. Wildenthal, B.H.: Prog. Part. Nucl. Phys. 11, 5 (1984)
11. Weaver, T.A. Woosely, S.E., Fuller, G.M., Numerical Astrophysics, editors J. Centrella, J. LeBlanc and R. Bowers (Portola: Science Book International, 1985), p. 374
12. Aufderheide, M.B., Fushiki, I., Woosley, S.E., Hartmann, D.H.: Ap. J. Supp. 91, 389 (1994)
13. Kar, K., Sarkar, S., Ray, A.: Phys. Lett. B 261, 217 (1991)
14. Kota, V.K.B., Majumdar, D.: Z. Phys. A 351, 365 (1995), this is paper III of this series
15. French, J.B., Kota, V.K.B., University of Rochester Report UR-1116 (1989); French, J.B., Kota, V.K.B., Smith, J.F., University of Rochester Report UR-1122 (1989); Kota, V.K.B., in Capture gamma-ray spectroscopy and related topics, edited by R.W. Hoff (AIP conference proceedings 238, N.Y., 1991), p. 685; French, J.B., Haq, R.U., Kota, V.K.B., to be submitted for publication (1995)
16. Kota, V.K.B., Kar, K.: Pramana – J. Phys. 32, 647 (1989)
17. McGrory, J.B., Wildenthal, B.H., Halbert, E.C.: Phys. Rev. C2, 186 (1970); Sharma, S.K., Bhatt, K.H.: Phys. Rev. Lett. 30, 623 (1973)
18. Richter, W.A., Vander Merwe, M.G., Julies, R.E., Brown, B.A.: Nucl. Phys. A 523, 325 (1991)
19. Sarkar, S.: Ph.D. Thesis, (University of Calcutta, India, 1993)
20. Brussaard, P.J., Glaudemans, P.W.M., Shell Model Applications in Nuclear Spectroscopy. (North Holland, N.Y., 1977)
21. Wu, C.S., Moszkowski, S.A., Beta Decay (Interscience Publishers, N.Y., 1966)
22. Proceedings of the international conference on Electromgnetic properties of Atomic Nuclei, editors H. Honie and H. Ohuma (Tokyo Institute of Technology Press, Japan, 1983)
23. Schenter, G.K., Vogel, P.: Nucl. Sci. Eng. 83, 393 (1983)
24. Ray, A., Kar, K., Sarkar, S., Chakravarti, S. in Nuclei in Cosmos, edited by F. Kappler and K. Wisshak, (IOP publishing, London, 1992), p. 363
25. Kota, V.K.B., Majumdar, D., Haq, R.U.: in preparation
26. Dilg, W., Schantl, W., Vonach, H., Uhl, M.: Nucl. Phys. A 217, 269 (1973)
27. Table of isotopes, VIIth Edn., edited by C.M. Lederer and V.S. Shirley (John Wiley & Sons INC., N.Y., 1978)
28. Macfarlane, M.H.: Phys. Lett. 182, 265 (1986)
29. Bloom, S.D., Fuller, G.M.: Nucl. Phys. A 440, 511 (1985)
30. Rapaport, J., Taddeucci, T., Welch, T.P., Gaarde, C., Larsen, J., Horen, D.J., Sugarbaker, E., Koncz, P., Foster, C.C., Goodman, C.D., Goulding, C.A., Masterson, T.: Nucl. Phys. A 410, 371 (1983)
31. Kar, K.: Private communication to V.K.B. Kota, Kar, K., Sarkar, S., Ray, A.: Ap. J. 434, 662 (1994)
32. Haq, R.U., Parikh, J.C.: Nucl. Phys. A 273, 410 (1976)
33. Oda, T. in Particles and nuclei, ed. H. Terazawa (World Scientific, Singapore, 1986), p. 255
34. French J.B., Kota, V.K.B.: Pramana – J. Phys. 32, 439 (1989)

ELSEVIER

Nuclear Physics A 604 (1996) 129–162

NUCLEAR PHYSICS A

Application of spectral averaging theory in large shell model spaces: Analysis of level density data of fp-shell nuclei

V.K.B. Kota, D. Majumdar [1]

Physical Research Laboratory, Ahmedabad 380 009, India

Received 13 July 1995; revised 11 March 1996

Abstract

In the spectral averaging theory extended to large shell model spaces (i) using a significant unitary group decomposition under which a one- plus two-body nuclear hamiltonian H decomposes into an orthogonal effective one-body part $\mathbf{h}$ and an irreducible two-body part $\mathbf{V}$ $(H \to \mathbf{h}+\mathbf{V})$, (ii) applying the Central Limit Theorems locally and (iii) decomposing the shell model space into distant non-interacting S-subspaces (for light nuclei S denotes $\hbar\omega$ excitation), the state densities $I^H(E)$ take a convolution form, $I^H(E) = \sum_S I^{\mathbf{h},S} \otimes \rho_G^{\mathbf{V},S}[E]$, where $I^{\mathbf{h}}$ is the renormalized non-interacting particle state density and ρ_G is a normalized spreading Gaussian due to $\mathbf{V}$. A similar convolution form is available for spin-cutoff densities that give spin-cutoff factors. For a first systematic analysis, using the extended spectral averaging theory, of experimental level density data in a given region of the periodic table, fp-shell nuclei are chosen as an example. Calculations, using surface delta interaction, are carried out by including configurations up to $2\hbar\omega$ excitations and by employing eight spherical orbits (ds, fp, $1g_{9/2}$) for the five nuclei ^{55}Mn, ^{56}Fe, ^{59}Co, ^{60}Co and ^{60}Ni and by extending for obvious reasons to ten spherical orbits (ds, fp, $1g_{9/2}$, $2d_{5/2}$, $1g_{7/2}$) for the three $A > 60$ nuclei ^{62}Ni, ^{63}Cu and ^{65}Cu. In general, spectral averaging theory is seen to provide a good representation of the observed total level densities and spin-cutoff factors.

1. Introduction

Recognizing that statistical features extend throughout the spectrum of a complex nucleus [1], the scope of statistical approach to nuclear structure was enlarged by French [2–6], by combining shell model ideas with Bethe and Wigner's treatment

[1] Present Address: Institute of Physics, Bhubaneswar 751 005, India.

PII S0375-9474(96)00148-0

[7,8] of statistical properties of nuclei, by introducing and developing the subject of statistical nuclear spectroscopy. Statistical spectroscopy provides a path for understanding the deeper connections between symmetries, statistical behavior, quantum chaos and information propagation and it divides into two parts, spectral averages being one part and fluctuations (around the averages) being the other with little communication between the two. The origin and nature of the separation are well understood [1,9–11]. Moreover, the Dyson–Mehta [12] spectral rigidity ensures that spectral fluctuations are small (they essentially carry no information except for symmetries) and hence they can be ignored for most purposes. Thus spectral averages (smoothed forms with respect to energy) carry essentially all the significant physical information about spectroscopic observables. The theory that allows one to deal with spectral averages is called spectral averaging theory (SAT). The fundamental result based on which SAT is developed is that there are central limit theorems (CLT's) operating in nuclear shell model spaces. The basic CLT results are: (i) state densities $I(E)$ [2] take a Gaussian ($\mathcal{G}$) form [9]; (ii) for positive definite operators (for example, K is $\mathcal{O}^\dagger\mathcal{O}$ type; $\mathcal{O}$ is a transition operator) expectation value densities $I_K(E)$ [3] take a Gaussian form [13] (see also [14]) and thus expectation values can be written as a ratio of two Gaussians; (iii) transition strength densities $\mathbf{I}_\mathcal{O}(E,E')$ [4] take a bivariate Gaussian form [15]. These CLT results are in general valid only in strongly interacting shell model subspaces, i.e. essentially in $0\hbar\omega$ spaces (for example: $(ds)^8$ for ^{24}Mg and $(fp)^{16}$ for ^{56}Fe etc.). The large number of studies that are carried out in $0\hbar\omega$ spaces, essentially up to 1983, are well documented [5,6,14,16–18]. Extensions of these CLT results to indefinitely large shell model spaces led to, what one may call, spectral averaging theory in large shell model spaces (SAT-LSS) and with this it is possible to address problems in statistical nuclear physics, i.e. calculation of level densities and spin-cutoff factors, construction of strength functions of various kinds etc. Broadly speaking, in SAT-LSS smoothed forms with *interactions* are derived and applied by using unitary group decompositions (of hamiltonians and the spectroscopic spaces), CLT's locally, and convolutions [13,15,19–21]; it is seen that the essential role of interactions is to produce local Gaussian spreadings of the non-interacting particle (NIP) densities.

With details of the theory (SAT-LSS) for state and spin-cutoff densities are now becoming available [13,20,21], the natural question that arises is how the theory works in practice in the analysis of experimental data. Therefore, as a first example, level density and spin-cutoff factor data for a series of fp-shell nuclei are analyzed using SAT-LSS and the results are reported in this paper. We will now give a preview.

A brief overview of the present form of SAT-LSS for state and spin-cutoff densities is given in Section 2. Spin-cutoff densities yield spin-cutoff factors that decompose state densities into observable level densities. Using the formalism given by SAT-LSS, systematic analysis of level density data is carried out for eight fp-shell nuclei – ^{55}Mn,

[2] $I(E)$ counts the number of states in the interval E and $E+\Delta E$; the normalized version of $I(E)$ is $\rho(E)$.
[3] $I_K(E) = \langle K\rangle^E I(E)$, where $\langle K\rangle^E$ is the expectation value of K.
[4] $\mathbf{I}_\mathcal{O}(E,E') = I(E')|\langle E'|\mathcal{O}|E\rangle|^2 I(E)$.

^{56}Fe, ^{59}Co, ^{60}Co and ^{60}Ni isotopes by performing calculations in an 8-orbit and for ^{62}Ni, ^{63}Cu and ^{65}Cu isotopes in a 10-orbit shell model space. All the calculations include up to $2\hbar\omega$ excitations. The 8- and 10-orbit results are given in Sections 3 and 4, respectively; Section 3 also gives calculational details. Finally some concluding remarks are given in Section 5.

2. Convolution forms for state and spin-cutoff densities

2.1. Preliminaries

In order to extend the applicability of CLT results to large shell model spaces, it is essential to deal with partitioning of the m-particle spaces to subspaces. The partitioning of relevance here is the partitioning of m-particle spaces according to spherical and unitary configurations and in order to define them, we begin with m-particles distributed over spherical shell model j-orbits (s.p. orbits) denoted by α, β, etc. (j_α being the angular momentum and $N_\alpha = 2j_\alpha + 1$ is the degeneracy of the orbit α). The m-particle spectroscopic space can be decomposed into spherical configurations $\mathbf{m} = (m_\alpha, m_\beta, \ldots)$, where m_α is the number of particles in the orbit α and $m = \sum_\alpha m_\alpha$. The dimensionality $d(\mathbf{m})$ of the configuration $\mathbf{m}$ is $d(\mathbf{m}) = \prod_\alpha \binom{N_\alpha}{m_\alpha}$. A unitary orbit $\boldsymbol{\alpha}$ is defined as a set of spherical orbits. As above, a unitary configuration $[\mathbf{m}] = (m_{\boldsymbol{\alpha}}, m_{\boldsymbol{\beta}}, \ldots)$; $m = \sum m_{\boldsymbol{\alpha}}$. Using the convention that the spherical orbits α belong to the unitary orbit $\boldsymbol{\alpha}$ and similarly the orbits β belong to $\boldsymbol{\beta}$, etc., the number of s.p. states in a unitary orbit $\boldsymbol{\alpha}$ is $N_{\boldsymbol{\alpha}} = \sum_{\alpha\in\boldsymbol{\alpha}} N_\alpha$; total number of s.p. states $N = \sum N_\alpha = \sum N_{\boldsymbol{\alpha}}$. The dimensionality of the configuration $[\mathbf{m}]$ is $d([\mathbf{m}]) = \prod_{\boldsymbol{\alpha}} \binom{N_{\boldsymbol{\alpha}}}{m_{\boldsymbol{\alpha}}}$. With spherical and unitary configurations, the decomposition of m-particle space is $m \to \sum [\mathbf{m}]$; $[\mathbf{m}] \to \sum \mathbf{m}$. It is worth remarking that this decomposition automatically produces fixed-parity subspaces by choosing the unitary orbits $\boldsymbol{\alpha}$ to be the set of spherical orbits with all of them having the same parity.

In the analysis given in Sections 3 and 4, the proton–neutron (pn) formalism (but not the isospin formalism) is used. Towards this end, in the remainder of the paper the following conventions are adopted. Let us assume that there are k_p proton orbits and k_n neutron orbits. Then the orbit label α takes values $\alpha = 1, 2, \ldots, k_p + k_n$ with $\alpha = 1, 2, \ldots, k_p$ the proton orbits $\alpha(p)$ and $\alpha = k_p + 1, k_p + 2, \ldots, k_p + k_n$ the neutron orbits $\alpha(n)$. A pn configuration $(\mathbf{m}_p, \mathbf{m}_n) = (m_p^1, m_p^2, \ldots, m_p^{k_p}, m_n^1, m_n^2, \ldots, m_n^{k_n})$ then corresponds to the configuration $\mathbf{m} = (m_1, m_2, \ldots, m_{k_p+k_n})$ with $m_1 = m_p^1$, $m_2 = m_p^2, \ldots, m_{k_p} = m_p^{k_p}$, $m_{k_p+1} = m_n^1$, $m_{k_p+2} = m_n^2, \ldots$, $m_{k_p+k_n} = m_n^{k_n}$, where $m_p^{\alpha(p)}$ and $m_n^{\alpha(n)}$ are the number of protons and neutrons in orbits $\alpha(p)$ and $\alpha(n)$, respectively, and m_α is the number of particles in the orbit α. Similarly defined are the orbit degeneracies N_α, dimensionalities $d(\mathbf{m})$, etc. All these conventions apply directly to the case of unitary orbits and unitary configurations. With these conventions, the NIP hamiltonian takes the simple form $h = \sum \epsilon_\alpha n_\alpha$, where ϵ_α are single-particle energies (SPE) and the summation is over the k_p proton and k_n neutron spherical orbits α.

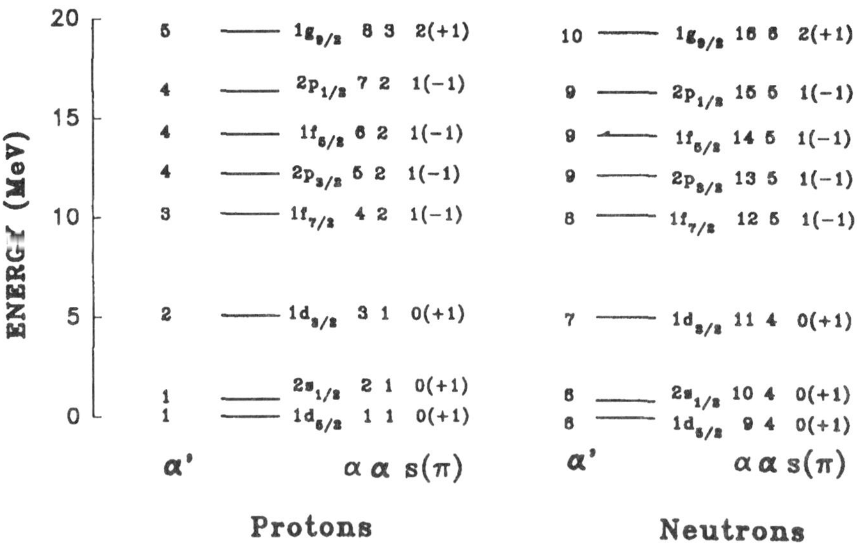

Fig. 1. Classification of $(ds, fp, g_{9/2})$ spherical orbits in the pn case in terms of unitary orbits and the corresponding s labels. The parity and all other quantum numbers for the orbits are shown. For example $1d_{5/2}$ implies $n = 0$, $\ell = 2$, $j = 5/2$ and the major oscillator number $N = 2n + \ell = 2$ and similarly $2p_{1/2}$ corresponds to $n = 1$, $\ell = 1$, $j = 1/2$, $N = 3$, etc. Given the s^π quantum numbers for the spherical orbits, the largest possible unitary orbits correspond to $\boldsymbol{\alpha}$ in the figure. The unitary orbits $\boldsymbol{\alpha}'$ are finer than $\boldsymbol{\alpha}$. For the pn case shown in the figure, the conventions defined in Section 2.1 have been adopted.

Starting with the decomposition of m-particle spaces into unitary and spherical configurations, a further decomposition is possible by attaching an s_α label to each spherical orbit α, where s_α for lighter nuclei denotes the $s\hbar\omega$ excitation value; for heavy nuclei the definition of s has to be somewhat modified. The purpose of the s quantum number is to catalog distant configurations. Fig. 1 gives an example. Given s^π values for the spherical orbits α, one can define the unitary orbits $\boldsymbol{\alpha}$ such that all the spherical orbits with the same s^π values belong to a given $\boldsymbol{\alpha}$ (in Fig. 1, the unitary orbits $\boldsymbol{\alpha}'$ do not satisfy this criterion). Then the corresponding unitary configurations $[\mathbf{m}]$ define uniquely the S-subspaces of the m-particle space. With this, the m-particle space decomposes into S-subspaces as follows:

$$m \to \sum S^\pi; \quad S^\pi \to \sum [\mathbf{m}]; \quad [\mathbf{m}] \to \sum \mathbf{m}; \quad S = \sum m_\alpha s_\alpha . \qquad (1)$$

A similar decomposition of the one- plus two-body hamiltonian $H = h + V$ is possible (V is usually given in terms of the two-body matrix elements $V^J_{\alpha\beta\gamma\delta}$). The NIP hamiltonian h by definition is S-preserving. The interaction V can be decomposed into a part ($V_{S\text{-pres}}$) that preserves S and a part ($V_{S\text{-mix}}$) that mixes different S-subspaces. With the S-decomposition of the m-particle space, the state densities $I^{H,m}(E) = \langle\langle \delta(H-E) \rangle\rangle^m$ decompose into partial $I^{H,S}(E) = \langle\langle \delta(H-E) \rangle\rangle^S$ densities; note that $\langle\langle \ \rangle\rangle$ denotes a trace and hereafter $\langle \ \rangle$ denotes average. It is important to recognize [3] that the fixed-S variance $\sigma^2(S)$ of $I^{H,S}$ decomposes into partial variances; $\sigma^2(S) = \sigma^2(S \to S) + \sum_{S \neq S'} \sigma^2(S \to S')$. The partial variance $\sigma^2(S \to S')$, which is due to $V_{S\text{-mix}}$, represents admixing between distant configurations (at least $2\hbar\omega$ away from each other) that belong to S and S' subspaces respectively and this leads to multimodal forms for densities unlike the unimodal forms given by the CLT results. A schematic model to understand the origin of the multimodal forms was introduced in [19,22] and tested in [23,24]. In large shell model spaces either one should derive a good method for incorporating the $\sigma^2(S \to S')$ corrections (i.e. $V_{S\text{-mix}}$ effects) to the CLT results or, to the extent they are not harmful, just ignore them. The latter is what is adopted in the remainder of this paper (the same approximation is made in [25] where large space shell model diagonalization calculations are carried out for some neutron excess ds-shell nuclei). Thus $H \to h + V_{S\text{-pres}} = h + V$. With this, H is S-preserving and $I(E)$ decomposes into non-interacting partial $I^{H,S}(E)$ densities,

$$I^{H,m}(E) = \sum_S I^{H,S}(E) = \sum_S \sum_{[\mathbf{m}]\in S} I^{H,[\mathbf{m}]}(E)\,,$$

$$I^{H,S} \Leftrightarrow M_p(S) = \left[\sum_{[\mathbf{m}]\in S} d([\mathbf{m}])\right]^{-1} \sum_{[\mathbf{m}]\in S} d([\mathbf{m}])M_p([\mathbf{m}])\,. \tag{2}$$

In (2) $M_p(S)$ is the pth moment of $I^{H,S}$ and similarly $M_p([\mathbf{m}])$ for $I^{H,[\mathbf{m}]}$ density; the fixed unitary configuration density $I^{H,[\mathbf{m}]}(E) = \langle\langle\delta(H-E)\rangle\rangle^{[\mathbf{m}]}$. Before going further, three points are worth mentioning: (i) as unitary configurations $[\mathbf{m}]$ are defined by a group (see Appendix A), the $M_p([\mathbf{m}])$ moments can be calculated via trace propagation equations [3,5,17,18] and then $M_p(S)$ follow via (2); (ii) the $V_{S\text{-pres}}$ in general admixes different unitary configurations $[\mathbf{m}]$ belonging to a given S and this admixing must be included while constructing the $I^{H,[\mathbf{m}]}$ densities appearing in (2); (iii) Eq. (2) extends directly to the case of expectation value densities.

2.2. *Convolution form for IP densities*

In the absence of interactions, $H = h$ will produce spikes corresponding to each spherical configuration $\mathbf{m}$ at energies $\sum \epsilon_\alpha m_\alpha$ with degeneracy $d(\mathbf{m})$. Now, with respect to the spherical configuration group decomposition, $V(= V_{S\text{-pres}})$ splits into two parts, $V = V^{[0]} + \mathbf{V}$ as given in Appendix A and similarly $h \to h^{[0]}$. The group theoretical meaning of $[0]$ in $V^{[0]}$ and $h^{[0]}$ is explained briefly in Appendix A. By decomposing $V^{[0]}$ with respect to the unitary configurations group that defines the S quantum number, H takes the form

$$H = \left[h^{[0][0]} + h^{[0][1]} + V^{[0][0]} + V^{[0][1]} + V^{[0][2]}\right] + \mathbf{V}$$

and the tensorial ranks $[0][\boldsymbol{\nu}]$; $\boldsymbol{\nu} = \mathbf{0}, \mathbf{1}, \mathbf{2}$ are described in Appendix A. To the extent that one can ignore the $V^{[0][2]}$ part, one has $H = \mathbf{h} + \mathbf{V}$ with the effective one-body part $\mathbf{h} = h^{[0][0]} + h^{[0][1]} + V^{[0][0]} + V^{[0][1]}$ and the irreducible (see Appendix A) two-body part $\mathbf{V}$ being orthogonal to each other [3,4]. The smallness of $V^{[0][2]}$ is well verified for large number of interactions all across the periodic table [13,19,21,24]. With this one has the following convolution form for state densities [13,19–21]:

$$
\begin{aligned}
I^H(E) &= \sum_S I^{\mathbf{h},S} \otimes \rho_{\mathcal{G}}^{\mathbf{V},S}[E] \\
&= \sum_S \left\{ \sum_{[\mathbf{m}]\in S} I^{\mathbf{h},[\mathbf{m}]} \otimes \rho_{\mathcal{G}}^{\mathbf{V},[\mathbf{m}]}[E] \right\},
\end{aligned}
$$

$$
\rho_{\mathcal{G}}^{\mathbf{V},[\mathbf{m}]} \Leftrightarrow \sigma_{\mathbf{V}}([\mathbf{m}]) = \left\{ \langle \mathbf{V}^2 \rangle^{[\mathbf{m}]} \right\}^{1/2},
$$

$$
\rho_{\mathcal{G}}^{\mathbf{V},S} \Leftrightarrow \sigma_{\mathbf{V}}(S) = \left[\left\{ \sum_{[\mathbf{m}]\in S} d([\mathbf{m}]) \right\}^{-1} \sum_{[\mathbf{m}]\in S} d([\mathbf{m}]) \langle \mathbf{V}^2 \rangle^{[\mathbf{m}]} \right]^{1/2}. \tag{3}
$$

The result in (3) is that, in each S-subspace, I^H is a convolution of the NIP density $I^{\mathbf{h}}$ with the spreading Gaussian produced by $\mathbf{V}$; note that the S-subspaces are non-interacting. An additional fact, not mentioned before but it is used in (3), is the feature of the constancy of the fixed-$\mathbf{m}$ (hence fixed-$[\mathbf{m}]$) widths. This is a well-known result in SAT (Refs. [17,18] give examples) and in Refs. [13,19–21] it is explicitly verified that the spreading widths for $\mathbf{m}$'s that belong to fixed-S subspaces fluctuate very little from each other (the fluctuation is $\lesssim 3\%$) and its origin lies in an ensemble representation of $\mathbf{V}$ [13,20,21]. The convolution result given by (3) tacitly assumes that the internal structure of $\mathbf{V}$ does not play any significant role in generating smoothed state densities. All it does is to produce a *scale* for each S subspace which is the fixed-S spreading width. This is consistent with the recent claim of Zuker and collaborators [26] who argued that as long as $V^{[0]}$ is taken care of properly, saturation properties of nuclei can be described by any realistic interaction. Eq. (3) for state densities extends to spin-cutoff densities $I^H_{J_Z^2}(E)$ [13,20,21],

$$
\begin{aligned}
I^H_{J_Z^2}(E) &= \langle J_Z^2 \rangle^E I(E) = \langle\langle J_Z^2 \delta(H-E) \rangle\rangle^m \\
&= \sum_S I^{\mathbf{h},S}_{J_Z^2} \otimes \rho^{\mathbf{V},S}_{J_Z^2:\mathcal{G}}[E] \\
&= \sum_S \left\{ \sum_{[\mathbf{m}]\in S} I^{\mathbf{h},[\mathbf{m}]}_{J_Z^2} \otimes \rho^{\mathbf{V},[\mathbf{m}]}_{J_Z^2:\mathcal{G}}[E] \right\},
\end{aligned}
$$

$$
\rho^{\mathbf{V},[\mathbf{m}]}_{J_Z^2}(y) = \langle J_Z^2 \delta(\mathbf{V}-y) \rangle^{[\mathbf{m}]} / \langle J_Z^2 \rangle^{[\mathbf{m}]},
$$

$$
\rho^{\mathbf{V},[\mathbf{m}]}_{J_Z^2:\mathcal{G}} \Leftrightarrow \left\{ \begin{aligned} \epsilon_{\mathbf{V}:J_Z^2}([\mathbf{m}]) &= \langle J_Z^2 \mathbf{V} \rangle^{[\mathbf{m}]} / \langle J_Z^2 \rangle^{[\mathbf{m}]} \\ \sigma_{\mathbf{V}:J_Z^2}([\mathbf{m}]) &= \left[\langle J_Z^2 \mathbf{V}^2 \rangle^{[\mathbf{m}]} / \langle J_Z^2 \rangle^{[\mathbf{m}]} - \left\{ \epsilon_{\mathbf{V}:J_Z^2}([\mathbf{m}]) \right\}^2 \right]^{1/2} \end{aligned} \right. . \tag{4}
$$

Modifications of (4) in the cases $\langle J_Z^2 \rangle^{[\mathbf{m}]} = 0$ is straightforward. Dividing the spin-cutoff density $I^H_{J_Z^2}(E)$ by the state density $I^H(E)$ gives spin-cutoff factors $\sigma_J^2(E)$ and they decompose $I(E)$ into fixed-J level densities $I_\ell(E,J)$, level densities $I_\ell(E, J_1 : J_2)$ in the spin window $J_1 - J_2$ and total level densities $I_\ell(E)$ (with or without fixed parity) as given in Appendix B. In Sections 3 and 4 data for $I_\ell(E)$ and $\sigma_J(E)$ are analyzed.

In order to apply the convolution forms given by (3), (4), one needs methods for calculating the spreading variances σ^2_{V} and $\sigma^2_{\mathrm{V}:J^2_Z}$ and the centroid shifts $\epsilon_{\mathrm{V}:J^2_Z}$. Trace propagation formulas for these are given in detail in [13,21]; see also Refs. [3,18]. A striking result here is that the spin-cutoff variances $\sigma^2_{\mathrm{V}:J^2_Z}$ are consistently smaller than the state density variances and the difference is $\sim 10\%$ [5]; an equivalent statement is that the fluctuations in the k_{22} cumulant of the bivariate $\rho(E,M)$, M is J_Z eigenvalue, distribution over spherical configurations that belong to a given S-space are negligible and the average value is essentially independent of S. This result is used [24] in Sections 3 and 4. In addition it is essential to have good methods for constructing the NIP densities $I^{\mathbf{h}}(E)$, $I^{\mathbf{h}}_{J^2_Z}(E)$ and their S-decompositions. These densities can be constructed exactly by using spherical orbits and spherical configurations. However, this method does not give any explanation to the connection between the CLT Gaussian form and the Bethe's exponential $(\exp 2\sqrt{aE})$ form for the NIP densities. Moment methods cure this problem [21,24,27,29]. Without loss of generality by putting $h = \mathbf{h}$, $I^{h,m}(E)$ can be written as (with $\mathcal{L}$ number of unitary orbits),

$$I^{h,m}(E) = \langle\langle \delta(h-E)\rangle\rangle^m = \sum_S \left\{ \sum_{[\mathbf{m}]\in S} I^{h,[\mathbf{m}]}(E) \right\} = \sum_S I^{h,S}(E)\,,$$

$$I^{h,[\mathbf{m}]}(E) = I^{h_{\alpha_1}}_{N_{\alpha_1}} \otimes I^{h_{\alpha_2}}_{N_{\alpha_2}} \otimes \ldots \otimes I^{h_{\alpha_{\mathcal{L}}}}_{N_{\alpha_{\mathcal{L}}}}[E]\,. \tag{5}$$

Eq. (5) follows by using the decomposition $h = \sum_\alpha \epsilon_\alpha n_\alpha = \sum_\alpha h_\alpha = \sum_{\alpha} h_\alpha = \sum_{\alpha}\{\sum_{\alpha\in\alpha} h_\alpha\}$. The convolution form in (5) implies that the cumulants of $I^{h,[\mathbf{m}]}$ are the sum of the cumulants of the single unitary orbit state densities $I^{h_\alpha}_{N_\alpha}$ and the methods for calculating the latter are given in detail in [13,21]. Applying CLT, the Edgeworth-corrected Gaussian form (see Eq. (8) ahead) for $I^{h,[\mathbf{m}]}$ is constructed and hence $I^{h,S}(E)$ and $I^{h,m}(E)$. Extension of (5) for J^2_Z density is immediate,

$$I^{h,m}_{J^2_Z}(E) = \langle\langle J^2_Z\delta(h-E)\rangle\rangle^m = \sum_S I^{h,S}_{J^2_Z}(E) = \sum_S \left[\sum_{[\mathbf{m}]\in S} I^{h,[\mathbf{m}]}_{J^2_Z}(E) \right]\,. \tag{6}$$

The moments of $I^{h,[\mathbf{m}]}_{J^2_Z}(E)$ follow from (5) with $h \to h_\lambda = h + \lambda J_Z$ and then carrying out a double parametric differentiation of the moments of $I^{h_\lambda,[\mathbf{m}]}$. These then define Edgeworth-corrected Gaussian representation for $I^{h,[\mathbf{m}]}_{J^2_Z}$; details are given in [13,20,21].

[5] Calculation of state density spreading variances σ^2_{V} is simple (so is the case with the centroid shifts $\epsilon_{\mathrm{V}:J^2_Z}$) while the calculation of spin-cutoff variances $\sigma^2_{\mathrm{V}:J^2_Z}$, though straightforward [13,21], is time consuming. However, from the previous level density calculations for ^{28}Si, ^{168}Er, ^{181}W and ^{234}U nuclei [20,21,27,28] and also from various ds- and fp-shell calculations [20,21,24] it is known that the ratio $R = \sigma^2_{\mathrm{V}:J^2_Z}/\sigma^2_{\mathrm{V}}$ for unitary configurations is ~ 0.9. For example, in the case of ^{56}Fe, employing the spectroscopic space defined ahead in Section 3, R takes values 0.9, 0.92 and 0.93 for configurations in $S = 0$, 1 and 2 spaces, respectively, and they are explicitly verified [24] to be within 1% of the exact results. Complete details of the method of using the ratio R in calculating $\sigma^2_{\mathrm{V}:J^2_Z}$ are given in [24].

The goodness of the moment methods is well tested in many numerical examples [13,21,24,27,29] and they are employed in the analysis reported in Sections 3 and 4.

2.3. Summary

The convolution forms (3), (4) for state and spin-cutoff densities as given by SAT-LSS provide a framework for analyzing experimental level density data with a (1+2)-body interaction. As described in Sections 2.1 and 2.2 these results are based on: (i) ignoring the $V_{S\text{-mix}}$ part of the interaction; (ii) neglecting $V^{[0][2]}$ part of $V_{S\text{-pres}}$; (iii) constancy of fixed-**m** variances; (iv) the spreadings due to interactions are Gaussian in nature; (v) NIP densities are constructed via moment methods using unitary orbits; (vi) the ratio $\sigma^2_{V:J_z^2}/\sigma^2_V \simeq 0.9$. In practice (ii), (iii) and (v) can easily be relaxed but then the compact convolution forms will be lost. In a variety of calculations it is well tested [13,20,21,24] that these and (vi) will not affect in any significant manner the calculated level densities. Shape (Edgeworth) corrections to Gaussian spreadings ((iv) above) involve calculation of higher-order traces and in principle it is possible to include the skewness and excess corrections. Here it is useful to point out that these shape corrections are significant near the ground state and this problem is cured by using the so called reference energy method (see Section 3.2.2 ahead). Thus Gaussian spreadings give reliable values for level densities above the reference energy. The questions related to (i) above are discussed in Sections 3.2.3 and 5. Incorporating effects due to (i) is beyond the scope of the present article.

3. Analysis of level density data for *fp*-shell nuclei: 8-orbit calculations

3.1. Preview

Level density calculations for 8 *fp*-shell nuclei $\{^{55}_{25}Mn_{30}, ^{56}_{26}Fe_{30}, ^{59}_{27}Co_{32}, ^{60}_{27}Co_{33}, ^{60}_{28}Ni_{32}, ^{62}_{28}Ni_{34}, ^{63}_{29}Cu_{34}, ^{65}_{29}Cu_{36}\}$ are carried out using (3), (4) and the results are described in this section and in Section 4. The various conditions that led to the selection of the above eight nuclei are given in Section 3.2.1. In order to proceed with the calculations, one needs to have a reference energy, a set of s.p. orbits, SPE and a two-body interaction and they are discussed in Sections 3.2.2 and 3.2.3; all the calculations are carried out using surface delta interaction (SDI). The theoretical calculations involve evaluating the spreading variances and the moments for constructing NIP densities. For these two, it is necessary to group the spherical orbits into unitary orbits and the choice of unitary orbits for the above two calculations need not be the same. In general the latter should be finer than the former. The unitary configurations in the former case define the classes [6] as described in [20,21] and using the classes and (3), (4), the

[6] Classes are defined by the unitary configurations in (3), (4). These unitary configurations are generated by the largest possible unitary orbits (orbits α in Fig. 1) consistent with the definition of the S-quantum

state and spin-cutoff densities are constructed. The choice of unitary orbits and the corresponding classes are given in Section 3.2.4. The calculational procedure is outlined in Section 3.2.5. The results of the calculations of the level densities and spin-cutoff factors and their comparison with the experimental data are given in Section 3.3 for the six nuclei ^{55}Mn, ^{56}Fe, ^{59}Co, ^{60}Co, ^{60}Ni, ^{62}Ni and ^{63}Cu employing the space spanned over the eight spherical orbits ($1d_{5/2}$, $2s_{1/2}$, $1d_{3/2}$, $1f_{7/2}$, $2p_{3/2}$, $1f_{5/2}$, $2p_{1/2}$, $1g_{9/2}$). Section 3.4 gives a summary of the 8-orbit results. Using ^{63}Cu nucleus as an example, it is clearly demonstrated in Section 3.3.6 that for $A > 60$ nuclei it is essential to go beyond the eight orbit space and therefore in the calculations, for ^{62}Ni, ^{63}Cu and ^{65}Cu nuclei, presented in Section 4, the space of ten spherical orbits is employed.

3.2. Selection of nuclei, choice of orbits and the hamiltonian and the procedure for calculations

3.2.1. Choice of fp-shell nuclei

The experimental data for level densities and spin-cutoff factors come from four different sources and they are

(1) direct counting of levels;
(2) neutron/proton resonance experiments;
(3) charged particle spectra;
(4) Ericson fluctuation measurements.

Theoretical calculations need a reference energy (Section 3.2.2) and there is also a free parameter (G; see Section 3.2.3) in the interaction. The low-energy data is used in fixing the reference energy and then the resonance data gives a value for G. The purpose here is not to derive a value for G but mainly to verify whether one can extrapolate the predictions much beyond the resonance energy (i.e. beyond say 10 MeV and up to say 20–25 MeV excitation) and interpolate well between the reference energy and the resonance energy. To verify this one needs data from sources (3) and/or (4). Thus, in the selection of the nuclei, for the analysis with the theory given in Section 2.2 we impose the restriction that the nucleus under question should have level density data from {(1) and/or (2)} and {(3) and/or (4)}. With this restriction, there are only eight fp-shell nuclei, as given in Table 1, that can be analyzed and the data for these eight nuclei are compiled in [24]; see also Appendix B.

3.2.2. Reference energy

The calculated state density yields a smoothed spectrum via the distribution function $F(E) = \int_{-\infty}^{E} I(x)dx$. Given a sequence of levels with angular momenta $J_1, J_2, J_3, \ldots, J_r$, inverting the equation

number. In fact, it is the fluctuation in the spreading widths of spherical configurations that belong to a given class that is small but not for the configurations that belong to a given S. It should be noted that the class variances involve partial variances unlike the fixed-S variances. Classes are introduced in [20,21] as they appear to provide a basis for the constant width approximation in terms of an ensemble representation of the hamiltonian.

Table 1
Sources of data for fp-shell nuclei. The sources (1)–(4) are defined in the text

Nucleus	Sources of data
^{55}Mn	(1), (2), (4)
^{56}Fe	(1), (3), (4)
^{59}Co	(1), (2), (3)
^{60}Co	(2), (4)
^{60}Ni	(1), (2), (3), (4)
^{62}Ni	(1), (2), (3)
^{63}Cu	(1), (2), (3)
^{65}Cu	(1), (2), (3)

$$F(E) = \sum_{k=1}^{r-1} (2J_k + 1) + (2J_r + 1)/2$$

gives the energy E_r of the rth level. With $N_r = \sum_{k=1}^{r} (2J_k + 1)$ denoting total number of states including and up to the rth level, one has to solve the equation [30]

$$\left[N_r - \frac{(2J_r + 1)}{2} \right] = \int_{-\infty}^{\overline{E_r}} I(E)\, dE \tag{7}$$

to obtain the smoothed energy $\overline{E_r}$. For sequence of levels with fixed parity (say π), extension of (7) is straightforward by replacing $I(E)$ by $I^{\pi}(E)$, N_r by N_r^{π} and J_r by J_r^{π}. Using this method one can calculate the ground-state energy and then measure all the energies with respect to this energy. Then comparison of calculated and experimental densities is immediate. However, as the fluctuations in low-energy domain are not "quantum chaotic", one should match the energy of a high-lying level with the corresponding calculated energy [3,30] and then measure all the energies with respect to this "reference energy". Thus the reference energy in some manner marks the region of onset of chaos [20]. There are several variants of the reference energy selection [1,20,21]. Thus in order to proceed to the level density calculations, a complete set of low-lying levels (up to sufficiently high energy) and the J^{π}_{ref} value of the last level (reference level) in the sequence (with energy $E_{\text{ref}}(J^{\pi}_{\text{ref}})$) are needed. These data for the eight nuclei mentioned in Table 1 are given in Sections 3.3, 4.2. Using

$$I(E) = \sum_{[\mathbf{m}]} I^{[\mathbf{m}]}(E) = \sum_{[\mathbf{m}]} d([\mathbf{m}])\rho^{[\mathbf{m}]}(E)$$

with $\rho^{[\mathbf{m}]}(E)$ taken to be a Edgeworth-corrected Gaussian, leads to (with N_{ref} giving the number of states up to and including the reference level),

$$\left[N_{\text{ref}} - \frac{1}{2}(2J_{\text{ref}} + 1) \right] = \sum_{[\mathbf{m}]} d([\mathbf{m}]) \int_{-\infty}^{\overline{E}_{\text{ref}}} \rho^{[\mathbf{m}]}(E)\, dE,$$

$$\rho^{[\mathbf{m}]}(E) = [\sigma([\mathbf{m}])]^{-1}\eta_{\mathcal{G}}(\widehat{E})\left\{1 + \frac{\gamma_1([\mathbf{m}])}{6}\mathrm{He}_3(\widehat{E}) + \left[\frac{\gamma_2([\mathbf{m}])}{24}\mathrm{He}_4(\widehat{E}) + \frac{[\gamma_1([\mathbf{m}])]^2}{72}\mathrm{He}_6(\widehat{E})\right]\right\},$$

$$\eta_{\mathcal{G}}(\widehat{E}) = \frac{1}{\sqrt{2\pi}}\exp\left\{-\frac{\widehat{E}^2}{2}\right\}, \qquad \widehat{E} = (E - \epsilon([\mathbf{m}]))/\sigma([\mathbf{m}]),$$

$$\int_{-\infty}^{\overline{E}_{\mathrm{ref}}} \rho^{[\mathbf{m}]}(E)\,dE = \frac{1}{2}\mathrm{erfc}\left[\frac{\epsilon([\mathbf{m}]) - \overline{E}_{\mathrm{ref}}}{\sqrt{2}\sigma([\mathbf{m}])}\right] - \eta_{\mathcal{G}}(\widehat{\overline{E}}_{\mathrm{ref}})\left\{\left[\frac{\gamma_1([\mathbf{m}])}{6}\mathrm{He}_2(\widehat{\overline{E}}_{\mathrm{ref}})\right] + \left[\frac{\gamma_2([\mathbf{m}])}{24}\mathrm{He}_3(\widehat{\overline{E}}_{\mathrm{ref}}) + \frac{[\gamma_1([\mathbf{m}])]^2}{72}\mathrm{He}_5(\widehat{\overline{E}}_{\mathrm{ref}})\right]\right\}. \tag{8}$$

In (8) γ_1 and γ_2 are skewness and excess parameters and He_ν are Hermite polynomials. Given the values of $(N^{\pi}_{\mathrm{ref}}, J^{\pi}_{\mathrm{ref}})$ one can solve (8) to obtain $\overline{E}_{\mathrm{ref}}$ and match it with the experimental energy $E_{\mathrm{ref}}(J^{\pi}_{\mathrm{ref}})$.

3.2.3. Choice of the single-particle orbits, single-particle energies and the two-body interaction

For the calculation of the level densities, eight single-particle shell model orbits are chosen (for both protons and neutrons) and they are $1d_{5/2}$, $2s_{1/2}$, $1d_{3/2}$, $1f_{7/2}$, $2p_{3/2}$, $1f_{5/2}$, $2p_{1/2}$ and $1g_{9/2}$. Thus (s^4p^{12}) is taken as the closed core and the space consists of ds-shell, fp-shell and $g_{9/2}$ orbit. The SPE given in [32] are used in the calculations and for the above eight orbits they are (in MeV) (−4.15, −3.28, 0.93, 1.55, 3.57, 7.74, 5.58, 8.74) respectively. They give the ds, fp and $g_{9/2}$ centroid energies to be −2.311, 4.214, and 8.74 MeV, respectively. The $ds - fp$ and $fp - g_{9/2}$ separations then are $\Delta_{ds-fp} = 6.53$ MeV and $\Delta_{fp-g_{9/2}} = 4.53$ MeV. However, the interaction V produces a further renormalization of Δ_{ds-fp} and $\Delta_{fp-g_{9/2}}$ as we are considering s^4p^{12} to be a closed core; the renormalization arises [20,21] due to the interaction of a particle in (ds), (fp) and $g_{9/2}$ shells with the core. The $ds - fp$ and $fp - g_{9/2}$ separations due to renormalization are calculated [20,21] using Kuo 15-orbit interaction [31] and they are $\delta\Delta^{\mathrm{Kuo}}_{sd-fp} = 5.66$ MeV and $\delta\Delta^{\mathrm{Kuo}}_{fp-g_{9/2}} = 3.38$ MeV. However, as surface delta interaction (SDI) is used for V in the present calculations (see ahead), it is plausible that one need not use the above values for $\delta\Delta$. A number of level density calculations are carried out to test the effects due to variations in $\delta\Delta$ (Section 3.2.5 gives an important discussion regarding these calculations) for all the eight fp-shell nuclei and found that it is best to use $\delta\Delta_{ds-fp} = 4.5$ MeV and $\delta\Delta_{fp-g_{9/2}} = 2$ MeV. The resulting zero centered SPE and the separations $\overline{\Delta} = \Delta + \delta\Delta$, used in the calculations, are given in Table 2. The two-body interaction that has been chosen for the present calculations is the SDI with one strength parameter G (i.e. $A_0 = A_1 = G$ in Eq. (6.44) of [33]). One important property of SDI is that it will not renormalize SPE. It is known from earlier calculations [20,21,27–29] for ^{168}Er and ^{234}U that $I(E)$ varies rapidly with G. Thus G can be determined using level density data. In other words the calculations determine a magnitude parameter

Table 2
Quantum numbers and energies of s.p. orbits

Orbit #	s	n	ℓ	j	ϵ_j (MeV) [a]
1	0	0	2	5/2	−1.838
2	0	1	0	1/2	−0.968
3	0	0	2	3/2	3.242
4	1	0	3	7/2	−2.664
5	1	1	1	3/2	[illegible]
6	1	0	3	5/2	3.526
7	1	1	1	1/2	1.366
8	2	0	4	9/2	0.000

[a] Given are the traceless sd, fp and $g_{9/2}$ energies; $\overline{\Delta}_{ds-fp} = 11.03$ MeV, $\overline{\Delta}_{fp-g_{9/2}} = 6.53$ MeV.

(spreading variance of say $S = 0$ configurations) and SDI is then used to fix the relative variances of other (S) configurations. It should be mentioned that in principle calculations can be carried out with realistic interactions defined over large shell model spaces. Such an exercise was carried out for the ^{28}Si nucleus by including up to $2\hbar\omega$ excitations with (sd) and (fp) shells [28] and it is seen that one has to adjust the $V^{[0]}$ part of the interaction in order to describe the data. This is entirely consistent with the claim of Zuker et al. [26] that effective interactions in general do not have a proper $V^{[0]}$ part and in addition they have also argued that all effective interactions will have an essentially identical **V** part. The agreements obtained in the large space calculations with a realistic interaction for ^{28}Si confirm that $V_{S\text{-mix}}$ effects in general should be small. In the event that they are large, one loses shell model stability.

3.2.4. Unitary orbits for spreading variances and NIP calculations

For calculating state and spin-cutoff density variances, the eight spherical orbits given in Table 2 for protons (and neutrons) are divided into three unitary orbits (hereafter called UOV orbits) $(d_{5/2}, s_{1/2}, d_{3/2})$, $(f_{7/2}, p_{3/2}, f_{5/2}, p_{1/2})$ and $(g_{9/2})$ denoted by Λ_{-1}, Λ_0 and Λ_1 with s = 0, 1 and 2, respectively. Given a fp-shell nucleus, the proton and neutron numbers (m_p, m_n) outside the s^4p^{12} core can be easily counted. The smallest total S value allowed for the nucleus is given by $S_{\min} = m_p + m_n - 24$. Given a unitary configuration $[\mathbf{m}_p, \mathbf{m}_n]$ defined over the three UOV orbits, the total S value $S_{\mathrm{TOT}}([\mathbf{m}_p, \mathbf{m}_n]) = \sum_{i=1}^{3}(\mathbf{m}_p)_i s_i + \sum_{i=1}^{3}(\mathbf{m}_n)_i s_i$. Then the $S\hbar\omega$ excitations are given by $S = S([\mathbf{m}_p, \mathbf{m}_n]) = S_{\mathrm{TOT}}([\mathbf{m}_p, \mathbf{m}_n]) - S_{\min}$. The calculations in this paper are restricted to $2\hbar\omega$ excitations, i.e. $S \leqslant 2$. As already mentioned before, the classes are defined by $[\mathbf{m}_p, \mathbf{m}_n]$. For a given S value, it is easy to enumerate the corresponding classes; for example $(\Lambda_0)^{m_p}(\Lambda_0)^{m_n}$ belongs to $S = 0$, $(\Lambda_{-1})^{m_p-1}(\Lambda_1)^{1_p}(\Lambda_0)^{m_n}$ belongs to $S = 1$ and $(\Lambda_0)^{m_p-1}(\Lambda_1)^{1_p}(\Lambda_0)^{m_n-1}(\Lambda_1)^{1_n}$ belongs to $S = 2$, etc. There are 15 classes in all – one $S = 0$, four $S = 1$ and ten $S = 2$. Therefore the level density calculations involve 15 spreading variances. For the calculation of the NIP densities $I^{\mathbf{h}}(E)$ and $I^{\mathbf{h}}_{j_z^2}(E)$, the three UOV unitary orbits are further divided into five unitary orbits (hereafter called UON orbits) $(d_{5/2}, s_{1/2})$, $(d_{3/2})$, $(f_{7/2})$, $(p_{3/2}, f_{5/2}, p_{1/2})$ and $(g_{9/2})$ with s = 0, 0, 1, 1 and

2, respectively. For each nucleus, the number of unitary configurations defined by UON unitary orbits for each of the 15 classes are easy to enumerate. For NIP calculations and in the final sum (using (3), (4)) involving convolution of NIP and spreading Gaussian densities there are $\sim$ 1500 configurations for both parities. However, with spherical orbits for both UOV and UON orbits, there will be $\sim 10^4$ configurations and then one has to calculate $\sim 10^4$ spreading variances unlike only 15 with UOV orbits.

3.2.5. Procedure for calculations

Following the theory outlined in Section 2, for calculating level densities and spin-cutoff factors, first the unitary decomposition (with respect to the direct sum group defined by UOV orbits) of the hamiltonian and J^2 operator are carried out. Then, for each of the fifteen classes, the spreading variances $\sigma^2_{\rm V}([\mathbf{m}])$, $\sigma^2_{{\rm V}:J_z^2}([\mathbf{m}])$ and the centroid shifts $\epsilon_{{\rm V}:J_z^2}([\mathbf{m}])$ are calculated. Generating the configurations $\{\mathbf{m}\}$ defined by UON orbits for each of the 15 classes (defined by UOV orbits), the first four NIP density cumulants $\epsilon_{\rm NIP}(\{\mathbf{m}\})$, $\sigma^2_{\rm NIP}(\{\mathbf{m}\})$, $\gamma_1^{\rm NIP}(\{\mathbf{m}\})$ and $\gamma_2^{\rm NIP}(\{\mathbf{m}\})$ are calculated both for NIP state and spin-cutoff densities. The $I^{\rm h} \otimes \rho^{\rm V}$ convolution, for each $\{\mathbf{m}\} \in [\mathbf{m}]$, is carried out by adding the $I^{\rm h}$ and $\rho^{\rm V}$ cumulants (in the present exercise $\rho^{\rm V}$ is a Gaussian). Using the first four cumulants (ϵ, σ^2, γ_1, γ_2) thus obtained, the fixed $\{\mathbf{m}\}$ densities are written as Edgeworth-corrected Gaussian densities (see Eq. (8)). This exercise is carried out for positive parity configurations in the case of even–even and odd–odd nuclei and for negative-parity configurations in the case of even–odd and odd–even nuclei in our set. All the calculations use $S = 0 \oplus S = 2$ space. The ratio of these fixed-parity spin-cutoff and state densities gives the fixed-parity spin-cutoff factors. This information is then converted into $I_\ell(E)$ and $\sigma^2_J(E)$ by using the equations given in Appendix B.

In the actual calculations, starting with the reference energy data ($E_{\rm ref}$, $J^\pi_{\rm ref}$, $N^\pi_{\rm ref}$) and a value for the SDI strength G ($G \simeq 25/A$ MeV), $N^\pi(E)$ is calculated at various energies E and then (8) is used to obtain $\overline{E}_{\rm ref}$. Then the ground-state (g.s.) energy in the calculation is $E_{\rm g.s.} = \overline{E}_{\rm ref} - E_{\rm ref}$. Now $I^\pi(E)$ is calculated at around 10 MeV excitation where resonance or some other data for $I_\ell(E)$ is available (we call it $I_{\ell;{\rm href}}(E)$). The $I^\pi(E)$ is converted into $I_\ell(E)$ and compared with $I_{\ell;{\rm href}}(E)$. The value of G is varied until the calculated $I_\ell(E)$ is within say 5% of the $I_{\ell;{\rm href}}(E)$. Using the G value thus obtained, $I_\ell(E)$ and $\sigma_J(E)$ are calculated for E from 4 to 25 MeV excitation and compared with data.

Good agreements with data alone do not mean that the calculations are proper. It is essential that (i) in the low-energy domain (say $E \lesssim 6$ MeV) the $S = 0$ densities should be much larger than the $S = 2$ densities ($I(E) = I_{S=0}(E) + I_{S=2}(E)$) and (ii) the value of G should be in a meaningful domain (low-energy spectroscopic calculations yield $G \simeq 25/A$ MeV). In fact, the Δ_{ds-fp} and $\Delta_{fp-g_{9/2}}$ separations (one set for all nuclei) given in Section 3.2.3 are chosen such that for all the fp-shell nuclei under consideration the condition (i) above is satisfied and good agreement with data is obtained using $G \simeq 20/A$ MeV. Table 3 shows the S-decomposition of the state densities at various

Table 3
S-decomposition of state densities. For a given nucleus and a given energy, the first entry is for $I^{\pi}_{\ell,S=0}(E)$ and the second for $I^{\pi}_{\ell,S=2}(E)$. The parity $\pi = +$ for even–even and odd–odd nuclei and $\pi = -$ for even–odd and odd–even nuclei, respectively

E (MeV)	^{55}Mn	^{56}Fe	^{59}Co	^{60}Co	^{60}Ni	^{63}Cu (a) [a]	^{63}Cu (b) [a]
4.0	24	9	26	258	14	26	6
	7	2	4	47	1	6	25
[illegible]	[illegible]	[illegible]	[illegible]	[illegible]	[illegible]	[illegible]	[illegible]
	144	42	97	988	23	136	513
12.0	3100	1479	3244	18546	2256	3002	1021
	2109	692	1613	14322	511	2304	7379
16.0	20121	11171	22046	97501	17057	19813	7934
	23607	8694	19914	150780	8015	28196	78174
20.0	96461	27442	113900	393590	97154	99534	45869
	204080	28022	187940	1201100	92063	260010	631190

[a] ^{63}Cu (a) and ^{63}Cu (b) calculations are described in Section 3.3.6 ahead.

energies for the six nuclei {^{55}Mn, ^{56}Fe, ^{59}Co, ^{60}Co, ^{60}Ni, ^{63}Cu}. In the case of ^{63}Cu using densities corresponding to ^{63}Cu (b) in Table 3, one gets excellent agreement with data as described in Section 3.3.6 ahead. However, here, in the low-energy domain the $S = 2$ densities are much larger than $S = 0$ densities. Thus these calculations cannot be considered to be good. This is not the case with the ^{63}Cu (a) calculations though here the agreements are only satisfactory as shown in Fig. 4 ahead.

Before going further it should be mentioned that the c.p. (charged particle) data, the data from Dilg et al., Iljinov et al., Katsanos et al., Lu et al. and the resonance data that are used in the analysis (see Figs. 2–4 ahead) are defined and explained in Appendix B.

3.3. 8-orbit results for ^{55}Mn, ^{56}Fe, ^{59}Co, ^{60}Co, ^{60}Ni and ^{63}Cu

3.3.1. ^{55}Mn

The low-energy data which is used in fixing the reference energy is obtained from [34,35] and it gives $E_{\text{ref}} = 1.528$ MeV, $J^{\pi}_{\text{ref}} = 3/2^-$ and $N^{\pi}_{\text{ref}} = 44$. With these values, the g.s. energy is calculated, with a trial value of G, using the method outlined in Section 3.2.2. The value of G is varied such that the total level density $I_\ell(E)$ at $E =$ 10 MeV becomes close to the values obtained from the experimental (proton resonance) data compiled by Iljinov et al. which is 3000 ± 800 MeV^{-1}. With $G = 0.365$ MeV the calculated value of $I_\ell(E)$ at 10 MeV is 3200 MeV^{-1}. The values of σ^2_{V}, $\sigma^2_{\text{V}:J_z^2}$ and $\epsilon_{\text{V}:J_z^2}$ calculated with $G = 0.365$ MeV are 17.2 MeV^2, 15.5 MeV^2 and 0.015 MeV respectively for $S = 0$ class, ~ 25 MeV^2, ~ 23 MeV^2 and ~ 0.01 MeV respectively for $S = 1$ classes, ~ 33 MeV^2, ~ 30 MeV^2 and ~ 0.02 MeV respectively for $S = 2$ classes. With these class variances and centroid shifts, $I(E)$, $I_\ell(E)$ and $\sigma_J(E)$ are calculated at various energies using the procedures given in Section 3.2.5 and the results for $I_\ell(E)$ and $\sigma_J(E)$ are compared with data in Figs. 2 and 3. From Fig. 2a, it is seen that the agreement between data and calculations is good up to ~ 16 MeV and the departures

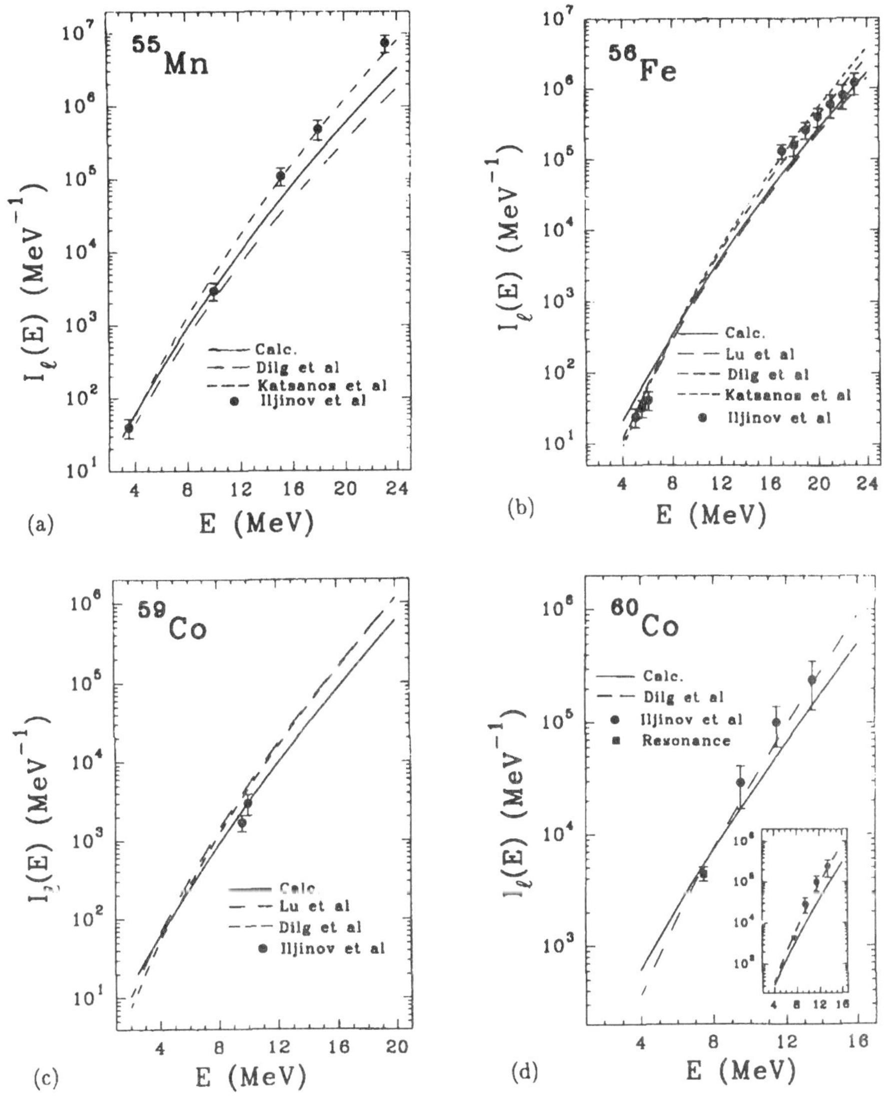

Fig. 2. Total level density $I_\ell(E)$ versus E: (a) ^{55}Mn; (b) ^{56}Fe; (c) ^{59}Co; (d) ^{60}Co; (e) ^{60}Ni; (f) ^{62}Ni; (g) ^{63}Cu; (h) ^{65}Cu. The calculations corresponding to the inset figures in case of ^{60}Co and ^{65}Cu are described in the text. The data from Dilg et al., Katsanos et al., Lu et al., Iljinov et al. and the resonance data are explained in Appendix B and the actual data compilation is given in [24].

appear to be significant beyond say 22 MeV. The calculated spin-cutoff factors are in excellent agreement with the data of Katsanos et al. The calculated state density $I(E)$ is fitted to the Lang LeCouteur (LLC) and back shifted Fermi gas forms (B.1)–(B.3) and the parameters (a, Δ) obtained from best fit are given in Table 4. The calculated densities are well represented by the Fermi gas forms; the Fermi gas fits are in general good for all the eight nuclei considered and the detailed fits are given in [24]. As can

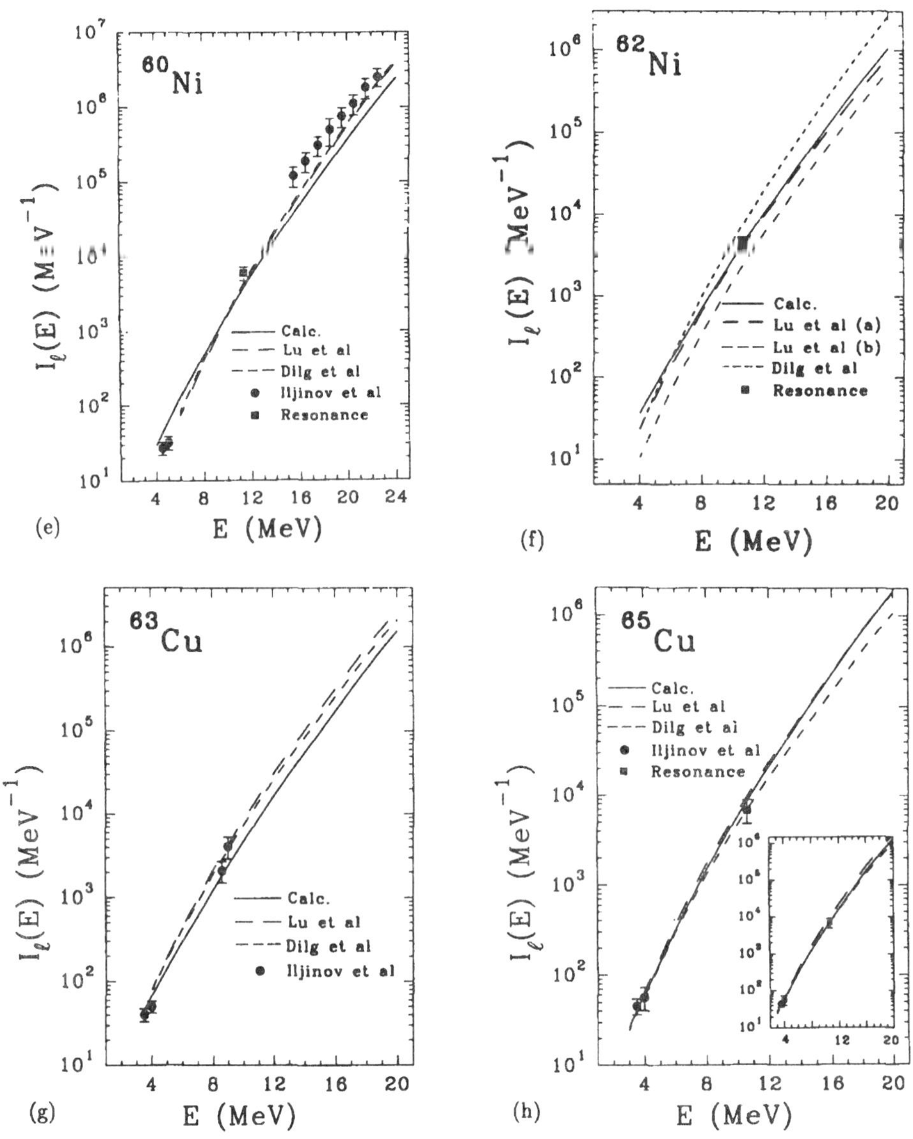

Fig. 2 — continued.

be seen from Table 4 the deduced (a, Δ) values are compatible more with the values of Dilg et al. than those of Katsanos et al. (these authors use the LLC form).

3.3.2. ^{56}Fe

The low-energy data which is used in fixing the reference energy is obtained from [35,36] and it gives $E_{\text{ref}} = 4.401$ MeV, $J^{\pi}_{\text{ref}} = 2^+$ and $N^{\pi}_{\text{ref}} = 147$. With these values, the g.s. energy is calculated with a trial value of G using the method outlined in Subsection 3.2.2. The value of G is obtained such that it fits reasonably well with (there

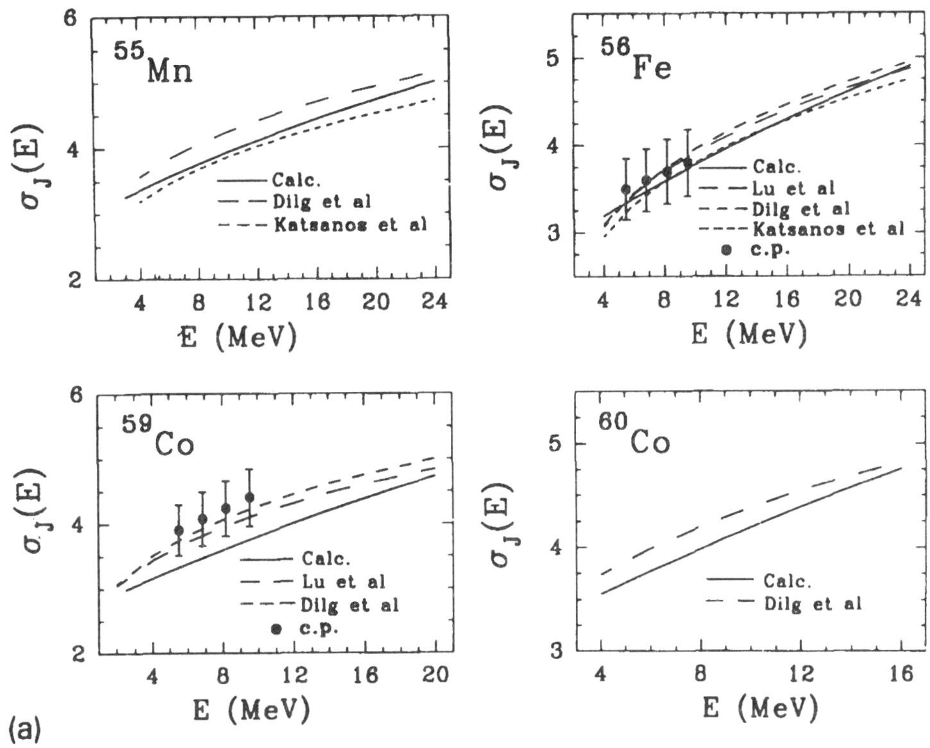

Fig. 3. Square root of spin-cutoff factor $\sigma_J(E)$ versus E for ^{55}Mn, ^{56}Fe, ^{59}Co, ^{60}Co, ^{60}Ni, ^{62}Ni, ^{63}Cu, ^{65}Cu nuclei. The c.p. (charged particle) data, the data from Dilg et al., Katsanos et al., Lu et al., Lu et al. (a) and Lu et al. (b) are explained in Appendix B and the actual data compilation is given in [24].

are no resonance data for ^{56}Fe) the data from Dilg et al. (1385 MeV^{-1}), Lu et al. (1064 MeV^{-1}) and Katsanos et al. (1476 MeV^{-1}) for $I_\ell(E)$ at 10 MeV excitation. At this energy the calculated $I_\ell(E)$ = 1304 MeV^{-1} for $G = 0.36$ MeV. For example, $\sigma^2_{\rm V}$, $\sigma^2_{{\rm V}:J_z^2}$ and $\epsilon_{{\rm V}:J_z^2}$ calculated with $G = 0.36$ MeV are 18.7 MeV^2, 16.9 MeV^2 and 0.006 MeV respectively for $S = 0$ class and $\sim$ 34 MeV^2, $\sim$ 32 MeV^2 and $\sim$ 0.02 MeV respectively for $S = 2$ classes. With these variances and centroid shifts $I(E)$, $I_\ell(E)$ and $\sigma_J(E)$ are calculated up to 24 MeV using the procedure outlined in Section 3.2.5 and the results for $I_\ell(E)$ and $\sigma_J(E)$ are compared with data in Figs. 2 and 3. Fig. 2b shows that the calculations are in excellent agreement with data right up to 24 MeV. The low-lying data (up to 6 MeV) from direct counting and high-lying data from Ericson fluctuations (17–23 MeV) are well described. Similarly, as can be seen from Fig. 3 the spin-cutoff factors are also well described; $\sigma_J(E) \sim$ 3.2–4.5 for E = 4–20 MeV. The deduced (a, Δ) values from the Fermi gas fits are given in Table 4 and they are in close agreement with values due to Lu et al. who used the charged particle data.

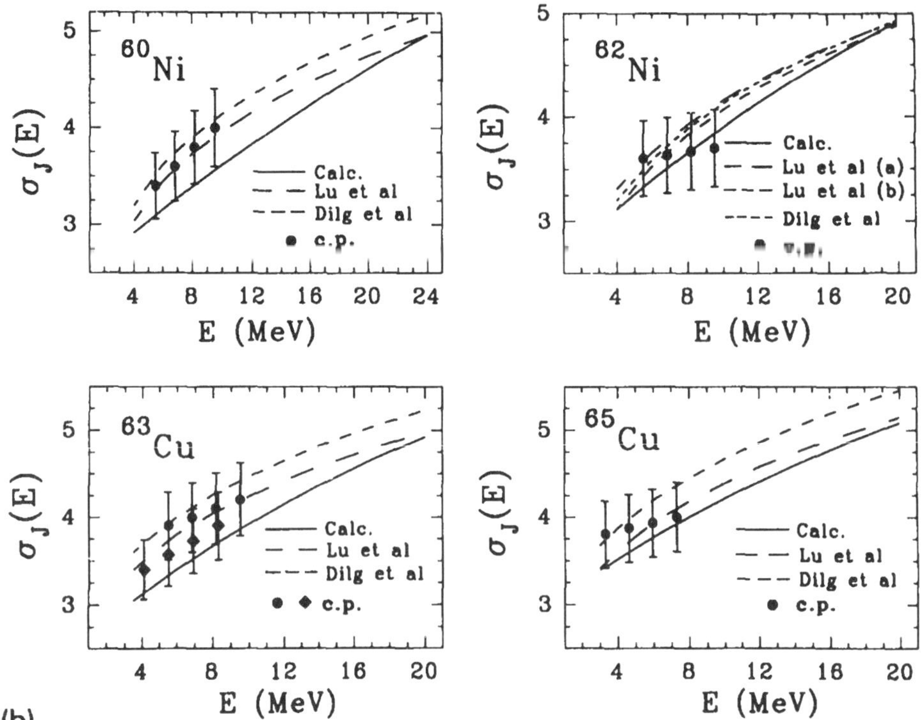

(b)

Fig. 3 — continued.

3.3.3. ^{59}Co

The low-energy data which is used in fixing the reference energy is obtained from [35,37] and it gives E_{ref} = 1.528 MeV, $J^{\pi}_{ref} = 3/2^-$ and $N^{\pi}_{ref} = 44$. With these values, the g.s. energy is calculated, with a trial value of G, using the method outlined in Section 3.2.2. The value of G is varied such that the total level density $I_{\ell}(E)$ at E = 10 MeV becomes close to the values obtained from the experimental (proton resonance) data compiled by Iljinov et al. which is 3000 ± 900 MeV^{-1}. With $G = 0.339$ MeV the calculated value of $I_{\ell}(E)$ at 10 MeV is 3045 MeV^{-1}. The state and spin-cutoff variances (σ^2_{V}, $\sigma^2_{\mathrm{V}:J^2_z}$) and the centroid shifts $\epsilon_{\mathrm{V}:J^2_z}$ are calculated with $G = 0.339$ MeV and then $I(E)$, $I_{\ell}(E)$ and $\sigma_J(E)$ are constructed using the procedure given in Section 3.2.5. The results for $I_{\ell}(E)$ and $\sigma_J(E)$ are compared with data in Figs. 2, 3. It is seen from Fig. 2c that the calculations extrapolate the densities from 10 MeV well up to 20 MeV. The calculated spin-cutoff factors (Fig. 3) are somewhat smaller (say by 20%) than the experimental values. The calculated $I(E)$ is well represented by LLC Fermi gas form with (a, Δ) = (5.84 MeV^{-1}, −0.74 MeV) and the deduced a values (see Table 4) are somewhat smaller than the values of Lu et al. and Dilg et al.

3.3.4. ^{60}Co

Complete low-energy data for ^{60}Co nucleus is available only up to 0.79 MeV [38]. As this energy is extremely low, it is not good to consider this energy to be the

Table 4
Fermi gas parameters. See Appendix B for the data from Dilg et al., Katsanos et al., Lu et al., Lu et al. (a), Lu et al. (b). Actual data compilation is given in [24].

Nucleus	Data values		Present calculation	
	a (MeV^{-1})	Δ (MeV)	a (MeV^{-1})	Δ (MeV)
^{55}Mn	5.45 [a]	−1.20 [a]	5.52 [b]	−1.15 [b]
	6.30 [c]	−0.50 [c]	5.53 [d]	−0.85 [d]
^{56}Fe	5.70 [e]	0.70 [a]	5.70 [b]	0.35 [b]
	6.15 [a]	1.10 [a]	5.72 [d]	0.64 [d]
	6.40 [c]	1.50 [c]		
^{59}Co	6.20 [e]	−0.80 [e]	5.84 [b]	−0.74 [b]
	6.31 [a]	−0.47 [a]	5.77 [d]	−0.74 [d]
^{60}Co	6.86 [a]	−2.16 [a]	5.98 [b]	−3.79 [b]
			5.99 [d]	−3.53 [d]
^{60}Ni	6.40 [e]	1.30 [e]	5.82 [b]	0.22 [b]
	6.42 [a]	1.14 [a]	5.84 [d]	0.51 [d]
^{62}Ni	6.40 [f]	0.50 [f]	6.44 [b]	0.40 [b]
	6.40 [g]	1.30 [g]	6.45 [d]	0.67 [d]
	7.27 [a]	1.07 [a]		
^{63}Cu	6.80 [e]	−0.50 [e]	6.44 [b]	−0.32 [b]
	6.63 [a]	−0.67 [a]	6.45 [d]	−0.06 [d]
^{65}Cu	6.60 [e]	−0.50 [e]	6.58 [b]	−0.38 [b]
	6.24 [a]	−0.77 [a]	6.60 [d]	−0.13 [d]

[a] Dilg et al.
[b] Lang LeCouteur Fermi gas form.
[c] Katsanos et al.
[d] Back shifted Fermi gas form.
[e] Lu et al.
[f] Lu et al. (a).
[g] Lu et al. (b).
[d] Back shifted Fermi gas form.

reference energy. Therefore in this case a different prescription (as compared to the method outlined in Section 3.2.2) is used for fixing the reference energy. Using the data from Dilg et al. for level densities, at 4 MeV the total number of positive parity states $N^+(E_{ref} = 4 \text{ MeV})$ is determined to be 1796. Starting with a value of G, the reference energy ($\overline{E}_{ref}$) is determined using $N^+(E_{ref})$ value. The value of G is then varied to reproduce the $I_\ell(E)$ value of Dilg et al. at 8 MeV excitation ($I_\ell(E=8 \text{ MeV}) = 7337 \text{ MeV}^{-1}$) and also the neutron resonance density at the resonance energy 7.491 MeV ($I_\ell(E=7.491 \text{ MeV}) = 4389 \pm 627 \text{ MeV}^{-1}$). To get agreement it is found that a rather low value (compared to $20/A$ MeV) of G has to be used. As this is not proper (G should be normally $25/A$ MeV), the value of G is fixed to be $G = 0.31$ MeV. Then with the above described method of fixing the reference energy, the $I_\ell(E)$ at $E = 8$ MeV and 7.491 MeV are found to be smaller by a factor 2 and at higher energies by a factor of 3–4 (see the inset figure in Fig. 2d). In order to produce satisfactory agreements, the calculated ground state has to be moved up by 1 MeV. The results obtained for $I_\ell(E)$ and $\sigma_J(E)$ with $G = 0.31$ MeV by the above mentioned correction for g.s. location, are shown in Figs. 2 and 3. A source of error in properly locating the g.s. could be that there is an extra back shifting as we are using SDI for an odd–odd nucleus. Alternatively this may be due to the fact that level densities at low energies are high for odd–odd nuclei compared to its even–even or odd-A neighbor and with SDI it may not be possible to reproduce these densities. It is worth pointing out that ^{60}Co is the only odd–odd nucleus that is so far analyzed using SAT-LSS [7]. The problems associated with odd–odd nuclei should be explored further. It can be seen from Figs. 2d and 3 the level densities and spin-cutoff factors for ^{60}Co are well described by the IP theory (level densities are calculated up to 16 MeV). The deduced (a, Δ) values from Fermi gas fits are not compatible, as shown in Table 4, with the values of Dilg et al. which are obtained by fitting the low-energy data and the neutron resonance data.

3.3.5. ^{60}Ni

For the ^{60}Ni nucleus there is complete low-energy data [35] up to 4.007 MeV, however there are several levels between 3.32 and 4.01 MeV for which the spins and parities are not assigned. In principle one can use the 3.318 MeV level as reference energy as all levels below are identified with J^π values assigned. As we are dealing with an even–even nucleus, for these level densities at low energies being low, it is always better to use as high a reference energy as possible. Therefore all the levels up to 4.007 MeV are used but with degeneracies assigned to levels with no J^π assignments by using the so-called information theory method [29]. According to information theory the total number of states $I^\pi(E)\,dE$ at energy E is given by

$$
\begin{aligned}
I^{\overline{\pi}}(E)\,dE &= \frac{\sum_{J\ (\text{known } J\text{'s or all } J\text{'s}),\pi=\overline{\pi}}(2J+1)P_{J\pi}}{\sum_{J\ (\text{known } J\text{'s or all } J\text{'s}),\pi}P_{J\pi}}\,,\\
P_{J\pi} &= C_J P_\pi\,,\\
C_J &= \frac{(2J+1)}{\sqrt{8\pi\sigma_J^3}}\exp\left[-\frac{(J+1/2)^2}{2\sigma_J^2}\right],\\
P_\pi &= 1\,, \qquad \text{desired parity}\,,\\
P_\pi &= \tfrac{1}{2}\,, \qquad \text{for unknown parity}\,,\\
P_\pi &= 0\,, \qquad \text{for opposite parity}\,.
\end{aligned} \tag{9}
$$

In applying (9), it should be kept in mind that the J summation is over the known J's if some J assignments are known and otherwise it is over all J's (i.e. $J = 0$ to ∞). Eq. (9) is applied with $\sigma_J = 3$ which is a typical value for ^{60}Ni at low energies as can be seen from Fig. 3. This gives $E_{\rm ref} = 4.007$ MeV, $J^\pi_{\rm ref} = 2^+$ and $N^\pi_{\rm ref} = 133.5$. With this reference energy the g.s. is calculated using a trial value for G and the method outlined in Section 3.2.2. The neutron resonance data is found to give a value of $G < 0.3$ (much lower than $20/A$ MeV) and therefore a search for a value of G (not much lower

[7] Behkami and Huizenga [39], in their analysis of fp-shell level density and spin-cutoff factor data using Fermi gas (NIP) theory with pairing, also considered the same set of nuclei listed in Table 1 but 'without' the ^{60}Co nucleus.

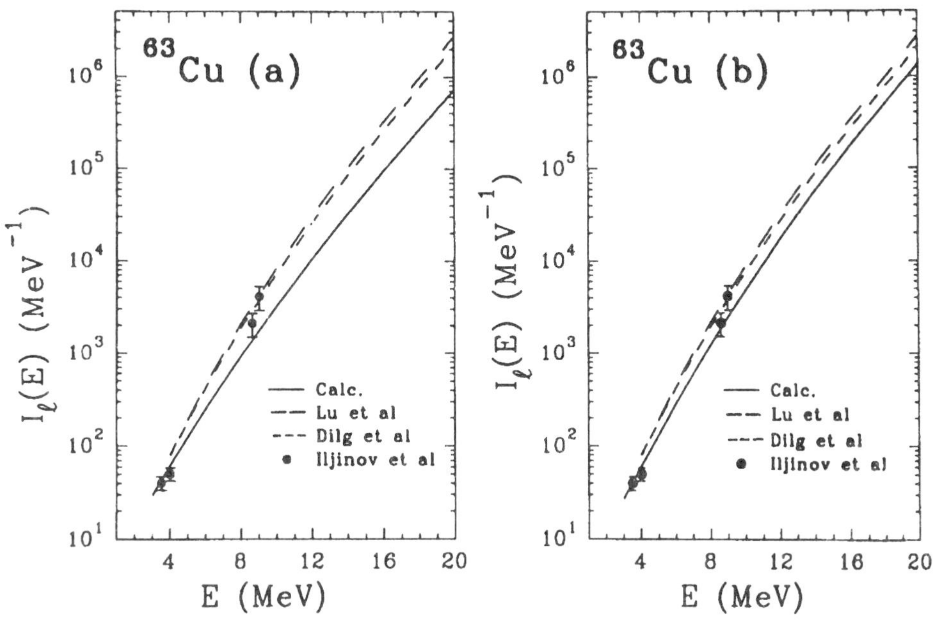

Fig. 4. Total level density $I_\ell(E)$ versus E for ^{63}Cu in 8-orbit calculations. The ^{63}Cu (a) and ^{63}Cu (b) calculations are described in the text.

than $20/A$ MeV) is made which gives an overall good description of $I_\ell(E)$ data from 4–24 MeV and the value found is $G = 0.31$ MeV. The calculated $I_\ell(E)$ and $\sigma_J(E)$ are compared with data in Figs. 2 and 3. The ^{60}Ni nucleus is the only nucleus in our set that has level density data from all the four sources mentioned in Section 3.2.1. As can be seen from Fig. 2e, the calculations with $G = 0.31$ MeV describe the data in the 4–24 MeV energy range. The deviations at energies > 16 MeV are not more than a factor 2. The calculated spin-cutoff factor values (see Fig. 3) are somewhat smaller than data values. The results for ^{60}Ni confirm that the IP theory describes the data in a wide energy range. The deduced (a, Δ) values are somewhat smaller than the values of Lu et al. (from charge particle spectra) and the values of Dilg et al. as shown in Table 4 but yet the calculated $I(E)$ is found [24] to be quite close to the curves from Lu et al. and Dilg et al.

3.3.6. ^{63}Cu

The low-energy data for fixing the reference energy is taken from [40] and it gives $E_{\text{ref}} = 2.337$ MeV, $J^{\pi}_{\text{ref}} = 5/2^-$ and $N^{\pi}_{\text{ref}} = 74$. With these data and a trial value of G the g.s. energy is determined. Then G is varied to get the best possible description of the data. An attempt to fix the value of G using the proton resonance data $I_\ell(E) = 2100 \pm 600$ MeV^{-1} at $E = 8.6$ MeV gives a value $G < 0.27$ MeV (note that $25/A = 0.397$ for ^{63}Cu) which is not in the acceptable range (this value of G violates the conditions laid down in Section 3.2.5 for the goodness of a calculation) and also with $G < 0.27$ MeV the $I_\ell(E)$ data above 10 MeV is not well described. With $G = 0.317$ MeV a reasonable description of data is obtained and the corresponding results are shown in Fig. 4 (labelled ^{63}Cu (a)). However, if the resonance data has to be well represented and also that the description of the $I_\ell(E)$ data beyond 10 MeV has to be good then one has to change the $\overline{\Delta}_{fp-g_{9/2}}$ spacing. By decreasing the $\overline{\Delta}_{fp-g_{9/2}}$ spacing from 6.53 to 4.53 MeV (given in Table 2) but still using G = 0.317 MeV, one gets excellent agreement with data as shown in Fig. 4 (labelled ^{63}Cu (b)). The decomposition of density into $S = 0$ and $S = 2$ parts clearly shows (Table 3) that the $S = 2$ densities dominate over the $S = 0$ densities in this case and therefore ^{63}Cu (b) calculations cannot be considered to be proper calculations. From the ^{63}Cu (b) results one can argue that the decrease in $\overline{\Delta}_{fp-g_{9/2}}$ may be a renormalization effect as we are restricting to 8-orbits. Thus the ^{63}Cu example shows the deficiencies of the choices made regarding s.p. orbits, their s values and $\overline{\Delta}$ spacings for $A > 60$ nuclei.

3.4. Summary of 8-orbit calculations

For the five fp-shell nuclei (^{55}Mn, ^{56}Fe, ^{59}Co, ^{60}Co, ^{60}Ni) 8-orbit calculations for total level densities, state densities and spin-cutoff factors are performed using SAT-LSS (which is a IP theory [8]) and compared with data in Sections 3.3.1–3.3.6. It is seen that the description of the data is excellent for the three nuclei {^{55}Mn, ^{56}Fe, ^{59}Co} and satisfactory for the two nuclei {^{60}Co, ^{60}Ni}. From the calculations the SDI strength 'G' is deduced to be $G \sim 20/A$ MeV (this value is compatible with the values deduced from low-energy spectroscopy). The ^{63}Cu example (also similar calculations for ^{62}Ni and ^{65}Cu) demonstrates that the 8-orbit calculations have problems when applied to $A > 60$ nuclei, i.e. in order to get the densities within a factor of 2–3, one has to necessarily violate the conditions (i) and (ii) given in Section 3.2.5. The solution appears to be that the spectroscopic space has to be enlarged by opening up ($2d_{5/2}$, $1g_{7/2}$) orbits. Because of this, in the present section, 8-orbit results for {^{62}Ni, ^{65}Cu} nuclei are not

[8] Starting from Bethe's $\exp(2\sqrt{aE})$ Fermi gas form, there is a large growth in the development of NIP theories for calculating level densities. These include: (i) modifications of the Bethe form (see Appendix B) mainly via various parametrizations of the level density parameter 'a' and back shifting parameter 'Δ' [41,42]; (ii) inclusion of pairing via a temperature-dependent gap equation [39,43]; (iii) collective effects via a convolution which gives an enhancement of the level densities by a collective enhancement factor given by the partition function of the collective hamiltonians [41,44,45]. On the other hand, there are attempts to develop an interacting particle level density theory. These studies include: (i) via Casimir operators of groups as done in the case of Wigner SU(4) by Bloch [46], Elliott $SU(3)$ by Kanestrom [47] and single j-shell pairing [48]; (ii) employing the CLT results of SAT as described in Section 1, a preliminary attempt in this direction is due to Ayik and Ginnochio [49] and a more elaborate study is due to Grimes et al. [50]. In addition there is [51] the phenomenological version of the theory given in Section 2.2; (iii) development of a theory analogous to what was described in Section 2.2 but employing a GOE to represent interactions (i.e. V) and this work is due to Sato and Yashida [52]. This theory suffers from the fact that GOE is not a model for generating nuclear level densities; (iv) the so-called static path approximation (SPA) due to Bertsch and others [53,54] which makes use of the Hubbard–Stratonovich transformation and is ideally suited for studies with pairing plus quadrupole–quadrupole interaction at higher temperatures. Though this approach is promising, so far no data analysis is carried out using this method; (v) the Monte Carlo (path integral) method, which generalizes SPA for the shell model as being developed by Koonin and collaborators [55,56]. The studies using this method are so far restricted to $0\hbar\omega$ spaces.

presented although they are in the selected set given in Table 1. The results of the 10-orbit calculations for ^{62}Ni and ^{63,65}Cu nuclei are described in the following section. Before proceeding further it has to be mentioned that 10-orbit calculations leave the 8-orbit results essentially unaltered. For example, for ^{56}Fe the total level density at 10 MeV differ by $\sim$ 1%, at 20 MeV by $\sim$ 4% and at 24 MeV by $\sim$ 7%. Similarly for ^{60}Ni the deviations are $\sim$ 1%, $\sim$ 6% and $\sim$ 9%, respectively. In fact with these deviations the agreements shown in Fig. 2 become even better at higher excitations. Furthermore the spin-cutoff values are found to remain essentially unaltered. In all these calculations, the value of 'G' determined by 8-orbit calculations is employed. Thus to within $\sim$ 10%, 8-orbit calculations are adequate for the five nuclei ^{55}Mn, ^{56}Fe, ^{59}Co, ^{60}Co, ^{60}Ni – the results are stable against expansion of the spectroscopic space.

4. 10-orbit results for ^{62}Ni, ^{63}Cu and ^{65}Cu nuclei

4.1. Choice of s values, single-particle energies, centroid separations and other calculational details

The orbits $(2d_{5/2}, 1g_{7/2})$ are added so that the density at $E > 10$ MeV grows and with this, with a reasonable value of G (strength of SDI) and keeping the $S = 0$ description to be dominant up to and above $E_{\rm ref}$, the data are well explained for the three nuclei ^{62}Ni, ^{63,65}Cu as can be seen from Sections 4.2.1–4.2.3 ahead. The important question that one has to face is that of the s value for the $(2d_{5/2}, 1g_{7/2})$ orbits. One possible choice is to assign $\bar{s} = 2$ for the three orbits $(1g_{9/2}, 2d_{5/2}, 1g_{7/2})$ with $\bar{s} = 0$ and 1 for (ds) and (fp) orbits respectively as in the previous section. Let us denote ds, fp and $(1g_{9/2}, 2d_{5/2}, 1g_{7/2})$ orbits as unitary orbits $\overline{\Lambda}_{-1}$, $\overline{\Lambda}_0$, $\overline{\Lambda}_1$, respectively. With this choice, the calculated density is not found to be acceptable as the $\bar{S} = 2$ intensity even at $E_{\rm ref}$ and below is nearly 100%. The reason for this is that the $S = 2$ class variances become larger by about 20% compared to the corresponding variances without $(2d_{5/2}, 1g_{7/2})$ orbits. This then leads to the natural choice of $s = 2$ for $1g_{9/2}$ as before and $s = 3$ for $(2d_{5/2}, 1g_{7/2})$ orbits. This is consistent with the fact that the $1g_{9/2}$ orbit is well separated from the $(2d_{5/2}, 1g_{7/2})$ orbits to form the magic number 50. With this choice there are four UOV orbits $\Lambda_{-1} = (d_{5/2}, s_{1/2}, d_{3/2})$, $\Lambda_0 = (f_{7/2}, p_{3/2}, f_{5/2}, p_{1/2})$, $\Lambda_1 = (g_{9/2})$, $\Lambda_2 = (d_{5/2}, g_{7/2})$ with s = 0, 1, 2, 3, respectively. Then for example, an $S = 2$ class

$$(\overline{\Lambda}_0)^{-2p}(\overline{\Lambda}_1)^{2p} \to (\Lambda_0)^{-2p}(\Lambda_1)^{2p} \oplus (\Lambda_0)^{-2p}(\Lambda_1)^{1p}(\Lambda_2)^{1p} \oplus (\Lambda_0)^{-2p}(\Lambda_2)^{2p}$$

with S = 2, 3 and 4, respectively. Thus we are led to $S = 4$ calculations in the 10-orbit case. There are one $S = 0$, ten $S = 2$, eight $S = 3$ and three $S = 4$ classes in this case.

The SPE for the ds, fp and $1g_{9/2}$ orbits are same as in the 8-orbit calculations. The $1g_{7/2}$ orbit energy is chosen to be 4 MeV above the $1g_{9/2}$ orbit and the $2d_{5/2} - 1g_{7/2}$ separation is chosen to be 0.68 MeV as in ^{116}Sn calculations [57]. The resulting traceless SPE and centroid separations $\overline{\Delta}^{(1)}$ are given in Table 5; the values of another set of

Table 5
Quantum numbers and energies of s.p. orbits

Orbit #	s	n	ℓ	j	ϵ_j (MeV) [a]
1	0	0	2	5/2	−1.838
2	0	1	0	1/2	−0.968
3	0	0	2	3/2	3.242
4	1	0	3	7/2	−2.664
5	1	1	1	3/2	0.611
6	1	0	3	5/2	3.526
7	1	1	1	1/2	1.366
8	2	0	4	9/2	0.000
9	3	1	2	5/2	0.389
10	3	0	4	7/2	−0.291

[a] Given are the traceless ds, fp, $g_{9/2}$ and $(d_{5/2}, g_{7/2})$ energies. Note that $\overline{\Delta}^{(1)}_{sd-fp} = 11.03$ MeV, $\overline{\Delta}^{(1)}_{fp-g_{9/2}} = 6.53$ MeV and $\overline{\Delta}^{(1)}_{g_{9/2}-\{d_{5/2},g_{7/2}\}} = 4.29$ MeV. Similarly $\overline{\Delta}^{(2)}_{sd-fp} = 11.03$ MeV, $\overline{\Delta}^{(2)}_{fp-g_{9/2}} = 5.78$ MeV and $\overline{\Delta}^{(2)}_{g_{9/2}-\{d_{5/2},g_{7/2}\}} = 3.29$ MeV.

centroid separations (denoted by $\overline{\Delta}^{(2)}$) are also given in the table. With the centroid separations $\overline{\Delta}^{(1)}$ given in Table 5, it is found that for ^{62}Ni and ^{63}Cu the calculated densities are usually smaller by a factor 2 and therefore the $\overline{\Delta}^{(1)}_{fp-g_{9/2}}$ is decreased by 0.75 MeV and $\overline{\Delta}^{(1)}_{g_{9/2}-\{d_{5/2},g_{7/2}\}}$ by 1 MeV and with this choice (denoted as $\overline{\Delta}^{(2)}$ in Table 5), together with a reasonable value for G, good agreement with the data is obtained and also a dominant $S = 0$ description up to and above $E_{\rm ref}$; in the case of ^{65}Cu, $\overline{\Delta}^{(1)}$ choice is found to be good. The S decomposition of the densities is given in Table 6. For the calculation of NIP densities, the UON orbits are chosen to be $(d_{5/2}, s_{1/2})$, $(d_{3/2})$, $(f_{7/2})$, $(p_{3/2}, f_{5/2}, p_{1/2})$, $(g_{9/2})$, $(d_{5/2}, g_{7/2})$ with $s = 0, 0, 1, 1, 2, 3$, respectively. All other calculational procedures are same as in the case of 8-orbit calculations (described in Section 3).

4.2. Results

4.2.1. ^{62}Ni

For ^{62}Ni nucleus there is complete low-energy data [58] up to 4.055 MeV, however there are several levels between 3.258 and 4.055 MeV for which the spins and parities are not assigned. In the present calculations all the levels up to 4.055 MeV are used but with degeneracies assigned to levels with no J and/or π assignments by using information theory method described in Section 3.3.5. In applying (9) the value of spin-cutoff factor has been taken to be 3.0. The final $(E_{\rm ref}, N^{\pi}_{\rm ref}, J^{\pi}_{\rm ref})$ values are (4.055 MeV, 167.2, 4^+), respectively. With this reference energy data and the unitary orbit centroid separation as given by $\overline{\Delta}^{(2)}$ in Table 5, the g.s. is calculated using a trial value for G. The level density (4552 MeV^{-1}) at the neutron resonance energy 10.597 MeV is reproduced with $G = 0.305$ MeV. The calculated $I_\ell(E)$ and $\sigma_J(E)$ are compared with data in Figs. 2 and 3. As can be seen from Fig. 2f, the calculations with $G = 0.305$ MeV describe

Table 6
S-decomposition of state densities. For a given nucleus and a given energy there are four entries and they are (in MeV^{-1}) for $I^{\pi}_{\ell;S=0}(E)$, $I^{\pi}_{\ell;S=2}(E)$, $I^{\pi}_{\ell;S=3}(E)$ and $I^{\pi}_{\ell;S=4}(E)$, respectively. The parity $\pi = +$ for ^{62}Ni and $\pi = -$ for ^{63,65}Cu

E (MeV)	^{62}Ni	^{63}Cu	^{65}Cu (a) [a]	^{65}Cu (b) [a]
E_{ref} [b]	13	6	2	6
	5	3	5	2
	1	0	1	0
	0	0	0	0
E_{res} [b]	915	422	611	1259
	982	408	2808	2034
	220	92	915	258
	7	3	43	3
16.0	14263	18962	8444	15021
	33048	50868	69794	56811
	10863	18678	31741	12108
	549	1115	2362	342

[a] ^{65}Cu (a) and ^{65}Cu (b) calculations are described in Section 4.2.3.
[b] For ^{62}Ni, $E_{\text{ref}} = 4.055$ MeV, $E_{\text{res}} = 10.597$ MeV; For ^{63}Cu, $E_{\text{ref}} = 2.337$ MeV, $E_{\text{res}} = 8.6$ MeV; For ^{65}Cu, $E_{\text{ref}} = 2.329$ MeV, $E_{\text{res}} = 10.65$ MeV.

the data in the 4–20 MeV energy range (the curve of Lu et al. (b) is consistently low compared to the calculated curve). The calculated spin-cutoff factor values (Fig. 3) are somewhat smaller than data values at low energies. The results for ^{62}Ni confirm that the IP theory describes the data in a wide energy range. The calculated state density $I(E)$ is well represented by the Fermi gas forms and the deduced (a, Δ) values are given in Table 4.

4.2.2. ^{63}Cu

The low-energy data given in Section 3.3.6, centroid separations given by $\overline{\Delta}^{(2)}$ in Table 5 and the proton resonance data ($E_{\text{res}} = 8.6$ MeV, $I_{\ell}(E) = 2100\ \text{MeV}^{-1}$), determine the value of G to be 0.305 MeV. Using this deduced value of G, $I_{\ell}(E)$ and $\sigma_J(E)$ are constructed and compared with data in Figs. 2 and 3. Beyond the resonance energy of 8.6 MeV, the calculated values are found to be lower, by about 40%, compared to data. The spin-cutoff factors change from $\sigma_J(E) = 3$ to 4 as E changes from 4 to 12 MeV while the data values are around 4 with an error of 10–20%. The deduced values of (a, Δ), from Fermi gas fits, are given in Table 4.

Thus ^{62}Ni and ^{63}Cu data are reasonably well described by using the centroids separation as given by $\overline{\Delta}^{(2)}$ in Table 5. The $\overline{\Delta}^{(1)}$ set is found to give densities to be too low, compared to data, to be acceptable.

4.2.3. ^{65}Cu

The low-energy data for fixing the reference energy is taken from [59] which gives $E_{\text{ref}} = 2.329$ MeV, $J^{\pi}_{\text{ref}} = 3/2^-$ and $N^{\pi}_{\text{ref}} = 58$. Two sets of calculations are performed for ^{65}Cu. In the first set of calculations, the centroid separations are chosen to be same as in the ^{62}Ni and ^{63}Cu cases, i.e. the $\overline{\Delta}^{(2)}$ set given in Table 5 is used. This set (the corresponding calculation is called ^{65}Cu (a)), the low-energy data given above and the resonance data (E_{res} = 10.65 MeV, $I_\ell(E_{res})$ = 6854 ± 2056 MeV^{-1}) are used to determine the value of G. The deduced value is G = 0.315 MeV and even with this high value of G (compared to 0.305 MeV used in ^{62}Ni and ^{63}Cu cases), the resonance density is predicted to be $\sim$ 9000 MeV^{-1}, which is just within the error bar of the resonance density 6854 ± 2056 MeV^{-1}. As a consequence, at $E \gtrsim 16$ MeV the calculated $I_\ell(E)$ differ from the data of Dilg et al. by a factor 2, however it is in close agreement with the data of Lu et al. In order to reproduce the resonance density value of 6800 MeV^{-1}, one has to use a much larger G value and also the $S = 2$ intensities become much larger than $S = 0$ intensities at low energies, which cannot be accepted. However, using a different set (the corresponding calculation is called ^{65}Cu (b)) of centroid separations, as given by the set $\overline{\Delta}^{(1)}$ in Table 5, the value of G is determined to be 0.3 MeV. With this the resonance density is reproduced rather well as also the data of Dilg et al. up to 20 MeV. However, here the deviations from the data of Lu et al. are by about 40% as shown in the inset figure of Fig. 2h. To be consistent with ^{62}Ni and ^{63}Cu calculations, the ^{65}Cu (a) results are accepted to be the final results. Therefore for ^{65}Cu (a) calculation, the results for $\sigma_J(E)$ versus E are shown in Fig. 3. The deduced values of (a, Δ), from Fermi gas fits, are given in Table 4.

4.3. Summary

The 8-orbit calculations for the five nuclei ^{55}Mn, ^{56}Fe, ^{59}Co, ^{60}Co and ^{60}Ni (described in Section 3.3) and 10-orbit calculations for the three nuclei ^{62}Ni, ^{63}Cu and ^{65}Cu (described in 4.2) show that SAT-LSS (given by (3), (4)) provides good extrapolation of level densities from the resonance energy ($\sim$ 8 MeV) to $\sim$ 20 MeV excitation and interpolation between the reference energy ($\sim$ 3–4 MeV) and the resonance energy. The calculated spin-cutoff factors and their energy variations (deriving from $J_Z^2 - \mathbf{V}^2$ correlations) are in good accord with the values deduced from data. Here it is appropriate to recall that (as can be seen from Sections 3.1–3.3, 4.1 and 4.2) in implementing the theory given by (3), (4), in practice one has to make choices regarding: (i) single-particle energies; (ii) centroid separations; (iii) effective interaction; (iv) reference energy selection; (v) selection of unitary orbits; (vi) additional experimental data for fixing free parameters in the interaction if any. In the present calculations not much freedom is exercised with regard to (i) as the s.p.e. are taken from data. As already mentioned, there are problems associated with (ii), i.e. with the choice of $V^{[0]}$. It is beyond the scope of the present paper to address this problem in any detail. In the absence of a good theory for deriving $V^{[0]}$, it may be worthwhile to find semi-empirical methods (as in [26]) for fixing $V^{[0]}$ in large shell model spaces. Once (ii) is taken care of properly, then (iii) is no longer a serious problem. A theory for the onset of chaos in nuclei will remove (iv). An empirical method for the same comes from experiments that guarantee completeness of levels up to a high enough energy [60]. The latter approach is adopted in Sections 3.3 and 4.2. A large number of studies [20,21] showed that (v) is not of any real consequence. In the present calculations as SDI is used, (vi) is relevant. This is a limitation of the present calculations (not of SAT-LSS). However, there is only one free parameter ('G') and it is able to extrapolate the results way beyond the data position that determined its value.

5. Conclusions

First systematic calculations of level densities and spin-cutoff factors, for a set of nuclei in a given region of the periodic table, within the SAT-LSS framework are presented in this paper with fp-shell nuclei as an example. Notwithstanding the various approximations that led to the theory and various options exercised in implementing the formalism (see Sections 2.3, 3.2 and 4.3), from the results shown in Figs. 2 and 3 it is plausible to conclude that SAT-LSS gives a good framework for calculating, with interactions, level densities and spin-cutoff factors. Full microscopic calculations using SAT-LSS require: (i) good theory/prescription for fixing the $V^{[0]}$ part of the effective interactions; (ii) a viable method/model, for incorporating $S = 0 \oplus 2 \oplus 4 \oplus \ldots$ mixing effects into the convolution forms for state and spin-cutoff densities. It appears that by pursuing further the work of Zuker et al. [26] might yield solutions to (i) and this belongs to an effective interaction theory. Similarly, the centroid and variance modified GOE model (and its extensions) studied by Pluhar and Widenmuller [61] might yield a solution to (ii). These two problems are being pursued and the results will be reported in a future publication. Finally, the results of the present paper together with the recent applications to nuclear astrophysics problems [62] and the earlier applications to the problem of TRNI in nucleon–nucleon force [15] and single-particle transfer strengths [63] show that SAT-LSS is a viable tool for analyzing and applying statistical properties of atomic nuclei.

Acknowledgements

The work presented in this paper has grown out of many discussions and correspondence one of the authors (V.K.B.K.) has with J.B. French in the last eight years and out of an ongoing project he is carrying with J.B. French and R.U. Haq. Thanks are due to V. Potbhare for his interest in the work presented in this paper. The authors thank K. Jain for help in the initial stages of the analysis presented in Section 3. Finally one of the authors (V.K.B.K.) thanks R.U. Haq for facilities at Laurentian University (Canada) where a draft of the paper has been prepared.

Appendix A. Unitary group decompositions

With m identical fermions distributed over N single-particle states, one can recognize the appearance of the $U(N)$ group, which is generated by the N^2 operators $a^{\dagger}_{j_r m_r} a_{j_s m_s}$; $r, s = 1, 2, \ldots, N$. Then the $\binom{N}{m}$ antisymmetric states form an irreducible representation (irrep) of the group $U(N)$, usually denoted by the Young shape $\{1^m\}$. One can in general seek a decomposition of a given operator into tensor operators (belonging to a definite irrep of $U(N)$) with respect to the $U(N)$ group. Given the one- plus two-body hamiltonian $H = h+V$, with respect to $U(N)$ group the operators h and V decompose as $h = \sum_{\nu=0,1} h^{\nu}$ and $V = \sum_{\nu=0,1,2} V^{\nu}$, respectively. The scalar part h^0 of h is a first-order polynomial in the number operator n and the remaining piece $h^1 = h - h^0$ is the unitary one-body part of h. Similarly, V^0 is a second-order polynomial in n (hence a $U(N)$ scalar), V^1 is the effective (n-dependent) one-body operator and the remaining piece is an irreducible two-body part. Details regarding Young columnar shapes associated with each ν and the explicit forms for the various tensor parts are given in [3].

The unitary group $U(N_\alpha)$ acting in each spherical orbit α generates m_α of spherical configurations $\mathbf{m}$; i.e. $\mathbf{m}$ behaves as $\{1^{m_\alpha}\} \otimes \{1^{m_\beta}\} \otimes \ldots$ with respect to the direct sum subgroup $U(N_\alpha) \oplus U(N_\beta) \oplus \ldots$ of $U(N)$. Thus in the spherical configurations space, the scalar operators are n_α's. Therefore it should be obvious that the NIP hamiltonian h is a scalar with respect to the spherical configuration group, $h = \sum \epsilon_\alpha n_\alpha = h^{[0]}$; $h^{[0]} = h^{[0,0,\ldots]}$ denotes that it is a scalar with respect to each spherical orbit. The $V^{[0]}$, the scalar part of V with respect to spherical configurations, is given by $V^{[0]} = \sum_{\alpha \geq \beta} V_{\alpha\beta} n_{\alpha\beta}$; $V_{\alpha\beta}$ are average two-particle matrix elements with the two particles in the spherical orbits α and β, respectively and $n_{\alpha\beta} = n_\alpha (n_\beta - \delta_{\alpha\beta})(1+\delta_{\alpha\beta})^{-1}$ As a one-body hamiltonian will be a scalar with respect to spherical configuration group (see Ref. [3] for exceptions), there cannot be an effective one-body part of V with respect to the spherical configuration group. Thus $\mathbf{V} = V - V^{[0]}$ should be like the $\nu = 2$ part $V^{\nu=2}$ introduced in the $U(N)$ case. Just as $V^{[0]}$ is scalar with respect to each of the $U(N_\alpha)$ groups (hence the notation $[0] = [0, 0, \ldots]$), $\mathbf{V}$ behaves as $V^{[2]} = V^{[\nu_\alpha, \nu_\beta, \ldots]}$; $\sum \nu_\alpha = 2$. The method for enumerating and constructing the various $V^{[\nu_\alpha, \nu_\beta, \ldots]}$; $\sum \nu_\alpha = 2$, is given in [3].

The unitary group $U(N_\alpha)$ acting in each unitary orbit α generates m_α of a unitary configuration $[\mathbf{m}]$; i.e. $[\mathbf{m}]$ behaves as $\{1^{m_\alpha}\} \otimes \{1^{m_\beta}\} \otimes \ldots$ with respect to the direct sum subgroup $U(N_\alpha) \oplus U(N_\beta) \oplus \ldots$. Thus in the unitary configurations space, the scalar operators are n_α's. One can carry out the tensor decomposition of H with respect to the group $U(N_\alpha) \oplus U(N_\beta) \oplus \ldots$. However, the only decomposition that is of relevance for the present paper is the decomposition of $H^{[0]} = h^{[0]} + V^{[0]}$. The decompositions are $h^{[0]} = h^{[0][0]} + h^{[0][1]}$, $V^{[0]} = V^{[0][0]} + V^{[0][1]} + V^{[0][2]}$ and $H^{[0]} = H^{[0][0]} + H^{[0][1]} + H^{[0][2]}$ where $H^{[0][0]} = h^{[0][0]} + V^{[0][0]}$, $H^{[0][1]} = h^{[0][1]} + V^{[0][1]}$ and $H^{[0][2]} = V^{[0][2]}$. It should be noted that: (i) $[\nu] = [\nu_\alpha, \nu_\beta, \ldots]$, $\sum \nu_\alpha = \nu$ and ν_α are integers; (ii) the part that behaves as $[\mathbf{0}] = [0, 0, \ldots]$ is a polynomial in m_α (first order for $h^{[0][0]}$ and second order for $V^{[0][0]}$); (iii) subtraction of $h^{[0][0]}$ from $h^{[0]}$ gives $h^{[0][1]}$; (iv) the $V^{[0][1]}$ part of a $V_{\alpha\beta}$ term will be of the form $(n_\beta - \delta_{\alpha\beta})F^1$ where F^1 is a one-body operator that behaves as h for the orbit α and its SPE depend on $\boldsymbol{\beta}$; (v) subtraction of $V^{[0][0]}$ and $V^{[0][1]}$ from $V^{[0]}$ gives $V^{[0][2]}$. Explicit forms for the various $[\nu]$ parts are given in [13,21]. The unitary configuration decomposition applies directly to the pn case with the conventions defined in Section 2.1.

Appendix B. Fermi gas forms and experimental data

Experimental data for level densities (also for spin-cutoff factors) are normally available in two forms: (i) numerical values at some select energies; (ii) values for the level density parameters assuming a form for the level densities. In the later case usually the NIP Fermi gas forms are used and this is always the case with the data available in literature for fp-shell nuclei. Before giving a brief description of empirical data, the relevant Fermi gas forms are given below.

B.1. Fermi gas forms

The explicit back shifted Fermi gas form for state density $I(E)$, in terms of the parameters a and Δ, is [7,64]

$$I(E) = \frac{\sqrt{\pi}}{12a^{1/4}(E-\Delta)^{5/4}} \exp\left(2\sqrt{a(E-\Delta)}\right). \tag{B.1}$$

The Lang LeCouteur (LLC) level density formula [65], which is commonly used in data analysis, is given by (it involves the temperature "t"),

$$I(E) = \frac{\sqrt{\pi}}{12a^{1/4}(E-\Delta+t)^{5/4}} \exp\left(2\sqrt{a(E-\Delta)}\right),$$

$$U = E - \Delta = at^2 - t; \tag{B.2}$$

$$t = \frac{1 \pm \sqrt{4a(E-\Delta)+1}}{2a} \simeq \sqrt{\frac{E-\Delta}{a}}. \tag{B.3}$$

The result in (B.3) that $t \simeq \sqrt{(E-\Delta)/a}$ also follows from (B.1) and the definition of nuclear temperature T, $T^{-1} = \partial \ln(I(E))/\partial E$. In order to analyze level density data it is necessary to decompose state densities into fixed-J level densities. Starting with fixed (E, M) bivariate densities $\rho(E,M)$ where M is a eigenvalue of the J_Z operator and recognizing that $\rho(E,M)$ can be written as $\rho(E,M) = \rho(E)\rho(M|E)$, then applying CLT for the conditional $\rho(M|E)$ density gives the J-decomposition of $I(E)$. In terms of the spin-cutoff factors $\sigma_J^2(E) = \langle J_Z^2 \rangle^E$, the formulas for fixed-J level densities $I_\ell(E,J)$, for the level densities $I_\ell(E, J_1 : J_2)$ in the spin window $J_1 - J_2$ ($I_\ell(E, J_1 : J_2)$ gives number of levels per MeV at energy E with spins between J_1 and J_2) and total level density $I_\ell(E)$ (which gives number of levels per MeV at energy E irrespective of the J value) are,

$$I_\ell(E,J) = \frac{1}{\sqrt{2\pi}\sigma_J(E)} \left[\exp\left\{-\frac{J^2}{2\sigma_J^2(E)}\right\} - \exp\left\{-\frac{(J+1)^2}{2\sigma_J^2(E)}\right\}\right] I(E)$$

$$\begin{aligned}
&\simeq -\frac{\partial}{\partial M} I(E,M)\Big|_{M=J+1/2} \\
&= \frac{2J+1}{\sqrt{8\pi\sigma_J^3(E)}} \exp\left\{-\frac{(J+1/2)^2}{2\sigma_J^2(E)}\right\} I(E) \\
&= C_J(E) I(E),
\end{aligned} \tag{B.4}$$

$$I_\ell(E,J) = (2J+1)\exp\left\{-\frac{(J+1/2)^2}{2\sigma_J^2}\right\} I_\ell(E,0);$$

$$I_\ell(E,0) = \frac{1}{\sqrt{8\pi\sigma_J^3}} I(E), \tag{B.5}$$

$$\begin{aligned}
I_\ell(E,J_1:J_2) &= \frac{1}{\sqrt{2\pi}\sigma_J(E)}\left[\exp\left\{-\frac{(J_1)^2}{2\sigma_J^2(E)}\right\} - \exp\left\{-\frac{(J_2+1)^2}{2\sigma_J^2(E)}\right\}\right] I(E) \\
&\simeq \left[\sum_{J=J_1}^{J_2} C_J(E)\right] I(E) \\
&= C_{J_1:J_2}(E) I(E),
\end{aligned} \tag{B.6}$$

$$\begin{aligned}
I_\ell(E) &= \sum_J I_\ell(E,J) = I_\ell(E,0:\infty) = C_{0:\infty} I(E) \\
&= \frac{1}{\sqrt{2\pi}\sigma_J(E)} I(E) \\
&= [\sqrt{2\pi}\sigma_J(E) C_{J_1:J_2}(E)]^{-1} I_\ell(E,J_1:J_2) = 2\sigma_J^2(E) I_\ell(E,0).
\end{aligned} \tag{B.7}$$

The spin-cutoff factors can be written in terms of a and Δ and rigid body moment of inertia $\mathcal{I}_{\rm rig}$ for a "A" nucleon system [66],

$$\sigma_J^2(E) = \frac{\chi \mathcal{I}_{\rm rig}}{\hbar^2}\sqrt{\frac{E-\Delta}{a}} \simeq \frac{\chi I_{\rm rig}}{\hbar^2} t. \tag{B.8}$$

With $\mathcal{I}_{\rm rig} = \frac{2}{5} m A R^2$ and $R = r_0 A^{1/3}$ fm ($r_0 \simeq 1.2$), a simpler formula is

$$\sigma_J^2(E) = 0.0096(\chi)(r_0^2) A^{5/3} t. \tag{B.9}$$

Unless otherwise specified, $\chi = 1$ in (B.8), (B.9) from now onwards. A simpler expression in terms of mass number A is [67]

$$\sigma_J^2(E) = 0.0888\sqrt{a(E-\Delta)}A^{2/3}. \tag{B.10}$$

The state density is sum of fixed parity (π) densities $I^\pi(E)$ and assuming equilibration (as commonly used in data analysis) of $I^\pi(E)$ densities gives,

$$\begin{aligned}
&I(E) = I^+(E) + I^-(E), \\
&I^+(E) \simeq I^-(E) \simeq \tfrac{1}{2} I(E).
\end{aligned} \tag{B.11}$$

Similar expressions for $I_\ell^\pi(E,J)$, $I_\ell^\pi(E,J_1:J_2)$ and $I_\ell^\pi(E)$ can be written down in terms of $I_\ell(E,J)$, $I_\ell(E,J_1:J_2)$ and $I_\ell(E)$ respectively. Given fixed parity spin-cutoff factors $\sigma_J^2(E;\pi)$, following (B.11) it is assumed that,

$$\sigma_J^2(E) = \tfrac{1}{2}\{\sigma_J^2(E;+) + \sigma_J^2(E;-)\};$$
$$\sigma_J^2(E;+) \simeq \sigma_J^2(E;-). \tag{B.12}$$

B.2. Experimental data

The experimental data from direct counting of levels gives level densities at few (2 or 3) low-lying energies (say $E \sim$ 2–4 MeV), neutron/proton resonance experiments give the level density at resonance energy (proton resonance experiments give densities at say 2 nearby energies), the charged particle spectra give values for level density parameters (a, Δ) say valid upto 20–25 MeV excitation and finally Ericson fluctuations measurements give level densities at a few (3 or 4) energies at high excitation energy (say around 20 MeV). The most recent exercise of selection and compilation of level density data from direct counting, proton/neutron resonances and Ericson fluctuations is due to Iljinov et al. [68] and the data taken from this compilation is referred to as data of "Iljinov et al.". The neutron resonance data is available for the nuclei (^{60}Co, ^{60}Ni, ^{62}Ni), the proton resonance data is available for the odd nuclei (^{55}Mn, ^{59}Co, ^{63,65}Cu) and data from Ericson fluctuations experiments is available for (^{55}Mn, ^{56}Fe, ^{60}Co, ^{60}Ni) nuclei. The charged particle experiments (for example ^{56}Fe$(\alpha,\alpha')^{56}$Fe, ^{63}Cu$(p,\alpha)^{60}$Ni etc.) are due to Lu et al. [69] and they are carried out for the nuclei ^{56}Fe, ^{59}Co, ^{60,62}Ni and ^{63,65}Cu; the authors deduced, using the LLC form for $I_\ell(E)$ (B.2)–(B.7), (B.9) one set of (a, Δ) values for the nuclei ^{56}Fe, ^{59}Co, ^{60}Ni, ^{63,65}Cu and two sets of (a, Δ) values for ^{62}Ni. The level densities and spin-cutoff factors calculated at different energies using the deduced (a, Δ) parameters and the smoothed NIP forms (B.2)–(B.7), (B.9) with $r_0 = 1.2$ are treated as experimental data. This data is called data of "Lu et al." (for ^{62}Ni there are data of "Lu et al. (a)" and "Lu et al. (b)" corresponding to the two sets of (a, Δ) parameters deduced for this nucleus). By fitting the low energy level density data and the data from neutron/proton resonance experiments, using the formulas (B.2), (B.3) and (B.4)–(B.7), (B.9) with $r_0 = 1.25$, Dilg et al. [70] deduced the values for the level density parameters (a, Δ) for about 200 nuclei. The values of Dilg et al. (a, Δ) are available for the six nuclei ^{55}Mn, ^{59,60}Co, ^{62}Ni, ^{63,65}Cu. Using (a, Δ) values that are deduced for $\sim$ 200 nuclei, Dilg et al. constructed smoothed formulas for (a, Δ) as a function of the mass number A. The smoothed formulas are

$$\begin{aligned} a(\mathrm{MeV}^{-1}) &= 2.40 + 0.067A, \\ \Delta(\mathrm{MeV}) &= P - 130/A, \\ P &= 25.6A^{-1/2} \text{ for even–even nuclei}, \\ &= 12.8A^{-1/2} \text{ for odd-mass nuclei}, \\ &= 29.4A^{-1} \text{ for odd–odd nuclei}. \end{aligned} \tag{B.13}$$

Using (B.13), the (a, Δ) values for ^{56}Fe and ^{60}Ni in the set of nuclei considered in Section 3 are calculated and they are called values of Dilg et al. (a, Δ) for these two nuclei. The level densities and spin-cutoff factors calculated at different energies using the parameters of Dilg et al. (a, Δ) and the smoothed NIP forms (B.2)-(B.7), (B.9) with $r_0 = 1.2$ are treated as experimental data. This data is called data of "Dilg et al." Katsanos et al. [71] fitted the Ericson fluctuations data (they measured the same) at certain energies (around 20 MeV excitation) and the data at low energies obtained by direct counting, using (B.2)-(B.7), (B.9) with $r_0 = 1.2$ and deduced (a, Δ) values. The level densities and spin-cutoff factors calculated at different energies using the deduced (a, Δ) parameters and the smoothed NIP forms (B.2)-(B.7), (B.9) with $r_0 = 1.2$ are treated as experimental data. This data is called data of "Katsanos et al.".

The measurement of spin-cutoff factors $\sigma_J^2(E)$, the quantity which characterizes the spin distribution of nuclear levels, cannot be carried out by a commonly available method. However information about $\sigma_J(E)$ basically come from three kinds of methods. They are: (i) isomeric ratio measurements [72]; (ii) the measurement of angular distribution of particles from charged particle reactions; (iii) using the analytical formulas (B.2)-(B.3), (B.8)-(B.10), (B.12) and the deduced values of a and Δ . For the eight fp-shell nuclei of interest to us, there is no $\sigma_J(E)$ data from isomeric ratio measurements. Lu et al. [73] carried out charged particle experiments for fp-shell nuclei and deduced $\sigma_J(E)$ values in an energy window E_1 to E_2 with σ_J between σ_1 and σ_2. As the energy changes from E_1 to E_2, the $\sigma_J(E)$ changes from σ_1 to σ_2 and it is assumed that at $E = E_1$, $E_1 + \delta$, $E_1 + 2\delta$ and E_2, $\sigma_J(E) = \sigma_1$, $\sigma_1 + \delta'$, $\sigma_1 + 2\delta'$ and σ_2 where $\delta = (E_2 - E_1)/3$ and $\delta' = (\sigma_2 - \sigma_1)/3$; this is similar to what was done in Ref. [39] for ^{56}Fe nucleus. The spin-cutoff values $\sigma_J(E)$ thus obtained at these four energies are referred to as "charged particle (c.p.)" data (though experiments give only spin-cutoff datum value). There are analytical formulas available for $\sigma_J(E)$ as given by (B.8)-(B.10), (B.3). Knowing (a, Δ) parameters and $\mathcal{I}_{\rm rig}$ or r_0 value, σ_J at different energies can be calculated. Using (B.9) and the a and Δ values given by Dilg et al. [70], Lu et al. [69] and Katsanos et al. [71], $\sigma_J(E)$ values are calculated and they are used in Fig. 3.

The compilation of data, from all the above sources, for the eight fp-shell nuclei $^{55}_{25}$Mn$_{30}$, $^{56}_{26}$Fe$_{30}$, $^{59}_{27}$Co$_{32}$, $^{60}_{27}$Co$_{33}$, $^{60}_{28}$Ni$_{32}$, $^{62}_{28}$Ni$_{34}$, $^{63}_{29}$Cu$_{34}$ and $^{65}_{29}$Cu$_{36}$ is given in [24]. In addition to data of Dilg et al., Lu et al., Lu et al. (a), Lu et al. (b), Katsanos et al. and charged-particle (referred as c.p. in Fig. 4), for some nuclei there is also resonance data coming from different sources [74-77] (see also [24]) than those given above. This data is referred to as "Resonance" in Fig. 2.

References

[1] T.A. Brody, J.F. Flores, J.B. French, P.A. Mello, A. Pandey and S.S.M. Wong, Rev. Mod. Phys. 53 (1981) 385.

[2] J.B. French, in: Nuclear Structure, eds. A. Hussain, Harun-ar-Rashid and M. Islam (North-Holland, Amsterdam, 1967) p. 85.

[3] F.S. Chang, J.B. French and T.H. Thio, Ann. Phys. (N.Y.) 66 (1971) 137.

[4] J.B. French, in: Dynamic Structure of Nuclear States, eds. D.J. Rowe, L.E.H. Trainor, S.S.M. Wong and T.W. Donnelly (University of Toronto Press, Toronto, 1972) p. 154.

[5] Moment Methods in Many Fermion Systems, eds. D.J. Dalton, S.M. Grimes, J.P. Vary and S.A. Williams (Plenum, New York, 1980).

[6] J.B. French and V.K.B. Kota, Annu. Rev. Nucl. Part. Sci. 32 (1982) 35.
[7] H.A. Bethe, Phys. Rev. 50 (1936) 332; Rev. Mod. Phys. 9 (1937) 69.
[8] E.P. Wigner, Ann. Math. 62 (1955) 548.
[9] K.K. Mon and J.B. French, Ann. Phys.(N.Y.) 95 (1975) 90.
[10] G.J.H. Labarge and R.U. Haq, Can. J. Phys. 68 (1990) 301.
[11] Y.D. Devi, R.U. Haq and V.K.B. Kota, Physical Research Laboratory Report No. PRL-TH-94/27 (1994).
[12] F.J. Dyson and M.L. Mehta, J. Math. Phys. 4 (1963) 701.
[13] J.B. French and V.K.B. Kota, University of Rochester Report No. UR-1116 (1989);
J.B. French, V.K.B. Kota and J.F. Smith III, University of Rochester Report No. UR-1122 (1989).
[14] V.K.B. Kota and K. Kar, Pramana-J. Phys. 32 (1989) 647.
[15] J.B. French, V.K.B. Kota, A. Pandey and S. Tomsovic, Ann. Phys. (N.Y.) 181 (1988) 198, 235.
[16] J.P. Draayer, J.B. French and S.S.M. Wong, Ann. Phys. (N.Y.) 106 (1977) 472, 503.
[17] J.C. Parikh, Group Symmetries in Nuclear Structure (Plenum, New York, 1978).
[18] S.S.M. Wong, Nuclear Statistical Spectroscopy (Oxford Univ. Press, Oxford, 1986).
[19] J.B. French and V.K.B. Kota, Phys. Rev. Lett. 51 (1983) 2183.
[20] J.B. French, R.U. Haq, V.K.B. Kota and J.F. Smith III, to be submitted for publication (1996).
[21] J.B. French, R.U. Haq and V.K.B. Kota, to be submitted for publication (1996).
[22] J.B. French, in: Mathematical and Computational Methods in Nuclear Physics, eds. J.S. Dehesa, J.M.G. Gomez and A. Polls (Springer, Berlin, 1984) p. 394.
[23] R.J. Leclair, M.Sc. Thesis, Laurentian University, Canada (1994).
[24] D. Majumdar, Ph.D. Thesis, M.S. University of Baroda (India) 1995.
[25] E.K. Warburton, J.A. Becker and B.A. Brown, Phys. Rev. C 41 (1990) 1147.
[26] A. Abzouzi, E. Caurier and A.P. Zuker, Phys. Rev. Lett. 66 (1991) 1134;
M. Dufour and A.P. Zuker, CRN Report No. 93-29, CNRS, Strassburg (1993).
[27] V.K.B. Kota, in: Capture Gamma-ray Spectroscopy and Related Topics, ed. R.W. Hoff, AIP Conf. Proc., Vol. 238 (AIP, New York, 1991) p. 685.
[28] J.B. French, A. Pandey and J.F. Smith III, in: Tests of Time Reversal Invariance in Neutron Physics, eds. N.R. Robertson, C.R. Gould and J.D. Bowman (World Scientific, Singapore, 1987) p. 80;
V.K.B. Kota, to appear in Recent Trends in Nuclear Physics, Proc. of the symposium held at M.S. University, Tirunelveli, India, 15–18 March 1993;
R.U. Haq, in: From Spectroscopy to Chaos, eds. D. Koltun and A. Das (World Scientific, Singapore, 1995) p. 89.
[29] J.F. Smith III, Ph.D Thesis, University of Rochester (1987).
[30] K.F. Ratcliff, Phys. Rev. C 3 (1971) 117.
[31] T.T.S. Kuo, Nucl. Phys. A 103 (1967) 71.
[32] B.M. Spicer and J.M. Eisenberg, Nucl. Phys. 63 (1965) 520.
[33] P.J. Brussaard and P.W.M. Glaudemans, Shell Model Application in Nuclear Spectroscopy (North-Holland, Amsterdan, 1977).
[34] Nuclear Data Sheets 44 (1985) 529.
[35] Table of Isotopes, VIIth Ed., eds. C.M. Lederer and V.S. Shirley (Wiley, New York, 1978).
[36] Nuclear Data Sheets 20 (1977) 286.
[37] Nuclear Data Sheets 39 (1983) 687.
[38] Nuclear Data Sheets 48 (1986) 284.
[39] A.N. Behkami and J.R. Huizenga, Nucl. Phys. A 217 (1973) 78.
[40] Nuclear Data Sheets 64 (1991) 834.
[41] A.V. Ignatyuk, in: The Statistical Properties of the Excited Atomic Nuclei (Energoatamitzdat, Moscow, 1983).
[42] S.K. Kataria and V.S. Ramamurthy, Nucl. Phys. A 349 (1980) 10;
S.K. Kataria, V.S. Ramamurthy and S.S. Kapoor, Phys. Rev. C 18 (1978) 549.
[43] J.R. Huizenga and L.G. Moretto, Annu. Rev. Nucl. Sci. 22 (1972) 427.
[44] A. D'Arrigo, G. Giardina and A. Taccone, Phys. Lett. B 262 (1991) 1.
[45] V.K.B. Kota, Europhys. Lett. 23 (1993) 481;
G. Maino, A. Mengoni and A. Ventura, Phys. Rev. C 42 (1990) 988;
A. Mengoni, G. Maino, A. Ventura and Y. Nakajima, in: Perspectives for the Interacting Boson Model, eds. R.F. Casten, A. Vitturi, A.B. Balantekin, B.R. Barrett, J.N. Ginocchio, G. Maino and T. Otsuka (1994) p. 421.

[46] C. Bloch, Phys. Rev. 9 (1954) 1094.
[47] I. Kanestrom, Nucl. Phys. 83 (1966) 380;
G. Hansen and A.S. Jensen, Nucl. Phys. A 406 (1983) 236.
[48] V. Paar, D.K. Sunko, S. Brant, M.G. Mustafa and R.G. Lanier, Z. Phys. A 345 (1993) 343.
[49] S. Ayik and J.N. Ginocchio, Nucl. Phys. A 234 (1974) 13.
[50] S.M. Grimes, S.D. Bloom, H.K. Vonach and R.F. Hausman Jr., Phys. Rev. C 27 (1983) 2893;
B. Strohmaier, S.M. Grimes and H. Satyanarayana, Phys. Rev. C 36 (1987) 1604;
B. Strohmaier and S.M. Grimes, Z. Phys. A 329 (1988) 431.
[51] R.U. Haq and S.S.M. Wong, Can. J. Phys. 60 (1982) 1502.
[52] K. Sato and S. Yoshida, Z. Phys. A 327 (1987) 421; A 330 (1988) 265; A 333 (1989) 141;
K. Sato, Y. Takahashi and S. Yoshida, Z. Phys. A 339 (1991) 129.
[53] B. Lauritzen, P. Arve and G.F. Bertsch, Phys. Rev. Lett. 61 (1988) 2835;
B. Lauritzen and G.F. Bertsch, Phys. Rev. C 39 (1989) 2412;
G. Puddu, P.F. Bortignon and R.A. Broglia, Ann. Phys. (N.Y.) 206 (1991) 409.
[54] Y. Alhassid and B.W. Bush, Nucl. Phys. A 565 (1993) 399;
R. Rossignoli and P. Ring, Ann. Phys. (N.Y.) 235 (1994) 350;
B.K. Agrawal and A. Ansari, Nucl. Phys. A 567 (1994) 1.
[55] S.E. Koonin, Nucl. Phys. A 574 (1994) 1c.
[56] C.W. Johnson, S.E. Koonin, G.H. Lang and W.E. Ormand, Phys. Rev. Lett. 69 (1992) 3157;
Y. Alhaasid, D.J. Dean, S.E. Koonin, G. Lang and W.E. Ormand, Phys. Rev. Lett. 72 (1994) 613;
D.J. Dean, S.E. Koonin, K. Langanke, P.B. Radha and Y. Alhassid, Phys. Rev. Lett. 74 (1995) 2909.
[57] G. Bonsignori, M. Savoia, K. Allaart, A. Van Egmond and G. Te Velde, Nucl. Phys. A 432 (1985) 389.
[58] Nuclear Data Sheets 60 (1990) 366.
[59] Nuclear Data Sheets 47 (1986) 178.
[60] T. Von Egidy, A.N. Behkami and H.H. Schimdt, Nucl. Phys. A 454 (1986) 109;
T. Von Egidy, H.H. Schimdt and A. N. Behkami, Nucl. Phys. A 481 (1988) 189;
A.N. Behkami and Z. Kargar, J. Phys. G 18 (1992) 1023.
[61] Z. Pluhar and H.A. Weidenmüller, Phys. Rev. C 38 (1988) 1046; Z. Phys. A 352 (1995) 257.
[62] V.K.B. Kota and D. Majumdar, Z. Phys. A 351 (1995) 377;
K. Kar, S. Sarkar and A. Ray, Ap. J. 434 (1994) 662.
[63] V. Potbhare and N. Tresseler, Nucl. Phys. A 530 (1991) 171.
[64] H. Vonach and M. Hille, Nucl. Phys. A 127 (1969) 289.
[65] J.M.B. Lang and K.J. LeCouteur, Proc. Phys. Soc. (London) A 67 (1954) 585.
[66] S.E. Woosley, in: Moment Methods in Many Fermion Systems eds. B.J. Dalton, S.M. Grimes, J.P. Vary and S.A. Williams (Plenum, New York, 1980) p. 61.
[67] A. Gilbert and A.G.W. Cameron, Can. J. Phys. 43 (1965) 1446.
[68] A.S. Iljinov, M.V. Mebel, N. Bianchi, E. De Sanctis, C. Guareldo, V. Lucherini, V. Muccifora, E. Polli, A.R. Reolon and P. Rossi, Nucl. Phys. A 543 (1992) 517.
[69] C.C. Lu, L.C. Vaz and J.R. Huizenga, Nucl. Phys. A 190 (1972) 229.
[70] W. Dilg, W. Schantl, H. Vonach and M. Uhl, Nucl. Phys. A 217 (1973) 269.
[71] A.A. Katsanos, R.W. Shaw Jr., R. Vandenbosch and D. Chamberlin, Phys. Rev. C 1 (1970) 594.
[72] R. Vandenbosch and J.R. Huizenga, Phys. Rev. 120 (1960) 1313.
[73] C.C. Lu, L.C. Vaz and J.R. Huizenga, Nucl. Phys. A 197 (1972) 321.
[74] J. Browne, H.W. Newson, E.G. Bilpuch and E. Mitchell, Nucl. Phys. A 153 (1970) 481.
[75] S.F. Mughabghab, M. Divadeenam and N.E. Holden, Neutron Cross Sections , Vol. 1, Part A ($Z = 1-60$) (Academic Press, New York, 1981).
[76] S.F. Mughabghab, Neutron Cross Sections, Vol. 1, Part B ($Z = 61 - 100$) (Academic Press, New York, 1984).
[77] G. Rohr, Level density data based on neutron resonances (Private communications to V.K.B. Kota, 1992);
G. Rohr, L. Maisano and R. Shelly, in: Specialists Meeting on Neutron Cross Sections of Fission Product Nuclei (Bologna, Italy, 1979) p. 197.

17 October 1996

PHYSICS LETTERS B

ELSEVIER

Physics Letters B 387 (1996) 227–232

Gamow–Teller strengths in f_p shell

S. Sarkar, K. Kar

Saha Institute of Nuclear Physics, Theory Group, Block-AF, Sector I, Bidhan Nagar, Calcutta 700 064, India

Received 19 January 1996; revised manuscript received 4 June 1996
Editor: W. Haxton

Abstract

We develop a simple parametrization based on the spectral distribution theory for the total Gamow–Teller (GT_+) strength in the f_p shell, which reproduces the sum of the observed $B(GT)$ of (n,p) experiments well and is useful for nuclear structure and astrophysical applications.

PACS: 23.40Hc; 25.45Kv; 97.60Bw
Keywords: Beta decays; Nuclear structure; Charge-exchange reactions; Supernovae

Recent studies of the Gamow–Teller (GT) strength in the GT_+ direction in the f_p shell are motivated primarily by two objectives. Firstly, experimental measurements of GT_+ strengths for quite a few nuclei in the f_p shell are now available and there is considerable advance made [1,2] in the shell model calculations involving the full f_p space. This may help in obtaining a better understanding for the long-standing problem of quenching of GT strength and in general the spin–isospin response of nuclei in the intermediate energy.

Secondly, the importance of weak interactions at the late stages of Si burning of pre-supernova (preSN) in the form of electron capture (EC) as well as beta decay is now well-established [3]. The rates of these interactions determine the electron-to-nucleon ratio, Y_e, of the system and during the early stages of gravitational collapse of type II supernova (SN) the electron capture on nuclei heavier than ^{56}Fe reduces this ratio further until the nuclei have neutron number greater than 40 and allowed captures get neutron-shell blocked [4,5]. As the hydrodynamic shock, launched at a later stage of the evolution and responsible for the eventual explosion, has its energy sensitively dependent on Y_e a careful evaluation of the weak interaction rates is essential for the SN problem. The nuclei involved in this preSN and SN stages are the f_p shell nuclei with $20 \leq Z \leq 40$ and $20 \leq N \leq 40$.

The allowed beta strength is dominated by the Gamow–Teller component, and the laboratory β^- decay or β^+ decay/electron capture on these nuclei can normally reach only the tail of the giant resonance due to energy constraints. On the other hand reaction studies through intermediate energy (p,n) and (n,p) reactions give information about the full resonance [6,7] for the $\beta^-(GT_-)$ and $\beta^+/EC(GT_+)$ directions respectively. The CHARGEX Collaboration at TRIUMF measured the total strengths (denoted as $\sum B(GT_+)$) up to excitation energy about 8–9 MeV in the daughter. Experimental strengths beyond 8–9 MeV are also available now. A recent (n,p) measurement [8] on $^{60,62,64}Ni$ isotopes give total strengths ($\sum B(GT_+)$) summed up to 30 MeV excitation energy in daughter nuclei. But though this energy range

PII S0370-2693(96)00966-5

is large, the estimates above 8 MeV are subjected to systematic uncertainties which may be quite large. Earlier in the absence of full f_p shell calculations the theoretical estimates commonly used were different truncated space shell model results and the Ikeda form [9] $S_{\beta^-} - S_{\beta^+} = 3(N-Z)$ where $S_{\beta\pm}$ are the GT $\beta^{\pm}$ total strength or sum rules. This relation is model independent in the nucleonic level but can only be used in the GT$_-$ direction for nuclei for which β_+ transition is completely blocked. The recent shell model studies in the full fp space have started probing how much of the quenching is due to truncation of space [10,1]. On the other hand there are efforts [11] to develop approximate forms for the GT sum rule strength which with very few parameters can give observed quenched strength for many nuclei in the fp shell. When these simple parametrized forms for the total GT$_+$ strength have good predictability they are very useful for astrophysical applications as well as for estimates in nuclear physics without making detailed time-consuming computations. The parameters are determined by fitting the forms to available observed total strengths of nuclei.

In this work we construct a form for the total GT$_+$ strength based on theoretical methods, known as the spectral distribution theory, which reproduces the averaged shell model values and relies on an asymptotic form for the density of energy levels in the large finite dimensional shell model space. We suggest improvements on a previous attempt [11] to develop empirical forms for the total GT$_+$ strength ($\sum B(\text{GT}_+)$ and show how the predictive power of these forms can be improved considerably.

The single particle estimate for the sum rule which assumes pure shell model configurations for the initial $|i\rangle$ and final $|f\rangle$ states and no neutron–proton correlations is given by,

$$\sum B(\text{GT}_+) = \sum_{r,s} \frac{n_s^h(\nu)\, n_r^p(\pi)\, |\langle s|\sigma\tau_+|r\rangle|^2}{(2j_r+1)(2j_s+1)} \tag{1}$$

where $n_r^p(\pi)$ is the number of proton particles in the orbit r of spin j_r of the parent ground state and $n_s^h(\nu)$ is the number of neutron holes in the daughter final orbit s of spin j_s, and the sum is over all the proton and neutron orbits. $\sigma\tau_+$ is the one-body Gamow–Teller operator.

In spectral distribution theory one observes that with large shell model spaces where the number of valence particles is not too small (≥ 4), the density of states as a function of energy goes asymptotically towards a Gaussian. The expectation value of an operator K in the state $|E\rangle$ of energy E can be expanded in terms of a polynomial in energy [12,13].

$$\begin{aligned} K(E) &= \langle E|K|E\rangle = \sum_i \langle i|K\delta(H-E)|i\rangle / d(m) \\ &\quad\times \sum_\nu \langle K P_\nu(H)\rangle^m P_\nu(E) \\ &= \langle K\rangle^m + \left\langle K\frac{(H-\epsilon)}{\sigma}\right\rangle^m \frac{(E-\epsilon)}{\sigma} + \ldots \\ &= \langle K\rangle^m + \zeta_{H-K}\sigma_K(E-\epsilon)/\sigma + \ldots \end{aligned} \tag{2}$$

Here $d(m)$ is the dimensionality of the space of m valence particles in N single particle states ($d(m) \stackrel{N}{=} C_m$), $\langle K\rangle^m$ stands for the average of the operator in the space ($\langle K\rangle^m = \sum\langle i|K|i\rangle/d(m)$), $P_\nu(x)$ are polynomials with the density of states as the weight function ($P_0(x) = 1$, $P_1(x) = (x-\epsilon)/\sigma, \ldots$). In the above expression ϵ and σ are the centroid and width of the Hamiltonian, σ_K is the width of the operator K in the m-particle space. One also observes that if under the transformation $H \to H + \alpha K$ the shape of the density of states $\rho(E)$ does not change for $\alpha \to 0$, then $K(E)$ is linear in energy, i.e., in the Eq. (2) only the first two terms survive [12]). In the final form of RHS in Eq. (2) ζ_{K-H} is the correlation coefficient of the operator K with H. If the correlation coefficient is very small the second term has small contribution compared to the first.

When one uses $K = O_{\text{GT}}^\dagger O_{\text{GT}}$ where $O_{\text{GT}} = \frac{1}{2}\sum_{i=1}^A \sigma(i)\tau_\pm(i)$ for the Gamow–Teller (GT$_\pm$) operator for the $\beta^\pm$ decay, it is found that its correlation coefficient with most (1 + 2)-body realistic Hamiltonians in the fp shell are indeed very small. Thus an expansion with only the first two terms in Eq. (2) is a good approximation to the shell model values [14]. But in the sd shell the correlation coefficient of K with (1 + 2)-body realistic Hamiltonians is large (about 0.5 or greater) and there the terms beyond the second in the expansion is needed [15]. This is one of the reasons for which the simple 1- or 2-parameter forms considered by Koonin and Langanke [11] fail in the sd shell.

In spectral distribution theory, the result in the m-particle space is called a scalar result. To make contact with real nuclei one should work in (m,T) spaces where T is the isospin of the states in the space considered or one works in (m_n, m_p) spaces where m_n/m_p stand for the valence neutron/proton number. These can also be extended to configuration-isospin $(\tilde{m},T)$ and configuration proton-neutron number $\tilde{m}^n,\tilde{m}^p)$ spaces. The notation $(\tilde{m}^n,\tilde{m}^p)$ stands for $m_1, m_2, m_3, \ldots, m_r$ (neutron, proton) particles in orbits numbered as $1,2,3,\ldots,r$.

We choose the scalar neutron-proton space (m_n, m_p) and write the expression for the GT sum rule strength as a function of the energy E of the initial state. For our purposes $E = E_{g.s.}$, the ground state energy of the target nucleus. So with $K = O_{GT_+}{}^{\dagger} O_{GT_+}$ Eq. (2) takes the form

$$K(E_{g.s.}) = a_1 \frac{m_p(N_n - m_n)}{N_p N_n} + b_1 \frac{m_n(N_n - m_n) m_p(N_p - m_p)}{N_n(N_n-1)N_p(N_p-1)} \cdot \frac{(E_{g.s.} - \epsilon)}{\sigma} \tag{3}$$

where a_1 is the average of the product of two one-body matrix elements of the GT operator and b_1 is the average of the product of these with the Hamiltonian two-body matrix element in the scalar neutron-proton space. N_n and N_p are the total number of protons and neutrons that can be accommodated in the shell. So for fp shell $N_n = N_p = 20$. It is seen that $(E_{g.s.} - \epsilon)/\sigma$ in spectral distribution theory lies mostly in the range -3.5 to -5.0 [13,14] for the nuclei in the mid-fp shell and for our purposes we assume it to be a constant. Then Eq. (3) can be written as

$$K(E_{g.s.}) = a m_p(20 - m_n) + b m_p(20 - m_p) m_n(20 - m_n) \tag{4}$$

or to put it in the form of Koonin-Langanke (KL) [11], who use Z_{val} and N_{val} instead of m_p and m_n one gets,

$$\frac{\sum B(GT_+)}{Z_{val}(20 - N_{val})} = a + b(20 - Z_{val}) N_{val} \tag{5}$$

If we put $b = 0$ then we get back the KL form exactly. Thus the spectral distributions give a theory which in its simplest form recovers the KL expression, and is also able to provide a correction term involving one extra parameter. We shall see that this correction term is indeed necessary in order to improve the predictive power throughout the fp shell.

Table 1
Total Gamow-Teller strength $\sum B(GT_+)$ extracted from (n,p) data for 12 nuclei along with the corresponding single particle estimate (Eq. (1)). Column (a) gives the $\sum B(GT_+)$ used in Ref. [11]. Both column (a) and (b) give $\sum B(GT_+)$ below 9 MeV except for the nuclei (^{51}V, ^{59}Co) and ^{48}Ti for which column (a) gives strengths summed up to 12 and 15 MeV respectively. Slightly different values for the $\sum B(GT_+)$ in column (b) than that in (a) for the 4 nuclei ^{51}V, ^{54,56}Fe and ^{58}Ni are taken from most recent references as quoted in column (b) (see text also).

No.	Nucleus	$\sum B(GT_+)$ (a)	$\sum B(GT_+)$ (b)	Single particle estimate, Eq. (1)
1	$^{45}_{21}Sc_{24}$	2.1	0.46 ± 0.05 [17]	33/14
2	$^{51}_{23}V_{28}$	1.48 ± 0.16	1.20 ± 0.16 [18]	36/7
3	$^{55}_{25}Mn_{30}$	1.72 ± 0.20	1.70 ± 0.20 [19]	60/7
4	$^{59}_{27}Co_{32}$	2.39 ± 0.25	1.90 ± 0.21 [18]	84/7
5	$^{48}_{22}Ti_{26}$	1.31 ± 0.20	0.91 ± 0.22 [20]	57/14
6	$^{54}_{26}Fe_{28}$	3.10 ± 0.60	3.30 ± 0.51 [21]	72/7
7	$^{56}_{26}Fe_{30}$	2.85 ± 0.30	2.90 ± 0.30 [19]	72/7
8	$^{58}_{28}Ni_{30}$	3.76 ± 0.40	3.80 ± 0.40 [19]	96/7
9	$^{60}_{28}Ni_{32}$	-	3.11 ± 0.32 [8]	96/7
10	$^{62}_{28}Ni_{34}$	-	2.53 ± 0.26 [8]	64/7
11	$^{64}_{28}Ni_{36}$	-	1.72 ± 0.19 [8]	32/7
12	$^{70}_{32}Ge_{38}$	0.84 ± 0.13	0.84 ± 0.13 [22]	16/3

In Table 1 we give $\sum B(GT_+)$ values for a set of fp shell nuclei along with the single particle estimates using Eq. (1). Column (a) of Table 1 is identical to the $\sum B(GT_+)$ values used by Koonin and Langanke [11] (given by their Table 1, column 3) except for the nuclei ^{51}V and ^{59}Co. For ^{51}V and ^{59}Co they give two values for each, but we quote only one value for each, the ones which were actually used for their fits. But we feel that going up to 12 and 15 MeV excitation energy for only three nuclei (^{51}V, ^{59}Co and ^{48}Ti respectively) is not proper. A better procedure should be adding up to the same excitation energy for all nuclei and so looking up the references of the experimental papers we construct $\sum B(GT_+)$ up to excitation of 8–9 MeV and display the values in column (b) of Table 1. For

Table 2
Description of fitting to Eq. (5) and KL1 form (i.e. with only the first term of Eq. (5)) with the set of $\sum B(GT_+)$ values (column (a)/(b) of Table 1) corresponding to various sets of nuclei constructed out of the 12 nuclei given in Table 1.

Description of fitted set of nuclei			Fitted parameter of		
			KL1 form	Eq. (5)	
Set No.	Description of the set	$\sum B(GT_+)$ values used	a (Error) $\times 10^{-2}$	a (Error) $\times 10^{-2}$	b (Error) $\times 10^{-4}$
I	8 nuclei of Table 1, column (a), ^{45}Sc excluded	(a)	4.14 (0.18)	6.75 (1.70)	−1.89 (1.22)
II	same 8 nuclei as in I	(b)	3.79 (0.17)	6.76 (1.63)	−2.13 (1.16)
III	all 12 nuclei of Table 1	(b)	3.88 (0.14)	1.84 (0.57)	1.54 (0.42)
IV	all odd–even nuclei	(b)	3.20 (0.19)	2.36 (0.67)	0.68 (0.53)
V	all even–even nuclei	(b)	4.64 (0.20)	2.51 (1.19)	1.48 (0.82)
VI	all even–even nuclei excluding ^{48}Ti & ^{70}Ge	(b)	4.95 (0.22)	3.48 (1.31)	1.00 (0.88)

example for ^{48}Ti the quoted $\sum B(GT_+)$ value in column (a) is 1.31 ± 0.20 which actually is the sum up to an excitation energy of 15 MeV whereas when we restrict the sum up to 9 MeV the value is 0.91 ± 0.22. For column (b) we add a systematic uncertainty of 10% of total $B(GT_+)$ along with the statistical uncertainty when only the later is quoted. $\sum B(GT_+)$ value for ^{45}Sc is misquoted in Ref. [11] as 2.1. But this value is 0.70 when strength is summed up to $\simeq 11$ MeV and is 0.46 ± 0.05 up to 9 MeV of daughter excitation. For ^{45}Sc we take only the systematic uncertainty as no statistical uncertainty is quoted.

We first do a parameter fit of Eq. (5) for the same set of 8 nuclei as used by Koonin and Langanke [11] using the same set of experimental values of $\sum B(GT_+)$ (column (a) of Table 1). The best fit values of the parameters "a" and "b" using Eq. (5) and also KL's one-parameter form (i.e. keeping only the first term of Eq. (5), denoted henceforth as KL1 form) are given in the first row of Table 2 and is called set I by us. The χ^2 per degree of freedom for the two fits are found to be 0.92 and 1.13 respectively. The KL two-parameter form (denoted throughout as KL2 form) is given as

$$\sum B(GT_+) = aZ_{val}(b - N_{val}) \tag{6}$$

The parameter "a" for the KL1 fit comes out as $a = (4.14 \pm 0.18) \times 10^{-2}$ whereas for KL2 the fit gives $a = (4.39 \pm 0.27) \times 10^{-2}$, $b = 19.60 \pm 0.19$ with a $\chi^2 = 1.05$. (Ref. [11] gives $a = 0.0429 \pm 0.0015$ with $\chi^2 = 1.1$ per degree of freedom for KL1 form. Also for KL2 form it gives $a = 0.0455 \pm 0.025$ and $b = 19.54 \pm 0.32$ with $\chi^2 = 1.0$. The differences of our results with the values given in Ref. [11] are due to the fact that Ref. [11] used rounded values for the quantity $B(GT_+)/Z_{val}$ in their fits [16]).

It is to be noted that Eq. (5) and the KL1 form are constrained to be within the fp space whereas in the KL2 form the number of single particle states itself is a parameter and can, in principle, involve orbits beyond the fp shell. That is why we do not include KL2 for the comparison in Table 2 but discuss it separately at a later stage.

Table 2 also gives best fit parameters of the spectral distribution form (Eq. (5)) and KL1 form using 8 nuclei from column (b) of Table 1 (set II, see Table 2) as well as using all 12 nuclei for which data is now available (set III) and then making a separation of odd Z-even N (o-e) (set IV) and even Z-even N (e-e) (sets V and VI). Fig. 1 gives the difference

$$\sum B(GT_+)^{\text{fitted}} - \sum B(GT_+)^{\text{exp.}}$$

plotted for the six sets of Table 2 along with the corresponding χ^2 for KL1 form and the spectral distribution form (Eq. (5)).

Because of the effect of pairing in the low lying region the strength distribution for the e-e and o-e nuclei are different and the systematics of e-e and o-e nuclei are not similar as observed by Koonin and

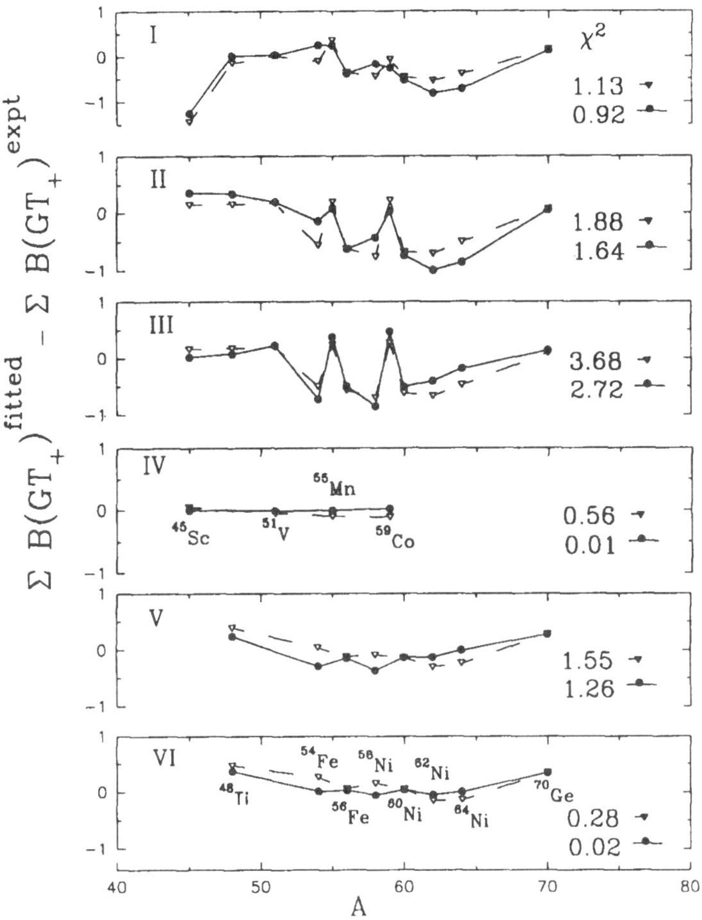

Fig. 1. The difference between the fitted and experimental $\sum B(\mathrm{GT}_+)$ is plotted against mass number A, showing the deviation of the predicted total strength using Eq. (5) (black dots) and KL1 form (open triangles) from the measured value for different nuclei. Also given are the χ^2 (per degree of freedom) values and the corresponding parameters are given in Table 2.

Langanke [11]. So to make further improvement of the predictive power of the spectral distribution form one has to treat these two types of nuclei separately. Doing that one sees that the χ^2 values come down to 1.26 and 1.55 for Eq. (5) and KL1 form with e-e nuclei whereas for o-e nuclei the χ^2 values are 0.01 and 0.56 respectively and it remains to be seen whether the fit stays so good when one encounters a larger set of nuclei in future. But if one reduces the set for fitting the e-e nuclei by excluding ^{48}Ti and ^{70}Ge (set VI of Table 2), the χ^2 values get reduced drastically and the prediction with this set for ^{48}Ti and ^{70}Ge also turn out to be good (see Fig. 1). So we suggest that for other e-e isotopes of Fe and Ni one should use Eq. (5) with set VI to predict the strength sums. But the drawback of this separation at this stage is the small sample sizes which increases the error in the parameters (the error in "b" has values 60% to 90% of the values of "b"). But we feel that the fitting with more data in future will reduce the errors and establish this separation of e-e and o-e nuclei on a firmer footing.

We now discuss briefly the results of fitting the KL2 form and compare it with fits to KL1.

KL2 form fits better than the KL1 form for both the e-e and o-e sets of nuclei. For the o-e set IV, KL2 form gives $\chi^2 = 0.21$ and the corresponding $a = (2.433 \pm$

0.710) $\times 10^{-2}$ and $b = 23.50 \pm 3.92$. For the set VI of six e-e nuclei the fit gives a $\chi^2 = 0.05$ and $a = (4.32 \pm 0.61) \times 10^{-2}$ and $b = 21.10 \pm 1.02$. This gives for the $\sum B(\mathrm{GT}_+;\, E_x \leq 9$ MeV) values as 1.30, 3.40, 2.88, 3.84, 3.14, 2.45, 1.76, 1.61 for the 8 e-e nuclei given in the same order as in Table 1. Thus the KL2 form reproduces the $\sum B(\mathrm{GT}_+)$ values very well within the fitting set but fails for the remaining two nuclei, Ti and Ge. One can note that the uncertainty associated with the two parameters of the KL2 form is also large similar to Eq. (5). It is seen in some cases that "b" is somewhat larger than 20 for KL2 form, indicating the participation of single particle states from the next higher orbits.

Finally for an overall comparison of the spectral distribution form of Eq. (5) with KL1 one should consider the largest sample of 12 nuclei (set III) which has the smallest errors. Then we find that the addition of the correction term involving the parameter "b" brings the χ^2 down substantially from the value of 3.68 to 2.72. The inclusion of this term becomes progressively important as one goes to more neutron-rich nuclei. The Gamow–Teller strength sum for such neutron-rich nuclei are considered important for electron captures both in the presupernova and supernova evolution.

To conclude, we have constructed here a phenomenological form for the $\beta^+/(n,p)$ Gamow–Teller strength sum in the fp shell which is:

(i) based on a formal theory of statistically averaged shell model results, and

(ii) include correction term to the form used by Koonin and Langanke.

We assert that this corrected form based on the spectral distribution theory should be consistently used for all nuclei through the fp shell, in particular the neutron-rich ones. Applications of the spectral distribution theory with configuration averaged forms is being looked into. Also the stellar weak interaction rates depend sensitively on the centroid of the Gamow–Teller resonances and an evaluation of the centroid using averaged strength distribution by the spectral distribution theory is being pursued. The evaluation of sum rule strengths for the excited states of the parent nuclei using spectral distribution theory is done by Kar et al. in Ref. [3] and needs follow-up work.

The authors would like to thank Alak Ray and F.K. Sutaria (TIFR, India) for useful discussions and valuable criticisms throughout this work, and Soumya Chakravarti (USA) for encouragement. One of the authors (S.S.) would like to thank the Theoretical Astrophysics Group of Tata Institute of Fundamental Research, Bombay, India, for their hospitality where part of this work was done. We also thank M. Saha Sarkar and A. Bhattacharya for critically reading the manuscript.

References

[1] E. Caurier, G. Martinez-Pinedo, A. Poves and A.P. Zuker, Phys. Rev. C 52 (1995) R1736, and references therein.
[2] K. Langanke, D.J. Dean, P.B. Radha, Y. Alhassid and S.E. Koonin, Phys. Rev. C 52 (1995) 718, and references therein.
[3] H.A. Bethe, G.E. Brown, J. Applegate and J.M. Lattimer, Nucl. Phys. A 324 (1979) 487;
H.A. Bethe, Rev. Mod. Phys. 62 (1990) 801;
K. Kar, A. Ray and S. Sarkar, Astrophys. J. 434 (1994) 662;
M.B. Aufderheide, I. Fushiki, G.M. Fuller and T.A. Weaver, Astrophys. J. 424 (1994) 257;
M.B. Aufderheide, I. Fushiki, S.E. Woosely and D.H. Hartmann, Astrophys. J. (Suppl.) 91 (1994) 389.
[4] G. Fuller, Astrophys. J. 252 (1982) 741.
[5] A. Ray, S.M. Chitre and K. Kar, Astrophys. J. 285 (1984) 766.
[6] D.J. Horen et al., Phys. Lett. B 99 (1981) 383.
[7] C. Gaarde, Nucl. Phys. A 396 (1983) 127.
[8] A.L. Williams et al., Phys. Rev. C 51 (1995) 1144.
[9] I. Ikeda, Prog. Theor. Phys. 31 (1964) 434.
[10] E. Caurier, A.P. Zuker, A. Poves and G. Martinez-Pinedo, Phys. Rev. C 50 (1994) 225.
[11] S.E. Koonin and K. Langanke, Phys. Lett. B 326 (1994) 5.
[12] J.P. Draayer, J.B. French and S.S.M. Wong, Ann. Phys. (N.Y.) 106 (1977) 472; 106 (1977) 503.
[13] J.B. French and V.K.B. Kota, Ann. Rev. Nucl. Part. Sci. 32 (1982) 35;
V.K.B. Kota and K. Kar, Pramana - J. Phys. 32 (1989) 647.
[14] S. Sarkar and K. Kar, J. Phys. G 14 (1988) L123.
[15] S. Sarkar and K. Kar, Phys. Rev. C 40 (1989) 1826.
[16] S.E. Koonin and K. Langanke, private communication.
[17] W.P. Alford et al., Nucl. Phys. A 531 (1991) 97.
[18] W.P. Alford et al., Phys. Rev. C 48 (1993) 2818.
[19] S. El-Kateb et al., Phys. Rev. C 49 (1994) 3128.
[20] W.P. Alford et al., Nucl. Phys. A 514 (1990) 49.
[21] M.C. Vetterli et al., Phys. Rev. C 40 (1989) 559;
T. Ronnqvist et al., Nucl. Phys. A 563 (1993) 225;
we take the average of the $\sum B(\mathrm{GT}_+;\, E_x)$ strengths given in these two references.
[22] M.C. Vetterli et al., Phys. Rev. C 45 (1992) 997.

PHYSICAL REVIEW C VOLUME 54, NUMBER 4 OCTOBER 1996

Strength functions and spreading widths of simple shell model configurations

Njema Frazier,[1,2] B. Alex Brown,[1,2] and Vladimir Zelevinsky[1,2,3]
[1]*National Superconducting Cyclotron Laboratory, Michigan State University, East Lansing, Michigan 48824-1321*
[2]*Department of Physics and Astronomy, Michigan State University, East Lansing, Michigan 48824-1116*
[3]*Budker Institute of Nuclear Physics, Novosibirsk 630090, Russia*

(Received 25 March 1996)

The exact solution of the many-body problem in the framework of the nuclear shell model with a realistic residual Hamiltonian makes it possible to study the fragmentation of simple configurations as a function of excitation energy and interaction strength. The analysis is performed for 839 states with quantum numbers $J^{\pi}T=0^{+}0$ in a system of 12 valence particles within the *sd* shell. Our statistics allow us to establish the generic shape of the strength function in the region of strong mixing. For the realistic interaction, the strength function is close to Gaussian in the central part and has exponential wings. The spreading width is larger than predicted by the standard golden rule. At the artificially suppressed interaction strength, we recover the Breit-Wigner shape and the golden rule for the spreading width. The transition between these regimes agrees with theoretical considerations based on the idea of chaotic dynamics. [S0556-2813(96)03609-6]

PACS number(s): 21.10.Ma, 21.10.Pc, 21.60.Cs

I. INTRODUCTION

The shell model with semiempirical residual interactions [1,2] is, to date, the most reliable approach to microscopic calculations of nuclear properties. It is especially successful in relatively light nuclei. With the matrix elements of the two-body interactions fitted by the known low-lying levels, it turns out to be possible to reproduce numerous observable quantities in the *sd*-shell nuclei. Supported by this success, we can move to the region of higher excitation energy and dense level spectrum in order to study the complicated nuclear states which are not accessible individually by experiment. Moreover, it is tempting to extrapolate our findings to other many-body quantum systems with strong interactions between constituents.

As excitation energy and level density increase, the stationary many-body states become exceedingly complicated superpositions of original "simple" shell model configurations. The only quantum numbers characterizing a given state are those of exact integrals of motion, angular momentum, parity, and isospin, $J^{\pi}T$, in our case. In such a situation we expect the new statistical regularities typical for quantum chaos to dominate the dynamics. As we have shown earlier [3–5], the signatures of quantum chaos in the level statistics [6–8] are seen already at weak residual interaction (0.2–0.3 of the actual strength). The nearest level spacing distribution $P(s)$ rapidly reaches the Wigner form. The spectral rigidity $\Delta(L)$ follows the predictions of the Gaussian Orthogonal Ensemble (GOE) up to $L\approx200$ where it upbends, probably due to the finite bandwidth of the original Hamiltonian matrix related to the selection rules of the two-body interaction. Similar studies have been carried out for heavy atoms [9] as well as for other modifications of the nuclear shell model [10–12].

Local level statistics give the simplest signatures of chaotic dynamics which are not sensitive enough to reveal the deviations from complete chaos in the structure of the many-body eigenfunctions. Using the representation-dependent criteria of information entropy [3,13], we quantified the complexity of the wave functions, discovering that the degree of complexity smoothly evolves along the spectrum, being in fact a weakly fluctuating function of excitation energy. Therefore, it is possible to relate information entropy calculated in the shell model basis to thermodynamic entropy and to the properties of the equilibrium thermal ensemble [4,5]. In an energy window which includes many levels but has an approximately constant level density, the generic wave functions "look the same" [14] which justifies the concept of statistical equilibrium.

One of the most important characteristics of the highly excited states is given by the strength function of simple modes. An external field, for example, of electromagnetic nature, acts by a simple one-body operator and excites, in the independent particle shell model, one-particle–one-hole states. In reality, such an excited state is a wave packet of many close stationary states. Each component carries a fraction of the strength of the original simple mode. An experiment with a resolution insufficient for the analysis of the dense fine structure spectrum displays a strength function as an envelope of the strength distribution. Using the language of time evolution, this is interpreted as damping of the simple mode [15], via its decay into complex stationary states. With reasonable statistical assumptions about the nearest level spacing distribution and the strength distribution among the invisible fine structure states, it is possible [16] to reconstruct their level densities and to recover the strength missing in the experiment with poor resolution.

Considering the background of fine structure states as a continuum, one would expect the exponential decay of the original excitation, and, therefore, a Breit-Wigner shape for the strength distribution as a function of the energy distance from the centroid. As known from quantum decay theory, the Breit-Wigner shape which has an infinite second moment cannot be exact [17]. However, it can be a good approximation, except maybe for the extreme wings of the distribution, as shown, for example, for the neutron resonances and giant resonances built on the ground state [18]. A microscopic mechanism for the coupling between the simple mode and its chaotic environment leading to the Breit-Wigner strength function is explained in detail in [18]. Under assumptions made in this derivation, the width of the distribution is given by the "golden rule" in terms of the mean coupling matrix elements and the background level density. We will call this theory a "standard model."

At some excitation energy, the spreading width saturates at a level determined by the original matrix elements for the coupling of the simple mode with the doorway states of the next degree of complexity. The saturation of the spreading width is known from the thorough systematics for the isobaric analog resonances [19]. A qualitative explanation of this behavior, using the concepts of chaotic wave functions, was given in [20,21]. Experimental data concerning the dipole giant resonance built on the compound states [22,23] also support the saturation of the spreading width.

The assumptions of the standard model break down when the width Γ_s obtained by the golden rule grows larger than the energy interval ΔE where the background level density and/or the coupling matrix elements can be considered as approximately constant. The existence of this limit of "strong coupling" was recognized long ago by Wigner [24] and discussed in the banded random matrix models [25]. The deviations from the standard model are responsible [26,27] for the narrow width of double giant resonances [28]. The formulation of the general approach which contains the standard model and the strong coupling as particular limiting cases was presented in [29].

The goal of the present paper is to study the strength function and the spreading width of simple nuclear configurations in the framework of the realistic shell model. Here we are able to check in detail various statistical hypotheses and to trace the transition between the weak coupling (standard model) and strong coupling situations. At this stage we consider the present work as a numerical experiment testing current theoretical ideas in realistic conditions. We start with a brief review of the standard model and its generalization in Sec. II. The results for the strength function obtained in our shell model are shown in Sec. III. At the realistic interaction strength, the generic shape of the strength function is close to the Gaussian in the central part. Up to high accuracy, the wings are exponential. The spreading width exceeds considerably the golden rule value which can be found with the help of the procedure of excluding a single basis state and considering its coupling to the rest of the system (Sec. IV). The strength function evolves as the interaction is artificially suppressed. In the weak coupling limit, we return to the domain of validity of the standard model and recover the Breit-Wigner shape. Accordingly, the spreading width dependence on the interaction strength changes from linear to quadratic as one proceeds from strong to weak coupling. We conclude with summarizing our results.

II. STRENGTH FUNCTION

A. Definitions

We consider a quantal system governed by the Hamiltonian H and considered in a truncated space spanned by the finite set of the basis states, $|k\rangle$. In this basis, which can be thought of as a traditional basis of the independent particle shell model, the Hamiltonian

$$H=H_0+H' \tag{1}$$

contains the diagonal configuration energies (which includes both the single-particle and diagonal two-body contributions), given by H_0, and the off-diagonal residual interaction H'. In the actual nuclear diagonalization, the integrals of motion such as total angular momentum (J), parity (π), and isospin (T) are exactly preserved by the projection of the simple shell model Slater determinants (m scheme; see below) so that all states under consideration have the same exact quantum numbers $J^\pi T$.

It is convenient to include the diagonal part of the residual interaction into the unperturbed Hamiltonian. Its eigenfunctions satisfy

$$H_0|k\rangle=\overline{E_k}|k\rangle, \tag{2}$$

where $\overline{E_k}=H_{kk}$. The strict degeneracy of the pure shell model configurations corresponding to a single partition is thereby removed.

By the diagonalization of the full Hamiltonian matrix including the off-diagonal matrix elements H'_{kl}, we obtain the eigenstates $|\alpha\rangle$ and their energies E_α,

$$H|\alpha\rangle=E_\alpha|\alpha\rangle. \tag{3}$$

The eigenstates are the complicated superpositions

$$|\alpha\rangle=\sum_k C_k^\alpha|k\rangle \tag{4}$$

of the basis states. The fragmentation of basis states, $|k\rangle=\Sigma_\alpha C_k^\alpha|\alpha\rangle$, is described by the same transformation coefficients C_k^α which can be taken as real in our case of time reversal invariance. This fragmentation is the object of our studies.

The average characteristics of the fragmentation can be expressed directly in terms of the matrix elements of the Hamiltonian (1). For a given basis state $|k\rangle$, the centroid of the strength distribution coincides with the unperturbed energy (2),

$$\sum_\alpha (C_k^\alpha)^2E_\alpha=\overline{E_k}. \tag{5}$$

The second moment of the strength distribution is determined by the sum of all off-diagonal matrix elements (starting at a given basis state) squared,

$$\sigma_k^2\equiv\sum_\alpha (C_k^\alpha)^2(E_\alpha-\overline{E_k})^2=\sum_{l\neq k}(H'_{kl})^2. \tag{6}$$

Closely related to, but different from this, is the effective local bandwidth ω_k,

$$\omega_k^2=\frac{1}{\sigma_k^2}\sum_{l\neq k}(\overline{E_k}-\overline{E_l})^2(H'_{kl})^2. \tag{7}$$

In contrast to the GOE, the actual shell model Hamiltonian matrix does not couple a state with all others; the two-body interaction leads to the specific selection rules which allow the matrix elements only between configurations which differ in orbits of not more than two particles. This determines the effective bandwidth (7) and brings the matrix closer to those described by the banded random matrix ensembles (BRME's) [25]. However, our many-body matrix elements are determined by a much smaller number of independent two-body matrix elements and therefore cannot be considered as uncorrelated which is the case in the BRME's. The ensembles based on two-body random interactions in a many-body systems were discussed in [6,30].

All the moments of the strength distribution can be found from the strength function

$$F_k(E)=\sum_{\alpha}(C_k^{\alpha})^2\delta(E-E_{\alpha}). \tag{8}$$

As compared to the full density of states

$$\rho(E)=\sum_{\alpha}\delta(E-E_{\alpha}), \tag{9}$$

the strength function (8) is frequently called the "local density of states." It determines the contribution of the basis state $|k\rangle$ to $\rho(E)$ at $E=E_{\alpha}$,

$$F_k(E)=\rho(E)\langle (C_k^{\alpha})^2\rangle_{E_{\alpha}=E}. \tag{10}$$

Equation (10) assumes that the strength function is approximated by the histogram where we sum in Eq. (8) over the eigenstates within a narrow energy bin which contains, due to the high level density, many states of close degree of complexity.

Our functions are normalized according to

$$\int dE\rho(E)=N, \quad \int dEF_k(E)=1. \tag{11}$$

Here N is the total dimension of Hilbert space.

B. Standard model

In the standard derivation of the strength function [18], one singles out a state $|k\rangle$ which is initially removed from the Hamiltonian matrix. The diagonalization of the remaining $(N-1)\times(N-1)$ matrix gives the intermediate "background" states $|\nu\rangle$, their eigenvalues e_{ν}, and wave functions $|\nu\rangle=\Sigma_{k'\neq k}\langle k'|\nu\rangle|k'\rangle$. The full matrix expressed in the basis $(|k\rangle,\{|\nu\rangle\})$ has E_k and intermediate energies e_{ν} on the main diagonal and off-diagonal matrix elements

$$V_{k\nu}=V_{\nu k}=\sum_{k'\neq k}H'_{kk'}\langle k'|\nu\rangle \tag{12}$$

due to the coupling between the single state $|k\rangle$ and the background. The advantage of this approach is that the omission of a single state cannot change significantly the statistical properties of the dense spectrum. We can expect that the level density of the background is the same as in the exact solution including all N states.

The problem of the interaction of the single state with the background can be easily solved. The eigenstates $|\alpha\rangle$ are the combinations

$$|\alpha\rangle=C_k^{\alpha}|k\rangle+\sum_{\nu}B_{\nu}^{\alpha}|\nu\rangle, \tag{13}$$

where the amplitudes C_k^{α} coincide with those in Eq. (4). Eliminating the coefficients B_{ν}^{α} with the use of the Schrödinger equation and normalizing the wave function, we obtain

$$(C_k^{\alpha})^2=\left[1+\sum_{\nu}\frac{V_{k\nu}^2}{(E_{\alpha}-e_{\nu})^2}\right]^{-1}. \tag{14}$$

The exact eigenvalues E_{α} are the poles of the Green function $G(E)=(E-H)^{-1}$ or the roots of the secular equation

$$E_{\alpha}-\overline{E}_k-\sum_{\nu}\frac{V_{k\nu}^2}{E_{\alpha}-e_{\nu}}=0, \tag{15}$$

whereas the weights (14) are the corresponding residues of $G(E)$. The roots (15) do not depend on the original choice of the excluded state $|k\rangle$. These equations do not contain any approximation.

The results of the standard model are based on additional assumptions.

(i) The background spectrum is dense and rigid, and so one can consider instead an equidistant sequence of levels with the mean spacing D.

(ii) The coupling intensities $V_{k\nu}^2$ are uncorrelated with the energies e_{ν} of the background states and weakly fluctuate around the mean value $\langle V^2\rangle$.

(iii) The mixing is sufficiently strong, $\langle V^2\rangle/D^2>1$.

Under these assumptions, it is easy to calculate the sums over the intermediate states ν and to obtain the strength function (10) of Breit-Wigner shape,

$$F_k(E)=\frac{1}{2\pi}\frac{\Gamma_s}{(E-E_k)^2+\Gamma_s^2/4}, \tag{16}$$

with the spreading width Γ_s given by the golden rule

$$\Gamma_s=2\pi\frac{\langle V^2\rangle}{D}. \tag{17}$$

Expressions (16) and (17) are sometimes taken for granted although they are valid under above-mentioned assumptions only. We note also that we use here the term "golden rule" for the spreading width (17) with no relation to perturbation theory. This result is valid in the domain of validity of the standard model.

C. General description and strong coupling limit

As discussed in [5], the assumptions of the standard model are correct as long as the resulting standard spreading width is relatively small. Γ_s should be compared with the energy range ΔE within which the level density and the coupling matrix elements can be considered as approximately constant. In the framework of the shell model, the two-body interactions are capable of admixing the close configurations of gradually increasing level of complexity. They serve as the doorway states for the further process of stochastization. The doorway states have their own spreading width which determines the energy range for effective mixing of the original state. The finiteness of this energy interval plays no role if $\Gamma_s < \Delta E$. The formal limit of $\Delta E \to \infty$ corresponds to the standard model.

As the coupling strength and the spreading width increase, the finite size of the doorway strength interval becomes important. One can estimate what happens in this limit assuming that the coupling $\overline{V_{k\nu}^2}$ suddenly disappears at some finite distance $\Delta E = e_\nu - \overline{E_k}$ from the centroid $\overline{E_k}$. A mean value of this interval can be inferred from the "bandwidth" ω_k, Eq. (7). The strength distribution is then determined by the outer roots of the secular equation (15) which are located well outside of the bandwidth at the distance

$$(E_\alpha - \overline{E_k})^2 = \sum_\nu V_{k\nu}^2 = \sigma_k^2, \tag{18}$$

where the last equality follows from the definition (6), expression (12) for the transformed coupling matrix elements, and the completeness of the set of the intermediate states $|\nu\rangle$ in space of dimension $(N-1)$.

Thus, the result (18) allows one to give a simple estimate

$$\Gamma_k \approx 2\sigma_k \tag{19}$$

for the spreading width in the limit of complete mixing. Written as Eq. (19), the expression for the spreading width does not refer to the specific form of decrease of the coupling matrix elements $V_{k\nu}$ as one moves away from the centroid $\overline{E_k}$. Using the idea of similarity of complicated states in a given energy region [29,5], we expect small variations of the width for various states $|k\rangle$ in this limit. Indeed, the quantity σ_k found for the sd-shell model is roughly constant for nearly all states.

The new and important property of the strong coupling limit (19) is that the quadratic dependence of the spreading width on the interaction strength is replaced by the linear dependence. Such a prediction was first made on qualitative grounds [26] in relation to the problem of the damping width of the double giant resonances. In the harmonic approximation for the giant mode, the matrix elements $V_{k\nu}$ of the coupling between the collective n-phonon state $|k\rangle$ and the compound states $|\nu\rangle$ scale as $\sqrt{n}$. Therefore, in agreement with data [28], the ratio $\Gamma(n)/\Gamma(1)$ of the widths of the multiple and single excitations should increase $\propto \sqrt{n}$ in the strong coupling limit (19) rather than $\propto n$ as predicted by the golden rule (17) with the quadratic dependence on the coupling matrix elements. The transition is nicely reproduced in numerical simulations [27].

To understand the underlying mechanism leading to the linear dependence of the spreading width, one can refer to the N scaling of the matrix elements [31,32] occurring as one proceeds to the mixed states at high excitation energy and high density. At a stage of the mixing process when a complicated state (4) contains N significant simple basis components $|k\rangle$, the matrix elements V between this state and one of simple states are reduced in average by factor $N^{-1/2}$ compared to typical matrix elements v between original simple states. The coupling intensity $\langle V^2 \rangle$ at this degree of complexity can be estimated as v^2/N. The quantity ND in the denominator of Eq. (17) gives the order of magnitude of the energy range covered by the fragmented strength of the original state. Thus, it is proportional to the spreading width, and the self-consistent consideration of Eq. (17) gives $\Gamma \propto v$, i.e., the linear dependence on the interaction strength. Beyond this point, the regime of strong coupling takes place where the saturation of the spreading width as a function of excitation energy is expected.

The detailed behavior of the spreading width as a function of the interaction strength depends on the explicit dependence of the level density and the coupling matrix elements on the energy of the background states. The general theory which incorporates this behavior as input and then predicts the strength function and the value of the spreading width was developed in [29]; see also [5]. When going from weak to strong coupling, the strength function evolves from the Breit-Wigner form, Eq. (16), to the nonuniversal shape with a finite second moment. The spreading width [defined as the full width at half maximum (FWHM)] undergoes a transition from the quadratic dependence of the standard golden rule to the linear dependence, similar to our estimate (19). However, the form of theory suggested in [29] assumes the uniformity of the statistical properties of all complicated states. This assumption is not completely valid in our case because the wide strength functions at the realistic interaction stength cover the regions with quite different degrees of complexity.

Since the specific features of the results for the shell model residual interaction should be shown below, at this point we limit ourselves with the simple interpolation formula for the spreading width as a function of the strength of the residual interaction λ,

$$\Gamma = \frac{\gamma\lambda^2}{1 + y\lambda}. \tag{20}$$

Foreseeing that in the region of chaotic dynamics one can define a generic strength function of the basic states, we omit here the basis state label k. Of course, there can be fluctuations of the spreading width from one basis state to another one. The case of the collective state with a nongeneric strength function was studied in [29]. The scale λ is defined as a common factor in front of the residual off-diagonal interaction H' in (1); $\lambda = 1$ corresponds to the realistic strength.

The parameters γ and y of Eq. (20) are related to the weak and strong coupling limits. For the weak interaction case, the comparison of Eqs. (17) and (20) determines

$$\gamma = \frac{2\pi}{D_0} \frac{\langle V^2(\lambda) \rangle}{\lambda^2}, \tag{21}$$

where D_0 is the mean level spacing for the unperturbed system. The strong coupling limit determines, according to Eq. (19),

$$1 + y = \frac{\gamma}{2\overline{\sigma}} \approx \frac{\pi \langle V^2 \rangle}{D_0 \overline{\sigma}}, \tag{22}$$

where $\langle V^2 \rangle$ and the mean value $\overline{\sigma}$ of the quantity (6) are taken at $\lambda = 1$ under the assumption that the actual value of the interaction strength belongs to the chaotic regime.

III. STRENGTH FUNCTIONS OF SHELL MODEL STATES

A. Model

The modern nuclear shell model provides an exact solution of the quantum many-body problem within a truncated space of several nucleon configurations. The space dimensions of the order of 10^3 are easy to handle. On the other hand, they are sufficiently large to make the results statistically reliable. We use the Wildenthal Hamiltonian [1] which defines the single-particle energies and the interaction between the valence particles by fitting more than 400 binding energies and excitation energies for the *sd*-shell nuclei. In this paper we carry out calculations exclusively for the $J^{\pi}T=0^{+}0$ states of 12 particles above the inert core of ^{16}O. The system can be considered as a model for the ^{28}Si nucleus in the approximation when the interaction across the major shells is neglected. Examples of other shell model applications for similar purposes can be found in [5].

The $0^{+}0$ class contains $N=839$ states. They are partitioned into shell model configurations according to the occupation numbers of the spherical orbits $0d_{5/2}, 0d_{3/2}$, and $1s_{1/2}$. The calculations are carried out with the OXBASH code [2]. The original configurations are described in the *m* scheme. This requires, as a prerequisite for the diagonalization, the construction, with the help of the projection operators, of the proper linear combinations of the *m* states within a partition which have the desired values $J^{\pi}T$. These superpositions are the basis states $|k\rangle$ of Eq. (2). As discussed in [5], the projection procedure already creates fairly complex nondeterminantal states which still have very close bare energies $\overline{E_k}$.

There are 63 independent two-body interaction matrix elements which define all many-body matrix elements. The latter therefore cannot be considered uncorrelated as in the GOE or BRME's. The global properties of the mixing interaction are seen [5] by inspection of the matrix H' prior to the actual diagonalization. The second moment σ_k, Eq. (6), and the effective bandwidth ω_k, Eq. (7), both stay essentially constant for the majority of the states $|k\rangle$. Their fluctuations are presumably of statistical character. The mean values of these quantities for the $0^{+}0$ states are $\overline{\sigma}\approx 10$ MeV and $\overline{\omega}\approx 8$ MeV. The constancy of σ_k is one of the manifestations of the N scaling [31,32]: As the mixing proceeds involving the increasing number N of fine structure states, each coupling matrix element diminishes $\propto 1/\sqrt{N}$, keeping the sum (6) constant. The same phenomenon is responsible for the saturation of the damping width of giant resonances [22,23] and isobaric analog resonances [20,21].

The direct diagonalization of the Hamiltonian matrix results in the energies E_α of the eigenstates $|\alpha\rangle$ and their wave functions expressed as the superpositions (4) of the basis states. The level density (9) has a Gaussian shape with the variance which is equal to $\sigma_E^0=8$ MeV without the off-diagonal interaction ($\lambda=0$). The variance broadens up to $\sigma_E=13$ MeV at $\lambda=1$. The mean level spacings near the middle of the spectrum are $D_0=24$ keV and $D=40$ keV for $\lambda=0$ and $\lambda=1$, respectively. It is known [6,33] that the Gaussian shape occurs mainly due to combinatorial reasons for the two-body interaction in the many-body system, irrespective of the random or deterministic (as in our case) nature of the two-body matrix elements. In contrast, the Wigner semicircle level density is obtained for the uncorrelated many-body matrix elements in a full or banded (but sufficiently wide) matrix.

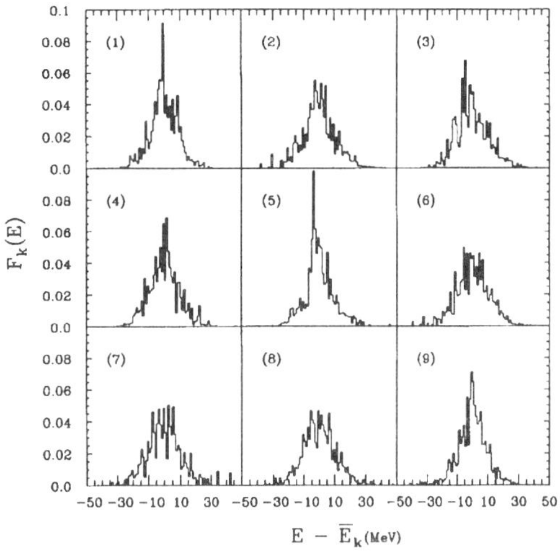

FIG. 1. The strength functions for nine individual $0^{+}0$ basis states $|k\rangle$ in the middle of the spectrum (histograms) vs the energy distance $E-\overline{E_k}$ from the centroid of the unperturbed state $|k\rangle$, panels 1–9. The bin size is 1 MeV.

We refer the reader to the review paper [5] for a detailed study of the chaotic properties of the $0^{+}0$ states. The local level statistics (the nearest level spacing distribution and the spectral rigidity) display an onset of chaos already for the weak interaction, $\lambda\approx 0.2$. The complexity of the eigenstates and related localization length measured in the original basis continues to regularly evolve beyond this point. This manifests the greater sensitivity of the wave functions to the deviations from the chaotic limit. Below we concentrate on the structure of the individual eigenstates which will allow us to extract the generic strength function.

B. Shape of the strength function

One of the main conclusions of the analysis given in [5] was that of the similarity of individual wave functions in the chaotic region, in accordance with Percival's conjecture [14]. The degree of complexity of the eigenstates, measured by information entropy and/or by the moments of the distribution function of the amplitudes C_k^α was found to saturate in the central part of the spectrum. The average properties of the eigenstates with approximately the same degree of complexity can be related to thermodynamical entropy and temperature of the equilibrium thermal ensemble [4]. These properties can be extracted by averaging out the fluctuations.

We expect that the closely related problem of the fragmentation of the basis states over the eigenstates can be solved similarly. Figure 1 shows the "empirical" strength functions $\underline{F_k}(E)$, Eq. (10), for nine basis states $|k\rangle$ with the centroids $\overline{E_k}$ located in the middle of the spectrum. The histograms obtained with the bin size of 1 MeV are plotted as a function of the energy distance from the corresponding centroid. Taking the basis states in this high-density mid-energy

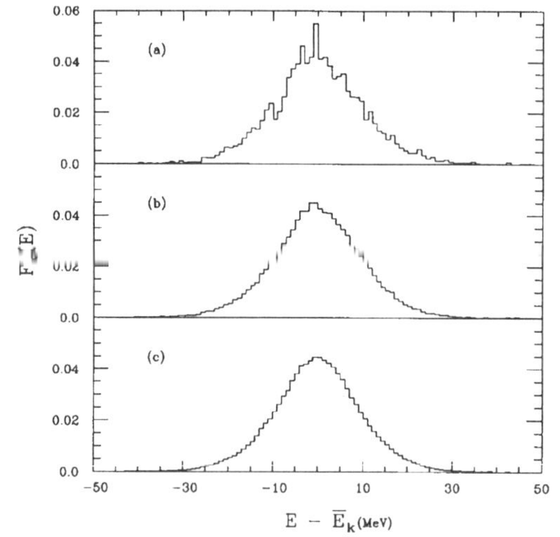

FIG. 2. The strength function averaged over 10, 100, and 400 0^+0 states in the middle of the spectrum, panels (a)–(c), respectively. The bin size is 1 MeV.

region as members of a statistical ensemble, we superimpose their strength functions in order to reduce the statistical fluctuations and produce a smooth, ''generic'' strength function. In Fig. 2 this averaging, with the same bin size, is performed over 10, 100, and 400 basis states, parts (a), (b), and (c), respectively.

The resulting strength function is already very smooth at the step (b). The wings of the curve are decreasing much faster than expected for the Breit-Wigner distribution (16). For the detailed fit we use the histograms with the finer bin size of 100 keV. Figure 3 demonstrates that the Breit-Wigner

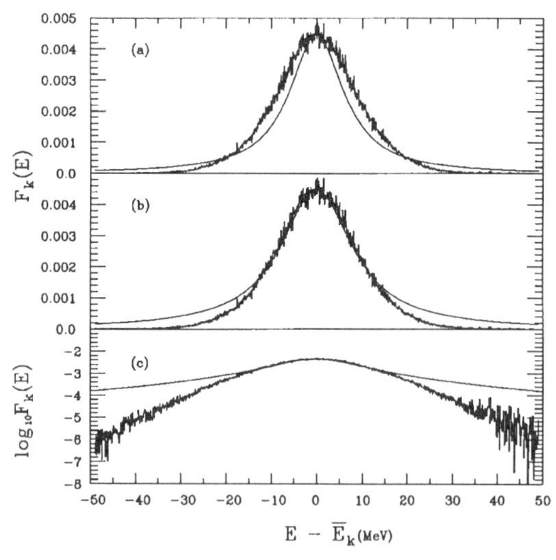

FIG. 3. The overall Breit-Wigner fit (solid lines) to the strength function of Fig. 2(c) (histograms), panel (a), to the central part of the strength function of Fig. 2(c) (histograms), panel (b), and the same fit on the logarithmic scale, panel (c). The bin size is 100 keV.

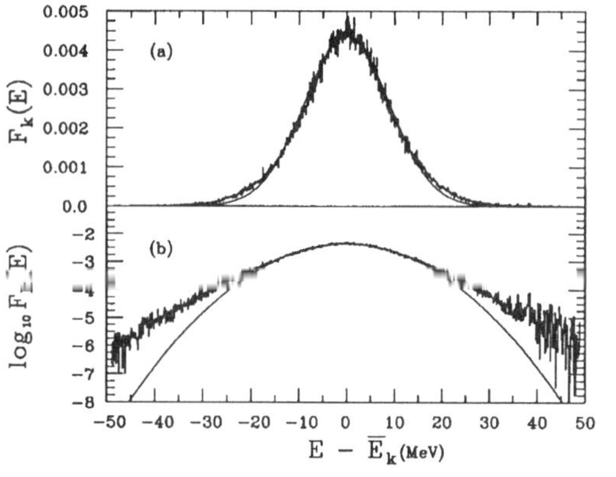

FIG. 4. The Gaussian fit (solid lines) to the strength function of Fig. 2(c) (histograms), panel (a), and the same fit on the logarithmic scale, panel (b). The bin size is 100 keV.

curve does not fit particularly well in any region [Fig. 3(a)]. Our attempt, Fig. 3(b), to fit only the central region results in an extremely poor fit to the tails, Fig. 3(c). Together with the width exceeding the quantity $\overline{\omega}$, this means that the strong coupling case is closer to reality for the states in the middle of the spectrum at the actual interaction strength.

The overall Gaussian fit, Fig. 4(a), is better. With the variance obtained from the Gaussian fit, $\sigma_F = 8.9 \pm 0.7$ MeV, the typical spreading width (FWHM) of the central 0^+0 states equals $\Gamma = (8\ln 2)^{1/2}\sigma_F = 21$ MeV. This agrees with the estimate (19), $\Gamma = 2\overline{\sigma} = 20$ MeV, made for the strong coupling limit. We again observe deviations from the Gaussian shape in the tails of the strength function, Fig. 4(b).

Because of the breakdown of both the Gaussian and the Breit-Wigner fits for the tails of the strength function, it becomes necessary to examine both the central region and the tails of the distribution more closely. According to Eq. (10), the strength function histograms actually involve two factors, the average value $\langle (C_k^\alpha)^2 \rangle$ of the eigenfunction components corresponding to the basis state $|k\rangle$ and the level density $\rho(E)$. The level density itself is described [5] by the Gaussian curve of Fig. 5. This is typical [6] for many-body systems with two-body interactions in the finite Hilbert space. The level density effects dominate the central part of the strength function. Eliminating the level density from the strength function, we come to the ''pure'' weight function $\langle (C_k^\alpha)^2 \rangle$ of Fig. 6.

We see that the Breit-Wigner shape that fits poorly to the strength function [Fig. 3(a)] is noticeably improved in the central region for the pure weight function of Fig. 6(a). The tail of the weight distribution on a logarithmic scale shows a linear behavior, Fig. 6(b). It is clear from Fig. 7, especially from the logarithmic plot of Fig. 7(b), that such an exponential fit does, in fact, represent the tails of the strength distribution. The final form of this fit is

$$F_k(E-\overline{E_k}) \approx F_0 \exp\left(-\frac{E-\overline{E_k}}{E_l}\right), \tag{23}$$

where the energy localization length is $E_l = 5$ MeV. Of course, the expression (23) is valid only for the tails of the

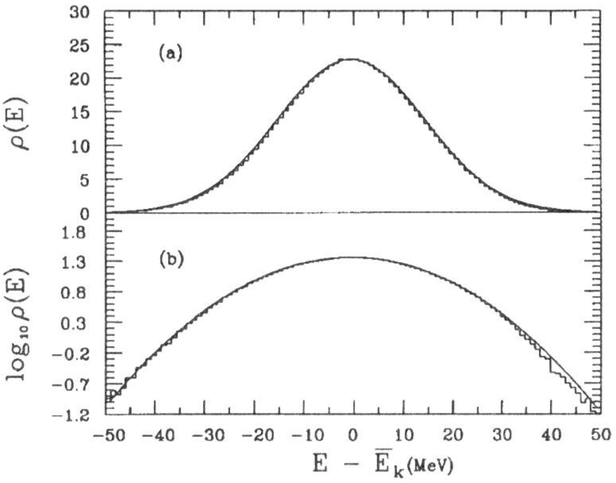

FIG. 5. Gaussian fit (solid lines) to the level density, $\rho(E)$, for the strength function averaged over 400 states, panel (a), and the same fit on the logarithmic scale, panel (b). The bin size is 1 MeV.

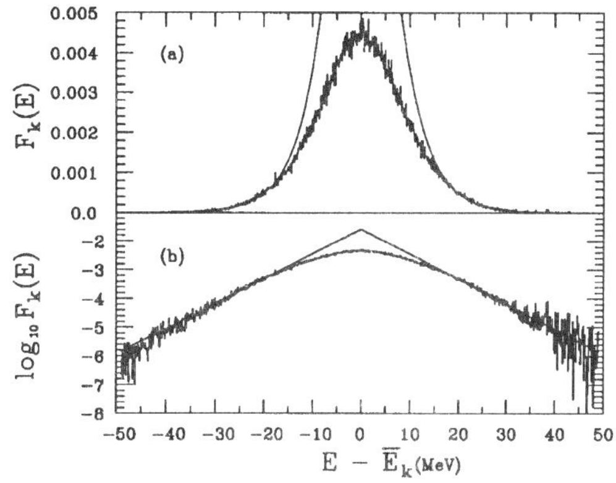

FIG. 7. The exponential fit of Eq. (23), solid lines, to the wings of the strength function of Fig. 2(c) (histograms), panel (a), and the same fit on the logarithmic scale, panel (b). The bin size is 100 keV.

distribution, $(E-\overline{E_k})>\Gamma$. The exponential behavior holds for at least three orders of magnitude.

A strength function falling off much faster than the Breit-Wigner distribution was also seen, with larger numerical uncertainty, in a study of atomic levels [9] where several interpolating formulas connecting the middle and the wings of the curve were suggested. The exponential localization is the part of the folklore accompanying studies of complicated wave functions [13]. To the best of our knowledge, the general proof of this notion is missing. The exponential localization of the wave functions in real space is known [34] in disordered solids. The situation in this case is different because the localized functions of nearly the same energy do not overlap if the distance between their centroids exceeds the spatial size of the wave function. In the limit of large N, these functions do not interact, which implies level statistics of the Poisson type. In our case, the exponential wings of the strength function coexist with the perfect chaotic level statistics.

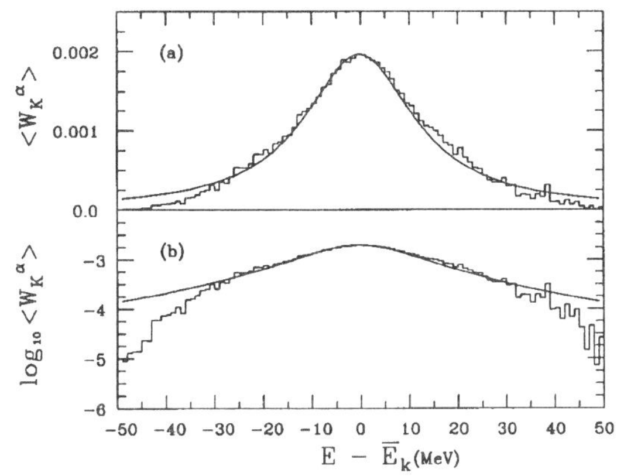

FIG. 6. Breit-Wigner fit (solid lines) to the average weight, $\langle W_k^\alpha\rangle=\langle (C_k^\alpha)^2\rangle$, for 400 states, panel (a), and the same fit on the logarithmic scale, panel (b). The bin size is 1 MeV.

IV. SPREADING WIDTH

A. Two-step diagonalization and the spectral function

In order to observe the spreading widths of individual shell model states, we performed the procedure of the two-step diagonalization described earlier in relation to the standard model. Taking out an arbitrary basis state $|k\rangle$ and performing the diagonalization of the remaining matrix we obtain the intermediate basis $(|k\rangle,\{|\nu\rangle\})$ with energies $(\overline{E_k},\{e_\nu\})$ and coupling matrix elements $V_{k\nu}$, Eq. (12), between the single excluded state and intermediate fine structure states. As we discussed, in the realistic situation these matrix elements are correlated with the energy distance $\overline{E_k}-e_\nu$. To characterize the distribution function of the coupling matrix elements, we introduce, for each removed state $|k\rangle$, the spectral function (form factor)

$$g_k(\overline{E_k},\overline{E_k}+\omega)=\sum_\nu |V_{k\nu}|^2\delta(\overline{E_k}-e_\nu+\omega). \tag{24}$$

This function can be presented by a histogram with the help of the density $\overline{\rho_k}(e_\nu)$ of the intermediate states available for the mixing with the single state $|k\rangle$. Introducing the average coupling intensity $\langle |V_{k\nu}|^2\rangle$, we get, for the spectral form factor (24),

$$g_k(\overline{E_k},\overline{E_k}+\omega)\approx\overline{\rho_k}(\overline{E_k}+\omega)\langle |V_{k\nu}|^2\rangle. \tag{25}$$

According to Eq. (18), this function is normalized as

$$\int d\omega g_k(\overline{E_k},\overline{E_k}+\omega)=\sigma_k^2. \tag{26}$$

Using the same arguments of uniform complexity, we expect to be able to extract the "generic" spectral function $\overline{g(\omega)}$, depending on the transition frequency ω only, from the superposition of the form factors g_k derived for different original states $|k\rangle$.

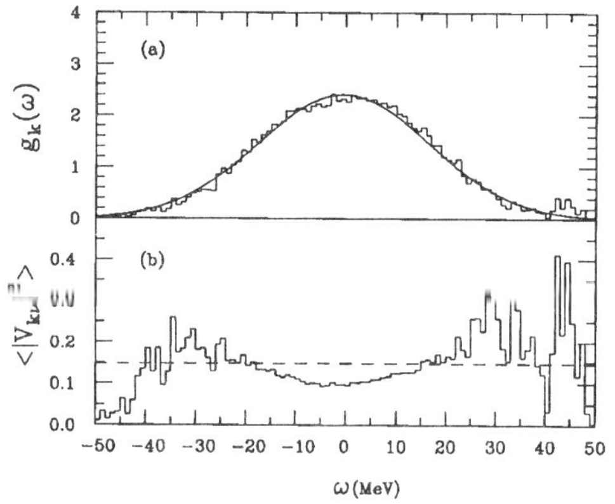

FIG. 8. (a) The spectral form factor $g(\omega)$ averaged over 100 0^+0 basis states in the middle of the spectrum, (histograms); Gaussian fit (solid line). (b) The coupling intensity $\langle V_{k\nu}^2\rangle$ for the same basis states $|k\rangle$ in the middle of the spectrum (histogram), as a function of $\omega = e_\nu - \overline{E}_k$; dashes correspond to the constant value of $\langle V^2\rangle = 0.149$ MeV^2.

As the interval of averaging increases, the average form factor rapidly evolves into the Gaussian shape. The form factor averaged over 100 close states is shown in Fig. 8(a) together with the Gaussian fit. The dispersion of the Gaussian is equal to $\sigma_g = 17$ MeV. The fitted normalization leads to the integral $\int d\omega g(\omega) = 104$ MeV^2 which agrees [see Eq. (26)], with the average value of $\sigma_k \approx \overline{\sigma} \approx 10$ MeV.

Dividing out the level density of the intermediate basis states, we determine, Fig. 8(b), the average coupling intensity $\langle |V_{k\nu}|^2\rangle$, Eq. (25). Except for the excess corresponding to the highest and, less pronounced, lowest states $|\nu\rangle$, the matrix elements are nearly constant on the level $\langle |V|^2\rangle \simeq 0.149$ MeV^2. This means that the realistic coupling is strong and it involves the remote parts of the spectrum with a different level density which mainly determines the shape of the form factor.

The value of the average coupling strength substituted into the golden rule (17) along with the average level spacing ~ 100 keV would result in the spreading width $\Gamma \approx 9.4$ MeV which is less than the value of 21 MeV found directly from the data by a factor of 2.2. The evident disagreement demonstrates that the golden rule estimate based on the standard model is not reliable when dealing with fragmentation and spreading widths which are of the order or larger than the scale of the change of the level density. Taking, instead of the average level density, the significantly larger level density in the center of the spectrum, $D \approx 40$ keV, we would get the golden rule width of 23 MeV, which is closer to the actual value.

Finally, we show in Fig. 9 that, in the intermediate basis, the wings of the strength function $F_k(E)$ can be actually calculated by perturbation theory using in Eq. (10) the weight coefficients

$$\langle (C_k^\alpha)^2\rangle \to (C_k^\nu)^2 \equiv \left(\frac{V_{k\nu}}{\overline{E}_k - e_\nu}\right)^2. \qquad (27)$$

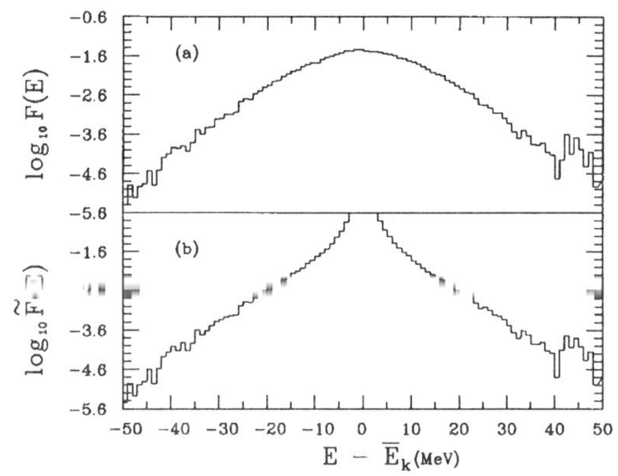

FIG. 9. Strength function $F_k(E)$, panel (a), and perturbative result $\tilde{F}_k(E)$, panel (b), on a logarithmic scale. Bin size is 1 MeV.

The agreement with the original calculation in the peripheral region is seen by coincidence of the two logarithmic plots.

B. Dependence on the interaction strength

Here we study the evolution of the strength function and the spreading width as a function of the strength of the residual interaction λ. As we mentioned, the level statistics reveal standard signatures of quantum chaos already at $\lambda \approx 0.2$. The mixing of the wave functions, the growth of the degree of complexity, and the evolution of the strength function are parallel aspects of the stochastization process.

In agreement with the general trends discussed in Sec. II, at weak interaction strength the shape of the strength function becomes closer to the Breit-Wigner one. Figure 10 clearly shows the Breit-Wigner behavior for $\lambda = 0.1, 0.2$, and 0.3, panels (a), (b), and (c), respectively. The whole evolution pattern is seen in the logarithmic plots of Figs. 11 and

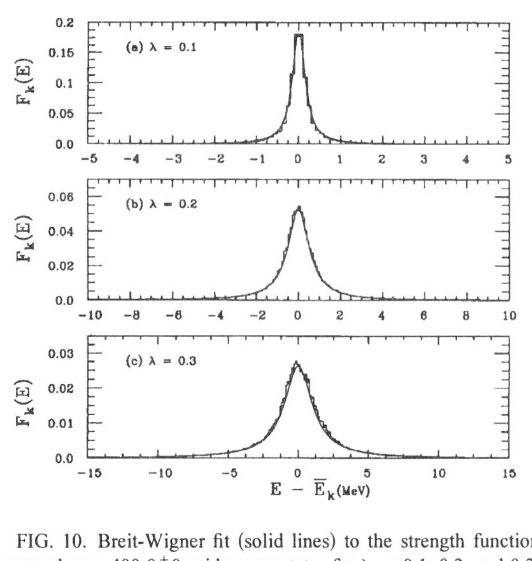

FIG. 10. Breit-Wigner fit (solid lines) to the strength function averaged over 400 0^+0 midenergy states for $\lambda = 0.1$, 0.2, and 0.3, panels (a), (b), and (c), respectively. The bin size is 100 keV.

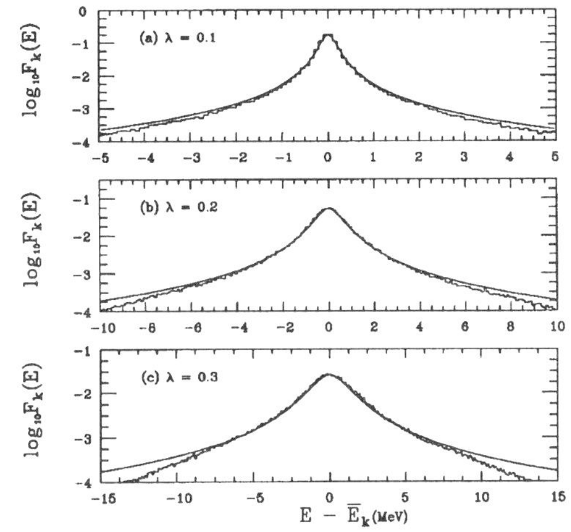

FIG. 11. Breit-Wigner fit (solid lines) to the strength function averaged over 400 0^+0 mid-energy states for $\lambda = 0.1$, 0.2, and 0.3, panels (a), (b), and (c), respectively, on a logarithmic scale. The bin size is 100 keV.

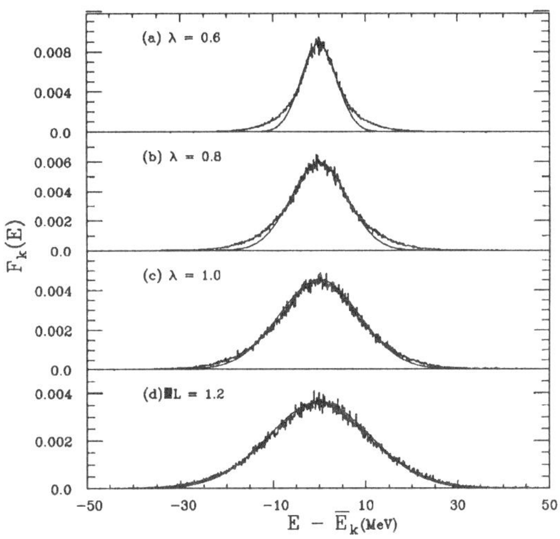

FIG. 13. Gaussian fit to the strength function averaged over 400 0^+0 mid-energy states for $\lambda = 0.6$, 0.8, 1.0, and 1.2, panels (a), (b), (c), and (d), respectively.

12. The curve evolves in the direction of the Gaussian in the central part with exponential tails. The Breit-Wigner description of the main part of the strength function can be considered as satisfactory up to $\lambda \approx 0.4$. In this region the narrow strength function is not strongly influenced by the change of the level density. As seen from Fig. 13, the quality of the Gaussian fit clearly improves as one goes to the strong coupling limit; the last panel, (d), corresponds to $\lambda = 1.2$.

Using "empirical" generic strength functions, we can trace the evolution of the spreading width (FWHM) as a function of the intensity of the residual interaction. Figure 14 shows the results for the values of λ between 0 and 1.2. The dependence of the spreading width on the interaction changes from quadratic in the weak coupling limit to linear at strong coupling. The results are nicely described by the simple interpolation (20) with parameters $\gamma \approx 44.9$ MeV and $y \approx 1.32$. Our estimates (21) and (22), using the value of the average level spacing at weak interaction $D_0 \approx 21$ keV, predict for these parameters $\gamma = 44.6$ MeV and $y = 1.23$.

V. CONCLUSION

The typical strength function for stationary nuclear states was, for the first time, extracted from the exact solution of the many-body problem in the truncated Hilbert space of

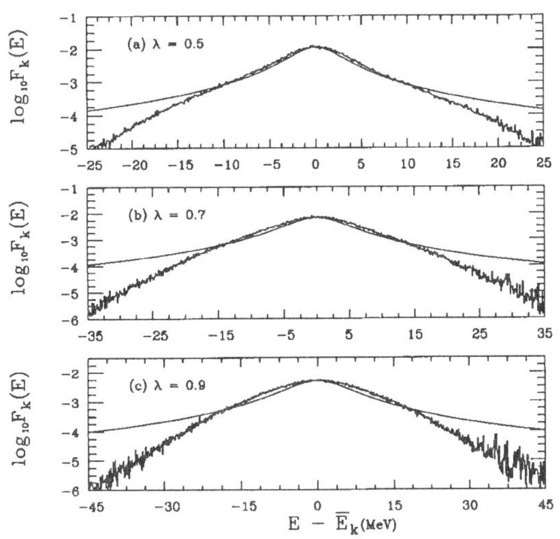

FIG. 12. Breit-Wigner fit (solid lines) to the strength function averaged over 400 0^+0 mid-energy states for $\lambda = 0.5$, 0.7, and 0.9, panels (a), (b), and (c), respectively, on a logarithmic scale. The bin size is 100 keV.

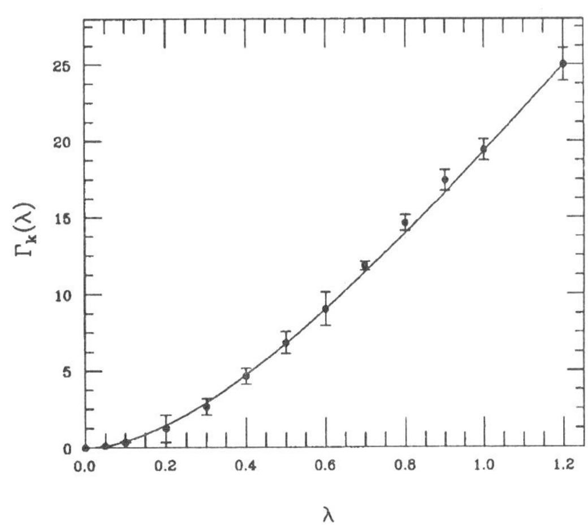

FIG. 14. Spreading width Γ_k of the basis states as a function of the interaction strength λ for 400 middle 0^+0 states. The solid line corresponds to Eq. (20) with $\gamma = 44.9$ MeV and $y = 1.32$. The error bars are defined by the deviations of the fit from the calculated data.

shell model configurations. The realistic interaction strength corresponds to the strong coupling case which cannot be satisfactorily described by the standard model of the strength function. The generic shape of the compound states at high excitation energy is close to a Gaussian but with exponential wings. At this point we cannot explain the factor responsible for the exponential localization length. In order to better understand this we should investigate the strength function evolution along the spectrum; here, we studied the most complicated states near the middle which could be considered as uniform in their properties including the shape and the width of the strength function.

The transition to the weak coupling case at the artificially suppressed strength of the residual interaction shows how the shape of the strength function regularly changes from the Gaussian to the "normal" Breit-Wigner. This is accompanied by the reduction of the spreading width and the smooth transition from the linear interaction dependence characteristic for the strong coupling limit to the ordinary quadratic dependence predicted by the golden rule. This transition is seen for the first time in realistic calculations.

New effects will enter when the presence of real decay into continuum is taken into account. The competition of the internal mixing, external decay, and interaction through common decay channels makes the whole problem more complicated. This interplay of various mechanisms will determine the physics of compound states at high excitation energy in stable nuclei and at lower excitation energy in weakly bound nuclei.

The authors acknowledge support from NSF Grants Nos. 94-03666 and 95-12831.

[1] B.A. Brown and B.H. Wildenthal, Annu. Rev. Nucl. Part. Sci. **38**, 29 (1988).

[2] B.A. Brown *et al.*, OXBASH, MSUNSCL Report No. 524, 1988.

[3] V. Zelevinsky, M. Horoi, and B.A. Brown, Phys. Lett. B **350**, 141 (1995).

[4] M. Horoi, V. Zelevinsky, and B.A. Brown, Phys. Rev. Lett. **74**, 5194 (1995).

[5] V. Zelevinsky, B.A. Brown, N. Frazier, and M Horoi, Phys. Rep. (to be published).

[6] T.A. Brody, J. Flores, J.B. French, P.A. Mello, A. Pandey, and S.S.M. Wong, Rev. Mod. Phys. **53**, 385 (1981).

[7] O. Bohigas and H.A. Weidenmüller, Annu. Rev. Nucl. Part. Sci. **38**, 421 (1988).

[8] F. Haake, *Quantum Signatures of Chaos* (Springer, New York, 1991).

[9] V.V. Flambaum, A.A. Gribakina, G.F. Gribakin, and M.G. Kozlov, Phys. Rev. A **50**, 267 (1994).

[10] B.A. Brown and G.F. Bertsch, Phys. Lett. **148B**, 5 (1984).

[11] W.E. Ormand and R.A. Broglia, Phys. Rev. C **46**, 1710 (1992).

[12] S. Drożdż, S. Nishizaki, J. Speth, and J. Wambach, Phys. Rev. C **49**, 867 (1994).

[13] F.M. Izrailev, Phys. Rep. **196**, 299 (1990).

[14] I.C. Percival, J. Phys. B **6**, L229 (1973).

[15] G.F. Bertsch, P.F. Bortignon, and R.A. Broglia, Rev. Mod. Phys. **55**, 287 (1985).

[16] G. Kilgus *et al.*, Z. Phys. A **326**, 41 (1987).

[17] P.M. Gopych *et al.*, Sov. J. Part. Nucl. **19**, 338 (1988).

[18] A. Bohr and B. Mottelson, *Nuclear Structure* (Benjamin, New York, 1969), Vol. 1.

[19] J. Reiter and H.L. Harney, Z. Phys. A **337**, 121 (1990).

[20] H.L. Harney, A. Richter, and H.A. Weidenmüller, Rev. Mod. Phys. **58**, 607 (1986).

[21] V.G. Zelevinsky and P. von Brentano, Nucl. Phys. **A529**, 141 (1991).

[22] A. Bracco *et al.*, Phys. Rev. Lett. **62**, 2080 (1989); G. Enders *et al.*, *ibid.* **69**, 249 (1992); J.J. Gaardhøje, Annu. Rev. Nucl. Part. Sci. **42**, 483 (1992).

[23] A. Bracco *et al.*, Phys. Rev. Lett. **74**, 3748 (1995).

[24] E.P. Wigner, Ann. Math. **62**, 548 (1955).

[25] M. Feingold, A. Gioletta, F.M. Izrailev, and L. Molinari, Phys. Rev. Lett. **70**, 2936 (1993).

[26] C.A. Bertulani and V.G. Zelevinsky, Phys. Rev. Lett. **71**, 967 (1993); Nucl. Phys. **A568**, 931 (1994).

[27] C.H. Lewenkopf and V.G. Zelevinsky, Nucl. Phys. **A569**, 183 (1994).

[28] S. Mordechai *et al.*, Phys. Rev. C **41**, 202 (1990); J. Ritman *et al.*, Phys. Rev. Lett. **70**, 533 (1993); R. Schmidt *et al.*, *ibid.* **70**, 1767 (1993); T. Aumann *et al.*, Phys. Rev. C **47**, 1728 (1993).

[29] B. Lauritzen, P.F. Bortignon, R.A. Broglia, and V. Zelevinsky, Phys. Rev. Lett. **74**, 5190 (1995).

[30] V.V. Flambaum, G.F. Gribakin, and F.M. Izrailev, Phys. Rev. E **53**, 5729 (1996).

[31] O.P. Sushkov and V.V. Flambaum, Pis'ma Zh. Éksp. Teor. Fiz. **32**, 377 (1980) [JETP Lett. **32**, 353 (1980)]; Usp. Fiz. Nauk **136**, 3 (1982) [Sov. Phys. Usp. **25**, 1 (1982)].

[32] V.G. Zelevinsky, Nucl. Phys. **A553**, 125c (1993); **A570**, 411c (1994).

[33] K.K. Mon and J.B. French, Ann. Phys. (N.Y.) **95**, 90 (1975).

[34] I.M. Lifshits, S.A. Gredescul, and L.A. Pastur, *Introduction to the Theory of Disordered Systems* (Wiley, New York, 1988).

893

Shell-model tests of the bimodal partial state densities in a 2 × 2 partitioned embedded random matrix ensemble

V.K.B. Kota, D. Majumdar, R. Haq, and R.J. Leclair

Abstract: The mixing of well-separated subspaces of an interacting many-particle system, such as a nucleus with active nucleons distributed in more than one major shell, can be studied using partitioned embedded ensembles of random matrices. The bimodal form of partial state densities (one-point functions) predicted earlier for a 2 × 2 partitioned embedded ensemble, which may be regarded as a model for the mixing of two well-separated degenerate subspaces, is tested using nuclear shell-model calculations in the $[(ds)^6 \oplus (ds)^4(f_{7/2})^2]^{J=0,T=0}$ space. The theoretical forms predicted by the binary correlation approximation theory are in good agreement with the shell-model results. This suggests that with suitable extensions it might be feasible to use the binary correlation method to deal with several interacting subspaces involving multimodal distributions.

PACS Nos.: 05.30.-d, 05.45.Mt, 24.60.Lz

Résumé : On peut utiliser des redistributions de matrices aléatoires pour étudier le mélange de sous-espaces d'états bien séparés de systèmes à N-corps en interaction, tels les noyaux dont les nucléons de valence seraient répartis sur plus d'une couche. Utilisant des calculs du modèle en couches dans l'espace $[(ds)^6 \oplus (ds)^4(f_{7/2})^2]^{J=0,T=0}$, nous vérifions la forme bimodale des densités partielles d'états (fonctions à un point) prédites plus tôt pour un ensemble redistribué de 2 × 2, qui peut être regardé comme un modèle du mélange de deux sous espaces dégénérés bien séparés. Les formes théoriques prédites dans l'approximation des corrélations binaires sont en bon accord avec les résultats du modèle en couches. Ceci suggère qu'une extension appropriée permettrait d'utiliser la méthode des corrélations binaires pour décrire plusieurs sous-espaces en interaction impliquant des distributions multimodales.
[Traduit par la rédaction]

1. Introduction

Random matrix theories in quantum physics along with their deeper connections to quantum chaos have been studied with increasing interest in the last 15 years or so [1]. The ideas and methods used in these two domains have contributed a great deal to the development of a statistical theory for isolated finite interacting many-particle systems. In these theories, embedded random matrix ensembles play a

Received March 25, 1999. Accepted December 8, 1999.

V.K.B. Kota and D. Majumdar.[1] Physical Research Laboratory, Ahmedabad 380 009, India.
R. Haq and R.J. Leclair. Department of Physics and Astronomy, Laurentian University, Sudbury, ON P3E 2C6, Canada.

[1] Present address: Department of Physics, University of Calcutta, Calcutta 700 009, India.

significant role [2] and have been the subject of intense study recently [3–8]. The embedded Gaussian orthogonal ensemble of random matrices (EGOE(k) or simply EGOE), introduced by French and Wong [9] and by Bohigas and Flores [10], incorporates in the classical GOE introduced by Wigner [11,12][2] the important fact that in the Hamiltonian describing physical systems, the rank (k) of the interaction is usually much smaller than the number of interacting particles (m). These embedded ensembles are generally studied numerically

(*i*) by constructing the EGOE matrices in the occupation number representation and using the Monte Carlo method;

(*ii*) by shell-model (in nuclear and atomic cases) calculations with realistic interactions and assuming that the shell-model Hamiltonian is a typical member of the EGOE; and

(*iii*) by using the shell model with the Monte Carlo method. Exact analytical results being difficult to obtain and often intractable, further insight is provided by the binary correlation approximation [2,13,14].

Interest in embedded ensembles has been heightened recently because large-scale nuclear shell-model results (matrix dimensionality ~3000) [15], spectroscopic calculations for the Ce atom [16], and the description of nuclear shell-model results for new measures of complexity and chaos [7] such as the number of principal components (NPC) and information entropy (S^{info}) have substantiated many of the results of EGOE. This has opened up the possibility of applications of embedded ensembles for other interacting particle systems such as atomic clusters, isolated quantum dots, etc.; see, for example, ref. 8. Besides EGOE, more general embedded ensembles, as modified versions of EGOE, have been introduced by many authors; they are

(i) embedded ensembles of (1+2)-body interactions with a varying strength parameter for the two-body part [4,15];

(ii) EGOE(2) in good angular momentum and isospin (JT) spaces with particle–hole symmetry [6];

(iii) $\mathcal{K}$ + EGOE(2) ensemble where $\mathcal{K}$ is a fixed operator [17]; and

(iv) partitioned EGOE (p-EGOE) [18,19]. The present paper deals with the p-EGOE random matrix ensembles.

To incorporate in ab-initio calculations of nuclear state densities the effects due to mixing between distant shell-model configurations (involving excitations across the major shells), a 2×2 p-EGOE random matrix ensemble is constructed, see refs. 18 and 19 and a plausible solution to the one-point function is given. Our purpose in this paper is to present nuclear shell-model tests of this solution. It is expected that these results will prompt further detailed investigations of p-EGOE (it should be mentioned here that there has been recent interest in p-EGOE in the study of the quasiparticle spectrum of a diffusive quantum dot [8]).

In Sect. 2, the p-EGOE ensemble is defined. Section 3 gives the bimodal form of the partial state density in a 2×2 p-EGOE representing mixing between two well-separated degenerate spaces. Results of nuclear shell-model tests of the bimodal forms are given in Sect. 4. Finally, Sect. 5 gives some concluding remarks.

[2] The Gaussian orthogonal ensemble (GOE) of random matrices is an ensemble of real symmetric matrices H with the matrix elements H_{ij} chosen to be independent zero-centered Gaussian random variables with variance $\overline{H_{ij}^2} = v^2(1+\delta_{ij})$; with v a constant.

2. Partitioned embedded random matrix ensembles

It is well known, as seen in nuclear $0 - \hbar\omega$ shell-model calculations, that EGOE generates a Gaussian state density whereas classical ensembles give a semicircle distribution of states [2,13,14]. However, in many nuclear structure studies such as level densities, Gamow–Teller strength distributions, gaint dipole resonance strengths and widths, etc., it is important to include multi-$\hbar\omega$ excitations in the shell-model spaces. In such nuclear structure calculations it is important to remember that shell-model stability ensures that the mixing between the distant multi-$\hbar\omega$ configurations is weak (i.e., the intensity mixing parameter τ defined ahead in (4) is small). Therefore, in general, the Hamiltonian H matrix divides into block-structured matrices with each diagonal block denoting the distant configurations (for light nuclei, these correspond to $0\hbar\omega$, $2\hbar\omega$, $4\hbar\omega$, ... for states having the same parity as the ground state) and the off-diagonal blocks representing the mixing between these distant configurations. It is this particular structure of the H-matrix that led to the introduction of p-EGOE.

In a two-particle space, p-EGOE(2) is defined by a partitioned GOE where the centroids of the diagonal blocks and the variances of the two-body matrix elements in each diagonal as well as off-diagonal blocks is different. As an example, consider N single particle states, the lowest N_1 states belonging to say $s = 0$ and the remaining $N_2 = N - N_1$ states belong to $s = 1$ space. Then in the two-particle space, three configurations $(2,0) \oplus (1,1) \oplus (0,2)$ with $S = \sum_i s_i = 0$, 1, and 2, respectively, are possible. The two-particle Hamiltonian matrix $H(2)$ is therefore a 3×3 block matrix with the blocks denoted by $H_{SS'}(2)$, $S, S' = 0, 1, 2$. The dimensions of the three diagonal blocks are $(N_1)(N_1-1)/2$, $(N_1)(N_2)$, and $(N_2)(N_2-1)/2$, respectively. The $H_{SS}(2)$ are represented by GOE with v_{SS}^2 being the variance of the matrix elements (note that for the diagonal matrix elements the variance is $2v_{SS}^2$) and they are all zero centered except that the diagonal matrix elements have centroid Δ_{SS}. The matrix elements of the off-diagonal $H_{SS'}(2)$, $S \neq S'$ blocks are independent zero-centred Gaussian variables with variance $v_{SS'}^2$ (i.e., they are off-diagonal GOE). Finally, it should be noted that $H(2)$ is real symmetric. In m (many) particle spaces, it is straightforward to construct $H(m)$ by writing the matrix elmements of p-EGOE in occupation number representation. In the situation that the parity of the single-particle states with a given s value is given by $(-1)^s$, the H matrix for the many-particle positive parity states, for parity preserving interactions, divides into $S = 0 \oplus 2 \oplus 4 \oplus \ldots$ subspaces. In the simple situation that $v_{SS'}^2$, the variances of the off-diagonal GOEs in two-particle space, are zero, the p-EGOE densities are a superposition of the EGOE densities for each of the S subspaces. For example, the state density (a one-point function) $I^m(E) = \langle\langle\delta(H-E)\rangle\rangle^m = \sum_S \langle\langle\delta(H-E)\rangle\rangle^{m,S} = \sum_S I^{m,S}(E)$; note that here the ensemble average is not shown explicitly. Our aim is to study the effects of $v_{SS'}^2$, $S \neq S'$, on $I^{m,S}(E)$ or equivalently on $I(E)$; it should be stressed that $I^m(E) = \sum_S I^{m,S}(E)$ is always valid. The solution to this p-EGOE problem appears to be quite complicated (see ref. 20 for the solution to the corresponding GOE).

3. Partial state densities in a 2 x 2 p-EGOE

For insight into the one-point functions (state densities) generated by p-EGOE, a simpler model dealing with only the $S = 0 \oplus 2$ subspace of the m-particle space and, in addition, putting the diagonal GOEs in the two-particle space to zero has been considered in refs. 18 and 19. For this two-spike model, the m-particle Hamiltonian is a 2×2 block matrix,

$$H(m) = \begin{pmatrix} A & X \\ \widetilde{X} & B \end{pmatrix} \longrightarrow \begin{pmatrix} 0 & X \\ \widetilde{X} & \Delta\, I \end{pmatrix} \tag{1}$$

In (1), (*i*) the dimensionality of matrix A is d_A, of B is d_B and one can assume, without loss of generality, that $d_A \leq d_B$; (*ii*) $A_{ij} = 0$ for all i, j and similarly $B_{ij} = \Delta\delta_{ij}$; (*iii*) the matrix X represents an off-diagonal EGOE.

Table 1. Shell-model (SM) results compared with the p-EGOE predictions from (5) for ϵ_M/Δ and σ_M^2/σ_{12}^2.

τ	ϵ_M/Δ		σ_M^2/σ_{12}^2	
	p-EGOE	SM	p-EGOE	SM
0.1	−0.09	−0.08	0.06	0.04
0.2	−0.16	−0.15	0.09	0.07
0.3	−0.22	−0.21	0.11	0.08
0.5	−0.33	−0.31	0.13	0.10
1.0	−0.55	−0.52	0.16	0.12
10.0	−2.37	−2.33	0.20	0.16

With α and β denoting the quantum numbers of the basis states defining the matrices A and B, respectively, the eigenstates take the form $|E\rangle = \sum_{\alpha=1}^{d_A} C_{\alpha(A),E}\ |\alpha\rangle + \sum_{\beta=1}^{d_B} C_{\beta(B),E}\ |\beta\rangle$. The state density $I_\Delta(E)$ and its decomposition into partial densities $I_\Delta^A(E)$ and $I_\Delta^B(E)$ is given by

$$I_\Delta(E) = \langle\langle\delta(H-E)\rangle\rangle = I_\Delta^A(E) + I_\Delta^B(E)$$

$$I_\Delta^A(E) = \langle\langle\delta(H-E)\rangle\rangle^A = \sum_{\alpha=1}^{d_A} \left|C_{\alpha(A),E}\right|^2, \qquad \alpha = 1,2,3,...d_A \tag{2}$$

$$I_\Delta^B(E) = \langle\langle\delta(H-E)\rangle\rangle^B = \sum_{\beta=1}^{d_B} |C_{\beta(B),E}|^2, \qquad \beta = 1,2,3,...d_B$$

The moments of $I_\Delta^A(E)$ are

$$M_p(I_\Delta^A) = \langle H^P\rangle^A = \frac{1}{d_A}\sum_E \sum_{\alpha=1}^{d_A} |C_{\alpha(A),E}|^2 E^p = \frac{1}{d_A}\sum_E I_\Delta^A(E)E^p$$

Then it is easily seen that $M_0(I_\Delta^A) = 1$, $M_1(I_\Delta^A) = 0$, and $M_2(I_\Delta^A) = \sigma_{12}^2$ where the partial variance σ_{12}^2 is defined by

$$\sigma_{12}^2 = (d_A)^{-1} \sum_{i,j} X_{ij}^2 \tag{3}$$

The shape of the densities in (2) depends on the the mixing parameter τ,

$$\tau = \sigma_{12}^2/\Delta^2 \tag{4}$$

Note that the mixing increases with increasing τ. The solution to $I_\Delta^A(E)$ is obtained by writing it first as a convolution integral involving $I_0^A(x)$ and then solving for $I_0^A(x)$. The convolution form is easy to recognize and, following ref. 19, we outline briefly the derivation for $I_0^A(x)$.

This density is generated by the matrix X, which corresponds to $S = 0 \rightarrow S = 2$ excitations generated by the defining two-body interaction (which is an off-diagonal GOE for $m = 2$). The binary correlation approximation [13] gives a counting theorem with which one can easily see that for EGOE the moments $M_{2\nu} =< H^{2\nu} >$ are Gaussian moments; the rule being that the H's in $< H^{2\nu} >$, which must be paired for nonvanishing contributions, can be paired in all possible ways so that $M_{2\nu} = (2\nu-1)!!\ (M_2)^\nu$. In the present problem however, we have $M_{2\nu}(I_0^A) =< (X\tilde{X})^\nu >^A$ involving alternate $A \rightarrow B$ and $B \rightarrow A$ excitations. Arranging the ν $X\tilde{X}$ terms in all possible ways gives $M_{2\nu}(I_0^A) = \nu!\ (\sigma_{12}^2)^\nu$. Then the corresponding I_0^A and I_Δ^A are given by [18,19],

$$I_0^A(E) = d_A|E|(\sigma_{12}^2)^{-1} \exp\left(\frac{-E^2}{\sigma_{12}^2}\right), \qquad -\infty \leq E \leq +\infty$$

$$I_\Delta^A(E) = |\frac{E-\Delta}{E}|^{1/2} I_0^A(\sqrt{E(E-\Delta)}), \qquad E \leq 0, \quad E \geq \Delta \tag{5}$$

$$I_\Delta^A(E) = 0, \qquad 0 \leq E \leq \Delta$$

Thus, $I_\Delta^A(E)$ is bimodal and using (5), one can calculate the modal centroid ϵ_M and variance σ_M^2 for the lower mode; see Table 1. Thus, $I_\Delta^A(E)$ a spike for $\tau = 0$, becomes bimodal as τ increases due to the mixing of space A states with the far off space B states. The derivation leading to $I_0^A(E)$, as given in (5), implicitly assumes that the $s = 0$ and 1 single-particle states are sufficiently large in number. However, even for moderate N_1 and N_2, the matrix dimensions d_A and d_B become prohibitively large, for example, (*i*) for $N_1 = 11$, $N_2 = 3$, and $m = 7$, we have $d_A = 330$ and $d_B = 1386$; (*ii*) $N_1 = 12$, $N_2 = 6$, and $m = 6$ gives $d_A = 924$ and $d_B = 7425$. Because of the large dimensionalities of the matrices involved, it is difficult to test (5) and also the results for ϵ_M and σ_M obtained through p-EGOE calculations (a limited comparison of analytic and numerical p-EGOE results is given in ref. 21). A good test of the theoretical predictions, with N_1 and N_2 sufficiently large, is provided by the shell model.

For completeness we mention that in the case the m-particle matrix X in (1) is represented by an off-diagonal GOE with variance of matrix elements given by v^2, the form of $I_0^A(E)$ is given by [19],

$$I_0^A(E) = (2\pi v^2|E|)^{-1}\sqrt{[(R_+^2 - E^2)(E^2 - R_-^2)]}, \qquad R_- \leq |E| \leq R_+$$

$$I_0^A(E) = 0, \qquad |E| < R_-\,,\ |E| > R_+ \tag{6}$$

$$R_\pm = v|\sqrt{d_B} \pm \sqrt{d_A}|$$

For GOE then, the characteristic form of the density is a semicircle for $d_A = d_B$ (Fig. 1*a*) and two separated deformed semicircles when $d_B > d_A$ (Fig. 1*b*). The partial densities $I_\Delta^A(E)$ for these two cases are shown in Fig. 2 for $\tau = 0.5$ (weak mixing) and 10 (strong admixing). The shape dependence of the partial densities on the mixing parameter is evident by comparing these shapes with those in Fig. 1 ($\tau = \infty$). It is important to mention that GOE models are unrealistic for densities since the independence of the many-body matrix elements implies multibody interactions.

Fig. 1. The partial density for space A, $I_o^A(E)$, as a function of energy E for the cases (*a*) and (*b*) where the m-particle matrix X is an off-diagonal GOE and (*c*) where X represents the many-particle matrix generated by the $sd \to f_{7/2}$ two-body matrix elements. The histogram obtained through numerical computation is compared with the analytic expression given in (6) for (*a*) and (*b*) and in (5) for (*c*). Note that $d_A = d_B$ in (*a*) and $d_A < d_B$ in (*b*).

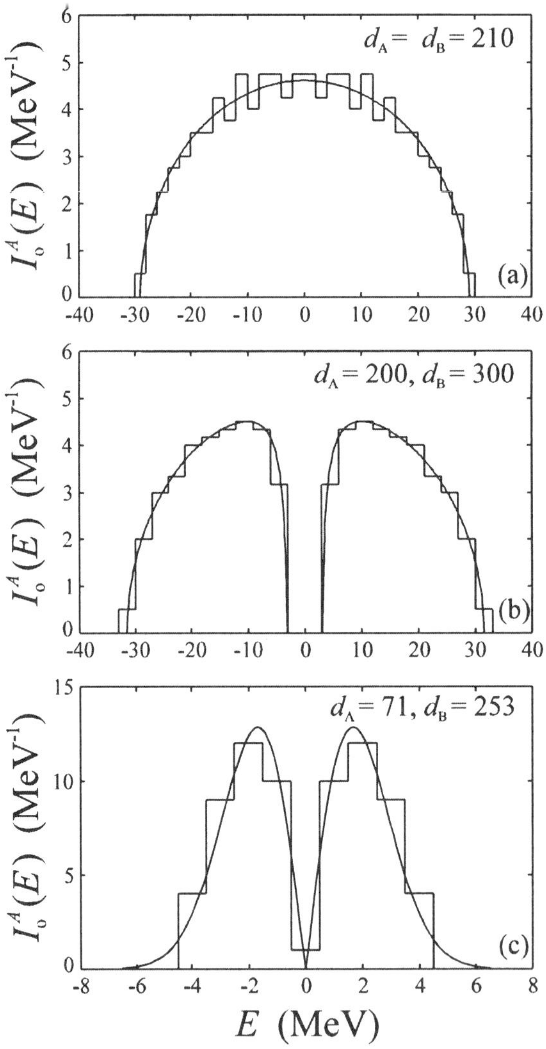

4. Shell-model results

For testing (5), a six-particle nuclear shell-model calculation is performed. The nucleons are assumed to be occupying single-particle orbits $({}^1d_{5/2}{}^2s_{1/2}{}^1d_{3/2})$ with $s = 0$ and ${}^1f_{7/2}$ with $s = 1$. Using the Kuo interaction, as an example, in the above space (see ref. 22), the H matrix is constructed for six particles carrying total angular momentum $J = 0$ and isospin $T = 0$ in $S = 0 \oplus 2$ space, i.e., in the $(ds)^{6,J=0,T=0;S=0} \oplus \left[(ds)^4 \times (f_{7/2})^2\right]^{J=0,T=0;S=2}$ space. The dimensions of the $S = 0$ (A

Fig. 2. The partial GOE density for space A, $I_{\Delta}^{A}(E)$, as a function of energy E for two representative values of the mixing parameter τ; $\tau = 0.5$ representing weak mixing and $\tau = 10$ representing strong mixing. The histograms are numerical results and the continuous curves represent the expressions given in (6).

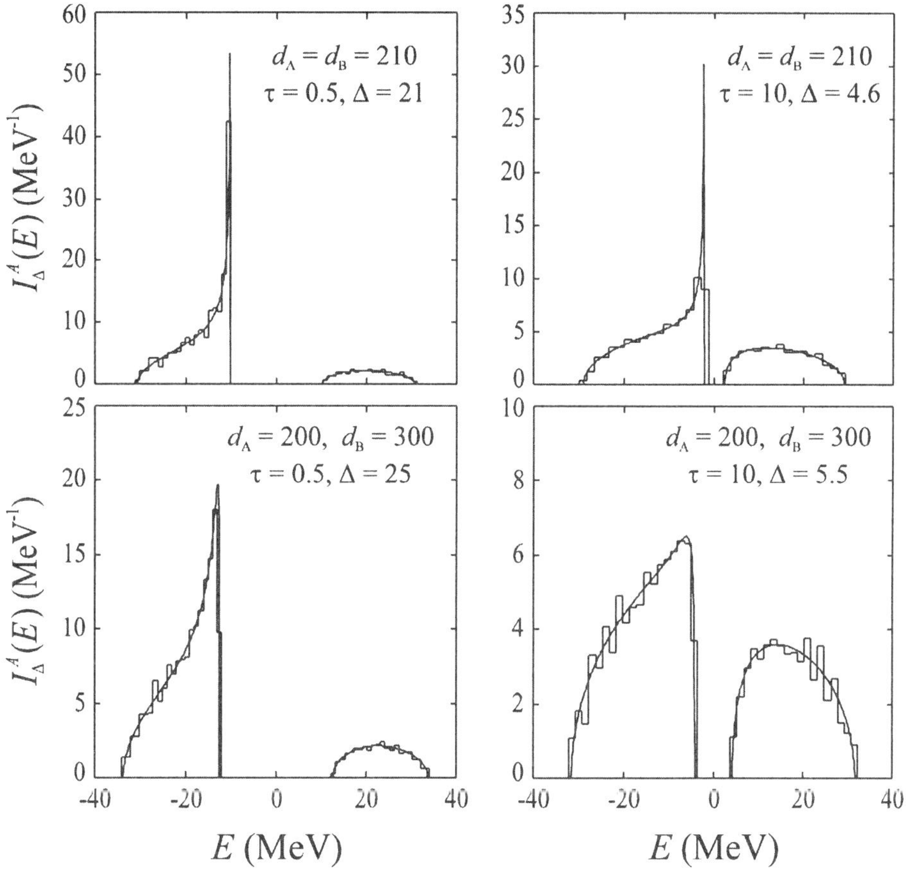

block) and $S = 2$ (B block) spaces are $d_A = 71$ and $d_B = 253$. The matrix dimensionality of H is $d = d_A + d_B = 324$. To generate a typical member of the two-spike p-EGOE matrix in (1), the A and B matrices are modified such that $A_{ij} = 0$ and $B_{ij} = (\Delta)\delta_{ij}$; the X matrix is generated by $sd \to f_{7/2}$ two-body Kuo matrix elements. In the present example, the partial variance $\sigma_{12}^2 = 5.6$ MeV^2. In the calculations, a desired value of τ is obtained by choosing Δ appropriately. The density $I_0^A(E)$ is shown in Fig. 1c. Using (2), the shell-model histograms for $I_{\Delta}^{A}(E)$ are constructed for $\tau = 0.05, 0.1, 0.3, 0.5$, 1.0, and 10 and compared with (5) in Fig. 3. In addition, the ϵ_M/Δ and σ_M^2/σ_{12}^2 for the lower mode of I_{Δ}^{A} are compared with the predictions from (5) and the results are given in Table 1. These shell-model results clearly confirm the form (5) for the one-point function $I_{\Delta}^{A}(E)$ given by the binary correlation approximation theory for p-EGOE. In pratical applications it is expected that $\tau \leq 0.3$ and in this case it is seen from Table 1 that only $\lesssim 10\%$ of the partial variance σ_{12}^2 contributes to the variance of the lower mode and the rest goes into the separation of the two modes; for all pratical purposes the lower mode can be neglected. Therefore, by switching on the diagonal A and B matrices in (1), it is expected that in the first approximation (by convolution argument) the variance of the lower mode of $I_{\Delta}^{A}(E)$, i.e., $\lesssim \sigma_{12}^2/10$, will add to the variance of the unperturbed $I^A(E)$.

Fig. 3. The partial EGOE density $I_{\Delta}^{A}(E)$ as a function of E for various values of the mixing parameter τ. Shell-model results (histograms) are compared with the bimodal form (5) given by p-EGOE (continuous curves). In all the shell-model calculations σ_{12}^{2} is fixed and the value of Δ is varied to obtain the desired τ value.

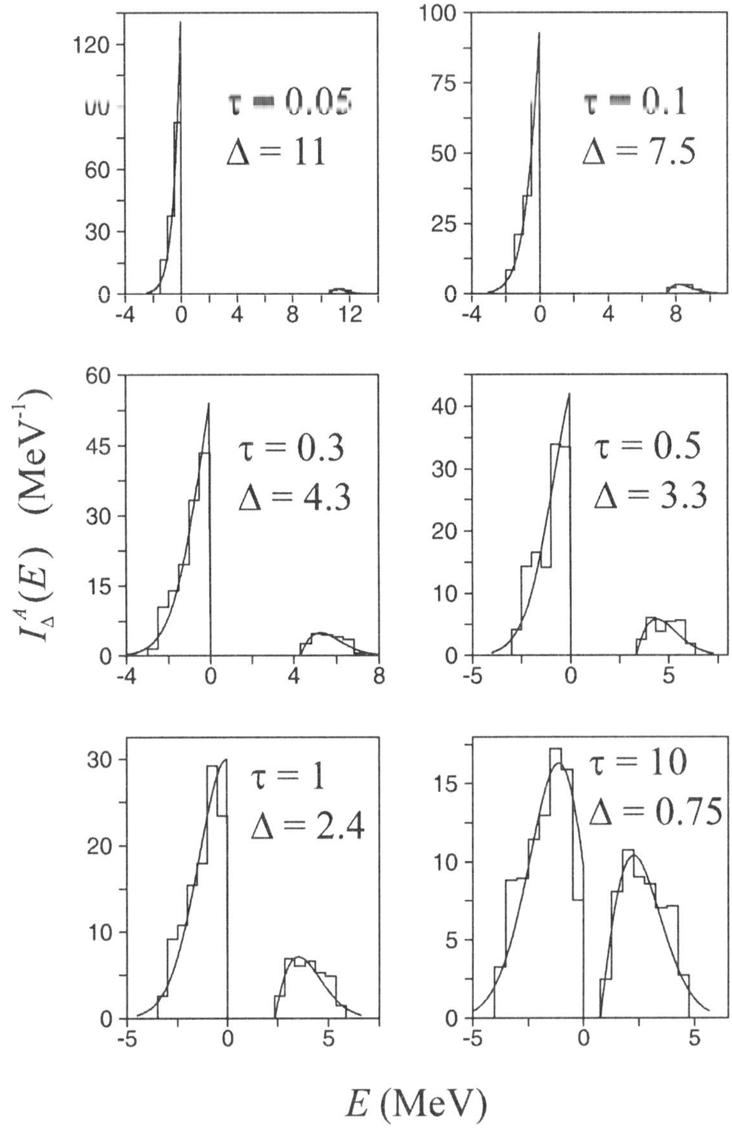

5. Conclusions and future outlook

From the present example, it is clear that mixing with distant configurations gives bimodal and will, in general, lead to multimodal one-point functions (partial state densities). Further, this important basic result extends to strength functions (or local density of states [15]) since partial state densities (I^A and I^B in (2)) may be regarded as strength functions. Recently, it has been recognized that strength functions, as predicted by EGOE, are Gaussian in shape in the chaotic domain for interacting particle systems and, in general, exhibit a transition from Breit–Wigner to Gaussian form [5,15,23]. Although we considered only the simple two-spike 2×2 p-EGOE given by (1), for practical applications one has to go beyond this and solve: (*i*) the two-space problem for general A and B matrices in (1) (the restrictive condition that they be spikes should be relaxed); (*ii*) the general case of mixing of $S = 0 \oplus 2 \oplus 4 \ldots$ spaces, i.e., go beyond the 2×2 block matrix case; and (*iii*) in large spaces, the corresponding problems for bivariate strength densities [14] generated by transition operators that connect distant configurations. The shell-model comparisons shown here and the good agreement obtained indicate that the binary correlation method that produced (5) may possibly lead to solutions for (*i*)–(*iii*).

Finally, we mention that current interest in embedded random matrix ensembles in generating a statistical mechanics for isolated finite interacting particle systems and their applications to mesoscopic systems in particular has prompted us to investigate p-EGOE in detail. Results of this paper constitute a first step in this direction.

Acknowledgements

The work presented here has grown out of a collaboration on statistical spectroscopy of two of the authors V.K.B. Kota and R. Haq with J.B. French.

References

1. T. Guhr, A. Muller-Groeling, and H.A. Weidenmüller. Phys. Rep. **299**, 189 (1998).
2. T.A. Brody, J. Flores, J.B. French, P.A. Mello, A. Pandey, and S.S.M. Wong. Rev. Mod. Phys. **53**, 385 (1981).
3. Ph. Jacquod and D.L. Shepelyansky. Phys. Rev. Lett. **79**, 1837 (1997); B. Georgeot and D.L. Shepelyansky. Phys. Rev. Lett. **81**, 5129 (1998).
4. V.V. Flambaum, G.F. Gribakin, and F.M. Izrailev. Phys. Rev. E: Stat. Phys. Plasmas Fluids Relat. Interdiscip. Top. **53**, 5729 (1996); V.V. Flambaum, F.M. Izrailev, and G. Casati. Phys. Rev. E: Stat. Phys. Plasmas Fluids Relat. Interdiscip. Top. **54**, 2136 (1996).
5. V.V. Flambaum and F.M. Izrailev. Phys. Rev. E: Stat. Phys. Plasmas Fluids Relat. Interdiscip. Top. **56**, 5144 (1997).
6. C.W. Johnson, G.F. Bertsch, and D.J. Dean. Phys. Rev. Lett. **30**, 2749 (1998).
7. V.K.B. Kota and R. Sahu. Phys. Lett. **429B**, 1 (1998).
8. C. Mejia-Monasterio, J. Richert, T. Rupp, and H.A. Weidenmüller. Phys. Rev. Lett. **81**, 5189 (1998); X. Leyronars, J. Tworzydlo, and C.W.J. Beenakker. Phys. Rev. Lett. **82**, 4894 (1999).
9. J.B. French and S.S.M. Wong. Phys. Lett. **33B**, 447 (1970); **35B**, 5 (1971).
10. O. Bohigas and J. Flores. Phys. Lett. **34B**, 261 (1971); **35B**, 383 (1971).
11. E.P. Wigner. Ann. Math. **67**, 325 (1958).
12. C.E. Porter (*Editor*). Statistical theories of spectra: fluctuations. Academic Press Inc. New York. (1965).
13. K.K. Mon and J.B. French. Ann. Phys. (N.Y.), **95**, 90 (1975).
14. J.B. French, V.K.B. Kota, A. Pandey, and S. Tomsovic. Ann. Phys. (N.Y.), **181**, 235 (1988).
15. V. Zelevinsky, B.A. Brown, N. Frazier, and M. Horoi. Phys. Rep. **276**, 85 (1996).
16. V.V. Flambaum, A.A. Gribakina, G.F. Gribakin, and M.G. Kozlov. Phys. Rev. A: At. Mol. Opt. Phys. **50**, 267 (1994); **58**, 230 (1998).
17. A. Pandey and J.B. French. J. Phys. A: At. Mol. Opt. Phys. **12**, L83 (1979); A. Cortes, R.U. Haq, and A.P. Zuker. Phys. Lett. **115B**, 1 (1982).
18. J.B. French and V.K.B. Kota. Phys. Rev. Lett. **51**, 2183 (1983).
19. J.B. French. Lect. Notes Phys. **209**, 100 (1984).
20. Z. Pluhar and H.A. Weidenmüller. Phys. Rev. C: Nucl. Phys. **38**, 1046 (1988); A. Pandey. Ann. Phys. (N.Y.), **134**, 110 (1981).
21. R.J. Leclair. M.Sc. Thesis. Laurentian University, Ont. Canada. 1994.
22. K.F. Ratcliff. Phys. Rev. C: Nucl. Phys. **3**, 117 (1971).
23. W. Wang, F.M. Izrailev, and G. Casati. Phys. Rev. E: Stat. Phys. Plasmas Fluids Relat. Interdiscip. Top. **57**, 323 (1998).

PHYSICAL REVIEW C, VOLUME 62, 024002

Level densities for nuclei with $20 \leq A \leq 41$

Po-lin Huang,* S. M. Grimes, and T. N. Massey
Ohio University, Athens, Ohio 45701
(Received 24 June 1999; revised manuscript received 5 January 2000; published 19 July 2000)

Some commonly used level density compilations are compared with one another and with data in the mass range $20 \leq A \leq 41$. A new formula is proposed which appears to be consistent with both low energy ($U < 5$ MeV) and high energy ($U \geq 20$ MeV) data. This new compilation is used as a benchmark for testing theoretical calculations of level densities utilizing the two-body interaction with moment methods.

PACS number(s): 21.10.Hw, 21.60.Cs, 27.30.+t

I. INTRODUCTION

Level densities are important in nuclear physics for a number of reasons. From a basic standpoint, our ability to predict the energy distribution of all of the excited levels of a nucleus provides a severe challenge to our theoretical understanding, given the size of the basis and the dependence on all facets of the interaction. A number of related areas of physics and technology are also dependent on our studies of the level density. These include nucleosynthesis studies in astrophysics, medical applications of nuclear physics, and fission and fusion reactor design.

The largest body of level density data is the set based on counting neutron resonances at low bombarding energy. Such studies yield a level density at an energy slightly above the neutron binding energy in the compound nucleus. There are some difficulties with this technique. For light nuclei, the level density is not large at this energy, if we assume that the levels observed are reached through $l=0$ absorption. At low energies this is expected to be the case, which means that a total level density can be inferred only if the parity ratio (fraction of levels with a specific parity) and the spin cutoff parameter are known. The combination of problems caused by the small number of levels of the appropriate spin and parity in the energy "window" at the binding energy, possible missing levels, and the need to know both the parity ratio and the spin cutoff parameter causes difficulty with compilations based solely on resonance parameters. Most of the sets [1–8] of level density parameters have been based on resonance counting at the binding energy.

To carry out a fit to these data, it has often been necessary to make an assumption to eliminate one of the two parameters of the Fermi gas model. Most often, it has been assumed that the energy shift δ in the Fermi gas formula was simply that due to pairing. This implicitly puts all of the shell effects in the level density parameter a. At high energies, one expects shell effects to anneal out of the level density, but this does not happen if there are shell effects in a. The Fermi gas level density form has

$$\rho(U)=\frac{1}{12\sqrt{2}}\frac{1}{\sigma a^{1/4}}\frac{\exp[2\sqrt{a(U-\delta)}]}{(U-\delta)^{5/4}}. \tag{1}$$

The parameter σ is the spin cutoff parameter and is $\langle J_z^2\rangle^{1/2}$ for levels at that energy. The density of levels with spin J at the energy U is then

$$\rho(U,J)=\frac{(2J+1)}{2\sigma^2}\exp\left[\frac{-\left(J+\frac{1}{2}\right)^2}{2\sigma^2}\right]\rho(U). \tag{2}$$

In general, σ will be energy dependent. For changes in a or δ, we obtain

$$\frac{d\rho}{\rho}=\sqrt{\frac{(U-\delta)}{a}}da-\sqrt{\frac{a}{(U-\delta)}}d\delta+\frac{3}{2(U-\delta)}d\delta. \tag{3}$$

Equation (3) follows from Eq. (1) if we adopt the usual U dependence for $\sigma(\propto U^{1/4})$. At high energies, $d\rho/\rho$ is $[\sqrt{a/(U-\delta)}]d\delta$ for shifts in energy but is $[\sqrt{(U-\delta)/a}]da$ for changes in a. Thus, level densities at high energies are very sensitive to a and are not so sensitive to δ, while the reverse is true at low energies.

To obtain the level density for levels of specified J and parity p ($=+1$ or -1), the parity ratio is introduced:

$$\Pi(U)=\frac{\rho^+(U)}{\rho^+(U)+\rho^-(U)}. \tag{4}$$

To deduce level densities from resonance counting at the binding energy, it has been necessary to make assumptions about $\Pi(U)$ and $\sigma(U)$. Ericson [9] has shown that under very general assumptions the total number of levels of positive parity is nearly equal to the number of negative parity levels, but data and shell model calculations agree in indicating that at low energies there is a tendency for levels of one parity to dominate. For even-even nuclei, the positive parity density is usually larger, while for odd A nuclei either even or odd parity levels dominate depending on the parity of the single particle orbital nearest the Fermi level. The mass range of the present study ($20 \leq A \leq 41$) is predominately in the sd shell region; for all nuclei with odd A except for $A=41$ it is expected that positive parity levels will dominate. Relatively little effort has been invested in determining the excitation energy at which the two parity level densities are equal, although it is expected that this point is reached at lower energy as A is increased.

*Present address: 94-2 Rao-an Road, Yung-an Hsiang, Kaohsiung County, Taiwan, ROC.

There are two commonly used paths to obtaining spin cutoff parameters. Ericson [9] has shown that the spin cutoff parameter should be

$$\sigma^2 = gt\langle m^2\rangle,$$

where g is the density of single particle states, t is the temperature, and $\langle m^2\rangle$ is the average of the square of the spin projection on the Z axis for single particle states near the Fermi level. g is connected to the level density parameter a through the equation

$$a = \frac{\pi^2}{6} g.$$

An early estimate of $\langle m^2\rangle$ was [10]

$$\langle m^2\rangle = 0.146A^{2/3}, \tag{5}$$

but it was later realized [11] that this estimate was unrealistic because it included all levels and not just those at the Fermi level. The earlier estimate was utilized in one of the most frequently used tabulations [1], however. It has been proposed [5] that the expression

$$\langle m^2\rangle = 0.24A^{2/3} \tag{6}$$

is a better representation of the average over levels at the Fermi energy.

The difference between expressions (5) and (6) is about 50%, which would translate into a similar discrepancy in level densities derived from resonance counting.

An alternative method of determining σ^2 is to assume that the nucleus is a rigid sphere. In this case, an assumed value for the radius of $R(A) = 1.25A^{1/3}$ fm yields a value for σ^2 of

$$\sigma^2 = 0.0145A^{5/3}t, \tag{7}$$

where the temperature is given by

$$t = \sqrt{\frac{U-\delta}{a}}, \tag{8}$$

so the forms for σ^2 become

$$\sigma^2 = \frac{6}{\pi^2} a \sqrt{\frac{U-\delta}{a}} 0.146A^{2/3} = 0.0888\sqrt{a(U-\delta)}A^{2/3}, \tag{9}$$

$$\sigma^2 = 0.1461\sqrt{a(U-\delta)}A^{2/3}, \tag{10}$$

$$\sigma^2 = 0.0145A^{5/3}\sqrt{\frac{U-\delta}{a}}. \tag{11}$$

Equation (10) results if Eq. (6) is used instead of Eq. (5), while Eq. (11) is based on the use of Eq. (7).

For values of a and δ typically used in this mass region, the σ^2 values estimated from Eq. (11) are between those predicted by Eqs. (9) and (10), but are closer to the latter.

Other databases have also been used to determine level density parameters. One analysis [12] has been based on the use of evaporation spectra to derive level density parameters for a wide range of nuclei, and other studies [13–17] have used this technique to determine parameters for a more limited set of nuclei. A good source of information about level densities at high energies is Ericson fluctuations; one compilation [18] has been based entirely on this type of information. Others [19,20] have derived sets of parameters in this mass region which incorporate information from three different types of data and similar analyses [21,22] have been prepared for specific nuclei. A recent compilation by Iljinov *et al.* [23] has utilized a very extensive body of data over a wide range in A to derive a set of level density parameters. In general, despite many years' work, the consistency of these compiled values is not as good as would be desired and many of them do not extrapolate correctly to the level density at low energy.

II. COMPILATION OF LEVEL DENSITY PARAMETERS

In general the compilations in widest use do not include information about the level density at more than one energy. It was decided to formulate a new set of parameters which would include more information in the fit. The fit of Braga-Marcazzan and Milazzo-Colli [18] (BMMC) and that of Rohr [4] were taken as starting points. The former is one of the few which is based on information from Ericson fluctuations and hence should be reliable at 20 MeV. BMMC report a values deduced individually for a number of nuclei based on the assumption that the Gilbert-Cameron [1] pairing energies could be used as the energy shift.

The Rohr compilation is based on resonance counting and so should be reliable at 7 MeV. In this case the energy shifts were also taken to be those of the Gilbert-Cameron compilation but the a values were fit with the function $a = \alpha A + C$, where A is the mass number, a is the level density parameter, and α and C are fitting constants. The use of a smooth form reduces the effect of level statistics in evaluating the level density for a particular nucleus and prevents local shell effects from appearing in a.

We have used a similar fitting function to represent the individual a values compiled by BMMC. The constants derived are slightly different from those of Rohr. While Rohr based his fit on data from more nuclei, the BMMC results, because they are at higher energy, are less likely to be biased by the choice of energy shift, since the effects of changing the energy shift vanish asymptotically as the energy increases.

It was found that using the Rohr or BMMC a values produced poor fits to the level density at low energies [24] based on comparisons with known levels for excitations below 5 MeV. This is perhaps not surprising, since the role of δ at low energies according to Eq. (3) is magnified relative to that of a.

Figures 1–3 show typical results for the Rohr and BMMC parameters compared with level counting at low energy. Note that the consistent tendency is for the predictions to underestimate the data.

The a values were, therefore, fixed at the BMMC magnitudes and a least squares fit was conducted for the δ value

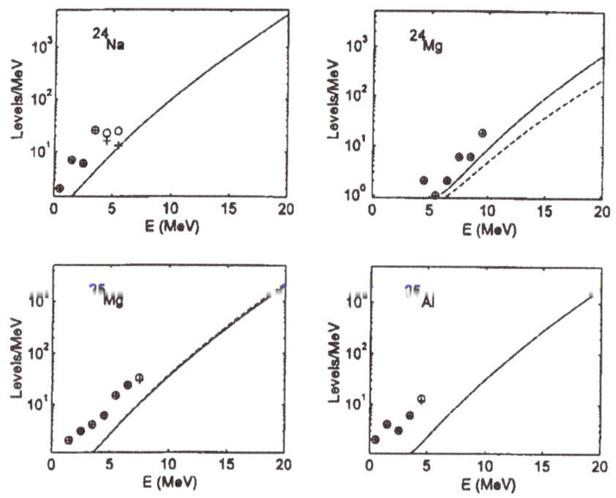

FIG. 1. Level densities for ^{24}Na, ^{24}Mg, ^{25}Mg, and ^{25}Al. In each panel, the + symbols denote the level density based on counting resolved levels of known spin and parity, the O symbols denote the level density based on counting all known levels, the solid line indicates the level density predicted by Rohr, and the dashed line is the prediction based on the systematics of Braga-Marcazzan and Milazzo-Colli.

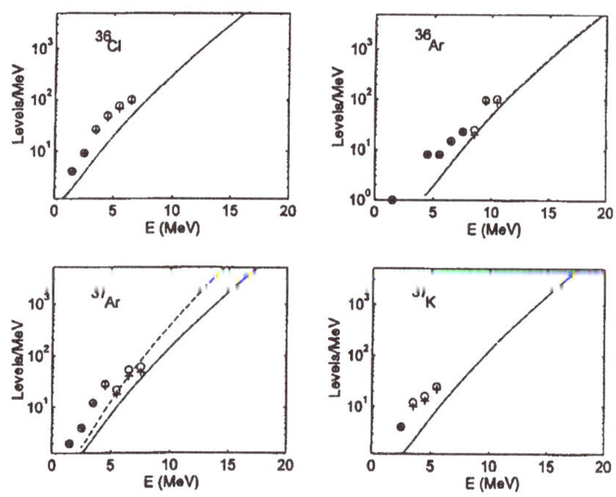

FIG. 3. Same as Fig. 1, except that the nuclei are ^{36}Cl, ^{36}Ar, ^{37}Ar, and ^{37}K.

needed to optimize the fit to the known levels at low energy. In all cases, reasonable fits could be achieved by modifying only the δ parameter. An effort to improve the agreement by fixing the δ values at their original magnitude and varying the a values was less successful, with numerous fits showing inappropriate slopes.

In retrospect, this result is not surprising. If, as indicated by Eq. (1), level densities at high energies have less sensitivity to δ than to a, an assumed value for δ would not have biased the a value significantly if the fitted point were at 20 MeV. The corrections to the values at 20 MeV caused by adjusting the δ values were about 30%, which is just within the estimated 30% error of the Ericson analysis. The best fit values for δ are listed in Table I. Also shown are the original

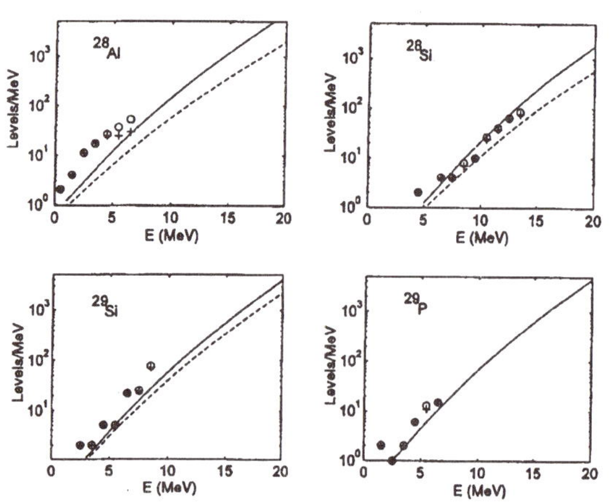

FIG. 2. Same as Fig. 1, except that the nuclei are ^{28}Al, ^{28}Si, ^{29}Si, and ^{29}P.

pairing shift values used by Rohr and BMMC; these were the pairing shifts proposed by Gilbert and Cameron and are labeled GC.

To make the formula for a useful in predicting level densities, it is necessary to find a formula for predicting δ as well. Two approaches were tried. The first is based on the use of the semiempirical mass formula. If the terms corresponding to pairing and shell corrections are dropped, the expression for the nuclear binding energy is

$$B(N,Z)=a_vA-a_sA^{2/3}-a_c\frac{Z^2}{A^{1/3}}-a_a\frac{(N-Z)^2}{A}, \quad (12)$$

where N, Z, and A are the neutron number, proton number, and mass number of the nucleus, respectively. a_v, a_s, a_c, and a_a are fitting constants. If the actual binding energy of the nucleus is compared with the value predicted by this formula, the difference should be the shift in the ground state energy caused by pairing and shell effects. Reasonable agreement was found between the predictions for δ using this formula with the parameters of Myers and Swiatecki [25]. Somewhat better results were obtained with the parameters deduced by Grimes [26], which are also listed in Table II. The latter parameters were deduced by fitting the parameters of this equation to nuclear masses *only* in the region 20≤A≤42.

The energy shifts in Table I under "Myers-Swiatecki" and "Grimes" should be compared with those under "Best fit"; the deviations are slightly smaller for the parameters of Grimes than for Myers and Swiatecki.

An alternative approach is based not on experimental mass values but on the systematics of the two-body interaction. The normal assumption used in pairing models is that the only interaction is between identical nucleons in states in which they couple to 0^+. This would yield a value of 0 for the energy shift of an odd-odd nucleus. The fits gave results which were normally not close to zero for odd-odd nuclei. This phenomenon has been observed previously and is the

TABLE I. Level density energy shifts.

	Myers-Swiatecki	Grimes	GC	Best fit	AIM
^{20}F	−4.39	−3.04	0.00	−2.43	−2.44
^{20}Ne	2.07	2.44	4.87	3.80	2.44
^{21}Ne	−1.84	−1.07	2.10	−0.10	−0.70
^{21}Na	−0.86	−0.96	2.77	−0.93	−0.92
^{22}Na	−2.80	−2.57	0.00	−1.95	−2.57
^{22}Ne	−0.17	1.04	4.77	1.80	2.39
^{23}Na	−1.30	−0.68	2.67	−0.10	−0.89
^{23}Mg	−0.50	−0.75	2.46	−0.70	−0.82
^{24}Na	−3.44	−2.39	0.00	−2.48	−2.57
^{24}Mg	2.97	3.06	5.13	3.60	2.57
^{25}Mg	−0.84	−0.37	2.46	−0.30	−0.82
^{25}Al	0.06	−0.33	2.67	−0.89	−0.89
^{26}Mg	0.80	1.70	4.26	1.70	2.13
^{26}Al	−1.69	−1.74	0.00	−1.50	−1.95
^{27}Al	0.02	0.36	1.80	0.50	−0.60
^{27}Si	0.70	0.19	2.09	0.00	−0.70
^{28}Al	−2.01	−1.26	0.00	−1.94	−1.95
^{28}Si	4.72	4.54	3.89	3.85	1.95
^{29}Si	1.67	1.87	2.09	0.80	−0.70
^{29}P	2.52	1.88	1.80	0.70	−0.60
^{30}P	0.63	0.32	0.00	−0.50	−1.65
^{30}Si	2.24	2.85	3.76	1.88	1.88
^{31}P	1.26	1.32	1.67	0.65	−0.56
^{31}S	1.96	1.18	1.62	0.50	−0.54
^{32}P	−1.10	−0.62	0.00	−1.35	−1.65
^{32}S	3.73	3.29	3.29	2.30	1.65
^{33}S	0.54	0.48	1.62	0.10	−0.54
^{33}Cl	1.34	0.46	1.67	0.00	−0.56
^{34}S	1.44	1.79	3.48	1.40	1.74
^{34}Cl	−0.62	−1.02	0.00	−1.25	−1.74
^{35}Cl	0.21	0.04	1.86	−0.10	−0.62
^{35}Ar	0.92	−0.08	1.62	−0.54	−0.54
^{36}Cl	−1.93	−1.70	0.00	−1.74	−1.74
^{36}Ar	2.80	2.14	3.48	1.70	1.74
^{37}Ar	−0.50	−0.79	1.62	−0.30	−0.54
^{37}K	0.30	−0.79	1.86	−0.35	−0.62
^{38}Ar	0.43	0.55	3.66	1.40	1.83
^{38}K	−1.05	−1.80	0.00	−1.20	−1.94
^{39}K	−0.18	−0.56	2.04	0.30	−0.68
^{39}Ca	0.52	−0.66	1.83	−0.20	−0.61
^{40}K	−3.46	−3.44	0.00	−1.80	−1.94
^{40}Ca	2.69	1.85	3.87	2.20	1.94
^{41}Ca	−1.26	−1.73	1.83	−2.20	−0.77
^{41}Sc	−0.25	−1.52	2.04	−0.30	−0.68

basis for the "backshifted" Fermi gas model [3,7,19]. This model includes an additional energy shift (beyond conventional pairing) which produces an enhancement in the level density. This change has been based on empirical systematics but is probably best justified by noting that an energy shift of this magnitude and sign is possibly serving to produce the level density enhancement that collective states produce.

TABLE II. Parameters for level density fit. Set A: $a=0.086A+1.18$. δ = binding energy $-a_vA+a_sA^{2/3}+a_cz^2/A^{1/3}+a_a(N-Z)^2/A$ Set B: $a=0.086A+0.23$. $\delta=$ binding energy $-a_vA+a_sA^{2/3}+a_cZ^2/A^{1/3}+a_a(N-Z)^2/A-2.7+0.025A$.

	Myers-Swiatecki (MeV)	Grimes (MeV)
a_v	15.56	15.598
a_s	17.23	17.97
a_c	0.697	$(0.8176-0.6457)A^{-1/3}$
a_a	23.285	24.1

It should be noted that the conventional pairing shift, which is 0 for odd-odd nuclei, $\Delta_p(\Delta_n)$ for odd-A nuclei with even $Z(N)$ and $\Delta_p+\Delta_n$ for even-even nuclei, is also "backshifted." The BCS formulation [12] has a condensation energy term in the solution; if this term is ignored, it essentially represents a "backshift" for the conventional pairing shift model and an additional "backshift" for the conventional backshifted model.

Comparison of the observed "best-fit" energy shifts with the predictions of the simple pairing model suggested an alternative to the semiempirical mass formula for obtaining the correct energy shift. Since the shifts would be such as to make the correction for even-even and odd-odd nuclei about equal in magnitude and opposite in sign, the two cases were shifted in size by half of the traditional correction. Instead of using $\Delta_p+\Delta_n$ for even-even nuclei and 0 for odd-odd, the shifts were set to $\frac{1}{2}(\Delta_p+\Delta_n)$ and $-\frac{1}{2}(\Delta_p+\Delta_n)$, respectively, where Δ_p and Δ_n were taken from Gilbert and Cameron. To be consistent with the simple pairing model, the energy shift for even Z-odd N and odd Z-even N nuclei should be set to $\frac{1}{2}(\Delta_p-\Delta_n)$ and $\frac{1}{2}(\Delta_n-\Delta_p)$, respectively. It is found that better agreement with the observed shifts occurs if this shift is set to $-\frac{1}{3}\Delta_p$ and $-\frac{1}{3}\Delta_n$, respectively. These values are also tabulated in Table I. Note that this average interaction method (AIM) gives results which are roughly as good as those obtained with the semiempirical mass equation.

The results of the fits with modified δ values are shown in Figs. 4–6 for typical nuclei. Note the considerably better fits shown in Figs. 4–6 compared to those in Figs. 1–3.

In order to test the effect of the modified δ values on the level densities at 20 MeV, a second set of level density parameters, denoted set B, was also deduced. For set B, the parameters of set A were adjusted so as to preserve the level density value at 5 MeV while restoring the value at 20 MeV to that of the Braga-Marcazzan parameters before the energy shift was changed. This fit has consistently lower a values and slightly lower (more negative) energy shifts as well. Table II lists the parameters of set B as well as those of set A.

III. COMPARISON WITH THEORY

The compilations are based on the macroscopic Fermi-gas model. This model was introduced by Bethe and utilized the

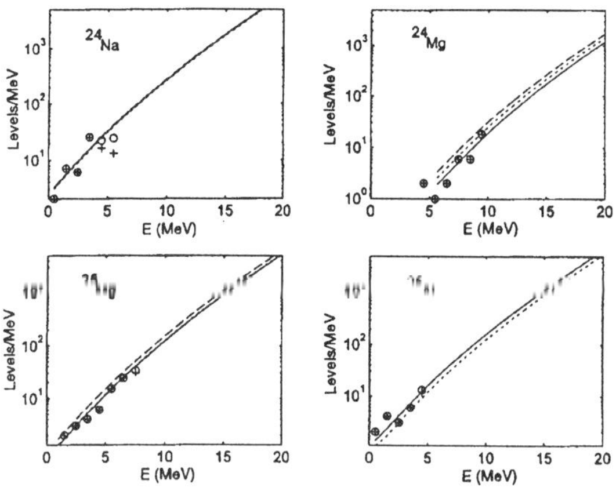

FIG. 4. Level densities of ^{24}Na, ^{24}Mg, ^{25}Mg, and ^{25}Al. The + symbols and ○ symbols are as explained in Fig. 1. The solid line is the best-fit value of δ used with the a value systematics of Table II, while the short dashed and long dashed curves use the same a value but are based on the systematics of Grimes and the AIM method, respectively.

assumption that the single-particle state density is constant with energy.

This assumption can be relaxed and the partition function evaluated for a "realistic" set of single-particle levels. This model is called the microscopic Fermi-gas model. It has the virtue that shell effects should automatically be included in the calculation. Further, by using the BCS [27] formalism, the effects of pairing can also be included.

Single-particle energy sets proposed by Nilsson [28], Seeger and Perisho [29], and Seeger and Howard [30] were used in these calculations. The level density values predicted by the model with the single-particle energies of Nilsson were too high for all of the nuclei studied. For nuclei with A below 28, the level scheme of Seeger-Perisho gave the best results. Near this value for A, both Seeger-Perisho and Seeger-

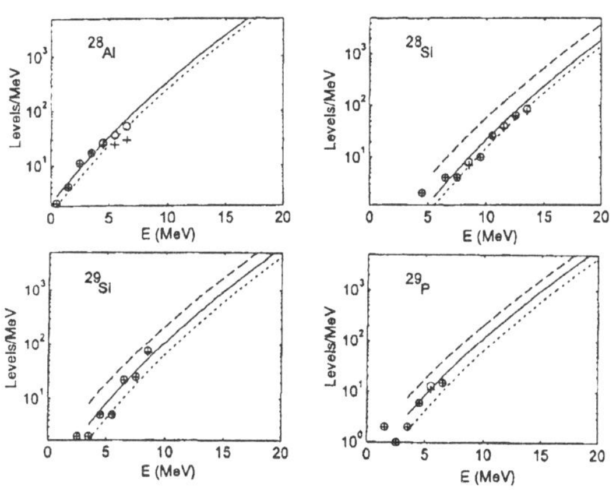

FIG. 5. Level densities of ^{28}Al, ^{28}Si, ^{29}Si, and ^{29}P. The significance of the curves and symbols is as indicated in the caption of Fig. 4.

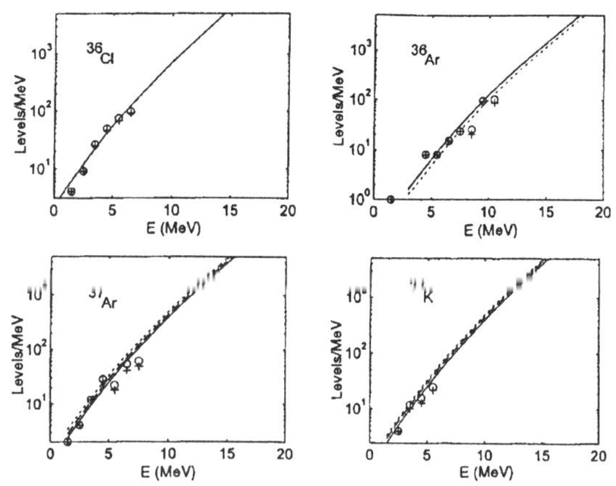

FIG. 6. Level densities of ^{36}Cl, ^{36}Ar, ^{37}Ar, and ^{37}K. The significance of the curves and symbols is as indicated in the caption of Fig. 4.

Howard levels provided a good representation of the data. Beyond $A=30$, the predictions of the model with the levels of Seeger and Howard were superior to those of Seeger and Perisho. Comparisons are shown in Figs. 7–12.

These single-particle schemes also provide reasonable representations of the spin-cutoff parameter. As has already been indicated, this quantity is especially important. It is used in Hauser-Feshbach calculations and is also necessary in interpreting the level density information provided by neutron resonance counting. A conclusive test of the spin-cutoff parameter predictions is difficult because the database for this parameter is somewhat sparse.

One additional parameter is also of interest. The parity ratio represents the fraction of the levels at a given excitation energy which have positive parity. This parameter is expected to be 0.5 at high energy but approaches 0 or 1 (depending on the orbital parity of the level closest to the Fermi level and whether the nucleus is even or odd) at low energy. As is the case for the spin-cutoff parameter, knowledge of this parameter is important in extracting information about level densities from neutron resonance counting. It has been pointed out [31] that the statistical mechanical approach, though reliable for level densities and spin-cutoff parameters, is not accurate in calculating parity ratios. This is because the formalism in evaluating the partition function produces the correct average N and Z but does have some dispersion in these quantities. Thus, the level density is an average of the level density for the nucleus with (N,Z) and small fractions of the corresponding densities for $(N-1,Z+1)$ and $(N+1,Z-1)$. Since the level densities are similar, it is possible to obtain [31] accurate values for the level density despite this averaging. The consequences of the averaging [31] are much more dramatic for the parity ratio and result in an energy dependence for the parity ratio that cannot overshoot 0.5, in contrast to more realistic calculations.

The two primary drawbacks to the Fermi-gas model are that no interactions (other than pairing) are included and that parity ratios are not reliable. This latter parameter is of par-

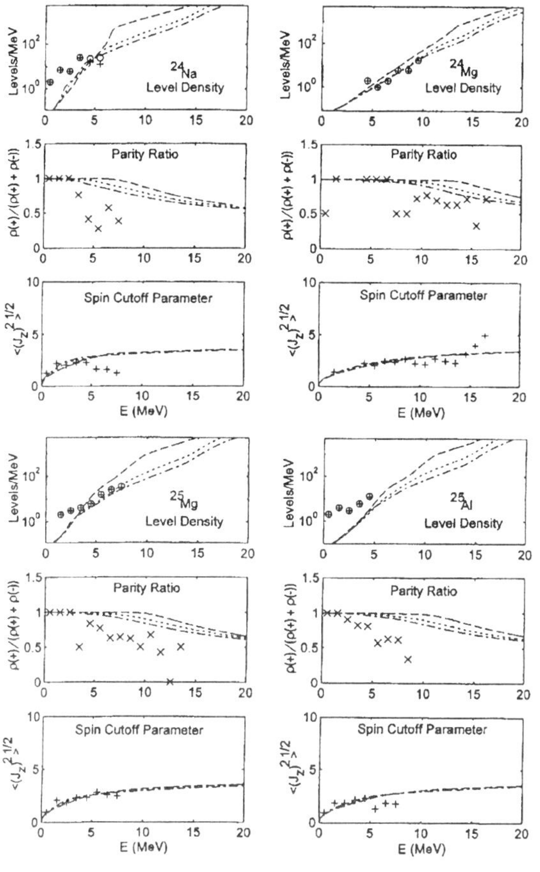

FIG. 7. Level density, parity ratio, and spin-cutoff parameter for ^{24}Na and ^{24}Mg. The long-dashed curve is the calculation based on levels proposed by Nilsson, the short-dashed curve is based on levels proposed by Seeger and Perisho, and the dot-dashed line is the result based on levels proposed by Seeger and Howard.

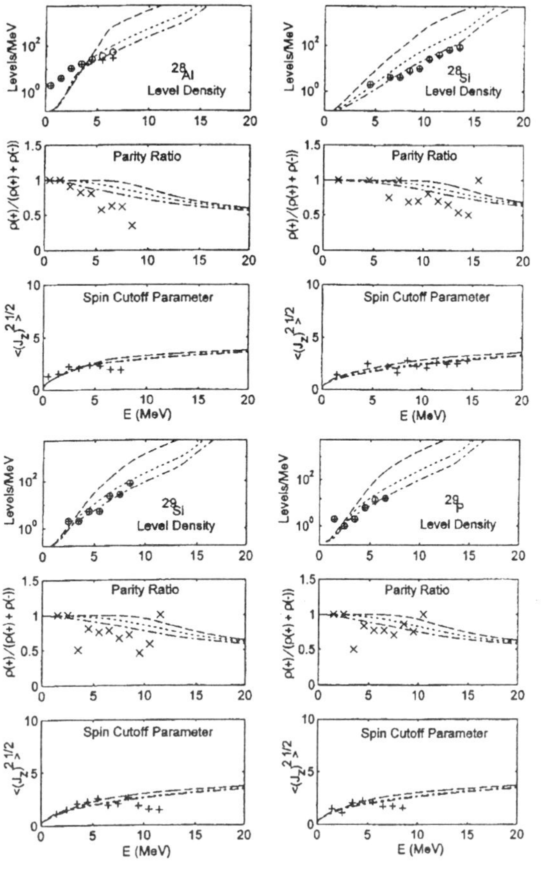

FIG. 8. Same as Fig. 7, except that the nuclei are ^{25}Mg and ^{25}Al.

ticular interest for light nuclei, since there is considerable evidence that it has not reached its asymptotic value of 0.5 at energies at which neutron resonance counts are made.

A formalism has been developed which allows full inclusion of the two-body force in such calculations. The first such calculations included the expectation values of $\langle H\rangle$ and $\langle H^2\rangle$ in an expansion of the level density [32–36], but more recent work has also included moments as high as $\langle H^4\rangle$ [37–39]. These calculations utilize the result of unitarity to allow calculation of moments of the Hamiltonian in the nondiagonal shell model basis. With values for these moments, an expansion for the level density itself can be calculated.

One approach to making such calculations is to use Hermite polynomial expansions. These allow a quick and efficient calculation of expansion coefficients and are consistent with asymptotic Gaussian behavior. Unfortunately, they also produce distributions which are not positive definite. Often, the negative excursions come in the low-excitation-energy region, making it impossible to use the calculations in the region of most physical interest.

One possible solution to this problem is to expand the moment information in the form of delta functions [40–42]. These obviously do not have negative excursions but have the drawback of being discrete instead of continuous. Nonetheless, the use of many Slater determinants to make the expansion produces a quasicontinuous distribution. The resulting distribution is, however, somewhat too narrow in the tail region.

A recent paper [43] has suggested an alternative procedure. If the Gaussian function is replaced by an exponential with polynomial argument, the resulting distribution is both positive definite and continuous. The present results are based on the use of the exponential with polynomial argument (EPA) technique.

Input parameters needed for such calculations include single-particle energies and two-body matrix elements. Finding suitable two-body matrix elements is somewhat complicated. In some cases, sets of two-body matrix elements are available which have been derived from fitting shell model calculations to low-lying levels. These are not suited to the present calculations because they normally do not have a procedure for extending the calculation to neighboring orbits.

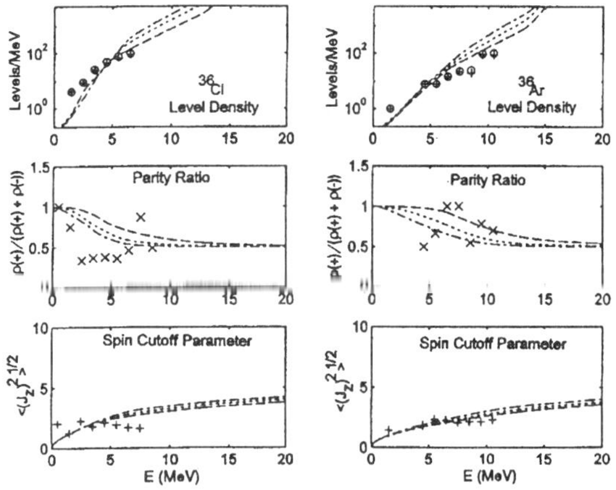

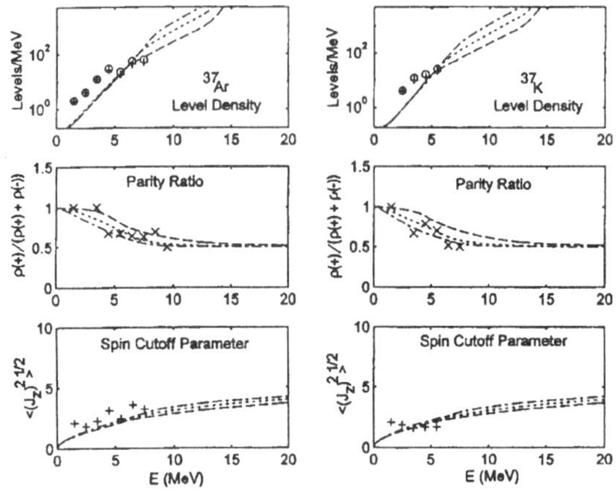

FIG. 9. Same as Fig. 7, except that the nuclei are ^{28}Al and ^{28}Si.

A set of matrix elements based on specified parameters for a potential is required. The Petrovich-McManus-Madsen-Atkinson (PMMA) set proposed by Petrovich *et al.* [44] was chosen. It was found that they had a high correlation coefficient with the empirically derived set of Warburton and Brown [45]. The PMMA parameters are fit to Yukawa potentials in the even states only (the triplet and singlet even potentials are nonzero; the triplet and singlet odd potentials are set to zero). The PMMA potential has been used with some success in shell model calculations [46].

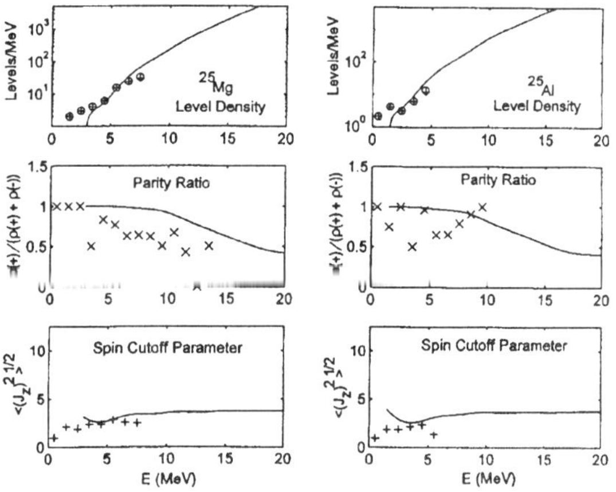

FIG. 11. Same as Fig. 7, except that the nuclei are ^{36}Cl and ^{36}Ar.

There are three particular challenges in finding the appropriate input for these calculations. Single-particle energies are supposed to include the effects of the interaction with "core" particles and the kinetic energy; for these calculations, the core is much smaller than in conventional shell model calculations. This would argue for single-particle energies which are not as spread out as those in a conventional calculation. The authors of a similar study [47] to this one carried out in the mass region near $A=60$ have found it necessary to make this kind of adjustment. A second complication is in the strength of the two-body force. Since our basis is much larger than that in a conventional shell model calculation, the interaction strength should be somewhat weaker. This change is expected because the interaction in a

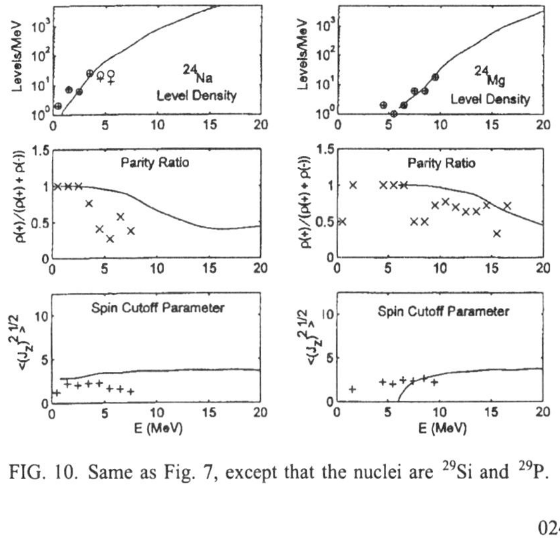

FIG. 10. Same as Fig. 7, except that the nuclei are ^{29}Si and ^{29}P.

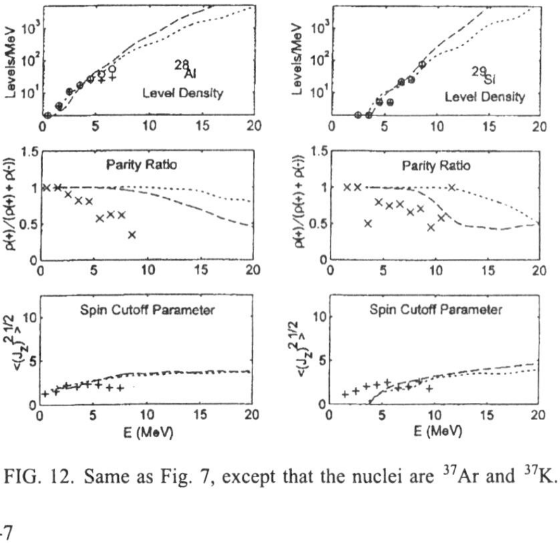

FIG. 12. Same as Fig. 7, except that the nuclei are ^{37}Ar and ^{37}K.

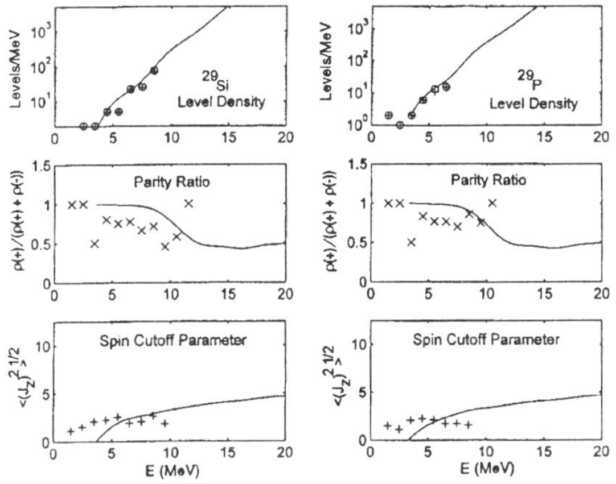

FIG. 13. Level density, parity ratio, and spin-cutoff factors for ^{24}Mg and ^{24}Na. The calculated values curves are based on use of a two-body force with the moment method; experimental points are based on known levels and their spin and parity values.

small space must be renormalized to compensate for missing orbits. To obtain proper fragmentation of states, the interaction needs to be strengthened compared to the appropriate strength in a large basis. We must, therefore, undo this renormalization. Finally, most shell model calculations have focused on calculating levels of one parity. This leaves out some information about the interaction.

Initial calculations with the PMMA force found that the force was too strong. A reduction factor of 0.7 was found to be appropriate. This reduction still did not resolve all problems in achieving a fit to the data. A simultaneous fit to the level density and spin cutoff parameter required adjustment of the single-particle energies as well as the PMMA force. Even these adjustments did not bring the parity ratio into agreement with data. This parameter was not included in the study of Kar and Majumdar [47], although the interaction used by these authors was not the PMMA interaction.

To obtain a simultaneous fit to the level density, spin-cutoff factor, and parity ratio, it was necessary to add a component of the two-body force in odd states. This magnitude was set to $\frac{3}{7}$ of the force strength of the even-state interaction and it was given the opposite sign.

Results of these calculations are shown in Figs. 13–17. A generally good representation of the level density, parity ratio, and spin cutoff factor is obtained. Table III lists the input parameters.

As has been pointed out in the Introduction, an understanding of the variation of both the spin-cutoff factor and parity ratio with energy and with A is central to a level density study based on neutron resonance counting. Both the spectral distribution method and the microscopic Fermi-gas formalism seem to give reasonable results for the level density and spin-cutoff factor for these nuclei. Only the spectral distribution method can give reliable predictions for the parity ratio. Comparison of the present calculations with the compiled values discussed in Sec. II shows reasonable agreement. The calculated level densities are normally within 35%

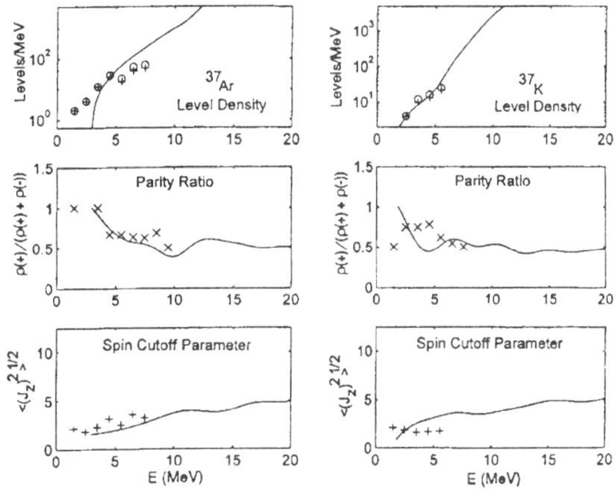

FIG. 14. Same as Fig. 13, except that the nuclei are ^{25}Mg and ^{25}Al.

of the predicted values over the energy range 3–15 MeV when compared with the set A parametrization. The predictions of this compiled set are typically above results of the moment method level density calculation. set B parameters (Table II) give very similar results to those of set A for energies below 10 MeV but show a less rapid rise above this energy. They show a slightly better agreement with the moment method calculation over the 3–15 MeV range than does set A.

IV. COMPARISON WITH OTHER COMPILATIONS

As discussed in Sec. I, the most extensively used compilations do not always predict level densities which are consistent for particular nuclei. Two of the leading compilations, Gilbert and Cameron [1] and Rohr [4], made different assumptions about the A dependence of σ, while both assumed that the value of $\Pi(U)$ was 0.5 at the binding energy. To see how these assumptions affected their analysis, a correction

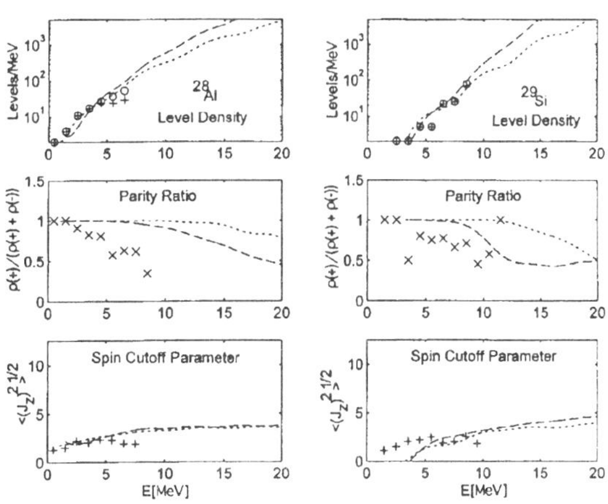

FIG. 15. Same as Fig. 13, except that the nuclei are ^{28}Al and ^{28}Si.

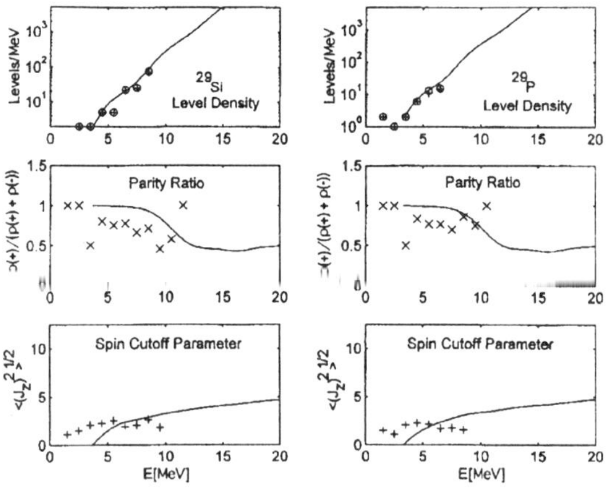

FIG. 16. Same as Fig. 13, except that the nuclei are ^{29}Si and ^{29}P.

factor was calculated based on the calculated σ and Π at the binding energy based on the moment method results. These factors are tabulated in Table IV and indicate that the predictions of Gilbert and Cameron should be increased by about 30% and those of Rohr decreased by 10% in this mass region. Other differences with compilations based on neutron resonances, including assumptions about energy shifts and poor statistics on level counting, cannot be corrected in as direct a fashion.

In Table V we present a comparison of level densities predicted by a number of compilations for specific nuclei. The column labeled "present results" lists level densities (set B); the other columns give the ratio of the level density predicted by that compilation to the present results. Note that in many cases deviations of 50% or more are common.

Examination of the comparison in Table V indicates that the present results (set B) give a consistently slower increase of level density with energy than the results of Rohr. This is consistent with the fact that the present results are "backshifted." A correlated observation is that Rohr did not include information about level densities above 10 MeV in his fit. Comparison with Gilbert and Cameron is challenging, since those authors present a set of "systematic" parameters which vary between spherical and deformed nuclei. In addition, they also present specific a and δ values for some nuclei, which are not consistent with the "systematics." The values indicated as Gilbert-Cameron 1 are from systematics; those labeled Gilbert-Cameron 2 are specifically listed for that nucleus. Again, because the Gilbert-Cameron values are not "backshifted," they predict level densities which rise more rapidly than the present values. The predictions of Beckerman [20] agree better with the present results, perhaps because high-energy data were also included in the fit.

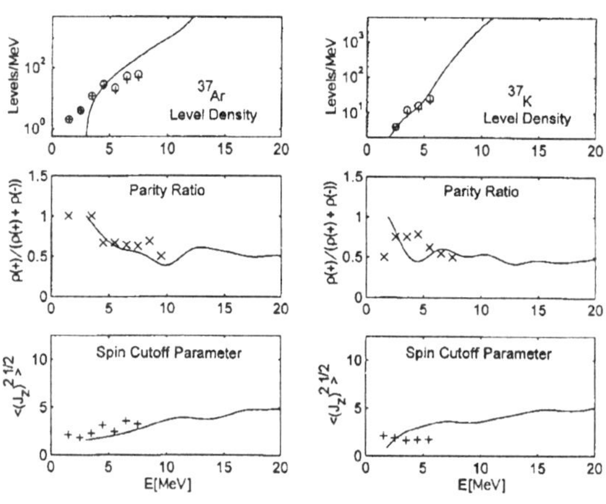

FIG. 17. Same as Fig. 13, except that the nuclei are ^{37}Ar and ^{37}K.

TABLE III. Single-particle energies.

$$E(n,L)=\frac{17.4}{A^{1/3}}(n+1)+0.4L-24.7+\frac{2.82}{A^{1/3}}$$
$$\times(-L) \text{ for } J=L+\frac{1}{2}$$
$$\times(+L+1) \text{ for } J=L-\frac{1}{2}$$

$n=2$ for $p_{1/2}\,p_{3/2}$ shell
$n=3$ for $d_{5/2}\,s_{1/2}\,d_{3/2}$ shell
$n=4$ for $f_{7/2}\,p_{3/2}\,p_{1/2}\,f_{5/2}$ shell

Not shown in the table but in better agreement with the present results are level densities predicted by Gadioli and Zetta [19] and Iljinov *et al.* [23]. The former is slightly above the present results at energies above 12 MeV. Iljinov *et al.* have lower values for a but include a rotational enhancement. The net effect of these two changes is to produce level densities which are generally within 20% of the present results.

Recent compilations have included specific correction factors for level density enhancements due to collective effects [23,48]. Many earlier compilations did not. It may seem surprising that although the enhancement factors are substantial, the fits without the factors can also describe the data. Indeed, one of the compilations [23] presents fits both with and without enhancement factors.

Although the enhancement factors are substantial, they apparently do not vary rapidly with energy. A study of the rotational enhancement factor [49] gives a very slow reduction in magnitude as the energy increases, with the enhancement vanishing at energies above 40 MeV. The nearly constant enhancement can be represented by increasing a or reducing δ, if knowledge of the detailed energy dependence

TABLE IV. Correction factors.

A	Gilbert-Cameron	Rohr
20	1.3	0.9
24	1.2	0.85
28	1.15	0.8
32	1.3	0.9
36	1.35	0.95
40	1.3	0.9

TABLE V. Comparison of level density compilations.

Nucleus	Excitation energy (MeV)	Level density present results	Ratio to present results			
			Rohr	Beckerman	Gilbert-Cameron 1	Gilbert-Cameron 2
^{24}Na	5	22	0.4	1.4	1	–
	10	140	0.7	1.6	2	–
	15	690	1.0	2.0	3.8	–
	20	3020	1.4	2.2	6.1	–
^{24}Mg	5	1.5	–	0.8	–	–
	10	14	0.6	1.2	1.1	0.7
	15	94	1.0	1.35	1.9	1.2
	20	500	1.3	1.4	3.3	1.8
^{25}Mg	5	10	0.3	1.0	0.3	0.3
	10	75	0.5	1.2	1.0	1.0
	15	460	0.7	1.3	1.6	1.6
	20	2040	1.0	1.6	2.7	3.7
^{25}Al	5	13	0.2	0.8	0.2	–
	10	88	0.4	1.0	0.5	–
	15	515	0.6	1.1	0.7	–
	20	2406	0.8	1.3	1.0	–
^{30}Si	5	4.2	0.4	1.22	0.4	0.3
	10	55	0.5	1.0	0.6	0.6
	15	420	0.8	1.0	0.8	1.0
	20	2530	1.0	1.0	1.1	1.5
^{30}P	5	24	0.6	0.6	0.4	0.5
	10	210	0.8	0.6	0.4	0.8
	15	1360	1.1	0.8	0.45	1.0
	20	7270	1.4	0.9	0.5	1.3
^{31}P	5	10	0.5	0.9	0.5	0.6
	10	106	0.8	1.0	0.6	1.2
	15	780	1.1	1.1	1.0	1.8
	20	4620	1.3	1.4	1.3	2.4
^{32}S	5	4	0.5	0.8	0.5	0.4
	10	56	0.7	0.8	0.5	0.5
	15	470	1.0	0.9	0.5	0.6
	20	3070	1.2	0.8	0.5	0.7

of the level density is not extensive. Early research on level densities in this mass region found a need to "backshift" the level density in order to match both the magnitude of the level density at the binding energy and the slope and magnitude of the level density at lower energies; a similar result was found in the present analysis. It is plausible that collective effects are the cause of this shift. In addition to the fact that collective effects enhance the level density, which is the same direction of change caused by the energy shift, the energy shift is correlated with ground state binding energies. One expects the shifts in these numbers to enhance the level density between closed shells relative to that at closed shells; the same behavior is expected for collective effects.

An additional approach has been taken to deal with shell effects. Ignatyuk *et al.* [48] have proposed the use of an a which varies with energy:

$$a=a_0\left(1-\frac{\delta}{E}(1-e^{-\gamma E})\right). \tag{13}$$

In this expression a is the (energy-dependent) level density parameter, a_0 is the asymptotic value of the level density parameter at high energies, δ is a shell shift, and γ is a convergence parameter ($\gamma\approx 0.05$). For typical values of the energy shift ($|\delta|\leq 2$ MeV), this form yields a value for a which is reduced (increased) about 10% for δ positive (negative) at low energy. It has moved about halfway to the asymptotic value at $E=25$ MeV. Thus, this approach yields a very slowly energy dependent a. As is the case for collective enhancement, the resulting energy dependence is similar to that of the conventional Fermi gas.

V. SUMMARY

An examination of the assumptions of some level density compilations has led to the development of a new compilation. This set of parameters has as its basis level density values derived from Ericson fluctuation measurements at energies above 20 MeV, but parameter adjustments produced a reasonable description of data below 5 MeV as well. The values for level density parameters from the present compilation were used in Hauser-Feshbach calculations which were compared with data by Bateman *et al.* [50]; reasonable agreement was found.

This set of parameters is compared with calculations based on BCS quasiparticles and one including the two-body force. In both cases, additional information helpful in testing the calculation included spin-cutoff parameters and parity ratios. Although these parameters are also useful in interpreting experimental information and in making Hauser-Feshbach calculations, the database is not as extensive as would be desirable.

Reasonable fits to the data were achieved with the two-body force calculations. It was possible to use the comparison with data to refine the input parameters.

The present results confirm a previous analysis which showed that including two-body forces in a large basis required a "compression" of the single-particle spectrum. Furthermore, comparison of parity ratio predictions with data gave indication of a need for an odd component to the two-body force.

More extensive study of this formalism will be needed in order to make it useful in *a priori* predictions.

[1] A. Gilbert and A. G. W. Cameron, Can. J. Phys. **43**, 1446 (1965).

[2] W. Dilg, W. Schantl, H. Vonach, and M. Uhl, Nucl. Phys. **A217**, 269 (1973).

[3] H. Vonach and M. Hille, Nucl. Phys. **A127**, 289 (1969).

[4] G. Rohr, Z. Phys. A **318**, 299 (1984).

[5] U. Facchini and E. Saetta-Menichella, Energ. Nucl. (Milan) **15**, 54 (1967).

[6] A. V. Ignatyuk, G. N. Smirenkin, and A. S. Tishin, Yad. Fiz. **21**, 485 (1975) [Sov. J. Nucl. Phys. **21**, 255 (1975)].

[7] J. R. Huizenga, A. N. Behkami, J. S. Swintek, and R. W. Atcher, Nucl. Phys. **A223**, 577 (1974).

[8] J. R. Huizenga, A. N. Behkami, R. W. Atcher, J. S. Swintek, H. C. Britt, and H. Freiesleben, Nucl. Phys. **A223**, 589 (1974).

[9] T. Ericson, Adv. Phys. **9**, 425 (1960).

[10] J. H. D. Jensen and J. M. Luttinger, Phys. Rev. **86**, 907 (1952).

[11] D. W. Lang, Nucl. Phys. **77**, 545 (1966).

[12] S. M. Grimes, J. D. Anderson, J. W. McClure, B. A. Pohl, and C. Wong, Phys. Rev. C **10**, 2373 (1974).

[13] J. Benveniste, G. Merkel, and A. Mitchell, Phys. Rev. **174**, 1357 (1968).

[14] C. C. Lu, L. C. Vaz, and J. R. Huizenga, Nucl. Phys. **A190**, 229 (1972).

[15] L. C. Vaz, C. C. Lu, and J. R. Huizenga, Phys. Rev. C **5**, 463 (1972).

[16] J. C. Pacer, J. Wiley, C. R. Lux, and N. T. Porile, Nucl. Phys. **A226**, 413 (1974).

[17] J. Wiley, J. C. Pacer, C. R. Lux, and N. T. Porile, Nucl. Phys. **A212**, 1 (1973).

[18] G. Braga-Marcazzan and L. Milazzo-Colli, Energ. Nucl. (Milan) **15**, 186 (1968).

[19] E. Gadioli and L. Zetta, Phys. Rev. **167**, 1016 (1968).

[20] M. Beckerman, Nucl. Phys. **A278**, 333 (1977).

[21] V. Mishra, N. Boukharouba, C. E. Brient, S. M. Grimes, and R. S. Pedroni, Phys. Rev. C **49**, 750 (1984).

[22] F. B. Bateman, S. M. Grimes, N. Boukharouba, V. Mishra, C. E. Brient, R. S. Pedroni, and T. N. Massey, Phys. Rev. C **55**, 133 (1997).

[23] A. S. Iljinov, M. V. Mebel, N. Bianchi, E. Di Sanctis, C. Guaraldo, V. Lucherini, V. Muccifora, E. Polli, A. R. Reolon, and P. Rossi, Nucl. Phys. **A543**, 517 (1992).

[24] The level density at low energies was deduced by summing the number of levels listed in P. M. Endt, Nucl. Phys. **A521**, 1 (1990), in 1 MeV bins. This same database was used to calculate $\sigma(=\langle J_z^2\rangle^{1/2})$ and Π in each bin.

[25] W. D. Myers and W. J. Swiatecki, Nucl. Phys. **81**, 1 (1966).

[26] S. M. Grimes (private communication).

[27] J. N. Bardeen, L. N. Cooper, and J. R. Schrieffer, Phys. Rev. **108**, 1125 (1957).

[28] S. G. Nilsson, K. Dan. Vidensk. Selsk. Mat. Fys. Medd. **29**, 16 (1955).

[29] P. A. Seeger and R. C. Perisho, Los Alamos Scientific Laboratory Report No. LA-3751, 1967.

[30] P. A. Seeger and W. M. Howard, Nucl. Phys. **A238**, 491 (1975).

[31] S. M. Grimes, Phys. Rev. C **38**, 2362 (1988).

[32] H. Banerjee and J. B. French, Phys. Lett. **23**, 245 (1966).

[33] J. B. French, Phys. Lett. **23**, 248 (1966).

[34] F. S. Chang, J. B. French, and K. F. Ratcliff, Phys. Lett. **23**, 251 (1966).

[35] J. B. French and K. F. Ratcliff, Phys. Rev. C **3**, 94 (1971).

[36] K. F. Ratcliff, Phys. Rev. C **3**, 117 (1971).

[37] G. D. Lougheed and S. S. M. Wong, Nucl. Phys. **A243**, 215 (1975).

[38] S. M. Grimes, S. D. Bloom, R. F. Hausman, Jr., and B. J. Dalton, Phys. Rev. C **19**, 2378 (1979).

[39] S. M. Grimes, S. D. Bloom, H. K. Vonach, and R. F. Hausman, Phys. Rev. C **27**, 2893 (1983).

[40] B. Strohmaier, S. M. Grimes, and H. Satyanarayana, Phys. Rev. C **36**, 1604 (1987).

[41] B. Strohmaier, S. M. Grimes, and H. Satyanarayana, Phys. Rev. C **36**, 1604 (1987).

[42] B. Strohmaier and S. M. Grimes, Z. Phys. A **329**, 431 (1988).

[43] S. M. Grimes and T. N. Massey, Phys. Rev. C **51**, 606 (1995).

[44] F. Petrovich, H. McManus, V. A. Madsen, and A. J. Atkinson, Phys. Rev. Lett. **22**, 895 (1969).

[45] E. K. Warburton and B. A. Brown, Phys. Rev. C **46**, 923 (1992).

[46] C. Wong, S. D. Bloom, S. M. Grimes, R. F. Hausman, Jr., and V. A. Madsen, Phys. Rev. C **18**, 2052 (1978).

[47] V. K. B. Kota and D. Majumdar, Nucl. Phys. **A604**, 129 (1996).
[48] A. V. Ignatyuk, K. K. Istekov, and G. N. Smirenkin, Yad. Fiz. **29**, 875 (1979) [Sov. J. Nucl. Phys. **29**, 450 (1979)].
[49] G. Hansen and A.S. Jensen, in "Proceedings of the IAEA Advisory Group Meeting on Basic and Applied Problems of Nuclear Level Densities," Brookhaven National Laboratory, Upton, NY, 1983, Brookhaven National Lab Report No. BNL-NCS-51694, 1983, p. 161.
[50] F. B. Bateman, R. C. Haight, M. B. Chadwick, S. M. Sterbenz, S. M. Grimes, and H. K. Vonach, Phys. Rev. C **60**, 064609 (1998).

PHYSICAL REVIEW C, VOLUME 62, 054607

Statistical theory of parity nonconservation in compound nuclei

S. Tomsovic
Department of Physics, Washington State University, Pullman, Washington 99164

Mikkel B. Johnson, A. C. Hayes, and J. D. Bowman
Los Alamos National Laboratory, Los Alamos, New Mexico 87545
(Received 22 November 1999; published 10 October 2000)

We present the first application of statistical spectroscopy to study the root-mean-square value of the parity nonconserving (PNC) interaction matrix element M determined experimentally by scattering longitudinally polarized neutrons from compound nuclei. Our effective PNC interaction consists of a standard two-body meson-exchange piece and a "doorway" term to account for 0^- spin-flip excitations. Strength functions are calculated using realistic single-particle energies and a residual strong interaction adjusted to fit the experimental density of states for the targets, ^{238}U for $A \sim 230$ and ^{104,105,106,108}Pd for $A \sim 100$. Using the standard Desplanques-Donoghue-Holstein estimates of the weak PNC meson-nucleon coupling constants, we find that M is about a factor of 3 smaller than the experimental value for ^{238}U and about a factor of 1.7 smaller for Pd. The significance of this result for refining the empirical determination of the weak-coupling constants is discussed.

PACS number(s): 24.60.Dr, 24.80.+y

I. INTRODUCTION

Studies of the weak interaction in strongly interacting systems have been of great interest in nuclear physics for a number of reasons. First of all, on the experimental side, observables may be obtained cleanly from measurements because (1) the symmetries respected by the weak interaction are different from those respected by the strong interaction, allowing the weak interaction sector to be uniquely disentangled from the purely strong interaction sector, and (2) the strength of the weak-interaction is sufficiently small that its matrix element between definite eigenstates of the strong interaction Hamiltonian may be extracted perturbatively. On the theoretical side, knowledge of the weak-interaction observables is a rich source of insights into the behavior of many-body systems. In those cases where the weak interaction is known, the experimental observable provides an opportunity to learn about the nuclear many-body wave function, giving an often unique glimpse at a specific feature of the eigenstate of the nuclear Hamiltonian. If, on the other hand, the weak interaction is not completely known (as is the case for the hadronic weak interaction), the measurements offer the possibility of learning about the weak interaction itself. To realize this goal, the main theoretical task is to calculate the properties of the nuclear many-body system as accurately as possible with a Hamiltonian that is known to describe the initial and final states in the matrix element.

Experimental information likely to shed light on the nature of the hadronic weak interaction is available in both light and heavy nuclei. Measurements of weak-interaction observables in these cases are driving the development of theoretical formulations of the many-body problem along two principal lines distinguished by the significance placed on the interpretability of individual matrix elements. The shell model formulation of the many-body problem is an example of a nonstatistical approach, and here one envisions individual matrix elements, such as those of operators describing transitions between discrete nuclear levels, to be calculable and hence meaningful. In light nuclei [1] the data have been available for many years, and these data are being analyzed using such a shell-model approach.

Our particular interest is in heavy nuclei. In this case [2], measurements based on longitudinally polarized neutron scattering and focusing on the hadronic parity nonconserving (PNC) interaction have become available only in the last 5–10 years, and development of the microscopic statistical approach is being stimulated by the availability of this data. The main focus of the latest experiments [3] has been the strength function of the effective PNC interaction in nuclei in the mass $A \sim 230$ and mass $A \sim 100$ regions of the periodic table. The root-mean-square (rms) matrix element M of this interaction is the physically meaningful quantity, as evidenced by the fact that the nuclear spacing between the resonances involved is on the order of 10–100 eV; note that it is not possible to obtain individual many-body matrix elements from this data. The values of M have been obtained, and these provide a new means to study the parity violating interaction in nuclei based on statistical theory [2].

The particular formulation on which we base our paper is that of statistical spectroscopy [4]. Statistical spectroscopy has proved very successful for calculations of strength functions of positive parity operators, but it has not been developed as fully for negative parity operators, especially for the case of parity violation. In this paper we will develop the theory in a study of parity violation for the targets measured in Ref. [3], namely, ^{238}U for $A \sim 230$ and ^{104,105,106,108}Pd for $A \sim 100$. The statistical densities that we will need are calculated using realistic single-particle energies and a residual strong interaction adjusted to fit the experimental density of states in the two elements of interest here. The determination of the smoothed strength and level densities is well established, and the theory has been shown to be well justified through numerical and theoretical evaluations in the early work [4]. The status of statistical spectroscopy in the other nuclei measured in Ref. [3] will be explored more fully in a later paper [5]. On the other hand, the application of statistical spectroscopy to parity violation raises new issues, and it is the main purpose of this paper to identify the new issues and show how statistical spectroscopy may be implemented in view of them.

We discuss the statistical observables in Sec. II, emphasizing their advantage for study of the weak PNC interaction in compound nuclei. In Sec. III and the Appendix, we emphasize the microscopic foundation of the statistical theory and also collect ingredients that will be used in this paper for specific calculations. The limited aim of our calculations of M, presented in Sec. IV, is to benchmark the theory, i.e., provide reference points for future calculations based on standard π plus ρ meson exchange interactions with strengths taken from the existing literature (specifically we use the "DDH best" set; see Ref. [1]). Since we are not adjusting parameters to fit data, we leave open the question of whether the data prefer some alternative parameter set [5].

In Sec. V we discuss sources of uncertainty in such a calculation and indicate additional studies that would establish a more refined theoretical error analysis. Finally, in Sec. VI we present a summary of our paper and draw our conclusions.

II. STATISTICAL AND NONSTATISTICAL LIMITS

Experiments are often able to determine individual matrix elements $\langle\Psi_I|O|\Psi_K\rangle$ of some operator O between stationary states $|\Psi_I\rangle$ and $|\Psi_K\rangle$ of the nucleus. Microscopic, nonstatistical calculations of such matrix elements are often quite successful. However, such calculations of individual matrix elements are not guaranteed to be stable. For example, a basic problem arises in calculating weak-interaction operators O (e.g., the axial charge operator or the hadronic PNC interaction) within the shell-model framework. This is because the matrix elements

$$\langle\Psi_I|O|\Psi_K\rangle=\sum_{i,k} a_I^{*i}\langle\phi_i|O|\phi_k\rangle a_K^k, \tag{1}$$

which we have expressed in terms of the expansion coefficients a_I^i of the nuclear eigenstates $|\Psi_I\rangle$ in a many-body basis $|\phi_i\rangle$,

$$|\Psi_I\rangle=\sum_i a_I^i|\phi_i\rangle, \tag{2}$$

often involve strong cancellations, so the predictions are sensitive to small changes in the Hamiltonian of the strong interaction.

It is instructive to see how this sensitive interplay comes about. In order to avoid the problem of diagonalizing the complete many-body Hamiltonian, one assumes that the important pieces of the nuclear wave functions are the low-lying components, which one isolates by making a truncation into a "model space." Because the components in the model space correspond to just a piece of the wave function, one must make a correction, either by applying a renormalization to the hadronic weak interaction to account for the excitations out of the model space not accounted for by the effective interaction or by enlarging the model space. In the shell-model theory of Ref. [6], this was done by enlarging the space. Enlarging the space was accompanied by corresponding changes in the coefficients a_I^i and a_K^k, leading to dramatic changes in the matrix elements, illustrated very clearly in Ref. [6]. It is the conclusion of this and other investigations that because of the sensitive interplay among various components, the nuclear wave function must be calculated very precisely, especially by including multiple particle-hole (p-h) admixtures and carefully eliminating spurious center-of-mass motion. To avoid this strong model dependence, it was pointed out that the PNC matrix element could be calculated from the measured analog first-forbidden decay of ^{18}Ne.

With more computer power the individual matrix elements may of course be calculated in light nuclei with sufficient stability. However, for heavy nuclei, where one has many more degrees of freedom, chaotic behavior may occur, and then the problem becomes much more serious. In this case eigenstates of the Hamiltonian have so many components, with amplitudes roughly equal in size and with phases that occur in a nearly random fashion that the individual matrix elements become impossible to interpret in principle. Even small changes in the Hamiltonian (or approximation scheme) can dramatically affect the wave function and hence the expectation value of an observable [7]. (The effective dimensionality, or number of significant Slater determinant components in the eigenstates, is far more stable.) Although a large amount of information is irretrievably lost in the transition from simple to complex systems in this fashion, significant physically meaningful information does remain. Indeed, it was the original motivation in developing the theory of statistical spectroscopy to identify this information and develop methods to extract it. The chaotic behavior of the nuclear wave function suggests application of the central limit theorem for this purpose and leads to the notion that matrix elements of operators may be thought of as elements of a statistical distribution.

For chaotic nuclei, one is thus led to identify as the observable the square of the matrix element of the PNC interaction averaged over N_I and N_K states of appropriate symmetry occurring in a small window of energy,

$$M^2\equiv\frac{1}{N_IN_K}\sum_{I,K}|\langle\Psi_I|O|\Psi_K\rangle|^2, \tag{3}$$

which can be cast into a form very similar to Eq. (1) using the statistical properties of the wave function. To see this, first use Eq. (2) to rewrite Eq. (3) as follows:

$$M^2=\sum_{i_1,k_1}\sum_{i_2,k_2}\left(\frac{1}{N_I}\sum_I a_I^{*i_2}a_I^{i_1}\right)\langle\phi_{i_1}|O|\phi_{k_1}\rangle \times\langle\phi_{i_2}|O|\phi_{k_2}\rangle^*\left(\frac{1}{N_K}\sum_K a_K^{*k_2}a_K^{k_1}\right). \tag{4}$$

Since the phases of the components are random when the statistical approach is valid, we may write

$$\frac{1}{N_K}\sum_K a_K^{k_2^*} a_K^{k_1} = \delta(k_1,k_2)\frac{1}{N_K}\sum_K |a_K^{k_2}|^2, \tag{5}$$

and if the a_K^k $(1 \leq k \leq n$, where n is the effective dimensionality) in the averaging interval are additionally assumed to be elements of an ensemble of Gaussian distributed random variables, we may write, for each k_2,

$$\frac{1}{N_K}\sum_K |a_K^{k_2}|^2 = \frac{1}{n}. \tag{6}$$

Typically, N_I and N_K are rather small in comparison to the effective dimensionality n. Equation (6) is the stronger assumption. In Ref. [8], the space is partitioned with the different partitions being described by separate ensembles. If we adopt the weaker statistical assumption of Eq. (5), Eq. (4) becomes

$$M^2 = \frac{1}{N_I N_K}\sum_{I,K}\sum_{i,k} |a_I^{*i}\langle \phi_i | O | \phi_k\rangle a_K^k|^2. \tag{7}$$

Comparing Eq. (7) to Eq. (1), we see that now the different terms in the sum over i,k occur with the *same* sign, so that the problem with delicate cancellations that would prevent a meaningful calculation of an individual matrix element is avoided. Thus, observables amenable to a statistical analysis are robust against approximations that do not change the overall scale characterizing the distribution of the a_I^i.

In actual situations where the statistical theory is applicable, only the low-lying components a_K^k of the many-body wave function are statistical in the sense of Eq. (5). We look at a more systematic method for separating the statistical and nonstatistical behavior in Sec. III below, and find that when a more careful separation is made, we arrive at the same conclusion.

III. FORMULATION OF STATISTICAL THEORY FOR PARITY VIOLATION

The specific problem we will be addressing in trying to understand the TRIPLE data of Ref. [3] is how to calculate M when the operator O in Eq. (7) is the interaction V^{PNC}, corresponding to the hadronic parity-violating (PV) interaction between two particles in free space; we base our numerical results in this paper on a meson-exchange form for the weak PNC interaction with the coupling strengths presented by Desplanques, Donoghue, and Holstein (DDH) [9]. Although these coupling strengths were estimated by DDH from empirical nonleptonic hyperon decay data, they also incorporated some estimates of various contributions such as the strong-interaction enhancements to the symmetry breaking; as a result, the couplings still contain a considerable amount of uncertainty. Because of these uncertainites and the fact that the DDH interaction has not been fitted to nuclear PNC data, one does not expect to be able to explain the TRIPLE measurements without some further refinement of the coupling parameters.

In our formulation, we will be using the fact that statistical methods [4] provide a relatively straightforward and model-independent means to average over the important components (i,k) in Eq. (7), namely, those that mix chaotically among themselves by the action of the strong interaction. These components correspond to the independent-particle configurations formed by placing valence particles in single-particle orbitals that lie within a few MeV of each other (within the spreading width) outside a closed shell, in other words, to components within a model space. The central limit theorem is used to express the strength functions in terms of bivariate Gaussian functions characterized by a few parameters as described in Sec. III A below. Statistical mixing of single-particle states lying farther than a typical spreading width (a few MeV) away from these is weak [10] and naturally suppressed in the statistical formulation.

However, statistical mixing is unfortunately only part of the problem of understanding the TRIPLE data. Nonstatistical mixing of states lying outside the model space is expected to occur (see Sec. III B 3), and the real question at issue in applying statistical spectroscopy to parity violation is how to deal in a satisfactory way with important components of the wave function that lie just $1\hbar\omega$ outside the model space, for example, the giant resonances. One solution to this problem is provided by effective interaction theory [11].

Effective interaction theory provides bookkeeping techniques for building the contribution of the (i,k) components lying outside the model space in Eq. (7) into an *effective* operator $\overline{O}$ acting entirely within the model space. More generally, it prescribes not only how one can calculate microscopically various effective operators $\overline{O}$ corresponding to actual observables O, but also the effective Hamiltonian $\overline{H}$ whose eigenfunctions $|\overline{\Psi}_K\rangle$,

$$\overline{H}|\overline{\Psi}_K\rangle = E_K|\overline{\Psi}_K\rangle, \tag{8}$$

correspond to the true stationary states $|\Psi_K\rangle$. Once the model space has been set up, matrix elements of the Hermitian, energy-independent operators $\overline{H}$ and $\overline{O}$ may be calculated diagrammatically according to the rules of degenerate many-body perturbation theory [12]. The connection to the observable given in Eq. (1) is

$$\langle \overline{\Psi}_I|\overline{O}|\overline{\Psi}_K\rangle = \langle \Psi_I|O|\Psi_K\rangle. \tag{9}$$

An application of effective interaction theory for the calculation of PV observables in the sense outlined here was followed in Ref. [13].

Once one has obtained the effective operator $\overline{V}^{PNC}$ corresponding to V^{PNC} by making use of this theory, one may apply the powerful methods of statistical spectroscopy directly to $\overline{V}^{PNC}$. The expression for M^2 in Eq. (7) may be expressed in the model space by expanding $|\overline{\Psi}_K\rangle$ as in Eq. (2) to obtain

$$|\overline{\Psi}_K\rangle = \sum_k \overline{a}_K^k|\phi_k\rangle, \tag{10}$$

where the sum now runs over independent-particle configurations k lying purely in the model space. (Note that the coefficients $\overline{a_K^k}$ do not necessarily have a simple connection to a_K^k.) By substituting Eqs. (9) and (10) into Eq. (1), we find

$$\langle\Psi_I|O|\Psi_K\rangle=\langle\overline{\Psi}_I|\overline{O}|\overline{\Psi}_K\rangle=\sum_{i,k}\overline{a_I^{*i}}\langle\phi_i|\overline{O}|\phi_k\rangle\overline{a_K^k}. \tag{11}$$

Similarly, the model-space counterpart to Eq. (7) is found by substituting Eq. (11) into Eq. (3) and following the analogous reasoning leading from Eq. (4) to Eq. (7),

$$M^2=\frac{1}{N_I N_K}\sum_{I,K}\sum_{i,k}|\overline{a_I^{*i}}\langle\phi_i|\overline{O}|\phi_k\rangle\overline{a_K^k}|^2. \tag{12}$$

The connection between M in Eq. (12) measured in a region of excitation energy E, the effective parity-violating interaction, and the strength function $S_{\overline{O}}$ defined below in Sec. III A (with $\overline{O}=\overline{V}^{PNC}$) is

$$\begin{aligned}M^2&\equiv M^2(E_1,E_2,\Gamma_1,\Gamma_2)\\&=D(E_1,\Gamma_1)D(E_2,\Gamma_2)S_{\overline{O}}(E_1,\Gamma_1,E_2,\Gamma_2),\end{aligned} \tag{13}$$

where $D(E,\Gamma)$ is the mean spacing between neutron resonances and Γ is the set of conserved quantum numbers labeling the states of interest (i.e., angular momentum, parity, isospin, etc.).

As we have emphasized, in Eqs. (11) and (12) the sums run only over independent-particle configurations i,k lying within the model space. By comparing these equations we arrive at a conclusion similar to that found at the end of Sec. II, namely, that the delicate cancellations that occur among amplitudes $\overline{a_I^{*i}}\langle\phi_i|\overline{O}|\phi_k\rangle\overline{a_K^k}$ in calculations of individual matrix elements are absent when one calculates the mean-square matrix elements in effective interaction theory. However, we also see that some residual interferences may creep back into the statistical problem through the calculation of $\overline{O}$, whose various terms may have fluctuating signs. Since these signs do not arise from statistical behavior, one may expect that this method of calculating mean-square matrix elements is the more robust procedure when statistical behavior is indicated.

Note how the statistical and nonstatistical aspects of the many-body physics cooperate to produce physical observables in this formulation: the statistical behavior is confined to the model space and handled within statistical spectroscopy, whereas the nonstatistical behavior (including any $1\hbar\omega$ collectivity that may arise from giant resonances, for example) is concentrated in the effective operator terms $\overline{O}-O$. These two aspects are interwoven at the end when M is evaluated in Eq. (12) or Eq. (13).

A. Statistical spectroscopy of French and collaborators

From a statistical point of view, spectroscopy in many-body systems reduces to two largely independent physical features: fluctuations and the secular behavior of system-dependent properties such as level densities, expectation values, and strength densities. Random matrix theories, first introduced by Wigner [14], and their growing number of generalizations especially with regards to mesoscopic, many-body electron systems [15], address the physics of fluctuation properties. This is the most widely recognized aspect of a statistical approach, but is not our concern here. Instead, we shall only describe an implementation of statistical spectroscopy that describes system-dependent large-scale secular behaviors that arise in the construction of the strength function of $\overline{V}^{PNC}$ in chaotic regimes of heavy nuclei. Just as Weyl (see Ref. [16]) was able to show that to leading order in wave number the only surviving information in the count of modes in a cavity was its volume, generalized central limit theorems (CLTs) and moment methods, as developed by French and co-workers over the past 30 years [4], identify the analogous surviving information in a microscopic approach to many-body spectroscopy and provide the means to calculate secular behaviors. In this context, convergence of the CLTs improves with the dilute limit of many valence particles occupying a much larger number of available single-particle states. It should be noted also that experience has shown good convergence to the CLT forms even in ds-shell-model examples [4] that involve spaces much smaller and less dilute than those appropriate for medium and heavy nuclei.

The matrix element M defined in Eq. (3) is an example of a strength density $S_{\overline{O}}(E_1,\Gamma_1;E_2,\Gamma_2)$, which may be expressed as a trace. In statistical spectroscopy, we consider the trace decomposed into its sum over particle configurations, $\mathbf{M}=(m_1,m_2,m_3,\ldots)$ where each m_i is the number of particles in a subset of single particle levels and $\mathbf{N}=(N_1,N_2,N_3,\ldots)$. The grouping is not restricted to a particular scheme, but for the present discussion each subset is taken to be a j orbital.

In general, strength densities are bivariate functions of energy. Using mainly the notation of French and co-workers [4]—i.e., $\langle\langle\ \rangle\rangle$ indicates the trace and $\langle\ \rangle$ indicates the expectation value ($\langle\langle\ \rangle\rangle/N$)—the expression is

$$S_{\overline{O}}(E_1,\Gamma_1;E_2,\Gamma_2)=\langle\langle\overline{O}^\dagger P_{\Gamma_2}\delta(E_2-\overline{H})\overline{O}P_{\Gamma_1}\delta(E_1-\overline{H})\rangle\rangle, \tag{14}$$

where $\overline{H}$ is the effective nuclear Hamiltonian, P_Γ is the projector onto the Γ subspace, and $\overline{O}$ is, for example, the effective parity-violating interaction. It is useful to perform a partitioning decomposition of Eq. (14):

$$\begin{aligned}S_{\overline{O}}&(E_1,\Gamma_1;E_2,\Gamma_2)\\&=\sum_{\mathbf{M},\mathbf{M}'}\langle\langle\overline{O}^\dagger P_{\Gamma_2}\delta(E_2-\overline{H})P_{\mathbf{M}'}\overline{O}P_{\Gamma_1}\delta(E_1-\overline{H})P_{\mathbf{M}}\rangle\rangle\\&=\sum_{\mathbf{M},\mathbf{M}'}S_{\overline{O}}(E_1,\Gamma_1;E_2,\Gamma_2;\mathbf{M},\mathbf{M}').\end{aligned} \tag{15}$$

Parity is taken care of trivially since the configurations respect it and one can just restrict the $\mathbf{M},\mathbf{M}'$ sums to the appropriate parities. For the model spaces of the medium and heavy nuclei considered in this paper, only the angular momentum decomposition remains to be taken into account. If $\overline{O}$ were an electromagnetic transition operator that coupled different angular momenta, we would have to modify the angular momentum decomposition given below. However, for the case we are interested in here, $\overline{O}$ is a parity-violating interaction and is a J scalar. In Eq. (14), P_{Γ_2} commutes with $\overline{O},\overline{H}$ and can be translated next to P_{Γ_1}. The strength expression vanishes unless $\Gamma_2=\Gamma_1=\Gamma$. There are two additional simplifications. For the purposes of the angular momentum decomposition relevant to the parity violation calculations, $E_1=E_2=E$. Finally, as a consequence of statistical spectroscopy, the spreading widths are nearly constant. Therefore, $S_{\overline{O}}(E,E,\Gamma)$ is proportional to the J-decomposed density of states, where the constant of proportionality $C(j|E)$ is the conditional probability density of finding the angular momentum j given E. We find

$$S_{\overline{O}}(E_1,\Gamma_1;E_2,\Gamma_2)=S_{\overline{O}}(E_1,E_2;\Gamma)\approx C(j|E)S_{\overline{O}}(E,E)$$
$$=C(j|E)\sum_{\mathbf{M},\mathbf{M}'}S_{\overline{O}}(E,E;\mathbf{M},\mathbf{M}'). \quad (16)$$

To the level of our statistical approximations, $C(j|E)$ is well approximated by a partitioned version of Bethe's form [17] with an energy-dependent spin cutoff factor $\sigma_j^2(E,\mathbf{M})$ evaluated as

$$C(j|E)=\begin{pmatrix}\sum N_i\\ \sum m_i\end{pmatrix}^{-1}\sum_{\mathbf{M}}C(J|E;\mathbf{M})\prod_i\begin{pmatrix}N_i\\ m_i\end{pmatrix},$$

$$C(j|E;\mathbf{M})=\left(\frac{(2j+1)^2}{8\pi\sigma_j^6(E,\mathbf{M})}\right)^{1/2}\exp\left(-\frac{(2j+1)^2}{8\sigma_j^2(E,\mathbf{M})}\right),$$

$$3\sigma_j^2(E,\mathbf{M})\approx\langle J^2\rangle^{\mathbf{M}}+\langle J^2(H-E_{\mathbf{M}})\rangle^{\mathbf{M}}\frac{E-E_{\mathbf{M}}}{\sigma_{\mathbf{M}}^2}+\cdots, \quad (17)$$

where

$$E_{\mathbf{M}}=\sum_i m_i\epsilon_i \quad (18)$$

and

$$\sigma_{\mathbf{M}}^2=\langle V^2\rangle^{\mathbf{M}}=\sum_{\substack{r\leq s\\ t\leq u}}\frac{(N_r-m_r)(N_s-m_s-\delta_{rs})m_t(m_u-\delta_{tu})}{(N_r-\delta_{rt}-\delta_{ru})(N_s-\delta_{st}-\delta_{su}-\delta_{rs})N_t(N_u-\delta_{tu})}\sum_\Gamma[\Gamma]|V_{rstu}^\Gamma|^2 \quad (19)$$

(below we define V_{rstu}^Γ), with ϵ_i the single-particle energy of the ith orbital including any contribution to the energy from the part of the two-body interaction which transforms like a one-body operator. It is further supposed that the single-particle basis diagonalizes the full one-body part of the Hamiltonian. The V_{rstu}^Γ are the matrix elements of the remaining two-body interaction. The counting factor $[\Gamma]$ reduces to $2j+1$ above.

In addition, only those terms survive in Eq. (16) in which the fundamental matrix elements of $\overline{O}$ appear as absolute squares. We thus have

$$S_{\overline{O}}(E_1,E_2;\mathbf{M},\mathbf{M}')$$
$$=\prod_{i,j}\begin{pmatrix}N_i\\ m_i\end{pmatrix}\begin{pmatrix}N_j\\ m_j\end{pmatrix}|\langle\mathbf{M}'|\overline{O}|\mathbf{M}\rangle|^2\rho(E_1,E_2;\mathbf{M},\mathbf{M}') \quad (20)$$

to evaluate. If $\overline{O}$ is a two-body operator, the matrix element can be deduced from Eq. (19) with the $\{r,s,t,u\}$ sums suppressed and, if not, from the appropriate analogous equations for other rank operators. The theoretical expression for $\rho(E_1,E_2;\mathbf{M},\mathbf{M}')$, which can be derived assuming statistical properties of many-body wave functions and has been confirmed in numerical tests, is that of a unit normalized bivariate Gaussian,

$$\rho(E_1,E_2;\mathbf{M},\mathbf{M}')=\frac{1}{2\pi\sigma_{\mathbf{M}}\sigma_{\mathbf{M}'}\sqrt{1-\xi^2}}\times\exp\left(-\frac{1}{2(1-\xi^2)}\left\{\frac{(E_1-E_{\mathbf{M}})^2}{\sigma_{\mathbf{M}}^2}-\frac{2\xi(E_1-E_{\mathbf{M}})(E_2-E_{\mathbf{M}'})}{\sigma_{\mathbf{M}}\sigma_{\mathbf{M}'}}+\frac{(E_2-E_{\mathbf{M}'})^2}{\sigma_{\mathbf{M}'}^2}\right\}\right), \quad (21)$$

which to specify completely requires the evaluation of two centroids $E_{\mathbf{M}},E_{\mathbf{M}'}$, two variances $\sigma_{\mathbf{M}}^2,\sigma_{\mathbf{M}'}^2$, and a normalized correlation coefficient ξ. In principle, these moments are given by

$$S_{00}=\langle\langle\overline{O}^\dagger P_{\mathbf{M}'}\overline{O}P_{\mathbf{M}}\rangle\rangle,$$

$$\xi=\langle\langle\overline{O}^\dagger\overline{H}P_{\mathbf{M}'}\overline{O}\overline{H}P_{\mathbf{M}}\rangle\rangle/(S_{00}\sigma_{\mathbf{M}}\sigma_{\mathbf{M}'}),$$

$$E_{\mathbf{M}}=\langle\langle\overline{O}^\dagger P_{\mathbf{M}'}\overline{O}\overline{H}P_{\mathbf{M}}\rangle\rangle/S_{00},$$

$$E_{\mathbf{M}'}=\langle\langle\overline{O}^\dagger\overline{H}P_{\mathbf{M}'}\overline{O}P_{\mathbf{M}}\rangle\rangle/S_{00},$$

$$\sigma_{\mathbf{M}}^2=\langle\langle \overline{O}^\dagger P_{\mathbf{M}'}\overline{O}\overline{H}^2 P_{\mathbf{M}}\rangle\rangle/S_{00},$$

$$\sigma_{\mathbf{M}'}^2=\langle\langle \overline{O}^\dagger \overline{H}^2 P_{\mathbf{M}'}\overline{O}P_{\mathbf{M}}\rangle\rangle/S_{00}. \tag{22}$$

However, the centroids and variances can be replaced by the simpler expressions given in Eqs. (18) and (19) with almost no loss in accuracy of the approximation.

The correlation coefficient ξ is quite important. As it approaches zero, strength is distributed uniformly in the sense that the many-body mean-square matrix element is a constant independent of (E_1,E_2). In the opposite limit of $\xi\to 1$, the mean-square matrix elements of $\overline{O}$ are concentrated along the diagonal and their behavior approaches being inversely proportional to the level density; see below. Although ξ depends on $(\mathbf{M},\mathbf{M}')$, the fluctuations from configuration to configuration are small and the errors made in using a single averaged value negligible compared to other approximations. In fact, we use Gaussian ensembles with operators of fixed particle rank to evaluate ξ since the strength density is not extremely sensitive to small variations. In the dilute limit of N (number of single-particle states) $\to\infty$, $\overline{H}$ a k-body operator, $\overline{O}$ a k'-body operator, and m valence particles,

$$\xi=\binom{m-k'}{k}\Big/\binom{m}{k}\approx 1-kk'/m. \tag{23}$$

The finite N expression is more complicated but analytically known for a Gaussian ensemble. It is this expression that we use and which can be found in Ref. [8]. It is not sufficiently illuminating to give the expression here. Typical values for ξ in medium and heavy nuclei are $\sim$0.7–0.9.

The criterion for choosing an appropriate model space is more complicated in the strength forms than for the level density since it depends on the joint contribution of $(\mathbf{M},\mathbf{M}')$ and the matrix elements connecting configurations. Nevertheless, one uses a similar logic and the most important configurations are the same as in the level density calculation, i.e., the single-particle orbitals strongly mixed by the residual two-body interaction. If O connects to important configurations outside the model space, statistical methods must be extended to incorporate perturbative effects as discussed in the following subsection.

B. Model space and effective interactions

Having made these general remarks, we next discuss the choice of model space and present the contributions to the effective one-body and two-body parity-violating interactions that we will use in this work. In the theory, there exist also three- and higher-body contributions to the effective operator $\overline{O}$, but there have been no calculations of these terms, so we will omit them.

1. Model space and strong interaction

There are two practical issues with regard to the choice of the strong interaction in statistical spectroscopy. One is the specification of the one-body part of the interaction, which specifies the model space and single-particle energies. The other is the choice of the spreading interaction that is crucial for getting the level density correct. Once these two elements are specified, then the partitioning of the model space and the calculation of the bivariate densities in Sec. III A are completely specified.

By "spreading interaction," one means that part of $\overline{H}$ that spreads the configurations across the eigenstates. To determine it from a given $\overline{H}$ requires removing those parts responsible for shifting the centroids of levels. In the language of Ref. [4], the spreading interaction is the irreducible rank-2 part of $\overline{H}$. The shell-model or mean-field contribution, which determines the single-particle spectrum, is the rank-1 contribution of $\overline{H}$. Note that a somewhat different definition of the one- and two-body parts of an effective interaction is used in microscopic shell-model studies, where the single-particle energies are often taken to be those appropriate to the beginning or end of the shell. The residual interaction for multiple numbers of particles in a shell can then both spread levels and shift their centroids. One must keep this distinction in mind when comparing the shell-model and statistical approaches to avoid confusion.

In our applications of statistical spectroscopy, we choose both the mean-field and spreading interactions phenomenologically. By so doing, we thus depart from a purely theoretical implementation of the theory. The mean-field part is identified with the shell-model potential. By taking the single-particle energies to be those corresponding to such a phenomenological potential, a smooth A dependence is imposed. The spreading interaction is adjusted phenomenologically to reproduce the level density in the excitation region of nuclei of interest where the weak spreading width is to be calculated. The energy dependence of the level densities for nuclei measured by TRIPLE have been extensively studied and experimentally tabulated [18], and improved level densities have been obtained also in the region of interest for M in Ref. [3]. We used these results for our numerical studies.

The choice of model space is determined mainly by the considerations discussed in Sec. III, namely, that chaotic mixing occurs within the model space. Since the strong spreading width of states is on the order of 2 MeV, a model space spanned by single-particle levels lying within 2 or 3 MeV of each other is expected to be sufficient for those nuclei in which there are at least two or three particles (or holes) in the valence space for both neutrons and protons.

For the nuclei of interest in this paper, it is a relatively straightforward matter to choose the model space subject to these criteria by looking at shell-model level schemes. In spherical nuclei, the single-particle orbitals will correspond to eigenstates in a spherical potential and in deformed nuclei to Nilsson levels. Since statistical spectroscopy has always been applied (even in regions of deformation, for example, the $A\sim 230$ region) exclusively in the spherical basis, we will do our present calculations in this basis as well, in accordance with our goal to benchmark the theory. We estimate the importance of deformation by including it through an effective one-body PNC interaction (see Sec. V B).

TABLE I. $0\hbar\omega$ model space for mass region $A\sim 230$. Note that we assume neutron (N) closure at $N=126$ and proton (P) closure at $Z=82$. The single-particle energies are given for the daughter ^{239}U.

Particle	Orbit	(n,l)	Parity
P	$h_{9/2}$	(0,5)	−
P	$i_{13/2}$	(0,6)	+
P	$f_{7/2}$	(1,3)	−
N	$i_{11/2}$	(0,6)	+
N	$j_{15/2}$	(0,7)	−
N	$g_{9/2}$	(1,4)	+
N	$d_{5/2}$	(2,2)	+

In our work, the single-particle energies of the active orbits for ^{239}U, given in Table I, were taken from Ref. [8] and for Pd were calculated [19] using a Woods-Saxon potential,

$$U=Vf(r)+V_{ls}\mathbf{l}\cdot\mathbf{s}\frac{r_0^2}{r}\frac{d}{dr}f(r)$$

$$f(r)=\left[1+\exp\left(\frac{r-R}{a}\right)\right]^{-1}, \tag{24}$$

with realistic parameters [20]

$$R=r_0A^{1/3},\quad r_0=1.27\ \text{fm},\quad a=0.67\ \text{fm},$$

$$V=\left(-51+33\tau_z\frac{N-Z}{A}\right)\ \text{MeV},\quad V_{ls}=-0.44V. \tag{25}$$

The resulting energies for the daughter ^{107}Pd are given in Table II. The model space is spanned by the set of all such configurations (N in number) that can be formed by putting m valence particles in a few (active) single-particle orbits lying in the immediate neighborhood of the nearest closed neutron and proton subshells (we refer to this as a $0\hbar\omega$ space). Note that there are both positive and negative parity levels included. The appearance of the intruder states of opposite parity is a feature particular to heavy nuclei: for the light nuclei for which parity violation has been considered, there are no such intruder states.

TABLE II. $0\hbar\omega$ model space for mass region $A\sim 100$. Note that we assume neutron (N) closure at $N=50$ and proton (P) closure at $Z=28$. The single-particle energies are given for the daughter ^{107}Pd.

Particle	Orbit	(n,l)	Parity	Orbital energy (MeV)
P	$g_{9/2}$	(0,4)	+	−7.94
P	$p_{1/2}$	(1,1)	−	−9.04
P	$f_{5/2}$	(0,3)	−	−11.2
P	$p_{3/2}$	(1,1)	−	−10.7
N	$d_{5/2}$	(1,2)	+	−8.44
N	$g_{7/2}$	(0,4)	+	−7.34
N	$s_{1/2}$	(2,0)	+	−6.63
N	$d_{3/2}$	(1,2)	+	−5.96
N	$h_{11/2}$	(0,5)	−	-4.85

In the following two subsections, Secs. III B 2 and III B 3, we discuss three different pieces of $\overline{V}^{PNC}$, a standard one-body piece, a standard two-body piece, and a higher-order correction. The latter will be seen to be peculiar to heavy nuclei, where the relative importance of the standard one-body and two-body PNC interactions is different from what it is in light nuclei.

2. Lowest-order effective one- and two-body PNC potentials

The standard two-body piece of the effective PNC interaction in the nucleus $V_{Std}^{PNC(2)}$ that we use is related to the free-space two-body PNC interaction as discussed in Ref. [1] and in the Appendix. The standard one-body interaction $V_{Std}^{PNC(1)}$ is, in physical terms, that part of the effective PNC interaction acting on nucleons outside the core (the valence nucleons) that originates from the core nucleons. Since $V_{Std}^{PNC(1)}$ is just an average of $V_{Std}^{PNC(2)}$ over the core of the nucleus [1], we have a relatively simple connection between the two. Referring to the Fermi gas model [see Eq. (20) of Ref. [1]], we obtain

$$V_{Std}^{PNC(1)}=\frac{\rho}{\rho_0}\sum_i (C_0+C_1\tau_{zi})\boldsymbol{\sigma}_i\cdot\mathbf{p}_i, \tag{26}$$

where ρ is the density of the core and ρ_0 is the central density in nuclei. The coefficients C_0 and C_1 are given in Eq. (A3) of the Appendix. For $V_{Std}^{PNC(1)}$ as well as $V_{Std}^{PNC(2)}$ we retain only the contributions from F_π and F_0, the weak π-NN and (isoscalar) ρ-NN coupling constants, respectively, since these are the dominant contributions from the DDH parameter set.

Next, we want to discuss the higher-order pieces of this interaction. The situation can become quite involved, since higher-order pieces of the effective one-body and two-body parity violating interactions can be formed from both $V_{Std}^{PNC(1)}$ and $V_{Std}^{PNC(2)}$. It is our aim in this paper to limit our attention to issues that have been discussed already in the literature, so we will not enumerate the other terms in any systematic fashion. This limits us to only one type of higher-order term, discussed in the next subsection.

3. Higher-order contributions arising from $V_{Std}^{PNC(1)}$

One of the differences between parity violation in light and heavy nuclei is the relative importance of single-particle transitions in PNC observables. Large shell-model calculations (see, for example, Refs. [21,22]) indicate that the PNC transitions in light nuclei are largely single particle in nature [1,23]. However, in heavy nuclei, because of the appearance of intruder states of opposite parity in the $0\hbar\omega$ model space, $\overline{V}^{PNC(2)}$ can connect states within this space, and many-body mixing plays a much more important role. The fact that there are generally no active single-particle states that are coupled by $V_{Std}^{PNC(1)}$ (i.e., that there are no opposite-parity single-particle levels with the same total angular momentum inside the $0\hbar\omega$ model space in a heavy spherical nucleus, even though there are intruder states) means that $V_{Std}^{PNC(1)}$ acts indirectly. Its contribution to the effective PNC interaction may be enumerated in second- and higher-order perturbation theory; in the theory of this paper, these are the major contributors to the difference of the operators $O-\overline{O}$ defined in Eq. (9) and accounts fully for the effect of $V_{Std}^{PNC(1)}$ on M. In a deformed nucleus the situation is somewhat different; see Sec. V.

The leading perturbative contribution of $V_{Std}^{PNC(1)}$ to $\overline{V}^{PNC}$ has been discussed in Ref. [24]. In an effort to use physical ideas to identify the important correction, Auerbach [25] developed the idea of "doorway" states for the PNC interaction. In this approach, the mixing of $V_{Std}^{PNC(1)}$ is mediated by the isoscalar and isovector 0^- spin-flip giant resonances. The disadvantage of Auerbach's formulation is that it is highly phenomenological. A microscopic expression $V_{Dwy}^{PNC(2)}$ for the doorway contribution to the effective two-body PNC interaction was obtained in Ref. [13] by identifying the doorway state as a collective one-particle one-hole 0^- phonon in the Tamm-Dancoff approximation (TDA). The result brought together the results of Refs. [24,25], showing that the effect of the doorway is to renormalize the perturbative contribution of Ref. [24] by a factor (ω_0/ω), where ω_0 is the energy of the unperturbed 0^- state, and ω is the energy of the TDA phonon. The result for $V_{Dwy}^{PNC(2)}$ consists of one isoscalar contribution, given in Eq. (4.9) of Ref. [13], and two isovector contributions, given in Eqs. (4.10) and (4.11) of that paper. The renormalization amounts to a suppression of the isovector 0^- part $V_{Dwy}^{PNC(2)}$ by about a factor of about 3 with little change in the isoscalar 0^- part (which was found to be much smaller for other reasons).

Desplanques has extended the study by considering the 0^- resonances in the random-phase approximation (RPA) [27] (see also Ref. [26]). This result shows that the multiparticle multihole correlations are very important. To get a nonvanishing contribution from the collective excitations in the RPA, it is necessary to include nonlocal terms in the residual strong interaction. The combined effect of these two considerations is to undo most of the suppression arising from the one-particle one-hole TDA correlations found in Ref. [13]. Desplanques found, essentially, that the doorway result changes by the following substitutions:

$$\omega_0/\omega \rightarrow \frac{1}{1+\lambda_p}, \quad \text{isoscalar},$$

$$\omega_0/\omega \rightarrow \frac{1}{1+\lambda_p'}, \quad \text{isosvector}, \tag{27}$$

where λ_p and λ_p' are defined in terms of the Fermi-liquid parameters as

$$\lambda_p = \frac{1}{3}G_1 - \frac{10}{3}H_0 + \frac{4}{3}H_1 - \frac{2}{15}H_2,$$

$$\lambda_p' = \frac{1}{3}G_1' - \frac{10}{3}H_0' + \frac{4}{3}H_1' - \frac{2}{15}H_2', \tag{28}$$

and where the effective mass of the nucleon M^* is set to its free-space value (the dominant momenta in the phonon are momenta above the Fermi momentum p_F). Using the Fermi-liquid parameters of Ref. [28], we calculate that $\lambda_p = -0.21$ and $\lambda_p' = 0.27$.

We use here the RPA treatment [27] of the O^- resonances rather than the TDA treatment. This is implemented by using Eq. (A2) (see the Appendix) with the replacements in Eq. (27). The isovector interaction, which is the only significant piece to this doorway result, continues to be suppressed in the RPA, but the renormalization is about a factor of 2 smaller than it was in Ref. [13].

IV. CALCULATIONS

In this section we describe our calculation of M and compare the results to the values of M deduced from scattering longitudinally polarized neutrons from the target nuclei ^{238}U and ^{104,105,106,108}Pd. These nuclei are good first cases to consider, because our calculation is relatively straightforward for proton and neutron midshell occupancy, and because the empirical values of the weak spreading width Γ_w are rather well determined [3] for these cases. The relationship between Γ_w and M is most precisely stated if we affix a subscript J on M to specify the total angular momentum of the CN states over which the PNC interaction is averaged and introduce D_J, the theoretical level spacing for s-wave levels of total angular momentum J in the region of the spectrum where the measurement has been performed. Then,

$$\Gamma_w = 2\pi |M_J|^2 / D_J, \tag{29}$$

where the label J has been omitted from Γ_w because of the expectation that this quantity is insensitive to the details of the distribution of CN levels. The resulting values for M_{expt} are given in Table III. Also shown in the table is the spacing D_0 of all s-wave levels, which is the same as D_J only for $J=0$ targets. In the case of the daughter ^{106}Pd there are two relevant intermediate states, of $J=2$ and 3, and in this case the quantity D_0 is the combined level density for the two states. To determine value of M_{expt} in this case required making an assumption about the relationship between D_0 and the two values of D_J.

We report the results of our calculation of M based on the theory developed in earlier sections of this paper in Table IV. The contributions to M arise from two sources: the standard two-body parity-violating interaction $V_{Std}^{PNC(2)}$ (Sec. III A 2) and the doorway piece $V_{Dwy}^{PNC(2)}$ (Sec. III A 3). To evaluate the matrix element M corresponding to $V_{Dwy}^{PNC(2)}$, we replace $\omega_0/\omega_{iv} \rightarrow 0.78$ in Table VII in accordance with Eq. (27). Details of our calculation of the matrix elements of $V_{Dwy}^{PNC(2)}$ and $V_{Std}^{PNC(2)}$ are given in the Appendix. The values of M corresponding to these two pieces of the interaction separately are given in the second and third columns of Table IV, respectively. In the case of the daughter ^{106}Pd, there are two different sets of s-wave resonances corresponding to $J=2$ and $J=3$ as discussed earlier; the individual values of M_J differ by less than 10%, so we have given only the average value M in this case in Table IV. The small value of M for ^{106}Pd reflects a smaller level spacing seen in Table III and the near constancy of Γ_m in the Pd isotopes.

TABLE III. The empirical weak spreading width Γ_w^{expt} [3] (preliminary) for the daughters ^{239}U and ^{105,106,107,109}Pd. Also shown are the neutron number N, the proton number Z, the theoretical level spacings D, and the value of M calculated according to Eq. (29).

Nucleus	(N,Z)	Γ_w^{expt} (10^7 eV)	D_0 (eV)	M_{expt} (meV)
^{239}U	(146,92)	$1.35^{+0.97}_{-0.64}$	20.75 (20.9)	$0.67^{+0.24}_{-0.16}$
^{105}Pd	(59,46)	$1.40^{+6.91}_{-0.99}$	165 (165±61)	$2.2^{+2.4}_{-0.9}$
^{106}Pd	(60,46)	$0.086^{+0.098}_{-0.042}$	13.1 (13.3±0.7)	$0.20^{+0.10}_{-0.07}$
^{107}Pd	(61,46)	$0.18^{+0.91}_{[illegible]}$	159 (159±24)	$0.79^{+0.88}_{[illegible]}$
^{109}Pd	(63,46)	$0.81^{+4.45}_{-0.62}$	159 (159±24)	$1.6^{+2.0}_{-0.7}$

The fourth column of Table IV contains the value of M corresponding to the sum of Dwy and Std. These results must be added in quadrature to obtain the combined value of M. Note that in doing this one must allow for the possibility that the two pieces can interfere. We can easily verify from the values given in Table VII that in practice there is only a small interference between the Std and Dwy contributions. We find the interference is constructive for $A \sim 230$ but destructive for $A \sim 100$.

In the last column of Table IV we show the empirical value of M taken from Table III. We see that for $A \sim 100$ the experimental values of M fluctuate around the theoretical values. As the weak spreading widths are expected to be much more stable than M (the theoretical spreading widths are stable to within about $\pm 10\%$), we compare in Table V the theoretical and experimental spreading widths in the two mass regions considered. Given that the spreading widths are quadratic in M, we see from Table VI that the M in the mass $A \sim 230$ region is about a factor of 3 smaller than the empirical value and in the mass $A \sim 100$ region about a factor of 1.7 smaller. The agreement between theory and experiment is remarkably close and encourages further applications with the goal of refining the empirical determination of the weak coupling parameters.

Earlier calculations [29,24,13] in the mass region $A \sim 230$ share some features in common with the present approach. Reference [29] described the nucleus as a chaotic Fermi gas; i.e., the basis states of the nucleons were taken to be plane waves. The framework of statistical spectroscopy was used to calculate M with the strength function of the PNC interaction normalized to the strength function calculated in a finite nucleus [8]. When plane waves are used, the selection rules are much weaker than with shell-model states, and for this reason both $V_{Std}^{PNC(2)}$ and $V_{Std}^{PNC(1)}$ contributed *directly* to M. In the present work $V_{Std}^{PNC(1)}$ contributes directly only if the nuclear core is deformed (see Sec. V B). Otherwise, $V_{Std}^{PNC(1)}$ contributes indirectly through spin-flip excitations. As discussed in Sec. III B 3, $V_{Std}^{PNC(2)}$ contributes directly because of the presence of opposite-parity intruder states. In Ref. [29], M was found to be several times larger than the empirical value (see also Ref. [13]), whereas in the present work it appears to be several times smaller. It would thus appear that the nuclear matter treatment overestimates the result of statistical spectroscopy in a finite nucleus for the contribution of our one-body PNC interaction for the daughter ^{239}U.

Reference [24] employed a $0\hbar\omega$ model space like ours, so $V_{Std}^{PNC(1)}$ contributed indirectly, through what can be thought of as the perturbative piece of the doorway process. The average over states was defined by a Lorentzian function approximating the spreading width and normalized to an empirical estimate of the number of "principal" components of the nuclear wave function. Good agreement between measured and calculated values of M were reported. In Ref. [24] the indirect contribution of $V_{Std}^{PNC(1)}$ dominated the contribution of $V_{Std}^{PNC(2)}$, and the latter was neglected. In our work, the matrix elements of $V_{Std}^{PNC(2)}$ are similar to those in Ref. [24] (note that a different representation of the two-body PNC interaction was used in our work). However, in contrast to Ref. [24], the matrix elements of $V_{Std}^{PNC(2)}$ are comparable to the sizes of the matrix elements of our $V_{Dwy}^{PNC(2)}$, as can be inferred from Table IV. Although we have used different parametrizations of the effective strong interaction, we do not believe that this can fully account for the large differences we find.

It was suggested in Ref. [24] that the contribution of $V_{Std}^{PNC(1)}$ to M, relative to that of $V_{Std}^{PNC(2)}$, would follow an

TABLE IV. Theoretical values of M for the effective parity-violating interaction. Contributions are shown separately for the standard (Std) and doorway (Dwy) pieces of the two-body interaction. A comparison of the experimental value of M given in Table III is also shown.

Nucleus	M_{Std} (meV)	M_{Dwy} (meV)	$M_{Std+Dwy}$ (meV)	M_{expt} (meV)
^{239}U	0.116	0.177	0.218	$0.67^{+0.24}_{-0.16}$
^{105}Pd	0.70	0.79	1.03	$2.2^{+2.4}_{-0.9}$
^{106}Pd	0.304	0.357	0.44	$0.20^{+0.10}_{-0.07}$
^{107}Pd	0.698	0.728	0.968	$0.79^{+0.88}_{-0.36}$
^{109}Pd	0.73	0.72	0.97	$1.6^{+2.0}_{-0.7}$

TABLE V. Comparison of theoretical weak spreading width to experiment. The value of Γ_w^{expt} for Pd is obtained by statistically combining the results for all values given in Table III; the theoretical value of Γ_w^{the} is, to within a few percent, the same for all Pd isotopes.

Nucleus	$\Gamma_w^{the}(10^7\ \text{eV})$	$\Gamma_w^{expt}(10^7\ \text{eV})$
^{239}U	0.143	$1.35^{+0.97}_{-0.64}$
Pd	0.40	$1.73^{+1.65}_{-0.84}$

$A^{1/3}$ law. We see from Table IV that the relative value of the doorway contribution does increase with A as predicted in Ref. [24], but a more detailed estimate suggests that it increases by more like a power of $A^{1/2}$. The existence of the different rate of increase of the standard and doorway contributions shows that the A dependence of the measurements may provide an empirical means to determine the size of the doorway contribution.

V. DISCUSSION

In this section we will address the implementation of statistical spectroscopy and various sources of uncertainty in the calculations of the previous section. Most of the uncertainties we discuss are amenable to further analysis in large-basis shell-model studies, which would be valuable to gain additional confidence in the theoretical results.

The theory of level densities and strength distributions within statistical spectroscopy applies to individual systems and operators, and as such is not inherently an ensemble theory. The central limit theorem results and correction expansions are all based on having many strongly interacting particles dilutely spread among even more single-particle states. The inputs required for all calculations are low-order operator moments. If it were possible to calculate exact moments up to products of approximately, say, six operators decomposed for the symmetries involved, we would have an essentially "perfect" implementation of statistical spectroscopy. It would be extremely accurate and would work even far out in the tails of the distributions. Instead, we have made a number of practical compromises in order to complete this first implementation for the parity-violation problem. Some of the compromises can and should be eliminated in future, planned, improved calculations.

Most of the compromises we have chosen are motivated by the following considerations: (i) the necessary angular momentum decomposed moments cannot be written down analytically; (ii) even the nondecomposed moments (scalar moments) are not worked out beyond a product of four operators—the expressions for products of three or four operators, though they exist, are almost prohibitively unwieldy; and, (iii) the exact operators may not be known in the model spaces—for example, the exact, residual, strong operator in a model space appropriate for heavy nuclei. We have addressed these points by (i) treating the angular momentum as a statistical variable in the spirit of Bethe's ansatz for the spin cutoff factor; (ii) using partitioning of the moments to improve accuracy instead of incorporating polynomial corrections to the lowest-order central limit theorems—this alleviates much of the need for higher moments; and (iii) supplementing exact moments with ensemble calculations and nuclear modeling—the idea being that in chaotically behaving systems whole classes of operators deviate very little from certain ensemble results (subject to very few restrictions). This also reduces the need for higher-order moments.

To be more specific about ensemble results, we have in mind two-body random operator ensembles which are a subclass of embedded Gaussian ensembles [30]. The basic ensemble would have random k-body matrix elements embedded in an m-particle space distributed over N single-particle levels; the two-body matrix element ($k=2$) is the most important one. Some of the moment simplifications that we are making in this paper [see the comments after Eq. (22)] are ensemble results. Expectation value relations such as

$$\langle OH\rangle=\langle O\rangle\langle H\rangle,$$

$$\langle O^2H^2\rangle=\langle O^2\rangle\langle H^2\rangle, \tag{30}$$

which express a factoring approximation true in systems that behave statistically, have been found to hold to a considerable accuracy (see Refs. [4,8]). Operators O or their higher powers, which effectively contain a piece proportional to H (or a function of H), will not satisfy relations of this kind.

In the following subsections, we address issues related to the nuclear modeling we have done, specifically the size of the model space, deformation, correlations between $\overline{O}$ and $\overline{H}$ including spurious center of mass motion, and the choice of weak coupling parameters. We critique each uncertainty below in Secs. V A–V D. In Sec. V E we discuss how the shell model could be used to gain additional confidence in the models. All of the statistical spectroscopy compromises that were discussed above could (and should) be incorporated into the shell-model testing.

A. Size of the model space

The physical motivation for choosing a $0\hbar\omega$ model space was discussed in Sec. III. However, to establish that the optimal selection of active orbits has been made, it must be verified that the amount of parity mixing as determined by our statistical methods is stable under enlargements of the model space. For this purpose, we added a spherical $1f_{5/2}$ neutron and $1g_{7/2}$ proton orbit to the model space for ^{239}U. Adding these two orbitals, we found that the calculation of M decreased by about 10% when the residual strong and weak interactions were chosen as described above. Note that according to the theory of Sec. III, the effective interactions should be changed as the model space is enlarged. Although we have not made such compensating changes, we expect that the effect is small for the small enlargements considered here. Our results thus indicate that the estimates based on the model space given in Table I and Table II are sufficiently accurate for the purposes of comparing to experiment, where the errors are greater than 10%.

The reason for the stability is that the neutron and proton valence orbits are not close to being either fully occupied or completely empty. In the opposite situation, choosing the optimal model space becomes a more delicate issue. There are a number of cases measured in Ref. [3] in which this is true for one of the shells (for example, Nb and Cs), and in these cases the choice of the optimal model space requires more effort. The data show that the spreading width is anomalously small in these cases, and it will be an interesting test of the theory to be able to reproduce this trend.

D. Deformation

Deformation contributes in two possible ways: if the nucleus has a permanent deformation, the one-body PNC potential develops a nonspherical component arising from the interaction of a valence nucleon with the deformed core. Otherwise, the two-body effective interaction may have a contribution that arises as one of the nucleons interacts with a core nucleon, exciting it and causing a core-deformation fluctuation. A second nucleon may then experience a long-range interaction with the first through the resulting core polarization. We estimate the effect of permanent deformation in this section.

1. Permanent deformation

The fact that the empirical and theoretical values of M more nearly agree for $A\sim 100$ than they do in the region of $A\sim 230$ could point to an important contribution of permanent deformation, since ^{238}U is a highly deformed nucleus but Pd is not. It is easy to show that there can be nonvanishing matrix elements of the deformed piece of $V_{Std}^{PNC(1)}$ for opposite-parity spherical orbits in which the j differ by zero or *two* units (and for opposite-parity deformed orbits in which the z-projection quantum numbers Ω are the same). For the spherical orbits in Table I there are pairs of orbits for both protons and neutrons that satisfy this requirement and have rather large matrix elements. Thus, for our calculations in the $A\sim 230$ region we will have direct contributions from $V_{Std}^{PNC(1)}$ even in the $0\hbar\omega$ model space, similar to the situation in nuclear matter [29].

We find an estimate for the contribution to M arising from the deformation by using a one-body potential similar to that given in Eq. (26) and using a harmonic oscillator basis, replacing ρ by a density having a deformed Woods-Saxon shape with radius R and diffuseness a. The contribution of the one-body PV potential arising from deformation is found by expanding the density to lowest order in the deformation parameter β_2, leading to the following shape,

$$\delta\rho(r)/\rho_0 = -\frac{R\beta_2}{a}g(r)Y_{20}(\hat{r}), \tag{31}$$

where

$$g(r)=\frac{\rho(r)}{\rho_0}\frac{e^{(r-R)/a}}{1+e^{(r-R)/a}}. \tag{32}$$

Taking matrix elements between the single-particle orbitals, we obtain

$$\begin{aligned}\langle n_f l_f j_f m_i|\delta V^{PNC(1)}|n_i l_i j_i m_i\rangle &= \frac{-im\omega_0 R\beta_2}{a}(C_0+C_1, C_0-C_1)\\ &\times\langle n_f l_f j_f m_i|\frac{1}{2}[g(r)\sigma\cdot p+\sigma\cdot p g(r)]\\ &\times Y_{20}(\hat{r})|n_i l_i j_i m_i\rangle,\end{aligned} \tag{33}$$

where at $\rho=\rho_0$, $C_0+C_1=3.53\times 10^{-8}$ and $C_0-C_1=-3.15\times 10^{-9}$ for (proton,neutron) orbits, and where the radial quantum number n is the number nodes in the radial wave function, l is the orbital angular momentum, j is the total angular momentum, m is the projection of the total angular momentum along the z axis, and

$$\begin{aligned}&\langle n_f l_f j_f^{\mp} m_i|\frac{1}{2}[g(r)\sigma\cdot p+\sigma\cdot p g(r)]Y_{20}(\hat{r})|n_i l_i j_i m_i\rangle\\ &=(-1)^{m_i-3/2}\sqrt{(2j_f+1)(2j_i+1)}\begin{Bmatrix} j_i & 2 & j_f\\ l_f\pm 1 & \frac{1}{2} & l_i\end{Bmatrix}\sqrt{\frac{5(2l_i+1)}{4\pi}}\begin{pmatrix} j_i & 2 & j_f\\ m_i & 0 & -m_i\end{pmatrix}\sqrt{2(l_f\pm 1)+1}\begin{pmatrix} l_i & l_f\pm 1 & 2\\ 0 & 0 & 0\end{pmatrix}\\ &\times\langle n_f l_f|\frac{1}{2}[g(r)\sigma\cdot p+\sigma\cdot p g(r)]|n_i l_i\pm 1\rangle,\end{aligned} \tag{34}$$

with $j_f^{\mp}\equiv l_f\pm\frac{1}{2}$.

Matrix elements of Eq. (33) do not vanish in the model space of Table I. There are two nonvanishing matrix elements: one for the pair proton orbits $(h_{9/2}, i_{13/2})$ and one for the pair of neutron orbits $(j_{15/2}, i_{11/2})$. The matrix element of $\delta V^{PNC(1)}$ of Eq. (33) is about 2.4 times smaller for neutrons than for protons. Averaging the square of Eq. (33) over the quantum number m_i (noting that there are $2j_<+1$ nonvanishing matrix elements), we find that the magnitude of the average proton matrix element is 0.011 eV.

We have estimated M corresponding to these values bynormalizing to the calculation in Ref. [8] along the lines explained in Ref. [29], which gives

$$M_{Def}^2(\text{keV}^2) = 2.6\alpha_p^2, \tag{35}$$

where

$$\alpha_p^2 = \frac{(1.2)^2}{A_v} \frac{\text{Tr}[(\delta V^{PNC(1)})^2]}{\text{Tr}[(U_2)^2]}. \tag{36}$$

The factor of $1/A_v$ arises because the trace is taken over the squares of the one- and two-body matrix elements within the model space; the square of the matrix elements is weighted by the relative number of valence nucleon pairs, and the factor of 1.2^2 represents the relative normalization of the strength function for a one- and two-body operator. The two-body interaction U_2 has no diagonal matrix elements but is otherwise modeled after the surface delta-function interaction [31]. We estimated in this fashion that in ^{238}U, $M_{Def} = 0.004$ meV.

Compared to the values of M arising from other sources in Table IV, we see that the contribution of deformation to M is completely negligible. A more quantitative calculation would entail a calculation using a deformed model space of Nilsson orbits, which would require no new theoretical development because statistical spectroscopy as formulated in Sec. III B applies in either a spherical or deformed basis.

2. *Core polarization*

Core polarization as described at the beginning of this section has been studied phenomenologically, and its contribution to the effective nucleon-nucleon interaction is known to act dominantly in $T=1$ and be largely independent of spin [32]. Just as in the case of the doorway model [13] the effective parity-violating interaction results from the commutator with the residual strong interaction and $\boldsymbol{\sigma}\cdot\mathbf{r}$. This commutator does not vanish for the core-polarization piece of the effective nucleon-nucleon interaction, and the resulting contribution to the effective PNC interaction deserves quantitative study in the future.

C. Correlations between O and H

There are several classes of correlations that we wish to mention. One arises when O couples strongly to a mode of motion generated by H lying predominantly outside the model space. In such a case, one would miss important physics by averaging O over only those states within the model space. To deal with this situation properly, one must use effective interaction theory and consider instead of O its effective operator counterpart $\bar{O}$. The bookkeeping rules of the theory ensure that corrections necessary to represent the effect of this correlation are included in $\bar{O}$.

Such a correlation is quite important for parity violation, where the isoscalar and isovector 0^- spin-flip resonances couple strongly to the one-body piece of the parity-violating interaction, so in this case O is identified with $V_{Std}^{PNC(1)}$. In order to deal with this situation in the context of parity violation, the doorway model was introduced, as discussed in Sec. III A 3. We showed that the effect of the 0^- spin-flip excitations is quite important.

Yet another type of significant correlation occurs when the operator O and Hamiltonian H commute with each other or where both commute with a third operator R. Let us consider the latter case, so that H and O may be characterized by the same quantum numbers Γ that are the good quantum numbers of R. Values of the mean-square average of O over distinct ensembles of states, each characterized by its own value of Γ, would in general be different. Thus, it would be misguided to average O over an ensemble of states with mixed Γ, since H does not produce such mixtures. If a statistical ensemble of mixed Γ were formed, either because one was not aware of the symmetry R or because of some unfortunate approximation, this ensemble would fail to properly describe the system.

One place where such a consideration is relevant is in the description of spurious center-of-mass motion. Generally, the eigenstates of H may be grouped into sets distinguished by quantum numbers specifying the center-of-mass motion. Suppose that each such set is spanned by distinct individual-particle basis states. Then, any mean-square average of O corresponding to one set would differ from that of any other, because the corresponding statistical ensembles are different. For example, if one compares two ensemble averages of O, one in which the center-of-mass motion is pure and one in which it is a mixture of several modes, one may find different results. Fortunately, this problem turns out not to be a serious one for us. As can be seen from Tables I and II, our $0\hbar\omega$ basis does not permit any spurious center-of-mass excitations. Thus, all matrix elements of $V^{PNC(2)}$ are completely nonspurious. As is generally the case in applying perturbation theory to include excitations outside the model space, some level of spuriosity is unavoidably introduced in calculating $V^{PNC(1)}$. However, the nature of statistical calculations, where the observable is a mean-squared matrix element insensitive to relative phases, considerably reduces the magnitude of the problem compared to the standard shell-model case.

D. Choice of weak-coupling parameters

It is important to bear in mind that the relative sizes of M corresponding to $V_{Dwy}^{PNC(2)}$ and $V_{Std}^{PNC(2)}$ depend upon the choice of weak-coupling constants. The values of these coupling constants are not certain, so the relative sizes of M for the two contributions may actually be different (perhaps substantially so) than the values reported in Tables III–V when more refined choices of the coupling constants are made. As we have indicated, we use the DDH parameter set for our calculations given in these tables. A more detailed breakdown of the operators involved as well as a careful statistical study of the additional empirical information available from TRIPLE and from other sources should allow extraction of a consistent set of values for F_π and F_0 from experiment. The application of our statistical theory to the remaining nuclei of the TRIPLE data set requires a more elaborate study because, as we have remarked, in cases not examined here the valence shells contain fewer nucleons and are therefore more sensitive to details requiring further refinement, such as the size of the model space and possibly the single-particle energies.

Note that the size of the Dwy contribution in our calculations is determined predominantly by our choice of the weak pion-nucleon coupling constant. If instead of the DDH value for this coupling parameter we would rely on empirical values inferred from ^{18}F [1], Dwy would be considerably smaller. Likewise, the value we found for the Std component of the force may change substantially. For example, if the measured anapole moment of ^{133}Cs is taken to determine the weak π and ρ coupling constants, with F_π constrained by the measurement in ^{18}F, the value of the weak ρ coupling constant would be substantially larger [33], as would our value of the Std contribution in Table IV. The interplay between the role of the Std and Dwy contributions clearly depends on the region in F_π-F_0 space of interest, the details of which will be reported elsewhere.

E. Shell-model tests

There are a number of useful tests of the validity of the statistical model for parity violation that could be carried out within large-basis standard shell-model calculations. We give one example: how to model parity mixing, as manifested in M, as it develops between states within the $0\hbar\omega$ model space and those lying outside it. We have assumed in this paper that the mixing occurs through the doorway mechanism as modified by Desplanques [27]. The assumption can be evaluated in a toy model (based on large but finite spaces and schematic interactions) as follows: On the one hand, the chosen interaction may be diagonalized numerically (thus obtaining a model-exact result) using a large-basis shell-model code. This same problem may then be examined in a model space applying a version of effective interaction theory. Various questions may be answered by comparing results obtained from the exact diagonalization and from effective interaction theory. Limiting cases would be instructive to consider.

For example, if excited states are initially well separated from the model space by a large gap, the leading terms in the perturbative expansions discussed in Sec. III will dominate. In the opposite limit of a very small gap, which corresponds to the well-defined case of statistical mixing in the full space, the calculation again becomes simple. The more difficult question of theoretical interest is how to handle the intermediate cases, where the gap is neither small nor large. It is in this regime where the doorway model has been conjectured to be applicable. It would be interesting to (1) confirm the doorway model in specific numerical cases; it may even be possible to motivate extensions of it valid over larger gap sizes by making selective summations over specific classes of diagrams to account for multiple p-h excitations; (2) investigate the relative importance of effective three-body PNC interactions, which have so far been neglected in all theoretical studies. In the most favorable outcome, the doorway result we have already evaluated describes this mixing over gaps comparable to shell spacings. Well-posed questions in such a model could lead to a better understanding of the theoretical uncertainties in applying statistical spectroscopy to PNC.

VI. SUMMARY AND CONCLUSION

In this paper, we have made the first extension of statistical spectroscopy to the case of parity violation and the weak spreading width in nuclei as measured by the TRIPLE Collaboration [3]. We have emphasized that statistical strength-function methods are advantageous in this case because the observables are expressed in terms of matrix elements of the effective PNC interaction averaged over *squares* of wave function components in a basis of independent-particle model states. We have also stressed that the corresponding theoretical results are less sensitive to theoretical uncertainties in familiar shell-model calculations even though the nuclear wave functions are very complicated.

In extending statistical spectroscopy to parity violation, we find that it is essential to include corrections to the underlying parity violating interaction to account for nonstatistical correlations. Of particular importance are the spin-flip correlations between states separated by $n\hbar\omega$, with $n \geq 1$. In this regard, effective interaction theory is particularly useful, and the $n\hbar\omega$ corrections have been incorporated through an effective operator. In calculating the effective operator of the PNC interaction we have implemented in the RPA approximation suggested by Desplanques [27].

We have evaluated the weak-spreading width for $A \sim 230$ in the ^{238}U target and for $A \sim 100$ in the ^{104,105,106,108}Pd targets using the standard estimates for the weak-coupling parameters of Desplanques, Donoghue, and Holstein [9]. The theoretical results are in qualitative agreement with the experimental results: the measured M in the Pd isotopes differs by about a factor of 1.7 and in U by a factor of 3. This observation supports the hope that ambiguities present in the values of the underlying weak meson-nucleon coupling parameters may be settled by using the weak-spreading width measured via neutron scattering from compound nuclei [3].

We have investigated the sensitivity of our results to various improvements in the theory, such as the sensitivity to enlargements of the model space and to permanent deformation. Permanent deformation seems to be a small correction, and our estimates indicate that results for M are stable to 10% accuracy under enlargements of the space. The reason for the stability is that these nuclei are in regions of the periodic table where large numbers of neutrons and protons occupy the shells. When, on the other hand, shells are nearly empty (or nearly full), choosing the optimal model space becomes a more delicate issue. We have indicated that the explanation of the anomalously small spreading width in the Nb and Cs cases [3] will constitute an interesting test of the theory. Because of the extended space required, these calculations require considerably more effort and have therefore not been considered in this paper. We have suggested additional toy-model calculations that might provide insight and improve confidence in the theoretical results.

ACKNOWLEDGMENTS

The work of M.B.J. and J.D.B. was supported in part by the U.S. Department of Energy under Contract No. W-7405-ENG-36 and that of S.T. by National Science Foundation Grant No. PHY-9800106. S.T. expresses his gratitude for the hospitality extended by the Physics Division of Los Alamos National Laboratory.

APPENDIX: MATRIX ELEMENTS OF THE EFFECTIVE PARITY-VIOLATING INTERACTION

In this appendix we give explicit expressions for the effective two-body PNC interaction and their matrix elements in an harmonic oscillator basis that have been used for our numerical calculations.

We may represent V^{PNC} in general as

$$V^{PNC}=\sum_{\alpha} V_{\alpha}^{PNC}, \tag{A1}$$

where V_{α}^{PNC} are terms depending on spin operators $\theta^{S(\alpha)}$ and isospin operators $\theta^{T(\alpha)}$. The free-space PNC interaction in the meson-exchange model is given in Table VI. Although there have been numerous estimates of the strengths of the weak meson-nucleon coupling constants in the literature, for our numerical work we use the DDH "best" values of F_{π} and F_0 (corresponding to the values $f_{\pi}=0.454\times 10^{-6}$ and $h_{\rho}^{0}=-1.14\times 10^{-6}$). As we stated earlier, we omit the other coupling constants, since they are quite small in the DDH analysis.

The interplay between the two-body strong and weak interaction is approximately described [1] by the two-body correlation function $f(r)$ given by Ref. [34]. The modification in $V_{Std}^{PNC(2)}$ arising from the short-range correlation function occurs by multiplying the initial and final nuclear wave function by $1+f(r)$, where $f(r)$ describes the suppression of the relative wave function at short distances due to the action of the nucleon-nucleon interaction. Otherwise, the representation of $V_{Std}^{PNC(2)}$ is the same as that for the free-space interaction as given in Eq. (A1) and Table VI.

Our representation of $V_{Dwy}^{PNC(2)}$ is

$$V_{Dwy}^{PNC(2)}=\sum_{\alpha=7,9,11,13}\theta^{S(\alpha)}\cdot\mathbf{v}_{1}^{\alpha}(r)V_{0}^{\alpha}(R)\theta^{T(\alpha)} + \sum_{\alpha=15,17}\theta^{S(\alpha)}\cdot\mathbf{V}_{1}^{\alpha}(R)v_{0}^{\alpha}(r)\theta^{T(\alpha)}, \tag{A2}$$

where a dependence on R, the center-of-mass position of the two nucleons, is now possible because of the presence of a third body (the nucleus). The definition of the various quantities is given in Table VII. The pieces associated with the isovector 0^{-} resonance are proportional to C_1, which is dominated by F_{π}. These, in particular Eq. (4.10) of Ref. [13], are much more important than the isoscalar piece associated with the isoscalar 0^{-} resonance, which is proportional to C_0 and dominated by F_0 [the numerical values of C_{π} and C_{ρ} in Eq. (4.10) of Ref. [13] are misquoted there and should rather be identified with the values of c_{ρ}^{s} and c_{π}^{v} given in Eq. (A3), given next]. These constants are given in the local density approximation by

$$C_0=c_{\rho}^{s}F_0+c_{\pi}^{s}F_{\pi}\frac{N_c-Z_c}{A_c},$$

$$C_1=c_{\pi}^{v}F_{\pi}+c_{\rho}^{v}F_0\frac{N_c-Z_c}{A_c}, \tag{A3}$$

where $c_{\rho}^{s}=7.61\times 10^{-3}$, $c_{\pi}^{s}=1.74\times 10^{-2}$, $c_{\pi}^{v}=1.74\times 10^{-2}$, and $c_{\rho}^{v}=1.3\times 10^{-3}$, and where N_c and Z_c are the neutron and proton numbers of the core ($A_c=N_c+Z_c$). In our evaluation of $V_{Dwy}^{PNC(2)}$ in this paper, we have taken $\rho=\rho_0$.

Next, we give an explicit expression for the matrix elements of $V_{Std}^{PNC(2)}$ and $V_{Dwy}^{PNC(2)}$. We express the matrix element of the effective interaction $\overline{V}^{PNC(2)}$ in second-quantized notation as

$$\overline{V}^{PNC(2)}=\frac{1}{4}\sum_{\beta j\beta' j' J}(2J+1)^{1/2} \times\langle\beta_1 j_1;\beta_2 j_2\|\tilde{V}_J^{PNC(2)}\|\beta_1' j_1';\beta_2' j_2'\rangle \times\{(a_{\beta_1 j_1}^{\dagger}a_{\beta_2 j_2}^{\dagger})^{J}(\tilde{a}_{\beta_2' j_2'}\tilde{a}_{\beta_1' j_1'})^{J}\}_{0}^{0}, \tag{A4}$$

where the sum runs over all quantum numbers (β,j), where β corresponds to the radial quantum number n, the orbital angular momentum, the spin, and the isospin of the nucleon. Here $a_{\beta_1 j_1}^{\dagger}$ is a creation operator, and $\tilde{a}_{\beta_1 j_1}$ is related to the corresponding annihilation operator by $\tilde{a}_{jm}=(-1)^{j+m}a_{j,m}$. The density matrix corresponds to initial and final pairs of particles (of total angular momentum J) coupled to $\Delta J=0$. The matrix element is antisymmetrized,

$$\langle\beta_1 j_1;\beta_2 j_2\|\tilde{V}_J^{PNC(2)}\|\beta_1' j_1';\beta_2' j_2'\rangle =\langle\beta_1 j_1;\beta_2 j_2\|\overline{V}_J^{PNC(2)}\|\beta_1' j_1';\beta_2' j_2'\rangle -(-1)^{j_1+j_2-J}\langle\beta_2 j_2;\beta_1 j_1\|\overline{V}_J^{PNC(2)}\|\beta_1' j_1';\beta_2' j_2'\rangle, \tag{A5}$$

and reduced in angular momentum only (we calculate separately matrix elements for neutron pairs, proton pairs, and neutron-proton pairs).

A general expression for $\overline{V}_J^{PNC(2)}$ is

$$
\begin{aligned}
\langle \beta_1 j_1;\beta_2 j_2 \| \overline{V}_J^{PNC(2)} \| \beta_1' j_1';\beta_2' j_2'\rangle = & \sum_{\alpha} \langle N_{\beta_1} N_{\beta_2} | T^{(\alpha_T)} | N_{\beta_{1'}} N_{\beta_{2'}}\rangle \sum_{\substack{nl\\NL}} \sum_{\substack{n'l'\\N'L'}} \sum_{S\mathcal{L}} \sum_{S'\mathcal{L}'} [(2j_1+1)(2j_2+1)(2j_1'+1)(2j_2'+1) \\
& \times (2\mathcal{L}+1)^2(2S+1)^2(2\mathcal{L}'+1)(2S'+1)]^{1/2} \begin{Bmatrix} l_1 & l_2 & \mathcal{L} \\ \frac{1}{2} & \frac{1}{2} & S \\ j_1 & j_2 & J \end{Bmatrix} \begin{Bmatrix} l_1' & l_2' & \mathcal{L}' \\ \frac{1}{2} & \frac{1}{2} & S' \\ j_1' & j_2' & J \end{Bmatrix} \\
& \times \langle n'l',N'L',\mathcal{L}' | n_1'l_1',n_2'l_2',\mathcal{L}'\rangle \langle nl,NL,\mathcal{L} | n_1 l_1,n_2 l_2,\mathcal{L}\rangle (-)^{J+\mathcal{L}'+S} \begin{Bmatrix} \mathcal{L} & \mathcal{L}' & 1 \\ S' & S & J \end{Bmatrix} \\
& \times [3(2l+1)(2L+1)]^{1/2} \begin{Bmatrix} \mathcal{L} & \mathcal{L}' & 1 \\ l & l' & k_1 \\ L & L' & k_2 \end{Bmatrix} \langle S \| S_\lambda^{S(\alpha_S)} \| S'\rangle \langle nl \| v^{\alpha}_{k_1} \| n'l'\rangle \langle NL \| V^{\alpha}_{k_2} \| N'L'\rangle,
\end{aligned}
\tag{A6}
$$

where the functional dependence of α_S, α_T, k_1, and k_2 on α as well as the values of v^{α}_{k1}, V^{α}_{k2}, $S^{(\alpha_S)}$, and $T^{(\alpha_T)}$ are given in Tables VI and VII for the standard and doorway pieces of the effective interaction, and we have made a Moshinsky transfor-

TABLE VI. Free-space parity-violating interaction of Desplanques, Donaghue, and Holstein [10].

(a) Definition of $V_{\alpha}^{PNC} \equiv \theta^{T(\alpha)} \boldsymbol{\theta} S(\alpha) \cdot \mathbf{v}^{\alpha}$ for the standard PNC interaction.

α	α_S	α_T	$\theta^{T(\alpha)} = T^{(\alpha_T)}$	$\theta^{S(\alpha)} = \mathbf{S}^{(\alpha_S)}$	v^{α}
1	2	3	$t_{10}^{(3)}$	$\mathbf{S}^{(2)}$	$\frac{\sqrt{2}}{M}(F_\pi \mathbf{u}_\pi^{(-)} + H_1 \mathbf{u}_\rho^{(-)})$
2	1	4	$F_0 t_{00}^{(4)} - \frac{1}{2\sqrt{3}} F_1 t_{10}^{(2)} - \frac{1}{2\sqrt{3}} F_2 t_{20}^{(6)}$	$\mathbf{S}^{(1)}$	$-\frac{\sqrt{3}}{M}\mathbf{u}_\rho^{(+)}$
3	3	4	$F_0 t_{00}^{(4)} - \frac{1}{2\sqrt{3}} F_1 t_{10}^{(2)} - \frac{1}{2\sqrt{3}} F_2 t_{20}^{(6)}$	$\mathbf{S}^{(3)}$	$-\frac{\sqrt{6}}{M}(1+\mu_V)\mathbf{u}_\rho^{(-)}$
4	1	2	$G_0 t_{00}^{(0)} + \frac{1}{2} G_1 t_{10}^{(2)}$	$\mathbf{S}^{(1)}$	$\frac{1}{M}\mathbf{u}_\omega^{(+)}$
5	3	2	$G_0 t_{00}^{(0)} + \frac{1}{2} G_1 t_{10}^{(2)}$	$\mathbf{S}^{(3)}$	$\frac{\sqrt{2}}{M}(1+\mu_S)\mathbf{u}_\omega^{(-)}$
6	2	1	$t_{10}^{(1)}$	$\mathbf{S}^{(2)}$	$\frac{1}{2M}(G_1 \mathbf{u}_\omega^{(+)} - F_1 \mathbf{u}_\rho^{(+)})$

(b) Definition of operators in Table VI(a). Note that the momentum operator $\mathbf{p}$ is defined as twice the relative momentum, $\mathbf{p} = \mathbf{p}_1 - \mathbf{p}_2$.

Operator		Operator	
Spin		Isospin	
$\mathbf{S}^{(1)}$	$\boldsymbol{\sigma}(1) - \boldsymbol{\sigma}(2)$	$t_{00}^{(0)}$	1
$\mathbf{S}^{(2)}$	$\boldsymbol{\sigma}(1) + \boldsymbol{\sigma}(2)$	$t_{10}^{(1)}$	$\tau_z(1) - \tau_z(2)$
$\mathbf{S}^{(3)}$	$i\boldsymbol{\sigma}(1) \times \boldsymbol{\sigma}(2)/\sqrt{2}$	$t_{10}^{(2)}$	$\tau_z(1) + \tau_z(2)$
Space		$t_{10}^{(3)}$	$i[\boldsymbol{\tau}(1) \times \boldsymbol{\tau}(2)]_z/\sqrt{2}$
$\mathbf{u}(r)^{(-)}$	$[p, u_0]$	$t_{00}^{(4)}$	$-\boldsymbol{\tau}(1) \cdot \boldsymbol{\tau}(2)/\sqrt{3}$
$\mathbf{u}(r)^{(+)}$	$\{p, u_0\}$	$t_{20}^{(6)}$	$[3\tau_z(1)\tau_z(2) - \boldsymbol{\tau}(1) \cdot \boldsymbol{\tau}(2)]/\sqrt{6}$
$u_0(r)$	$e^{-mr/4\pi r}$		

TABLE VII. Effective interaction arising from the doorway contribution as formulated in Ref. [18]. Some of the notation is defined in Table VI. Note that the doorway contribution is defined for odd values of α only. $\lambda_{00}=-53.9$ MeV fm^3, $\lambda_{11}=239$ MeV fm^3, $\lambda_{01}=200$ MeV fm^3, $\lambda_{10}=59.8$ MeV fm^3, $M=4.758$ fm^{-1} (nucleon mass), and $\mu=3.897$ fm^{-1} (ρ meson mass).

α	α_S	α_T	k_1	k_2	$\theta^{T(\alpha)}=T^{(\alpha_T)}$	$\theta^{S(\alpha)}=\mathbf{S}^{(\alpha_S)}$	$v_{k1}^{\alpha}(r)$ (fm^{-1})	$V_{k2}^{\alpha}(R)$
7	3	4	1	0	$\frac{-\sqrt{6}M\mu\omega_0}{\omega_{is}\hbar c}\lambda_{11}C_0 t_{00}^{(4)}$	$S^{(3)}$	$\frac{i\mu}{4\pi}e^{-\mu r}\hat{r}$	1
9	3	2	1	0	$\frac{\sqrt{2}M\mu\omega_0}{\omega_{is}\hbar c}\lambda_{10}C_0 t_{00}^{(0)}$	$S^{(3)}$	$\frac{i\mu}{4\pi}e^{-\mu r}\hat{r}$	1
11	2	3	1	0	$\frac{\sqrt{2}M\mu\omega_0}{2\omega_{iv}\hbar c}(\lambda_{01}+\lambda_{11})C_1 t_{10}^{(3)}$	$S^{(2)}$	$\frac{i\mu}{4\pi}e^{-\mu r}\hat{r}$	1
13	3	4	1	0	$\frac{\sqrt{2}M\mu\omega_0}{2\omega_{iv}\hbar c}(\lambda_{10}+\lambda_{11})C_1 t_{10}^{(2)}$	$S^{(3)}$	$\frac{i\mu}{4\pi}e^{-\mu r}\hat{r}$	1
15	1	3	0	1	$\frac{\sqrt{2}M\mu\omega_0}{\omega_{iv}\hbar c}(\lambda_{01}-\lambda_{11})C_1 t_{10}^{(3)}$	$S^{(1)}$	$\frac{i}{4\pi r}e^{-\mu r}$	$\mu\vec{R}$
17	3	1	0	1	$\frac{\sqrt{2}M\mu\omega_0}{\omega_{iv}\hbar c}(\lambda_{10}-\lambda_{11})C_1 t_{10}^{(1)}$	$S^{(3)}$	$\frac{i}{4\pi r}e^{-\mu r}$	$\mu\vec{R}$

mation to the relative and center-of-mass coordinates ($\langle nl,NL,\mathcal{L}|n_1l_1,n_2l_2,\mathcal{L}\rangle$ is a Moshinsky bracket [35]). This is a generalization of the results given in Ref. [36]. Our angular momentum conventions follow Brink and Satchler [37]. We have checked our results against various special cases, and two codes have been written independently to check the calculation of $V_{Std}^{PNC(2)}$. The V_{rstu}^{Γ} of Eq. (19) is equal to the $\tilde{V}_J^{PNC(2)}$ in Eq. (A4) multiplied by $\sqrt{2}$ when the quantum numbers of particle 1 or 2 are equal in the initial or final states.

[1] E. G. Adelberger and W. C. Haxton, Annu. Rev. Nucl. Part. Sci. **35**, 35 (1985).

[2] J. D. Bowman, G. T. Garvey, M. B. Johnson, and G. E. Mitchell, Annu. Rev. Nucl. Part. Sci. **43**, 829 (1993).

[3] J. D. Bowman *et al.*, Phys. Rev. Lett. **65**, 1192 (1990); X. Zhu *et al.*, Phys. Rev. C **46**, 768 (1992); C. M. Frankle *et al.*, Phys. Rev. Lett. **46**, 1542 (1992); B. E. Crawford *et al.*, Phys. Rev. C **58**, 1225 (1998); J. D. Bowman *et al.* (unpublished).

[4] J. B. French and L. S. Hsu, Phys. Lett. **25B**, 75 (1967); F. S. Chang, J. B. French, and T. H. Thio, Ann. Phys. (N.Y.) **66**, 137 (1971); K. K. Mon and J. B. French, *ibid.* **95**, 90 (1975); J. P. Draayer, J. B. French, and S. S. M. Wong, *ibid.* **106**, 472 (1977); **106**, 503 (1977); J. B. French, P. A. Mello, and A. Pandey, *ibid.* **113**, 277 (1978); J. B. French and V. K. B. Kota, Phys. Rev. Lett. **51**, 2183 (1983); J. B. French, V. K. B. Kota, A. Pandey, and S. Tomsovic, Ann. Phys. (N.Y.) **181**, 198 (1988); **181**, 235 (1988).

[5] S. Tomsovic, M. B. Johnson, A. Hayes, and J. D. Bowman (unpublished).

[6] W. C. Haxton, Phys. Rev. Lett. **46**, 698 (1981).

[7] O. Bohigas and H. A. Weidenmueller, Annu. Rev. Nucl. Part. Sci. **38**, 421 (1988).

[8] S. Tomsovic, Ph.D. thesis, University of Rochester, Report No. 974, 1987 (unpublished).

[9] B. Desplanques, J. F. Donoghue, and B. R. Holstein, Ann. Phys. (N.Y.) **124**, 449 (1980).

[10] J. B. French and V. K. B. Kota, Phys. Rev. Lett. **51**, 2183 (1983).

[11] P. J. Ellis and E. Osnes, Rev. Mod. Phys. **49**, 777 (1976); B. R. Barrett and W. M. Kirson, in *Advances in Nuclear Physics 6*, edited by M. Baranger and E. Vogt (Plenum, New York, 1973), p. 219.

[12] M. B. Johnson and M. Baranger, Ann. Phys. (N.Y.) **62**, 172 (1971).

[13] M. B. Johnson and J. D. Bowman, Phys. Rev. C **51**, 999 (1995).

[14] E. P. Wigner, Ann. Math. **62**, 548 (1955); M. L. Mehta, *Random Matrices and the Statistical Theory of Energy Levels*, 2nd ed. (Academic Press, New York, 1990).

[15] See several of the courses in *Mesoscopic Quantum Physics*, Proceedings of the Les Houches Summer School of Theoretical Physics, Haute-Savoie, France, 1994, edited by E. Akkermans, G. Montambaux, J.-L. Pichard, and J. Zinn-Justin (Elsevier Science, New York, 1995), Session LXI.

[16] H. Baltes and E. R. Hilf, *Spectra of Finite Systems* (Bibliographisches Institut, Mannheim, 1976).

[17] H. A. Bethe, Phys. Rev. **50**, 332 (1936).

[18] W. Dilg, W. Schantl, H. Vonach, and M. Uhl, Nucl. Phys. **A217**, 269 (1973); A. Gilbert and A. G. W. Cameron, Can. J. Phys. **43**, 1446 (1965).

[19] S. Perez, FORTRAN code BDSTS, 1973.

[20] A. Bohr and B. Mottelson, *Nuclear Structure* (Benjamin, New York, 1969), Vol. I.

[21] W. C. Haxton, B. F. Gibson, and E. M. Henley, Phys. Rev. Lett. **45**, 1677 (1980).

[22] B. A. Brown, W. A. Richter, and N. S. Godwin, Phys. Rev. Lett. **45**, 1681 (1980).

[23] A. C. Hayes, in *Parity and Time Reversal Violation in Compound Nuclear States and Related Topics*, edited by N. Auerbach and J. D. Bowman (World Scientific, Singapore, 1996), p. 272.

[24] V. V. Flambaum and O. K. Vorov, Phys. Rev. Lett. **70**, 4051 (1993); Phys. Rev. C **51**, 1521 (1995).

[25] N. Auerbach, Phys. Rev. C **45**, R514 (1992).

[26] V. V. Flambaum and [illegible] (1994).

[27] B. Desplanques, Phys. Rep. **297**, 1 (1998).

[28] D. Pines, K. F. Quader, and J. Wambach, Nucl. Phys. **A477**, 365 (1988).

[29] M. B. Johnson, J. D. Bowman, and H. S. Yoo, Phys. Rev. Lett. **67**, 310 (1991); M. B. Johnson, S. H. Yoo, and D. J. Bowman (unpublished).

[30] J. B. French and S. S. M. Wong, Phys. Lett. **33B**, 449 (1970); O. Bohigas and J. Flores, *ibid.* **34B**, 261 (1971); T. A. Brody, J. Flores, J. B. French, P. A. Mello, A. Pandey, and S. S. M. Wong, Rev. Mod. Phys. **53**, 385 (1981).

[31] P. W. M. Glaudemans and P. J. Brussard, *Shell-Model Applications in Nuclear Spectroscopy* (North-Holland, Amsterdam, 1977).

[32] A. Molinari, M. B. Johnson, H. A. Bethe, and W. M. Alberico, Nucl. Phys. **A239**, 43 (1974).

[33] W. S. Wilburn and J. D. Bowman, Phys. Rev. C **57**, 3425 (1998).

[34] G. A. Miller and J. E. Spencer, Ann. Phys. (N.Y.) **100**, 562 (1976).

[35] T. A. Brody and M. Moshinsky, *Tables of Transformation Brackets for Nuclear Shell-Model Calculations* (Gordon and Breach, New York, 1967).

[36] B. Castel and I. S. Towner, *Modern Theories of Nuclear Moments* (Clarendon Press, Oxford, 1990).

[37] D. M. Brink and G. R. Satchler, *Angular Momentum* (Clarendon Press, Oxford, 1968).

PHYSICAL REVIEW E VOLUME 62, NUMBER 3 SEPTEMBER 2000

Theory for matrix elements of one-body transition operators in the quantum chaotic domain of interacting particle systems

V. K. B. Kota[1] and R. Sahu[1,2]
[1]Physical Research Laboratory, Ahmedabad 380 009, India
[2]Physics Department, Berhampur University, Berhampur 760 007, India

(Received 7 March 2000)

Demonstrating the equivalence between the recent theory of Flambaum and collaborators which is based on smoothed strength functions, with the much earlier formulation due to French and collaborators which is based on embedded random matrix ensembles and smoothed transition strength densities, we derive a theory for matrix elements of one-body transition operators in the quantum chaotic domain of isolated finite interacting particle systems with a mean-field and a chaos generating two-body interaction (V). The role of the bivariate correlation coefficient (ζ) arising out of the noncommutability of V and the transition operator (in the theory of Flambaum *et al.*, $\zeta=0$) is tested in numerical embedded ensemble calculations with a one- plus two-body Hamiltonian generating order-chaos transitions.

PACS number(s): 05.45.Mt, 05.30.−d, 21.10.Pc, 24.60.Lz

In the last few years the study of quantum chaos in isolated finite interacting particle systems has turned from spectral statistics to properties of eigenfunctions and transition strengths. For the former the classical random matrix ensembles [Gaussian orthogonal ensemble (GOE), Gaussian unitary ensemble (GUE), etc.] provide the predictions. For the latter, it has been recently recognized by a large number of research groups in atomic, molecular, nuclear, and mesoscopic physics [1–6] that embedded random matrix ensembles (EEs) and in particular EGOE(k), the embedded Gaussian orthogonal ensemble of random matrices of k-body interactions, are relevant.

EGOE(k) for many (m) fermion systems is generated by defining the Hamiltonian H, which is, say, k body, to be GOE in k-particle space and then propagating it to m-particle spaces by using the geometry of the m-particle spaces [7]. Here one assumes that the m-particle space is a direct product space, of single-particle states (say, N in number), for example, as in the nuclear shell model. In many situations Hamiltonians for interacting particle systems contain a mean-field producing part (one-body part h) and a two-body residual interaction V mixing the configurations built out of the distribution of particles in the mean-field single-particle states; h is defined by single-particle energies $\epsilon_i, i=1-N$, and V is defined by two-particle matrix elements. Then it is more realistic to use EGOE(1+2), the embedded ensemble of (1+2)-body Hamiltonians defined by $\{H\}=[h(1)]+\lambda\{V(2)\}$ where $\{V\}$ is EGOE(2) and $[h]$ is a fixed Hamiltonian or an ensemble with single-particle energies chosen random but following some distribution; $[h]$ and $\{V\}$ are independent. In the literature EGOE(1+2) [or the more general EE(1+2) where the matrix elements of $\{V\}$ in two-particle space may or may not be Gaussian distributed] is also called TBRIM (two-body random interaction model) [2]. It is to be expected that the generic features of the EGOE(1+2) approach those of EGOE(2) for sufficiently large values of λ and significant results emerge as λ is varied starting from $\lambda=0$; the first study with a λ variation [using EE(1+2) instead of EGOE(1+2)], for observables in rotating nuclei, is due to Åberg [8].

The nature of occupancies (of single-particle states), strength functions, information entropy (S^{info}), inverse participation ratio (IPR), transition strength sums (for example, Gamow-Teller strength sums in nuclei), matrix elements (transition strengths) of one-body transition operators, interaction-driven thermalization, Fock-space localization, etc., in the chaotic domain of interacting particle systems is being studied in several systems in an attempt to characterize quantum chaos in many particle systems [1–6,8–14]. EGOE results for S^{info} and IPR are reported in [3] and there is a newly emerging understanding, obtained via the study of occupancies and strength sums [4,11,14], that in the chaotic domain of isolated finite interacting many-particle systems smoothed densities (they include strength functions) define the statistical description of these systems and these densities follow from EEs. This paper deals with chaos in relation to matrix elements of one-body transition operators. In particular, for systems with a mean-field and a chaos generating two-body interaction V, the seemingly different formulations due to French, Kota, Pandey, Tomsovic, and Majumdar (FKPTM) [9,15–17] and Flambaum, Gribakina, Gribakin, Kozlov, and Ponomarev (FGGKP) [1,13,18] are analyzed, in this paper, in order to establish a theory for matrix elements of one-body transition operators in the quantum chaotic domain of isolated finite interacting particle systems. In addition numerical EGOE(1+2) calculations are presented for testing the theory. Let us begin with a brief discussion of the FKPTM formulation.

Given a Hamiltonian H and its m-particle eigenstates $|E\rangle$, the transition strengths or matrix elements generated by a transition operator $\mathcal{O}$ are $|\langle E_f|\mathcal{O}|E_i\rangle|^2$. As discussed in detail in [19], in general the state densities $I^{H,m}(E)=\langle\langle\delta(H-E)\rangle\rangle^m=d(m)\rho^{H,m}(E)$ for EGOE(k) take a Gaussian ($\mathcal{G}$) form, i.e., $I^{H,m}(E)\rightarrow I_{\mathcal{G}}^{H,m}(E)$, and similarly $\rho^{H,m}(E)$ which is normalized to unity. Note that $\langle\langle\;\rangle\rangle$ denotes trace, $d(m)$ is the dimension of the m-particle space, and $I_{\mathcal{G}}^{H,m}$ is defined by its centroid (ϵ) and width (σ). In addition, it is also known [9] that the bivariate strength densities (matrix elements of $\mathcal{O}$ weighted by the state densities at the initial and final energies) $I_{biv;\mathcal{O}}^{H;m_i m_f}(E_i,E_f)=\langle\langle\mathcal{O}^\dagger\delta(H-E_f)\mathcal{O}\delta(H-E_i)\rangle\rangle^{m_i}=\langle\langle\mathcal{O}^\dagger\mathcal{O}\rangle\rangle^{m_i}\rho_{biv;\mathcal{O}}^{H;m_i,m_f}(E_i,E_f)$ take bivariate Gaussian form for EGOE(k), i.e., $I_{biv;\mathcal{O}}^{H;m_i m_f}(E_i,E_f)\rightarrow I_{biv-\mathcal{G};\mathcal{O}}^{H;m_i m_f}(E_i,E_f)$, and similarly $\rho_{biv;\mathcal{O}}^{H;m_i,m_f}(E_i,E_f)$ which is normalized to unity. It should be pointed out that, for number nonconserving transition operators $\mathcal{O}$, the number of particles m_i and m_f in the initial and final states, respectively, will not be same. The bivariate Gaussian strength density $I_{biv-\mathcal{G};\mathcal{O}}^{H;m_i m_f}$ is defined by the centroids (ϵ_i, ϵ_f) and widths (σ_i, σ_f) of its two marginals and the bivariate correlation coefficient ζ:

$$\zeta=\left\langle\left\langle \mathcal{O}^{\dagger}\left(\frac{H-\epsilon_f}{\sigma_f}\right)\mathcal{O}\left(\frac{H-\epsilon_i}{\sigma_i}\right)\right\rangle\right\rangle^{m_i} \Big/ \langle\langle \mathcal{O}^{\dagger}\mathcal{O}\rangle\rangle^{m_i}.$$

From now on some or all the superscripts over the various densities and traces will be dropped at will when no confusion arises. The FKPTM [9] result for transition matrix elements starts with $H=h(1)+V(2)$ and the bivariate strength density $I_{biv;\mathcal{O}}^{h}(x_i,x_f)$ due to $h(1)$, the mean-field producing part of H. With $V(2)$ generating chaos and thus represented by EGOE(2), its role is to Gaussian spread (with constant widths) the spikes at the energies x_i and x_f in $I_{biv;\mathcal{O}}^{h}$. The spreadings, more importantly, are correlated; i.e., the spreading function $\rho_{biv;\mathcal{O}}^{V}$, in the convolution form $I_{biv;\mathcal{O}}^{H}(E_i,E_f)=I_{biv;\mathcal{O}}^{h}\otimes\rho_{biv;\mathcal{O}}^{V}[E_i,E_f]$ is a bivariate Gaussian $\rho_{biv-\mathcal{G};\mathcal{O}}^{V}(y_i,y_f;0,0,\sigma_i,\sigma_f,\zeta)$ with ζ arising out of the noncommutability of V and the transition operator $\mathcal{O}$:

$$\zeta\sim\langle\mathcal{O}^{\dagger}V\mathcal{O}V\rangle/\langle\mathcal{O}^{\dagger}\mathcal{O}\rangle\langle VV\rangle.$$

Decomposing the m-particle space into the subspaces $\mathbf{\Gamma}$ defined by $h(1)$ $[m\rightarrow\Sigma\mathbf{\Gamma}$ with $\mathbf{\Gamma}$ labeling the eigenstates of $h(1)]$, $I_{biv;\mathcal{O}}^{H}=I_{biv;\mathcal{O}}^{h}\otimes\rho_{biv-\mathcal{G};\mathcal{O}}^{V}$ can be rewritten as

$$|\langle E_f|\mathcal{O}|E_i\rangle|^2=\sum_{\mathbf{\Gamma}_i,\mathbf{\Gamma}_f}\frac{I_{\mathcal{G}}^{\mathbf{\Gamma}_i}(E_i)I_{\mathcal{G}}^{\mathbf{\Gamma}_f}(E_f)}{I_{\mathcal{G}}^{m_i}(E_i)I_{\mathcal{G}}^{m_f}(E_f)}|\langle\mathbf{\Gamma}_f|\mathcal{O}|\mathbf{\Gamma}_i\rangle|^2 \times\frac{\rho_{biv-\mathcal{G};\mathcal{O}}^{\mathbf{\Gamma}_i,\mathbf{\Gamma}_f}(E_i,E_f;\epsilon_i,\epsilon_f,\sigma_i,\sigma_f,\zeta)}{\rho_{\mathcal{G}}^{\mathbf{\Gamma}_i}(E_i)\rho_{\mathcal{G}}^{\mathbf{\Gamma}_f}(E_f)}, \tag{1}$$

$$|\langle\mathbf{\Gamma}_f|\mathcal{O}|\mathbf{\Gamma}_i\rangle|^2=[d(\mathbf{\Gamma}_i)d(\mathbf{\Gamma}_f)]^{-1}\sum_{\alpha\in\mathbf{\Gamma}_i,\beta\in\mathbf{\Gamma}_f}|\langle\mathbf{\Gamma}_f\beta|\mathcal{O}|\mathbf{\Gamma}_i\alpha\rangle|^2,$$

$$I^{\mathbf{\Gamma}_r}(E_r)=\langle\langle\delta(H-E_r)\rangle\rangle^{\mathbf{\Gamma}_r}=d(\mathbf{\Gamma}_r)\rho^{\mathbf{\Gamma}_r}(E_r),$$

$$\epsilon_r=\langle H\rangle^{\mathbf{\Gamma}_r},\quad \sigma_r^2=\langle(H-\epsilon_r)^2\rangle^{\mathbf{\Gamma}_r},\quad r=i,j.$$

In Eq. (1), $d(\mathbf{\Gamma}_r)$ are dimensions of the subspaces $\mathbf{\Gamma}_r$. Given a one-body transition operator $\mathcal{O}=\Sigma_{\alpha,\beta}\epsilon_{\alpha\beta}a_{\alpha}^{\dagger}a_{\beta}$ where $a_{\alpha}^{\dagger}$ creates a particle in the single-particle state α and a_{β} destroys a particle in state β, it is easy to write down the expression for $|\langle\mathbf{\Gamma}_f|\mathcal{O}|\mathbf{\Gamma}_i\rangle|^2$ in terms of $\epsilon_{\alpha\beta}$ and matrix elements of number operators. References [16,17] give the details of Eq. (1) for one-body operators with $\mathbf{\Gamma}$'s chosen not only to be eigenstates of $h(1)$ but also for their various groupings. Equation (1) is applied in nuclear structure problems with two-body transition operators in [9,15], one-body transition operators, but using the so-called unitary orbits (defined by grouping single-particle states), in [17,20], and one-particle transfer operators in [21]. Although $I_{biv;\mathcal{O}}^{H}=I_{biv;\mathcal{O}}^{h}\otimes\rho_{biv-\mathcal{G};\mathcal{O}}^{V}$ is more general, it is Eq. (1) when applied to one-body transition operators that bring out the relationship between FKPTM and FGGKP.

The FGGKP formulation starts with the transition matrix elements written in terms of matrix elements in the mean-field basis states $|k_i\rangle$ by using the expansion $|E_i\rangle=C_{k_i}^{E_i}|k_i\rangle$:

$$\begin{aligned}|\langle E_f|\mathcal{O}|E_i\rangle|^2&=\left\{\sum_{k_i,k_f}C_{k_i}^{E_i}C_{k_f}^{E_f}\langle k_f|\mathcal{O}|k_i\rangle\right\}^2\\&=\left[\sum_{k_i,k_f}|C_{k_i}^{E_i}|^2|C_{k_f}^{E_f}|^2|\langle k_f|\mathcal{O}|k_i\rangle|^2\right]\\&\quad+\left[\sum_{k_i\neq k_i',k_f\neq k_f'}C_{k_i}^{E_i}C_{k_i'}^{E_i}C_{k_f}^{E_f}C_{k_f'}^{E_f}\langle k_f|\mathcal{O}|k_i\rangle\times\langle k_f'|\mathcal{O}|k_i'\rangle\right]=[\text{diag}]+[\text{offdiag}].\end{aligned} \tag{2}$$

Assuming that the transitions between different pairs of mean-field basis states are uncorrelated, the "offdiag" term in Eq. (2) is neglected. In the "diag" term, for a given k_f and k_i only one $\epsilon_{\alpha\beta}$ in $\mathcal{O}$ will contribute and [16] $|\langle k_f|\mathcal{O}|k_i\rangle|^2=|\epsilon_{\alpha\beta}|^2\langle n_\beta(1-n_\alpha)\rangle^{\mathcal{E}_i}\delta(\mathcal{E}_f-[\mathcal{E}_i-\epsilon_\beta+\epsilon_\alpha])$; ϵ_α are energies of the single-particle states α and $\mathcal{E}_i$ are the energies $\langle k_i|H|k_i\rangle$ of the mean-field basis states $|k_i\rangle$. It is well verified by EGOE(1+2) calulations [2] that $\langle n_\beta(1-n_\alpha)\rangle^{\mathcal{E}_i}$ do not vary much over the basis states that contribute to the given initial (E_i) or final (E_f) state in the chaotic domain. With this, Eq. (2) simplifies to

$$\begin{aligned}|\langle E_f|\mathcal{O}|E_i\rangle|^2&\rightarrow|\langle E_f|\mathcal{O}|E_i\rangle|_{\text{diag}}^2\\&=\sum_{\alpha,\beta}|\epsilon_{\alpha\beta}|^2\langle n_\beta(1-n_\alpha)\rangle^{E_i}\\&\quad\times\left\{\sum_{\mathcal{E}_i}|C_{\mathcal{E}_i}^{E_i}|^2|C_{\mathcal{E}_f=\mathcal{E}_i-\epsilon_\beta+\epsilon_\alpha}^{E_f}|^2\right\}.\end{aligned} \tag{3}$$

The $|C|^2$'s in Eq. (3) are nothing but the strength functions $F_k(E)$, $F_k(E)=|C_k^E|^2/\overline{D(E)}$, where $\overline{D(E)}$ is mean spacing.

With EGOE, the strength functions $F_k(E)$ take a Gaussian form characterized by the spectral width σ_k while the standard form normally employed in many applications is the Breit-Wigner (BW) form $\Gamma_k/[2\pi\{(E-\mathcal{E}_k)^2+\Gamma_k^2/4\}]$ characterized by a spreading width Γ_k. Usually, given (N,m,Δ_{spe}) where Δ_{spe} is the average single-particle level spacing, for EGOE(1+2) the quantum chaotic regime is defined by a critical interaction strength λ_c that is necessary for the emergence of Wigner-Dyson level spacing statistics [5]. However, as is well demonstrated in a recent EGOE(1+2) calculation [22], when the interacting particle system is chaotic, there is yet another border defined by the interaction strength λ_{F_k} (calculations show that $\lambda_{F_k}>\lambda_c$) beyond which the strength functions make a transition from BW to Gaussian form; see also [10,12,14]. Thus the order-chaos transition implies BW form to the Gaussian transition for the $F_k(E)$, with the BW form extending beyond λ_c up to λ_{F_k} (earlier, Georgeot and Shepelyansky [3] also showed that the BW form extends to the chaotic domain).

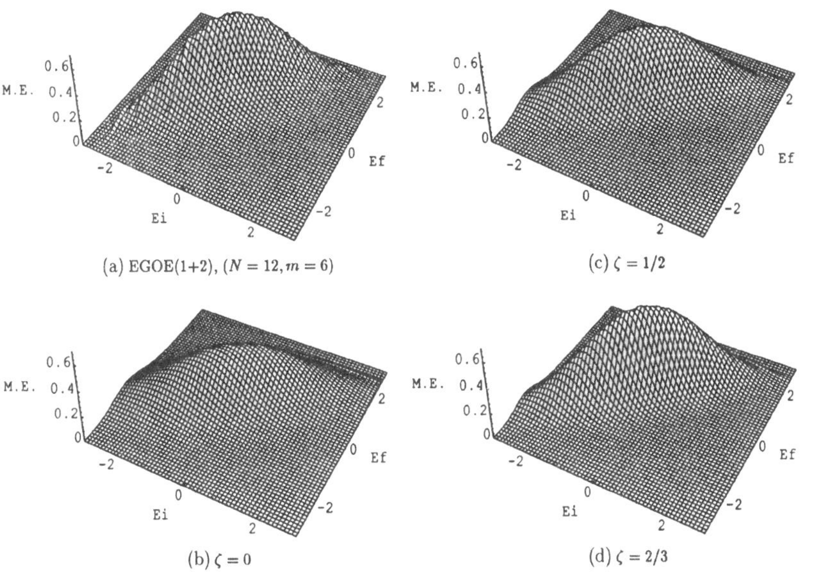

FIG. 1. Transition strengths $|\langle E_f|\mathcal{O}|E_i\rangle|^2$ vs (E_i,E_f). (a) Exact EGOE(1+2) strengths, (b) Eq. (6) with $\zeta=0$, (c) Eq. (6) with $\zeta=1/2$, and (d) Eq. (6) with $\zeta=2/3$. Here Ei$=\hat{E}_i=(E_i-\epsilon)/\sigma$ and Ef$=\hat{E}_f=(E_f-\epsilon)/\sigma$. Similarly M.E. stands for the strengths $|\langle E_f|\mathcal{O}|E_i\rangle|^2$. The EGOE(1+2) system and the one-body transition operator $\mathcal{O}$ are defined in the text. In all the calculations the strengths in the window, $\hat{E}_i\pm\Delta'/2$ and $\hat{E}_f\pm\Delta'/2$, are summed and plotted at $(\hat{E}_i,\ \hat{E}_f)$; Δ' is chosen to be 0.1. It should be noted that the total strength is 252. An enhancement in strengths due to the bivariate correlation coefficient ζ is clearly seen in (c) and (d) when compared to (b).

With the additional assumptions that (i) strength functions are only a function of $(E-\mathcal{E}_k)/s_k$ where s_k is a scale parameter (Γ_k for BW and σ_k for Gaussian) and (ii) the scale parameters s_k, in the chaotic domain, are constant (for example, $s_k\rightarrow\overline{s_i}$ for the initial many-particle basis states), the sum over $\mathcal{E}_i$ in Eq. (3) can be converted into integrals involving $F_k(E)$ (note that $\Sigma_{\mathcal{E}_i}[\,]\rightarrow\int[\,]d\mathcal{E}_i/\overline{D(E)}$). The final result, in terms of occupancies and the mean spacings, is

$$\begin{aligned}|\langle E_f|\mathcal{O}|E_i\rangle|^2&=\sum_{\alpha,\beta}|\epsilon_{\alpha\beta}|^2\langle n_\beta(1-n_\alpha)\rangle^{E_i}\overline{D(E_f)}\\&\quad\times\int F_{\mathcal{E}_i}(E_i)F_{\mathcal{E}_f=\mathcal{E}_i-\epsilon_\beta+\epsilon_\alpha}(E_f)d\mathcal{E}_i\\&=\sum_{\alpha,\beta}|\epsilon_{\alpha\beta}|^2\langle n_\beta(1-n_\alpha)\rangle^{E_i}\overline{D(E_f)}\mathcal{F}(\Delta,\overline{s}_i,\overline{s}_f),\\&\qquad\Delta=E_f-E_i+\epsilon_\beta-\epsilon_\alpha.\end{aligned}\tag{4}$$

Flambaum *et al.* [13] advocated use of Eq. (4) with BW form for F's which in turn gives a BW form for $\mathcal{F}$,

$$\mathcal{F}(\Delta,\overline{\Gamma}_i,\overline{\Gamma}_f)_{\rm BW}=\frac{1}{2\pi}\frac{\overline{\Gamma}_i+\overline{\Gamma}_f}{\Delta^2+(\overline{\Gamma}_i+\overline{\Gamma}_f)^2/4},\tag{5}$$

where $\overline{\Gamma}_i$ and $\overline{\Gamma}_f$ are the average BW spreading widths for the basis states over the initial and final many-particle states, respectively. The FGGKP theory is given by Eqs. (4) and (5) and it was subjected to EGOE(1+2) tests in [2] and also using dipole ($E1$) transitions in a Ce atom [13,18]. It is seen in the EGOE(1+2) calculations that there is always an enhancement (some times it is even by a factor of 2) around $\Delta=0$ in the matrix elements compared to the results given by Eqs. (4) and (5). In other words, the "offdiag" term in Eq. (2) in fact gives a coherent contribution. It is also seen that the enhancements grow with number of particles m; an estimate for the enhancements is given in [2]. Now we will show that Eq. (1) cures these problems.

Consider Eq. (1) by taking the subspace labels Γ_i and Γ_f to be the configurations defined by distributing the particles in single-particle states. For $\zeta=0$, the $\rho_{biv-\mathcal{G}}/\rho_{\mathcal{G}}\rho_{\mathcal{G}}$ will be unity and Eq. (1) will be identical to Eq. (3). Following the steps that led to Eq. (4), i.e., evaluating $|\langle\Gamma_f|\mathcal{O}|\Gamma_i\rangle|^2$ which gives the $\langle n_\beta(1-n_\alpha)\rangle^{\mathcal{E}_i}$ term, replacing it by $\langle n_\beta(1-n_\alpha)\rangle^{E_i}$, assuming constant spectral widths [i.e., σ_r^2 in Eq. (1) do not depend on Γ_r, $\sigma_i^2\rightarrow\overline{\sigma_i^2}$ and $\sigma_f^2\rightarrow\overline{\sigma_f^2}$], noting that $I^\Gamma(E)/I^m(E)$ are nothing but $|C|^2$, and converting the sum over $\Gamma_i(\mathcal{E}_i)$ into an integral give, directly,

$$\begin{aligned}|\langle E_f|\mathcal{O}|E_i\rangle|^2&=\sum_{\alpha,\beta}|\epsilon_{\alpha\beta}|^2\langle n_\beta(1-n_\alpha)\rangle^{E_i}\overline{D(E_f)}\\&\quad\times\int\rho_{biv-\mathcal{G};\mathcal{O}}(E_i,E_f;\mathcal{E}_i,\mathcal{E}_f=\mathcal{E}_i-\epsilon_\beta\\&\quad+\epsilon_\alpha,\overline{\sigma}_i,\overline{\sigma}_f,\zeta)d\mathcal{E}_i\\&=\sum_{\alpha,\beta}|\epsilon_{\alpha\beta}|^2\langle n_\beta(1-n_\alpha)\rangle^{E_i}\overline{D(E_f)}\\&\quad\times\mathcal{F}(\Delta=E_f-E_i+\epsilon_\beta-\epsilon_\alpha,\overline{\sigma}_i,\overline{\sigma}_f,\zeta)_{biv-\mathcal{G}},\end{aligned}\tag{6}$$

$$\begin{aligned}\mathcal{F}(\Delta,\overline{\sigma}_i,\overline{\sigma}_f,\zeta)_{biv-\mathcal{G}}&=\frac{1}{\sqrt{2\pi(\overline{\sigma_i^2}+\overline{\sigma_f^2}-2\zeta\overline{\sigma_i}\overline{\sigma_f})}}\\&\quad\times\exp\left(-\frac{\Delta^2}{2(\overline{\sigma_i^2}+\overline{\sigma_f^2}-2\zeta\overline{\sigma_i}\overline{\sigma_f})}\right).\end{aligned}$$

With Gaussian form for the strength functions, FGGKP theory given by Eq. (4) will coincide with Eq. (6) for $\zeta=0$. Thus, unlike FGGKP, FKPTM equation (6) includes correlated Gaussian spreadings and, as already pointed out, in the chaotic domain (i.e., for $\lambda>\lambda_{F_k}$) Gaussian spreadings are more appropriate than BW spreadings [22]. Second, Eq. (6) gives, for $\Delta=0$ and $\zeta \rightarrow 1$, an enhancement in the matrix elements compared to the diag($\zeta=0$) approximation; for $\overline{\sigma}_i \sim \overline{\sigma}_f$, the enhancement is $1/\sqrt{(1-\zeta)}$. The so-called binary correlation approximation [9] for EGOE gives $\zeta \sim 1-2/m$. Therefore ζ grows with m and hence the enhancements grow with m. Thus Eq. (6) reproduces all the peculiar results observed in the EGOE(1+2) calculations in [2].

For further confirmation that Eq. (6) is a proper theory, in the chaotic domain defined by $\lambda > \lambda_{F_k}$, for matrix elements of one-body transition operators, numerical calculations are carried out for various λ values using a 25-member EGOE(1+2) ensemble $\{H\}=h(1)+\lambda\{V(2)\}$ in 924-dimensional $N=12$, $m=6$ space; $h(1)$ is defined by the single-particle energies $\epsilon_i=(i)+(1/i)$, $i=1,2,\ldots,12$ just as in [2]. The one-body transition operator employed in the calculations is $\mathcal{O}=a_2^\dagger a_9$. Results for $\lambda=0.3$ are shown in Fig. 1. For the present EGOE(1+2) system, $\lambda_{F_k}=0.2$ [22]. Thus, for $\lambda=0.1$ the strength functions are close to BW form and comparing the EGOE(1+2) numerical results, for matrix elements of the one-body operator, with Eqs. (4) and (5) the disagreement between the two is found to be quite similar to the comparision between Figs. 1(a) and 1(b); i.e., it is essential to construct a theory with correlated BW forms. For $\lambda=0.3$, the strength functions are close to Gaussian and the bivariate Gaussian form is a good approximation for strength densities. Therefore for this case Eq. (6) applies as shown in Fig. 1. The comparisons in Fig. 1 clearly emphasize the role of the bivariate correlation coefficient ζ and without ζ it is not possible to get a meaningful description of the transition matrix elements. Numerical calculations give $\zeta=1/2$ and shown in the figure is also the plot for $\zeta=2/3$ as given by the binary correlation approximation. Here the agreement between theory and calculations is even better. This is encouraging as in practice it is often difficult [9,15,17] to calculate the exact values of ζ.

In summary, analyzing FKPTM and FGGKP formulations, a theory (6) for the matrix elements of one-body transition operators in the chaotic domain (with $\lambda > \lambda_{F_k}$) of isolated finite interacting particle systems is derived and tested. Further investigations of Eq. (6) in real systems are highly desirable.

This work has been partially supported by DST (India).

---

[1] V. V. Flambaum, A. A. Gribakina, G. F. Gribakin, and M. G. Kozlov, Phys. Rev. A **50**, 267 (1994).

[2] V. V. Flambaum, G. F. Gribakin, and F. M. Izrailev, Phys. Rev. E **53**, 5729 (1996).

[3] V. K. B. Kota and R. Sahu, Phys. Lett. B **429**, 1 (1998); B. Georgeot and D. L. Shepelyansky, Phys. Rev. Lett. **79**, 4365 (1997).

[4] V. K. B. Kota, R. Sahu, K. Kar, J. M. G. Gómez, and J. Retamosa, Phys. Rev. C **60**, 051 306 (1999); in *Perspectives in Nuclear Physics*, edited by C. R. Praharaj (Narosa, New Delhi, India, in press) (e-print nucl-th/0005066).

[5] Ph. Jacquod and D. L. Shepelyansky, Phys. Rev. Lett. **79**, 1837 (1997).

[6] B. L. Altshuler, Y. Gefen, A. Kamenev, and L. S. Levitov, Phys. Rev. Lett. **78**, 2803 (1997); C. Mejia-Monasterio, J. Richert, T. Rupp, and H. A. Weidenmüller, *ibid.* **81**, 5189 (1998); X. Leyronas, J. Tworzydlo, and C. W. J. Beenakker, *ibid.* **82**, 4894 (1999).

[7] T. A. Brody, J. Flores, J. B. French, P. A. Mello, A. Pandey, and S. S. M. Wong, Rev. Mod. Phys. **53**, 385 (1981).

[8] S. Åberg, Phys. Rev. Lett. **64**, 3119 (1990).

[9] J. B. French, V. K. B. Kota, A. Pandey, and S. Tomsovic, Phys. Rev. Lett. **58**, 2400 (1987); Ann. Phys. (N.Y.) **181**, 235 (1988).

[10] V. Zelevinsky, B. A. Brown, N. Frazier, and M. Horoi, Phys. Rep. **276**, 85 (1996).

[11] V. V. Flambaum and F. M. Izrailev, Phys. Rev. E **56**, 5144 (1997).

[12] W. Wang, F. M. Izrailev, and G. Casati, Phys. Rev. E **57**, 323 (1998); F. Borgonovi, I. Guarneri, and F. M. Izrailev, *ibid.* **57**, 5291 (1998).

[13] V. V. Flambaum, A. A. Gribakina, G. F. Gribakin, and I. V. Ponomarev, Physica D **131**, 205 (1999).

[14] V. K. B. Kota and R. Sahu, in *Dynamical Systems: Recent Developments*, edited by V. Srinivasan and P. K. Panigrahi (Allied Publishers, Hyderabad, India, in press).

[15] S. Tomsovic, Ph.D. thesis, University of Rochester, 1986.

[16] V. K. B. Kota and D. Majumdar, Z. Phys. A **351**, 365 (1995).

[17] V. K. B. Kota and D. Majumdar, Z. Phys. A **351**, 377 (1995).

[18] V. V. Flambaum, A. A. Gribakina, and G. F. Gribakin, Phys. Rev. A **58**, 230 (1998).

[19] K. K. Mon and J. B. French, Ann. Phys. (N.Y.) **95**, 90 (1975); J. B. French and V. K. B. Kota, Phys. Rev. Lett. **51**, 2183 (1983).

[20] D. Majumdar, K. Kar, and A. Ansari, J. Phys. G **24**, 2103 (1998).

[21] V. Potbhare and N. Tresseler, Nucl. Phys. A **530**, 171 (1991).

[22] For a 25-member EGOE(1+2) with $N=12$, $m=6$, and single-particle energies as used in generating Fig. 1, the strength functions $F_k(E)$ are constructed for $\lambda=0.01$, 0.05, 0.1, 0.2, 0.3, and 0.5. It is seen that, for $\lambda=0.01$, 0.05, and 0.1 only the BW form, for 0.2 both the BW and ED (ED is Edgeworth corrected Gaussian [23]) forms, and for 0.3 and 0.5 only the ED form are found to be good. Thus, clearly there is BW form to the Gaussian transition in $F_k(E)$ and the value of $\lambda=\lambda_{F_k}$ for the onset of this transition is $\lambda_{F_k}\approx 0.2$. This should be compared with $\lambda=\lambda_c\approx 0.08$, for the present EGOE(1+2), for the onset of chaos in level fluctuations, occupancies, and strength sums of one-body operators [4,5,14] (the numerically determined λ_c is in agreement with the estimate obtained from the theory given in [5]). In the nuclear shell model example considered in [10], $\lambda_c \sim 0.3$ and $\lambda_{F_k} \sim 0.6$. Let us point out that unlike in the case of λ_c, no theory is available yet for determining λ_{F_k} for EGOE(1+2) given (N,m,Δ_{spe}) and this problem is being addressed.

[23] M. G. Kendall and A. Stuart, *Advanced Theory of Statistics*, 3rd ed. (Hafner, New York, 1969), Vol. 1.

VOLUME 87, NUMBER 1 PHYSICAL REVIEW LETTERS 2 JULY 2001

Nonuniversal Behavior of the k-Body Embedded Gaussian Unitary Ensemble of Random Matrices

L. Benet, T. Rupp, and H. A. Weidenmüller
Max-Planck-Institut für Kernphysik, D-69029 Heidelberg, Germany
(Received 30 October 2000; published 15 June 2001)

Using a novel approach, we investigate the shape of the average spectrum and the spectral fluctuations of the k-body embedded unitary ensemble in the limit of a large matrix dimension. We identify the transition point between a semicircle and a Gaussian shape. The transition also affects the spectral fluctuations which deviate from the Wigner-Dyson form and become Poissonian in the limit $k \ll m \ll l$. Here m is the number of fermions and l the number of degenerate single-particle states.

DOI: 10.1103/PhysRevLett.87.010601 PACS numbers: 05.40.–a, 05.30.–d, 73.21.–b

Introduction.—The stochastic behavior displayed by spectra and wave functions of quantum many-body systems (atoms, molecules, atomic nuclei, quantum dots) is usually and successfully modeled in terms of canonical random-matrix theory (RMT) [1,2]. However, this type of modeling is not completely realistic: All the above-mentioned many-body systems are effectively governed by one- and two-body forces, while canonical RMT is tantamount to assuming many-body forces between the constituents. Thus, a stochastic modeling of the one- and two-body interactions would yield a much smaller number of independent random variables than used in RMT. For instance, the number of independent two-body matrix elements in a shell-model calculation in atoms or nuclei is typically much smaller than the dimension N of the matrices involved, while the number of independent random variables in RMT is of order N^2. This difference poses the question whether a more realistic stochastic modeling of many-body systems might yield results which differ from RMT predictions. The question was addressed in the 1970s with the help of numerical simulations using matrices of fairly small dimensions. The main results were: In a certain limit, the average level density does not have the shape of a semicircle but is Gaussian; the ensembles are neither stationary nor ergodic; unfolding of the spectra yields Wigner-Dyson spectral fluctuation properties; see Ref. [3]. Interest in model Hamiltonians with random two-body interactions has resurged in recent years in several areas of many-body physics (see Ref. [4], and references therein), and the question of possible further differences between such models and RMT has resurfaced. It is the purpose of this Letter to shed new light on this question. In particular, we will show that for certain parameter values, the spectral fluctuation properties of realistic models differ from those of RMT. We discuss the implications of this novel and surprising result.

Focusing attention on the case of unitary symmetry, we use a paradigmatic model, the k-body embedded Gaussian random ensemble EGUE(k) introduced by Mon and French [5]. For this ensemble, a number of analytical results have been obtained [1,5,6]. However, as emphasized in Ref. [1], page 413, a comprehensive analytical approach to the spectral fluctuation properties of the ensemble is still lacking. This is due to the fact that, in contrast to RMT, embedded ensembles do not possess the (orthogonal or unitary) invariance in Hilbert space which is so essential for the successful analytical treatment of RMT.

We use three different methods: The supersymmetry approach, the "binary correlation approximation" of Mon and French [5], and the construction of two "limiting ensembles." We also make use of two fundamental and novel results on the second moment: The eigenvector expansion and duality. We compare our results with those of the Gaussian unitary ensemble (GUE) of random matrices.

Definitions.—Our Hilbert space is spanned by $N = \binom{l}{m}$ Slater determinants $|\mu\rangle$, $\mu = 1, \ldots, N$, obtained by distributing m spinless fermions over l degenerate single-particle states. The ratio $f = m/l$ is the filling factor. Using standard creation and annihilation operators $a_j^\dagger$ and a_j, $j = 1, \ldots, l$, we have $|\mu\rangle = \prod_{s=1}^{m} a_{j_s}^\dagger |0\rangle$. Here $|0\rangle$ is the vacuum state. The creation operators are written in ascending order of the indices j_s. The k-body interaction V_k has the form ($k = 1, \ldots, m$)

$$V_k = \sum{}' v_{j_1,\ldots,j_k;i_1,\ldots,i_k} a_{j_1}^\dagger \cdots a_{j_k}^\dagger a_{i_k} \cdots a_{i_1} . \tag{1}$$

Here, and in Eqs. (2) and (9), the prime on the summation sign indicates that the sums over all indices run from 1 to l with the constraint that $j_1 < \cdots < j_k$ and $i_1 < \cdots < i_k$. The k-body interaction matrix $v_{j_1,\ldots,j_k;i_1,\ldots,i_k}$ is complex Hermitian. The independent matrix elements are uncorrelated Gaussian random variables with zero mean and a common second moment $\overline{v^2}$. The value of $\sqrt{\overline{v^2}}$ determines the overall energy scale and is set equal to unity without loss of generality. By taking matrix elements $\langle \nu | V_k | \mu \rangle$, we "embed" the random k-body interaction into an m-fermion system. This defines EGUE(k). The ensemble is invariant under unitary transformations of the single-particle states $1, \ldots, l$ and, for $k = m$, reduces to the standard GUE. As always in RMT, we are interested in the limit $N \to \infty$ ($l \to \infty$).

Second moment.—The Gaussian distribution of the matrix elements in Eq. (1) implies that the $\langle\nu|V_k|\mu\rangle$'s are also Gaussian distributed random variables with zero mean value. All spectral properties of EGUE(k) are, therefore, determined by the second moment

$$A^{(k)}_{\mu\nu,\rho\sigma} = \overline{\langle\mu|V_k|\sigma\rangle\langle\rho|V_k|\nu\rangle} = \sum{}' \langle\mu|a^\dagger_{j_1}\cdots a^\dagger_{j_k} a_{i_k}\cdots a_{i_1}|\sigma\rangle \times \langle\rho|a^\dagger_{i_1}\cdots a^\dagger_{i_k} a_{j_k}\cdots a_{j_1}|\nu\rangle. \quad (2)$$

The bar denotes the ensemble average. We have $A^{(k)}_{\mu\nu,\rho\sigma} = A^{(k)}_{\rho\sigma,\mu\nu}$ and $(A^{(k)}_{\mu\nu,\rho\sigma})^* = A^{(k)}_{\sigma\rho,\nu\mu} = A^{(k)}_{\mu\nu,\rho\sigma}$. The matrix $A^{(k)}$ is Hermitian in the pairs of indices (μ,ν) and (ρ,σ). Moreover, it is easy to prove the "duality" relation $A^{(k)}_{\mu\nu,\rho\sigma} = A^{(m-k)}_{\mu\sigma,\rho\nu}$ which connects the second moments of the k-body and the $(m-k)$-body interactions.

Eigenvector expansion.—We construct the eigenvectors $C^{(sa)}$ and eigenvalues $\Lambda^{(s)}(k)$ of the Hermitian matrix $A^{(k)}$ satisfying $\sum_{\rho\sigma} A^{(k)}_{\mu\nu,\rho\sigma} C^{(sa)}_{\sigma\rho} = \Lambda^{(s)}(k) C^{(sa)}_{\mu\nu}$. Here $s = 0,\ldots,m$ and a labels the degenerate eigenvectors. We are guided by the example of the GUE where the second moment of the Hamiltonian H reads $\overline{H_{\mu\sigma}H_{\rho\nu}} = (\lambda^2/N)\delta_{\mu\nu}\delta_{\rho\sigma}$. The two Kronecker symbols display the unitary invariance of the GUE. The matrix $\delta_{\sigma\rho}$ is an eigenfunction of the second moment with eigenvalue λ^2. All traceless unitary matrices are likewise eigenfunctions but belong to eigenvalue zero. In the present case, we use the ansatz $C^{(sa)}_{\mu\nu} = \langle\mu|a^\dagger_{j_1}\cdots a^\dagger_{j_s} a_{i_s}\cdots a_{i_1}|\nu\rangle$. The label a enumerates all possible distinct choices of the indices $j_1 < \cdots < j_s; i_1 < \cdots < i_s$. It is easy to check that $C^{(sa)}$ is an eigenvector of the matrix $A^{(k)}$ if no two indices $(j_r, i_{r'})$ are equal. The corresponding eigenvalue is

$$\Lambda^{(s)}(k) = \binom{m-s}{k}\binom{l-m+k-s}{k}. \quad (3)$$

When at least two indices $(j_r, i_{r'})$ are equal, $C^{(sa)}_{\mu\nu}$ is not an eigenfunction of $A^{(k)}$ but a linear combination of eigenfunctions with labels $s' \leq s$. This is because an eigenfunction $C^{(s'a)}_{\mu\nu}$ with eigenvalue $\Lambda^{(s')}(k)$, $s' < s$, looks like a member of class s when the operator defining $C^{(s'a)}$ is multiplied by $(\sum_p a^\dagger_p a_p)^{s-s'} = m^{s-s'}$. To remove the components belonging to lower s' values, we orthogonalize (in the sense of the trace) all matrices $C^{(sa)}_{\mu\nu}$ in which at least two indices $(j_r, i_{r'})$ are equal, to all matrices generated from classes $s' < s$ in the manner just described. The resulting matrices are eigenvectors with eigenvalue $\Lambda^{(s)}(k)$.

We choose Hermitian linear combinations of the degenerate eigenvectors which obey the orthonormality condition $\sum_{\mu\nu} C^{(sa)}_{\mu\nu} C^{(tb)}_{\nu\mu} = N\delta_{st}\delta_{ab}$. The number $D^{(s)}$ of linear independent eigenvectors in class s is given by $D^{(0)} = 1$ and

$$D^{(s)} = \binom{l}{s}^2 - \binom{l}{s-1}^2, \quad s \geq 1. \quad (4)$$

We have $\sum_{s=0}^{m} D^{(s)} = N^2$ showing that the eigenvectors form a complete orthonormal set. Hence, the matrix $A^{(k)}$ possesses an eigenvalue decomposition of the form

$$A^{(k)}_{\mu\nu,\rho\sigma} = \frac{1}{N}\sum_{s=0}^{m} \Lambda^{(s)}(k) \sum_a C^{(sa)}_{\mu\nu} C^{(sa)}_{\rho\sigma}. \quad (5)$$

We note that the eigenvectors do not depend on the rank k of the interaction, only the eigenvalues do. Equation (3) shows that the sum over s actually terminates at $s = m-k$. For $k = m$ only $s = 0$ contributes, and the result reduces to the GUE expression with $\lambda^2 = \Lambda^{(0)}(m)$. Conversely, Eq. (5) extends the GUE result to EGUE(k) and constitutes a central result of this paper.

Moments of V_k.—Using the eigenvector decomposition of $A^{(k)}$, duality, and the orthonormality of the $C^{(sa)}$'s, we calculate the low moments of V_k and the kurtosis κ for which we write $\kappa = 2 + Q(k,m,l)$. We recall that $\kappa = 2(3)$ for the semicircle (Gaussian, respectively). We find $Q(k,m,l) = (1/N)\sum_{s=0}^{\min(m-k,k)} \times \{[\Lambda^{(s)}(k)\Lambda^{(s)}(m-k)]/[\Lambda^{(0)}(k)]^2\}D^{(s)}$. For $l \to \infty$ and keeping both k and m fixed, we have $Q(k,m,l) \to 0$ if $2k > m$ while $Q(k,m,l) \to \binom{m-k}{k}/\binom{m}{k}$ for $2k \leq m$. This shows that the transition of the average spectrum from a semicircle to a Gaussian shape begins (with decreasing k) at $2k = m$. We ascribe the special role of the point $2k = m$ to duality. Likewise we can show that the relative fluctuations of the first and second moments of V_k and, thus, nonergodic features vanish for $l \to \infty$.

Supersymmetry.—We calculate spectrum shape and spectral fluctuation properties of EGUE(k) using the supersymmetry approach [7,8]. After averaging over the ensemble, the integrand of the generating functional contains an exponential whose argument depends linearly on the matrix $A^{(k)}$. With the eigenvalue decomposition (5), the argument of the exponential becomes a sum of squares of bilinear forms in the integration variables. This allows us to perform the Hubbard-Stratonovich transformation. For each pair (s,a) we introduce a supermatrix $\sigma^{(sa)}$ of composite variables. The resulting integral over the composite variables contains the factor $\exp(-\mathcal{L}_{\text{eff}})$. The effective Lagrangian $\mathcal{L}_{\text{eff}}$ is given by

$$\frac{N}{2}\sum_{sa} \text{trg}[\sigma^{(sa)}]^2 + \text{tr}_\mu\, \text{trg} \ln\left[\left(E - \frac{1}{2}\epsilon L\right)\delta_{\mu\nu} - \sum_{sa} \lambda^{(s)}(k)\sigma^{(sa)} C^{(sa)}_{\mu\nu} - J_{\mu\nu}\right]. \quad (6)$$

Here, $\lambda^{(s)}(k) = +[\Lambda^{(s)}(k)]^{1/2}$. The energy arguments E_1 and E_2 of the advanced and the retarded Green function define $E = (1/2)(E_1 + E_2)$ and $\epsilon = E_2 - E_1$, while J stands for the source terms. The diagonal supermatrix L distinguishes the retarded and advanced cases and is defined in Ref. [8]. The saddle-point equation has the solution $\sigma^{(sa)}_{\text{s.p.}} = \delta_{s0}\tau^{(0)}$

where $\tau^{(0)}$ is the standard GUE saddle-point solution. Thus, the saddle-point condition yields a semicircular spectrum and universal GUE spectral fluctuations. To determine the range of validity of this solution, we have calculated the first nonvanishing term in the loop expansion. The expansion is obtained by writing $\sigma^{(sa)} = \sigma^{(sa)}_{\text{s.p.}} + \delta\sigma^{(sa)}$ and expanding in powers of $\delta\sigma^{(sa)}$. We recall that for the GUE, each term of the loop expansion vanishes in the limit $N \to \infty$ with an inverse power of N. For the one-point function, we find that the loop correction is proportional to $Q(k, m, l)$. This is consistent with the result of the previous paragraph and reaffirms our conclusion that the transition from a semicircular to a Gaussian shape sets in at $2k = m$. It would be desirable to show that all higher terms of the loop expansion vanish likewise asymptotically for $2k > m$ but this proof is beyond our means. For the two-point function, the loop correction yields nonuniversal spectral fluctuations of the type first considered by Kravtsov and Mirlin [9]. The amplitude of this correction (given in units of the inverse mean level spacing) vanishes, however, for $l \to \infty$. It does so as N^{-2} for $k = m$ but only as $(\ln N)^{-2k}$ for both k and f fixed. We see that for $l \to \infty$, the supersymmetry approach does not yield a limit on the range of validity of the universal GUE spectral fluctuations of EGUE(k).

The case $k \ll m \ll l$.—For this case where the supersymmetry method does not yield relevant information on spectral fluctuations, we use a modification of the binary correlation approximation [1,5]. In the average two-point function $\overline{g(z_1)g(z_2)}$, we expand [6] both traced Green functions g in powers of V_k. We collect the terms containing equal powers of V_k. The ensemble average is taken by Wick-contracting pairs of V_k's in all possible ways. We evaluate all pairs located on the same Green function as in Ref. [5]. For the rest, we use simple counting arguments. We show that for $k \ll m \ll l$, terms where s pairs of V_k's are not on the same Green function, are smaller by at least a factor l^{-sk} than the terms with $s = 0$. Thus, the connected part of $\overline{g(z_1)g(z_2)}$ vanishes asymptotically compared to the disconnected part, and the two-point correlation function

$$R_2(z_1, z_2) = \frac{\overline{g(z_1)g(z_2)}}{\overline{g(z_1)} \cdot \overline{g(z_2)}} - 1 \tag{7}$$

approaches zero in the limit $k \ll m \ll l$. For similar reasons, all higher correlation functions also vanish in the same limit, and the spectral fluctuations become Poissonian.

The case $k = 1$ with $k \ll m \ll l$ illustrates this result. Taking first $m = k = 1$, we construct the eigenvalues ϵ_α and eigenfunctions Ψ_α of a given realization of EGUE(1) by diagonalization. The ϵ_α's obey GUE spectral statistics; the average spectrum has a semicircle shape, and the coefficients $U_{\alpha j}$ in the expansion $\Psi_\alpha = \sum_{j=1}^{l} U_{\alpha j} a_j^\dagger |0\rangle$ in terms of the single-particle basis are Gaussian distributed random variables. For $m > 1$ and every realization of the ensemble, the eigenfunctions are Slater determinants χ of the Ψ_α's, the eigenvalues are sums of the ϵ_α's, and the average spectrum is an m-fold convolution of the semicircle. This shows immediately that for $m \gg 1$, the spectrum has a Gaussian shape, and that the spectral fluctuations are Poissonian. We have also shown that the states χ are localized.

Regular graphs.—Further insight into the spectral properties of EGUE(k) is gained by using yet another approach involving regular graphs and limiting ensembles. A graphical representation of EGUE(k) is obtained by assigning to each Hilbert-space vector $|\mu\rangle$ a vertex μ, and to each nondiagonal element $\langle\nu|V_k|\mu\rangle$ which does not vanish identically, a link connecting the vertices ν and μ. The diagonal matrix elements $\langle\mu|V_k|\mu\rangle$ are represented by loops attached to the vertices μ. The number of vertices is N. The number M of links emanating from a given vertex is the same for all vertices and given by

$$M = \sum_{s=1}^{k} \binom{m}{s} \binom{l-m}{s}. \tag{8}$$

For $k < m$, we have $M < N - 1$ while $M = N - 1$ for $k = m$. The resulting graphical structure is called a "regular graph" in the mathematical literature. The total number P of links is given by $P = (1/2)MN$. The number K of uncorrelated matrix elements of EGUE(k) is given by $K = \binom{l}{k}^2$. It is interesting to study the ratio K/P. For fixed m and $f \leq 1/2$, K/P grows monotonically with k, starting out with very small values and reaching the limit $2N/(N-1)$ for $k = m$. This suggests that deviations of EGUE(k) from universal GUE behavior are caused by the fact that the number of independent random variables is too small in comparison with the number of links, making it impossible for the system to become thoroughly mixed.

Limiting ensembles.—To test this hypothesis, we have constructed and analyzed two limiting ensembles. The first, EGUE$_{\text{min}}(k)$, is given by the matrix elements $\langle\nu|V_k^{\text{min}}|\mu\rangle$ of the interaction $V_k^{\text{min}} = v\sum' a_{j_1}^\dagger \cdots a_{j_k}^\dagger a_{i_k} \cdots a_{i_1}$. The factor v is a Gaussian complex random variable. The ensemble EGUE$_{\text{min}}(k)$ has the same graphical representation as EGUE(k) but possesses only a single random variable. We succeeded in proving that for $k = 1$ and $k = m$, EGUE$_{\text{min}}(k)$ is fully integrable, has a Gaussian average spectrum, and spectral fluctuations which are not of a GUE type. We are convinced that these properties hold for all k. The second limiting ensemble, EGUE$_{\text{max}}(k)$, carries the maximum number of uncorrelated random variables consistent with the graph structure of EGUE(k) and has Hilbert-space matrix elements given by

$$v_{\nu\mu} \sum{}' \langle\nu|a_{j_1}^\dagger \cdots a_{j_k}^\dagger a_{i_k} \cdots a_{i_1}|\mu\rangle. \tag{9}$$

The matrix $v_{\nu\mu}$ is complex Hermitian. Elements not connected by symmetry are uncorrelated complex Gaussian random variables with mean value zero and variance

$\overline{v_{\nu\mu}v_{\nu'\mu'}} = \delta_{\mu\nu'}\delta_{\nu\mu'}$. Using supersymmetry, we have shown that $\mathrm{EGUE}_{\mathrm{max}}(k)$ has an average spectrum of semicircle shape, that the spectral fluctuations are of universal GUE type, and that for $l \to \infty$ the leading term of the loop expansion vanishes both for the one-point and for the two-point functions. We conclude that the spectral properties of $\mathrm{EGUE}_{\mathrm{max}}(k)$ coincide with those of GUE.

Conclusions.—We have studied the shape of the average spectrum and the eigenvalue fluctuations of the embedded ensemble EGUE(k) in the limit of infinite matrix dimension, attained by letting the number l of degenerate single-particle states go to infinity. We have shown that for sufficiently high rank k of the random interaction ($2k > m$ where m is the number of fermions), EGUE(k) behaves generically: The spectrum has a semicircle shape, and the eigenvalue fluctuations obey Wigner-Dyson statistics. This does not come as a surprise. A transition to a different regime takes place at or near $2k = m$. It has long been known that the average spectrum changes into a Gaussian shape, although the point of departure from the semicircle shape was not known previously. We have presented conclusive evidence that in addition—and contrary to general expectations—the level fluctuations also change and are not of a Wigner-Dyson type for $2k \lesssim m$. In the extreme case $k \ll m \ll l$, the spectral fluctuations are Poissonian, and the eigenfunctions are likely to display localization in Fock space. We cannot pin down precisely the k value where such change occurs nor can we penetrate deeply into the intermediate regime. This is not surprising as we do not know of any other case where such an aim would have been achieved. We cannot even say definitively whether the transition from Wigner-Dyson to Poissonian statistics is smooth or sudden. But we have circumstantial evidence for a smooth transition: (i) The nonuniversal fluctuations calculated from the loop correction set in smoothly. (ii) In the case $k = 1$, the transition from GUE (for $m = 1$) to Poisson behavior (for $k \ll m$) is smooth. (iii) The ratio K/P of the number of uncorrelated random variables over the number of links changes smoothly with k for fixed m.

Formally and using diagrammatic language, we ascribe the deviations from universal GUE behavior to the fact that with decreasing k, intersecting Wick contraction lines gain increased weight. Universal GUE results are obtained whenever such contributions are negligible. This apparently is the case for $2k > m$. We ascribe the special role of the transition point $2k = m$ to duality.

Physically, our results can be understood in terms of the ratio K/P of the number of uncorrelated random variables over the number of links. We have shown that if all links were to carry uncorrelated random variables, the ensemble would have GUE spectral fluctuations. Conversely, if all links were to carry the same random variable, the ensemble would be completely integrable and display Poissonian statistics. These statements hold for all values of k. The actual situation is located between these two limits. EGUE(1) is closest to the integrable case, and EGUE(m) corresponds to the GUE. This shows that deviations from GUE behavior are not caused by the number of zeros in the matrix representation of the interaction but are strictly due to correlations between the matrix elements carried by the links.

As mentioned in the introduction, numerical simulations have shown good agreement between the spectral fluctuations of EGUE(k) and those of GUE. We ascribe this result to the fact that the dimensions of the matrices used were quite small. It is easy to see that the ratio K/P is relatively close to unity for typical values such as $k = 2$, $m = 8$, and $l = 20$. It is only in the limit $l \gg 1$ that K/P becomes very small, resulting in significant deviations of EGUE(k) spectral fluctuation behavior from Wigner-Dyson statistics.

Details, further results, and the extension to EGOE(k), the orthogonal case, are given in Ref. [10].

We are grateful to O. Bohigas and T. H. Seligman. Both helped us with many questions and suggestions and were actively involved in some aspects of this work.

[1] T. A. Brody, J. Flores, J. B. French, P. A. Mello, A. Pandey, and S. S. M. Wong, Rev. Mod. Phys. **53**, 385 (1981).

[2] T. Guhr, A. Müller-Groeling, and H. A. Weidenmüller, Phys. Rep. **299**, 189 (1998).

[3] J. B. French and S. S. M. Wong, Phys. Lett. **35B**, 5 (1971); O. Bohigas and J. Flores, Phys. Lett. **34B**, 261 (1971); **35B**, 383 (1971); A. Gervois, Nucl. Phys. **A184**, 507 (1972); T. A. Brody, Lett. Nuovo Cimento **7**, 482 (1973); J. B. French, Rev. Mex. Fis. **22**, 221 (1973); T. A. Brody, E. Cota, J. Flores, and P. A. Mello, Nucl. Phys. **A259**, 87 (1976).

[4] V. K. B. Kota, Phys. Rep. (to be published).

[5] K. K. Mon and J. B. French, Ann. Phys. (N.Y.) **95**, 90 (1975).

[6] J. J. M. Verbaarschot and M. R. Zirnbauer, Ann. Phys. (N.Y.) **158**, 78 (1984).

[7] K. B. Efetov, Adv. Phys. **32**, 53 (1983).

[8] J. J. M. Verbaarschot, H. A. Weidenmüller, and M. R. Zirnbauer, Phys. Rep. **129**, 367 (1985).

[9] V. E. Kravtsov and A. D. Mirlin, Pis'ma Zh. Eksp. Teor. Fiz. **60**, 645 (1994) [JETP Lett. **60**, 656 (1994)].

[10] L. Benet, T. Rupp, and H. A. Weidenmüller, Ann. Phys. (N.Y.) (to be published); cond-mat/0010426.

ACADEMIC PRESS

Available online at www.sciencedirect.com

SCIENCE DIRECT®

Annals of Physics 306 (2003) 58–77

ANNALS of PHYSICS

www.elsevier.com/locate/aop

Convergence of moment expansions for expectation values with embedded random matrix ensembles and quantum chaos

V.K.B. Kota*

Physical Research Laboratory, Room 464 Theoretical Physics Division, Ahmedabad 380 009, India

Received 23 December 2002

Abstract

Smoothed forms for expectation values $\langle \mathbf{K} \rangle^E$ of positive definite operators **K** follow from the **K**-density moments either directly or in many other ways each giving a series expansion (involving polynomials in E). In large spectroscopic spaces one has to partition the many particle spaces into subspaces. Partitioning leads to new expansions for expectation values. It is shown that all the expansions converge to compact forms depending on the nature of the operator **K** and the operation of embedded random matrix ensembles and quantum chaos in many particle spaces. Explicit results are given for occupancies $\langle n_i \rangle^E$, spin-cutoff factors $\langle J_Z^2 \rangle^E$ and strength sums $\langle \mathcal{O}^\dagger \mathcal{O} \rangle^E$, where $\mathcal{O}$ is a one-body transition operator.

1. Introduction

Given a operator **K**, its expectation value $\langle \mathbf{K} \rangle^E$ is the diagonal matrix element of **K** in the Hamiltonian (H) diagonal basis (in the case that the eigenvalues E have a degeneracy, then it is an average over the degenerate states). For isolated finite interacting many particle systems like nuclei, atoms, quantum dots, etc. many observables can be written as expectation values. For example, the occupation numbers which determine the single particle/collective structures at low-energies and the thermodynamic behavior at higher energies are expectation values of the number operators n_i,

* Fax: +91-79-6301502.
E-mail address: vkbkota@prl.ernet.in.

doi:10.1016/S0003-4916(03)00058-7

where i denotes a single particle state or orbit (the word orbit signifies that it contains many states as happens for example when there is rotational symmetry, then a j-orbit will have $(2j+1)$ states). Similarly $\langle J_Z^2 \rangle^E$ gives spin-cutoff factors which are needed for spin decomposition of state densities (note that normally J-densities are, for examples, observables in nuclei and atoms). Another example is strength sums given by $\langle \mathcal{O}^\dagger \mathcal{O} \rangle^E$ with, for example, $\mathcal{O}$ being the Gamow–Teller operator is important in nuclear astrophysics and $\mathcal{O}$ being a dipole operator is important for atoms etc. Expectation values, just as state densities [1], can be decomposed into smoothed (with respect to E) and fluctuating parts. In the situation that the smoothed part carries most of the significant information, one can ignore the fluctuations (or deal with them separately) and consider only smoothed expectation values.

In the past several attempts are made by French and his collaborators [2–7] to derive a theory for smoothed expectation values and in the process developed several methods for calculating expectation values in terms of traces (moments); in the remaining part of this paper the word 'smoothed' is often dropped for the sake of brevity. All the moment methods for expectation values, as described in Section 3 ahead, result in series expansions in terms of lower order moments of the state and expectation value densities (expectation value density, as defined in Section 3 ahead, is the product of expectation values and state density). However very little is known about the convergence of these expansions. There is new interest in the subject of expectation values with growing emphasis on the role of two-body random matrix ensembles in finite interacting many particle systems in general (see [8,9] and references therein and also [10–23]) with clear applications in nuclei [8,18,19], atoms [20,21], quantum dots [22], interacting spin systems used in models of quantum computers [16,23], etc., due to the work of Izrailev, Borgonovi, Zelevinsky and others in quantum chaos studies of thermalization in interacting particle systems [11,14–17] and because of new interest in statistical nuclear [24–26] and atomic [20,27] spectroscopy. Therefore an immediate question to be answered is about the convergence properties of the moment expansions for expectation values. The purpose of the present paper is to discuss the convergence properties of the moment expansions as dictated, for a given operator $\mathbf{K}$, by the operation of one plus two-body random matrix ensembles (EGOE(1 + 2), see Section 2 ahead for the definition) and quantum chaos. In the context of EGOE(1 + 2), smoothed expectation values correspond to ensemble averaged expectation values. In this paper we restrict ourselves to expectation values of positive definite operators (eigenvalues are positive) and deal with $\langle n_i \rangle^E$, $\langle J_Z^2 \rangle^E$, and $\langle \mathcal{O}^\dagger \mathcal{O} \rangle^E$ with $\mathcal{O}$ being a one-body transition operator as examples. Now we will give a preview.

Section 2 gives the back ground material, for easy reading of the paper, consisting of: (i) moment methods for state densities; (ii) random matrix ensembles and quantum chaos generating generic forms for state densities and other related quantities. Section 3 gives a summary of four general methods for expectation values involving averages (moments) over total many particle spaces. Convergence properties of these expansions with the action of EGOE(1 + 2) is discussed in Section 4. Partitioning of the spectroscopic spaces and the associated expansions with convergences dictated by EGOE(1 + 2) is the topic of Section 5. It is established in Sections 4 and 5 that the physical basis for the convergence of the expansions for the expectation values of a given **K** depends on the structure of **K** and the operation of quantum chaos and one plus two-body random matrix ensembles in many particle spaces. Finally Section 6 gives some concluding remarks and future outlook.

2. Moments, state densities, random matrix ensembles, and quantum chaos

Let us consider a finite dimensional Hilbert space generated by distributing, say, m number of fermions over a finite N dimensional set of single particle states ψ_i to form a vector space of dimensionality $d(m) = \binom{N}{m}$ in which the Hamiltonian H operates. It is profitable to deal not with the spectrum itself, $\mathbf{I}(E) = \sum_j \delta(E - E_j)$, where $H\Psi_j = E_j\Psi_j$, but with the smoothed version of it, $I(E)$, which is generated by a set of lower order spectral moments. The moments $M_p(m)$ are density weighted integrals of E^p and equivalently they are also traces of H^p over the m particle space,

$$M_p(m) = d^{-1}(m) \int I(E)E^p\,\mathrm{d}E = \int \rho(E)E^p\,\mathrm{d}E = \langle H^p \rangle^m = d^{-1}(m)\langle\langle H^p \rangle\rangle^m. \quad (1)$$

It should be noted that the state density $I(E) = \langle\langle \delta(H - E)\rangle\rangle^m$ and the corresponding normalized (to unit integral) density $\rho(E) = \langle \delta(H - E)\rangle^m$; $I(E) = d(m)\,\rho(E)$. In (1) $\langle\langle\ \rangle\rangle$ denotes a trace and $\langle\ \rangle$ denotes average; we shall refer to $\langle H^p \rangle^m$ also as a "trace" or as an "average." The moments [28] are generated by the Fourier transform of $\rho(E)$, i.e., $\phi(t) = \int \mathrm{e}^{\mathrm{i}tE}\rho(E)\,\mathrm{d}E = \sum (it)^p (p!)^{-1} M_p$. The moments can be converted into central moments $\mathcal{M}_p = \langle (H - M_1)^p \rangle^m$. Moment description of densities is significant as we shall see ahead that we need only low-order moments (say $p \leqslant 4$) to represent the densities via corrections to Gaussian form. Besides the first two moments, centroid $\epsilon = M_1$ and variance $\sigma^2 = M_2 - M_1^2 = \mathcal{M}_2$, it is often more appropriate to consider cumulants than the moments M_p with $p \geqslant 3$. Cumulants are generated by the logarithm of $\phi(t)$, i.e., $\ln(\phi(t)) = \sum_{r=1}^{\infty} (it)^r (r!)^{-1} K_r$. For $r \leqslant 6$, $K_1 = M_1$, $K_2 = \mathcal{M}_2$, $K_3 = \mathcal{M}_3$, $K_4 = \mathcal{M}_4 - 3\mathcal{M}_2^2$, $K_5 = \mathcal{M}_5 - 10\mathcal{M}_3\mathcal{M}_2$, and $K_6 = \mathcal{M}_6 - 15\mathcal{M}_4\mathcal{M}_2 - 10\mathcal{M}_3^2 + 30\mathcal{M}_2^3$. It should be noted that the cumulants K_r with $r > 1$ depend only on the central moments $\mathcal{M}_p$. In order to make the moments and cumulants scale free, one defines reduced moments $\mu_p = \mathcal{M}_p/\sigma^p$ and the reduced cumulants $k_r = K_r/\sigma^r$; the $k_3 = \gamma_1$ is called the skewness and $k_4 = \gamma_2$ is the excess parameter. For a Gaussian density,

$$\rho(E)\,\mathrm{d}E = \frac{1}{\sqrt{2\pi}\sigma} \exp - \frac{(E - \epsilon)^2}{2\sigma^2}\,\mathrm{d}E,$$

where $k_r = 0$ for $r \geqslant 3$. Thus γ_2 and γ_2 measure deviations from Gaussian. For different methods of constructing Gaussians with corrections, see Appendix A.

Moment methods of constructing densities derive their importance from the fact that there are propagation theorems which evaluate many-particle traces in terms of few particle traces, the simplest being $\langle F(k)\rangle^m = \binom{m}{k}\langle F(k)\rangle^k$ (so that for a k-body operator, with n being the number operator, $\binom{n}{k}\langle F(k)\rangle^k$ is the "trace equivalent" of F). The trace propagation theorems extend to subspaces defined by a wide class of symmetry decompositions of both single and m-particle spaces and thereby of the densities; see Appendix E of [8]. Having understood that moments can be calculated using trace propagation without recourse to exact matrix diagonalization (i.e., without knowing exact eigenvalues and eigenfunctions of H in m particle spaces), the important issue remains is how to derive generic forms for the densities.

For non-interacting particle systems (NIP), the Hamiltonian $H \rightarrow h(1) = \sum_i \epsilon_i n_i$, where ϵ_i are the single particle energies and n_i is the number operator for the ith single particle state. Then, except for the Pauli blocking corrections which arise because no two particles can be in the same state, the m particle state density is simply an m-fold convolution product of the one-particle state density. As long as the one-particle variance $\sigma^2(1)$ is finite, the convolution product leads, by the action of the most elementary Central Limit Theorem (CLT), to a Gaussian density for large m. It is also easy to see that, except for singular one-particle spectra (in which a few states are far removed from the others), the Gaussian behavior does set in with only a few particles ($m \sim 6$). Since cumulants add under convolutions, CLT gives the m particle centroid to be $\epsilon(m) = m\epsilon(1)$ which is exact and $\sigma^2(m) = m\sigma^2(1)$ which is not exact; the exact result is $\sigma^2(m) = m(N-M)(N-1)^{-1}\sigma^2(1)$ which includes the blocking corrections. Similarly the exact expressions for $\gamma_1(m)$ and $\gamma_2(m)$ are,

$$\begin{aligned} \gamma_1(m) &= \left(\frac{N-2m}{N-2}\right)\left(\frac{N-1}{N-m}\right)^{1/2} m^{-1/2}\gamma_1(1), \\ \gamma_2(m) &= \frac{(N-1)}{(N-2)(N-3)m(N-m)}\left\{[N(N+1) - 6m(N-m)]\gamma_2(1) \right. \\ &\quad \left. + 6N\left[1 - \frac{m(N-m)}{N-1}\right]\right\}. \end{aligned} \tag{2}$$

For the uniform NIP spectrum with $\epsilon_i = i - 1$ and singular extreme pairing spectrum with $\epsilon_1 = 0$ and $\epsilon_i = 1$ for $i \neq 1$, it is easy to see the CLT action in the case of the former and the singular behavior in the later case by evaluating $\gamma_1(1)$ and $\gamma_2(1)$ and applying Eq. (2). Since interaction energies, unlike the mean-field (generated by $h(1)$) energies, are not additive, the NIP (CLT) argument does not extend directly to the case of interacting particle systems. At this stage it is important to remind that deriving physics via smoothed densities (defined by lower order moments) implicitly assumes that the fluctuations are well separated from the averages and they are small. In deriving forms for the interacting particle state densities and understanding the nature of level (and strength) fluctuations, one sees the need for dealing with random matrix ensembles and quantum chaos.

The Gaussian orthogonal ensemble (GOE) of random matrices [29], originally proposed by Wigner as a model for a complicated system with time reversal invariant H, is now (starting with the pioneering paper by Bohigas et al. [30]) universally regarded as a model for a corresponding chaotic system and it is applied in all most all branches of physics. For the GOE, the H matrix is real symmetric and the matrix elements H_{ij} are independent (except for the real symmetric nature) Gaussian random variables with zero center and same variance, $\overline{H_{ij}^2} = v^2(1 + \delta_{ij})$, where v^2 is a constant. The nearest neighbor spacing distribution $P(S)\,\mathrm{d}S$ for GOE, follows the Wigner's law

$$P(S)\,dS = \frac{\pi}{2} S \exp - \frac{\pi}{4} S^2 \, dS \tag{3}$$

and the Dyson–Mehta $\Delta_3(n)$ statistic (it is the root mean-square deviation of the distribution function, over a length of n mean spacings, from the best fit straight line) will be logarithmic, i.e., $\Delta_3(n) \sim \ln n$. These level fluctuation measures show that the GOE spectrum is rigid and exhibits level repulsion. Before going further it should be pointed out that 'quantum chaos' refers to quantum systems which in the classical limit show chaotic dynamics (examples are billiard systems, kicked top, kicked rotor, quartic oscillator, etc.). As quantum chaotic systems exhibit [29] random matrix fluctuations, quite often quantum chaos for interacting many particle quantum systems implies operation of random matrix ensembles and vice-versa. It is in this sense quantum chaos is used in this paper. The question of importance now is to identify the random matrix ensemble appropriate for finite interacting many particle systems.

State density for GOE is a semi-circle which is radically different from the Gaussian form observed in interacting particle systems such as the nuclear shell model examples. This difference shows up as the GOE is an ensemble of multi-body, not two-body, interactions (a GOE for m particle systems includes m-body interactions). This led, French and Wong and Bohigas and Flores [31] to introduce two-body random matrix ensembles (TBRE). For the GOE version of TBRE, the two-body interaction is taken as GOE and then the m particle Hamiltonian matrix is constructed with m-particle basis states being direct products of single particle states (antisymmetric for fermions and symmetric for bosons; in this paper all the discussion is restricted to fermion systems). Then it is called embedded GOE of two-body interactions or simply EGOE(2) as a GOE in 2-particle spaces is embedded in m ($m > 2$) particle spaces and similarly for k-body interactions EGOE(k) is defined (present paper deals only with EGOE(2)). It is well known that for EGOE(2) (and in general for EGOE(k)) in the dilute limit ($m \to \infty$, $N \to \infty$ and $m/N \to 0$) the ensemble averaged state densities approach Gaussian form [12,32] (similarly the transition strength densities approach bivariate Gaussian form, see Appendix B). Moreover, EGOE(2) gives GOE fluctuations. The GOE fluctuations are seen for large but finite N in many numerical examples [1,8]; there could be deviations from the GOE behavior in the $N \to \infty$ limit as argued recently by Benet et al. (see [9] and the second reference in [32]) but this is of no concern in this paper. Another important property of EGOE(2) is that it produces sharp separation between average density and level fluctuations with the former being a Gaussian with lower order shape (cumulants) corrections and the later determined by very high order density moments [1]. Ergodicity of EGOE(2) ensures that the Gaussian form applies to spectrally smoothed state densities for all most all two-body interactions; see [1,8] for more discussion on these basic results.

Hamiltonian for realistic systems is not just two-body but it consists of a mean-field one-body plus a complexity generating two-body interaction. The nature of the state densities generated by such a Hamiltonian is understood by assuming that H is representable by one plus two-body embedded Gaussian orthogonal ensemble of random matrices (EGOE(1 + 2)), $H \to \{H\} = h(1) + \lambda\{V(2)\}$, where $\{\ \}$ denotes an ensemble, $h(1)$ is a fixed one-body operator (or an ensemble) generating single particle spectrum with single particle energies ϵ_i having average spacing $\Delta = 1$, $\{V(2)\}$ being EGOE(2) (i.e., $V(2)$ in two-particle space is GOE) with the two-particle $V(2)$ matrix elements variance chosen to be unity and λ is the interaction strength. For EGOE(1 + 2) one can define two 'quantum chaos' markers λ_c and λ_{F_k} ($\lambda_c \propto 1/m^2N$ and $\lambda_{F_k} \propto 1/\sqrt{m}$) [8]. For $\lambda > \lambda_c$ there is chaos in the sense that the level fluctuations start coming close to GOE fluctuations [10]. The λ_{F_k} determines the nature of strength functions. Given the mean-field $h(1)$ basis states $|k\rangle = \sum_E C_k^E|E\rangle$, the strength function $F_k(E)$ is

$$F_k(E) = \sum_{\beta \in E} |C_k^{E,\beta}|^2 = \overline{|C_k^E|^2} I(E).$$

As λ is increasing, $F_k(E)$ changes from δ-function to Breit–Wigner (BW) to Gaussian form [8,18,33]; note that

$$F_{k;\mathrm{BW}}(E) = \frac{1}{2\pi} \frac{\Gamma_k}{(E - E_k)^2 + \frac{\Gamma_k^2}{4}},$$

where E_k is the centroid and Γ_k is the spreading width. For $\lambda \ll \lambda_c$ strength functions are close to δ-function and then the BW form sets in. For $\lambda_c \leqslant \lambda \leqslant \lambda_{F_k}$ they are BW and for $\lambda > \lambda_{F_k}$ they approach Gaussian form; the $\lambda > \lambda_{F_k}$ region is called the Gaussian domain and similarly the $\lambda_c \leqslant \lambda \leqslant \lambda_{F_k}$ region is called BW region. It should be emphasized that the local fluctuations in both BW and Gaussian domain are of GOE type. These chaos markers λ_c and λ_{F_k} are used in Sections 4 and 5 to derive proper theoretical forms for expectation values. Before proceeding to expectation values, another important point to be mentioned is that the state densities for EGOE(1 + 2) take Gaussian form (in the dilute limit) independent of the value of λ (also as $\lambda \to \infty$ EGOE(1 + 2) ensemble behaves as EGOE(2)). However for $\lambda < \lambda_c$ the fluctuations are of Poisson type. Similar results are valid even for the bivariate strength densities (see Appendix B and [34,35]).

3. General methods for expectation values

Given an operator $\mathbf{K}$, its expectation values $\langle \mathbf{K} \rangle^E$, taking into account degeneracies in the H spectrum, are defined by

$$\langle \mathbf{K} \rangle^E = [I(E)]^{-1} \sum_{\beta \in E} \langle E\beta \mid \mathbf{K} \mid E\beta \rangle = \langle\langle \mathbf{K}\delta(H - E)\rangle\rangle^m / \langle\langle \delta(H - E)\rangle\rangle^m, \tag{4}$$

where $|E, \beta\rangle$ are the eigenstates of H. Just as the state densities, one can define, for positive definite $\mathbf{K}$'s, expectation value densities (or simply $\mathbf{K}$-densities) $I_{\mathbf{K}}^m(E) = \langle\langle \mathbf{K}\delta(H - E)\rangle\rangle^m$ and its normalized counterpart $\rho_{\mathbf{K}}^m(E)$,

$$\rho_{\mathbf{K}}^m(E) = \langle\langle \mathbf{K}\delta(H - E)\rangle\rangle^m / \langle\langle \mathbf{K}\rangle\rangle^m = I_{\mathbf{K}}^m(E) / \langle\langle \mathbf{K}\rangle\rangle^m. \tag{5}$$

The positive definiteness of $\rho_{\mathbf{K}}(E)$ is seen by writing the trace $\langle\langle \mathbf{K}\delta(H - E)\rangle\rangle^m$ in the $\mathbf{K}$ eigenstates basis and then by expanding these states in the H eigen basis. The moments of $\rho_{\mathbf{K}}(E)$ are $M_p(\mathbf{K}) = \langle \mathbf{K}H^p\rangle^m / \langle \mathbf{K}\rangle^m$ and they can be converted into

$\mathcal{M}_p(\mathbf{K})$, $\mu_p(\mathbf{K})$, $K_p(\mathbf{K})$ and $k_p(\mathbf{K})$; for many purposes the moments $\hat{\mathcal{M}}_p(\mathbf{K}) = \langle\mathbf{K}(H-\epsilon)^p\rangle^m/\langle\mathbf{K}\rangle^m$ are also useful. It should be stressed that the centroids of $\rho(E)$ and $\rho_\mathbf{K}(E)$ are in general different and therefore one can choose only one of them to be zero. This aspect plays a significant role in the discussion ahead. Note that $\epsilon(\mathbf{K}) = \epsilon + \langle\mathbf{K}(H-\epsilon)\rangle^m/\langle\mathbf{K}\rangle^m$. Let us point out that sometimes the superscripts 'm' and 'H' are included in $I(E)$ (and also for $I_\mathbf{K}(E)$, $\rho(E)$, $\rho_\mathbf{K}(E)$), i.e., for example, $I(E)$ is written as $I^{m,H}(E)$ to point out that they are defined over m-particle spaces and the Hamiltonian involved is H. Usually we drop the superscripts m or H or both of them when there is no confusion. Similarly we write sometimes $\epsilon(m)$, $M_p(\mathbf{K}:m)$, etc. when we need to specify m. Having defined all the essential quantities, now we will briefly discuss four methods for calculating expectation values.

3.1. Moment method

In the moment method one uses the fact that the $\mathbf{K}$ densities are proper (positive definite) densities for positive definite operators and therefore they can be constructed directly in terms of their low-order moments assuming that the asymptotic form of these densities is known. Then $\langle\mathbf{K}\rangle^E$ is given by (see [6,7]),

$$\langle\mathbf{K}\rangle^E = \langle\mathbf{K}\rangle^m \rho^m_\mathbf{K}(E)/\rho^m(E). \tag{6}$$

There are many ways of expanding a density function given its cumulants and an asymptotic form [28]. Applying this to the two densities in (6), it can be recognized that Eq. (6) in principle gives expectation values as a ratio of two infinite series and they can be truncated depending on the form one adopts. Appendix A gives five different forms that are used for $\rho(E)$ (they can be used for $\rho_\mathbf{K}(E)$ as well) in literature. All of them are closely related to the Gaussian form and they involve only $(\epsilon, \sigma, \gamma_1, \gamma_2)$ parameters. There is some freedom in choosing any expansion one likes as it is often difficult to derive a unique theory for the corrections to the asymptotic Gaussian form.

3.2. Parametric derivative method

In the parametric derivative method the linear response of the density under the deformation $H \longrightarrow H_\zeta = H + \zeta\mathbf{K}$ is considered. Then it is easily seen that [3,36]

$$\langle\mathbf{K}\rangle^E = -[\rho^H(E)]^{-1} \lim_{\zeta\to 0} \int_{-\infty}^{E} \frac{\partial\rho^{H_\zeta}(x)}{\partial\zeta}\,\mathrm{d}x = -[\rho^H(E)]^{-1} \lim_{\zeta\to 0} \frac{\partial F^{H_\zeta}(E)}{\partial\zeta}. \tag{7}$$

With $\rho(E)$ is defined say by the parameters S_ν, Eq. (7) can be applied using the relation

$$\frac{\partial\rho^{H_\zeta}(x)}{\partial\zeta} = \sum_\nu \frac{\partial\rho^H(x)}{\partial s_\nu}\frac{\partial s_\nu}{\partial\zeta}.$$

Then it should be clear that Eq. (7) also gives expectation values as a ratio of two infinite series and they can be truncated for example using the forms given in Appendix A. It is useful to note that, with $F^1(x) = F(x)$ and $F^t(x) = \int_{-\infty}^x F^{t-1}(y)\,\mathrm{d}y$, the expectation value

$$\langle \mathbf{K}^t \rangle^E = (-1)^t [\rho^H(E)]^{-1} \lim_{\zeta \to 0} \frac{\partial F^{H_{\zeta:t}}(E)}{\partial \zeta^t} \quad \text{for } [H,K] = 0.$$

3.3. Polynomial expansion method

In the polynomial expansion method the δ-function in the numerator of Eq. (5) is expanded in terms of the ortho-normal polynomials $P_\mu(E)$ defined by the exact state density, $\delta(H - E) = \rho^H(E) \sum_\mu P_\mu(H) P_\mu(E)$. Then Eq. (5) gives expectation values as a series expansion involving P_μ's [4,5]

$$\langle \mathbf{K} \rangle^E = \sum_\mu \langle \mathbf{K} P_\mu^m(H) \rangle^m P_\mu^m(E). \tag{8}$$

Note that the polynomials P_μ^m are uniquely defined by the moments of $\rho^{m;H}(E)$ [5]. A significant difference between Eq. (8) and Eqs. (6) and (7) is that the later involves ratio of infinite series while the former involves only one series. However had we used the polynomials defined by some other density (say the asymptotic form of $\rho(E)$), then Eq. (8) also will involve ratio of two infinite series. Eq. (8) was used in many nuclear structure applications assuming that it converges to $\mu = 0 + 1$ terms in (8) for any $\mathbf{K}$; see for example [37].

3.4. Bivariate density method

In the special situation that H and $\mathbf{K}$ commute (a good example is $\mathbf{K} = J_z$ in nuclei and atoms), an alternative, for generating a moment expansion for $\langle \mathbf{K}^t \rangle^E$, to the previous three methods is to use the bivariate density method. Here the bivariate density $\rho(E, E_k)$ defined by the eigenvalues E and E_k of H and $\mathbf{K}$ is constructed and then the expectation values are given by [38],

$$\langle \mathbf{K}^t \rangle^E = \int_{-\infty}^{\infty} (E_k)^t \rho(E, E_k)\, \mathrm{d}E_k \Big/ \int_{-\infty}^{\infty} \rho(E, E_k)\, \mathrm{d}E_k \quad \text{for } [HK] = 0. \tag{9}$$

Just as in Appendix A, there are many ways one can construct the bivariate density $\rho(E, E_k)$ given its bivariate cumulants and an asymptotic form [28,39] and each of them can be used in (9); see Appendix C for an example and [38,39] for more details. An alternative to (9) is to use the tth parametric derivative method as explained below Eq. (7); one can always use (6), (7), or (8) with $\mathbf{K}$ replaced by $\mathbf{K}^t$.

4. Convergence of moment expansions with the action of EGOE(1 + 2)

All the four methods in Section 3 are known before and in principle they are mathematically exact. They differ in practice when the expansions given by each of them are truncated in terms of the moments of the state and $\mathbf{K}$-densities upto some maximum order. In fact truncation of the expansions is equivalent to assuming convergence to certain number of terms (this is also equivalent to smoothing [1]). Clearly, the expansions given by (6)–(9) will be useful only if their convergence properties are known. In the past forms given by (8) without regard to the nature of $\mathbf{K}$ are used by some groups and forms given by (6) are used by some others. In many numerical examples it is verified that each of them is reasonably accurate. Some general conditions for the convergence of the expansions given by (6)–(9) are known before [5,8,34]. The only important question left unanswered is which form (among (6)–(9)) has the proper physical basis for a given **K**. This will be answered now by assuming that H is representable by EGOE(1 + 2) and that one is in the $\lambda > \lambda_c$ chaotic domain (then $\rho^H(E)$ is close to a Gaussian and the fluctuations follow GOE).

For the number operator n_i an obvious choice is to consider $H \rightarrow H_\zeta = H + \zeta n_i$, then the single particle energy ϵ_i changes to $\epsilon_i + \zeta$. With H represented by EGOE(1 + 2), H_ζ for ζ small is also represented by EGOE(1 + 2) and therefore the shape of $\rho^H(E)$ will be unchanged (from the Gaussian form) under the ζ deformation. Using this and applying (7), gives

$$\langle n_i\rangle^E \overset{\mathrm{EGOE}(1+2)}{\rightarrow} \langle n_i\rangle^m + \langle n_i(H-\epsilon(m))\rangle^m (E-\epsilon(m))/\sigma^2(m). \tag{10}$$

The linear form (10), seen in many nuclear shell model calculations, was discussed in [5]; the two terms in (10) correspond to the first two terms in the polynomial expansion (8). Thus the EGOE action in the chaotic domain ($\lambda > \lambda_c$) gives Eq. (10) to be the form for the expectation values of single state (or orbit) number operators with correction given by the third term in (8) or by the terms given by (7) due to (γ_1, γ_2) corrections to the state density Gaussian. Recently Eq. (10) is used [17] to derive a compact form for the single particle entropy defined in terms of the occupancies $\langle n_i\rangle^E$.

For the J_Z^2 expectation values, for H preserving angular momentum (as it is the case for example with nuclear and atomic Hamiltonians), it is more appropriate to consider (9). Note that $\langle J_z^2\rangle^E = \frac{1}{3}\langle J^2\rangle^E$ and J^2 gives χ_3^2 distribution. From the EGOE(1 + 2) representation of H and the fact that J_z spectrum is uniform give the marginals of $\rho(E, J_Z)$ to be Gaussians. Also in this case the bivariate correlation coefficient is zero. This gives, via (9), $\langle \mathbf{K}\rangle^E = \langle \mathbf{K}\rangle^m$. Adding lower-order Edgeworth corrections (see (C.2)) to $\rho(E, J_Z)$, spin-cutoff factors are given by [38],

$$\langle J_z^2\rangle^E \overset{\mathrm{EGOE}(1+2)}{\rightarrow} \langle J_z^2\rangle^m \left\{1 + [k_{12}(m)\hat{E}] + \left[(k_{22}(m) - k_{12}(m)\gamma_1(m))(\hat{E}^2 - 1)/2\right]\right\}. \tag{11}$$

In (11), the bivariate cumulants $k_{12}(m) = \langle J_Z^2(H-\epsilon(m))\rangle^m/(\langle J_z^2\rangle^m \sigma(m)$ and $k_{22} = \langle J_Z^2(H-\epsilon(m))^2\rangle^m/\langle J_z^2\rangle^m \sigma^2(m) - 1$. The quadratic form given by (11) is seen in many nuclear shell model calculations. The expansion used by Johnson et al. [26] recently is closely related to (11).

Given a transition operator $\mathcal{O}$, the operator $\mathbf{K} = \mathcal{O}^\dagger\mathcal{O}$ will have a spectrum quite different from a Gaussian [40] (for example, with Q_μ^2 denoting electric dipole transition operator, $Q \cdot Q$ gives χ_5^2 distribution) and therefore the arguments used in deriving (10) can not be applied for $\langle \mathcal{O}^\dagger\mathcal{O}\rangle^E$. However here one can use the result that for EGOE(1 + 2) the bivariate strength density, in general (for non-singular one and two-body transition operators, see discussion on p. 253 of [34] and Appendix B) will be a bivariate Gaussian. Then $\rho_{\mathbf{K}}(E)$ being a marginal of this bivariate Gaussian, it will be a Gaussian. Therefore using (6) we get $\langle \mathcal{O}^\dagger\mathcal{O}\rangle^E$ to be a ratio of Gaussians,

$$\langle \mathbf{K}\rangle^E \overset{\mathrm{EGOE}(1+2)}{\rightarrow} \langle \mathbf{K}\rangle^m \rho^m_{\mathbf{K};\mathcal{G}}(E)/\rho^m_{\mathcal{G}}(E), \quad \mathbf{K} = \mathcal{O}^\dagger\mathcal{O}, \tag{12}$$

where $\mathcal{O}$ is at most two-body and $\mathcal{G}$ denotes Gaussian. A better approximation to Eq. (12) is to add, for example, Edgeworth corrections to $\rho_{\mathbf{K}}$ and ρ using (A.2). Numerical EGOE(1 + 2) tests of Eq. (12), for one-body transition operators, are given in [8]. Note that in principle one can apply Eqs. (7) and (8) for deriving a form for $\langle \mathcal{O}^\dagger \mathcal{O} \rangle^E$. An important point here is that (12) keeps the **K**-density centroids and widths to be the exact values unlike in the case with Eqs. (7) and (8) which give expansions involving corrections even to the centroids and variances. Such expansions are called non-standard expansions in statistics literature [38]; they arise as in general $\epsilon(\mathbf{K}:m) \neq \epsilon(m)$ and $\sigma(\mathbf{K}:m) \neq \sigma(m)$. Recently Eq. (12) is used successfully in analyzing Gamow–Teller (GT), electric quadrupole (E2) and magnetic dipole (M1) transition strength sums in large nuclear pf and sdpf spaces [41].

In summary, with EGOE(1 + 2) operating in the quantum chaotic domain (with $\lambda > \lambda_c$) of m-particle spaces, the convergence of the moment expansions for expectation values depend on the nature of the operator **K**. For example, Eqs. (10) and (12) with corrections will give the forms for $\langle n_i \rangle^E$ and $\langle \mathcal{O}^\dagger \mathcal{O} \rangle^E$ respectively ($\mathcal{O}$ is at most two-body) and their convergence is to within GOE fluctuations. Note that the EGOE(1 + 2) used here is an ensemble for spin-less fermion systems. Therefore, strictly speaking one can not use this ensemble for $\langle J_z^2 \rangle^E$ as here the Hamiltonian must be J invariant for spin-cutoff factors are to be physically meaningful. Therefore for $\langle J_z^2 \rangle^E$ case it is necessary to deal with EGOE(1 + 2)-J (for nuclei with EGOE(1 + 2)-JT, where T is isospin). Analytical treatment of the ensembles with J and JT symmetries is still not developed; see [8] for a discussion of these ensembles. In this sense, at present, Eq. (11) and its relatives can be used only as methods for spin-cutoff calculations. Finally it is important to point out that Eqs. (10)–(12) are valid even in the $\lambda < \lambda_c$ domain as here also the state and strength densities (in the dilute limit) are Gaussians (see [35,42] for examples in the $\lambda = 0$ case) but however the convergences are not meaningful as the fluctuations here are of Poisson type.

5. Expectation values with partitioning

Results given by (10)–(12) are based on the convergence of the state density to Gaussian form with GOE fluctuations, i.e., for EGOE(1 + 2) with $\lambda > \lambda_c$. However, the convergence to Gaussian being asymptotic, in general it will not be good in the wings. Therefore [2,42,43] in large spectroscopic spaces one has to partition the space to subspaces. With partitioning, as we shall see ahead the methods of Section 3 will apply, albeit in an indirect way. Let us consider m particle spaces partitioned into subspaces denoted by Γ each with dimension $d(\Gamma)$ so that $d(m) = \sum_\Gamma d(\Gamma)$. Then, with $m \to \sum \Gamma \oplus$, the state density decomposes into partial densities $I^\Gamma(E)$,

$$I(E) = \langle\langle \delta(H - E) \rangle\rangle^m = \sum_\Gamma \langle\langle \delta(H - E) \rangle\rangle^\Gamma = \sum_\Gamma I^\Gamma(E). \quad (13)$$

Note that the trace $\langle\langle \ \rangle\rangle^\Gamma$ means summing over all states with the same Γ label. It should be stressed that $I^\Gamma(E)$ is always positive definite and therefore it is a proper density function. In the situation that Γ are the mean-field $h(1)$ eigenstates, then $I^\Gamma(E)$ are nothing but strength functions or local density of states. From now on we refer to $I^{\Gamma}(E)$ (for any definition of Γ) as strength functions as this nominiclature is more familiar. For spin-less fermion systems, $\Gamma = \widetilde{m} = (m_1, m_2, \ldots, m_N)$, where m_i is the number of particles in the *ith* single particle state and $\sum_i m_i = m$. Note that $m_i = 0$ or 1 and $d(\widetilde{m}) = 1$ in this case. For systems with spin, in general $d(\widetilde{m}) \neq 1$ and $m_i > 1$ is possible with i's denoting single particle orbits. Just as the case with state densities, expectation values also decompose into traces over Γ subspaces (this is seen easily from Eq. (4)),

$$\langle \mathbf{K} \rangle^E = [I(E)]^{-1} \sum_{\Gamma} \langle\langle \mathbf{K}\delta(H-E)\rangle\rangle^{\Gamma} = [I(E)]^{-1} \sum_{\Gamma} I^{\Gamma}_{\mathbf{K}}(E). \tag{14}$$

An essential complication now is that $I^{\Gamma}_{\mathbf{K}}(E)$ is not in general positive definite even if $\mathbf{K}$ is positive definite. Usually the $m \to \Gamma$ decomposition is carried out using a group symmetry in the m particle spaces [2,7,44]. Then Γ are the irreducible representations (irreps) of the groups in the group–subgroup chain involved. If the Casimir invariants of the groups are C_i's, the $I^{\Gamma}_{\mathbf{K}}(E)$ depend on the correlations between H, $\mathbf{K}$, and C_i's (also on (m, N)). There is no general theory yet for understanding these correlations [7]. Therefore to proceed further, one should consider situations where Eq. (14) reduces to a simpler form depending on K and due to the action of quantum chaos and EGOE(1+2). As an aside let us point out that the polynomial method (Eq. (8)) can be applied to Eq. (14) by using the identity $\delta(H-E) = \rho^{\Gamma}(E) \sum_{\mu} P^{\Gamma}_{\mu}(H) P^{\Gamma}_{\mu}(E)$, where P^{Γ}_{μ} are the polynomials defined by the normalized partial densities $\rho^{\Gamma}(E) = (d(\Gamma))^{-1} I^{\Gamma}(E)$. This form was used in the past by assuming that it converges to $\mu \leqslant 2$ terms in general [5,44,45].

5.1. Convergence with partitioning according to K subspaces

A major simplification of Eq. (14) arises if we choose Γ to be the eigen subspaces of the operator $\mathbf{K}$, then $I^{\Gamma}_{\mathbf{K}}(E) = \langle \mathbf{K} \rangle^{\Gamma} I^{\Gamma}(E)$. This gives,

$$\langle \mathbf{K} \rangle^E = \sum_{\Gamma} \frac{I^{\Gamma}(E)}{I(E)} \langle \mathbf{K} \rangle^{\Gamma}, \quad \text{for } \mathbf{K}|\Gamma\alpha\rangle = f(\Gamma)|\Gamma\alpha\rangle. \tag{15}$$

In (15) α are additional labels needed for complete specification of the basis states with a given Γ. In order to apply (15) the form of I^{Γ} has to be known. For $\mathbf{K} = n_i$ the Γ's are $\widetilde{m}$'s defined earlier. As pointed out in Section 2, for EGOE(1+2) in the chaotic domain, the strength functions $I^{m}(E)$ take Gaussian form for $\lambda > \lambda_{F_k}$ and BW form in the $\lambda_c \leqslant \lambda \leqslant \lambda_{F_k}$ domain. Then,

$$\begin{aligned} \langle n_i \rangle^E &\equiv \sum_{\widetilde{m}} \frac{I^{\widetilde{m}}(E)}{I(E)} \langle n_i \rangle^{\widetilde{m}}, \\ &\xrightarrow{\text{EGOE}(1+2):\lambda_c \leqslant \lambda \leqslant \lambda_{F_k}} \sum_{\widetilde{m}} \frac{I^{\widetilde{m}}_{\text{BW}}(E)}{I(E)} \langle n_i \rangle^{\widetilde{m}}, \\ &\xrightarrow{\text{EGOE}(1+2):\lambda > \lambda_{F_k}} \sum_{\widetilde{m}} \frac{I^{\widetilde{m}}_{\mathcal{G}}(E)}{I(E)} \langle n_i \rangle^{\widetilde{m}}. \end{aligned} \tag{16}$$

Note that: (i) in a recent work, the first form of (16) is applied with the smoothed $I^m(E)$ constructed using a semi-classical theory for interacting spin systems [46]; (ii) the second form of (16) is applicable in the BW domain of EGOE(1 + 2) and this is used by Flambaum for atoms [20]; (iii) the third form in (16), i.e., the Gaussian domain result was used (see [42] and the first two references in [37]) in the past for nuclei by assuming (with tests via nuclear shell model) that it is valid for nuclear Hamiltonians; (iv) it is easy to add (γ_1, γ_2) corrections (Appendix A) to the Gaussian partial densities in (16); (v) Eq. (16) is used for J_z^2 expectation values by many groups [26,47] although in this case the first equality itself is not exact. To some extent justification for such a procedure for $\langle J_z^2 \rangle^E$ comes from Eq. (22) and its extension discussed ahead.

A second situation where (15) applies is for K's built from the generators of a group G_1. To be more precise, consider $G_1 \supset G_2$ group–subgroup chain within the $U(N)$ group defining the (N, m) space (see [2,13,18,42] for details regarding the $U(N)$ group). Say $\mathcal{O}$ is a generator of G_1 (then it will be a one-body operator) with tensorial character $\mathcal{O}^{\Gamma_1^0 \Gamma_2^0 \alpha_0}$ with respect to $G_1 \supset G_2$, i.e., $\mathcal{O}$ behaves as Γ_1^0 irrep of G_1, Γ_2^0 irrep of G_2 and α_0 gives the labels of the subgroups of G_2 [47]. Now consider the transition strength sum operator

$$\mathbf{K} = \mathcal{O}^{\Gamma_1^0 \Gamma_2^0} \cdot \mathcal{O}^{\Gamma_1^0 \Gamma_2^0} = \sum_{\alpha_0} \left(\mathcal{O}^{\Gamma_1^0 \Gamma_2^0 \alpha_0} \right)^\dagger \left(\mathcal{O}^{\Gamma_1^0 \Gamma_2^0 \alpha_0} \right).$$

Then, the Γ in (15) goes to $(\Gamma_1 \Gamma_2)$ with $\Gamma_1 \in G_1$, $\Gamma_2 \in G_2$, and $\mathcal{O}^{\Gamma_1^0 \Gamma_2^0} \cdot \mathcal{O}^{\Gamma_1^0 \Gamma_2^0} |\Gamma_1 \Gamma_2 \alpha\rangle = f(\Gamma_1, \Gamma_2) |\Gamma_1 \Gamma_2 \alpha\rangle$ as $\mathcal{O}^{\Gamma_1^0 \Gamma_2^0} \cdot \mathcal{O}^{\Gamma_1^0 \Gamma_2^0}$ is a G_2 scalar in G_1. This gives,

$$\left\langle \mathcal{O}^{\Gamma_1^0 \Gamma_2^0} \cdot \mathcal{O}^{\Gamma_1^0 \Gamma_2^0} \right\rangle^E = \sum_{\Gamma_1, \Gamma_2} \frac{I^{(\Gamma_1 \Gamma_2)}(E)}{I(E)} \left\langle \mathcal{O}^{\Gamma_1^0 \Gamma_2^0} \cdot \mathcal{O}^{\Gamma_1^0 \Gamma_2^0} \right\rangle^{(\Gamma_1 \Gamma_2)}. \tag{17}$$

In order to apply (17), the shape of $I^{(\Gamma_1 \Gamma_2)}(E)$ need to be known. In principle this is a well defined group theoretical problem. However it is inconceivable that a general solution to this problem can be worked out. To proceed further we make use of the relationship between chaos and thermalization in finite interacting particle systems. Recently it is shown that [17], given the EGOE(1 + 2) Hamiltonian $\{H\} = h(1) + \lambda\{V(2)\}$, different definitions of entropy (for example, thermodynamic entropy defined by the state density, information entropy in wavefunctions expanded in the mean-field basis, single particle entropy defined by occupation numbers) will coincide for $\lambda \sim \lambda_t$, where λ_t is defined by the condition that the spreading variance generated by $h(1)$ equals the variance generated by $\lambda\{V(2)\}$. The λ_t value defines (note that $\lambda_t > \lambda_{F_k}$) the thermodynamic region for a chaotic ($\lambda > \lambda_{F_k}$) many particle system. For $\lambda \sim \lambda_t$ the mean-field ($h(1)$) basis states are completely mixed by $V(2)$ and the dual $V(2)$ basis [48] is also completely mixed by $h(1)$. Thus in the thermodynamic region there is no preferential basis. Then it is plausible to argue, following Percival conjecture [49] for wavefunctions in the chaotic domain, that the shape of the strength functions defined by any group–subgroup chain (i.e., $I^{(\Gamma_1 \Gamma_2)}(E)$) will be Gaussian just as the decomposition with respect to the mean-field basis (note that $F_k(E)$ are, as discussed in Section 2, Gaussian for $\lambda > \lambda_{F_k}$ where $|k\rangle$ are mean-field basis states). Then in the thermodynamic region,

$$\left\langle \mathcal{O}^{\Gamma_1^0\Gamma_2^0}\cdot\mathcal{O}^{\Gamma_1^0\Gamma_2^0}\right\rangle^{E} \stackrel{\text{EGOE}(1+2):\lambda_c\sim\lambda_t}{\longrightarrow} \sum_{\Gamma_1,\Gamma_2}\frac{I_{\mathcal{G}}^{(\Gamma_1\Gamma_2)}(E)}{I(E)}\left\langle \mathcal{O}^{\Gamma_1^0\Gamma_2^0}\cdot\mathcal{O}^{\Gamma_1^0\Gamma_2^0}\right\rangle^{(\Gamma_1\Gamma_2)}. \tag{18}$$

For (2s1d) shell nuclei it is well verified that, for realistic nuclear Hamiltonians, $\lambda\sim\lambda_t$ [14,17]. For GT strength sums in nuclei, with GT operator being a $SU(4)$ generator, Eq. (18) is used [50] with Wigner's spin–isospin $SU(4)\supset SU_S(2)\otimes SU_T(2)$ irreps $\{f\}ST$; i.e., $\Gamma_1=\{f\}_{SU(4)}$ and $\Gamma_2=(ST)$. Similarly one can use (18) for electric quadrupole strength sums in nuclei using say Elliott's $SU(3)\supset O(3)$ [44,51] and for electric dipole strength sums in atoms using $O(4)\supset O(3)$ [52].

A third situation where a simplification occurs is when we consider one-body transition operators with defining matrix elements representable by independent random variables. Then, for a one-body transition operator $\mathcal{O}=\sum_{i\neq j}\epsilon_{ij}a_i^\dagger a_j$, we have $\overline{\epsilon_{ij}\epsilon_{kl}}=\overline{\epsilon_{ij}^2}\delta_{(ij),(kl)}$. The average here is with respect to the ensemble defining ϵ_{ij}'s. Then the ensemble averaged transition strength sum generated by $\mathcal{O}^\dagger\mathcal{O}$ can be written in terms of number operators,

$$\overline{\mathcal{O}^\dagger\mathcal{O}}\simeq\sum_{i\neq j}\overline{\epsilon_{ij}^2}n_j(1-n_i). \tag{19}$$

Using (19) and applying the considerations that led to (16) give,

$$\begin{aligned}\overline{\langle\mathcal{O}^\dagger\mathcal{O}\rangle^E} &\stackrel{\text{EGOE}(1+2):\lambda_c\leqslant\lambda\leqslant\lambda_{F_k}}{\longrightarrow}\sum_{i\neq j}\overline{\epsilon_{ij}^2}\sum_{\widetilde{m}}\frac{I_{\text{BW}}^{\widetilde{m}}(E)}{I(E)}\langle n_j(1-n_i)\rangle^{\widetilde{m}}.\\ &\stackrel{\text{EGOE}(1+2):\lambda>\lambda_{F_k}}{\longrightarrow}\sum_{i\neq j}\overline{\epsilon_{ij}^2}\sum_{\widetilde{m}}\frac{I_{\mathcal{G}}^{\widetilde{m}}(E)}{I(E)}\langle n_j(1-n_i)\rangle^{\widetilde{m}}.\end{aligned} \tag{20}$$

In (20) the average is with respect to the ensemble defining ϵ_{ij}. Thus in deriving (20), EGOE representation is used both for H and $\mathcal{O}$. A close examination reveals that these two ensembles are assumed to be independent. In fact Eq. (12) is also based on such a consideration [34]. Eq. (20) in BW domain [53] for atoms and Gaussian domain for nuclei [54] are derived from quite different considerations in the past and its possible extensions in the Gaussian domain are discussed in [55,56]; see also Eq. (23) ahead.

Before turning to other situations that give a theory for expectation values, let us mention that if the symmetries defining Γ subspaces are exact symmetries of H, then $\langle\mathbf{K}\rangle^E$ will be $\langle\mathbf{K}\rangle^\Gamma$, i.e., they are fixed symmetry averages and their evaluation is a well defined group theoretical problem [2,7,40,44].

5.2. Convolution factoring

Given a $H=h(1)+V(2)$ represented by EGOE(1 + 2), it is now well understood (see [8,18,57] and references therein) that in the Gaussian domain, the widths of the strength functions $I^{\widetilde{m}}(E)$ in (16) are nearly constant (independent of $\widetilde{m}$), $\sigma^2(\widetilde{m})\sim\langle(\widetilde{V(2)})^2\rangle^m$, where $\widetilde{V(2)}$ is obtained from $V(2)$ after subtracting the part that contributes to the centroids of $I^{\widetilde{m}}(E)$. Tacitly one assumes that this part is already added to the mean-field generating $h(1)$; see the Appendix in [18]. Therefore the centroids of $\widetilde{I^m}(E)$ are generated by $h(1)$ and the constant variance by $V(2)$. Then it is easy to see that the sum in (16) gets simplified to give a convolution structure for occupation numbers,

$$\langle n_i \rangle^E \stackrel{\text{EGOE}(1+2)}{\rightarrow} [I^H(E)]^{-1} I^h_{n_i} \otimes \rho^V_{\mathcal{G}}[E]. \tag{21}$$

In (21) $\otimes$ denotes convolution. In fact, for EGOE(1 + 2), $I^H(E) = \sum_{\widetilde{m}} \widetilde{I^m}(E) \sim I^h \otimes \rho^V_{\mathcal{G}}[E]$. This together with the parametric derivative method of Eq. (7) gives (21) with $H \rightarrow H + \zeta n_i = [h(1) + \zeta n_i] + V(2)$. Similarly applying double parametric differentiation (see the discussion after Eq. (7)) with $H \rightarrow H + \zeta J_Z = [h(1) + \zeta J_Z] + V(2)$ gives,

$$\langle J_Z^2 \rangle^E \stackrel{\text{EGOE}(1+2)}{\rightarrow} [I^H(E)]^{-1} I^h_{J_Z^2} \otimes \rho^V_{\mathcal{G}}[E]. \tag{22}$$

In the approximations leading to Eqs. (21) and (22) the $\mathbf{K} - V$ correlations are neglected. A simple prescription for putting back some of these correlations is to replace $\rho^V_{\mathcal{G}}$ in Eqs. (21) and (22) with $\rho^V_{\mathbf{K};\mathcal{G}}$. Together with this, Eq. (22) was used in some earlier studies of nuclear level densities [58]. Let us add that the construction of the $I^h_{\mathbf{K}}$ densities in Eqs. (21) and (22) is simple. It should be pointed out that some of the results discussed in [11] for occupancies follow from Eq. (21). Similarly Eq. (21) is also employed more recently in [59] in the study of the classical statistical mechanics of a few-body interacting spin model.

In the case of $\langle \mathcal{O}^\dagger \mathcal{O} \rangle^E$, a form similar to ((21) and (22)) can be derived by starting with the bivariate convolution form (with the two convoluting factors generated by h and V respectively) [34] for the bivariate strength densities $I^H_{\mathcal{O}}(E_i, E_f)$ generated by the transition operator $\mathcal{O}$. It should be stressed that this form is derived by assuming that H is represented by EGOE(1 + 2) and the transition operator $\mathcal{O}$, a t-body operator with $t \leqslant 2$, by EGOE(t). Integrating over the final energies E_f in $I^H_{\mathcal{O}}(E_i, E_f)$ gives,

$$\langle \mathcal{O}^\dagger \mathcal{O} \rangle^E \stackrel{\text{EGOE}(1+2)}{\rightarrow} [I^H(E)]^{-1} I^h_{\mathcal{O}^\dagger \mathcal{O}} \otimes \rho^V_{\mathcal{O}^\dagger \mathcal{O}:\mathcal{G}}[E], \tag{23}$$

where the correction to $\rho^V_{\mathcal{O}^\dagger \mathcal{O}:\mathcal{G}}$ follows via the moment method and $I^h_{\mathcal{O}^\dagger \mathcal{O}}$ construction is simple. Note the close relationship between the forms given by (23), (21) and (22) although they are derived from quite different approaches.

Convolution forms given by Eqs. (21)–(23) are based on the strong condition that the variances of the strength functions are constant. As this result is verified for nuclear Hamiltonians [8,44] and, as pointed out in Section 5.1, $\lambda \sim \lambda_t$ for nuclear Hamiltonians, makes it plausible that Eqs. (21)–(23) may be valid only in the thermodynamic ($\lambda \sim \lambda_t$) regime of the spectrum. A formal proof of this result is not yet available.

6. Conclusions and future outlook

Moment expansions, for expectation values of a given operator $\mathbf{K}$, that are developed in the last 30 years are brought to one place in this paper. More importantly, it is clearly established that the action of EGOE(1 + 2) and quantum chaos, defined by the markers λ_c, λ_{F_k} and λ_t, dictate the convergences of these expansions depending on the nature of **K**. Eqs. (10), (16), and (21) and (12), (18), (20), (23), respectively give the EGOE(1 + 2) theory for occupancies $\langle n_i \rangle^E$ and transition strength sums given by $\langle \mathcal{O}^\dagger \mathcal{O} \rangle^E$ or $\langle \mathcal{O} \cdot \mathcal{O} \rangle^E$ with $\mathcal{O}$ being a non-singular one-body transition operator (in some situations it can be two-body; see Appendix B and [34]). However it is seen that the spin-cutoff theory given by Eqs. (21) and (22) is not complete and this requires detailed study of EGOE(1 + 2)-J. Let us point out that there are many applications of the results in Sections 4 and 5 in nuclear structure and nuclear astrophysics [26,37,44,55,58]. The results for strength sums may be useful in the study of line distributions in atoms (see for example [27]). Similarly, the results for occupancies may find applications in interacting spin systems used to model quantum computers core [16,23].

All the convergences discussed in this paper are to within GOE fluctuations. However in general, near the spectrum ends, for EGOE(1 + 2), the fluctuations may not take GOE form [1] and this is re-emphasized recently by Flores et al. [60] who argued that they may be close to semi-Poisson type (then the nearest neighbor spacing distribution $P(S)\,dS = 4S e^{-2S}\,dS$ as opposed to the Wigner form given in Eq. (3)). Therefore chaos will not set in uniformly throughout the spectrum of a finite interacting many particle system. Thus there is a need to study the energy dependence of the chaos markers λ_c, λ_{F_k}, and λ_t. Such a study for the λ_c parameter was carried out by Jacquod and Shepelyansky [10]. Further examination of their result and its extensions are clearly needed for a more complete theory for smoothed expectation values.

Finally let us mention that considerations based on group symmetries in the (N, m) space are used in deriving Eq. (18) and the need to study EGOE-J is seen in the context of spin-cutoff factors. Another situation where group symmetries enter explicitly is in the case of expectation values of operators with well defined tensorial character [47] with respect to a group symmetry of the hamiltonian. Let us consider the m particle basis states $|m\beta\Gamma\alpha\rangle$, where Γ are irreps of a group G which is a symmetry of H, α are the irrep labels of the subgroups of G and β are the additional labels required for completely specifying the eigenstates. With G a symmetry of H, all eigenvalues will be degenerate with respect to α and the total degeneracy (number of α labels) is denoted by $[\Gamma]$. For example with H having J (angular momentum) symmetry, $G = O(3)$ and $[\Gamma] = (2J + 1)$. Now, for a transition operator $\mathcal{O} \to \mathcal{O}^{\Gamma_0\alpha_0}$ with respect to the symmetry group G where Γ_0 is a irrep of G and α_0 are its subgroup labels, one has to deal with reduced expectation values $\langle || \mathcal{O}^{\Gamma_0} || \rangle^{E,\Gamma}$ as the eigenstates are of the type $|E\beta'\Gamma\alpha\rangle$. Then Eq. (4) is modified as follows,

$$\begin{aligned} \langle || \mathcal{O}^{\Gamma_0} || \rangle^{m,E,\Gamma} &= \left[I_\ell^{m,\Gamma}(E)\right]^{-1} \langle\langle || \mathcal{O}^{\Gamma_0}\delta(H-E) || \rangle\rangle^{m,\Gamma}, \\ \langle\langle || \mathcal{O}^{\Gamma_0}\delta(H-E) || \rangle\rangle^{m,\Gamma} &= \sum_{\beta' \in E} \langle m, E\beta'\Gamma || \mathcal{O}^{\Gamma_0} || m, E\beta'\Gamma \rangle, \\ I_\ell^{m,\Gamma}(E) &= [\Gamma]^{-1} \langle\langle \delta(H-E) \rangle\rangle^{m,\Gamma} \end{aligned} \tag{24}$$

Note that the generalized Wigner coefficient $\langle \Gamma\alpha\Gamma_0\alpha_0 \mid \Gamma\alpha \rangle$ is factored out, from the matrix elements of $\mathcal{O}^{\Gamma_0\alpha_0}$, in writing Eq. (24) using the Wigner–Eckart theorem [47]. Good examples for applications of (24) are static moments, electric quadrupole and magnetic dipole in nuclei and electric dipole in atoms where $G = O(3)$ and $\Gamma = J$.

In the past the methods of Sections 3 and 5 are applied to the reduced expectations values defined in Eq. (24) [5,61]. From Sections 4 and 5 it is clear that the conditions for the convergence of moment expansions for reduced expectation values can be derived only by studying EGOE with group symmetries and the associated chaos markers (which will depend on Γ). It should be pointed out that the significance of EGOE with group symmetries is recognized only recently [8,13,62]. Clearly much of the future lies in their study.

Acknowledgments

This paper is dedicated to the memory of J.B. French and many results in this paper followed from the discussions the author has had with him.

Appendix A

Let us say that the centroid and width of a normalized density $\rho(x)$ are (ϵ, σ). Then in terms of the standardized variable $\hat{x} = (x - \epsilon)/\sigma$ one has

$$\rho(x)\,\mathrm{d}x = \eta(\hat{x})\,\mathrm{d}\hat{x}.$$

For a Gaussian, $\eta_{\mathcal{G}}(\hat{x}) = (2\pi)^{-1/2} \exp -(\hat{x})^2/2$. Then the Gram–Charlier (GC), Edgeworth (Ed), and Cornish–Fisher (CF) expansions for $\eta(\hat{x})$ in terms of $\eta_{\mathcal{G}}(\hat{x})$ including only γ_1 and γ_2 corrections are [28,63],

$$\eta_{\mathrm{GC}}(\hat{x}) = \eta_{\mathcal{G}}(\hat{x}) \left\{ 1 + \frac{\gamma_1}{6}(\hat{x}^3 - 3\hat{x}) + \frac{\gamma_2}{24}(\hat{x}^4 - 6\hat{x}^2 + 3) \right\}, \tag{A.1}$$

$$\eta_{\mathrm{Ed}}(\hat{x}) = \eta_{\mathcal{G}}(\hat{x}) \left\{ 1 + \frac{\gamma_1}{6}(\hat{x}^3 - 3\hat{x}) + \frac{\gamma_2}{24}(\hat{x}^4 - 6\hat{x}^2 + 3) + \frac{\gamma_1^2}{72}(\hat{x}^6 - 15\hat{x}^4 + 45\hat{x}^2 - 15) \right\}, \tag{A.2}$$

$$\eta_{\mathrm{CF}}(\hat{x}) = \frac{1}{(2\pi)^{1/2}} \left| 1 - \frac{\gamma_1}{3}\hat{x} - \frac{\gamma_2}{8}(\hat{x}^2 - 1) + \frac{\gamma_1^2}{36}(12\hat{x}^2 - 7) \right| \times \exp\left\{ -\frac{1}{2}\left(\hat{x} - \frac{\gamma_1}{6}(\hat{x}^2 - 1) - \frac{\gamma_2}{24}(\hat{x}^3 - 3\hat{x}) + \frac{\gamma_1^2}{36}(4\hat{x}^3 - 7\hat{x}) \right)^2 \right\}. \tag{A.3}$$

In the maximum entropy (maxent) method, as advocated by Grimes and Massey [64], the form for $\rho(x)$ is,

$$\rho_{\mathrm{maxent}}(x) = \exp(a + bx + cx^2 + dx^3 + ex^4) \tag{A.4}$$

and the parameters (a, b, c, d, e) in (A.4) are fixed (with the help of an appropriate least square fit procedure) by the normalization condition and the first four moments.

With the binomial form introduced by Zuker [24,26], the $I(x)$ is written as a continuous binomial (with $Z-b$ denoting Zuker's binomial),

$$I_{Z-b}(x) = \sqrt{\frac{8}{\pi n}} \exp\left\{ -(n-1)\left[\frac{x}{x_{\max}}\ln\frac{x}{x_{\max}} + \frac{x_{\max}-x}{x_{\max}}\ln\frac{x_{\max}-x}{x_{\max}}\right] + n\frac{x}{x_{\max}}\ln\lambda\right\}. \tag{A.5}$$

In (A.5) x is the excitation energy (i.e., the energy measured with respect to the ground state) and $x_{\max}$ is the maximum excitation energy. Solving $d=(1+\lambda)^n$ with d is the dimension and $\gamma_1 = (1-\lambda)/\sqrt{n\lambda}$ one can obtain (n,λ) in (A.5). Alternatively solving $\gamma_1=(1-\lambda)/\sqrt{n\lambda}$ and $\gamma_2=(1+\lambda^2-4\lambda)/n\lambda$ one can obtain (n,λ) and then multiplying $I(x)$ with a constant gives the correct dimension. Finally the exact centroid and width are taken into account by moving the center and scaling. In the limit $n\to\infty$ and $\lambda\to 1$, $I_{Z-b}(x)$ goes to a Gaussian.

Appendix B

Given a Hamiltonian H and its m-particle eigenstates $|E\rangle$, the transition strengths or matrix elements generated by a transition operator $\mathcal{O}$ are $|\langle E_{\rm f} \mid \mathcal{O} \mid E_i\rangle|^2$. The bivariate strength densities (matrix elements of $\mathcal{O}$ weighted by the state densities at the initial and final energies)

$$\begin{aligned} I^{H;m_i\,m_{\rm f}}_{{\rm biv};\mathcal{O}}(E_i,E_{\rm f}) &= \langle\langle \mathcal{O}^\dagger\delta(H-E_{\rm f})\mathcal{O}\delta(H-E_i)\rangle\rangle^{m_i} \\ &= \langle\langle \mathcal{O}^\dagger\mathcal{O}\rangle\rangle^{m_i}\rho^{H;m_i,m_{\rm f}}_{{\rm biv};\mathcal{O}}(E_i,E_{\rm f}) \end{aligned} \tag{B.1}$$

take [34] bivariate Gaussian form for EGOE(1+2) for $\mathcal{O}$ at most two-body, i.e.,

$$I^{H;m_i\,m_{\rm f}}_{{\rm biv};\mathcal{O}}(E_i,E_{\rm f}) \to I^{H;m_i\,m_{\rm f}}_{{\rm biv}\text{-}\mathcal{G};\mathcal{O}}(E_i,E_{\rm f})$$

and similarly $\rho^{H;m_i,m_{\rm f}}_{{\rm biv};\mathcal{O}}(E_i,E_{\rm f})$ which is normalized to unity. This result assumes that $\mathcal{O}$ is also representable by EGOE and it is independent of H but for the non-commutability of $\mathcal{O}$ and H. With H a k-body and $\mathcal{O}$ a t-body operator, the bivariate Gaussian form is good for $|k(2t-3)/m| \lesssim 0.3$ [34]. Therefore for $k=1$ or 2, t has to be maximum 2 (for $m\geqslant 6$). Thus, with a EGOE(1+2), t can be at most 2-body. The bivariate Gaussian strength density $I^{H;m_i m_{\rm f}}_{{\rm biv}\text{-}\mathcal{G};\mathcal{O}}$ is defined by the centroids $(\epsilon_i,\ \epsilon_{\rm f})$ and widths $(\sigma_i,\ \sigma_{\rm f})$ of its two marginals and the bivariate correlation coefficient ζ,

$$\zeta = \left\langle \mathcal{O}^\dagger\left(\frac{H-\epsilon_{\rm f}}{\sigma_{\rm f}}\right)\mathcal{O}\left(\frac{H-\epsilon_i}{\sigma_i}\right)\right\rangle^{m_i} / \langle\langle \mathcal{O}^\dagger\mathcal{O}\rangle\rangle^{m_i}. \tag{B.2}$$

One and two-body transition operators $\mathcal{O}$ for which the bivariate Gaussian form is not valid, can be called singular transition operators.

Appendix C

Given the bivariate Gaussian (with correlation coefficient ζ),

$$\eta_{\mathcal{G}}(\widehat{x}_1,\widehat{x}_2)=\frac{1}{2\pi\sqrt{(1-\zeta^2)}}\exp\left\{-\frac{\widehat{x}_1^2-2\zeta\widehat{x}_1\widehat{x}_2+\widehat{x}_2^2}{2(1-\zeta^2)}\right\} \tag{C.1}$$

the bivariate Edgeworth expansion including bivariate cumulants k_{rs} corrections with $r+s \leqslant 4$ is [28,38]

$$\begin{aligned}\eta_{\text{biv-Ed}}(\widehat{x}_1,\widehat{x}_2)=&\left\{1+\sum_{p+q=3,4}\frac{k_{pq}}{p!q!}He_{p,q}(\widehat{x}_1,\widehat{x}_2)\right.\\&+\sum_{p_1+q_1=3,p_2+q_2=3,p_1\geqslant p_2}\frac{k_{p_1q_1}k_{p_2q_2}}{p_1!p_2!q_1!q_2!(1+\delta_{p_1p_2}\delta_{q_1q_2})}\\&\left.\times He_{p_1+p_2,q_1+q_2}(\widehat{x}_1,\widehat{x}_2)\right\}\eta_{\mathcal{G}}(\widehat{x}_1,\widehat{x}_2).\end{aligned} \tag{C.2}$$

In (C.2) the p's and q's are all positive integers. The bivariate Hermite polynomials $He_{m_1,m_2}(\widehat{x}_1,\widehat{x}_2)$ in (C.2) satisfy the recursion relation,

$$\begin{aligned}(1-\zeta^2)He_{m_1+1,m_2}(\widehat{x}_1,\widehat{x}_2)&=(\widehat{x}_1-\zeta\widehat{x}_2)He_{m_1,m_2}(\widehat{x}_1,\widehat{x}_2)-m_1He_{m_1-1,m_2}(\widehat{x}_1,\widehat{x}_2)\\&\quad+m_2\zeta He_{m_1,m_2-1}(\widehat{x}_1,\widehat{x}_2),\\He_{0,0}(\widehat{x}_1,\widehat{x}_2)&=1,\quad He_{1,0}(\widehat{x}_1,\widehat{x}_2)=(\widehat{x}_1-\zeta\widehat{x}_2)/(1-\zeta^2),\\He_{0,1}(\widehat{x}_1,\widehat{x}_2)&=(\widehat{x}_2-\zeta\widehat{x}_1)/(1-\zeta^2),\\He_{1,1}(\widehat{x}_1,\widehat{x}_2)&=\frac{(\widehat{x}_1-\zeta\widehat{x}_2)(\widehat{x}_2-\zeta\widehat{x}_1)}{(1-\zeta^2)^2}+\frac{\zeta}{1-\zeta^2}.\end{aligned} \tag{C.3}$$

References

[1] T.A. Brody, J. Flores, J.B. French, P.A. Mello, A. Pandey, S.S.M. Wong, Rev. Mod. Phys. 53 (1981) 385;
K. Patel, M.S. Desai, V. Potbhare, V.K.B. Kota, Phys. Lett. A 275 (2000) 329.

[2] J.B. French, V.K.B. Kota, Ann. Rev. Nucl. Part. Sci. 32 (1982) 35.

[3] F.S. Chang, J.B. French, Phys. Lett. B 44 (1973) 131;
Phys. Lett. B 44 (1973) 135.

[4] J.P. Draayer, J.B. French, V. Potbhare, S.S.M. Wong, Phys. Lett. B 55 (1975) 263.

[5] J.P. Draayer, J.B. French, S.S.M. Wong, Ann. Phys. (N. Y.) 106 (1977) 472.

[6] J.B. French, V.K.B. Kota, J.F. Smith, University of Rochester Report ER-40425-245/UR1122 (1989).

[7] V.K.B. Kota, K. Kar, Pramana-J. Phys. 32 (1989) 35.

[8] V.K.B. Kota, Phys. Rep. 347 (2001) 223.

[9] L. Benet, H.A. Weidenmüller, arXiv: cond-mat/0207656 (2002), to be published.

[10] Ph. Jacquod, D.L. Shepelyansky, Phys. Rev. Lett. 79 (1997) 1837.

[11] V.V. Flambaum, F.M. Izrailev, Phys. Rev. E 56 (1997) 5144.

[12] L. Benet, T. Rupp, H.A. Weidenmüller, Phys. Rev. Lett. 87 (2001) 010601.

[13] V.K.B. Kota, K. Kar, Phys. Rev. E 65 (2002) 026130.
[14] M. Horoi, V. Zelevinsky, B.A. Brown, Phys. Rev. Lett. 74 (1995) 5194.
[15] F. Borgonovi, I. Guarneri, F.M. Izrailev, G. Casati, Phys. Lett. A 247 (1998) 140.
[16] G. Benenti, G. Casati, D.L. Shepelyansky, Eur. Phys. J. D 17 (2001) 265.
[17] V.K.B. Kota, R. Sahu, Phys. Rev. E 66 (2002) 037103.
[18] V.K.B. Kota, R. Sahu, Phys. Rev. E 64 (2001) 016219.
[19] V. Velazquez, A.P. Zuker, Phys. Rev. Lett. 88 (2002) 072502.
[20] V.V. Flambaum, A.A. Gribakina, G.F. Gribakin, I.V. Ponomarev, Physica D 131 (1999) 205.
[21] D. Angom, V.K.B. Kota, Phys. Rev. A, in press.
[22] Y. Alhassid, Ph. Jacquod, A. Wobst, Phys. Rev. B 61 (2000) R13357;
Ph. Jacquod, A.D. Stone, Phys. Rev. B 64 (2001) 214416;
Y. Alhassid, A. Wobst, Phys. Rev. B 65 (2002) 041304;
T. Papenbrock, L. Kaplan, G.F. Bertsch, Phys. Rev. B 65 (2002) 235120.
[23] B. Georgeot, D.L. Shepelyansky, Phys. Rev. E 62 (2000) 3504;
B. Georgeot, D.L. Shepelyansky, Phys. Rev. E 62 (2000) 6366;
G.P. Berman, F. Borgonovi, F.M. Izrailev, V.I. Tsifrinovich, Phys. Rev. E 65 (2001) 015204;
V.V. Flambaum, F.M. Izrailev, Phys. Rev. E 64 (2001) 026124.
[24] A. Zuker, Phys. Rev. C 64 (2001) 021303.
[25] J.U. Nabi, C.W. Johnson, W.E. Ormand, arXive:nucl-th/0111067 (2001).
[26] C.W. Johnson, J.U. Nabi, W.E. Ormand, arXive:nucl-th/0111068 (2001).
[27] G. O'Sullivan, P.K. Carroll, P. Dunne, R. Faulkner, C. McGuinness, N. Murphy, J. Phys. B 32 (1999) 1893.
[28] M.G. Kendall, A. Stuart, Advanced Theory of Statistics, vol. 1, third ed., Hafner Publishing Company, New York, 1969.
[29] T. Guhr, A. Müller-Groeling, H.A. Weidenmüller, Phys. Rep. 299 (1998) 189;
F. Haake, Quantum Signatures of Chaos, second ed., Springer, New York, 2000.
[30] O. Bohigas, M.J. Giannoni, C. Schmit, Phys. Rev. Lett. 52 (1984) 1.
[31] J.B. French, S.S.M. Wong, Phys. Lett. B 33 (1970) 447;
Phys. Lett. B 35 (1971) 5;
O. Bohigas, J. Flores, Phys. Lett. B 34 (1971) 261;
Phys. Lett. B 35 (1971) 383.
[32] K.K. Mon, J.B. French, Ann. Phys. (N. Y.) 95 (1975) 90;
L. Benet, T. Rupp, H.A. Weidenmüller, Ann. Phys. (N. Y.) 292 (2001) 67.
[33] V.V. Flambaum, F.M. Izrailev, Phys. Rev. E 61 (2000) 2539.
[34] J.B. French, V.K.B. Kota, A. Pandey, S. Tomsovic, Ann. Phys. (N. Y.) 181 (1988) 235.
[35] V.K.B. Kota, D. Majumdar, Z. Phys. A 351 (1995) 365;
D. Majumdar, Ph.D. Thesis, M.S. University (Gujarat, India, 1995), unpublished.
[36] T.R. Halemane, J. Math. Phys. 22 (1981) 2961;
V.K.B. Kota, V. Potbhare, P. Shenoy, N. Tressler, Pramana-J. Phys. 32 (1989) 507.
[37] V. Potbhare, S.P. Pandya, Nucl. Phys. A 256 (1976) 253;
V.K.B. Kota, V. Potbhare, Nucl. Phys. A 331 (1979) 93;
T.R. Halemane, J.B. French, Phys. Rev. C 25 (1982) 2029;
S. Sarkar, K. Kar, Phys. Lett. B 387 (1996) 227.
[38] V.K.B. Kota, Z. Phys. A 315 (1984) 91.
[39] T.P. Hutchinson, C.D. Lai, Continuous Bivariate Distributions, Emphasizing Applications, Rumsby Scientific Publishing, Adelaide, South Australia, 1990;
D.D. Mari, S. Kotz, Correlations and Dependence, Imperial College Press, London, 2001.
[40] J.B. French, J.P. Draayer, in: W. Beiglbock, et al. (Eds.), Group Theoretical Methods in Physics, Springer, Berlin, 1979, p. 394.
[41] J.M.G. Gómez, K. Kar, V.K.B. Kota, J. Retamosa, R. Sahu, Phys. Rev. C 64 (2001) 034305.
[42] F.S. Chang, J.B. French, T.H. Thio, Ann. Phys. (N. Y.) 66 (1971) 137.
[43] J.B. French, V.K.B. Kota, Phys. Rev. Lett. 51 (1983) 2183.

[44] J.C. Parikh, Group Symmetries in Nuclear Structure, Plenum, New York, 1978;
S.S.M. Wong, Nuclear Statistical Spectroscopy, Oxford University Press, New York, 1986.
[45] K. Kar, Nucl. Phys. A 368 (1981) 285.
[46] F. Borgonovi, G. Celardo, F.M. Izrailev, G. Casati, Phys. Rev. Lett. 88 (2002) 054101.
[47] P.H. Butler, Phil. Trans. R. Soc. Lond. A 277 (1975) 546.
[48] Ph. Jacquod, I. Varga, Phys. Rev. Lett. 89 (2002) 134101.
[49] I.C. Percival, J. Phys. B 6 (1973) L229.
[50] K. Kar, Phys. Rev. C 28 (1983) 2446;
J. Phys. G 9 (1983) 735.
[51] J.P. Elliott, Proc. R. Soc. Lond. A 245 (1958) 128.
[52] B.G. Wybourne, Classical Groups for Physicists, Wiley, New York, 1974.
[53] V.V. Flambaum, A.A. Gribakina, G.F. Gribakin, M.G. Kozlov, Phys. Rev. A 50 (1994) 267.
[54] K. Kar, S. Sarkar, A. Ray, Phys. Lett. B 261 (1991) 217;
M.H. Macfarlane, Phys. Lett. B 182 (1986) 265.
[55] V.K.B. Kota, D. Majumdar, Z. Phys. A 351 (1995) 377.
[56] V.K.B. Kota, R. Sahu, Phys. Rev. E 62 (2000) 3568.
[57] V.V. Flambaum, G.F. Gribakin, F.M. Izrailev, Phys. Rev. E 53 (1996) 5729.
[58] V.K.B. Kota, D. Majumdar, Nucl. Phys. A 604 (1996) 129.
[59] F. Borgonovi, F.M. Izrailev, Phys. Rev. E 62 (2000) 6475.
[60] J. Flores, M. Horoi, M. Mueller, T.H. Seligman, Phys. Rev. E 63 (2000) 026204.
[61] J.J.M. Verbaarschot, P.J. Brussaard, Nucl. Phys. A 423 (1984) 77.
[62] Z. Pluhar, H.A. Weidenmüller, Ann. Phys. (N. Y.) 297 (2002) 344.
[63] V.K.B. Kota, V. Potbhare, P. Shenoy, Phys. Rev. C 34 (1986) 2330.
[64] S.M. Grimes, T.N. Massey, Phys. Rev. C 51 (1995) 606.

PHYSICAL REVIEW C **67**, 054309 (2003)

Spin- and parity-dependent nuclear level densities and the exponential convergence method

Mihai Horoi,[1] Joshua Kaiser,[1,2] and Vladimir Zelevinsky[2,3]
[1]*Physics Department, Central Michigan University, Mount Pleasant, Michigan 48859*
[2]*Department of Physics and Astronomy, Michigan State University, East Lansing, Michigan 48824*
[3]*National Superconducting Cyclotron Laboratory, East Lansing, Michigan 48824*
(Received 26 December 2001; revised manuscript received 28 January 2003; published 21 May 2003)

The knowledge of the nuclear spin- and parity-dependent level densities $\rho(E,J,\pi)$ is important for understanding statistical properties of nuclei and reaction rates. We show that the earlier suggested method of exponential extrapolation, which allows one to find the yrast energy with high accuracy even for very large shell model Hamiltonian matrices, can be combined with the formalism of statistical spectroscopy in order to obtain the reliable level density.

DOI: 10.1103/PhysRevC.67.054309 PACS number(s): 21.10.Ma, 21.10.Dr, 21.10.Hw, 21.60.Cs

The knowledge of the level density is an important element of the description of highly excited many-body systems. In application to nuclei, as well as other finite systems such as atomic clusters, the many-body level density determines all statistical properties, both for the discrete spectrum and for reactions. This was recognized more than sixty-five years ago, see Refs. [1,2] and historic references therein, when the statistical approach based on thermodynamic analogies and the Fermi-gas model of the nucleus was developed [3]. The importance of the level densities is obvious, especially in connection to neutron resonances in heavy nuclei and related applications, nuclear reactions in general and advances far from stability in particular, and the astrophysical problem of nucleosynthesis.

The most common semiphenomenological approaches to the problem of the nuclear level density use basically the noninteracting Fermi-gas model of a nucleus with corrections that are needed to take into account the average effects of the shell structure, pairing correlations, and collective excitations associated mainly with the static or dynamical deformation of the mean field [1,4–8]. The residual interaction is fully taken into account in modern large-scale shell model calculations based on a direct diagonalization or, for example, with the use of the Monte Carlo techniques [9–11]; a comparative analysis of these methods was performed in Ref. [12]. One can hope that at not very high excitation energy the finiteness of model space is not crucial because of a relatively low statistical weight of intruder configurations. Furthermore, one can try to smoothly match the exact shell model results to the improved Fermi-gas predictions at higher energies.

As the dimensions of the shell model Hamiltonian matrices grow beyond the directly tractable limits, the approach using the methods of statistical spectroscopy [13–17] is still possible. Moreover, the statistical approach is the most appropriate for answering the questions formulated in statistical terms. Here one needs to build the Hamiltonian matrix and calculate the traces of its moments without an explicit diagonalization. According to the central limit theorem, the distributions of level densities, strength functions, and related quantities are converging to the Gaussian so that the lowest moments are expected to provide sufficient information [17]. The underlying physics can be understood in terms of many-body quantum chaos. Two-body interactions of the fermions in the region of high level density are effectively strong and act essentially at random, as follows from the general analysis [18,19] and from detailed calculations for complex atoms [20] and nuclei [21]. The standard partitioning of Hilbert space according to the particle configurations, which is a usual tool in the construction of the shell model basis, is very convenient here as well since it facilitates the calculation of the traces. The traces can be found for few-particle states and then extended by a combinatorial "propagation" to many-body states. Unfortunately, the propagation procedure for the traces including the projection operators for additional quantum numbers is much more complicated [22]. Meanwhile, reliable information on the level density $\rho(E,J^{\pi},T)$ as a function of excitation energy, total spin J, parity, and isospin is highly desirable.

Taking into account only the first two scalar moments of total density seems to be insufficient in cases where the inclusion of many major shells and other shell effects distort the shape of the level density. Recently, Zuker [23] showed that small asymmetries of the level density can be addressed by including the third or/and the fourth static moment(s). In cases where the fluctuations are more pronounced, it was recently shown [24] that after decomposing the shell model space in partitions according to the arrangement of nucleons in single-particle orbits, the total density of states can be found as a sum of the Gaussians G_p on the set of partitions $p=1,2,\ldots,P$ of A valence particles in s single-particle states (e.g., $[n_1(p),n_2(p),\ldots,n_s(p)]$, $A=\Sigma_{i=1}^{s}n_i(p)$),

$$\rho(E)=\sum_{p=1}^{P} D_p G_p(E+E_0-\overline{E}_p,\overline{\sigma}_p), \tag{1}$$

provided that the ground state energy E_0 can be accurately calculated, for example, with the aid of the exponential convergence method [25]. Previously, similar procedures were used to describe the shape of the density, see, e.g., the title of Sec. 3 of Ref. [26]. In Eq. (1), D_p is the total number of states in the partition p, E is the excitation energy, E_0 is the ground state energy, $\overline{E}_p$ and $\overline{\sigma}_p$ are the first (centroid) and second (width) moments, respectively, for each partition. The partition moments $\overline{E}_p$ and $\overline{\sigma}_p$ can be calculated using a

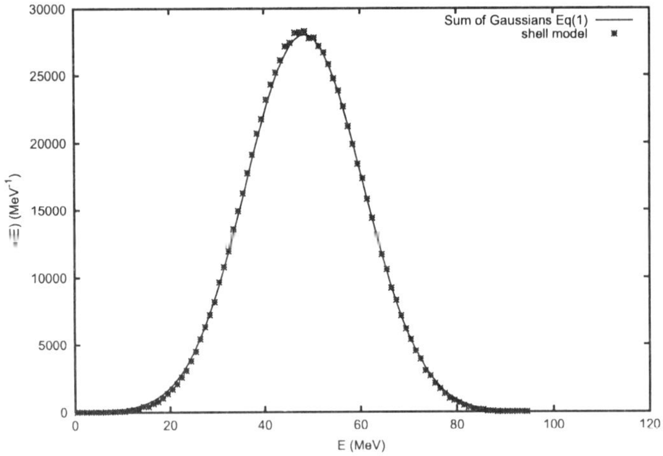

FIG. 1. Shell model density of states compared with the sum of Gaussians approach, Eq. (1), based on partition [illegible] of statistical spectroscopy.

straightforward method introduced in Ref. [14], see below. The precise knowledge of the ground state energy E_0 is crucial for the application of the partition summation. Figure 1 compares the results of Eq. (1) with full shell model calculations available for the case of $A=12$ valence particles in the sd shell. Partitions have an important advantage of forming an invariant subspace with respect to rotational symmetry and parity. All the results shown in Figs. 1–5 are obtained using the USD interaction of Ref. [27]. In all cases the number of protons equals the number of neutrons and states of all isospins are included.

Previous work on level density mostly stressed only the description of the shape of the density, leaving aside the role of the starting energy. Papers based on Monte Carlo techniques [9–11] use (without emphasizing) the ground state energy, however, they can only include very limited interactions to avoid the sign problem that induces high errors in the value of ground state energy [28]. Kota and Majumbar [29] recognize the importance of knowing the location of the ground state energy. In practice, their approach uses, besides theoretical results related to the shape of the density, set(s) of experimental data, when available.

The total density of states $\rho(E)$ is normally calculated as the first step in many other approaches, including the Fermi-gas model and Monte Carlo method. To obtain the density of levels $\rho(E,J)$ for a given spin J, one usually employs a rescaling procedure based on the statistical distribution of spins, which we describe below. However, we show that this procedure is not sufficiently accurate, especially for the lowest and highest spins.

The approach developed by Jacquemin *et al.* [22,26,30] allows one to calculate exactly shell model centroids $\overline{E}(J)$ and widths $\sigma(J)$ for the level densities at fixed spin J. In principle, the knowledge of the cumulative level number $\mathcal{N}(E,J)=\int_{-\infty}^{E}dE'\,\rho(E',J)$ makes it possible to reproduce the positions of the discrete levels [14]. However, in the lowest part of the spectra this procedure would not work because even the corrected Gaussian approximation does not describe the lowest part of the spectrum that is very far from the centroid. In this situation it is natural to enrich the methods of statistical spectroscopy with information based on the generic properties of complicated ("chaotic") many-body wave functions found in the analysis of the exact shell model diagonalization [21] for the cases of smaller dimensions.

The classical derivation of the level density for given spin [2,3] is based on the idea of random coupling of individual particle spins into the total spin of the many-body system. This "geometric chaoticity" originates from the presence of many possible coupling schemes described by the complicated $3nj$ symbols rather than from specific features of nuclear forces. The random-walk process of spin coupling results in the Gaussian level density as a function of the total spin projection $J_z=M$,

$$\rho_M(E)=\frac{\rho(E)}{\sqrt{2\pi\sigma^2}}e^{-M^2/2\sigma^2}, \tag{2}$$

where $\rho(E)$ is the total density of states at excitation energy E, and σ^2 is the average value $\langle M^2\rangle$ in the shell model space. Assuming Eq. (2), the level density $\rho(E,J)$ for spin J can be obtained with no further approximations in a conventional way,

$$\rho(E,J)=\rho_{M=J}(E)-\rho_{M=J+1}(E), \tag{3}$$

which leads to the standard expression

$$\rho(E,J)=\frac{\rho(E)}{\sqrt{8\pi\sigma^3}}(2J+1)e^{-(J+1/2)^2/2\sigma^2}. \tag{4}$$

The next steps usually taken are the Fermi-gas approximation for the total density of states $\rho(E)$, the backward shift of the excitation energy, $E\rightarrow U=E-\Delta$, where $\Delta(Z,N)$ reflects the pairing threshold for two-quasiparticle excitations (in even-even nuclei), and the expression of σ in terms of the statistical (rigid body) moment of inertia.

One can notice that the random walk approximation (2)–(4) works much better for the total number of states, dimension $d(J)$ in Hilbert space, the quantity of a pure geometrical nature, than for the energy-dependent density $\rho(E,J)$ that is

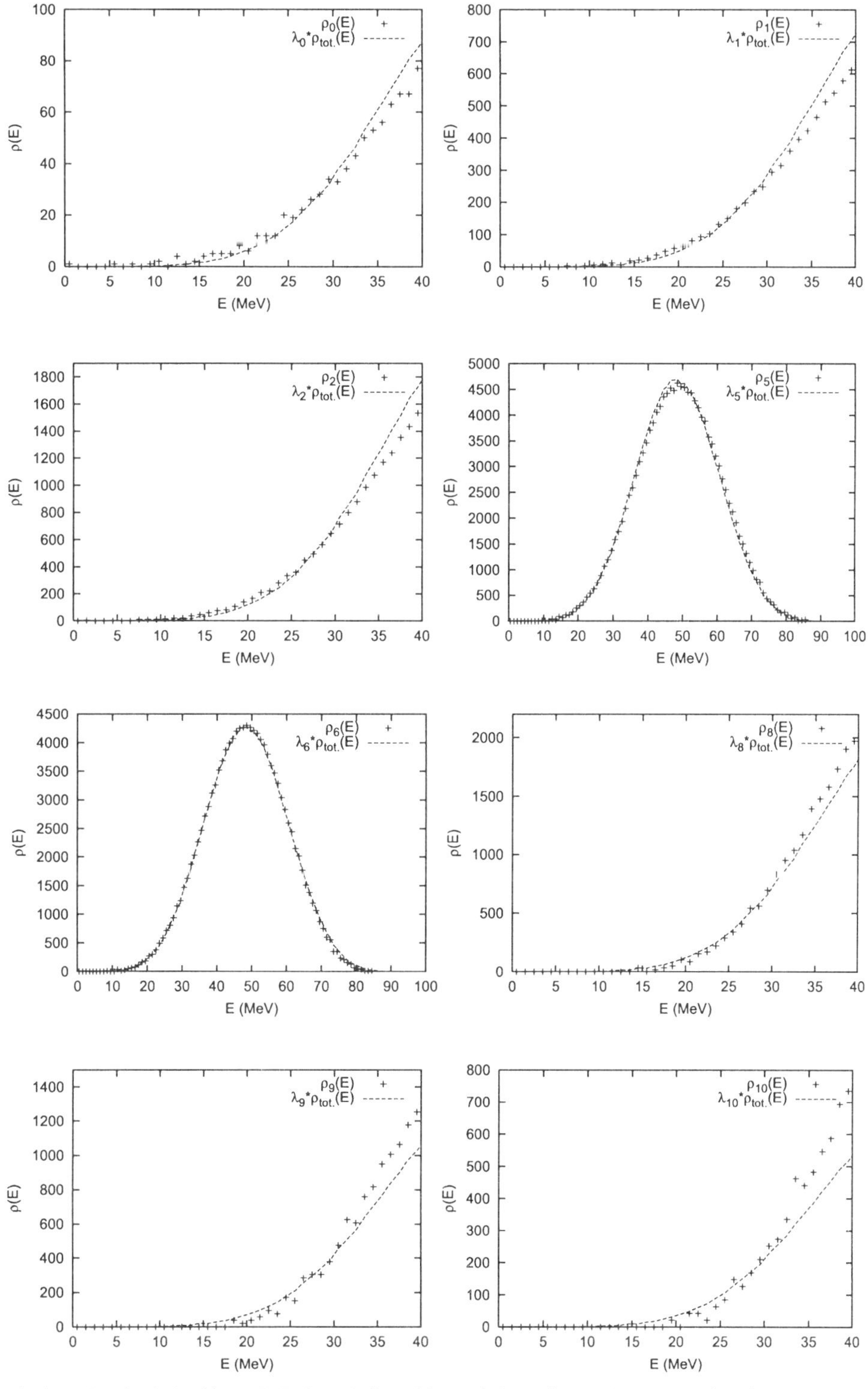

FIG. 2. Spin-dependent level densities $\rho(E,J)$ from shell model calculations (times $2J+1$) (crosses) and rescaled densities (broken lines) for 12 particles in the sd-shell.

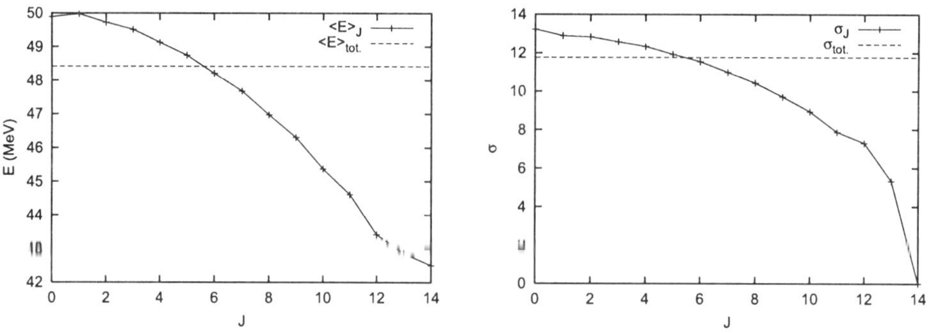

FIG. 3. Shell model scalar centroids and widths of the partial level density $\rho(E,J)$ as a function of total spin J for 12 particles in the sd-shell. Broken horizontal lines show the same quantities for the total density $\rho(E)$.

determined by the interaction dynamics. If one tries to rescale the partial level densities $\rho(E,J)$ from the total level density $\rho(E)$ of Eq. (1) with the aid of the scaling factors $\lambda_J = d(J)/d_{tot}$, the deviations turn out to be much more pronounced (see Fig. 2), even for the most important low spins. The physical reason for this discrepancy is that the centroids and widths found according to the recipes of statistical spectroscopy [22,26] depend on spin, Fig. 3. Similar results for the J dependence of centroids and widths were reported in Refs. [31,32]. For $J=5$ and 6, the J-dependent scalar centroids and widths coincide with those of the total density (Fig. 3), and the rescaling scheme describes reasonably well the J-dependent level density, see Fig. 2. In fact, as a rule,

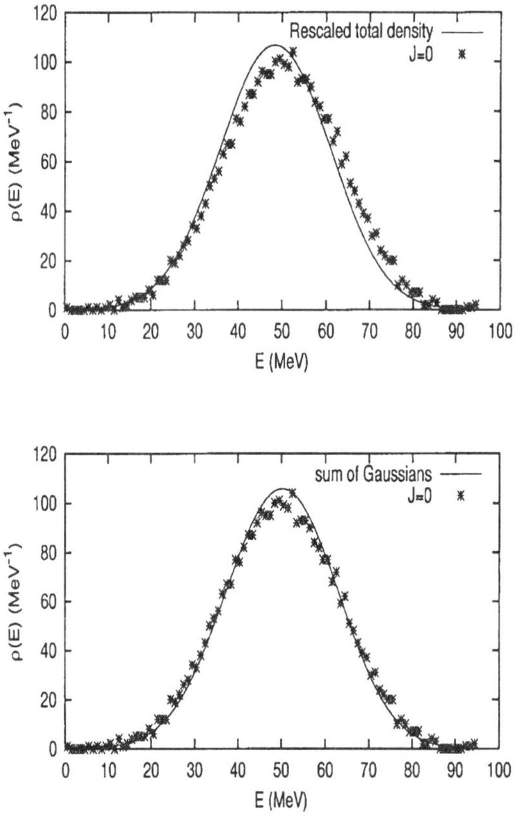

FIG. 4. $J=0$ shell model level density for 12 particles in the sd shell model (stars), compared with the rescaled total density $\lambda_0\rho(E)$ (upper panel), and fixed-J Gaussian sum of Eq. (9) (lower panel).

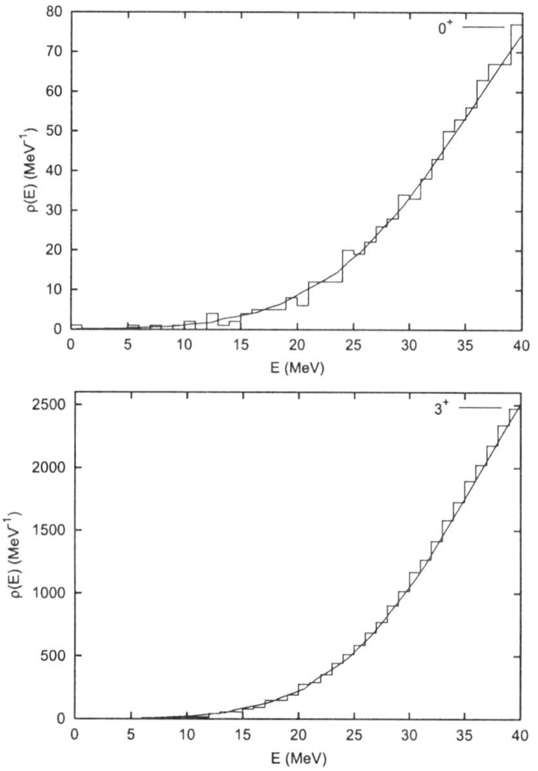

FIG. 5. Low energy part of $J=0$ (upper panel) and $J=3$ (lower panel) shell model level density times $2J+1$ for ^{28}Si (histogram) compared with the fixed-J sum of finite range Gaussians, Eq. (11).

the variance $\sigma_c(J)$ of the centroid distributions makes the main contribution to the total width $\sigma(J)$ larger than the average width of individual partitions weighted with their dimensions [22].

As a practical solution for the problem of calculation of shell-model level densities without full diagonalization we can suggest the following. For each shell model partition $p=1,\ldots,P$, the fixed-J centroid $\bar{E}_p(J)$ is calculated in accordance with the prescriptions of statistical spectroscopy [22,17], together with the fixed-J width $\sigma_p(J)$. Since the results for the level density obtained from the spectral moments are sensitive to the position E_0 of the ground state, we apply the method of exponential extrapolation to find the ground state energy. This method, based on the generic behavior of small admixtures of remote basis states to the wave functions of low-lying states, was suggested, tested, and justified by shell-model examples, random matrix theory, and exactly solvable models [25]. It was later successfully used for ground state energies of fp nuclei [37] and the superdeformed state in ^{56}Ni as well as for single-particle occupancies [33].

We remind the reader the main justification of the exponential extrapolation. The typical properties of complicated many-body states are revealed, for example, by a detailed analysis in the framework of the nuclear shell model [21]. Very similar results for complex atoms [20] show that these properties are indeed generic for many-body quantum chaos. It is practically important that the determination of the relevant global parameters does not require the diagonalization of the huge Hamiltonian matrix. The distribution of the components C_k^α of a many-body state $|\alpha\rangle$ in the basis of $J^\pi T$-projected independent particle configuration states $|k\rangle$ is close to the Gaussian predicted by the random matrix theory [1,18,19]. The same is valid for the strength functions

$$F_k(E)=\sum_\alpha |C_k^\alpha|^2\delta(E-E_\alpha) \qquad (5)$$

of the shell model basis states $|k\rangle$ for the realistic residual interaction strength [34]. The centroids $\bar{E}_k$ of these distributions are given by the diagonal matrix elements of the full Hamiltonian H,

$$\bar{E}_k\equiv\sum_\alpha |C_k^\alpha|^2E_\alpha=H_{kk}. \qquad (6)$$

The width σ_k of the strength function (5) is also known prior to the diagonalization as the sum of the off-diagonal matrix elements squared,

$$\sigma_k^2\equiv\sum_\alpha |C_k^\alpha|^2(E_\alpha^2-\bar{E}_k^2)=\sum_{l(\neq k)} H_{kl}^2. \qquad (7)$$

An illustration of these distributions for the case of $J^\pi T=0^+0$ states of $A=12$ particles in the sd-shell model, where the USD interaction [27] was used, can be found in Figs. 1 and 2 of Ref. [35]. The results are qualitatively the same for different effective interactions; the fp-shell model provides also quite similar results. We need to stress that no diagonalization is required here. As was shown for the first time in Ref. [21], the widths σ_k are nearly constant, $\sigma_k\approx\bar{\sigma}$. The mechanism of this equilibration is provided by the above mentioned geometric chaoticity that is inevitably growing in the process of constructing $J^\pi T$-projected states [21,36]. As discussed earlier, the same chaoticity of almost random coupling of individual spins into total spin J is in the core of the standard derivation [1–3] of the spin-dependent level density. The fact of constancy of the widths σ_k is essential for establishing a practical boundary for the shell model truncation [35] on the level of $(3-4)\sigma$. This truncation gives a first approximation to the energy eigenvalues.

The further precise definition of energies of low-lying states can be achieved using the exponential extrapolation [25,37]. It was established [34] that the remote tails of the generic strength function are exponential. This can be translated into exponentially small admixtures of highly excited states, located beyond threshold of $(3-4)\sigma$ from the centroid, to the exact wave functions of low-lying states. The exponential convergence was rigorously proved in some models and tested by the full matrix diagonalization when possible. A practical algorithm was worked out [37] to establish the onset of the exponential regime by several progressive truncations keeping the partitions in their entirety. Using this algorithm, it is usually possible to locate the ground and low-lying states very close (better than 200–300 keV) to the exact energy value using shell model spaces of dimensions that are less than 1% of the full dimension [37,33].

With the knowledge of the ground state position we determine the set of the Gaussians,

$$G_p(E,J)\equiv G[E+E_0-\bar{E}_p(J),\sigma_p(J)], \qquad (8)$$

where E is the excitation energy from the ground state of energy E_0. The superposition of the contributions of the partitions weighted with their dimensions $d_p(J)$ leads to the full level density

$$\rho(E,J)=\sum_p^P d_p(J)G_p(E,J). \qquad (9)$$

In this approach each partition has a definite number of particles and parity; no special effort for conserving number of particles and parity is required, as in the case of a model that cannot select subspaces with correct symmetries, such as shell model Monte Carlo [38,39]. The generalization to the fixed-isospin level density is straightforward. $\bar{E}_p(J)$ and $\sigma_p(J)$ are calculated using techniques similar to those described in Ref. [22].

A part of the present approach was sometimes mentioned in the literature as a method to determine the shape, but not the J-dependent level density $\rho(E,J)$. A novel feature here is the opportunity of estimating the position of the ground state with very good accuracy [25,37]; the method works if the ground state energy is known with an accuracy not worse than 100–200 keV. But this is the precision one can achieve with the aid of the exponential convergence [37]. The exponential convergence method was recently proven to work for many major shells, and many $\hbar\omega$ excitations [40]. Figure 4 shows the results for this approach applied to the case of 12 valence particles in the sd-shell model using the USD interaction of Ref. [27].

Recently, a method proposed by Zuker [23], which uses four scalar moments, was generalized by Johnson and collaborators to the sum of configurations [41–43]. Due to the complexity of the calculations, they could not obtain all J-dependent moments, as in Eq. (9). Therefore, they relied on a rescaling procedure based on Eq. (4), but using an energy dependent spin-cutoff factor σ. This approach attempts to calculate more accurately the shape of the total density of states, but introduces two new approximations, Eq. (4) and the approximate calculation of $\langle \hat{J}^2\rangle(E)$, that could affect the overall accuracy. Figs. 2–4 of Ref. [43] do not prove that the inclusion of the third and fourth moments improves the accuracy of the results.

One can understand why our sum using only the first and second moments, Eq. (9), turns out to be so reliable. Although the higher moments describing skewness and excess are significant, in adding contributions from many configurations, the asymmetries have a tendency to average out. As Fig. 1 of Ref. [35] suggests, the only configurations for which the asymmetry may survive are those with the centroids close to the edges of the spectrum. In addition, the ranges of those distributions are finite [typically, the ranges of the spectrum are about $(12–16)\sigma$]. To take into account these effects, we replaced the Gaussians in Eq. (9) by the finite range Gaussians (FRGs) that span over a range of at most $\pm 3\sigma$, and do not extend lower than the energy of the yrast state for each spin,

$$G_p(E,J)\mapsto(\mathrm{FRG})_p(E,J,E_J), \tag{10}$$

where

$$(\mathrm{FRG})_p(E,J,E_J)=\sqrt{\frac{2}{\pi}}\frac{e^{-t^2}}{\sigma[\mathrm{erf}(t_-)+\mathrm{erf}(t_+)]} \tag{11}$$

if $t_- < t < t_+$, and 0 otherwise. Here

$$t=\frac{E+E_0-\bar{E}_p(J)}{\sqrt{2}\sigma}, \tag{12}$$

t_- corresponds to either the $\bar{E}_p(J)-3\sigma$ or the yrast state energy (whichever is higher), and, similarly, $t_+=\bar{E}_p(J)+3\sigma$. The choice of $\pm 3\sigma$ for the FRGs is based on the observation of Ref. [35] (see, e.g., Fig. 3 of Ref. [35]) that, for the low-lying states, the configurations located within $\pm 3\sigma$ to the centroid provide more than 90% of the overlap with the exact wave function. An accurate calculation of the yrast state energy can be obtained using the exponential convergence method with the same effort as for the ground state energy [25,37].

The results obtained with the use of this procedure are shown in Fig. 5 for the lowest part of spectrum in the cases of $J=0$ and $J=3$. Figure 6 shows the density of 0^+ and 0^- states for six particles in psd model space using the PSDMK interaction [44] (no attempt was made to remove the center-of-mass spurious components; the effect of the center-of-mass spurious motion on the nuclear level density is still under investigation [43,48]). The comparison with exact shell model calculations proves to be very good.

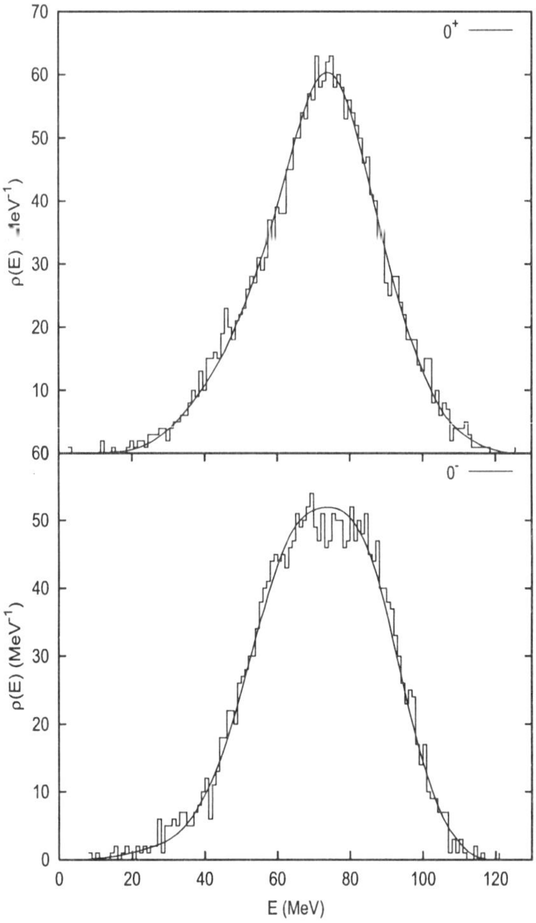

FIG. 6. $J=0$ shell model level density for 6 particles in psd model space (histograms) compared with the fixed-J sum of finite range Gaussians, Eq. (11); upper panel shows the density of positive parity states, while lower panel shows the negative parity states.

In conclusion, we proposed a new practical method of calculating the total density of states using the procedure of statistical spectroscopy for the centroids and widths for each partition in a given shell model space. We show that the rescaling scheme based on Eq. (4) for the total density of states, sometimes used with the Fermi-gas and Monte Carlo methods [28], is not accurate enough in describing shell model J-dependent densities, especially for low spins J that are of special interest in applications. However, a sum of Gaussians over partitions with the J-dependent centroids and widths, Eqs. (8)–(12), can successfully describe spin- and parity-dependent partial level densities; the essential ingredient is an accurate estimate of the ground state energy of the system, which is possible with the aid of the exponential convergence method.

We acknowledge support from the National Science Foundation, Grant Nos. PHY-0070911 and DMR-9977582.

[1] A. Bohr and B. Mottelson, *Nuclear Structure* (Benjamin, New York, 1969), Vols. 1 and 2.
[2] T. Ericson, Adv. Phys. **9**, 425 (1960).
[3] H.A. Bethe, Phys. Rev. **50**, 332 (1936).
[4] J.R. Huizenga and L.G. Moretto, Annu. Rev. Nucl. Sci. **22**, 427 (1972).
[5] W. Dilg, W. Schantl, H. Vonach, and M. Uhl, Nucl. Phys. **A217**, 269 (1973).
[6] A.V. Ignatyuk, A.V. Smirenkin, and A.S. Tishin, Sov. J. Nucl. Phys. **21**, 255 (1975).
[7] S. Goriely, Nucl. Phys. **A605**, 28 (1996).
[8] P. Demetriou and S. Goriely, Nucl. Phys. **A695**, 95 (2001).
[9] N. Cerf, Phys. Rev. C **49**, 852 (1994); **50**, 836 (1994).
[10] H. Nakada and Y. Allhassid, Phys. Rev. Lett. **79**, 2939 (1997).
[11] K. Langanke, Phys. Lett. B **438**, 235 (1998).
[12] W.E. Ormand, Phys. Rev. C **56**, R1678 (1997).
[13] F.S. Chang, J.B. French, and T.H. Thio, Ann. Phys. (N.Y.) **66**, 137 (1971).
[14] J.B. French and K.F. Ratcliff, Phys. Rev. C **3**, 94 (1971); K.F. Ratcliff, *ibid.* **3**, 117 (1971).
[15] J.P. Draayer, J.B. French, and S.S.M. Wong, Ann. Phys. (N.Y.) **106**, 472 (1977); **106**, 503 (1977).
[16] J.B. French, V.K.B. Kota, A. Pandey, and S. Tomsovic, Ann. Phys. San Diego **181**, 198 (1988); **181**, 235 (1988).
[17] S. S. M. Wong, *Nuclear Statistical Spectroscopy* (Oxford University Press, New York, 1986).
[18] T.A. Brody, J. Flores, J.B. French, P.A. Mello, A. Pandey, and S.S.M. Wong, Rev. Mod. Phys. **53**, 385 (1981).
[19] V.K.B. Kota, Phys. Rep. **347**, 223 (2001).
[20] V.V. Flambaum, A.A. Gribakina, G.F. Gribakin, and M.G. Kozlov, Phys. Rev. A **50**, 267 (1994).
[21] V. Zelevinsky, B.A. Brown, M. Horoi, and N. Frazier, Phys. Rep. **276**, 85 (1996).
[22] C. Jacquemin and S. Spitz, J. Phys. G **5**, L95 (1979); Z. Phys. A **290**, 251 (1979).
[23] A.P. Zuker, Phys. Rev. C **64**, 021303(R) (2001).
[24] M. Horoi, Bull. Am. Phys. Soc. **45**, 155 (2002).
[25] M. Horoi, A. Volya, and V. Zelevinsky, Phys. Rev. Lett. **82**, 2064 (1999).
[26] C. Jacquemin, Z. Phys. A **303**, 135 (1981); C. Jacquemin, G. Auger, and C. Quesne, *ibid.* **309**, 55 (1982).
[27] B.A. Brown and B.H. Wildenthal, Annu. Rev. Nucl. Part. Sci. **38**, 29 (1988).
[28] H. Nakada (private communication), about Fig. 3 of Ref. [10].
[29] V.K.B. Kota and D. Majumbar, Nucl. Phys. **A604**, 129 (1996).
[30] C. Jacquemin and S.K. Kataria, Z. Phys. A **324**, 261 (1986).
[31] J.J.M. Verbaarschot, Phys. Lett. **102B**, 201 (1982).
[32] V.K.B. Kota, Z. Phys. A **315**, 91 (1984).
[33] M. Horoi, B.A. Brown, and V. Zelevinsky, Phys. Rev. C **67**, 034303 (2003). Based on the present capabilities of our shell model code, we can calculate level densities for shell model cases with full dimension of the order 10^{10}–10^{11}.
[34] N. Frazier, B.A. Brown, and V. Zelevinsky, Phys. Rev. C **54**, 1665 (1996).
[35] M. Horoi, B.A. Brown, and V. Zelevinsky, Phys. Rev. C **50**, R2274 (1994).
[36] V. Zelevinsky, Annu. Rev. Nucl. Part. Sci. **46**, 237 (1996).
[37] M. Horoi, B.A. Brown, and V. Zelevinsky, Phys. Rev. C **65**, 027303 (2002).
[38] Y. Allhassid, S. Liu, and H. Nakada, Phys. Rev. Lett. **83**, 4265 (1999).
[39] Y. Allhassid, G.F. Bertsch, S. Liu, and H. Nakada, Phys. Rev. Lett. **84**, 4313 (2000).
[40] M. Horoi and V. Zelevinsky, Bull. Am. Phys. Soc. **47**(2), 165 (2002).
[41] C.W. Johnson, J-U. Nabi, and W. E Ormand, nucl-th/0105041.
[42] J-U. Nabi, C.W. Johnson, and W. E Ormand, nucl-th/0111067.
[43] C.W. Johnson, J-U. Nabi, and W. E Ormand, nucl-th/0111068.
[44] The PSDMK interaction combines the Cohen-Kurath interaction [45] for the p space, Preedom-Wildenthal interaction [46] for the sd space, and Millener-Kurath interaction [47] for the coupling between the p and sd spaces.
[45] S. Cohen and D. Kurath, Nucl. Phys. **A73**, 1 (1965).
[46] O. Preedom and B.H. Wildenthal, Phys. Rev. C **6**, 1633 (1972).
[47] D.J. Millener and D. Kurath, Nucl. Phys. **A255**, 315 (1975).
[48] M. Horoi and V. Zelevinsky (unpublished).

RAPID COMMUNICATIONS

PHYSICAL REVIEW C **71**, 041304(R) (2005)

Regularities with random interactions in energy centroids defined by group symmetries

V. K. B. Kota

Physical Research Laboratory, Ahmedabad 380 009, India

(Received 14 December 2004; published 29 April 2005)

Regular structures generated by random interactions in energy centroids defined over irreducible representations (irreps) of some of the group symmetries of the interacting boson models *sd*IBM, *sdg*IBM, *sd*IBM-*T*, and *sd*IBM-*ST* are studied by deriving trace propagation equations for the centroids. It is found that, with random interactions, the lowest and highest group irreps in general carry most of the probability for the corresponding centroids to be lowest in energy. This generalizes the result known earlier, via numerical diagonalization, for the more complicated fixed spin *J* centroids where simple trace propagation is not possible.

DOI: 10.1103/PhysRevC.71.041304 PACS number(s): 24.60.Lz, 21.60.Fw, 21.10.Re, 05.45.−a

Johnson, Bertsch, and Dean in 1998 [1], using the nuclear shell model, found that random two-body interactions lead to ground states for even-even nuclei, having spin 0^+ with very high probability. Similarly, Bijker and Frank [2], using the interacting boson model (IBM) with *s* and *d* bosons (*sd*IBM), showed that random interactions generate vibrational and rotational structures with high probability. Further studies using the shell model, fermions in one or two *j* orbits, *sd*, *sp*, and *sdg* IBM's, bosons in a single ℓ orbit, etc., revealed statistical predominance of odd-even staggering in binding energies, $0^+, 2^+, 4^+, \ldots$ yrast sequence, regularities in ground states in parity distributions, occupation numbers, and so on; see [3–10] and references therein. Notably, Zelevinsky *et al.* [4] introduced the idea of geometric chaos as a basis for the regularities observed in shell model studies. Similarly, Zhao *et al.* [8] developed a prescription based on sampling of the corners of the parameter space, and Bijker, Frank and Kota [7,9] employed mean-field methods. The unexpected results for regularities with random interactions are reviewed in [6,10]. As Zhao *et al.* stated [10], "a more fundamental understanding of the robustness of $0^+_{\text{g.s.}}$ dominance is still out of reach." Therefore, going beyond the ground states and near yrast levels, energy centroids, spectral widths, and correlations among them are also being investigated by several groups [4,5,11–13] as they are expected to give new insights into regularities generated by random interactions. For example, Zhao *et al.* [11,12] initiated the study of energy centroids and analyzed fixed-*L* (fixed-*J*, *JT*) centroids in IBM (in shell model) spaces. They found that $L_{\min}$ (or $J_{\min}$) and $L_{\max}$ (or $J_{\max}$) will be lowest with largest probabilities, and others appear with negligible probability. Similarly Papenbrock and Weidenmüller [13] recently analyzed the structure of fixed-*J* spectral widths for fermions in a single-*j* shell.

An interesting and important question is the extension of the spin-zero ground state dominance (and also other regular structures seen in shell model and IBM studies) to group theoretical models with Hamiltonians preserving a symmetry higher than *J* (or *L*). Similarly, one may consider centroids and variances defined over good or broken symmetry subspaces. They open a new window to the regularities of many-body systems in the presence of random forces. Initiating work in this direction [9], we recently used random one- plus two-body Hamiltonians invariant with respect to $O(\mathcal{N}_1) \oplus O(\mathcal{N}_2)$ symmetry of a variety of interacting boson models to investigate the probability of occurrence of a given $(\omega_1\omega_2)$ irreducible representation (irrep) to be the ground state in even-even nuclei; $[\omega_1]$ and $[\omega_2]$ are symmetric irreps of $O(\mathcal{N}_1)$ and $O(\mathcal{N}_2)$ respectively. We found that the 0^+ dominance observed in ground states of even-even nuclei extends to group irreps. The purpose of this paper and others to follow is to go beyond this and study regularities generated by random interactions in energy centroids, variances, etc. defined over group irreps. Reported in this rapid communication are the results of a first analysis of energy centroids with examples from *sd*IBM, *sdg*IBM, *sd*IBM-*T* with the bosons carrying isospin (*T*), and *sd*IBM-*ST* with the bosons carrying spin-isospin (*ST*) degrees of freedom. Before proceeding further, it is important to stress that energy centroids (also variances) can be calculated as a function of particle number *m* and the quantum numbers labeling the group irreps, without recourse to the construction of the Hamiltonian matrix. The principle used here is trace propagation, a subject introduced in the context of statistical nuclear spectroscopy by French [14,15]. Readers not interested in the details of group algebra and derivation of trace propagation equations for the energy centroids [given by Eqs. (5)–(8)], may jump ahead to the discussion of results starting just after Eq. (8).

Let us begin with the spectrum generating algebra (SGA), say G_1, of a group theoretical model with all the many particle states in the model belonging to the irrep Γ_1 of G_1. For example, the SGA G_1 for *sd*IBM is U(6). Now the average of an operator $\mathcal{O}(k)$ of maximum body rank *k* over the irreps Γ_2 of a subalgebra G_2 of G_1 (G_2 in general denotes a set of subalgebras contained in G_1 and Γ_2 denotes all their irreps) is defined by

$$\langle \mathcal{O}(k)\rangle^{\Gamma_1,\Gamma_2} = \sum_{\beta}\sum_{\alpha\in\Gamma_2} \langle \Gamma_1\beta\Gamma_2\alpha|\mathcal{O}(k)|\Gamma_1\beta\Gamma_2\alpha\rangle \Big/ \left[\sum_{\beta}\sum_{\alpha\in\Gamma_2}\langle \Gamma_1\beta\Gamma_2\alpha|1|\Gamma_1\beta\Gamma_2\alpha\rangle\right] \quad (1)$$

In Eq. (1), β labels the multiple occurrence (multiplicity) of Γ_2 in a given Γ_1 irrep (i.e., in the reduction of Γ_1 to Γ_2). Removing the denominator in Eq. (1) gives the trace over (Γ_1, Γ_2) space, i.e., $\text{tr}[\mathcal{O}(k)]^{\Gamma_1,\Gamma_2}$. General theory for propagation of traces of operators over irreps of group symmetries was developed in Refs. [16–19]. In particular, Quesne [16] showed that for $G_1 \supset G_2$, trace propagation over the irreps Γ_1 and Γ_2 of G_1 and G_2 algebras is related to the so-called integrity basis of G_2 in G_1 which gives the minimal set of G_2 scalars in G_1. As discussed in Refs. [17,18], it is seen that in general the multiplicity of Γ_2 in a given Γ_1 irrep results in the propagation of the matrix of traces $\mathrm{tr}[\mathcal{O}(k)]^{\Gamma_1\Gamma_2;\beta\beta'} = \sum_\alpha \langle\Gamma_1\beta\Gamma_2\alpha|\mathcal{O}(k)|\Gamma_1\beta'\Gamma_2\alpha\rangle$. However, quite often the trace of this trace matrix or its average, as given by Eq. (1), which is important in applications, may not propagate in a simple manner. There are approximate methods for propagating trace of the trace matrix, and they are significant, particularly when the integrity basis contains far too many operators [18]. A very important example here is fixed-L averages in IBM's or fixed-J (and JT) averages in the shell model. For these, it is not possible to write a simple propagation equation in terms of the defining space averages. On the other hand, traces over irreps of group symmetries (higher than J symmetry) can be propagated in many situations using Casimir invariants. French and Draayer [19] showed that by simple counting of irreps of G_2 in G_1 and the scalars one can construct in terms of the Casimir invariants of G_1 and G_2 will immediately confirm if propagation via Casimir invariants is possible; in this situation, the integrity basis reduces to Casimir operators of G_1 and G_2. In this paper we restrict ourselves to examples in IBM's where this result applies; Refs. [20,21] give the first IBM examples.

For IBM's the SGA, called G_1 above, is $U(\mathcal{N})$, with $\mathcal{N}=6$ for sdIBM, 15 for sdgIBM, etc, and its irreps Γ_1 are labeled uniquely by the boson number m as all m boson states are symmetric with respect to $U(\mathcal{N})$. Now, consider the average of an operator $\mathcal{O}(k)$ over the irreps (m,Γ_2) with Γ_2 being the irreps of a subalgebra G_2 of $U(\mathcal{N})$. Say the number of (m,Γ_2), called Γ^i hereafter, for $m \leqslant k$ is r. Also assume that there are r number of invariants $\hat{C}_i$, $i=1,2,\ldots,r$, of maximum body rank k constructed out of the products of m and the Casimir invariants of G_2. Then, for any irrep Γ^0, clearly $\langle\mathcal{O}\rangle^{\Gamma^0} = \sum_{i=1}^r a_i\langle\hat{C}_i\rangle^{\Gamma^0}$, where a_i are constants. The a_i can be determined by assuming that the averages $\langle\mathcal{O}\rangle^{\Gamma^j}$ are known for the irreps Γ^j, $j=1,2,\ldots,r$. For example, Γ_j can be chosen to be the irreps (m,Γ_2) for $m\leqslant k$. With this, defining the row matrices $[C]$ and $[\mathcal{O}_{\rm inp}]$ and the $r\times r$ matrix $[X]$ as

$$
\begin{aligned}
[C] &\Leftrightarrow C_i = \langle\hat{C}_i\rangle^{\Gamma^0},\\
[\mathcal{O}_{\rm inp}] &\Leftrightarrow \mathcal{O}_{{\rm inp}:i} = \langle\mathcal{O}\rangle^{\Gamma^i},\\
[X] &\Leftrightarrow X_{ij} = \langle\hat{C}_j\rangle^{\Gamma^i},
\end{aligned} \tag{2}
$$

the propagation equation is

$$
\langle\mathcal{O}\rangle^{\Gamma^0} = [C]\,[X]^{-1}\,[\widetilde{\mathcal{O}_{\rm inp}}]. \tag{3}
$$

As the eigenvalues of the Casimir invariants of the algebras $U(\mathcal{N})$, $O(\mathcal{N})$, etc. are known, construction of $[C]$ and $[X]$ is easy. In the reminder of this paper, the H is assumed to be $(1+2)$-body. As an example, let us consider SU(3) centroids in sdIBM. Here $G_1=U(6)$ and $G_2=SU(3)$. Simple counting of scalar in terms of the number operator $\hat{n}$ and the quadratic Casimir operator $\hat{C}_2$ and the cubic Casimir operator $\hat{C}_3$ of SU(3) confirms that they exhaust all the scalars needed for propagating $\langle\mathcal{O}(k)\rangle^{m,(\lambda\mu)}$ for any k ... O denotes SU(3) irreps. Propagation eq... c centroids over SU(3) irreps can be wr... $a_0 + a_1 m + a_2 m^2 + a_3 C_2(\lambda\mu)$, where

$$
C_2(\lambda\mu) = \langle(\lambda\mu)\alpha|\hat{C}_2|(\lambda\mu)\alpha\rangle = [\lambda^2+\mu^2+\lambda\ldots
$$

Using Eqs. (3) and (4), the propagation equation, the energy centroids for $m\leqslant 2$, is [21]

$$
\begin{aligned}
\langle H\rangle^{m,(\lambda\mu)} &= \tfrac{1}{2}(2-3m+m^2)\langle H\rangle^{0,(00)} + (2m-m^2)\langle H\ldots\\
&\quad + \left[-\tfrac{5}{6}m + \tfrac{5}{18}m^2 + \tfrac{1}{18}C_2(\lambda\mu)\right]\langle H\rangle^{2,(40)}\\
&\quad + \left[\tfrac{1}{3}m + \tfrac{2}{9}m^2 - \tfrac{1}{18}C_2(\lambda\mu)\right]\langle H\rangle^{2,(02)}. \qquad (\ldots
\end{aligned}
$$

Equation (5) extends easily to the SU(3) limit of pfIBM with U(10) SGA but not to sdg, $sdgpf$, etc., IBM's. Now we will derive three new propagation equations for energy centroids.

In the $U(\mathcal{N})\supset\sum_i\,[U(\mathcal{N}_i)\supset O(\mathcal{N}_i)]\oplus$ symmetry limits of IBM's, with the bosons carrying angular momenta $\ell_1,\ell_2,\ldots$ so that $\mathcal{N}_i=(2\ell_i+1)$ and $\mathcal{N}=\sum_i\mathcal{N}_i$, for a given ith orbit, $U(\mathcal{N}_i)$ generates number of particles m_i in the orbit and $O(\mathcal{N}_i)$ generates the corresponding seniority quantum number ω_i. The number operators $\hat{n}_i$ of $U(\mathcal{N}_i)$ and the quadratic Casimir operators of $O(\mathcal{N}_i)$ or the corresponding pairing operators $\hat{P}_2(O(\mathcal{N}_i))$ suffice to give fixed $\widetilde{m\omega}=(m_1\omega_1,m_2\omega_2,\ldots)$ averages of H. Appendix A in Ref. [22] gives the explicit form of $\hat{P}_2(O(\mathcal{N}_i))$ for a general situation. Fixed-$\widetilde{m\omega}$ centroids of H can be written as $\langle H\rangle^{\widetilde{m\omega}} = \sum_i m_i\epsilon_i + \sum_{i\geqslant j} a_{ij}m_i(m_j-\delta_{ij}) + \sum_i c_i\langle\hat{P}_2(O(\mathcal{N}_i))\rangle^{m_i\omega_i}$. Solving for a_{ij} and c_i in terms of the centroids for $m\leqslant 2$, the final propagation equation, for IBM's with no internal degrees of freedom, is

$$
\begin{aligned}
\langle H\rangle^{\widetilde{m\omega}} &= \sum_i m_i\epsilon_i + \sum_{i>j}\overline{V_{ij}}\,m_im_j\\
&\quad + \sum_i \frac{m_i(m_i-1)}{2}\,\langle V\rangle^{m_i=2,\omega_i=2}\\
&\quad + \sum_i \frac{\langle V\rangle^{m_i=2,\omega_i=0}-\langle V\rangle^{m_i=2,\omega_i=2}}{2\mathcal{N}_i}\\
&\quad \times (m_i-\omega_i)(m_i+\omega_i+\mathcal{N}_i-2);\\
\overline{V_{ij}} &= \{[\mathcal{N}_i(\mathcal{N}_j+\delta_{ij})]/(1+\delta_{ij})\}^{-1}\\
&\quad \times \sum_L V^L_{\ell_i\ell_j\ell_i\ell_j}(2L+1),\\
\langle V\rangle^{m_i=2,\omega_i=0} &= \langle(\ell_i\ell_i)L_i=0|V|(\ell_i\ell_i)L_i=0\rangle,\\
\langle V\rangle^{m_i=2,\omega_i=2} &= \left[\frac{\mathcal{N}_i(\mathcal{N}_i+1)}{2}\,\overline{V_{ii}} - \langle V\rangle^{m_i=2,\omega_i=0}\right]\\
&\quad \Bigg/\left[\frac{\mathcal{N}_i(\mathcal{N}_i+1)}{2}-1\right].
\end{aligned} \tag{6}
$$

Note that in (6), ϵ_i are energies of the single- particle levels with angular momentum ℓ_i and $V^L_{\ell_i\ell_j\ell_i\ell_j} = \langle(\ell_i\ell_j)L|V|(\ell_i\ell_j)L\rangle$ are two-particle matrix elements of the two-body part of H. Also in Eq. (6), for s orbit, $m_s=2$ and $\omega_s=2$ and there will be no two-boson state with $\omega_s=0$. Equation (6) for sdgIBM is given first in [23], i.e., for averages over the irreps of the algebras in the chain $U_{sdg}(15)\supset U_s(1)\oplus[U_d(5)\supset O_d(5)]\oplus[U_g(9)\supset$

(9)]. Similarly Eq. (6) gives H averages over the irreps $U_{sd}(6) \supset U_d(5) \supset O_d(5)$ of sdIBM, $U_{sdpf}(16) \supset [U_d(5) \supset O_d(5)] \oplus [U_p(3) \supset O_p(3)] \oplus [U_f(7) \supset O_f(7)]$ of $sdpf$IBM, etc. Moreover, this extends easily (this will be discussed elsewhere) to IBM's with internal degrees of freedom. Let us add that it is also possible to write propagation equations for the variances $\langle [H - \langle H \rangle^{m\tilde{\omega}}]^2 \rangle^{m\tilde{\omega}}$ using the results in [20,24].

In IBM-T with $U(3\mathcal{N}) \supset U(\mathcal{N}) \otimes [SU_T(3) \supset O_T(3)]$ where $U(\mathcal{N})$ gives the spatial part (for sd, sdg, $sdpf$, etc.) and $O_T(3)$ generating isospin [25], it is possible to propagate the centroids $\langle H \rangle^{m,\{f\},T} \equiv \langle H \rangle^{m,(\lambda\mu),T}$. Note that the $U(\mathcal{N})$ irreps are labeled by $\{f\} = \{f_1, f_2, f_3\}$ where $f_1 \geqslant f_2 \geqslant f_3 \geqslant 0$ and $m = f_1 + f_2 + f_3$. The corresponding $SU_T(3)$ irreps are $(\lambda, \mu) = (f_1 - f_2, f_2 - f_3)$. The $SU_T(3)$ to $O_T(3)$ reductions follow from the formulas given by Elliott [26,27]. The scalars $1, \hat{n}, \hat{n}^2, \hat{C}_2(SU_T(3))$ and $\hat{T}^2$ and the energy centroids for $m \leqslant 2$, via Eqs. (2)–(4), give

$$\begin{aligned}\langle H \rangle^{m,(\lambda\mu),T} &= \left[1 - \tfrac{3}{2}m + \tfrac{m^2}{2}\right] \langle H \rangle^{0,(00),0} + [2m - m^2] \\ &\times \langle H \rangle^{1,(10),1} + \left[-\tfrac{1}{6}m + \tfrac{1}{18}m^2 + \tfrac{1}{9}C_2(\lambda\mu) \right. \\ &\left. - \tfrac{1}{6}T(T+1)\right] \langle H \rangle^{2,(20),0} + \left[-\tfrac{5}{6}m + \tfrac{5}{18}m^2 \right. \\ &\left. + \tfrac{1}{18}C_2(\lambda\mu) + \tfrac{1}{6}T(T+1)\right] \langle H \rangle^{2,(20),2} \\ &+ \left[\tfrac{1}{2}m + \tfrac{1}{6}m^2 - \tfrac{1}{6}C_2(\lambda\mu)\right] \langle H \rangle^{2,(01),1}. \qquad (7)\end{aligned}$$

For sdIBM-T, starting with the general Hamiltonian given in Appendix A of [27] which contains the s and d boson energies and 17 two-particle matrix elements $V^{L,t}_{\ell_1\ell_2\ell_3\ell_4}$, it is easy to write $\langle H \rangle^{m,(\lambda\mu),T}$ for $m \leqslant 2$; for $m = 2$, the two-boson isospins t uniquely define the corresponding $SU_T(3)$ irreps. Thus Eq. (7) for $\langle H \rangle^{m,(\lambda\mu),T}$ is easy to apply for any m.

In IBM-ST, a group chain of interest is [28] $U(6\mathcal{N}) \supset U(\mathcal{N}) \otimes [SU_{ST}(6) \supset O_{ST}(6)]$, with $U(\mathcal{N})$ generating the spatial part and $SU_{ST}(6)$ [or $U_{ST}(6)$] generating the spin-isospin part; note that the Wigner's spin-isospin super-multiplet algebra $SU_{ST}(4)$ is isomorphic to $O_{ST}(6)$. Just as before, it is possible to propagate the centroids $\langle H \rangle^{m,\{f\},[\sigma]}$. Here $\{f\}$'s are the irreps of $U(\mathcal{N})$ or equivalently $U_{ST}(6)$ and $\{f\} = \{f_1, f_2, f_3, f_4, f_5, f_6\}$, where $\sum_i f_i = m$ and $f_i \geqslant f_{i+1} \geqslant 0$. The $O_{ST}(6)$ irreps are labeled by $[\sigma] = [\sigma_1, \sigma_2, \sigma_3]$, and the $\{f\}$ to $[\sigma]$ reductions, needed for the results discussed ahead, follow from the analytical formulas given in [27] and the tabulations in [29]. Equations (2) and (3) give, using the quadratic Casimir invariants $\hat{C}_2$ of $U_{ST}(6)$ and $O_{ST}(6)$,

$$\begin{aligned}\langle H \rangle^{m,\{f\},[\sigma]} &= \left[1 - \tfrac{3}{2}m + \tfrac{m^2}{2}\right] \langle H \rangle^{0,\{0\},[0]} + [2m - m^2] \\ &\times \langle H \rangle^{1,\{1\},[1]} + \left[-\tfrac{5}{3}m + \tfrac{1}{4}m^2 + \tfrac{1}{6}C_2(\{f\}) \right. \\ &\left. + \tfrac{1}{12}C_2([\sigma])\right] \langle H \rangle^{2,\{2\},[2]} + \left[-\tfrac{1}{12}m \right. \\ &\left. + \tfrac{1}{12}C_2(\{f\}) - \tfrac{1}{12}C_2([\sigma])\right] \langle H \rangle^{2,\{2\},[0]} \\ &+ \left[\tfrac{5}{4}m + \tfrac{1}{4}m^2 - \tfrac{1}{4}C_2(\{f\})\right] \langle H \rangle^{2,\{1^2\},[1^2]}, \qquad (8)\end{aligned}$$

where $C_2(\{f\}) = \langle \hat{C}_2[U_{ST}(6)] \rangle^{\{f\}} = \sum_{i=1}^{6} f_i(f_i + 7 - 2i)$ and $C_2([\sigma]) = \langle \hat{C}_2[O_{ST}(6)] \rangle^{[\sigma]} = \sum_{i=1}^{3} \sigma_i(\sigma_i + 6 - 2i)$. Diagonalizing $\hat{C}_2(O_{ST}(6))$ in the $|(\ell_1\ell_2)LST\rangle$ basis and applying the resulting unitary transformation to the H matrix in this basis will give the input averages in Eq. (8).

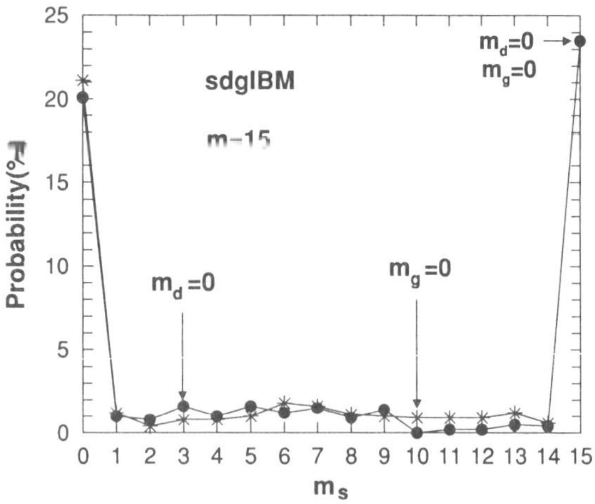

FIG. 1. Probabilities for sdgIBM fixed-$(m_s, m_d, v_d, m_g, v_g)$ centroid energies to be lowest in energy vs m_s for a system of 15 bosons ($m = 15$). For each m_s, the probability shown is the sum of the probabilities for the irreps with the seniority quantum number lowest [$v_\ell = \pi(m_\ell)$] and highest ($v_\ell = m_\ell$). Filled circles and stars are for configurations with $m_d = 0$ and $m_g = 0$, respectively; they are joined by lines to guide the eye. Note that for $m_s = 15$, both $m_d = 0$ and $m_g = 0$.

Now we will apply Eqs. (5)–(8) to study regularities generated by random interactions in energy centroids. In all the calculations used are independent Gaussian random variables with zero center and unit variance and a 1000-member ensemble. We begin with the simplest example of sdIBM centroids. The highest SU(3) irrep for a given m is $(2m, 0)$ and Eq. (5) gives $\langle H \rangle^{m,(\lambda\mu)} - \langle H \rangle^{m,(2m,0)} = [C_2(\lambda\mu) - C_2(2m, 0)]\Delta/18$ with $\Delta = \langle H \rangle^{2,(40)} - \langle H \rangle^{2,(02)}$. Therefore, the probability of finding Δ to be positive or negative will simply give the probability for finding the highest or lowest m particle SU(3) irrep to be lowest in energy. With the two-particle matrix elements chosen to be Gaussian variables (with zero center and unit variance), Δ itself will be a Gaussian variable with zero center. For $m = 3k, 3k+1$, and $3k+2$, k being a positive integer, the lowest SU(3) irreps are (00), (20), and (02), respectively. They will be lowest in energy with 50% and the $(2m, 0)$ irrep will be lowest in energy with 50% probability. Thus, it is easy to understand the regularities in centroids defined over fixed SU(3) irreps in sdIBM with one- plus two-body Hamiltonians, without constructing the many boson Hamiltonian matrix but just by using the propagation equation (5).

In sdgIBM, regularities in fixed-$(m_s, m_d, v_d, m_g, v_g)$ centroids are studied using the propagation equation (6). Choosing the three single-particle energies $(\epsilon_s, \epsilon_d, \epsilon_g)$ and the 16 diagonal two-particle matrix elements $V^L_{\ell_1\ell_2\ell_1\ell_2}$, with $\ell_i = 0, 2$, and 4 to be Gaussian variables, the probability for the centroid of a given $(m_s, m_d, v_d, m_g, v_g)$ configuration to be lowest is calculated for $m = 6 - 25$, and the results are shown in Fig. 1 for $m = 15$. To maintain proper scaling, the ϵ are divided by m and the V^L by $m(m-1)$ just as in [2]. For the discussion

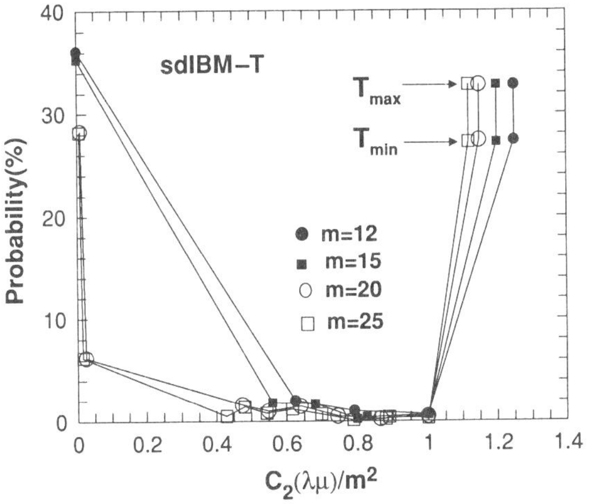

FIG. 2. Probabilities for the sdIBM-T's $(\lambda\mu)T$ centroid energies to be lowest in energy vs $C_2(\lambda\mu)/m^2$ for boson systems with $m = 12, 15, 20$, and 25. Except for the highest $(\lambda\mu)$, for all other $(\lambda\mu)$ shown in the figure, the probabilities are for T_{max} if $\lambda \neq 0$ and $\mu \neq 0$ and they are for T_{min} if $\lambda = 0$ or $\mu = 0$. For the irreps not shown in the figure, the probability is <0.1%. All the points for a given m are joined by lines to guide the eye.

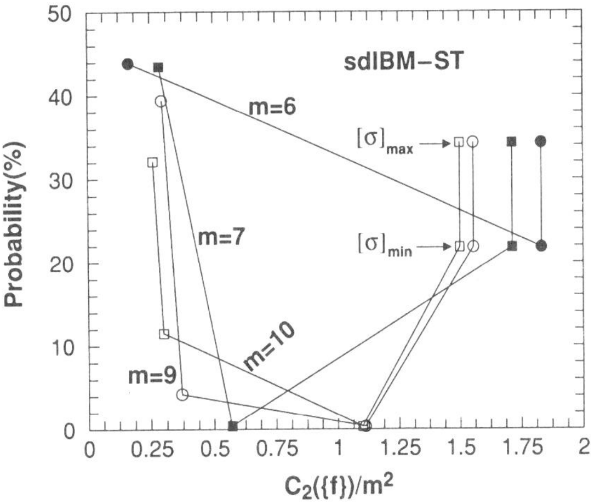

FIG. 3. Probabilities for the sdIBM-ST $\{f\}[\sigma]$ centroid energies to be lowest in energy vs $C_2(\{f\})/m^2$ for boson systems with $m = 6, 7, 9$, and 10. For the structure of the irreps with probability >2%, see text. For $m = 7, 9$, and 10 there is one additional irrep with ~0.3% probability. For the irreps not shown in the figure, the probability is <0.1%. All the points for a given m are joined by lines to guide the eye.

of the results, we define $\pi(x)$ such that $\pi(x) = 0$ for x even and $\pi(x) = 1$ for x odd. It is seen from Fig. 1, and also valid for any m, that the configurations $(m_s, m_d = v_d = m - m_s, m_g = v_g = 0)$, $[m_s, m_d = m - m_s, v_d = \pi(m_d), m_g = v_g = 0]$, $(m_s, m_d = v_d = 0, m_g = v_g = m - m_s)$ and $[m_s, m_d = v_d = 0, m_g = m - m_s, v_g = \pi(m_g)]$ exhaust about 91% probability. Moreover, the configurations with $m_s = m_d = 0$ carry ~20%, $m_s = m_g = 0$ carry ~21%, $m_s = m$ carries ~24%, and $m_s \neq 0$ but $m_d = 0$ or $m_g = 0$ carry ~26% probability. Thus the $m_s = m$ configuration and the four configurations with $m_s = 0$ are most probable to be lowest in energy. However, other configurations with $m_s \neq 0, m$ (they are 49 out of 1195 configurations in the $m = 15$ example) give nonnegligible probability for being lowest. Thus, about ~4% of the $(m_s, m_d, v_d, m_g, v_g)$ configurations will have probability to be lowest with random interactions.

For sdIBM-T, it is easily seen from Eq. (7) that the one-body part of H will not play any role in the study of fixed-$(\lambda, \mu)T$ centroids. Choosing $V^{L,t}_{\ell_1\ell_2\ell_1\ell_2}$ to be Gaussian variables, the centroids are generated, using Eq. (7), for $m = 10 - 25$ and for all allowed $(\lambda, \mu)T$. Some typical results for the regularities are shown in Fig. 2. First, for a given m the highest $SU_T(3)$ irrep is $(m, 0)$ with $T_{max} = m$ and $T_{min} = \pi(m)$. For $m = 3k, 3k + 1$, and $3k + 2$, with k being a positive integer, the lowest $SU_T(3)$ irreps are (00), (10), and (01) with $T = 0, 1$, and 1, respectively; for the latter two situations, the next lowest irreps are (02) and (20), respectively, with $T_{min} = 0$. For $m = 3k$, Fig. 2 shows that the lowest $SU_T(3)$ irrep centroid (here T is unique) is lowest with ~35% probability. Similarly the highest irrep centroid is lowest with ~60% probability and this splits into ~30% each for the lowest and highest T's. For $m = 3k + 1$ and $3k + 2$, the probability for the centroid of the highest irrep to be lowest in energy is the same as for $m = 3k$. However, for the centroid of the lowest irrep, the probability is ~29%, and the next lowest irrep appears with ~6%. Thus, in general, the centroids of the highest and the lowest (for $m = 3k + 1$ and $3k + 2$, the lowest two) $SU_T(3)$ irreps exhaust about 95% of the probability for being lowest in energy. As the two-particle centroids $X^t = \langle H\rangle^{m=2,t}$ are linear combinations of V, it can be seen that they themselves are Gaussian variables. Note that $\langle H\rangle^{m,(\lambda\mu)T} - \langle H\rangle^{m,(m,0)m} = [\mathcal{C}_2(\lambda\mu) - \mathcal{C}_2(m, 0)]\Delta_1 + [T(T+1) - m(m+1)]\Delta_2$, where $\Delta_1 = \frac{1}{9}X^0 + \frac{1}{18}X^2 - \frac{1}{6}X^1$ and $\Delta_2 = \frac{1}{6}(X^2 - X^0)$. Calculations with X^t taken as Gaussian variables with the same variance (actually, the variance of X^0 and X^2 are the same, and that of X^1 is ~20% higher) are carried out, and it is seen that they give almost the same results as in Fig. 2.

For sdIBM-ST, as seen from Eq. (8), the energy centroids $\langle H\rangle^{m,\{f\},[\sigma]}$ are determined by the two-particle averages $\langle H\rangle^{2,\{2\},[2]}$, $\langle H\rangle^{2,\{2\},[0]}$, and $\langle H\rangle^{2,\{1^2\},[1^2]}$, and they are linear combinations of the two-particle matrix elements V^{LST} in the $|(\ell_1\ell_2)LST\rangle$ basis. Instead of choosing V^{LST} to be Gaussian variables, we have chosen, using the result found in the sdIBM-T examples, the three two-particle averages to be Gaussian variables. Using this, the probabilities are calculated for various m values, and some of the results are shown in Fig. 3. First, for a given m, the highest $\{f\}$ is $\{m\}$. The corresponding highest and lowest $[\sigma]$ are $[m]$ and $[\pi(m)]$. For all m, the centroid of the highest $U_{ST}(6)$ irrep is lowest with ~56% probability, and this splits into ~34% and ~22% for the highest and lowest $O_{ST}(6)$ irreps. For $m = 6k, 6k \pm 1, 6k \pm 2$, and $6k + 3$, with k a positive integer, the lowest $U_{ST}(6)$ irreps are those that can be reduced to the irreps $\{0\}$, $\{1\}$, $(\{2\}, \{1^2\})$, and $(\{1^3\}, \{21\})$, respectively. These irreps with the corresponding lowest $[\sigma]$ are lowest, with probability $\sim$43%.

In conclusion, with random interactions, the lowest and highest group irreps (i.e., irreps of G_2 in $G_1 \supset G_2$) carry most of the probability for the corresponding centroids to be lowest in energy. With the inclusion of a subalgebra ($G_1 \supset G_2 \supset G_3$), these probabilities split into the probabilities for the corresponding lowest and highest irreps of the subalgebra. This is indeed the situation for all the examples discussed in this paper. Continuing with the process of embedding subalgebras, the O(3) algebra generating L can be reached (with generalization for systems with LT, LST, or JT). Then, clearly, the energy centroids of highest and lowest L should be most probable, and this is found to be true numerically in [11,12]. An important aspect of the energy centroids is that they propagate via Casimir invariants in many situations. New propagation equations are derived in this paper [Eqs. (6)–(8)]. In fact, there are many other situations where such equations can be derived; an example is the centroids over the irreps $[m_{sd}(\lambda_{sd}\mu_{sd}); m_{pf}(\lambda_{pf}\mu_{pf})]$ of $[\mathrm{U}_{sd}(6) \supset \mathrm{SU}_{sd}(3)] \oplus [\mathrm{U}_{pf}(10) \supset \mathrm{SU}_{pf}(3)]$ algebra of *sdpf*IBM [30]. These will be discussed in a longer paper along with extensions of the present work to spectral variances and also to shell model symmetries. Finally, an important observation is that the propagators carry information about $G_1 \supset G_2$ geometry (i.e., $G_1 \supset G_2$ reduced Wigner coefficients and G_2 Racah coefficients); thus, it is plausible that propagation equations may be useful in quantifying geometric chaos. This is being investigated, and it should be remarked that only recently has the role of Wigner-Racah algebra in two-body random matrix ensembles been established [31].

Thanks are due to Y. M. Zhao for useful correspondence and for making available Ref. [12] before it was submitted for publication.

[1] C. W. Johnson, G. F. Bertsch, and D. J. Dean, Phys. Rev. Lett. **80**, 2749 (1998).
[2] R. Bijker and A. Frank, Phys. Rev. Lett. **84**, 420 (2000).
[3] V. Velázquez, J. G. Hirsch, A. Frank, and A. P. Zuker, Phys. Rev. C **67**, 034311 (2003); Y. M. Zhao, A. Arima, N. Shimizu, K. Ogawa, N. Yoshinaga, and O. Scholten, *ibid.* **70**, 054322 (2004).
[4] D. Mulhall, A. Volya, and V. Zelevinsky, Phys. Rev. Lett. **85**, 4016 (2000).
[5] V. K. B. Kota and K. Kar, Phys. Rev. E **65**, 026130 (2002).
[6] V. Zelevinsky and A. Volya, Phys. Rep. **391**, 311 (2004).
[7] R. Bijker and A. Frank, Phys. Rev. C **64**, 061303(R) (2001).
[8] Y. M. Zhao, A. Arima, and N. Yoshinaga, Phys. Rev. C **66**, 034302 (2002).
[9] V. K. B. Kota, High Energy Phys. Nucl. Phys. (China) **28**, 1307 (2004); nucl-th/0401038.
[10] Y. M. Zhao, A. Arima, and N. Yoshinaga, Phys. Rep. **400**, 1 (2004).
[11] Y. M. Zhao, A. Arima, and N. Yoshinaga, Phys. Rev. C **66**, 064323 (2002).
[12] Y. M. Zhao, A. Arima, and K. Ogawa, Phys. Rev. C **71**, 017304 (2005).
[13] T. Papenbrock and H. A. Weidenmüller, Phys. Rev. Lett. **93**, 132503 (2004).
[14] J. B. French, in *Isospin in Nuclear Physics*, edited by D. H. Wilkinson (North Holland, Amsterdam, 1969), p. 259; F. S. Chang, J. B. French, and T. H. Thio, Ann. Phys. (NY) **66**, 137 (1971).
[15] J. C. Parikh, *Group Symmetries in Nuclear Structure* (Plenum, New York, 1978); S. S. M. Wong, *Nuclear Statistical Spectroscopy* (Oxford University Press, New York, 1986).
[16] C. Quesne, J. Math. Phys. **16**, 2427 (1975).
[17] J. P. Draayer and G. Rosensteel, Nucl. Phys. **A439**, 61 (1985).
[18] V. K. B. Kota and K. Kar, Pramana J. Phys. **32**, 647 (1989).
[19] J. B. French and J. P. Draayer, in *Group Theoretical Methods in Physics*, edited by W. Beiglbock *et al.* (Springer, Berlin, 1979), p. 394.
[20] V. K. B. Kota, J. Physique-LETTRES **40**, L579 (1979); **41**, L143 (1980).
[21] V. K. B. Kota, Ann. Phys. (NY) **134**, 221 (1981).
[22] V. K. B. Kota, Ann. Phys. (NY) **280**, 1 (2000).
[23] Y. D. Devi and V. K. B. Kota, Physical Research Laboratory (Ahmedabad, India) Technical Report PRL-TN-90-68, 1990 (unpublished).
[24] C. Quesne and S. Spitz, Ann. Phys. (NY) **112**, 304 (1978).
[25] J. P. Elliott and A. P. White, Phys. Lett. **B97**, 169 (1980).
[26] J. P. Elliott, Proc. R. Soc. London Sec. A **245**, 128 (1958).
[27] V. K. B. Kota, Ann. Phys. (NY) **265**, 101 (1998).
[28] J. P. Elliott and J. A. Evans, Phys. Lett. **B101**, 216 (1981).
[29] B. G. Wybourne, Symmetry Principles and Atomic Spectroscopy (Wiley Interscience, New York, 1970).
[30] J. Engel and F. Iachello, Phys. Rev. Lett. **54**, 1126 (1985).
[31] V. K. B. Kota, J. Math. Phys. **46**, 033514 (2005); Z. Pluhar and H. A. Weidenmüller, Ann. Phys. (NY) **297**, 344 (2002).

JOURNAL OF MATHEMATICAL PHYSICS **46**, 033514 (2005)

SU(*N*) Wigner–Racah algebra for the matrix of second moments of embedded Gaussian unitary ensemble of random matrices

V. K. B. Kota
Physical Research Laboratory, Ahmedabad 380 009, India

(Received 20 July 2004; accepted 3 November 2004; published online 16 February 2005)

Recently Pluhar and Weidenmüller [Ann. Phys. (N.Y.) **297**, 344 (2002)] showed that the eigenvectors of the matrix of second moments of embedded Gaussian unitary ensemble of random matrices generated by k-body interactions [EGUE(k)] for m fermions in N single particle states are SU(N) Wigner coefficients and derived also an expression for the eigenvalues. Going beyond this work, we will show that the eigenvalues of this matrix are square of a SU(N) Racah coefficient and thus the matrix of second moments of EGUE(k) is solved completely by SU(N) Wigner–Racah algebra. © *2005 American Institute of Physics.*
[DOI: 10.1063/1.1850179]

I. INTRODUCTION

Interacting finite quantum systems such as nuclei, atoms, quantum dots, nanometer-scale metallic grains, etc., are governed by Hamiltonians of low (compared to the number of particles in the system) particle rank. Therefore, for many purposes, the random matrix models appropriate for these systems are embedded random matrix ensembles of k-body interactions originally introduced, via nuclear shell model calculations, by French and Wong and Bohigas and Flores.[1] For a system of m spinless fermions in N single particle states (we will use fermions throughout this paper and turn briefly to bosons at the end) the embedded Gaussian unitary ensemble of random matrices of k-body interactions [EGUE(k)] is generated by defining the Hamiltonian H, which is given to be k-body, to be GUE in the k-particle spaces and then propagating it to the m particle spaces by using the geometry (direct product structure) of the m-particle spaces. Just as EGUE(k), the EGOE(k) and other embedded ensembles are defined.[2] With m particle space dimension given by $N_m=\binom{N}{m}$, one has the unitary groups SU(N), U(N_k), and U(N_m) with EGUE(k) invariant under U(N_k) and the embedding in m-particle spaces is defined by SU(N); note that a GUE in m particle spaces is invariant under U(N_m) but not the EGUE(k), $k<m$. Very early, using the so-called binary correlation approximation, Mon and French[3] and later French *et al.*[4] derived some analytical properties of embedded ensembles valid in the dilute limit [defined by $(N,m,k)\to\infty$, $m/N\to 0$ and $k/m\to 0$]. However only recently rigorous analytical results, valid for any (N,m,k) are derived for these ensembles by Benet *et al.*[5] and very soon Pluhar and Weidenmüller (hereafter called PW) demonstrated[6] that these results indeed follow from considerations based on the SU(N) embedding algebra. With all the m-particle matrix elements being linear combinations of the k-particle matrix elements [see Eq. (9) ahead], the joint distribution for the matrix elements will be a multivariate Gaussian. Thus all the information about EGUE(k) is in the covariance matrix or the matrix of second moments [Eq. (10) ahead]. PW have shown that the eigenvectors of this matrix are SU(N) Wigner [or Clebsch–Gordon (CG)] coefficients and derived the expression for their eigenvalues using a duality relation for EGUE(k). The purpose of this paper is to show that the eigenvalues can be written as SU(N) Racah coefficients and thus the matrix of second moments is solved completely by SU(N) Wigner–Racah algebra. To this end results for SU(N) Racah coefficients given in Refs. 7–9 are used. We will start with some basic results given in the PW paper.

II. BASIC DEFINITIONS AND RESULTS

Let us begin with m particles in N single particle states (unfortunately in PW ℓ is used in place of N but to keep the notations the same as in our earlier papers,[2,10] we use N). The single particle (sp) creation operator $a_i^\dagger$ for any ith sp state transforms as the irreducible representation $\{1\}$ of U(N) and similarly a product of r creation operators transform, as we have fermions, as the irrep $\{1^r\}$ in Young tableaux notation. Let us add that a U(N) irrep $\{\lambda_1,\lambda_2,\ldots,\lambda_N\}$ defines the corresponding SU(N) irrep as $\{\lambda_1-\lambda_N,\lambda_2-\lambda_N,\ldots,\lambda_{N-1}-\lambda_N\}$ with $N-1$ rows [there are also other equivalent ways of defining SU(N) irreps given a U(N) irrep (Ref. 8)]. This U(N)$\leftrightarrow$SU(N) correspondence is used throughout and therefore we use U(N) and SU(N) interchangeably. In PW, $\{1^r\}$ is denoted by f_r and we will follow this notation from now on. With v_r denoting irreps (and other multiplicity labels) of the groups in a subgroup chain of U(N) that supply the labels needed for a complete specification of any m-particle state (for the purpose of the present paper the subgroup chain need not be specified), the operator $\Pi_{i=1}^{r} a_i^\dagger$ and a normalized r-particle creation operator $A^\dagger(f_r v_r)$ behave as the SU(N) tensors $T^{f_r v_r}$ and $(1/\sqrt{r!})T^{f_r v_r}$, respectively. Using the composition formula,

$$T^{f_m v_m} = \sum_{v_k, v_s} C^{f_m v_m}_{f_k v_k\, f_s v_s} T^{f_k v_k} T^{f_s v_s}, \quad s = m-k, \tag{1}$$

where $C^{f_m v_m}_{f_k v_k f_s v_s}$ is a SU(N) CG coefficient, a m-particle state $|f_m v_m\rangle = A^\dagger(f_m v_m)|0\rangle$ can be written as a product of k and $s=m-k$ particle states as

$$|f_m v_m\rangle = \binom{m}{k}^{-1/2} \sum_{v_k v_s} A^\dagger(f_k v_k)|f_s v_s\rangle C^{f_m v_m}_{f_k v_k\, f_s v_s}. \tag{2}$$

Some properties of the CG coefficients, used later in simplifications, are

$$C^{f_{ab}v_{ab}}_{f_a v_a f_b v_b} = (-1)^{\phi(f_a,f_b,f_{ab})} C^{f_{ab}v_{ab}}_{f_b v_b\, f_a v_a}, \quad C^{f_{ab}v_{ab}}_{f_a v_a\, f_b v_b} = C^{\overline{f_{ab}v_{ab}}}_{\overline{f_a v_a}\,\overline{f_b v_b}},$$

$$C^{f_{ab}v_{ab}}_{f_a v_a\, f_b v_b} = (-1)^{\phi(f_a,f_b,f_{ab})} \sqrt{\frac{d(f_{ab})}{d(f_a)}}\, C^{f_a v_a}_{f_{ab}v_{ab}\, \overline{f_b v_b}},$$

$$C^{f_a v_a}_{00\, f_a v_a} = 1, \quad C^{00}_{f_a v_a\, \overline{f_a v_a}} = \frac{1}{\sqrt{d(f_a)}}, \quad \left(C^{f_{ab}v_{ab}}_{f_a v_a\, \overline{f_a v_b}}\right)^* = C^{f_{ab}v_{ab}}_{f_a v_b\, \overline{f_a v_a}},$$

$$\sum_{v_a, v_b} \left(C^{f_{ab}v_{ab}}_{f_a v_a\, \overline{f_a v_b}}\right)^* C^{f'_{ab}v'_{ab}}_{f_a v_a\, \overline{f_a v_b}} = \delta_{f_{ab} f'_{ab}}\, \delta_{v_{ab} v'_{ab}}. \tag{3}$$

In (3), ϕ is a function that defines the phase for the $1\leftrightarrow 2$ interchange in the CG coefficients, $d(f)$ is the dimension of the irrep f and $\overline{f}$ is the irrep conjugate to f. For $f_r=\{1^r\}$, $\overline{f_r}=\{1^{N-r}\}$ and it also contains a phase factor as given in Eq. (53) of Ref. 7 [this is also seen easily by comparing the second and third equalities in (3) with the corresponding relations for the standard CG coefficients for angular momentum[11]]. Similarly with $f_a=f_k$, $f_b=\overline{f_k}$, one has

$$f_{ab} = g_\nu = \{2^\nu 1^{N-2\nu}\}, \quad \nu = 0,1,\ldots,k.$$

Note that $g_0=\{0\}$ for SU(N) and also $g_\nu=\overline{g_\nu}$. The function ϕ in Eq. (3) is of the form $\phi(\lambda_1,\lambda_2,\lambda_3)=F(\lambda_1)+F(\lambda_2)+F(\lambda_3)$ where F is some function with $F(\lambda)$ an integer and $F(\lambda)=F(\overline{\lambda})$; for SU($N$) irreps that appear in this paper these results are valid as can be seen from Eq. (60) of Ref. 7 except that there can be an overall N dependent factor which will not change any of the final results.

SU(N) irreducible tensors $B_k(g_\nu \omega_\nu)$ constructed out of $A^\dagger(f_k v_k)A(f_k v'_k)$ are defined by

$$B_k(g_\nu\omega_\nu) = \sum_{v_k,v'_k} A^\dagger(f_k v_k) A(f_k v'_k) C^{g_\nu\omega_\nu}_{f_k v_k\, \overline{f_k v'_k}}. \tag{4}$$

It is useful to note that the tensors B's in (4) multiplied by $k!$ are, to within a phase factor, the same as the tensors defined in Eq. (48) of Ref. 7. Wigner–Eckart theorem decomposes the matrix elements of $B_k(g_\nu\omega_\nu)$ in m-particle spaces into a reduced matrix element $\langle\| \,\|\rangle$ and a CG coefficient,

$$\langle f_m v_m | B_k(g_\nu\omega_\nu) | f_m v'_m\rangle = \langle f_m \| B_k(g_\nu) \| f_m\rangle C^{g_\nu\omega_\nu}_{f_m v_m\, \overline{f_m v'_m}}. \tag{5}$$

Two important properties of $B_k(g_\nu\omega_\nu)$ are

$$\langle\langle B_k(g_\nu\omega_\nu)\rangle\rangle^k = \sqrt{N_k}\,\delta_{g_\nu,\{0\}},$$

$$\langle\langle B_k(g_\nu\omega_\nu) B_k(g_\mu\omega_\mu)\rangle\rangle^k = \delta_{g_\nu g_\mu}\delta_{\omega_\nu\omega_\mu}. \tag{6}$$

In (6), $\langle\langle\ \rangle\rangle^k$ denotes trace over the k-particle spaces. The first equality in Eq. (6) easily follows from the fact that here $g_\nu=\{0\}$ as traces are scalars with respect to SU(N) and then applying the fifth equality in Eq. (3). Similarly the second equality in Eq. (6) follows from the fact that only $g_\nu=g_\mu$ will give a scalar. Now we will turn to EGOE(k).

Consider a k-body Hamiltonian,

$$H(k) = \sum_{v_a,v_b} V_{v_a v_b}(k) A^\dagger(f_k v_a) A(f_k v_b), \tag{7}$$

where $V_{v_a v_b}(k)$ are matrix elements of $H(k)$ in k-particle space and form a GUE, i.e., $V_{v_a v_b}(k)$ are independent Gaussian variables with zero center and variance given by

$$\overline{V_{v_a v_b}(k) V_{v_c v_d}(k)} = \frac{\lambda^2}{N_k}\delta_{v_a v_d}\delta_{v_b v_c}. \tag{8}$$

In Eq. (8) the overline indicates ensemble average and λ^2 is ensemble averaged variance of $H(k)$ in k-particle space. The m-particle matrix elements of $H(k)$ are, with $s=m-k$,

$$H_{v^1_m v^2_m}(k) = \langle f_m v^1_m | H(k) | f_m v^2_m\rangle = \binom{m}{k} \sum_{v_a,v_b,v_s} (C^{f_m v^1_m}_{f_k v_a f_s v_s})^* C^{f_m v^2_m}_{f_k v_b f_s v_s} V_{v_a v_b}(k). \tag{9}$$

Equation (9) is obtained easily by substituting the definition (7) for $H(k)$, then inserting the complete set of states between $A^\dagger$ and A operators and applying Eq. (2). The EGUE(k) in $m \geq k$ spaces is defined by Eqs. (7)–(9). Now it is clear that for any analysis of EGUE(k) all one needs to know is the covariance between any two m-particle matrix elements $H_{v^1_m v^2_m}(k)$ and this defines the matrix of second moments,

$$A_{v^1_m v^4_m : v^3_m v^2_m} = \overline{H_{v^1_m v^2_m}(k) H_{v^3_m v^4_m}(k)}. \tag{10}$$

As stressed by PW, the most important step in EGUE(k) analysis is to derive a "generalized eigenvalue expansion" of A defined by $A_{ij}=\Sigma_k C_{ik} E_k C_{jk}$ with E_k the eigenvalues and C_{jk} the eigenvectors such that E_k are positive and C's Hermitian. To this end, it is useful to consider the unitary decomposition of $H(k)$ in terms of the SU(N) tensors $B_k(g_\nu\omega_\nu)$,

$$H(k) = \sum_{g_\nu,\omega_\nu} B_k(g_\nu\omega_\nu) W_{g_\nu\omega_\nu}(k). \tag{11}$$

The expansion coefficients $W_{g_\nu\omega_\nu}(k)$ are easily given by

$$W_{g_\nu\omega_\nu}(k) = \langle\langle H(k) B_k(g_\nu\omega_\nu)\rangle\rangle^k \tag{12}$$

and this follows by using the definition (11) and Eq. (6). The most significant property of the W coefficients is that they are independent Gaussian variables with zero center and variance given by [derived using Eqs. (11), (4), (7), and (8) in that order and using the orthonormal properties of the CG coefficients]

$$\overline{W_{g_\nu\omega_\nu}(k) W_{g_\mu\omega_\mu}(k)} = \frac{\lambda^2}{N_k} \delta_{g_\nu g_\mu} \delta_{\omega_\nu\omega_\mu}. \tag{13}$$

III. MATRIX OF SECOND MOMENTS

First we will derive an expression for the covariance $\overline{H_{v^1_m v^2_m}(k) H_{v^3_m v^4_m}(k)}$ in terms of SU(N) CG coefficients and Racah coefficients and then turn to the eigenvalues and eigenvectors of A, the matrix of second moments. Applying Eqs. (10), (11), and (5) in that order gives

$$\begin{aligned}
\overline{H_{v^1_m v^2_m}(k) H_{v^3_m v^4_m}(k)} &= \sum_{g_\nu,\omega_\nu,g_\mu,\omega_\mu} \overline{\langle f_m v^1_m | B_k(g_\nu\omega_\nu) W_{g_\nu\omega_\nu}(k) | f_m v^2_m\rangle \langle f_m v^3_m | B_k(g_\mu\omega_\mu) W_{g_\mu\omega_\mu}(k) | f_m v^4_m\rangle} \\
&= \sum_{g_\nu,\omega_\nu,g_\mu,\omega_\mu} \overline{W_{g_\nu\omega_\nu}(k) W_{g_\mu\omega_\mu}(k)} \langle f_m v^1_m | B_k(g_\nu\omega_\nu) | f_m v^2_m\rangle \langle f_m v^3_m | B_k(g_\mu\omega_\mu) | f_m v^4_m\rangle \\
&= \frac{\lambda^2}{N_k} \sum_{g_\nu,\omega_\nu} \langle f_m v^1_m | B_k(g_\nu\omega_\nu) | f_m v^2_m\rangle \langle f_m v^3_m | B_k(g_\nu\omega_\nu) | f_m v^4_m\rangle \\
&= \frac{\lambda^2}{N_k} \sum_{g_\nu,\omega_\nu} |\langle f_m \| B_k(g_\nu) \| f_m\rangle|^2 C^{g_\nu\omega_\nu}_{f_m v^1_m \overline{f_m v^2_m}} C^{g_\nu\omega_\nu}_{f_m v^3_m \overline{f_m v^4_m}}.
\end{aligned} \tag{14}$$

Equations (55) and (56) of Ref. 7 together with Eqs. (3) [see also the remark just after Eq. (4)] allows one to write the reduced matrix element in Eq. (14) as a SU(N) Racah or U-coefficient,

$$|\langle f_m \| B_k(g_\nu) \| f_m\rangle|^2 = \frac{(N_m)^2 \binom{m}{k}^2}{d(g_\nu)(N_{m-k})} [\mathrm{U}(f_m f_{N-k} f_m f_k; f_{m-k} g_\nu)]^2. \tag{15}$$

With $\{2^\nu 1^{N-2\nu}\} = \{1^\nu\} \otimes \{1^{N-\nu}\} - \{1^{\nu-1}\} \otimes \{1^{N-\nu+1}\}$ where $\otimes$ denotes Kronecker product, the dimension $d(g_\nu)$ is given by

$$d(g_\nu) = d(\nu) = (N_\nu)^2 - (N_{\nu-1})^2 = \frac{(N!)^2 (N+1)(N-2\nu+1)}{(\nu!)^2 (N-\nu+1!)^2}. \tag{16}$$

Before going further let us define, for a given (m,N), a function $\Lambda^\nu(k)$,

$$\Lambda^\nu(k) = \binom{m-\nu}{k} \binom{N-m+k-\nu}{k}. \tag{17}$$

Now, using Eq. (16) and substituting the expression given by Eq. (61) of Ref. 7 for the U-coefficient in (15), it is seen that

$$|\langle f_m||B_k(g_\nu)||f_m\rangle|^2=\Lambda^\nu(m-k),\quad \nu=0,1,2,\ldots,k. \tag{18}$$

Combining (14) with (18) yields an expression for the covariance between H matrix elements in m-particle spaces,

$$\begin{aligned}\overline{H_{v_m^1v_m^2}(k)H_{v_m^3v_m^4}(k)}&=\frac{\lambda^2}{N_k}\sum_{\nu=0,1,\ldots,k;\omega_\nu}\frac{(N_m)^2\binom{m}{k}^2}{d(g_\nu)(N_{m-k})}[\mathrm{U}(f_mf_{N-k}f_mf_k;f_{m-k}g_\nu)]^2C^{g_\nu\omega_\nu}_{f_mv_m^1\overline{f_m}\overline{v_m^2}}C^{g_\nu\omega_\nu}_{f_mv_m^3\overline{f_m}\overline{v_m^4}}\\&=\frac{\lambda^2}{N_k}\sum_{\nu=0,1,\ldots,k;\omega_\nu}\{\Lambda^\nu(m-k)\}C^{g_\nu\omega_\nu}_{f_mv_m^1\overline{f_m}\overline{v_m^2}}C^{g_\nu\omega_\nu}_{f_mv_m^3\overline{f_m}\overline{v_m^4}}.\end{aligned} \tag{19}$$

Equation (19) will be useful in deriving expressions for the moments of H spectrum, i.e., $\overline{\langle H^p\rangle^m}$. However the disadvantage of (19) is that it is not in a proper form to give the eigenvalues and eigenvectors of the matrix A in Eq. (10). In order to obtain them, the CG coefficients in Eq. (14) should be changed to $C^{g_\mu\omega_\mu}_{f_mv_m^1\overline{f_m}\overline{v_m^4}}C^{g_\mu\omega_\mu}_{f_mv_m^3\overline{f_m}\overline{v_m^2}}$ and this can be accomplished by a SU(N) Racah transform. Using Eq. (3.2.17) of Ref. 9 one has, for example,

$$\sum_{\omega_\nu}C^{g_\nu\omega_\nu}_{f_mv_m^1\overline{f_m}\overline{v_m^2}}C^{f_mv_m^4}_{g_\nu\omega_\nu f_mv_m^3}=\sum_{g_\mu\omega_\mu}\mathrm{U}(f_m\overline{f_m}f_mf_m;g_\nu g_\mu)C^{g_\mu\omega_\mu}_{\overline{f_m}\overline{v_m^2}f_mv_m^3}C^{f_mv_m^4}_{f_mv_m^1g_\mu\omega_\mu}. \tag{20}$$

Then Eqs. (20) and (3) will give

$$\sum_{\omega_\nu}C^{g_\nu\omega_\nu}_{f_mv_m^1\overline{f_m}\overline{v_m^2}}C^{g_\nu\omega_\nu}_{f_mv_m^3\overline{f_m}\overline{v_m^4}}=\sum_{g_\mu\omega_\mu}\sqrt{\frac{d(g_\nu)}{d(g_\mu)}}\mathrm{U}(f_m\overline{f_m}f_mf_m;g_\nu g_\mu)C^{g_\mu\omega_\mu}_{f_mv_m^1\overline{f_m}\overline{v_m^4}}C^{g_\mu\omega_\mu}_{f_mv_m^3\overline{f_m}\overline{v_m^2}}. \tag{21}$$

Finally Eqs. (21) and (15) combined with (14) produce the generalized eigenvalue expansion of the matrix of second moments A,

$$\begin{aligned}A_{v_m^1v_m^4:v_m^3v_m^2}=\overline{H_{v_m^1v_m^2}(k)H_{v_m^3v_m^4}(k)}=\sum_{g_\mu\omega_\mu}C^{g_\mu\omega_\mu}_{f_mv_m^1\overline{f_m}\overline{v_m^4}}C^{g_\mu\omega_\mu}_{f_mv_m^3\overline{f_m}\overline{v_m^2}}\Bigg\{\lambda^2\frac{(N_m)^2\binom{m}{k}^2}{N_k\,(N_{m-k})}\Bigg[\sum_{g_\nu}\sqrt{\frac{1}{d(g_\nu)d(g_\mu)}}\\\times[\mathrm{U}(f_mf_{N-k}f_mf_k;f_{m-k}g_\nu)]^2\mathrm{U}(f_m\overline{f_m}f_mf_m;g_\nu g_\mu)\Bigg]\Bigg\}.\end{aligned} \tag{22}$$

Obviously the quantity in the curly brackets in Eq. (22) gives the eigenvalues of A and the C's are eigenvectors.

IV. EIGENVALUES AS SU(N) RACAH COEFFICIENTS AND THEIR APPLICATIONS

In order to proceed further, it is useful to consider $6j$ symbols of SU(N) and they are defined by [see Eq. (3.2.18) of Ref. 9],

$$\mathrm{U}(\lambda_1\lambda_2\lambda\lambda_3;\lambda_{12}\lambda_{23})=\sqrt{d(\lambda_{12})d(\lambda_{23})}(-1)^{\phi(\lambda_2,\overline{\lambda_2},0)+\phi(\lambda_{12},\lambda_3,\overline{\lambda})+\phi(\lambda_1,\lambda_2,\overline{\lambda_{12}})}\begin{Bmatrix}\lambda_1&\lambda_{23}&\overline{\lambda}\\\overline{\lambda_3}&\lambda_{12}&\lambda_2\end{Bmatrix}. \tag{23}$$

In (23) λ's are SU(N) irreps and the four couplings involved in the U-coefficient are assumed to be multiplicity free (for the applications in the present paper this assumption is always valid). Symmetry properties of the $6j$-symbol appearing on the right-hand side of (23) are well known.[8,9] In the present analysis, the Biedenharn–Elliott sum rule extended to SU(N) (Refs. 8 and 9) plays a central role. This sum rule relates a product of three Racah coefficients (weighted appropriately by dimension factors and phase factors with the irreps in the Racah coefficients appearing in some particular order) with sum over a common irrep label to a product of two Racah coefficients. After converting the Racah coefficients in (22) into $6j$ symbols of SU(N) using Eq. (23) and then applying the symmetry properties of the $6j$ symbols, it is seen that the sum in the square brackets in Eq. (22) is exactly in the required form. Applying the Biedenharn–Elliott sum rule, the sum then simplifies to

$$\frac{N_{m-k}}{N_k d(g_\mu)} U^2(f_m f_{N-m+k} f_m f_{m-k}; f_k g_\mu).$$

Here $\mu=0,1,\ldots,m-k$. Now the eigenvalues of the matrix A, in terms of the U(N) Racah coefficients is given by

$$E_\mu = \frac{\lambda^2}{N_k} \frac{(N_m)^2 \binom{m}{k}^2}{d(g_\mu)(N_k)} [\mathrm{U}(f_m f_{N-m+k} f_m f_{m-k}; f_k g_\mu)]^2, \quad \mu = 0,1,\ldots,m-k \tag{24}$$

with degeneracy $d(g_\mu)$ [see Eq. (16)]. Equation (24) is the central result of this paper. With this, the matrix A is completely specified by the U(N) Wigner and Racah coefficients. Now substituting the formula [Eq. (61) of Ref. 7] for the U-coefficients in Eq. (24) produces the result of PW,

$$A_{v_m^1 v_m^4 : v_m^3 v_m^2} = \overline{H_{v_m^1 v_m^2}(k) H_{v_m^3 v_m^4}(k)} = \sum_{g_\mu \omega_\mu} C^{g_\mu \omega_\mu}_{f_m v_m^1 \overline{f_m} \overline{v_m^4}} C^{g_\mu \omega_\mu}_{f_m v_m^3 \overline{f_m} \overline{v_m^2}} E_\mu, \tag{25}$$

where

$$E_\mu = \frac{\lambda^2}{N_k} \Lambda^\mu(k), \quad \mu = 0,1,\ldots,m-k. \tag{26}$$

Note that the function $\Lambda^\mu(k)$ is defined by Eq. (17).

Information about EGUE(k) is contained in the ensemble averaged moments $M_p=\overline{\langle H^p \rangle^m}$ and the bivariate moments $\Sigma_{pq}=\overline{\langle H^p \rangle^m \langle H^q \rangle^m}$. In deriving the formulas for the lower order moments, we will show the usefulness of Eq. (19). Obviously, ensemble averaged centroid is zero and the variance is

$$\overline{\langle H^2 \rangle^m} = \frac{1}{N_m} \sum_{v_m^i, v_m^j} \overline{H_{v_m^i v_m^j} H_{v_m^j v_m^i}} = \frac{1}{N_m} \sum_{g_\mu, \omega_\mu} \Lambda^\mu(k) \sum_{v_m^i, v_m^j} C^{g_\mu \omega_\mu}_{f_m v_m^i \overline{f_m} \overline{v_m^i}} C^{g_\mu \omega_\mu}_{f_m v_m^j \overline{f_m} \overline{v_m^j}} = \Lambda^0(k). \tag{27}$$

The second equality follows from (25) and the final result follows by applying (3); note that $\Sigma_{v_m^i} C^{g_\mu \omega_\mu}_{f_m v_m^i \overline{f_m} \overline{v_m^i}} = \sqrt{N_m} \delta_{\mu,0}$. The variance in (27) is in λ^2/N_k units and this factor is dropped as all the quantities we consider from now on are all scaled with respect to $\{\overline{\langle H^2 \rangle^m}\}^{1/2}$. As the third moment is zero, we will turn to the fourth moment,

$$\begin{aligned}
\overline{\langle H^4 \rangle^m} &= \frac{1}{N_m} \sum_{v_m^i, v_m^j, v_m^{k'}, v_m^l} \overline{H_{v_m^i v_m^j} H_{v_m^j v_m^{k'}} H_{v_m^{k'} v_m^l} H_{v_m^l v_m^i}} \\
&= \frac{1}{N_m} \sum_{v_m^i, v_m^j, v_m^{k'}, v_m^l} \Bigg\{ 2 \Bigg[\sum_{g_\nu, \omega_\nu} \langle f_m v_m^i | B_k(g_\nu \omega_\nu) | f_m v_m^j \rangle \langle f_m v_m^j | B_k(g_\nu \omega_\nu) | f_m v_m^{k'} \rangle \Bigg] \\
&\quad \times \Bigg[\sum_{g_\mu, \omega_\mu} \langle f_m v_m^{k'} | B_k(g_\mu \omega_\mu) | f_m v_m^l \rangle \langle f_m v_m^l | B_k(g_\mu \omega_\mu) | f_m v_m^i \rangle \Bigg] + \Bigg[\sum_{g_\nu, \omega_\nu} \langle f_m v_m^i | B_k(g_\nu \omega_\nu) | f_m v_m^j \rangle \\
&\quad \times \langle f_m v_m^{k'} | B_k(g_\nu \omega_\nu) | f_m v_m^l \rangle \Bigg] \Bigg[\sum_{g_\mu, \omega_\mu} \langle f_m v_m^j | B_k(g_\mu \omega_\mu) | f_m v_m^{k'} \rangle \langle f_m v_m^l | B_k(g_\mu \omega_\mu) | f_m v_m^i \rangle \Bigg] \Bigg\}
\end{aligned}$$

$$= 2[\Lambda^0(k)]^2 + \frac{1}{N_m} \sum_{v_m^i, v_m^j, v_m^{k'}, v_m^l} \left\{ \sum_{\nu=0,1,\ldots,k;\omega_\nu} \{\Lambda^\nu(m-k)\} C^{g_\nu \omega_\nu}_{f_m v_m^i \overline{f_m} \overline{v_m^j}} C^{g_\nu \omega_\nu}_{f_m v_m^{k'} \overline{f_m} \overline{v_m^l}} \right\}$$

$$\times \left\{ \sum_{\mu=0,1,\ldots,m-k;\omega_\mu} \{\Lambda^\mu(k)\} C^{g_\mu \omega_\mu}_{f_m v_m^j \overline{f_m} \overline{v_m^i}} C^{g_\mu \omega_\mu}_{f_m v_m^l \overline{f_m} \overline{v_m^{k'}}} \right\}$$

$$= 2[\Lambda^0(k)]^2 + \frac{1}{N_m} \sum_{\nu=0}^{\min\{k,m-k\}} \Lambda^\nu(m-k)\Lambda^\nu(k)d(\nu). \tag{28}$$

The second equality in Eq. (28) follows by applying Eqs. (11) and (13). In the third equality, it is easy to recognize the first term. The second term follows by applying Eqs. (19) and (25) to the two pieces in the corresponding term in the second equality. The final result follows by applying the orthonormality of the CG coefficients. Equations (27) and (28) will give the excess (γ_2) parameter of the density of eigenvalues of EGUE(k),

$$\gamma_2 = \frac{\overline{\langle H^4 \rangle^m}}{[\overline{\langle H^2 \rangle^m}]^2} - 3 = \left[\frac{1}{N_m} \sum_{\nu=0}^{\min\{k,m-k\}} \frac{\Lambda^\nu(m-k)\Lambda^\nu(k)d(\nu)}{[\Lambda^0(k)]^2} \right] - 1. \tag{29}$$

Turning now to the lowest bivariate moment Σ_{11}, it is easily seen that

$$\Sigma_{11} = \overline{\langle H \rangle^m \langle H \rangle^m} = \frac{1}{(N_m)^2} \sum_{v_m^i, v_m^j} \overline{H_{v_m^i v_m^i} H_{v_m^j v_m^j}} = \frac{1}{N_m} \Lambda^0(m-k). \tag{30}$$

Applying (19) and recognizing that only $\nu=0$ will contribute to the traces give immediately Eq. (30). However Eq. (25) generates a different formula for Σ_{11} and equating it to (30) gives the identity (derived in PW using the duality transformation),

$$\frac{1}{N_m} \sum_{\nu=0}^{m-k} \Lambda^\nu(k)d(\nu) = \Lambda^0(m-k).$$

The variance of the distribution of centroids of the H spectra over the EGUE(k) ensemble is

$$\hat{\Sigma}_{11} = \frac{\Sigma_{11}}{\overline{\langle H^2 \rangle^m}} = \frac{1}{N_m} \frac{\Lambda^0(m-k)}{\Lambda^0(k)}. \tag{31}$$

Finally Σ_{22} is given by

$$\Sigma_{22} = \overline{\langle H^2 \rangle^m \langle H^2 \rangle^m} = \frac{1}{(N_m)^2} \sum_{v_m^i, v_m^j, v_m^{k'}, v_m^l} \overline{|H_{v_m^i v_m^j}|^2 |H_{v_m^{k'} v_m^l}|^2}$$

$$= \frac{1}{(N_m)^2} \sum_{v_m^i, v_m^j, v_m^{k'}, v_m^l} \overline{|H_{v_m^i v_m^j}|^2}\, \overline{|H_{v_m^{k'} v_m^l}|^2} + \frac{2}{(N_m)^2} \sum_{v_m^i, v_m^j, v_m^{k'}, v_m^l} [\overline{H_{v_m^i v_m^j} H_{v_m^{k'} v_m^l}}]^2$$

$$= [\Lambda^0(k)]^2 + \frac{2}{(N_m)^2} \sum_{\nu=0}^{m-k} [\Lambda^\nu(k)]^2 d(\nu). \tag{32}$$

Here in the second equality used is the property $\overline{x^2y^2} = (\overline{x^2})(\overline{y^2}) + 2[\overline{xy}]^2$ of Gaussian variables x and y. Similarly the final result follows by applying (25) to the second term in the second equality and simplifying the CG coefficients using Eq. (3). Now, the variance of the distribution of the variances of the H spectra over the EGUE(k) ensemble is

$$\hat{\Sigma}_{22}=\frac{\Sigma_{22}}{[\langle H^2\rangle^m]^2}-1=\frac{2}{(N_m)^2}\sum_{\nu=0}^{m-k}\left[\frac{\Lambda^\nu(k)}{\Lambda^0(k)}\right]^2 d(\nu). \tag{33}$$

Equations (27), (29), (31), and (33) are also given by Benet *et al.*;[5] this paper neither gives details of the derivations nor uses SU(N) Racah coefficients. Also in this work, $\hat{\Sigma}_{11}$, $\hat{\Sigma}_{22}$, and γ_2+1 are denoted by S, R, and Q, respectively, while we have followed Ref. 3.

V. CONCLUSIONS

Going beyond PW, matrix elements of the matrix of second moments are written explicitly in terms of SU(N) Wigner and Racah coefficients and this result is obtained by recognizing that the reduced matrix elements of $B_k(g_\nu)$ are SU(N) Racah coefficients. With this one has Eq. (19) and this is converted into the generalized eigenvalue expansion form by first applying a SU(N) Racah transform and then applying the Biedenharn–Elliott sum rule extended to SU(N). This gives the eigenvalues of the matrix of second moments explicitly in terms of SU(N) Racah coefficients [Eq. (24)]. The two different forms given by Eqs. (19) and (25) for the covariances of m-particle H matrix elements, give in a simple manner the formulas for the low order moments M_p that define the state density and the bivariate moments Σ_{pq} that give information about fluctuations.

Although EGUE(k) for only fermions is considered in this paper, all the results in fact translate to those of EGUE(k) for bosons by using the well known $N\to -N$ symmetry,[12,13] i.e., in the fermion results replace N by $-N$ and then take the absolute value of the final result. For example, the m boson space dimension is $d(m)=|\binom{-N}{m}|=\{(N-m+1)_m\}$. More importantly the eigenvalues of the matrix of the second moments are

$$\Lambda_B^\nu(k)\to\left|\binom{m-\nu}{k}\binom{-N-m+k-\nu}{k}\right|=\binom{m-\nu}{k}\binom{N+m+\nu-1}{k}. \tag{34}$$

This result was explicitly derived in Ref. 14; see Eq. (14) of this paper. Moreover for bosons, $\{k\}\otimes\{k^{N-1}\}\to g_\nu=\{2\nu,\nu^{N-2}\}$, $\nu=0,1,\ldots,k$. Also, the $N\to -N$ symmetry and Eq. (16) give $d(g_\nu)=\{(N+\nu-1)_\nu\}^2-\{(N+\nu-2)_{\nu-1}\}^2$ and this is same as Eq. (15) of Ref. 14. Similarly Eqs. (27), (29), (31), and (33) extend directly to the boson EGUE(k) with $\Lambda^\nu(k)$ replaced by $\Lambda_B^\nu(k)$ defined in Eq. (34). In addition, for fermions to bosons there is also a $m\leftrightarrow N$ symmetry and this connects fermion results (say for M_p and Σ_{pq}) in the dilute limit to boson results in the dense limit.[13]

Recently there is considerable interest in mesoscopic physics to study EGUE(k) for fermions with spin[15] and here the embedding algebra is $U(2N)\supset U(N)\otimes SU(2)$ with SU(2) generating spin. The approach presented in Secs. II–IV is being applied to this system; some useful results for the $U(2N)\supset U(N)\otimes SU(2)$ Wigner–Racah algebra are available in Refs. 16 and 17. Finally, Wigner–Racah algebra analysis of embedded ensembles with more general group symmetries (see Refs. 2, 6, 17, and 18 for examples) should be possible in the future, thus opening up a new direction in random matrix theory.

ACKNOWLEDGMENT

Thanks are due to H.A. Weidenmüller for reading the first draft of the paper and for his valuable comments.

[1] J. B. French and S. S. M. Wong, Phys. Lett. **33B**, 449 (1970); **35B**, 5 (1971); O. Bohigas and J. Flores, *ibid.* **34B**, 261 (1971); **35B**, 383 (1971).
[2] V. K. B. Kota, Phys. Rep. **347**, 223 (2001); L. Benet and H. A. Weidenmüller, J. Phys. A **36**, 3569 (2003).
[3] K. K. Mon and J. B. French, Ann. Phys. (N.Y.) **95**, 90 (1975); T. A. Brody, J. Flores, J. B. French, P. A. Mello, A. Pandey, and S. S. M. Wong, Rev. Mod. Phys. **53**, 385 (1981).
[4] J. B. French, V. K. B. Kota, A. Pandey, and S. Tomsovic, Ann. Phys. (N.Y.) **181**, 235 (1988).
[5] L. Benet, T. Rupp, and H. A. Weidenmüller, Phys. Rev. Lett. **87**, 010601 (2001); Ann. Phys. (N.Y.) **292**, 67 (2001).
[6] Z. Pluhar and H. A. Weidenmüller, Ann. Phys. (N.Y.) **297**, 344 (2002).
[7] K. T. Hecht, Commun. Math. Phys. **41**, 135 (1975).
[8] J. R. Derome and W. T. Sharp, J. Math. Phys. **6**, 1584 (1965); P. H. Butler, Philos. Trans. R. Soc. London **277**, 545

(1975); P. H. Butler and B. G. Wybourne, Int. J. Quantum Chem. **10**, 581 (1976).

[9] P. H. Butler, *Point Group Symmetry Applications: Methods and Tables* (Plenum, New York, 1981).

[10] V. K. B. Kota, Ann. Phys. (N.Y.) **306**, 58 (2003); V. K. B. Kota and R. Sahu, Phys. Rev. E **64**, 016219 (2001).

[11] A. R. Edmonds, *Angular Momentum in Quantum Mechanics*, 3rd ed. (Princeton University Press, Princeton, NJ, 1974).

[12] M. Nomura, Ann. Phys. (N.Y.) **91**, 83 (1975); V. K. B. Kota, *ibid.* **134**, 221 (1981); P. Cvitanovic and A. D. Kennedy, Phys. Scr. **26**, 5 (1982).

[13] V. K. B. Kota and V. Potbhare, Phys. Rev. C **21**, 2637 (1980); K. Patel, M. S. Desai, V. Potbhare, and V. K. B. Kota, Phys. Lett. A **275**, 329 (2000).

[14] T. Asaga, L. Benet, T. Rupp, and H. A. Weidenmüller, Ann. Phys. (N.Y.) **298**, 229 (2002).

[15] Ph. Jacquod and A. D. Stone, Phys. Rev. Lett. **84**, 3938 (2000); Phys. Rev. B **64**, 214416 (2001); T. Papenbrock, L. Kaplan, and G. F. Bertsch, *ibid.* **65**, 235120 (2002).

[16] K. T. Hecht and J. P. Draayer, Nucl. Phys. A **223**, 285 (1974).

[17] V. K. B. Kota and K. Kar, Phys. Rev. E **65**, 026130 (2002).

[18] T. Papenbrock and H. A. Weidenmüller, Phys. Rev. Lett. **93**, 132503 (2004).

PHYSICAL REVIEW C **73**, 034324 (2006)

Underlying symmetries of realistic interactions and the nuclear many-body problem

K. D. Sviratcheva,[1] J. P. Draayer,[1] and J. P. Vary[2]

[1]*Department of Physics and Astronomy, Louisiana State University, Baton Rouge, Louisiana 70803, USA*

[2]*Department of Physics and Astronomy, Iowa State University, Ames, Iowa 50011, USA*

(Received 5 December 2005; published 30 March 2006)

The present study brings forward important information, within the framework of spectral distribution theory, about the types of forces that dominate three realistic interactions, CD-Bonn, CD-Bonn+3terms, and GXPF1, in nuclei and their ability to account for many-particle effects such as the formation of correlated nucleon pairs and enhanced quadrupole collective modes. Like-particle and proton-neutron isovector pairing correlations are described microscopically by a model interaction with $\mathfrak{sp}(4)$ dynamical symmetry, which is extended to include an additional quadrupole-quadrupole interaction. The analysis of the results for the $1f_{7/2}$ level shows that both CD-Bonn+3terms and GXPF1 exhibit a well-developed pairing character compared to CD-Bonn, while the latter appears to build up more (less) rotational isovector $T = 1$ (isoscalar $T = 0$) collective features. Furthermore, the three realistic interactions are in general found to correlate strongly with the pairing+quadrupole model interaction, especially for the highest possible isospin group of states where the model interaction can be used to provide a reasonable description of the corresponding energy spectra.

DOI: 10.1103/PhysRevC.73.034324 PACS number(s): 21.60.Fw, 21.30.Fe, 21.10.Re, 21.60.Cs

I. INTRODUCTION

A unified microscopic description of light to heavy nuclei requires a comprehensive understanding of the strong interaction and how it manifests itself in the nuclear medium. Effective interaction theory attempts to model the essence of this strong interaction in terms of one-, two-, and sometimes higher (three- or even four-) body interactions for the purpose of supporting microscopic shell-model calculations that target reproducing striking features of nuclei. Such are strong pairing correlations found near closed shells that yield to collective rotational motion as one moves away from shell closure as well as more subtle effects that must be understood at a deeper level to reproduce, for example, nuclear abundances as realized through fast or slow decay processes within and between nuclear species. While good progress is being made toward understanding the strong force, especially through the recent work on lattice quantum chromodynamics studies, much work remains to be done.

Short such a comprehensive understanding of the strong interactions, one way to gain insight into the principal characteristics of various microscopic interactions is to perform a detailed study of their matrix elements or common quantities they generate, such as eigenvectors or eigenvalues. For instance, in the same environment two very similar interactions are expected to yield similar patterns of their matrix elements (see, for example, Refs. [1–3]), as well to yield large overlaps of corresponding eigenstates (e.g., Refs. [4,5]) and close energy spectra (e.g., Refs. [6,7]). On the other hand, a complementary comparison that is based on the theory of spectral distributions [8,9] that invokes overall correlations of two interactions offers a broader view on their global behavior and universal properties [10–14].

The theory of spectral distributions of French and collaborators is an alternative approach for studying effective interactions [8,15,16] and continues to be a powerful concept with recent applications in quantum chaos and nuclear astrophysics including studies on nuclear level densities, transition strength densities, and parity/time-reversal violation (for example, see Refs. [17–22]). The significance of the method is related to the fact that low-order energy moments over a certain domain of single-particle states, such as the energy centroid of an interaction (its average expectation value) and the deviation from that average, yield valuable information about the interaction that is of fundamental importance [11,16,23–29] without the need for carrying out large-dimensional matrix diagonalization and with little to no limitations due to the dimensionality of the vector space. Within this general framework, a simple and elegant global comparison of pairs of interactions can be carried in a unified way regardless of how the interactions are built or of the models that adopt them. It is based on their correlation, which is a measure that is independent of the centroids of the interactions. This correlation measure turns on a comparison of the one- and two-body parts of the interactions, and in so doing probes beyond the mean-field potential.

In addition, the group-theoretical foundation of the spectral distribution methods naturally establishes a propagation of information from nuclear systems of two particles to many-fermion nuclei [8,10] and makes the approach especially suitable for studies of the goodness/breaking of symmetries in complex many-particle systems [10,11,26]. Such studies can likewise help reduce the dimensionality of a model space to a tractable size by detecting the optimal subspace domain for a particular many-body problem where microscopic calculations become feasible [9,28,30].

In this paper we employ the theory of spectral distributions to compare three realistic interactions, namely CD-Bonn [31], CD-Bonn+3terms [32], and GXPF1 [2] and two pairing+quadrupole model interactions [33,34]. Such a study is important for understanding the types of forces that dominate a realistic interaction and its ability to describe correlated and collective phenomena. Most significant are the formation of nucleon pairs and quadrupole collective excitations that possess a clear symplectic algebraic structure, $\mathfrak{sp}(4)$ and $\mathfrak{sp}(3,\mathbb{R})$, respectively.

The Sp(4) dynamical symmetry of like-particle and proton-neutron pairing correlations [33,35,36] between nucleons occupying the same major shell has been found to provide for a reasonable microscopic description of the pairing-governed isobaric analog 0^+ states in light and medium mass nuclei [34,37]. Currently, these nuclei have a significant impact in astrophysical studies [38,39] and have been extensively explored after the advent of radioactive beams. The comparison with realistic interactions can determine the extend to which the significantly simpler Sp(4) model Hamiltonian can readily be used to obtain an approximate, yet very good description of low-lying nuclear structure and in turn, one can apply the model to larger model spaces that are otherwise prohibitive in size. Furthermore, we introduce a possible Sp(4) symmetry breaking by an additional quadrupole-quadrupole interaction and examine the capacity of the extended model interaction to imitate realistic interactions. This, in turn, provides a further step towards gaining a better understanding of the underlying foundation of the microscopic interactions.

II. SYMPLECTIC $\mathfrak{sp}(4)$ PAIRING MODEL INTERACTION

The close interplay of like-particle and proton-neutron isovector (isospin = 1) pairing correlations have long been recognized as a major driver that shapes nuclear systems with valence protons and neutrons occupying the same major shell. While like-particle pairing interactions are known to dominate far from the $N = Z$ line, closer to it strong proton-neutron pair correlations are also very important. Hence, isovector pairing plays a crucial role in understanding the microscopic structure of light and medium mass nuclei around as well as far off the valley of stability. A group-theoretical microscopic description of isovector pairing, based on the fermion realization of the $\mathfrak{so}(5)$ algebra [35] [isomorphic to $\mathfrak{sp}(4)$], was successfully applied to the structure of fp-shell $N = Z$ nuclei [36]. These algebraic results have since been confirmed through pairing-plus-quadrupole shell model work [40]. Indeed, some recent results show that the symplectic Sp(4) dynamical symmetry is fundamental to the nuclear interaction that governs fully-paired isobaric analog 0^+ states of light and medium mass even-A nuclei with valence protons and neutrons occupying the same shell [34].

The general model Hamiltonian with Sp(4) dynamical symmetry for a system of n valence nucleons in a 4Ω-dimensional space consists of one- and two-body terms and can be expressed through the Sp(4) group generators,

$$H_{\mathfrak{sp}(4)} = -G\sum_{i=-1}^{1}\hat{A}_i^\dagger\hat{A}_i - F\hat{A}_0^\dagger\hat{A}_0 - \frac{E}{2\Omega}\left(\hat{T}^2 - \frac{3\hat{N}}{4}\right) - D\left(\hat{T}_0^2 - \frac{\hat{N}}{4}\right) - C\frac{\hat{N}(\hat{N}-1)}{2} - \epsilon\hat{N}, \quad (1)$$

where $\hat{N}$ counts the total number of valence particles, $\hat{T}^2 = \Omega\{\hat{T}_+,\hat{T}_-\} + \hat{T}_0^2$ is the isospin operator, $\hat{A}^\dagger_{0,+1,-1}$ creates a proton-neutron (pn) pair, a proton-proton (pp) pair or a neutron-neutron (nn) pair of total angular momentum $J^\pi = 0^+$ and isospin $T = 1$, G, F, E, D, and C are interaction strength parameters and $\epsilon > 0$ is the Fermi level energy. This Hamiltonian, which is rotationally invariant, conserves the number of particles and the third projection (T_0) of the isospin, while it includes scattering of a pp pair and a nn pair into two pn pairs and vice versa, along with a J-independent isoscalar ($T = 0$) pn force. The significant interplay between isovector and isoscalar interactions is evident in the low-lying structure of $N = Z$ odd-odd nuclei with valence protons and neutrons filling the same major shell.

Estimates for the interaction strength parameters in Eq. (1) were found [33,34] as a result from an optimal reproduction of the Coulomb corrected [41] experimental energies [42,43] of the lowest isobaric analog 0^+ states of even-A nuclei with valence nucleons occupying the $1f_{7/2}$ orbit or the $1f_{5/2}2p_{1/2}2p_{3/2}1g_{9/2}$ major shell.[1] For the $1f_{7/2}$ level with a ^{40}Ca core the interaction strengths were estimated to be, $G/\Omega = 0.453$, $F/\Omega = 0.072$, $C = 0.473$, $D = 0.149$, $E/(2\Omega) = -1.120$, $\epsilon = 9.359$. The analysis revealed that the model interaction with Sp(4) dynamical symmetry accounts quite well for the available experimental energies of isobaric analog 0^+ states for a total of 149 nuclei [34] and in addition for the observed detailed structure beyond mean-field effects such as the $N = Z$ anomalies, isovector pairing gaps and staggering effects [37]. This in turn allowed us to interpret the main driving force that defines the properties of the states under consideration and to provide a reasonable description of these states, while retaining the physical validity and the proper limits of the strengths of interactions available in literature.

An important feature of our algebraic Hamiltonian (1) is that it arises naturally within a microscopic picture. Because of this, the Sp(4) interaction can be compared to realistic interactions and, as well, the physical nature of the model interaction and its strength can be realized. From a microscopic perspective, the pair-creation operators, $\hat{A}^{(\dagger)}$, and their annihilation counter parts, $\hat{A}$, are realized in terms of creation $c^\dagger_{jm\sigma}$ and annihilation $c_{jm\sigma}$ single-fermion operators with the standard anticommutation relations $\{c_{jm\sigma}, c^\dagger_{j'm'\sigma'}\} = \delta_{j,j'}\delta_{m,m'}\delta_{\sigma,\sigma'}$, where these operators create (annihilate) a particle of type $\sigma = \pm 1/2$ (proton/neutron) in a state of total angular momentum j (half integer) with projection m in a finite space $2\Omega = \Sigma_j(2j+1)$. There are ten independent scalar products (zero total angular momentum) of the fermion operators:

$$\begin{aligned}
\hat{A}^\dagger_\mu &= \frac{1}{\sqrt{2\Omega(1+\delta_{\sigma\sigma'})}}\sum_{jm}(-1)^{j-m}c^\dagger_{jm\sigma}c^\dagger_{j,-m,\sigma'},\\
\hat{A}_\mu &= (\hat{A}^\dagger_\mu)^\dagger, \quad (\mu = \sigma + \sigma'),\\
\hat{T}_\pm &= \frac{1}{\sqrt{2\Omega}}\sum_{jm}c^\dagger_{jm,\pm1/2}c_{jm,\mp1/2},\\
\hat{N} &= \sum_{\sigma jm}c^\dagger_{jm\sigma}c_{jm\sigma}, \quad \hat{T}_0 = \sum_\sigma \sigma\sum_{jm}c^\dagger_{jm\sigma}c_{jm\sigma},
\end{aligned} \quad (2)$$

[1]The lowest isobaric analog 0^+ states of odd-odd nuclei have the isospin of the ground state of the even-even same-mass neighbor with a larger difference in proton and neutron numbers. These states are ground states for even-even nuclei and only some [$N \approx Z$] odd-odd nuclei.

which form a fermion realization of the symplectic $\mathfrak{sp}(4)$ Lie algebra. Such an algebraic structure is exactly the one needed to describe isovector (like-particle plus *pn*) pairing correlations and isospin symmetry in nuclear isobaric analog 0^+ states.

Using relations (2), the one- and two-body interaction (1) can be rewritten in standard second quantized form in terms of fermion creation $a^{\dagger}_{jm(1/2)\sigma} = c^{\dagger}_{jm(1/2)\sigma}$ and annihilation $a_{jm(1/2)\sigma} = (-1)^{j+m+1/2+\sigma} c_{j-m(1/2)-\sigma}$ tensor operators,

$$\begin{aligned} H = & - \sum_{\substack{r \leqslant s \\ \{r=(j_r,\frac{1}{2})\}}} \sqrt{[r]}\varepsilon_{\mathrm{rs}}\{a^{\dagger}_r \otimes a_s\}^{(00)} \\ & - \sum_{\substack{r \leqslant s \\ t \leqslant u \\ \Gamma=(J,T)}} \frac{\sqrt{[\Gamma]}}{\sqrt{(1+\delta_{\mathrm{rs}})(1+\delta_{tu})}} \\ & \times W^{\Gamma}_{\mathrm{rstu}}\{\{a^{\dagger}_r \otimes a^{\dagger}_s\}^{\Gamma} \otimes \{a_t \otimes a_u\}^{\Gamma}\}^{(00)} \\ = & - \sum_{r \leqslant s} \sqrt{[r]}\varepsilon_{\mathrm{rs}}\{a^{\dagger}_r \otimes a_s\}^{(00)} \\ & - \frac{1}{4} \sum_{\substack{\mathrm{rstu} \\ \Gamma}} \sqrt{(1+\delta_{\mathrm{rs}})(1+\delta_{tu})[\Gamma]} \\ & \times W^{\Gamma}_{\mathrm{rstu}}\{\{a^{\dagger}_r \otimes a^{\dagger}_s\}^{\Gamma} \otimes \{a_t \otimes a_u\}^{\Gamma}\}^{(00)}, \end{aligned} \tag{3}$$

with $[r] = 2(2j_r + 1)$ and $[\Gamma] = (2J+1)(2T+1)$, where $\varepsilon_{\mathrm{rs}}$ is the single-particle energy and W^{JT}_{rstu} is the two-body antisymmetric matrix element in the *JT*-coupled scheme $[W^{\Gamma}_{\mathrm{rstu}} = -(-)^{r+s-\Gamma} W^{\Gamma}_{srtu} = -(-)^{t+u-\Gamma} W^{\Gamma}_{\mathrm{rsut}} = (-)^{r+s-t-u} W^{\Gamma}_{\mathrm{srut}} = W^{\Gamma}_{\mathrm{turs}}]$. For an isospin nonconserving two-body interaction of isospin rank $\mathcal{T}$, the coupling of fermion operators is as follows, $\{\{a^{\dagger}_r \otimes a^{\dagger}_s\}^{JT} \otimes \{a_t \otimes a_u\}^{JT}\}^{(0\mathcal{T})}$, with $W^{(\mathcal{T})JT}_{\mathrm{rstu}}$ matrix elements. The latter are expressed through the parameters of the model interaction for isospin rank 0 and 2 of $H_{\mathfrak{sp}(4)}$ and $\{r \leqslant (s,t); t \leqslant u\}$ orbits as follows:

$$\begin{aligned} W^{(0)JT}_{\mathrm{rstu}} & \equiv W^{JT}_{\mathrm{rstu}} = \langle rsJTMT_0|H^{(0)}|tuJTMT_0\rangle \\ & = -\left(G + \frac{F}{3}\right)\frac{\sqrt{\Omega_r\Omega_t}}{\Omega}\delta_{(JT),(01)}\delta_{\mathrm{rs}}\delta_{tu} \\ & \quad - \left\{-\left(\frac{E}{2\Omega} + \frac{D}{3}\right)\left[(-)^T + \frac{1}{2}\right] + C\right\}\delta_{rt}\delta_{su} \end{aligned} \tag{4}$$

$$\begin{aligned} W^{(2)JT}_{\mathrm{rstu}} & = \langle rsJTMT_0|H^{(2)}|tuJTMT_0\rangle \\ & = \frac{\sqrt{2}}{3}\left(F\frac{\sqrt{\Omega_r\Omega_t}}{\Omega}\delta_{J0}\delta_{\mathrm{rs}}\delta_{tu} - D\delta_{rt}\delta_{su}\right)\delta_{T1}. \end{aligned} \tag{5}$$

The isotensor part (5) of the model interaction introduces isospin dynamical symmetry through the *D*-term (retaining *T* as a good quantum number and splitting the energy degeneracy along the third projection of the isospin) and as well a plausible, but very weak, isospin mixing (*F*-term) [44].

For the purposes of this paper, we will use only the isoscalar part of our model Hamiltonian (4) and set all the orbits equal to $j = 7/2$ $(r = s = t = u)$ because we choose to focus on a study of nuclei in the single $1f_{7/2}$ level. In addition, these matrix elements correspond to the pure nuclear interaction and do not include Coulomb repulsion because its effect is corrected in the experimental energies themselves by applying an empirical formula deduced in [41]. This may result in slightly more bound states predicted by our model when compared to estimates of realistic interactions.

Within the isospin-invariant picture, the two-body matrix elements of the model Hamiltonian W^{JT}_{rstu} (4) depend only on three parameters, $G_0 = G + \frac{F}{3}$, $E_0 = (\frac{E}{2\Omega} + \frac{D}{3})$ and C,

$$W^{JT}_{\frac{7}{2}\frac{7}{2}\frac{7}{2}\frac{7}{2}} = -G_0\delta_{(JT),(01)} - \left\{-E_0\left[(-)^T + \tfrac{1}{2}\right] + C\right\}. \tag{6}$$

The two-body matrix elements reflect the microscopic aspect of the model interaction, which is *J*-independent for all but $J = 0$. Hence $H_{\mathfrak{sp}(4)}$ describes the average behavior of higher-*J* states, while it distinguishes between $T = 0$ and $T = 1$ groups of states. The smaller the magnitude of E_0 (< 0), the smaller the separation of these groups. As expected, the pairing correlations contribute only to the $(J = 0, T = 1)$ state and they are absent for higher-*J* states where both particles are uncoupled. Relative to the 0^+ $T = 1$ state, the bigger the G_0 pairing strength, the larger the energy gap to the higher-*J* states.

The role of the Sp(4) dynamical symmetry in generating the energy spectrum of the $1f_{7/2}$ nuclei can be further understood by comparing the Sp(4) interaction to the CD-Bonn [31], CD-Bonn+3terms [32], and GXPF1 [2] realistic interactions. CD-Bonn is a charge-dependent one-boson-exchange nucleon-nucleon (*NN*) potential that is one of the most accurate in reproducing the world proton-proton and neutron-proton scattering data. In addition, the CD-Bonn+3terms interaction introduces phenomenological isospin-dependent central terms plus a tensor force with strengths and ranges determined in no-core $0\hbar\omega$ shell model calculations to achieve an improved description of the $A = 48$ Ca, Sc and Ti isobars. The GXPF1 effective interaction is obtained from a realistic G-matrix interaction based on the Bonn-C potential [50] by adding empirical corrections determined through systematic fitting to experimental energy data in the *fp* shell.

III. THEORY OF SPECTRAL DISTRIBUTIONS

Group theory underpins spectral distribution theory [8–10, 12,45]. The model space is partitioned according to particular group symmetries and each subsequent subgroup partitioning yields finer and more detailed spectral estimates. For *n* particles distributed over $\mathcal{N}$ single-particle states, a *scalar* distribution (denoted by "*n*" in the formulae) is called the spectral distribution averaged over all *n*-particle states associated with the $\mathrm{U}(\mathcal{N} = 4\Omega)$ group structure and an *isospin-scalar* distribution (denoted by "n, T") is averaged over the ensemble of all *n*-particle states of isospin *T* associated with $\mathrm{U}(\mathcal{N} = 2\Omega) \otimes \mathrm{U}(2)_T$.

For a spectral distribution α (α is *n* or n, T), the correlation coefficient between two Hamiltonian operators, H and H', is defined as

$$\zeta^{\alpha}_{H,H'} = \frac{\langle (H^{\dagger} - \langle H^{\dagger}\rangle^{\alpha})(H' - \langle H'\rangle^{\alpha})\rangle^{\alpha}}{\sigma_H \sigma_{H'}} \tag{7}$$

$$= \frac{\langle H^{\dagger}H'\rangle^{\alpha} - \langle H^{\dagger}\rangle^{\alpha}\langle H'\rangle^{\alpha}}{\sigma_H \sigma_{H'}}, \tag{8}$$

where the "width" of the distribution is the positive square root of the variance,

$$\left(\sigma_H^\alpha\right)^2 = \langle (H - \langle H\rangle^\alpha)^2\rangle^\alpha = \langle H^2\rangle^\alpha - (\langle H\rangle^\alpha)^2. \quad (9)$$

The average values, related to the trace of an operator divided by the dimensionality of the space, are given in terms of the ensemble considered. In the (isospin-)scalar case, the correlation will be denoted by ζ^n ($\zeta^{n,T}$) or simply ζ (ζ^T) for $n = 2$.

The steps for computing the ζ^α correlation coefficient and the σ^α variance[2] are given in Refs. [8–10,12,24] (see also computational codes [12,46]) and take on the simple form for a single-j level:

$$\langle H^\dagger H'\rangle^\alpha - \langle H^\dagger\rangle^\alpha\langle H'\rangle^\alpha = \sum_\tau p_2(\alpha,\tau)\frac{1}{\sum_\Delta[\Delta]}\sum_J[\Delta] \times W^{J\tau}_{rrrr}(2)W'^{J\tau}_{rrrr}(2), \quad (10)$$

$$W^{J\tau}_{rrrr}(2) = W^{J\tau}_{rrrr} - W_c^{(\tau)}, \quad (11)$$

$$W_c^{(\tau)} = \frac{1}{\sum_\Delta[\Delta]}\sum_\Delta[\Delta]W^{J\tau}_{rrrr}, \quad (12)$$

where $\tau = \{0 \text{ or } 1\}$ is the isospin label of the two-body matrix elements, $W^{J\tau}_{rrrr}(2)$ is the traceless *pure two-body* interaction and $W_c^{(\tau)}$ is the monopole moment or centroid in the (isospin-) scalar case, that is the average expectation value of (the isospin-τ part of) the two-body interaction for a two-particle system $n = 2$. In the scalar ($\alpha = n$) case the following holds, $\Delta = \Gamma = (J,\tau)$, $\sum_\Gamma[\Gamma] = \binom{\mathcal{N}}{2}$, $\mathcal{N} = 4\Omega = 2(2j_r+1)$, and the τ-independent propagator is

$$p_2(n,\tau) = \frac{n(n-1)(\mathcal{N}-n)(\mathcal{N}-n-1)}{2(\mathcal{N}-2)(\mathcal{N}-3)}. \quad (13)$$

In the isospin-scalar ($\alpha = n, T$) case: $\Delta = J$, $\sum_J[J] = \frac{\mathcal{N}(\mathcal{N}+(-1)^\tau)}{2}$, $\mathcal{N} = 2\Omega$, and the propagator functions are [10,47]

$$p_2(n,T,\tau=0) = \frac{[n(n+2)-4T(T+1)]\left[\left(\mathcal{N}-\frac{n}{2}\right)\left(\mathcal{N}-\frac{n}{2}+1\right)-T(T+1)\right]}{8\mathcal{N}(\mathcal{N}-1)} \quad (14)$$

$$\begin{aligned} p_2(n,T,\tau=1) = {} & \frac{1}{\mathcal{N}(\mathcal{N}+1)(\mathcal{N}-2)(\mathcal{N}-3)} \\ & \times\left\{\frac{1}{2}T^2(T+1)^2(3\mathcal{N}^2-7\mathcal{N}+6)\right. \\ & +\frac{3}{8}n(n-2)\left(\mathcal{N}-\frac{n}{2}\right)\left(\mathcal{N}-\frac{n}{2}+1\right) \\ & \times(\mathcal{N}+1)(\mathcal{N}+2)+\frac{1}{2}T(T+1) \\ & \times[(5\mathcal{N}-3)(\mathcal{N}+2)n\left(\frac{n}{2}-\mathcal{N}\right) \\ & \left.+\mathcal{N}(\mathcal{N}-1)(\mathcal{N}+1)(\mathcal{N}+6)]\right\}. \end{aligned} \quad (15)$$

In terms of a geometrical picture, the correlation coefficient ζ defines the angle between two vectors (H and H') of length $\sigma_{H^{(\prime)}}$ (9) and hence its square gives a normalized measure (percentage) of one of the vectors, e.g., the Sp(4) interaction, that is contained in the other, such as a realistic interaction. The correlation coefficient is a measure that is independent of the averages of the interactions. Clearly these averages, though an interesting measure, are irrelevant when the focus is on detailed property-defining two-body interaction beyond strong mean-field effects.

For the Sp(4) interaction, the average two-body interaction is expressed in terms of the model parameters in the scalar case as

$$W_c = -\frac{3G_0}{\binom{\mathcal{N}}{2}} + \frac{3E_0}{2(\mathcal{N}-1)} - C \quad (16)$$

and in the isospin-scalar case as

$$W_c^T = -\frac{G_0}{\binom{\mathcal{N}}{2}}\delta_{T1} + E_0\left[(-1)^T + \frac{1}{2}\right] - C. \quad (17)$$

Hence the pure two-body $W^{JT}_{\frac{7}{2}\frac{7}{2}\frac{7}{2}\frac{7}{2}}(2)$ matrix elements (6), and consequently the correlation coefficients involving $H_{\mathrm{sp}(4)}$, are independent of the C (and E_0) parameter(s) in the (isospin-) scalar case.

IV. UNDERLYING SYMMETRIES OF REALISTIC INTERACTIONS

We now use statistical concepts to probe the nature of the CD-Bonn [31], CD-Bonn+3terms [32], and GXPF1 [2] realistic interactions, hereafter referred as H_R. Specifically, we will compare these interactions to the symplectic pairing and quadrupole interactions through their mutual correlations. Clearly, if two interactions have similar matrix elements they will be strongly correlated and any pattern that is observed in the behavior of one will be reflected in the other. This can be made quantitative by evoking measures from statistical spectroscopy, namely, the closer the correlation coefficient

[2]This follows from the decomposition of the one($k=1$)- and two ($k=2$)-body interaction H into definite particle rank terms [irreducible tensors $\mathcal{H}_k(\nu)$ of rank $\nu = 0,1,2$], that is into a collection of pure zero-, one- and two-body interactions. For example, in the scalar single-j case for n particles, the Hamiltonian can be rendered, $H = n\mathcal{H}_1(0) + \binom{n}{2}\mathcal{H}_2(0) + \mathcal{H}_1(1) + (n-1)\mathcal{H}_2(1) + \mathcal{H}_2(2) = -n\varepsilon - \binom{n}{2}W_c - \frac{1}{2}\sum_\Gamma\sqrt{[\Gamma]}W^\Gamma_{rrrr}(2)\{\{a_r^\dagger\otimes a_r^\dagger\}^\Gamma\otimes\{a_r\otimes a_r\}^\Gamma\}^{(00)}$, where ε is the single-particle energy and for a single-j level the pure one-body part is trivially zero.

between two interactions is to unity the more similar their spectra with the two coinciding within a rescaling factor when the correlation coefficient is unity.

In a similar manner, the projection onto a model Hamiltonian that describes collective rotational excitations or/and pairing correlations can be used to probe the rotational and pairing characteristics of a microscopic interaction [11,26,48,49]. The dynamical symmetry of the pairing (or quadrupole-quadrupole) interaction sets a specific relation between the matrix elements of the Hamiltonian that models it. If this relation is found in a realistic interaction, that is, the model and realistic interactions are strongly correlated, then the latter possesses the underlying symmetry and will reflect the characteristic properties of the pairing (quadrupole) Hamiltonian. It should be clear that the complement is also true, namely, if a model interaction is strongly correlated with a realistic one, the associated model calculations can be used to investigate the behavior of physical systems.

A. The sp(4) model and pairing character

An interesting feature of any interaction is its trace-equivalent part. If the latter is found dominant then only the underlying group scalars are enough to provide for an approximate and yet reasonably good solution [9]. The greatest advantage in this case is the simplicity of the many-body problem and the tractable size of the model space. In the isospin-scalar case, the centroid of a Hamiltonian expressed through the ε single-particle energy and the $W_c^{0,1}$ monopole moments (12) is [47]

$$\langle H^{\dagger}\rangle^{n,T} = -n\varepsilon - \binom{n}{2}\frac{W_c^0+3W_c^1}{4} - \left[T(T+1)-\frac{3}{4}n\right] \times \frac{W_c^1-W_c^0}{2}. \tag{18}$$

For a Hamiltonian with symplectic dynamical symmetry, $H_{sp(4)}$, the trace-equivalent part in the isospin-scalar distribution includes the E-, C- andε-terms of Eq. (1). When applied to the lowest isobaric analog 0^+ states of the nuclei in the $1f_{7/2}$ orbit, it reproduces their energy within 1% of the experimental value for about a third of the nuclei. While for these states the centroid is sufficient to achieve a good description, its difference with experiment goes up to 7% compared to only 0.4% achieved by the whole Sp(4) Hamiltonian. In addition, a model with a trace-equivalent Sp(4) Hamiltonian will not be capable of explaining the fine nuclear structure where $H_{sp(4)}$ (1) succeeded [37] and will not correlate with any of the realistic interactions. The latter indicates an inadequate reproduction of the entire energy spectrum. Indeed, while such an interaction was found insufficient for a description of ds shell nuclei when compared to several effective interactions, a drastic improvement was achieved with the inclusion of pure two-body residual interactions of the spin-orbit and quadrupole types [11] as well as pairing correlations [26]. In summary, the Sp(4) symmetric Hamiltonian (1) provides for a more accurate description of nuclear structure by adding to an average interaction suitable for the isobaric analog 0^+ states in $1f_{7/2}$ a significant isovector pairing part.

Furthermore, the Sp(4) dynamical symmetry allows the model Hamiltonian to reflect on the charge dependence of the nuclear interaction, which is evident from experiments and present in almost all of the modern realistic interactions (e.g., CD-Bonn). While the small isospin admixture found in the $1f_{7/2}$ isobaric analog 0^+ states has been directly estimated through the $H_{sp(4)}$ eigenstates [44], the theory of spectral distributions provides a further estimate of isospin symmetry breaking throughout the entire spectrum [47] based on the Sp(4) isotensor interaction (7). Using Eqs. (4), (17), and (25) in Ref. [10] the isospin $T+2$ admixture into an average T state is found in $1f_{7/2}$ to be, as expected, much smaller (on average less than 0.0001%) than the one detected among the 0^+ seniority-zero states [44]. As expected, it is also much smaller than the measure calculated in Ref. [10] for the two-body Coulomb interaction in the $1f_{7/2}$ shell (with a maximum value of 0.009%), because the latter corresponds to the stronger $\Delta T = 1$ admixture. The quite small isospin symmetry breaking that the isotensor Sp(4) model interaction introduces allows us to carry the present study without its consideration.

The extent to which the Sp(4) dynamical symmetry governs the H_R realistic interactions within a certain domain of states is represented by the correlation coefficients between $H_{sp(4)}$ and H_R (Table I).

In the **scalar** distribution, where the outcome is averaged over the isospin values, the analysis of the results shows that all of the realistic interactions correlate between themselves to a high degree (Table I, upper cell for each pair of interactions). In comparison, each of them has a correlation with the Sp(4) symmetric interaction of order of 0.6–0.8, which is typically regarded as a good one [13]. This implies that the realistic interactions possess around 40–60% of the dynamical symplectic symmetry of $H_{sp(4)}$ $[(\zeta_{H_{sp(4)},H_R})^2]$ and hence 0.4–0.6 portion of H_R is dynamically symmetric under Sp(4) transformations. Equivalently, the $H_{sp(4)}$ model interaction contains 40–60% of the realistic interactions under consideration. This is a very interesting result, and definitely valuable concerning the restrictions the symplectic model is subject to.

A much more interesting scenario occurs when the **isospin-scalar** case is considered. This is because in this case the space is divided into two regions specified by their isospin values

TABLE I. Correlation coefficients for a two-nucleon system, $n = 2$, in the scalar (ζ) and isospin-scalar (ζ^T) distributions.

$\zeta(\zeta^{T=0}, \zeta^{T=1})$	CD-Bonn	CD-Bonn+3terms	GXPF1
$H_{sp(4)}$	0.66	0.64	0.76
	(–, 0.61)	(–, 0.85)	(–, 0.71)
CD-Bonn		0.95	0.96
		(0.99, 0.94)	(0.98, 0.99)
CD-Bonn+3terms			0.97
			(0.99, 0.97)
$H_Q^{\perp}(2)$	0.47	0.60	0.53
	(0.60, 0.73)	(0.68, 0.50)	(0.74, 0.65)
H_M	0.81	0.87	0.93
	(0.60, 0.95)	(0.68, 0.98)	(0.74, 0.96)

with a view towards a more detailed examination of the nature of the interactions under consideration. Indeed the centroids of both $T = 0$ and $T = 1$ regions are considerably separate as is observed in the realistic and Sp(4) interactions and as well confirmed by experiment. In addition, the important pairing correlations that are described in the symplectic model enter in the $T = 1$ channel, where the tendency towards pair formation of realistic interactions can be detected.

As in the scalar case, all of the realistic interactions are quite strongly correlated in both the $T = 0$ and $T = 1$ channels (Table I, lower cell for each pair of models). The $T = 1$ correlation coefficients between $H_{\mathfrak{sp}(4)}$ and the H_R realistic interactions do not depend on any of the parameters in $H_{\mathfrak{sp}(4)}$ including the pairing strength itself[3] and show that $H_{\mathfrak{sp}(4)}$ correlates strongly with the realistic interactions. Among the three H_R, the $T = 1$ part of the CD-Bonn+3terms interaction possesses the closest similarity to the pn and like-particle $J = 0$ pairing correlations. This is indicated by its large projection of 72% $[(\zeta^1_{H_{\mathfrak{sp}(4)},H_R})^2$ in Table I] onto the $T = 1$ $H_{\mathfrak{sp}(4)}$ pairing interaction. Hence, the CD-Bonn+3terms interaction is expected to describe quite well phenomena of a pairing character.

The individual pairing strength associated with each realistic interaction is typically invoked for purposes of comparison. Compared to CD-Bonn, the $J = 0$ isovector pairing strength estimate turns out to be stronger for both CD-Bonn+3terms and GXPF1 with a relatively weaker coupling observed in the latter. Indeed, GXPF1 was shown to tend towards smaller $J = 0$ pairing strength [2] when compared to two other effective interactions, namely the G interaction based on the Bonn-C potential [50] and the KB3G interaction [51]. However, pairing effects, with strong or weak coupling, may be fully or partially reflected in H_R considerably depending on the strength of the overall interaction. It is the correlation coefficient between a realistic (H_R) and symmetry-holding (as $H_{\mathfrak{sp}(4)}$) interactions that manifests what part of H_R is ruled by the symmetry. In this sense, we can identify that pairing features are more fully developed in CD-Bonn+3terms, then in GXPF1 and the least in CD-Bonn ($\zeta^1_{H_{\mathfrak{sp}(4)},H_R}$ in Table I). In the latter, other types of interaction compete stronger with pair formation than in the former two interactions and hence suppress pairing coherence.

The large $J = 0$ coherence and its strong coupling observed in the CD-Bonn+3terms interaction should not be surprising because it reproduces (by an optimal fit) the energy difference between the ground state and the first 2^+ state of ^{48}Ca. Such an observable is believed to be directly affected by the formation of correlated pairs in the ground state of the spherical core of ^{48}Ca and the pairing gap that occurs below the first excited state of a broken pair. It is interesting to point out that while the addition of three phenomenological terms to the CD-Bonn interaction to obtain CD-Bonn+3terms keeps the close similarity between both interactions [$\zeta^1 = 0.94$ (Table I)], it causes the correlations with isovector pairing interaction to double in strength $[(\zeta^1_{H_{\mathfrak{sp}(4)},H_R})^2]$. In short, the analysis shows that the simple Sp(4) model interaction can reproduce reasonably the $T = 1$ low-lying energy spectra generated by the CD-Bonn+3terms realistic interaction for a system of two nucleons in the $1f_{7/2}$ orbit and for this reason can be used as a good approximation.

Another result in favor of the algebraic $\mathfrak{sp}(4)$ model follows from a comparison of the lowest isobaric analog 0^+ states in the $A = 42$ isobars. These are precisely the ($J = 0, T = 1$) states, which are expected to be shaped by strong proton-neutron and like-particle pairing correlations [52] and are well described by the $\mathfrak{sp}(4)$ model [34]. The outcome reveals a very close similarity between the estimate of the two-body ($J = 0, T = 1$) matrix element for the symplectic interaction (−1.85 MeV) and both CD-Bonn+3terms (−2.06 MeV) and GXPF1 (−2.44 MeV) realistic interactions. In addition, the energy differences between the first 2^+ state and the 0^+ ground state for the different effective interactions are also very close, namely, 1.91 MeV (for $H_{\mathfrak{sp}(4)}$), 2.00 MeV (for CD-Bonn+3terms) and 1.50 MeV (for GXPF1). All these estimates are rather different from CD-Bonn with 0.48 MeV 2^+ to 0^+ energy difference but very close to the experimental energy gap for the $A = 42$ isobars, namely, 1.56 MeV for ^{42}Ti, 1.59 MeV for ^{42}Sc and 1.52 MeV for ^{42}Ca.

B. Pairing+quadrupole model interaction

While the pairing-governed isobaric analog 0^+ state energies are well determined within the framework of the Sp(4) model, the nuclear spectrum as described by the Sp(4)-symmetric Hamiltonian contains degenerate higher-J states averaged for a given isospin value as can be clearly seen from its microscopic structure (4). Nonetheless, the correlation of $H_{\mathfrak{sp}(4)}$ with realistic interactions for the $1f_{7/2}$ level turns out to be reasonably strong. A question one can pose concerns the role of other significant interactions in nuclei such as the quadrupole-quadrupole interaction ($Q \cdot Q$). As our results indicate, answers to such questions can be found within the framework of statistical measures.

The pairing model based on the $\mathfrak{sp}(4)$ algebra [incorporating like-particles pairing through an $\mathfrak{su}(2)$ subalgebra] is commonly considered to be inappropriate for two reasons. The first reason is related to the degeneracy of the single-particle levels, which is not a problem for the $1f_{7/2}$ shell considered as a single orbit well-separated from the ds shell and the upper fp shell. The second reason is the lack of the $Q \cdot Q$ interaction. This is because one usually neglects the fact that the pairing interaction contains in itself a part of the quadrupole-quadrupole interaction. This part is not negligible with the correlation being typically between 0.4 to 0.6 depending on the distribution considered: 15% when the whole space is considered and 35% in the $T = 1$ region. This is probably one of the reason why the Sp(4) model interaction turns out to work rather well despite an explicit appearance of the quadrupole-quadrupole interaction.

Because of the fact that the $Q \cdot Q$ interaction is already present in the $\mathfrak{sp}(4)$ Hamiltonian, its additional influence can be studied following the construction prescribed in Ref. [11]. In short, we add a $Q \cdot Q$ term to the symplectic $H_{\mathfrak{sp}(4)}$

[3]For the isospin-scalar distribution, the Sp(4) $T = 0$ model interaction contains only a trace-equivalent part and hence the correlation coefficients cannot be determined.

Hamiltonian (1) in a way that this term is only the part of the pure two-body quadrupole-quadrupole interaction that is not contained in Sp(4), or in the vector algebra terminology we add only the part that is orthogonal to the pure two-body Sp(4) Hamiltonian [16],

$$H_M = H_{\mathfrak{sp}(4)} + H_Q^{\perp}(2), H_Q = -\frac{\chi}{2} Q \cdot Q. \quad (19)$$

Such a Hamiltonian does not affect the centroid of $H_{\mathfrak{sp}(4)}$ because $H_Q^{\perp}(2)$ is traceless. In this way this collective interaction preserves the shell structure that is built into $H_{\mathfrak{sp}(4)}$ and established by a harmonic oscillator potential and as a result is favored in many studies [11,26,30,53].

Compared to the pairing $H_{\mathfrak{sp}(4)}$ Hamiltonian, the additional collective interaction, $H_Q^{\perp}(2)$, has a lower correlation with H_R for all the cases except for $T = 1$ CD-Bonn and where $\zeta_{H_{\mathfrak{sp}(4)},H_R}$ cannot be determined (Table I). The realistic interactions contain the Sp(4) interaction by 5% to 50% more than they contain $H_Q^{\perp}(2)$, the largest value being for the $T = 1$ CD-Bonn+3terms interaction. We should emphasize that this outcome does not imply that the realistic interactions correlate better with the pairing interaction than they do with the quadrupole-quadrupole interaction nor that their pairing character is dominant. This is because the $H_Q^{\perp}(2)$ interaction represents only that part of the rotational interaction that is not included in the Sp(4) interaction and the entire $Q \cdot Q$ collective mode affects both $\zeta^{(T)}_{H_Q^{\perp}(2),H_R}$ and $\zeta^{(T)}_{H_{\mathfrak{sp}(4)},H_R}$ correlations. The outcome only implies that a comparatively larger part of the overall correlations is already accounted for solely by the symplectic $\mathfrak{sp}(4)$ algebraic model interaction.

In our study, we vary only χ, the quadrupole strength parameter in Eq. (19), to find its optimal value (which is an exact solution) by maximizing the correlation coefficient ζ between H_M and H_R [54]. We do not alter the parameters of the Sp(4) model, which have already been shown in an appropriate domain of states to be valid for reproducing various quantities (such as binding energies and pairing gaps) and are in agreement with estimates available in literature [34,37]. This implies that the σ "width" of $H_{\mathfrak{sp}(4)}$ (9) does not change. The minimization procedure is performed for H_M compared to each realistic interaction and in the isospin-scalar case, for each isospin value (Tables I and II).

In both scalar and isospin-scalar cases, the addition of the quadrupole-quadrupole interaction definitely improves the $\zeta^{(T)}_{H_M,H_R}$ correlation (Table I), which is associated with the angle between the two-body effective interaction and its projection on the plane spanned by two orthogonal vectors, two-body $H_{\mathfrak{sp}(4)}$ and $H_Q^{\perp}(2)$, in an abstract operator space (Fig. 1). In the same representation, the angles between the realistic interactions and both axes give the $\zeta^{(T)}_{H_Q^{\perp}(2),H_R}$ and $\zeta^{(T)}_{H_{\mathfrak{sp}(4)},H_R}$ correlations, and the length of each vector is specified by $\sigma_H^{(T)}$ (9). Therefore, while enhanced quadrupole effects rotate the projection of an interaction closer to the $H_Q^{\perp}(2)$ axis, greater influence of additional interactions neglected in H_M pushes H_R away from the H_M horizontal plane.

The GXPF1 interaction is found to correlate best with the model H_M Hamiltonian for the scalar distribution (86%) and in the $T = 0$ case compared to the other interactions (Table I). More than 50% of the $T = 0$ GXPF1 interaction is accounted by the isoscalar model interaction. The $T = 0$ correlation coefficients between the model H_M Hamiltonian and H_R are reasonably good and determined solely by the $Q \cdot Q$ interaction independent of its strength, χ (19).

In the $T = 1$ region, all the realistic interactions considered are reproduced to the 90%–97% level by such a pairing+quadrupole model interaction. Other interactions

TABLE II. First and second energy moments [the centroid W_c (12) and the "width" σ (9)] of realistic and model interactions for a two-nucleon system, $n = 2$, in the $1f_{7/2}$ level. H_M is determined by an estimate for the quadrupole-quadrupole strength χ for each realistic interaction; its centroid energies coincide with the ones of $H_{\mathfrak{sp}(4)}$ for a given distribution.

	CD-Bonn	CD-Bonn+3terms	GXPF1
	Scalar distribution		
W_c	0.36	0.19	−0.06
σ	0.59	0.80	0.98
H_M, W_c	−0.63	−0.63	−0.63
H_M, σ	1.23	1.36	1.21
H_M, χ	0.096	0.124	0.092
	Isospin-scalar distribution, $T = 1$		
W_c^1	0.54	0.46	−0.17
σ^1	0.27	0.57	0.62
H_M, W_c^1	−0.01	−0.01	−0.01
H_M, σ^1	0.55	0.41	0.48
H_M, χ	0.071	0.036	0.055

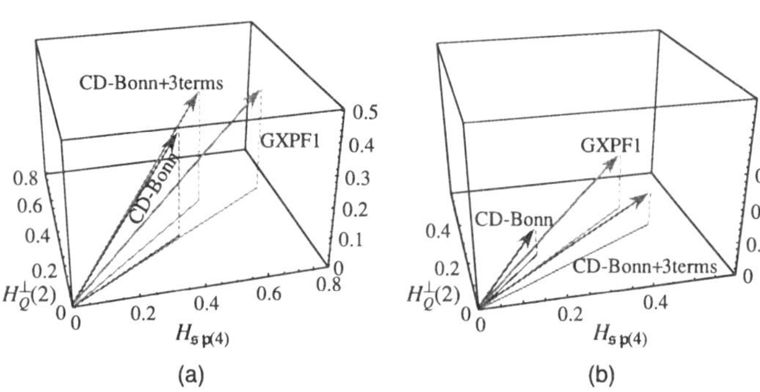

FIG. 1. (Color online) Geometrical representation of the realistic interactions, CD-Bonn (light blue), CD-Bonn+3terms (red) and GXPF1 (green), in an abstract operator space, where the horizontal plane is spanned by the orthogonal linear operators, the pure two-body $H_{\mathfrak{sp}(4)}$ and $H_Q^{\perp}(2)$ model Hamiltonians, both linearly independent of additional operators represented by the vertical axis. (a) Scalar distribution. (b) Isospin-scalar distribution, $T = 1$. The orientation of the vectors remains the same for any particle number $n \geqslant 2$ in (a) and for all $T = n/2$ cases in (b).

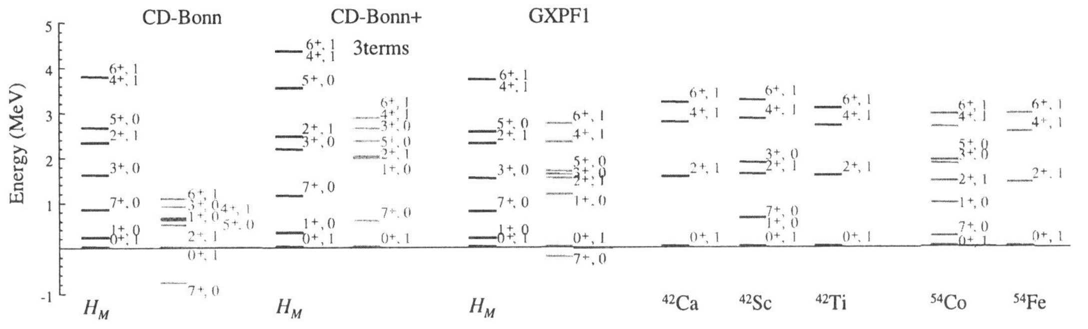

FIG. 2. (Color online) Energy spectra of two-particle states in the $1f_{7/2}$ level predicted by the CD-Bonn (light blue), CD-Bonn+3terms (red), and GXPF1 (green) realistic interactions. Each is compared to the model Hamiltonian H_M (black) with $\chi = 0.096, 0.124$ and 0.092, respectively (Table II). For comparison, the available experimental energy spectra of the $A = 42$ Ca, Sc, Ti isobars (blue) and $A = 54$ Co and Fe isobars (magenta) are also shown.

contribute almost negligibly as is clearly seen from Fig. 1(b) with their contribution being least for the CD-Bonn+3terms interaction. In addition, within the $T = 1$ distribution, the smallest χ value (Table II) is found for the CD-Bonn+3terms realistic interaction as expected due to its reasonable correlation with the Sp(4) model interaction. In summary, the results once again prove that the pairing and quadrupole-quadrupole interactions are significant in shaping nuclear structure and are dominant for the $T = 1$ two-body nuclear interaction for the $1f_{7/2}$ orbit.

From the point of view of the model interaction (19) we adopted, the rotational character of the three realistic interactions may appear to be obscure because a strong correlation to the entire H_Q quadrupole-quadrupole interaction (not only to its projection $H_Q^{\perp}(2)$), $\zeta^{(T)}_{H_Q,H_R}$, is needed. However, development of rotational features turns out to follow qualitatively the $\zeta^{(T)}_{H_Q^{\perp}(2),H_R}$ correlation coefficients (Table I). This is because H_Q is already present in the H_M model Hamiltonian and one can visualize H_Q in the scalar (isospin-scalar, $T = 1$) case in Fig. 1 as an axis that lies in the horizontal plane around 67° (54°) counterclockwise from the $H_{\mathrm{sp}(4)}$ axis and hence it lies furthest away from $H_{\mathrm{sp}(4)}$ compared to all three realistic interactions. Therefore, the comparatively largest collectivity is attained within CD-Bonn+3terms in the scalar case, within GXPF1 for the $T = 0$ domain of states (with an exact quantitative measure $\zeta^0_{H_Q,H_R} = \zeta^0_{H_Q^{\perp}(2),H_R}$), and within CD-Bonn in the $T = 1$ case. In contrast, the $T = 1$ part of the CD-Bonn interaction shows the smallest individual quadrupole strength, which is an illustrative example of a very prominent rotational behavior (detected via ζ) but of a weak strength (depending on σ). In general, the individual quadrupole strength associated with the GXPF1 interaction is very similar to, yet slightly stronger than, the one of CD-Bonn+3terms with the same trend observed for GXPF1 [2] with respect to KBG3.

C. Energy spectrum

The scalar and isospin-scalar $T = 1$ distributions show strong correlations of the pairing+quadrupole model interaction with the realistic interactions and hence a similar pattern of energy states is expected.

While correlation coefficients (8) prove useful in studies of nuclear properties shaped by the residual pure two-body interaction, the discrete energies of a quantum-mechanical system are additionally influenced by the centroid, $W_c^{(T)}$ (12), and the overall interaction strength related to $\sigma^{(T)}$ (9). The centroid of the H_M pairing+quadrupole model interaction, which coincides with the one for Sp(4) ($W^{(T)}_{c,H_M} = W^{(T)}_{c,H_{\mathrm{sp}(4)}}$) (Table II), is very close to that for GXPF1 and both differ from the other two realistic interactions. However, this quantity is irrelevant for the energy spectra relative to the ground state within a given distribution.

For the scalar distribution, the H_M model Hamiltonian generates energy spectra that are comparatively more spread out, especially with respect to CD-Bonn (see Fig. 2 and σ in Table II). A renormalization of H_M will push the higher-lying states down and will establish an energy pattern very much like the ones observed in GXPF1, CD-Bonn+3terms and experiment (except for the 1^+ state). Such a renormalization, however, is not done because the scalar distribution itself introduces averaging over isospin values. Instead we turn to the more detailed $T = 1$ spectral distribution.

The $T = 1$ part of the pure two-body realistic interactions is reproduced quite well by the model Hamiltonian. This in turn yields a similar energy spectra (Fig. 3) as predicted by the model interaction H_M and both the CD-Bonn+3terms and GXPF1 realistic interactions. They also agree well with the available experimental data for the $A = 42$ and $A = 54$ isobars (the latter refer to systems of two holes). Here again, the agreement between H_M and CD-Bonn is not as good as for the other two interactions, especially in reproducing the spreading of the states, which is smaller for CD-Bonn relative to the experimental energy spectra. Compared to the Sp(4) model interaction, the energy of the first 2^+ state is brought slightly lower by the addition of $H_Q^{\perp}(2)$, to 1.47 MeV (when χ is determined in comparison to CD-Bonn+3terms) and 1.22 MeV (to GXPF1). Such values are a bit closer to the experimental results, namely, 1.52–1.59 MeV for the $A = 42$ isobars and 1.41–1.44 MeV for the $A = 54$ isobars in the $1f_{7/2}$ level. While the influence of the quadrupole-quadrupole

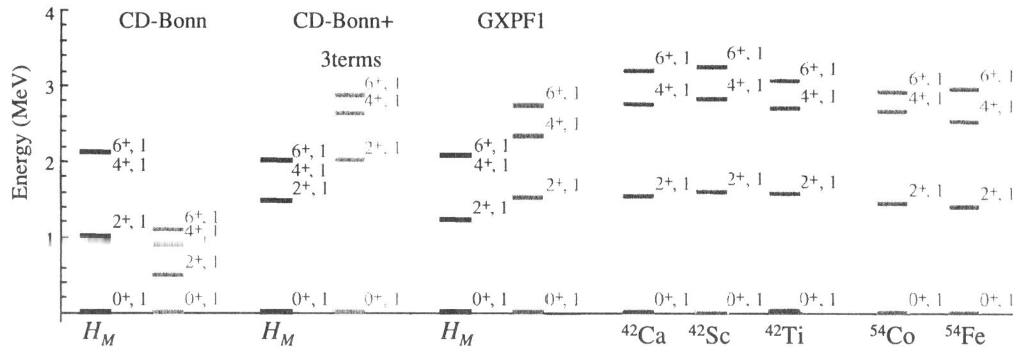

FIG. 3. (Color online) Energy spectra of $T = 1$ states predicted by the CD-Bonn (light blue), CD-Bonn+3terms (red) and GXPF1 (green) realistic interactions. Each is compared to the model Hamiltonian H_M (black) with $\chi = 0.071, 0.036$ and 0.055, respectively (Table II). For comparison, the experimental $T = 1$ energy spectra of the $A = 42$ Ca, Sc, Ti isobars (blue) and $A = 54$ Co and Fe isobars (magenta) are also shown.

interaction is significant in the 2^+ $T = 1$ states, it does not affect the estimate for the Sp(4) parameters because they were determined with regard to the nuclear isobaric analog 0^+ states.

An interesting result is that in the detailed case of isospin-scalar distribution, the $(J = 0, T = 1)$ two-body matrix element is not affected at all by the $Q \cdot Q$ interaction added to $H_{\mathrm{sp}(4)}$ because $H_Q^{\perp}(2)$ has a zero contribution to this state. In this way, the addition of a quadrupole-quadrupole interaction does not alter the 0^+ $T = 1$ ground state of the $A = 42$ isobars, which are precisely the lowest isobaric analog 0^+ states where the Sp(4) model has been applied. Moreover, the average effect of this two-body collective interaction throughout the entire shell is zero [see Eq. (19)]. Even though this property of $H_Q^{\perp}(2)$ is imposed by construction, such a form (as already mentioned) is typically preferred so as to preserve the shell structure. In short, the domain of states where the Sp(4) model was applied is not influenced by the inclusion of the quadrupole degree of freedom to the pairing model. The Sp(4) model interaction itself actually accounts for all the effects, small or large, due to the influence of the quadrupole-quadrupole interaction on these states.

D. Correlations between interactions for nuclear systems with more than two nucleons

An important feature of spectral distribution theory is that the correlation coefficient concept can be propagated beyond the defining two-nucleon system to derivative systems with larger numbers of nucleons [8] and in the isospin-scalar case, for higher values of isospin [10]. The propagation formulas (10) determine how the averages extracted from the two-nucleon matrix elements in the two-nucleon system get carried forward into many-nucleon systems. This propagation of information is model-independent. In this way one can track the similarity of pairing/rotational characteristics between different interactions in many-nucleon systems [49].

In the scalar case the correlations between the interactions retain their values as given in Table I. For the isospin-scalar distribution, the correlation coefficients between the realistic interactions and the H_M model interaction decrease for $n > 2$ and higher-T values compared to the $n = 2$ $T = 1$ case and increase when compared to the $n = 2$ $T = 0$ estimates (Fig. 4).

For given n and T, the ζ values can again be found as the maximum correlation for an optimal value of the quadrupole-quadrupole strength χ, which is related to the angle between the geometrically represented H_R and its projection onto the H_M plane. When $T = n/2$ (Fig. 4), which corresponds to the highest possible isospin states (including $n = 2$, $T = 1$) in 41 nuclei with valence nucleons occupying the $1f_{7/2}$ orbit, the realistic interactions continue to be strongly correlated with the H_M model Hamiltonian, namely, $\zeta^{(T)}_{H_M,H_R}$ is 0.95, 0.98 and 0.96 for CD-Bonn, CD-Bonn+3terms, and GXPF1, respectively. For these states, the other types of interactions are negligible and constitute only 3% of CD-Bonn+3terms, 8% of GXPF1 and 9% of CD-Bonn [Fig. 1(b)].

For all the cases with $T \neq n/2$ throughout the entire shell, the correlation coefficients are found to retain almost the same value (Fig. 4), namely, $\zeta^{T}_{H_M,H_R}$ is around $0.63 - 0.70$, $0.76 - 0.84$ and $0.80 - 0.86$ for CD-Bonn, CD-Bonn+3terms, and GXPF1, respectively, with the corresponding optimal strength χ of the quadrupole-quadrupole interaction in the intervals, $0.087 - 0.096$, $0.042 - 0.047$, and $0.057 - 0.058$. The smaller χ, the weaker the Sp(4) symmetry breaking resulting from the additional quadrupole-quadrupole interaction. Again, the least strength is observed when H_M is compared to CD-Bonn+3terms. In the case of GXPF1, χ remains almost the same for all the states, with $T \neq n/2$ as well as with $T = n/2$ (Table II, last row). In addition, the relative contribution of other types of interaction, which are not accounted for by H_M (19), is somewhat greater for CD-Bonn+3terms and CD-Bonn compared to GXPF1 because of their comparatively smaller $\zeta^{T\neq n/2}_{H_M,H_R}$ (Fig. 4. In short, among the three realistic interactions, when T differs from $n/2$ the Sp(4) dynamical symmetry continues to be reflected the most in CD-Bonn+3terms, while the extended pairing+quadrupole model interaction correlates the best with GXPF1.

An interesting result is that among the $T \neq n/2$ cases the highest $\zeta^{T}_{H_M,H_R}$ correlations for each realistic interaction

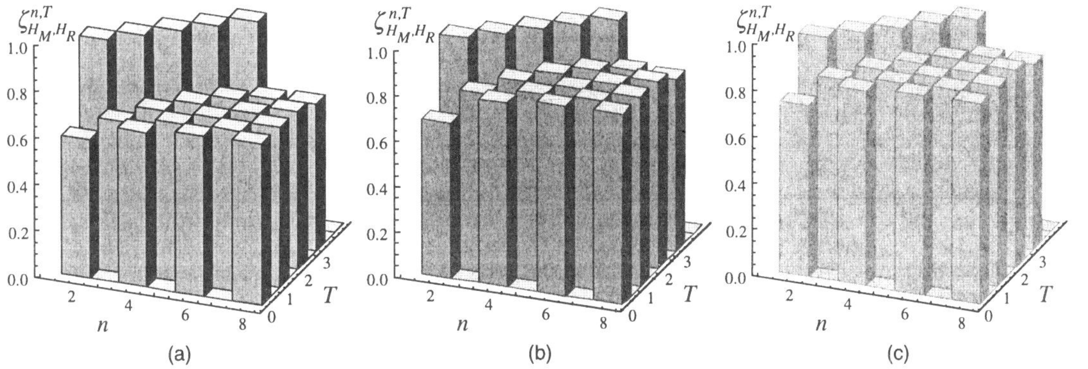

FIG. 4. (Color online) Correlation coefficients, $\zeta^{n,T}_{H_M,H_R}$, of the pairing+quadrupole model interaction with (a) CD-Bonn, (b) CD-Bonn+3terms, and (c) GXPF1 as a function of n and T for the $1f_{7/2}$ orbit. The representation is symmetric with respect to the sign of $n-2\Omega$.

are observed for the low-T mid-shell nuclear states, where the χ strength is relatively smaller. This suggests that for these states the other kinds of interactions not present in H_M grow weaker and the realistic interactions are comparatively closer in behavior to the symplectic pairing $H_{\mathfrak{sp}(4)}$. The lowest correlation is observed for the mid-shell nuclei with isospin next to the highest, where χ is the largest. For these cases, both quadrupole-quadrupole and other types of interactions accounted for by the realistic interactions increase in relative importance.

In summary, for all of the (n, T) distributions the model pairing+quadrupole interaction accounts on average for about 59%, 77%, and 78% of the CD-Bonn, CD-Bonn+3terms, and GXPF1 realistic interactions, respectively, and up to 91%, 97%, and 92% of those in the highest possible isospin states for a given n, where the H_M model interaction can be used to provide a reliable description.

Although, the aforementioned results refer to a single-j orbit ($1f_{7/2}$), they represent a first step towards a generalization to multi-j major shells (such as the fp shell or even a set of several major shells). Such an extension is certainly feasible and not space limited from the perspective of spectral distribution theory. Even though the statistics for the $1f_{7/2}$ level are not large compared to the fp shell, it is a natural choice. Specifically, the $1f_{7/2}$ orbital is comparatively far away from both the neighboring ds and upper fp shells which means there is a preponderance of a single-j coherence over configuration mixing in the low-lying nuclear states for such nuclei. For this reason it is an interesting example in its simplicity and provides a quite clear view of the pairing/rotational foundation of the nuclear interaction that is free of competing configuration mixing effects. In addition, it does represents a partitioning of the fp-space and as such it provides for more detailed spectral measures that may reflect important fine effects that are otherwise averaged out when the entire fp major shell is taken into account. It will be also important to augment this single-j shell study with a follow-on, complementary multi-shell analysis of similar type. The multi-shell case will be the topic of a future publication.

In addition to the above, we note for completeness that the estimates for the parameters of the model Sp(4) interaction are not really relevant to the primary objective of the present study. The reason is that the correlation coefficients in the detailed isospin-scalar case are independent of the interaction strength parameters, which therefore do not affect the correlation measures of the model interaction with realistic ones or the pairing and rotational characteristics of the latter.

V. CONCLUSIONS

With a view towards a broader study within multi-j shells, we compared three realistic interactions and two model pairing and quadrupole interactions for the $1f_{7/2}$ orbit by means of the theory of spectral distributions.

In the more detailed case of isospin-scalar distribution, the CD-Bonn, CD-Bonn+3terms, and GXPF1 realistic interactions were found to contain on average 59%, 77%, and 78%, respectively, of the pairing+quadrupole interaction. Moreover, this percentage goes up to 91%, 97%, and 92%, respectively, for the highest possible isospin group of states for all the nuclei with valence protons and neutrons occupying the $1f_{7/2}$ shell. For these states, the strongest correlation was observed between the CD-Bonn+3terms and the pairing+quadrupole model interaction, where other types of interaction accounted in the realistic interactions represent only 3% of it. They constitute 8% of the GXPF1 realistic interaction, and 9% of CD-Bonn. For these cases, the pairing+quadrupole model interaction has been shown to be a very good approximation that provides a reasonable description of the energy spectra of the nuclei in the $1f_{7/2}$ level. While both interactions, CD-Bonn+3terms and GXPF1, exhibit a well-developed pairing character compared to CD-Bonn, the latter appears to build up more (less) rotational collective features that are outside of the scope of the $T=1$ ($T=0$) Sp(4) interaction.

The major advantage of the $\mathfrak{sp}(4)$ algebraic model, which focuses on the isovector pairing correlations and also includes a certain portion of the quadrupole-quadrupole interaction, is that it provides an elegant solution for describing the pairing-governed isobaric analog 0^+ states in light and medium mass nuclei. In addition, while it correlates to a reasonably good extent with the realistic interactions, the description of the low-lying nuclear energy spectrum of higher-J states is improved with the inclusion of the quadrupole-quadrupole interaction that being symmetric under SU(3) breaks the Sp(4) symmetry and removes degeneracies. Nevertheless, we found that for the isospin-scalar distribution the Sp(4) model interaction accounts for a large part of the CD-Bonn + 3terms realistic interaction. It also includes between 15% to 35% of the rotational collective interaction and typically accounts for a rather large portion of the overall correlation of the realistic interactions with the pairing+quadrupole interaction. Moreover, the additional quadrupole degree of freedom incorporated in the symplectic model interaction does not affect the domain of states where the Sp(4) model was applied and hence introduces no errors in the estimates of the parameters of the symplectic interaction. These results confirm the conclusion that the Sp(4) interaction can provide for an approximate pattern of the nuclear energy spectra and, above all, can be accepted as a very reasonable approximation to describe the pairing-governed isobaric analog 0^+ states in the nuclei in the $1f_{7/2}$ orbit.

ACKNOWLEDGMENTS

This work was supported by the U.S. National Science Foundation, Grant No. 0140300 and the Southeastern Universities Research Association.

[1] K. Klingenbeck, W. Knüpfer, M. G. Huber, and P. W. M. Glaudemans, Phys. Rev. C **15**, 1483 (1977).
[2] M. Honma, T. Otsuka, B. A. Brown, and T. Mizusaki, Phys. Rev. C **69**, 034335 (2004).
[3] A. G. Negoita, J. P. Vary, S. Stoica, P. Navratil, and S. Popescu, to be published.
[4] P. Halse, Phys. Rev. C **39**, 1104 (1989).
[5] M. Horoi, B. A. Brown, and V. Zelevinsky, Phys. Rev. Lett. **87**, 062501 (2001).
[6] J. P. Vary and S. N. Yang, Phys. Rev. C **15**, 1545 (1977).
[7] A. Poves, J. Sánchez-Solano, E. Caurier, and F. Nowacki, Nucl. Phys. **A694**, 157 (2001).
[8] F. S. Chang, J. B. French, and T. H. Thio, Ann. Phys. (NY) **66**, 137 (1971).
[9] J. B. French and K. F. Ratcliff, Phys. Rev. C **3**, 94 (1971).
[10] K. T. Hecht and J. P. Draayer, Nucl. Phys. **A223**, 285 (1974).
[11] T. R. Halemane, K. Kar, and J. P. Draayer, Nucl. Phys. **A311**, 301 (1978).
[12] V. K. B. Kota, Phys. Rev. C **20**, 347 (1979); Fortran Programs for Statistical Spectroscopy Calculations.
[13] G. Rosensteel, Nucl. Phys. **A341**, 397 (1980).
[14] J. P. Draayer and G. Rosensteel, Nucl. Phys. **A386**, 189 (1982).
[15] J. P. Draayer, J. B. French, V. Potbhare, and S. S. M. Wong, Phys. Lett. **B55**, 263 (1975); J. P. Draayer, J. B. French, M. Prasad, V. Potbhare, and S. S. M. Wong, *ibid.* **57**, 130 (1975); J. P. Draayer, J. B. French, and S. S. M. Wong, Ann. Phys. **106**, 472 (1977); B. D. Chang and J. P. Draayer, Phys. Rev. C **20**, 2387 (1979).
[16] V. Potbhare, Nucl. Phys. **A289**, 373 (1977).
[17] J. B. French, V. K. B. Kota, A. Pandey, and S. Tomsovic, Ann. Phys. (NY) **181**, 235 (1988).
[18] V. K. B. Kota and D. Majumdar, Z. Phys. A **351**, 365 (1995); **351**, 377 (1995).
[19] S. Tomsovic, M. B. Johnson, A. Hayes, and J. D. Bowman, Phys. Rev. C **62**, 054607 (2000).
[20] J. M. G. Gomez, K. Kar, V. K. B. Kota, R. A. Molina, and J. Retamosa, Phys. Lett. **B567**, 251 (2003).
[21] M. Horoi, M. Ghita, and V. Zelevinsky, Phys. Rev. C **69**, 041307(R) (2004); M. Horoi, J. Kaiser, and V. Zelevinsky, *ibid.* **67**, 054309 (2003).
[22] V. K. B. Kota, Phys. Rev. C **71**, 041304(R) (2005).
[23] K. F. Ratcliff, Phys. Rev. C **3**, 117 (1971).
[24] J. B. French, in *Dynamic Structure of Nuclear States*, edited by D. J. Rowe *et al.* (University of Toronto Press, Toronto, 1972), p. 154.
[25] B. J. Dalton, W. J. Baldridge, and J. P. Vary, Phys. Rev. C **20**, 1908 (1979).
[26] C. R. Countee, J. P. Draayer, T. R. Halemane, and K. Kar, Nucl. Phys. **A356**, 1 (1981).
[27] J. B. French, Nucl. Phys. **A396**, 87c (1983).
[28] J. P. Draayer and G. Rosensteel Phys. Lett. **B124**, 281 (1983); G. Rosensteel and J. P. Draayer, Nucl. Phys. **A436**, 445 (1985).
[29] S. Sarkar, K. Kar, and V. K. B. Kota, Phys. Rev. C **36**, 2700 (1987).
[30] J. P. Draayer and G. Rosensteel, Phys. Lett. **B125**, 237 (1983).
[31] R. Machleidt, F. Sammarruca, and Y. Song, Phys. Rev. C **53**, R1483 (1996); R. Machleidt, *ibid.* **63**, 024001 (2001).
[32] S. Popescu, S. Stoica, J. P. Vary, and P. Navratil, to be published.
[33] K. D. Sviratcheva, A. I. Georgieva, and J. P. Draayer, J. Phys. G: Nucl. Part. Phys. **29**, 1281 (2003).
[34] K. D. Sviratcheva, A. I. Georgieva, and J. P. Draayer, Phys. Rev. C **70**, 064302 (2004).
[35] K. T. Hecht, Nucl. Phys. **63**, 177 (1965); Phys. Rev. **139**, B794 (1965); Nucl. Phys. **A102**, 11 (1967); J. N. Ginocchio, Nucl. Phys. **74**, 321 (1965).
[36] J. Engel, K. Langanke, and P. Vogel, Phys. Lett. **B389**, 211 (1996).
[37] K. D. Sviratcheva, A. I. Georgieva, and J. P. Draayer, Phys. Rev. C **69**, 024313 (2004).
[38] K. Langanke, Nucl. Phys. **A630**, 368c (1998).
[39] P. T. Hosmer *et al.*, Phys. Rev. Lett. **94**, 112501 (2005).
[40] K. Kaneko, M. Hasegawa, and J. Y. Zhang, Phys. Rev. C **59**, 740 (1999).
[41] J. Retamosa, E. Caurier, F. Nowacki, and A. Poves, Phys. Rev. C **55**, 1266 (1997).
[42] G. Audi and A. H. Wapstra, Nucl. Phys. **A595**, 409 (1995).
[43] R. B. Firestone and C. M. Baglin, *Table of Isotopes*, 8th ed. (Wiley, New York , 1998).
[44] K. D. Sviratcheva, A. I. Georgieva, and J. P. Draayer, Phys. Rev. C **72**, 054302 (2005).
[45] J. C. Parikh, *Group Symmetries in Nuclear Structure* (Plenum, New York, 1978).
[46] B. D. Chang, J. P. Draayer, and S. S. M. Wong, Comput. Phys. Commun. **28**, 41 (1982).

[47] J. B. French, in *Isospin in Nuclear Physics*, edited by D. H. Wilkinson (North Holland, Amsterdam, 1969), p. 259.
[48] J. P. Draayer, Nucl. Phys. **A216**, 457 (1973).
[49] V. K. B. Kota, S. P. Pandya, and V. Potbhare, Nucl. Phys. **A349**, 397 (1980).
[50] M. Hjorth-Jensen, T. T. S. Kuo, and E. Osnes, Phys. Rep. **261**, 125 (1995).
[51] A. Poves, J. Sánchez-Solano, E. Caurier, and F. Nowacki, Nucl. Phys. **A694**, 157 (2001).
[52] A. Bohr, B. R. Mottelson, and D. Pines, Phys. Rev. **110**, 936 (1958).
[53] C. Bahri, J. P. Draayer, O. Castanos, and G. Rosensteel Phys. Lett. **B234**, 430 (1990).
[54] B. D. Chang, Nucl. Phys. **A304**, 127 (1978).

677

Nuclear spectroscopy in the chaotic domain: level densities

J.B. French, S. Rab, J.F. Smith, R.U. Haq, and V.K.B. Kota

Abstract: In heavy nuclei, the structure generating the slow-neutron resonance spectrum extends downward in energy to ~(1–2) MeV excitation and, of course, upward as well until particle emission becomes significant, thereby generating an Embedded Gaussian Orthogonal Ensemble (EGOE) spectrum built on a secular mean-density function. In this extended chaotic domain, principles and methods for the calculation of one-point quantities (e.g., level densities, spin-cutoff factors, occupancies, etc.,) have been well developed during the last several years. The economy and the resultant generic forms follow from the dominance of unitary symmetries, central limit theorems, and quantum chaos. In this paper, techniques used for level densities are illustrated by a detailed study of several heavy nuclei, the input data being taken from the observed low-lying spectrum and the far-separated neutron-resonance spectrum, this in itself saying much about long-range spectral rigidity. Explicit forms for the interacting particle state densities, expectation values, and expectation-value densities of operators in Hamiltonian eigenstates are given. Extension of the formalism to two-point functions that deal with spectral fluctuations, transition strengths, and analysis of measures for broken symmetries and which involve the same formal structure is indicated; higher order correlation functions are of little immediate interest because they define quantities only rarely measurable.

PACS Nos.: 21.10Ma, 21.60Cs, 24.60−k, 24.60.Lz

Résumé : Dans les noyaux lourds, la structure responsable du spectre de résonance de neutrons lents s'étend vers le bas jusqu'à des énergie d'excitation de 1–2 MeV et vers le haut jusqu'à ce que l'émission de particules devienne significative, générant ainsi un spectre « Ensemble Gaussien Orthogonal Inclus (EGOE) » construit sur une fonction séculaire de densité moyenne. Dans ce domaine chaotique étendu, les principes et les méthodes pour le calcul de quantités ponctuelles (densités de niveaux, facteurs de coupure de spin, occupations, etc.,) ont été bien développés ces dernières années. L'économie et les formes génériques résultantes découlent de l'influence dominante des symétries unitaires, des théorèmes à valeur centrale limite et du chaos quantique. Nous illustrerons ici les techniques de calcul des

Received 23 December 2004. Accepted 20 March 2006. Published on the NRC Research Press Web site at http://cjp.nrc.ca/ on 15 September 2006.

J.B. French,[1] **S. Rab,**[2] **and J.F. Smith.**[3] Department of Physics and Astronomy, University of Rochester, Rochester, NY 14627, USA.
R.U. Haq.[4] Department of Physics and Astronomy, Laurentian University, Sudbury, ON P3E 2C6, Canada.
V.K.B. Kota. Physical Research Laboratory, Ahmedabad, 380 009, India.

[1]Deceased
[2]Present Address: Department of Physics and Astronomy, SUNY Stony Brook, NY 11794, USA.
[3]Present Address: Naval Research Laboratory, Code 5741, Washington DC 20375-5320, USA.
[4]Corresponding author (e-mail: rhaq@nickel.laurentian.ca).

 doi: 10.1139/P06-047

densités de niveau dans plusieurs noyaux lourds, les données de départ provenant des spectres observés de basse énergie et du spectre largement séparé de résonance neutronique, ce dernier point donnant beaucoup d'information sur la rigidité spectrale à longue portée. Nous présentons des résultats explicites pour les densités de particules en interaction, ainsi que les valeurs attendues et les densités de valeur attendues d'opérateurs dans des états propres du Hamiltonien. Nous suggérons comment étendre le formalisme aux fonctions à deux points qui incluent les fluctuations spectrales, aux forces de transition et comment développer des analyses de mesure pour les symétries brisées qui impliquent la même structure formelle. Les fonctions de corrélation d'ordre supérieur sont de faible intérêt, parce qu'elles définissent des quantités rarement mesurables.

[Traduit par la Rédaction]

1. Introduction

The spectrum of a heavy nucleus gradually changes its nature with increasing energy, from non-chaotic at low energy and dominated then by details of the Hamiltonian, to chaotic at higher energy (below and some distance above the resonance region) with a strong secular variation of the density and *locally* dominated by a many-particle extension, the embedded ensemble (EGOE) of the standard Gaussian Orthogonal Ensemble (GOE) with statistical structure (level repulsion and spectral rigidity) for each fixed angular momentum and parity (J, π)-set [1, 2]. Depending on circumstances such a set, or a fixed-π collection of them, forms the "chaotic domain" [3]. Our aim, is to construct, *in this chaotic domain*, the state density as a function of the Hamiltonian parameters (equivalent, therefore, to the partition function) extending upward close to the energy at which particle emission begins and connecting at low energy with the known spectrum, as in Fig. 1. We further carry out the J (or (J, π)) decomposition of the state densities (giving level densities) so as to be able to deal with the observed neutron-resonance spacings.

It is noteworthy that studies of the observable neutron-resonance spectra in heavy nuclei have given birth to the subject of "quantum chaos" with large consequences for a broad class of quantum systems. Our ultimate goal is to "open up" this spectral region in a nucleus that begins with the onset of chaos and extends upward in the spectrum to somewhat above the neutron-resonance (or proton-resonance) energy for a detailed examination of transition strengths, and of symmetry breakings both fundamental (parity, time reversal) and otherwise (including measures for collectivities and (or) localization [4–7].

We proceed, not by conventional matrix methods (hopeless for large dimensions; model spaces for ^{234}U and ^{168}Er extending up to the resonance region are of dimension 10^{19} and 10^{14}, respectively) but via statistical methods that take into account ergodic behavior and in which the generic forms largely given by the (EGOE) of random matrices follow from the dominance of unitary symmetries, central limit theorems, and quantum chaos. We also take advantage of the algebraic simplicities that arise in the narrow neutron-resonance domain to calculate the smoothed state density. Parenthetically, it is good to keep in mind that the GOE, originally proposed by Wigner [8] as the model for a complicated system with a time-reversal invariant Hamiltonian H, and now almost universally regarded as a model for a corresponding chaotic system [3], is an ensemble of multibody, not two-body interactions [9]. This difference shows up both in the one-point and two-point functions; the GOE density is "semi-circular" while that of the EGOE(2) (the embedded ensemble of two-body interactions) is Gaussian; the low-order two-point fluctuations are also radically different [1, 10, 11] though this difference seems not to have been experimentally demonstrated. It is of consequence that the embedded ensembles give a sharp separation between shape fluctuations of the density and intrinsic fluctuations [1, 11, 12] and consequently define a proper density for a realistic many-particle system.

Statistical spectroscopy gives rise to smoothed forms (in the region where level fluctuations are close to GOE) for level densities, expectation values such as occupation numbers, spin-cutoff factors etc., strength sums such as M1, Gamow–Teller (GT) and transition strengths (for example E2, GT etc.,).

Fig. 1. The experimental distribution function (staircase) $F_{ex}(E)$ compared with its theoretical counterpart (continuous curve) $F_{th}(E; G)$ for $^{234}U^{+}$ treated as 10 protons and 16 neutrons distributed over the *spe* spectra shown. The SDI interaction is used. The optimal G value ($= 0.129 \pm 0.001$ MeV) follows by matching the two distribution functions (experimental and theoretical) at $E_r^* = 1.354$ MeV and the two partial level densities (1.8×10^6/MeV) at $E_R = 6.843$ MeV where the total π^+ state density is 7.01×10^8 states/MeV. The spreading width is 2.2 MeV; should we drop the $d_{5/2}(n)$ orbit, we would find $G = 0.141$ MeV and $\sigma_A = 2.0$ MeV. The statistical ground state as indicated by the arrow falls 0.54 MeV below the exact one.

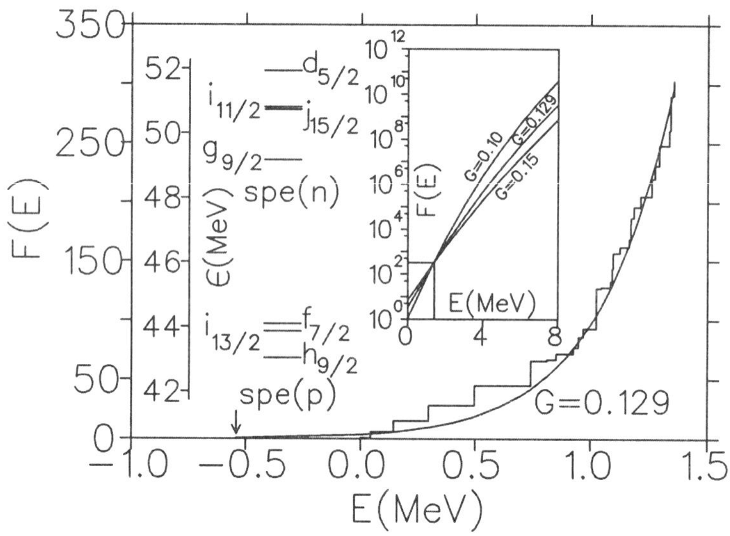

Statistical methods are of course not new [4, 13] but a new understanding of the laws that justify them gives rise to a convolution factoring for densities, simplifies their application and greatly extends the usable model space and energy domains [6, 14]. Statistical and shell-model methods start with the same input and, in principle, treat the same problems. But the statistical laws are least valid in the ground-state domain where shell model diagonalizations (limited however in huge spaces) or more complicated techniques are most successful. Thus, for many purposes the methods are complementary, statistical methods making an extension to indefinitely large spaces of results which originate in relatively small ones, joining the two being in principle achieved by matching numerical values of the spectral distribution function.

Since statistical methods do not extend to the ground state (gs) region, therefore, a definite parity being always understood, our "base-point" will not be the gs but rather the "dominant reference level", the highest level (assumed to be in the chaotic domain) up to which the spectrum is complete, all J values being known. This level and all lower "chaotic" levels we label by E_r (the dominant one by E_r^* when necessary and the approximate minimal one by $\mathcal{E}$); in the ground-state region, as data shows, fluctuations carry information and, therefore, they are of Poisson or semi-Poisson type (this is also a property of EGOE as demonstrated recently [15]). An essential fact now is that, because GOE fluctuations scale with the smoothed density, an EGOE subspectrum with fixed-(J, π) yields, apart from that density function, only a single piece of information that we may take as the mean level spacing in the neighborhood of any one point in the spectrum. Essentially, the same is true for a mixture of EGOEs with different angular momenta.

If, as is very often the case, it is not clear where the chaotic domain begins, we might have to settle for $\mathcal{E}$. It is never satisfactory to choose the ground state as reference; the analysis will not be valid there and the fluctuations that scale with the spacing, will be large in energy units and this of course propagates up the spectrum, a single spacing error at low energy then showing up as thousands of spacings in the resonance region. But it is known from EGOE theory and also from large shell-model calculations that the natural fluctuation unit is one exact symmetry spacing and that measurable fluctuations are of that order, increasing only logarithmically with the effective dimensionality of the space. Thus, the spectral rigidity ensures an accuracy that would be lost should we use the ground-state reference. We note that we stand to learn much about the behaviour of complex systems (in particular about the transition between nonchaotic and chaotic energy domains) as the experimental reference energy is moved up by new experiments.

Now via parametric differentiation on the state density or otherwise we construct angular-momentum distributions, single-state occupancies, and other one-point ($k = 1$) quantities, all as smooth functions of the excitation energy expressible in terms of generic forms which, for polynomial operators in particular, may have extensions into the nonchaotic domain. The techniques used extend to two-point ($k = 2$) correlation functions such as those needed for the study of spectral fluctuations, transitions strengths, and the analysis of measures for broken symmetries (see, for example, refs. 5 and 6) and, in principle to ($k \geq 3$)-point functions which, being essentially unmeasurable, are of little present interest. Thus, a $k = (1 + 2)$ treatment is essentially complete [16]; see however Sect. 6. Moreover there is evidence as in ref. 17 and Fig. 1 that, in accord with the exponential increase of the state density with energy, chaotic behavior, at least in its aspects of interest to us, is established quite rapidly, say at 1–2 MeV excitation (i.e., $\mathcal{E} \sim 2$ MeV) in heavy nuclei whose low-lying spectra are dominated by collectivity (as with the 33 heavy nuclei we have studied in detail). It is expected that statistical spectroscopy becomes operational around 2 MeV excitation energy and should be fully effective at somewhat higher excitation energy (perhaps defined by the region of four quasiparticle excitations for even–even and three quasiparticle excitations for odd-A heavy nuclei).

It should be obvious however that, without dealing successfully with the extremely dense spectrum in the slow-neutron resonance domain, our analysis will be hopelessly uncertain; hence the importance of the slow-neutron spectrum that supplies for us the important "marker" viz the mean spacing of the observed resonances (corrected in lighter nuclei for any secular variation) when the target (t) nucleus is even–even so that $J_t^\pi = 0^+$ and $J^\pi = (1/2)^+$ (since a slow neutron carries only spin angular momentum into the nucleus). For odd-target nuclei J_t is nonzero and we have two ($J_t \pm 1/2, \pi_t$) sequences but in practice only a combination of these two will be identified. In both cases we "see" only a small fraction of the final states. Hence, information from the neutron-resonance region will fix for us an essential parameter of the lower lying state density.

The method of calculating densities essentially consists of three parts. The smoothed chaotic state density is first expressed in terms of the Hamiltonian parameters through an elementary unitary group decomposition of the interaction coupled with the use of spectral-averaging methods as described in Sect. 2. Secondly, the spin-cutoff densities are constructed in much the same way and thereby the spin-cutoff factors for the J-decomposition are produced as described in Sect. 3. Finally, for an accurate determination of the density at high energy, the Hamiltonian parameters are derived from a nontrivial fitting of the theory with the neutron-resonance spectrum (which, for ^{234}U, begins at 6.843 MeV, about 3×10^8 states above the ground state). In general, in heavy nuclei, the observed resonance spectrum (whose lowest energy we label as E_R) may contain up to a few hundred levels spanning a few keV at an excitation energy usually $\sim$(5–8) MeV; below this there may be, say, 10^8 states, almost all being individually unobservable. The state density $I^H(E)$ and the distribution function $F^H(E) = \int_{-\infty}^{E} I^H(X)\,\mathrm{d}X$ (which counts the states up to energy E) then emerge as explicit functions of the Hamiltonian matrix elements. It is a striking aspect of "spectral rigidity" that we can successfully combine "input" information from these two far-separated regions. Indeed the replacement of *extrapolation* (from the shell-model domain) by *interpolation* (between the two observable domains) is an essential feature of the procedure. Note incidentally that the slow-neutron resonances are mostly observable when the parity of the resonance states is the same as that of the target (see references to neutron resonance data in ref. 2). It is of no concern if these functions do not accurately extend much below the (1–2) MeV region. The method is illustrated for the nucleus ^{234}U, treated as an example in Sect. 4. Results for heavy nuclei are described in Sect. 5. A brief discussion of extensions to strength distributions and two-point functions is given in Sect. 6 and concluding remarks are presented in Sect. 7.

2. State densities: unitary group decompositions and convolutions

State densities and their decompositions, by conserved symmetries, which give rise to level densities, by weakly broken symmetries; and by configurations (distribution of the "active" particles among the single-particle (sp) orbits) are important because they are in some cases measurable, and because they enter, for example, into the calculation of transition-strength densities and reaction cross sections. Moreover, when given as functions of the Hamiltonian parameters, they produce a smoothed partition function and determine much about the structure of the nucleus.

We are dealing with a finite-dimensional space, one in which we distribute a certain number, say m, of active nucleons over a finite N-dimensional set of single-particle states to form a space of dimensionality (N, m) in which the Hamiltonian operates. We partition the (sp) model space according to harmonic-oscillator (ho) orbits, the higher orbits being modified as usual for spin-orbit effects. The many-particle states are described first in terms of ho configurations $\boldsymbol{m}$ (distributions of m particles over the ho orbits), which are assigned an "excitation-parameter" s where $s = 0$ labels the ground-state ho configuration while $s = 1$ indicates a $1\hbar\omega$ excitation, etc., Thus, labeling and ordering the ho orbits by Λ_r (with $r = 0, \pm 1, \pm 2, ...$), Λ_0 is so defined that Λ_0^m is the m-particle ground-state configuration with $s = 0$ while $(\Lambda_{-1})^{-1}\Lambda_0^{m+1}$ and $\Lambda_0^{m-1}(\Lambda_1)^1$ have $s = 1$ and in general $s = \sum r\boldsymbol{m}_r$ where $\boldsymbol{m}_r$ counts particles when $r > 0$ and holes when $r < 0$; we take $m = 0$ for a ground-state configuration with only filled ho orbits (for an example, see Table 1). For light nuclei and small s the configurations will have a definite parity, otherwise a parity splitting into two parts labeled by $s^{\pm}$, which we treat separately (often ignoring the superscript). In heavy nuclei (our main interest) all many-particle ho configurations will have essentially equal number of positive-parity and negative-parity states.

The problem is that of calculating the density from the Hamiltonian parameters or, conversely, of extracting Hamiltonian information from the observed density. We consider not the spectrum itself but a smoothed version of it generated by calculating the finite set of low-order spectral moments and using them to fit the densities. In general terms, our results will tell us new things about the propagation of information up through the spectrum, as represented by the relationship between spectra in the ground-state domain and densities at high excitation. It will be useful to divide the interaction into three parts; the first, shifts but does not renormalize the spectrum, and for heavy nuclei at least, is well represented by a "surface-delta" interaction; the second gives rise to the renormalization; and the third is significant only for relatively high energies.

Following ref. 14 (see also refs. 18–20 for a preliminary account and some details), we write the interaction as $(V + \hat{V})$ with $V = V_{t=0}$ and $\hat{V} = \sum_{t>0} V_t$ where V_t (with t integral) connects only configurations differing by $t\hbar\omega$ in the ho energy. But $V_{t>0}$ will mostly connect well-separated states generating a relatively weak admixing and resulting in mainly multimodal densities [21], which (though this is not essential) we can, in the present context, ignore. Then s becomes a good quantum number, the interaction becomes V, which generates close-to-Gaussian density distributions (CLT action due to EGOE(2)), and the Hamiltonian becomes

$$H = C(0) + h(1) + V \tag{1}$$

a sum of (0, 1, and 2)-body parts. The zero-body operator $C(0)$ measures the internal energy of the inactive nucleons and along with the one-body $h(1) = \sum \epsilon_i n_i$ defines the primary mean field in which the active nucleons find themselves.

Table 1. For ^{28}Si we list first (for $s = 0, 1, 2$) the variance classes along with their dimensionalities, and then the spreading widths RMS-averaged over all (dimensionality-weighted) fixed-s configurations as well as the corresponding averages and dimensionalities for each variance class (defined by a single *ho* configuration, and having necessarily a fixed s). Note that $\Lambda_{-2} = ({}^1s_{1/2})$, $\Lambda_{-1} = ({}^1p_{3/2}, {}^1p_{1/2})$, $\Lambda_0 = ({}^1d_{5/2}, {}^2s_{1/2}, {}^1d_{3/2})$, and $\Lambda_1 = ({}^1f_{7/2}, {}^2p_{3/2}, {}^2p_{1/2}, {}^1f_{5/2})$. The decrease for $s = 2$ (and similarly for $s = 1$) in the fractional width fluctuation from 8% to the canonical 3% is due to the nonstatistical structure described in the text. Note incidentally that in averaging, we must count twice each configuration (all but three of them) that is not invariant under $p \leftrightarrow n$ since we have not displayed these separately.

s	Class	Dimensionality	$\sigma_A(s)\,\{1 \pm \delta\}$	$\sigma_A(s;c)\,\{1 \pm \delta\}$
0	$(\Lambda_0)^{6p}(\Lambda_0)^{6n}$	853 776	8.6(1 ± 0.03)	8.6(1 ± 0.03)
1	$(\Lambda_{-1})^{-1p}(\Lambda_0)^{7p}(\Lambda_0)^{6n}$	4 390 848	11.0(1 ± 0.08)	12.5(1 ± 0.03)
	$(\Lambda_0)^{5p}(\Lambda_0)^{6n}(\Lambda_1)^{1p}$	14 636 160		10.6(1 ± 0.03)
2	$(\Lambda_{-2})^{-1p}(\Lambda_0)^{7p}(\Lambda_0)^{6n}$	1 463 616	13.2(1 ± 0.09)	16.8(1 ± 0.03)
	$(\Lambda_{-1})^{-1p}(\Lambda_{-1})^{1p}$	102 453 120		13.9(1 ± 0.03)
	$(\Lambda_{-1})^{-2p}(\Lambda_0)^{8p}(\Lambda_0)^{6n}$	6 860 700		15.8(1 ± 0.02)
	$(\Lambda_{-1})^{-1p}(\Lambda_{-1})^{-1n}(\Lambda_0)^{7p}(\Lambda_0)^{7n}$	22 581 504		15.6(1 ± 0.02)
	$(\Lambda_0)^{4p}(\Lambda_0)^{6n}(\Lambda_1)^{2p}$	86 902 200		12.3(1 ± 0.04)
	$(\Lambda_0)^{5p}(\Lambda_0)^{5n}(\Lambda_1)^{1p}(\Lambda_1)^{1n}$	250 905 600		12.2(1 ± 0.04)
	$(\Lambda_{-1})^{-1p}(\Lambda_0)^{7p}(\Lambda_0)^{5n}(\Lambda_1)^{1n}$	75 271 680		14.0(1 ± 0.03)

We next construct the two-body operator $V^{0,0,0\cdots} = V^{[0]}$ say (a function of the *spherical-orbit* two-body number operators, $n_{\alpha\beta} = n_\alpha(n_\beta - \delta_{\alpha\beta})(1 + \delta_{\alpha\beta})^{-1}$) so normalized that its eigenvalues yield, for every particle number m, all the spherical-configuration centroids. With $N_{\alpha\beta}$ and $m_{\alpha\beta}$ defined similarly by orbit dimensionalities N_α, N_β, with occupancies m_α, m_β, and with $[J] = (2J + 1)$ we have for the $[J]$-weighted diagonal average of V, given explicitly in terms of its two-body matrix elements

$$V_{\alpha\beta} = N_{\alpha\beta}^{-1} \sum_J [J] V^J_{\alpha\beta\alpha\beta}, \qquad V^{[0]} = \sum_{\alpha \le \beta} V_{\alpha\beta} n_{\alpha\beta} \tag{2}$$

Here $V_{\alpha\beta}$ is the $(2J+1)$-weighted average interaction energy for the two particles, one each in spherical orbits α, β which may coincide. Information is of course lost in the J-averaging; see Sect. 3 for a way to recover it.

Now with $N = N_s$, the dimensionality of the single-particle space for the specified s (the parity being understood), we make a $U(N)$ decomposition of $V^{[0]}$ into its three irreducible parts [13] corresponding to the Young columnar shapes $[N - \nu, \nu]$ with $\nu = 0, 1, 2$. Thus, $V^{[0]} = \sum_\nu V^{[0],\nu}$ where the $\nu = 0$ term shifts the entire fixed-s spectrum, giving then a mass-correction term; $\nu = 1$ renormalizes the single-particle energies (*spe*), changing the spectrum (and the ground-state mass) without shifting the centroid; $\nu = 2$ is always negligible. Thus, $V^{[0]}$ simply modifies both parts of the mean field. Finally, with

$$\boldsymbol{C} = C(0) + h^0(1) + V^{[0],0}, \qquad \boldsymbol{h} = h^1(1) + V^{[0],1} \tag{3}$$

we have

$$H = \{\boldsymbol{C} + \boldsymbol{h} + \boldsymbol{V}\} \tag{4}$$

where $\boldsymbol{V} = V - V^{[0]}$, easily found to be $U(N)$-irreducible with $\nu = 2$, generates the spreading and admixing of the configurations.

When $\boldsymbol{V} = 0$, we have an effective $(0 + 1)$-body Hamiltonian (i.e., one which for fixed particle number, is not distinguishable from a true $(0 + 1)$-body operator) whose spectrum is a set of "spikes" with the degeneracies of the spherical configurations and the energies and parities of $(\boldsymbol{C}+\boldsymbol{h})$. When $\boldsymbol{V}$ is "turned on" the spikes spread and admix without moving their centroids (this because, by construction $\boldsymbol{V}$ makes no contribution to any centroid) and thus form, for each particle number, a rigid lattice. For our immediate purpose, it is adequate to take the spreadings as Gaussian; then we need only calculate the configuration (spreading) variances $< \boldsymbol{V}^2 >^{\boldsymbol{m}}$ and combine them with the dimensionalities and centroids to produce the final $[J]$-averaged densities. It is worth noting that a similar decomposition of the interaction into a monopole and a multipole (non-monopole) part has been stressed by Zuker [22] in the context of detailed spectroscopy via the Shell Model.

To make clear the roles played by different parts of the interaction and to indicate the explicit convolution factoring used in the neutron-resonance data analysis of the state density, we proceed as follows: with $d(m)$ and $d(\boldsymbol{m})$ the total and configuration dimensionalities, with $\langle\langle \mathcal{G} \rangle\rangle^m = d(m)\,\langle \mathcal{G} \rangle^m$ (similarly for $d(\boldsymbol{m})\,\langle \mathcal{G} \rangle^{\boldsymbol{m}}$) the traces of an operator $\mathcal{G}$, with $< \boldsymbol{h} >^{\boldsymbol{m}} = \epsilon(\boldsymbol{m})$ the corresponding configuration centroids (here and from now onwards, the constant $\boldsymbol{C}$ in (4) is absorbed, without loss of generality, in $\boldsymbol{h}$), with $A \otimes B[E]$ the convolution, evaluated at E, of functions A, B, we have first for fixed s, and then for the total density,

$$
\begin{aligned}
\rho^{\boldsymbol{h}+\boldsymbol{V},m}(E) &= \langle \delta(\boldsymbol{h}+\boldsymbol{V}-E) \rangle^m \\
&= d(m)^{-1} \sum_{\boldsymbol{m}} d(\boldsymbol{m})\, \langle \delta(\boldsymbol{h}+\boldsymbol{V}-E) \rangle^{\boldsymbol{m}} \\
&\rightarrow d(m)^{-1} \sum_{\boldsymbol{m}} d(\boldsymbol{m})\, \langle \delta(\epsilon(\boldsymbol{m})+\boldsymbol{V}-E) \rangle^{\boldsymbol{m}} \\
&= d(m)^{-1} \sum_{\boldsymbol{m}} d(\boldsymbol{m}) \int \mathrm{d}z\, \langle \delta(\boldsymbol{h}-z) \rangle^{\boldsymbol{m}}\, \langle \delta(\boldsymbol{V}-E+z) \rangle^{\boldsymbol{m}} \\
&\rightarrow d(m)^{-1} \sum_{\boldsymbol{m}} d(\boldsymbol{m}) \rho^{\boldsymbol{h},\boldsymbol{m}} \otimes \rho^{\boldsymbol{V},m}[E] \\
&= \rho^{\boldsymbol{h},m} \otimes \rho^{\boldsymbol{V},m}[E]
\end{aligned}
\tag{5}
$$

$$
\begin{aligned}
I^H(E) &= \sum_s I^{\boldsymbol{h}+\boldsymbol{V},s}(E) = \sum_s I^{\boldsymbol{h},s} \otimes \rho^{\boldsymbol{V},s}[E] \\
I^{\boldsymbol{h},s}(E) &= \sum_{\boldsymbol{m}} d(\boldsymbol{m},s)\delta(E-\epsilon(\boldsymbol{m},s)), \qquad \epsilon(\boldsymbol{m},s) = < \boldsymbol{h} >^{\boldsymbol{m},s}
\end{aligned}
\tag{6}
$$

The form following the first arrow in (5) derives from the Gaussian nature of $\rho^{\boldsymbol{h}+\boldsymbol{V},\boldsymbol{m}}$ which, if strictly valid, would imply that, for every $\boldsymbol{m}$, all moments may be expressed in terms of the first two, $< (\boldsymbol{h}+\boldsymbol{V}) >^{\boldsymbol{m}}$ and $< (\boldsymbol{h}+\boldsymbol{V})^2 >^{\boldsymbol{m}}$, in both of which $\boldsymbol{h} = \epsilon(\boldsymbol{m})$; we may say that in these moments $\boldsymbol{h}$, $\boldsymbol{V}$ commute to within the fluctuations described by departures from Gaussian. In the form following the second arrow, we have assumed (also as discussed and verified in ref. 14) that the fluctuations in the spreading variances for configurations of the same s may be ignored, i.e., that $\rho^{\boldsymbol{V},\boldsymbol{m}} = \rho^{\boldsymbol{V},m}$. Note that the decoupling in $\int \mathrm{d}z$ which gives rise to the final form derives from the fact that the $\boldsymbol{m}$ are degenerate eigenspaces of $\boldsymbol{h}$; the spreading densities $\rho^{\boldsymbol{V},\boldsymbol{m}}$ are all zero-centered and of unit integral. For completeness, we remark that generic $\rho^{\boldsymbol{V}}$ densities may have shape corrections (especially near the ground state and provided that cumulants of order ≤ 4 have fairly small fluctuations, for which there is some evidence).

It is possible to take into account shape corrections to the Gaussians via the skewness and excess parameters in a number of different (canonical as in Appendix A or noncanonical) ways [10, 23]; for the most recent attempt in this direction see ref. 24. But still the asymptotic CLT action and the constancy of variances are the things that really count. Note also that we can use the isospin formalism, which treats nucleons (η) as identical, or the (p,n) formalism, as is done here, which treats p and n as distinct. Before proceeding further two remarks are in order. First, in applying (6), as the number of configurations $\boldsymbol{m}$ will be in general large (e.g., 21 282 in the $s = 0^+$ space for ^{234}U), we may reduce it by combining sets of close-lying spherical orbits into "unitary" orbits. Secondly, the s summation in (6) is better replaced by unitary configurations defining various "classes" and we will now briefly discuss the "class" structure.

Spreading variances are determined by transitions generated by $\boldsymbol{V}$, necessarily between states of the same s^π, which, when $s > 0$, may or may not belong to the same "class" as defined by the *ho*-orbits involved. "Variance classes" arise because the active harmonic oscillator orbits have different radial functions and angular momenta thereby imposing on the system a nonstatistical structure, specifically in giving different variances for the various harmonic oscillator configurations. A striking result is the essential constancy of the spreading variances in the same class; for example, with ^{28}Si, described in terms of protons and neutrons distributed over the lowest four *ho* (10 spherical) orbits, we find for $s = 0, 1$ (respectively with 1, 2 classes), the spreading widths $\sigma_A = 8.6(1 \pm 0.03)$, $12.5(1 \pm 0.03)$, and $10.6(1 \pm 0.03)$ MeV as given by a Kuo interaction [25] RMS-averaged over the final configurations; similarly for $s = 2$ the widths are different for the seven classes, with $12.2 \le \sigma_A \le 16.8$ MeV but those of each class are constant with a $3 \pm 1)\%$ fluctuation; without class partitioning the fluctuation is 8%. See Table 1 for further details. Incidentally, the $s = 0$ dimensionality for ^{28}Si is 853 776 and that for the largest $s = 2$ class is 250 905 600 (for a $s = 0 \oplus 2$ analysis of ^{28}Si level density data using (6) and the results in Sect. 3, see ref. 19). The spreading widths for heavy nuclei are smaller than above, ~2.2 MeV for ^{234}U$^+$ (i.e., for the π^+ (positive parity) states of ^{234}U) with $s = 0$; see Table 4 ahead. Finally, it should be added that constancies (of the variances) derive from the nature of the admixings generated by CLT's; this is an aspect of quantum chaos [26] and a EGOE basis for this result is given in ref. 27.

It should be remarked that that the unitary-group convolution factoring of the general type $\rho^H = \rho^C \otimes \rho^h \otimes \rho^V$ is a result of considerable significance making clear the role played by the conventional noninteracting particle (NIP) theory [28, 29], which treats the nucleus as an ideal Fermi system and which has been used for half a century, with empirical determinations of the parameters and minor ad hoc modifications of the form; for reviews and extensions see ref. 30. The *form* of the NIP density is more or less preserved in the interacting-particle (IP) case, i.e., when we take account of the *residual* interaction (it being understood that a field-forming interaction is already taken into account). The principal effects of the interaction are then a (constant) shift and a centroid-preserving renormalization of the NIP Hamiltonian (i.e. a change in the *sp* energies representing modifications of the mean field) along with a local spreading and admixing of the NIP states. The spreading effects are representable by a convolution since the spreading width varies slowly with energy. We stress that the spreading, though narrow, necessarily has a profound effect on the ground-state energy, on the structure of the states, and on the state-density parameter that appears in Bethe's equation.

In light nuclei, say through the *ds* shell (for which the dozen or so interactions proposed have essentially the same norms for $\boldsymbol{V}$, effectively differing only in the induced *spe*) and perhaps well into *fp* shell, the interaction is known well enough to yield good renormalizations of the *spe*. This is not true for the heavier nuclei for which it is more or less standard to use the surface-delta interaction, the SDI of refs. 31 and 32.

$$V^{\mathrm{SDI}}(1,2) = -4\pi G D^{-1} \delta\left(\boldsymbol{r}(1) - \boldsymbol{r}(2)\right) \delta\left(r(1) - R_0\right) \tag{7}$$

where R_0 is a radius parameter, D a radial integral, and G the strength-parameter (to be determined ahead). This interaction does *not* renormalize the primary $\nu = 1$ *spe*. Now instead of seeking a badly known renormalization of a badly known primary *sp* spectrum we simply take (for the *spe* of the heavy nuclei) the zero-deformation limit of the Nilsson deformed-orbit spectrum [33], a procedure closely connected with experiment.

Table 2. Propagating polynomials $R_{[p]}$ for the central moments $\mu_p(m)$, $p \geq 2$. Note that $\mu_p(m) = \sum_{[p]=[p_1,p_2,\ldots]} R_{[p]} \prod_i \mu_{p_i}(1)$ where $[p]$ is a partition of p and the partitions with any $p_i = 1$ will not appear in the summation. In this table $g(p) = m(N-m)[(N-1)(N-2)\ldots(N-p+1)]^{-1}$.

p	$[p]$	$R_{[p]}$
2	[2]	$g(2)$
3	[3]	$g(3)(N-2m)$
4	[4]	$g(4)[N(N+1)-6m(N-m)]$
	[22]	$g(4)[3(m-1)(N-m-1)N]$
5	[5]	$g(5)\{(N-2m)[N(N+5)-12m(N-m)]\}$
	[32]	$g(5)[(N-2m)10N(m-1)(N-m-1)]$
6	[6]	$g(6)[120m^2(N-m)^2-30N(N+3)m(N-m)+N(N+1)(N^2+15N-4)]$
	[42]	$g(6)\{15N(m-1)(N-m-1)[(N+4)(N-1)-6m(N-m)]\}$
	[33]	$g(6)\{10N(m-1)(N-m-1)[(N^2-N+4)-4m(N-m)]\}$
	[222]	$g(6)[15N^2(m-1)(m-2)(N-m-1)(N-m-2)]$

We make use of (6) and (7) to construct state densities for heavy nuclei. The spreading variances defining ρ^V depend only on the G parameter of the SDI interaction and they follow easily from the formulation given in ref. 13. The $\boldsymbol{h}$ (or $\boldsymbol{C}+\boldsymbol{h}$) is determined by both G and the primary sp energies and one can construct ρ^h either exactly or by using unitary orbits approximately. We follow the later approach as it is more economical. Grouping the single-particle states into unitary orbits i (with degeneracy N_i), which are say L in number, the number of single-particle states N will be a sum of N_i and similarly h will be a sum of h_i. Note that with unitary orbits, parity decompositin of I^h is immediate. Now it is easy to see that I^h is a sum of unitary configuration densities. Let us say that $[m_1, m_2, \ldots, m_L]$ is a unitary configuration, then an important theorem is that $I^{[m_1,m_2,\ldots,m_L]}$ is a convolution of densities in each unitary orbit,

$$
\begin{aligned}
I^{h,m}(E) &= \sum_{[m_1,m_2,\ldots,m_L]} I^{h,[m_1,m_2,\ldots,m_L]}(E) \\
I^{h,[m_1,m_2,\ldots,m_L]}(E) &= I^{h_1,m_1} \otimes I^{h_2,m_2} \otimes \ldots \otimes I^{h_L,m_L}(E)
\end{aligned}
\tag{8}
$$

The convolution form implies that the cumulants of $I^{h,[m_1,m_2,\ldots,m_L]}(E)$ are sum of the cumulants of the single-orbit density cumulants [34]. Using this, $I^{h,[m_1,m_2,\ldots,m_L]}(E)$ are constructed in terms of their first four cumulants as Edgeworth-corrected Gaussians (Appendix A) and then $I^{h,m}$ as a sum of these as given by the first equality in (8). For completeness we show in Table 2 explicit forms of the first six central moments of single orbit NIP state densities.

3. Spin-cutoff densities and spin-cutoff factors

The parity decomposition of the state density, with spherical or unitary orbits, is immediate. We need also the J-decomposition for the neutron-resonance marker. For the fixed-(J,π) partial densities, we introduce σ_J, Bethe's "spin-cutoff factor" whose square is the locally smoothed expectation value, $< E\pi|J_z^2|E\pi >$; for brevity from now on we drop the parity label π. An often used but inadequate method simply sums the J_z^2-expectation values over the separate spherical configurations, thereby ignoring the J^2V and the larger J^2V^2 correlations that result from the interaction between configurations. In the lower inset of Fig. 3 ahead we see the very large errors, especially at low energy, which result from ignoring them. The spin-cutoff factor combines with the state density $I(E)$ via Bethe's equation [28]

$$
I_l(E,J) \simeq I(E)(2J+1)(2\sigma_J^2(E))^{-3/2}\pi^{-1/2}\exp\{-(J+1/2)^2/2\sigma_J^2(E)\} \tag{9}
$$

to give the calculated *level* density. We discuss next the methods of calculating the energy-dependent spin-cutoff factor $\sigma_J(E)$ and this will complete the formalism for the use of neutron resonance data in constructing the density function.

Taking degeneracies in the spectrum into account, a theory for $\sigma_J^2(E)$ should start with spin-cutoff densities $I_{J_z^2}^{H,m}(E)$

$$I_{J_z^2}^{H,m}(E) = \left\langle\left\langle J_z^2\delta(H-E)\right\rangle\right\rangle^m, \qquad \int I_{J_z^2}^{H,m}(E)\,\mathrm{d}E = \left\langle\left\langle J_z^2\right\rangle\right\rangle^m$$

$$\rho_{J_z^2}^{H,m}(E) = \left[\left\langle J_z^2\right\rangle^m\right]^{-1}\left\langle J_z^2\delta(H-E)\right\rangle^m, \qquad \int \rho_{J_z^2}^{H,m}(E)\,\mathrm{d}E = 1, \qquad M_p(J_z^2) = \left[\left\langle J_z^2\right\rangle^m\right]^{-1}\left\langle J_z^2H^p\right\rangle^m$$

$$\left\langle J_z^2\right\rangle^E = \frac{I_{J_z^2}^{H,m}(E)}{I^{H,m}(E)} \tag{10}$$

To derive the forms for $I_{J_z^2}^{H,m}(E)$ we begin with the $s=0$ case. Starting with $H_\lambda = H + \lambda J_z$ and the corresponding deformed density $\rho^\lambda = \rho^{H+\lambda J_z}$ and recognizing that the J_z operator commutes with H, the spin-cutoff factor follows ref. 35 from the double parametric derivative of ρ^λ

$$\left\langle J_z^2\right\rangle^E = [\rho(E)]^{-1}\lim_{\lambda\to 0}\frac{\partial^2}{\partial\lambda^2}\int_{-\infty}^{E}\mathrm{d}y\int_{-\infty}^{y}\rho^{H+\lambda J_z}(x)\,\mathrm{d}x \tag{11}$$

Starting with the Edgeworth expansion of ρ^λ (see Appendix A)(11) gives a calculable series expansion for the spin-cutoff factors. This result based on the assumption that ρ^λ is close to a Gaussian, a condition well-satisfied as long as ρ is close to Gaussian; note that the J_z spectrun is uniform and it adds to the one-body part of H. It should be remarked that the result obtained here is seen to be identical to the form obtained with the bivariate Edgeworth representation for fixed-(E, M) densities [36]; M's being the eigenvalues of the J_z operator. An alternative to the parametric differentiation is to use the polynomials defined by the state identity $\rho(E)$ for expanding the δ-function in (10); see ref. 4 for details of the polynomial expansion. A third method is to represent $\rho_{J_z^2}^{H,m}(E)$ directly as an Edgeworth-corrected Gaussian (Appendix A) using its exact first four cumulants. This "moment method" is seen to give an excellent representation for spin-cutoff factors. An example is shown in Fig. 2. It should be remarked that the double parametric derivative method is equivalent to using a nonstandard Edgeworth form [34] for $\rho_{J_z^2}^{H,m}(E)$, and hence, we understand why the moment method gives a better representation. A final remark is that the moments of spin-cutoff density can be written as double parametric derivatives of the state-density moments,

$$(p+2)(p+1)\left\langle J_z^2H^p\right\rangle^m = \lim_{\lambda\to 0}\frac{\partial^2}{\partial\lambda^2}\left\langle (H+\lambda J_z)^{p+2}\right\rangle^m \tag{12}$$

Therefore, the results in Table 2 immediately give the first four NIP spin-cutoff moments, a result we shall use later.

We now extend our formalism to take account of the convolution factoring (5,6) of the state density. With $H = \boldsymbol{h} + V$ and J_z being strictly a $\nu = 1$ operator, the $H_\lambda = H + \lambda J_z \to (\boldsymbol{h} + \lambda J_z) + V$. Then the convolution form for the deformed density $\rho^{H_\lambda} = \rho^{\boldsymbol{h}+\lambda J_z} \otimes \rho^V$ with a double parametric derivative as in (11) gives,

$$\rho_{J_z^2}^{H}(E) = \rho_{J_z^2}^{\boldsymbol{h}} \otimes \rho^V[E] \tag{13}$$

a particularly simple form, the second convolution factor being the state-density spreading function while the first involves only the NIP density (six moments as given in Table 2 are needed, as follows

Fig. 2. Square of the spin-cutoff factor $\sigma_{J_z}^2(E) = \langle J_z^2 \rangle^E$ versus E. Results for the moment method (continuous curve), parametric derivative approach (dot-dash), and the polynomial expansion (broken line) are compared with exact shell model results (open squares) in $(2s1d)^{m=5,T=1/2}$ space. In the exact results, averaging is done over a bin size of $0.3\sigma \simeq 2.3$ MeV; σ is the spectrum width. For the state density (with fixed T) the first four cumulants (ϵ, σ, γ_1, γ_2) are (−25.36 MeV, 8.01 MeV, 0.173, −0.031). Similarly, for the spin-cutoff density the parameters are (−26.49 MeV, 7.23 MeV, 0.233, −0.01). Finally, the calculations use the Kuo interaction [25] with ^{17}O sp energies $(\epsilon_{d_{5/2}}, \epsilon_{s_{1/2}}, \epsilon_{d_{3/2}}) = (-4.15, 0.93, -3.28)$ MeV. It is seen from this figure that the first two methods give excellent agreement with the shell model over the entire spectrum but the polynomial method fails near both ends, a result previously demonstrated by Grimes et al. [37]. The agreement with the moment method shows that the spin-cutoff density is well represented by an Edgeworth-corrected Gaussian.

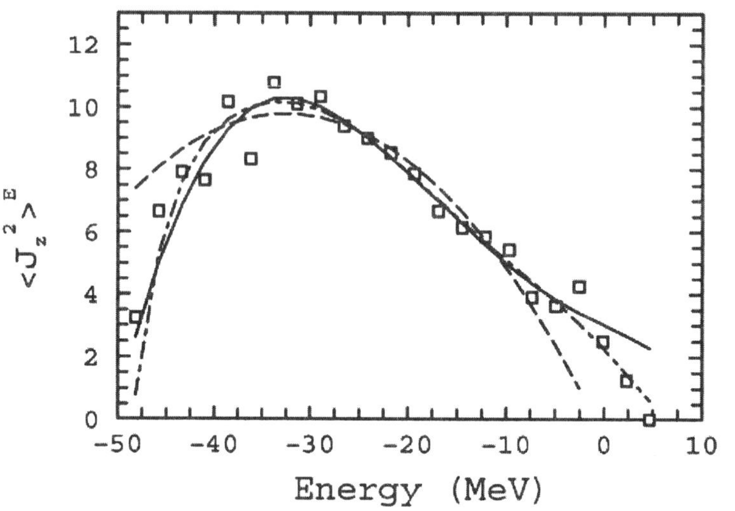

from (12), to produce a four-moment (Edgeworth) $\rho_{J_z^2}^{h}$ density). A form improved over (13) replaces ρ^V by $\rho_{J_z^2}^{V}$,

$$\rho_{J_z^2}^{H}(E) = \rho_{J_z^2}^{h} \otimes \rho_{J_z^2}^{V}[E] \tag{14}$$

The point here being that, with configuration partitioning the partial $\rho_{J_z^2}^{H,m}(E)$ densities are to a good approximation, except for the centroids, well-represented by $\rho_{J_z^2}^{V}$. This requires that the configuration spin-cutoff density variances be nearly constant and equal to the variance $\sigma_{V,J_z^2}^2$ of $\rho_{J_z^2}^{V}$. Table 3 gives some tests of this result. The centroids $\langle J_z^2 H \rangle^m / \langle J_z^2 \rangle^m = \langle H \rangle^m + \Delta(\boldsymbol{m}) = \langle \boldsymbol{h} \rangle^m + \Delta(\boldsymbol{m})$ and the $\Delta(\boldsymbol{m})$, giving the centroid shift, are well approximated by $\langle J_z^2 V \rangle^m / \langle J_z^2 \rangle^m$. With these we have (14), which takes into account the $J^2 V$ and $J^2 V^2$ correlations that are missing in (13). We adopt (14) with the NIP part constructed as an Edgeworth-corrected Gaussian following the procedure defined by (8) and (12); in applying (8) and (12), h is replaced by $h + \lambda J_z$. The centroids and variances of $\rho_{J_z^2}^{V}$ are generated following the trace propagation results given in refs. 18 and 39; see also the second reference in ref. 10. Finally, (14) with extension to include summation over subspaces defined by s-values and the related class structure just as the case with (6), gives spin-cutoff factors via (10).

Fig. 3. For ^{234}U we give, with the optimal G value as well as for $G = 0.15$ and 0.10 MeV (*a*) the total state density; (*b*) the partial level density for $J = (2, 3)^+$ (since ^{233}U has a $J = 5/2^+$ ground state); (*c*) the spin-cutoff factor for $G = 0.129$ when the J^2V and J^2V^2 correlations are ignored; (*d*) when only J^2V correlations are treated; (*e*) when both types of correlations are treated. Note the large correlation effects, especially at low energies.

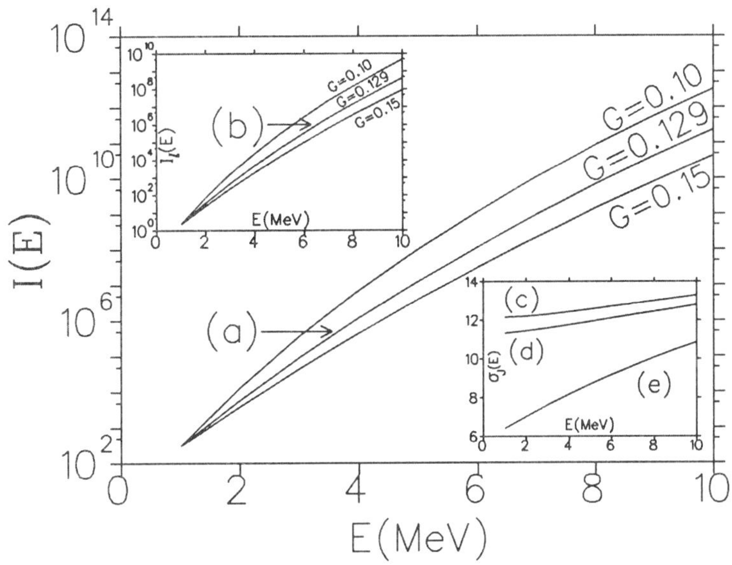

4. An example: the nucleus ^{234}U

To illustrate the method we consider ^{234}U as an example. The procedure is outlined in the following steps.

1. Define the spectroscopic space in terms of the number of active nucleons distributed over a set of active spherical orbits, the size of the set being determined by, among other considerations, the maximum excitation energy of interest. For ^{234}U treated as 10 protons and 16 neutrons and for excitation energy $\leq$8 MeV, a reasonable first choice of orbital sets is $(h_{9/2}, i_{13/2}, f_{7/2})$ for protons and $(g_{9/2}, j_{15/2}, i_{11/2}, d_{5/2})$ for neutrons.

2. Choose the single-particle energies, ϵ_i, thereby defining the one-body Hamiltonian $h(1)$. As mentioned earlier, for many heavy nuclei these will derive from the zero-deformation limit of the Nilsson Hamiltonian, a spherical harmonic oscillator with a spin-orbit and an l^2 term [33].

3. Choose the two-body interaction V. We use the surface-delta interaction (SDI) with one free parameter $G = G_{\mathrm{pp}} = G_{\mathrm{nn}} = G_{\mathrm{pn}}$ to be determined using experimental data.

4. Assign s-values for the orbits. We take $s = 0$, which is appropriate for most purposes up to 6–8 MeV excitation. Beyond that we need $s > 0$ (see item 15 ahead).

5. Carry out the $V = V^{[0]} + V$ decomposition observing that $V^{[0]}$ modifies only the effective NIP structure and V only the spreading variances. For the SDI with $G = 1$ the spherical configuration generator, $V^{[0]}$, is defined by the average two-particle matrix elements. Then, with $G = 1$, $V_{\alpha\beta} = -N_\alpha/2(N_\alpha - \delta_{\alpha\beta})$ for pp and nn and $G = -1$ for pn.

Table 3. Average widths in $(2s1d)^m$ space. For a given m, the results in the first row are for the Kuo interaction [25] and second row are for the Chung–Wildenthal interaction [38]. The quantity σ_A^2 is the average spherical-orbit configuration $(\boldsymbol{m})$ variance for the state density and σ^2_{A,J_z^2} is the corresponding variance for the spin-cutoff densities. Similarly σ^2_{V,J_z^2} is the variance of $\rho^V_{J_z^2}$. Note that the widths shown in this table have particle hole symmetry, $\sigma(m) = \sigma(24 - m)$. The last column gives the percentage deviation of the widths of $\rho^{H,\boldsymbol{m}}_{J_z^2}$ from σ_{A,J_z^2} (i.e., $\sigma(\rho^{H,\boldsymbol{m}}_{J_z^2}) \simeq \sigma_{A,J_z^2}\{1 \pm \delta/100\}$). The results in columns 3–5 clearly show that $\sigma_{A,J_z^2} \sim \sigma_{V,J_z^2}$ and the fluctuations in the variances are small as required by (14).

m	σ_A	σ_{V,J_z^2}	σ_{A,J_z^2}	$\delta(\%)$
4	4.87	4.59	4.54	5
	4.98	4.70	4.64	4
6	6.91	6.49	6.41	4
	7.07	6.65	6.56	4
8	8.37	7.84	7.74	4
	8.55	8.03	7.93	4
12	9.53	8.92	8.80	3
	9.74	9.14	9.01	4

6. Combine close-lying spherical orbits with the same (s, π) to define larger, not necessarily maximal "unitary" orbits. The minimum number of unitary orbits in our example is $(2 + 2)$, thus $h_{9/2} \otimes f_{7/2}$, $i_{13/2}$ for protons and $g_{9/2} \otimes i_{11/2} \otimes d_{5/2}$, $j_{15/2}$ for neutrons, these arrangements of course being dominated by parity considerations.

7. Generate appropriate unitary configurations and calculate the corresponding spreading variances produced by V. There are 94 positive parity and 93 negative parity unitary configurations which should be compared to 21 282 positive and 21 117 negative parity configurations had we used spherical orbits. The general result is that, from proper unitary configurations, we may expect a large reduction in complexity with only a relatively small decrease in the accuracy. Finally, the dimensionality weighted average variance, σ_V^2, with $G = 1$ MeV is 294 MeV^2 for both positive and negative parity configurations.

8. With respect to the unitary orbital space defined above, decompose $h(1)$ and $V^{[0]}$ into unitary tensor parts to obtain the renormalized (occupancy-dependent) one-body Hamiltonian h which determines the "effective" NIP density expressed in terms of configuration densities, each one of which is represented by a four-moment Edgeworth expansion.

9. Construct the state density by convoluting the NIP configuration densities with the spreading Gaussian due to V and adding these to obtain the theoretical state density $I_{\rm th}(W)$ and, by integration, the distribution function $F_{\rm th}(W)$.

10. Following the same procedure, construct the spin-cutoff density $I_J(W)$ (the function which when divided by the state density, gives the spin-cutoff factor, σ_J^2 as a function of energy). The NIP part of the J-density is expressed by an Edgeworth expansion in terms of moments $h^p(J^2)^q$ and the spreading function involves correlations $< J^2V >$ and $< J^2V^2 >$. The average centroid shift and variance of the spin-cutoff density with $G = 1$ is 0.35 MeV and 280 MeV^2, respectively.

11. The known fixed-parity spectrum of ^{234}U [40] has of course two segments. For ^{234}U$^+$ the ground-state segment extends to $E_{\rm r}^* = 1.354$ MeV where $J = 6$. It consists of 29 levels, the largest J-value being 14 (at 1.341 MeV) and the total number of states is 303. The experimental distribution function $F_{\rm ex}(E)$, (a staircase with step-heights determined by the level degeneracies and widths by the level spacings) counts all states with energy $< E$ and half of the states at E if E is an H-eigenvalue. Thus, $F_{\rm ex}(E_{\rm r}^*) = 296.5$. Well above this segment the density of $J^+ = (2+3)^+$ levels (^{233}U has ground state $5/2^+$) is measurable via the observable resonance spectrum that begins at 6.843 MeV where the partial *level* density is 1.8×10^6/MeV [41].

12. We are left now with three problems, determining G, filling in the (1.354–6.843) MeV gap, and extending the experimental resonance subspectrum to all J values (and both parities). For the first we construct, in terms of the interacting Gaussian partial densities and for a reasonable range of the interaction parameter G, the theoretical state density $I_{\rm th}^H(E; G)$ and, by integration, the smoothed distribution function $F_{\rm th}^H(E; G)$. For the optimal G ($= 0.129 \pm .001$) MeV where the error derives from a 10% uncertainty in the resonance density) we have proceeded as in the Fig. 1 caption but (to confirm stability) have averaged over a narrow segment proceeding downward from $E_{\rm r}^*$. The question then is whether this choice of G will result in the two spectra tracking well in the chaotic domain starting *downward* from $E_{\rm r}^*$ (but eventually deviating, at energy $\mathcal{E}$, when the statistical assumptions can no longer reproduce the dilute low-lying rigid structure that in ^{234}U is rotational). That this requirement is remarkably well met in ^{234}U is obvious from Fig. 1, which above $\mathcal{E} \sim 1.0$ MeV confirms the accuracy of our smoothed calculations and indicates the downward extension of the chaotic domain.

13. There is no a priori matching requirement for the experimental and theoretical *ground-state* domains since chaotic behavior determines the second but, near the ground state, not the first. There is however evidence (see the arrow in Fig. 1) that the statistical ground state falls *below* the exact one by an amount $\Delta \sim 0.5$ MeV (0.54 MeV in ^{234}U$^+$), a deviation that rapidly disappears as we go up in the spectrum. In general, Δ varies from 0.3 MeV for ^{150}Sm to 0.9 MeV for ^{235}U for the 33 heavy nuclei studied. It would then seem reasonable to regard this as a general rule that could be used (with the resonance spectrum) for nuclei whose low-lying spectrum is little known.

14. Figure 3 shows the total state density and partial level density for the resonance spins $(2, 3)^+$. The J decomposition of the state density is carried out using the convolution form (14). The inset figure showing $\sigma_J(E)$ versus E exemplifies two general "spin" phenomena, the (not surprising) increase of the spin-cutoff factor, σ_J, with energy, and the reduction in σ_J that arises predominately from the (J^2V^2) correlation or, equivalently, from the fact that the J_z^2 density has a width that is ~4% less (for example, for ^{234}U) than the state density width. The latter is significant; under $\sigma \to \sigma + \partial\sigma$ for a zero-centered Gaussian density, we find that, at energy E, $\delta\rho(E)/\rho(E) = \{(E/\sigma)^2 - 1\}\delta\sigma/\sigma$; actual state densities are dominated by Gaussian densities well removed from their centroids. This is clearly seen in Fig. 3, which shows a correlation-effect reduction of σ_J, for example, by

$\sim$25% at 4 MeV and $\sim$15% at 8 MeV. *We conclude that all elementary spin-cutoff calculations are invalid since they necessarily omit the correlation*; moreover, since computing the (J^2V^2) correlation in large spaces takes much time, accepting the $\sim$4% width reduction as applying in general, which is probably reasonable, would greatly simplify calculation of the J distribution; see also ref. 20.

15. Finally, to check accuracy we expanded the spectroscopic space and treated ^{234}U as 16 protons and 24 neutrons distributed in the orbital sets $(d_{3/2}, s_{1/2}, h_{9/2}, i_{13/2}, f_{7/2}, f_{5/2}$, and $p_{3/2})$ and $(f_{5/2}, p_{1/2}, g_{9/2}, j_{15/2}, i_{11/2}, d_{5/2}, g_{7/2}, s_{1/2}$, and $d_{3/2})$, respectively. For the decomposition $V = V_{t=0} + \hat{V}$ of the interaction, we group and label the *ho* orbits by Λ_r as follows: $\Lambda_{-1}(p) \equiv (d_{3/2}, s_{1/2}), \Lambda_0(p) \equiv (h_{9/2}, i_{13/2}, f_{7/2}), \Lambda_1(p) \equiv (f_{5/2}, p_{3/2}), \Lambda_{-1}(n) \equiv (f_{5/2}, p_{1/2}), \Lambda_0(n) \equiv (g_{9/2}, j_{15/2}, i_{11/2}, d_{5/2})$, and $\Lambda_1(n) \equiv (g_{7/2}, s_{1/2}, d_{3/2})$. The ground-state configuration $(\Lambda_{-1}(p))^6(\Lambda_0(p))^{10}(\Lambda_{-1}(n))^8(\Lambda_0(n))^{16}$ has $s = 0$. Note that the orbital grouping is by energy, the maximum s values are 16 and 22 for protons and neutrons and the total s-excitation is 38. Whereas, it is possible to consider this huge space of all s-excitations, a smaller subspace with the lowest few s values may be adequate. We, therefore, consider the $s = 0 + 1 + 2$ subspace, which should be appropriate for most purposes up to 10 MeV excitation. The dimensionalities of the s = 0, 1, and 2 subspaces are 1.34×10^{19}, 5.01×10^{20}, and 8.98×10^{21}, respectively. For positive parity, we obtain, with $G = 1$, σ_V^2 as 294, 333, 371 MeV^2 for s = 0, 1, and 2, respectively. The average centroid shift and variance of spin-cutoff densities with $G = 1$ is (0.35, 280), (0.31, 318), (0.26, 356) for the three s values. We find that $G = 0.131$ MeV reproduces the experimental level density. A study of the density as a function of excitation energy indicates that the relative contributions of s = 1 and 2 are small in the region of interest (at E_R, the density is 88% $s = 0$ and 12% $s = 1$) and small even somewhat beyond (at 10 MeV excitation, the numbers are 76%, 20%, 2% while at 20 Mev the contributions are 27%, 55%, and 27%). These results indicate that an $s = 0$ calculation should usually suffice. Indeed, the change in G for the truncated space ($s = 0$ only) is 1.5%.

5. Results for heavy nuclei

For the 33 heavy nuclei (A = 150–240; ^{150}Sm–^{240}Pu) studied the results are given in Table 4. This set is essentially the same as was used for the level fluctuation analysis [2, 16] and represents nuclei for which adequate data are available both in the ground-state and neutron-resonance region. We have taken the *sp* space as belonging to a single harmonic oscillator orbit (along with the intruder opposite parity state). The model space dimensionalities vary from 10^{14} for ^{167}Er to 10^{20} for ^{240}Pu. The orbital sets for protons and neutrons contain orbits of both parities. This will be particularly important in dealing with nuclei (for example, ^{150}Sm, ^{156}Gd) where the parity of the s-wave resonances is opposite to the parity of the low-lying spectrum and in those rare cases (^{207}Pb, ^{239}U) where, in addition, p-wave resonances have been observed. In the majority of cases, the parity of the low-lying spectrum coincides with that of the neutron-resonance spectrum, as in ^{234}U, this because the "wrong-parity" resonance spectrum (p-wave resonances) is scarcely observable. If we assume, as in (7), that the G-parameter is indeed independent of parity, or if we have a rule connecting $G^{\pm}$, we can deal with the entire chaotic domain and a correspondingly enlarged range of phenomena. Moreover, if we admit two or more independent interaction parameters we could (if the data were adequate) fix *them* by considering two or more close-lying nuclei. For the nuclei studied, the spreading variances range from 1.67 MeV^2 for ^{171}Yb to 5.93 MeV^2 for ^{240}Pu and the strength parameter G from 0.09 MeV for ^{171}Yb to 0.15 MeV for ^{168}Er.

Table 4. Results for heavy nuclei. For each nucleus the resonance parameters are listed in the first column: the energy E_R, the spin-parity J^{π}_{R} and the mean level spacing D of the resonances. The second and third columns give details about the spectroscopic space: active number of protons m_p, active number of neutrons m_n and the fixed parity dimensionality d^{π}; the *sp* orbits and the corresponding *spe*'s from zero deformed Nilsson potential with $\sum_i \ell_i^2$ and $\sum_i \ell_i \cdot s_i$ strengths are as given by Bohr and Mottelson [33]. The fourth column gives reference level information: the reference energy E_r, the spin J^{π}_{r} of the reference level and the total number of states $F(E_r)$ up to E_r where for the reference level we use only half the degeneracy. The calculated strength parameter G of SDI, the state density spreading variance σ^2_V, the spin-cutoff density variance $\sigma^2_{V:J_z^2}$ and centroid shift $\epsilon_{V:J_z^2}$, and the spin-cutoff factor $\sigma_J(E_R)$ at the resonance energy are given in column 5. Also given, for some nuclei, the energy difference Δ between the exact and the statistical ground state. The (†) in the table indicates that a shift of 0.5 MeV has been used in fixing the ground state. The resonance data are taken from Mughabghab's compilation [41]. The reference energy data are from Nuclear Data Sheets 53–78.

Nucleus E_R (MeV) J^{π}_{R} D (eV)	m_p m_n d^{π}	Orbits *spe* (MeV)	E_r (MeV) J^{π}_{r} $F(E_r)$	G (MeV) σ^2_V (MeV^2) σ^2_{V,J_z^2} (MeV^2) ϵ_{V,J_z^2} (MeV) $\sigma_J(E_R)$	Δ
^{150}Sm	12	$7/2^+, 5/2^+, 11/2^-$	1.279	0.115	0.3
7.985	6	(43.122, 43.819, 44.759)	6^+	3.44	
$(3,4)^-$	7.96×10^{14}	$3/2^+, 1/2^+$	33.5	3.25	
2.2		(46.269, 46.567)		0.014	
		$7/2^-, 9/2^-, 13/2^+$		7.45	
		(50.342, 51.029, 51.832)			
		$3/2^-, 5/2^-, 1/2^-$			
		(53.390, 53.772, 54.860)			
^{153}Sm	12	$7/2^+, 5/2^+, 11/2^-$	0.066	0.108	1.0
5.868	9	(42.838, 43.531, 44.465)	$9/2^+$	4.21	
$1/2^+$	8.00×10^{18}	$3/2^+, 1/2^+$	33	3.99	
51.8		(45.964, 46.261)		0.014	
		$7/2^-, 9/2^-, 13/2^+$		7.38	
		(50.010, 50.693, 51.491)			
		$3/2^-, 5/2^-, 1/2^-$			
		(53.038, 53.418, 54.499)			
^{155}Sm	12	$7/2^+, 5/2^+, 11/2^-$	0.359	0.112	0.7
5.814	11	(42.653, 43.343, 44.273)	$13/2^+$	5.28	
$1/2^+$	8.66×10^{17}	$3/2^+, 1/2^+$	43	5.03	
115		(45.766, 46.061)		0.016	
		$7/2^-, 9/2^-, 13/2^+$		7.41	
		(49.794, 50.474, 51.268)			
		$3/2^-, 5/2^-, 1/2^-$			
		(52.809, 53.187, 54.263)			
^{155}Gd	14	$7/2^+, 5/2^+, 11/2^-$	0.423	0.103	0.9
6.438	9	(42.653, 43.343, 44.273)	$7/2^+$	3.96	
$1/2^+$	1.67×10^{17}	$3/2^+, 1/2^+$	84	3.76	
14.5		(45.766, 46.060)		0.014	
		$7/2^-, 9/2^-, 13/2^+$		7.72	
		(49.794, 50.474, 51.268)			
		$3/2^-, 5/2^-, 1/2^-$			
		(52.809, 53.187, 54.263)			

Table 4. (*Continued*).

Nucleus E_{R} (MeV) J^{π}_{R} D (eV)	m_{p} m_{n} d^{π}	Orbits *spe* (MeV)	E_{r} (MeV) J^{π}_{r} $F(E_{\mathrm{r}})$	G (MeV) σ^2_V (MeV^2) σ^2_{V,J^2_z} (MeV^2) ϵ_{V,J^2_z} (MeV) $\sigma_J(E_{\mathrm{R}})$	Δ
^{156}Gd 8.536 $(1,2)^-$ 1.8	14 10 5.85×10^{17}	$7/2^+, 5/2^+, 11/2^-$ (42.562, 43.250, 44.178) $3/2^+, 1/2^+$ (45.668, 45.963) $7/2^-, 9/2^-, 13/2^+$ (49.688, 50.366, 51.159) $3/2^-, 5/2^-, 1/2^-$ (52.696, 53.073, 54.147)	1.924 12^+ 308.5	0.118 5.66 5.38 0.017 8.02	0.7
^{157}Gd 6.360 $1/2^+$ 37.8	14 11 1.81×10^{18}	$7/2^+, 5/2^+, 11/2^-$ (42.471, 43.158, 44.084) $3/2^+, 1/2^+$ (45.571, 45.865) $7/2^-, 9/2^-, 13/2^+$ (49.582, 50.259, 51.049) $3/2^-, 5/2^-, 1/2^-$ (52.584, 52.960, 54.032)	0.276 $11/2^+$ 30	0.104 4.74 4.51 0.015 7.84	0.7
^{158}Gd 7.937 $(1,2)^-$ 4.9	14 12 4.97×10^{18}	$7/2^+, 5/2^+, 11/2^-$ (42.382, 43.066, 43.991) $3/2^+, 1/2^+$ (45.474, 45.768) $7/2^-, 9/2^-, 13/2^+$ (49.477, 50.153, 50.942) $3/2^-, 5/2^-, 1/2^-$ (52.473, 52.848, 53.918)	1.407 4^+ 106.5	0.118 6.47 6.17 0.018 8.02	0.8
^{159}Gd 5.943 $1/2^+$ 85	14 13 1.22×10^{19}	$7/2^+, 5/2^+, 11/2^-$ (42.293, 42.976, 43.899) $3/2^+, 1/2^+$ (45.379, 45.672) $7/2^-, 9/2^-, 13/2^+$ (49.373, 50.048, 50.835) $3/2^-, 5/2^-, 1/2^-$ (52.363, 53.737, 53.804)	0.373 $13/2^+$ 23	0.097 4.65 4.44 0.015 7.89	0.5
^{162}Dy(+ve) 8.197 $(2,3)^+$ 2.67	16 14 3.45×10^{19}	$7/2^+, 5/2^+, 11/2^-$ (42.030, 42.709, 43.626) $3/2^+, 1/2^+$ (45.097, 45.388) $7/2^-, 9/2^-, 13/2^+$ (49.067, 49.737, 50.519) $3/2^-, 5/2^-, 1/2^-$ (52.037, 52.410, 53.470)	1.635 5^+ 155.5	0.101 5.35 5.12 0.016 8.62	

Table 4. (*Continued*).

Nucleus E_R (MeV) J_R^π D (eV)	m_p m_n d^π	Orbits *spe* (MeV)	E_r (MeV) J_r^π $F(E_r)$	G (MeV) σ_V^2 (MeV^2) σ^2_{V,J_z^2} (MeV^2) ϵ_{V,J_z^2} (MeV) $\sigma_J(E_R)$	Δ
^{162}Dy(-ve)	16	$7/2^+, 5/2^+, 11/2^-$	1.959	0.100	
8.197	14	(42.030, 42.709, 43.626)	9^-	5.26	
$(2, 3)^+$	3.45×10^{19}	$3/2^+, 1/2^+$	315.5	5.03	
2.67		(45.097, 45.388)		0.016	
		$7/2^-, 9/2^-, 13/2^+$		8.62	
		(49.067, 49.737, 50.519)			
		$3/2^-, 5/2^-, 1/2^-$			
		(52.037, 52.410, 53.470)			
^{167}Er	10	$5/2^+, 11/2^-, 3/2^+$	0.0	0.144	0.5†
6.436	9	(42.279,43.186,44.642)	$7/2^+$	3.51	
$1/2^+$	9.23×10^{13}	$1/2^+$	4	3.32	
38.28		(44.931)		0.044	
		$9/2^-, 13/2^+, 3/2^-$		8.83	
		(49.235, 50.010, 51.513)			
		$5/2^-, 1/2^-$			
		(51.881, 52.931)			
^{168}Er	10	$5/2^+, 11/2^-, 3/2^+$	1.943	0.152	0.4
7.771	10	(42.195,43.100,44.554)	12^+	4.24	
$(3, 4)^+$	2.49×10^{14}	$1/2^+$	278.5	4.01	
4		(44.84)		0.044	
		$9/2^-, 13/2^+, 3/2^-$		8.87	
		(49.138, 49.910, 51.410)			
		$5/2^-, 1/2^-$			
		(51.778, 52.826)			
^{169}Er	10	$5/2^+, 11/2^-, 3/2^+$	0.243	0.138	0.5†
6.003	11	(42.111,43.015,44.466)	$7/2^+$	3.72	
$1/2^+$	5.89×10^{14}	$1/2^+$	4	3.51	
100.26		(44.753)		0.039	
		$9/2^-, 13/2^+, 3/2^-$		8.76	
		(49.040, 49.812, 51.309)			
		$5/2^-, 1/2^-$			
		(51.676, 52.722)			
^{170}Yb(+ve)	12	$5/2^+, 11/2^-, 3/2^+$	1.670	0.143	
8.47	10	(42.028, 42.931, 44.378)	6^+	3.84	
$(3, 4)^+$	4.36×10^{15}	$1/2^+$	186.5	3.63	
2.2		(44.665)		0.044	
		$9/2^-, 13/2^+, 3/2^-$		9.21	
		(48.944, 49.714, 51.208)			
		$5/2^-, 1/2^-$			
		(51.574, 52.618)			

Table 4. (*Continued*).

Nucleus E_R (MeV) J_R^π D (eV)	m_p m_n d^π	Orbits *spe* (MeV)	E_r (MeV) J_r^π $F(E_r)$	G (MeV) σ_V^2 (MeV^2) σ^2_{V,J_z^2} (MeV^2) ϵ_{V,J_z^2} (MeV) $\sigma_J(E_R)$	Δ
^{170}Yb(−ve)	12	$5/2^+, 11/2^-, 3/2^+$	1.903	0.135	
[illegible]	10	[illegible]	[illegible]	3.45	
$(3, 4)^+$	4.36×10^{15}	$1/2^+$	191.5	3.26	
2.2		(44.665)		0.042	
		$9/2^-, 13/2^+, 3/2^-$		9.21	
		(48.94399, 49.713716, 51.208014)			
		$5/2^-, 1/2^-$			
		(51.574373, 52.617939)			
^{171}Yb(+ve)	12	$5/2^+, 11/2^-, 3/2^+$	0.369	0.094	
6.615	11	(41.946, 42.847, 44.292)	$13/2^+$	1.79	
$1/2^+$	4.88×10^{15}	$1/2^+$	37	1.69	
37		(44.577)		0.028	
		$9/2^-, 13/2^+, 3/2^-$		9.74	
		(48.848, 49.617, 51.108)			
		$5/2^-, 1/2^-$			
		(51.474, 52.515)			
^{171}Yb(−ve)	12	$5/2^+, 11/2^-, 3/2^+$	0.981	0.091	
6.615	11	(41.946, 42.847, 44.292)	$11/2^-$	1.67	
$1/2^+$	4.88×10^{15}	$1/2^+$	214	1.58	
37		(44.577)		0.027	
		$9/2^-, 13/2^+, 3/2^-$		9.74	
		(48.848, 49.617, 51.108)			
		$5/2^-, 1/2^-$			
		(51.474, 52.515)			
^{172}Yb(+ve)	12	$5/2^+, 11/2^-, 3/2^+$	1.863	0.134	
8.020	12	(41.865, 42.764, 44.206)	5^+	3.84	
$(0, 1)^-$	1.69×10^{15}	$1/2^+$	287.5	3.64	
5.8		(44.491)		0.039	
		$9/2^-, 13/2^+, 3/2^-$		9.30	
		(48.754, 49.520, 51.009)			
		$5/2^-, 1/2^-$			
		(51.374, 52.413)			
^{172}Yb(−ve)	12	$5/2^+, 11/2^-, 3/2^+$	1.840	0.129	
8.020	12	(41.865, 42.764, 44.206)	9^-	3.54	
$(0, 1)^-$	1.69×10^{15}	$1/2^+$	179.5	3.36	
5.8		(44.491)		0.037	
		$9/2^-, 13/2^+, 3/2^-$		9.30	
		(48.754, 49.520, 51.009)			
		$5/2^-, 1/2^-$			
		(51.374, 52.413)			

Table 4. (*Continued*).

Nucleus E_R (MeV) J_R^π D (eV)	m_p m_n d^π	Orbits *spe* (MeV)	E_r (MeV) J_r^π $F(E_r)$	G (MeV) σ_V^2 (MeV^2) $\sigma_{V,J_z^2}^2$ (MeV^2) ϵ_{V,J_z^2} (MeV) $\sigma_J(E_R)$	Δ
^{173}Yb(+ve)	12	$5/2^+, 11/2^-, 3/2^+$	0.603	0.128	
6.368	13	(41.784, 42.681, 44.120)	$13/2^+$	3.63	
$1/2^+$	3.12×10^{15}	$1/2^+$	25	3.44	
70		(44.405)		0.036	
		$9/2^-, 13/2^+, 3/2^-$		9.02	
		(48.659, 49.425, 50.910)			
		$5/2^-, 1/2^-$			
		(51.275, 52.312)			
^{173}Yb(−ve)	12	$5/2^+, 11/2^-, 3/2^+$	1.122	0.135	
6.368	13	(41.784, 42.681, 44.120)	$5/2^-$	4.04	
$1/2^+$	3.12×10^{15}	$1/2^+$	173	3.83	
70		(44.405)		0.038	
		$9/2^-, 13/2^+, 3/2^-$		9.02	
		(48.659, 49.425, 50.910)			
		$5/2^-, 1/2^-$			
		(51.275, 52.312)			
^{174}Yb(−ve)	12	$5/2^+, 11/2^-, 3/2^+$	1.885	0.103	
7.465	14	(41.704, 42.599, 44.036)	5^-	2.48	
$(2, 3)^-$	5.13×10^{15}	$1/2^+$	54.5	2.35	
7.8		(44.320)		0.028	
		$9/2^-, 13/2^+, 3/2^-$		9.30	
		(48.566, 49.330, 50.813)			
		$5/2^-, 1/2^-$			
		(51.176, 52.212)			
^{174}Yb(+ve)	12	$5/2^+, 11/2^-, 3/2^+$	1.926	0.117	
8.020	14	(41.704, 42.599, 44.036)	5^+	3.18	
$(2, 3)^-$	5.13×10^{15}	$1/2^+$	212.5	3.02	
7.8		(44.320)		0.032	
		$9/2^-, 13/2^+, 3/2^-$		9.30	
		(48.566, 49.330, 50.813)			
		$5/2^-, 1/2^-$			
		(51.176, 52.212)			
^{175}Yb(+ve)	12	$5/2^+, 11/2^-, 3/2^+$	0.522	0.127	
5.822	15	(41.624, 42.518, 43.951)	$13/2^+$	3.83	
$1/2^+$	7.53×10^{15}	$1/2^+$	29	3.64	
162		(44.235)		0.034	
		$9/2^-, 13/2^+, 3/2^-$		8.87	
		(48.473, 49.236, 50.716)			
		$5/2^-, 1/2^-$			
		(51.078, 52.112)			

Table 4. (*Continued*).

Nucleus E_R (MeV) J_R^π D (eV)	m_p m_n d^π	Orbits *spe* (MeV)	E_r (MeV) J_r^π $F(E_r)$	G (MeV) σ_V^2 (MeV^2) $\sigma_{V,J_z^2}^2$ (MeV^2) ϵ_{V,J_z^2} (MeV) $\sigma_J(E_R)$	Δ
^{175}Yb(−ve)	12	$5/2^+, 11/2^-, 3/2^+$	1.009	0.127	
5.822	13	(41.624, 42.518, 43.951)	3/2	3.84	
$1/2^+$	7.53×10^{15}	$1/2^+$	123	3.65	
162		(44.235)		0.034	
		$9/2^-, 13/2^+, 3/2^-$		8.87	
		(48.473, 49.236, 50.716)			
		$5/2^-, 1/2^-$			
		(51.078, 52.112)			
^{175}Hf(+ve)	14	$5/2^+, 11/2^-, 3/2^+,$	0.897	0.119	0.6
6.709	13	(41.624, 42.518, 43.951)	$19/2^+$	3.06	
$1/2^+$	2.66×10^{15}	$1/2^+$	124	2.90	
21		(44.235)		0.031	
		$9/2^-, 13/2^+, 3/2^-$		9.23	
		(48.473, 49.236, 50.715)			
		$5/2^-, 1/2^-$			
		(51.078, 52.112)			
^{175}Hf(−ve)	14	$5/2^+, 11/2^-, 3/2^+,$	0.819	0.125	0.6
6.709	13	(41.624, 42.518, 43.951)	$17/2^-$	3.38	
$1/2^+$	2.66×10^{15}	$1/2^+$	161	3.20	
21		(44.235)		0.031	
		$9/2^-, 13/2^+, 3/2^-$		9.23	
		(48.473, 49.236, 50.715)			
		$5/2^-, 1/2^-$			
		(51.078, 52.112)			
^{178}Hf	14	$5/2^+, 11/2^-, 3/2^+,$	1.803	0.115	
7.626	16	(41.389, 42.277, 43.703)	3^-	3.13	
$(3, 4)^-$		$1/2^+$	221.5	2.97	
2.4		(43.985)		0.029	
		$9/2^-, 13/2^+, 3/2^-$		9.24	
		(48.199, 48.957, 50.429)			
		$5/2^-, 1/2^-$			
		(50.790, 51.818)			
^{178}Hf	14	$5/2^+, 11/2^-, 3/2^+,$	1.803	0.115	
7.626	16	(41.389, 42.277, 43.703)	3^-	3.12	
3^-		$1/2^+$	221.5	2.96	
5.3		(43.985)		0.029	
		$9/2^-, 13/2^+, 3/2^-$		9.24	
		(48.199, 48.957, 50.429)			
		$5/2^-, 1/2^-$			
		(50.790, 51.818)			

Table 4. (*Continued*).

Nucleus E_R (MeV) J_R^π D (eV)	m_p m_n d^π	Orbits *spe* (MeV)	E_r (MeV) J_r^π $F(E_r)$	G (MeV) σ_V^2 (MeV^2) σ^2_{V,J_z^2} (MeV^2) ϵ_{V,J_z^2} (MeV) $\sigma_J(E_R)$	Δ
^{178}Hf	14	$5/2^+, 11/2^-, 3/2^+,$	1.803	0.115	
7.626	16	(41.389, 42.277, 43.703)	3^-	3.15	
4^-		$1/2^+$	221.5	2.98	
4.4		(43.985)		0.029	
		$9/2^-, 13/2^+, 3/2^-$		9.24	
		(48.199, 48.957, 50.429)			
		$5/2^-, 1/2^-$			
		(50.790, 51.818)			
^{179}Hf	14	$5/2^+, 11/2^-, 3/2^+,$	0.631	0.111	
6.099	17	(41.312, 42.199, 43.622)	$17/2^+$	2.95	
$1/2^+$	8.43×10^{15}	$1/2^+$	61	2.80	
62		(43.903)		0.028	
		$9/2^-, 13/2^+, 3/2^-$		9.02	
		(48.109, 48.866, 50.335)			
		$5/2^-, 1/2^-$			
		(50.695, 51.721)			
^{180}Hf	14	$5/2^+, 11/2^-, 3/2^+,$	1.291	0.123	
7.388	18	(41.235, 42.120, 43.541)	3^+	3.65	
$(4, 5)^+$		$1/2^+$	78.5	3.46	
4.4		(43.822)		0.030	
		$9/2^-, 13/2^+, 3/2^-$		8.88	
		(48.020, 48.775, 50.242)			
		$5/2^-, 1/2^-$			
		(50.601, 51.625)			
^{180}Hf	14	$5/2^+, 11/2^-, 3/2^+,$	1.291	0.121	
7.388	18	(41.235, 42.120, 43.541)	3^+	3.54	
4^+		$1/2^+$	78.5	3.36	
8.7		(43.822)		0.030	
		$9/2^-, 13/2^+, 3/2^-$		8.90	
		(48.020, 48.775, 50.242)			
		$5/2^-, 1/2^-$			
		(50.601, 51.625)			
^{180}Hf	14	$5/2^+, 11/2^-, 3/2^+,$	1.291	0.121	
7.388	18	(41.235, 42.120, 43.541)	3^+	3.55	
5^+		$1/2^+$	78.5	3.37	
7.7		(43.822)		0.030	
		$9/2^-, 13/2^+, 3/2^-$		8.91	
		(48.020, 48.775, 50.242)			
		$5/2^-, 1/2^-$			
		(50.601, 51.625)			

Table 4. (*Continued*).

Nucleus E_R (MeV) J^π_R D (eV)	m_p m_n d^π	Orbits *spe* (MeV)	E_r (MeV) J^π_r $F(E_r)$	G (MeV) σ^2_V (MeV^2) σ^2_{V,J^2_z} (MeV^2) ϵ_{V,J^2_z} (MeV) $\sigma_J(E_R)$	Δ
^{181}W 0.686 $1/2^+$ 23	16 17 3.16×10^{15}	$5/2^+, 11/2^-, 3/2^+$ (41.139, 42.043, 43.460) 1/2+ (43.741) $9/2^-, 13/2^+, 3/2^-$ (47.932, 48.686, 50.149) $5/2^-, 1/2^-$ (50.508, 51.530)	0.675 11/2 86	0.104 2.43 2.30 0.022 9.04	0.3
^{207}Pb 6.739 $1/2^+$ 37100	8 11 9.13×10^{17}	$3/2^+, 1/2^+, 9/2^-$ (41.559,41.827,45.044) $13/2^+, 7/2^-$, (45.370,45.849) $3/2^-, 5/2^-, 1/2^-$, (47.955,48.298,49.275) $9/2^+, 15/2^-, 11/2^+$ (51.522,51.932,52.277)	2.340 $7/2^-$ 16	0.124 5.01 4.76 0.018 5.88	
^{207}Pb 6.739 $(1/2, 3/2)^-$ 5000	8 11 9.13×10^{17}	$3/2^+, 1/2^+, 9/2^-$ (41.559,41.827,45.044) $13/2^+, 7/2^-$, (45.370,45.849) $3/2^-, 5/2^-, 1/2^-$, (47.955,48.298,49.275) $9/2^+, 15/2^-, 11/2^+$ (51.522,51.932,52.277)	2.340 $7/2^-$ 16	0.124 5.01 4.76 0.018 5.88	
^{207}Pb 6.739 $(3/2, 5/2)^+$ 7400	8 11 9.13×10^{17}	$3/2^+, 1/2^+, 9/2^-$ (41.559,41.827,45.044) $13/2^+, 7/2^-$, (45.370,45.849) $3/2^-, 5/2^-, 1/2^-$, (47.955,48.298,49.275) $9/2^+, 15/2^-, 11/2^+$ (51.522,51.932,52.277)	2.340 $7/2^-$ 16	0.124 5.01 4.76 0.018 5.88	
^{233}U 5.753 $1/2^+$ 4.6	10 15 7.42×10^{18}	$9/2^-, 13/2^+, 7/2^-$, (43.11,43.92,44.16) $9/2^+, 15/2^-, 11/2^+$ (49.24,50.79,50.87) $5/2^+$ (52.01)	0.153 $11/2^+$ 30	0.115 3.76 3.57 0.039 9.49	0.47
^{234}U 6.843 $(2, 3)^+$ 0.55	10 16 1.34×10^{19}	$9/2^-, 13/2^+, 7/2^-$ (43.048,43.863,44.096) $9/2^+, 15/2^-, 11/2^+$ (49.170, 50.718, 50.802) $5/2^+$ (51.934)	1.354 6^+ 296.5	0.129 4.87 4.64 0.045 9.56	0.5

Table 4. (*Continued*).

Nucleus E_{R} (MeV) J^{π}_{R} D (eV)	m_{p} m_{n} d^{π}	Orbits *spe* (MeV)	E_{r} (MeV) J^{π}_{r} $F(E_{\mathrm{r}})$	G (MeV) σ^2_V (MeV^2) σ^2_{V,J_z^2} (MeV^2) ϵ_{V,J_z^2} (MeV) $\sigma_J(E_{\mathrm{R}})$	Δ
^{235}U 5.298 $1/2^+$ 10.6	10 17 2.21×10^{19}	$9/2^-, 13/2^+, 7/2^-$ (42.987,43.800,44.033) $9/2^+, 15/2^-, 11/2^+$ (49.100, 50.646, 50.730) $5/2^+$ (51.860)	0.225 $9/2^+$ 61	0.124 4.67 4.45 0.045 9.38	0.9
^{236}U 6.546 $(3,4)^-$ 0.44	10 18 3.32×10^{19}	$9/2^-, 13/2^+, 7/2^-$ (42.926,43.738,43.971) $9/2^+, 15/2^-, 11/2^+$ (49.031, 50.575, 50.658) $5/2^+$ (51.787)	1.127 5^+ 150.5	0.127 4.99 4.76 0.047 9.71	0.4
^{237}U 5.126 $1/2^+$ 14.7	10 19 4.54×10^{19}	$9/2^-, 13/2^+, 7/2^-$ (42.865,43.677,43.909) $9/2^+, 15/2^-, 11/2^+$ (48.962, 50.504, 50.587) $5/2^+$ (51.714)	0.327 $11/2^+$ 72	0.127 5.08 4.85 0.047 9.39	0.8
^{238}U 6.153 $(0,1)^+$ 20.9	10 20 5.68×10^{19}	$9/2^-, 13/2^+, 7/2^-$ (42.80, 43.62, 43.85) $9/2^+, 15/2^-, 11/2^+$ (48.89, 50.43, 50.52) $5/2^+$ (51.64)	1.127 4^+ 135.5	0.129 5.35 5.11 0.049 9.60	0.3
^{239}U 4.806 $1/2^+$ 20.9	10 21 6.49×10^{19}	$9/2^-, 13/2^+, 7/2^-$ (42.745,43.555,43.786) $9/2^+, 15/2^-, 11/2^+$ (48.82, 50.36, 50.44) $5/2^+$ (51.57)	0.373 $13/2^+$ 103	0.128 5.28 5.04 0.049 9.36	0.5
^{239}U 4.806 $(1/2, 3/2)^-$ 7.2	10 21 6.49×10^{19}	$9/2^-, 13/2^+, 7/2^-$ (42.74, 43.55, 43.79) $9/2^+, 15/2^-, 11/2^+$ (48.82, 50.36, 50.44) $5/2^+$ (51.57)	0.373 $13/2^+$ 103	0.128 5.28 5.04 0.049 9.36	
^{233}Th 4.786 $1/2^+$ 16.81	8 17 3.61×10^{18}	$9/2^-, 13/2^+, 7/2^-$ (43.109, 43.925, 44.159) 9/2+, 15/2−, 11/2+ (49.240, 50.791, 50.875) $5/2^+$ (52.008)	0.278 $7/2^+$ 88	0.118 3.82 3.63 0.042 9.34	0.8

Table 4. (*Concluded*).

Nucleus E_R (MeV) J_R^π D (eV)	m_p m_n d^π	Orbits *spe* (MeV)	E_r (MeV) J_r^π $F(E_r)$	G (MeV) σ_V^2 (MeV^2) σ^2_{V,J_z^2} (MeV^2) ϵ_{V,J_z^2} (MeV) $\sigma_J(E_R)$	Δ
^{240}Pu 6.534 $(0, 1)^+$ 2.3	12 20 1.99×10^{20}	$9/2^-, 13/2^+, 7/2^-$ (42.606,43.494,43.723) $9/2^+, 15/2^-, 11/2^+$ (48.757, 50.292, 50.375) $5/2^+$ (51.498)	1.031 3^+ 84.5	0.131 3.93 5.66 0.050 9.87	0.3

For the "wrong-parity" case, consider the nucleus ^{239}U [41, 42]. The resonance energy $E_R = 4.806$ MeV and the observed resonance spacings $D_0 = 20.9$ eV for the s-wave with $J^\pi = 1/2^+$, and $D_1 = 7.2$ eV for the p-wave resonances with $J^\pi = (1/2^-, 3/2^-)$. The low-lying spectrum up to 0.373 MeV has 13 positive parity levels. With $\mathcal{E} = 0.373$ MeV ($J^\pi = 13/2^+$) as the reference state, we determine $G^+ = 0.128$ MeV by matching the experimental distribution function $F_{ex}(\mathcal{E}) = 103$ and the observed s-wave density 4.66×10^4/MeV with the corresponding theoretical ones. Assuming now that $G^- = G^+$ we calculate the p-wave resonance density and find remarkable agreement with the observed value 1.39×10^5/MeV.

As a different example of the wrong-parity spectra consider the nucleus ^{156}Gd [41, 43]. The angular momentum of the target nucleus (^{155}Gd) is $3/2^-$ leading to $J^\pi = (1^-, 2^-)$ resonances at $E_R = 8.536$ MeV with a resonance level density 5.55×10^5/MeV. The low-lying spectrum contains 35 positive parity levels up to $E_R^* = 1.924$ MeV. Once again assuming that $G^+ = G^-$, we determine the optimal G value to be 0.118 MeV. As shown in Fig. 4, the smoothed distribution function agrees well with the experimental staircase down to $\mathcal{E} = 1$ MeV below the reference state of 1.924 MeV. The statistical ground state was found to be 0.7 MeV below the experimental ground state. In several nuclei (^{162}Dy, $^{170-175}$Yb, ^{175}Hf), we consider reference energies with positive parity and negative parity and in both cases the deduced G values (or equivalently the spreading variances) are, as expected, quite close.

The nucleus ^{207}Pb in Table 4 is unique since not only s-wave ($1/2^+$) and p-wave ($(1/2, 3/2)^-$) but also d-wave ($(3/2, 5/2)^+$) resonances are observed. It is pleasing to note that a single G value is able to reproduce the observed resonance spacings in all three cases to within a factor of two.

Finally, consider the Hf isotopes. The ground state of ^{177}Hf is $7/2^-$ leading to the resonance spins $(3^-, 4^-)$ in ^{178}Hf [41, 44] with a resonance spacing of 2.4 eV at the resonance energy $E_R = 7.626$ MeV. Using a reference state at 1.803 MeV, we find $G = 0.115$ MeV. Since the spins have been identified in the resonance spectrum, we can also determine G using the spacings of the 3^- and 4^- separately. The values are $G = 0.114$ and 0.115 MeV using the 3^- and 4^- subspectrum spacings. Similarly, for ^{180}Hf [41, 45], we obtain $G = 0.123, 0.121$, and 0.120 MeV using the resonance spacings corresponding to the spins $(4, 5)^+$, 4^+, and 5^+, respectively. Keeping in mind the extreme sensitivity of the density on G, as in Fig. 1 where a 1% change in G leads to a 10% change in $I(E)$, the near constancy of the G values determined for the Hf isotopes reflects the stability and self-consistency of the method. For details regarding the deduced values of σ_V^2, σ_J, and G see Table 4. With this particular strength, SDI can then be used for studies of smoothed transition strengths, strength sums etc., within the framework of statistical spectroscopy.

Fig. 4. The distribution function $F(E)$ for ^{156}Gd with $G^{-} = G^{+} = 0.118$ MeV determined using a reference energy at 1.924 MeV and the observed density at the resonance energy $E_{\mathrm{R}} = 8.534$ MeV. The agreement extending below up to ~1 MeV may be regarded as an indication of the onset of chaos.

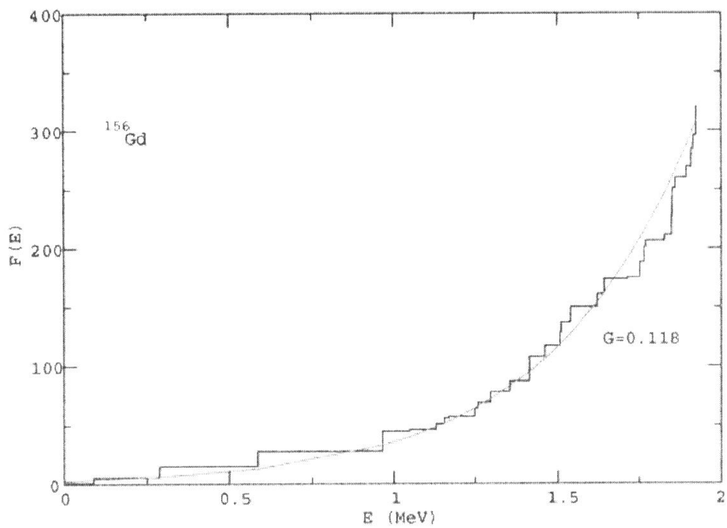

6. Strength distributions and two-point functions

We turn briefly to two-point quantities that essentially complete the spectroscopy of the chaotic domain. Examples are smoothed transition strengths and measures for weak symmetry breaking. The techniques used for the one-point function have "obvious" extensions to the two-point functions. We give here a brief introduction leaving a detailed discussion and applications for the future.

Strengths are absolute squares of matrix elements of a transition operator T between H-eigenstates, the variables being the initial and final energies and exact-symmetry labels. This then leads to bivariate densities, reduced cumulants, convolutions etc., as well as new one-point functions (marginal densities for example) as defined in ref. 6. Once again we start with $V = 0$, which gives for each particle number (or pair of such if particle number is not conserved) an exactly solvable noninteracting particle problem. Then to introduce V we rely on the fairly well-tested proposition that when a pair (initial and final) of h-eigenstates spreads among the H-eigenstates as V is turned on, the initial strength connecting them spreads in proportion to the product of admixing intensities.

For a transition operator T, the exact evaluation of the NIP bivariate density is simple, both by inspection and by a formal procedure. In the dilute limit, $N >> m >> 1$, the density is describable as a "weakly singular" bivariate Gaussian. Proceeding now to the interacting particle (IP) case we should assume Gaussian spreadings induced by V and carry out the double convolution, keeping in mind that in large (N, m) spaces partitioning of the sp space is essential (and straightforward), and that a correlation between the two V-spreadings must be allowed, as in the case of time-reversal symmetry breaking where the correlation coefficients are close to unity [6]. The method, used already [6] for bounds on time-reversal noninvariance and more recently [5] for β-decay rates relevant for pre-supernova evolution, and in the experimental data analysis of parity violation in compound nuclei [46], leads to a bivariate Gaussian strength distribution; it can be verified (or improved upon) by admitting low-order Edgeworth corrections; nontrivial problems of trace evaluation may be encountered here and extensions would be needed for $s > 0$ and for operators of mixed parity. Also, there are additional complexities in the case where the interaction is weak as compared to the mean-field [47].

7. Conclusions

We have given the interacting-particle (IP) nuclear state density as a function of the Hamiltonian parameters in a simple calculable form that contains the original Bethe (NIP) density for noninteracting particles as a convolution factor. Our major interest has been in energies above an experimentally observable reference energy that for heavy nuclei will often be ~2 MeV. Our procedure has been to fix the centroids of the interacting-particle spherical orbit configurations and then to incorporate, by convolution, the spreading and admixing effects of the residual interaction. Formally, this has involved unitary-group decompositions of the interaction, the recognition that much of nuclear structure is dominated by central limit theorems, and the fact that the nucleus, even at relatively low excitation is in a well-defined sense, a set of chaotic subsystems. A striking consequence is that the results (deriving, as they do, from low-order traces of the Hamiltonian) are essentially independent of the phase relations between the spreading-interaction matrix elements. This indicates that most of the information carried by the interaction is in fact lost except for a small set of variances of the spreading interaction V, if we restrict ourselves to one-point functions. A little more information (some of it however of fundamental importance) derives from the study of symmetries, via fluctuations described by two-point functions. But beyond that there is little indeed. With regard to bivariate strength densities for transition operators, the issues involved are complex and their detailed discussion is left for the future.

The neutron-resonance data analysis indicates that for heavy nuclei the spreading widths are ~2 MeV and the spin-cutoff widths are ~4% smaller. The significance of J^2V^2 correlations in producing the energy variation of spin-cutoff factors is pointed out. The SDI interaction has been used in the analysis as the primary purpose here was to make clear the structures and procedures. The present analysis along with the previous $(1f2p)$-shell nuclei analysis validate the theory for one-point quantities.

The one essential approximation is the neglect of interactions connecting configurations, which on the average, are separated by $2\hbar\omega$ or more; this approximation could be improved upon by introducing few-parameter forms for the bimodal or multimodal densities, which would be generated by such interactions and also by exploring further the so-called partitioned EGOE [21]. Much of the future, in applying the theory, lies in generating realistic effective interactions for heavy nuclei and efforts in this direction (see, ref. 48 for a recent attempt) are called for. With this, a complete ab-initio approach to nuclear level densities is possible and we are now close to solving this long-standing problem in nuclear structure. A systematic survey of fixed-(JT) level densities over the periodic table should be possible. After the present work was completed, Horoi et al. reported on statistical spectroscopy calculations for level densities for light nuclei [49] by evaluating exact fixed-JT centroids and variances and using a different method for fixing the ground state.

Finally, since the state density is given in terms of Hamiltonian parameters, playing the same role, therefore, as the partition function (its Laplace transform), it easily gives occupancies and we shall deal in future papers with the Fermi–Dirac representation of occupancies; see, ref. 50 for recent interest in Fermi occupancies for isolated finite interacting particle systems in the context of quantum chaos and thermalization. We shall also make the extension to two-point functions (briefly described here), which are encountered, for example, in the study of transition strengths and weakly broken symmetries such as time-reversal, isospin and parity.

Acknowledgments

This work has been supported in part by the U.S. Department of Energy Grant DE-FG02-88ER40425; the Natural Sciences and Engineering Research Council (NSERC), Canada; and the Physical Research Laboratory, Ahmedabad, India.

References

1. T.A. Brody, J. Flores, J.B. French, P.A. Mello, A. Pandey, and S.S.M. Wong. Rev. Mod. Phys. **53**, 385 (1981).
2. R.U. Haq, A. Pandey, and O. Bohigas. Phys. Rev. Lett. **48**, 1086 (1982).
3. O. Bohigas, M.J. Giannoni, and C. Schmit. Phys. Rev. Lett. **52**, 1 (1984); O. Bohigas and M.J. Giannoni. Lect. Notes Phys. **209**, 1 (1984); O. Bohigas and H. Weidenmüller. Ann. Rev. Nucl. Part. Sci. **38**, 421 (1988).
4. J.P. Draayer, J.B. French, and S.S.M. Wong. Ann. Phys. (N.Y.), **106**, 503 (1977).
5. V.K.B. Kota and D. Majumdar. Z. Phys. A, **351**, 365, 377 (1995).
6. J.B. French, V.K.B. Kota, A. Pandey, and S. Tomsovic. Ann. Phys. (N.Y.), **181**, 235 (1988).
7. V.K.B. Kota and R. Sahu. Phys. Lett. **429B**, 1 (1998); Phys. Rev. E, **64**, 016219 (2001).
8. E.P. Wigner. *In* Essays in honour of Michael Polanyi (Routledge and Kegan Paul, London, 1961). Reprinted in Symmetries and reflections. *Edited by* W.J. Moore and M. Scriven. Indiana University Press, Bloomington, Ind., USA. 1967; Ox Bow Press, Woodbridge, Conn., USA. 1979. p. 203.
9. J.B. French and S.S.M. Wong. Phys. Lett. **33B**, 449 (1970); O. Bohigas and J. Flores. Phys. Lett. **34B**, 261 (1971); S.S.M. Wong and J.B. French. Nucl. Phys. **A198**, 188 (1972).
10. K.K. Mon and J.B. French. Ann. Phys. (N.Y.), **95**, 90 (1975); V.K.B. Kota. Phys. Rep. **347**, 223 (2001).
11. G.J.H. Laberge and R.U. Haq. Can. J. Phys. **68**, 301 (1990); K. Patel, M.S. Desai, V. Potbhare, and V.K.B. Kota. Phys. Lett. **275A**, 329 (2000).
12. K.F. Ratcliff. Phys. Rev. C, **3**, 117 (1971).
13. F.S. Chang, J.B. French, and T.H. Thio. Ann. Phys. (N.Y.), **66**, 137 (1971).
14. J.B. French and V.K.B. Kota. Phys. Rev. Lett. **51**, 2183 (1983); J.B. French. Lect. Notes Phys. **209**, 100 (1984).
15. J. Flores, M. Horoi, M. Müller, and T.H. Seligman. Phys. Rev. E, **63**, 026204 (2000).
16. O. Bohigas, R.U. Haq, and A. Pandey. Phys. Rev. Lett. **54**, 1645 (1985).
17. E. Cota, J. Flores, P.A. Mello, and E. Yépez. Phys. Lett. **53B**, 32 (1974); A.Y. Abul-Magd, H.L. Harney, M.H. Simbel, and H.A. Wiedenmüller. Phys. Lett. **579B**, 278 (2004).
18. J.B. French, V.K.B. Kota, and J.F. Smith. University of Rochester Report ER40425-245/UR1122 (1989).
19. V.K.B. Kota. *In* Capture gamma-ray spectroscopy and related topics. *Edited by* R.W. Hoff. AIP conference proceedings 238, New York. 1991. p. 685; R.U. Haq. *In* From spectroscopy to chaos. *Edited by* D. Koltun and A. Das. World Scientific, Singapore. 1995. p. 89; J.B. French. *In* A gift of prophecy - essays in celebration of the life of R.E. Marshak. *Edited by* E.C.G. Sudarshan. World Scientific, Singapore. 1995. p. 156; V.K.B. Kota. *In* Recent trends in nuclear structure physics. *Edited by* G. Shanmugam and N. Arunachalam. M.S. University Publisher, Tirunelveli, India. 1997. p. I2-1.
20. V.K.B. Kota and D. Majumdar. Nucl. Phys. **A604**, 129 (1996).
21. V.K.B. Kota, D. Majumdar, R.U. Haq, and R.J. Leclair. Can. J. Phys. **77**, 893 (1999).
22. A.P. Zuker. Lect. Notes Phys. **209**, 157 (1984); A. Abzouzi, E. Caurier, and A.P. Zuker. Phys. Rev. Lett. **66**, 1134 (1991); A.P. Zuker. Nucl. Phys. **A576**, 65 (1994); M. Dufour and A.P. Zuker. Phys. Rev. C, **54**, 1641 (1996); J. Dufflo and A.P. Zuker. Phys. Rev. C, **59**, R2347 (1999).
23. V.K.B. Kota, V. Potbhare, and P. Shenoy. Phys. Rev. C, **34**, 2330 (1986); S.M. Grimes and T.N. Massey. Phys. Rev. C, **51**, 606 (1995).
24. A.P. Zuker. Phys. Rev. C, **64**, 021303 (2001); C.W. Johnson, J. Nabi, and W.E. Ormand. e-print nucl-th/0105041 (2001).
25. T.T.S. Kuo. Nucl. Phys. **A103**, 71 (1967).
26. V. Zelevinsky, B.A. Brown, N. Frazier, and M. Horoi. Phys. Rep. **276**, 85 (1996).
27. V.V. Flambaum, G.F. Gribakin, and F.M. Izrailev. Phys. Rev. E, **53**, 5729 (1996).
28. H.A. Bethe. Phys. Rev. **50**, 332, (1936); H.A. Bethe. Rev. Mod. Phys. **9**, 69 (1937).
29. C. Van Lier and G.E. Uhlenbeck. Physica, **4**, 531 (1937).
30. T. Ericson. Adv. Phys. **9**, 425 (1960); A. Gilbert and A.G.W Cameron. Can. J. Phys. **43**, 1446 (1965); J. Huizenga and L.G. Moretto. Ann. Rev. Nucl. Sci. **22**, 427 (1972); see also separate reviews by S.M. Grimes, S.E. Woosley, G. Reffo, and M. Beckerman. *In* Moment methods in many-fermion systems. *Edited by* B.J. Dalton, S.M. Grimes, J.P. Vary, and S.A. Williams. Plenum, NY, USA. 1980. For papers which make closer connection with the present work see also R.U. Haq and S.S.M. Wong. Can. J. Phys.

60, 1502 (1982); K. Sato, Y. Takahashi, and N. Yoshida. Z. Phys. **A339**, 129 (1991) and earlier papers in the same journal; Z. Pluhar and H. Weidenmüller. Phys. Rev. C, **38**, 1046 (1988).
31. I.M. Green and S.A. Moszkowski. Phys. Rev. B, **139**, 790 (1965).
32. P.J. Brussaard and P.W.M. Glaudemans. Shell-model applications in nuclear spectroscopy. North Holland, New York. 1977.
33. S.G. Nilsson. Dan. Mat. Fys. Medd. **29**(16), (1955); A. Bohr and B.R. Mottelson. Nuclear structure. Vol. II. Nuclear deformations. Benjamin, Mass. 1975. p. 219.
34. M.G. Kendall and A. Stuart. Advanced theory of statistics. Vol. 1. 3rd ed. Hafner Publishing Co., New York. 1969.
35. F.S. Chang and J.B. French. Phys. Lett. **44B**, 131, 135 (1973).
36. V.K.B. Kota. Z. Phys. A, **315**, 91 (1984).
37. S.M. Grimes, S.D. Bloom, R.F. Hausman, Jr., and B.J. Dalton. Phys. Rev. C, **19**, 2378 (1979).
38. J.P.Vary and S.N. Yang. Phys. Rev. C, **15**, 1545 (1977).
39. S.S.M. Wong. Nuclear statistical spectroscopy. Oxford University Press, New York. 1986.
40. Y.A. Akovali. Nucl. Data. Sheets, **71**, 181 (1994).
41. S.F. Mughabghab. Neutron cross sections. Vol. 1B. Academic Press, New York. 1984.
42. M.R. Schmorak. Nucl. Data Sheets, **66**, 839 (1992).
43. R.G. Helmer. Nucl. Data Sheets, **65**, 65 (1992).
44. E. Browne. Nucl. Data Sheets, **72**, 221 (1994).
45. E. Browne. Nucl. Data Sheets, **71**, 81 (1994).
46. S. Tomsovic, M.B. Johnson, A.C. Hayes, and J.D. Bowman. Phys. Rev. C, **62**, 054607 (2000).
47. V.V. Flambaum, A.A. Gribakina, G.F. Gribakin, and I.V. Ponomarev. Physica D, **131**, 205 (1999); V.K.B. Kota and R. Sahu. Phys. Rev. E, **62**, 3568 (2000); M. Horoi, B.A. Brown, and V. Zelevinsky. Phys. Rev. Lett. **87**, 062501 (2001).
48. L. Losano, H. Dias, and T.T.S. Kuo. Phys. Lett. **384B**, 20 (1996); A. Holt, T. Engeland, M. Hjorth-Jensen, and E. Osnes. Phys. Rev. C, **61**, 4318 (2000); D.J. Dean, T. Engeland, M. Hjorth-Jensen, M.P. Kartamyshev, and E. Osnes. Prog. Part. Nucl. Phys. **53**, 419 (2004).
49. M. Horoi, J. Kaiser, and V. Zelevinsky. Phys. Rev. C, **67**, 054309 (2003); M. Horoi, M. Ghita, and V. Zelevinsky. Phys. Rev. C, **69**, 041307(R) (2004).
50. M. Horoi, V. Zelevinsky, and B.A. Brown. Phys. Rev. Lett. **74**, 5194 (1995); F. Borgonovi and F.M. Izrailev. Phys. Rev. E, **62**, 6475 (2000); F. Borgonovi, G. Celardo, F.M. Izrailev, and G. Casati. Phys. Rev. Lett. **88**, 054101 (2002); G. Benenti, G. Casati, and D.L. Shepelyansky. Euro. Phys. J. **D17**, 265 (2001).

Appendix A.

The low-order cumulants of $\rho(E)$, which add under convolution and vanish for Gaussian when $p > 2$, are $K_1 = M_1$, $K_2 = \mu_2 = \sigma^2$, $K_3 = \mu_3 = \gamma_1\sigma^3$, $K_4 = \mu_4 - 3(\mu_2)^2 = \gamma_2\sigma^4$ (where γ_1, γ_2 are, respectively, the "skewness" and "excess"), $K_5 = \mu_5 - 10\mu_3\mu_2$, and $K_6 = \mu_6 - 15\mu_4\mu_2 - 10\mu_3^2 + 30\mu_2^3$; note that μ_p is the pth central moment. In terms of the "reduced cumulants" $k_p = K_p/\sigma^p$ and Hermite polynomials He_p we have then for the fourth-order Edgeworth expansion [34] of the density normalized to $\int \rho^{[m]}(E)\,\mathrm{d}E = \int \eta(z)\,\mathrm{d}z = 1$, so that $\rho^{[m]}(E) = \sigma^{-1}\eta(z)$

$$\eta(z) = \eta_G(z)\left\{1 + \frac{1}{6}k_3He_3(z) + \left[\frac{1}{24}k_4He_4(z) + \frac{1}{72}(k_3)^2He_6(z)\right]\right\}$$

$$z = (E - \epsilon)/\sigma$$

$$\eta_G(z) = (2\pi)^{-1/2}\exp\left(-z^2/2\right)$$

$$He_p(z) = z^p\sum_{\nu=0}^{[\frac{p}{2}]}(-1)^\nu\binom{p}{2\nu}(2\nu-1)!!z^{-2\nu}$$

$$\int_{-\infty}^{\infty} \eta_G(z) He_p(z) He_q(z)\, \mathrm{d}z = \delta_{pq}\, p! \tag{A.1}$$

In this "standard" Edgeworth form, in contrast to the "non-standard" one, the centroid and variance are carried by the Gaussian function so that the lowest corrections are of third order. This will be of consequence for expectation values. The same is true in the closely related Gram–Charlier expansions, identical to Edgeworth to fourth order except that the $(k_3)^2$ term is missing.